FROM ORDER TO CHAOS II

Essays: Critical, Chaotic and Otherwise

WORLD SCIENTIFIC SERIES ON NONLINEAR SCIENCE

Editor: Leon O. Chua
University of California, Berkeley

Series A. MONOGRAPHS AND TREATISES

Published Titles

Volume 17: The Analysis of Complex Nonlinear Mechanical Systems: A Computer Algebra Assisted Approach
M. Lesser

Volume 18: Propagation of Waves in Shear Flows
A. L. Fabrikant and Yu. A. Stepanyants

Volume 19: Continuum Mechanics via Problems and Exercises
Edited by M. E. Eglit and D. H. Hodges

Volume 20: Chaotic Dynamics
C. Mira, L. Gardini, A. Barugola and J.-C. Cathala

Volume 21: Hopf Bifurcation Analysis: A Frequency Domain Approach
G. Chen and J. L. Moiola

Volume 22: Chaos and Complexity in Nonlinear Electronic Circuits
M. J. Ogorzalek

Volume 23: Nonlinear Dynamics in Particle Accelerators
R. Dilão and R. Alves-Pires

Volume 24: From Chaos to Order
G. Chen and X. Dong

Volume 25: Chaotic Dynamics in Hamiltonian Systems
H. Dankowicz

Volume 30: Quasi-Conservative Systems: Cycles, Resonances and Chaos
A. D. Morozov

Volume 31: CNN: A Paradigm for Complexity
L. O. Chua

Volume 32: From Order to Chaos II
L. P. Kadanoff

Volume 33: Lectures in Synergetics
V. I. Sugakov

Volume 35: Introduction to Control of Oscillations and Chaos
A. L. Fradkov and A. Yu. Pogromsky

Forthcoming Titles

Volume 27: Thermomechanics of Nonlinear Irreversible Behaviours
G. A. Maugin

Volume 34: Introduction to Nonlinear Dynamics: Chua's Circuit
L. Kocarev

WORLD SCIENTIFIC SERIES ON NONLINEAR SCIENCE

Series A Vol. 32

Series Editor: Leon O. Chua

FROM ORDER TO CHAOS II

Essays: Critical, Chaotic and Otherwise

Leo P. Kadanoff

University of Chicago

World Scientific

Singapore • New Jersey • London • Hong Kong

Published by

World Scientific Publishing Co. Pte. Ltd.

P O Box 128, Farrer Road, Singapore 912805

USA office: Suite 1B, 1060 Main Street, River Edge, NJ 07661

UK office: 57 Shelton Street, Covent Garden, London WC2H 9HE

British Library Cataloguing-in-Publication Data
A catalogue record for this book is available from the British Library.

The author and publisher would like to thank the following publishers of the various journals and books for their assistance and permission to include the selected reprints found in this volume:

Academic Press
American Association of Physics Teachers
American Institute of Physics
American Physical Society
Council of Turkey (TUBITAK)
CS Publications
Elsevier Science Publishers
Encyclopaedia Britannica
European Physical Society
Gordon and Breach Scientific Publishers
IEEE
La Recherche
Longman Group, UK
National Academy of Sciences
Royal Swedish Academy of Sciences
Società Italiana di Fisica
Society for Computer Simulation
Springer Verlag

FROM ORDER TO CHAOS II

ISBN 981-02-3433-3
ISBN 981-02-3434-1 (pbk)

Printed in Singapore.

This book is an outcome of the exchanges of information with my professors, coworkers, and students. They have all been my teachers.

Table of Contents

Publications of Leo P. Kadanoff

1. (with P. C. Martin) Knight shift in superconductors, *Phys. Rev. Lett.* **3**, 322 (1959).
2. (with P. C. Martin) Transport properties of superconductors, *Bull. Am. Phys. Soc.* **5**, 13 (1960).
3. Radiative transport within an ablating body, *Trans. ASME: Series* **C3**, 215 (1961).
4. (with V. Ambegaokar) Electromagnetic properties of superconductors, *Nuovo Cimento* **22**, 914 (1961).
5. (with G. Baym) Conservation laws and correlation functions, *Phys. Rev.* **124**, 237 (1961).
6. (with P. C. Martin) Theory of many-particle systems II: Superconductivity, *Phys. Rev.* **124**, 670 (1961).
7. (with G. Baym) Quantum statistical mechanics (W. A. Benjamin, New York, 1962), p. 203. Also translated into Russian, Japanese, and Chinese.
8. (with H. Hidalgo) Comparison between theory and flight ablation data, *Amer. Inst. of Aeronautics and Astronautics J.* **1**, 41 (1963).
9. Boltzmann equation for polarons, *Phys. Rev.* **130**, 1364 (1963).
10. (with D. Markowitz) Effect of impurities upon critical temperature of anisotropic superconductors, *Phys. Rev.* **131**, 563 (1963).
11. (with P. C. Martin) Hydrodynamic equations and correlation functions, *Ann. Phys.* **24**, 419 (1963).
12. Failure of the electronic quasiparticle picture for nuclear spin relaxation in metals, *Phys. Rev.* **132**, 2073 (1963).
13. (with D. C. Langreth) Perturbation theoretic calculation of polaron mobility, *Phys. Rev.* **133**, A1070 (1964).
14. (with R. E. Prange) Transport theory for electron-phonon interactions in metals, *Phys. Rev.* **134**, A566 (1964).
15. (with M. Revzen) Green's function formulation of the Feynman model of the polaron, *Nuovo Cimento* **33**, 397 (1964).
16. (with I. I. Falko) Ultrasonic attenuation in superconductors containing magnetic impurities, *Phys. Rev.* **136**, A1170 (1964).
17. The electron-phonon interaction in normal and superconducting metals, in *Lectures on the Many-Body Problem*, Vol. 2 (Academic Press, New York and London, 1964), p. 787.
18. On the problem of ultrasonic attenuation in superconductors containing magnetic impurities, in *Proc. 9th Int. Conf. on Low Temp. Physics* (Plenum Press, New York, 1965), Part A, pp. 378-80.
19. (with J. W. Kane) Green's functions and superfluid hydrodynamics, *J. Math. Phys.* **6**, 1902 (1965).
20. (with A. B. Pippard) Ultrasonic attenuation in superconductors, *Proc. Roy. Soc.* **A292**, 299 (1966).

21. Scaling laws for Ising models near critical points, *Proc. 1966 Midwest Conf. on Theor. Phys.* (Bloomington, Ind.).
22. Scaling laws for Ising models near Tc, *Physics* **2**, 263 (1966).
23. Spin-spin correlation in the two-dimensional Ising model, *Nuovo Cimento* **44**, 276 (1966).
24. Basic principles of physics: Electricity, magnetism, and heat (W. A. Benjamin, New York, 1967).
25. (with J. W. Kane) Long range order in superfluid helium, *Phys. Rev.* **155**, 80 (1967).
26. (with W. Gotze, D. Hamblen, R. Hecht, E. A. S. Lewis, V. V. Palciauskas, M. Rayl, J. Swift, D. Aspnes and J. W. Kane) Static phenomena near critical points: Theory and experiment, *Rev. Mod. Phys.* **39**, 395 (1967).
27. (with J. Swift) Transport coefficients near the critical point: A master-equation approach, *Phys. Rev.* **165**, 310 (1968).
28. (with J. Swift) Transport coefficients near the liquid-gas critical point, *Phys. Rev.* **166**, 89 (1968).
29. Wave function fluctuations in finite superconductors, *Comments on Solid State Phys.* **1**, 2 (1968).
30. Transport coefficients near critical points, *Comments on Solid State Phys.* **1**, 5 (1968).
31. (with G. Laramore) Anomalous electrical conductivity above the superconducting transition, *Phys. Rev.* **175**, 579 (1968).
32. (with J. Swift) Transport coefficients near the L transition of helium, *Ann. Phys.* **50**, 312 (1968).
33. Operator algebra and the determination of critical indices, *Phys. Rev. Lett.* **23**, 1430 (1969).
34. Correlations along a line in the two-dimensional Ising model, *Phys. Rev.* **188**, 859 (1969).
35. (with J. R. Voss and W. J. Bouknight) A city grows before your eyes, *Computer Decisions*, Dec., 1969.
36. (with G. E. Laramore) Anomalous ultrasonic attenuation above the magnetic critical point, *Phys. Rev.* **187**, 619 (1969).
37. (with J. R. Voss and W. J. Bouknight) Computer display and analysis through time and space, *Technological Forecasting and Social Change* **2**, 77 (1970).
38. The droplet model and scaling, in *Critical Phenomena, Proceedings of the Int. School of Physics, "Enrico Fermi"*, Course LI, ed. M. S. Green (Academic Press, New York, 1971), p. 118.
39. Critical behavior universality and scaling, in *Critical Phenomena, Proceedings of the Int. School of Physics, "Enrico Fermi"*, Course LI, ed. M. S. Green (Academic Press, New York, 1971), p. 101.
40. Studying and displaying urban growth patterns, *Aerospace and Electronic Systems* **7**, No. 3 (1971).
41. (with H. Ceva) Determination of an operator algebra for the two-dimensional Ising model, *Phys. Rev.* **B3**, 3918 (1971).
42. From simulation model to public policy: An examination of Forrester's urban

dynamics, Simulation **16**, 261 (1971). Reprinted in *Best Computer Papers of 1971*, ed. O. Petrocelli (Anerbach, New York, 1971), *American Scientist* **60**, 74 (1972).

43. (with F. J. Wegner) Some critical properties of the eight-vertex model, *Phys. Rev.* **B4**, 2909 (1971).
44. A modified Forrester model of the United States as a group of metropolitan areas, in *Urban Dynamics: Extensions and Reflections* (San Francisco Press, 1972).
45. (with H. Weinblatt) Public policy conclusions from urban growth models, *Urban Dynamics: Extensions and Reflections* (San Francisco Press, (1972).
46. (with H. Weinblatt) Public policy conclusions from urban growth models, *IEEE Trans. Systems, Man, and Cybernetics* **SMC-2**, 159 (1972).
47. (with M. K. Grover and F. J. Wegner) Critical exponents for the Heisenberg model, *Phys. Rev.* **B6**, 311 (1972).
48. (with B. Chinitz, G. Crampton, S. Jacobs, J. Tucker and H. Weinblatt) The Brown University national metropolitan models, in *Socio-Economic Systems and Principles*, ed. W. Vogt *et al.* (University of Pittsburgh, Pennsylvania, 1973). Also published in *Synergetics*, ed. A. Haken (B. G. Taubner, Stuttgart, 1973).
49. Renormalization equations: Conceptual basis and a simple example. in *Renormalization Group in Critical Phenomena and Quantum Field Theory*, ed. J. D. Gunton and M. S. Green (Temple University Press, Philadelphia, 1973), p. 21.
50. Renormalization group techniques on a lattice, in *Cooperative Phenomena*, ed. H. Haken (North Holland, 1974).
51. (with B. Harrison and B. Chinitz) A simulation model of urban labor markets and development policy, in *Urban Development Models*, ed. R. Baxter *et al.* (The Construction Press, 1975).
52. (with A. Houghton) Numerical evaluations of the critical properties of the two-dimensional Ising model, *Phys. Rev.* **B11**, 377 (1975).
53. Variational principles and approximate renormalization group calculations, *Phys. Rev. Lett.* **16**, 1005 (1975).
54. Scaling, universality, and operator algebras, in *Phase Transitions and Critical Phenomena*, ed. C. Domb and M. S. Green, Vol. 5A (Academic Press, New York, 1976), p. 1.
55. From simulation model to public policy, *Amer. Scientist* **60**, 74 (1972).
56. (with A. Houghton and M. C. Yalabik) Variational approximations for renormalization group transformations, *J. Stat. Phys.* **14**, 171 (1976). Also published in *Proc. IUPAP Statistical Mechanics Conference* (Hungarian Academy of Sciences, 1976).
57. Notes on Migdal's recursion formulas, *Ann. Phys.* **100**, 359 (1976).
58 The application of renormalization group techniques to quarks and strings, *Rev. Mod. Phys.* **49**, 267 (1977). Also published in *Lecture Notes in Physics* **54** (Springer-Verlag), (1976), p. 276.
59. (with J. V. Jose, S. Kirkpatrick and D. Nelson) Renormalization, vortices, and symmetry breaking perturbations in the two-dimensional planar model, *Phys. Rev.* **B16**, 1217 (1977).
60. Connections between the critical behavior of the planar model and that of the eight-vertex model, *Phys. Rev. Lett.* **39**, 903 (1977).

61. A model for interacting quarks and strings, in *Proc. IUPAP Statistical Mechanics Conference*, Haifa, 1977.
62. (with H. Maris) Teaching the renormalization group, *Amer. J. Phys.* **46**, 625 (1978).
63. Lattice coulomb gas representation of two-dimensional problems, *J. Phys.* **A11**, 1399 (1978).
64. (with S. J. Shenker and A. Pruisken) A variational real space renormalization group transformation based on the cumulant expansion, *J. Phys.* **A12**, 91 (1979).
65. Multicritical behavior at the Kosterlitz-Thouless critical point, *Ann. Phys.* **120**, 39 (1979).
66. (with R. Ditzian) High temperature series expansion methods for Ising systems with quenched impurities, *Phys. Rev.* **19**, 4531 (1979).
67. (with A. Brown) Correlation functions on the critical lines of the Baxter and Ashkin-Teller models, *Ann. Phys.* **121**, 319 (1979).
68. (with R. Ditzian) Series studies of the four state Potts model, *J. Phys.* **A12**, L229 (1979).
69. (with M. Kohmoto) SMJ's analysis of Ising model correlation functions, *Ann. Phys.* **A126**, 371 (1980).
70. (with R. Ditzian, J. R. Banavar and G. S. Grest) Phase diagram for the Ashkin-Teller model in three dimensions, *Phys. Rev.* **22**, 2542 (1980).
71. Singularities near the bifurcation point of the Ashkin-Teller model, *Phys. Rev.* **B2**, 1405 (1980).
72. (with A. M. M. Pruisken) Marginality, universality and expansion techniques for critical lines in two dimension, *Phys. Rev.* **B22**, 5154 (1980).
73. (with M. Kohmoto) Lower bound RSRG approximation for a large η system, *J. Phys.* **A13**, 3339 (1980).
74. (with E. Fradkin) Disorder variables and para-fermions in two-dimensional statistical mechanics, *Nucl. Phys.* **B170**, 1 (1980).
75. (with A. N. Berker) Ground state entropy and algebraic order at low temperatures, *J. Phys.* **A13**, L259 (1980).
76. (with A. Zisook) Planar model correlation functions, *J. Phys.* **A13**, L379 (1980).
77. (with A. Zisook) Correlation functions on the critical line of the two-dimensional planar model: Logarithms and correlations to scaling, *Nucl. Phys.* **B180**, 61 (1981).
78. (with M. Kohmoto) Quantum mechanical ground states, non-linear Schrödinger equations, and linked cluster expansions, *J. Phys.* **A14**, 1291 (1981).
79. (with S. J. Shenker) Band to band hopping in one-dimensional maps, *J. Phys.* **A14**, L23 (1981).
80. (with M. Kohmoto and M. den Nijs) Hamiltonian studies of the $d = 2$ Ashkin-Teller model, *Phys. Rev.* **B24**, 5229 (1981).
81. (with M. Kohmoto and M. P. M. den Nijs) Connections among different phase transition problems in two dimensions, *Physica* **106A**, 122 (1981).
82. (with M. Kohmoto) Disorder variables for a non-abelian symmetry group, *Nucl. Phys.* **B190**, 671 (1981).
83. Many point correlation functions in a modified Ising model, *Phys. Rev.* **B24**, 5382 (1981).

84. (with S. J. Shenker) Critical behavior of a KAM surface: I. Empirical results, *J. Stat. Phys.* **27**, 631 (1982).
85. Scaling for a critical Kolmogorov-Arnold-Moser (KAM) trajectory, *Phys. Rev. Lett.* **47**, 1641 (1981).
86. Critical behavior of a KAM surface: II. Renormalization approach, in *Melting, Localization, and Chaos*, eds. R. K. Kalia and P. Vashista (North-Holland, New York, 1982).
87. (with M. Widom) Renormalization group analysis of bifurcations in area preserving maps, *Physica* **D5**, 287 (1982).
88. (with M. Feigenbaum and S. J. Shenker) Quasiperiodicity in dissipative systems: A renormalization group analysis, *Physica* **D5**, 370 (1982).
89. (with S. Howes and M. den Nijs) Quantum model for commensurate-incommensurate transitions, *Nucl. Phys.* **B215**, 169 (1983).
90. (with E. Domany, S. Alexander and D. Bensimon) Solutions to the Schrödinger equation on some fractal lattices, *Phys. Rev.* **B28**, 3110 (1983).
91. Supercritical behavior of an ordered trajectory, *J. Stat. Phys.* **31**, 1 (1983).
92. (with M. Kohmoto and C. Tang) Localization problem in one dimension: Mapping and escape, *Phys. Rev. Lett.* **50**, 1870 (1983).
93. (with M. Widom, D. Bensimon and S. J. Shenker) Strange objects in the complex plane, *J. Stat. Phys.* **32**, 443 (1983).
94. Roads to chaos, *Physics Today* **30**, 46 (1983).
95. Analysis of cycles for a volume preserving map, unpublished.
96. (with D. Bensimon) Extended chaos and disappearance of KAM trajectories, *Physica* **D13**, 82 (1984).
97. (with C. Tang) Escape from strange repellers, *Proc. Nat. Acad. Sci.* **81**, 1276 (1984).
98. Applications of scaling ideas to dynamics, in *Proc. Erice Summer School on Dynamics* (Springer-Verlag, 1984).
99. (with D. Bensimon and B. Shraiman) Mean-field theories for a ballistic model of aggregation, *Proc. Kinetics, Aggregation and Gellation Conf.*, Athens, Georgia, 1984).
100. From periodic motion to unbounded chaos: Investigations of the simple pendulum, *Physica Scripta* **T9**, 5 (1985).
101. (with J. R. Banavar and A. A. M. Pruisken) Energy spectrum for a fractal lattice in a magnetic field, *Phys. Rev.* **B31**, 1388 (1985).
102. Simulating hydrodynamics: A pedestrian model, *J. Stat. Phys.* **39**, 267 (1985).
103. (with S. Liang) Scaling in a ballistic aggregation model, *Phys. Rev.* **A31**, 2628 (1985).
104. (with P. Cvitanovic, M. H. Jensen and I. Procaccia) Renormalization, unstable manifolds and the fractal structure of mode locking, *Phys. Rev. Lett.* **55**, 343 (1985).
105. Fractal singularities in a measure and how to measure singularities on fractal, Cargese School (1985).
106. (with P. Cvitanovic, M. H. Jensen and I. Procaccia) Circle maps in the complex plane, *Proc. 6th Int. Symposium on 'Fractals in Physics'*, eds. L. Pietronero and E. Tosatti (North-Holland, 1985).

107. (with M. H. Jensen, A. Libchaber, I. Procaccia and J. Stavans) Global universality at the onset of chaos: Results of a forced Rayleigh-Bénard experiment, *Phys. Rev. Lett.* **55**, 2798 (1985).
108. (with D. Bensimon, S. Liang, B. I. Shraiman and C. Tang) Complex analytic methods for viscous flows in two dimensions, in *Proc. Mexican Summer School, 1985.* Also published in *Directions in Condensed Matter Physics*, eds. G. Grinstein and G. Mazenko (World Scientific, 1986), p. 51.
109. (with D. Bensimon, T. C. Halsey, M. H. Jensen, A. Libchaber, B. I. Shraiman and J. Stavans) More on microcanonical paradigms, Suppl. to *J. Irrep. Events* **A17**, 1597 (1985).
110. (with T. C. Halsey, M. H. Jensen, I. Procaccia and B. I. Shraiman) Fractal measures and their singularities: The characterization of strange sets, *Phys. Rev.* **A33**, 1141 (1986). See also erratum: *Phys. Rev.* **A34**, 1601E (1986).
111. Fractals: Where's the physics, *Phys. Today* **39**, 6 (1986).
112. (with D. Bensimon, S. Liang, B. I. Shraiman and C. Tang) Viscous flows in two dimensions, *Rev. Mex. Fisica* **32**, S101 (1986).
113. (with D. Bensimon and M. H. Jensen) Renormalization-group analysis of the global structure of the period-doubling attractor, *Phys. Rev.* **A33**, 3622 (1986).
114. Chaos: A view of complexity in the physical sciences, in *The Great Ideas Today* (Encyclopedia Brittanica, Chicago, 1986), p. 86.
115. On two levels, *Phys. Today* **39**, 7 (1986).
116. Computational physics: Pluses and minuses, *Phys. Today* **39**, 7 (1986).
117. Saffman-Taylor bubble problem, unpublished.
118. Cathedrals and other edifices, *Phys. Today* **39**, (1986).
119. Renormalization group analysis of the global properties of a strange attractor, *J. Stat. Phys.* **43**, 395 (1986).
120. On complexity, *Phys. Today* March (1987).
121. Dimensional calculations for Julia sets, in *Proceedings of European Condensed Matter Physics Congress*, *Physica Scripta* **T19**, 19-22 (1987).
122. (with D. Bensimon) The breakdown of KAM trajectories, published in Chaotic Phenomena in Astrophysics, *Annals of the New York Academy of Sciences* **497**, 110-16 (1987).
123. (with M. H. Jensen and I. Procaccia) Scaling structure and thermodynamics of strange sets, *Phys. Rev.* **A36**, 1409 (1987).
124. (with G. Zocchi, B. Shaw and A. Libchaber) Finger narrowing under local perturbations in the Saffman-Taylor problem, *Phys. Rev.* **A36**, 1894 (1987).
125. From neutrinos to quasiparticles, *Physics Today* 7-9 (1987).
126. (with G. McNamara and G. Zanetti) A poiseuille viscometer for lattice gas automata, *Complex Systems* **1**, 791-803 (1987).
127. (with M. Feingold and O. Piro) Diffusion of passive scalars in fluid flows: Maps in three dimensions, *A Way to Connect Fluid Dynamics to Dynamical Systems: Passive Scalars*, in *Fractal Aspects of Materials: Disordered Systems*, p. 203, eds. A. J. Hurd, D. A. Weitz and B. B. Mandelbrot (Materials Research Society, Pittsburgh, 1987). Also appears in *Universalities in Condensed Matter*, eds. R. Jullian, L. Peliti,

R. Rammal and N. Boccara, Springer Proc. Phys. (Springer, Berlin 1988). Transport of passive scalars: KAM surfaces and diffusion in three-dimensional Liouvillean maps. Instabilities and nonequilibrium structures, eds. E. Tirapequi and D. Villarroel (D. Reidel, Dordrecht), p. 37 (1989).

128. The big, the bad and the beautiful, *Physics Today* **9** (1988).
129. (with M. Feingold and O. Piro) Passive scalars, 3D volume preserving maps and chaos, *J. Stat. Phys.* **50**, 529 (1988).
130. Interactive computation for undergraduates, *Phys. Today* **41**, 9-11 [Reference Frame] (1988).
131. (with G. McNamara and G. Zanetti) From automata to fluid flow: Comparisons of simulation and theory, *Phys. Rev.* **A40**, 4527 (1989).
132. (with B. Castaing, G. Gunaratne, F. Heslot, A. Libchaber, S. Thomae, X.-Z. Wu, S. Zaleski and G. Zanetti) Scaling of hard thermal turbulence in Rayleigh Bénard convection, *Fluid Mechanics* **209**, 1-30 (1989).
133. (with S. R. Nagel, L. Wu and S.-M. Zhou) Scaling and universality in avalanches, *Phys. Rev.* **A39**, 6524-6537 (1989).
134. Fractals and multifractals in avalanche models, *Physica* **D38**, 213-214 (1989).
135. Scaling and structures in the hard turbulence region of Rayleigh Bénard convection, *Proceedings of Newport Conference on Turbulence* (1989), *Proceedings of Europhysics Conference on Turbulence* (Moscow, 1989).
136. Scaling and universality in statistical physics, *Physica* **A163**, 1-14 (1990).
137. (with W.-S. Dai and S.-M. Zhou) Singularities in complex interface dynamics, *Proceedings of Cargese Conference on Nonlinear Flow Problems* (1990).
138. (with X.-Z. Wu, A. Libchaber and M. Sano) Frequency power spectrum of temperature fluctuations in free convection, *Phys. Rev. Lett.* **64**, 2140-2144 (1990).
139. (with P. Constantin) Singularities in complex interfaces, *Phil. Trans. R. Soc. London* **A333**, 379-389 (1990) .
140. Exact solutions for the Saffman-Taylor problem with surface tension, *Phys. Rev. Lett.* **65**, 2986-2988 (1990).
141. Scaling and multiscaling (fractals and multifractals), *Proceedings of Second Latin-American Workshop on Nonlinear Phenomena* (1990).
142. Chaos and complexity: The results of non-linear processes in the physical world — Springer proceedings in physics, Volume 57, *Evolutionary Trends in the Physical Sciences*, eds. M. Suzuki and R. Kubo (Springer-Verlag, 1991).
143. (with P. Constantin) Dynamics of a complex interface, *Physica* **D47**, 450-460 (1991)
144. (with W.-S. Dai and S.-M. Zhou) Interface dynamics and the motion of complex singularities, *Phys. Rev.* **A43**, 6672-6682 (1991).
145. (with Sue Copppersmith, M. J. Vinson, A. J. Kolan and Scott Wunsch) Chaos, computers, and physics, Laboratory Notes for Physics 251 (1987, 1990, 1991,1997).
146. (with I. Procaccia, E. Ching, P. Constantin, A. Libchaber and X.-Z. Wu) Transitions in convective turbulence: The role of thermal plumes, *Phys. Rev.* **A44**, 8091 (1991).
147. (with A. Libchaber, E. Moses and G. Zocchi) Turbulence in a box, *La Recherche* **22**, 628-638 (1991).

148. (with G. Vasconcelos) Stationary solutions for the Saffman-Taylor problem with surface tension, *Phys. Rev.* **A44**, 6490-6495 (1991).
149. Complex structures from simple systems, *Physics Today* **9**, March (1991).
150. Scaling and multiscaling: Fractals and multifractals (Review), *Chinese Journal of Physics* **29**, 613-635 (1991).
151. Scaling in hydrodynamics, to be published in *Proceedings of Cargese Summer School* (Plenum, 1991).
152. (with C. Amick, S. C. E. Ching and V. Rom-Kedar) Beyond all orders: Singular perturbations in a mapping, *J. Nonlinear Science* **2**, 9-68 (1992)
153. (with E. S. C. Ching, A. Libchaber and X.-Z. Wu) Turbulent convection in helium gas, *Physica* **D58**, 414-422 (1992).
154. Hard times, *Physics Today* (October 1992).
155. (with A. B. Chhabra, A. J. Kolan, M. J. Feigenbaum and I. Procaccia) Critical indices for singular diffusion, *Phys. Rev.* **A45**, 6095-6098 (1992).
156. (with P. Constantin, T. F. Dupont, R. E. Goldstein, M. Shelley and S.-M. Zhou) Droplet breakup in a model of the Hele-Shaw cell, *Phys. Rev.* **E47**, 4169-4181 (1993).
157. (with A. B. Chhabra, M. J. Feigenbaum, A. J. Kolan and I. Procaccia) Sandpiles, avalanches and the statistical mechanics of non-equilibrium stationary states, *Phys. Rev.* **E47**, 3099-3121 (1993).
158. (with M. J. Vinson and A. J. Kolan) Chaos, computers, and physics, *Laboratory Notes for Physical Science* 114, (1993).
159. *From Order to Chaos: Essays Critical Chaotic and Otherwise* (World Scientific, 1993).
160. (with T. F. Dupont, R. E. Goldstein and S.-M. Zhou) Finite-time singularity formation in Hele-Shaw systems, *Phys. Rev.* **E47**(6), 4182-4196 (1993).
161. (with V. Rom-Kedar, E. S. C. Ching and C. Amick) The break-up of a heteroclinic connection in a volume preserving mapping, *Physica* **D62**, 51-65 (1993).
162. (with S. Boatto and P. Olla) Traveling-wave solutions to thin-film equations, *Phys. Rev.* **E48**(6), 4423-4431 (1993).
163. (with A. Bertozzi, M. Brenner and T. F. Dupont) Singularities and similarities in interface flows, in *L. Sirovitch Editor, Trends and Perspectives in Applied Mathatematics*, Springer-Verlag Applied Math Series, Volume 100 (1994), pp. 155-208.
164. Conditional averages in convective turbulence, *Physica* **A204**, 341-345 (1994).
165. (with D. H. Rothman) Bubble, Bubble, Boil, and Trouble, *Computations in Physics* **8**, 199-204 (1994).
166. Greats, reference frame, *Physics Today* April (1994).
167. (with Y. Du and H. Li) Breakdown of hydrodynamics in a one dimensional system of inelastic particles, *Phys. Rev. Lett.* **74**(8), 1268-1271 (1995).
168. (with D. Lohse, J. Wang and R. Benzi) Scaling and dissipation in the GOY shell model, *Phys. Fluids* **7**(3), 617-629 (1995).
169. A model of turbulence, reference frame, *Physics Today*, September (1995).
170. (with N. Schorghofer and D. Lohse) How the viscous subrange determines inertial range properties in turbulence shell models, *Physica* **D88**, 40-54 (1995).
171. (with D. Lohse and N. Schörghofer) Scaling and linear response in the GOY turbulence model, *Physica* **D100**, 165-186 (1997).

172. (with T. Zhou) Inelastic collapse of three particles, *Phys. Rev.* **E54**, 623 (1996).
173. (with V. Emsellem, D. Lohse, P. Tabeling and J. Wang) Transitions and probes in helium turbulence, *Phys. Rev.* **E55(3)**, 2672-2681 (1997).
174. Turbulent excursions, fluid motion, news and views, *Nature*, Vol. 382, July 11 (1996).
175. Cascade models of turbulence and mixing, *Tr. J. of Physics* **21**, 1-14 (1997).
176. (with S. N. Coppersmith, T. C. Jones, A. Levin, J. P. McCarten, S. R. Nagel, S. C. Venkataramani and X. Wu) Self-organized short-term memories, *Phys. Rev. Lett.* **78**(21), 3983-3986 (1997).
177. (with S. Wunsch and T. Zhou) Scaling properties of passive scalars in one dimension, *Physica* **A244**, 190-212 (1997).
178. (with N. Goldenfeld) Can you tell the difference between a stock market crash and a black hole? for *NumeriX www site*, unpublished (1997).
179. Built upon sand: Theoretical ideas inspired by the flow of granular materials, submitted (1998).
180. Blowups and singularities, reference frames, *Physics Today*, September, pp. 11-12 (1997).
181. (with M. Brenner, P. Constantin, A. Schenkel and S. C. Venkataramani) Diffusion, attraction, and collapse, in preparation (1998).

General Introduction: The Worlds of Science

This book is a selection of my research and popular essays, with particular emphasis on works which review or discuss in a general way some scientific or technical question. The papers are all about the world of science, or rather about the different worlds in which a scientist works. In my own work I can see at least four different kinds of things which might be meant when one talks about the worlds of a scientist.

First, I might point to a little society or social grouping composed of scientists and a few associates. This **social world** defines the group in which we work and exchange ideas. This is the audience for our papers, the source of our applause, and our critics and competition. A scientist can go to different places all over the world and see mostly just the usual group of associates. A scientist can be thrown into a new little group, formed in an allied field of scientific endeavor, and immediately recognize the society and the social norms. This little world is close and closed. It defines our successes and failures.

But there is in addition a more intimate social group which defines our work. Much scientific work is done in direct collaboration with other scientists. Many of the papers in this volume have several authors. Typically each author brings a slightly different experience and point of view to the joint effort, so that the eventual product is much better than would have been produced by any single person.

This fact came home to me at the very beginning of my career as an 'independent' scientist. Gordon Baym and I had both been trained at Harvard, he under Schwinger and myself under Paul Martin and Roy Glauber. He had learned how to apply variational methods to the derivation of Green's function approximations. I was working on the development of approximations which built in some thermodynamic and conservation laws. With huge effort over a period of months, I had derived one or two approximations which fit my criteria. A day after I had described to him what I had done, he showed me how to construct an infinite number of new approximations which fit into the general scheme. The results appeared in part in our book *Quantum Statistical Mechanics* and in part in our paper, which appears as #5 in my publication list.[1] Two heads had done a lot more than one.

My scientific life has contained many other very fruitful collaborations. I describe some of these in the introductory essays which head the various sections of this book.

But, we scientists also work in a very different kind of tight little world, the artificial little world constructed by our ideas. Some of my recent work has been related to the development of models for the behavior of avalanches or sand slides. To construct this **model world**, we considered a simplified example in which square or cubic grains of sand were stacked in neat piles. If a given pile overtopped its neighbors by more than a specified amount, then several grains would fall onto the neighboring stacks. Clearly, this model represented a totally artificial oversimplification of any picture of the behavior of real sand. Nonetheless, a whole group of us threw ourselves quite wholeheartedly into the study of this little model. For several years at a time, we took this artificial example and pretended

[1] The publication list can be found at the end of this book. In general, items in this list will be given by a number prefixed by the sign '#'.

it was the whole universe.[2] We examined this tiny world with the same seriousness that one might examine the history of the British Empire, or a science of the human mind. Our goal was to develop and understand the laws which governed behavior in this tiny closed-off cosmos.

Why should serious people study such inadequate toys? Clearly these toys cannot accurately represent the third type of scientific world, the **real world** in which we work and live. Nonetheless it might be profitable to study such hermetic little model worlds because perhaps the experience developed in the little world can be extended and applied to our real world. Maybe there is something in our model avalanche which can be carried over and give some deep insight into how avalanches work. Perhaps these ideas might even have some practical use in the protection of Swiss mountain villages or to the design of particle detectors. Maybe not. Probably not. But one cannot tell what might be carried to the real world until the model world is examined and understood.

Science also sees another version of the real world, the world of people. People and the society support science. Naturally some return is demanded. One demand is that science generates ideas and concepts which can be meaningful to the public at large and can catch its imagination or satisfy its curiosity. To realize this goal, we scientists must be teachers in the broadest sense Another demand is that we, from time to time, satisfy the aspirations of society for better technology or for a better understanding of the applications or limits of technology. We can only occasionally help the society in this direction, but our help can be quite crucial. We have served in the development of weapons, of communications, and of health care. We cannot be sure where we will be needed in the future, but we are required to be alert to ways in which we can serve.

The fourth version of the world of the scientist is the most important and the one which we scientists most vividly experience. What is it that one really transfers from the oversimplified model world to the complex world of reality? Clearly the medium of exchange is ideas. One carries some concept of how an avalanche must work. Some idea. And then one takes the idea and asks how well that idea agrees with mundane reality. (But reality is in most cases richer, more beautiful, and more fertile than our imaginations. So in most cases, this comparison enriches rather than just checks our ideas.) The results of all the model building, all the comparison with reality, and all the back and forth of scientific exchange is a set of concepts which can then be applied to other situations. Our final outcome then is something which can be added to the **world of ideas**.

The series of essays in this volume relates in some degree to all four worlds.

I cannot imagine anyone who would wish to go through all this material from beginning to end. So let me take the reader for a walking tour through this material so that he or she might plan a particular path which might prove pleasing or useful.

This book is divided into five parts, which I shall describe in inverse order. The last part, **Complex Patterns** is concerned with the source and nature of complexity in physical systems. This subject is in the process of development. However, some of the important ideas in it have already become apparent. In one view, this subject starts from a question: Given that the laws of physics are simple and predictive, how can we have a world which

[2]In this case, the collaborative team included Ashvin B. Chhabra, Mitchell Feigenbaum, Amy Kolan, Sidney Nagel, Itamar Procaccia, Lei Wu, and Su-Min Zhou.

is so complex and apparently unpredictable. The question is clearly in the world of ideas. To answer it, one turns to the development of mathematical models and of real physical systems, in both cases looking for simultaneous simplicity and complexity.

The work described in the previous study arose naturally from previous studies of turbulence and chaos in fluid systems, as described in the section **Turbulence and Chaos**. This section is largely concerned with the development of mathematical theories of chaos. These theories are intended to live in the real world, since they are designed to confront real experimental data. So the somewhat philosophical questions about complexity are intended to be answered in a context in which one is informed by actual observations of nature.

This volume's third section describes my introduction to complexity. During the late 1960s, it became fashionable for scientists to look away from the traditional applications of their research to military systems, and to focus instead upon problems which might be relevant to the broader needs of society. In the U.S. National Science Foundation, this relevance boom even gave rise to a new program RANN, Research Applied to National Needs.[3] In any case, the same social forces which pushed the NSF toward RANN pushed me toward studying the complexity of forces which shape the physical and social environment of our urban areas. This part of the book, **Simulations**, **Urban Studies**, and **Social Systems**, includes the outcome of this effort. It also includes some editorial pieces written for *Physics Today* which report upon the health and decay in another kind of social system, the one of physics itself.

The second section in this book come from my best and most important contribution to science, my work on the understanding of **Scaling and Phase Transitions**. Here I played a part in the invention of a tiny world, the critical system, and then devoted considerable effort to studying the detailed properties of that world. In this period, many different and very intelligent people focused an amazing amount of effort upon a very closed and partial model of reality. Despite the limited focus of the work, it has had its consequences. For more details about what happened, I ask you to look at the introduction to that section.

The book's first section, **Fundamentals Issues in Hydrodynamics, Condensed Matter and Field Theory**, is devoted to describing the relationship among different models, either in general or in particular examples. Physics contains many different models, which might describe different aspects of the very same physical system. Clearly, one should ask how the different realities caught by the different models fit together. Some of the essays in this section directly confront this general issue, others ask how it is resolved in a particular physical example. A major theme of this section is that physics is really about how models, which give different levels of physical description, may fit together.

Each section in the book is headed by a specific introduction which goes into more details on the questions mentioned here, and outlines some of the contents of the papers.

There is an overall theme to the whole book. Each of the first three sections is devoted to 'old' scientific questions, questions which have been asked and mostly answered. One can

[3]This program was in large measure designed and put into place by Joel Snow. Our recent elected officials have had many unkind words for civil servants. In my experience, I have found governmental science administrators to be thoughtful, hard-working people. As a group, they have contributed a lot to science without getting much thanks.

expect that each field of science starts from questions built upon small piece of the world, and that in time these questions will become answered as well as the times and means available permit. Then, the subfield gets mined out. Naturally, the scientist must then direct his or her work away from these particular aspects of reality. Naturally, also, when that happens there is some temptation to see and bemoan 'the end of science'. Perhaps this is particularly tempting right now for a physicist since some of our most exciting problems have either been solved or run into major technical barriers.

However, in the introduction to the last section, I shall argue that we have in front of us a mostly uncharted territory, concerning the development of complexity in the world. The ideas in this part of science is connected with understanding the relationship and linkage among the worlds we have explored. Right now, I see not an end for physics but a beginning. Now is an exciting time to be working in physics.

Section A

Fundamental Issues in Hydrodynamics, Condensed Matter and Field Theory

From Level to Level

This section contains a collection of essays about apparently disconnected subjects in field theory, condensed matter physics, and hydrodynamics. There is, however, a thread of connection among all these different essays. In each case, we ask some kind of question about the relation between different levels of description of the physical world. This is a very natural question for a person trained as I am, in the area of condensed matter physics. My thesis advisers, Paul Martin and Roy Glauber, continually directed my attention to the relation between a microscopic description of reality and a macroscopic description. Thus a gas is composed of molecules, but it also obeys the laws of fluid mechanics. A microwave cavity contains not only photons but also an electric field. Or again, a fluid near its critical point is a bunch of molecules, but they also be described by a scale-invariant field theory.

In my imagination I see every physics problem as a kind of little world. Each world has its own rules, which apply to the description at that level. This idea is brought out in the paper[4] A1, *On Two Levels*, which looks for a more macroscopic, hydrodynamic, level of description within a computer game in which simplified particles go through a kind of dance. Naturally, one sees the dancers by looking very closely, while a more lumped description shows the hydrodynamic flow. In this paper the logic is one in which the more lumped description is built up from the more microscopic one. This same point is made again, backwards, in the next essay. Here Paul Martin and I start from the equations of motion of hydrodynamics, and look at local fluctuations to gain a more microscopic description of fluid flow. In this case, one and the same physical description covers two quite different ranges of physical size and physical phenomena.

There is something more to say about this paper. A parallel computation was done by Landau and Placzek, long before Paul Martin and I wrote our piece. I do not recall ever having seen this parallel effort. Nonetheless its existence serves as a reminder that rarely do we produce something completely new in science. Every piece is based upon predecessors, and if we did not do the work, it still probably would be produced by others in substantially similar form. This point was brought home to me in my work on electrons and phonons. Essay A3 in this volume is a review piece, based upon the research described in paper #14 on my publication list. The latter is a joint work by Richard Prange and myself. It arose in that form because Prange and I did substantially identical, independent research on this topic. His work was written up first, and appeared one morning in my mailbox. After I called him and described my own thinking about this subject, he generously suggested that we write it up together. During these many years, I have felt myself in his debt for this fine courtesy.

[4]Each reference to a paper which appears in this volume is given as a letter followed by a number. The letter denotes the section in this volume while the following number gives the placement of the paper in the section.

Paper A4 is partially a reprise of the ideas of A3 but now applied to superfluid motion rather than the normal state.[5] It is illustrative of an important idea about the relation between different levels of experience. The basic microscopic forces and interactions are quite the same in normal materials as in superfluids. However, the nature of the physical state of the two systems is vastly different. The superfluid forms a condensate in which all the particles in the system cooperate to produce a single quantum state. This condensation extends across the entire spatial extent of the system and produces a change in behavior which spans both the micro and the macro levels of description. The superfluid is thus essentially different from its normal state counterpart in both worlds of description.

A qualitative change in the collective behavior of a large group of particles is called a phase transition. Often a phase transition is accompanied by a change in the symmetries shown by the system. The paper on superfluidity shows in one example how phase transitions may manifest themselves equally in the microscopic and macroscopic domains. Paper A5 reviews Kenneth Wilson's work on symmetry changes and phase transitions in systems of quarks and strings. This world in strong interactions is described by a theory, *quantum chromodynamics*, in which quarks are the fundamental particles. The basic question here is how can one have a description of quantum chromodynamics in terms of almost free quarks, and yet not directly see quarks in our world. The answer is that the confinement is a concept which describes the macro level, the world of ordinary nuclear physics, while the micro-level is correctly given by quarks only weakly coupled by gluons. This work that Wilson started has grown into the subbranch of particle and nuclear physics called lattice gauge theory.

This review, A5, is one of many papers I have written to describe and perhaps help explain the work of other physicists. I have always felt enlarged when I could work with the beautiful and deep ideas of others. In fact, many of my papers were intended to be in part or in whole explanations or descriptions of beautiful ideas of others. Thus, for example, my paper #57 on my publication list extends the work of A. A. Migdal,[6] #59, written with J. Jose, S. Kirkpatrick, and D. Nelson, is partially built to convince the reader of the correctness of a theory due to Kosterlitz and Thouless,[7] #133 addresses the ideas of Bak, Tang, and Wiesenfeld,[8] while in #69 Mahito Kohmoto and I review the work of Sato, Miwa, and Jimbo. Of course, this recognition of others is not entirely unselfish. I have always held the opinion that people are more likely to recognize your work if your recognize theirs.

In paper A6, Eduardo Fradkin and I explain ideas from the theory of critical phenomena and phase transitions in a way which has proved fruitful in a variety of other fields. This paper shows how 'particles' with very peculiar behavior arises from a treatment of two

[5]Another important thread of my research effort in the earlier part of my career was concerned with understanding the relationship between the microscopic and the macroscopic properties of superconductors and superfluids. These include, for example, the collaborative works with P. C. Martin (#1, #2, and #6) on my publication list, with Vinay Ambegaokar (#4), with Brian Pippard (#20) and with my students D. Markowitz (#10), Igor I. Falko (#16), G. Laramore (#31), and Jack Swift (#32).

[6]A. A. Migdal, *Zh. Eksp. Teoret. Fiz.* **69**, 810, 1475 (1975).

[7]J. M. Kosterlitz and D. Thouless, *J. Phys.* **C6**, 1181 (1973); J. M. Kosterlitz, *J. Phys.* **C7**, 1046 (1974). See also the work of V. L. Berezinskii, *JETP* **32**, 493 (1971) and *JETP* **34**, 610 (1971).

[8]P. Bak, C. Tang, and K. Wiesenfeld, *Phys. Rev. Lett.* **59**, 381 (1987); *Phys. Rev.* **A38**, 364 (1988).

coupled Ising models. It is one of the first treatments of fractional statistics particles, the so-called anyons, which have had a vogue in condensed matter physics and field theory. This paper has perhaps received a bit less attention than it deserved. It is part of a long series of papers aimed at describing the mathematical structure of two-dimensional statistical mechanics. These will be discussed in more detail in the introductory essay for Section B.

In my work, as in the work of most physicists, a major goal has been the building of bridges between different worlds of experience. Science has two traditional windows for looking at the world, through experimental study or theoretical construction. In recent decades a third window has opened, which approaches reality through the construction of computer models. Paper A7 is an attempt to describe some of the great accomplishments of computer-based physics. It argues that the importance of this relatively new tool is that it can be effectively used in conjunction with the older tools of analysis and experiment. If I were writing this column now, I would say the same thing in a more negative way by pointing out that computer-based physics can be ineffective when it constructs totally closed worlds, which then cannot be related to the richer perspectives of theory and of experiment. However, now as then, I believe that this new tool can be properly used to either explore some totally new area of behavior, or alternatively to check a precisely formulated idea.

From the earliest period of my work I have made use of computers. For example, an applied paper[9] on heat transfer in semi-transparent materials contained a solution to a radiation transfer equation obtained with the aid of a computer. Later on Jack Swift and I, see paper B4, described a computer algorithm which might be used to simulate the approach to equilibrium in a system obeying statistical mechanics.[10] Abdullah Sadiq later implemented this idea in his PhD thesis. In more recent years, my students, my coworkers and I have made a major effort to use the computer as a device to generate ideas which could be checked against theory and experimental reality. We have concentrated on efforts on small computers, because we found these to be truly flexible tools for studying a tiny portion of reality.

Paper A7 was written for the Reference Frames column in *Physics Today*. Gloria Lubkin, the editor, had the idea of such a column, describing personal views of the worlds of physics. She and I worked together in producing these columns. I have had much good feeling about them. I think that they are rather successful, in large measure because of her enthusiasm and good sense. This section ends with two columns, A8 and A9, intended to assess some of the worlds constructed by scientists and how they are related to one another. The first does this by looking at physics as a whole; the second looks at an example from hydrodynamics, condensed matter physics, and elementary particle experiment.

[9]This joint work with Henry Hidalgo is #8 on my publication list.

[10]This approach use the method of *cellular automata*. Later on, this approach was very much improved by others. See paper A1 for a discussion of some of its implications. After a while I returned to this subject in work done jointly with McNamara and G. Zanetti, #126 and #131. We contributed to the further development and checking of the applicability of such models to describe equilibrium and nonequilibrium situations. The avalanche work already mentioned uses another kind of automaton.

reference frame

On two levels

Leo P. Kadanoff

Some of the most interesting situations in physics, and indeed in other sciences as well, concern the connections between two "levels of reality." How does the presumed world of strings connect with the more observable world of quarks and gluons? How do quantum problems "go to" their classical limits? How does the irreversibility of the macroscopic world connect with the known time-reversibility of microscopic description? In each of these cases, there is a tension between two levels of description. For each situation, different laws, formulations, conceptualizations, theories and experiments apply at each of the two levels.

The motion of particles in a fluid gives an interesting example of such a "two-level system." In this column I will discuss this motion using a variant of a model originally proposed by J. Hardy, O. de Pazzis and Yves Pomeau. (For some recent papers on the version of this model discussed here, see the references on page 9.) On level 1, the rules are that particles hop in a simple deterministic fashion upon a two-dimensional lattice. One obtains level 2 by averaging the velocities of many particles in a given region, thereby getting a continuum description of the system in terms of an averaged flow velocity $\mathbf{u}(\mathbf{r},t)$. In moving back and forth between the levels, one might wish to know how to derive the continuum equations from the rules or, conversely, how to derive the simplest rules that yield a prescribed continuum theory.

Real flows can be very rich and beautiful. Figure 1 shows a laboratory experiment involving the flow of a fluid past a cylinder. The flow is made visible by the injection of smoke particles into the fluid. In the depicted flow the fluid moves at a relatively low velocity. At higher speeds the pattern becomes even more intricate.

Figure 2 depicts a different world, containing particles in motion. The particles in this "hexagonal lattice gas" are distinguished by the six possible directions of their "momenta." The momenta are vectors of equal length directed parallel to the basic axes of the triangular lattice. One calculates the total momentum (or average velocity) for a group of particles by finding the sum (or average) of these momentum vectors for the individual particles. The model is an algorithm that tells you how the particles move about and change their momenta. The time development comes about through two types of steps. In the first type of step, which occurs between 2a and 2b, each particle moves one lattice constant in a direction determined by the direction of its momentum. This kind of step is immediately followed by a step of the next type, in which particles undergo collisions that conserve momentum and the number of particles. In the version depicted, the particles collide when the total momentum of the particles on a given site is zero. Then, as one can see by comparing figures 2b and 2c, these particles change their momenta by a counterclockwise rotation through 60°. The algorithm simply repeats the two steps again and again.

Of course, I am about to argue that the particle system of figure 2 will, on the continuum level, generate flows like the one in figure 1. How do I know? The references give both theoretical and "experimental" support for this belief. On the theoretical side, one knows real fluids are described by a set of relations called the Navier–Stokes

continued on page 9

Leo Kadanoff, a condensed-matter theorist, is John D. MacArthur Professor of Physics at the University of Chicago.

Fluid flow behind a circular cylinder (in a wind tunnel) is made visible by smoke filaments; the Reynolds number is about 300. Photograph by Peter Bradshaw (Imperial College, London), reproduced in Milton Van Dyke, *An Album of Fluid Motion* (Parabolic Press, Stanford, Calif., 1982). Figure 1

reference frame

equations. These express the space–time dependence of the local fluid velocity **u** and pressure p in the form

$$\partial \mathbf{u}/\partial t = -(\mathbf{u}\cdot\nabla)\mathbf{u} - \rho^{-1}\nabla p + \nu\nabla^2\mathbf{u}$$
$$\nabla\cdot\mathbf{u} = 0$$

The first of these equations is essentially an expression of local momentum conservation (or $\mathbf{F} = m\mathbf{a}$) for a situation in which the fluid's mass density ρ and its viscosity ν are constant. According to fluid mechanics, the density is held constant whenever flow velocities are very small in comparison with the speed of sound. The zero-divergence condition is then simply an expression of local mass conservation. The Navier–Stokes equations thus arise as a consequence of the real fluid's conservation laws. Because the "right" conservation laws are built into the lattice-gas model, in the appropriate limit the model should also obey the Navier–Stokes equations. (The symmetries are important too, but the hexagonal symmetry gives the correct continuum properties.)

The other evidence for the connection between real fluids and the lattice gas comes from computer simulations. For example, figure 3 shows the result of a computer simulation of roughly the same physical situation as that depicted in figure 1. The two pictures show qualitatively similar patterns of swirls behind the moving cylinder. Work in progress is aimed at quantitative comparisons.

I find it fantastic and beautiful that the tiny, trivial world of the lattice gas can give rise to the intricate structures of hydrodynamic flow. The physical universe is also wonderfully simple at some levels, but overpoweringly rich in others.

The exact connections between the two levels of this model system will be worked out in the next few years. In addition, the approach may turn out to be of some engineering and technical importance. Knowing flow patterns is necessary for designing an airplane, missile, ship or chemical plant, or for controlling air, water or soil pollution. Most predictions of complex flows will come from simulations. (Real experiments will also be quite necessary because they probably provide the best atmosphere for exploratory work.) It is possible that lattice-gas models might be an efficient tool for performing the required calculations. The relative efficiency of this method and of more standard methods of solving nonlinear partial differential equations is a matter for present debate and future investigation. The outcome of this competition among calculational methods may well depend upon future developments in parallel-processing hardware and software and in programming technique, and perhaps upon discovering more about the basic physics of flow processes.

Some recent references

- U. Frisch, B. Hasslacher, Y. Pomeau, Phys. Rev. Lett. **56**, 1505 (1986).
- S. Orszag, V. Yakhot, Phys. Rev. Lett. **56**, 1691 (1986).
- N. Margolis, T. Toffoli, G. Vichniac, Phys. Rev. Lett. **56**, 1694 (1986).
- J. Salem, S. Wolfram, in *Theory and Applications of Cellular Automata*, S. Wolfram, ed., World Scientific, Singapore (1986), p. 362. □

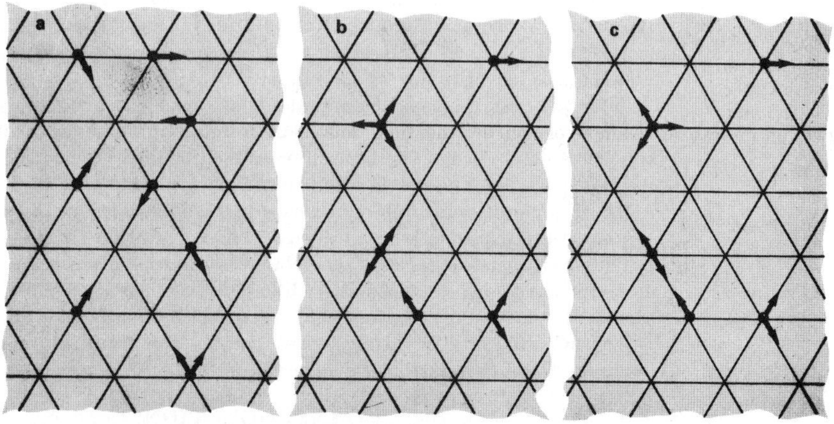

A lattice-gas model. Parts a, b and c show successive stages in the history of the gas; the arrows show the directions of the momenta. Note that more than one particle can occupy a lattice site. Figure 2

Flow behind a moving cylinder in the lattice-gas model of figure 2. The arrows show averaged local velocities; the Reynolds number is about 100. Simulation ©1986 by Bruce Nemnich, J. Salem and Stephen Wolfram. Figure 3

ANNALS OF PHYSICS: **24**, 419–469 (1963)

Hydrodynamic Equations and Correlation Functions

LEO P. KADANOFF

Physics Department, University of Illinois, Urbana, Illinois

AND

PAUL C. MARTIN

Lyman Laboratory of Physics, Harvard University, Cambridge, Massachusetts

The response of a system to an external disturbance can always be expressed in terms of time dependent correlation functions of the undisturbed system. More particularly the linear response of a system disturbed slightly from equilibrium is characterized by the expectation value in the equilibrium ensemble, of a product of two space- and time-dependent operators. When a disturbance leads to a very slow variation in space and time of all physical quantities, the response may alternatively be described by the linearized hydrodynamic equations. The purpose of this paper is to exhibit the complicated structure the correlation functions must have in order that these descriptions coincide. From the hydrodynamic equations the slowly varying part of the expectation values of correlations of densities of conserved quantities is inferred. Two illustrative examples are considered: spin diffusion and transport in an ordinary one-component fluid.

Since the descriptions are equivalent, all transport processes which occur in the nonequilibrium system must be exhibited in the equilibrium correlation functions. Thus, when the hydrodynamic equations predict the existence of a diffusion process, the correlation functions will include a part which satisfies a diffusion equation. Similarly when sound waves occur in the nonequilibrium system, they will also be contained in the correlation functions.

The description in terms of correlation functions leads naturally to expressions for the transport coefficients like those discussed by Kubo. The analysis also leads to a number of sum rules relating the dissipative linear coefficients to thermodynamic derivatives. It elucidates the peculiarly singular limiting behavior these correlations must have.

I. INTRODUCTION

Although the general nonequilibrium behavior of a many-particle system is exceedingly complex, there exists a well-developed—and relatively simple—theory of nonequilibrium behavior for situations in which physical quantities vary extremely slowly in space and time. The best known example of such a theory is ordinary fluid mechanics. The full nonequilibrium problem involves virtually infinite complexity; on the other hand, the hydrodynamical limit is simply characterized by five partial differential equations.[1]

The simplification occurs because when all physical quantities vary slowly in space and time each portion of the system is almost in thermodynamic equilibrium. Under these conditions, the variation in the system is completely described by local values of the various thermodynamic variables—for example, by giving the pressure, density, and velocity as a function of space and time. The basis of fluid mechanics is the partial differential equations satisfied by these local thermodynamic quantities.

In these hydrodynamic equations, there appear a variety of parameters whose values are not given by fluid mechanics. These parameters fall into two categories. First, there are the thermodynamic derivatives which arise because changes in the various local variables are related by thermodynamic identities. Second, there are the transport coefficients like viscosity and thermal conductivity which enter because the fluxes of thermodynamic quantities contain terms proportional to the gradients of the local variables. To find the values of the transport coefficients and thermodynamic derivatives, we must turn to a more fundamental theory than fluid mechanics.

Recently, it has been appreciated that time-dependent correlation functions afford a powerful theoretical tool for investigating nonequilibrium behavior. Indeed a wide variety of nonequilibrium phenomena are described by thermodynamically averaged expectation values of products of pairs of densities of conserved quantities at different space-time points. In particular these correlation functions completely describe the nonequilibrium behavior of a system in which the deviation from equilibrium is small. Since, in principle, we know how to compute these equilibrium-averaged (*2*, *3*) time-dependent correlation functions, we are in principle able to completely determine the behavior of a system slightly disturbed from equilibrium. Specifically, a calculation of the time dependent correlation functions must lead both to the hydrodynamic equations and the numerical values of all the thermodynamic derivatives and transport coefficients.

In practice the computational difficulties involved in evaluating correlation functions are nontrivial. Indeed, the part of the correlation function which varies slowly in space and time and reflects the hydrodynamic equations is the most difficult part to compute.

The reason for this difficulty is easy to see. The hydrodynamic equations refer to a system in local thermodynamic equilibrium. This local equilibrium is produced and enforced by the frequent collisions between particles.[2] So, the hydrodynamic equations refer to a situation in which the behavior of the system is

[1] This simplification is called a contraction of the description. It is discussed in ref. *1*.

[2] A description of how this occurs is given in ref. *1* and also in ref. *4*.

dominated by collisions. On the other hand, the conventional methods (*2*, *3*) for computing correlation functions are based, in one sense or another, on a expansion in some parameter describing the number of collisions in the system. This parameter is most often the strength of the interparticle potential. Since the hydrodynamic equations only appear when the behavior is dominated by the secular effects of collisions, the most straightforward techniques for determining the correlation functions cannot be successfully applied to the prediction of hydrodynamic phenomena.[3]

In this paper, which is largely pedagogical, we shall be primarily concerned with using the hydrodynamic equations to learn about the correlation functions. Our analysis will bear on the inverse problem, the derivation of the hydrodynamical equations, mainly in a negative way. We shall see that the correlation functions must exhibit complicated singular behavior at long wavelengths and low frequencies. This behavior, which does not result in each order of perturbation theory, indicates the necessity for determining successive approximations through iterative integral equations, or equivalently through extensive resummation of perturbation expansions.

We first consider the simplest example of a transport process: spin diffusion. In this case the only hydrodynamic equation is a diffusion equation for the spin magnetization. From this hydrodynamic equation, we determine the form of the slowly varying part of the magnetization-magnetization correlation function. The hydrodynamic description which involves the spin susceptibility (a thermodynamic derivative) and the spin diffusion coefficient (a transport coefficient) enters into the correlation function. By comparing this result with the correlation function description we find how the correlation function determines both the thermodynamic derivative and the transport coefficient.

A very similar type of analysis is applied to the algebraically more complex case of transport in a single component fluid. Here, the linearized form of the usual equations of fluid mechanics serve as the hydrodynamic equations. The form of the correlation functions composed of the densities of conserved operators (number, energy, and momentum) are again determined from these hydrodynamic equations. In particular, it is shown how the correlation functions yield the various thermodynamic derivatives and the formulas discussed by Kubo (*6*) and many other people (*7*) for the relevant transport coefficients: the viscosity, the bulk viscosity, and the thermal conductivity.

The expressions derived are useful in calculating quantities which appear in the transport equations. They are also interesting for the converse purpose. The correlation functions themselves are of direct experimental interest. Inelastic neutron scattering, for example, directly measures the density-density correla-

[3] One possible correlation function approach has been discussed by Baym and Kadanoff (*5*).

tion function (*8*). By incorporating information about the form of the correlation function and the thermodynamic derivatives and transport coefficients which occur in it, we may attempt to interpret this kind of experimental data (*9*). In fact, a similar but more heuristic analysis (*10*) is already in use in this connection.

II. SPIN DIFFUSION

A. Hydrodynamic Description

As a concrete example of the simplest kind of transport process possible we consider a fluid composed of uncharged particles with spin ½. The particles interact through a velocity- and spin-independent force. This situation is realized to an excellent approximation in at least one system of current interest, liquid He^3.

In describing spin transport, we choose a specific direction of spin quantization. If, at a given point in space, the spin of the particles were just as likely to point antiparallel to the direction of quantization as parallel to it, the spin magnetization would vanish there. However, if there were an imbalance between the densities of particles pointing in the two directions, there would be a magnetization proportional to the difference in densities. We shall represent the magnetization in the direction of quantization at the space-time point $\mathbf{r}$, t by the symbol $M(\mathbf{r}, t)$.

An essential feature of the discussion of spin transport will be the assumption that the total magnetization is conserved, that is,

$$\frac{d}{dt}\int d\mathbf{r} M(\mathbf{r}, t) = 0. \tag{1}$$

This neglects, for example, any coupling of the electron spins with magnetic impurities or nuclear spins. The conservation law (1) follows from the fact that the total magnetization is proportional to the total spin of the entire system, and this total spin is a constant of the motion. The conservation law also has a differential form, a continuity equation for the magnetization

$$\frac{\partial}{\partial t} M(\mathbf{r}, t) + \nabla \cdot \mathbf{j}^M(r, t) = 0. \tag{2}$$

Here, $\mathbf{j}^M(\mathbf{r}, t)$ is the magnetization current. We can write expressions for these quantities in terms of the quantum mechanical operators which describe the individual particles in the system. Let the νth particle have position $\mathbf{r}_\nu(t)$, momentum $\mathbf{p}_\nu(t)$, and spin in the direction of quantization $s_\nu(t)$. Let m and γ be the mass and spin magnetic moment of all the particles. Then, the magnetization and magnetization currents are given by

$$\begin{aligned} M(\mathbf{r}, t) &= \textstyle\sum_\nu \gamma s_\nu(t)\delta(\mathbf{r} - \mathbf{r}_\nu(t)) \\ \mathbf{j}^M(\mathbf{r}, t) &= \textstyle\sum_\nu \gamma s_\nu(t)\{\mathbf{p}_\nu(t), \delta(\mathbf{r} - \mathbf{r}_\nu(t))/2m\}. \end{aligned} \tag{3}$$

Here, the curly brackets represent the anticommutator

$$\{A, B\} = AB + BA.$$

For this system the hydrodynamic equation is extremely simple. When all the properties of the system vary slowly in space and time[4]

$$\langle \mathbf{j}^M(\mathbf{r}, t)\rangle = -D\nabla\langle M(\mathbf{r}, t)\rangle. \tag{4}$$

The transport coefficient, D, is called the spin diffusion coefficient. By combining (2) and (4), we get a diffusion equation for the magnetization,

$$\frac{\partial}{\partial t}\langle M(\mathbf{r}, t)\rangle = D\nabla^2\langle M(\mathbf{r}, t)\rangle. \tag{5}$$

So far, we have not asked about how the system came to be disturbed from full thermodynamic equilibrium. Of course, Eq. (5) is correct, whenever the variation in space and time is sufficiently slow, independent of the type of initial disturbance. Nevertheless, it is useful for us to consider a specific mechanism for producing the deviation from equilibrium.

The simplest such mechanism is a magnetic field $H(\mathbf{r}, t)$ pointing in the direction of quantization. Let us suppose that a spatially varying magnetic field has been adiabatically applied and is suddenly turned off at time $t = 0$, so that

$$\begin{aligned} H(\mathbf{r}, t) &= H(\mathbf{r})e^{\epsilon t} \qquad & t < 0 \\ &= 0 & t > 0, \end{aligned} \tag{6}$$

where ϵ is an infinitesimal positive number. Of course, a magnetic field which is independent of time and varies slowly in space will induce a magnetization of the form

$$\langle M(\mathbf{r})\rangle = \chi H(\mathbf{r}). \tag{7}$$

The coefficient, χ, is called the spin susceptibility. It is the thermodynamic derivative

$$\chi = \left.\frac{\partial M}{\partial H}\right|_{H=0}. \tag{8}$$

Now, we have a complete description of the response to the disturbance (6). While the magnetic field is applied, the magnetization must satisfy (7); after it is turned off, $M(\mathbf{r}, t)$ will satisfy (5). In order to represent the relaxation behavior in a convenient form, we define a quantity $M(\mathbf{k}, z)$ which is the Fourier

[4] This relation was first proposed by N. Bloembergen (*11a*) for spins on nuclei fixed on a lattice. See also (*11b*). It was derived for particles which are free to move by Torrey (*11c*) and also by Hart (*11d*).

transform of the induced magnetization in space and effectively the Laplace transform in time. That is, we write

$$M(\mathbf{k}, z) = \int d\mathbf{r} e^{-i\mathbf{k}\cdot\mathbf{r}} \int_0^\infty dt e^{izt} \langle M(\mathbf{r}, t) \rangle. \tag{9}$$

In Eq. (9), z is a complex number. It must lie in the upper half of the complex plane for the time integral to converge.

It is quite easy to calculate $M(\mathbf{k}, z)$. We perform the transformation indicated in (9) upon (5), finding

$$0 = \int d\mathbf{r} e^{-i\mathbf{k}\cdot\mathbf{r}} \int_0^\infty dt e^{izt} \left[\frac{\partial}{\partial t} \langle M(\mathbf{r}, t) \rangle - D\nabla^2 \langle M(\mathbf{r}, t) \rangle \right].$$

After this equation is integrated by parts, it becomes

$$0 = \int d\mathbf{r} e^{-i\mathbf{k}\cdot\mathbf{r}} \int_0^\infty dt e^{izt} (-iz + Dk^2) \langle M(\mathbf{r}, t) \rangle - \int d\mathbf{r} e^{-i\mathbf{k}\cdot\mathbf{r}} \langle M(\mathbf{r}, 0) \rangle.$$

According to (7), the spatial Fourier transform of the magnetization at time zero is χ times the Fourier transform of the magnetic field. Thus, we have

$$0 = (-iz + Dk^2) M(\mathbf{k}, z) - \int d\mathbf{r} \chi H(\mathbf{r}) e^{-i\mathbf{k}\cdot\mathbf{r}}.$$

We use the symbol $H(\mathbf{k})$ to denote the Fourier transform of the magnetic field at time zero and find

$$M(\mathbf{k}, z) = \frac{\chi H(\mathbf{k})}{-iz + Dk^2} \cdot \tag{10}$$

Equation (10) is a simple representation of the information contained in the hydrodynamic equations for spin diffusion. Notice that the existence of a diffusion process is reflected in the pole in (10) at $z = -iDk^2$.

We shall use the evaluation (10) of $M(\mathbf{k}, z)$ to determine the magnetization-magnetization correlation function.

B. Correlation Function Description

In order to develop the correlation function description of spin diffusion, we notice that an external magnetic field can be represented by an extra time-dependent term added to the Hamiltonian of the system

$$\delta\mathcal{H}(t) = -\int d\mathbf{r} M(\mathbf{r}, t) H(\mathbf{r}, t). \tag{11}$$

According to the standard techniques of quantum mechanical perturbation theory, the linear change in the average of any operator, $A(\mathbf{r}, t)$, induced by an extra term in the Hamiltonian is

$$\delta\langle A(\mathbf{r}, t)\rangle = -i \int_{-\infty}^{t} dt' \langle [A(\mathbf{r}, t), \delta\mathcal{H}(t')]\rangle_{\text{eq.}} . \tag{12}$$

Equation (12) applies to a system which was in complete thermal equilibrium at time minus infinity; the expectation value on the right hand side, $\langle\ \rangle_{\text{eq.}}$, is the expectation value in the equilibrium ensemble. This result is discussed in some detail in Appendix A.

We apply (12) to a discussion of the induced magnetization by using the change in the Hamiltonian given by (6) and (11). The induced magnetization is given by

$$\begin{aligned} \langle M(\mathbf{r}, t)\rangle &= i \int_{-\infty}^{t} dt' e^{\epsilon t'} \int d\mathbf{r}' H(\mathbf{r}') \langle [M(\mathbf{r}, t), M(\mathbf{r}', t')]\rangle_{\text{eq.}} \quad t \leqq 0, \\ &= i \int_{-\infty}^{0} dt' e^{\epsilon t'} \int d\mathbf{r}' H(\mathbf{r}') \langle [M(\mathbf{r}, t), M(\mathbf{r}', t')]\rangle_{\text{eq.}} \quad t \geqq 0. \end{aligned} \tag{13}$$

In order to compare (13) with the result of our hydrodynamic discussion, we introduce an integral representation for the commutator of the magnetization at different space-time points. Because of the space-time translational invariance of the equilibrium system we may write

$$\langle [M(\mathbf{r}, t), M(\mathbf{r}', t')]\rangle_{\text{eq.}} = \int \frac{d\omega}{\pi} \int \frac{d\mathbf{k}}{(2\pi)^3} \chi''(\mathbf{k}, \omega) e^{i\mathbf{k}\cdot(\mathbf{r}-\mathbf{r}') - i\omega(t-t')} . \tag{14}$$

We shall call $\chi''(\mathbf{k}, \omega)$ the absorptive part of the dynamic susceptibility. Because of the rotational invariance of the system, $\chi''(\mathbf{k}, \omega)$ depends only upon the magnitude of $\mathbf{k}$—not its direction. Because $M(\mathbf{r}, t)$ is a Hermitian operator, $\chi''(k, \omega)$ is real and an odd function of the frequency, ω.

Equation (13) now becomes

$$\langle M(\mathbf{r}, t)\rangle = \int \frac{d\mathbf{k}}{(2\pi)^3} H(\mathbf{k}) e^{i\mathbf{k}\cdot\mathbf{r}} \int \frac{d\omega}{\pi} \frac{\chi''(k, \omega)}{\omega} \qquad \text{for } t \leqq 0, \tag{15}$$

$$= \int \frac{d\mathbf{k}}{(2\pi)^3} H(\mathbf{k}) e^{i\mathbf{k}\cdot\mathbf{r}} \int \frac{d\omega}{\pi} \frac{\chi''(k, \omega)}{\omega} e^{-i\omega t} \qquad \text{for } t \geqq 0. \tag{16}$$

We convert Eq. (16) into an expression for $M(\mathbf{k}, z)$ by employing the definition (9) of this Laplace-Fourier transform. In this way we find

$$M(\mathbf{k}, z) = \int \frac{d\omega'}{\pi i} \frac{\chi''(k, \omega')}{\omega'(\omega' - z)} H(\mathbf{k}). \tag{17}$$

Equation (10) gives an expression for $M(\mathbf{k}, z)$ which is appropriate in the limit of small k; Eq. (17) gives an expression for $M(\mathbf{k}, z)$ in terms of $\chi''(k, \omega)$.

We can therefore solve for $\chi''(k, \omega)$ in the long wavelength limit. We notice that when $\chi''(k, \omega')$ is a smooth function of ω', we can use the identity

$$\lim_{\epsilon\to 0} \frac{1}{\omega' - \omega - i\epsilon} = \mathcal{P}\frac{1}{\omega' - \omega} + \pi i\delta(\omega - \omega') \tag{18}$$

where $\mathcal{P}$ stands for the principal value. Thus, when z lies just above the real axis, $z = \omega + i\epsilon$,

$$\text{Re}[M(\mathbf{k}, \omega + i\epsilon)/H(\mathbf{k})] = \chi''(k, \omega)/\omega. \tag{19}$$

Equation (10) yields the expression

$$\chi''(k, \omega) = \frac{\chi Dk^2\omega}{\omega^2 + (Dk^2)^2}. \tag{20}$$

Because the hydrodynamic equations are valid for slowly varying disturbances, Eq. (19) gives the correct expression for χ'' at small k. Notice that at long wavelengths and low frequencies the value of $\chi''(k, \omega)$ depends sensitively on the relative magnitude of ω and k^2. When $\omega \ll Dk^2$

$$\chi''(k, \omega) \cong \omega\chi/Dk^2$$

while when $Dk^2 \ll \omega$

$$\chi''(k, \omega) \cong \chi Dk^2/\omega.$$

Equation (20) for the dynamic spin susceptibility contains the same information as the hydrodynamic equations from which it was derived. The fact that the magnetization satisfies a diffusion equation is reflected in the poles of (20) at frequencies $\pm\, iDk^2$. The magnetization-magnetization commutator deduced from (20) and (14),

$$\langle[M(\mathbf{r}, t), M(\mathbf{r}', t')]\rangle_{\text{eq.}} = -\,i\chi D \int \frac{d\mathbf{k}}{(2\pi)^3}\, k^2 e^{i\mathbf{k}\cdot(\mathbf{r}-\mathbf{r}') - Dk^2(t-t')} \qquad \text{for } t > t'$$

exhibits this diffusive character.

C. Sum Rules for $\chi''(k, \omega)$

So far, we have only made use of $\chi''(k, \omega)$, the absorptive part of the dynamic spin susceptibility. In our further work, it will be convenient to use the complex dynamic susceptibility

$$\chi(k, z) = \int \frac{d\omega'}{\pi} \frac{\chi''(k, \omega')}{\omega' - z}. \tag{21}$$

When z lies just above the real axis $\chi(k, z)$ may be split into its real and imaginary parts

$$\chi(k, \omega + i\epsilon) = \chi'(k, \omega) + i\chi''(k, \omega),$$

the identity (18) yielding

$$\chi'(k, \omega) = \mathcal{P} \int \frac{d\omega'}{\pi} \frac{\chi''(k, \omega')}{\omega' - \omega}. \tag{22a}$$

Equation (22a) is ordinarily called a Kramers-Kronig relation. There are two such relations, which give the real part of the response in terms of the imaginary part and vice versa. The other Kramers-Kronig relation is

$$\chi''(k, \omega) = -\mathcal{P} \int \frac{d\omega'}{\pi} \frac{\chi'(k, \omega')}{\omega' - \omega}. \tag{22b}$$

Equations (22a) and (22b) may be derived from one another by using the relation

$$\mathcal{P} \int \frac{d\bar{\omega}}{\pi} \frac{1}{\omega - \bar{\omega}} \frac{1}{\omega' - \bar{\omega}} = \pi\delta(\omega - \omega').$$

Notice that $M(\mathbf{k}, z)$ can be expressed in terms of $\chi(k, z)$. From (17) we deduce

$$\frac{-izM(\mathbf{k}, z)}{H(\mathbf{k})} = \int \frac{d\omega'}{\pi} \frac{\chi''(k, \omega')}{\omega'} - \chi(k, z)$$

and hence,

$$\frac{M(\mathbf{k}, z)}{H(\mathbf{k})} = \frac{-\chi(k, 0) + \chi(k, z)}{iz}. \tag{23}$$

The quantity $\chi(k, 0)$ will prove to be particularly important in all that follows. Its importance is illustrated by taking the Fourier transform of Eq. (15), which gives

$$M(\mathbf{k}) = \int d\mathbf{r} e^{-i\mathbf{k}\cdot\mathbf{r}} \langle M(\mathbf{r}, t = 0) \rangle$$

$$M(\mathbf{k}) = \int \frac{d\omega'}{\pi} \frac{\chi''(k, \omega')}{\omega'} H(\mathbf{k}) \tag{24}$$

$$= \chi'(k, 0)\, H(\mathbf{k})$$

$$= \chi(k, 0)\, H(\mathbf{k}).$$

Since the response at time zero is a response to an adiabatically applied disturbance, $\chi(k, 0)$ is the static, wave-number dependent, magnetic susceptibility. Henceforth, we shall use the conventional abbreviation $\chi(k) \equiv \chi(k, 0)$. From

(24) we have

$$\chi(k) = \int \frac{d\omega}{\pi} \frac{\chi''(k, \omega)}{\omega}. \tag{25}$$

Equation (24) gives the exact response to an adiabatically magnetic field. However, according to Eq. (7), when the field varies very slowly in space

$$M(\mathbf{k}) = \chi H(\mathbf{k}). \tag{7}$$

Thus, it follows that

$$\chi = \lim_{k\to 0} \chi(k) = \lim_{k\to 0} \int \frac{d\omega}{\pi} \frac{\chi''(k, \omega)}{\omega}. \tag{26}$$

In general, we may view (25) as a sum rule which expresses the static susceptibility in terms of an integral of $\chi''(k, \omega)$. Eq. (25) is just an application of one of the Kramers-Kronig relations at zero frequency. In the long wavelength limit, the value of the sum rule is the thermodynamic derivative,

$$\chi = \left.\frac{\partial M}{\partial H}\right|_{H=0}.$$

It is instructive to compare (26) with a more familiar type of sum rule which expresses moments of $\chi''(k, \omega)$ in terms of equal time commutators. The first nontrivial example of this kind of sum rule is obtained by taking the time derivative of Eq. (14) and applying the conservation law (2). This leads to the identity

$$\frac{\partial}{\partial t} \langle [M(\mathbf{r}, t), M(\mathbf{r}', t')] \rangle_{\text{eq.}} = -\langle [\nabla \cdot \mathbf{j}^M(\mathbf{r}, t), M(\mathbf{r}', t')] \rangle_{\text{eq.}}$$

$$= -i \int \frac{d\omega}{\pi} \int \frac{d\mathbf{k}}{(2\pi)^3} \omega \chi''(k, \omega) e^{i\mathbf{k}\cdot(\mathbf{r}-\mathbf{r}') - i\omega(t-t')}.$$

We can easily compute the equal time commutator of the magnetization and the magnetization current by using the definitions (3) of these quantities and the canonical commutation relations. The result

$$\langle [\mathbf{j}^M(\mathbf{r}, t), M(\mathbf{r}', t')] \rangle_{\text{eq.}} |_{t'=t} = (\gamma^2/4m) i \nabla' \delta(\mathbf{r} - \mathbf{r}') \langle n(\mathbf{r}, t) \rangle_{\text{eq.}} \tag{27}$$

is a very disguised version of the fundamental statement that the commutator of the position and the momentum is i. Here, $n(\mathbf{r}, t)$ is the density of particles at the space-time point r, t. Equation (27) implies

$$-\int \frac{d\omega}{\pi} \int \frac{d\mathbf{k}}{(2\pi)^3} \omega \chi''(k, \omega) e^{i\mathbf{k}\cdot(\mathbf{r}-\mathbf{r}')} = \frac{\gamma^2 \langle n \rangle_{\text{eq.}}}{4m} \nabla^2 \delta(\mathbf{r} - \mathbf{r}')$$

or

$$\int \frac{d\omega}{\pi} \omega\chi''(k, \omega) = \frac{n}{m}\frac{\gamma^2}{4} k^2. \tag{28}$$

Equation (28) is the spin analog of the longitudinal f-sum rule which has been extensively discussed in the literature. This sum rule is exact for all values of k as is the static Kramers-Kronig relation (25). However the latter has an independently computable thermodynamic value only for small k. The sum rules also differ in that Eq. (28) expresses a moment of χ'' in terms of an equal time commutator while Eq. (25) gives the value of the time integral of a commutator.

These sum rule statements can be incorporated in Eq. (17) for $M(\mathbf{k}, z)/H(\mathbf{k})$ by performing an expansion for large values of z. In particular, Eqs. (23) and (21) may be rewritten as

$$\frac{M(\mathbf{k}, z)}{H(\mathbf{k})} = \frac{i}{z}\chi(k) + \frac{i}{z^2}\int \frac{d\omega}{\pi} \chi''(k, \omega) + \frac{i}{z^3}\int \frac{d\omega}{\pi} \omega\chi''(k, \omega) + \frac{i}{z^3}\int \frac{d\omega}{\pi} \frac{\omega^2}{z - \omega} \chi''(k, \omega). \tag{29}$$

The coefficient of i/z^2 vanishes because $\chi''(k, \omega)$ is an odd function of the frequency. According to the sum rule (27), the coefficient of i/z^3 is $\frac{1}{4}nk^2\gamma^2/m$. Therefore, for small k,

$$\lim_{z\to\infty} \frac{M(\mathbf{k}, z)}{H(\mathbf{k})} = -\frac{\chi}{iz} - \frac{n\gamma^2 k^2}{4iz^3 m} + O\left(\frac{1}{z^4}\right). \tag{30}$$

Using the hydrodynamic equations, we found that for small values of k and z

$$\frac{M(\mathbf{k}, z)}{H(\mathbf{k})} = \frac{\chi}{-iz + Dk^2} \tag{10}$$

and

$$\chi''(k, \omega) = \frac{\chi Dk^2\omega}{\omega^2 + (Dk^2)^2}. \tag{20}$$

Let us observe now that the hydrodynamic analysis agrees with the sum rule (25) but completely fails to satisfy the rule (28). The easiest way of verifying both properties is to notice that Eq. (10) agrees with Eq. (30) at large values of z only to order i/z.

We might have anticipated that the second sum rule was not satisfied by the expression obtained from the hydrodynamic approximation (10) since that sum rule gives a result of order k^2 while (10) is only expected to be appropriate for the smallest values of k^2. We can understand phenomenologically how the sum rule (28) is satisfied by extending the hydrodynamic description to include the effect of a collision time.

D. Introduction of Relaxation Time

The main reason why the function $\chi''(k, \omega)$ deduced from the hydrodynamic equations fails to satisfy the commutation sum rule can be traced to the assumption that the current responds instantly to changes in the magnetization according to

$$\langle \mathbf{j}^M(\mathbf{r}, t)\rangle = -D\nabla\langle M(\mathbf{r}, t)\rangle \qquad \text{for } t \geqq 0. \quad (4)$$

Actually, there must be some lag in the response of the magnetic current to rapid changes in the magnetization. Let us suppose that this response lag is described by a single relaxation time, τ, according to the equation

$$\frac{\partial}{\partial t}\langle \mathbf{j}^M(\mathbf{r}, t)\rangle = -\frac{1}{\tau}\left[\langle \mathbf{j}^M(\mathbf{r}, t)\rangle + D\nabla\langle M(\mathbf{r}, t)\rangle\right] \qquad \text{for } t \geqq 0.$$

We may substitute this form for the current into the conservation law and find

$$\left[\frac{\partial^2}{\partial t^2} + \frac{1}{\tau}\left(\frac{\partial}{\partial t} - D\nabla^2\right)\right]\langle M(\mathbf{r}, t)\rangle = 0 \qquad \text{for } t \geqq 0. \quad (31)$$

We again find $M(\mathbf{k}, z)$ by Laplace transforming in time and Fourier transforming in space the equation of motion. After using the initial conditions

$$\langle M(\mathbf{r}, 0)\rangle = \chi H(\mathbf{r}) \qquad \frac{\partial}{\partial t}\langle M(\mathbf{r}, t)\rangle\,|_{t=0} = 0$$

we obtain

$$\frac{M(\mathbf{k}, z)}{H(\mathbf{k})} = \frac{\chi(1 - iz\tau)}{-iz + Dk^2 - \tau z^2}. \quad (32)$$

To see whether (32) agrees with our sum rules, we expand for large z obtaining

$$\frac{M(\mathbf{k}, z)}{H(\mathbf{k})} = \frac{i\chi}{z} + \frac{iDk^2\chi}{\tau z^3} + \mathcal{O}\left(\frac{1}{z^4}\right).$$

Hence, all the requirements including (30) can be satisfied by taking a relaxation time which satisfies

$$D = \frac{n\gamma^2}{4m}\frac{\tau}{\chi}. \quad (33)$$

Equation (33) has been used by D. Hone (*12a*) to achieve a semiquantitative understanding of the experimental value of D in liquid He^3 at very low temperatures. It has also been used by T. Moriya (*12b*) in discussing the spin correlation function in ferromagnets.

According to Eq. (19), the dynamic susceptibility can be determined as $\text{Re}[M(\mathbf{k}, \omega)/H(\mathbf{k})]\omega$. Therefore, from Eqs. (32) and (33) we find in the single collision time approximation

$$\chi''(k, \omega) = \frac{\chi D k^2 \omega}{\omega^2 + D^2[k^2 - (4\omega^2 \chi m/n\gamma^2)]^2}. \tag{34}$$

Also, from (32) and (33) we find

$$\chi(k, z) = \chi + \frac{iM(k, z)z}{H(k)} = \chi\left(1 - \frac{iz}{Dk^2} - \frac{4z^2\chi m}{n\gamma^2 k^2}\right)^{-1}. \tag{35}$$

E. Dispersion Relation Representations for Susceptibility

It should be emphasized that Eqs. (34) and (35) are in no sense exact. In this section, we shall generalize the phenomenological discussion in an exact form.

In order to derive this generalization, we first examine the analytic properties of $\chi(k, z)$. We note that $\chi(k, z)$ is an analytic function of the complex variable z whose singularities lie entirely on the real z axis. From Eqs. (35) and (23), we see that, in the constant collision time approximation,

$$\frac{1}{\chi(k, z)} - \frac{1}{\chi} = -\frac{4mz^2}{n\gamma^2 k^2} - \frac{iz}{D\chi k^2}. \tag{36}$$

Consequently, we might guess that the quantity

$$\frac{1}{z^2}\left[\frac{1}{\chi(k, z)} - \frac{1}{\chi(k)}\right] + \frac{4m}{n\gamma^2 k^2}$$

has a relatively simple analytic structure.

To justify this inference we examine the zeros of $\chi(k, z)$. The important observation to be made is that in a thermodynamically stable system the quantity $\omega\chi''(k, \omega)$, which measures the difference between the energy fed into the system and the energy given up, by a weakly applied field, must be positive definite. For a canonical ensemble this positive definiteness may be directly verified by expanding the commutator in terms of matrix elements and using the fluctuation dissipation theorem. Both of these statements are proven in the appendices. Using the oddness of $\chi''(k, \omega)$ we may write

$$\chi(k, z) = \int \frac{d\omega}{\pi} \frac{\omega}{\omega^2 - z^2} \chi''(k, \omega).$$

Consequently, if $z^2 = x + iy$, we have

$$\text{Im}\, \chi(k, (x + iy)^{1/2}) = \int \frac{d\omega}{\pi} \frac{\omega y}{(\omega^2 - x)^2 + y^2} \chi''(k, \omega).$$

We see that $\text{Im}\chi(k, z)$ only vanishes for real z^2—that is for z either purely real or purely imaginary. Moreover when z^2 is negative $\chi(k, z)$ is real and positive.

Thus the only possible zeros of $\chi(k, z)$ appear for real z. Since the zeros of $\chi(k, z)$ are poles of $1/\chi(k, z)$, the only poles of $1/\chi(k, z)$ lie on the real frequency axis. Finally we recall that in the limit of large z

$$\chi(k, z) = -\int \frac{d\omega}{\pi} \frac{\omega}{z^2} \chi''(k, \omega) = -\frac{n\gamma^2 k^2}{4mz^2}.$$

From these properties we deduce the spectral form

$$\frac{1}{z^2}\left[\frac{1}{\chi(k, z)} - \frac{1}{\chi(k)}\right] + \frac{4m}{nk^2\gamma^2} = -\frac{1}{\chi(k)} \int \frac{d\omega}{\pi} \frac{f(k, \omega)}{\omega z(\omega - z)}. \tag{37}$$

To interpret the spectral weight function, $f(k, \omega)$, we compare this result with Eq. (36). In Eq. (36), the variable z is restricted to lie in the upper half of the complex plane. When $z = \omega + i\epsilon$ the spectral representation becomes

$$\frac{\chi(k)}{\chi(k, \omega + i\epsilon)} - 1 = -\frac{4m\omega^2\chi(k)}{n\gamma^2 k^2} - if(k, \omega) - \mathcal{P} \int \frac{d\omega'}{\pi} \frac{f(k, \omega')\omega}{\omega'(\omega' - \omega)}. \tag{38}$$

Since $2if(k, \omega)$ is equal to the discontinuity in the function $\chi^{-1}(k, z)$ across the real axis and that discontinuity is imaginary, the function $f(k, \omega)$ is a real odd function of the frequency. If the function ωf were independent of frequency, the last term in Eq. (38) would vanish and Eq. (38) would be identical with Eq. (36) with $f = \omega/Dk^2$. Therefore, it is reasonable to write

$$f(k, \omega) = \omega/D(k, \omega)k^2 \tag{39}$$

with the knowledge that in the limit small wave number and very small frequency $D(k, \omega)$ reduces to the spin diffusivity, at least when $\omega^2 \cong -(Dk^2)^2$.

This leads to an exact spectral representation for $\chi(k, z)$

$$\frac{\chi(k, z)}{\chi(k)k^2} = \left[k^2 - \frac{4mz^2}{n\gamma^2} \chi(k) + \int \frac{d\omega}{\pi} \frac{z}{z - \omega} \frac{1}{D(k\omega)}\right]^{-1} \tag{40}$$

and, from Eq. (19)

$$\chi''(k, \omega) = \frac{\chi(k)D(k, \omega)k^2\omega}{\omega^2 + [\;\;]^2}$$

$$[\;\;] \equiv D(k, \omega)k^2 - \frac{4m\omega^2}{n\gamma^2} D(k, \omega)\chi(k) + \mathcal{P} \int \frac{d\omega'}{\pi} \frac{\omega}{\omega - \omega'} \frac{D(k, \omega)}{D(k, \omega')} \tag{41}$$

Of course, Eqs. (40) and (41) give only one of many possible spectral representations for the response. The virtue of this particular representation, however, is that the low frequency and low wave number limit of all the quantities appearing on the right hand side of these equations will be regular. This knowledge is quite useful both for theoretically estimating $\chi(k)$ and $D(k, \omega)$ and for interpreting experimental results on $\chi''(k, \omega)$ in terms of χ and D. Note that if the function $D(k, \omega)$ were constant (40) and (41) would be identical with (35) and (34). This gives a precise meaning to the single collision time approximation.

Equations (40) and (41) are easily modified to describe other self-diffusion processes. In our analysis, the spin on the particles just serves as a kind of label. If the label were somewhat different; for example, if the system contained particles painted red and green or identically interacting isotopes, the results would be unaltered.

Another useful representation is obtained by observing that the function

$$-\frac{4z^2m}{nk^2\gamma^2}\frac{\chi(k,z)}{\chi(k)}$$

is analytic off the real axis and approaches unity at infinity. Its logarithm is therefore analytic for complex z and its real part is continuous for real z. It is therefore possible to represent the function as

$$\frac{\chi(k,z)}{\chi(k)} = -\frac{n\gamma^2k^2}{4mz^2}\exp\int\frac{d\omega'}{\pi}\frac{\delta(k,\omega')}{\omega'-z}$$

where δ is a real function. As ω approaches the real axis we find

$$\chi'(k, \omega) + i\chi''(k, \omega) = \lambda(k, \omega)\chi(k)e^{i\delta(k, \omega)}$$

where

$$\lambda(k, \omega) = -\frac{n\gamma^2k^2}{4m\omega^2}\exp \mathcal{P}\int\frac{d\omega'}{\pi}\frac{\delta(k, \omega')}{\omega'-\omega}.$$

Moreover the discontinuity in $\chi^{-1}(k, z)/\chi^{-1}(k)$ is given by

$$f(k, \omega) = \lambda^{-1}(k, \omega)\sin\delta(k, \omega).$$

This permits us to identify the argument of $\chi^{-1}(k, z)$, that is, δ, as

$$\omega\cot\delta(k, \omega) = Dk^2$$

in a first approximation, and as

$$\omega\cot\delta(k, \omega) = D\left(k^2 - \frac{4\omega^2\chi m}{n\gamma^2}\right)$$

in the next. We might therefore introduce D as the constant term in an effective range expansion.

F. Expressions Entailing the Fluctuation-Dissipation Theorem

One may carry the analysis a bit further by employing the fluctuation-dissipation theorem. This theorem relates the canonically or grand canonically averaged commutator and anticommutator of any pair of hermitian operators $A_i(\mathbf{r}, t)$, each of which commutes with the number operator N and transforms in time according to

$$i\dot{A}_i(\mathbf{r}, t) = [\mathcal{H}, A_i(r, t)].$$

This is to say, the operators are assumed to have no explicit time dependence and the usual implicit time dependence of Heisenberg representation operators. The theorem states that if the commutator of two such operators is given by

$$\langle [A_i(\mathbf{r}, t), A_j(\mathbf{r}', t')] \rangle_{\text{eq.}} \equiv \int \frac{d\omega}{\pi} \int \frac{d\mathbf{k}}{(2\pi)^3} e^{i\mathbf{k}\cdot(\mathbf{r}-\mathbf{r}')-i\omega(t-t')} \chi''_{ij}(\mathbf{k}, \omega) \tag{42a}$$

their anticommutator is given by

$$\langle \{ A_i(\mathbf{r}, t) - \langle A_i \rangle_{\text{eq.}}, A_j(\mathbf{r}', t') - \langle A_j \rangle_{\text{eq.}} \} \rangle$$
$$= \int \frac{d\omega}{\pi} \int \frac{d\mathbf{k}}{(2\pi)^3} e^{i\mathbf{k}\cdot(\mathbf{r}-\mathbf{r}')-i\omega(t-t')} \coth \frac{\beta\omega}{2} \chi''_{ij}(\mathbf{k}, \omega). \tag{42b}$$

This relationship between the commutator and the anticommutator is called a fluctuation-dissipation theorem because the anticommutator expresses the time dependent correlations or fluctuations in the system and, as we have seen, the commutator describes the transport coefficient or dissipation.[5]

In particular, the magnetization anticommutator is

$$\frac{\beta}{2} \langle \{ M(\mathbf{r}, t), M(\mathbf{r}', t') \} \rangle = \int \frac{d\omega}{\pi} \int \frac{d\mathbf{k}}{(2\pi)^3} \frac{\beta}{2} \coth \frac{\beta\omega}{2} \chi''(k, \omega) e^{i\mathbf{k}\cdot(\mathbf{r}-\mathbf{r}')-i\omega(t-t')}. \tag{43}$$

In the remainder of this section we shall continue to omit the subscripts since we are only considering one operator, the magnetization. From Eq. (20), we see that for large values of r and t, or small values of k and ω, the anticommutator is

$$\frac{\beta}{2} \langle \{ M(\mathbf{r}, t), M(\mathbf{0}, 0) \} \rangle_{\text{eq.}} = \chi \int \frac{d\mathbf{k}}{(2\pi)^3} e^{i\mathbf{k}\cdot\mathbf{r} - Dk^2|t|}.$$

Therefore, the anticommutator also has a part which satisfies the diffusion equation.

Of course, it is hardly surprising that this correlation function behaves in the same way as a response to an external disturbance. The correlation function re-

[5] The fluctuation-dissipation theorem was first derived by H. Nyquist (*13a*) who related the random noise in an electrical circuit (the fluctuations) to the response of the circuit to an applied voltage (the dissipation). H. B. Callen and T. R. Welton (*13b*) recognized the importance of Nyquist's idea and generalized it somewhat. The fluctuation-dissipation theorem lies at the very heart of much recent work in many particle physics. It is, for example, the "boundary condition" utilized by Martin and Schwinger (*2*). For a discussion of the fluctuation-dissipation theorem which is close to the spirit and purpose of this article, see (*6*).

flects how the natural fluctuations in the system die out, while the hydrodynamic equations describe how externally induced deviations from equilibrium disappear. But the system should not really have any way of knowing whether a particular deviation from equilibrium was produced by a natural fluctuation or an external disturbance. Therefore, the same transport processes which appear in the hydrodynamic equations should also manifest themselves in correlation functions.

The fluctuation-dissipation theorem can be used to obtain a frequently quoted expression for the spin diffusion coefficient D. From (20)

$$D\chi = \lim_{\omega\to 0}\left[\lim_{k\to 0}\frac{\omega}{k^2}\chi''(k,\omega)\right].$$

Using (43), $D\chi$ can be expressed in terms of the magnetization anticommutator as

$$D\chi = \lim_{\omega\to 0}\left[\lim_{k\to 0}\int d\mathbf{r}\int dt\, e^{-i\mathbf{k}\cdot(\mathbf{r}-\mathbf{r}')+i\omega(t-t')}\frac{\omega^2}{k^2}\frac{\beta}{4}\langle\{M(\mathbf{r},t),M(\mathbf{r}',t')\}\rangle_{\text{eq.}}\right].$$

The differential conservation law, $(\partial M/\partial t) + \nabla\cdot\mathbf{j}^M = 0$, now gives

$$D\chi = \lim_{\omega\to 0}\left[\lim_{k\to 0}\int d\mathbf{r}\int dt\, e^{-i\mathbf{k}\cdot(\mathbf{r}-\mathbf{r}')+i\omega(t-t')}\frac{\beta}{4k^2}\langle\{\mathbf{k}\cdot\mathbf{j}^M(\mathbf{r},t),\mathbf{k}\cdot\mathbf{j}^M(\mathbf{r}',t')\}\rangle\right].$$

Since the direction of $\mathbf{k}$ is now quite irrelevant, we can replace $\mathbf{k}\cdot\mathbf{j}^M/\mathbf{k}$ by, say, the x component of $\mathbf{j}^M$. Thus, we finally find

$$D\chi = \lim_{\omega\to 0}\int d\mathbf{r}\, dt\, e^{i\omega(t-t')}\frac{\beta}{4}\langle\{\mathbf{j}_x^M(\mathbf{r},t),\mathbf{j}_x^M(\mathbf{r}',t')\}\rangle. \tag{44}$$

This type of expression, in which the transport coefficient is given in terms of the anticommutator of the currents, has been much discussed in the literature (*6, 7*).

In addition to relating fluctuations to dissipation as in Eqs. (43) and (44) we may use the identity (42) to make another inference. For this purpose we observe that the susceptibility is the thermodynamic derivative,

$$\chi = \left.\frac{\partial M}{\partial H}\right|_{H=0}. \tag{8}$$

For a system in thermal equilibrium in the presence of a static uniform field the magnetization can be calculated in the grand canonical ensemble where the expectation value of any operator is defined by

$$\langle A\rangle \equiv \text{tr}\,[\rho A]$$
$$\rho \equiv \frac{\exp(-\beta[\mathcal{H}-\mu\mathfrak{N}])}{\text{tr}\,[\exp(-\beta[\mathcal{H}-\mu\mathfrak{N}])]}. \tag{45}$$

Here, $\mathcal{H}$ and $\mathfrak{N}$ are the Hamiltonian and number operators. The trace is a diagonal sum over all states of the system with all possible values of the energy and the particle number. The parameters μ and β are respectively the chemical potential and the inverse temperature (in energy units).

In calculating the effect of a magnetic field which is independent of r, we apply Eq. (45) to the case in which the Hamiltonian is the Hamiltonian in the absence of the magnetic field, $\mathcal{H}_0$, plus the magnetic energy $-H \int d\mathbf{r}M(\mathbf{r})$. The density matrix ρ, can be expanded to first order in the magnetization since the total magnetization commutes with both $\mathcal{H}_0$ and $\mathfrak{N}$. In this manner we obtain

$$\rho \cong \rho_{\text{eq.}} + \rho_{\text{eq.}}\, \beta H \int d\mathbf{r}\, \{M(\mathbf{r}) - \text{tr}\, [\rho_{\text{eq.}}\, M(\mathbf{r})]\}$$

where $\rho_{\text{eq.}}$ is the density matrix with no magnetic field

$$\rho_{\text{eq.}} = \frac{\exp\,(-\beta[\mathcal{H}_0 - \mu\mathfrak{N}])}{\text{tr}\,[\exp\,(-\beta[\mathcal{H}_0 - \mu\mathfrak{N}])]}.$$

Since the magnetization must vanish when there is no field,

$$\langle M \rangle = \beta H\, \text{tr} \left[\rho_{\text{eq.}}\, M(\mathbf{r}) \int d\mathbf{r}'\, M(\mathbf{r}') \right] = \beta H \left\langle M(\mathbf{r}) \int d\mathbf{r}'\, M(\mathbf{r}') \right\rangle_{\text{eq.}},$$

the susceptibility is[6]

$$\frac{\partial M}{\partial H} = \chi = \frac{\beta}{2} \int d\mathbf{r}'\, \langle \{M(\mathbf{r}), M(\mathbf{r}')\} \rangle_{\text{eq.}}. \tag{46a}$$

(In writing Eq. (46a) we have taken advantage of the fact that, for equal times, the magnetization at one point will commute with the magnetization at another to replace $M(\mathbf{r}')M(\mathbf{r})$ by the more symmetrical combination $\frac{1}{2}\{M(\mathbf{r}), M(\mathbf{r}')\}$.) Using the fluctuation-dissipation theorem in the form (43), we may rewrite Eq. (46a) as

$$\chi = \int \frac{d\omega}{\pi} \frac{\beta}{2} \coth \frac{\beta\omega}{2}\, \chi''(k, \omega)|_{k=0}. \tag{46b}$$

Let us compare this thermodynamic form for the susceptibility with our previous result

$$\chi = \lim_{k \to 0} \int \frac{d\omega}{\pi} \frac{\chi''(k, \omega)}{\omega}. \tag{26}$$

The positive definite integrand in (46b) is greater than or equal to the positive definite integrand of (26) and the two expressions are equal only at $\omega = 0$. It

[6] This expression was first discussed by J. Kirkwood (*14*).

therefore follows that $\chi''(k, \omega)/\omega$ must be very sharply peaked about zero frequency.[7] This sharp peaking of χ''/ω is predicted by Eq. (20) according to which χ''/ω becomes a delta function at zero frequency as k goes to zero. Moreover the integral conservation law

$$\frac{d}{dt} \int d\mathbf{r}\, M(\mathbf{r}, t) = 0$$

implies that the $\mathbf{k} = \mathbf{0}$ part of the anticommutator can only contain zero frequency components. Thus, the frequency integral in (46b) can only contribute at exactly zero frequency and expressions (46b) and (26) are completely consistent with one another.

III. TRANSPORT IN A FLUID

A. Hydrodynamic Equations

Spin transport is particularly simple because it is described, in the hydrodynamic limit, by a simple diffusion equation. For most systems, however, the hydrodynamic equations are more complex. This is because there is one transport equation connected with each differential conservation law. For a one-component fluid, for example, there is a conservation law and transport equation for the density of particles, $n(\mathbf{r}, t)$, the momentum density, $\mathbf{g}(\mathbf{r}, t)$, and the energy density, $\epsilon(\mathbf{r}, t)$. These conservation laws can be written as

$$\frac{\partial}{\partial t} n(\mathbf{r}, t) + \nabla \cdot \mathbf{g} \frac{(\mathbf{r}, t)}{m} = 0 \qquad \text{number conservation,} \quad (47a)$$

$$\frac{\partial}{\partial t} g(\mathbf{r}, t) + \nabla \cdot \tau(\mathbf{r}, t) = 0 \qquad \text{momentum conservation,} \quad (47b)$$

$$\frac{\partial}{\partial t} \epsilon(\mathbf{r}, t) + \nabla \cdot \mathbf{j}^{\epsilon}(\mathbf{r}, t) = 0 \qquad \text{energy conservation.} \quad (47c)$$

Here, j^{ϵ} is the energy current density and τ is the stress tensor, which serves as a momentum current.

Of course Eqs. (47) are incomplete in themselves. They must be supplemented with the assumption that when all variations in space and time are slow, the system can be treated as if it is in thermodynamic equilibrium locally. Since the state of the fluid in equilibrium is characterized by the five conserved variables or five associated intensive variables we expect local equilibrium to be characterized either by the local densities of the conserved variables or by related spatially

[7] There are a few cases known in which the limit as k goes to zero of an integral like (46) is not equal to its value at zero wave number. The most notable example of this pathological behavior is a system of particles interacting through a Coulomb force, in which the long-ranged interaction makes the limit of small wave numbers peculiar.

and temporally varying intensive quantities. Conventionally these are chosen to be the temperature, pressure, and average velocity.

We define an average velocity by writing the momentum density as

$$\langle \mathbf{g}(\mathbf{r}, t)\rangle = \langle n(\mathbf{r}, t)\rangle m\mathbf{v}(\mathbf{r}, t).$$

We shall consider the case in which the deviation from complete equilibrium is small. We may further suppose that the complete equilibrium system is taken to be at rest and uniform. We may then write to first order

$$\langle \mathbf{g}(\mathbf{r}, t)\rangle = nm\mathbf{v}(\mathbf{r}, t) \tag{48a}$$

where n is the equilibrium density of particles. For a system of particles in complete equilibrium, moving with uniform velocity, $\mathbf{v}$, Galilean invariance implies an energy current

$$\mathbf{j}^{\epsilon} = (\epsilon + p)\mathbf{v}.$$

When the system is in local equilibrium the energy current will generally contain a term of this form. However, if there is a temperature gradient in the system, there is an extra flow of energy from hot regions to cold regions. These two effects lead to an energy current of the form

$$\mathbf{j}^{\epsilon}(\mathbf{r}, t) = (\epsilon + p)\mathbf{v}(\mathbf{r}, t) - \kappa\nabla T(\mathbf{r}, t) \tag{48b}$$

where ϵ and p are the equilibrium parameters.[8] The coefficient κ is called the thermal conductivity.

Of course, the temperature which appears in (48b) is not independent of the other variables. Because the system is in local thermodynamic equilibrium, variations in the intensive parameters satisfy the usual thermodynamic relations. Thus, a change in the temperature is related to changes in the density and energy density by

$$\nabla T(\mathbf{r}, t) = \left.\frac{\partial T}{\partial n}\right|_{\epsilon} \nabla n(\mathbf{r}, t) + \left.\frac{\partial T}{\partial \epsilon}\right|_{n} \nabla \epsilon(\mathbf{r}, t).$$

[8] However, in a superfluid there exists more general modes of motion than this mode in which the fluid appears locally to be moving as a whole. The superfluid can sustain without appreciable decay the relative motion of its excitations (the normal fluid) against a sea of its condensed state (the superfluid). This extra freedom results in the local equilibrium situation being described by not one but two velocities: the condensed mode velocity v_s and the normal mode velocity v_n . In this case, the energy current, neglecting dissipation, is

$$\mathbf{j}^{\epsilon}(\mathbf{r}, t) = (\epsilon + p)\mathbf{v}_n(\mathbf{r}, t) + \mu\rho_s(\mathbf{v}_s(\mathbf{r}, t) - \mathbf{v}_n(\mathbf{r}, t))$$

where ρ_s is the density of the superfluid (condensed) component. The inclusion of this extra degree of freedom of the superfluid changes the hydrodynamic equations and this invalidates all the main conclusions of the present work.

To complete the set of equations (48), it is necessary to specify the stress tensor, τ. For a fluid at rest in complete equilibrium the stress tensor takes the form

$$\tau_{ij} = \delta_{ij}p$$

where p is the pressure. When the fluid is disturbed from equilibrium, extra stresses are produced as a result of viscous forces in the fluid. These forces are proportional to gradients of the velocity so that the full stress tensor may be written as

$$\tau_{ij}(\mathbf{r}, t) = \delta_{ij}p(\mathbf{r}, t) - \eta\left(\frac{\partial v_i(\mathbf{r}, t)}{\partial r_j} + \frac{\partial v_j(\mathbf{r}, t)}{\partial r_i}\right) - \delta_{ij}\, \nabla\cdot\mathbf{v}(\mathbf{r}, t)\left(\zeta - \frac{2}{3}\eta\right). \tag{48c}$$

Here η is called the viscosity and ζ the second viscosity or bulk viscosity. Again there may be additional terms in a superfluid. We shall henceforth ignore this possibility, restricting ourselves to normal fluids. Changes in the pressure, p, are, in a normal fluid, related to changes in the density, energy density, and temperature through the usual thermodynamic relations. Therefore, Eqs. (47) together with Eqs. (48) form a complete description of the fluid. In fact, they are the linearized form of the usual equations of fluid mechanics.[9]

Now, we recombine these equations in a form which is convenient for our purposes. With the aid of (48a) and (48c), the momentum conservation law may be written as

$$\frac{\partial}{\partial t}\langle\mathbf{g}(\mathbf{r}, t)\rangle + \nabla p(\mathbf{r}, t) - \frac{\eta}{mn}\nabla^2\langle\mathbf{g}(\mathbf{r}, t)\rangle - \frac{\zeta + \frac{1}{3}\eta}{mn}\nabla\nabla\cdot\langle\mathbf{g}(\mathbf{r}, t)\rangle = 0. \tag{49}$$

It is convenient to divide the momentum density into longitudinal and transverse parts, that is, to write

$$\mathbf{g}(\mathbf{r}, t) = \mathbf{g}_l(\mathbf{r}, t) + \mathbf{g}_t(\mathbf{r}, t)$$

where

$$\nabla\cdot\mathbf{g}_t(\mathbf{r}, t) = 0$$

$$\nabla\times\mathbf{g}_l(\mathbf{r}, t) = 0.$$

With these definitions, the transverse part of the momentum satisfies the diffusion equation

$$\frac{\partial}{\partial t}\langle\mathbf{g}_t(\mathbf{r}, t)\rangle = \frac{\eta}{mn}\nabla^2\langle\mathbf{g}_t(\mathbf{r}, t)\rangle. \tag{50a}$$

[9] See, for example, ref. *15*.

To get the remaining hydrodynamic equations, we take the divergence of (49) and use the number conservation law (47a) to eliminate $\mathbf{g}(\mathbf{r}, t)$. We then find

$$\left[-m \frac{\partial^2}{\partial t^2} + \frac{\frac{4}{3}\eta + \zeta}{n} \frac{\partial}{\partial t} \nabla^2\right] \langle n(\mathbf{r}, t)\rangle + \nabla^2 p(\mathbf{r}, t) = 0. \tag{50b}$$

The momentum density may also be eliminated from the energy conservation law which results from substituting (48b) into (47c). In this way, we find

$$\frac{\partial}{\partial t}\left[\langle \epsilon(\mathbf{r}, t)\rangle - \frac{\epsilon + p}{n} \langle n(\mathbf{r}, t)\rangle\right] - \kappa\nabla^2 T(\mathbf{r}, t) = 0. \tag{50c}$$

The analysis of the diffusion equation (50a) follows along exactly the same lines as the analysis of spin diffusion given earlier. We suppose $\mathbf{g}_t(\mathbf{r}, t)$ dies off at large distances so that we may define

$$\mathbf{g}_t(\mathbf{k}, z) = \int d\mathbf{r} \int_0^\infty dt\, e^{-i\mathbf{k}\cdot\mathbf{r}+izt} \langle \mathbf{g}_t(\mathbf{r}, t)\rangle$$

$$\mathbf{g}_t(\mathbf{k}) = \int d\mathbf{r}\, e^{-i\mathbf{k}\cdot\mathbf{r}} \langle \mathbf{g}_t(\mathbf{r}, 0)\rangle = mn\mathbf{v}_t(\mathbf{k}).$$

We then find

$$\mathbf{g}_t(\mathbf{k}, z) = \frac{mn\mathbf{v}_t(\mathbf{k})}{-iz + (\eta k^2/mn)}. \tag{51}$$

Equations (50b) and (50c) can be analyzed in a very similar way. We define

$$n(\mathbf{k}, z) = \int d\mathbf{r} \int_0^\infty dt\, e^{-i\mathbf{k}\cdot\mathbf{r}+izt} \langle n(\mathbf{r}, t)\rangle$$

$$p(\mathbf{k}, z) = \int d\mathbf{r} \int_0^\infty dt\, e^{-i\mathbf{k}\cdot\mathbf{r}+izt} p(\mathbf{r}, t)$$

$$n(\mathbf{k}) = \int d\mathbf{r}\, e^{-i\mathbf{k}\cdot\mathbf{r}} \langle n(\mathbf{r}, 0)\rangle, \text{ etc.}$$

We notice that we can guarantee that $\langle \partial n(\mathbf{r}, t)/\partial t\rangle_{t=0} = 0$ by taking the longitudinal part of $v(\mathbf{k})$ to be zero initially. With this additional requirement, the transform of Eqs. (50b) and (50c) become

$$imz(-iz + D_l k^2)\, n(\mathbf{k}, z) - k^2 p(\mathbf{k}, z)$$

$$= -m(-iz + D_l k^2)\, n(\mathbf{k}) - iz\left[\epsilon(\mathbf{k}, z) - \frac{\epsilon + p}{n} n(\mathbf{k}, z)\right] + \kappa k^2 T(\mathbf{k}, z) \tag{52a}$$

$$= -\left[\epsilon(\mathbf{k}) - \frac{\epsilon + p}{n} n(\mathbf{k})\right] \tag{52b}$$

where we have introduced the abbreviation $D_l = (\frac{4}{3}\eta + \zeta)/mn$ for the "longitudinal" diffusion coefficient.

Notice that (52b) involves the quantity

$$q(\mathbf{k}, z) = \epsilon(\mathbf{k}, z) - \frac{\epsilon + p}{n} n(\mathbf{k}, z) \tag{53}$$

which is the change in the energy density minus the enthalpy per particle times the change in the number density. The response $q(\mathbf{k}, z)$ and the corresponding operator,

$$q(\mathbf{r}, t) = \epsilon(\mathbf{r}, t) - \frac{\epsilon + p}{n} n(\mathbf{r}, t)$$

will play an important role in all that follows. To understand q, we recall the thermodynamic relation

$$TdS = dE + pdV$$

which holds at constant particle number, N. If $dN = 0$, we have the additional identities

$$-dV/V = dn/n$$

and

$$dE = d(\epsilon V) = Vd\epsilon + \epsilon dV = V[d\epsilon - (\epsilon/n)dn].$$

Thus, at constant N

$$\frac{T}{V} dS = d\epsilon - \frac{\epsilon + p}{n} dn.$$

This permits us to identify $q(\mathbf{r}, t)$ as an operator whose changes represent T times the change in the entropy density. Thus, we shall call $q(\mathbf{r}, t)$ the density of heat energy.

We are, of course, permitted to use any convenient set of variables in analyzing Eqs. (52). It will prove convenient to use the matter density $n(\mathbf{k}, z)$ and the heat energy density $q(\mathbf{k}, z)$. Because the system is in local thermodynamic equilibrium the temperature and pressure can be written as[10]

$$T(\mathbf{k}, z) = \left.\frac{\partial T}{\partial n}\right|_S n(\mathbf{k}, z) + \frac{V}{T} \left.\frac{\partial T}{\partial S}\right|_n q(\mathbf{k}, z)$$

$$p(\mathbf{k}, z) = \left.\frac{\partial p}{\partial n}\right|_S n(\mathbf{k}, z) + \frac{V}{T} \left.\frac{\partial p}{\partial S}\right|_n q(\mathbf{k}, z).$$

[10] Because our identification of q was made at fixed N, all the thermodynamic derivatives here and below must be taken at fixed N.

For the variables which characterize the initial conditions, it is convenient to use not $q(\mathbf{k})$ and $n(\mathbf{k})$ but the pressure and temperature defined by

$$n(\mathbf{k}) = \frac{\partial n}{\partial p}\bigg|_T p(\mathbf{k}) + \frac{\partial n}{\partial T}\bigg|_p T(\mathbf{k})$$

$$q(\mathbf{k}) = \frac{T}{V}\frac{\partial S}{\partial p}\bigg|_T p(\mathbf{k}) + \frac{T}{V}\frac{\partial S}{\partial T}\bigg|_p T(\mathbf{k}).$$

Written in terms of these new variables, (52) becomes

$$\left[izm(-iz + D_l k^2) - k^2 \frac{\partial p}{\partial n}\bigg|_S\right] n(\mathbf{k}, z) - k^2 \frac{V}{T}\frac{\partial p}{\partial S}\bigg|_n q(\mathbf{k}, z)$$
$$= -m(-iz + D_l k^2)\left[\frac{\partial n}{\partial p}\bigg|_T p(\mathbf{k}) + \frac{\partial n}{\partial T}\bigg|_p T(\mathbf{k})\right] \qquad (54a)$$

$$\left[-iz + \kappa k^2 \frac{V}{T}\frac{\partial T}{\partial S}\bigg|_n\right] q(\mathbf{k}, z) + \kappa k^2 \frac{\partial T}{\partial n}\bigg|_S n(\mathbf{k}, z)$$
$$= \frac{T}{V}\frac{\partial S}{\partial T}\bigg|_p T(\mathbf{k}) + \frac{T}{V}\frac{\partial S}{\partial p}\bigg|_T p(\mathbf{k}). \qquad (54b)$$

Before discussing the general solution to (54), we consider a particular simplification which occurs at very low temperatures. As the temperature goes to zero the coupling between the mechanical variables (the pressure and the density of particles) and the thermal variables (the temperature and entropy) becomes very weak. The pressure becomes a function of the density not the temperature. Hence the thermodynamic derivatives coupling $q(\mathbf{k}, z)$ and $T(\mathbf{k})$ into Eq. (54a) disappear. Similarly, the entropy depends more sensitively on the temperature than on the density or pressure; consequently the thermodynamic derivatives which couple $n(\mathbf{k}, z)$ and $p(\mathbf{k})$ into (54b) vanish. In this case the solution to the equations is

$$n(\mathbf{k}, z) = -(-iz + D_l k^2)\frac{\partial n}{\partial p} p(\mathbf{k})\left[z^2 - \frac{1}{m}\frac{dp}{dn}k^2 + izD_l k^2\right]^{-1} \qquad (55a)$$

and

$$q(\mathbf{k}, z) = \frac{T}{V}\frac{dS}{dT} T(\mathbf{k})\left[-iz + \kappa k^2 \frac{V}{T}\frac{dT}{dS}\right]^{-1}. \qquad (55b)$$

Equation (55b) states that the temperature satisfies a diffusion equation

$$\frac{\partial}{\partial t} T(\mathbf{r}, t) = D_T \nabla^2 T(\mathbf{r}, t)$$

for $t > 0$, with the thermal diffusivity given by

$$D_T = \kappa \Big/ \frac{T}{V}\frac{dS}{dT}.$$

Equation (55a) states that the density and pressure satisfy a damped sound wave equation

$$\left(\frac{\partial^2}{\partial t^2} - c^2\nabla^2 - \Gamma\frac{\partial}{\partial t}\nabla^2\right) p(\mathbf{r}, t) = 0$$

for $t > 0$, with the sound velocity c given by

$$mc^2 = dp/dn$$

and the sound wave damping constant

$$\Gamma = D_l\,.$$

By examining the solutions to Eqs. (54) in the general case, we see that sound propagation can be isolated from heat diffusion whenever k is so small that

$$(D_T k^2)^2 \ll c^2 k^2.$$

For these wavelengths the solutions to (54), omitting only terms of order $(D_T k/c)^2$, are

$$\begin{aligned} n(\mathbf{k}, z) &= p(\mathbf{k}) \left.\frac{\partial n}{\partial p}\right|_T \left(1 - \frac{c_v}{c_p}\right)[-iz + D_T k^2]^{-1} \\ &\quad - p(\mathbf{k}) \left.\frac{\partial n}{\partial p}\right|_T \left[\Gamma\frac{c_v}{c_p}k^2 + D_T\left(1 - \frac{c_v}{c_p}\right)k^2 - iz\frac{c_v}{c_p}\right][z^2 - c^2k^2 + iz\Gamma k^2]^{-1} \qquad (56) \\ &\quad + T(\mathbf{k}) \left.\frac{\partial n}{\partial T}\right|_p [-iz + D_T k^2]^{-1} - T(\mathbf{k}) \left.\frac{\partial n}{\partial T}\right|_p D_T k^2[z^2 - c^2k^2 + iz\Gamma k^2]^{-1} \end{aligned}$$

and

$$\begin{aligned} q(\mathbf{k}, z) &= T(\mathbf{k}) mnc_p[-iz + D_T k^2]^{-1} + p(\mathbf{k}) \frac{T}{V}\left.\frac{\partial S}{\partial p}\right|_T [-iz + D_T k^2]^{-1} \\ &\quad - p(\mathbf{k}) \frac{T}{V}\left.\frac{\partial S}{\partial p}\right|_T D_T k^2[z^2 - c^2k^2 + iz\Gamma k^2]^{-1} \end{aligned} \qquad (57)$$

where c_v and c_p are the specific heats at constant volume and pressure

$$mnc_p = \frac{T}{V}\left.\frac{\partial S}{\partial T}\right|_p$$

$$mnc_v = \frac{T}{V}\left.\frac{\partial S}{\partial T}\right|_n.$$

In this more general situation, the sound velocity is given by

$$mc^2 = \left.\frac{\partial p}{\partial n}\right|_S = \left.\frac{c_p}{c_v}\frac{\partial p}{\partial n}\right|_T \tag{58}$$

while the thermal diffusivity and the sound wave damping constants are

$$D_T = \kappa/mnc_p \tag{59}$$

$$\Gamma = D_l + D_T[(c_p/c_v) - 1]. \tag{60}$$

B. Construction of Disturbance for Correlation Function Description

Before we can use the solution for the hydrodynamic equations that we have just derived, we must look into the following conceptual problem. We wish to compare the previous description with a description in which we mechanically displace a system from equilibrium in such a way that all variations in time and space are slow. In our discussion of spin diffusion there was a very natural mechanism by which this deviation from complete equilibrium could be mechanically induced. The spin magnetic moment could be altered by applying an external magnetic field. There exists no such handle for the molecules in a fluid. In particular, the mechanical forces by which a heat conduction process is set up are rather subtle.

Now almost any force which disturbs the system from equilibrium will set up heat conduction and sound propagation processes, and if we wait long enough, these will be the only modes we will generally find. However if we are to infer the form of the correlation functions from the hydrodynamic equations, which are only true when the system is in local equilibrium, we must apply a disturbance which guarantees that the system is in local equilibrium at all times, not just for long times. That is to say, we must select an interaction Hamiltonian which disturbs the system in such a way that the system is even in local equilibrium initially.

To aid us in choosing such a mechanical disturbance, we recall the method for computing the average value of an operator, $A(\mathbf{r}, t)$, in a system in full equilibrium. If the system is moving with a velocity $\mathbf{v}$, the average of A in the grand canonical ensemble is

$$\begin{aligned} \langle A(\mathbf{r}, t)\rangle &= \operatorname{tr}[\rho A(\mathbf{r}, t)] \\ \rho &= \exp \Xi\, [\operatorname{Tr} \exp \Xi]^{-1} \\ \Xi &= -\beta\left[\mathcal{H}_0 - \mu\mathfrak{N} + \tfrac{1}{2} mv^2\mathfrak{N} - \int d\mathbf{r}\mathbf{g}(\mathbf{r})\cdot\mathbf{v}\right]. \end{aligned} \tag{61}$$

The thermodynamic state of the system is described by μ, β, and $\mathbf{v}$. If the velocity is small, the v^2 term in (61) may be neglected.

We note that it is possible to represent a situation in which the chemical potential changes from μ to $\mu + \delta\mu$, the temperature changes from T to $T + \delta T$, and the velocity goes from zero to $\delta\mathbf{v}$ by writing the density matrix in the complete equilibrium form

$$\rho = \exp\left[-\beta(\mathcal{H} - \mu\mathfrak{N})\right]\{\mathrm{Tr}\exp\left[-\beta(\mathcal{H} - \mu\mathfrak{N})\right]\}^{-1}$$

with a modified Hamiltonian, $\mathcal{H}_0 + \delta\mathcal{H}$, where

$$\delta\mathcal{H} = -\int d\mathbf{r}\left\{\frac{\delta T}{T}\left[\epsilon(\mathbf{r}) - \mu n(\mathbf{r})\right] + \delta\mu n(\mathbf{r}) + \delta\mathbf{v}\cdot g(\mathbf{r})\right\}. \tag{62}$$

In analogy with (62) it is appealing to use the interaction Hamiltonian

$$\begin{aligned}\delta\mathcal{H}(t) = -\int d\mathbf{r}\left\{\frac{\delta T(\mathbf{r})}{T}\left[\epsilon(\mathbf{r}, t) - \mu n(\mathbf{r}, t)\right] + \delta\mu(r)n(\mathbf{r}, t)\right.&\\ \left. + \,\delta\mathbf{v}(r)\cdot\mathbf{g}(\mathbf{r}, t)\right\} e^{\epsilon t} &\quad \text{for } t < 0, \qquad (63)\\ = 0 &\quad \text{for } t > 0.\end{aligned}$$

to represent a situation in which the system is in local thermodynamic equilibrium for all times less than zero. We would, of course, guess that the local velocity would be $\delta\mathbf{v}(\mathbf{r})$, the local temperature $T + \delta T(\mathbf{r})$ and the local chemical potential $\mu + \delta\mu(\mathbf{r})$. If we can show that the system is in local thermodynamic equilibrium for times less than zero, then we can use (63) as an interaction Hamiltonian for producing hydrodynamic flow.

To justify the use of (63), we must prove that for all times less than zero, the average of any operator $A(r, t)$ changes from its complete equilibrium value by the amount

$$\delta\langle A(\mathbf{r}, t)\rangle = \left.\frac{\partial A}{\partial\mu}\right|_{T,\mathbf{v}}\delta\mu(\mathbf{r}) + \left.\frac{\partial A}{\partial T}\right|_{\mu,\mathbf{v}}\delta T(\mathbf{r}) + \left.\frac{\partial A}{\partial\mathbf{v}}\right|_{\mu,T}\cdot\delta\mathbf{v}(\mathbf{r}) \quad \text{for } t < 0. \tag{64}$$

The derivatives indicated in (64) are, of course, thermodynamic derivatives.

The proof of (64) is essentially identical with the proof (in Section II, C) that $\lim_{k\to 0}\chi(k)$ is the thermodynamic derivative dM/dH. For simplicity, we consider the case in which $\delta\beta = \delta\mathbf{v} = 0$. We write a spectral form for the $A - n$ commutator

$$\langle[A(\mathbf{r}, t), n(\mathbf{r}', t')]\rangle = \int\frac{d\omega}{\pi}\int\frac{d\mathbf{k}}{(2\pi)^3}e^{i\mathbf{k}\cdot(\mathbf{r}-\mathbf{r}')-i\omega(t-t')}\chi''_{A,n}(\mathbf{k}, \omega).$$

According to the fluctuation-dissipation theorem (Eq. (43)) the $A - n$ anticommutator is

$$\langle\{A(\mathbf{r},t) - \langle A\rangle_{\text{eq.}}, n(\mathbf{r}',t') - \langle n\rangle_{\text{eq.}}\}\rangle_{\text{eq.}}$$

$$= \int \frac{d\omega}{\pi} \int \frac{d\mathbf{k}}{(2\pi)^3} e^{i\mathbf{k}\cdot(\mathbf{r}-\mathbf{r}')-i\omega(t-t')} \chi''_{An}(\mathbf{k},\omega) \coth\frac{\beta\omega}{2}.$$

We can calculate the thermodynamic derivative $\partial A/\partial\mu$ from Eq. (61), obtaining

$$\left.\frac{\partial A}{\partial\mu}\right|_{T,\mathbf{v}} = \frac{\beta}{2}\int d\mathbf{r}'[\langle\{A(\mathbf{r},t), n(\mathbf{r}',t')\}\rangle_{\text{eq.}} - \langle A\rangle_{\text{eq.}}\langle n\rangle_{\text{eq.}}].$$

It is of course implicit in writing this equation that the static correlation function has no long range part or that the integral converges. This is not the case with $\partial A/\partial v$, or $\{A, v\}$ in a superfluid. Apart from such exceptional situations, we may write

$$\left.\frac{\partial A}{\partial\mu}\right|_{T,v} = \int \frac{d\omega}{\pi}\frac{\beta}{2}\coth\frac{\beta\omega}{2}\,\chi''_{A,n}(\mathbf{0},\omega).$$

Since the total number of particles is independent of time $\chi''_{A,n}(0,\omega)/\omega$ must be just a delta function at zero frequency. Therefore, just as before, we can make the replacement

$$(\beta\omega/2)\coth(\beta\omega/2)\chi''_{An}(\mathbf{0},\omega) = \chi''_{An}(\mathbf{0},\omega)$$

and find

$$\left.\frac{\partial A}{\partial\mu}\right|_{T,v} = \int \frac{d\omega}{\pi}\frac{\chi''_{An}(\mathbf{0},\omega)}{\omega}. \tag{65}$$

We can use Eq. (12) to calculate the response to the time-dependent disturbance (63). Then, in just the same way as we obtained Eq. (15), we find

$$\delta\langle A(\mathbf{r},t)\rangle = \int \frac{d\mathbf{k}}{(2\pi)^3}\mu(\mathbf{k})e^{i\mathbf{k}\cdot\mathbf{r}}\int\frac{d\omega}{\pi}\frac{\chi''_{An}(\mathbf{k},\omega)}{\omega} \qquad \text{for } t \leqq 0$$

where $\mu(\mathbf{k})$ is the Fourier transform of $\delta\mu(\mathbf{r})$. Thus $A(\mathbf{k})$, the Fourier transform of $\delta\langle A(\mathbf{r},0)\rangle$, is

$$A(\mathbf{k}) = \mu(\mathbf{k})\int\frac{d\omega}{\pi}\frac{\chi''_{An}(\mathbf{k},\omega)}{\omega}. \tag{66}$$

If $\delta\mu(\mathbf{r})$ contains only very small wave numbers (or equivalently, varies slowly in space) then the $\mathbf{k}$ which appears in (66) may be replaced by $\mathbf{0}$. A comparison of (65) and (66) indicates that

$$A(\mathbf{k}) = \left.\frac{\partial A}{\partial\mu}\right|_{T,\mathbf{v}}\mu(\mathbf{k})$$

or

$$\delta\langle A(\mathbf{r},t)\rangle = \left.\frac{\partial A}{\partial \mu}\right|_{T,\mathbf{v}} \delta\mu(\mathbf{r}) \qquad \text{for } t < 0.$$

In this way we can verify that Eq. (64) is valid whenever $\delta\mu(\mathbf{r})$, $\delta T(\mathbf{r})$, and $\delta\mathbf{v}(\mathbf{r})$ vary sufficiently slowly in space. In the limit of slow variation, the system appears to be in local thermodynamic equilibrium and $\mu + \delta\mu(\mathbf{r})$, $T + \delta T(\mathbf{r})$, $\delta\mathbf{v}(\mathbf{r})$ are just the local chemical potential, temperature, and velocity.

This is, however, a weak link in this derivation of (64), namely, our assumption that

$$\lim_{\mathbf{k}\to 0} \int \frac{d\omega}{\pi} \frac{\chi''_{ij}(\mathbf{k},\omega)}{\omega} = \int \frac{d\omega}{\pi} \frac{\chi''_{ij}(\mathbf{0},\omega)}{\omega}.$$

There are situations in which $\mathbf{k} = \mathbf{0}$ is quite different from all $\mathbf{k} \neq \mathbf{0}$. This difference will appear whenever there are infinitely long-ranged correlations. These correlations tend to affect $\mathbf{k} = \mathbf{0}$ modes very differently from $\mathbf{k} \neq \mathbf{0}$ ones. Thus, for example, in a Coulomb system, the exact shape and nature of the surface will determine the behavior of the plasma oscillation at $\mathbf{k} = \mathbf{0}$. Also, in a superfluid, the surfaces and past history of the body will determine the relative proportions of superfluid and normal flow at $\mathbf{k} = \mathbf{0}$. This effect appears because of the infinitely long-ranged correlations in the superfluid component. Thus, in these cases, Eq. (64) fails to be correct.

For the purposes of the above argument, the chemical potential, the temperature, and the velocity were a convenient complete set of variables. However, the chemical potential does not have any direct physical meaning in the one-component system. Consequently, it is more convenient to eliminate the local chemical potential in favor of the local pressure by using the thermodynamic relation

$$dp = n\,d\mu + (S/V)\,dT = n\,d\mu + (\epsilon + p - \mu n)\,dT/T \tag{67a}$$

to define

$$p(\mathbf{k}) = n\mu(\mathbf{k}) + (\epsilon + p - \mu n)T(\mathbf{k})/T. \tag{67b}$$

To see that $p(\mathbf{k})$ has the significance of a change in the pressure in the limit of slow spatial variation, it is only necessary to use the thermodynamic relation (67a) to rewrite (64) as

$$A(\mathbf{r},t) = \left[n\delta\mu(\mathbf{r}) + (\epsilon + p - \mu n)\frac{\delta T(\mathbf{r})}{T}\right]\left.\frac{\partial A}{\partial p}\right|_{T,\mathbf{v}} + \delta T(\mathbf{r})\left.\frac{\partial A}{\partial T}\right|_{p,\mathbf{v}} + \mathbf{v}(\mathbf{r})\cdot\left.\frac{\partial A}{\partial \mathbf{v}}\right|_{p,T}.$$

Then, if we use (67b) to define

$$\delta p(\mathbf{r}) = n\delta\mu(\mathbf{r}) + (\epsilon + p - \mu n)\delta T(\mathbf{r})/T$$
$$\equiv \int \frac{d\mathbf{k}}{(2\pi)^3} e^{i\mathbf{k}\cdot\mathbf{r}} p(\mathbf{k}),$$

we have

$$\delta\langle A(\mathbf{r}, t)\rangle = \delta p(\mathbf{r}) \left.\frac{\partial A}{\partial p}\right|_{T,\mathbf{v}} + \delta T(\mathbf{r}) \left.\frac{\partial A}{\partial T}\right|_{p,\mathbf{v}} + \delta\mathbf{v}(\mathbf{r}) \cdot \left.\frac{\partial A}{\partial \mathbf{v}}\right|_{T,p} \tag{68}$$

so that $\delta p(\mathbf{r})$ does indeed have the meaning of a change in the pressure.

Finally, we eliminate $\delta\mu(\mathbf{r})$ from the disturbance (63) by making use of (67b). With this substitution Eq. (63) becomes

$$\delta\mathcal{H}(t) = -\int d\mathbf{r} \left[\frac{\delta p(\mathbf{r})}{\langle n\rangle} n(\mathbf{r}, t) + \frac{\delta T(\mathbf{r})}{T} q(\mathbf{r}, t) + \mathbf{v}(\mathbf{r})\cdot\mathbf{g}(\mathbf{r}, t)\right] e^{\epsilon t} \quad \text{for } t < 0, \tag{69}$$
$$= 0 \quad \text{for } t > 0,$$

where $q(\mathbf{r}, t)$ is the operator previously encountered which represents changes in the density of heat energy

$$q(\mathbf{r}, t) = \epsilon(\mathbf{r}, t) - \frac{\langle\epsilon\rangle + p}{\langle n\rangle} n(\mathbf{r}, t).$$

C. Representation of the Commutators

We can now write the response of the system to the disturbance (69) as

$$\delta\langle A(\mathbf{r}, t)\rangle = \int_{-\infty}^{t} dt' \int d\mathbf{r}' e^{\epsilon t'} \{\ \} \quad \text{for } t < 0,$$
$$= \int_{-\infty}^{0} dt' \int d\mathbf{r} e^{\epsilon t'} \{\ \} \quad \text{for } t > 0$$

where

$$\{\ \} = \langle[A(\mathbf{r}, t), n(\mathbf{r}', t')]\rangle_{\text{eq}}.\delta p(\mathbf{r}')/n + \langle[A(\mathbf{r}, t), q(\mathbf{r}', t')]\rangle_{\text{eq}}.\delta T(\mathbf{r}')/T + \langle[A(\mathbf{r}, t), \mathbf{g}(\mathbf{r}', t')]\rangle_{\text{eq}}\cdot\mathbf{v}(\mathbf{r}').$$

We introduce the representation (42a) for the commutators, the A_i being q, n, and the components of $\mathbf{g}$. For times less than zero, we then have

$$A(\mathbf{k}) = \int \delta\langle A(\mathbf{r}, t)\rangle e^{-i\mathbf{k}\cdot\mathbf{r}}\, d\mathbf{r}$$
$$= \int \frac{d\omega}{\pi} \frac{\chi''_{A,n}(\mathbf{k}, \omega)}{\omega} \frac{p(\mathbf{k})}{n} + \int \frac{d\omega}{\pi} \frac{\chi''_{A,q}(\mathbf{k}, \omega)}{\omega} \frac{T(\mathbf{k})}{T} + \int \frac{d\omega}{\pi} \frac{\chi''_{A,\mathbf{g}}(\mathbf{k}, \omega)}{\omega} \cdot \mathbf{v}(\mathbf{k}). \tag{70}$$

For times greater than zero, the response may be represented by

$$A(\mathbf{k}, z) = \int_0^\infty dte^{izt} \int d\mathbf{r}e^{-i\mathbf{k}\cdot\mathbf{r}}\delta\langle A(\mathbf{r}, t)\rangle$$

$$= \int \frac{d\omega}{\pi i} \frac{\chi''_{A,n}(\mathbf{k}, \omega)}{\omega(\omega - z)} \frac{p(\mathbf{k})}{n} + \int \frac{d\omega}{\pi i} \frac{\chi''_{A,q}(\mathbf{k}, \omega)}{\omega(\omega - z)} \frac{T(\mathbf{k})}{T} \qquad (71)$$

$$+ \int \frac{d\pi}{\pi i} \frac{\chi''_{A,\mathbf{g}}(\mathbf{k}, \omega)}{\omega(\omega - z)} \cdot \mathbf{v}(\mathbf{k}).$$

We are interested in the cases in which $A(\mathbf{r}, t)$ is $n(\mathbf{r}, t)$, $q(\mathbf{r}, t)$, or $\mathbf{g}(\mathbf{r}, t)$. Therefore, we shall briefly discuss the properties of the Fourier transforms of the commutators formed from these conserved operators. By using time-reversal invariance, rotational invariance, and the Hermitian nature of the operators, one can show that $\chi''_{n,n}$, $\chi''_{q,q}$, $\chi''_{n,q}$ and $\chi''_{q,n}$ are each real odd functions of ω, and that

$$\chi''_{n,q}(k, \omega) = \chi''_{q,n}(k, \omega). \qquad (72)$$

Equation (72) expresses a reciprocity which was first discussed by Onsager (*16*). From (71) and (72) it follows that the response of the density to a change in the temperature (at constant pressure) differs by only a factor of n from the change in the entropy density induced by a change in the pressure (at constant temperature). In more complex situations than those we shall consider here, this reciprocity leads to a connection between transport coefficients which would otherwise have no obvious relation with one another.

The Fourier transform of the momentum-momentum commutator is a tensor, since it is an average of a direct product of two vectors. However, the only tensor quantities of which $\chi''_{g_i,g_j}(\mathbf{k}, \omega)$ could be composed, in the absence of long-range correlations, are the direct product k_ik_j and the unit matrix $\delta_{i,j}$. We find it convenient to express χ''_{g_i,g_j} in terms of linear combinations of these as

$$\chi''_{g_i,g_j} = (k_ik_j/k^2)\chi''_l(k, \omega) + (\delta_{ij} - k_ik_j/k^2)\chi''_t(k, \omega). \qquad (73)$$

Here the l and t stand for longitudinal and transverse since the splitting that we have indicated in (73) divides the tensor into two parts, one with components in the direction of k, the other whose dot product with k is zero. Both parts are real functions, odd in the frequency variable.

The conservation law

$$\frac{\partial n}{\partial t} + \frac{1}{m} \nabla\cdot\mathbf{g} = 0$$

enables us to express $\chi''_{\mathbf{g},n}$ in terms of χ''_l as

$$\chi''_{n,\mathbf{g}}(\mathbf{k}, \omega) = \chi''_{\mathbf{g},n}(\mathbf{k}, \omega) = \frac{\mathbf{k}}{m\omega} \chi''_l(k, \omega) \qquad (74)$$

while a double application of this law gives

$$\chi''_{n,n}(k, \omega) = \frac{k^2}{m^2\omega^2} \chi''_l(k, \omega). \tag{75}$$

One more result of the number conservation law is

$$\chi''_{\mathbf{g},q}(\mathbf{k}, \omega) = \chi''_{q,\mathbf{g}}(\mathbf{k}, \omega) = \frac{\mathbf{k}}{m\omega} \chi''_{q,n}(k, \omega). \tag{76}$$

D. Sum Rules

By comparing Eqs. (70) and (68) we can deduce a variety of Kramers-Kronig relation sum rules analogous to Eq. (26) for the integrals of the various commutators. For example, we may take $A(\mathbf{r}, t) = n(\mathbf{r}, t)$. Then Eq. (68) gives

$$n(\mathbf{k}) = \left.\frac{\partial n}{\partial p}\right|_T p(\mathbf{k}) + \left.\frac{\partial n}{\partial T}\right|_p T(\mathbf{k})$$

so that Eq. (70) implies

$$\lim_{k\to 0} \int \frac{d\omega}{\pi} \frac{\chi''_{n,n}(k, \omega)}{\omega} = \lim_{k\to 0} \chi_{nn}(k) = n \left.\frac{\partial n}{\partial p}\right|_T \tag{77a}$$

$$\lim_{k\to 0} \int \frac{d\omega}{\pi} \frac{\chi''_{n,q}(k, \omega)}{\omega} = T \left.\frac{\partial n}{\partial T}\right|_p . \tag{77b}$$

For $A(\mathbf{r}, t) = q(\mathbf{r}, t)$ we find

$$\lim_{k\to 0} \int \frac{d\omega}{\pi} \frac{\chi''_{n,q}(k, \omega)}{\omega} = n \frac{T}{V} \left.\frac{\partial S}{\partial p}\right|_T = T \left.\frac{\partial n}{\partial T}\right|_p \tag{77c}$$

$$\lim_{k\to 0} \int \frac{d\omega}{\pi} \frac{\chi''_{q,q}(k, \omega)}{\omega} = \frac{T^2}{V} \left.\frac{\partial S}{\partial T}\right|_p = mnc_p T. \tag{77d}$$

For $A(\mathbf{r}, t) = \mathbf{g}(\mathbf{r}, t)$

$$A(\mathbf{k}) = mn\mathbf{v}(\mathbf{k})$$

so that Eq. (70) implies

$$\lim_{k\to 0} \int \frac{d\omega}{\pi} \frac{\chi''_{g_i g_j}(k, \omega)}{\omega} = \delta_{ij}\, mn$$

which may be written

$$\lim_{k\to 0} \int \frac{d\omega}{\pi} \frac{\chi''_t(k, \omega)}{\omega} = mn. \tag{77e}$$

and

$$\int \frac{d\omega}{\pi} \frac{\chi''_l(k, \omega)}{\omega} = \int \frac{d\omega}{\pi} \frac{\omega^2 m^2}{k^2} \frac{\chi''_{nn}(k, \omega)}{\omega} = mn. \tag{77f}$$

Finally, by using the fact that the heat current is zero even when the system is in motion, it is possible to show that

$$\lim_{k\to 0} \int \frac{d\omega}{\pi} \chi_{j^q,g}(k, \omega) = 0.$$

By using the conservation laws, we can derive from this relation the sum rule

$$\lim_{k\to 0} \int \frac{d\omega}{\pi} \frac{\omega^2}{mk^2} \chi''_{q,n}(k, \omega) = 0. \tag{77g}$$

Since Eq. (68) is only valid for small k, the sum rules (77) need only be valid in this limit. However, not all of the identities are really subject to this restriction. Equation (77f), which expresses a sum rule on the density-density correlation function is, in fact, valid for all k. This sum rule can be derived from an argument identical to the one that we used to get (27). That is to say, this result is a consequence of the exact commutation relation

$$[n(\mathbf{r}), \mathbf{g}(\mathbf{r}')] = -i\nabla[n(\mathbf{r})\delta(\mathbf{r} - \mathbf{r}')]$$

which holds whenever there are velocity-independent forces. In fact, the sum rule (77f) is a famous result. In solid state physics, it is usually referred to as the longitudinal f-sum rule. It has played a very important role in the discussion of the BCS theory of superconductivity.[11] In neutron scattering studies, it is known as the Placzek sum rule (*9*).

In the classical limit, subject to the existence of the velocity correlation function Fourier transform or the absence of long-range order, (77e) is also exact for all k. It is a statement of the van Leeuwen theorem that the orbital magnetic susceptibility of a classical system vanishes. Landau (*18*) has discussed how diamagnetic susceptibility can, in fact, appear in a quantum mechanical system. In our language, this is a consequence of the fact that

$$\chi_l(k) - \chi_t(k) = \mathcal{O}(k^2)$$

can contain a term of order k^2 in the limit of small k^2.

The quantum effects are even more drastic in a superfluid. In a superfluid (77e) is not even satisfied in the limit $k^2 \to 0$. This failure of the sum rule is reflected in the anomalous electromagnetic properties of the superconductor, the Meissner effect and the persistence of supercurrents and in the corresponding properties of liquid helium. The source of this failure has been indicated. In superfluids, correlation functions which involve the momentum die off too slowly in space to permit the analysis we have employed.

The sum rule (77a) is particularly interesting since it is an additional sum rule

[11] See, for example, ref. *17*.

on the density-density correlation function, a function which is very important both experimentally and theoretically.[12]

An alternative, and very useful, expression of these sum rules is given by taking the limit as z goes to infinity in Eq. (71). Then if A is n, q, and $\mathbf{g}$, we find

$$\lim_{z\to\infty} -izn(\mathbf{k},z) = p(\mathbf{k})\left[\frac{\partial n}{\partial p}\bigg|_T + \frac{k^2}{mz^2} + \mathcal{O}\left(\frac{1}{z^3}\right)\right] + T(\mathbf{k})\left[\frac{\partial n}{\partial T}\bigg|_p + \mathcal{O}\left(\frac{1}{z^3}\right)\right] + \mathbf{k}\cdot\mathbf{v}(\mathbf{k})\left[\frac{1}{z} + \mathcal{O}\left(\frac{1}{z^2}\right)\right] \tag{78a}$$

$$\lim_{z\to\infty} -izq(\mathbf{k},z) = p(\mathbf{k})\left[\frac{T}{V}\frac{\partial S}{\partial p}\bigg|_T + \mathcal{O}\left(\frac{1}{z^3}\right)\right] + T(\mathbf{k})\left[mnc_p + \mathcal{O}\left(\frac{1}{z}\right)\right] + \mathbf{k}\cdot\mathbf{v}(\mathbf{k})\mathcal{O}\left(\frac{1}{z^2}\right) \tag{78b}$$

$$\lim_{z\to\infty} -iz\mathbf{g}(\mathbf{k},z) = \mathbf{v}(\mathbf{k})[mn + \mathcal{O}(1/z)] + \mathbf{k}p(\mathbf{k})[1 + \mathcal{O}(1/z)] + \mathbf{k}T(\mathbf{k})\mathcal{O}(1/z). \tag{78c}$$

Notice that these expressions agree, as they must, with the results of our hydrodynamic analysis as given in Eqs. (51), (56), and (57).

E. Low Temperature Forms for Correlation Functions

The hydrodynamic analysis led to particularly simple forms for the correlation functions in the low temperature limit. (See Eqs. (55a) and (55b).) The response $q(\mathbf{k}, z)$ had no term proportional to $p(\mathbf{k})$ and was proportional to $T(\mathbf{k})$. By comparing (55b) and (71), we see that for z in the upper half of the complex plane

$$\int \frac{d\omega'}{\pi i}\frac{\chi''_{q,q}(k,\omega')}{\omega'(\omega'-z)} = \frac{T^2}{V}\frac{dS}{dT}\left[-iz + \kappa k^2 \frac{V}{T}\frac{dT}{dS}\right]^{-1}. \tag{79}$$

Since

$$\chi''_{q,q}(k,\omega) = \mathrm{Re}\int \frac{d\omega'}{\pi i}\frac{\chi''_{q,q}(k,\omega')}{\omega'(\omega'-\omega-i\epsilon)}$$

we find the same diffusion structure for this heat-energy correlation function as

[12] This sum rule, of course, expresses information about both the commutator and the anticommutator of the density. In its anticommutator form, this result was used by Ornstein and Zernicke in their classical work on critical fluctuations. It has been more recently employed by J. M. Ziman (*19a*). The commutator form has been known to the authors for quite some time. It was discussed by N. D. Mermin (*19b*). It has more recently been stressed by D. Pines.

we found for the spin density correlation function, namely,

$$\chi''_{qq}(k, \omega) = \kappa T k^2 \omega \left[\omega^2 + \left(\kappa k^2 \frac{V}{T} \frac{dT}{dS} \right)^2 \right]^{-1}. \tag{80}$$

(See, for comparison, Eq. (19).)

In the low temperature case, the response of the density is also simple since it contains, according to (55a), no term proportional to $T(k)$ but only a term proportional to $p(k)$. By comparing (55a) and (71) we see

$$\int \frac{d\omega}{\pi i} \frac{\chi''_{n\, n}(k, \omega)}{\omega(\omega - z)} = -\frac{1}{iz} n \frac{\partial n}{\partial p} \left[1 + \frac{1}{m} \frac{\partial p}{\partial n} k^2 \left(z^2 - \frac{1}{m} \frac{\partial p}{\partial n} k^2 + i D_l z k^2 \right)^{-1} \right] \tag{81}$$

and consequently

$$\chi''_{n,n}(k, \omega) = \frac{n k^4 \Gamma \omega}{m} \left[\left(\omega^2 - \frac{1}{m} \frac{dp}{dn} k^2 \right)^2 + (\Gamma \omega k^2)^2 \right] \tag{82}$$

where $\Gamma = D_l$.

Equations (81) and (82) give the density response in the limit of small k. But, because the density correlation function is such an important quantity both experimentally and theoretically, it is worthwhile for us to examine some of the general properties of this function. Our analysis in this case closely parallels the establishment of a dispersion relation for $M(\mathbf{k}, z)$ in section E of the chapter on spin response. In analogy with this work we define

$$\chi_{n,n}(k, z) = \int \frac{d\omega}{\pi} \frac{1}{\omega - z} \chi''_{n,n}(k, \omega)$$

and

$$\chi_{n,n}(k) = \chi_{n,n}(k, 0) \tag{83}$$

and notice that the facts

$$\lim_{z \to \infty} \chi_{n,n}(k, z) = -\frac{1}{z^2} \int \frac{d\omega}{\pi} \omega \chi''_{n,n}(k, \omega) = -\frac{k^2 n}{m z^2}$$

$$\omega \chi''_{n,n}(k, \omega) \geqq 0$$

imply the dispersion relation

$$\frac{1}{z^2} \left[\frac{\chi_{n,n}(k)}{\chi_{n,n}(k, z)} - 1 \right] = -\frac{m \chi_{n,n}(k)}{n k^2} - \int \frac{d\omega'}{\pi} \frac{f_{n,n}(k, \omega')}{\omega' z(\omega' - z)}.$$

When z lies just above the real axis, we have

$$\frac{\chi_{M,M}(k)}{\chi_{M,M}(k, \omega + i\epsilon)} - 1 = -\frac{m\omega^2}{nk^2} \chi_{n,n}(k) - i f_{n,n}(k, \omega) + \mathcal{P} \int \frac{d\omega'}{\pi} \frac{\omega}{\omega'} \frac{f_{n,n}(k, \omega')}{(\omega - \omega')} \tag{84}$$

This result should be compared with Eq. (38) which gave the spin response as

$$\frac{\chi_{M,M}(k)}{\chi_{M,M}(k, \omega + i\epsilon)} - 1 = -\frac{4m\omega^2\chi(k)}{nk^2\gamma^2} - if_{M,M}(k, \omega) + \mathcal{P}\int \frac{d\omega'}{\pi}\frac{\omega}{\omega'}\frac{f_{M,M}(k, \omega')}{(\omega - \omega')}. \tag{38}$$

In the spin case, $f(k, \omega)$ reduced to ω/k^2D_M in the limit as the wave number and the frequency went to zero. Therefore, we decided to define a frequency- and wave number-dependent diffusivity by

$$f_{M,M}(k, \omega) = \omega/k^2D_M(k, \omega). \tag{39}$$

In this sound wave propagation situation, Eq. (81) implies that for small k^2,

$$\frac{1}{\chi_{n,n}(k, \omega + i\epsilon)} - \frac{1}{\chi_{n,n}(k)} = \frac{1}{\chi_{n,n}(k, \omega + i\epsilon)} - \frac{1}{n}\frac{dp}{dn} = -\frac{m\omega^2}{nk^2} - i\omega\Gamma\frac{m}{n}.$$

Therefore, we can write the spectral weight function which appears in (84) as

$$f_{n,n}(k, \omega) = \chi_{n,n}(k)\,\Gamma(k, \omega)\,\frac{m}{n}\,\omega$$

with the knowledge that for low temperature systems $\Gamma(k, \omega)$ reduces to the sound wave damping constant, $\Gamma = D_l$, in the limit as the frequency and wave number go to zero. With this definition the spectral function of (84) becomes

$$\chi_{n,n}^{-1}(k, \omega + i\epsilon) - \chi_{n,n}^{-1}(k) = -\frac{m\omega^2}{nk^2} - \frac{m}{n}i\omega\Gamma(k, \omega) + \frac{m\omega}{n}\mathcal{P}\int\frac{d\omega'}{\pi}\frac{\Gamma(k, \omega')}{\omega - \omega'} \tag{85}$$

Consequently, an exact form for $n(k, z)/p(k)$ is

$$\left.\frac{n(k, z)}{p(k)}\right|_{z=\omega+i\epsilon} = -\frac{1}{\omega}\left\{\chi_{n,n}(k) - \left[\chi_{n,n}^{-1}(k) - \frac{m\omega^2}{nk^2} - i\omega\Gamma(k, \omega)\frac{m}{n} + \frac{m\omega}{n}\mathcal{P}\int\frac{d\omega'}{\pi}\frac{\Gamma(k, \omega')}{\omega - \omega'}\right]^{-1}\right\}$$

and

$$\chi''_{n,n}(k, \omega) = \frac{(n/m)\Gamma(k, \omega)k^4\omega}{\left(\omega^2 - \dfrac{n}{m}\dfrac{k^2}{\chi_{n,n}^{-1}(k)} - \omega^2k^2\,\mathcal{P}\displaystyle\int\frac{d\omega'}{\pi}\frac{\Gamma(k, \omega)}{\omega^2 - \omega'^2}\right)^2 + [\omega k^2\Gamma(k, \omega)]^2} \tag{86}$$

We should emphasize that the equations (85) and (86) are exact. However, whether or not these equations are useful depends critically upon the simplicity, or lack thereof, of the function $\Gamma(k, \omega)$. In the particular case of a low temperature ordinary fluid, we have seen that the equations of fluid mechanics imply that $\Gamma(k, \omega)$ goes to the constant Γ for low frequencies and small wave numbers with $\omega \sim ck$ and also that $\chi_{n,n}(k) \rightarrow n(dp/dn)^{-1}$. However, at nonvanishing temperature $\Gamma(k, \omega)$ depends on the manner in which ω and k approach zero, since $\chi''_{n,n}(k, \omega)$ must include the thermal diffusion process indicated in (56). That is to say, Γ, like χ'', has different limits, depending on the ratio of ω and k as they both approach zero. Specifically it can be shown that $f(k, \omega)$ has a term which behaves like

$$f \cong \left(\frac{c_p}{c_v} - 1\right) D_T k^2 \omega \left[\omega^2 + D_T^2 \left(\frac{c_p}{c_v}\right)^2 k^4\right]^{-1}$$

for small k and ω.

We next comment further on the significance of the function $f(k, \omega)$. We observe that the part to which we have just referred has poles at $\omega = \pm i D_T (c_p/c_v) k^2 = \pm i(\kappa/mn)k^2$ corresponding to relaxation as a result of thermal conduction. In addition, the function $f(k, \omega)$ contained the term which at low frequencies became $m\omega\Gamma\chi/n$. This term vanished at high frequencies. Both terms are consistent with a form

$$f(k, \omega) = \int_0^\infty \frac{d\tau}{\tau} \frac{\phi(k, \tau)}{\omega^2 + (1/\tau)^2} \frac{\omega m \chi}{n} \quad \text{or} \quad \sum_i \frac{\phi_i(k)\omega}{\tau_i(k)\,(\omega^2 + (1/\tau_i(k))^2} \frac{m\chi}{n}.$$

Conversely, such a form would give rise to an expression for $\chi^{-1}(k, z)$

$$\chi^{-1}(k, z) = -\frac{m}{n}\frac{z^2}{k^2} + \chi^{-1}(k) + \chi^{-1}(k) z \mathcal{P} \int \frac{d\omega'}{\pi} \frac{f(k, \omega')}{\omega'(z - \omega')}$$

which could be rewritten for z in the upper half-plane as

$$= -\frac{m}{n}\left\{\frac{z^2}{k^2} - c_\infty^2 - \int_0^\infty d\tau \frac{1}{1 - iz\tau} \phi(k, \tau)\right\}$$

or

$$= -\frac{m}{n}\left\{\frac{z^2}{k^2} - c_\infty^2 - \sum_i \frac{\phi_i(k)}{1 - iz\tau_i(k)}\right\}$$

where

$$c_\infty^2 \equiv \frac{n}{m\chi} + \int_0^\infty d\tau \phi(k, \tau), \quad \text{or} \quad \frac{n}{m\chi} + \sum_i \phi_i(k).$$

The form we have hypothesized therefore corresponds to a weighted distribution of relaxation times, τ. Experimental evidence (*20*) seems to indicate that at least this complicated a form is required. Such a form would be obtained if the function $f(k, \omega)$ could be calculated and if its analytic continuation into the lower half ω-plane only possessed poles, or a branch line on the negative imaginary axis. Generally, however, there may be both real and imaginary parts to the singularities of f, and therefore of continuations of $\chi^{-1}(k, z)$ into the lower half plane. Under these circumstances f contains a distribution of resonant frequencies as well as a distribution of relaxation times.

Finally we remark without exhibiting details that the phase representation, which is convenient for carrying out stationary phase asymptotic evaluations of $\chi^{-1}(k, z)$, in the general case reduces to

$$\omega \cot \delta(k, \omega) \cong \frac{c_v}{c_p} \frac{(\omega^2 - c^2k^2)\omega^2}{Dk^2[1 - (c_v/c_p)](\omega^2 - c^2k^2) - (c_v/c_p)\Gamma k^2\omega^2}$$

and in the low temperature limit to

$$\omega \cot \delta(k, \omega) \cong (c^2k^2 - \omega^2)/\Gamma k^2.$$

F. Evaluation of the Absorptive Susceptibility

At higher temperatures, $n(\mathbf{k}, z)$ and $q(\mathbf{k}, z)$ as given in Eqs. (56) and (57) contain both sound wave and diffusion poles. By using the same device as before, we can calculate the absorptive susceptibilities which describe the long wavelength, low frequency response in this more general case. From (56), (57), and (71) we find, for example, that for $T(\mathbf{k})$ and $\mathbf{v}(\mathbf{k})$ equal to zero in the ordinary fluid

$$\begin{aligned}\chi''_{nn}(k, \omega) &= \omega \operatorname{Re} [n(\mathbf{k}, z)/p(\mathbf{k})]_{z=\omega+i\epsilon} \\ &= n \left.\frac{\partial n}{\partial p}\right|_T \frac{[1 - (c_v/c_p)]\, D_T k^2\omega}{\omega^2 + (D_T k^2)^2} + n \left.\frac{\partial n}{\partial p}\right|_T \frac{\omega c^2 k^4 \Gamma (c_v/c_p)}{(\omega^2 - c^2k^2)^2 + (\omega k^2\Gamma)^2} \\ &\quad - n \left.\frac{\partial n}{\partial p}\right|_T \frac{D_T[1 - (c_v/c_p)]\,(\omega^2 - c^2k^2)\,\omega k^2}{(\omega^2 - c^2k^2)^2 + (\omega k^2\Gamma)^2}.\end{aligned} \tag{87a}$$

Similarly we find that

$$\chi''_{q,q}(k, \omega) = \frac{nmc_pT\, D_T\, k^2\omega}{\omega^2 + (D_T k^2)^2} \tag{87b}$$

and

$$\chi''_{n,q}(k, \omega) = \left.\frac{\partial n}{\partial T}\right|_p \left[\frac{D_T k^2\omega}{\omega^2 + (D_T k^2)^2} - \frac{D_T k^2 \omega(\omega^2 - c^2k^2)}{(\omega^2 - c^2k^2)^2 + (\omega\Gamma k^2)^2}\right]. \tag{87c}$$

Finally, from (51), it follows that

$$\chi_t''(k, \omega) = \frac{\eta k^2\omega}{\omega^2 + (\eta k^2/mn)^2}. \tag{87d}$$

We can sum up our results as follows: In the low wave number, low frequency limit, the correlation function composed of the transverse component of the momentum exhibits a diffusion structure with diffusivity, $D_t = \eta/mn$, given by the viscosity divided by the mass density. The heat energy-heat energy correlation function also has a diffusion structure but here the diffusivity is the thermal diffusivity, $D_T = \kappa/mnc_p$. The density-density correlation function exhibits both this diffusion process and a damped sound wave propagation. The total weight of $\chi''_{n,n}/\omega$ is $n(\partial p/\partial n)_T$ of which a proportion $(1 - c_v/c_p)$ comes from the diffusion process and a proportion c_v/c_p comes from the sound propagation. The heat energy-density correlation function also reflects both processes but the sound propagation contributes negligible weight to $\chi''_{n,q}/\omega$.

G. Expressions for the Transport Coefficients

In this section, we derive expressions for the transport coefficients: the thermal conductivity, κ, the viscosity, η, and the bulk viscosity, ζ. The expressions we derive are Kubo-type formulas in that they relate the transport coefficients to correlation functions formed of the currents of the conserved operators. The argument that we use is essentially identical to that used in deriving Eq. (44). In that situation, we started from the fact that the spin diffusion coefficient obeyed

$$\lim_{\omega\to 0}\left[\lim_{k\to 0}\frac{\omega}{k^2}\chi''_{M,M}(k,\omega)\right] = D_M\,\chi_{M,M}.$$

We then applied the spin conservation law to find

$$D_M\chi_{M,M} = \frac{\beta}{4}\lim_{\omega\to 0}\left[\lim_{k\to 0}\int d\mathbf{r}\int dt\, e^{-i\mathbf{k}\cdot\mathbf{r}+i\omega t}\,\langle\{j_x^M(\mathbf{r},t), j_x^M(\mathbf{0},0)\}\rangle\right].$$

According to Eq. (87b),

$$\lim_{\omega\to 0}\left[\lim_{k\to 0}\frac{\omega}{k^2}\chi''_{qq}(k,\omega)\right] = mnc_p\,TD_T = T\kappa.$$

Therefore, the thermal conductivity can be expressed as

$$\kappa = \frac{\beta}{4T}\lim_{\omega\to 0}\left[\lim_{k\to 0}\int d\mathbf{r}\int dt\, e^{-i\mathbf{k}\cdot\mathbf{r}+i\omega t}\langle\{j_x^q(\mathbf{r},t), j_x^q(\mathbf{0},0)\}\rangle\right]. \tag{88a}$$

However, according to (87a),

$$\lim_{\omega\to 0}\left[\lim_{k\to 0}\frac{\omega}{k^2}\chi''_{n,n}(k,\omega)\right] = 0$$

so that

$$0 = \lim_{\omega\to 0}\left[\lim_{k\to 0}\int d\mathbf{r}\int dt\, e^{-i\mathbf{k}\cdot\mathbf{r}+i\omega t}\langle\{g_x(\mathbf{r},t), g_x(\mathbf{0},0)\}\rangle\right],$$

and also, according to (87c),

$$\lim_{\omega \to 0} \left[\lim_{k \to 0} \frac{\omega}{k^2} \chi''_{n,q}(k, \omega) \right] = 0$$

so that

$$0 = \lim_{\omega \to 0} \left[\lim_{k \to 0} \int d\mathbf{r} \int dt\, e^{-i\mathbf{k}\cdot\mathbf{r}+i\omega t} \langle \{ j_x^q(\mathbf{r}, t), g_x(\mathbf{0}, 0) \} \rangle \right].$$

Therefore, not only is the thermal conductivity given by expression (88a); it is also given by the much more general expression

$$\kappa = \frac{\beta}{4T} \lim_{\omega \to 0} \left[\lim_{k \to 0} \int d\mathbf{r} \int dt\, e^{-i\mathbf{k}\cdot\mathbf{r}+i\omega t} \times \langle \{ j_x^q(\mathbf{r}, t) + \lambda g_x(\mathbf{r}, t), j_x^q(\mathbf{0}, 0) + \lambda g_x(\mathbf{0}, 0) \} \rangle \right] \tag{88b}$$

where λ is any constant. The choice $\lambda = (\epsilon + p)/(mn)$ is particularly instructive since

$$\mathbf{j}^q(\mathbf{r}, t) + \frac{\epsilon + p}{mn} \mathbf{g}(\mathbf{r}, t) = \mathbf{j}^\epsilon(\mathbf{r}, t)$$

where $\mathbf{j}^\epsilon$ is just the energy current. With this choice of λ, the thermal conductivity can be expressed as

$$\kappa = \frac{\beta}{4T} \lim_{\omega \to 0} \left[\lim_{k \to 0} \int d\mathbf{r} \int dt\, e^{-i\mathbf{k}\cdot\mathbf{r}+i\omega t} \langle \{ j_x^\epsilon(\mathbf{r}, t), j_x^\epsilon(\mathbf{0}, 0) \} \rangle \right]. \tag{88c}$$

The fact that the thermal conductivity can be represented by either expression (88a) or expression (88c) sheds some light on an apparently puzzling relation between the work of H. Mori and M. S. Green (*21*). Both authors worked with $k = 0$ from the very beginning of their calculation. Mori used the grand canonical ensemble and found a result of the form of (88a) in which the thermal conductivity is expressed in terms of a correlation function formed with $\mathbf{j}^q$'s. Green used the microcanonical ensemble and found that the thermal conductivity could be represented in the form (88c), in which the correlation function was formed from energy currents. The difference between their two results and Green's explanation of it is rather disturbing since it seems peculiar to ascribe significance to a correlation function whose value depends upon the ensemble used.

Our result complements the results of these two authors. We do not begin with the case in which k is truly set equal to zero since, in this case, not all correlation functions are well defined, and it is true that some depend on the ensemble. The hydrodynamic equations manifest this ensemble dependence through their strong dependence on initial conditions in time and boundary conditions at the edges of the container in which the system is enclosed. However, the hydrodynamic equations indicate that so long as $1/k$ is much smaller than a linear dimension of the container, these complications are irrelevant. As we might expect physically, we can unambiguously associate a transport coefficient with a spontaneous fluctuation function whenever the spontaneous fluctuation function is physically well defined and ensemble independent. This is the case whenever $1/k$ is much smaller than a container dimension but larger than any possible microscopic length. The evaluation of the physical functions in the limit $k \to 0$ is therefore assumed to take place after the container walls have receded to infinity. The limit as $k \to 0$, when the volume is kept finite, is ensemble dependent in a manner which we can understand from the hydrodynamic equations both mathematically and physically. The choice of a correct ensemble and current when $k \to 0$ first is in fact dictated by the requirement that the ensemble and current yield a result in agreement with the ensemble independent limit appropriate for $V \to \infty$ and then $k \to 0$. Under these circumstances, our analysis of the hydrodynamic equations indicates that the thermal conductivity can be expressed in terms of correlations of either $\mathbf{j}^{\epsilon}$ or $\mathbf{j}^{q}$. Mori's and Green's discussions each present an ensemble and a current for which no error results from the unphysical limiting process.

Finally, we indicate that we can obtain the standard Kubo-type expressions for the viscosity by employing the facts that, from (87d)

$$\lim_{\omega\to 0}\left[\lim_{k\to 0}\frac{\omega}{k^2}\chi_t''(k,\omega)\right] = \eta \tag{89a}$$

and from (87a)

$$\lim_{\omega\to 0}\left[\lim_{k\to 0}\frac{\omega}{k^2}\chi_l''(k,\omega)\right] = \lim_{\omega\to 0}\left[\lim_{k\to 0}\frac{m^2\omega^2}{k^4}\chi_{n,n}''(k,\omega)\right] = \frac{4}{3}\eta + \zeta. \tag{89b}$$

Thus,

$$\frac{4}{3}\eta + \zeta = \frac{\beta}{4}\lim_{\omega\to 0}\left[\lim_{k\to 0}\int d\mathbf{r}\int dt\frac{\omega^2}{k^4}\langle\{\mathbf{k}\cdot\mathbf{g}(\mathbf{r},t),\mathbf{k}\cdot\mathbf{g}(\mathbf{0},0)\}\rangle e^{-i\mathbf{k}\cdot\mathbf{r}+i\omega t}\right].$$

Applying the same arguments as before, we find that the viscosity and bulk viscosity may be obtained from the well-known correlation function expression

$$\eta\left(\delta_{ij} - \frac{1}{3}\frac{k_ik_j}{k^2}\right) + \zeta\frac{k_ik_j}{k^2} = \frac{\beta}{4}\lim_{\omega\to 0}\left[\lim_{k\to 0}\int d\mathbf{r}\int dt e^{-i\mathbf{k}\cdot\mathbf{r}+i\omega t} \times \sum_{nm}\frac{k_nk_m}{k^2}\langle\{\tau_{im}(\mathbf{r},t),\tau_{jn}(\mathbf{0},0)\}\rangle\right] \tag{90}$$

where τ is the stress tensor.

APPENDIX A. PERTURBATION THEORY

In this appendix we remind the reader of the expression which results from using time dependent perturbation theory to describe the effect of an external disturbance. We suppose that prior to time t_0 the state Ψ is a stationary state of the time independent Hamiltonian H. Subsequent to t_0 an external disturbance is applied which couples to the observable properties, $A_j(\mathbf{r}, t)$, of the system. We describe this disturbance by an additional term in the Hamiltonian

$$\mathcal{H}_{\text{ext}} = \int d\mathbf{r} \sum_j A_j(\mathbf{r}, t)\, a_j(\mathbf{r}, t) .$$

The functions $a_j(\mathbf{r}, t)$ represent the generalized external forces. For example, the observables might include components of the magnetization, in which case the corresponding force a_j would be the components of the external magnetic field. To calculate the expectation value at time t of the observable A_i we must calculate

$$\langle \Psi(t), A_i^{\mathrm{S}}(\mathbf{r}, t)\Psi(t)\rangle, \tag{A.1}$$

where $\Psi(t)$ is the Schroedinger wave function which was equal to Ψ for $t < t_0$ and $A_i^{\mathrm{S}}(\mathbf{r}, t)$ is the operator in the Schroedinger representation which characterizes the observable. Time dependent perturbation theory may be generated by introducing a wave function $\Phi(t)$ at time t_0 which would have become $\Psi(t)$ if no external perturbation had been applied, that is,

$$\Psi(t) \equiv e^{-iH(t-t_0)}\Phi(t).$$

From the Schroedinger equation we obtain

$$\begin{aligned} i\partial\Phi/\partial t &= e^{iH(t-t_0)} H_{\text{ext}}^{\mathrm{S}}(t) e^{-iH(t-t_0)}\Phi(t) \\ \Phi(t) &= \Psi(t_0) - i\int_{t_0}^{t} dt' H_{\text{ext}}^{\mathrm{I}}(t')\Phi(t') \end{aligned} \tag{A.2}$$

where for any operator $O^{\mathrm{S}}(t)$, the corresponding interaction representation operator is defined by

$$O^{\mathrm{I}}(t) = e^{iH(t-t_0)} O^{\mathrm{S}}(t) e^{-iH(t-t_0)}.$$

The formal solution of Eq. (A.2) is

$$\Psi(t) = e^{-iH(t-t_0)}\left(\exp\left[-i\int_{t_0}^{t} dt' H_{\text{ext}}^{\mathrm{I}}(t')\right]\right)_+ \Psi(t_0) \tag{A.3}$$

where the formal expression, the ordered product in brackets, is defined by the power series generated by iterating Eq. (A.2). For later purposes we note that direct integration of the equation for Ψ yields the alternative equivalent expression,

$$\exp\left[-i\int_{t_0}^{t}(H+H_{\text{ext}}^{\text{S}}(t'))\,dt'\right]_+ = e^{-iH(t-t_0)}\exp\left[-i\int_{t_0}^{t}H_{\text{ext}}^{\text{I}}(t')\,dt'\right]_+ . \tag{A.4}$$

Indeed, it is for this purpose that we have generated the perturbation series in terms of states instead of developing it for the density matrix directly. Substituting into (A.1) we obtain

$$\langle\Psi(t), A_i^{\text{S}}(\mathbf{r},t)\,\Psi(t)\rangle = \left\langle\Psi(t_0), \exp\left[i\int_{t_0}^{t}H_{\text{ext}}^{\text{I}}(t')\,dt'\right]_- \cdot A_j^{\text{I}}(\mathbf{r},t)\exp\left[-i\int_{t_0}^{t}H_{\text{ext}}^{\text{I}}(t')\,dt'\right]_+\Psi(t_0)\right\rangle.$$

By expanding the exponential we obtain the result to any desired order. In particular, if we expand to first order, and denote by a bracket the average over an ensemble of stationary states at time t_0, we deduce

$$\langle A_i(\mathbf{r},t)\rangle = \langle A_i^{\text{I}}(\mathbf{r},t)\rangle + i\sum_j\int d\mathbf{r}'\int_{t_0}^{t}dt' \cdot\langle[A_i^{\text{I}}(\mathbf{r},t), A_j^{\text{I}}(\mathbf{r}',t')]\rangle\, a_j(\mathbf{r}',t'). \tag{A.5}$$

If the observables are not explicitly time dependent in the Schroedinger representation, the operators $A^{\text{I}}(\mathbf{r}, t)$ are the Heisenberg operators for the Hamiltonian H. We shall henceforth assume that this is the case and omit the superscripts I. In terms of the absorptive susceptibility, defined by

$$\tilde{\chi}_{ij}''(r,r';t-t') \equiv \tfrac{1}{2}\langle[A_i(\mathbf{r},t), A_j(\mathbf{r}',t')]\rangle = \int\frac{d\omega}{2\pi}\,e^{-i\omega(t-t')}\chi_{ij}''(\mathbf{r},\mathbf{r}';\omega), \tag{A.6}$$

and the integral representation of the step function

$$\eta(t-t') = \lim_{\epsilon\to0} i\int\frac{d\omega}{2\pi}\frac{e^{-i\omega(t-t')}}{\omega+i\epsilon} = 0 \qquad \text{for } t<t'$$

$$= 1 \qquad \text{for } t>t',$$

we may write

$$\delta\langle A_i(\mathbf{r},t)\rangle = \lim_{\epsilon\to0}\sum_j 2i\int dr'\int_{-\infty}^{\infty}dt'\int\frac{d\bar\omega}{2\pi}\frac{ie^{-i\bar\omega(t-t')}}{\bar\omega+i\epsilon} \cdot\int\frac{d\omega}{2\pi}\chi_{ij}''(\mathbf{r},\mathbf{r}';\omega)e^{-i\omega(t-t')}a_j(\mathbf{r}',t').$$

We finally obtain

$$\delta\langle A_i(\mathbf{r}, t)\rangle = \sum_j \int_{-\infty}^{\infty} dt' \int d\mathbf{r}' \tilde{\chi}_{ij}(\mathbf{r}, \mathbf{r}'; t - t') a_j(\mathbf{r}', t'), \tag{A.7}$$

where $\tilde{\chi}_{ij}(\mathbf{r}, \mathbf{r}'; t - t')$ is the Fourier transform of the complex susceptibility $\chi_{ij}(\mathbf{r}, \mathbf{r}'; \omega)$ and

$$\chi_{ij}(\mathbf{r}, \mathbf{r}'; \omega) = \chi'_{ij}(\mathbf{r}, \mathbf{r}'; \omega) + i\chi''_{ij}(\mathbf{r}, \mathbf{r}'; \omega) \tag{A.8}$$

is the boundary value as z approaches ω on the real axis from above, of the analytic function of z

$$\chi_{ij}(\mathbf{r}, \mathbf{r}'; z) = \int \frac{d\bar{\omega}}{\pi} \frac{\chi''_{ij}(\mathbf{r}, \mathbf{r}'; \bar{\omega})}{\bar{\omega} - z}. \tag{A.9}$$

APPENDIX B. SOME PROPERTIES OF THE COMPLEX SUSCEPTIBILITY

1. Symmetries

In the text we noted the symmetry properties of χ''_{ij} for a spatially invariant system (that is for a system invariant under rotations, translations, and inversions) when A_i was the same as A_j. We summarize here the more general symmetry properties.[13]

(a) Since $\tilde{\chi}''_{ij}$ is a commutator, it is antisymmetric under interchange of $\mathbf{r}$ with $\mathbf{r}'$, i with j, and t with t'. We therefore have

$$\begin{aligned} \tilde{\chi}''_{ij}(\mathbf{r}, \mathbf{r}'; t - t') &= -\tilde{\chi}''_{ji}(\mathbf{r}', \mathbf{r}; t' - t) \\ \chi''_{ij}(\mathbf{r}, \mathbf{r}'; \omega) &= -\chi''_{ji}(\mathbf{r}', \mathbf{r}; -\omega). \end{aligned} \tag{B.1}$$

(b) The fact that $\tilde{\chi}''_{ij}$ is the commutator of hermitian operators leads to the identity

$$\begin{aligned} [\tilde{\chi}''_{ij}(\mathbf{r}, \mathbf{r}'; t - t')]^* &= -[\tilde{\chi}''_{ij}(\mathbf{r}, \mathbf{r}'; t - t')] \\ \chi''_{ij}(\mathbf{r}, \mathbf{r}'; \omega) &= -\chi''^*_{ij}(\mathbf{r}, \mathbf{r}'; -\omega) = \chi''^*_{ji}(\mathbf{r}, \mathbf{r}'; \omega). \end{aligned} \tag{B.2}$$

Thus the part of $\chi''_{ij}(\mathbf{r}, \mathbf{r}'; \omega)$ which is symmetric under interchange of i with j and $\mathbf{r}$ with $\mathbf{r}'$ is both real and odd in ω while the antisymmetric part is imaginary and even in ω. These statements imply in particular that if $\chi_{ii}(\mathbf{r}, \mathbf{r}'; \omega)$ is spatially invariant it is real and odd in the frequency.

(c) A similar result applicable to different operators follows from time re-

[13] It should be noted that all properties of the system are defined in terms of commutators. The corresponding classical relations can be obtained by using the equivalence of the commutator with the Poisson bracket multiplied by i in the correspondence limit. Obviously, this leaves the various symmetry properties unaltered.

versal. Since the time reversal operator, T, has the property

$$(T\Psi, T\Phi) = (\Phi, \Psi),$$

and for hermitian operators, $A_i(\mathbf{r}, t)$

$$TA_i(\mathbf{r}, t)T^{-1} = \epsilon_i A_i(\mathbf{r}, -t),$$

where ϵ_i is the signature of the operator A_i under time reversal, we have

$$(\Psi, [A_i(\mathbf{r}, t), A_j(\mathbf{r}', t')]\Psi) = \epsilon_i\epsilon_j(T\Psi, [A_j(\mathbf{r}', -t'), A_i(\mathbf{r}, -t)]T\Psi).$$

Consequently whenever the Hamiltonian and the ensemble of states are invariant under time reversal

$$\begin{aligned} \tilde{\chi}''_{ij}(\mathbf{r}, \mathbf{r}'; t - t') &= \epsilon_i\epsilon_j\tilde{\chi}''_{ji}(\mathbf{r}', \mathbf{r}; t - t') \\ \chi''_{ij}(\mathbf{r}, \mathbf{r}'; \omega) &= \epsilon_i\epsilon_j\chi''_{ji}(\mathbf{r}', \mathbf{r}; \omega). \end{aligned} \tag{B.3}$$

This means that if A_i and A_j have the same signature under time reversal $\chi''_{ij}(\mathbf{r}, \mathbf{r}'; \omega)$ is odd in ω, real, and symmetric under interchange of i with j and $\mathbf{r}$ with $\mathbf{r}'$. If they have opposite signature, $\chi''_{ij}(\mathbf{r}, \mathbf{r}'; \omega)$ is even, imaginary, and antisymmetric.

If the Hamiltonian and ensemble involve a magnetic field or some other property which changes sign under time reversal, then the relation

$$\chi''_{ij}(\mathbf{r}, \mathbf{r}'; \omega; \mathbf{B}) = \epsilon_i\epsilon_j\chi''_{ji}(\mathbf{r}', \mathbf{r}; \omega; -\mathbf{B}) \tag{B.3$'$}$$

is obtained. Hence for two operators with the same signature under time reversal there will be an additional part of $\chi''_{ij}(\mathbf{r}, \mathbf{r}'; \omega)$ which is odd in the field, $\mathbf{B}$, even in ω, imaginary, and antisymmetric in i, $\mathbf{r}$ and j, $\mathbf{r}'$.

The symmetry properties of $\chi'_{ij}(\mathbf{r}, \mathbf{r}'; \omega; \mathbf{B})$ are determined from the relation

$$\chi'_{ij}(\mathbf{r}, \mathbf{r}'; \omega; \mathbf{B}) = \mathcal{P}\int \frac{d\bar{\omega}}{\pi} \frac{\chi''_{ij}(\mathbf{r}, \mathbf{r}'; \bar{\omega}; \mathbf{B})(\bar{\omega} + \omega)}{\bar{\omega}^2 - \omega^2}. \tag{B.4}$$

which means that they are identical apart from the interchange of evenness and oddness in ω.

2. Sum Rules and Moment Expansions

The moment sum rule discussed in the text is the first of a sequence of statements

$$\left\langle\left[\frac{d^n A_i(\mathbf{r}, t)}{dt^n}, A_j(\mathbf{r}', t')\right]\right\rangle\Bigg|_{t'=t} = \int \frac{d\bar{\omega}}{\pi}(-i\bar{\omega})^n\chi''_{ij}(\mathbf{r}, \mathbf{r}'; \bar{\omega}). \tag{B.5}$$

These statements and the Kramers-Kronig relation (B.4) gives rise to a moment expansion, valid at high frequencies,

$$\chi'_{ij}(\mathbf{r}, \mathbf{r}'; \omega) = -\sum_n \int \frac{d\bar{\omega}}{\pi} \left(\frac{\bar{\omega}}{\omega}\right)^n \frac{\chi''_{ij}(\mathbf{r}, \mathbf{r}'; \omega)}{\omega}$$

$$= \sum_n \frac{i^n}{\omega^{n+1}} \left\langle \left[-\frac{d^n A_i(\mathbf{r}, t)}{dit^n}, A_j(\mathbf{r}, t') \right] \right\rangle \bigg|_{t=t'}. \tag{B.6}$$

3. Identification of χ'' with Dissipation

The rate at which mechanical work is done on a system by an external force is equal to the explicit rate of change in the Hamiltonian

$$\frac{dW}{dt} = -\sum_{i,j} \int \dot{a}_i(\mathbf{r}, t)\tilde{\chi}_{ij}(\mathbf{r}, \mathbf{r}'; t - t')a_j(\mathbf{r}', t')\, dt'\, dr\, dr'.$$

The mechanical dissipation (which is equal to the entire dissipation at constant entropy) is obtained by integrating this expression over time.

$$\int dW = \sum_{i,j} \int d\mathbf{r} \int d\mathbf{r}' a_i(\mathbf{r}, t) \left[\frac{\partial}{\partial t} \chi_{ij}(\mathbf{r}, \mathbf{r}'; t - t') \right] a_j(\mathbf{r}', t')\, dt\, dt'. \tag{B.7}$$

Since $\partial\chi'/\partial t$ is antisymmetrical in time and $\partial\chi''/\partial t$ is symmetrical, only $\partial\chi''/\partial t$ contributes to the dissipation. Alternatively, for a single frequency of applied field

$$H_{\text{ext}}(t) = -\int d\mathbf{r} \sum_i A_i(\mathbf{r}, t) \operatorname{Re} a_i(\mathbf{r})e^{-i\omega t}$$

and the mean value of the work done is

$$\overline{\frac{dW}{dt}} = -\frac{1}{2} \operatorname{Re} \sum_{ij} \int d\mathbf{r}\, d\mathbf{r}' a_i^*(\mathbf{r}) i\omega\chi_{ij}(\mathbf{r}, \mathbf{r}'; \omega)a_j(\mathbf{r}').$$

Since χ' and χ'' are both hermitian the average comes only from χ''_{ij}

$$\overline{\frac{dW}{dt}} = \frac{1}{2} \sum_{ij} \int d\mathbf{r}\, d\mathbf{r}' a_i^*(\mathbf{r})\chi''_{ij}(\mathbf{r}, \mathbf{r}'; \omega)a_j(\mathbf{r}')\omega. \tag{B.8}$$

We may also write this expression for the rate of energy exchange as the energy, ω, times the difference between transition probabilities for absorption and emission

$$= \frac{\omega}{4} \sum_{ij} \int d\mathbf{r}\, d\mathbf{r}' \int dte^{i\omega t} a_i^*(\mathbf{r})\langle[A_i(\mathbf{r}, t), A_j(\mathbf{r}', 0)]\rangle a_j(\mathbf{r}')$$

$$= \omega \int dte^{i\omega t} \int dEw(E) \int dE'\rho(E')$$

$$\cdot \left\{ \left| \left\langle E \left| \frac{1}{2} \int d\mathbf{r} \sum_i a_i(\mathbf{r})A_i(\mathbf{r}, 0) \right| E' \right\rangle \right|^2 (e^{i(E-E')t} - e^{-i(E-E')t}) \right\}$$

where $\rho(E')$ is the density of states of the Hamiltonian, and $w(E)$ the normalized weighting of states of the stationary ensemble. In writing this equation and the subsequent equations (B.9) and (B.14), we have assumed that the states may be labeled by the energy alone, that is, that there is no degeneracy. When the states are degenerate, these equations should include averages over states of identical energy. This extra averaging process changes none of our conclusions. We therefore obtain

$$
\begin{aligned}
= \omega\Big\{2\pi \int dEw(E)\left|\left\langle E\left|\frac{1}{2}\sum_i \int d\mathbf{r}a_i(\mathbf{r})A(\mathbf{r},0)\right|E+\omega\right\rangle\right|^2 \rho(E+\omega) \\
- 2\pi \int dEw(E)\left|\left\langle E\left|\frac{1}{2}\sum_i \int d\mathbf{r}a_i(\mathbf{r})A_i(\mathbf{r},0)\right|E-\omega\right\rangle\right|^2 \rho(E-\omega)\Big\}.
\end{aligned} \tag{B.9}
$$

The rate of change of mechanical energy may of course be associated with the rate of change of free energy in an ensemble at constant temperature. Since the matrix $\chi''_{ij\omega}$ describes the dissipation it must be positive definite in any stable system. This positive definiteness of $\omega\chi''_{ij}(\mathbf{r}, \mathbf{r}'; \omega)$ has implications for $\chi'_{ij}(\mathbf{r}, \mathbf{r}'; \omega)$. In particular it follows from Eq. (B.4) that $\chi'_{ij}(\mathbf{r}, \mathbf{r}'; \omega)$ is a nonnegative matrix at vanishing frequency. Hence, for example, the static electric polarizability must be positive. Likewise, for the one-component fluid discussed in the text, this requirement reduces in the long wavelength limit to the familiar thermodynamic stability conditions $(dp/dn)_S > 0$, $c_p > 0$, and $(dp/dn)_T > 0$.

Note also that at large frequencies the "sign" of the matrix χ'_{ij} is always negative. This behavior is just what we expect for an oscillator bound by a restoring force $n\omega_0^2$ and perturbed by an external force of frequency, ω. Its displacement will be 180° out of phase with the force when $\omega \gg \omega_0$ and the absorption is sufficiently small so that there is oscillation (that is, when we have an oscillator damped less than critically); also the absorption will be maximum at an intermediate frequency when the displacement is out of phase with the force by 90°. Although we shall not pursue the point, it should be clear that in the representation in which $\chi_{ij}(\mathbf{r}, \mathbf{r}'; \omega)$ is diagonalized, its logarithm gives a natural definition for frequency-dependent phase shifts in precise analogy with the above description and the phase shift representation discussed in connection with the dispersion relation for magnetic susceptibility.

4. Fluctuation Dissipation Theorem

The time translation property of the weighting factor for a canonical ensemble and the cyclical property of the trace imply the identities

$$
\begin{aligned}
\mathrm{Tr}\, e^{-\beta H}A_i(\mathbf{r}, t)A_j(\mathbf{r}', t') &= \mathrm{Tr}\, A_i(\mathbf{r}, t+i\beta)e^{-\beta H}A_j(\mathbf{r}', t') \\
&= \mathrm{Tr}\, e^{-\beta H}A_j(\mathbf{r}', t')A_i(\mathbf{r}, t+i\beta).
\end{aligned} \tag{B.10}
$$

Moreover Tr [exp $(-\beta H)A(\mathbf{r}, t)$] is independent of time. Consequently, provided the time Fourier transform

$$\tfrac{1}{2}\langle[A_i(\mathbf{r},t)-\langle A_i(\mathbf{r},t)\rangle][A_j(\mathbf{r},t)-\langle A_j(\mathbf{r},t)\rangle]\rangle=\int\frac{d\omega}{2\pi}f_{ij}(\mathbf{r},\mathbf{r}';\omega)e^{-i\omega(t-t')}$$

exists, it satisfies

$$f_{ij}(\mathbf{r},\mathbf{r}';\omega)=f_{ji}(\mathbf{r}',\mathbf{r};-\omega)e^{\beta\omega}$$

and therefore,[14]

$$\chi''_{ij}(\mathbf{r},\mathbf{r}';\omega)=(1-e^{-\beta\omega})f_{ij}(\mathbf{r},\mathbf{r}';\omega)=(e^{\beta\omega}-1)f_{ji}(\mathbf{r}',\mathbf{r};-\omega). \tag{B.11}$$

Likewise the transform of the symmetrized product

$$\tfrac{1}{2}\langle\{[A_i(\mathbf{r},t)-\langle A_i(\mathbf{r},t)\rangle],[A_j(\mathbf{r}',t')-\langle A_j(\mathbf{r}',t')\rangle]\}\rangle$$
$$=\int\frac{d\omega}{2\pi}s_{ij}(\mathbf{r},\mathbf{r}';\omega)e^{-i\omega(t-t')} \tag{B.12}$$

satisfies the identity

$$s_{ij}(\mathbf{r},\mathbf{r}';\omega)=(1+e^{\beta\omega})f_{ij}(\mathbf{r},\mathbf{r}';\omega)$$

and the fluctuation dissipation theorem[15]

$$\begin{aligned}\frac{1}{2}s_{ij}(\mathbf{r},\mathbf{r}';\omega)&=\omega\left[\frac{1}{2}+\frac{1}{e^{\beta\omega}-1}\right]\frac{\chi''_{ij}(\mathbf{r},\mathbf{r}';\omega)}{\omega}\\&=\frac{1}{2}\coth\frac{\beta\omega}{2}\,\chi''_{ij}(\mathbf{r},\mathbf{r}';\omega).\end{aligned} \tag{B.13}$$

We may use the expression (B.8) for the dissipation and (B.13) to demonstrate that our statement of stability

$$W=\frac{1}{2}\sum_{ij}\int d\mathbf{r}\,d\mathbf{r}'a_i^*(\mathbf{r})\chi''_{ij}(\mathbf{r},\mathbf{r}';\omega)a_j(\mathbf{r})\omega\geqq 0$$

is satisfied by the canonical ensemble. For this purpose, with the aid of Eq. (B.13) it is only necessary to show that the fluctuations at a single frequency are positive definite. By introducing a set of intermediate states we obtain

$$\begin{aligned}W=\tfrac{1}{2}\omega\tanh\tfrac{1}{2}\beta\omega\int dEw_{\text{e.q.}}(E)\int dE'\rho(E')[\delta(E'-E+\omega)+\delta(E'-E-\omega)]\\\times 2\pi\left|\left\langle E\left|\sum_i\int d\mathbf{r}a_i(\mathbf{r})[A_i(\mathbf{r},0)-\langle A_i(\mathbf{r},0)\rangle]\right|E'\right\rangle\right|^2\geqq 0,\end{aligned} \tag{B.14}$$

where $w_{\text{eq.}}(E)$ is the normalized distribution for the equilibrium ensemble.

[14] This is the first relation in the appendices which depends on canonical averaging. We shall subsequently assume this particular density matrix.

[15] Had Poisson brackets and classical mechanics been employed, this relation would have involved $2/\beta\omega$ instead of coth $(\beta\omega/2)$.

APPENDIX C. RELAXATION

In this final appendix we make explicit the connection of our discussion with the widely quoted peculiar looking dissipation function of Kubo. We also demonstrate how the latter arises directly in a discussion of relaxation and leads alternatively to the expression for relaxation discussed in the text. We recall a familiar parallel. In classical electromagnetic theory we use the retarded Green's function for the wave equation to determine the radiation emitted by charges moving along prescribed trajectories. We also use this function to find the behavior of radiation in free space in terms of the radiation present at an initial time. The former corresponds to the characterization of the response to externally applied forces. The latter corresponds to the relaxation problem we shall now discuss. It would clearly be possible to consider simultaneously emission of radiation by charges undergoing prescribed motions and the propagation and absorption of incident radiation present initially. Likewise it would be possible to discuss a system relaxing to equilibrium and simultaneously subjected to external forces. Since there is no really new effect we shall confine ourselves to free relaxation.

We suppose that initially the system is characterized by a disturbed density matrix

$$\rho = \Xi\,[\mathrm{Tr}\,\Xi]^{-1}$$

where

$$\begin{aligned}\Xi &\equiv \exp\left[-\beta H + \beta \sum_j \int d\mathbf{r}' A_j(\mathbf{r}')a_j(\mathbf{r}')\right] \\ &\equiv \exp\,[-\beta(H + H_{\text{ext}})]\end{aligned} \tag{C.1}$$

Note particularly that although we have again used the symbol H_{ext} it here describes an initial condition and does not depend on the time in any way. For times $t > 0$, a property A_i of the system transforms according to

$$A_i(\mathbf{r}, t) = e^{iHt}A_i(\mathbf{r})e^{-iHt}.$$

Thus we have

$$\langle A_i(\mathbf{r}, t)\rangle_{\text{noneq.}} = \mathrm{Tr}\,[A_i(\mathbf{r}, t)\rho]. \tag{C.2}$$

We now employ a special form of the identity (A.4)

$$\begin{aligned}\exp\,[-\beta(H + H_{\text{ext}})] &= \exp\,(-\beta H)\exp\left(-\int_0^\beta H^{\mathrm{I}}_{\text{ext}}(-i\beta')\,d\beta'\right)_+ \\ H^{\mathrm{I}}_{\text{ext}}(-i\beta') &= \exp\,(\beta' H)H_{\text{ext}}\exp\,(-\beta' H)\end{aligned} \tag{C.3}$$

To first order in H_{ext}, we obtain from Eqs. (C.1–3)

$$\delta\langle A_i(\mathbf{r}, t)\rangle = \sum_j \int d\mathbf{r}' \int_0^\beta d\beta' [\langle A_j(\mathbf{r}', -i\beta')A_i(\mathbf{r}, t)\rangle_{eq.} - \langle A_j(\mathbf{r}', -i\beta)\rangle_{eq.}\langle A_i(\mathbf{r}, t)\rangle_{eq.}]a_j(\mathbf{r}'). \tag{C.4}$$

Taking into account the relaxation between the unordered products and the commutator (B.11) we obtain

$$\begin{aligned}\delta\langle A_i(\mathbf{r}, t)\rangle &= \sum_j \int d\mathbf{r}' \int_0^\beta d\beta' \int \frac{d\omega}{\pi} \frac{\chi''_{ij}(\mathbf{r}, \mathbf{r}'; \omega)}{e^{\beta\omega} - 1} e^{-i\omega(t+i\beta')} a_j(\mathbf{r}') \\ &= \sum_j \int d\mathbf{r}' \int \frac{d\omega}{\pi} \frac{e^{-i\omega t}}{\omega} \chi''_{ij}(\mathbf{r}, \mathbf{r}'; \omega) a_j(\mathbf{r}').\end{aligned} \tag{C.5}$$

Thus the Kubo expression (C.4) is a peculiar way of writing the ensemble averaged commutator, that is to say

$$\int_0^\beta \langle \dot{A}_i(\mathbf{r}, t)A_j(\mathbf{r}, i\beta')\rangle_{eq.}\, d\beta' = -i\langle[A_i(\mathbf{r}, t), A_j(\mathbf{r}', 0)]\rangle_{el.} \tag{C.6}$$

Since these expressions apply for $t > 0$ it is natural to introduce one-sided Fourier transforms as in the text and write

$$A_i(\mathbf{r}, z) = -\frac{1}{iz} \sum_j \int d\mathbf{r}' \left[\int \frac{d\omega}{\pi} \frac{\chi''_{ij}(\mathbf{r}, \mathbf{r}'; \omega)}{\omega} - \chi_{ij}(\mathbf{r}, \mathbf{r}'; z) \right] a_j(\mathbf{r}'). \tag{C.7}$$

REFERENCES

1. G. E. Uhlenbeck, "Lectures in Applied Mathematics," Appendix. Interscience, New York, 1959.

2. P. C. Martin and J. Schwinger, *Phys. Rev.* **115**, 1342 (1959).

3. E. W. Montroll and J. C. Ward, *Phys. Fluids* **1**, 55 (1958); J. M. Luttinger and J. C. Ward, *Phys. Rev.* **118**, 1417 (1960).

4. S. Chapman and T. G. Cowling, "Mathematical Theory of Non-Uniform Gases." Cambridge Univ. Press, London, 1939.

5. G. Baym and L. P. Kadanoff, *Phys. Rev.* **124**, 287 (1962).

6. R. Kubo, *J. Phys. Soc. Japan* **12**, 570 (1957). See also the article by Kubo in "Lectures in Theoretical Physics," Vol. I, Chapter 4. Interscience, New York, 1959.

7. M. S. Green, *J. Chem. Phys.* **22**, 398 (1954); *Phys. Rev.* **119**, 829 (1960); M. Lax, *Phys. Rev.* **109**, 1921 (1958); H. Mori, *Phys. Rev.* **112**, 1829 (1958); and many others.

8. L. Van Hove, *Phys. Rev.* **95**, 249 (1959).

9. "Inelastic Scattering of Neutrons in Solids and Liquids." International Atomic Energy Agency, Vienna, 1961.

10. G. H. Vineyard, *Phys. Rev.* **110**, 999 (1959).

11a. N. Bloembergen, *Physica* **15**, 386 (1949).

11b. A. M. Portis, *Phys. Rev.* **104**, 584 (1956).

11c. H. C. Torrey, *Phys. Rev.* **104**, 536 (1956).

11d. H. R. Hart, Jr., Ph.D. Thesis, University of Illinois, Urbana, Illinois, 1960 (unpublished).

12a. D. Hone, *Phys. Rev.* **121,** 669 (1961).
12b. T. Moriya, *Progr. Theoret. Phys.* (*Kyoto*) **28,** 371 (1962)
13a. H. Nyquist, *Phys. Rev.* **32,** 110 (1928).
13b. H. B. Callen and T. R. Welton, *Phys. Rev.* **83,** 34 (1951).
14. J. Kirkwood, *J. Chem. Phys.* **7,** 911 (1939).
15. L. D. Landau and E. M. Lifshitz, "Fluid Mechanics." Addison-Wesley, Reading, Mass., 1959.
16. L. Onsager, *Phys. Rev.* **37,** 405 (1931); *ibid.* **38,** 2265 (1931).
17. P. W. Anderson, *Phys. Rev.* **112,** 1900 (1958).
18. L. Landau, *Z. Physik* **64,** 629 (1930).
19a. J. M. Ziman, *Phil. Mag.* **6,** 1013 (1961).
19b. N. D. Mermin, *Ann. Phys.* (*N. Y.*) **18,** 421 (1962).
20. K. Herzfeld and T. Litovitz, "Absorption and Dispersion of Ultrasonic Waves," Sec. 105. Academic Press, New York, 1959.
21. H. Mori, *Phys. Rev.* **112,** 1829 (1958); M. S. Green, *J. Chem. Phys.* **22,** 398 (1954).

The Electron-Phonon Interaction in Normal and Superconducting Metals[1]

LEO P. KADANOFF

Department of Physics, University of Illinois, Urbana, Illinois

I. Correlation Function Approach

This series of lectures has two purposes:

(a) To report on recent developments in the theory of the electron-phonon interaction in metals.

(b) To serve as an introduction to the use of Green's function methods in many particle physics.[2]

We begin by noticing that one of the very simplest things you can do to a many-particle system is add a particle to it or pull one out. For this reason, it is very convenient to describe the properties of the system in terms of the creation and annihilation operators:

$$C_p^+(t') = \text{creation operator}$$

$$C_p(t) = \text{annihilation operator.}$$

When these operators act to the right on a state of the system, they respectively add a particle with momentum p to this state at the time t' and pull one out at the time t.

[1] A series of lectures given at the 1963 Spring School of Physics at Ravello, Italy. This work was supported in part by the U.S. Army Ordinance DA-ARO(D)-31-124-G340.

[2] The Green's function methods that I shall employ can be traced back to the work of Martin and Schwinger (*1*). These methods are discussed in detail in a book by Baym and Kadanoff (*2*).

Reprinted from *Lectures on the Many-Body Problem*, Vol. 2, ed. E. R. Caianiello, Academic Press.

We shall also make use of the wave field operators $\psi^+(r', t')$ and $\psi(r, t)$ which respectively add and substract a particle at the space-time points r', t' and r, t. If one quantizes in a box of unit volume with periodic boundary conditions the two sets of operators are related by

$$\begin{aligned} \psi^+(r', t') &= \sum_p \exp(-ip \cdot r')\, C_p^+(t') \\ \psi(r, t) &= \sum_p \exp(ip \cdot r)\, C_p(t) \end{aligned} \tag{1.1}$$

where the momentum sums run over all the allowed momenta in the box. (We use units in which $\hbar = 1$.)

These are all Heisenberg representation operators and, as such, they have the time depencence

$$A(t) = \exp(iH't) \quad A(0) \quad \exp(-iH't) \tag{1.2}$$

Of course, we may choose the zero point of energy at our convenience. We choose our zero point by using in (1.2)

$$H' = H - \mu N \tag{1.3}$$

where H is the standard Hamiltonian, μ the chemical potential, and N the number operator to be the basic operator which gives the time dependence of our operators.

The two basic correlation functions which we shall discuss are

$$\begin{aligned} G^>(r, t;\, r', t') &= \langle \psi(r, t)\psi^+(r,' t')\rangle \\ G^<(r, t;\, r', t') &= \langle \psi^+(r', t')\psi(r, t)\rangle \end{aligned} \tag{1.4}$$

where $\langle\ \rangle$ stands for both a quantum-mechanical average and a statistical average according to the averaging procedure defined by the grand canonical ensemble of statistical mechanics.

Because of the equilibrium nature of the system all physical quantities are independent of the time. Furthermore, our periodic boundary condition in space guarentees the spatial homogeneity of the system. Therefore, $G^>$ and $G^<$ cannot depend upon r and r' or t and t' individually but only upon the difference variables $r - r'$, $t - t'$. This fact is conveniently included by using the Fourier representations of these quantities as

$$G^{>}(r, t; r', t') = G^{>}(r - r'; t - t') \tag{1.5a}$$

$$= \sum_p \int \frac{d\omega}{2\pi} \exp[ip \cdot (r - r') - i\omega(t - t')]\, G^{>}(p, \omega)$$

$$= \sum_p \exp[ip \cdot (r - r')] \langle C_p(t)\, C_p^{+}(t') \rangle$$

while

$$G^{<}(r - r'; t - t') = \sum_p \exp[ip \cdot (r - r')] \langle C_p^{+}(t')\, C_p(t) \rangle \tag{1.5b}$$

$$\sum_p = \int \frac{d\omega}{2\pi} \exp[ip \cdot (r - r') - i\omega(t - t')]\, G^{<}(p, \omega)\,.$$

$G^{>}(p, \omega)$ and $G^{<}(p, \omega)$ have a very direct physical interpretation. Consider $G^{<}(p, \omega)$. We know that $C_p(t)$ is an operator which removes a particle with momentum at the time t. In the usual quantum-mechanical way, the time Fourier transform of $C_p(t)$ removes a particle of momentum p and energy ω. But, if you are going to remove such a particle, there must be a particle with this energy and momentum present initially. For this reason,

$$G^{<}(p, \omega) = \text{density of particles with momentum } p \text{ and energy } \omega. \tag{1.6}$$

Because the time Fourier transform of $C_p^{+}(t')$ adds a particle with momentum p and energy ω, we can see that $G^{>}(p, \omega)$ measures the system's ability to accept a particle with this energy and momentum. That is,

$$G^{>}(p, \omega) = \text{effective density of states in } p, \omega \tag{1.7}$$

or, if you prefer, this may be termed a density of holes.

These statements (1.6) and (1.7) will be very important for our future discussions. Therefore, we examine them in a little detail. From (1.5b), the total number of particles with momentum p, $N_p = C_p^{+} C_p$, is given by

$$\langle N_p \rangle = \langle C_p^{+} C_p \rangle = \int \frac{d\omega}{2\pi} G^{<}(p, \omega) \tag{1.8}$$

which is the integral over all energies of the density as function of energy. Also the total effective density of states is, from (1.5b)

$$\int \frac{d\omega}{2\pi} G^{>}(p, \omega) = \langle C_p C_p^{+} \rangle .$$

However, the creation and annihilation operators satisfy the equal time commutation or anticommutation relation

$$C_p(t) C_p^{+}(t) \mp C_p^{+}(t) C_p(t) = 1 \tag{1.9}$$

where the upper sign here, and in what follows, is appropriate when the particle obey Bose statistics and the lower sign is for Fermi statistics. Therefore, the total density of states is

$$\int \frac{d\omega}{2\pi} G^{>}(p, \omega) = 1 \pm \langle N_p \rangle . \tag{1.10}$$

For fermions as $\langle N_p \rangle$ increases, the effective density of states decreases while for bosons, which are very gregarious, the larger the occupation of a state, the larger is the effective density of states.

To illustrate the meaning of $G^{>}$ and $G^{<}$, let us consider a tunneling experiment. In its simplest form this is an experiment in which two

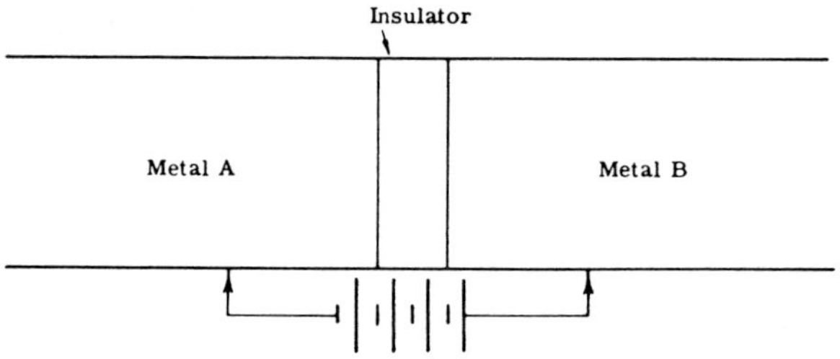

FIG. 1

conductors are separated by a thin insulating layer and a DC voltage is applied across the gap. The induced current can then be measured as a function of voltage.

Electrons cross the gap via a quantum-mechanical tunneling process. Following many recent authors (*3*; *4*) (see the chapter by R. Prange in this volume) we can describe the tunneling by a tunneling amplitude $T(p', p)$, which is the probability amplitude for an electron with momentum p on one side of the gap reappearing with momentum p' on the other. (This is essentially the overlap between the wave functions on opposite sides of the insulator.)

We calculate the tunneling rate from A to B by making use of the golden rule, which states that a transition rate is proportional to a matrix element squared [here, $| T(p, p') |^2$], an energy conserving δ-function, a density of initial and final states, summed over all initial and all final states. Thus,

$$\left.\frac{dN}{dt}\right|_{A\to B} = \sum_{pp'} \int \frac{d\omega}{2\pi} \int \frac{d\omega'}{2\pi} \, | T(p, p') |^2 \qquad (1.11a)$$
$$2\pi\delta(\omega - \omega' + eV) \; G_A^{<}(p, \omega) \; G_B^{>}(p', \omega') \, .$$

Here the δ-function requires that the energy change on traversing the barrier be eV. $G_A^{<}$ and $G_B^{>}$, of course, represent the density of particles in A and the density of states in B.

The rate of tunneling in the opposite sense is given by a result identical to (1.11a) except for the appearance of the density of particles in B and the density of states in A, that is

$$\left.\frac{dN}{dt}\right|_{B\to A} = \sum_{pp'} \int \frac{d\omega}{2\pi} \int \frac{d\omega'}{2\pi} \, | T(p, p') |^2 \qquad (11.1b)$$
$$2\pi\delta(\omega - \omega' + eV) \; G_A^{>}(p, \omega) \; G_B^{<}(p', \omega') \, .$$

To obtain the net tunneling rate we substract (1.11b) from (1.11a) to find

$$\text{net} \, \frac{dN}{dt} = \sum_{pp'} \int \frac{d\omega}{2\pi} \int \frac{d\omega'}{2\pi} \, | T(p, p') |^2 \, 2\pi\delta(\omega - \omega' + eV) \qquad (1.12)$$
$$[G_A^{<}(p, \omega) \; G_B^{>}(p', \omega') - G_A^{>}(p, \omega) \; G_B^{<}(p', \omega)].$$

Equation (1.12) may be used to predict the tunneling rate in all cases except that in which both conductors are in a superconducting state. In that case, some very exceptional behavior results as first predicted by Josephson (*5*) and experimentally verified by Anderson and Rowell (*6*).

This kind of tunneling is discussed in detail by P. W. Anderson in this volume.

In all but this exceptional case, one expects that, as $V \to 0$, the tunneling current should vanish. However, if A and B are different materials so that $G_A^< \neq G_B^<$, Eq. (1.12) does not appear to predict the vanishing of the current at $V \to 0$.

In order to see how this vanishing does in fact occur, we must make use of a kind of detailed balancing condition which relates the density of particles and the density of states. It is a characteristic feature of a system in equilibrium at temperature T that the relative occupation of a state with energy ω is proportional to $\exp(-\beta\omega)$, where $\beta = (kT)^{-1}$, and k is the Boltzmann constant. This characteristic feature shows up as a restriction upon $G^<$ and $G^>$ that, for a system in equilibrium,

$$G^<(p, \omega) = \exp(-\beta\omega)\ G^>(p, \omega) \tag{1.13}$$

$$(\text{density of particles}) = \exp(-\beta\omega)\ (\text{density of states}).$$

We shall prove this condition in a moment. For now let us see its consequences in the tunneling rate given by (1.12). We use (1.13) to eliminate $G_A^<$ in favor of $G_A^>$ and $G_B^<$ in favor of $G_B^>$. Then (1.12) becomes

$$\text{net}\ \frac{dN}{dt} = \sum \int \int |T|^2\ 2\pi\delta(\omega - \omega' + eV)$$

$$G_A^>(p, \omega)\ G_B^>(p', \omega')\ [\exp(-\beta\omega) - \exp(-\beta\omega')].$$

Clearly now as $V \to 0$, so does the current.

Equation (1.13) will be quite crucial for us so we consider its proof in some detail. In the ground canonical ensemble, the expectation value of any operator is given by

$$\begin{aligned} \langle X \rangle &= \sum_\xi \exp[-\beta(E_\xi - \mu N_\xi)]\ \langle \xi | X | \xi \rangle / Z_G \\ &= \mathrm{Tr}\ [\exp(-\beta H')\ X]/Z_G \end{aligned} \tag{1.14}$$

where the sum runs over all possible states with any number of particles and where

$$Z_G = \sum_\xi \exp[-\beta(E_\xi - \mu N_\xi)] = \mathrm{Tr}\ [\exp(-\beta H')]$$

is the grand partition function.

Now consider the two functions

$$F^{>}(t) = \langle A(t)\, X \rangle$$
$$F^{<}(t) = \langle X\, A(t) \rangle\,. \tag{1.15}$$

If $A(t)$ is a Heisenberg representation operator with no explicit time dependence

$$A(t) = \exp(iH't) \quad A(0) \quad \exp(-iH't)\,. \tag{1.2}$$

Equation (1.2) can be used to define $A(t)$ for complex values of the time. Now consider, in particular,

$$\begin{aligned} F^{>}(t)\,|_{t=-i\beta} &= \langle \exp(+\beta H')\, A(0) \exp(-\beta H')\, X \rangle \\ &= \mathrm{Tr}\,[\exp(-\beta H') \exp(\beta H')\, A(0) \exp(-\beta H')\, X]/Z_G \\ &= \mathrm{Tr}\,[A(0) \exp(-\beta H')\, X]/Z_G \\ &= \mathrm{Tr}\,[\exp(-\beta H')\, X\, A(0)]/Z_G = \langle X\, A(0) \rangle\,. \end{aligned}$$

Consequently,

$$F^{>}(t)\,|_{t=-i\beta} = F^{<}(t)\,|_{t=0}\,. \tag{1.16}$$

Let us apply this theorem to $G^{>}$ and $G^{<}$. Equation (1.16) implies that

$$\begin{aligned} G^{>}(r, t; r', t')\,|_{t=-i\beta} &= G^{<}(r, t; r', t')\,|_{t=0} \\ \int \frac{d\omega}{2\pi} \exp[-i\omega(t-t')\, G^{>}(p, \omega)\,|_{t=-i\beta} & \\ = \int \frac{d\omega}{2\pi} \exp[-i\omega(t-t')]\, G^{<}(p, \omega)\,|_{t=0} & \end{aligned} \tag{1.17}$$

or

$$\int \frac{d\omega}{2\pi} \exp(i\omega t') \exp(-\beta\omega)\, G^{>}(p, \omega) = \int \frac{d\omega}{2\pi} \exp(i\omega t')\, G^{<}(p, \omega),$$

which immediately leads us back to (1.13).

The functions $G^{>}(p, \omega)$ and $G^{<}(p, \omega)$ contain a tremendous amount of useful information. They describe, as we have seen, the density of particles and the density of states, and they describe all the thermodynamic properties of the system. Consequently, it is quite worthwhile to know these functions.

In actual calculations, it is very convenient to make use of the spectral weight function

$$A(p,\omega) = G^{>}(p,\omega) \mp G^{<}(p,\omega) \tag{1.18}$$
$$= \text{Fourier transform } \langle C_p(t)C_p^{+}(t') \mp C_p^{+}(t')C_p(t)\rangle .$$

The usefulness of $A(p, \omega)$ is derived from the equal time commutation relation (1.7), which implies the sum rule

$$\int \frac{d\omega}{2\pi} A(p,\omega) = 1 . \tag{1.19}$$

By making use of the detailed balancing condition (1.13) together with the definition (1.17) we can express $G^{>}$ and $G^{<}$ in terms of A as

$$G^{>}(p,\omega) = [1 \pm f(\omega)]\, A(p,\omega) \tag{1.20}$$
$$G^{<}(p,\omega) = f(\omega)\, A(p,\omega)$$

where

$$f(\omega) = \frac{1}{\exp(\beta\omega) \mp 1} \tag{1.21}$$

is the a priori probability of observing a particle with energy ω in the grand canonical ensemble.

Now we must face the hard job of determining $G^{>}$ and $G^{<}$. To see how this goes consider just the trivial case of free particles. In general, ω represents the total energy of a particle added to the system, its kinetic energy plus the interaction energy that it aims through its interaction with all the other particles in the system. However, for noninteracting particles ω must be just the kinetic energy

$$\varepsilon_p = p^2/2m - \mu . \tag{1.22}$$

Therefore $G^{>}$, $G^{<}$, and A are all proportional to $\delta(\omega - \varepsilon_p)$. However, for $A(p, \omega)$, we have a sum rule (1.18) which tells us the constant of proportionality, so that

$$A(p,\omega) = 2\pi\delta(\omega - \varepsilon_p) . \tag{1.23}$$

Using (1.22), we can easily calculate the density of particles with momentum p as

$$\begin{aligned}\langle N_p \rangle &= \int \frac{d\omega}{2\pi} G^<(p, \omega) = \int \frac{d\omega}{2\pi} f(\omega)\, A(p, \omega) \\ &= f(\varepsilon_p) \\ &= \frac{1}{\exp\left[\beta(p^2/2m - \mu)\right] \mp 1},\end{aligned}$$

the familiar result for a system of noninteracting fermions or bosons.

In generalizing these free particle results, it is convenient for us to define a function of a complex variable z:

$$G(p, z) = \int \frac{d\omega}{2\pi} \frac{A(p, \omega)}{z - \omega}. \tag{1.24}$$

For the free-particle system:

$$G(p, z) = \frac{1}{z - \varepsilon_p}. \tag{1.25}$$

In the general case,

$$G(p, z) = \frac{1}{z - \varepsilon_p - \Sigma\,(p, z)}. \tag{1.26}$$

Since the variable ω represents the total energy of an added particle we can interpret $\Sigma\,(p, \omega)$ [that is, $\Sigma\,(p, z)$ for real z] as the extra interaction energy that would be produced by the addition of a hypothetical particle of momentum p and energy ω. For this reason Σ is termed the self-energy. If the hypothetical energy is to be a realizable energy difference ω must satisfy the dispersion relation

$$\omega = \varepsilon_p + \Sigma\,(p, \omega). \tag{1.27}$$

Notice that these energy levels give the position of the poles of $G(p, z)$ and hence of δ-function singularities in $A(p, \omega)$.

To illustrate how $\Sigma\,(p, \omega)$ may be calculated, let us consider the interaction of electrons with longitudinal phonons. In the simplest model, this interaction may be represented by the interaction Hamiltonian

$$H_{ep} = \sum_q [v_q a_q \int dr \exp(iq \cdot r)\, \psi^+(r)\psi(r) + v_q^* a_q^+ \int dr \exp(-iq \cdot r)\, \psi^+(r)\psi(r)]. \tag{1.28}$$

Here a_q^+ and a_q are phonon creation and annihilation operators, and v_q is a matrix element which measures the coupling strength. We can rewrite (1.27) as

$$H_{ep} = \sum_{qp} (v_q a_q\, C_{p+q}^+ C_p + v_q^* a_q^+ C_{p,q}^+ C_p) \tag{1.29}$$

The first term in (1.29) describes a process in which an electron with momentum p absorbs a phonon with momentum q and hence scatters into the momentum state $p + q$. The second term describes a similar process in which a phonon is emitted.

As a preliminary to the calculation of $\Sigma(p, \omega)$, we calculate $\langle H_{ep} \rangle$ in second order perturbation theory. As usual, the second order perturbation theory gives a sum over all initial and the final states of a matrix element squared divided by an energy denominator which is the difference in energy between the initial and the final state. In this case

$$\langle H_{ep} \rangle = \int \frac{d\omega}{2\pi} \int \frac{d\omega}{2\pi} \sum_{p,p'} G^<(p, \omega)\, G^>(p', \omega')\, |\, v_q\, |^2 \left[\frac{N_q}{\omega + \omega_q - \omega'} + \frac{N_q + 1}{\omega - \omega_q - \omega'} \right]_{q=|p-p'|}. \tag{1.30}$$

Here ω_q is the energy of a phonon with wave vector q, and

$$N_q = \frac{1}{\exp(\beta\omega_q) - 1}$$

is the equilibrium number of phonons with wave vector q. The first term in the square bracket of (1.30) describes a process in which an electron with (p, ω) absorbs a phonon and scatters into the state (p', ω'). The energy difference between the initial state an the final state is $\omega + \omega_q - \omega'$. The density of initial states is given by the density of electrons, $G^<(p, \omega)$ times the density of phonons, N_q, and the density of final states if $G^>(p', \omega')$. In just the same way, the other term in the square bracket gives the energy shift due to phonon emission processes.

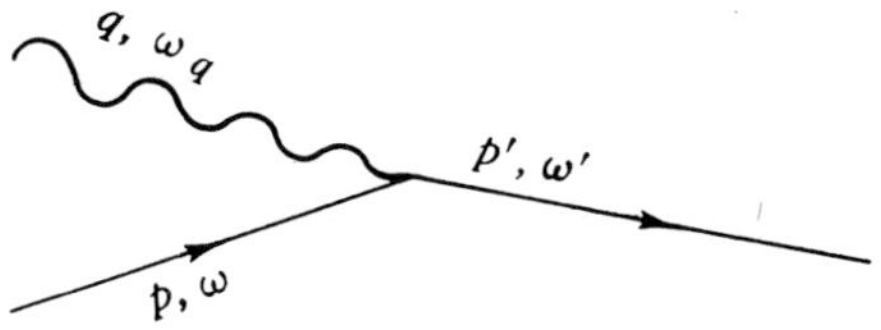

FIG. 2

We calculate the self-energy as the change in the interaction energy which occurs upon the addition to the system of a particle with momentum p and energy ω. This addition may be represented by changing the density of particles according to

$$G^<(p', \omega') \to G^<(p', \omega') + 2\pi\delta(\omega - \omega')\, \delta_{p,p'}. \tag{1.31a}$$

The added particle also changes the density of states. Since an electron is a fermion, the change is a reduction in the density of states according to

$$G^>(p', \omega') \to G^>(p', \omega') - 2\pi\delta(\omega - \omega')\, \delta_{p,p'}. \tag{1.31b}$$

The self-energy is then defined as

$$\Sigma(p, \omega) = \delta\langle H_{ep}\rangle \tag{1.32}$$

$$= \int \frac{d\omega'}{2\pi} \sum_{p'} |v_q|^2 \left\{ G^>(p', \omega') \left[\frac{N_q}{\omega + \omega_q - \omega'} + \frac{N_q + 1}{\omega - \omega_q - \omega'} \right] \right.$$

$$\left. + G^<(p', \omega') \left[\frac{N_q}{\omega - \omega_q - \omega'} + \frac{N_q + 1}{\omega + \omega_q - \omega'} \right] \right\}_{q=|p-p'|}.$$

This result can again be understood in terms of second order perturbation theory.[3]

[3] The argument given here may be extended, but only with considerable care, to obtain Σ as a variational derivative of an object related to $\langle H_{ep}\rangle$ with respect to G. A rigorous version of this argument is presented by Baym (*7*).

Before we go any further, we must face a serious difficulty. $\Sigma(p, \omega)$ is, strictly speaking, undefined at a fantastic number of different points on the real axis: all those points at which the energy denominator may vanish. The exact position of these singularities depends in great detail upon the exact size and shape of the system. We seek a method of eliminating this unwanted detail and retaining only that part of the information contained in (1.32) which pertains to the properties of a very large system.

There is a very simple method for accomplishing this elimination. We simply replace the ω which appears in (1.32) by the complex variable z. Then (1.32) becomes

$$\Sigma(p, z) = \int \frac{d\omega'}{2\pi} \sum_{p'} | v_q |^2 \left\{ G^{>}(p', \omega') \left[\frac{N_q}{z + \omega_q - \omega'} + \frac{N_q + 1}{z - \omega_q - \omega'} \right] + G^{<}(p', \omega') \left[\frac{N_q}{z - \omega_q - \omega'} + \frac{N_q + 1}{z + \omega_q - \omega'} \right] \right\}_{q = | p - p' |} . \tag{1.33}$$

By taking z to have an imaginary part we stand far enough away from the singularities so that we may consider them to be essentially continuously distributed along the real axis. In fact, we may allow them to become continuously distributed by taking the limit as the volume of the system goes to infinity and

$$\sum_{p'} \to \int \frac{d^3p'}{(2\pi)^3} . \tag{1.34}$$

Then, $\Sigma(p, z)$ has the form:

$$\Sigma(p, z) = \int \frac{d\omega}{2\pi} \frac{\Gamma(p, \omega)}{z - \omega} \tag{1.35}$$

with $\Gamma(p, \omega)$ now being a continuous function of ω.

II. Green's Function Approach

At this point, we shall consider how the results we have just obtained may be rederived by using a more formal Green's function technique. To do this, we define

$$G(r, t;\ r', t') = \frac{1}{i}\langle T[\psi(r, t)\psi^+(r', t')]\rangle \tag{2.1}$$

where the time variables are limited to be on the imaginary axis with

$$0 < it < \beta\,; \qquad 0 < it' < \beta\,.$$

Here T is a Wick time ordering symbol which tells you to order the operators according to the sense of the "time" it: the operator with the larger value of it appearing on the left. Also, for fermions, t contains a factor of $(-i)$ for each permutation of the operators from their standard order. Thus,

$$G(r, t;\ r', t') = \begin{cases} \dfrac{1}{i}\, G^{>}(r, t;\ r', t') & \text{for} \quad it > it' \\ \pm \dfrac{1}{i}\, G^{<}(r, t;\ r', t') & \text{for} \quad it < it'\,. \end{cases} \tag{2.2}$$

The main reason for defining G for imaginary times lies in the fact that we can relate the values of G at the two end points of its region of definition by using Eq. (1.16). This implies

$$G(r, t;\ r'\ t')\,|_{t=0} = \pm\, G(r, t;\ r', t')\,|_{t=-i\beta}\,. \tag{2.3}$$

The boundary condition (2.3) is most conveniently represented by writing G as a Fourier series in its time variable as

$$G(r, t;\ r', t') = \int \frac{d^3p}{(2\pi)^3}\,\frac{1}{-i\beta}\sum_{2} \exp\,[ip\cdot(r-r') - iz_{\nu}(t-t')]\; G_2(p)\,. \tag{2.4}$$

This form for G will necessarily satisfy the boundary condition if

$$z_{\nu} = \frac{\pi\nu}{-i\beta}$$

$$\nu = \begin{cases} \text{even integer for bosons} \\ \text{odd integer for fermions.} \end{cases}$$

By inverting the Fourier series, one can easily show that the Fourier coefficient is

$$G_\nu(p) = G(p, z_\nu) = \int \frac{d\omega}{2\pi} \frac{A(p, \omega)}{z_\nu - \omega} . \qquad (2.5)$$

A useful technique for evaluating G involves the equation of motion of the annihilation operator:

$$i \frac{\partial \psi(r, t)}{\partial t} = - [H', \psi(r, t)] .$$

For free particles this equation of motion is

$$\left[i \frac{\partial}{\partial t} + \frac{\nabla^2}{2m} + \mu \right] \psi(r, t) = 0 .$$

In calculating the time derivative of $G(r, t; r', t')$, we use Eq. (2.4) together with the time derivative of the discontinuity at $it = it'$ produced by the time ordering. This discontinuity is

$$\frac{1}{i} \langle [\psi(r, t) \psi^+(r', t) \mp \psi^+(r', t) \psi(r, t) \rangle = \frac{1}{i} \delta(r - r') .$$

Thus for free particles G obeys:

$$\left[i \frac{\partial}{\partial t} + \frac{2}{2m} + \mu \right] G(r, t; r', t') = \delta(r - r') \, \delta(ti - it')i . \qquad (2.6)$$

To find $G(p, z)$ from (2.6), we multiply by $\exp [- ip \cdot (r - r') + iz(t - t')]$ and integrate over all space and all time in the interval $[0, - i\beta]$. Then, we find

$$[z_\nu - \varepsilon_p] \, G(p, z_\nu) = 1$$

or

$$G(p, z_\nu) = \frac{1}{z_\nu - \varepsilon_p} = \int \frac{a\omega}{2\pi} \frac{A(p, \omega)}{z_\nu - \omega} .$$

There exists a theorem which states that if an equation that is $F(z)$ and $G(z)$ are two analytic functions of z such that $F(z) = G(z)$ holds for all

$z_\nu = (\pi\nu/-i\beta)$ for even or odd integral ν and if $F(z)$ and $G(z)$ have no essential singularity at infinity then

$$F(z) = G(z)$$

for all z. Therefore (2.6) implies

$$\int \frac{d\omega}{2\pi} \frac{A(p,\omega)}{z-\omega} = \frac{1}{z-\varepsilon_p}$$

for all z and we find

$$A(p,\omega) = 2\pi\delta(\omega-\varepsilon_p)$$

just as before.

To illustrate the application of Green's function techniques, we consider the interaction of electrons with longitudinal phonon, which may be represented by the Hamiltonian

$$H' = H_e + H_p + H_{ep} \tag{2.7a}$$

$$H_e = \int dr\, \psi^+(r)\left[\frac{\nabla^2}{2m} - \mu\right]\psi(r) \tag{2.7b}$$

$$H_p = \sum_q \omega_q a_q^+ a_q \tag{2.7c}$$

$$\begin{aligned} H_{ep} = & \sum_q v_q a_q \int dr \exp(+iq\cdot r)\,\psi^+(r)\psi(r) \\ & + \sum_q v_q^* a_q^+ \int dr \exp(-iq\cdot r)\,\psi^+(r)\psi(r)\,. \end{aligned} \tag{2.7d}$$

Here $\psi^+(r)$ and $\psi(r)$ are electron wave field creation and annihilation operators, a_q^+ and a_q are phonon creation and annihilation operators, ω_q is the phonon energy. For simplicity, we ignore Umklapp processes, the possibility of several types of phonons, and the spin of the electrons.

We see immediately that the various operators obey the equations of motion

$$\begin{aligned} & \left[i\frac{\partial}{\partial t} + \frac{\nabla^2}{2m} + \mu\right]\psi(r,t) \\ & \quad = \sum_q \left[v_q a_q(t)\exp(iq\cdot r) + v_q^* a_q^+(t)\exp(-iq\cdot r)\right]\psi(r,t) \end{aligned} \tag{2.8a}$$

$$\left(i\frac{\partial}{\partial t} - \omega_q\right) a_q(t) = \int dr\, v_q{}^* \exp(iq \cdot r)\, \psi^+(r, t)\, \psi(r, t) \tag{2.8b}$$

$$\left(i\frac{\partial}{\partial t} + \omega_q\right) a_q{}^+(t) = -\int dr\, v_q \exp(-iq \cdot r)\, \psi^+(t, r)\, \psi(r, t)\,. \tag{2.8c}$$

It is quite simple to compute the equation of motion of the electronic Green's function. We have:

$$\begin{aligned}\left[i\frac{\partial}{\partial t} + \frac{\nabla^2}{2m} + \mu\right] G(r, t; r', t') &= \delta(r - r')\,\delta(t - t') \\ &+ \frac{1}{i}\langle T\{\sum_q [v_q a_q(t) \exp(iq \cdot r) \\ &+ v_q{}^* a_q{}^+(t) \exp(-iq \cdot r)] \\ &\psi(r, t)\,\psi^+(r', t')\}\rangle\,.\end{aligned} \tag{2.9}$$

Therefore, in order to find $G(l, l')$ we need to know

$$\langle T[a_q(t)\, \psi(1)\, \psi^+(1')]\rangle$$

and

$$\langle T[a_q{}^+(t)\, \psi(1)\, \psi^+(1')]\rangle$$

where 1 stands for r, t, and $1'$ for r', t'. These functions are not known, but we can use the equation of motion for $a_q(t)$ and $a_q(t')$ to eliminate the phonon variables from these expressions. If we are to make use of these equations of motion, we must allow the time in $a_q(t)$ and $a_q{}^+(t')$ to be different from t_1, so that we may differentiate with respect to t. For this reason, we define:

$$F_q(t, 1, 1') = \langle T[a_q(t)\, \psi(1)\, \psi^+(1)]\rangle \tag{2.10a}$$

$$\tilde{F}_q(t, 1, 1') = \langle T[a_q{}^+(t)\, \psi(1)\, \psi^+(1')]\rangle\,. \tag{2.10b}$$

From (2.3b) and (2.3c) we find

$$\left[i\frac{\partial}{\partial t} - \omega_q\right] F_q(t, 1, 1') \tag{2.11a}$$

$$= v_q{}^* \langle T[\int d\bar{r} \exp(-iq \cdot \bar{r})\, \psi^+(\bar{r}, t)\, \psi(\bar{r}, t)\, \psi(1)\, \psi^+(1')]\rangle$$

and

$$\left[i\frac{\partial}{\partial t}+\omega_q\right]\tilde{F}_q(t,1,1') \tag{2.11b}$$

$$=-v_q\langle T[\int d\bar{r}\exp(iq\cdot\bar{r})\,\psi^+(\bar{r},t)\,\psi(\bar{r},t)\,\psi(1)\,\psi^+(1')]\rangle\,.$$

There are no discontinuity terms coming from differentiating T because the equal time commutator of a_q and a_q^+ with ψ and ψ^+ vanishes.

Equations (2.11) are to be solved with the standard 0 to $-i\beta$ boundary condition:

$$F_q(t,1,1')\,|_{t=0}=F_q(t,1,1')\,|_{t=-i\beta} \tag{1.16}$$

and similarly for $\tilde{F}$. In obtaining the solution to (2.11) it is convenient to use the function $d_\omega(t,t')$, which is defined by the differential equation

$$\left[i\frac{\partial}{\partial t}-\omega_q\right]d_{\omega_q}(t,t')=\delta(t-t')\quad\text{for}\quad\begin{matrix}0<it<\beta\\0<it'<\beta\end{matrix}$$

and the boundary condition:

$$d_{\omega_q}(0,t')=d_{\omega_q}(-i\beta,t')\,.$$

The solution to this equation for d_ω may be written down at once. We split d_ω into two parts in just the same way as we divided $G(1,1')$ into two parts by writing

$$id_{\omega_q}(t_1,t_2)=\begin{cases}d^>_{\omega_q}(t_1,t_2) & \text{for}\quad it_1>it_2\\ d^<_{\omega_q}(t_1,t_2) & \text{for}\quad it_1<it_2\,.\end{cases} \tag{2.12}$$

We can see immediately that $d^>_{\omega_q}$ and $d^<_{\omega_q}$ are exactly analogous to $G^>(p,t_1-t_2)$ and $G^<(p,t_1-t_2)$ for free bosons if the replacement

$$\left(\frac{p^2}{2m}-\mu\right)\rightarrow\omega$$

is made. Therefore, we have

$$\begin{aligned}d^>_{\omega_q}(t_1-t_2)&=\exp[-i\omega_q(t_1-t_2)]\,(N_q+1)\\ d^<_{\omega_q}(t_1-t_2)&=\exp[-i\omega_q(t_1-t_2)]\,N_q\end{aligned} \tag{2.13a}$$

where N_q is the equilibrium number of phonons with wave vector **q**.

$$N_q = \frac{1}{\exp(\beta\omega_q) - 1} .$$

Also, we see

$$\begin{aligned} d^{>}_{-\omega_q}(t_1 - t_2) &= - \exp [i\omega_q(t_1 - t_2)]\, N_q \\ d^{<}_{-\omega_q}(t_1 - t_2) &= - \exp [i\omega_q(t_1 - t_2)]\, (N_q + 1) \end{aligned} \tag{2.13b}$$

since

$$\frac{1}{\exp(-\beta\omega_q) - 1} = - \frac{1}{\exp(\beta\omega)_q - 1} - 1 = -(1 + N_q) .$$

We can see immediately that solution to Eq. (2.11) with the boundary condition (1.16) may be expressed in terms of d as:

$$F_q(t, 1, 1') = \int d\bar{r} \int_0^{-i\beta} d\bar{t}\, v_q{}^* \exp(-iq \cdot \bar{r})\, d_{\omega_q}(t, \bar{t})$$
$$\langle T[\psi^+(\bar{r}, \bar{t})\, \psi(\bar{r}, \bar{t})\, \psi(1)\, \psi^+(1')] \rangle \tag{2.14a}$$

$$\tilde{F}_q(t, 1, 1') = -\int d\bar{r} \int_0^{-i\beta} d\bar{t}\, v_q \exp(iq \cdot \bar{r})\, d_{\omega_q}(t, \bar{t})$$
$$\langle T[\psi^+(\bar{r}, \bar{t})\, \psi(\bar{r}, \bar{t})\, \psi(1)\, \psi^+(1')] \rangle . \tag{2.14b}$$

Notice the great utility of the boundary condition and the $[0, -i\beta]$ approach. This approach enables us to eliminate all reference to the phonon variables in expressions like F_q and $\tilde{F}_q$ and reduce these to expressions which involve only electron operators. Of course, we have to pay a price for this elimination.. This price may be seen from the result of substituting (2.14) and back into the equation of motion of (2.6):

$$\left[i\frac{\partial}{\partial t} + \frac{\nabla_1^2}{2m} + \mu \right] G(1, 1') = \delta(1 - 1')$$
$$+ \frac{1}{i} \sum_q | v_q |^2 \int d\bar{r} \int_0^{-i\beta} d\bar{t}$$
$$[\exp(iq \cdot (r - \bar{r}))\, d_{\omega_q}(t, \bar{t})$$
$$- \exp(-iq \cdot (\bar{r} - \bar{r}))\, d_{\omega_q}(\bar{t}, \bar{t})$$
$$\langle T[\psi(1)]\, \psi^+(1')\, \psi^+(\bar{r}, \bar{t})\, \psi(\bar{r}, \bar{t})] \rangle. \tag{2.15}$$

It is convenient to rewrite this expression in terms of

$$G_2(1, 2;\ 1', 2') = \left(\frac{1}{i}\right)^2 \langle T[\psi(1)\,\psi(2)\,\psi^+(2')\,\psi^+(1')]\rangle \tag{2.16}$$

as

$$\left[i\frac{\partial}{\partial t} + \frac{\nabla_1^2}{2m} + \mu\right] G(1, 1') = \delta(1 - 1') \pm i\int dr_2 \int_0^{-i\beta} dt_2\, V(1 - 2)\; G_2(1, 2;\ 1', 2) \tag{2.17}$$

where

$$V(1 - 2) = \sum_q |\, v_q\,|^2\, [\exp\,[iq\cdot(r_1 - r_2)]\; d_{\omega_q}(t_1, t_2)\; \exp\,[-\,iq\cdot(r_1 - r_2)]\; d_{-\omega_q}(t_1, t_2)]\,. \tag{2.18}$$

The reason that Eq. (2.17) is so instructive is that it is quite similar in structure to the equation of motion which would emerge from an ordinary electron-electron interaction. If the ordinary interaction can be represented by a central potential $v(|\,\mathbf{r}_1 - \mathbf{r}_2\,|)$, then the equation of motion for G again takes the form (2.13) except for this case:

$$V(1, 2) = \delta(t_1 - t_2)\; v(|\,\mathbf{r}_1 - \mathbf{r}_2\,|)\,. \tag{2.19}$$

Therefore the phonons act to produce an affective electron-electron potential. This potential may be considered to result from the fact that an electron at one point in space and time may change the phonon field by emitting or absorbing phonons. This change in the phonon field can at a later time, scatter electrons at another point in space so that the scattering looks like it is produced by a retarded interaction between electrons.

To find G from Eq. (2.17), we must make some approximation for the $G_2(1, 2;\ 1', 2')$ which appears in that equation. This two-particle Green's function, of course, describes the correlated motion of two electrons added to the system at $1'$ and $2'$ and removed at 1 and 2. The simplest possible approximation for this G_2 is to come from the assumption that the added electrons move quite independently of one another. This assumption may be expressed mathematically as

$$G_2(1, 2;\ 1', 2') \approx G(1, 1')\quad G(2, 2') - G(1, 2')\quad G(2, 1')\,. \tag{2.20}$$

The appearance of two terms in this expressions is a result of Fermi statistics obeyed by the electrons which requires that

$$G_2(1, 2;\ 1', 2') = -\ G_2(2, 1;\ 1', 2')\,.$$

When we substitute (2.20) in (2.17) we find

$$\left[i\frac{\partial}{\partial t} + \frac{\nabla_1^2}{2m} + \mu\right] G(1, 1') = \delta(1 - 1') \pm i[\int V(1 - 2)\, G(1, 2)\, d2\, G(1, 1') + i\int d2\, V(1 - 2)\, G(1, 2)\, G(2, 1') \tag{2.21}$$

the middle term on the right-hand side of (2.21) should be neglected for two reasons:

(a) This term is proportional to

$$\int d2\, V(1 - 2)\, \langle n(2)\rangle = \langle n\rangle \int dr \int_0^{-i\beta} dt\, V(r, t)\,.$$

After the r_2 integral is performed, this expression becomes proportional to

$$|\, v_q\,|^2\,|_{q=0}\,.$$

However, a $q = 0$ phonon is pure nonsense. This "phonon" represents the effect of picking up and bodily displacing the whole crystal. Therefore this whole term is physically meaningless.

(b) This term is proportional to the average potential field produced by all the electrons in the system. As such, it has actually been included in the original definition of the band structure which underlies our original Hamiltonian. Therefore, this term has already been counted and it must now be left out.

After this term is thrown out Eq. (2.21) may be rewritten as:

$$\left[i\frac{\partial}{\partial t} + \frac{\nabla_1^2}{2m} + \mu\right] G(1, 1') = \delta(1 - 1') + \int d2\, \Sigma(1, 2)\, G(2, 1') \tag{2.22}$$

where the self-energy Σ is

$$\Sigma(1, 1') = iV(1 - 1')\, G(1, 1')\,. \tag{2.23}$$

It is a trivial matter to verify that $\Sigma(1, 1')$ obeys the same boundary conditions as G, i.e.,

$$\Sigma(1, 1')\,|_{t_1=0} = -\Sigma(1, 1')\,|_{t_1 i=\beta} \tag{2.24}$$

since $V(1, 1')$ obeys the boson boundary condition

$$V(1, 1')\,|_{t_1=0} = V(1, 1')\,|_{t_1=-i\beta}.$$

Since Σ obeys the same boundary condition as G, its formal properties are very similar to the formal properties of G. For example, when $\Sigma(1, 1')$ is split into two parts as

$$i\,\Sigma(1, 1') = \begin{cases} \Sigma^{>}(1-1') = \displaystyle\int \frac{d^3p}{(2\pi)^3}\frac{d\omega}{2\pi} \\ \quad \cdot \exp\,[ip\cdot(r_1 - r_2) - i\omega(t_1 - t_1')]\,\Sigma^{>}(p, \omega) \quad \text{for } it_1 > it_1' \\ -\Sigma^{<}(1-1') = -\displaystyle\int \frac{d^3p}{(2\pi)^3}\frac{d\omega}{2\pi} \\ \quad \cdot \exp\,[ip\cdot(r_1 - r_1') - i\omega(t_1 - t_1')]\,\Sigma^{<}(p, \omega) \quad \text{for } it_1 < it_1' \end{cases} \tag{2.25}$$

then $\Sigma^{>}(p, \omega)$ and $\Sigma^{<}(p, \omega)$ obey the detailed balancing relation

$$\Sigma^{<}(p, \omega) = \exp\,(-\beta\omega)\,\Sigma^{<}(p, \omega)\;;$$

also, Σ may be written as a Fourier series

$$\Sigma(1, 1') = \int \frac{d^3p}{(2\pi)^3} \exp\,[ip\cdot(r_1 - r_1')]\,\frac{1}{-i\beta} \cdot \sum_\nu \exp\,[-iz_\nu(t_1 - t_1')]\,\Sigma_\nu(p) \tag{2.26}$$

with the Fourier coefficient given by

$$\Sigma_\nu(p) = \Sigma(p, z_\nu) = \int \frac{d\omega}{2\pi}\,\frac{\Sigma^{>}(p, \omega) + \Sigma^{<}(p, \omega)}{z_\nu - \omega} = \int \frac{d\omega}{2\pi}\,\frac{\Gamma(p, \omega)}{z_\nu - \omega}. \tag{2.27}$$

This Fourier series representation of $\Sigma(1, 1')$ is very convenient because it enables us to solve the differential Equation (2.22) quite directly.

If we multiply this equation by $\exp[-ip\cdot(r_1-r_1')+iz_\nu(t_1-t_1')]$ and then integrate over all r_1 and all t_1 in the interval $0<it_1<\beta$, then (2.22) becomes

$$[z_\nu-\varepsilon_p-\Sigma(p,z_\nu)]\,G(p,z_\nu)=1\,. \qquad (2.28)$$

Equation (2.28) is a relation between the analytic functions

$$\Sigma(p,z)=\int\frac{d\omega}{2\pi}\,\frac{\Sigma^>(p,\omega)+\Sigma^<(p,\omega)}{z-\omega}=\int\frac{d\omega}{2\pi}\,\frac{\Gamma(p,\omega)}{z-\omega} \qquad (2.29a)$$

$$G(p,z)=\int\frac{d\omega}{2\pi}\,\frac{G^>(p,\omega)+G^<(p,\omega)}{z-\omega}=\int\frac{d\omega}{2\pi}\,\frac{A(p,\omega)}{z-\omega} \qquad (2.29b)$$

which is known to hold on all the points

$$z=z_\nu=\frac{\pi\nu}{-i\beta}\qquad \nu=\text{odd integer.}$$

From the fact that this relation holds on this limited set of points and the fact that neither $\Sigma(p,z)$ nor $G(p,z)$ has an essential singularity at ∞ it follows that (2.28) holds for all z, i.e., that

$$G(p,z)=\frac{1}{z-\varepsilon_p-\Sigma(p,z)}\,. \qquad (2.30)$$

In order to make use of Eq. (2.30) we use (2.18) and (2.23) to write

$$\begin{aligned}\Sigma^>(1,1')&=V^>(1,1')\,G^>(1,1')\\&=\sum_q|v_q|^2\exp[iq\cdot(r_1-r_1')]\,d_{\omega_q}(t_1-t_1')\\&\quad-\exp[-iq\cdot(r_1-r_1')]\,d_{-\omega_q}(t_1-t_1')]\,G^>(1,1')\,.\end{aligned} \qquad (2.31)$$

The Fourier transform of (2.31) is

$$\begin{aligned}\Sigma^>(p,\omega)=\int\frac{d^3p'}{(2\pi)^3}\,\frac{d\omega'}{2\pi}\,G^>(p',\omega')\,\{|v_q|^2\,[N_q\,2\pi\delta(\omega+\omega_q-\omega')\\+(N_q+1)\,2\pi\delta(\omega-\omega_q-\omega')]\}\qquad q=|p-p'|\,.\end{aligned} \qquad (2.32a)$$

A similar evaluation gives

$$\Sigma^<(p, \omega) = \int \frac{d^3p'}{(2\pi)^3} \frac{d\omega'}{2\pi} G^<(p', \omega') \{| v_q |^2 [(N_q + 1)2\pi\delta(\omega + \omega_q - \omega')$$

$$+ N_q 2\pi\delta(\omega - \omega_q - \omega')]\} \qquad q = | p - p' | . \tag{2.32b}$$

The substitution of (2.32) into (2.29a) leads back to our earlier result (1.33):

$$\Sigma(p, z) = \int \frac{d\omega'}{2\pi} \int \frac{d^3p'}{(2\pi)^3} | v_q |^2$$

$$\cdot \left\{ G^>(p', \omega') \left[\frac{N_q}{z + \omega_q - \omega'} + \frac{N_q + i}{z - \omega_q - \omega'} \right] \right. \tag{2.33}$$

$$\left. + G^<(p', \omega') \left[\frac{N_q}{z - \omega_q - \omega'} + \frac{N_q + 1}{z + \omega_q - \omega'} \right] \right\} \qquad q = | p - p' | .$$

It is now quite simple to derive an expression for $A(p, \omega)$ since $iA(p, \omega)$ is the discontinuity in $G(p, z)$ as z crosses the real axis. From (2.30),

$$A(p, \omega) = \frac{\Gamma(p, \omega)}{[\omega - \varepsilon_p - \operatorname{Re} \Sigma(p, \omega)]^2 + [\Gamma(p, \omega)/2]^2} . \tag{2.34}$$

Here $\operatorname{Re} \Sigma(p, \omega)$ is the part of $\Sigma(p, z)$ which is continuous as z crosses the real axis. It is given by an expression identical with (2.33) except that

$$\frac{1}{z \pm \omega_q - \omega'} \rightarrow P \frac{1}{\omega \pm \omega_q - \omega'}$$

where P stands for principal value. $\Gamma(p, \omega)$ is the discontinuity in (2.33). In Γ, the energy denominators in (2.33) are replaced by δ-functions:

$$\frac{1}{z \pm \omega_q - \omega'} \rightarrow 2\pi\delta(\omega \pm \omega_q - \omega') .$$

Since $A(p, \omega)$ is defined in terms of $G^>$ and $G^<$ by (2.29b) and $\Sigma(p, z)$ is defined in terms of $G^>$ and $G^<$ by (2.33), Eq. (2.34) gives us one relation between $G^>$ and $G^<$. The boundary condition

$$G^<(p, \omega) = \exp(-\beta\omega)\, G^>(p, \omega) \tag{2.35}$$

provides another relation between $G^>$ and $G^<$. Thus, we have two equations in two unknowns and we may solve for $G^>$ and $G^<$.

III. Specialization to a Metal

At this point, I would like to quote two rather surprising facts about the approximation described above:

(a) This approximation is exact in a normal (nonsuperconducting) metal if the correct phonon energy spectrum is employed and the electron-electron interactions are neglected.

(b) The highly nonlinear equations we have written down are exactly soluble for a metal.

These statements have been proved by Midgal (*8*). Before outlining Midgal's proof, I should point out that these statements are very closely related to the well-known fact that the adiabatic approximation (*9*) gives a correct description of the electron-phonon interaction in metals Therefore, I and II really tell us that our Green's function approximation is a correct restatement of the adiabatic approximation.

In order to see the essence of Migdal's arguments, let us consider the diagrammatic expansion for the self-energy in a power series in G and V. The terms we have considered in the self-energy correspond to the diagrams

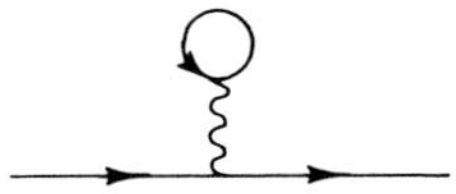

FIG. 3

$$\Sigma_1(1, 1') = \delta(1 - 1') \left[-i \int d2V(1 - 2)\, G(2, 2)\right] \tag{3.1}$$

and

FIG. 4

$$\Sigma_2(1, 1') = iV(1, 1')\, G(1, 1')\,. \tag{3.2}$$

Here the $\longrightarrow$ line stands for a G and $\sim\!\sim$ for the interaction V. The first of these terms vanishes and the next is the one we have taken into account.

The next order terms, which we have neglected are:

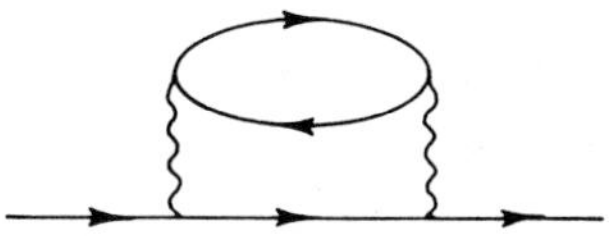

FIG. 5

$$\Sigma_3(1, 1') = i\delta V(1, 1')\, G(1, 1') \tag{3.3}$$

and

$$\delta V(1, 1') = -i \int V(1, 2)\, G(2, 2')\, G(2', 2)\, V(2', 1')\, d2'\, d2$$

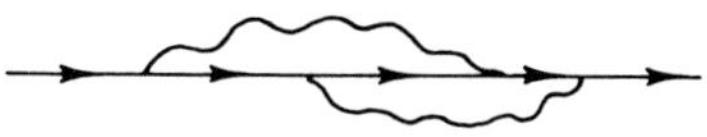

FIG. 6

$$\Sigma_4(1, 1') = i^2 \int d2\, d2'\, V(1, 2)\, G(1, 2')\, G(2', 2)\, G(2, 1')\, V(2, 1')\,. \tag{3.4}$$

The substance of point I in the Migdal paper is that terms like Σ_4 are negligible while terms like Σ_3 serve to shift the phonon energies in the $V(1, 1')$ of Σ_2 from the bare phonon energies to the energies of the fully interacting phonon system. The manner in which Σ_3 serves to shift the energies is, I think, clear. However, some further discussion is necessary in order to see why Σ_4 is small.

A crucial fact in this discussion is that only electrons with very small values of $\varepsilon_p (\varepsilon_p \ll \mu)$ play any role in the electron-phonon interaction. The reason for this limitation lies in the fact that the phonon energies are all quite small, in fact,

$$(\omega_q/\mu) \sim \sqrt{\frac{m}{M}} \ll 1 \tag{3.5}$$

where m is the electronic mass and M the ionic mass. Therefore, an electron which initially lies quite close to the Fermi surface must always remain quite close to $\varepsilon_p \approx 0$. If all electronic momenta must be very close to

$$p = p_F = \sqrt{2m\mu}\,.$$

This puts considerable restrictions on the momentum integrals which are to be done in evaluating Σ_4. In fact, this phase space restriction leads to Σ_4 being smaller than Σ_2 by a factor of order $\sqrt{m/M}$. (We assume that the dimensionless coupling constant $[|\, v_q\,|^2\, mp_F/\omega_q]$ is of order unity.)

In this way, Migdal proved that the approximation that we discussed in the last two sections was an exact perturbation theoretic expansion of the self-energy to order $(m/M)^{1/2}$. Migdal apparently did not publish this proof for several years because it obviously contained a grave defect—it appears to prove that the electron-phonon interaction has nothing to do with superconductivity. The trouble is that, for a superconductor, perturbation theory does not converge. Hence this whole perturbation theoretic argument is quite wrong for a superconductor.

We shall return to the discussion of the superconductor later. For now let us consider Migdal's second point. We can solve (2.34) for $A(p, \omega)$. To effect this solution we examine (2.33) and notice that $\Sigma(p, z)$ is a very slowly varying function of p, but a rapidly varying function of z. The rate of variation in z is measured by ω_q but the rate of variation in $\varepsilon_p = (p^2/2m - \mu)$, is measured by μ. Because of his slow variation in p, we can simply set $p = p_F$ in Eq. (2.33) and write (2.34) as

$$A(p, \omega) = \frac{\Gamma(\omega)}{[\omega - \varepsilon_p - \operatorname{Re}\Sigma(\omega)]^2 + [\Gamma(\omega)/2]^2} \tag{3.6}$$

with

$$\operatorname{Re}\Sigma(\omega) = \operatorname{Re}\Sigma(p_F, \omega)$$

$$\Gamma(\omega) = \operatorname{Re}\Gamma(p_F, \omega)\,.$$

Let us observe that (3.6) implies

$$\int \frac{d\varepsilon}{2\pi}\, A(p, \omega) = 1\,. \tag{3.7}$$

Equation (3.7) is crucially important because a huge variety of physical quantities depend upon the integral (3.7). Because this integral is independent of the strength of the electron-phonon interaction, a whole host of calculations are immensely simplified. The physical meaning of (3.7) can be roughly appreciated if we notice that the electron-phonon interaction serves to scatter electrons only over the very narrow range of momenta

$$\left| \frac{p^2}{2m} - \mu \right| \lesssim \omega_q \, .$$

That is to say the interaction produces a quasi-particle state by redistributing the electron momenta and superposing states of slightly different electronic momentum. However, when we exmaine a sum over all p (or equivalently, all ε) all this redistribution becomes irrelevant so that the integral in (3.7) is quite independent of the strength of electron-phonon interaction.

Equation (3.7) can be used to simplify the expression (2.33) for $\Sigma(p, z)$. This simplification is effected by writing

$$\int \frac{d^3p}{(2\pi)^3} = \int d\varepsilon_p \int \frac{d\Omega}{4\pi} \frac{mp}{2\pi^2} \approx \frac{mp_F}{2\pi^2} \int d\varepsilon \int \frac{d\Omega}{4\pi} \tag{3.8}$$

where

$$\Omega = \vec{p}/|\vec{p}| \approx p/p_F \, .$$

We shall employ the abbreviation

$$N(0) = \frac{mp_F}{2\pi^2} \tag{3.9}$$

for the (unperturbed) density of states in energy at the Fermi surface. Then by using the relations

$$G^>(p, \omega) = [1 + f(\omega)] \, A(p, \omega)$$

$$G^<(p, \omega) = f(\omega) \, A(p, \omega) \, ,$$

we can rewrite Eq. (2.33) as

$$\Sigma(z) = N(0) \int d\omega' \int \frac{d\Omega'}{4\pi} \, |v_q|^2 \cdot \left\{ [1 + f(\omega')] \left[\frac{N_q}{z + \omega_q - \omega'} + \frac{N_q + 1}{z - \omega_q - \omega'} \right] \right. \tag{3.10}$$

$$+f(\omega')\left[\frac{N_q}{z-\omega_q-\omega'}+\frac{N_q+1}{z+\omega_q-\omega'}\right]\Bigg\}$$

$$\cdot\int\frac{d\varepsilon'}{2\pi}A(\varepsilon',\omega')\,.\qquad q=p_F\,|\,\Omega-\Omega'\,| \tag{3.10 cont.}$$

Since the ε' integral of A is unity according to (3.7), we have reduced the evaluation of $\Sigma(z)$ and hence $A(p, \omega)$ to quadratures. Schrieffer and Englesberg (*10*) have done these integrals at zero temperature and examined the form of A.

However, for most purposes it is not even necessary to know A: the sum rule (3.7) is quite sufficient for obtaining a variety of useful results. For example, consider a tunneling experiment: We can rewrite expression (1.12) for the tunneling rate as

$$\begin{aligned}\text{net}\ \frac{dN}{dt} &= 2\int\frac{d\omega}{2\pi}\int\frac{d\omega'}{2\pi}\int\frac{d^3p}{(2\pi)^3}\int\frac{d^3p'}{(2\pi)^3}\\ &\quad\cdot|\,T(p,p')\,|^2\,A_A(p,\omega)\,A_B(p',\omega')\\ &\quad\cdot 2\pi\delta(\omega-\omega'+eV)\,[f(\omega)-f(\omega')]\,.\end{aligned} \tag{3.11}$$

(The factor 2 is a result of the sum over electron spins.) Then we can use the fact that $T(p, p')$ depends only very weakly upon ε and ε' to do the ε and ε' integrals in (3.11). The result

$$\begin{aligned}\text{net}\ \frac{dN}{dt} &= 2[N(0)]^2\,2\pi\int\frac{d\Omega'}{4\pi}\int\frac{d\Omega'}{4\pi}\,|\,T(p_F\Omega,p_F\Omega')\,|^2\\ &\quad\cdot\int d\omega\int d\omega'\,\delta(\omega-\omega'+eV)\,[f(\omega)-f(\omega')]\\ &= 2[N(0)]^2\,2\pi\int\frac{d\Omega}{4\pi}\int\frac{d\Omega'}{4\pi}\,|\,T(p_F\Omega,p_F\Omega')\,|^2\,eV\end{aligned} \tag{3.12}$$

is very interesting because all effects of the electron-phonon interaction have completely cancelled out.

Notice, incidentally, the complete failure of the conventional quasi-particle picture which gives the effective density of states in energy as

$$N(0)\,\frac{d\varepsilon}{dE} \tag{3.13}$$

with E given by

$$E_p = \varepsilon_p + \Sigma(p, E_p) . \tag{3.14}$$

Even for regions of E in which $\Sigma(p, E)$ is completely real so that the quasi particles are infinitely long-lived expression (3.13) is irrelevant. The relevant density of states for tunneling is the simpler expression $N(0) = mp_F/2\pi^2$.

IV. The Superconductor

All of the arguments of the previous lectures can be generalized to the case of the superconductor. This generalization requires the use of the additional propagators

$$F(1, 2) = \frac{1}{i} \langle T[\psi_\uparrow(1)\, \psi_\downarrow(2)]\rangle \tag{4.1}$$

and

$$F^+(1, 2) = \frac{1}{i} \langle T[\psi_\downarrow{}^+(1)\, \psi_\uparrow{}^+(2)]\rangle , \tag{4.2}$$

which respectively describe processes in which a bound pair breaks up so that two particles may be removed from the system and processes in which two particles added to the system form a bound pair. (The formulation given here is appropriate tin the case in which the pairs are bound in an S state of zero total spin. In more complex situations, more correlation functions need be considered. See the work of R. Balian in this volume.)

It is important that F and F^+ be considered handled in exactly the same way as G. Nambu[4] (*11*) has introduced a simple trick for handling G, F, and F^+ at the same time. He begins with the spinor fields

$$\Psi(1) = \begin{pmatrix} \psi_\uparrow(1) \\ \psi_\downarrow{}^+(1) \end{pmatrix} ; \qquad \Psi^+(1) = \overbrace{\underbrace{\psi_\uparrow{}^+(1')\, \psi_\downarrow(1')}} \tag{4.3}$$

and notices that the basic Hamiltonian may be rewritten in terms of

[4] Nambu's formalism provides a very convenient technique for extending the B.C.S. theory (*12*) to the case of retarded interactions.

these fields. For example, the kinetic energy is:

$$K.E. = \int dr \left\{ \left(\frac{\nabla \cdot \nabla'}{2m} - \mu \right) [\psi_\uparrow^+(r')\, \psi_\uparrow(r) + \psi_\downarrow^+(r')\, \psi_\downarrow(r)] \right\}_{r'=r}$$

$$= \int dr \left\{ \left(\frac{\nabla \cdot \nabla'}{2m} - \mu \right) [\psi_\uparrow^+(r')\, \psi_\uparrow(r) - \psi_\downarrow(r')\, \psi_\downarrow(r) + \delta(r - r')] \right\}_{r'=r}$$

The δ-function term is quite irrelevent since it is just a number added to the Hamiltonian. When this term is neglected the kinetic energy can be written in terms of an inner product of the spinors

$$K.E. = \int dr \left\{ \left(\frac{\nabla \cdot \nabla'}{2m} - \mu \right) [\Psi^+(r')\tau_3 \Psi(r)] \right\}_{r'=r} \tag{4.4}$$

Here τ_3 is the standard Dirac spin matrix

$$\tau_3 = \begin{pmatrix} 1 & 0 \\ 0 & -1 \end{pmatrix}.$$

Similarly, the interaction term can be written in terms of Ψ^+ and Ψ as

$$\begin{aligned} H_{ep} &= \int dr\, [v_q a_q \exp(iq \cdot r) + v_q^* a_q^+ \exp(-iq \cdot r)] \\ &\qquad [\psi_\uparrow^+(r)\, \psi_\uparrow(r) + \psi_\downarrow^+(r)\, \psi_\downarrow(r)] \qquad (4.5) \\ &\rightarrow \int dr\, [v_q a_q \exp(iq \cdot r) + v_q^* a_q^+ \exp(-iq \cdot r)]\, [\Psi^+(r)\tau_3 \Psi(r)]. \end{aligned}$$

The properties of the superconductor are described by a spinor Green's function

$$\begin{aligned} g(1, 1') &= \frac{1}{i} \langle T[\Psi(1)\Psi^+(1')] \rangle \\ &= \begin{pmatrix} G(1, 1') & F(1, 1') \\ F^+(1, 1') & -G(1', 1) \end{pmatrix}. \end{aligned} \tag{4.6}$$

Since this spinor function obeys the same boundary condition as G, it has exactly the same formal properties as G. In particular, we can notice that it can be expanded in Fourier series with matrix Fourier coefficients $g(p, z_\nu)$ where

$$g(p, z) = \int \frac{d\omega}{2\pi} \frac{a(p, \omega)}{z - \omega} = \int \frac{d\omega}{2\pi} \frac{g^{>}(p, \omega) + g^{<}(p, \omega)}{z - \omega}. \tag{4.7}$$

The two pieces of g are again related by

$$g^{<}(p, \omega) = \exp(-\beta\omega)\, g^{<}(p, \omega). \tag{4.8}$$

Equivalently, we can write

$$\begin{aligned} g^{<}(p, \omega) &= f(\omega)\, a(p, \omega) \\ g^{<}(p, \omega) &= [1 - f(\omega)]\, a(p, \omega), \end{aligned} \tag{4.9}$$

where $a(p, \omega)$ is now a matrix.

We can express

$$g(p, z) = \frac{1}{z - \varepsilon_p - \sigma(p, z)} = \int \frac{d\omega}{2\pi} \frac{a(p, \omega)}{z - \omega} \tag{4.10}$$

where the matrix $\sigma(p, z)$ is expandable in exaclty the same kind of perturbation theory as in the normal state. The only difference lies in the fact that the matrix g replaces the scalar G and a factor of τ_3 appears at each vertex. Thus, in lowest order

$$\sigma(1, 1') = \qquad = i\, V(1 - 1')\, \tau_3\, g(1, 1') \tau_3 \tag{4.11}$$

or

$$\begin{aligned} \sigma(p, z) = \int \frac{d^3 p'}{(2\pi)^3} \frac{d\omega'}{2\pi} \, | v_q |^2 \\ \cdot \left\{ \tau_3\, g^{>}(p', \omega')\, \tau_3 \left[\frac{N_q}{z + \omega_q - \omega'} + \frac{N_q + 1}{z - \omega_q - \omega'} \right] \right. \\ \left. + \tau_3\, g(p', \omega')\, \tau_3 \left[\frac{N_q}{z - \omega_q - \omega'} + \frac{N_q + 1}{z + \omega_q - \omega'} \right] \right\}. \end{aligned} \tag{4.12}$$

Before we go any further, we should notice that, as in the normal metal, $\sigma(p, z)$ can be considered to be independent of p. Hence, we write $\sigma(p, z) \approx \sigma(p_F, z) = \sigma(z)$.

In order to solve $g(p, z)$ we guess the form of the answer for $\sigma(p, z)$.

Our guess, which we verify below, is that $\sigma(p, z)$ contains no contribution proportional to τ_3 and none proportional to τ_2, but only terms proportional to the unit matrix

$$1 = \begin{pmatrix} 1 & 0 \\ 0 & 1 \end{pmatrix}$$

and

$$\tau_1 = \begin{pmatrix} 0 & 1 \\ 1 & 0 \end{pmatrix}.$$

Thus we write

$$\sigma(z) = \sigma_0(z)\, 1 + \sigma_1(z)\, \tau_1 \,. \tag{4.13}$$

Actually it is slightly more convenient to express $\sigma(p, z)$ in terms of the variables

$$\begin{aligned} Z(z) &= [1 - \sigma_0(z)/z] \\ Z(z)\, \Delta(z) &= \sigma_1(z) \end{aligned} \tag{4.14}$$

with this redefunction of the variables, Eq. (4.10) becomes

$$\begin{aligned} g(p, z) &= \frac{1}{Z(z)z - \varepsilon\tau_3 - \Delta(z)Z(z)\tau_1} \\ &= \frac{zZ(z) + \varepsilon\tau_3 + \Delta(z)Z(z)\tau_1}{Z^2(z)\,[z^2 - \Delta^2(z)] - \varepsilon^2} \end{aligned} \tag{4.15}$$

and[5]

$$\int \frac{d\varepsilon}{2\pi}\, g(p, z) = \frac{z + \Delta(z)\tau_1}{\sqrt{-z^2 + [\Delta(z)]^2}} = \int \frac{d\varepsilon}{2\pi} \int \frac{d\omega}{2\pi}\, \frac{a(p, \omega)}{z - \omega} \,. \tag{4.16}$$

Here, the sign of the square root is choosen so that a $z \to \infty$ below the real axis $\sqrt{-z^2} \to -iz$, while as $z \to \infty$ above the real axis $\sqrt{-z^2} \to +iz$. There is a branch line in the definition of the square root along the real axis. The ε integral of $a(p, \omega)$ is now easily computed in terms of the discontinuity of Eq. (4.16) as z crosses the real axis. This gives

$$\int \frac{d\varepsilon}{2\pi}\, a(p, \omega) = \mathrm{Re}\, \frac{\omega + \Delta(\omega)\,\tau_1}{\sqrt{\omega^2 - [\Delta(\omega)]^2}} \tag{4.17}$$

[5] The first application of this ε integration trick to the superconductor appeared in P. Morel and P. W. Anderson (*13*).

where the square root is defined by the statement

$$\lim_{\omega\to\infty} \left\{\omega \Big/ \sqrt{\omega^2 - [\Delta(\omega)]^2}\right\} = 1 . \tag{4.12}$$

Now we return to Eq. (4.12). We make the substitution $g^{>} = (1 - f)a$, $g^{<} = fa$, and write the momentum integral as

$$\int \frac{d^3p'}{(2\pi)^3} \to N(0) \int d\varepsilon' \int \frac{d\Omega'}{4\pi} ,$$

after the ε' integral is performed with the aid of Eq. (4.17), we find

$$\begin{aligned} \sigma(z) &= z[1 - Z(z)]\, 1 + Z(z)\, \Delta(z)\, \tau_1 \\ &= \int d\omega' \operatorname{Re}\left[\frac{\omega'\, 1 - \Delta(\omega')\, \tau_1}{\sqrt{\omega'^2 - [\Delta(\omega')]^2}}\right] K(z, \omega') \end{aligned} \tag{4.18}$$

where

$$\begin{aligned} K(z, \omega') = N(0) \int \frac{d\Omega'}{4\pi} \, | v_q |^2 & \\ \left\{ \frac{N_q[1 - f(\omega')] + N_q + 1) f(\omega')}{z + \omega_q - \omega'} \right. & \\ \left. + \frac{(N_q + 1)\, [1 - f(\omega')] + N_q f(\omega')}{z - \omega_q - \omega'} \right\}_{q = p_F |\Omega - \Omega'|} & \end{aligned} \tag{4.19}$$

We see that our original assumption, Eq. (4.13) about the form of $\sigma(z)$ is indeed justified.

It is interesting to note that Eq. (4.18) may be reduced to an integral equation for Δ alone since

$$Z(z) = 1 - \frac{1}{z} \int d\omega'\, \omega' \operatorname{Re} \frac{1}{\sqrt{\omega'^2 - [\Delta(\omega')]^2}} \, F(z, \omega') \tag{4.20}$$

so that

$$\begin{aligned} \Delta(z) = - \int d\omega' \, [K(z, \omega')] & \\ \left\{ \operatorname{Re} \frac{\Delta(\omega')}{\sqrt{\omega'^2 - \Delta(\omega')^2}} - \frac{\Delta(z)\omega'}{z} \operatorname{Re} \frac{1}{\sqrt{\omega'^2 - [\Delta(\omega')]^2}} \right\} . & \end{aligned} \tag{4.21}$$

An equation essentially similar to this one has been solved by J.R. Schrieffer *et al.* (*14*) in the zero temperature limit. The only modification in this work beyond the formulas indicated here is the appearance of a pseudopotential in $K(z, \omega')$ which is introduced as a representation of the coulomb potential between electrons.

This solution can be used to describe the tunneling characteristics of a junction between a normal metal and a superconductor. We again use Eq. (3.11)

$$\left.\frac{dN}{dt}\right|_{A\to B} = 2[N(0)] \int \frac{d\Omega}{4\pi} \int \frac{d\Omega'}{4\pi} \,|\, T(p, p') \,|^2 \quad (3.11)$$

$$\int \frac{d\omega}{2\pi} \int \frac{d\omega'}{2\pi} 2\pi\delta(\omega - \omega' + eV) \,[f(\omega) - f(\omega')]$$

$$\int d\varepsilon\, A_A(\varepsilon, \omega) \int d\varepsilon'\, A_B(\varepsilon', \omega')$$

for the case in which A is a superconductor and B a normal metal. Just as before, we employ

$$\int d\varepsilon'\, A_B(\varepsilon', \omega') = 2\pi\,.$$

For the superconductor, Eq. (4.16) implies

$$\int d\varepsilon\, a(\varepsilon, \omega) = 2\pi \,\mathrm{Re} \left\{ \frac{z + \Delta(z)\, \tau_1}{\sqrt{z^2 [\Delta(z)]^2}} \right\}_{z=\omega}. \quad (4.22)$$

Since the scalar $A(p, \omega)$ is the upper left-hand corner of matrix $a(p, \omega)$

$$\int d\varepsilon\, A_A(\varepsilon, \omega) = 2\pi \,\mathrm{Re} \frac{\omega}{\sqrt{\omega^2 - [\Delta(\omega)]^2}}\,. \quad (4.23)$$

[This expression for the density of states was first derived by Schieffer *et al.* (*14*).]

In this way, we find that the tunneling rate is given by

$$\left.\frac{dN}{dt}\right|_{A\to B} = 4\pi[N(0)]^2 \int \frac{d\Omega}{4\pi} \int \frac{d\Omega'}{4\pi} \,|\, T(p, p') \,|^2 \quad (4.24)$$

$$\int d\omega \int d\omega'\, \delta(\omega - \omega' + eV) \,[f(\omega) - f(\omega')]$$

$$\mathrm{Re} \frac{\omega}{\sqrt{\omega^2 - [\Delta(\omega)]^2}}$$

at zero temperature

$$f(\omega) = \begin{cases} 1 & \text{for } \omega < 0\,, \\ 0 & \text{for } \omega > 0 \end{cases}$$

so that

$$\left.\frac{dN}{dt}\right|_{A\to B} \sim \int_{-eV}^{0} d\omega \,\mathrm{Re}\, \frac{\omega}{\sqrt{\omega^2 - [\Delta(\omega)]^2}}\,. \tag{4.25}$$

Therefore, we see that the tunneling rate, expressed as a function of voltage, directly measures

$$\mathrm{Re}\, \frac{\omega}{\sqrt{\omega^2 - [\Delta(\omega)]^2}}\,. \tag{4.26}$$

In ref. (*14*) this quantity is evaluated theoretically and compared with the experiment of J. M. Rowell *et al.* (*15*) which measures the tunneling rate in a junction composed of superconducting AlPb and normal Al. The agreement between theory and experiment is quite impressive.

It should be noticed that the effective density of states for tunneling is

$$N(0)\,\mathrm{Re}\, \frac{\omega}{\sqrt{\omega^2 - [\Delta(\omega)]^2}} = \frac{mp_F}{2\pi^2}\,\mathrm{Re}\, \frac{\omega}{\sqrt{\omega^2 - [\Delta(\omega)]^2}}\,. \tag{4.27}$$

This is very different from the expression for the density of states which emerges from the quasi-particle picture. The latter is

$$N(0)\,\frac{d\varepsilon}{dE}$$

where the quasi-particle energy is given by

$$E^2 = \varepsilon^2 + [\Delta(E)]^2$$

and hence

$$\frac{d\varepsilon}{dE} = \frac{1}{\varepsilon}\left[E - \Delta(E)\,\frac{d\Delta(E)}{dE}\right]$$

$$= \frac{E}{\sqrt{E^2 - [\Delta(E)]^2}} - \frac{\Delta(E)\,\dfrac{d\Delta(E)}{dE}}{\sqrt{E^2 - [\Delta(E)]^2}}\,.$$

Consequently, the quasi-particle picture gives an effective tunneling density of states as

$$N(0)\left[\frac{E}{\sqrt{E^2-[\Delta(E)]^2}}-\frac{\Delta(E)\dfrac{d\Delta(E)}{dE}}{\sqrt{E^2-[\Delta(E)]^2}}\right]. \tag{4.28}$$

Whenever $\Delta(E)$ depends upon E the quasi-particle picture gives a density of states different from the exact expression, Eq. (4.27). Therefore, the quasi-particle picture fails whenever $\Delta(E)$ depends upon E.

References

1. P. Martin and J. Schwinger, *Phys. Rev.* **113**, 1352 (1959).
2. G. Baym and L. Kadanoff, "Quantum Statistical Mechanics" W. A. Benjamin, New York (1962).
3. J. Bardeen, *Phys. Rev. Letters* **9**, 147; *Phys. Rev. Letters* **6**, 57 (1961).
4. M. H. Cohen, L. M. Falicov, and D. C. Phillips, *Phys. Rev. Letters* **8**, 316 (1962).
5. B. D. Josephson, *Phys. Rev. Letters* **1**, 251 (1962).
6. P. W. Anderson, and J. M. Rowell, *Phys. Rev. Letters* **10**, 230 (1963).
7. G. Baym, *Phys. Rev.* **127**, 1391 (1962).
8. A. B. Midgal, J. *Exptl. Theoret. Phys. USSR* **34**, 996 (1958).
9. J. M. Ziman, "Electrons and Phonons", pp. 175-177; 212-219. Oxford Univ. Press (Clarenden), London (1962).
10. J. R. Schrieffer and S. Englesberg, *Phys. Rev.* (to be published).
11. Y. Nambu, *Phys. Rev.* **117**, 648 (1968).
12. J. Bardeen, L. Cooper, and J. R. Schrieffer, *Phys. Rev.* **108**, 1175 (1957).
13. P. Morel and P. W. Anderson, *Phys. Rev.* **125**, 1263 (1962).
14. J. R. Schrieffer, K. J. Scalapino, and J. W. Wilkins, *Phys. Rev. Letters* **2**, 336 (1963).
15. J. M. Rowell, P. W. Anderson, and D. E. Thomas, *Phys. Rev.* **10**, 324 (1963).

Comments on Solid State Physics (1) 1 (1968).

Wave-Function Fluctuations in Finite Superconductors

The basis of our microscopic picture of superconductivity is the use of a single wave function $\langle\psi(\mathbf{r})\rangle$ to describe the quantum-mechanical state of a macroscopic number of Cooper pairs.[1] This wave function was originally introduced in the phenomenological theory of Landau and Ginzburg.[2] After the pairing idea was derived by Bardeen, Cooper, and Schrieffer (BCS),[3] one could see that the phase of this wave function was related to the "superfluid" velocity and the vector potential by

$$\begin{aligned} \langle\psi(\mathbf{r})\rangle &= \sqrt{n_0(\mathbf{r})}\, e^{i\phi(\mathbf{r})} , \\ m v_s(\mathbf{r}) &= \hbar\nabla\phi(\mathbf{r}) - 2eA(\mathbf{r})/c . \end{aligned} \tag{1}$$

This picture was capped by Gor'kov's proof[4] that the Landau-Ginzburg theory of the behavior of $\langle\psi(\mathbf{r})\rangle$ was derivable from the BCS theory.

Much recent effort in superconductivity theory has been devoted to solving the various equations for this condensate wave function. Particularly notable advances in this direction were produced by Abrikosov[5] and by de Gennes and collaborators.[6] Now, we are seeing a second generation of effort. People have gotten over their surprise that a quantum system—even a fermion system—could show a behavior dominated by a single average wave function, $\langle\psi(\mathbf{r})\rangle$, and are now beginning to look into fluctuations in this wave function.

Yang[7] pointed out that in the BCS theory of superconductivity, the superconducting state differs from the normal state because

$$\langle\psi^*(\mathbf{r})\psi(\mathbf{r}')\rangle \tag{2}$$

is nonzero even when r and r' are very far apart. This long-range order can be reduced by fluctuations. In fact, there is a theorem that no system which is infinite in only one dimension, such as an infinitely long cable of superconducting lead, can ever exhibit long-range order.[8] Ferrell[9] pointed out how the BCS theory cannot be internally consistent in any one-dimensional system because the fluctuations in the usual state are sufficient to radically change the nature of this state. T. M. Rice[10] extended this calculation by using the fluctuation formula

$$\langle \psi^*(\mathbf{r})\psi(\mathbf{r}')\rangle = n_0 \exp\{-\langle[\phi(\mathbf{r}) - \phi(\mathbf{r}')]^2\rangle/2\} , \tag{3}$$

which is exact for Gaussian fluctuations in the phase. He employed the Landau-Ginzburg free energy

$$G = \int d\mathbf{r}\{a|[\psi(\mathbf{r})]|^2 + b|\psi(\mathbf{r})|^4 + c|[\nabla\psi(\mathbf{r})]|^2\}$$

as an effective Hamiltonian for a kind of "classical statistical mechanics" in which instead of averaging over the momentum-coordinate phase space, one averages over all possible values of the wave function $\psi(\mathbf{r})$. Therefore, the average wave function has now become a variable in a fluctuation theory.

This new idea led to a very impressive conclusion: In any system which is finite in one of its dimensions, for example, in a plate of lead one meter thick in the z direction but infinite in the x and y directions, $\langle\phi(\mathbf{r})\rangle^2 \to \infty$. Therefore, the infinite fluctuations in the phase preclude the possibility of long-range order in the function (2). Consequently, the usual BCS superconductivity theory cannot be rigorously true in any such finite system. Later on, Hohenberg[11] pointed the way to a proof that Rice's result was rigorously true while Kane and Kadanoff[12] showed how it could be connected with the two-fluid hydrodynamics.

If the BCS theory is wrong in a finite system, what is the behavior of the system at low temperatures? Bychkov, Gor'kov, and Dzyaloshinskii, in a very controversial paper,[13] proposed that superconductivity could persist even in a one-dimensional system. According to their point of view, the superconductivity remains but the Landau-Ginzburg equation is wrong. In this way they were able to circumvent the results of the fluctuation analysis. However, the one-dimensional low-temperature state is so complex that one cannot be sure that their calculation is right. In the two-dimensional case, Dyson[14] has suggested that fluctuations might produce domains of all sizes in the region below T_c. It is not clear that Dyson's proposed phase transition can exist or even if it does exist that it will have infinite electrical conductivity.

In the one-dimensional case, there might well be no phase transition. If there is no real superconducting phase transition in a lead cable, the thermodynamic properties must vary continuously in temperature. In particular, its electrical resistance must be a continuous function of temperature and cannot go to zero at any finite value of T. But we certainly know that a very thick lead cable can show an electrical resistance which is, for all practical purposes, infinitesimally small.

Various authors[15,16,17] have addressed themselves to the problem of computing the resistance of a cable of superconducting material as a function of thickness.[18] All three groups assume that $\psi(\mathbf{r})$ must fluctuate to zero, at least at a point, to produce electrical resistance. Langer and Ambegaokar[16] follow Rice in using the full form of the Landau-Ginzburg free energy as an effective Hamiltonian. They give a particularly beautiful and compelling calculation of this effect by computing the rate at which the fluctuations can pass over a free energy barrier.

Unfortunately, this calculation does not seem to agree with experiment.[17,19] Instead the data are analyzable by assuming that resistance between two points x and x' in the cable will be produced whenever

$$\langle[\phi(x) - \phi(x')]^2\rangle \sim 1 \ .$$

For long cables that is a much weaker condition than demanding that ψ vanish somewhere between x and x'. To my knowledge, there is no theoretical explanation of why this phase condition works. This phase condition also seems to fit the resistance of thin films of superconducting material.[20]

We are faced with a real puzzle. It is true that the experiments are difficult ones. Perhaps, as suggested in Ref. 16, nucleation effects are producing these results. Perhaps the Rice fluctuation theory is invalidated by the very fluctuations it predicts, so that one must base calculations of electrical resistance upon complex states like those discussed in Refs. 13 and 14. Perhaps an unexpected resistive mechanism is operative. More information will be necessary before we can decide.

Leo P. Kadanoff

Note added in proof: The experimental work of W. W. Webb and R. J. Warburton [Phys. Rev. Letters 20, 461 (1968)] on electrical resistance in thin superconducting tin whiskers agrees with the Langer-Ambegaokar fluctuation theory (Ref. 16). Consequently, this theory now appears right for clean superconductors, but it still appears to disagree with experiment for dirty superconductors, as studied in Refs. 17, 19, and 20.

References

1. L. N. Cooper, Phys. Rev. **104**, 1189 (1956).
2. V. L. Ginzburg and L. D. Landau, Zh. Eksperim.i Teor. Fiz. **20**, 1064 (1950).
3. J. Bardeen, L. N. Cooper, and J. R. Schrieffer, Phys. Rev. **108**, 1175 (1957).
4. L. P. Gor'kov, Soviet Phys.—JETP **9**, 1364 (1959).
5. A. A. Abrikosov, Soviet Phys.—JETP **5**, 1174 (1957).
6. Much of this work is summarized in P. G. de Gennes, *Superconductivity of Metals and Alloys* (W. A. Benjamin, New York, 1966).
7. C. N. Yang, Rev. Mod. Phys. **34**, 694 (1962).
8. L. D. Landau and I. M. Lifshitz, *Statistical Physics* (Addison-Wesley, 1958), p. 482.
9. R. A. Ferrell, Phys. Rev. Letters **13**, 330 (1964).
10. T. M. Rice, Phys. Rev. **140**, A1889 (1965).
11. P. Hohenberg, Phys. Rev. **158**, 383 (1967).
12. J. Kane and L. P. Kadanoff, Phys. Rev. **155**, 80 (1967).
13. Yu. A. Bychkov, L. P. Gor'kov, and I. E. Dzyaloshinskii, Soviet Phys.—JETP **23**, 489 (1966).
14. F. Dyson, in *Statistical Physics, Phase Transitions and Superfluidity*, 1966 Brandeis Summer Institute (Gordon and Breach, New York, 1968), Vol. I.
15. W. A. Little, Phys. Rev. **156**, 396 (1967).
16. J. S. Langer and V. Ambegaokar, Phys. Rev. **164**, 498 (1967).
17. R. D. Parks and R. P. Groff, Phys. Rev. Letters **18**, 342 (1967).
18. This is a very important practical point, particularly in view of Little's suggestion [Phys. Rev. **134**, A1416 (1964)] that a chain organic molecule might be a room-temperature superconductor.
19. R. P. Groff, S. Marcelja, W. E. Masker, and R. D. Parks, Phys. Rev. Letters **19**, 1328 (1967).
20. M. Strongin, O. F. Kammerer, S. Marcelja, R. D. Parks, and D. H. Douglass (private communication).

The application of renormalization group techniques to quarks and strings

Leo P. Kadanoff

Department of Physics, Brown University, Providence, Rhode Island 02912

Recent work by K. G. Wilson, A. A. Migdal and others has led to a statistical mechanical treatment of systems of interaction quarks and strings. This work is summarized here. The major topics discussed include boson and fermion variables in statistical mechanics; descriptions of local and gauge symmetries; exact solutions of one-dimensional problems with nearest-neighbor interactions; exact solutions of two dimensional problems with plaquette interactions; Wilson's model of quarks and strings; asymptotic freedom and trapping for this model; the effect of a phase transition in this system; approximate recursion relations of the Migdal form. Finally, all this is put together to give a partial argument for the simultaneous existence of asymptotic freedom and trapping O_2 in the quark-string case. Arguments are developed which distinguish this case from the superficially analogous example of quantum electrodynamics.

CONTENTS

INTRODUCTION

This paper describes how modern renormalization techniques may be applied to models of elementary particle phenomena in which quarks and strings are placed upon a lattice.

We concentrate on two points: (1) how the theory may be made consistent with the apparently contradictory concepts of asymptotic freedom and quark trapping, and (2) how one might begin to approach actual renormalization calculations for these problems.

Section I describes the formulation of lattice problems in statistical physics. Some statistical variables are described, including the standard Gaussian "boson" random variables and anticommuting "fermion" random variables. The discussion of variable types is continued in Sec. II, which is essentially a description of the relationship between symmetries and variable types. Local and global symmetries and the roles of statistical variables as bases for representations of the symmetry are discussed. Finally, K. G. Wilson's model (1974, 1975a, 1976a) of quarks and strings on a lattice is explicitly written down.

In Sec. III renormalization group techniques are introduced as a method of solving one-dimensional problems. Particular attention is given to the fermion variables, which can serve as a representation of quarks, and to the case in which the statistical variables define a homogeneous space for the symmetry group. Following Migdal (1975a), we describe how the solution of the one-dimensional problem for this case can be converted into a solution of a two-dimensional problem with a local (gauge) symmetry. In this section we also describe how the renormalization method can be used to calculate Green's functions.

Section IV describes the qualitative properties of Wilson's lattice model. The quarks are represented by fermion random variables and the strings by homogeneous variables. This formulation can include asymptotic freedom and the unobservability of free quarks. However, Lorentz invariance is not an automatic consequence of the theory. Instead, it can only arise because the theory is near a critical point. The renormalization group technique is then suggested as a natural way of attacking and viewing these near-critical problems. Furthermore, the simultaneous existence of asymptotic freedom and trapped quarks is explained as a consequence of the nonexistence of a phase transition in the string variables.

Section V introduces approximation techniques for attacking lattice problems. The potential moving method of Kadanoff (1975) and Kadanoff, Houghton, and Yalabik (1976) is used to "derive" Migdal's (1975a, 1975b) approximate recursion relations.

Section VI discusses the phenomenology of fixed points with particular reference to the critical dimensionalities at which phase transitions become unstable or disappear. Finally, Section VII applies the Migdal approximation to the quark–string model.

I. STATISTICAL SUMS AND PARTICLE PHYSICS

A. Introduction

In recent years, a whole body of knowledge has been developed about the connection between problems in quantum field theory and those in classical statistical physics. A first point of contact is the similarity between the diagrammatic expansions employed in the two areas. The mathematical similarity between these areas was then further exploited by K. G. Wilson's [(1970, 1971, 1972) see also Wilson and Fisher (1972) and the review of Wilson and Kogut (1974)] wedding of the renormalization ideas in particle physics (Gell-Mann and Low, 1954) with the concepts of universality and scaling which were current in phase transition physics. This extended view of the renormalization group then provided a microscopic theory of phase transitions as well as many insights into the structure of particle theory.

Constructive field theorists have deepened our understanding of these heuristically derived connections by showing how problems in "classical" (i.e., commuting-variable) statistical physics in Euclidean space of dimensionality d could be connected to quantum mechanics problems in $d-1$ spatial dimensions. [See, for example, Osterwalder and Schrader (1973a, b, 1975), Nelson (1973), Simon (1974), and Jaffe and Glimm (1976).] Roughly speaking, the two kinds of problems can be linked together by comparing their Green's functions. One goes from one kind of problem to the other by making an analytic continuation of one of the coordinates—the "time"—according to

$$t \to \pm it \tag{1.1}$$

This "Wick rotation" of the coordinate can, in practice, usually be carried out very simply. Imagine that the problem in statistical physics were described in terms of field variables $\sigma(x)$ and $\bar{\sigma}(x)$ and that the simplest correlation function constructed from these was the one-particle propagator

$$G(x,x') = \langle\sigma(x)\bar{\sigma}(x')\rangle \ .$$

Here $\langle\ \rangle$ means some sort of statistical average, and x and x' are points in a d-dimensional Euclidean space. In the most elementary example, $G(x,x')$ will have a Fourier transform with a simple pole, i.e.,

$$G(x,x') = \int \frac{dp}{(2\pi)^d}\,\frac{e^{ip(x-x')}}{p^2+m^2} \ . \tag{1.2}$$

The "momentum" integral, $\int dp$, is an integral over a Euclidean momentum vector. The net result is that $G(x,x')$ will depend upon a distance in Euclidean space

$$r = \left(\sum_{\alpha=0}^{d-1} (x_\alpha - x'_\alpha)^2\right)^{1/2}$$

and will have an asymptotic form for large separations,

$$G(r) \sim e^{-mr} \ .$$

Under the Wick rotation, x_0 and x'_0 become pure imaginary. Thence r can become real for spacelike separations and imaginary for timelike separations. This analytic continuation of $G(r)$ yields the typical time-ordered Green's functions of field theory. In particular, the correlation function (1.2) gives, when continued, the usual propagator for a noninteracting spin-zero particle, including the typical oscillatory structure for timelike separations. However, in one sense, the whole continuation is quite unnecessary since whenever we see a structure like (1.2) we can simply recognize it as the Euclidean reflection of a spin-zero particle with mass m.

Thus it is formally possible to move back and forth between quantum field theory and classical problems involving statistical fields. However, the classical statistical mechanics of fields is itself a subtle subject, involving all kinds of potential divergences. To eliminate the ultraviolet (short-distance) divergences, one can replace the continuum problem by a lattice problem in which the basic "fields" are only defined at the set of lattice points

$$x = (x_1, x_2, \ldots, x_d) = a_0(n_1, n_2, \ldots, n_d) \ .$$

Here the n's are integers, and the length a_0 is called a lattice constant. The entire ultraviolet divergence problem is then reduced to defining a suitable limit, $a_0 \to 0$. This limit should, for example, yield a Euclidean rotational invariance of the statistical problem so that the related quantum field theory can have a Lorentz invariant structure.

This limiting process must be rather sophisticated. For example, the basic Green's function $G(x,x')$ may be considered to depend upon the separation vector $\Delta x = x - x'$ the lattice constant a_0, and a set of coupling parameters $\mathbf{K}$ which appear in the Hamiltonian for the statistical problem. Hence we write $G(x,x')$ as $G(\Delta x, a_0, \mathbf{K})$. The limit $a_0 \to 0$ is defined first of all by keeping the coordinates x, x', and Δx fixed. To get a physically meaningful result, one must then vary the coupling parameters $\mathbf{K}$ with the cutoff a_0 so as to keep physical quantities like the mass of Eq. (1.2) fixed. The renormalization group (Gell-Mann and Low, 1954; Callen, 1972; Symanzik, 1971) is exactly a method of discussing how physical parameters vary with the change in couplings and cutoffs. Hence the renormalization group is an absolutely essential ingredient in deriving physical information from a lattice theory.

Alternatively, one can make contact with the standard theory of critical phenomena on a lattice by saying that the mass m is exactly the reverse coherence length ξ^{-1}. For fixed couplings K the coherence length is a fixed multiple M^{-1} of the lattice constant a_0. Therefore we can write

$$m = (a_0/\xi)a_0^{-1} = M(\mathrm{K})a_0^{-1} \ . \tag{1.3}$$

As $a_0 \to 0$ with fixed couplings, $m \to \infty$. In order to get finite masses, one must adjust the couplings K as a function of a_0 so that $M(\mathrm{K})$ decreases linearly with a_0. A decreasing dimensionless mass M is evidence that the problem is coming closer and closer to a critical point. Thus the physical limit of a lattice theory is necessarily one in which the lattice problem shows near-critical behavior.

B. Formulation of lattice problems in statistical physics

To describe a problem in $d-1$ dimensions of space and one time dimension, define a d-dimensional hypercubic lattice with the set of lattice points separated by the distance a_0. On each lattice point define a statistical variable $\sigma(x)$. Since there may indeed be several variables at each point, $\sigma(x)$ can be considered to be a vector of variables $\sigma_i(x)$, with an internal index $i = 1, 2 \ldots$. One then defines a statistical problem by giving a meaning to statistical sums and averages.

The basic summation operation is a sum over a single variable at a particular point in space x, which we denote by $\mathrm{tr}_{\sigma(x)}$. The full statistical sum is then written as

$$\mathrm{Tr}_\sigma = \prod_x \mathrm{tr}_{\sigma(x)} \ . \tag{1.4}$$

For example, in the Ising model, each $\sigma(x)$ takes on the values ± 1 and

$$\mathrm{tr}_\sigma f(\sigma) = f(1) + f(-1) \ .$$

Next, one weighs the sum (1.4) with a factor

$$\rho(\sigma) = (\exp A[K,\sigma]) \div Z[K] \ . \tag{1.5}$$

Here $A[K,\sigma]$ is an action which depends on the variables σ and a set of coupling functions or parameters K. (In standard statistical physics this action is replaced by minus the Hamiltonian divided by Boltzmann constant times temperature.) For example, a nearest-neighbor coupling problem is described by a coupling function $K(\sigma,\sigma')$ and has

$$A[K,\sigma] = \sum_{\langle xx'\rangle} K(\sigma(x),\sigma(x')) \ ,$$

where $\langle xx'\rangle$ is used as a notation for a sum over all nearest-neighbor pairs on the lattice.

Finally, the partition function $Z[K]$ is defined to give a proper normalization to the probability function (1.5). It is

$$Z[K] = \mathrm{Tr}_\sigma e^{A[K,\sigma]} \ . \tag{1.6}$$

The density matrix defined by Eq. (1.5) is used to define all the statistical averages. Given any function of the statistical variables $Q(\sigma)$, the average of Q is defined to be

$$\langle Q\rangle = \mathrm{Tr}_\sigma(\sigma)Q(\sigma) \ . \tag{1.7}$$

In particular, then, the basic Green's function is defined as an average $\langle\sigma(x)\bar\sigma(x')\rangle$, where $\bar\sigma(x)$ is a variable conjugate to $\sigma(x)$. In the spin-zero case, $\sigma(x)$ may be complex, and $\bar\sigma$ will then be σ^*; for the spin one-half case $\sigma(x)$ will become a spinor $\psi(x)$, and $\bar\sigma(x)$ will be

$$\bar\psi(x) = \psi^\dagger(x)\gamma_0 \ ,$$

where γ_0 is one of the usual gamma matrices.

C. Variable types

In this paper we shall use local spinor variables $\psi_i(x)$ and $\bar\psi_i(x)$ to represent quarks. The strings will, however, have a slightly more complex representation via statistical variables $U_{ij}(x,x')$ with two spatial indices and two internal indices. The basic set of U's are defined when x and x' are nearest neighbors. In this case, we shall say that $U_{ij}(x,x')$ describes a *string bit.* To handle a finite length of string one forms the matrix product of a succession of connected string bits. Thus, if $x_0, x_1, x_2, \ldots, x_n$ are a succession of points such that x_i and x_{i+1} are always nearest neighbors, a piece of string extending from x_0 to x_n is described by a variable

$$U_\gamma(x_0,x_n) = U(x_0,x_1)U(x_1,x_2)\cdots U(x_{n-2},x_{n-1})U(x_{n-1},x_n) \ . \tag{1.8a}$$

In this definition, matrix multiplication over the internal indices of the U's is implied. The subscript γ on $U_\gamma(x_0,x_n)$ is intended as a reminder that this product variable depends upon the path $\gamma = (x_0, x_1, x_2, \ldots, x_n)$ followed by the string bits in going from x_0 to x_n.

The case $x_0 = x_n$ is especially important. In this case, the piece of string is closed. This string loop is represented here by writing a capital Γ instead of a small γ, to remind ourselves that the path is closed, and taking

$$U_\Gamma = U_\gamma(x_0,x_0) \ . \tag{1.8b}$$

The string loop variable (1.8b) has two internal indices like any other string variable. For a depiction of these different variable types, see Fig. 1.

For the moment, let us put these string variables aside and focus our attention upon the more usual vari-

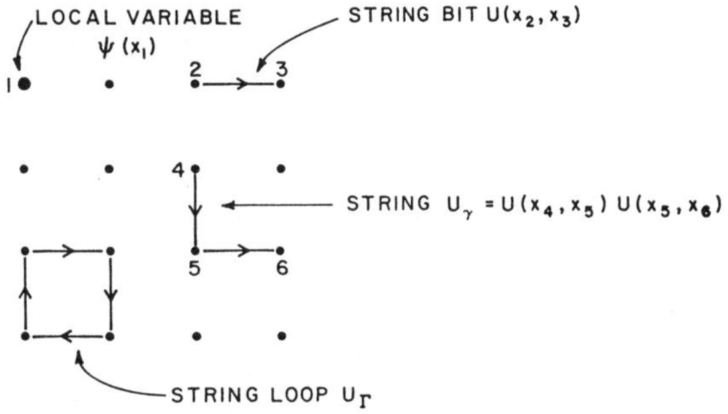

FIG. 1. Type of variables.

ables which have but a single spatial index. The simplest statistical variable is an Ising variable for which $\sigma(x)$ takes on the values ± 1. These Ising variables may then be fully defined by two algebraic statements. The first is a definition of the square of the variables

$$[\sigma(x)]^2 = 1 \ . \tag{1.9a}$$

The second is a definition of the basic trace operation for a single Ising variable

$$\begin{aligned} \mathrm{tr}_\sigma 1 &= 1 \\ \mathrm{tr}_\sigma \sigma &= 0 \ . \end{aligned} \tag{1.9b}$$

In general, we can define the behavior of statistical variables $\sigma(x)$ by giving:

(1) algebraic rules for adding and multiplying the variables and,

(2) a definition of the fundamental statistical sum tr_σ. We shall use these algebraic rules for constructing variables to represent bosons and fermions. In either case, let (1) stand for the statistical variable at the space-time point x_1 with internal index i_1. For the boson case, $\sigma(1)$ is a complex number $\phi(1)$ and the conjugate variable $\bar{\sigma}(1)$ is just the complex conjugate of $\varphi(1)$, i.e., $\varphi^*(1)$. Thus the boson variables are added and multiplied as complex numbers. The basic trace operation is

$$\mathrm{tr}_\varphi = \int \frac{d(\mathrm{Re}\varphi) d(\mathrm{Im}\varphi)}{2\pi} \ .$$

Fermions have a more complex statistical representation [see Berezin (1966) and Abers and Lee (1973]. The basic objects are statistical variables $\psi(1)$ and $\bar{\psi}(1)$ which have typical fermion anticommutation properties

$$\{\psi(1), \psi(2)\} = \{\bar{\psi}(1), \bar{\psi}(2)\} = \{\psi(1), \bar{\psi}(2)\} = 0 \ . \tag{1.10}$$

Therefore, for a particular position and value of the internal indices, there are only four possible quantities which can be formed

$$1, \psi(1), \bar{\psi}(1), \bar{\psi}(1)\psi(1) = -\psi(1)\bar{\psi}(1) \ .$$

The basic trace operation is defined by giving the trace of these objects according to

$$\mathrm{tr}_\psi 1 = \mathrm{tr}_\psi \psi = \mathrm{tr}_\psi \bar{\psi} = 0 \tag{1.11a}$$

but

$$\mathrm{tr}_\psi \bar{\psi}\psi = -\mathrm{tr}_\psi \psi\bar{\psi} = 1 \ . \tag{1.11b}$$

Thus the only terms which contribute to $\mathrm{Tr} e^{A[\psi]}$ are then products over all spatial indices and internal indices of $\bar{\psi}(1)\psi(1)$. We shall see that this rather peculiar representation generates the standard Fermi Green's functions.

Notice that the summation operations described by Eqs. (1.11) are not representable as sums with positive weights. For this reason, some of the standard theorems of statistical mechanics will fail for the ψ's.

D. Free bosons and fermions

To handle both bosons and fermions at one time, we introduce the pair of variables $\sigma(1)$ and $\bar{\sigma}(1)$ to stand for $\psi(1)$ and $\bar{\psi}(1)$, in the Fermi case, and $\phi(1)$ and $\phi^*(1)$ in the Bose. The action for a non-interacting system may be written as

$$A[\sigma] = -\bar{\sigma}(\bar{1})\sigma(\bar{1}) + \bar{\sigma}(\bar{1})\Sigma(\bar{1}, \bar{2})\sigma(\bar{2}) \ , \tag{1.12}$$

where we employ the convention that one should sum over any repeated barred index. We wish to calculate the partition function and the set of Green's functions

$$G(12\cdots; 1'2'\cdots) = \langle \sigma(1)\sigma(2)\cdots\bar{\sigma}(2')\bar{\sigma}(1')\rangle \ . \tag{1.13}$$

To calculate this sum, we isolate the part of the action which depends upon the variables $\sigma(1)$ and $\bar{\sigma}(1)$ as

$$A_1[\sigma] = -\bar{\sigma}(1)\sigma(1) + \bar{\sigma}(1)\eta(1) + \bar{\eta}(1)\sigma(1) \tag{1.14}$$

with

$$\begin{aligned} \eta(1) &= \Sigma(1, \bar{2})\sigma(\bar{2}) \\ \bar{\eta}(1) &= \bar{\sigma}(\bar{2})\Sigma(\bar{2}, 1) \end{aligned} \tag{1.15}$$

[we take $\Sigma(1,1)$ to be zero]. This part of the action may be used to analyze the behavior of averages of product terms like $\sigma(1)$ or $\bar{\sigma}(1)$ or $\sigma(1)\bar{\sigma}(1)$ inside an average such as the one defined by Eq. (1.13).

The calculation starts from the fact that we can calculate the trace of $A_1[\sigma]$ as

$$z = \mathrm{tr}_{\sigma(1)} e^{A_1[\sigma]} = \pm e^{\bar{\eta}(1)\eta(1)} \ . \tag{1.16}$$

Here the upper sign refers to bosons, the lower to fermions. For the Fermi case, the result (1.16) is derived by using the fact that

$$[\sigma(1)] = [\bar{\sigma}(1)]^2 = 0$$

so that e^{A_1} may be written

$$\begin{aligned} e^{A_1[\sigma]} = 1 &- \bar{\sigma}(1)\sigma(1) + \bar{\sigma}(1)\eta(1) \\ &+ \bar{\eta}(1)\sigma(1) + \bar{\sigma}(1)\eta(1)\bar{\eta}(1)\sigma(1) \ . \end{aligned}$$

Then Eq. (1.16) follows directly from Eq. (1.11). The calculation then proceeds by noting that

$$\frac{1}{z} \mathrm{tr}_{\sigma(1)} e^{A_1[\sigma]}\sigma(1) = \eta(1) \tag{1.17}$$

$$\frac{1}{z} \mathrm{tr}_{\sigma(1)} e^{A_1[\sigma]}\sigma(1)\bar{\sigma}(1) = \eta(1)\bar{\eta}(1) + 1 \ . \tag{1.18}$$

Equation (1.17) serves as an equation of motion for σ since it says that if the index value 1 appears only once in $G(12\cdots; 1'2'\cdots)$ one can make the replacement

$$\sigma(1) \to \eta(1) = \Sigma(1, \bar{2})\sigma(\bar{2})$$

under the average (1.13). On the other hand, Eq. (1.18) describes the modification in this replacement which is required if a σ and $\bar{\sigma}$ have the same index. This modification is, then, essentially identical with a Wick contraction.

When applied to $G(1,1')$ these two rules give

$$G(1, 1') = \delta(1, 1') + \Sigma(1, \bar{1})G(\bar{1}, 1') \ , \tag{1.19}$$

which is the standard Dyson equation for the one-particle Green's function. For the two-particle function, we find

$$G(12; 1'2') = G(1, 1')G(2, 2') \pm G(1, 2)G(2, 1') \ . \tag{1.20}$$

Therefore this and all higher-order Green's functions appear in the standard form appropriate for noninteracting bosons and fermions.

Finally, we write the partition function. As $\Sigma(1,2)$ is changed infinitesimally, the partition function changes according to

$$\delta \ln Z = \pm G(\bar{1},\bar{2})\delta\Sigma(\bar{2},\bar{1}) \ .$$

If we imagine that $G(1,2)$ is a matrix in its indices so that we can write a formal solution to Eq. (1.19) as

$$G = 1/(1-\Sigma) \ ,$$

then the partition function obeys

$$\delta(\ln Z) = \delta[\pm \operatorname{trace} \ln(1-\Sigma)] \ .$$

Here trace is a diagonal sum over the indices of G. This equation can then be integrated to read

$$\ln Z = \pm \operatorname{trace} \ln(\pm G) \ . \tag{1.21}$$

E. Specific examples

To see the possible relationship between these Green's functions and the propagators of high-energy physics, we shall look at two specific examples. Start from a "scalar" example in which the fields have no internal indices. Then a nearest-neighbor interaction with coupling constant K can be represented by choosing

$$\Sigma(x_1,x_2) = \sum_\alpha K[\delta(x_1-x_2-\hat{e}_\alpha a_0)+\delta(x_1-x_2+\hat{e}_\alpha a_0)] \ . \tag{1.22}$$

Here $\hat{e}_\alpha$ are the set of d lattice vectors

$$\hat{e}_1=(1,0,0,\ldots) \ ,$$
$$\hat{e}_2=(0,1,0,\ldots) \ , \tag{1.23}$$
$$\hat{e}_3=(0,0,1,\ldots) \ .$$

Then Eq. (1.19) is immediately solved by Fourier transformation, which gives

$$G(p) = 1/[1-\Sigma(p)] \tag{1.24}$$

with

$$\Sigma(p) = K\sum_{\hat{e}} e^{ip\cdot\hat{e}a_0}$$
$$=2K\sum_{\alpha=1}^{d} \cos p_\alpha a_0 \ . \tag{1.25}$$

In turn then the coordinate space Green's function takes the form

$$G(x,x') = \int \frac{dp\, a_0^d}{(2\pi)^d} \frac{e^{ip\cdot(x-x')}}{1-2K\sum_\alpha \cos p_\alpha a_0} \ . \tag{1.26}$$

We wish to focus on the limit in which $a_0 \to 0$ while $x-x'$ remains fixed. In order that G not vanish in this limit, we require that $1-2dK$ be very small. In particular, we write

$$1-2dK = \frac{a_0^2}{2d} m^2 \ll 1 \ . \tag{1.27}$$

Then the denominator in Eq. (1.26) may be expanded in a power series in a_0 to give

$$G(x,x') = \int \frac{dp}{(2\pi)^d} \frac{e^{ip\cdot(x-x')}}{m^2+p^2} \frac{2d}{a_0^2} a_0^d \ . \tag{1.28}$$

Except for the multiplicative factor $2da_0^{d-2}$, Eq. (1.28) is almost the same as the boson propagator in field theory. The major difference is between the Euclidean nature of the metric in Eq. (1.28) and the Minkowski metric in high-energy physics. However, the difference may be eliminated by an appropriate analytic continuation.

The second example involves $\psi(1)$ and $\bar{\psi}(1)$ with spinor internal indices. These indices appear in matrix multiplication of γ matrices γ_α with $\alpha=1,2,\ldots,d$. These matrices obey the Euclidean version of the standard anticommutation relations, i.e.,

$$\{\gamma_\alpha,\gamma_\beta\} = 2\delta_{\alpha,\beta} \ . \tag{1.29}$$

We then follow Wilson (1975a) and choose an interaction which, instead of (1.22), has the structure

$$\Sigma(x_1,x_2) = \sum_\alpha K[(1-\hat{e}_\alpha\gamma)\delta(x_1-x_2-\hat{e}_\alpha a_0)$$
$$+(1+\hat{e}_\alpha\gamma)\delta(x_1-x_2+\hat{e}_\alpha a_0)] \ , \tag{1.30}$$

and find that Eq. (1.24) holds true once more but, instead of (1.25), $\Sigma(p)$ obeys

$$\Sigma(p) = 2K\sum_\alpha [\cos p_\alpha a_0 - i\gamma_\alpha \sin p_\alpha a_0] \ . \tag{1.31}$$

Once again we focus on the possibility of having a finite $G(x,x')$ in the limit $a_0 \to 0$. To achieve this result, we pick K to be very close to $(2d)^{-1}$ and define a mass m by

$$1-2dK = ma_0/d \ . \tag{1.32}$$

Thence, by the same line of reasoning as before, we find the (almost) standard result for a spin-one-half particle

$$G(x,x') = \int \frac{dp}{(2\pi)^d} \frac{e^{ip(x-x')}}{m-i\gamma p} da_0^{d-1}$$
$$= \int \frac{dp}{(2\pi)^d} \frac{e^{ip\cdot(x-x')}}{m^2+p^2} (m+i\gamma p) da_0^{d-1} \ . \tag{1.33}$$

II. SYMMETRIES

In setting up a quark–string model we shall eventually choose an action $A(\sigma)$ which depends upon two kinds of variables, quark creation and annihilation variables $\bar{\psi}(1)$ and $\psi(1)$, and string variables $U(1,2)$. These statistical quantities will contain internal indices reflecting the basic symmetries of the problem. The quark variables contain spinor indices, color indices, and flavor indices, while $U(1,2)$ is a matrix in its color indices. The color symmetry needed for this problem is of a very special nature. It is a gauge symmetry in which there is an independent symmetry operation at each point in space. But, before discussing gauge symmetries, let us look at the simpler case of a global symmetry in which there is a single set of symmetry operations for the entire action.

A. Global symmetry

Consider an action A which depends on two sets of variables $\sigma_i(x)$ and $\bar{\sigma}_i(x)$. Here $\sigma(x)$ and $\bar{\sigma}(x)$ may be identical, as in the Ising model, or different as in the case in which σ and $\bar{\sigma}$ are ψ and $\bar{\psi}$. We say that σ and $\bar{\sigma}$ form a basis for a representation of the symmetry group $\{G\}$ if there exists a set of matrices G^α (with the different matrices labeled by the index α) such that

$$A[\sigma,\bar{\sigma}]=A[\sigma',\bar{\sigma}'] \tag{2.1}$$

whenever the new variables σ' are defined as linear combinations of the old variables by

$$\sigma_i'(x)=\sum_j G_{ij}^{\alpha}\sigma_j(x)$$
$$\bar{\sigma}_i'(x)=\sum_j (G_{ij}^{\alpha})^*\bar{\sigma}_j(x). \tag{2.2}$$

We limit ourselves to unitary representations of the group, i.e., those which obey

$$\sum_j G_{ij}^{\alpha}(G_{kj}^{\alpha})^*=\delta_{ik}.$$

For example, if we take a nearest-neighbor interaction

$$A[\sigma]=\sum_{\langle xx'\rangle}K(\sigma(x),\sigma(x')),$$

Eq. (2.1) is equivalent to

$$K(\sigma_1,\sigma_2)=K(G^{\alpha}\sigma_1,G^{\alpha}\sigma_2).$$

Of course, Eq. (2.1) is not sufficient to guarantee the invariance of the final problem. We must have the basic statistical sum Tr also be invariant under the change of variables (2.2). This additional condition will be met if tr_σ is an invariant sum; i.e., if it obeys

$$\mathrm{tr}_\sigma f(\sigma)=\mathrm{tr}_\sigma f(G^{\alpha}\sigma) \tag{2.3}$$

for all group elements G^{α} and all possible functions f. Then, tr_σ is said to be an *invariant sum*.

Equations (2.1) and (2.2) are the condition that the action be invariant under group operations, while Eq. (2.3) is the requirement that the basic sums be so invariant. The action will certainly be invariant if $A\{\sigma\}$ is a function of scalars like

$$\sum_i \bar{\sigma}_i(x)\sigma_i(x')=\bar{\sigma}(x)\sigma(x).$$

Thus, for example, the conventional two-dimensional rotation symmetry of the XY model is obtained by taking $\sigma_i(x)$ to be a two-component vector and by choosing the coupling function K to depend upon the combination

$$\sigma_1(x)\sigma_1(x')+\sigma_2(x)\sigma_2(x')=\boldsymbol{\sigma}(x)\cdot\boldsymbol{\sigma}(x').$$

This action will then be invariant under two-dimensional rotations with the rotation matrix

$$G^{\alpha}=\begin{pmatrix}\cos\alpha & \sin\alpha\\ -\sin\alpha & \cos\alpha\end{pmatrix}.$$

The full statistical problem will be invariant if $\mathrm{tr}_{\sigma(x)}$ takes the form

$$\mathrm{tr}_{\sigma(x)}=\int d\sigma_1 d\sigma_2\, w(\sigma_1^2+\sigma_2^2) \tag{2.4}$$

where w is any weight function.

B. Transitive representations

In this way, we can make the σ's a basis for a representation of a symmetry group. Let us give a few examples. For the Ising case, $\sigma(x)=\pm1$, an action which is bilinear in the σ's, e.g.,

$$K\sum_{\langle xx'\rangle}\sigma(x)\sigma(x')$$

has two symmetry operations,

$$\sigma(x)\to\sigma(x)$$

and

$$\sigma(x)\to-\sigma(x). \tag{2.5}$$

Hence the representation has two elements $G^0=1$ and $G'=-1$. The abstract group so represented is called Z_2.

Another rather analogous case is one in which $\sigma(x)$ runs over all real values between $-\infty$ and ∞. Then

$$\mathrm{tr}_\sigma=\int_{-\infty}^{\infty}d\sigma.$$

If the action is even in σ, the transformations (2.5) are once again the basic symmetry operations. But there is a significant difference between the two cases. In the Ising case, the symmetry operations

$$\sigma\to G^{\alpha}\sigma$$

define all the possible variations in the basic variables. Thus the basic Ising sum can be written as

$$\mathrm{tr}_\sigma f(\sigma)=\sum_\alpha f(G^{\alpha}\sigma). \tag{2.6}$$

If G^{α} are the two transformations of Z_2, Eq. (2.6) gives a correct representation of the Ising sum, but certainly not of a sum over all real numbers. Whenever Eq. (2.6) is satisfied we say that $\sigma(x)$ forms a homogeneous space for the symmetry group or that $\sigma(x)$ is a transitive representation of the symmetry. We shall find that transitive representations are particularly simple and useful representations of the symmetry, since for these representations the group theory will give all possible information about the variation of $\sigma(x)$.

For the XY model described above, the $\sigma_i(x)$ form a transitive representation only if the weight $w(\sigma_1^2+\sigma_2^2)$ in Eq. (2.4) is a delta function which limits the magnitude of $\sigma_1^2+\sigma_2^2$ to a particular value (say 1).

The group SU_n can be represented by choosing $\sigma(x)$ to be an n-component complex vector of unit length

$$\sum_i \sigma_i(x)\sigma_i^*(x)=1.$$

The basic symmetry is the transformation

$$\sigma\to\sigma'=u\sigma,$$

where the u is a unitary matrix ($uu^{\dagger}=1$) which is required to have a unit determinant. This representation is also transitive if we choose tr_σ to be a proper invariant sum. Thus

$$\mathrm{tr}_\sigma f(\sigma)=\int_u f(u\sigma).$$

Here the invariant sum $\int_u$ can be defined in the following way (Murnaghan, 1938). Since u is an n by n complex matrix it has n^2 different complex components. Call the integral over the complex plane of all these components $\int du$. The conditions on u are then inserted as simple delta function conditions under the integral.

Hence

$$\int_u = \int du\, \delta(\det u - 1)\, \delta(1 - u^\dagger u)\,. \tag{2.7}$$

This invariant integral obeys

$$\int_u f(u) = \int_u f(uu_1) \tag{2.8}$$

for any unitary u_1 with determinant equal to 1.

We shall not make direct use of Eq. (2.7) for finding averages over SU_3 transformations. Instead, we shall simply list some averages defined by

$$\langle x \rangle = \int_u x(u) \Big/ \int_u 1\,. \tag{2.9}$$

We have

$$\begin{aligned}
&\langle 1 \rangle = 1\,,\\
&\langle u_{ij} \rangle = \langle u^\dagger_{ij} \rangle = 0\,,\\
&\langle u_{ij}\, u_{kl} \rangle = \langle u^\dagger_{ij}\, u^\dagger_{kl} \rangle = 0\,,\\
&\langle u_{ij} (u^\dagger)_{kl} \rangle = \tfrac{1}{3}\delta_{il}\,\delta_{jk}\,,\\
&\langle u_{ij}\, u_{jk}\, u^\dagger_{mn} \rangle = 0\,,\\
&\langle u_{ij}\, u_{kl}\, u_{mn} \rangle = \tfrac{1}{6}\epsilon_{ikm}\,\epsilon_{jln}\,.
\end{aligned} \tag{2.10}$$

Here ϵ_{ijk} is the completely antisymmetric tensor of rank 3.

For instance $\langle u_{ij} \rangle = 0$ can be proved as follows: $\int_u$ is invariant under the transformation $u \to u' = ue^{\pm 2\pi i/3}$, because of Eq. (2.8). Then $\langle u_{ij} \rangle$ is an average over the three cubic roots of unity, and hence it must vanish.

Later on there will be considerable use made of various transitive representations.

C. String variables and local symmetries

For the global symmetry, the entire action $A\{\sigma\}$ is transformed with the aid of the same matrix G^α_{ij}. In particular, the new variables are

$$\psi'_i(x) = \sum_j G^\alpha_{ij}\, \psi_j(x)\,,$$

$$\bar{\psi}'_i(x) = \sum_j (G^\alpha)^*_{ij}\, \bar{\psi}_j(x)\,.$$

If there is a string variable $U_{ij}(x,x')$ defined for each ordered pair of nearest neighbors xx', it transforms according to

$$U'(x,x') = G^\alpha U(x,x') (G^\alpha)^\dagger\,.$$

The color symmetry assumed in high-energy physics is a far richer symmetry than this. Under transformations, $A\{\psi, \bar{\psi}, U\}$ is required to be invariant even when the transformation is different at every point in space. Thus, if $u(x)$ is a color transformation matrix which depends upon the space–time point x, A is required to be unchanged under the transformation,

$$\begin{aligned}
&\psi(x) \to \psi'(x) = u(x)\, \psi(x)\,,\\
&\psi(x) \to \bar{\psi}'(x) = \bar{\psi}(x)\, [u(x)]^\dagger\,,\\
&U(x,x') \to U'(x,x') = u(x)\, U(x,x')\, [u(x')]^\dagger\,.
\end{aligned} \tag{2.11}$$

(We omit here the ij indices.)

This new kind of invariance is easily constructed. Just let A depend upon terms like

$$\sum_{ij} \bar{\psi}_i(x)\, U_{ij}(x,x')\, \psi_j(x') = \bar{\psi}(x)\, U(x,x')\, \psi(x')\,. \tag{2.12a}$$

This kind of term is certainly invariant under the local color symmetry. It represents the motion of a quark from x' to x, with the aid of the bit of string $U_{ij}(x,x')$. Other invariant terms which involve no motion are

$$\sum_i \bar{\psi}_i(x)\, \psi_i(x) = \bar{\psi}(x)\, \psi(x) \tag{2.12b}$$

and

$$\sum_{ij} U_{ij}(x,x')\, U_{ji}(x',x)\,. \tag{2.12c}$$

One additional term is needed: A term which provides string–string interactions. Imagine that we construct a closed path Γ from x_1 to x_2 to $\ldots x_n$ to x_1 and construct the combination U_Γ which is a product of string bits along the path as in Eq. (1.8). Then any trace in the 3 by 3 color space of a power of U_Γ

$$\mathrm{trace}(U_\Gamma)^p \tag{2.12d}$$

is also a scalar. We now construct an action from these invariant pieces.

D. The Wilson model

Now, we can put together the pieces and write the Wilson model (1975a, 1976a) of quarks and strings. Each $\psi(1)$ and $\bar{\psi}(1)$ is a four-component spinor with an additional flavor index f and a color index $i = 1, 2, 3$. The color symmetry is a SU_3 symmetry and is an exact local symmetry. The flavor index takes on the values f = up, down, strange, and (perhaps) charmed. The first three values represent the broken SU_3 symmetry of Gell-Mann and Ne'eman. Thus we write $\bar{\psi}_{i,f}(x)$ and $\psi_{i,f}(x)$. The spinor indices are never written explicitly but represented by the γ matrices of Eq. (1.29).

Each ordered pair of nearest-neighbor sites on the lattice defines a string bit variable $U_{ij}(x, x')$. Here i and j are color indices. They run from 1 to 3. We make $U_{ij}(x, x')$ itself a unitary matrix with unit determinant. To limit the number of variables, we choose

$$U(x',x) = [U(x,x')]^\dagger = [U(x,x')]^{-1}\,. \tag{2.13}$$

For each x, x' and j, $U_{ij}(x,x')$ is a vector (with index i) which is a basis function for a transitive representation of the SU_3 symmetry, so that, again, the trace will be a sum over group elements.

Now we will set up an action which includes all the symmetries mentioned so far. This action is a sum of two terms

$$A = A_\psi + A_U\,. \tag{2.14}$$

Here A_U depends only on the string variables, while A_ψ depends on both quark and string variables. Visualize a d-dimensional simple "cubic" lattice. On this lattice the nearest neighbors are connected by vectors $\pm\hat{e}_\mu a_0$ where $\hat{e}_\mu$ are the lattice vectors given by Eq. (1.23). These $\hat{e}_\mu$ enable one to construct a closed path over a basic square (called a plaquette). The path starts at x, proceeds to $x + \hat{e}_\mu a_0$, then to $x + (\hat{e}_\mu + \hat{e}_\nu) a_0$, to $x + \hat{e}_\nu a_0$, and finally returns to x. This closed path is denoted by $\Gamma_{\mu\nu}(x)$. Then, from Eq. (1.8b), one defines a product of string bits around this path as

$$U_{\Gamma_{\mu\nu}(x)} = U(x, x+\hat{e}_\mu a_0)\, U(x+\hat{e}_\mu a_0, x+\hat{e}_\mu a_0+\hat{e}_\nu a_0) \times U(x+\hat{e}_\mu a_0+\hat{e}_\nu a_0, x+\hat{e}_\nu a_0)\, U(x+\hat{e}_\nu a_0, x). \tag{2.15}$$

The basic string–string coupling in the lattice can be written in terms of these closed loop variables as

$$A_U = \sum_{x,\mu,\nu} J(U_{\Gamma_{\mu\nu}}(x)). \tag{2.16}$$

This structure will be color symmetric if the coupling function $J(U)$ is only a function of the trace of U and powers of U. A structure of this kind was employed by Wilson (1975a, 1976b) as a generalization of an action first used by Wegner (1971a) and then analyzed by Balian, Drouffe, and Itzykson (1974, 1975a, 1975b). This form is

$$J(U) = \tfrac{1}{2} J \,\mathrm{trace}(U + U^\dagger). \tag{2.17}$$

Here J is the statistical version of the string–string coupling constant.

The action (2.17) is the simplest color symmetric structure which can be constructed from the U's. Notice that terms like (2.12c) cannot usefully be included because Eq. (2.13) implies that these terms are simply unity.

Next consider A_ψ. We would like to make this part of the action as close as possible to the form of the free-fermion action (1.12). To do this we write

$$A_\psi = -\sum_{x,i,f} \bar{\psi}_{if}(x)\psi_{if}(x) + \sum_{\langle xx'\rangle}\sum_{fij} \bar{\psi}_{if}(x)\Sigma_{0f}(x,x')\,U_{ij}(x,x')\,\psi_{jf}(x'). \tag{2.18}$$

The first term simply sets the normalization of the quark variables; the second is a nearest-neighbor sum which describes the hopping of quarks from one site to neighboring sites with the aid of the bit of string U and the hopping amplitude Σ_0. This nearest-neighbor hopping amplitude is written exactly as in Eq. (1.30) as

$$\Sigma_{0f}(x,x') = K_f\left(1 - \gamma\cdot\frac{x-x'}{a_0}\right). \tag{2.19}$$

This action maintains the basic color symmetry (2.11). Notice that the action could contain additional color singlet terms which could be built from the color singlet combinations

$$\sum_i \psi_{if}(x)\bar{\psi}_{if'}(x), \tag{2.20a}$$

$$\sum_{ijk} \epsilon_{ijk}\psi_{if_1}(x)\psi_{jf_2}(x)\psi_{kf_3}(x). \tag{2.20b}$$

The combination (2.20a) can be connected with a SU_6 meson multiplet of dimension 35, and an SU_6 singlet. The combination (2.20b) is connected with a 56-dimensional representation of SU_6 including the nucleons. These physical objects do not explicitly appear in our starting action but they should be generated by the action of the renormalization group. As we shall see, the exclusion of these terms from our action in the limit $a_0\to 0$ is closely related to the free quark behavior generated by the theory in this limit.

E. Connection with field theory

In the limit $a_0\to 0$, the theory just outlined should reduce to the field theory description of a set of Fermi particles interacting with a gauge field. To make this connection, we write $U(x,x')$ in terms of a line integral of a vector potential $A_\mu(x)$ as

$$U(x,x') = \exp\left[ig\int_x^{x'} d\mathbf{x}''\cdot \mathbf{A}(x'')\right]. \tag{2.21}$$

Here g will turn out to be the coupling constant of the gauge field theory. Since x and x' will be separated by one lattice constant, one can expand the exponential in Eq. (2.21) in a power series in a_0. When this expansion is applied to the right-hand side of Eq. (2.16) one finds a part of the action

$$A_u = -\tfrac{1}{2}\,\mathrm{trace}\sum_{\mu\nu}\sum_x a_0^4 F_{\mu\nu}F_{\mu\nu}g^2 J_{\mu\nu}$$

with

$$F_{\mu\nu} = \partial_\mu A_\nu(x) - \partial_\nu A_\mu(x) - ig[A_\mu(x), A_\nu(x)]. \tag{2.22}$$

The standard coupling term (cf. Abers and Lee, 1973) is then recovered if we replace the sum over x by an integral and choose $J_{\mu\nu}$ as

$$J_{\mu\nu} = 1/2g^2. \tag{2.23}$$

Then A_U becomes simply

$$A_U = -\frac{1}{4}\sum_{\mu\nu}\int dx\, F_{\mu\nu}F_{\mu\nu}. \tag{2.24}$$

Notice one very important point. To the statistical mechanic, weak coupling means $J\to 0$; to the field theorist, weak coupling means $g\to 0$. Therefore, according to Eq. (2.23), the field theorist and statistical mechanic have exactly opposite views of weak and strong coupling.

The fermion term A_ψ can be handled in a similar fashion. Take the coupling in Eq. (2.19) to be of the form

$$K_F = (1/2d)(1 - m_f a_0/d), \tag{2.25}$$

where m_f is physically the mass of a particle with flavor f. Then an expansion of Eq. (2.18) in a_0 yields

$$A_\psi = \frac{1}{a_0^3}\int dx \sum_f \bar{\psi}_f[m_f - \gamma_\mu(\partial_\mu - igA_\mu)]\psi_f. \tag{2.26}$$

This is then the standard structure of a fermion term in a gauge field theory. [See, for example, Abers and Lee (1973).]

III. RENORMALIZATION THEORY: A POINT OF VIEW AND A CALCULATIONAL METHOD

The quark–string theory involves a formulation in which the variables $\psi(1)$ and $U(x,x')$ appear on a lattice with lattice constant much smaller than one fermi. Clearly this lattice is only a formulational and calculational tool. It must disappear from all the final results of the theory.

The renormalization theory can be viewed [see Wilson (1976b)] as a description of how we can simultaneously change the lattice constant and the basic action but nonetheless leave all physical results of the theory entirely unchanged. This approach, then, permits us to visualize how it might be possible for the original lattice to drop out of any physical end results of the theory.

At the same time as the renormalization method provides an important insight into the formulation of the problem, it can also provide a very useful calculational tool. In statistical physics, a variety of problems which do not yield to perturbation theory or any other "classical" analytic tool were attacked with considerable success by approximate renormalization methods.

In this section, we describe the general formulation of these methods and their particular application to one- and two-dimensional problems.

A. Formulation

Given a set of statistical variables σ on a lattice with lattice constant a_0 and an action $A[K,\sigma]$ which depends upon a set of coupling constants or coupling functions K, we can, in principle, compute the partition function and all the correlation functions via Eqs. (1.4)–(1.7). Now imagine that we have another set of variables, μ, on a lattice embedded in the original lattice (see Fig. 2). This new lattice has lattice constant λa_0. We view these new variables as providing an alternative description of the original problem.

To make the conversion from one description to the other, we define a function $T(\mu,\sigma)$. At the start, T is arbitrary; in the end, we shall make very specific choices of T to achieve calculational convenience. This T is used to define a new effective action via

$$e^{A'(\mu)} = \mathrm{Tr}_\sigma e^{T(\mu,\sigma)+A[K,\sigma]}. \tag{3.1}$$

We demand the one condition that the transformation leave the partition function invariant, i.e., that

$$\mathrm{Tr}_\mu e^{T(\mu,\sigma)} = 1. \tag{3.2}$$

Then the partition function generated by $A'(\mu)$, i.e.,

$$Z' = \mathrm{Tr}_\mu e^{A'(\mu)}$$

will be identical with the partition function generated by $A[K,\sigma]$.

The new action $A'(\mu)$ will, in general, be very complex. If the transformation obeys translational (Galilean) invariance, $A'(\mu)$ can be characterized by the same kind of coupling functions which might appear in a very general $A[K,\sigma]$. These coupling functions, K, would describe two-body, three-body, ... interactions. Thus, we can write $A'(\mu)$ as $A[K',\mu]$, where K' is the set of new couplings generated by the transformation.

In summary the change of description $\sigma\to\mu$ has two effects. The lattice constant changes from a_0 to λa_0; the couplings change from K to K'. If we could calculate the sum in Eq. (3.1), we could find out how the new couplings K' depend upon the old. In general, we write this dependence as

$$K' = R^\lambda[K]. \tag{3.3}$$

Equation (3.3) defines a renormalization transformation. It is also possible to define a composition rule for these transforms. If R^{λ_1} represents the change $a_0\to\lambda_1 a_0$, and R^{λ_2} describes the change $a_0\to\lambda_2 a_0$, then the function R^λ, defined by

$$K' = R^\lambda[K] = R^{\lambda_2}[R^{\lambda_1}[K]],$$

defines a new transformation with a change in lattice constant

$$\lambda = \lambda_1\lambda_2.$$

We can express this type of relation formally as the composition rule

$$R^{\lambda^n} = \{R^\lambda\}^n. \tag{3.4}$$

This renormalization transform has several important invariance properties. Since the partition function is left invariant, we see that

$$Z[K] = Z[R^\lambda[K]]. \tag{3.5}$$

Equation (3.5) is very useful for describing the "phases" of the statistical system. The different phases of the system are regions in the space of possible couplings distinguished by different singularity structures in $Z[K]$. For example, Fig. 3 shows a piece of the coupling space defined by an action for an Ising system [$\sigma(x)=\pm 1$]

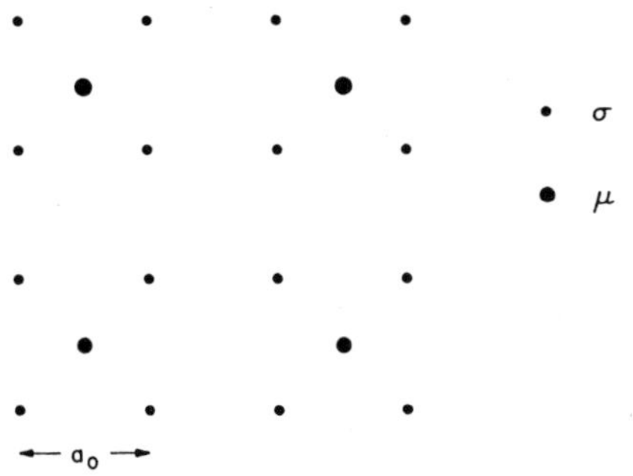

FIG. 2. A renormalization transform with $\lambda=2$.

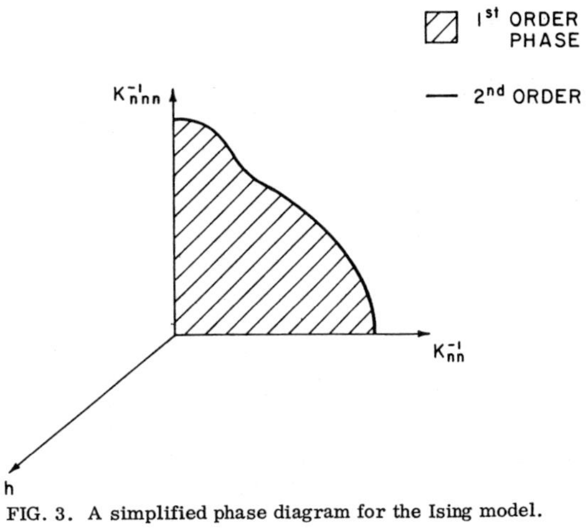

FIG. 3. A simplified phase diagram for the Ising model.

with nearest-neighbor couplings K_{nn} and next-neighbor couplings K_{nnn}, together with a magnetic field h. The action is then

$$A[K,\sigma]=\sum_x h\sigma(x)+\sum_{\langle xx'\rangle} K_{nn}\sigma(x)\sigma(x')$$
$$+\sum_{nnn} K_{nnn}\sigma(x)\sigma(x')\,.$$

We depict three different phases:

1. A first-order region at $h=0$ and sufficiently strong K_{nn} and K_{nnn}. In this region $\partial(\ln Z)/\partial h$, the magnetization, has a discontinuity at $h=0$.

2. A second-order line at the boundary of the first-order region. On this line, $\partial^2(\ln Z)/\partial h^2$ is infinite and the "mass" $M[K]$ goes to infinity.

3. The remainder of the space shown, where the system is in the high-temperature phase, which is characterized by having no singularities at all.

The renormalization transform (3.3) is assumed to have the property that $R^\lambda[K]$ is a totally nonsingular function of K. If Eq. (3.4) is to hold the singularities on both sides of the equation must come totally from Z. Hence we conclude that, *when K' and K are connected by a renormalization transformation, they lie in the very same phase.*

One can make much more quantitative statements too. We assume that $\exp T[\mu,\sigma]$ is a short-ranged function which induces correlations between $\mu(x)$ and the $\sigma(x')$ lying near x. Then the range, ξ, of the $\mu-\mu$ and the $\sigma-\sigma$ correlation function must be the same. From Eq. (1.3) we find that if $m=\xi^{-1}$ is to remain invariant, the dimensionless mass $M[K]=\xi/a_0$ must obey

$$M[R^\lambda[K]]=\lambda M[K]\,. \tag{3.6}$$

Now let us follow the consequences of this point of view. Imagine that we started with couplings K_0 and constructed the couplings

$$K_\alpha=R^\lambda[K_{\alpha-1}]$$

for $\lambda>1$ and $\alpha=1,2,\ldots$. If we could arrange for the coupling to stay in the space of K_{nn} and K_{nnn}, we might imagine a picture of successive transformations like those shown in Fig. 4. The flow lines shown depict how different couplings may be connected while always flowing within the same phase.

Notice how the flow lines converge on fixed points. These points are not special parts of the phase diagram, but are instead determined by the particular renormalization transformation. Nonetheless, it is very helpful to analyze the flow patterns with the aid of these fixed points. To do this, turn to Fig. 5a, where we draw a slice of Fig. 4 including all three fixed points.

The flow lines leave the critical fixed point and flow toward the other two. As we shall see this instability (flow away from the fixed point) is a necessary concomitant of a critical point characterized by an infinite correlation length, i.e., $M[K]=0$. All fixed points are by definition places where there is scale invariance; but this invariance is possible for $M=0$ and for $M=\infty$. The critical fixed points are those with zero mass. Since we wish $m=M/a_0$ to approach finite values as $a_0\to 0$, we are aiming at a theory which approaches such a

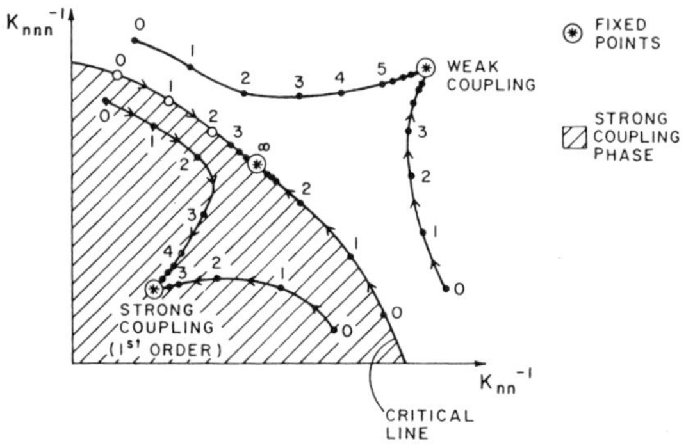

FIG. 4. Structure of the fixed points for the Ising model.

zero-mass fixed point.

The discussion in this section is a very simplified picture of the fixed point theory introduced by Wilson [(1970); see also Wilson and Kogut (1974)] and then given further mathematical form by Wegner (1972 and 1973). In Sec. VI we shall further develop this discussion. For now, however, let us turn to specific examples.

B. Example: Decimation in one dimension

All of this can be made very explicit in a one-dimensional example. [See Houghton and Kadanoff (1973) and Nelson and Fisher (1975).] Physically, the one dimension will become time after the Wick rotation. Hence, for the particle physicist, our "one-dimensional" example includes no spatial variable whatsoever.

Let the x in $\sigma(x)$ be simply na_0, where n is an integer. Let the basic action be

$$A[\sigma,K]=\sum_n K\left(\sigma(na_0),\,\sigma((n+1)a_0)\right)\,. \tag{3.7}$$

Let the new variables $\mu(x)$ be defined at the points shown in Fig. 6

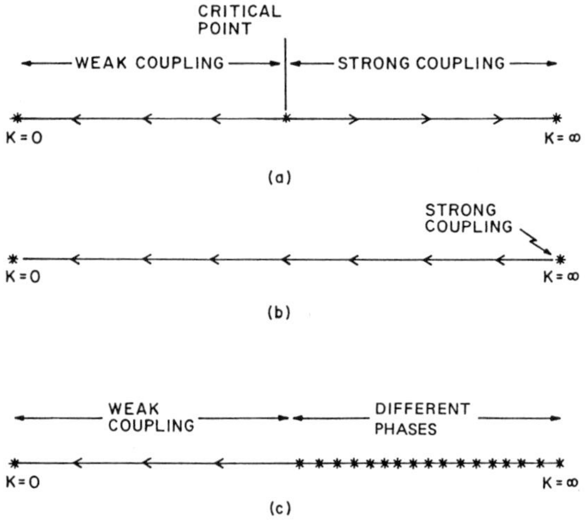

FIG. 5. Different phase diagrams. The stars are fixed points; the arrowheads in the diagrams show the direction of flow of the couplings as the lattice constant is increased.

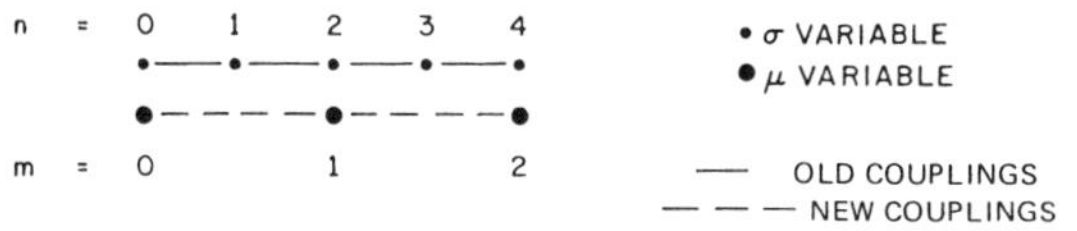

FIG. 6. Decimation applied to a one-dimensional problem.

$$X_m = \lambda m a_0 = 2 m a_0 .$$

Hence we have a change in lattice constant by a factor of 2. We choose these new variables to be essentially identical with the old ones at the same point by writing

$$e^{T(\mu,\sigma)} = \prod_m \delta(\mu(X_m) - \sigma(X_m)) \tag{3.8}$$

where δ is the delta function. The net result of the transformation function (3.1) is to decimate or thin out the statistical variables, leaving us with as many variables in the new problem as in the old. Each new variable is equal to an old variable.

In one dimension the new action $A'(\mu)$ is easy to calculate from Eq. (3.1). The result is exactly of the nearest-neighbor form (3.7),

$$A'(\mu) = \sum_m K'\big(\mu(m a_0 \lambda),\ \mu((m+1) a_0 \lambda)\big) \tag{3.9}$$

with a new coupling function K', which is given by

$$e^{K'(\mu,\mu')} = \mathrm{tr}_\sigma e^{K(\mu,\sigma)+K(\sigma,\mu')} . \tag{3.10}$$

Equation (3.10) then serves as an explicit construction of the dependence of K' upon K. This construction gives an explicit definition of the function $R_0^2[K]$. Here the superscript "2" indicates a change in lattice constant by a factor of 2, while the subscript "0" indicates that we have done the renormalization in the trivial, one-dimensional, example.

One has found $R_0^2[K]$ via Eq. (3.10); one can apply Eq. (3.4) to find R_0^λ via

$$R_0^\lambda = \{R_0^2\}^{\log_2 \lambda} .$$

After that, one can rather easily find all the correlation functions for the problem. For example consider the correlation function

$$G(x - x', K, a_0) = \langle \sigma(x) \bar{\sigma}(x') \rangle . \tag{3.11}$$

Since $\mu(x)$ is $\sigma(x)$, this correlation function can just as easily be calculated in terms of an average of μ's, at least for the case in which $(x - x')/a_0$ is an even integer. Thence we find the identity, applicable to the decimation transform

$$G(x - x', R_0^2[K], 2a_0) = G(x - x', K, a_0) .$$

The successive application of this rule gives

$$G(x - x', K, a_0) = G(x - x', R_0^\lambda[K], \lambda a_0) . \tag{3.12}$$

C. Fermions in one dimension

A simple example of this analysis is the fermion case for which $K(\psi, \psi')$ has the form

$$K(\psi,\psi') = -\tfrac{1}{2}\bar{\psi}\psi - \tfrac{1}{2}\bar{\psi}'\psi' + K\bar{\psi}(1-\gamma_1)\psi' + K\bar{\psi}'(1+\gamma_1)\psi, \tag{3.13}$$

including one coupling constant, K. In this case, Eq. (1.24) implies that

$$G(p) = \frac{1}{1 - 2Ke^{-ip\gamma_1 a_0}} .$$

The inversion of this Fourier transform gives, for $x \neq x'$,

$$G(x - x') = \tfrac{1}{2}[\,\mathrm{sgn}(1 - 2K) + \gamma_1 \,\mathrm{sgn}(x - x')]e^{-m|x-x'|} . \tag{3.14}$$

In Eq. (3.14) sgn is the sign function, which is plus or minus one, depending upon the sign of its argument. The mass m appears in the form (1.3), i.e., $m = M/a_0$, with the dimensionless mass being

$$M(K) = |\ln 2K| . \tag{3.15}$$

Notice the singularities in (3.14) and (3.15) which appear at

$$K = K^* = \tfrac{1}{2} .$$

At this critical point, the mass passes through zero and the roles of holes and particles are interchanged.

The trace (3.10) is calculated in Appendix A. There it is shown that the new coupling has the same form as (3.13) with the parametric change

$$K \to K' = \tfrac{1}{2}(2K)^2 = R_0^2(K).$$

Equation (3.4) then implies that the recursion relation for general λ is given by

$$K \to K' = R_0^\lambda(K) = \tfrac{1}{2}(2K)^\lambda . \tag{3.16}$$

Equation (3.16) implies that there are three fixed points: a weak coupling point $K^* = 0$; a strong coupling point $K^* = \infty$; and a critical point $K^* = \frac{1}{2}$. At these fixed points the mass is infinite ($K^* = 0, \infty$) or zero ($K^* = \frac{1}{2}$). Hence these fixed points are—like all fixed points—places where the theory is scale invariant. This series of fixed points appears in the same form as shown in Fig. 5a. Notice also the pattern of flows. As λ increases, K' moves away from the fixed point at $K^* = \frac{1}{2}$ and toward the other two fixed points. Hence the direction of the arrows in Fig. 5a.

In general, to give a quantitative meaning to this flow away from the fixed point, one can follow Wegner (1972, 1973) and define a variable $h(K)$ which is an analytic function of K near the fixed point, vanishes at the fixed point, and has the simple recursion relation

$$h[K'] = \lambda^y h[K] . \tag{3.17}$$

Then y is termed a scaling index (or critical index) for h. Notice that the scaling property (3.6) of the dimensionless mass implies that

$$M(h) \sim h^\nu , \tag{3.18a}$$

where the coherence length index ν is given by

$$\nu = 1/y . \tag{3.18b}$$

In this fermion example, the Wegner variable is

$$h = -\ln 2K \tag{3.19a}$$

with associated index

$$y = 1. \tag{3.19b}$$

Notice that this variable is essentially the same as the dimensionless mass, M. The scale invariant quantity $|h/a_0|$ is the physical mass, m.

We can use the recursion relations to calculate the Green's functions if we happen to know them anywhere in a given phase. For example, at small K, first-order perturbation theory gives the nearest-neighbor Green's function as

$$G(x-x'=\pm a_0, K, a_0) = (1\pm\gamma_1)K. \qquad (3.20)$$

We apply the recursion formula (3.12) for the Green's function to the case $\lambda = |x-x'|a_0$. Then Eq. (3.12) reads

$$G(x-x', K, a_0) = G(\mathrm{sgn}(x-x')\lambda a_0, R^\lambda(K), \lambda a_0). \qquad (3.21)$$

For $K' = R^\lambda[K] \ll 1$, we can evaluate the Green's function on the right-hand side of (3.21) as the nearest-neighbor result (3.20) to find

$$G(x-x', K, a_0) = [1+\mathrm{sgn}(x-x')\gamma_1]R^\lambda(K).$$

If we substitute the appropriate values of λ and R^λ we find

$$G(x-x', K, a_0) = \frac{1+\mathrm{sgn}(x-x')\gamma_1}{2}(2K)^{|x-x'|/a_0} \qquad (3.22)$$

which is the exact Green's function in the weak coupling phase. Hence recursion calculations can indeed give us Green's functions and the values of masses.

In the end, we are interested in interacting quarks and strings. The analysis of this section can be extended to the case in which string variables are included in the basic coupling in a term of the form (2.12a). Then the basic nearest-neighbor coupling takes the form

$$K(\psi, \psi', U) = -\tfrac{1}{2}\bar\psi\psi - \tfrac{1}{2}\bar\psi'\psi' + K\bar\psi U(1-\gamma_1)\psi' + K\bar\psi' U^\dagger(1+\gamma_1)\psi$$

instead of (3.13). Here we have assumed that $\psi' = \psi(x')$, $\psi = \psi(x)$, $U = U(x, x')$, and $x' > x$. Once again we calculate the trace (3.9) as in Appendix A. Just as before, we find a new coupling function of exactly the same form as the old one, with a new coupling parameter given by Eq. (3.16). The only difference is that the new coupling invokes a new string variable which is the matrix product of the old string variables. Thus if X and X' are nearest-neighbor sites on the new lattice, and x is the point between them, the new string variable is

$$U'(X, X')$$

with

$$U'_{ij}(X, X') = \sum_k U_{ik}(X, x)U_{kj}(x, X'). \qquad (3.23)$$

Thus the one-dimensional recursion relation (3.16) will remain valid in the presence of strings. However, the Green's functions will be changed by the strings. In fact, the nearest-neighbor Green's function (3.20) will be proportional to an average of $U(x, x')$ and this average will vanish. For this reason

$$G(x-x', K, a_0) = 0 \quad \text{for } x \neq x'$$

in the presence of strings.

D. Other examples

The Ising case has a variable $\sigma = \pm 1$ and a coupling of the form

$$K(\sigma, \sigma') = K_0 + K\sigma\sigma'. \qquad (3.24)$$

The basic sum (3.10) can be calculated quite easily if we write

$$e^{K(\sigma,\sigma')} \sim 1 + \sigma\sigma' \tanh K.$$

Then we find a new coupling of the form (3.24) with a new coupling parameter K' given by

$$\tanh K' = (\tanh K)^2. \qquad (3.25a)$$

Equation (3.25a) represents the effect of changing the lattice constant by a factor of 2. More generally, if the change is by a factor of λ, the new coupling is $K(\lambda a_0)$ given by

$$\tanh K(\lambda a_0) = [\tanh K(a_0)]^\lambda. \qquad (3.25b)$$

As λ increases $K(\lambda a_0)$ decreases. This situation is represented by the phase diagram shown in Fig. 5b. There are two fixed points at $K^* = 0$ and $K^* = \infty$. This phase diagram shows no critical point for any finite value of K. There is, after all, no phase transition for the one-dimensional Ising model. However, as K goes to infinity the model does show a quasicritical structure. [See Houghton and Kadanoff (1973) and Nelson and Fisher (1975).] For example, for large K, the recursion relation (3.25b) reads

$$K(\lambda a_0) = K(a_0) - \frac{1}{2}\ln\lambda \qquad (3.26)$$

so that K is close to a fixed point, in that it is changing only logarithmically slowly with λ. Alternatively one can look at the correlation function, which has the form

$$\langle\sigma(x)\sigma(x')\rangle = (\tanh K)^{|x-x'|/a_0},$$

and see that the mass is given by

$$m = -(\ln\tanh K)/a_0 = M(K)/a_0.$$

For large K, the dimensionless mass M becomes very small, i.e.,

$$M(K) \approx 2e^{-2K}. \qquad (3.27)$$

But even though this mass can be small, the system shows no phase transition and all its properties are analytic functions in the region $-\infty < K < \infty$.

Most of the systems analyzed in statistical physics show phase diagrams which bear a qualitative similarity to either Fig. 5a or 5b. We could hope that the systems of interest for particle physics might fall into one of these two classes. However, more complex situations are conceivable. For example, let

$$K' = 1 + (K-1)[1 - \mathrm{Re}\sqrt{1-K}\,]^{\lambda-1}. \qquad (3.28)$$

Then all points for $K < 1$ are in a weak coupling phase while each point for $K > 1$ is a fixed point. Hence each of these $K > 1$ points is a separate phase. We believe that such a possible existence of an infinite number of different phases is characteristic of the Baxter [see Baxter (1971 and 1972) and also Kadanoff and Wegner (1971)] model for $d = 2$. It is also probably the kind of

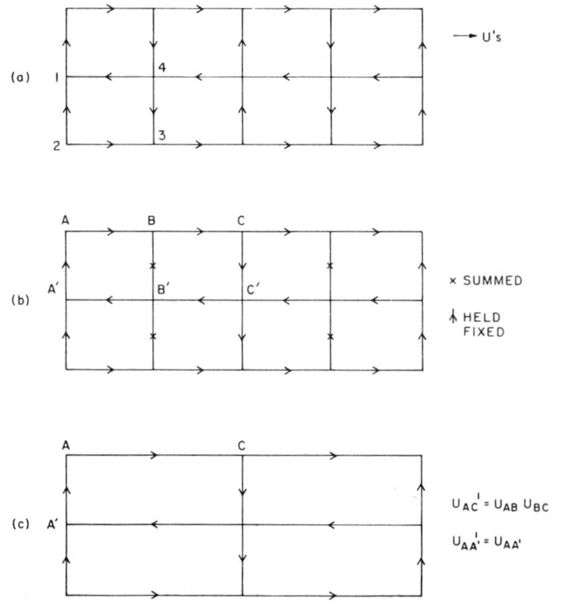

FIG. 7. Decimation on a two-dimensional gauge lattice.

behavior shown by the XY model at $d=2$. This phase diagram is shown in Fig. 5c.

E. Recursion relations for gauge symmetries at $d=2$

Consider a two-dimensional problem on the square lattice drawn in Fig. 7a. Each arrow represents a string bit $U(x,x')$ labeled by the coordinates of its end points. Each $U(x,x')$ is an n by n matrix which obeys

$$U(x,x')=[U(x',x)]^{-1}=[U(x',x)]^{\dagger}.$$

The gauge invariance is represented by the transform

$$U(x,x')\rightarrow u(x)U(x,x')u^{-1}(x').$$

Here $U(x)$ is a matrix in the representation of the symmetry group. In principle, then, U and u are entirely different kinds of n by n matrices. However, if we demand that U forms a basis for a transitive representation of the symmetry, then we can always pick the set of possible U's to be identical with the set of transformation matrices, u.

In this case of a transitive representation, one can find the recursion relations for a gauge interaction on the basic plaquettes. We assume a gauge invariant action which is of the form of a sum over squares, i.e.,

$$A[U]=\sum_{\Gamma}J(U_\Gamma), \tag{3.29}$$

where the string loop U_Γ is defined as in Eq. (1.8). For example, the numbered square in Fig. 7a has

$$U_\Gamma=U_{\Gamma_{1234}}=U(x_1,x_2)U(x_2,x_3)U(x_3,x_4)U(x_4,x_1). \tag{3.30}$$

If we insist that J obeys

$$J(U)=J(U^{-1}) \tag{3.31a}$$

as well as the gauge invariance condition

$$J(U)=J(uUu^{-1}), \tag{3.31b}$$

then $J(U_\Gamma)$ will be independent of the particular method of tracing the path Γ. In addition, the basic sum obeys the invariance conditions

$$\text{tr}_u f(U)=\text{tr}_u f(Uu)=\text{tr}_u f(uU) \tag{3.32}$$

for any function, f. Since U and u are exactly the same kinds of matrices, $\text{tr}_u f(uU)$ will have exactly the same meaning as $\text{tr}_U f(U)$.

To find a recursion relation, sum over the string bits defined by the crosses in Fig. 7b. This sum will lead us to a new problem on the lattice shown in Fig. 7c. Since the latter has a lattice constant in the x-direction which is twice as large as the former, we describe the summation indicated as an x decimation. If we call the summation variables U_x, the new action is given by

$$e^{A'[U]}=\text{Tr}_{U_x}e^{A[U]}. \tag{3.34}$$

Notice that each summation is completely independent of all of the others. Thus we can look at a basic pair of squares like the one with the labeled vertices in Fig. 7b and calculate the summation as

$$e^{J'[U]}=\text{tr}_{U_{BB'}}\exp[J(U_{\Gamma_{ABB'A'}})+J(U_{\Gamma_{BB'C'C}})] \tag{3.35}$$

with the U_Γ's defined as in Eq. (3.30).

Migdal (1975a) made two important observations regarding the sum (3.35). The first is that since $J'[U]$ is a gauge invariant quantity, it can only depend upon the one gauge invariant product that one can form from the U's in the boxes. After $U_{BB'}$ is integrated out, this one product is

$$U_{\Gamma_{ABCC'B'A'}}=U_\Gamma.$$

Therefore J' must depend upon this product alone. Without loss of generality, we may evaluate (3.33) by setting all U's except U_{AB} and $U_{BB'}$ equal to the unit matrix. Thence we get a relatively simple recursion relation

$$e^{J'(U_{AB})}=\text{tr}_{U_{BB'}}e^{J(U_{AB}U_{BB'})}e^{J(U_{BB'})}. \tag{3.36}$$

Migdal's second observation involved the connection between Eq. (3.36) and a special form of the one-dimensional recursion relation. Imagine a nearest-neighbor problem at $d=1$ with the matrices U as the basic statistical variables. Pick a coupling with the special form

$$K(U_1,U_2)=J(U_1U_2^{\dagger}) \tag{3.37}$$

and a $J(U)$ which obeys Eqs. (3.31). For this problem, the nearest-neighbor recursion relation Eq. (3.10) will read

$$e^{K'(U_1,U_2)}=\text{tr}_U e^{J(U_1U)+J(U_2U)}. \tag{3.38}$$

By changing the variable of integration in (3.38) according to

$$U\rightarrow U_2^{-1}U$$

we can see that K' is also of the form (3.37). Therefore we can set $U_2=1$ and write the nearest-neighbor recursion relation as

$$e^{J'(U_1)}=\text{tr}_U e^{J(U_1U)+J(U)}. \tag{3.39}$$

But, Eq. (3.39) is identical to the gauge theory recursion relation of Eq. (3.36)!

We conclude that *the one-dimensional recursion relation applies equally well to the two-dimensional gauge case, if we specialize the one-dimensional theory to couplings which obey Eqs. (3.37) and (3.31).* [For work which led up to Migdal's and included the special case $U=\pm 1$, i.e., the symmetry group Z_2, see Balian *et al.* (1974, 1975a, 1975b) and also Wegner (1971a).] Thus we can say that a change of lattice constant in the x-direction leads to a new coupling which is determined by exactly our previous recursion function

$$J' = R_0^\lambda[J]. \tag{3.40}$$

If we change the lattice constant in both directions we must compose two such transforms. According to Eq. (3.4) this composition is given by

$$J(\lambda a_0) = R_0^{\lambda^2}[J(a_0)]. \tag{3.41}$$

Equation (3.41) represents a complete formal solution of the two-dimensional gauge problem. Since no one-dimensional nearest-neighbor problems involving bounded variables and couplings have a phase transition, none of the $d=2$ gauge problems for compact symmetry groups will show a phase transition. They will all have a phase diagram like Fig. 5b.

For future reference, we list two results for the case in which the basic symmetry group is Z_2. In that case $U(x,x')$ is the Ising variable $U=\pm 1$ and the coupling takes the form (3.24)

$$J(U) = J_0 + JU. \tag{3.42}$$

Then, according to Eq. (3.25), the recursion relation takes the form

$$J(\lambda a_0) = (\tanh^{-1})[\tanh J(a_0)]^{\lambda^2} \tag{3.43}$$

which for large J reduces to

$$J(\lambda a_0) = J(a_0) - \ln\lambda. \tag{3.44}$$

Finally, if U_Γ is a product of U's which surround an area A, the average of U_Γ is

$$\langle U_\Gamma\rangle = \exp(-CA) \tag{3.45}$$

where the constant is

$$C = |\ln\tanh J|/a_0^2. \tag{3.46}$$

This rule that the average of U_Γ decreases exponentially with the area will be very important in what follows.

IV. PROPERTIES OF THE QUARK-STRING MODEL

This section is devoted to describing how the quark-string theory can lead to two different pictures of elementary particle phenomena, both of which seem to have a good experimental basis. These pictures are:

1. *Asymptotic Freedom.* This picture arose in some measure from the parton concept (Feynman, 1970; and Bjorken and Paschos, 1969) in which the properties of elementary particles are explained by treating them as weakly interacting Han–Nambu (1965) quarks. It was extended to the suggestion (Gross and Wilczek, 1973a; Politzer, 1973) that in the high-momentum transfer limit quark–quark interactions renormalized to zero.

2. *The infrared trap.* The experimental point is the fact that no free quarks have ever been observed. A picture which explains this fact is that quarks are bound, with infinite binding energy, into color singlet combinations [see 't Hooft (1972, 1974); Gross and Wilczek (1973b); Weinberg (1973)].

The renormalization group point of view suggests a way out of this dilemma presented by the apparent incompatibility of these two statements. How can quarks be at once weakly interacting and unobservable? They can be so if the qualitative nature of the couplings change in the different energy ranges (length scales) so that a free quark picture, which is asymptotically correct for small distances, becomes vastly wrong for large distances.

Thus our physical discussion will proceed in three stages. First, the trapping will be derived in the $J\to 0$ limit. Next, the freedom will be shown to be a consequence of the theory in the $J\to\infty$ limit. Finally, we shall argue that renormalizations can effectively connect these two limiting cases if only there is no critical fixed point for $0\le J<\infty$.

A. Arguments for trapping

All quark Green's functions must be invariant under gauge transformations. Thus, for example, the one-quark propagator $G(ifx;jf'x')$ must be unchanged under the transformation

$$G\to u_{ik}(x)G(kfx;k',f',x')(u^{-1})_{k'j}. \tag{4.1}$$

This invariance will only be possible if G is diagonal in its color indices and in x, i.e.,

$$G(ifx;jf'x') = \delta_{x,x'}\delta_{i,j}F(f,f'). \tag{4.2}$$

Hence this quark "propagator" in essence says that free quarks will not propagate, but instead will just disappear an instant after one tries to create them.

Similarly the gauge invariance implies that the two- and three-quark propagators only describe the motion of color singlet combinations in the form described, respectively, by Eqs. (2.20a) and (2.20b). For example, $G_3(123;1'2'3')$ has (for x_1 unequal to x_1' or x_2' or x_3') a piece which describes the propagation of the singlet combination which has the quantum numbers of the baryons. This piece only appears for

$$\begin{aligned} &x_1 = x_2 = x_3,\\ &x_{1'} = x_{2'} = x_{3'}, \end{aligned} \tag{4.3}$$

and has the form

$$G(123;1'2'3') = \epsilon_{i_1,i_2,i_3}\epsilon_{i_{1'},i_{2'},i_{3'}} \times F(f_1f_2f_3;f_{1'}f_{2'}f_{3'};x_1-x_{1'}) \tag{4.4}$$

In fact, the flavor indices were introduced with exactly the purpose of allowing such baryon propagation.

It is tempting, but wrong, to assert that results like (4.2) and (4.4) mean that free quarks are unobservable. A separated quark–antiquark pair might be observed if our measuring apparatus detected something different from these Green's functions. A more careful argument would have to be built upon the possibility of observing the energy carried by a free quark with some local detector even if its color was not observable. The necessity for constructing this more careful argument becomes obvious if we notice that the same argument which gives the vanishing of $G(x_1, x_{1'})$ for $x_1 \neq x_{1'}$ can also be applied to electrodynamics for which G is the electron propagator. Hence if we take the above argument really literally, we would conclude that electrons were not observable either!

To make a more careful analysis, we consider a process in which a quark–antiquark pair is produced at the space-time point $x_0 = (0, 0, 0, 0)$ annihilated at $x = (0, 0, 0, t)$. In the meantime a quark is observed at the spatial point $r = (0, 0, \frac{1}{2}z)$ and an antiquark at $r = (0, 0, -\frac{1}{2}z)$. A heuristic picture of such a process is shown in Fig. 8a and is redrawn on the lattice in Fig. 8b.

These pictures suggest that it is not impossible to observe a separated quark and antiquark. But, is it likely? We follow Wilson (1975a, 1976a) in showing that for small J and K this process is so unlikely as to be unobservable when the quarks are separated over an appreciable distance in space (z) and time (t).

Let us examine the probability for the process in question. Each step in the path involves moving the quark via a term $K(1 - e\gamma)U$. Therefore to lowest order in K the probability takes the form

$$\mathrm{Prob} \sim \frac{\mathrm{Tr}_U\, e^{A_U} U_\Gamma (K)^{(2z+2t)/a_0}}{\mathrm{Tr}_U\, e^{A_U}}. \tag{4.5}$$

Here U_Γ is the product of the U's over the closed path shown in Fig. 8. Since the average of each individual U vanishes

$$\mathrm{Tr}_{U(x,x')}\, U(x, x') = 0,$$

one must expand A_U in Eq. (4.5) in order to get a nonzero result. The nonzero terms arise from structures like $\mathrm{tr}_U UU^\dagger$ which are indeed nonzero.

The first nonzero result in perturbation theory arises

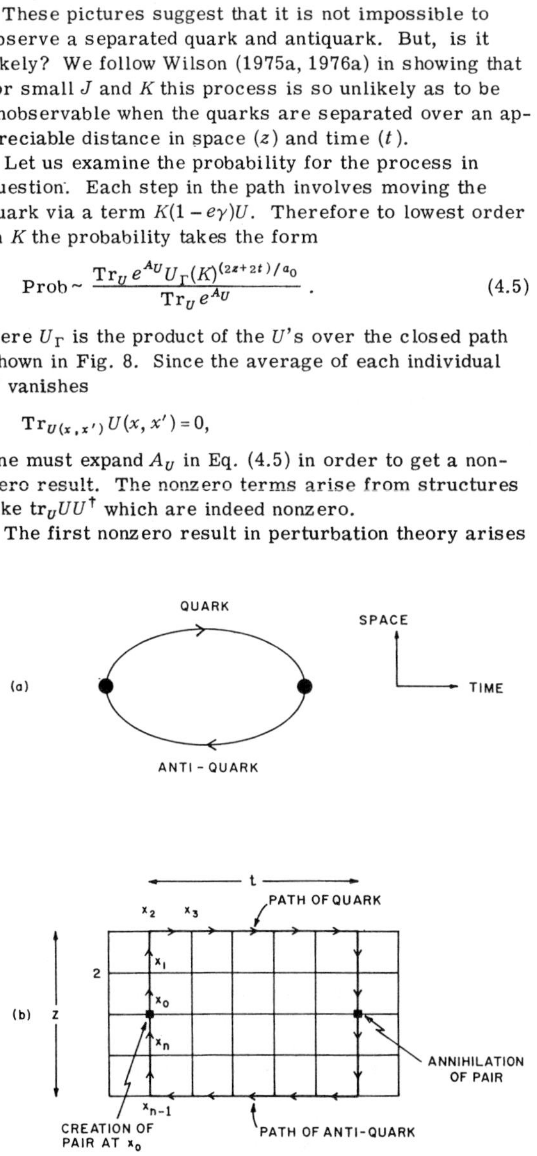

FIG. 8. Separation of a quark–antiquark pair.

when one takes a term in perturbation theory

$$JU^\dagger(1, 2)U^\dagger(2, 3)U^\dagger(3, 4)U^\dagger(4, 1)$$

for each and every square contained in the path of Fig. 8. Thence, for small J, we estimate the probability of this kind of process as

$$\mathrm{Prob} \sim J^{zt/a_0^2} = e^{-(zt/a_0^2)|\ln J|}. \tag{4.6}$$

To interpret Eq. (4.6), remember that in the Euclidean space the probability of a process is proportional to the exponential of $-\Delta E \times \Delta t$, where ΔE is the energy of the process and Δt is the time during which that energy is available. We then see from Eq. (4.6) that the energy of separating two quarks by a distance z is proportional to the separation distance

$$\Delta E \sim z|\ln J|/a_0^2. \tag{4.7}$$

If the energy grows linearly with the separation distance then clearly the quarks cannot become unbound.

Thus we have established the infinitely strong binding of quarks in lowest-order perturbation theory. It has arisen because $\langle U_\Gamma \rangle$, where U_Γ is defined as a product over a closed loop, is of the form

$$\ln \langle U_\Gamma \rangle \sim -\text{Area of loop} \tag{4.8}$$

for large loops.

But, Eq. (4.8) is only known to be true in perturbation theory. If Eq. (4.8) remains true in the exact theory then quarks will be trapped with an infinite binding energy. If this result of perturbation theory disappears in the exact theory quarks can become unbound. What will happen?

Our experience in statistical mechanics indicates that qualitative results of perturbation theory, like Eq. (4.8), will remain true for some range of couplings J, whenever J is too weak to produce a phase transition. Thus, if J is lower than some critical value J_c, we might expect Eq. (4.8) to remain true so that free quarks will be unobservable. We follow W. Bardeen and call this situation a baryon phase. In particular, in the baryon phase for large loops of linear dimension L (see Balian *et al.*, 1975a)

$$\ln \langle U_\Gamma \rangle \sim -L^2 \quad \text{(baryon phase)}. \tag{4.9a}$$

In the opposite limit of large J one can do an expansion in $1/J$ and find, to lowest order, that

$$\ln \langle U_\Gamma \rangle \sim -L \quad \text{(quark phase)}. \tag{4.9b}$$

As we shall see, in this situation, free quarks are definitely observable. More generally, one can imagine that

$$\ln \langle U_\Gamma \rangle \sim -L^{\delta(J)} \quad \text{(complex phase)} \tag{4.9c}$$

where perhaps δ depends continuously upon J.

To study the observability of quarks, we are then impelled to understand the phase transitions of the system as a function of J. [See Migdal (1975a).]

B. Asymptotic freedom

There is a host of theoretical work in which the behavior of elementary particles at high energy is described by assuming that particles are made up of noninteracting or weakly interacting quarks. The renormalization group point of view and the quark–string model provide a beautiful description of how this might occur.

In the renormalization group picture, one can have different forms of the action to describe different energy ranges. In particular, many renormalizations are required to move from high-energy phenomena to lower-energy phenomena. As the renormalization proceeds, the effective interaction can change. Let us assume that the effective interaction which describes high-energy phenomena includes a very strong four-string interaction J. This strong interaction will tend to suppress fluctuations in $U(x,x')$. If U cannot fluctuate, then the quark interaction A_ψ is a pure two-body term. There is no higher-order quark interaction. Hence the quarks behave as free particles.

Wilson (1975a) has shown how to make this conceptual framework more explicit. The string–string interaction is

$$A_U = \frac{1}{2}\sum_{\text{plaquettes}} J\,\text{trace}(U_\Gamma + U_\Gamma^\dagger). \tag{4.10}$$

For very large positive J, we would like to make the trace as large as possible. Since U_Γ is a unitary matrix, the largest possible value of the trace is achieved when U_Γ equals the unit matrix. A general form of $U(x,x')$ which will lead to $U_\Gamma = 1$ is

$$U(x,x') = u(x)u^{-1}(x') \tag{4.11}$$

for arbitrary gauge matrices $u(x)$. When J is large, this form of U might be expected to describe all short-ranged correlation phenomena reasonably well. However, we cannot expect to apply (4.11) to long-ranged correlations, e.g., to closed-path U_Γ's in which there are a very large number of links in the path. According to (4.11), U_Γ will be the unit matrix for all paths. But, in each step of the product, there will be some error arising from the imperfection in the approximation (4.11). If there is no phase transition, (that is, if we are not in the quark phase) these errors will accumulate after a large number of steps. Thus we expect to find that there is a characteristic distance ξ_0 and a characteristic number of steps n such that

$$\begin{aligned} U_\Gamma &\approx 1 \quad \text{for } na_0 \ll \xi_0, \\ &\approx 0 \quad \text{for } na_0 \gg \xi_0. \end{aligned} \tag{4.12}$$

For larger distances than ξ_0, fluctuations in U_Γ will be very important; for smaller distances they will be unimportant. For the small distances, we can expect to use a theory based upon Eq. (4.11).

This theory then has as its action

$$\begin{aligned} A = A_\infty = &-\sum_{xf} \bar\psi_f(x)\psi_f(x) \\ &+ \sum_{\substack{\langle xx'\rangle \\ f}} \bar\psi_f(x)u(x)\Sigma_{0f}(x,x')u^{-1}(x')\psi_f(x') \end{aligned} \tag{4.13}$$

where Σ_{0f} is given by Eq. (2.19). The sum over $U(x,x')$'s has been performed by setting them equal to the special values (4.11), but a sum over all u's, i.e., all gauges, and a sum over all ψ's remain to be calculated. Hence the remaining sums can be represented by

$$\text{Tr}_U \text{Tr}_\psi \to \text{Tr}_u \text{Tr}_\psi. \tag{4.14}$$

However, the remaining sums are trivial. They all can be performed by making the gauge transformation

$$\begin{aligned} &\psi(x) \to \psi'(x) = u(x)\psi(x), \\ &\text{Tr}_\psi = \text{Tr}_{\psi'}. \end{aligned} \tag{4.15}$$

This transformation completely eliminates the u's from the action (4.13) and leaves one with a pure free fermion calculation. Hence the partition function and all other gauge invariant averages are exactly the same in the limit $J \to \infty$ in the free fermion case. In short for all gauge invariant quantities the quarks behave as free particles.

What about gauge dependent quantities, e.g., $G(1,1')$, which is, according to Eq. (4.13)

$$G(1,1') = \frac{\text{Tr}_u \text{Tr}_\psi e^{A_\infty[\psi,u]}\psi(1)\bar\psi(1')}{\text{Tr}_u \text{Tr}_\psi e^{A_\infty[\psi,u]}}. \tag{4.16}$$

After the transformation (4.15), we find

$$G(xf, x'f') = \frac{\text{Tr}_u \text{Tr}_\psi e^{A_\infty[\psi,1]} u(x)\psi_f(x)\bar\psi_f(x')u^\dagger(x')}{\text{Tr}_u \text{Tr}_\psi e^{A_\infty[\psi,1]}}. \tag{4.17}$$

The remaining trace over u vanishes unless $x = x'$. Thence only the gauge invariant piece at $x = x'$ is left in G, and this—once again—can be evaluated by the free fermion theory.

The end result is that the limit $J \to \infty$ gives the average over gauges of free fermion behavior.

C. Renormalization effects

So far this chapter has made two essential arguments:

(1) If J is sufficiently small, i.e., J is smaller than a critical coupling at which a phase transition occurs, then quarks will be bound together with an infinite binding energy. Call this critical value of the coupling J^*.

(2) In the limit $J \to \infty$, the quarks will show almost free particle behavior, except when they are separated by a very large distance. The larger the value of J, the greater the distance (measured in lattice constants) over which free particle behavior will be seen.

In short, trapping is characteristic of a theory with sufficiently weak couplings; freedom is characteristic of a theory with very strong couplings J. In nature, freedom and trapping are both observed, but in different regions of energy, i.e., on different distance scales.

These contrasting observations can be made to agree within the context of the very simplest renormalization group point of view. Imagine that under successive increases of lattice constant $a_0 \to \lambda a_0$, J continually decreases, and that after many renormalizations J approaches zero. Thus, no matter how large J is initially, a sufficient number of renormalizations will bring it close to zero. This kind of behavior is characteristic of systems which show no phase transition. (See Fig. 5b.) Then, for small distance scales, we can have an action with very large J—i.e., asymptotic freedom—while we always remain in the baryon "phase." There is no quark "phase," so unbound quarks cannot be observed.

Hence our contrasting observations of freedom and trapping will be consistent if there is no phase transition in the four-dimensional system of quarks and strings, no matter how large J might be.

These observations can be expressed graphically by redoing Fig. 5 as in Fig. 9. In the latter, we have indicated the physical nature of the different phases which arise. The arrows on the lines show the directions of charge of the couplings when a_0 *decreases*. An arrow going toward $g=0$ (or $J=\infty$) shows asymptotic freedom. Only the diagram without a phase transition (9b) is consistent with asymptotic freedom. In this diagram all values of the coupling (save $g=0$) put the system in the baryon phase, and therefore show quark trapping. In the other two cases, the existence of quark trapping is a function of coupling and only occurs for sufficiently small J.

If the gauge theory has the same kind of recursion equations as the Z_2 theory in two dimensions, then both freedom and trapping would be possible. Equation (3.43) implies that the phase diagram is the same as in Fig. 9b while Eq. (3.45) shows a correlation function like Eq. (4.9a) and hence trapping. In fact, Gross and Wilczek (1973) and Politzer (1973) have analyzed the renormalization structure of the coupled quark–string theory. They used the continuum form described in Sec. II.E and concluded that at $d=4$ if there were not too many quark flavors (fewer than 17) there would be no phase transition near $J=\infty$.

We shall try to follow a similar line of argument for the lattice version of the theory.

V. APPROXIMATION METHODS FOR LATTICE SYSTEMS

To make further progress, one needs approximation techniques. In this section, we will describe some approximation techniques borrowed from statistical mechanics.[1] They are all designed to produce effective calculations of a "free energy" $F[A]$ or partition function $Z[A]$ defined via

$$-F[A]=\ln Z[A]=\ln \mathrm{Tr}_\sigma e^{A[K,\sigma]} \quad . \tag{5.1}$$

In statistical applications, one finds that the approximations described here are especially useful and accurate in calculations of critical indices. They have not yet been extensively employed for correlation functions. Hence we have no experience which would inform us about how accurate mass calculations might be.

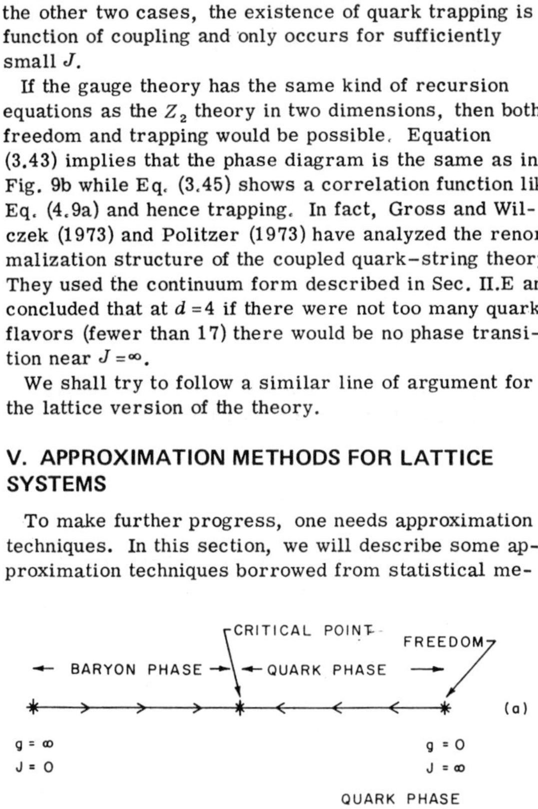

FIG. 9. Possible structures of phases for the string system. The arrows show the flow of couplings for decreasing lattice constant. We hope case *b* appears in the quark–string theory.

A. Lower bound approximations

Start from the exact recursion relation for the action defined by

$$e^{A'(\mu)}=e^{A[K',\mu]}=\mathrm{Tr}_\sigma e^{T(\mu,\sigma)+A[K,\sigma]} \quad . \tag{5.2}$$

Here, $T(\mu,\sigma)$ is normalized so that

$$\mathrm{Tr}_\mu e^{T(\mu,\sigma)}=1 \quad . \tag{5.3}$$

Hence the free energy or partition function defined from A' is identical to that defined from A, i.e.,

$$F[A']=F[A] \quad . \tag{5.4}$$

Unfortunately, one cannot calculate the sum in Eq. (5.2). To circumvent this difficulty, we define an approximate calculation that we can indeed perform. We add to the exponent in (5.2) an error term $\Delta(\mu,\sigma)$ which makes the sum calculable. Then we find an approximate recursion equation

$$A'_a(\mu)=\ln \mathrm{Tr}_\sigma e^{T(\mu,\sigma)+A[K,\sigma]}e^{\Delta(\mu,\sigma)} \quad . \tag{5.5}$$

This approximate action can be described in terms of some new coupling functions K', which are some functional of the coupling functions. This form of the approximate recursion relation is then written

$$K'=R_a[K] \quad . \tag{5.6}$$

How do we choose a good approximation of this nature? More specifically, given several possible choices of $\Delta(\mu,\sigma)$, how can we choose the one which is "smallest" and thus generates the smallest possible error?

One guide comes from a variational principle—or rather an inequality. This inequality requires the following conditions:

(a) $\mathrm{Tr}_\sigma e^{T(\mu,\sigma)+A[K,\sigma]}$ is a sum with positive semidefinite weights.

(b) $\Delta(\mu,\sigma)$ is real.

(c) The average of Δ is zero, i.e.,

$$\mathrm{Tr}_\mu \, \mathrm{Tr}_\sigma e^{T(\mu,\sigma)+A[K,\sigma]}\Delta(\mu,\sigma)=0 \quad . \tag{5.7}$$

Under these conditions, the free energy generated from A'_a is smaller than the true free energy, i.e., instead of

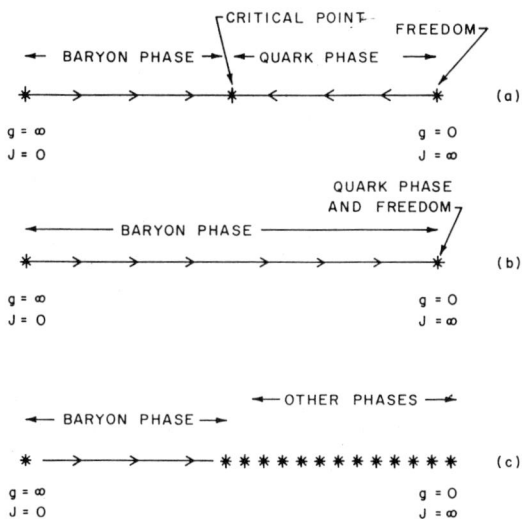

[1]The particular style of lattice renormalization approximation which we shall describe in this chapter, the potential-moving method, was developed by Kadanoff (1975) and Kadanoff, Houghton, and Yalibik (1976). Earlier lattice calculations include Niemeijer and van Leeuwen (1973 and 1974), Wilson (1975b), Nauenberg and Nienhuis (1974, 1975), and Houghten and Kadanoff (1973).

Eq. (5.4) we have

$$F[A_a] < F[A'] = F[A] \ . \tag{5.8}$$

To prove Eq. (5.8), one defines

$$F(\lambda) = -\ln \mathrm{Tr}_\mu \, \mathrm{Tr}_\sigma e^{T(\mu,\sigma)+A[K,\sigma]+\lambda\Delta(\mu,\sigma)} \ . \tag{5.9}$$

Then $F(0)$ is the exact free energy, $F(1) = F[A_a']$. By virtue of Eq. (5.7)

$$\frac{dF}{d\lambda}(\lambda)\bigg|_{\lambda=0} = 0 \ . \tag{5.10}$$

Also,

$$\frac{d^2F}{d\lambda^2} = -\langle(\Delta - \langle\Delta\rangle_\lambda)^2\rangle_\lambda \tag{5.11}$$

where $\langle\ \rangle_\lambda$ is an average with weight $\exp(T + A + \lambda\Delta)$. If the weight is positive, the second derivative is negative and the theorem is proved.

Thus, from all possible Δ's, we choose that Δ which maximizes the approximately calculated free energy and we then get the "best" possible answer. In the meantime, the average of the squared fluctuations in Δ, as defined by integrating Eq. (5.11) over λ between 0 and 1, has been minimized.

The first problem is to find a Δ which obeys Eq. (5.7). To do this imagine any set of local variables $a_i(x)$. For example, $a_i(x)$ might be $\sigma(x)\sigma(x+\hat{e}_1 a_0)$. The labels i on $a_i(x)$ distinguish among different kinds of inequivalent variables; the labels x describe equivalent variables at different positions. Then if

$$\Delta(\mu,\sigma) = \sum_i c_i(x) a_i(x) \ , \tag{5.12}$$

where $c_i(x)$ is some set of coefficients independent of μ and σ which obeys

$$\sum_x c_i(x) = 0 \quad \text{for all } i \ , \tag{5.13}$$

then Eq. (5.10) will certainly be satisfied because, at $\lambda = 0$, the average of $a_i(x)$ is independent of x.

If we consider $a_i(x)$ to be in effect bits and pieces of the action $A[K,\sigma]$, then the net effect of $\Delta(\mu,\sigma)$ is to add something to the action at some points and subtract something at others. The condition (5.13) says that we are allowed to add and subtract such couplings within the variational constraint if we just demand that for every bit of strength we add at one set of points we make sure we subtract an equivalent total strength at other points.

More simply stated: the variational principle allows us to move potential terms from one set of bonds in the lattice to equivalent bonds but not to increase or decrease the total amount of any type of bond.

This potential moving method will permit us to conduct approximate recursion calculations in a controllable fashion. The basic technique is to use the potential moving to move hard-to-handle bonds into a location where their effect may be taken into account.

B. Migdal approximation

To see this approximation technique in its simplest form, we follow Kadanoff (1976) and consider the derivation of an approximation similar to that employed by Migdal (1975a, 1975b). We start with variables $\sigma(x)$ and nearest-neighbor bonds in the $x, y, z, \ldots$ directions. Thus the action is

$$A[K,\sigma] = \sum_{x\alpha} K_\alpha(\sigma(x), \sigma(x+\hat{e}_\alpha a_0)) \ .$$

The label α on K_α distinguishes the different bonds in the different directions.

Now we employ a recursion calculation in which the new variables $\mu(x)$ are defined to be exactly the same as the old variables $\sigma(x)$ on a fraction $1/\lambda$ of the lattice sites; i.e.,

$$\mu(x) = \sigma(x) \quad \text{for } x = (\lambda n_1, n_2, n_3, \ldots) a_0 \ .$$

The remaining σ's are summation variables. (See Figure 10.)

To make the summation possible we pick $\Delta[\sigma]$ to be

$$\Delta(\sigma) = \sum_x a_\alpha(x) K(\sigma(x), \sigma(x+\hat{e}_\alpha a_0 1)) \tag{5.14}$$

and take a_α to be exactly zero if $\alpha = 1$. For the remaining bonds, we choose

$$a_\alpha(x) = -1 \tag{5.15a}$$

when $\sigma(x)$ and $\sigma(x+\hat{e}_\alpha a_0)$ are summation variables and

$$a_\alpha(x) = (\lambda - 1) \tag{5.15b}$$

for the bonds which connect two μ variables. The net effect of (5.15) is to move $(\lambda-1)K_\alpha$ bonds ($\alpha = 2, 3, \ldots$) from summation bonds to the bonds between two μ variables.

Now the summation over σ is easy to perform. We sum as before and find a new x coupling

$$\mathrm{K}_1' = R_0^\lambda(\mathrm{K}_1) \ . \tag{5.16a}$$

The other bonds are the sum of the bond that was always present and the $\lambda - 1$ bonds that were moved

$$\mathrm{K}_\alpha' = \lambda K_\alpha, \quad \alpha = 2, 3, \ldots \ . \tag{5.16b}$$

Migdal's result now emerges if we consider the effect of successive $x, y, z, \ldots$ decimations. All these decimations together change the lattice constant from a_0 to λa_0. Successive applications of Eqs. (5.16) imply that the x coupling constant after the change is

$$\mathrm{K}_1(\lambda a_0) = \lambda^{d-1} R_0^\lambda(\mathrm{K}_1(a_0)) \tag{5.17a}$$

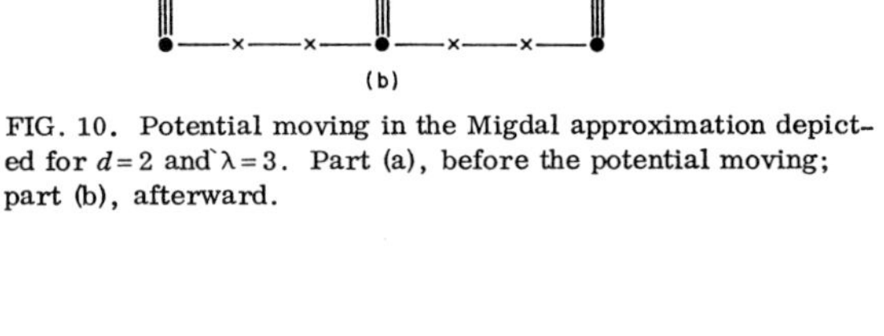

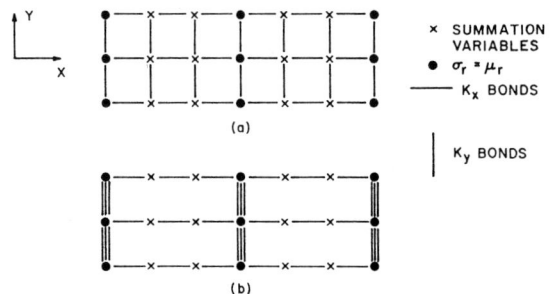

FIG. 10. Potential moving in the Migdal approximation depicted for $d = 2$ and $\lambda = 3$. Part (a), before the potential moving; part (b), afterward.

while all the other coupling constants obey

$$K_\alpha(\lambda a_0) = \lambda^{d-\alpha} R_0^\lambda(\lambda^{\alpha-1} K_\alpha(a_0)) \quad . \tag{5.17b}$$

Here $\alpha = 2, 3, 4 \ldots$ is the index which describes the coupling constant in the $y, z, t, \ldots$ direction.

Equation (5.17a) is exactly the same as Migdal's result. We shall discuss the consequences of Eq. (5.17) below.

The recursions (5.16) and (5.17) will generate a lower bound on the free energy for all problems with positive semidefinite statistical weights. Unfortunately, we need to apply them to fermion systems in which the positivity is lost. Hence one of our main controls on the accuracy of the approximation is lost too.

When is this kind of approximation likely to be accurate? There are three limits in which we might expect reasonable results from Eqs. (5.16):

(1) As $d \to 1$. It is exact at $d = 1$ and the number of bonds to be moved goes to zero as $d \to 1$.

(2) For weak couplings $K_\alpha(\sigma, \sigma')$ for all $\alpha > 1$. Then the errors in Z must be of order K_α^2.

(3) For very strong couplings $K(\sigma, \sigma')$ which force $\sigma \approx \sigma'$ with a very high probability. Then the donor and the recipient sites are likely to have very closely the same values of σ and σ' so that the effect of the motion is quite small.

C. Migdal approximations for the gauge system

The same general scheme can be applied to the gauge system, with interaction $J(U_\Gamma)$ on square plaquettes. A three-dimensional version of this situation is shown in Fig. 11. The recursion is a decimation in which the lattice constant in the "1" direction changes by a factor of $\lambda = 3$. The new variables μ are shown on the figure.

By using the same calculational method as in the two-dimensional case, we could do the sums if only the couplings J_{12} and J_{13} were present. However, couplings J_{23}, like those in plaquette D, prevent us from calculating the correlated summations of the neighboring variables. We therefore move all couplings like D which link together summation variables to plaquettes like C where they give no difficulty. The net result is a new coupling on the plaquettes C of the form

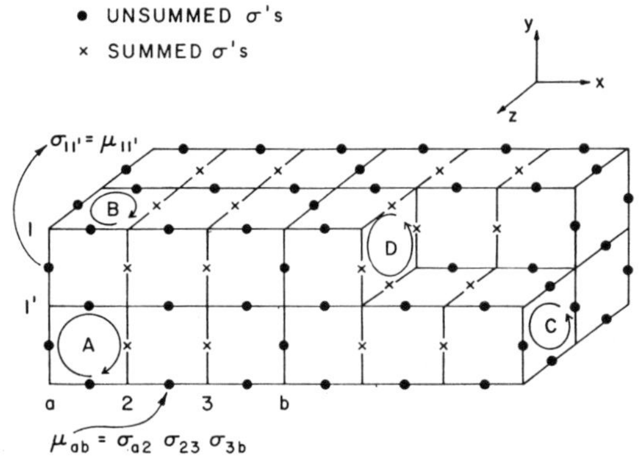

FIG. 11. The three-dimensional Migdal recursion for a gauge system.

$$J'_{\alpha\beta} = \lambda J_{\alpha\beta} \quad \text{for } \alpha \neq 1 \text{ and } \beta \neq 1 \quad . \tag{5.18a}$$

Now we are able to do the summations. Just as in two dimensions these sums can be calculated as a special case of the one-dimensional nearest-neighbor recursion. In direct analogy to Eq. (3.40), we find

$$J'_{\alpha\beta} = R_0^\lambda[J_{\alpha\beta}] \quad \text{for } \alpha = 1 \text{ or } \beta = 1 \ . \tag{5.18b}$$

Once again, the recursions may be presumed to be accurate if the couplings (5.18a) are weak, if the couplings (5.18b) are strong, or if $d \to 2$.

To derive the full consequences of this approximation, consider the effect of successive decimations in the directions $1, 2, \ldots, d$ upon $J_{\alpha\beta}$. Take the spatial indices $\alpha\beta$ to be such that $\alpha < \beta$ and $\beta = 2, 3, \ldots, d$. Then Eqs. (5.18) imply the Migdal-style recursion relations

$$J_{\alpha\beta}(\lambda a_0) = \lambda^{d-\beta} R_0^\lambda[\lambda^{\beta-\alpha-1} R_0^\lambda[\lambda^{\alpha-1} J_{\alpha\beta}(a_0)]] \quad . \tag{5.19}$$

An especially interesting example of this recursion occurs if we take $\alpha = 1, \beta = d/2 + 1$. In that case, the recursion (5.19) is a composition of two identical steps. Each step can be described by an effective recursion

$$J \to J' = R_1^\lambda[J] = \lambda^{d/2-1} R_0^\lambda[J] \quad . \tag{5.20}$$

In terms of this effective recursion function R_1^λ, Eq. (5.19) may be written as

$$J_{\alpha\beta}(\lambda a_0) = R_1^\lambda[R_1^\lambda[J_{\alpha\beta}(a_0)]] \, . \tag{5.21}$$

But, if we view the change in lattice constant $a_0 \to \lambda a_0$ as taking place in two steps

$$a_0 \to \sqrt{\lambda}\, a_0 \to \sqrt{\lambda}\,(\sqrt{\lambda}\, a_0) \ ,$$

then we can consider R_1^λ to be the recursion function for a single step. In this way, we reinterpret (5.21) as

$$\begin{aligned} J_{\alpha\beta}(\sqrt{\lambda}\, a_0) &= R_1^\lambda[J_{\alpha\beta}(a_0)] \\ &= \lambda^{d/2-1} R_0^\lambda[J_{\alpha\beta}(a_0)] \quad . \end{aligned} \tag{5.22}$$

The net result of this argument is that a particular coupling function $J_{1,d/2+1}$ in the gauge case obeys exactly the same style of recursion as Eq. (5.17a)—which describes the nearest-neighbor case. Thus, for these particular couplings, *the recursion in the d-dimensional gauge case is just the same as the recursion in the d/2-dimensional nearest-neighbor situation* (Migdal, 1975a). For example, the four-dimensional gauge case has some recursions which are exactly the same (in the Migdal approximation) as the two-dimensional nearest-neighbor situation. Thus, if the Migdal approximation is accurate, the four-dimensional gauge case can be understood in terms of the much simpler problem of nearest-neighbor interactions in two dimensions.

But the Migdal approximation *is* accurate when the couplings are strong. And, it is exactly this strong-interaction limit which is significant for discussing whether or not the quark–string system has a phase transition. Therefore the Migdal approach can be very useful for determining whether the Wilson model does in fact give both asymptotic freedom and quark trapping.

D. The Ising example (Migdal, 1975b)

To illustrate the considerations of this chapter, consider a nearest-neighbor Ising model. In this case, the couplings K obey the one-dimensional recursion relation

$$K' = R_0^\lambda(K) ,$$

where, according to Eq. (3.25),

$$\tanh K' = (\tanh K)^\lambda .$$

Therefore, from Eq. (5.17a), the x-direction coupling for the system with nearest-neighbor interactions obeys

$$K_x(\lambda a_0) = \lambda^{d-1} \tanh^{-1}[\tanh K_x(a_0)]^\lambda . \tag{5.23}$$

Equation (5.23) expresses the result of one recursion in which the lattice constant increases by a factor of λ.

The case $d\to 1$ is especially interesting. In this situation, there is a fixed point for large values of K_x. When $K_x \gg 1$, Eq. (5.23) implies

$$K_x(\lambda a_0) = \lambda^{d-1}[K_x(a_0) - \ln\lambda/2] . \tag{5.24}$$

Then $d\to 1$, there is a fixed point at $K_x = K^*$, where K^* goes to infinity as $d\to 1$, in the form

$$2K^* = 1/(d-1) . \tag{5.25}$$

Notice also that the recursion relation (5.24) has a critical index

$$y = d-1 \tag{5.26}$$

which goes to zero.

Clearly one dimension is a very special limit of the Ising model. In this limit, the critical couplings go to infinity. For $d<1$, the critical point disappears entirely. We describe a value of the dimension at which the critical couplings go to infinity and then the phase transition disappears as a *lower critical dimension* d_L^*. For the nearest-neighbor Ising model $d_L^* = 1$.

A very similar analysis can be applied to the gauge-style coupling. In this case, for Ising variables $U(x,x') = \pm 1$, the basic coupling on a plaquette takes the form (3.42). For each pair of spatial indices (α,β) there is a single coupling $J_{\alpha\beta}$, which is directly analogous to the K_α described above.

Let us apply Eq. (5.22) to the case in which $J_{\alpha\beta}$ is very large. Then by using the same calculation which led from Eq. (5.23) to Eq. (5.24), we find a recursion relation for the gauge case

$$J_{\alpha\beta}(\lambda a_0) = \lambda^{d-2} J_{\alpha\beta}(a_0) - \tfrac{1}{2}\ln(\lambda^{d-\beta} + \lambda^{d-\alpha-1}) . \tag{5.27}$$

This recursion relation shows a lower critical dimensionality d_L^* equal to 2. Near the lower critical dimensionality there is a fixed point at very strong values of the coupling,

$$J_{\alpha\beta}^* = \frac{\lambda^{2-\beta} + \lambda^{1-\alpha}}{2(d-2)} , \tag{5.28}$$

and a critical index

$$y = d-2 . \tag{5.29}$$

We now make an analogy between these results for the group Z_2 near $d=2$ and the desired results for the group SU_3 at $d=4$. Assume for a moment that nature had one space and one time dimension and had a "color" symmetry Z_2. For $d\approx 2$, and strong coupling, the Migdal recursion relations of the section would be reliable. They would show a phase transition, i.e., the structure of Fig. 9a, for $d = 2+\epsilon$, with $\epsilon > 0$. However, at $\epsilon = 0$ there would be no phase transition and the phase diagram would look like Fig. 9b. Hence the theory would show both quark trapping and asymptotic freedom. The theory, however, would be far from trivial since a perturbation theory in g would only work in the quark phase. Since this phase only exists at $g=0$, it is very likely that the radius of convergence of this perturbation theory would be zero.

Two dimensions is special for a Z_2 gauge theory because $d_L^* = 2$ for this theory. According to the Migdal approximation, this value of the lower critical dimensionality is in turn derivable from the fact that $d_L^* = 1$ for the Z_2 nearest-neighbor coupling. In fact, the general result is that for a given representation of a given symmetry, the lower critical dimensionality of the gauge theory, (d_L^*) gauge, is related to the lower critical dimensionality of the corresponding nearest-neighbor theory, $(d_L^*)_{\text{global}}$:

$$(d_L^*)_{\text{gauge}} = 2(d_L^*)_{\text{global}} . \tag{5.30}$$

Now let us turn to the consideration of the interesting case, one in which $U(x,x')$ is the fundamental representation of a particular Lie group. According to Migdal (1975a) and to the more accurate calculations of Brezin and Zinn-Justin (1976) and of Polyakov (1975), the lower critical dimension for the global symmetry in this situation is $(d_L^*)_{\text{global}} = 2$. Therefore the gauge case shows a lower critical dimensionality at $d=4$!

In fact, one can make a slightly (but crucially) stronger statement. For all the regular representations of compact semisimple Lie groups (e.g., SU_3), according to calculational methods of Brezin and Zinn-Justin (1976), there will be no phase transition in the $d=2$ nearest-neighbor case and consequently no phase transition in the $d=4$ gauge case. Therefore, if the string-string interactions dominated the quark-string interaction, the strings would show no phase transition. In this case, we would indeed obtain the desired phase diagram, i.e., Fig. 9b. We would then have a theory containing as we wish both asymptotic freedom and trapping. It would be near-critical (since $d_L^* = 4$), so it might even be Lorentz invariant. What could be more satisfying!

But there is more. Quantum electrodynamics can be expressed in this same language, with a symmetry group U_1, an Abelian group. The corresponding nearest-neighbor problem is called the XY model. For this special case, all proofs of the non-existence (Polyakov, 1975; Brezin and Zinn-Justin, 1976) of a phase transition at d_L^* fail. In fact, there are plausible arguments which suggest the existence of a phase transition at d_L^*, perhaps of the nasty nature shown by the phase diagram 9c. But this phase transition is quite desirable from the point of view of experiment. It permits the observation of electrons and positrons and also permits the theory to not be asymptotically free.

However, the reader should be aware that all of the analysis of this section depends on the idea that the strings determine their own interactions with no help from the quarks. In the Migdal approximation, this idea is valid. Different couplings can be moved independently of one another in this approximation. But, in the real world, the quarks may enter into the string recursion relations in an essential way and thereby invalidate the reasoning outlined here. In fact, the analyses of Gross and Wilczek (1973) and Politzer (1973) show that as the number of flavors increases, the quarks in fact do produce a phase transition in the string system.

VI. FIXED POINT PHENOMENOLOGY

Before analyzing the Migdal-style recursion relations, we go through a digression in which we discuss the physical interpretation of recursion equations and their fixed points. We are particularly interested in surveying the meaning of a lower critical dimensionality and of a marginal critical index, i.e., an index y which goes to zero. The basis for this section is, of course, Wilson's (1970) recognition of the importance of fixed points and his concern (1975b) with marginal variables. The first explicit treatment of such variables is in Kadanoff and Wegner (1971). The beautiful mathematical formulation of fixed point behavior is largely due to Wegner (1972 and 1973).

A. A first example

For a system with the Ising model symmetry $\sigma_r \to -\sigma_r$ one can follow Wilson and Fisher (1972) and draw a plot of critical indices versus dimensionality. For example, of one looks at the critical index y_1 which appears in the recursion relation for the deviation of the coupling strength from its critical value

$$(K_1 - K^*)' = \lambda^{y_1}(K_1 - K^*) \ ,$$

one can draw a picture like Fig. 12. Here the physical quantity of interest is not really y_1 but instead $\nu = 1/y_1$ which is defined by the statement that the inverse coherence length (i.e., mass) behaves near critically as

$$\xi^{-1} = m \sim (K_1 - K^*)^{\nu} \ . \tag{6.1}$$

Figure 12 contains a description of two different fixed points. (There are actually many, many more). There are the nontrivial Ising-like fixed point and the Gaussian fixed point. Notice that these cross at an upper "critical" dimensionality (namely $d = 4$) and that the Ising fixed point disappears at the lower "critical" dimensionality ($d = 1$).

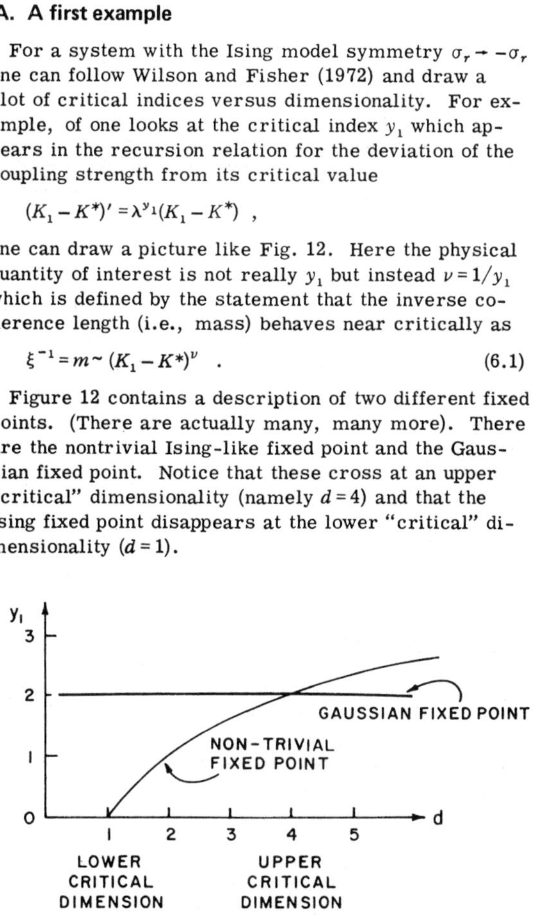

FIG. 12. Critical index y_1, plotted against dimensionality.

One of the major purposes of a phenomenology of fixed points is to explain and describe the behavior in the neighborhood of these critical dimensionalities.

B. Wegner variables, free energy, and scaling

To form this explanation, we begin from a recursion relation

$$K' = R^{\lambda}[K] \ .$$

This expresses the new coupling functions K' as a function of the old ones K, when the lattice constant is changed by a factor of λ, i.e.,

$$a_0 \to \lambda a_0 \ . \tag{6.2}$$

If we express the coupling functions K in terms of a set of coupling parameters $\mathbf{K}$ then the basic recursion for the parameters may be written as

$$\mathbf{K}' = R^{\lambda}(\mathbf{K}) \ . \tag{6.3}$$

Once we have picked the form of the transformation function $T(\mu,\sigma)$ the recursion relation (6.3) is well-defined. It may have one or several or a continuum of fixed points. We pick a given fixed point K^*, which obeys

$$K^* = R^{\lambda}(K^*) \ . \tag{6.4}$$

We expand about this fixed point by writing $K_\alpha = K_\alpha^* + h_\alpha$ with $h_\alpha \ll 1$. Then, to first order in h, Eq. (6.3) may be expanded as

$$h'_\alpha = \sum_\beta B_{\alpha\beta} h_\beta \tag{6.5}$$

where

$$B_{\alpha\beta} = \left. \frac{\partial R_\alpha^\lambda(K)}{\partial h_\beta} \right|_{K=K^*} \ . \tag{6.6}$$

Then we form the eigenvectors of Eq. (6.5) by writing the linear combinations

$$h'_i = \sum_\alpha U_{i\alpha} h_\alpha \tag{6.7}$$

which diagonalize $B_{\alpha\beta}$. These linear combinations then obey

$$h'_i = \lambda^{y_i} h_i \ . \tag{6.8}$$

Here we have written the eigenvalue of $B_{\alpha\beta}$ as λ^{y_i}. This approach then provides a full description of all first-order deviations from criticality and from the fixed point. But one can go further. One can construct the Wegner (1972) functions

$$h_i = h_i(\mathbf{K} - \mathbf{K}^*) \ ,$$

which have a power series expansion in the form

$$h_i = \sum_\alpha U_{i\alpha} h_\alpha + \sum_{\alpha\beta} U_{i\alpha\beta}^2 h_\alpha h_\beta + \cdots$$

when $U, U^2, \ldots$ are chosen correctly the Wegner functions obey Eq. (6.8) for all values of h_α. Wegner has described how to perform this construction. It will work whenever the y_i are incommensurate, i.e., whenever

$$\sum_i y_i m_i \neq 0 \tag{6.9}$$

for any set of positive or negative integers m_i.

The most significant kind of failure of Eq. (6.9) occurs when one of the y_i equals zero. But, for the moment, let us ignore this case of a marginal variable and all other failures of Eq. (6.9) and assume that Eq. (6.8) is true arbitrarily far away from the critical point. This assumption enables us to analyze the behavior of any quantity with a simple scale dependence. Consider, for example, an inverse coherence length or mass defined by

$$\xi^{-1} = m = M(K)/a_0 \ . \tag{6.10}$$

Express M as a function of the Wegner variables h_i. Then the scale invariance of the physical mass implies that $M(h)$ obeys

$$M(\lambda^{y_1}h_1, \lambda^{y_2}h_2, \ldots) = \lambda M(h_1, h_2, \ldots) \ . \tag{6.11}$$

Therefore M must obey the homogeneity requirement

$$M(h_1, h_2, \ldots) = h_1^{1/y_1}\bar{M}(h_2/h_1^{\delta_2}, h_3/h_1^{\delta_3}, \ldots) \tag{6.12a}$$

with

$$\delta_i = y_i/y_1 \ . \tag{6.12b}$$

Here $\bar{M}$ is called a scaling function.

Some of the h_i's satisfy Eq. (6.12) in a very simple way: they do not appear in this or any other quantity describing the critical behavior. These redundant variables may be safely ignored.

The remaining variables may be arranged into three categories according to the sign of y_i. For the three possible cases we say that

$$\text{if} \begin{cases} y_i > 0 \\ y_i = 0 \\ y_i < 0 \end{cases} \text{then } h_i \text{ is} \begin{cases} \text{relevant} \\ \text{marginal} \\ \text{irrelevant} \end{cases} .$$

For most purposes one can neglect the irrelevant variables. In statistical physics they are neglected because one is interested in long-ranged correlations which may be studied as $\lambda \to \infty$ and in this limit they vanish. In particle physics they are set equal to zero so that the action may have a well-defined limit as $\lambda \to 0$, i.e., for zero lattice constant.

If we throw away these variables, then the mass M depends upon only a few different variables, the small class for which $y_i \geq 0$. This mass is then a universal function of any of a few variables. This universality is a justification for studying model problems since these models can well be sufficient for establishing all the important functional dependence of such quantities as $\bar{M}$.

C. Universality failure and marginal variables

The analysis of the previous section fails whenever Eq. (6.9) fails. The most interesting case of this kind occurs when we have a marginal variable, i.e., one with $y_i = 0$.

This marginality may take several forms. The simplest case conceptually is the one in which the marginal Wegner variable, h_{mar} simply obeys

$$h'_{\text{mar}} = h_{\text{mar}} \tag{6.13}$$

for a whole range of values of h_{mar}. Then we have a case in which the fixed point K^*, the scaling function $\bar{M}$, and all the other critical indices y_i can be continuous functions of this one (or perhaps several) marginal variables. This case is depicted in Fig. 3.4c and is realized in Baxter's $d=2$ solution of the 8-vertex model. Luther and Scalapino (1976) have suggested that there are two marginal variables of this type in the two-dimensional XY model.

Another kind of failure occurs when the recursion equation takes the form

$$h'_{\text{mar}} = h_{\text{mar}} + b \ .$$

In one sense, this equation says that there is no fixed point. In another sense, we "almost" have a fixed point. If, for example, $\lambda = 2$ then a transition through n steps changes a_0 into $2^n a_0$ and gives

$$h_{\text{mar}}(2^n a_0) = h_{\text{mar}}(a_0) + bn$$

or

$$h_{\text{mar}}(\lambda a_0) = h_{\text{mar}}(a_0) + b \log_2\lambda \ . \tag{6.14}$$

This very slow motion away from a fixed point is characteristic of the behavior at the lower critical dimensionality at which the phase transition just barely disappears.

Notice that Eq. (6.14) exactly describes the Ising model recursion at its lower critical dimension, i.e., Eq. (3.26).

A third conceivable behavior is given by the recursion relation

$$h' = h + bh^2 \ .$$

If h is small, and the basic step is 2, this equation may be written as

$$h(2^{n+1}a_0) = h(2^n a_0) + b[h(2^n a_0)]^2 \ .$$

Let $h_n = h(2^{n+1}a_0)$ and assume h varies slowly with n. Then

$$\frac{dh_n}{dn} = bh_n^2 \ ,$$

so that

$$h_n = \frac{1}{h_0^{-1} - bn} \ ,$$

or

$$h(\lambda a_0) = \frac{h(a_0)}{1 - h(a_0)b \log_2\lambda} \ . \tag{6.15}$$

If $bh(a_0)$ is negative, then after many iterations $h(\lambda a_0)$ will very slowly approach zero. If, on the other hand, $bh(a_0)$ is positive, after a large number of iterations $h(\lambda a_0)$ will get very far from zero, and we will approach an entirely different coupling structure.

This type of behavior is characteristic of a situation in which there are two fixed points with very similar physical behavior which approach one another. For example, at dimensionality 4, the Gaussian fixed point and the Ising-type fixed point become essentially identical in all of the critical indices and critical behavior.

The variable which takes you from one of these fixed points to the other is h and its slow charge reflects the degeneracy between these two solutions. The logarithm in Eq. (6.15) is reflected in logarithms which show up in the thermodynamic behavior at these "upper" critical dimensionalities.

D. Stability

To see this behavior in more detail, we plot in analogy to Figure 12 the second largest critical index for a variable which is even in σ. This plot is Fig. 13.

From this figure it follows that the nontrivial fixed point has an extra thermodynamically relevant variable h_2 at dimensionalities above four, while the Gaussian fixed point has this extra variable at dimensionalities below four.

To see the consequences of these extra variables, we should turn to an examination of the standard relevant variables. These include:

(1) h_0, which is just a constant term in the action of the Hamiltonian and uninteresting.

(2) h_1, which physically represents a deviation of the coupling from its critical strength, i.e., is $T-T_c$ or $-(K-K^*)$. As a_0 grows to λa_0 the action for the system is pushed further and further from its critical point so that the coherence length—which is measured in units of a_0—may remain fixed. Thus renormalization increases the value of this and every other thermodynamically relevant variable.

(3) h_σ, a relevant variable, which is odd in the spin and represents a symmetry breaking term like a magnetic field. This symmetry breaking term also grows and forces one away from the critical point as one does successive renormalizations.

To get to the critical point, the experimentalist adjusts the temperature to set $h_1=0$. Either nature—or the experimentalist—adjusts the symmetry breaking term to zero. Then what happens to h_2? Assume that we are "near" a Gaussian fixed point at say $d=3$. Assume h_2 is small but not zero. Nonetheless $h_2 \neq 0$ means we are not at the Gaussian fixed point. How can we cal-calculate where we are? We observe the system on a large distance scale. To describe this observation, the theorist renormalizes with $\lambda > 1$. In this renormalization h_2 grows and grows and pushes the system toward

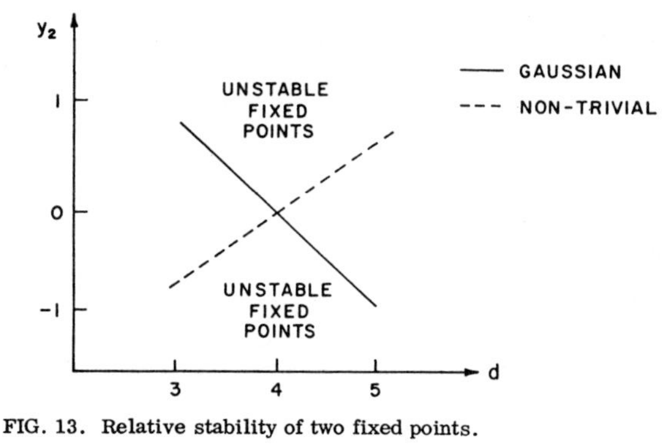

FIG. 13. Relative stability of two fixed points.

the competitive fixed point which is the nontrivial fixed point.

We conclude that in statistical systems we will naturally observe only that fixed point with the fewest relevant variables. These fixed points are termed the "most stable" fixed points. For $d<4$ the nontrivial fixed point is most stable. For $d>4$ the Gaussian fixed point is most stable.

VII. THE MIGDAL RECURSION FORMULAE APPLIED TO THE QUARK-STRING SYSTEM

The Migdal-style recursion formulae can be applied quite directly to the Wilson action defined by Eqs. (2.17), (2.18), and (2.19). The pieces of this calculation are already before us. The quarks have a nearest-neighbor coupling and can be attacked via the method of Sec. V.B—and especially with the aid of the recursion formula (5.17b). The strings have a gauge-style coupling and can be attacked via the method of Sec. V.C—with the key equation being (5.22) for this case.

Notice that both types of potential-moving approximation can be applied simultaneously. In the motion, the different types of potential bonds do not interfere with each other. Thus, in this first analysis, we can handle the different parts of the recursion problem quite independently of one another. [For descriptions of perturbation theories on lattices see Wilson (1975a) and Baluni and Willemsen (1976).]

A. Recursion relation for quarks

To handle the quarks, we write the quark part of the action in the form of a sum of nearest-neighbor coupling terms. In particular, we write the A_ψ of Eq. (2.18) as

$$A_\psi = \sum_{x,\mu} K_{\mu f}(\psi(x), \psi(x+\hat{e}_\mu a_0), U(x, x+\hat{e}_\mu a_0)). \tag{7.1}$$

To fit the form (2.18), we take the coupling function to be

$$K_{\mu f}(\psi, \psi', U) = -K^0_{\mu f} \sum_i (\bar{\psi}_{if}\psi_{if} + \bar{\psi}'_{if}\psi'_{if}) + K_{\mu f} \sum_{ij} [\bar{\psi}_{if}(1-\gamma_\mu)U_{ij}\psi'_{jf} + \bar{\psi}'_{if}(1+\gamma_\mu)(U^\dagger)_{ij}\psi_{jf}]. \tag{7.2}$$

Expression (7.2) contains two coupling constants: The hopping parameter $K_{\mu f}$, and a normalization parameter $K^0_{\mu f}$. We arrange these in a vector

$$\mathbf{K}_{\mu f} = (K^0_{\mu f}, K_{\mu f}). \tag{7.3}$$

Our starting point is a symmetrical situation in which $\mathbf{K}_{\mu f}$ is independent of μ. To make the calculation as symmetrical as possible we also choose $K^0_{\mu f}$ to be μ independent. Then, to match Eq. (2.26), we must choose the normalization parameter to be

$$K^0_{\mu f} = 1/2d. \tag{7.4}$$

Thus we begin from

$$\mathbf{K}_{\mu f} = (1/2d, K_f). \tag{7.5}$$

Now we apply a decimation in the x direction, $\mu=1$.

The couplings in the other directions simply increase by a factor of λ as in Eq. (5.16b), i.e.,

$$\mathbf{K}'_{\mu f}=\lambda(K^0_{\mu f},K_{\mu f}) \quad \text{for } \mu>1 . \tag{7.6}$$

On the other hand, the new coupling in the "1" direction is given by

$$e^{K'_{1f}(\psi,\psi',U')}=\operatorname{tr}_\eta e^{K_{1f}(\psi,\eta,U_1)+K_{1f}(\eta,\psi',U_2)}$$

for the particular case in which $\lambda=2$. Except for an additive constant, $K'_{1f}(\psi,\psi',U')$ turns out to be exactly of the form (7.2) with a set of new couplings

$$\mathbf{K}'_{1f}=(K^0_{1f},(K_{1f})^2/K^0_{1f}), \tag{7.7}$$

and a new U which is

$$U_{ij}=\sum_k (U_1)_{ik}(U_2)_{kj}. \tag{7.8}$$

Thus, as we would wish, the longer string is represented by a variable U' which is simply the matrix product of the variables representing its component parts.

Equation (7.7) describes the case $\lambda=2$. By composing $n=\log_2\lambda$ transformations of the form (7.7), we find the more general result

$$\mathbf{K}'_{1f}=(K^0_{1f},K^\lambda_{1f}/(K^0_{1f})^{\lambda-1}) \tag{7.9}$$

which represents the lattice constant in the x direction by a factor of λ.

Now imagine changing successively the lattice constants in the $1,2,\ldots,d$ direction by a factor of λ. Then we would have to apply $d-1$ transforms of the form (7.6) and one transform of the form (7.9). The net result is that for all μ the new coupling is

$$\mathbf{K}'_{\mu f}=\lambda^{d-1}(K^0_{\mu f},K^\lambda_{\mu f}/(K^0_{\mu f})^{\lambda-1}).$$

Given the starting point (7.5) we have

$$\mathbf{K}'_{\mu f}=\lambda^{d-1}(1/2d,(2dK_f)^\lambda/2d). \tag{7.10}$$

Equation (7.10) has one satisfactory feature and one unsatisfactory feature. The new couplings are independent of direction, as we would like. However, the normalization term K^0 is no longer given by Eq. (7.4), but instead by

$$K^0_{\mu f}=\lambda^{d-1}/2d .$$

We would like to recover the structure of our original action in which K^0 is $1/2d$. Fortunately we can very easily redefine the size of our couplings by making the replacement

$$\psi(1)\to Z_3\psi(1). \tag{7.11}$$

This transformation changes no correlation functions or equations of motion; it merely changes a constant additive term in the action. Under this new "renormalization" transformation all the couplings $K_{\mu f}$ change according to

$$\mathbf{K}_{\mu f}\to\mathbf{K}_{\mu f}Z_3^2 .$$

Hence we pick $Z_3=\lambda^{-(d-1)/2}$ and derive from (7.10) the new coupling

$$\mathbf{K}'_{\mu f}=(1/2d,(2dK_f)^\lambda/2d).$$

We have now constructed the renormalization so that the normalization term $K^0_{\mu f}$ is left invariant by the transformation, the only change being in K_f which is seen to obey the recursion relation

$$K_f(\lambda a_0)=[2dK_f(a_0)]^\lambda/2d . \tag{7.12}$$

Equation (7.12) is our desired result, a simple recursion relation for the hopping constant obtained from the Migdal scheme.

B. Interpretation of the quark recursion formula

Equation (7.16) has exactly the same structure as the one-dimensional recursion formula, (3.16)

$$K_f(\lambda a_0)=\tfrac{1}{2}[2K_f(a_0)]^\lambda .$$

This latter formula was derived for the special case $U(x,x')=1$, but it applies equally for arbitrary U. Hence Eq. (7.12) is exact, as it must be, when $d=1$.

Our new recursion formula has two major consequences: First, there is a fixed point at

$$K_f^*=1/2d . \tag{7.13}$$

Secondly, at this fixed point, there is a critical index

$$y=1 \tag{7.14}$$

just as in one dimension.

We can check these results against the solvable special case $U(x,x')=1$. In that case, the fixed point and critical index are exactly those of Eq. (7.13) and (7.14). Hence the consequences of Eq. (7.12) are right at least in this special limit. Since this limit is essentially the asymptotic freedom limit, we have set up a formalism which at least generates the known answers in that asymptotic situation.

But this result is certainly not new to us. Consider our discussion of the limit $a_0\to 0$ in Sec. II.E. Then we wrote

$$K_f=(1/2d)(1-m_f a_0/d) \tag{2.25}$$

and considered the mass m_f to be a scale invariant quantity. Equation (2.25) has an $a_0\to 0$ limit (a fixed point) at exactly the value (7.13). The correct term scales linearly with a_0, which is the statement (7.14).

Thus we have learned that the Migdal approach generates a simple recursion formula for K_f which is at least right in the limit $a_0\to 0$.

C. Recursion relation for the strings: One-dimensional recursion

The gauge theory is represented in terms of matrix variables U which form a homogeneous space for the symmetry group in question. In particular, each plaquette contains a coupling of the form $J(U_c)$, where U_c is the product of U's around the square. We choose $J(U_c)$ real and demand that J be parity independent, i.e.,

$$J(U_c)=J(U_c^\dagger). \tag{7.15}$$

In addition, the symmetry requires that

$$J(uU_cu^\dagger)=J(U_c) \tag{7.16}$$

for all u which are matrix representation of the symmetry group. The transformation matrix u will be of exactly the same type as the variables U and their product U_c. Finally, the homogeneity of the space demands that for any two matrices U and U' we can construct a u which converts U into U' via

$$U' = uU . \tag{7.17}$$

This gauge theory maps into a one-dimensional nearest-neighbor theory in which $K(U_1, U_2)$ has the structure

$$K(U_1, U_2) = J(U_1 U_2^\dagger) .$$

Hence the one-dimensional recursion formula can be written for $\lambda = 2$, in the form

$$e^{J'(U_1 U_2^\dagger)} = \mathrm{tr}_U e^{J(U_1 U^\dagger) + J(U U_2^\dagger)} . \tag{7.18}$$

The most convenient way of writing Eq. (7.18) comes from the choice of U_1 and U_2

$$U_1 = U_2^\dagger = U_T^{1/2} .$$

Then (7.18) takes the form

$$e^{J'(U_T)} = \mathrm{tr}_U e^{J(U_T^{1/2} U^\dagger) + J(U U_T^{-1/2})} \tag{7.19}$$

since U_T is unitary.

The usual way of parametrizing U starts from the generators Λ_a of the group representation. Then, since U is unitary, it may be written as

$$U = \exp i\theta_a \Lambda_a = \exp iX . \tag{7.20}$$

Here there is an implied sum over a. The parameters θ_a are all real.

We are interested in analyzing Eq. (7.19) in the strong coupling limit. In this limit, $J(U)$ is largest for small values of θ_a. Then it is reasonable to expand $J(U)$ in the form

$$J(U) = J_0 - \frac{J}{2!} \,\mathrm{trace}\, X^2 + \frac{J_4}{4!} \,\mathrm{trace}\, X^4 + \cdots . \tag{7.21}$$

For most of our purposes, we need not worry about the constant term J_0, while the higher-order term J_4 will prove to be negligible, when J is sufficiently large. The trace in Eq. (7.21) is, of course, a diagonal sum over the indices of the representation matrix X.

We shall need two properties of the generators Λ_a: their commutators which can be written in the form

$$[\Lambda_a, \Lambda_b] = i f_{abc} \Lambda_c \tag{7.22}$$

and the trace of their product, which we shall assume to be

$$\mathrm{trace}\Lambda_a \Lambda_b = \delta_{ab} C . \tag{7.23}$$

We take C to be positive so that the Lie group is compact. Finally, since θ_a is required to be small by the condition that the coupling constant J is large, one can replace the invariant sum over U in Eq. (7.19) by a simple integral over the θ_a, i.e.,

$$\mathrm{tr}_U \sim \int \prod_a d\theta_a = \int d\theta . \tag{7.24}$$

Now the calculation is all set up. We write

$$U_T = \exp(i\varphi_a \Lambda_a) = \exp iX_T .$$

Then Eq. (7.19) reads

$$\exp[-\tfrac{1}{2} J' C \varphi_a \varphi_a] \sim \int d\theta \exp[Y(\theta, \varphi)]$$

$$= \int d\theta \exp[J(U_T^{1/2} U^\dagger) + J(U U_T^{-1/2})] . \tag{7.25}$$

We expand the exponent on the right-hand side of Eq. (7.25) in a power series in φ_a and θ_a. If we employ only the coupling J and not J_4, we discover—after some algebra—that

$$Y(\theta, \varphi) = -\frac{JC}{2}\left(\frac{\varphi_a \varphi_a}{2} + 2\theta_a \theta_a\right)$$

$$+ \frac{JC}{48} \varphi_a \theta_b \varphi_{a'} \theta_{b'} f_{abc} f_{a'b'c'} .$$

The integrals in Eq. (7.25) can now be performed directly to give the result

$$e^{-J' C \varphi_a \varphi_a / 2} \sim e^{-(JC/2)\varphi_a \varphi_a / 2} \, e^{\varphi_a \varphi_{a'} f_{abc} f_{a'bc}/24} . \tag{7.26}$$

In one sense, Eq. (7.26) is a very satisfactory result. We made the assumption that we need include only the Gaussian term in J but not the term in J_4 in Eq. (7.21). Now (7.26) shows that only bilinear terms in φ_a emerge after the recursion. This result is then consistent in that sense. But the quadratic form on the left-hand side of (7.26) is diagonal and a-independent, while the right-hand side does not seem to be diagonal. This appears inconsistent. However, if the Lie group is compact, we can always choose the Λ_a so that the terms in f come to a diagonal form

$$\sum_{bc} f_{abc} f_{a'bc} = C_f \delta_{aa'}$$

with C_f being a positive constant independent of a. In this way we derive the one-dimensional strong-coupling recursion relation

$$J' = \frac{J}{2} - \frac{C_f}{C} \frac{1}{12}$$

for $\lambda = 2$. By doing $n = \log_2 \lambda$ such recursions, we find the general strong-coupling recursion formula

$$J' = \frac{J}{\lambda} - \frac{C_f}{C} \frac{(\lambda - 1)}{6\lambda} . \tag{7.27}$$

D. Nearest-neighbor interactions in 2 + ϵ dimensions

Equation (7.27) can be applied to describe the phase transition behavior for a nearest-neighbor coupling of variables which are the regular representation of the symmetry group. However, notice that most often in statistical physics we do not deal with this particular representation. For example, when Brezin and Zinn-Justin (1976) attacked the group of rotations in n-dimensional Euclidean space their variables were not rotation matrices—which is the case we just analyzed—but instead n-component unit vectors.

The only regular representation of a Lie group conventionally treated in statistical physics is the representation of U1. This Abelian group has $U = e^{i\varphi}$. Then one can take a nearest-neighbor coupling of the form

$$K(U_1, U_2) = J(e^{i(\varphi_1 - \varphi_2)}) .$$

In the particular case in which $J(e^{i\varphi})$ is $J\cos\varphi$, then the model under consideration is called the XY or planar model. It is equivalent to the interaction of two-component unit vectors. In this special case, $C_f = 0$.

Let us analyze the structure of the recursions for the nearest-neighbor case in d dimensions via the Migdal approximation. From Eqs. (7.27) and (5.17a) one finds that after a change in lattice constant by a factor of λ there is a new coupling in the x direction, $J_x(\lambda a_0)$, which obeys

$$J_x(\lambda a_0)=\lambda^{d-2}J_x(a_0)-\frac{C_f}{C}\frac{(\lambda-1)}{6}. \tag{7.28}$$

Consequently, there is a fixed point at

$$J_x^*=\frac{C_f}{6C}\frac{\lambda-1}{(d-2)\ln\lambda} \tag{7.29}$$

for d approximately equal to 2. Equation (7.28) also implies a critical index

$$y=d-2=\nu^{-1}. \tag{7.30}$$

Notice that as $d\to 2$, y becomes marginal. When $C_f>0$, i.e., when the group is non-Abelian, the recursion equation has the structure (6.14) at two dimensions. Then, step by step, each recursion slowly weakens the coupling. Hence in this case there is no phase transition at $d=2$. There is a phase transition for $d>2$. Thence $d_L^*=2$ for these examples of non-Abelian nearest-neighbor interactions.

But, for the elementary particle application, the most important fact is that at the lower critical dimensionality the phase transition disappears or is rather pushed to zero temperature.

The U1 case is different. Here the group is Abelian so that C_f vanishes. Then at two dimensions the recursion relation is one again marginal but is apparently of the form (6.13). [This form has been more carefully demonstrated by Zittartz (1976) and the correlation functions calculated by Berezinski (1971).] Thus we have very strong indications that at $d=2$ the XY problem has a line of fixed points and shows the structure plotted in Fig. 5c.[2]

E. Gauge interactions in 4 + ϵ dimensions

The Migdal-style analysis of the string interactions in $4+\epsilon$ dimensions is precisely similar to the line of argument we have just carried out for the nearest-neighbor case.

The one-dimensional recursion formula (7.27) can be written

$$R_0^\lambda(J)=\frac{1}{\lambda}\left(J-\frac{C_f}{C}\frac{\lambda-1}{6}\right). \tag{7.31}$$

From this formula and Eq. (5.22) we derive the Migdal-style recursion formula for the coupling $J_{\alpha\beta}$, which describes gauge plaquettes in the α-β plane. This recursion formula is

$$J_{\alpha\beta}(\lambda a_0)=\lambda^{d-4}\left[J_{\alpha\beta}(a_0)-\frac{C_f}{C}\frac{\lambda-1}{6}(\lambda^{1-\alpha}+\lambda^{3-\beta})\right]. \tag{7.32}$$

Consequently the fixed point is given by

$$J_{\alpha\beta}^*=\frac{C_f}{6C}\frac{\lambda-1}{d-4}\frac{\lambda^{1-\alpha}+\lambda^{3-\beta}}{\ln\lambda} \tag{7.33}$$

for dimensionality near 4, while the critical index is

$$y=d-4. \tag{7.34}$$

Now four dimensions is the lower critical dimensionality for this phase transition. Just as in the nearest-neighbor situation we must distinguish between two cases: The Abelian situation (the group U1) in which $C_f=0$ and the non-Abelian case when $C_f>0$.

The color carrying strings are representations of SU_3 symmetry and fall therefore into the non-Abelian case. For this situation, there is no phase transition at $d=4$. Of course, this is exactly the result we would have wanted since then the situation is described by the phase diagram of Fig. 9b. As indicated in Sec. IV, this phase diagram implies both asymptotic freedom and quark trapping.

On the other hand, electromagnetism is represented by strings which have Abelian U1 symmetry. Our line of argument implies the existence of a whole line of fixed points for this case, as depicted in Fig. 9c. There are two closely related ways of forming this conclusion. First, one can directly analyze the four-dimensional gauge theory as was done by Polyakov (1975). Alternatively, one can follow Migdal's (1975a, 1975b) line of argument to map the gauge problem into the XY model and then employ perturbation theory about $J=\infty$ to find [see Zittantz (1976), Berezenski (1971), and Wegner (1971b)] a line of fixed points. Either line of argument leaves gaps, so that we should not consider the picture of Fig. 9c to be necessarily meaningful. On the other hand, one can correctly say that *the lines of argument which imply that quarks are at once trapped and asymptotically free fail to go through for electrons coupled by electromagnetism*. Therefore, quantum electrodynamics does *not* provide a counter argument to the line of reasoning that we have applied to quarks.

F. An assessment

Are these conclusions about the behavior of a quark-string system reliable? How far can we trust this kind of application of the Migdal approximation? In my view, some parts of the argument are quite trustworthy, while other parts hide some very real opportunities for substantial errors.

The Migdal-style argument seems to me to be most reliable when it is used to assess the possibility of a phase transition for a pure system of interacting strings at four dimensions and when these strings are describable as representations of a compact, non-Abelian group. These arguments basically describe the sign of the β-function (i.e., the direction of the arrows in Fig. 9) near zero and infinite couplings. These signs can be accurately calculated, I believe, for this situation. Then if the values of the β-function at these points have the same sign (i.e., if the arrows point in the same direction) there will be an even number of fixed points in the interval between zero and infinite coupling. If they have opposite signs, there will be an odd number of fixed points. We have concluded that in the non-Abelian

---

[2]See Zittartz (1976), Jose (1976), Berezinskii (1971, 1972), Wegner (1971b), Kosterlitz and Thouless (1973), Kosterlitz (1974), and Luther and Scalapino (1976), who discuss the two-dimensional case. Adler (1972) suggests a rather similar picture for the U1 gauge theory at $d=4$.

situation the signs are the same. Hence there are an even number of fixed points.

We hope that this even number is zero. We can make some physical arguments [see Eqs. (4.9)] about one kind of phase transition which might occur, producing a change in correlation structure from Eq. (4.9a) to Eq. (4.9b). But, it is harder to imagine how two or four or six phase transitions might take place. Thus we do have some foundation for our hope that in the non-Abelian case there might be no phase transition at all.

On the other hand, in the Abelian four-dimensional case, the β function as defined by

$$\beta(J) = \frac{\partial}{\partial \ln \lambda} \ln J(\lambda a_0) \Big|_{\lambda=1}$$

vanishes at $J=\infty$. In my view, one cannot form a reliable conclusion about the phase transition behavior from this fact alone.

The quarks further complicate this story. The analysis of this paper has been carried out as if the quarks did not play an essential role in the phase transition. They were merely probes which enabled us to observe this transition or its absence. However, even though this point of view is consistent with the Migdal approximation scheme, it has a rather doubtful validity. Notice that the quarks are also undergoing a near phase transition as $a_0 \to 0$. When two phase transitions occur together, they may interfere in a very subtle and complex manner. Notice, for example, that the Baxter (1971, 1972) phase transition may be viewed as the result of the interaction between two Ising models which are both going critical at the same time (Kadanoff and Wegner, 1971). And the Baxter solution is much richer and more complex than the Onsager solution of the Ising model. Similarly, the quark–string phase transition may be much richer than the phase transition of either quarks or strings alone.

An additional weakness arises because the quarks are fermions. Our potential moving argument is partially predicted upon a "variational principle." The "variational principle" in turn gives lower bounds on the free energy whenever the basic probability function Tr is a sum with positive semidefinite weights. This positivity fails for fermion variables. The best we can say, then, is that our error is second order in the moved potential, but we cannot assess the sign of the error.

This problem is made even more serious because the fermion problem does not have a "strong" interaction. In this case one-half of our argument for the Migdal approximation—that part based upon the strength of the interaction—is absent. Thus our argument about phase transitions in the coupled quark–string system is far from watertight. In fact it is wrong when there are too many different flavors. To make the situation even worse, we have no experience with any analogous renormalization calculations in any statistical mechanical calculation involving fermions on lattices.

On the other hand, there is some calculational evidence that we are moving in the right direction. In recent calculations, W. Bardeen and R. B. Pearson (1976) and also Kogut and Suskind (1975) have formulated partial lattice versions of the theory outlined here. These groups, respectively, have 2 and 1 continuum coordinates and consequently $d-2$ and $d-1$ dimensional lattices. These partial lattice theories have the advantage that they can include chiral invariance ($\psi \to \gamma_5 \psi$; $\bar{\psi} \to -\bar{\psi}\gamma_5$) in a convenient manner. These groups then attempt calculations with a fair success in giving, respectively, the leading Regge trajectories, the structure of the two-dimensional theory, and the lowest hadron masses.[3]

G. For the future

It is likely that further developments from the point of view presented in this paper would do well to include chiral symmetry and perhaps to allow the theory to spontaneously break that symmetry. Thus the underlying Lagrangian used in this paper may well have to be modified in some crucial respects in order to make proper contact with the physics.

In addition some kinds of technical progress will be necessary in order to make any kind of lattice theory into a useful tool. At the moment, we have only a very slight acquaintance with renormalization calculations involving fermions. It would be very helpful if this experience were extended through approximate renormalization calculations (perhaps on lattices) of such problems as the Schwinger (1962) model, the massive Thirring (1958) model [see Glaser (1958), Johnson (1961), Summerfield (1963), and Luther (1976)], and the backward scattering model [see Luther and Emery (1974)]. Indeed even some experience with purely Gaussian models with a bilinear action would be useful.

Notice that most approximate renormalization calculations performed to date describe thermodynamic functions rather than correlation functions. For the elementary particle example, the correlation functions are crucial since these functions define both the masses and the scattering amplitudes. It would be very useful if the relevant approximations could be extended to the calculation of correlation functions and the results checked against known accurate calculations.

Kogut and Suskind have pointed the way to a worthwhile parallel development: the use of perturbation theory about strong coupling in conjunction with series expansion techniques. One can hope that this method will yield expressions for many of the quantities of physical interest.

The problems of quarks and strings are indeed more complex than those problems which have been attacked in statistical physics. Nonetheless, we can hope and expect that the basic methods of modern statistical renormalization theory can be applied to gain some qualitative and indeed quantitative picture of the consequences of a quark–string model of elementary particles.

APPENDIX A. RECURSION RELATION FOR QUARKS IN ONE DIMENSION

We would like to calculate the trace in Eq. (3.10) for the case in which the coupling function involves fermion

---

[3]See Bardeen and Pearson (1976), Kogut and Suskind (1974, 1975, 1976), Carrol *et al.* (1975), Banks *et al.* (1976).

variables and is of the form (3.13). The quantity to be calculated is

$$e^{K'(\psi,\psi')}=\operatorname{tr}_\sigma e^{-1/2\bar{\psi}\psi-1/2\bar{\psi}'\psi'-\bar{\sigma}\sigma}\, e^{K[\bar{\psi}(1-\gamma_1)\sigma+\bar{\sigma}(1-\gamma_1)\psi']}$$

$$\times e^{K[\bar{\psi}'(1+\gamma_1)\sigma+\bar{\sigma}(1+\gamma_1)\psi]}\,. \tag{A1}$$

Here σ, ψ, and ψ' are anticommuting fermion variables, and γ_1 is a matrix with eigenvalues ± 1.

To calculate the trace (A1), we decompose the spinor variable σ into parts which are eigenvectors of γ_1 with eigenvalue ± 1. In particular, we write

$$\sigma_\pm=\tfrac{1}{2}(1\pm\gamma_1)\sigma\,,$$
$$\bar{\sigma}_\pm=\tfrac{1}{2}\bar{\sigma}(1\pm\gamma_1) \tag{A2}$$

and notice, then, that the trace over σ can be decomposed into separate traces over σ_+ and σ_-:

$$\operatorname{tr}_\sigma=\operatorname{tr}_{\sigma+}\operatorname{tr}_{\sigma-}\,. \tag{A3}$$

Thus Eq. (A1) implies

$$K'(\psi,\psi')=-\tfrac{1}{2}\bar{\psi}\psi-\tfrac{1}{2}\bar{\psi}'\psi'+a_+ +a_-\,, \tag{A4}$$

where

$$a_+=\ln\operatorname{tr}_{\sigma+}e^{-\bar{\sigma}_+\sigma_+}e^{K\bar{\psi}'(1+\gamma_1)\sigma_+}e^{K\bar{\sigma}_+(1+\gamma_1)\psi} \tag{A5}$$

and

$$a_-=\ln\operatorname{tr}_{\sigma-}e^{-\bar{\sigma}_-\sigma_-}e^{K\bar{\psi}(1-\gamma_1)\sigma_-}e^{K\bar{\sigma}_-(1-\gamma_1)}\,. \tag{A6}$$

Because the fermion variables have squares which vanish

$$\psi^2=\bar{\psi}^2=0 \tag{A7}$$

[see Eq. (1.10)], any exponential or logarithm of an expression involving these variables is very simple. In particular, for any pair of components of the spinors $\bar{\psi}_\alpha, \psi_\beta$

$$e^{K\bar{\psi}_\alpha\psi_\beta}=1+K\bar{\psi}_\alpha\psi_\beta\,, \tag{A8}$$

$$\ln(a+b\bar{\psi}_\alpha\psi_\beta)=\ln a+b/a\bar{\psi}_\alpha\psi_\beta\,. \tag{A9}$$

Thus, if the projection operation (A2) leaves only a single component in σ_+ and in σ_-, expression (A5) may be expanded as

$$a_+=\ln\operatorname{tr}_{\sigma_+}(1-\bar{\sigma}_+\sigma_+)[1+K\bar{\psi}'(1+\gamma_1)\sigma_+]$$
$$\times[1+K\bar{\sigma}_+(1+\gamma_1)\psi]\,. \tag{A10}$$

Because of (A7), this may be further simplified to

$$a_+=\ln\operatorname{tr}_{\sigma_+}[1-\bar{\sigma}_+\sigma_+ +K\bar{\psi}'(1+\gamma_1)\sigma_+ +K\bar{\sigma}_+(1+\gamma_1)\psi$$
$$+K^2\bar{\psi}'(1+\gamma_1)\sigma_+\bar{\sigma}_+(1+\gamma_1)\psi]\,. \tag{A11}$$

Then Eq. (1.11) permits the evaluation of the trace in the form

$$a_+=\ln-1-K^2\bar{\psi}'(1+\gamma_1)^2\psi$$

so that Eq. (A9) implies

$$a_+=(\ln-1)+2K^2\bar{\psi}'(1+\gamma_1)\psi\,. \tag{A12}$$

If the projection (A2) left n independent components in σ_+, the result for a_+ would be just the same, except that $\ln-1$ would be replaced by $n\ln-1$. When a_+ and a_- are added together as in Eq. (A4), this trivial term cancels out leaving

$$K'(\psi,\psi')=-\tfrac{1}{2}\bar{\psi}\psi-\tfrac{1}{2}\bar{\psi}'\psi'$$
$$+2K^2\bar{\psi}'(1+\gamma_1)\psi+2K^2\bar{\psi}(1-\gamma_1)\psi'\,. \tag{A13}$$

Eq. (A13) is the result discussed in Sec. III.C of the text.

APPENDIX B. SOLVING THE TWO-DIMENSIONAL GAUGE THEORY VIA A GAUGE TRANSFORM

It is instructive to find the solution of the two-dimensional gauge theory by using a gauge transformation argument. The basic goal is to reduce the two-dimensional gauge problem to a nearest-neighbor problem with the aid of a more physical argument than that employed in Sec. III.E.

Start with the fragment of lattice shown in Fig. 14a. The lattice sites are labeled (n,m) with n and m being the x and y coordinates of the sites in units of the lattice constant. The basic action is expressed in terms of the variables $U_{n,m;\,n'm'}$ $[=(U_{n',m';\,nm})^{-1}]$ where (n,m) and (n',m') are nearest-neighbor sites. In particular it is a function of the product variables

$$U_\Gamma(n,m)=U_{n,m;\,n,m+1}U_{n,m+1;\,n+1,m+1}$$
$$\times U_{n+1,m+1;\,n+1,m}U_{m+1,m;\,n,m}\,. \tag{B1}$$

In terms of these product variables, the action takes the form

$$A[U]=\sum_{n,m}J(U_\Gamma(n,m))\,. \tag{B2}$$

Let U_Γ be representations of SU_n, specifically unit determinant unitary matrices. Then, in Eq. (B1), we imply matrix multiplication over the internal indices of the matrices. The trace is a diagonal sum over these matrix indices. Now, imagine a calculation in which we find the partition function

$$Z=\operatorname{Tr}_U e^{AU} \tag{B3}$$

by doing an invariant sum over all these SU_n matrices. This sum is invariant under the replacement of all the U's according to $U\to U'$ with

$$U'_{nm;\,n'm'}=u(n,m)\,U_{nm;\,n'm'}[u(n',m')]^{-1}\,. \tag{B4}$$

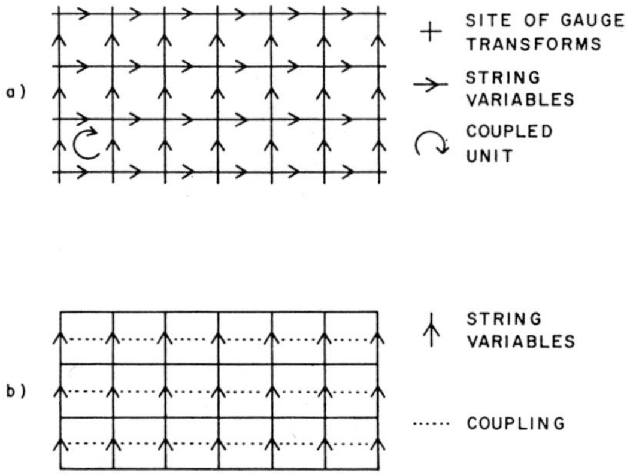

FIG. 14. Two-dimensional gauge problem before (a) and after (b) the choice of a special gauge.

In Eq. (B4), $u(n,m)$ is an n by n special unitary matrix, and matrix multiplication over the internal indices is understood.

For simplicity, assume that the summation in Eq. (B2) is over m from $-\infty$ to ∞ but only covers $n=1$ to ∞. Now choose a special gauge to perform the sum (B3). Basically, we wish to choose a gauge in which the vector potential in the x direction vanishes. To do this we take

$$u(1,m)=\underline{1} \tag{B5}$$

where $\underline{1}$ stands for the unit matrix, and

$$u(n+1,m)=[u(n,m)\,U_{n,m;\,n+1,m}]^{-1}. \tag{B6}$$

Given this special choice of u, the variables in the new gauge are very simple for all strings stretched in the x direction. In fact, according to Eqs. (B4) and (B6)

$$U'_{n,m;\,n\pm1,m}=\underline{1}. \tag{B7}$$

In this way we have "gauged away" one-half of the original variables. If we then choose the particular normalization

$$\mathrm{tr}_U 1=1,$$

the sum (B3) may be simplified to the form

$$Z=\mathrm{Tr}'_{U'}\,e^{A[U']}. \tag{B8}$$

Here the prime on Tr indicates that we sum over only y-direction strings. According to Eqs. (B1), (B2), and (B7), this new action still takes the form (B2)

$$A[U']=\sum_{n,m}J(U'_\Gamma(n,m)), \tag{B9}$$

but now the basic argument of the function J is

$$U'_\Gamma(n,m)=(U_{n,m;\,n,m+1}U_{n+1,m+1;\,n+1,m}). \tag{B10}$$

Notice that all strings point in the y direction. Nearest-neighbor bits of string separated by one unit in the x direction are coupled by Eqs. (B9) and (B10) *but* there is no coupling between strings with $U_{nm;\,nm'}$ with different values of their central point $(m+m')/2$. Then the summation (B8) decomposes into a group of uncorrelated individual sums over variables with different $(m+m')/2$ indices. The $d=2$ gauge problem is thereby reduced to a set of uncoupled $d=1$ nearest-neighbor problems QED.

ACKNOWLEDGMENTS

This material is a revised version of a set of lecture notes for a series of talks delivered at the University of Chicago. Several members of the staff, including the entire many-body group, helped with constructing and criticizing these presentations. I owe special thanks to G. Mazenko, who organized most of the show.

The work described here started at the IBM Research Laboratory in Zurich, was continued at Brown and Harvard, and was completed at Nordita. Particularly helpful criticism came to me during seminars delivered at Harvard, NAL, Princeton, and Urbana. In addition, I wish to thank G. Mazenko, A. Luther, A. Jaffe, S. Coleman, B. Lee, S. Weinberg, K. Wilson, W. Bardeen, R. Ditzian, and A. A. Migdal for their helpful discussions and private communication.

REFERENCES

Albers, E. S. and B. W. Lee, 1973, Phys. Rep. **9C**, 1.
Adler, S., 1972, Phys. Rev. D **5**, 3021.
Balian, R., J. Drouffe, and C. Itzykson, 1974, Phys. Rev. D **10**, 3376.
Balian, R., J. Drouffe, and C. Itzykson, 1975a, Phys. Rev. D **11**, 2098.
Balian, R., J. Drouffe, and C. Itzykson, 1975b, Phys. Rev. D **11**, 2104.
Baluni, V. and J. F. Willemsen, 1976, Phys. Rev. D **13**, 3342.
Banks, T., L. Suskind, and J. Kogut, 1976, Phys. Rev. D **13**, 1043.
Bardeen, W. and R. B. Pearson, 1976, Fermilab Preprint 76/24-THY 1976.
Baxter, R., 1971, Phys. Rev. Lett. **26**, 832.
Baxter, R., 1972, Ann. Phys. (N.Y.) **70**, 193.
Berezin, 1966, *The Method of Second Quantization* (Academic, New York).
Berezinski, V. L., 1971, Sov. Phys.–JETP **32**, 493.
Berezinskii, V. L., 1972, JETP **34**, 610.
Bjorken, J. D. and E. A. Paschos, 1969, Phys. Rev. **185**, 1975.
Brezin, E. and J. Zinn-Justin, 1976, Phys. Rev. Lett. **36**, 691.
Callan, C. G., Jr., 1972, Phys. Rev. D **5**, 3202.
Carrol, A., J. Kogut, D. K. Sinclair, and L. Suskind, 1975, Phys. Rev. D **13**, 2270.
Feynman, R. P., 1970, in *High Energy Physics*, edited by C. N. Yang *et al.* (Gordon and Breach, New York).
Gell-Mann, M. and F. E. Low, 1954, Phys. Rev. **95**, 1300.
Gross, D. and F. Wilczek, 1973a, Phys. Rev. Lett. **30**, 1346.
Gross, D. and F. Wilczek, 1973b, Phys. Rev. D **8**, 3633.
Han, M. Y. and Y. Nambu, 1965, Phys. Rev. **139B**, 1066.
Houghton, A. and L. Kadanoff, 1973, in *Renormalization Group in Critical Phenomena and Quantum Field Theory: Proceedings of a Conference, 1973*, edited by J. D. Gunton and M. S. Green (Dept. of Physics, Temple University, Philadelphia).
Jaffe, A. and J. Glimm, 1976, Lectures at Cargèse Summer School.
Jose, J., 1976, Phys. Rev. D to be published.
Kadanoff, L. P., 1975, Phys. Rev. Lett. **34**, 1005.
Kadanoff, L. P., 1976, Ann. Phys. (N.Y.), to be published.
Kadanoff, L. P., Wolfgang Gotze, D. Hamblen, R. Hecht, E. A. S. Lewis, V. V. Palciauskas, M. Rayl, J. Swift, D. Aspnes, and J. Kane, 1967, Rev. Mod. Phys. **39**, 395.
Kadanoff, L. P., A. Houghton, and M. C. Yalabik, 1976, J. Stat. Phys. **14**, 171.
Kadanoff, L. P. and F. Wegner, 1971, Phys. Rev. B **4**, 3989.
Kogut, J. and L. Suskind, 1974, Phys. Rev. D **9**, 3391.
Kogut, J. and L. Suskind, 1975, Phys. Rev. D **11**, 395.
Kogut, J. and L. Suskind, 1976, reported at the May 1976 Yeshiva Statistical Mechanics Meeting.
Kosterlitz, J. M., 1974, J. Phys. C **7**, 1046.
Kosterlitz, J. M. and D. J. Thouless, 1973, J. Phys. C **6**, 1181.
Luther, A. and D. Scalapino, 1976, reported by A. Luther at 1976 Cargèse Summer School.
Migdal, A. A., 1975a, Zh. Eksp. Teor. Fiz. **69**, 810.
Migdal, A. A., 1975b, Zh. Eksp. Teor. Fiz. **69**, 1457.
Murnaghan, F. D., 1938, *The Theory of Group Representations* (Reprinted 1963 by Dover, New York), Chap. 8.
Nauenberg, M. and B. Nienhuis, 1974, Phys. Rev. Lett. **33**, 944.
Nauenberg, M. and B. Nienhuis, 1975, Phys. Rev. B **11**, 4152.
Nelson, D. and M. Fisher, 1975, Ann. Phys. (N.Y.) **91**, 226.
Nelson, E., 1973, in *Constructive Quantum Field Theory*, edited by G. Velo and A. S. Wightman (Springer, Berlin), p. 94.
Niemeijer, Th. and J. M. J. van Leeuwen, 1973, Phys. Rev. Lett. **31**, 1411.
Niemeijer, Th. and J. M. J. van Leeuwen, 1974, Physica

(Utrecht) **71**, 17.
Osterwalder, K. and R. Schrader, 1973a, Commun. Math. Phys. **31**, 83.
Osterwalder, K. and R. Schrader, 1973, in *Lecture Notes in Physics*, edited by G. Velo and A. S. Wightman (Springer, Berlin).
Osterwalder, K. and R. Schrader, 1975, Commun. Math. Phys. **42**, 281.
Polyakov, A. M., 1975, Phys. Lett. B **59**, 79.
Schwinger, J., 1962, Phys. Rev. **128**, 2425.
Simon, B., 1974, *The* $p(\phi)_2$ *Euclideian* (*Quantum*) *Field Theory* (Princeton University, Princeton, N.J.).
Symanzik, K., 1971, Commun. Math. Phys. **23**, 49.
Thirring, W., 1958, Ann. Phys. (N.Y.) **3**, 91.
't Hooft, G., 1972, in Proceedings of Marseilles Conference on Gauge Theories (unpublished).
't Hooft, G., 1974, Nucl. Phys. B **72**, 461.
Wegner, F., 1971a, J. Math. Phys. **12**, 2259.
Wegner, F., 1971b, Z. Phys. **206**, 465.
Wegner, F., 1972, Phys. Rev. B **5**, 4529.
Wegner, F., 1973, Lecture Notes in Physics **37**, 171.
Weinberg, S., 1973, Phys. Rev. Lett. **31**, 494.
Wilson, K. G., 1970, Phys. Rev. D **2**, 1438.
Wilson, K. G., 1971, Phys. Rev. B **4**, 3174, 3184.
Wilson, K. G., 1972, Phys. Rev. Lett. **28**, 548.
Wilson, K. G., 1974, Phys. Rev. D **10**, 2445.
Wilson, K. G., 1975a, "Quarks and Strings on a Lattice," Erice Lecture Notes.
Wilson, K. G., 1975b, Rev. Mod. Phys. **47**, 773.
Wilson, K. G., 1976a, Phys. Rep. **23C**, 331.
Wilson, K. G., 1976b, "Relativistically Invariant Lattice Theories," presented at Coral Gables Conference.
Wilson, K. G. and M. E. Fisher, 1972, Phys. Rev. Lett. **28**, 240.
Wilson, K. G. and J. Kogut, 1974, Phys. Rep. **12C**, 75.
Zittartz, J., 1976, Z. Phys. B **23**, 55, 63.

Nuclear Physics B170 [FS1] (1980) 1–15

DISORDER VARIABLES AND PARA-FERMIONS IN TWO-DIMENSIONAL STATISTICAL MECHANICS

Eduardo FRADKIN

Department of Physics, University of Illinois at Urbana-Champaign, Urbana, IL 61801, USA

Leo P. KADANOFF

The James Franck Institute, 5640 S. Ellis Ave., The University of Chicago, Chicago, IL 60637, USA

Received 4 March 1980

It is shown that "clock" type models in two-dimensional statistical mechanics possess order and disorder variables ϕ_n and χ_m with n and m falling in the range $1, 2, \ldots, p$. These variables respectively describe abelian analogs to charged fields and the fields of 't Hooft monopoles with charges $q = n/p$ and topological quantum number m. They are related to one another by a dual symmetry. Products of these operators generate, via a short-distance expansion, para-fermion operators in which rotational symmetry and the internal symmetry group are tied together. The clock models in two dimensions are shown to be an ideal laboratory where these ideas have a very simple realization.

1. Introduction

It has become quite commonplace for concepts to move up and back between statistical physics and field theory. This paper is concerned with elaborating an example from statistical physics which might perhaps illuminate in a simple context some ideas which have been employed in particle physics. In particular, we study some fields which appear (at least) superficially similar to those describing (fractionally) charged particles and topological excitations like 't Hooft monopoles [1,2]. The underlying symmetry groups are abelian, and the context is a two-dimensional lattice statistical mechanics model. In this relatively simple context, we can make much progress in analyzing the correlations among the basic fields.

The example will be described in two contexts. The first, the gaussian model, can be solved quite completely to give all correlations among the interesting fields. The second, the clock model, is a generalization of the Ising model to a case in which the basic variable take on, instead of the two values of the Ising variables, p different values. We shall show that the same basic operators appear in the two contexts and play closely analogous roles. We also describe a continuous transformation which takes one from the solvable gaussian model to the "realistic"

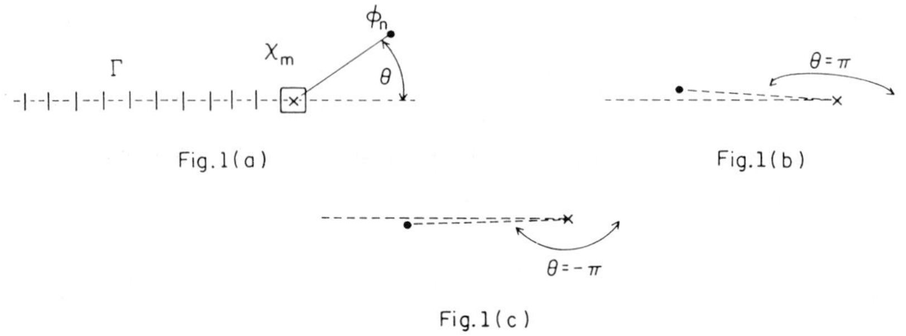

Fig. 1. (a) A disorder operator χ_m and an order operator ϕ_n. The disorder operator resides at the plaquette with the cross. The angle θ measures their relative position. The seam of vertical segments represents the set of bonds whose interactions is shifted by $2\pi m/p$. (b, c) Definition of the range of the angle θ.

statistical model. In this way, we argue that the properties found in the gaussian model are likely to reappear in the "realistic" model and that both models are likely to have counterparts in two-dimensional field theories.

The basic properties of the models are described here and then developed in detail in the succeeding sections. They are as follows.

(i) The models contain fields $\phi_n(\boldsymbol{r})$ $(n = 1, 2, \ldots, p)$ and a global abelian symmetry under which the fields transform as

$$\phi_n(\boldsymbol{r}) \to \phi_n(\boldsymbol{r})\, \mathrm{e}^{i\alpha_m n}\,, \tag{1.1}$$

where α_m is $(2\pi/p)$ times an integer, $m = 1, 2, \ldots, p$, describing the transformation.

(ii) From this symmetry there arise a set of dual fields $\chi_m(\boldsymbol{r})$, with m playing the role of a vortex quantum number of topological charge. There is a dual transformation—which leaves the hamiltonian invariant in form—which interchanges ϕ_n and χ_m. The field $\chi_m(\boldsymbol{r})$ may be identified as a field for topological excitations by the following operation. Take a correlation function containing a field $\phi_n(\boldsymbol{r}_1)$ and also $\chi_m(\boldsymbol{r}_2)$. Continuously vary $\boldsymbol{r}_1$ (see fig. 1) so that it moves in a complete circle about $\boldsymbol{r}_2$. Then the correlation function returns to its former value, except that it is multiplied by a phase factor $\exp(i\alpha_m n)$*. Hence this circuit effectively produces the symmetry operation of eq. (1.1).

(iii) Since the ϕ_n and χ_m contain the group symmetry in a very explicit fashion, their product generates, via the operator product expansion [4, 5] new operators $\psi_{nm}(\boldsymbol{r})$ which are generalized spinors or para-fermion variables. These operators are joint representations of the rotation symmetry of the two-dimensional space and also of the internal symmetry group. In particular, under 360° rotation these

* A similar situation was found in ref. [3] in the context of the two-dimensional Ising model.

operators transform as

$$\psi_{nm}(\boldsymbol{r}) \rightarrow e^{i\alpha_m n}\psi_{nm}(\boldsymbol{r}). \tag{1.2}$$

Hence they have an apparent angular momentum quantum number, which is given by

$$l_z = -2\pi \frac{mn}{p}. \tag{1.3}$$

Thus, for example, we shall see that a correlation function formed from two of these operators has an angular dependence which is given by

$$\langle \psi_{nm}(\boldsymbol{r}) \psi_{-n-m}(\boldsymbol{0}) \rangle \sim \left(\frac{x+iy}{x-iy} \right)^{i\alpha_m n}. \tag{1.4}$$

To develop these properties we use the following strategy. In sect. 2 we introduce fields $\phi_n(\boldsymbol{r})$ and $\chi_m(\boldsymbol{r})$ in the context of a trivially solvable gaussian model. In this context we develop explicit formulas for correlations among arbitrarily large numbers of ϕ_n's and χ_m's. Then we generalize the gaussian model by adding large numbers of excitations in which n/p and m/p are integral. The number of these excitations are controlled by two fugacities y_0 and y_p. In sect. 3, we show that the generalized model reduces at $y_0 = y_p = 1$ to the "clock model" or planar Potts model, which is one of the standard $d=2$ statistical models. The fields ϕ_n and χ_m are then shown to have natural generalizations in this context and also very natural physical interpretations. Finally, in sect. 4 we return to the gaussian model to gain information about operator product expansions and also explicit forms for the correlations among an arbitrary number of χ, ϕ and ψ fields.

2. The gaussian model

Consider a two-dimensional square lattice and a variable $\xi(\boldsymbol{r})$ at each site. The range of ξ is $[-\infty, +\infty]$. The gaussian model* is a system of such variables with a hamiltonian

$$\mathcal{H}_{\text{gaussian}} = \tfrac{1}{2} K \sum_{(\boldsymbol{r}\mu)} \left(\xi(\boldsymbol{r} + \hat{e}_\mu) - \xi(\boldsymbol{r}) \right)^2. \tag{2.1}$$

The partition function for the gaussian model is given by

$$Z_{\text{gaussian}} = \prod_{\boldsymbol{r}} \int_{-\infty}^{+\infty} d\xi(\boldsymbol{r}) \exp\left[-\mathcal{H}_{\text{gaussian}} \right]. \tag{2.2}$$

* We use the definition of the gaussian model as given in ref. [6].

Alternatively we can represent the gaussian model by a set of angle-like site variables $\theta(\boldsymbol{r})$ $(0 \leqslant \theta < 2\pi)$ and integer-valued link variables $l_\mu(\boldsymbol{r})$. In terms of these variables the partition function (2.2) reads [7]

$$Z_{\text{gaussian}} = \prod_{\boldsymbol{r}} \int_0^{2\pi} \frac{\mathrm{d}\theta(\boldsymbol{r})}{2\pi} \prod_{(\boldsymbol{r},\mu)} \sum_{l_\mu(\boldsymbol{r})=-\infty}^{+\infty} \left(\prod_{\boldsymbol{r}} \delta_{\varepsilon_{\mu\nu}\Delta_\nu l_\mu(\boldsymbol{r}),0} \right)$$
$$\times \exp\left\{ -\tfrac{1}{2}K \sum_{(\boldsymbol{r},\mu)} \left(\Delta_\mu\theta(\boldsymbol{r}) - 2\pi l_\mu(\boldsymbol{r})\right)^2 \right\}, \qquad (2.3)$$

where we have used the compact notation

$$\Delta_\mu\theta(\boldsymbol{r}) \equiv \theta(\boldsymbol{r}+\hat{e}_\mu) - \theta(\boldsymbol{r}),$$

$$\varepsilon_{\mu\nu}\Delta_\nu l_\mu(\boldsymbol{r}) \equiv l_\mu(\boldsymbol{r}) + l_\nu(\boldsymbol{r}+\hat{e}_\mu) - l_\mu(\boldsymbol{r}+\hat{e}_\nu) - l_\nu(\boldsymbol{r}), \qquad (2.4)$$

for the lattice gradient and the lattice curl, respectively.

It is trivial to compute both the partition function (2.2) and also the correlation function

$$\left\langle \left(\prod_{i=1}^{N_n} \phi_{n_i}(\boldsymbol{r}_i) \right) \left(\prod_{n=1}^{N_m} \chi_{m_j}(\boldsymbol{R}_j) \right) \right\rangle = \frac{Z[\{n\};\{m\}]}{Z_{\text{gaussian}}}, \qquad (2.5)$$

where

$$Z[\{n\};\{m\}] = \prod_{\boldsymbol{r}} \int_0^{2\pi} \frac{\mathrm{d}\theta(\boldsymbol{r})}{2\pi} \prod_{(\boldsymbol{r},\mu)} \sum_{l_\mu(\boldsymbol{r})=-\infty}^{+\infty} \left(\prod_{j=1}^{N_m} \delta_{\varepsilon_{\mu\nu}\Delta_\nu l_\mu(\boldsymbol{r}),m_j/p} \right)$$
$$\times \left(\prod_{i=1}^{N_n} \exp[i n_i \theta(\boldsymbol{r}_i)] \right) \times \exp\left\{ -\tfrac{1}{2}K \sum_{(\boldsymbol{r},\mu)} \left(\Delta_\mu\theta(\boldsymbol{r}) - 2\pi l_\mu(\boldsymbol{r})\right)^2 \right\}, \qquad (2.6)$$

whenever $m_i = 0 \pmod{p}$. If $m_i \neq 0 \pmod{p}$, $Z[nm]$ vanishes identically. Here $\boldsymbol{R}$ labels the sites of the dual lattice.

If we want the system to have a non-integer curl, m/p, at some point we must shift the l_μ variables residing at links crossed by a path like the one shown in fig. 1 by an amount equal to m/p. Once a path is chosen it is possible to compute the correlation function (2.5). The result is

$$\left\langle\left(\prod_{i=1}^{N_n}\phi_{n_i}(\boldsymbol{r}_i)\right)\left(\prod_{j=1}^{N_m}\chi_{m_j}(\boldsymbol{R}_j)\right)\right\rangle=\left(\prod_{i<j=1}^{N_n}\frac{1}{|\boldsymbol{r}_i-\boldsymbol{r}_j|^{n_in_j/2\pi K}}\right)$$
$$\times\left(\prod_{i<j=1}^{N_m}\frac{1}{|\boldsymbol{R}_i-\boldsymbol{R}_j|^{m_im_j2\pi K}}\right)$$
$$\times\left(\prod_{i=1}^{N_n}\prod_{j=1}^{N_m}\exp\left[in_im_j\theta(\boldsymbol{r}_i-\boldsymbol{R}_j)\right]\right), \qquad (2.7)$$

when the constraints

$$\sum_{i=1}^{N_n} n_i=0\,, \qquad \sum_{j=1}^{N_m} m_j=0 \qquad (2.8)$$

are satisfied. This correlation function is equal to zero otherwise. The angle $\theta(\boldsymbol{r}_i-\boldsymbol{R}_j)$, specifies the relative orientation of a charge n_i at $\boldsymbol{r}_i$ and a topological charge m_j at $\boldsymbol{R}_j$ and lies in the range $(-\pi,\pi)$ with the conventions shown in fig. 1. Notice that the angles are measured with a cut along the "path of shifted bonds" Γ. Thus each time ϕ_n crosses the path, it picks up a phase

$$\phi_n\to\phi_n\,\mathrm{e}^{i(2\pi/p)nm}\,. \qquad (2.9)$$

This shows that χ_m is a ladder operator which increases the phase by $2\pi m/p$.

Until now we have a gaussian model, which contains only symmetry breaking (or spin wave) excitations and fixed topological charges created by χ_m. Let us define a generalized gaussian model [6] whose excitations are spin waves, charges N and topological charges M. It is defined by the partition function

$$Z_{y_n,y_m,p}=\sum_{\{N_i\}}\sum_{\{M_j\}}Z[pN;pM]\,y_n^{\Sigma_iN_i^2}y_m^{\Sigma_jM_j^2}, \qquad (2.10)$$

where $Z[pN;\,pM]$ is given by eq. (2.6). The correlation functions of this model are

$$\left\langle\prod_i\phi_{n_i}(\boldsymbol{r}_i)\prod_j\chi_{m_j}(\boldsymbol{R}_j)\right\rangle=\frac{Z_{y_n,y_m,p}[\{n_i\},\{m_j\}]}{Z_{y_n,y_m,p}}\,, \qquad (2.11)$$

$$Z_{y_n,\,y_m,\,p}[\{n_i\},\,\{m_j\}]=\sum_{\{N_i\}}\sum_{\{M_j\}}Z[n+pN;\;m+pM]\,y_n^{\Sigma_iN_i^2}y_m^{\Sigma_jM_j^2}\,. \qquad (2.12)$$

Here the fugacities y_n and y_m control the number of charges N and topological charges M present in the system. It is important that the path-crossing property mentioned above is maintained in this generalized model. The quantum number m may be considered to be a textural singularity present in the system such that a complete circuit of ϕ_n about m is a symmetry operation of the system. In the language of spin systems, like the planar model, the textural singularity m is the endpoint of a domain wall (i.e., a magnetic dislocation). This wall favors a jump in the orientation of the spins equal to $2\pi m/p$.

3. The clock model

The clock model is a discrete version of the XY model in which the spin can point only along p different directions. The symmetry group of the clock model is the cyclic group of p elements Z_p. The partition function for the clock model is given by

$$Z_p^{\text{clock}} = \sum_{\{\theta(\boldsymbol{r})\}} \sum_{\{l_\mu(\boldsymbol{r})\}} \exp\left\{ -\tfrac{1}{2}K \sum_{(\boldsymbol{r},\mu)} \left(\Delta_\mu\theta(\boldsymbol{r}) - 2\pi l_\mu(\boldsymbol{r})\right)^2 \right\}, \tag{3.1}$$

where $\theta(\boldsymbol{r})$ is the (discrete) angle

$$\theta(\boldsymbol{r}) = \frac{2\pi}{p}\sigma(\boldsymbol{r}), \qquad \sigma = 0, \ldots, p-1. \tag{3.2}$$

The partition function of the p-state clock model relates very simply to that of the generalized gaussian model [7] (2.8)

$$Z_p^{\text{clock}} = \lim_{y_n, y_m \to 1} Z_{y_n, y_m}^{\text{gaussian}}. \tag{3.3}$$

The correlation functions of both models are also related through a similar expression.

As we have already pointed out in sect. 1, two kinds of operators can be defined in the clock model. Both contain the symmetry in a very explicit manner. The first type are the symmetry-breaking operators $\phi_n(\boldsymbol{r}) = \exp[in\theta(\boldsymbol{r})]$ which are order variables. The second kind, $\chi_m(\boldsymbol{R})$, are the disorder variables [3] and create vortex-like excitations. The disorder variable $\chi_m(\boldsymbol{R})$ located at a dual site $\boldsymbol{R}$, is introduced in the clock model by shifting the integer-valued bond variables $l_\mu(\boldsymbol{r})$ in (3.1) by a fractional amount $(q_\mu/p)(\boldsymbol{r})$ along a path Γ (see fig. 2). The integer-valued variables $q_\mu(\boldsymbol{r})$ satisfy the constraint

$$\varepsilon_{\mu\nu}\Delta_\nu q_\mu(\boldsymbol{r}) = m(\boldsymbol{R}). \tag{3.4}$$

Thus, the presence of disorder variables $\chi_m(\boldsymbol{R})$ introduces in the system fractional

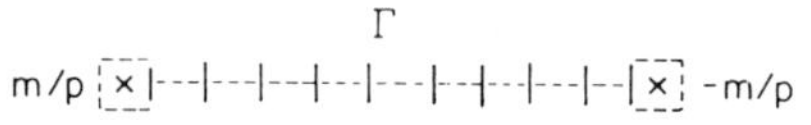

Fig. 2. Two disorder operator with topological charge m/p.

(a) (b)

Fig. 3. The order operator ϕ_n loops around the disorder operator χ_m by a 360° rotation. Note that the initial and final configurations differ just by the relative position of ϕ_n and the string Γ.

domain walls that favor a discontinuity in the orientation of the spins by an angle equal to $(2\pi/p)m$. If, for instance, two such disorder variables are present at dual sites **0** and $\boldsymbol{R}$, the domain wall will begin and end at these points. Notice by the way that, since this is a p-state model, the charge n of the order variable and the topological charge m of the disorder variable are only defined modulo p.

The clock model has the important property of self-duality*. Under a dual transformation the coupling constant K transforms like

$$\tilde{K} = \left(\frac{p}{2\pi}\right)^2 / K, \tag{3.5}$$

while order and disorder operators are transformed into each other: $\phi_n \leftrightarrow \chi_n$. For $p \leqslant 4$ (i.e., $p = 2, 3, 4$) the clock models have a continuous phase transition [6, 8] at the self-dual coupling variable $K = \tilde{K} = p/2\pi$. On the other hand, for $p > 4$ these models are supposed to have two phase transitions of a Kosterlitz-Thouless character at critical coupling values $K_{\max}(p)$ and $K_{\min}(p)$ related by eq. (3.5)

$$2\pi K_{\max}(p) = \frac{p^2}{2\pi K_{\min}(p)}. \tag{3.5'}$$

Let us show now that the fields ϕ_n and χ_m obey commutation relations like those pointed out in sect. 1. First of all let us see that the correlation function $\langle \Pi_r \phi_n(\boldsymbol{r}) \Pi_R \chi_m(\boldsymbol{R}) \rangle$ has a phase ambiguity as it did in the gaussian model. Consider a disorder variable $\chi_m(\boldsymbol{R})$ residing at the dual site $\boldsymbol{R}$ and its string of shifted bonds along a path Γ going to the left of $\boldsymbol{R}$. Consider now an order variable $\phi(\boldsymbol{r})$ at a site $\boldsymbol{r}$ neighbor to the path Γ but just under it. Imagine now performing a 360° rotation of the field $\phi_n(\boldsymbol{r})$ about the position of the disorder field $\chi_m(\boldsymbol{R})$ in a way that ends up just above the path Γ (fig. 3). Clearly the situation is almost

* Self-dual in the Kramers-Wannier sense. See ref. [9].

identical to the original one except for the fact that the path Γ is "misplaced" relative to ϕ_n. This problem can be easily solved just by performing a gauge transformation right at the location of ϕ_n as shown in fig. 2:

$$\phi_n(\boldsymbol{r}) \to \phi_n(\boldsymbol{r}) \exp\left[i \frac{2\pi}{p} mn \right],$$

$$q_\mu(\boldsymbol{r}) \to q_\mu(\boldsymbol{r}) + \Delta_\mu \alpha(\boldsymbol{r}),$$

$$\theta(\boldsymbol{r}) \to \theta(\boldsymbol{r}) + \frac{2\pi}{p} m, \tag{3.6}$$

where

$$\alpha(\boldsymbol{r}') = \begin{cases} m, & \boldsymbol{r}' \neq \boldsymbol{r}, \\ 0, & \text{otherwise.} \end{cases}$$

Thus after a 360° rotation the field ϕ_n has suffered a phase shift by $(2\pi/p)nm$ and its phase is thus ambiguous. This is so since ϕ_n represent a fractional charge equal to n/p residing at $\boldsymbol{r}$ and transforms like the nth irreducible representation of the Z_p group. Analogously if the disorder variable is rotated around the order variable we also end up with a phase shift. A way to understand this phase ambiguity is to say that the product of an order operator times a disorder operator requires an ordering prescription. This is precisely what we have done in fig. 1. Each time an order variable crosses a string attached to a disorder variable, it picks up a phase. Thus an order variable $\phi_n(\boldsymbol{r})$ and a disorder variable $\chi_m(\boldsymbol{R})$ do not commute with each other since the expectation value of products of these operators before and after a complete rotation by 360° differs by a phase $\exp(i(2\pi/p)mn)$.

We can summarize the result by stating that the fields ϕ_n and χ_m obey the commutation relations

$$\phi_n(\boldsymbol{r})\chi_m(\boldsymbol{R}) = \begin{cases} \exp\left[i \frac{2\pi}{p} mn \right] \chi_m(\boldsymbol{R})\phi_n(\boldsymbol{r}), & \text{(a)} \\ \chi_m(\boldsymbol{R})\phi_n(\boldsymbol{r}), & \text{(b)} \end{cases} \tag{3.7}$$

where alternative (a) is realized if the order variable $\phi_n(\boldsymbol{r})$ performs a complete rotation around the disorder variable $\chi_m(\boldsymbol{R})$, and alternative (b) is realized otherwise.

It is interesting to note that 't Hooft has found identical commutation relations in non-abelian gauge theories between Wilson loop operators and 't Hooft disorder operators [10]. Indeed this is not surprising since the 't Hooft commutation relations reflect the fact that the abelian group Z_p is the center of the non-abelian group $SU(p)$. More remarkable, however, is the close analogy between our disorder variables and the 't Hooft operators. These operators are like disorder operators in the sense that their behavior is dual of that of the Wilson loops [10]. Furthermore, the 't Hooft operators create currents of fractionally charged 't Hooft-Polyakov magnetic monopoles in the same manner than our disorder operators create fractional topological charges (vortices). In fact, the path of shifted bonds Γ is completely analogous to the Dirac string that is attached to the magnetic monopoles.

Let us consider now, in the framework of the clock model, the behavior of the composite operator $\phi_n(\boldsymbol{r})\chi_m(\boldsymbol{R})$, where $\boldsymbol{r}$ and $\boldsymbol{R}$ are nearby positions. It is clear that since both ϕ_n and χ_m contain the internal symmetry in a very explicit manner their product generates, via an operator product expansion, a set of new local operators $\psi_{n,m}(\boldsymbol{r})$ which are para-fermion variables. This can be seen very easily just by rotating an operator $\psi_{n,m}(\boldsymbol{r})$ around another $\psi_{n',m'}(\boldsymbol{r}')$ at $\boldsymbol{r}'$. The commutation relations (3.7) imply that the ψ's themselves satisfy

$$\psi_{n,m}(\boldsymbol{r})\psi_{n',m'}(\boldsymbol{r}') = \begin{cases} \exp\left\{i\dfrac{2\pi}{p}mn'\right\}\psi_{n',m'}(\boldsymbol{r}')\psi_{n,m}(\boldsymbol{r}), & \text{(a)} \\ \exp\left\{i\dfrac{2\pi}{p}m'n\right\}\psi_{n',m'}(\boldsymbol{r}')\psi_{n,m}(\boldsymbol{r}), & \text{(b)} \\ \psi_{n',m'}(\boldsymbol{r}')\psi_{n,m}(\boldsymbol{r}), & \text{(c)} \end{cases} \tag{3.8}$$

where in (a) $\psi_{n,m}$ performs a closed path around $\psi_{n',m'}$, (b) $\psi_{n',m'}$ performs a closed path around $\psi_{n,m}$ and (c) otherwise. As in eq. (3.7) these anomalous commutation relations originate in the fact that after a 360° rotation it is necessary to perform a gauge transformation to recover the original path configuration. Since the order variable $\phi_n(\boldsymbol{r})$ is not gauge invariant, a phase factor $\exp[i(2\pi/p)nm']$ arises. Therefore the composite operators $\psi_{n,m}(\boldsymbol{r})$ are joint representations of the (euclidean) rotations group in two-dimensional space and the internal symmetry group Z_p. In particular under at 360° rotation the operators $\psi_{n,m}$ transform as

$$\psi_{n,m}(\boldsymbol{r}) \to \psi_{n,m}(\boldsymbol{r}) \exp\left[i\frac{2\pi}{p}nm\right]. \tag{3.9}$$

Hence the operator $\psi_{n,m}$ creates an excitation that carries an internal quantum number l_z

$$l_z = -\frac{nm}{p} \tag{3.10}$$

which plays the role of an effective angular momentum.

4. Operator product expansion

To gain further insight into the properties of the para-fermion operators $\psi_{n,m}(\boldsymbol{r})$ we must give a local definition of them. This will be done by means of the operator product expansion [4, 5].

The basic philosophy behind the operator product expansion is that if the distance between the operators $\phi_n(\boldsymbol{r})$ and $\chi_m(\boldsymbol{R})$ is much smaller than the correlation length (although much bigger than the lattice spacing) then their product defines a local operator $\psi_{nm}(\boldsymbol{r})$:

$$\phi_n(\boldsymbol{r})\chi_m(\boldsymbol{R}) = C_{n,m}\left(\frac{\boldsymbol{R}-\boldsymbol{r}}{2}\right)\psi_{n,m}(\boldsymbol{r}) + \cdots . \tag{4.1}$$

It is very easy to evaluate the coefficients $C_{n,m}(\boldsymbol{r})$ in the gaussian model, but what we really want is to have an expansion like (4.1) for the clock model.

So what we would like to do is to understand how the clock model (described by $y_n = y_m = 1$) is related to the gaussian model (described by $y_n = y_m = 0$). Whenever these models lie in the same universality class, i.e., have the same critical behavior, the gaussian model analysis will serve to give us considerable additional information about the clock models. Thus, to make the connection between the two models we should investigate what happens when y_n and y_m are increased bit by bit, starting from the gaussian case in which both are zero.

The behavior is different depending upon whether $p > 4$, $p = 4$ or $p < 4$ [6]. For $p > 4$ there exists a range of gaussian model couplings, K_G,

$$4 < 2\pi K_G < \tfrac{1}{4}p^2 , \tag{4.2}$$

for which an infinitesimal increase in y_n and y_m will not change the structure of gaussian-model correlations. Technically speaking the corresponding excitations are irrelevant in the gaussian model. It is believed that this behavior persists in the generalized gaussian model for which $0 < y_n, y_m < 1$. Hence, the generalized model serves to give a smooth interpolation between the clock model and the essentially equivalent behavior of the gaussian model. That is, there exists a range of clock-model couplings between a minimum coupling $K_{\min}(p)$ and a maximum coupling $K_{\max}(p)$ such that if the clock-model coupling lies in the range

$$K_{\min}(p) < K_{\text{clock}} < K_{\max}(p), \tag{4.3}$$

then there exists a mapping between the gaussian model and the clock model,

$$K_G = F(K_{\text{clock}}, p), \tag{4.4}$$

such that when K_G and K_{clock} are related by eq. (4.4) the correlation functions of the two models are identical. In particular, the beginnings and ends of the two ranges (4.2) and (4.3) are connected by eq. (4.4).

At $p = 4$, according to eq. (4.2), the range of validity of the relationship between the two models shrink to the point $2\pi K_G = 4$. For $p < 4$, i.e., p equal 2 or 3, there is no region of validity of the mapping (4.4). That is to say, the gaussian model is never stable against $n = p$ and $m = p$ excitations. On the other hand, in the region

$$4 > 2\pi K_G > \tfrac{1}{4}p^2 , \tag{4.5}$$

the gaussian model is stable against these excitations.

Finally, for $p = 4$, there are a rich variety of connections between these clock-type models, called the Ashkin-Teller model or the 8-vertex model and gaussian model (see, e.g., Kadanoff and Brown [11], Luther and Peschel [12] and Coleman [13]). The exact connections are too complex and indeed too imperfectly known to be detailed here. The net result is a simple one: the gaussian model agrees with different aspects of clock-type models for all p so that a treatment of the gaussian case can give considerable insight into the behavior of more realistic models.

Thus, we return to eq. (4.1), which can be written for the gaussian model in the form

$$\phi_n(\boldsymbol{r})\chi_m(\boldsymbol{R}) = \exp\left[i\frac{nm}{p}\alpha(\boldsymbol{\delta})\right]\psi_{n,m}(\mathbf{r}) , \tag{4.6}$$

where $\boldsymbol{\delta} = \frac{1}{2}(\boldsymbol{R} - \boldsymbol{r})$ and $\alpha(\boldsymbol{\delta})$ is the angular position of the order variable relative to the disorder variable measured according to the convention shown in fig. 1.

Furthermore it is possible to compute the correlation function between $\psi_{n,m}$ variables:

$$G_{nm}(\boldsymbol{r}_2 - \boldsymbol{r}_1) = \langle \psi_{n,m}(\boldsymbol{r}_1)\psi_{-n,-m}(\boldsymbol{r}_2)\rangle = \frac{\exp[-i(2nm/p)\theta]}{|\boldsymbol{r}_2 - \boldsymbol{r}_1|^{2x_{n,m}}} , \tag{4.7}$$

where

$$2x_{n,m} = \frac{n^2}{2\pi K_G} + \frac{m^2}{p^2}2\pi K_G , \tag{4.8}$$

where K_G and K_{clock} are related by eq. (4.4) and θ is the angle shown in fig. 1. Thus we find that the correlation function $G_{n,m}$ has an explicit angular dependence. In particular, if we commute to para-fermion operators (i.e., $\theta = \pi$) the phase of the correlation function $G_{n,m}$ jumps by $2\pi nm/p$. Thus,

$$\langle \psi_{n,m}(\boldsymbol{r}_1)\psi_{-n,-m}(\boldsymbol{r}_2)\rangle = e^{-i2\pi nm/p}\langle \psi_{-n,m}(\boldsymbol{r}_2)\psi_{n,m}(\boldsymbol{r}_1)\rangle . \tag{4.9}$$

Therefore the operators $\psi_{n,m}(\boldsymbol{r}_1)$ do obey parastatistics.

It is also easy to get the following results:

$$\langle\psi_{n,m}(r_1)\psi_{-n,-m}(r_2)\rangle=\langle\psi_{-n,-m}(r_1)\psi_{n,m}(r_2)\rangle\,, \tag{4.10a}$$

$$\langle\psi_{n,-m}(r_1)\psi_{-n,m}(r_2)\rangle=\langle\psi_{-n,m}(r_1)\psi_{n,-m}(r_2)\rangle\,, \tag{4.10b}$$

$$\langle\psi_{n,-m}(r_1)\psi_{-n,m}(r_2)\rangle=\frac{\exp[+i(2nm/p)\theta]}{|r_2-r_1|^{2x_{n,m}}}\,, \tag{4.10c}$$

where θ and $x_{n,m}$ are the *same* quantities defined above. It is convenient to introduce a canonical* set of para-fermion operators whose correlations are real and simple. Two such canonical combinations are

$$\begin{aligned}\Psi^{(0)}_{n,m}(r)&=\tfrac{1}{2}[\psi_{n,m}(r)+\psi_{n,-m}(r)+\psi_{-n,-m}(r)+\psi_{n,m}(r)]\,,\\ \Psi^{(+)}_{n,m}(r)&=\tfrac{1}{2}e^{-i\pi/4}[\psi_{n,m}(r)+i\psi_{n,-m}(r)-\psi_{-n,-m}(r)-i\psi_{-n,m}(r)]\,.\end{aligned} \tag{4.11}$$

There are other combinations equally canonical but we will deal only with these two. The correlations between the canonical operators Ψ^0 and $\Psi^{(+)}$ are easy to compute:

$$\langle\Psi^{(0)}_{n,m}(r_1)\Psi^{(0)}_{n,m}(r_2)\rangle=\frac{\cos((2nm/p)\theta)}{|r_2-r_1|^{2x_{n,m}}}\,, \tag{4.12a}$$

$$\langle\Psi^{(+)}_{n,m}(r_1)\Psi^{(+)}_{n,m}(r_2)\rangle=\frac{\sin((2nm/p)\theta)}{|r_2-r_1|^{2x_{n,m}}}\,, \tag{4.12b}$$

$$\langle\Psi^{(0)}_{n,m}(r_1)\Psi^{(+)}_{n,m}(r_2)\rangle=0\,. \tag{4.12c}$$

Hence different canonical operators do not correlate with each other. Moreover, eqs. (4.12) have some properties worthwhile mentioning. First it is clear that different species of canonical para-fermions (i.e., Ψ^0, $\Psi^{(+)}$ and $\Psi^{(-)}$) do not correlate (4.12c). Moreover, for some geometries even operators of the same type are not correlated. This happens if

$$\theta=\frac{(2s+1)}{nm}\frac{1}{4}p\pi,\qquad s=0.1,\ldots\to\langle\Psi^{(0)}_{n,m}(1)\Psi^{(0)}_{n,m}(2)\rangle=0, \tag{4.13a}$$

and

$$\theta=\frac{sp\pi}{2nm},\qquad s=0,1,\ldots\to\langle\Psi^{(+)}_{n,m}(1)\Psi^{(+)}_{-n,m}(2)\rangle=0. \tag{4.13b}$$

* This operators are canonical in the sense that their correlations are real and simple. Notice that this definition of canonical operators is rather different from that used by Kadanoff [7] and Kadanoff and Brown [11].

For the operator $\Psi^{(0)}_{n,m}$ this situation occurs whenever $\theta = (2s+1)p\pi/4nm$ (s integer), while for $\Psi^{(+)}_{n,m}$ we must have $\theta = p\pi s/2nm$ (s integer). So in the case $p=4$ the correlation function $\langle \Psi^{(0)}_{1,1}(\boldsymbol{r}_1)\Psi^{(0)}_{1,1}(\boldsymbol{r}_2)\rangle$ vanishes if $\theta = \pi$. This means that $\boldsymbol{r}_2$ lies to the left of $\boldsymbol{r}_1$ along the x axis. Analogously, $\langle \Psi^{(+)}_{1,1}(\boldsymbol{r}_1)\Psi^{+}_{1,1}(\boldsymbol{r}_2)\rangle$ vanishes if $\theta = 0$, which is, in fact, the opposite ordering.

We can gain further insight into the properties of these operators by computing expectation values of several of them. Let us compute, as an example, the following three-point function

$$G^{(0)}_{(3)} = \langle \Psi^{(0)}_{n_1,m_1}(\boldsymbol{r}_1)\Psi^{(0)}_{n_2,m_2}(\boldsymbol{r}_2)\Psi^{(0)}_{n,m}(\boldsymbol{r}_3)\rangle\,, \tag{4.14}$$

where $n = n_1 + n_2$, $m = m_1 + m_2$ and $\{\boldsymbol{r}_1, \boldsymbol{r}_2, \boldsymbol{r}_3\}$ are three points on the plane. Using the fact that the only contributions to (4.14) come from those terms that satisfy $\Sigma^3_{i=1} n_i = \Sigma^3_{i=1} m_i = 0$ we obtain the result

$$G^{(3)}_{(0)}(\boldsymbol{r}_1, \boldsymbol{r}_2, \boldsymbol{r}_3) = \frac{1}{2}\frac{|\boldsymbol{r}_2 - \boldsymbol{r}_1|^{\alpha_{21}}}{|\boldsymbol{r}_3 - \boldsymbol{r}_1|^{\alpha_{31}}|\boldsymbol{r}_3 - \boldsymbol{r}_2|^{\alpha_{32}}}\cos\phi_{123}\,, \tag{4.15}$$

where

$$\alpha_{21} = \frac{n_1 n_2}{2\pi K_G} + \frac{m_1 m_2}{p^2} 2\pi K_G\,, \tag{4.16a}$$

$$\alpha_{31} = \frac{n n_1}{2\pi K_G} + \frac{m m_1}{p^2} 2\pi K_G\,, \tag{4.16b}$$

$$\alpha_{32} = \frac{n n_2}{2\pi K_G} + \frac{m m_2}{p^2} 2\pi K_G\,, \tag{4.16c}$$

$$\phi_{123} = \left(\frac{n_1 m_2 + n_2 m_1}{p}\right)\theta(\boldsymbol{r}_2 - \boldsymbol{r}_1) - \left(\frac{n_1 m + n m_1}{p}\right)\theta(\boldsymbol{r}_3 - \boldsymbol{r}_1)$$
$$- \left(\frac{n_2 m + n m_2}{p}\right)\theta(\boldsymbol{r}_3 - \boldsymbol{r}_2)\,. \tag{4.16d}$$

In the particular case when the three points fall on a line the phase ϕ depends only on the relative ordering of the operators. Thus if the ordering is (1, 2, 3), the phase ϕ is equal to zero since all the θ's are, by definition, zero in this case. But if the ordering is (1, 3, 2), we have

$$\theta(\boldsymbol{r}_2 - \boldsymbol{r}_1) = \theta(\boldsymbol{r}_3 - \boldsymbol{r}_1) = 0\,, \qquad \theta(\boldsymbol{r}_3 - \boldsymbol{r}_2) = \pi\,,$$

$$\phi = -\frac{\pi}{p}(n_2 m + n m_2)\,. \tag{4.17}$$

Thus

$$G^{(0)}_{123} = \frac{1}{2} \frac{|r_2 - r_1|^{\alpha_{21}}}{|r_3 - r_1|^{\alpha_{31}} |r_3 - r_2|^{\alpha_{32}}},$$

$$G^{(0)}_{132} = G^{(0)}_{123} \cos\left(\frac{\pi}{p}[n_2 m + n m_2]\right). \tag{4.18}$$

In particular if $p = 4$, $n_m = n_2 = m_1 = m_2 = 1$, we have $G^{(0)}_{132} = -G^{(0)}_{123}$. However, we may also consider the sequence (2, 3, 1) with the result

$$G^{(0)}_{231} = G^{(0)}_{123} \cos\left(\frac{2n_1 m_1}{p}\pi\right), \tag{4.19}$$

and, if $p = 4$ and $m_1 = n_1 = m_2 = n_2 = 1$, we find $G^{(0)}_{231} = 0$. We also get the results

$$\langle \Psi^{(+)}_{n_1 m_1}(r_1) \Psi^{(+)}_{n_2 m_2}(r_2) \Psi^{+}_{n,m}(r_3) \rangle = 0, \tag{4.20a}$$

$$\langle \Psi^{(+)}_{n_1 m_1}(r_1) \Psi^{(0)}_{n_2 m_2}(r_2) \Psi^{(0)}_{n,m}(r_3) \rangle = -\tfrac{1}{2} G^{(0)}_{123} \sin\phi, \tag{4.20b}$$

$$\langle \Psi^{(+)}_{n_1 m_1}(r_1) \Psi^{(0)}_{n_2 m_2}(r_2) \Psi^{(+)}_{nm}(r_3) \rangle = +\tfrac{1}{2} G^{(0)}_{123} \sin\phi, \tag{4.20c}$$

where ϕ is defined by eq. (4.16d).

Finally we quote results for the four-point functions

$$\langle \Psi^{(0)}_{1,1}(r_1) \Psi^{(0)}_{1,1}(r_2) \Psi^{(0)}_{1,1}(r_3) \Psi^{(0)}_{1,1}(r_4) \rangle = G^{(0)}_{1,1}(r_1, r_2, r_3, r_4), \tag{4.21}$$

$$G^{(0)}_{1,1}(r_1, r_2, r_3, r_4) = \frac{1}{8} \left[\frac{|r_2 - r_1| \, |r_4 - r_3|}{|r_3 - r_1| \, |r_3 - r_2| \, |r_4 - r_1| \, |r_4 - r_2|} \right]^{2x_{1,1}} \cos\phi(1,2,3,4)$$

$$+ 5 \text{ permutations}, \tag{4.22}$$

where

$$\phi(1,2,3,4) = \frac{2}{p} \{ \theta(r_2 - r_1) + \theta(r_4 - r_3) - \theta(r_3 - r_1)$$

$$- \theta(r_3 - r_2) - \theta(r_4 - r_1) - \theta(r_4 - r_2) \}, \tag{4.23a}$$

$$2x_{1,1} = \frac{1}{2\pi K_G} + \frac{2\pi K_G}{p^2}. \tag{4.23b}$$

Analogously, it is possible to obtain correlations between the other canonical combinations and higher point functions as well.

This work has been supported in part by the NSF through grants at the University of Chicago and at the University of Illinois NSF DMR78-21069.

References

[1] G. 't Hooft, Nucl. Phys. B79 (1974) 276
[2] A.M. Polyakov, JETP Lett. 20 (1974) 194
[3] L.P. Kadanoff and H. Ceva, Phys. Rev. B3 (1971) 3918
[4] K.G. Wilson, Phys. Rev. 174 (1969) 1499
[5] L.P. Kadanoff, Phys. Rev. Lett. 23 (1969) 1430
[6] J. Jose, L.P. Kadanoff, S. Kirkpatrick and D. Nelson, Phys. Rev. B16 (1977) 1217
[7] L.P. Kadanoff, Ann. of Phys. 120 (1979) 39
[8] S. Elitzur, R. Pearson and J. Shigemitsu, Phys. Rev. D19 (1979) 3698
[9] L.P. Kadanoff, J. Phys. A11 (1978) 1399
[10] G. 't Hooft, Nucl. Phys. B138 (1978) 1
[11] L.P. Kadanoff and A. Brown, Ann. of Phys. 121 (1979) 318
[12] A. Luther and I. Peschel, Phys. Rev. B12 (1975) 3908
[13] S. Coleman, Phys. Rev. D11 (1975) 2088

reference frame

Computational physics: pluses and minuses

Leo P. Kadanoff

In the last 50 years, a new approach to physics has arisen almost equal in importance to the two "old" branches of theory and experiment. This new type of effort is computational physics: It involves the use of computers large and small to simulate the behavior of physical systems and to work out the consequences of physical ideas, as expressed in mathematical form.

To judge the value of any effort in science, one should of course look at the ideas it produces and the impact of those ideas upon the intellectual and technical atmosphere throughout the world. By these standards one can identify several pieces of computational physics that easily bear comparison with the best work in theory and experiment.

Enrico Fermi, John Pasta and Stanislaw Ulam shook the roots of classical mechanics by finding an unexpected recurrence phenomenon in an anharmonic chain. Nicholas Metropolis, working with Arianna and Marshall Rosenbluth and Mitzi and Edward Teller, developed Monte Carlo methods that gave us rich new insights into the behavior of complex systems and materially aided the hydrogen-bomb project. Edward Lorenz has shown us how even very simple systems may be chaotic. Berni Alder and Thomas E. Wainwright discovered long-range particle correlations produced by the hydrodynamic flows induced by the particles.

Kenneth Wilson used a large, general-purpose computer to develop and realize his theory of the Kondo effect; this analysis in condensed-matter physics served as a very important example of the concept of asymptotic freedom in particle physics. In another very beautiful and important piece of work, Mitchell Feigenbaum used a personal-sized desk-top computer to elucidate the onset of chaos in simple dynamical systems.

Cellular automata are models of physical systems consisting of many simple, linked identical parts. Several different groups are using this tool from computer science for conceptual or simulational analysis. For example, recent calculations have suggested that an appropriate automaton might be used as an effective device for numerical calculations in hydrodynamics. Our understanding of spin dynamics has been materially aided by computer simulations, a recent example being the papers of Andrew Olgielski and his coworkers. Several of these calculations have used special-purpose computers designed for and sharply aimed at specific tasks.

Very many of us, seeing this exciting work being done, have modified our own research efforts to make the best use we can of these new tools. We have employed large computers and small ones, general-purpose computers and very specialized computers. As one might expect, some of the work in the field is of mixed quality, because we're only beginning to develop appropriate standards for this new class of endeavors.

Realizing the importance of this work, government sponsors have taken a considerable interest in advancing computational research and technique. For example, NSF has instituted five major supercomputer centers, intended to bring the best people together near excellent computers and to develop new methods for using those computers effectively. They are intended to be the nucleus of a national program aimed at more effective computation.

However, I see a threat to our joint endeavors, and particularly to the university side of this program. Apparently the government has decided to use visa restrictions to close access to the class of high-speed general-purpose computers called "supercomputers." Eastern European scientists would be forbidden to use these facilities without specific permission from the government. The proposed restriction would be enforced, as I hear it, by monthly reporting of all the users of each such facility to the government.

continued on page 9

But this is the feature we are the most proud of !

reference frame

continued from page 7

This proposed rule would force a major change in the way we do business at universities. At most universities, the facilities are open to all: faculty, students and visitors. Some universities have classified facilities, which are indeed closed to most scientific users. But to my knowledge the proposed rules would bring about the first major closure of unclassified facilities in the United States. How can we best respond to this proposed change?

The answer to this question is somewhat obscured for some of us by our inability to see the national-security need for the restriction. I for one cannot believe that the USSR could possibly mount a major (or even a minor) weapons-development effort based on time borrowed from America's supercomputers.

Another confusing question is, what makes supercomputers so special? These research tools are simply the latest in a long line of computers, each substantially more powerful than the last. If these devices are potentially useful to the developers of military equipment, so too is much of scientific instrumentation. Should we not expect to have soon similar restrictions placed upon the use of each and every piece of laboratory equipment, from a Macintosh to an oscilloscope?

Thus one might expect the restriction of supercomputers to be only the first step in a long process of closing off a substantial part of previously unclassified US science. This science is now open for a reason: Open science is more effective and more productive. If we close US science we risk a considerable loss in its quality.

In my view the open part of the university community should not participate in this closure process. The intellectual freedom we require to be effective scientists cannot survive in an atmosphere in which access to ideas and discussions is restricted. We need a system that sharply differentiates between those facts, ideas and facilities that are open and those that are restricted and closed. To maintain such a system, we should apply to closed facilities the same rules we apply to other classified facilities. Those institutions that feel it necessary to remain open should not offer institutional support to using or gaining access to these closed projects.

Computational physics has given us an exciting new technique for understanding nature. It would be ironic and unfortunate if it also gave us, through restrictions on supercomputers, a whole new technique for restricting and limiting the search for knowledge. □

reference frame

Cathedrals and other edifices

Leo P. Kadanoff

One argument often used to justify society's support of pure science is that contemporary science is producing great and enduring structures that will be passed on to future generations as a major portion of the legacy of our age. The analogy to medieval cathedral building is frequently pressed. In this column, I point to a few of the results of physics that are likely to endure and remain important, not to just a few specialists in physical law, but to people in general.

One lasting product is the grand idea that the entire universe is governed by a few simple laws and that these laws are within human understanding. One beginning for this point of view is the Newtonian insight that apples and moons obey the very same laws of force and acceleration. This start has led to our major effort to derive the most fundamental laws that govern all parts of the observable universe. Bit by bit, we have seen the unification of electricity and magnetism, of space and time, of quantum theory and electromagnetism, of these and the weak interactions, and now perhaps of all known forces via the proposed string theories. These advances have been paralleled by deeper insights into atomic physics and condensed-matter problems, into chemistry, biology and molecular biology (made in part by people trained in physics)—insights that have gradually led us to the view that every portion of the world can be understood as a manifestation of a small group of rationally comprehensible laws. Perhaps the pinnacle of that view is our gradually emerging understanding that living creatures, including even ourselves, are indeed ordinary parts of the natural world.

Another part of the grand structure is the opposite view—that natural laws are diverse. Different laws may apply at different levels of organization, partially because the objects studied at the various levels are not the same (see PHYSICS TODAY, September, page 7). Our own gross motions are described by classical mechanics; those of our blood, by hydrodynamics; the details of our hemoglobin (compared with that of other species), by the laws of genetics; the binding of oxygen with the hemoglobin, by quantum laws. One can even have situations in which laws at the different levels are in apparent contradiction: Microscopic laws of nature imply time-reversal invariance. This is roughly the statement that movies run backwards "look right." But thermodynamics says just the opposite. Our own perceptions fall within the thermodynamic world. Thus we see at each level of scale and organization within the world different rules, different generalizations, different natural laws.

Edward Purcell has produced a fine example of this "many worlds" view of nature by asking the question "What is the physics a bacterium needs to know?" The answer is that the bacterium lives in a high-viscosity world, with laws very different from those in a human-scale, low-viscosity situation. Another and perhaps even more striking example is the contrast between the quantum world and the classical one as revealed in the Einstein–Podolsky–Rosen *gedanken* experiment and John S. Bell's analysis of the classical interpretation of that experiment. In discussing this example, David Mermin has pointed out (PHYSICS TODAY, April 1985, page 38) that in a quantum world our ideas of logic and causation do not work out quite the way we might expect. And worse(?!) yet, the quantum world may impinge upon our own, leading to results that challenge our intuitive view of connection and causation.

Physics has enriched the discussion of many other topics that should interest any thinking individual. Consider time. Any person might wish to know the relation between the time we perceive and time as it "really is," and to do so that person might wish to think about the impossibility of communication across the light cone and the resulting loss of contact with regions inside black holes, and about the twin paradox (in which aging of different individuals depends upon the history of their motions), and about the Big Bang and the large-scale structure of space–time.

One more fundamental achievement of science is the prediction and control of the world around us. It is very pleasing to be able to answer such questions as why the sky is blue, why diamonds are hard and why snowflakes are so diverse. (We cannot fully answer the last one yet.) But predictability in natural situations should be more than just casually pleasing to us. This predictability demonstrates that the physical laws for the objects that surround us are within our grasp. From understanding can arise control. It is satisfying to see science and the engineering art working in concert to produce such useful advances as medical x rays, solid-state electronics and satellite communication.

The role of scientists in these types of engineering advances has changed considerably since the second world war. At that time there were very few engineers expressly trained in applying advanced technology, so physicists had to take much of the role now filled by engineers. No more. Now, most fields of technology have very highly trained applied people—some scientists, some engineers—who carry out the R&D process. The role of pure science is that of finding totally new phenomena and new areas where simple and elegant laws might apply. For example, physics and physicists are right now trying hard to ask what new laws might govern complex systems. It's very fashionable to ask what laws might apply to a large group of simple circuit elements, or a large group of neurons, or even a large number of molecules in random environments. Here we see

Leo Kadanoff, a condensed-matter theorist, is John D. MacArthur Professor of Physics at the University of Chicago.

reference frame

entirely new worlds to explore, but we recognize that the major problems in these areas have not yet been stated, much less solved. In another kind of advance, we are discovering that we can find or manufacture new systems that might yield exciting new generalizations. Manmade complex circuits, computer systems and complex (for example, layered) materials are new arenas for the application of scientific ideas. Even humble rocks have shown unexpected complexity, beauty and regularity. In studying these relatively new systems, we can expect new laws, new kinds of elegance and new beautiful science.

There's one more very important role that scientists can play in helping coordinate the research and design process. We scientists often have a reasonably reliable picture of what is feasible and what is impossible. For example, people cannot expect to communicate with their distant ancestors. A physical law stands in the way. One cannot achieve in any substantial measure the differential aging permitted by the twin paradox, because the accelerations required will be beyond our technology for the foreseeable future. For the same reason, we cannot expect any direct observation of strings. Scientists can and should point out these limitations about apparently feasible technology. Soviet agriculture and biology would have been saved many headaches had the government been willing to hear the voice of real science, rather than that of the party toady Trofim Lysenko. He made a career for himself by arguing that he could make advances in plant breeding via the inheritance of acquired characteristics, a goal that remained quite elusive. Here and today we are largely in a more favorable situation. Scientists can argue more effectively that there are limitations upon what technology can reliably achieve. We do not do extensive research in antigravity devices or faster-than-light communication because we have good reason for believing that these are impossible. We can also argue that some systems are too complex for effective control, given our present level of understanding. Thus, many scientists urge caution in activity related to the atmospheric release of fluorocarbons because we do not know the chemical and physical processes that relate these compounds to climate and to the level of uv radiation reaching the Earth. Ignorance of the processes involved and of the behavior of complex systems has effectively limited weather modification and has so far prevented reliable earthquake prediction. Earthquake modification seems only a distant dream. Similarly, many scientists have wondered whether any missile-defense system capable of protecting populations is feasible, given the weakness of our predictive powers for truly complex systems.

In our society, such questioning is considered a valuable activity. Scientific modes of thinking can be used to distinguish the correct from the incorrect and the practicable from the impractical. The possibility of making such distinctions is a kind of cathedral too. □

"THEY SAY THEY CAN'T UNDERSTAND WHY OUR LOOKING AT THE SAME OLD STARS SHOULD INTERFERE WITH A SHOPPING MALL PROMOTION."

SIDNEY HARRIS

reference frame

From neutrinos to quasiparticles

Leo P. Kadanoff

This column is about a particularly elegant idea for an experiment in neutrino physics. Of course any observation of neutrinos is a wonderful accomplishment. But the proposed experiment to be discussed here has the further charm that it makes essential use of the quantum mechanical properties of low-temperature matter. The matter, in this case, is liquid helium, and the experiment will rely on low-lying quantum excitations (called quasiparticles) as messengers to carry the energy transferred from the neutrinos to the helium.

A now classical experiment of Raymond Davis and coworkers sees solar neutrinos in a 600-ton vat of cleaning fluid via the reaction

$$\nu_e + Cl^{37} \rightarrow e^- + Ar^{37}$$

Roughly one argon atom is produced every two days. The Ar^{37}, having a halflife of 35 days, is essentially not present in nature and hence the few atoms that are produced can be detected. These experiments have measured a solar neutrino flux a factor of three lower than predicted, thereby stimulating a considerable reexamination of solar and weak-interaction theories.

The next generation of neutrino experiments, now being planned and proposed, will measure the neutrino arrival time much more precisely. The observation of neutrinos produced by the recent supernova has pointed up the importance of good time resolution (see the article by David Helfand on page 24). (The end products of neutrino interaction were seen in Cerenkov radiation detectors originally designed to study proton decay.) Arrival time data from these observations were used to estimate the neutrino mass. Unfortunately the chlorine conversions are unsuitable for a time-resolved experiment simply because the argon atoms produced cannot travel through the container rapidly enough. Their typical speed of about 300 meters per second would certainly be fast enough if they moved in a straight line. But they are in a liquid, so every few angstroms they bump into another molecule and change direction. The resulting tortuous path, called a random walk, then demands a very long time to achieve a reasonable displacement. For example, a time on the order of a month is required for the atom to move 1 meter away from its starting position. Given this time constant, the actual collection cannot utilize the natural motion of the argon atoms. Instead, at intervals, helium gas is bubbled through the fluid for 20 hours or so. The argon is caught in the helium and then detected. However, space and time resolution is lost in this collection process.

Several groups have been seeking a new approach to solar neutrino detection. One requirement is that the neutrinos produce an end product that will move directly and without substantial decay or delay to a detector. The proposal I'm writing about, by Robert Lanou, Humphrey Maris and George Seidel, is for a detection method based on the observation of rotons in liquid helium.[1] In a recent colloquium talk, I heard Blas Cabrera list eight different proposed detection schemes. I do not have any way of knowing which one is preferable. But I do wish to use this roton proposal as a device to discuss a fundamental idea from condensed matter physics, the idea of a quasiparticle.

Imagine any quantum system held at a temperature low enough that the energy of a typical excitation, kT, will be much lower than any characteristic interaction energy in the material. According to the ideas of Lev Landau and the theorists who followed him, such a system can be described in terms of elementary excitations called quasiparticles. In a material with a translational invariance, for example, a liquid or an ordered solid, each such excitation can be described by a position vector **q**, a momentum vector **p**, and a quasiparticle energy that depends on **p** and **q**. The quasiparticles move in a straight line, like classical particles, until they collide with another excitation. Landau continued this mode of thinking by using statistical mechanics to estimate the number of quasiparticle excitations in the system at low temperatures. One finds, quite naturally, that the number of excitations becomes smaller and smaller as the temperature goes to zero. And now comes the major point for our present purposes: If the temperature is low enough that there are few excitations, there will be few collisions and the quasiparticles will move for a long distance in a straight line. This straight line, or ballistic, motion is to be contrasted with the result of many scatterings: random walk, or diffusive, motion.

Now let us come back to the problem of neutrino detection. Visualize the neutrinos as they move through a large vat, perhaps 1 meter on a side, of helium at a few tenths of a kelvin. Occasionally a neutrino will interact with an electron or nucleon in the vat, transferring a respectable amount of energy to this particle. This energy will in turn soon cascade into lots and lots of excitations with lower energy. In liquid He^4, there are two kinds of excitations: phonons (which are the quanta of sound waves) and rotons. The latter are a kind of quasiparticle with a minimum energy of about 9 K. From one point of view, the roton is the quantum "ghost" of a vanishingly small vortex ring. A more mundane vision sees this excitation as an extra helium atom moving through the fluid, producing a swirling disturbance behind it.

Because the temperature is so low, the fluid is almost in its quantum mechanical ground state and contains very few thermally produced excitations. Hence according to the quasipar-

Leo Kadanoff, a condensed matter theorist, is John D. MacArthur Professor of Physics at the University of Chicago.

ticle concept, once the quanta are set into motion, they will find nothing much to bump into. They will move in straight lines until they reach the walls of the container or the surface of the fluid. From the velocity of the excitation, we can construct an estimate of the typical transit time of an excitation through the helium. That time is 10 milliseconds. Wonder of wonders: By simply changing materials, going from argon in cleaning fluid to rotons in low-temperature liquid helium, we change the transit time by a factor of 10^9 or so. This change is a reflection of the transition from diffusive to ballistic motion.

Furthermore, the quasiparticles have an energy just larger than the 8-K binding energy of a helium atom to a free helium surface at that temperature. In fact, roughly 30% of the rotons that hit the surface will knock a helium out of the liquid. These knocked-off heliums are detectable and thus form the basis of the proposed roton-based neutrino detection scheme.

I find something rather satisfying about the conception of this experiment. We start from Landau's theoretically derived idea that quantum excitations are indivisible wholes (called "quanta") and that they have labels like position, energy and momentum that can be given real meanings. We might end with a piece of apparatus that works in a way that would be totally unexpected from classical thinking. If constructed, it will give us knowledge about phenomena that lie at the center of natural laws, both those involving the neutrino and those of bulk matter.

Notice that many features of the proposed experiment have never been tested. For example, the mean free path of rotons has not been measured at the temperatures at which the experiment will be performed. Moreover, relevant roton–phonon branching ratios are unknown, and hence the total number of rotons produced is uncertain. Another example: One cannot know about the effects of possible radioactive decay of contaminants in the helium or the containers. In fact, there will be no detector with sufficient sensitivity to measure these backgrounds until a reasonably large prototype of the neutrino detector is itself built. To an outsider like myself, it would appear that planning experiments like the ones mentioned here takes both vision and guts.

Reference

1. R. E. Lanou, H. J. Maris, G. M. Seidel, Phys. Rev. Lett. **58**, 2498 (1987). □

Section B

Scaling and Phase Transitions

On the Joys of Creation

This section describes the development of ideas of scaling and universality as they relate to phase transitions and critical phenomena. In the late 1960s and early 1970s, a group of physicists and chemists changed the way scientists look at problems in statistical mechanics and related fields. Looking back at this work almost thirty years later, I still feel proud, pleased, and somewhat surprised that I could play a role in such an achievement.

My story starts when I was finishing up as a graduate student at Harvard in 1960. Kurt Gottfried and Paul Martin both pointed out to me that the problem of 'second order phase transitions' was quite interesting and not understood at all. Kurt and I even did a calculation, never published, in which we looked for critical fluctuations in three-dimensional superconductors. We got, and were discouraged by what is in retrospect the correct answer, which is that it is next to impossible to observe critical fluctuation in three-dimensional superconductors. Then, I put the problem aside for a while.

I came back to it during a nine-month period I was spending in the Cavendish laboratory in Cambridge University, invited by Neville Mott to take part in what he called a theoretical jamboree. This kind of jamboree permitted one to have lots of free time for research, so I began to struggle once again with the critical behavior which occurs very near second order phase transitions. Previously many scientists had believed that the Van der Walls-Weiss-Landau mean field theory of this transition was essentially correct. But, by this time, the experimental work of Sasha Voronel and others had shown that the mean field theory was wrong for the classical liquid gas phase transitions. C. F. Keller's experiment[11] and Brian Pippard's[12] analysis had shown that it did not work for the transitions in helium. On the theoretical side, Cyril Domb, Michael Fisher, and the King's College school had proven that the mean field theory did not apply to Ising models either. The time seemed ripe for a new approach.

I entered the problem by studying Lars Onsager's solution of the two-dimensional Ising model. The exact solution of this model of two-dimensional magnetism had been announced in the 1940s, but it had never been fully analyzed. Onsager and C. N. Yang had calculated some of the thermodynamic properties, but there was really no explanation of what physics might be demonstrated by Onsager's solution. Here was a tiny little world, just waiting to be explored and perhaps even captured. As far as I could tell, nobody had looked at the spatial correlations built into the Onsager solution. My own background pushed me toward looking at these correlations. The work on the connection between hydrodynamics and correlations (A2) were all about the relation between thermodynamic behavior and space-time correlations, especially in the long-wavelength limit. My Green's function studies and much of the other work described in Section A was aimed at understanding correlations over large regions in space and time. I was certainly ready to attack correlations in Ising models.

[11] See C. F. Kellers (Thesis, Duke University, 1960).

[12] A. B. Pippard, *Proc. Roy. Soc. (London)* **A216**, 547 (21953).

I began to calculate the spin-spin correlation function of the Ising model near its critical point. (See paper B1.) The calculation was long and difficult. It involved the Weiner-Hopf technique for solving integral equations, which made extensive use of complex variable techniques. Fortunately, I had received an excellent training in this method in graduate school with the applied math courses I had taken from George Carrier and Arthur Bryson and with the complex variable course I had taken from George Mackey. After about six months, I had a finished work which I sent off to the *Journal of Mathematical Physics.*

This paper was potentially very important for this subarea of physics. It contained the germs of much of what was to prove to be the correct theory of second order phase transitions, all worked out in a particular example. The paper also had defects. It was hard to read. The boundary conditions at the edge of the material were handled in a sloppy and incorrect fashion. (I did not believe that the boundary conditions were really relevant to the solution, so I did not inquire very deeply into their correct handling.) And the paper certainly did not proclaim why it was potentially important. It was rejected, twice. And in each rejection it crossed the Atlantic once or twice by slow boat. This paper was written before the days of extensive use of preprints. So, for a period of almost a year, there were only a few people who knew the contents of this paper. It was eventually published, after the second rejection, in *Nuovo Cimento.*

In the meantime, the field had progressed considerably. There was a conference on second order phase transitions, held at the U.S. National Bureau of Standards, in which the field was reviewed. Several people — notably Fisher, Ben Widom, and Michael Buckingham — described some conjectures about spatial correlations. Widom, in particular, had introduced some 'magic' (the word was used by Martin when he mentioned the work to me) relations among the critical indices. These critical indices are numbers used to describe the order of magnitudes or the various quantities which can be used to describe the near-critical behavior of a system near a second order phase transition. My spin correlation work also involved these critical indices. There were many of them, and I had a hard time remembering them all. So I had developed a mnemonic device, which involved expressing all orders of magnitudes in terms of two independent magnitudes, the natural fluctuations in the spin and in the energy density.

Then, in Christmas week of 1965 I had a sudden vision. A gift from the gods. I had a simple view of how these magnitude relations might be true and general. In modern terms, I had developed a scaling analysis of the critical behavior of Ising models based upon the idea of running coupling constants, i.e. couplings which depended upon the distance scale. (Some idea of this kind was also present in field theory in the work of Stückelberg and Peterman and of Murray Gell-Mann and Francis Low.[13] Unfortunately, I was only barely aware of this earlier work.) The awfully complex and convoluted extension of the Onsager solution which I had previously done could now be explained in terms of a few simple and appealing ideas. And better yet, these ideas could be extended to understanding most problems involving second order phase transitions.

Many people have written about the difficulties which they experienced in getting their ideas accepted by the scientific community. I have never had much problem in this direction.

[13] E. C. G. Stückelberg and A. Peterman, *Helv. Phys. Acta.* **26**, 499 (1953). Murray Gell-Mann and Francis Low, *Phys. Rev.* **95**, 1300 (1954).

I have always been an establishment figure, and most of my good work has gotten all the recognition it deserved. I had this 'Christmas vision' while I was on the faculty at Urbana. My colleagues there said they liked the work. My collaborator, Gordon Baym, had particularly warm words, while my seniors, John Bardeen and David Pines, nodded approval. I am told my preprint elicited seminars on the work at Harvard and Cornell. My joy was increased when I spoke to Mel Green, who told me about data that had been presented at the Bureau of Standards conference, which supported my conclusions. I also recalled Paul Martin's statement about 'magic' relations, and then discovered Ben Widom's papers which had just been published in the *Journal of Chemical Physics*.[14] Widom had developed scaling arguments which gave many of my conclusions. Later on I found out that Patashinskii and Pokrovskii had also gotten the right answers at about the same time.[15]

My own work was published in the journal *Physics*. It is paper B2 in this volume. My answers in this paper were the same as that of Widom and of Patashinskii and Pokrovskii, but my point of view was somewhat different. There were, in fact, quite a large number of contributors to the development of these new ideas. In hindsight, it is easy to see that we — all of us — had discovered and invented a wonderful new world. Near critical points, thermodynamic systems show a very special behavior. If you examine the system on distance scales which are much larger than the typical microscopic distances, you see correlations which extend over these very long distances. The system has to choose to fall into one or another thermodynamic phase, and is showing the vacillations in its decision-making. This process of choosing produces a beautiful, closed world with its own rules and its own, internally consistent explanations. The world had been penetrated and could be explained.

The next job was to see whether the ideas that had been developed all jibed with the experimental facts. To do this, a group of us got together at Urbana and ran a seminar aimed at looking at all the known experimental and theoretical data about critical phenomena, to see whether it fit into the newly proposed pattern. This group, W. Gotze, D. Hamblen, R. Hecht, E. A. S. Lewis, V. V. Palciauskas, M. Rayl, J. Swift, D. Aspnes, J. W. Kane and myself, looked at all the available literature and gave seminars on what we had learned. We convinced ourselves that all the data were consistent with the new point of view, and that much of it could be explained by the new scaling theory of critical phenomena. We reported our conclusions in the review paper, which is number B3 in this volume.

At some point in the development of this review paper, I am not sure when, a new idea was added to the concepts of scaling and running couplings. This idea has the modern name 'universality', which I first heard applied in a conversation with Sasha Polyakov and Sasha Migdal in a dollar bar in Moscow. Their use of the word came from descriptions of field theoretical solutions to problems of this kind. The basic idea is that there are only a few different solutions to near-critical problems. Using the solutions, one could group the problems into 'universality classes'. Many apparently different problems have the same solution and belong to the same class. In critical phenomena, the universality classes are in large measure defined by giving the dimensionality of the system and by describing the kind of information which the system must transfer over long distances. As a specific example,

[14] B. Widom, *J. Chem. Phys.* **43**, 3892, 3898 (1965).
[15] A. Z. Patashinskii and V. L. Pokrovskii, *Soviet Phys. JETP* **23**, 292 (1966).

liquid gas phase transitions and Ising-model magnets show exactly the same behavior near their respective critical points.

This universality idea provides a truly new way of looking at problems in statistical mechanics, field theory, hydrodynamics, etc. Instead of solving each problem, one looks for classes of problem, each class having a common solution. We essentially used the idea (before it had a name) in organizing our review paper. It had been explicitly used in the previous work on critical phenomena in helium by Pippard, and had been implicit in the work on Ising models of the King's College school. Bob Griffiths added considerably to the concept by pointing out its relation to the geometry of thermodynamic surfaces. By looking back on my Ising model work, I could see that the universality idea was clearly and closely related to the scaling concepts. The universality concept said that one could classify the different microscopic worlds produced by critical phenomena. This idea made critical behavior itself into a ideal world, beautiful and self-contained.

Over the next few years, I had a very considerable pleasure in exploring the corners of our new world. Paper B4, by Swift and myself, describes how to set up a cellular automaton so as to simulate hydrodynamic equations. A cellular automaton is a parallel processing computer set up to simulate the motion of many particles simultaneously. Such simulations have been much developed since the time that Jack and I worked together. (See, for example, papers E4 and E12, which make use of another automaton.) Paper B4 was interesting, maybe even important, but not of much use in critical phenomena. Paper B4 developed the idea that one could simulate hydrodynamic behavior by using methods based upon the motion of particles with discretely defined positions and velocities. This idea was much more fully developed by later workers. One outcome can be seen in papers E4 and E12, described below. The other paper with almost the same name, B5, is quite different. In this paper, Swift and I are developing a part of a theory of transport behavior near critical points. There were two elements of this new approach: mode coupling and scaling. As always, new ideas worked from solid older ones. Here the original mode coupling ideas, in perhaps slightly different form, were creations of Marshall Fixman and Kiozi Kawasaki. Some of the dynamic scaling ideas were in the work of Ferrell and collaborators,[16] and Halperin and Hohenberg.[17] My own development toward these ideas go back to work I had done on transport theory in my thesis (paper 6 with Paul Martin, on my publication list) and to a long series of papers I had done on transport with a wide variety of collaborators.[18]

Papers B6 and B7 explain and describe the world of scaling. In particular, paper B6 looks back to the older work of Fisher and of Buckingham and describes how their ideas can be incorporated in the new viewpoint. Paper B7 summarizes what we had learned about the phenomenological theory of critical systems, from a slightly more mature perspective than that of the 1967 review paper.

One important result in the theory of critical phenomena is what is called the operator product expansion or short distance expansion (see paper B8). This expansion, due

[16]R. A. Ferrell, N. Menyhárd, H. Schmidt, F. Schwabl, and P. Szepfalusy, *Phys. Rev. Lett.* **18**, 891 (1967).
[17]B. I. Halperin and P. C. Hohenberg, *Phys. Rev. Lett.* **19**, 700 (1967).
[18]During the early part of my career, I spent a lot of time working on transport properties of various many particle systems. All this work formed a basis for my understanding of the relation between the macro and micro levels of description. The resulting papers, especially #13 (with D. C. Langreth), #15 (with M. Revzen), and #25 (with J. W. Kane) formed an important part of my education.

to K. Wilson and myself, is based upon the idea that, at criticality, there are a limited number of different local fluctuating quantities. This idea permits one to express products of fluctuating quantities at criticality in terms of an expansion in the fundamental critical quantities. This idea provides a mechanism for understanding the critical symmetry in a rather deep way. In paper B9, H. Ceva and I explore this expansion for the two-dimensional Ising model.

Meanwhile the field had exploded. Ken Wilson had absorbed the new viewpoint, and magnificently extended it. He introduced two rich new concepts: the fixed point and the space of coupling constants. With these concepts the scaling and universality point of view became a theory: the renormalization theory of critical phenomena (and of much else). The remaining papers in this chapter trace out how the scaling-university theory and the renormalization group work together. Paper B11 attempts to be a formal review of my contributions to the whole area. I should also point back to paper A5, which describes how the running coupling constant concept found application in quantum chromodynamics. Paper B10, a joint effort with Humphrey Maris, describes how some of the new ideas could be presented at a level which is accessible to undergraduates.

I should emphasize once again my tremendous personal gratification in the accomplishments of the whole field. The story I have told has many actors and many excellent accomplishments, among them my own. In the end, we can say that a group of scientists created something beautiful, which was not there before. I saw it happen. I saw understanding and nature come into correspondence. It was great to have been there.

But science does not stand still to admire past accomplishments. Immediately after Wilson had invented the renormalization group, Franz Wegner showed how to phase the renormalization arguments into a formal algebra of coupling constants. Wilson and I independently (see papers B8 and B9) described the complementary algebra of fluctuating quantities near the critical point. My own work, which I did jointly with many collaborators[19] carried forward the understanding of these algebras in the context of two-dimensional critical behavior. Then Polyakov invented conformal field theory, which proved to be a tool which would enable one to understand all the behavior of almost any critical theory in two dimensions. The algebraic technique was refined and perfected by the field theorists. The scientific area of critical behavior exploded and indirectly contributed to the development of string theory.

In the meantime, with many collaborators, I contributed to the solidification of the gains that had been made in understanding critical phenomena. For example, Franz Wegner and I (see #43) developed concepts related to continuously varying critical indices.[20] N. Berker, R. Ditzian, Gary Grest, Michael Widom and I applied these concepts to other statistical mechanical systems. This work was subsequently carried forward and extended with many

[19] Among these coworkers are H. Ceva, Alan Brown, Michael Widom, Ad Pruisken, and Gary Grest.

[20] I lost a bet involving a bottle of bottle of Scotch (or Vodka) to A. A. Midgal on this issue. I had bet that all two-dimensional critical indices were rational numbers. I figured I could not lose, since it is usually very hard to prove that a number which arises in the context of some physical problem is not rational. But then Rodney Baxter proved that in the eight-vertex model, indices vary continuously. I paid off. The Midgal's junior (A. A.) and senior (A. B.) generously repaid with an 'irrational' vodka bottle which had been bent into an impossible shape in their home workshop.

other people.[21] In another series of papers, A. Houghton, M. C. Yalabik and I explored the consequences of the real space analysis of renormalization near critical points.

The understanding that our little community has generated, in the course of time, has spread far beyond the study of second order phase transitions. The scaling and correlation ideas form a crucial part of the phenomenology of particle physics and reappears in astrophysics in, for example, analyses of fractal distribution of matter in the universe. Classical Mechanics and Hydrodynamics contain many scalings, with the behavior in turbulence and near the onset of chaos being most like those in critical behavior. In materials science, since the work of P. de Gennes, B. Mandelbrot, and T. Witten, people are always looking for some sort of scaling or universality. We have come far by looking at the small corner of physical behavior very near critical points. In part, we have accomplished our trip by following Onsager's solution and squeezing out every generalization that could possibly be found in that rich reservoir.

Beyond the export of our specific technical methods, we have also exported a point of view, encompassing the way in which one might look at the structure of physics. One image of this structure is that each little world of phenomena is really based upon the physical laws which describe a more fundamental level of reality. This image leads one to a reductionist outlook. Then one would say that the true goal of physics should be to reach deeper and deeper toward the basic laws which describe the fundamental interactions in the world. However, the study of critical phenomena and other topics in condensed matter physics pushes one toward another and complementary image of nature. This view was expressed in P. W. Anderson's thoughtful paper 'More is Different'[22] in which he pointed out that every level of reality can have its own deep and fundamental laws. In studying a particular little model world, the goal of the scientist is threefold: First, to expose the fundamental laws in their most general form and to show how they work out in the specific system in hand. Second, to show how this particular level of experience is related to other closely connected parts of reality. And third, to take the ideas generated in the study of this one particular part of the world and apply them to other portions of the world. The study of critical phenomena is not in any way unique. However, it is a particularly successful and beautiful example of the generation of deep ideas about the simplified world of critical fluctuations, about how those fluctuations are defined by the microscopic behavior, and about how these ideas manifest themselves in macroscopic behavior. The ideas generated were exported widely and illuminated, in an unexpected fashion, many other areas of science.

[21] My collaborators during this period include most of a generation of Dutch Statistical Mechanics — Ad Pruisken, Marcel den Nijs, and Bernard Nienhuis — and other coworkers including J. Banavar and Morgan Grover.

[22] P. W. Anderson, *Science* **177**, 393 (1972).

L. P. Kadanoff
11 Agosto 1966
Il Nuovo Cimento.
Serie X. Vol. **44**, pag. 276-305

Spin-Spin Correlations in the Two-Dimensional Ising Model (*).

L. P. Kadanoff (**)

Department of Physics, University of Illinois - Urbana, Ill.

(ricevuto il 4 Gennaio 1966)

Summary. — The Onsager solution to the two-dimensional Ising model is phrased in the language of thermodynamic Green's functions. The description is quite closely analogous to a theory of noninteracting fermions, except that displacements along one of the lattice directions replace the time displacements of the standard fermion theory. This formulation is used to provide new derivations of the well-known results for the partition function and the zero-field magnetization. Aside from these formulational points, the main new result of this paper is an evaluation of the spin-spin correlation function in the limit of large, but not infinite, spatial separations between the spins. For the square lattice with nearest neighbor interactions and all coupling constants identical, the correlation functions respectively for T just greater than or just less than T_c are shown to be of the form $\langle[\sigma_{j_1,k_1}-\langle\sigma\rangle][\sigma_{j_2,k_2}-\langle\sigma\rangle]\rangle = \varepsilon^{\frac{1}{4}} f_{\gtrless}(\varepsilon R)$, where $\varepsilon = 4|K-K_c|$ is a measure of the distance from the critical temperature and $R=[(j_1-j_2)^2+(k_1-k_2)^2]^{\frac{1}{2}}$ is the spatial separation of the spin sites. This result holds when $\varepsilon\to 0$ but εR remains finite. The functions $f(x)$ are evaluated in the asymptotic limit of large x and shown to be $f_<(x) = e^{-2x}x^{-2}\pi^{-1}2^{-21/8}$ and $f_>(x) = e^{-x}(\pi x)^{-1/2}2^{-3/8}$.

1. – Introduction.

This paper has two purposes. The first is to state the Onsager solution [1-4] of the two-dimensional Ising model in Green's function language. Many of the well-known results are very conveniently derived using Green's functions. Since these funcions have proved to be powerful tools in perturbative calcu-

---

(*) Work supported in part by NSF Grant GP 4937.

(**) Alfred P. Sloan Foundation Fellow.

(1) L. Onsager: *Phys. Rev.*, **65**, 117 (1944).

(2) L. Onsager: *Suppl. Nuovo Cimento*, **6**, 261 (1949).

(3) B. Kaufman: *Phys. Rev.*, **76**, 1232 (1949).

(4) B. Kaufman and L. Onsager: *Phys. Rev.*, **76**, 1244 (1949).

5152

lations, it is hoped that these results can be used in describing other physical situations as perturbation expansions about the Onsager solution.

Section **2** is devoted to this translation into the Green's function language. In Sect. **3** results from **2** are applied to a rederivation of Onsager's result for the zero-field magnetization. The spin correlation functions for T just smaller than T_c and T just greater than T_c are discussed in Sect. **4** and **5** respectively.

2. – Mathematical formulation.

2·1. *Field-theoretic statement.* – Given a set of « spins », $\sigma_{j,k}$, on the two-dimensional square lattice labelled with x and y co-ordinates $j=1, 2, \dots m$, $k=1, 2, \dots n$, the nm variables $\sigma_{j,k}$ can each have the values ± 1. The Hamiltonian includes an interaction $-K\beta^{-1}$ between nearest neighbors separated by one unit in the x-direction and $-K'\beta^{-1}$ between nearest neighbors in the y-direction, where β is the inverse temperature in energy units. We take K and K' to be positive so that we may discuss the analog of the ferromagnetic transition. Cyclic boundary conditions are employed so that $\sigma_{j,k}=\sigma_{j+m,k}=\sigma_{j,k+n}$. Thus, the partition function may be written as

$$Z=\sum_{\sigma_{j,k}=\pm 1}\exp\Big[\sum_{j,k}(K\sigma_{j,k}\sigma_{j+1,k}+K'\sigma_{j,k}\sigma_{j,k+1})\Big]\,. \tag{2.1}$$

We shall attack the Ising model by using the transfer matrix method of Kramers and Wannier ([5]) and Montroll ([6]) which is carefully explained in the review of Newell and Montroll ([7]). Instead of deriving the method anew we merely borrow the relevent equations of N-M. The partition function is written as (N-M, eq. (2.3))

$$Z=\text{trace}\,\boldsymbol{P}^m. \tag{2.2}$$

Here $\boldsymbol{P}$ is a 2^n-dimensional matrix called the transfer matrix. It can be written in terms of $2n$ simpler matrices $\boldsymbol{s}_k$ and $\boldsymbol{C}_k$ which obey (N-M, pp. 355-356)

$$\begin{cases}[\boldsymbol{s}_k,\,\boldsymbol{s}_{k'}]=[\boldsymbol{C}_k,\,\boldsymbol{C}_{k'}]=0,\\ \boldsymbol{s}_k^2=\boldsymbol{C}_k^2=1\,,\\ [\boldsymbol{s}_k,\,\boldsymbol{C}_{k'}]=0\,,\ \text{for } k\neq k',\\ \{\boldsymbol{s}_k,\,\boldsymbol{C}_k\}=0\,.\end{cases} \tag{2.3}$$

([5]) H. A. Kramers and G. H. Wannier: *Phys. Rev.*, **60**, 252, 263 (1941).

([6]) E. Montroll: *Journ. Chem. Phys.*, **9**, 706 (1941).

([7]) G. F. Newell and E. W. Montroll: *Rev. Mod. Phys.*, **25**, 353 (1953), hereafter denoted as N-M. We use the notation of this reference as much as possible.

Hence the trace in (2.2) can be thought of as a diagonal sum over the 2^n eigenvectors of the complete set of commuting matrices $\boldsymbol{s}_k$. The transfer matrix is

$$\boldsymbol{P} = (2\sinh 2K)^{n/2}\boldsymbol{V}_2\boldsymbol{V}_1\,, \tag{2.4}$$

where (N-M (2.12) and (2.13))

$$\boldsymbol{V}_2 = \exp\left[K'\sum_{k=1}^{n}\boldsymbol{s}_k\,\boldsymbol{s}_{k+1}\right], \qquad \boldsymbol{V}_1 = \exp\left[K^*\sum_{k=1}^{n}\boldsymbol{C}_k\right]. \tag{2.5}$$

Here, $\boldsymbol{s}_{n+1} = \boldsymbol{s}_1$ and

$$\operatorname{tgh} K^* = e^{-2K}. \tag{2.6}$$

The matrix $\boldsymbol{s}_k$ can be thought of as an operator which represents the spin on the site $(j = 1, k)$. In fact, if we define an average in the matrix space as

$$\langle\!\langle \boldsymbol{\Lambda} \rangle\!\rangle = Z^{-1}\,\mathrm{trace}\,[(\boldsymbol{P})^m\boldsymbol{\Lambda}]\,, \tag{2.7}$$

then the spin-spin correlation function on the row $j = 1$ is

$$\langle \sigma_{j=1,k_1}\sigma_{j=1,k_2}\rangle = \langle\!\langle \boldsymbol{s}_{k_1}\boldsymbol{s}_{k_2}\rangle\!\rangle\,.$$

More general spin-spin correlation functions may be obtained from N-M eq. (6.7) which is, in the present notation,

$$\langle \sigma_{j_1,k_1}\sigma_{j_2,k_2}\rangle = \langle\!\langle (\boldsymbol{P})^{j_1-1}\boldsymbol{s}_{k_1}(\boldsymbol{P})^{-j_1+1}(\boldsymbol{P})^{j_2-1}\boldsymbol{s}_{k_2}(\boldsymbol{P})^{-j_2+1}\rangle\!\rangle$$

for $j_1 < j_2$. This result is most easily interpreted if we think of the matrix

$$\boldsymbol{s}_{j,k} = (\boldsymbol{P})^{j-1}\boldsymbol{s}_k(\boldsymbol{P})^{-(j-1)} \tag{2.8}$$

as an operator which represents the spin at the site j, k. Then we can write the spin-spin correlation function as

$$\langle \sigma_{j_1,k_1}\sigma_{j_2,k_2}\rangle = \langle\!\langle (\boldsymbol{s}_{j_1,k_1}\boldsymbol{s}_{j_2,k_2})_+\rangle\!\rangle\,, \tag{2.9}$$

where the $(\;)_+$ is an ordering symbol which instructs us to put operators with larger values of j to the right of those with smaller j.

This form of writing correlation functions is quite analogous to the standard field-theoretic statements of quantum statistical mechanics [8]. Think of j as

---

(8) See for example L. P. KADANOFF and G. BAYM: *Quantum Statistical Mechanics* (New York, 1962), expecially Chap. I.

being like the time variable and k as being like the space variable in the quantum theory. Then, $\boldsymbol{P}$ is like the time translation operator $e^{-i\mathscr{H}t}$ since from (2.8), $\boldsymbol{P}$ translates j from one row to the next. Equation (2.8) puts the j-dependence into the « Heisenberg representation ». From (2.2), we see that $(\boldsymbol{P})^m$ is analogous to $e^{-\beta\mathscr{H}}$.

2·2. *The variables.* – To utilize these definitions, we must diagonalize $\boldsymbol{P}$. This is partially achieved through the use of the variables ([9-11])

$$\begin{cases} \boldsymbol{b}_{k,+} = \dfrac{1}{\sqrt{2}}\, \boldsymbol{s}_{k-1} \prod_{r=k}^{n} \boldsymbol{C}_r\,, \\ \boldsymbol{b}_{k,-} = -\dfrac{i}{\sqrt{2}}\, \boldsymbol{s}_k \prod_{r=k}^{n} \boldsymbol{C}_r\,. \end{cases} \tag{2.10}$$

We use these definitions in the range $1<k<n$ where the b's are connected to the more physical variables by

$$\begin{cases} \boldsymbol{b}_{k,+}\boldsymbol{b}_{k,-} = \frac{1}{2} i \boldsymbol{s}_{k-1}\boldsymbol{s}_k\,, \\ \boldsymbol{b}_{k,+}\boldsymbol{b}_{k-1,-} = -\frac{1}{2} i \boldsymbol{C}_{k-1}\,. \end{cases} \tag{2.11}$$

The $\boldsymbol{b}$'s are introduced because they have simple commutation relations with $\boldsymbol{P}$. In fact, for $2<k<n$, $\boldsymbol{P}\boldsymbol{b}_{k,\pm}\boldsymbol{P}^{-1}$ is just a linear combination of $\boldsymbol{b}$'s at the sites k, $k+1$, and $k-1$. If we use the symbol $\boldsymbol{b}_{k,\tau}$ to denote the two components $\boldsymbol{b}_{k,+}$ and $\boldsymbol{b}_{k,-}$, the result is, for $2<k<n$,

$$\boldsymbol{P}\boldsymbol{b}_{k,\tau}\boldsymbol{P}^{-1} = \sum_{\tau'=-}^{+} \sum_{k'} q_{\tau,\tau'}(k-k')\,\boldsymbol{b}_{k',\tau'}\,, \tag{2.12}$$

where q happens to vanish unless $|k-k'|$ equals zero or one. To see how (2.12) comes about, consider the simpler operator

$$\boldsymbol{V}_1\boldsymbol{b}_{k,+}\boldsymbol{V}_1^{-1} = \exp\left[K^* \sum_{k'=1}^{n} \boldsymbol{C}_{k'}\right] \boldsymbol{b}_{k,+} \exp\left[-K^* \sum_{k''=1}^{n} \boldsymbol{C}_{k''}\right]. \tag{2.13}$$

For $1<k<n$, all the $\boldsymbol{C}_{k'}$ commute with $\boldsymbol{b}_{k,+}$ except for $\boldsymbol{C}_{k-1}$ which anticommutes. Thus, the only terms which really contribute to the right-hand side of (2.13)

([9]) These variables are identical to the variables P_j, Q_j of ref. ([3]).

([10]) T. D. SCHULTZ, D. C. MATTIS and E. H. LIEB [*Rev. Mod. Phys.*, **36**, 856 (1964)] describe an alternative field-theoretic approach to the one we utilize here. They use the variables $\boldsymbol{b}$ as fermion operators and apply them directly to the diagonalization of $\boldsymbol{P}$.

([11]) E. W. MONTROLL, R. B. POTTS and J. C. WARD: *Journ. Math. Phys.*, **4**, 308 (1963).

are $k' = k'' = k - 1$. This right-hand side is

$$\exp[2K^* \boldsymbol{C}_{k-1}] \boldsymbol{b}_{k+}$$

or, since $\boldsymbol{C}_k^2 = 1$,

$$\boldsymbol{V}_1 \boldsymbol{b}_{k,+} \boldsymbol{V}_1^{-1} = \cosh 2K^* \boldsymbol{b}_{k,+} + \sinh 2K^* \boldsymbol{C}_{k-1} \boldsymbol{b}_{k,+}$$

and finally, from (2.10),

$$\boldsymbol{V}_1 \boldsymbol{b}_{k,+} \boldsymbol{V}_1^{-1} = \cosh 2K^* \boldsymbol{b}_{k,+} - i \sinh 2K^* \boldsymbol{b}_{k-1,-} . \tag{2.14a}$$

A similar calculation for $b_{k,-}$ implies

$$\boldsymbol{V}_1 \boldsymbol{b}_{k,-} \boldsymbol{V}_1^{-1} = \cosh 2K^* \boldsymbol{b}_{k,-} + i \sinh 2K^* \boldsymbol{b}_{k+1,+} . \tag{2.14b}$$

When $\boldsymbol{V}_2 \boldsymbol{b}_{k,\tau} \boldsymbol{V}_2^{-1}$ is computed, this too turns out to be a linear combination of $\boldsymbol{b}_{k'\tau'}$'s and when all this is put together, the result (2.12) emerges together with specific expressions for the nonvanishing coefficients $q_{\tau'\tau'}(k - k')$.

In order to make use of (2.12), it is necessary to invent a compact notation for the expression of linear combinations of $\boldsymbol{b}$'s like those which appear on the right-hand side of (2.14). To do this, think of $\boldsymbol{b}_{k,+}$ and $\boldsymbol{b}_{k,-}$ as forming the two components of a spinor

$$\boldsymbol{b}_k = \begin{pmatrix} \boldsymbol{b}_{k+} \\ \boldsymbol{b}_{k-} \end{pmatrix} .$$

Define a set of three « Pauli matrices » τ_1, τ_2, τ_3 which have the effect of rearranging the components of the spinor $\boldsymbol{b}_k$, *i.e.*

$$\begin{cases} \tau_1 \boldsymbol{b}_k = \begin{pmatrix} \boldsymbol{b}_{k-} \\ \boldsymbol{b}_{k+} \end{pmatrix} , \\ \tau_2 \boldsymbol{b}_k = i \begin{pmatrix} -\boldsymbol{b}_{k-} \\ \boldsymbol{b}_{k+} \end{pmatrix} , \\ \tau_3 \boldsymbol{b}_k = \begin{pmatrix} \boldsymbol{b}_{k+} \\ -\boldsymbol{b}_{k-} \end{pmatrix} . \end{cases} \tag{2.15}$$

Then, eq. (2.14) may be written in this spinor notation as

$$\boldsymbol{V}_1 \boldsymbol{b}_k \boldsymbol{V}_1^{-1} = \sum_{k'} q_1(k - k') \boldsymbol{b}_{k'}$$

5156

with

$$q_1(k-k')=\int_0^{2\pi}\frac{dp_y}{2\pi}\exp[ip_y(k-k')]\exp[-ip_y\tau_3/2]\exp[2K^*\tau_2]\exp[ip_y\tau_3/2]\,.$$

A similar notation may be applied to (2.12), *i.e.*

$$\boldsymbol{P}\boldsymbol{b}_k\boldsymbol{P}^{-1}=\sum_{k'}q(k-k')\boldsymbol{b}_{k'} \tag{2.16}$$

with

$$q(k-k')=\int_0^{2\pi}\frac{dp_y}{2\pi}\exp[ip_y(k-k')]q(p_y)\,, \tag{2.17}$$

where

$$q(p_y)=\exp[-ip_y\tau_3/2]\exp[2K^*\tau_2]\exp[ip_y\tau_3/2]\exp[-2K'\tau_2]\,. \tag{2.18}$$

Since $\boldsymbol{b}_{j,k,\tau}$ is defined as $\boldsymbol{P}^{j-1}\boldsymbol{b}_{k,\tau}\boldsymbol{P}^{-(j-1)}$, eq. (2.16) may also be written as

$$\boldsymbol{b}_{j+1,k}=\sum_{k'}q(k-k')\boldsymbol{b}_{j,k'}\,. \tag{2.19}$$

The « Heisenberg equation of motion » (2.19) will form one of the keystones of our subsequent analyses. The other keystone will be the equal-j anticommutation relation

$$\{b_{j,k,\tau},b_{j,k',\tau'}\}=\delta_{k,k'}\delta_{\tau,\tau}1\,, \tag{2.20}$$

which again holds when k and k' are between unity and n.

2·3. *Green's functions.* – The equation of motion and commutation relation can be used in the determination of the Green's function

$$g(j,k,\tau;j'k',\tau')=\langle\!\langle(b_{j,k,\tau}b_{j'k'\tau'})_+\rangle\!\rangle\,, \tag{2.21}$$

where now the ordering symbol is designed to be appropriate for a set of operators which obey the « fermion » anticommutation relations, (2.20) *i.e.*

$$(b_{j,k,\tau}b_{j'k'\tau'})_+=\begin{cases}b_{j,k,\tau}b_{j'k'\tau'} & \text{for } j'\geqslant j\,,\\ -b_{j'k'\tau'}b_{j,k,\tau} & \text{for } j'<j\,.\end{cases} \tag{2.22}$$

Then, the Green's function obeys

$$g(j+1,k,j',k')-\sum_{\bar{k}}q(k-\bar{k})g(j,\bar{k};j'k')=-q(k-k')\delta_{j,j'}\,. \tag{2.23}$$

In writing (2.23), we have surpressed the explicit τ-dependence by considering g to be 2×2 matrix in the τ-space. The right-hand side of (2.23) reflects the re-ordering of b's (from (2.22)) as j passes through j'. This is the analog of the discontinuity term in the standard Green's function theory.

For some purposes, it is necessary to have an even more compact notation to express results like (2.23). In this notation $g(jk\tau; j'k'\tau')$ is considered to be a matrix in a 2 mn-dimensional space in which the basis vectors are labeled by j, k, and τ. Translation operators $\exp[\pm iP_x]$ and $\exp[\pm iP_y]$ are defined by taking

$$f = \exp[\pm iP_x]g\,, \qquad h = \exp[\pm iP_y]g\,, \tag{2.24a}$$

to imply

$$\begin{cases} f(jk\tau; j'k'\tau') = g(j\pm 1,\, k,\, \tau;\, j',\, k',\, \tau')\,, \\ h(jk\tau; j'k'\tau') = g(j,\, k\pm 1,\, \tau;\, j',\, k',\, \tau')\,. \end{cases} \tag{2.24b}$$

Thence, eq. (2.23) may be written in this more compact notation as

$$\exp[iP_x]\,g - Qg = -Q\,, \tag{2.25}$$

with, from (2.18),

$$Q = \exp[-iP_y\tau_3/2]\exp[2K^*\tau_2]\exp[iP_y\tau_3/2]\exp[-2K'\tau_2]\,. \tag{2.26}$$

Strictly speaking, eq. (2.25) does not determine g. A full determination requires this difference equation to be supplemented by an appropriate set of boundary conditions. The boundary condition in j is very easy: one sees without difficulty that

$$g(m+1,\, k;\, j'k') = -g(1k;\, j'k')\,. \tag{2.27}$$

The other boundary condition, at the boundaries $k=1$ and $k=n$ is much more difficult to construct. In fact, our basic equation of motion (2.19) fails near these boundaries so that, if the boundary conditions are really important, we cannot proceed further without considerable extra effort.

But, as the system becomes quite large, $n\to\infty$ and $m\to\infty$, what happens near the boundaries should not affect the behavior of g in the interior region $1\ll j\ll m$, $1\ll k\ll n$ just so long as no infinities are introduced near the boundaries. Thus, instead of introducing the correct spatial boundary condition, we immediately proceed to the limit of an infinite system and demand only that $g(jk;\, j'k')$ remain finite as j and k get very far from j' and k'. To enforce this demand and to include the fact that the large system must be translationally invariant, we write

$$(2.28)\qquad g(j,k;j',k')=\int_0^{2\pi}\frac{dp_x}{2\pi}\int_0^{2\pi}\frac{dp_y}{2\pi}\exp[ip_x(j-j')+ip_y(k-k')]g(p_x,p_y)$$

and from (2.24) that the 2×2 matrix $g(p_x, p_y)$ obeys

$$(2.29)\qquad g(p_x,p_y)=q(p_y)/[q(p_y)-\exp[ip_x]].$$

It is interesting to ask why this simpleminded neglect of boundary conditions produces unique answers. To study this point, we look back at our original definition of $\langle\!\langle\ \rangle\!\rangle$ given by (2.7) and see that it involves a weighting function $(\boldsymbol{P})^m$. Of course, as $m\to\infty$ this weighting factor becomes a projection operator for the eigenstate of P with maximum eigenvalue. We denote the eigenstate by $|\text{max}\rangle$ and the eigenvalue by P'_{max}. Thus, as $m\to\infty$

$$g(jk;j'k')=\begin{cases}\langle\text{max}|b_k(\boldsymbol{P}/P'_{\text{max}})^{j'-j}b_{k'}|\text{max}\rangle & \text{for } j\leqslant j',\\ -\langle\text{max}|b_{k'}(P/P'_{\text{max}})^{j-j'}b_k|\text{max}\rangle & \text{for } j'<j\,.\end{cases}$$

Because P/P'_{max} has eigenvalues all of which are less than or equal to unity, g has no infinities as $|j-j'|\to\infty$. If we tried to define our $\langle\!\langle\rangle\!\rangle$ by using any other weighting factor besides $\boldsymbol{P}^m$, the equations of motion, (2.25), will remain unaffected but the proof that g remains finite as $|j-j'|\to\infty$ will fail. Thus, in the absence of degeneracy of the maximum eigenvalue of $\boldsymbol{P}$ our boundary condition of regularity at infinity uniquely defines g in the limit of the large system. Unfortunately it is known that $\boldsymbol{P}$ does have a degeneracy in this eigenvalue below T_c. Nonetheless, we still expect that our precedure is designed to pick out the physically meaningful solution to eq. (2.25) and therefore that our solution is correct in the limit $n\to\infty$, $m\to\infty$.

2'4. *The partition function.* – From g, we can easily determine the first derivative of Z with respect to K or K' and hence the average of $\mathscr{H}$. From (2.2), (2.4), (2.5) and (2.8), if we make an infinitesimal change in K and K', the corresponding change in the partition function obeys

$$(2.30)\qquad \mathscr{D}\left[\log Z-\frac{mn}{2}\log(2\sinh 2K)\right]=\sum_{j,k}[(\mathscr{D}K')\langle\!\langle\boldsymbol{s}_{j,k}\boldsymbol{s}_{j,k+1}\rangle\!\rangle+(\mathscr{D}K^*)\langle\!\langle\boldsymbol{C}_{j,k}\rangle\!\rangle].$$

We use a script D for this differential to distinguish it from all others. Because of the translational invariance, the expectation values do not depend upon

j or k. From (2.11), (2.23), (2.24), and (2.28),

(2.31a) $$\langle\!\langle \boldsymbol{s}_{j,k}\boldsymbol{s}_{j,k+1}\rangle\!\rangle = -2i\langle\!\langle \boldsymbol{b}_{j,k,+}\boldsymbol{b}_{j,k,-}\rangle\!\rangle = 2i\langle\!\langle \boldsymbol{b}_{j,k,-}\boldsymbol{b}_{j,k,+}\rangle\!\rangle = -\int_0^{2\pi}\frac{\mathrm{d}p_x}{2\pi}\int_0^{2\pi}\frac{\mathrm{d}p_y}{2\pi}\,\mathrm{tr}\left[\tau_2 g(p_x, p_y)\right],$$

(2.31b) $$\langle\!\langle \boldsymbol{C}_{j,k}\rangle\!\rangle = 2i\langle\!\langle \boldsymbol{b}_{j,k,+}\boldsymbol{b}_{j,k-1,-}\rangle\!\rangle = -2i\langle\!\langle \boldsymbol{b}_{j,k-1,-}\boldsymbol{b}_{j,k,+}\rangle\!\rangle = \int_0^{2\pi}\frac{\mathrm{d}p_x}{2\pi}\int_0^{2\pi}\frac{\mathrm{d}p_y}{2\pi}\,\mathrm{tr}\left[\tau_2\exp\left[ip_y\tau_3\right]g(p_x, p_y)\right].$$

Here tr indicates a diagonal sum over the 2×2 τ-space. It is a matter of simple spin algebra to evaluate these traces and thus find the differential of the partition function.

With a view to our later analyses, we evaluate Z by a slightly different route, which looks more complicated but actually involves less algebra. By using (2.31), we can write (2.30) as a diagonal sum over the entire $2mn$-dimensional space of j, k, and τ. For example, the coefficient of $\mathscr{D}K'$ in (2.30) is

$$-\sum_{j=1}^{m}\sum_{k=1}^{n}\mathrm{tr}\left[\tau_2 g(jk; jk)\right] = -\mathrm{Tr}\left[g\tau_2\right],$$

where Tr denotes a diagonal sum over all the variables in g. With a similar expression for the other term, the right-hand side of (2.30) becomes

$$\mathrm{Tr}\left[-(\mathscr{D}K')g\tau_2 + \exp\left[-iP_y\tau_3\right]\tau_2(\mathscr{D}K^*)g\right]$$

when we use our most implicit matrix notation. In this notataion, the « solution » of (2.26) is

$$g = \frac{1}{Q - e^{iP_x}}\,Q = Q\,\frac{1}{Q - e^{iP_x}}$$

and, from (2.21), the differential of Q is

$$\mathscr{D}Q = -2Q\mathscr{D}K'\tau_2 + 2\exp\left[-iP_y\tau_3\right]\tau_2\mathscr{D}K^*Q\,.$$

Thence, the right-hand side of (2.30) is

$$\frac{1}{2}\,\mathrm{Tr}\left[\frac{1}{Q - e^{iP_x}}\,\mathscr{D}Q\right].$$

5160

But it is true for any operator X that

$$\mathscr{D}(\mathrm{Tr}\ \log X) = \mathrm{Tr}\,(X^{-1}\mathscr{D}X)$$

even if $\mathscr{D}X$ and X do not commute. Consequently the differential eq. (2.30) may be integrated in the form

$$\log Z - \frac{mn}{2}\log\,(2\sinh 2K) = +\frac{1}{4}\,\mathrm{Tr}\,[\log\,(Q - e^{iP_x})^2] + \mathrm{const}\,, \tag{2.32}$$

where the constant must be independent of K and K'. A brief examination of the low-temperature limit, $K'\to\infty$, $K\to\infty$, $K^*\to 0$ indicates that the constant in fact vanishes.

The trace in (2.32) is easily evaluated since Q is diagonal in momentum space. Thus, this term is

$$\frac{mn}{4}\int_0^{2\pi}\frac{\mathrm{d}p_x}{2\pi}\int_0^{2\pi}\frac{\mathrm{d}p_y}{2\pi}\,\mathrm{tr}\,\{\log\,[q(p_y) - e^{ip_x}]^2\}\,. \tag{2.33}$$

To find the remaining two-dimensional trace we must know the eigenvalues of the 2×2 matrix $q(p_y)$. Denote these by $q'(p_y)$ and $q''(p_y)$. From (2.18)

$$\begin{cases} \det q(p_y) = q'(p_y)\,q''(p_y) = 1\,, \\ \tfrac{1}{2}\,\mathrm{tr}\,q(p_y) = \tfrac{1}{2}[q'(p_y) + q''(p_y)] = \\ \qquad\qquad = \cosh 2K'\cosh 2K^* - \cos p_y \sinh 2K'\sinh 2K^*. \end{cases} \tag{2.34}$$

If we write $q'(p_y) = \exp\,[\gamma(p_y)]$, $q''(p_y) = \exp\,[-\gamma(p_y)]$ and pick $\gamma(p_y) > 0$, eq. (2.34) implies

$$\cosh\gamma(p_y) = \cosh 2K'\cosh 2K^* - \cos p_y\sinh 2K'\sinh 2K^*, \tag{2.35}$$

while the expression (2.33) becomes

$$\log Z = \frac{mn}{2}\left[\log\,(2\sinh 2K) + \int_0^{2\pi}\frac{\mathrm{d}p_y}{2\pi}\,\gamma(p_y)\right] \tag{2.36}$$

for the partition function of the two-dimensional Ising model. This partition function is regular except just at $K' = K^*$ where $\gamma(p_y)$ has a singularity at $p_y = 0$. When T is below T_c, *i.e.*, for $K' > K^*$, the system is in its ordered (« ferromagnetic ») phase; for $K' < K^*(T > T_c)$ the system is in its disordered phase.

3. – Spin correlation functions at infinite distances.

3·1. *Formulation.* – This paper is mainly concerned with the computation of the correlation function $\langle\langle(\boldsymbol{s}_{j_1,k_1}\boldsymbol{s}_{j_2,k_2})\rangle\rangle$. Since the $\boldsymbol{s}$'s are not simply expressible in terms of $\boldsymbol{b}$'s we must employ a trick. We use the familiar field-theoretic device of the generalization of the « Hamiltonian », here $\boldsymbol{P}^m$.

Let us define for the case $j_1 \neq j_2$

$$\boldsymbol{P}(j) = \begin{cases} \boldsymbol{s}_{k_1}\boldsymbol{P} & \text{for } j = j_1, \\ \boldsymbol{s}_{k_2}\boldsymbol{P} & \text{for } j = j_2, \\ \boldsymbol{P} & \text{otherwise,} \end{cases} \tag{3.1}$$

$$\boldsymbol{U}(j) = \Big(\prod_{r=1}^{j-1} \boldsymbol{P}(r)\Big)_+ , \tag{3.2}$$

and thereby generate a new partition function instead of (2.2). This is

$$Z' = \text{trace}\,[\boldsymbol{U}(m+1)] = Z\langle\sigma_{j_1,k_1}\sigma_{j_2,k_2}\rangle\,. \tag{3.3}$$

Thus the new partition function serves as a way of generating the average we wish to compute.

In employing the new ensemble, we define, in analogy to (2.7) and (2.8),

$$\langle\!\langle\boldsymbol{\Lambda}\rangle\!\rangle' = \text{trace}\,[\boldsymbol{\Lambda}\boldsymbol{U}(m+1)]/Z', \tag{3.4}$$

$$\boldsymbol{\Lambda}(j) = \boldsymbol{U}(j)\boldsymbol{\Lambda}[\boldsymbol{U}(j)]^{-1}, \tag{3.5}$$

and especially

$$g'(jk\tau; j'k'\tau') = \langle\!\langle(b_{j,k,\tau}b_{j',k',\tau'})_+\rangle\!\rangle'\,. \tag{3.6}$$

From (2.12), (3.1), and (3.2), the equation of motion for b is

$$b_{j+1,k,\tau} = \sum_{\bar{k}\bar{\tau}} q_{\tau,\bar{\tau}}(k-\bar{k}) \times \begin{cases} s_{j_1,k_1}b_{j,\bar{k},\bar{\tau}}s_{j_1,k_1} & \text{for } j = j_1, \\ s_{j_2,k_2}b_{j,\bar{k},\bar{\tau}}s_{j_2,k_2} & \text{for } j = j_2, \\ b_{j,\bar{k},\bar{\tau}} & \text{otherwise.} \end{cases} \tag{3.7}$$

From the definition, (2.10), of b it follows that the only effect of the factors s in (3.7) is to produce extra minus signs in this equations, that is

$$s_{j_1,k_1}b_{j_1,k,\tau}s_{j_1,k_1} = \begin{cases} b_{j_1,k,\tau} & \text{for } n > k > k_1 > 1, \\ -b_{j_1,k,\tau} & \text{for } 1 < k \leqslant k_1 < n\,. \end{cases}$$

5162

Thus, (3.7) may now be written as

$$b_{j+1,k,\tau} = \sum_{\bar{k},\bar{\tau}} q_{\tau,\bar{\tau}}(k-\bar{k})[1-2\eta_1(j,\bar{k})-2\eta_2(j,\bar{k})]b_{j,\bar{k},\bar{\tau}}\,. \tag{3.8}$$

Here the $\eta_1(j,k)$ and $\eta_2(j,k)$ have been inserted to produce the correct minus signs. They are defined by

$$\eta_1(j,k) = \begin{cases} 1 & \text{if } j=j_1 \textit{ and } n>k_1 \geqslant k>1, \\ 0 & \text{otherwise,} \end{cases}$$

and a similar expression for η_2.

Exactly the same logic which gave us (2.25) now implies that g' obeys

$$e^{iP_x}g' - Q'g' = -Q', \tag{3.9}$$

where Q' differs from Q because of the extra minus signs in (3.8). In this implicit notation we write

$$Q' = Q(1-2\eta_1-2\eta_2)\,, \tag{3.10a}$$

where the η's are projection operators diagonal in j, k, and τ, *e.g.*

$$\eta_1(jk\tau, j'k'\tau') = \delta_{j,j'}\delta_{k,k'}\delta_{\tau,\tau'}\eta_1(j,k)\,. \tag{3.10b}$$

We solve (3.9) by again using the boundary condition that $g'(jk\tau;j'k'\tau')$ does not diverge as $|j-j'|$ or $|k-k'|$ become large. Just so long as the points (j_1,k_1), (j_2,k_2), (j,k), and (j',k') all be very far from the boundaries, we should expect that the g' thereby determined will only differ infinitesimally from the true g'. The identical calculation which gave eq. (2.32) now gives

$$\log Z' = \frac{mn}{2}\log(2\sinh 2K) + \frac{1}{4}\,\mathrm{Tr}\,[\log(Q'-e^{iP_x})^2] + (\text{const})', \tag{3.11}$$

where (const)$'$ must be independent of K and K'. Thence, the subtraction of (2.32) from (3.11) gives

$$\log\langle\sigma_1\sigma_2\rangle = \log Z' - \log Z = \tfrac{1}{4}\,\mathrm{Tr}\,[\log(Q'-e^{iP_x})^2 - \log(Q-e^{iP_x})^2] + \\ + (\text{const})' = \tfrac{1}{4}\,\mathrm{Tr}\,\{\log[1-2g(\eta_1+\eta_2)]^2\} + (\text{const})'. \tag{3.12}$$

In (3.12), σ_1 and σ_2 are, of course, abbreviations for σ_{j_1,k_1} and σ_{j_2,k_2}. The last line of (3.12) is true because

$$\mathrm{Tr}\log X + \mathrm{Tr}\log Y = \mathrm{Tr}\log XY$$

even if X and Y do not commute. The constant in (3.12) must be independent of K and K'—below it is shown to be zero. If we assume for the moment that this is true, the fact that for any X

$$\operatorname{tr}\log X = \log\det X$$

implies that

$$\langle\sigma_1\sigma_2\rangle^4 = \det[1-2g(\eta_1+\eta_2)]^2.$$

Similar determinantal forms for the spin-spin correlation function have been derived by a variety of techniques in the previous literature, *e.g.*, ref. ([4,10,11]). The derivation outlined here is new, however.

From the fact that the η's are projection operators, it follows that eq. (3.12) can be restated as

$$\text{(3.13)} \qquad \log\langle\sigma_1\sigma_2\rangle = \tfrac{1}{4}\operatorname{Tr}\log[1-2(\eta_1+\eta_2)g(\eta_1+\eta_2)]^2 + (\text{const})'.$$

Equation (3.13) will be our basic tool for the investigation of correlation functions.

There is a sense in which eq. (3.13) is obviously wrong. Our basic expressions for g and g' are only correct when $1 \ll j \ll m$ and $1 \ll k \ll n$, *i.e.* when we look far from the boundaries of the system. However, the trace in (3.13) is a diagonal sum over *all* j and k. Since this sum includes regions in which g and g' are inaccurately determined we should be very seriously concerned about the correctness of (3.13). We attempt to rescue ourselves from this mathematical difficulty by a physical argument. It is reasonable to believe that the sum over j and k defined by the traces in eqs. (2.32), (3.11), and (3.13) is, in fact, a summing up of contributions from different regions of the lattice. The behavior of the spin-spin correlation function in (3.13) should be produced by regions of the lattice which are not infinitely far from the spin sites $((j_1, k_1)$ and $(j_2, k_2))$ just so long as the distance from any one spin to a boundary is infinitely greater than the distance between spins. By this physical argument, the only terms in the diagonal sum of (3.13) which should make any appreciable contribution to $\langle\sigma_1\sigma_2\rangle$ should have j (or k) infinitely greater than unity and infinitely less than m (or n). To insure that this physical requirement is satisfied, we insert a convergence factor in (3.13), writing this as

$$\text{(3.13a)} \qquad \log\langle\sigma_1\sigma_2\rangle = \tfrac{1}{4}\operatorname{Tr}\left\{I\log[1-2(\eta_1+\eta_2)g(\eta_1+\eta_2)]^2\right\} + (\text{const})'.$$

Here, I is an operator diagonal in j, k, and τ, which serves to eliminate the unwanted boundary regions. It is chosen to be equal to unity within the deep interior of the lattice, in particular in the neighborhood of (j_1, k_1) and (j_2, k_2), but to decay exponentially to zero toward the boundaries. We choose this decay to be sufficiently slow so that I essentially commutes with the argument of the log in (3.13*a*). With this choice, I will usually play no very essential role in our manipulations of the trace. In most of our calculations it simply serves as a reminder that we need not make any special calculations of contributions from boundary regions because these contributions have been eliminated by a physical argument. Henceforth, we do not write I explicitly, except when absolutely necessary.

3·2. *Properties of g.* – Before (3.13) is attacked, good strategy dictates a listing of relevant properties of g. For clarity, in this Section we exhibit the j, k indices of g explicitly; for compactness, we surpress the τ, τ' indices so that we are always working in a 2×2 τ-space.

We begin from $q(p_y)$, defined by eq. (2.18). This can be arranged to show all the τ-dependence explicitly as

$$q(p_y) = \tfrac{1}{2}\{1 + [\Phi(p_y)]^{\bar{\tau}_3}\tau_2\}\exp[\gamma(p_y)] + \tfrac{1}{2}\{1 - [\Phi(p_y)]^{\bar{\tau}_3}\tau_2\}\exp[-\gamma(p_y)] \tag{3.14}$$

with

$$\bar{\tau}_3 = \exp[K'\tau_2]\tau_3\exp[-K'\tau_2]\,. \tag{3.15}$$

Since $\bar{\tau}_3$ anticommutes with τ_2, $\frac{1}{2}(1\pm\Phi^{\bar{\tau}_3}\tau_2)$ are projection operators which pick out the eigenvalues $q'(p_y) = \exp[\pm\gamma(p_y)]$. The function $\Phi(p_y)$ is rather complicated. It is conveniently written in terms of the constants A and B which are, in the notation of YANG ([12]),

$$\begin{cases} A = \dfrac{\cosh 2K'\cosh 2K^* + \cosh 2K' + \cosh 2K^* + 1}{\sinh 2K'\sinh 2K^*}\,, \\[2ex] B = \dfrac{\cosh 2K'\cosh 2K^* - 1 + |\cosh 2K' - \cosh 2K^*|}{\sinh 2K'\sinh 2K^*}\,. \end{cases} \tag{3.16}$$

Notice that

$$A > B > 1\,. \tag{3.17}$$

For $T < T_c$, *i.e.*, for $K' > K^*$,

$$\Phi(p_y) = \left(\frac{B - e^{ip_y}}{A - e^{ip_y}}\cdot\frac{Ae^{ip_y} - 1}{Be^{ip_y} - 1}\right)^{\frac{1}{2}}, \tag{3.18}$$

while for $T > T_c$, *i.e.*, for $K' < K^*$,

$$\Phi(p_y) = \left[\frac{(B - e^{ip_y})\cdot(A - e^{ip_y})}{(Be^{ip_y} - 1)\cdot(Ae^{ip_y} - 1)}\right]^{\frac{1}{2}}. \tag{3.19}$$

([12]) C. N. YANG: *Phys. Rev.*, **85**, 808 (1952).

Inside the trace (3.13), one can perform a unitary transform which leaves the trace unchanged and only has the effect of converting $\bar{\tau}_3$ into τ_3. Consequently, we may safely ignore the difference between these two quantities and we shall henceforth write τ_3 instead of $\bar{\tau}_3$.

From (3.14) and (2.29)

$$g(p_x, p_y) = \frac{1}{2}\{1 + [\Phi(p_y)]^{\tau_3}\tau_2\}\frac{\exp[\gamma(p_y)]}{\exp[\gamma(p_y)] - \exp[iP_x]} + \\ + \frac{1}{2}\{1 - [\Phi(p_y)]^{\tau_3}\tau_2\}\frac{\exp[-\gamma(p_y)]}{\exp[-\gamma(p_y)] - \exp[iP_x]}\,. \tag{3.20}$$

Equation (3.20) permits the calculation of the p_x integral in (2.28) and yields g as

$$g(jk; j'k') = \int_0^{2\pi}\frac{dp_y}{2\pi}\exp[-\gamma(p_y)|j-j'| + ip_y(k-k')]\cdot \begin{cases} \frac{1}{2}\{1 + [\Phi(p_y)]^{\tau_3}\tau_2\} & \text{for } j \leqslant j', \\ -\frac{1}{2}\{1 - [\Phi(p_y)]^{\tau_3}\tau_2\} & \text{for } j > j'. \end{cases} \tag{3.21}$$

From eq. (3.21), it follows that g decreases exponentially as (j, k) and (j', k') become far from one another, except exactly at $T = T_c$. This exponential behavior is important because it helps to insure that traces like (3.13a) converge rapidly.

Since η_1 is a projection operator which picks out $j = j_1$, $\eta_1 g \eta_1$ involves only the form of g in (3.21) which results when we set $j = j'$. In particular, the jk, $j'k'$ matrix element of $\eta_1 g \eta_1$ is

$$\int_0^{2\pi}\frac{dp_y}{2\pi}\frac{1}{2}\{1 + [\Phi(p_y)]^{\tau_3}\tau_2\}\exp[ip_y(k-k')] \tag{3.22}$$

when $j = j' = j_1$, $k \leqslant k_1$, and $k' \leqslant k_1$. Otherwise this matrix element vanishes. This result may be written in the more implicit notation as

$$1 - 2\eta_1 g \eta_1 = 1 - \eta_1 - \eta_1[\Phi(P_y)]^{\tau_3}\tau_2\eta_1\,. \tag{3.23}$$

For our later analyses, it will be important to decompose $\Phi(p_y)$ into factors which have simple analytic properties in p_y. For $T < T_c$, (3.18) may be factorized to yield

$$(3.24)\qquad \begin{cases} \Phi(p_y) = c_0(p_y)\, c_i^{-1}(p_y)\,, \\[2mm] c_0(p_y) = \left(\dfrac{B - e^{iP_y}}{A - e^{iP_y}}\right)^{\frac{1}{2}}, \\[2mm] c_i(p_y) = \left(\dfrac{Be^{iP_y} - 1}{Ae^{iP_y} - 1}\right)^{\frac{1}{2}}, \end{cases}$$

while for $T < T_c$, (3.19) may be factorized as

$$(3.25)\qquad \begin{cases} \Phi(p_y) = e^{iP_y} d_0(p_y)\, d_i^{-1}(p_y)\,, \\ d_0(p_y) = [(A - e^{iP_y})(B - e^{iP_y})]^{-\frac{1}{2}}\,, \\ d_i(p_y) = e^{iP_y}[(Ae^{iP_y} - 1)(Be^{iP_y} - 1)]^{-\frac{1}{2}}. \end{cases}$$

The factorizations (3.24) and (3.25) have been chosen so that c_0, c_0^{-1}, d_0, and d_0^{-1} are each functions of $z = e^{ip_y}$ which have all their singularities outside the unit circle, $|z| = 1$. Inside this circle these functions are each analytic. Conversely c_i, c_i^{-1}, d_i, and d_i^{-1} have all the singularities inside the unit circle and are analytic everywhere outside this circle including the point at $z = \infty$. The usefulness of these analytic properties becomes apparent if the c's and d's are transformed into matrices in j and k. For example, write the matrix element

$$(3.26)\qquad [c_0^{\pm 1}]_{jk;j'k'} = \int_0^{2\pi} \frac{dp_y}{2\pi} \exp[ip_y(k-k')][c_0(p_y)]^{\pm 1}\delta_{j,j'} = \\ = \oint \frac{dz}{2\pi i}\, z^{(k-k'-1)} \left(\frac{B-z}{A-z}\right)^{\pm\frac{1}{2}} \delta_{j,j}\,,$$

where the contour is the unit circle. Because of the analytic properties of $c_0(p_y)$, $(c_0^{\pm 1})_{jk;j'k'}$ vanish when $k > k'$. We denote the matrices in $jk\tau$ space, whose matrix elements are given by (3.26), by $c_0^{\pm 1}(P_y)$ or, more simply, by $c_0^{\pm 1}$. Because η_1 is a projection operator onto ($j = j_1$, $k \leqslant k_1$) and because c_0 is diagonal in j, the vanishing of (3.26) for $k > k'$ implies that all matrix elements of $(1 - \eta_1)c_0^{\pm 1}\eta_1$ must vanish. Thus, the relations

$$(3.27)\qquad (1-\eta_1)c_0^{\pm 1}\eta_1 = (1-\eta_1)d_0^{\pm 1}\eta_1 = \eta_1 c_i^{\pm 1}(1-\eta_1) = \eta_1 d_i^{\pm 1}(1-\eta_1) = 0\,,$$

follow from the analytic structure of the c's and d's. Equation (3.27) will prove invaluable in the evaluation of the spin-spin correlation function.

5167

3·3. *Zero-field magnetization.* – When the constant of integration in (3.13) is evaluated, this equation will permit the calculation of the spin-spin correlation function in a variety of interesting cases. Since (const)′ is independent of K and K', it can be evaluated by a consideration of the zero-temperature situation in which all the spins are known to point in the same direction so that $\langle\sigma_1\sigma_2\rangle = 1$ and (3.13*a*) reads

$$(3.28)\qquad (\text{const})' = -\tfrac{1}{4}\,\text{Tr}\left\{\log[1-2(\eta_1+\eta_2)g(\eta_1+\eta_2)]^2\right\}.$$

But, at $T=0$, $K\to\infty$, $K'\to\infty$, $K^*\to 0$ so that γ, A, and B all become infinite. Then Φ reduces to unity and from (3.21)

$$g(jk; j'k') = \tfrac{1}{2}(1+\tau_2)\,\delta_{j,j'}\,\delta_{k,k'}$$

or in the more implicit notation

$$g = \tfrac{1}{2}(1+\tau_2)\,.$$

Since the η's commute with τ_2 and since $\eta_1^2=\eta_1$, $\eta_1\eta_2=0$ the argument of the log in (3.28) is simply the unit operator. Thence, (const)′ vanishes.

According to ref. [10,12], the zero-field magnetization is proportional to $\langle\sigma\rangle$, where $\langle\sigma\rangle$ is given as the positive number such that

$$(3.29)\qquad \langle\sigma\rangle = \left[\lim_{|j_1-j_2|\to\infty}\langle\sigma_{j_1,k_1}\sigma_{j_2,k_2}\rangle\right]^{\frac{1}{2}}.$$

However as j_1 and j_2 become infinitely distant $\eta_1 g\eta_2$ and $\eta_2 g\eta_1$ become exponentially small so that (3.13) implies

$$(3.30)\qquad \log\langle\sigma\rangle = \tfrac{1}{8}\,\text{Tr}\left\{\log[(1-2\eta_1 g\eta_1)^2(1-2\eta_2 g\eta_2)^2]\right\} =$$

$$= \tfrac{1}{8}\,\text{Tr}\left\{\log[(1-2\eta_1 g\eta_1)^2]\right\} + \tfrac{1}{8}\,\text{Tr}\left\{\log[(1-2\eta_2 g\eta_2)^2]\right\} = \tfrac{1}{4}\,\text{Tr}\left\{\log(1-2\eta_1 g\eta_1)^2\right\}.$$

The last line of (3.30) follows from the translational invariance which requires that the traces be independent of j_1, k_1, j_2 and k_2.

In ref. [12], YANG evaluated the trace of eq. (3.30) by finding all the eigenvalues of the operator inside the log and then performing the indicated summation. This calculation is possible because the difference equation which appears in this eigenvalue problem is soluble through the use of the Wiener-Hopf method [13]. To see a solution of this type notice that for $T<T_c$ from

[13] P. M. MORSE and H. FESHBACK: *Methods of Theoretical Physics* (New York, 1953), p. 978.

(3.23) and (3.24)

$$1-2\eta_1 g\eta_1 = 1-\eta_1-\eta_1 c_i^{-\tau_3} c_0^{\tau_3}\tau_2\eta_1 \,. \tag{3.31}$$

Then the Wiener-Hopf method gives the inverse of the matrix (3.31) as

$$(1-2\eta_1 g\eta_1)^{-1} = 1-\eta_1-c_0^{\tau_3}\eta_1 c_i^{-\tau_3}\tau_2 \,. \tag{3.32}$$

Equation (3.32) may be verified by direct matrix multiplication with the aid of (3.27).

We employ a slightly different technique to get the ONSAGER (2) and YANG (12) result for $\langle\sigma\rangle$. Take the differential of (3.30) with respect to K and K^*

$$\mathscr{D}[\log\langle\sigma\rangle] = -\tfrac{1}{2}\operatorname{Tr}[(1-2\eta_1 g\eta_1)^{-1}\mathscr{D}(2\eta_1 g\eta_1)] \,.$$

With the aid of (3.32), this yields

$$\mathscr{D}[\log\langle\sigma\rangle] = \tfrac{1}{2}\operatorname{Tr}[c_0^{\tau_3}\eta_1 c_i^{-\tau_3}\mathscr{D}(c_i^{\tau_3}c_0^{-\tau_3})\eta_1] \,. \tag{3.33}$$

Notice that the only terms which contribute in the diagonal sum of (3.33) are those for $j=j_1$ and $k\leqslant k_1$. Thence, (3.33) is really a sum over k and over $\tau_3=\pm 1$. Unfortunately, this sum is only conditionally convergent; it is necessary to do the τ_3 sum first in order to make the k sum converge for $k\ll k_1$. To make this sum unambiguous, we use the convergence factor I, which we here take to be a diagonal matrix in j, k, and τ, which is independent of τ, and is proportional to $\exp[-\delta|k-k_1|]$. If δ is infinitesimally small but $\delta k_1\gg 1$, this factor effects the removal of the boundary region from the sum in (3.23) and hence is exactly the type of convergence factor demanded earlier in this analysis.

Once I is inserted into (3.33), so that this becomes

$$\mathscr{D}[\log\langle\sigma\rangle] = \tfrac{1}{2}\operatorname{Tr}[Ic_0^{\tau_3}\eta_1 c_i^{-\tau_3}\eta_1\mathscr{D}(c_i^{\tau_3}c_0^{-\tau_3})\eta_1] \,, \tag{3.33a}$$

one can proceed to use all the tricks for evaluating traces without worrying about convergence difficulties. Equation (3.27) and the diagonality of I imply that the last two factors of η_1 in (3.33*a*) can be replaced by unity without affecting the value of the trace. Thence, (3.33*a*) becomes

$$\mathscr{D}[\log\langle\sigma\rangle] = \tfrac{1}{2}\operatorname{Tr} Ic_0^{\tau_3}\eta_1 c_i^{-\tau_3}\mathscr{D}[c_0^{-\tau_3}c_i^{\tau_3}] \,. \tag{3.34}$$

Since the c's are functions of the translation operators, e^{iP_y}, the sums over k in (3.34) are most easily performed in momentum space. With this end in view define

(3.35) $$I(p_y - p_y') = \sum_{kk'} \exp[-ip_y k] I_{j_1k;j_1k'} \exp[ip_y'k'] = \\ = \sum_{k=-\infty}^{+\infty} \exp[-i(p_y - p_y')k] \exp[-\delta|k-k_1|] .$$

As $\delta \to 0$ this becomes equal to 2π times the delta-function of $(p_y - p_y')$. Similarly, define

(3.36) $$\eta(p_y - p_y') = \sum_{kk'} \exp[-ip_y k] \eta_{1j_1k;j_1k'} \exp[ip_y k'] = \sum_{k=-\infty}^{k_1} \exp[-i(p_y - p_y')k] = \\ = \frac{\exp[-i(p_y - p_y')k_1]}{1 - \exp[i(p_y - p_y')]} .$$

In (3.36), we have neglected a term proportional to $\exp[-i(p_y - p_y')\infty]$ which oscillates too rapidly to have any effect.

Equation (3.34) may now be written in momentum space as

$$\mathscr{D}[\log \sigma] = \frac{1}{2}\int_0^{2\pi} \frac{dp_y}{2\pi} \int_0^{2\pi} \frac{dp_y'}{2\pi} I(p_y - p_y')\eta(p_y' - p_y) \sum_{\tau_3=\pm 1} [c_0(p_y')]^{\tau_3}[c_i(p_y)]^{-\tau_3} \cdot \\ \cdot \mathscr{D}\{[c_0(p_y)]^{-\tau_3}[c_i(p_y)]^{-\tau_3}\} .$$

Because of the δ-function behavior of I, this reduces to

(3.37) $$\mathscr{D}[\log \sigma] = \frac{1}{2}\int_0^{2\pi} \frac{dp_y}{2\pi} \lim_{p_y' \to p_y} \eta(p_y - p_y') \sum_{\tau_3=\pm 1} [c_0(p_y')]^{\tau_3}[c_i(p_y)]^{-\tau_3} \cdot \\ \cdot \mathscr{D}\{[c_0(p_y)]^{-\tau_3}[c_i(p_y)]^{\tau_3}\} .$$

From (3.36), near $p_y' = p_y$

$$\eta(p_y - p_y') = \frac{i}{p_y - p_y'} ,$$

while the τ_3 sum in (3.37) is, near $p_y' = p_y$,

$$-(p_y - p_y')\frac{\partial c_0(p_y)}{\partial p_y}[c_i(p_y)]^{-1}[c_0(p_y)]^{-1} c_i(p_y) - \\ -(p_y - p_y')\frac{\partial [c_0(p_y)]^{-1}}{\partial p_y} c_i(p_y) c_0(p_y)[c_i(p_y)]^{-1} .$$

5170

Thus, the limit on the right-hand side of (3.37) is well defined and, after a bit of algebra, this equation reduces to

$$\mathscr{D}[\log\langle\sigma\rangle] = -\int_0 \frac{dp_y}{2\pi i}\left[\frac{\partial}{\partial p_y}\log c_0(p_y)\right]\mathscr{D}[\log c_i(p_y)] = \tag{3.38}$$

$$= -\frac{1}{4}\int \frac{dz}{2\pi i}\left[\frac{1}{A-z}-\frac{1}{B-z}\right]\left[\frac{(\mathscr{D}B)z}{Bz-1}-\frac{(\mathscr{D}A)z}{Az-1}\right] =$$

$$= \frac{1}{8}\mathscr{D}\log\left[(A^2-1)(B^2-1)/(AB-1)^2\right].$$

Since $\langle\sigma\rangle = 1$ at $T = 0$, *i.e.*, $A \to B \to \infty$, (3.38) may be integrated to yield the well-known result

$$\langle\sigma\rangle = (A^2-1)^{\frac{1}{8}}(B^2-1)^{\frac{1}{8}}/(AB-1)^{\frac{1}{4}}. \tag{3.39}$$

Near $T = T_c$, $B \approx 1$, $A \approx [(\cosh 2K' + 1)/\sinh 2K']^2$ so that (3.39) becomes

$$\langle\sigma\rangle \approx [2(B-1)\cosh 2K']^{\frac{1}{8}}. \tag{3.40}$$

Since $(B-1)$ is proportional to $T_c - T$ near T_c, the magnetization is proportional to $(T_c - T)^{\frac{1}{8}}$.

3 4. $T > T_c$: *vanishing magnetization.* – Above the critical temperature, the analog of eqs. (3.30) and (3.31) reads

$$\log\langle\sigma\rangle = \tfrac{1}{4}\,\mathrm{Tr}\left\{\log[1-2\eta_1 g\eta_1]^2\right\} = \tag{3.41}$$

$$= \tfrac{1}{4}\,\mathrm{Tr}\left\{\log[1-\eta_1-\eta_1\exp[iP_y\tau_3]d_0^{\tau_3}d_i^{-\tau_3}\tau_2\eta_1]^2\right\}.$$

Since the magnetization must be zero above T_c, $1-2\eta_1 g\eta_1$ must have at least one vanishing eigenvalue ([11]). A bit of guesswork coupled with the use of eq. (3.23) indicates that the projection operator for the zero eigenstate is

$$|0\rangle\langle 0| = d_0\,\delta_1\tfrac{1}{2}(1+\tau_3)\,d_i/N. \tag{3.42}$$

In (3.42), δ_1 is a projection operator which is diagonal in j, k, and τ. It is equal to unity when $j = j_1$ and $k = k_1$ and zero otherwise. N is a normalizing factor,

$$N = \mathrm{Tr}\left[d_0\,\delta_1\frac{1}{2}(1+\tau_3)\,d_i\right] = \int_0^{2\pi}\frac{dp_y}{2\pi}\,d_0(p_y)\,d_i(p_y), \tag{3.43}$$

which makes $|0\rangle\langle 0|$ into a proper projection operator.

For our later work, it will be necessary to know the portion of the trace (3.41) in which the zero eigenstate is left out:

$$\log\langle\tilde{\sigma}\rangle = \tfrac{1}{4}\,\mathrm{Tr}\log[1-2\eta_1 g\eta_1 + |0\rangle\langle 0|]^2. \tag{3.44}$$

This trace is evaluable by exactly the same method we used for $\langle\sigma\rangle$. The first step is to calculate the inverse

$$[1-2\eta_1 g\eta_1 + |0\rangle\langle 0|]^{-1} = (1-|0\rangle\langle 0|)f_1(1-|0\rangle\langle 0|) + |0\rangle\langle 0|\,, \tag{3.45}$$

where

$$f_1 = 1-\eta_1 - \exp[iP_y(1+\tau_3)/2]\,d_0^{\tau_3}\eta_1 d_i^{-\tau_3}\exp[-iP_y(1-\tau_3)/2]\,. \tag{3.46}$$

The correctness of this result may be verified by matrix multiplication. Next, consider the change in $\log\langle\tilde{\sigma}\rangle$ produced by an infinitesimal change in the parameters A and B. From (3.44), this change is

$$\begin{aligned}\mathscr{D}[\log\tilde{\sigma}] &= \tfrac{1}{2}\,\mathrm{Tr}\,[1-2\eta_1 g\eta_1 + |0\rangle\langle 0|]^{-1}\mathscr{D}[1-2\eta_1 g\eta_1 + |0\rangle\langle 0|] = \\ &= \tfrac{1}{2}\,\mathrm{Tr}\,f\mathscr{D}[1-2\eta_1 g\eta_1] + \tfrac{1}{2}\langle 0|f(1-2\eta_1 g\eta_1)\mathscr{D}(|0\rangle) + \tfrac{1}{2}\mathscr{D}(\langle 0|)(1-2\eta_1 g\eta_1)f|0\rangle\,.\end{aligned} \tag{3.47}$$

The three terms which appear on the right-hand side of (3.47) are individually easily calculated. The first turns out to have a structure very much like (3.27), *i.e.*

$$\begin{aligned}\frac{1}{2}\mathrm{Tr}\,\{f\mathscr{D}[1-2\eta_1 g\eta_1]\} &= \frac{1}{2}\mathrm{Tr}\,[d_0^{\tau_3}\eta_1 d_i^{-\tau_3}\mathscr{D}(d_i^{\tau_3}d_0^{-\tau_3})\eta_1] = \\ &= -\int_0^{2\pi}\frac{\mathrm{d}p_y}{2\pi}\,\mathscr{D}[\log d_0(p_y)]\frac{\partial}{\partial p_y}[\log d_i(p_y)] = \\ &= \mathscr{D}\frac{1}{8}\log[(1-A^{-2})(1-B^{-2})(1-A^{-1}B^{-1})^2]\,.\end{aligned} \tag{3.48}$$

In deriving the last two lines of (3.48) one follows the same line of calculation as in Sect. **3·3**. The next terms in (3.47) are

$$\begin{aligned}\tfrac{1}{2}\langle 0|(1-d_0\,\delta d_0^{-1})\mathscr{D}|0\rangle + \tfrac{1}{2}(\mathscr{D}\langle 0|)(1-d_i^{-1}\,\delta d_i)|0\rangle &= \\ = \tfrac{1}{2}\mathscr{D}\{\log N - \log[d_0(z=0)] - \log[d_i(z=\infty)]\}\,.\end{aligned} \tag{3.49}$$

Finally, we integrate (3.47) to find

$$\langle\tilde{\sigma}\rangle = (A^2-1)^{\frac{1}{8}}(B^2-1)^{\frac{1}{8}}(AB-1)^{\frac{1}{4}}N^{\frac{1}{2}}\,. \tag{3.50}$$

5172

In writting (3.50), we have set a multiplicative constant of integration equal to unity at infinite temperature, $A = B = \infty$, $d_0 = d_i = A$, and (3.44) immediately implies $\langle\tilde{\sigma}\rangle = 1$. Since (3.50) has the same implication at infinite temperatures, the constant of integration is correct.

4. – Correlation function for $T < T_c$.

In this Section, we consider the asymptotic expansion of the correlation function for large $|j_1 - j_2|$. In Sect. **3˙3**, the correlation function was evaluated for infinite $|j_1 - j_2|$. To get the next term, we combine (3.14) and (3.25) to find

$$\log[\langle\sigma_{j_1,k_1}\sigma_{j_2,k_2}\rangle/\langle\sigma\rangle^2] = \tag{4.1}$$

$$= \tfrac{1}{4}\,\mathrm{Tr}\log[1-(1-\eta_1 2g\eta_1)^{-1}\eta_1 2\,g\eta_2-(1-\eta_2 2g\eta_2)^{-1}\eta_2 2g\eta_1]^2 = \tfrac{1}{4}\,\mathrm{Tr}\log[1-\Lambda]^2,$$

where

$$\Lambda = (1-\eta_1 2g\eta_1)^{-1}\eta_1 2g\eta_2(1-\eta_2 2g\eta_2)^{-1}\eta_2 2g\eta_1\,. \tag{4.2}$$

From (3.23) and (3.26)

$$(1-\eta_1 2g\eta_1)^{-1}\eta_1 = -\,c_0^{\tau_3}\eta_1 c_i^{-\tau_3}\tau_2\,. \tag{4.3}$$

Thus, when $j_1 > j_2$, Λ has the form

$$\Lambda = c_0^{\tau_3}\eta_1(c_0^{-\tau_3}-c_i^{-\tau_3}\tau_2)\exp[-\gamma|j_1-j_2|]\,c_0^{\tau_3}\exp[-iP_x(j_1-j_2)]\,\eta_2\cdot \tag{4.4}$$

$$\cdot\exp[iP_x(j_1-j_2)]\,(c_0^{-\tau_3}+c_i^{-\tau_3}\tau_2)\exp[-\gamma|j_1-j_2|]\,.$$

Here $\gamma = \gamma(P_y)$ is an operator whose matrix elements are defined in direct analogy with (3.26),

$$\big(\gamma(P_y)\big)_{jk;j'k'} = \delta_{k,k'}\int_0^{2\pi}\frac{dp_y}{2\pi}\,\gamma(p_y)\exp[iP_y(j-j')]\,.$$

4˙1. *Large spatial separation.* – As $|j_1 - j_2|$ becomes large, the logarithms on both the left-hand side and the right-hand side of (4.1) have arguments which approach unity. If $\gamma|j_1 - j_2| \gg 1$, these logarithms may be expanded to give

$$\frac{1}{\langle\sigma\rangle^2}\langle[\sigma_1-\langle\sigma\rangle][\sigma_2-\langle\sigma\rangle]\rangle = \frac{1}{2}\,\mathrm{Tr}\,\Lambda\,. \tag{4.5}$$

This trace is calculable in momentum space since

$$\langle p_x, p_y|\eta_1|p'_x, p'_y\rangle = \frac{\exp[ij_1(p'_x - p_x) + ik_1(p'_y - p_y)]}{1 - \exp[i(p_y - p'_y)]}, \tag{4.6}$$

so that (4.5) becomes

$$\langle[\sigma_1 - \langle\sigma\rangle][\sigma_2 - \langle\sigma\rangle]\rangle/\langle\sigma\rangle^2 = \tag{4.7}$$

$$= \frac{1}{8\pi^2}\int_{-\pi}^{\pi}\mathrm{d}p\int_{-\pi}^{\pi}\mathrm{d}p' \exp[i(p'-p)(k_1-k_2) + i(p+p') - [\gamma(p)+\gamma(p')]|j_1-j_2|]\cdot$$
$$\cdot\left[\frac{c_i(p)\,c_0(p) - c_i(p')\,c_0(p')}{e^{ip} - e^{ip'}}\right]^2 \frac{1}{c_0(p)\,c_i(p)\,c_0(p')\,c_i(p')}.$$

Equation (4.7) was derived under the assumption that, for all p_y,

$$\gamma(p_y)|j_1 - j_2| \gg 1. \tag{4.8}$$

In this case, the only consistent way of evaluating the integrals in (4.7) is to assume that the exponential is quite small. The interesting case is the one in which the system lies just below T_c, *i.e.*, K' is only very slightly greater than K^*. Then, from (3.19),

$$\begin{cases} A = (\mathrm{tgh}\, K')^{-2}, \\ B = 1 + \varepsilon, \\ \varepsilon = 2(K' - K^*)/\sinh 2K' \ll 1. \end{cases} \tag{4.9}$$

The main contribution to the integrals in (4.7) occurs when p and p' are of the order of ε. In this region

$$\begin{cases} \gamma(p) \approx \sinh 2K'[\varepsilon^2 + p^2]^{\frac{1}{2}}, \\ c_0(p)\,c_i(p) \approx \frac{1}{2}(\mathrm{tgh}\, K')\gamma(p). \end{cases} \tag{4.10}$$

Thus, the right-hand side of (4.7) can be approximated by

$$\frac{1}{8\pi^2}\int \mathrm{d}p\,\mathrm{d}p' \exp[-i(p-p')(k_1-k_2) - [\gamma(p)+\gamma(p')]|j_1-j_2|]\cdot \tag{4.11}$$
$$\cdot\left[\frac{\gamma(p)-\gamma(p')}{p-p'}\right]^2 \frac{1}{\gamma(p)\gamma(p')},$$

where the integrals are now allowed to run from $-\infty$ to ∞.

5174

We should expect the correlation function to show some hint of sperical symmetry in the case in which the coupling constants are the same in the x and y direction, *i.e.*, when $\sinh 2K' = 1$. However, (4.11) does not seem to exhibit this symmetry. To show that (4.11) is, in fact, rotationally invariant, we notice that the contour of the p and p' integrals may safely be translated up or down in the complex plane by any amount less than ε since the first singularity of the integral is the branch point at $\pm i\varepsilon$. We make the change of variables

$$\begin{cases} p = \varepsilon \sinh (u - i\theta)\,, \\ p' = \varepsilon \sinh (u' + i\theta)\,, \end{cases} \tag{4.12}$$

take u and u' to run between $-\infty$ and ∞, and pick

$$\theta = \mathrm{tg}^{-1}[(k_1 - k_2)/(\sinh 2K'|j_1 - j_2|)] \tag{4.13}$$

with $-\pi/2 < \theta < \pi/2$. After a bit of rearrangement, the integral (4.11) becomes

$$\frac{1}{8\pi^2}\int du\, du' \exp[-(\cosh u + \cosh u')\varepsilon R]\left[\mathrm{tgh}\,\frac{1}{2}(u + u')\right]^2, \tag{4.14}$$

where

$$R = [(j_1 - j_2)^2 \sinh^2 2K' + (k_1 - k_2)^2]^{\frac{1}{2}}\,. \tag{4.15}$$

The condition (4.8) for the validity of (4.7) requires that $\varepsilon R \gg 1$. This in turn implies that (4.14) contributes for $u \ll 1$ and $u' \ll 1$. After the hyperbolic functions are expanded for small u and u' the integral is easily calculated to yield

$$\begin{aligned} &\langle[\sigma_1 - \langle\sigma\rangle][\sigma_2 - \langle\sigma\rangle]\rangle = \\ &\qquad = [2\varepsilon \cosh 2K']^{\frac{1}{4}}[\exp[-2\varepsilon R](8\pi\varepsilon^2 R^2)^{-1} + O(\exp[-4\varepsilon R])]\,. \end{aligned} \tag{4.16}$$

4·2. *Smaller* εR. – Very near T_c, *i.e.*, for $\varepsilon \ll 1$, it is possible to see the form of the correlation function for large R but εR arbitrary. The result is

$$\langle[\sigma_1 - \langle\sigma\rangle][\sigma_2 - \langle\sigma\rangle]\rangle/\langle\sigma\rangle^2 = f_<(\varepsilon R)\,, \tag{4.17}$$

where $f_<$ depends only on εR and not on ε or R individually.

To see how (4.17) arises, consider the expansion of (4.1) in a power series in Λ. Under the consequent traces it is possible to make a unitary transfor-

5175

mation and replace Λ by

$$\Lambda' = \eta_1[(c_0 c_i)^{-\tau_3}\tau_2 - 1]\exp[-\gamma|j_1 - j_2|]\exp[-iP_x(j_1 - j_2)]\,\eta_2\cdot \\ \cdot\exp[iP_x(j_1 - j_2)][(c_0 c_i)^{-\tau_3}\tau_2 + 1]\exp[-\gamma|j_1 - j_2|]\,. \tag{4.18}$$

For notational convenience, we set $j_2 = k_2 = 0$.

A typical term in the trace (4.1) is the trace over the 2×2 « spin matrix » space of

$$\int_{-\infty}^{\infty} dk' \int_{-\infty}^{\infty} dk\, \langle k|\Lambda'|k'\rangle\, \langle k'|\Lambda'|k\rangle\,. \tag{4.19}$$

Here, we have replaced the sum over k by an integral because a large number of terms contribute to the summation. For each power of Λ', there is an integral over k. Hence, in our dimensional analysis of (4.1), we should consider $dk\langle k|\Lambda'|k'\rangle$ which is, for $k < k_1$,

$$d(\varepsilon k)\int du \int du' \cosh u \cosh u'\, i[\sinh u - \sinh u']^{-1}\cdot \\ \cdot\exp[i \sinh u(\varepsilon k) - i \sinh u'(\varepsilon k') - (\cosh u + \cosh u')\sinh 2K\varepsilon|j_1|]\cdot \\ \cdot\left[\left(\frac{\varepsilon}{2}\,\mathrm{tgh}\, K'\right)^{-\tau_3}(\cosh u)^{-\tau_3}\tau_2 - 1\right]\left[\left(\frac{\varepsilon}{2}\,\mathrm{tgh}\, K'\right)^{-\tau_3}(\cosh u')^{-\tau_3}\tau_2 + 1\right]. \tag{4.20}$$

A further simplification of (4.20) may be obtained if we make the unitary transform

$$\Lambda' \to \Lambda'' = (\tfrac{1}{2}\varepsilon\,\mathrm{tgh}\, K')^{\tau_3/2}\Lambda'(\tfrac{1}{2}\varepsilon\,\mathrm{tgh}\, K')^{-\tau_3/2}\,. \tag{4.21}$$

This transformation leaves the trace (4.1) unchanged but it has the effect of replacing the factor $((\varepsilon/2)\,\mathrm{tgh}\, K')^{\tau_3}$ in (4.20) by unity. Then the expression (4.20) only depends upon εk, $\varepsilon k'$, and εj_1 but has no separate dependence upon k, k', j_1, or ε. It follows that for small ε and large $|j_1 - j_2|$, the trace (4.20) depends only upon $\varepsilon|j_1 - j_2|\sinh 2K'$ and $\varepsilon|k_1 - k_2|$.

The only remaining step in the proof of (4.17) is the proof of the rotational invariance, *i.e.*, that there is no separate dependence of (4.1) upon $j_1 - j_2$ or $k_1 - k_2$ but only upon the combination R defined by (4.15). We shall not go into this in detail but merely point out that the analysis of the higher-order terms in the expansion is very similar to the logic which led from (4.11) to (4.14).

We now consider that (4.17) has been established. The correlation function just below T_c is thus given in terms of $f_<(\varepsilon R)$. For large εR, $f_<$ is given by (4.16). For $\varepsilon R \to 0$, KAUFMAN (3) has calculated $f_<$, the spin-spin correlation function along a diagonal and found a result which can be reduced to (14)

$$f_<(\varepsilon R) \to (\text{const})(\varepsilon R)^{-\frac{1}{4}} \tag{4.22}$$

if one assumes elliptical symmetry. Finally we should notice that the magnetic susceptibility is

$$\chi = \chi^0 \sum \langle [\sigma_1 - \langle\sigma\rangle][\sigma_2 - \langle\sigma\rangle] \rangle\,, \tag{4.23}$$

where χ_0 is the high-temperature susceptibility. According to (4.17), as $\varepsilon \to 0$,

$$\chi = \chi^0 [2\varepsilon \cosh 2K']^{\frac{1}{4}} [\varepsilon^2 \sinh 2K']^{-1} 2\pi \int_0^\infty \mathrm{d}Z Z f_<(Z)\,. \tag{4.24}$$

Since the final integral in (4.24) is independent of ε, this agrees with the result of ESSAM and FISHER (15) who predicted via the Padé approximant method that $\chi \sim \varepsilon^{-7/4}$.

5. – Correlation functions for $T > T_c$.

From (3.14), the correlation function for $T > T_c$ is given by

$$\log \langle \sigma_1 \sigma_2 \rangle = \tfrac{1}{4}\, \mathrm{Tr} \log [1 - 2(\eta_1 + \eta_2) g (\eta_1 + \eta_2)]^2 \tag{5.1}$$

as $|x - x'| \to \infty$, the cross terms $\eta_1 g \eta_2$ and $\eta_2 g \eta_1$ become very small. However, this does not imply that these terms are in any sense negligible. The remaining terms, $1 - 2\eta_1 g \eta_1 - 2\eta_2 g \eta_2$ have two eigenstates with zero eigenvalue $|0\rangle$ and $|0'\rangle$, where

$$|0\rangle\langle 0| = \tfrac{1}{2}(1 + \tau_3) d_0 \delta_1 d_1 / N\,, \qquad |0'\rangle\langle 0'| = \tfrac{1}{2}(1 + \tau_3) d_0 \delta_2 d_i / N\,. \tag{5.2}$$

Here δ_2 is a projection operator onto $j = j_2$ and $k = k_2$. When the cross terms are neglected, the right-hand side of (5.1) diverges. Thus, for large $|x - x'|$, these zero eigenstates dominate the behavior of the trace (5.1).

We handle these special eigenstates with some care by writing the operator in the bracket of (5.1) as

$$[1 - 2\eta_1 g \eta_1 + |0\rangle\langle 0|][1 - 2\eta_2 g \eta_2 + |0'\rangle\langle 0'|][X - |0\rangle\langle 0| - |0'\rangle\langle 0'|]\,, \tag{5.3}$$

---

(14) See a preprint by R. HARTWIG and J. STEPHENSEN for a further discussion of this point.

(15) J. W. ESSAM and M. E. FISHER: *Journ. Chem. Phys.*, **38**, 802 (1963).

where

(5.4) $X = 1 - [1 - 2\eta_1 g\eta_1 + |0\rangle\langle 0|]^{-1}\eta_1 2g\eta_2 - [1 - 2\eta_2 g\eta_2 + |0'\rangle\langle 0'|]^{-1}\eta_2 2g\eta_1$.

Because of the addition of $|0\rangle\langle 0|$, the first square bracket in (5.3) is a non-singular operator. In fact, its determinant has already been evaluated in Sect. 3·4 and was there denoted as $[\langle\tilde{\sigma}\rangle]^{\frac{1}{2}}$. With this notation (5.1) may be rewritten as

(5.5) $$\log[\langle\sigma_1\sigma_2\rangle/\langle\tilde{\sigma}\rangle^2] = \tfrac{1}{4}\,\mathrm{Tr}\log[X - |0\rangle\langle 0| - |0'\rangle\langle 0'|]^2 =$$
$$= \tfrac{1}{4}\,\mathrm{Tr}\,[\log X^2] + \tfrac{1}{4}\,\mathrm{Tr}\log[1 - X^{-1}|0\rangle\langle 0| - X^{-1}|0'\rangle\langle 0'|]^2 .$$

The last trace in (5.5) is the log of a simple 2×2 determinant. After this determinant is evaluated, (5.5) implies

(5.6) $$\log[\langle\sigma_1\sigma_2\rangle/\langle\tilde{\sigma}\rangle^2] = \tfrac{1}{4}\,\mathrm{Tr}[\log X^2] +$$
$$+ \tfrac{1}{4}\log\{\langle 0|\eta_1 2g\eta_2 X^{-1}|0\rangle\langle 0'|\eta_2 2g\eta_1 X^{-1}|0'\rangle - \langle 0|\eta_1 2g\eta_2 X^{-1}|0'\rangle\langle 0'|\eta_2 2g\eta_1 X^{-1}|0\rangle\}^2 ,$$

5·1. *Very large* $|j_1 - j_2|$. – As $|j_1 - j_2| \to \infty$, $X - 1$ goes to zero as $\exp[-\gamma|j_1 - j_2|]$. Then the trace in (5.6) goes to zero and may be neglected. Inside the curly bracket, the diagonal matrix elements are each proportional to $\exp[-2\gamma|j_1 - j_2|]$ and may be neglected relative to the off-diagonal matrix elements which are proportional to $\exp[-\gamma|j_1 - j_2|]$. Thus in this limit,

(5.7) $$\langle\sigma_1\sigma_2\rangle = \langle\tilde{\sigma}\rangle^2|\langle 0|\eta_1 2g\eta_2|0'\rangle\langle 0'|\eta_2 2g\eta_1|0\rangle| .$$

The matrix elements in question are easily evaluable. From (5.2) it follows that

(5.8) $$-\langle 0|\eta_1 2g\eta_2|0'\rangle = \langle 0'|\eta_2 2g\eta_1|0\rangle =$$
$$= \int_{-\pi}^{\pi}\frac{\mathrm{d}p}{2\pi}\exp[-ip(k_1 - k_2) - \gamma|j_1 - j_2|]\,d_0(p)\,d_i(p)/N .$$

Just above the critical temperature B is very close to unity. We again write $B = 1 + \varepsilon$, and evaluate the integral in the numerator (5.8) in the limit

(5.9) $$\begin{cases} \varepsilon = 2(\sinh 2K')^{-1}(K^* - K') \ll 1 , \\ R = \{[\sinh 2K'(j_1 - j_2)]^2 + (k_1 - k_2)^2\}^{\frac{1}{2}} \gg \dfrac{1}{\varepsilon} , \end{cases}$$

5178

just as in Sect. **4**·1. When this is coupled with eq. (3.50) for $\langle\tilde{\sigma}\rangle$, (5.7) yields the asymptotic form of the correlation function:

$$(5.10)\qquad \langle\sigma_1\sigma_2\rangle = (2\varepsilon)^{\frac{1}{4}}(A+1)^{\frac{1}{4}}(A-1)^{\frac{3}{4}}\cdot \cdot\int_{-\pi}^{\pi}\frac{dp}{2\pi}\, d_0(p)\, d_i(p)\exp\left[-\gamma(p)|j_1-j_2| + ip(k_1-k_2)\right],$$

so that

$$(5.11)\qquad \langle\sigma_1\sigma_2\rangle = (2\varepsilon\cosh 2K')^{\frac{1}{4}}\left[\frac{e^{-\varepsilon R}}{(2\pi\varepsilon R)^{\frac{1}{2}}} + O(e^{-3\varepsilon R})\right],$$

where $\varepsilon = B-1$ is again a measure of the deviation from the critical temperature and R measures the distance between the sites in question.

The R-dependence in eq. (5.11) has already been obtained by FISHER ([16]) by an argument rather simpler than that employed here. Equation (5.11) is a partial justification for the result of DOMB and SYKES ([17]) that the susceptibility, χ, is proportional to $\varepsilon^{-7/4}$ for $\varepsilon \ll 1$. Of course, (5.11) does not by itself justify this conclusion; it is also necessary to know the form of the correlation function for $\varepsilon R \sim 1$.

5·2. *Smaller* εR. – The result of DOMB and SYKES follows at once if we assume

$$(5.12)\qquad \langle\sigma_1\sigma_2\rangle = (2\varepsilon\cosh 2K')^{\frac{1}{4}} f_>(\varepsilon R)$$

for $\varepsilon \ll 1$ and $(j_1 - j_2) \gg 1$. In Sect. **4**·2 we were able to construct a result of this type for $T < T_c$ by describing the asymptotic form of the operators which appear inside our traces.

To construct a similar argument for $T > T_c$, we must examine the structure of the X defined by eq. (5.4). The inverses in X are given by (3.45) which has the form

$$(5.13)\qquad [1 - 2\eta_1 g\eta_1 + |0\rangle\langle 0|]^{-1} = (1 - |0\rangle\langle 0|) f_1 (1 - |0\rangle\langle 0|) + |0\rangle\langle 0|,$$

where f_1 is given by (3.46). If we substitute the resulting expression for X into eq. (5.6) and try to construct the analog of the arguments of Sect. **4**·2, we run into great difficulties with the terms resulting from the $|0\rangle\langle 0|$ projection oper-

(16) M. E. FISHER: *Physica*, **25**, 521 (1959).
(17) C. DOMB and M. F. SYKES: *Proc. Roy. Soc.* (*London*), A **240**, 214 (1957).

5179

ators in (5.13). In the denominators of these operators there appears

$$N = \int \mathrm{d}p\, d_0(p)\, d_i(p) \sim \log \varepsilon\,. \tag{5.14}$$

This extra ε-dependence considerably obscures all further arguments aboue the correlation function. In fact, it is just these questions about $\log \varepsilon$-dependenct which tends to complicate (and perhaps even invalidate) the work of PATASHINSKI and POKROVSKI (18) on the second-order phase transition.

In this Ising model case, the normalization factor (5.14) and the consequent complications may be eliminated from the evaluation of the correlation function. To effect this elimination we consider how the unwanted terms involving $|0\rangle\langle 0|$ enter into (5.6) by defining

$$\begin{aligned} X(\lambda) = 1 - [(1-\lambda|0\rangle\langle 0|)f_1(1-\lambda|0\rangle\langle 0|) + \lambda|0\rangle\langle 0|]\eta_1 2g\eta_2 - \\ - [(1-\lambda|0'\rangle\langle 0'|)f_2(1-\lambda|0'\rangle\langle 0'|) + \lambda|0'\rangle\langle 0'|]\eta_2 2g\eta_1\,. \end{aligned} \tag{5.15}$$

At $\lambda = 1$, $X(\lambda) = X$. As $\lambda \to 0$, the unwanted factors $|0\rangle\langle 0|$ disappear from (5.15). We wish to show that we can safely replace X by $X(0)$ in (5.6).

To do this, we define

$$\begin{aligned} F(\lambda) = \tfrac{1}{4}\,\mathrm{Tr}\log[X(\lambda)]^2 + \tfrac{1}{4}\log\{\langle 0|\eta_1 2g\eta_2[X(\lambda)]^{-1}|0\rangle\langle 0'|\eta_2 2g\eta_1[X(\lambda)]^{-1}|0'\rangle - \\ - \langle 0|\eta_1 2g\eta_2[X(\lambda)]^{-1}|0'\rangle\langle 0'|\eta_2 2g\eta_1[X(\lambda)]^{-1}|0\rangle\}^2 = \\ = \tfrac{1}{4}\,\mathrm{Tr}\log\{X(\lambda) - (|0\rangle\langle 0| + |0'\rangle\langle 0'|[X(\lambda) + \eta_1 2g\eta_2 + \eta_2 2g\eta_1]\}^2)\,. \end{aligned} \tag{5.16}$$

At $\lambda = 1$, $F(\lambda)$ is the left-hand side of (5.6). A brief calculation based upon the last line of (5.16) indicates that $\mathrm{d}F/\mathrm{d}\lambda = 0$. Since $F(\lambda = 1) = F(\lambda = 0)$, we can safely replace X by $X(0)$ in (5.6). Once the projection operators $|0\rangle\langle 0|$ and $|0'\rangle\langle 0'|$ are eliminated from the X's in (5.6) precisely the same arguments as used in Sect. **4**·2 can be constructed to yield the result (5.12).

* * *

I would like to thank Prof. Sir N. MOTT for the hospitality of the Cavendish Laboratory, where most of the work reported here was performed. Drs. G. BAYM and P. NOZIÈRES provided some very helpful criticism. I would also like to thank Dr. T. T. WU for a very helpful discussion and comparison with his work (19), which enabled me to discover a numerical error in my calculations.

(18) A. Z. PATASHINSKI and V. L. POKROVSKI: *Žurn. Èksp. Teor. Fiz.*, **46**, 994 (1963) [translation *Sov. Phys. JETP*, **19**, 677 (1964)].

(19) T. T. WU: *Phys. Rev.* (to be published).

Physics Vol. 2, No. 6, pp. 263-272, 1966. Physics Publishing Co.

SCALING LAWS FOR ISING MODELS NEAR T_c*

LEO P. KADANOFF †

Department of Physics, University of Illinois
Urbana, Illinois

(*Received* 3 *February* 1966)

Abstract

A model for describing the behavior of Ising models very near T_c is introduced. The description is based upon dividing the Ising model into cells which are microscopically large but much smaller than the coherence length and then using the total magnetization within each cell as a collective variable. The resulting calculation serves as a partial justification for Widom's conjecture about the homogeneity of the free energy and at the same time gives his result $s\nu' = \gamma' + 2\beta$.

1. Introduction

IN a recent paper [1] Widom has discussed the consequences of the assumption that the free energy in a system near a phase transition of second order is a homogeneous function of parameters which describe the deviation from the critical point and has shown that this assumption leads to consequences which roughly agree with our present numerical information [2] about the behavior of such systems. Another paper by Widom [3] written at the same time explores the consequences of an apparently quite independent idea: that the behavior of the interface separating droplets of the 'wrong phase' within a system just below a phase transition should be quite similar to the behavior of an interface separating a region of fluctuation in the order parameter from the surrounding medium [4]. Here again information is derived which agrees with numerical calculations and experiment [2].

Widom's ideas about interfaces are based upon physical plausibility arguments; his idea about the homogeneity of the singular part of the free energy is not given any very strong justification beyond the fact that it appears to work. In the present paper, the Ising model is analyzed in a manner which is designed to throw light upon how correlations between the order parameter in different regions of the lattice scale when the parameters describing the deviation from the critical point - in this case $T - T_c$ and the applied magnetic field - are changed. Widom's homogeneity condition upon the singular part of the free energy and some of his results for interfaces are then derived as a consequence of these scaling arguments based upon our model.

* This research was supported in part by the National Science Foundation under grant NSF-GP 4937.

† A.P. Sloan Foundation Fellow.

Although the argument is carried out in Ising model language, it is clear that the arguments could be generalized to other cases of second order phase transitions.

2. Description of Model

Consider an Ising Model in a weak magnetic field near T_c. This model can be described in terms of two parameters

$$\varepsilon = (T - T_c)/T_c, \tag{1}$$

which measures the deviation from the critical temperature, and h, a dimensionless magnetic field defined so that flipping a single spin gives a change in magnetic energy $2h/kT$. A full solution of the Ising model would be obtained if we knew $f(\varepsilon, h)$, the free energy per site in the presence of the magnetic field.

To get some feeling for the behavior of the Ising model, imagine that we divided the entire lattice into cells of L lattice sites on a side. Then each cell in this s-dimensional lattice contains L^s lattice sites. We take L to be large, but much smaller than the coherence length, ξ, which describes the range of spin correlations, measured in lattice constants. Since ξ goes to infinity at $\varepsilon = 0$ and $h = 0$, it is easy to find an L which satisfies this criterion.

To zeroth order, each cell can be considered to be isolated from the others and from the external magnetic field. Then to zeroth order $f(\varepsilon, h) = f_L(\varepsilon)$ where $f_L(\varepsilon)$ is a free energy per site of a lattice of side L in no magnetic field. Since the small size of the cell tends to eliminate the singularities from the phase transition, $f_L(\varepsilon)$ should be an analytic function of ε but not of L, i.e.

$$f_L(\varepsilon) = f_L^{(0)} + \varepsilon f_L^{(1)} + \varepsilon^2 f_L^{(2)} + \dots \tag{2}$$

where f_L^0, f'_L and f_L^2 should not all be analytic functions of L [4].

Next consider interactions of the cell α with the magnetic field. This gives a term in $-\beta\mathscr{H}$ of the form

$$h \sum_{\mathbf{r}\in\alpha} \sigma_{\mathbf{r}} \tag{3}$$

The basic assumption of our analysis is that within each cell, the spins tend to line up so that they mostly point either up or down. That is,

$$\sum_{\mathbf{r}\in\alpha} \sigma_{\mathbf{r}} = \mu_\alpha \langle \sigma \rangle_L L^s \tag{4}$$

where μ_α is either plus one or minus one. The average spin, $\langle\sigma\rangle_L$, defined in (4) is given by

$$\langle \sigma \rangle_L^2 = \sum_{\mathbf{r}\in\alpha} \sum_{\mathbf{r}'\in\alpha} \frac{\langle \sigma_{\mathbf{r}} \sigma_{\mathbf{r}'} \rangle}{L^{2s}} \tag{5}$$

Because of the "small" size of the cell, this depends strongly on only L, not upon ε or h. Thus, the interaction with the magnetic field takes the form of a term in $\exp[-\beta\mathscr{H}]$ of

$$\exp\Big(\sum_\alpha h\ \mu_\alpha\ L^s \langle \sigma \rangle_L\Big) \tag{6}$$

Next, consider the interaction among cells. The free energy will tend to be larger if the spins on neighboring cells are lined up. There will tend to be a smaller contribution if they are anti-parallel. Then, in net, this tends to make a contribution to the exponential $\exp[-\beta F]$

$$\exp \sum_{\alpha,\beta} \left\{ \mu_\alpha \mu_\beta \tilde{K}(\varepsilon, L) + f_{int}(\varepsilon, L) \right\}. \tag{7}$$

Here the sum extends over nearest neighbor cells, $\tilde{K} + f_{int}$ gives the contribution to the free energy when neighboring cells are aligned and, $-\tilde{K} + f_{int}$ gives the contribution to the free energy when they are out of step. Since the direct interactions between cells which produce f_{int} occur within a distance which is very short compared to the coherence length, we assume that f_{int} is, like $f_L(\varepsilon)$, a regular function of ε, but not necessarily a regular function of L. On the other hand, $\tilde{K}$ is perhaps a somewhat more subtle beast. This describes the extra free energy that it costs to put two cells out of step. This involves, then, the rather delicate difference between the ways cells can match up when they are in step and when they are out of step. Nonetheless, it seems reasonable to assume that $\tilde{K}(\varepsilon, L)$ is also a regular function of ε; but, we assert this with somewhat less confidence than our other statements relating to this model. In writing (7) we are asserting that the correlations among cells can be totally represented by these interactions among near neighbors and that there are no less direct interactions that we need include in (7) as long ranged interactions. This statement, together with the assertion that the cell can be represented by the double valued variable, μ_α, are the two basic assumptions of this model.

To find the correction to the zeroth order result (1), we sum the product of (6) and (7) over both possible orientations of each μ_α. This sum is, of course, just an Ising model calculation with coupling constant $\tilde{K}$ and effective magnetic field $\tilde{h}$. This then gives a contribution to the total free energy

$$\sum_\alpha s\, f_{int}(\varepsilon, L) + f(\tilde{\varepsilon}, \tilde{h}) \tag{8}$$

where $\tilde{\varepsilon}$ measures the deviation of the new coupling constant from the critical value, and where, from (6) the effective magnetic field in the cell problem is

$$\tilde{h}(h, L) = h \langle\sigma\rangle_L L^s \tag{9}$$

Since the original Ising model problem has within it correlations over many cells, we assert that $\tilde{\varepsilon}$ and $\tilde{h}$ must be sufficiently small so that the new Ising model problem retains long-ranged correlations in $\langle \mu_\alpha \mu_\beta \rangle$. This requires $\tilde{\varepsilon} \ll 1$, $\tilde{h} \ll 1$.

Since there are L^s sites per cell, equations (1) and (9) give the free energy as

$$f(\varepsilon, h) = f_L(\varepsilon) + sL^{-s} f_{int}(\varepsilon, L) + L^{-s} f(\tilde{\varepsilon}, \tilde{h}). \tag{10}$$

Equation (10) is one of the two basic relations in our analysis. The other is obtained by discussing the spin-spin correlation function $\langle \sigma_{\mathbf{r}} \sigma_{\mathbf{r}'} \rangle$ for the case in which the distance between spins $|\mathbf{r} - \mathbf{r}'|$, as measured in lattice constants, is much larger than L. Then each $\sigma_{\mathbf{r}}$ can be replaced by the average of $\sigma_{\mathbf{r}}$ over the cell in which it lies so that (4) may be used to rewrite the correlation function as

$$\langle \sigma_{\mathbf{r}} \sigma_{\mathbf{r}'} \rangle = g(\varepsilon, h, |\mathbf{r} - \mathbf{r}'|) = \langle\sigma\rangle_L^2 \langle \mu_\alpha \mu_{\alpha'} \rangle = \langle\sigma\rangle_L^2 g(\tilde{\varepsilon}, \tilde{h}, |\mathbf{r} - \mathbf{r}'|/L). \tag{11}$$

The last line of (11) follows because the μ's are described by an Ising model with effective distance from T_c, $\tilde{\epsilon}$, effective magnetic field $\tilde{h}$ and cell L lattice constants long.

3. Analysis of the Model

The free energy is a singular function of ϵ. For example if the specific heat diverges as $\epsilon^{-\alpha}$ [$(-\epsilon)^{-\alpha'}$] above [below] T_c, for small ϵ and $h = 0$,

$$f(\epsilon, 0) = \begin{cases} f_0 + f_1\epsilon + f_2\epsilon^{2-\alpha} + \ldots \text{ for } T > T_c \\ \\ f_0 + f_1\epsilon + f'_2(-\epsilon)^{2-\alpha'} + \ldots \text{ for } T < T_c \end{cases} \tag{12a}$$

In three dimensions (2) $0 \leqslant \alpha \leqslant 1$, $0 \leqslant \alpha' \leqslant 1$. Another possible behavior has the specific heat above and/or below T_c diverge logarithmically so that α and/or α' are zero but

$$f(\epsilon, 0) = f_0 + f_1\epsilon + 1/2\ \epsilon^2\ [B - A \log \epsilon] \text{ for } T > T_c \tag{12b}$$

and/or

$$f(\epsilon, 0) = f_0 + f_1\epsilon + 1/2\ \epsilon^2\ [B' - A' \log|\epsilon|] \text{ for } T < T_c \tag{12c}$$

We first attack the case in which the α's differ from zero and later consider the possibility of behaviors (12b) and/or (12c).

Consider first, at $h = 0$, the terms in equation (10) which are singular in ϵ and $\tilde{\epsilon}$. If we equate these singular terms above T_c, we find

$$\epsilon^{2-\alpha} = [\tilde{\epsilon}]^{2-\alpha}/L^s.$$

It follows then that

$$\begin{aligned} \tilde{\epsilon} &= \epsilon L^{1/\nu} \text{ above } T_c \\ \tilde{\epsilon} &= \epsilon L^{1/\nu'} \text{ below } T_c \end{aligned} \tag{13}$$

and

$$s\nu = (2 - \alpha) \tag{14a}$$

$$s\nu' = (2 - \alpha') \tag{14b}$$

Notice that if we assert that $\tilde{K}$ is a regular function of ϵ, $\tilde{\epsilon}$, which is $\tilde{K} - K_c$, cannot have a discontinuous first derivative with respect to ϵ. Therefore, our assertion about the regularity of K implies

$$\nu = \nu' \tag{15}$$

which in turn implies $\alpha = \alpha'$. It should be recognized that equation (15) is more dubious than the other assertions of this paper. Consequently, we hold this statement aside and do not use it in simplifying our further analysis.

More generally, when $h \neq 0$, we can equate the singular parts in ε and h on both sides of equation (10) and find

$$f_{sing}(\varepsilon, h) = \frac{1}{L^s} f_{sing}(\tilde{\varepsilon}, \tilde{h}). \tag{16}$$

This can only be true if h is proportional to some power of L, i.e. if

$$< \sigma > = L^{-\Psi} \tag{17}$$

and then (16) and (14) can hold true if the singular part of $f(\varepsilon, h)$ obeys

$$f_{sing}(\varepsilon, h) = \begin{cases} \varepsilon^{2-\alpha} F(\varepsilon^{(s-\psi)\nu}/h) \text{ for } T > T_c \\ \\ |\varepsilon|^{2-\alpha'} F'(|\varepsilon|^{(s-\psi)\nu'}/h) \text{ for } T < T_c \end{cases} \tag{18}$$

This is Widom's homogeneity assertion [1] which we have now derived from our model.

If $\alpha = 0$, and f has the form (12b), then the last term on the right-hand side of equation (10) has a singular part

$$\frac{\tilde{\varepsilon}^2}{L^s} [B - A \log \tilde{\varepsilon}] = \varepsilon^2 [B - A \log \varepsilon - A \log L^{1/\nu}] \tag{19}$$

if we again employ (13) for this case. The extra term in log $L^{1/\nu}$ must be cancelled out by some other term on the right-hand side of (10). But since this term is regular in ε, the compensating term can be obtained by taking $f_L{}^{(2)}$ in equation (2) to be

$$f_L^{(2)} = A \log L^{1/\nu} + \text{regular terms} \tag{20}$$

or by a similar structure in $sL^{-s}f_{int}(\varepsilon, L)$. When $\alpha' > 0$, this term in (20) does not contribute to the leading singularity in $f(\varepsilon, h)$ for $T < T_c$ and consequently the analysis which led to (18) can go through just as before except that a term in $\varepsilon^2 \log |\varepsilon|$ must be added to F and F'.

But, when α' is also zero, the same term, (20), must cancel out log L terms both above and below T_c. The coefficient in front of this term cannot change at T_c since f_L is, by hypothesis, regular in ε. Then when $\alpha = \alpha' = 0$, and hence $\nu = \nu'$ by (14), we conclude with Widom that

$$A = A' \tag{21}$$

It is relevant to notice that this equality holds in the Onsager solution of the two dimensional

Ising model [4]. It is interesting, but perhaps quite besides the point to notice that it is also true for the Λ-transition [5] of He^4 and in at least one anti-ferromagnetic transition [6].

The arguments of Widom are designed to use the homogeneity of the singular part of $f(\varepsilon, h)$, as exhibited in equation (18), to derive relations between the average magnetization,

$$M \sim \langle\sigma\rangle \sim \left.\frac{\partial f(\varepsilon,\ h)}{\partial h}\right|_\varepsilon \tag{22}$$

the spin susceptibility

$$\chi \sim \left.\frac{\partial\langle\sigma\rangle}{\partial h}\right|_\varepsilon \tag{23}$$

and the parameters, α, α', ν and ν' which we have already defined. If we define β, γ, γ' and δ in the conventional manner [1,2]

$$\begin{aligned} \langle\sigma\rangle &= |\varepsilon|^\beta \quad \text{for } T < T_c,\ h = 0 \\ \chi &= \varepsilon^{-\gamma} \quad \text{for } T > T_c,\ h = 0 \\ \chi &= |\varepsilon|^{-\gamma'} \quad \text{for } T < T_c,\ h = 0 \\ \langle\sigma\rangle &= |h|^{1/\delta} \quad \text{for } \varepsilon = 0 \end{aligned} \tag{24}$$

it follows from (18) that

$$\alpha' + \beta(1 + \delta) = 2 \tag{25a}$$

$$\gamma' + 2\beta + \alpha' = 2 \tag{25b}$$

$$\gamma'/\gamma = (2 - \alpha')/(2 - \alpha). \tag{25c}$$

Since the parameters in (25) are all experimental quantities, these relations could be checked if we could find an Ising model in nature.

In the course of this analysis, we also find that ψ can be written in terms of experimental quantities as

$$2\psi = s - \gamma/\nu. \tag{26}$$

This result is useful in the analysis in the spin-spin correlation function of equation (11), which equation can now be written as

$$g(\varepsilon, h, |r - r'|) = L^{-2\psi}\ g(\varepsilon L^{1/\nu}, hL^{s-\psi}, |r - r'|/L) \tag{27}$$

for $T > T_c$. The right-hand side of (27) will only be independent of L if g is a homogeneous function of its arguments of the form

$$g(\varepsilon,\ h, |r - r'|) = \begin{cases} \dfrac{1}{|r - r'|^{2\psi}} G(\varepsilon^{\nu(s-\psi)}/h, |r - r'|\varepsilon^{\nu}) & \text{for } T > T_c \\ \dfrac{1}{|r - r'|^{2\psi}} G'(\varepsilon^{\nu'(s-\psi)}/h, |r - r'|\varepsilon^{\nu'}) & \text{for } T < T_c. \end{cases} \quad (28)$$

Consequently, the coherence length that we discussed at the beginning of this paper must be at $h = 0$

$$\xi = \begin{cases} \varepsilon^{-\nu} & \text{for } T > T_c \\ |\varepsilon|^{-\nu'} & \text{for } T < T_c \end{cases} \quad (29)$$

and at $\varepsilon = 0$

$$\xi = |h|^{-(s-\psi)}. \quad (30)$$

The result (29) makes contact with the notation of Widom [3] and Fisher [2] and permits us to identify ν and ν' as experimental quantities. Therefore, equations (14a) and (14b) may be viewed as experimental relations.

There is one more experimental relation to be obtained. From (28) as ε and h go to zero, g is proportional to $|\mathbf{r} - \mathbf{r}'|^{-2\psi}$. This relation is conventionally [2] written as

$$g(\varepsilon = 0,\ h = 0, |r - r'|) = \frac{1}{|r - r'|^{s-2+\eta}} \quad (31)$$

so that, from (26), we can write

$$\gamma = (2 - \eta)\nu \quad (32)$$

which is our final experimental relation.

4. Discussion and Comparison with Other Theoretical Results

One can view the preceding work as developing relations among the nine "experimental" quantities α, α', β, γ, γ', δ, ν, ν' and η. Equations (14a), (14b), (25a), (25b), (25c) and (31) are the six such relations that we consider to be direct consequences of our model. The relation (15), $\nu = \nu'$, which implies $\alpha = \alpha'$, $\gamma = \gamma'$ is a somewhat more tenuous conclusion. With (15) included, we have seven relations among our nine parameters; if (15) is rejected there are only six such relations. In the first case, then, there are three free parameters; in the second two.

None of these relations are new. Essam and Fisher [7] argued for (25b) on the basis of homogeneity considerations not totally different from those of Widom [1] who made full use of the homogeneity conjecture to get (25a), (25b), (25c) and (31) as well as (15). Besides, (31) can be viewed as a tautology: a definition of the coherence length. In this sense, the relation

$$\gamma' = (2 - \eta)\nu'$$

which follows from (31), (25c), and (14) can also be viewed as trivial.

There exists an alternative derivation of equation (14a) due to Pippard [8] and Ginsberg [9]. Imagine that we were in a situation of zero magnetic field in which the temperature lay just above T_c, i.e. $\varepsilon > 0$. Then imagine that a temperature fluctuation occurred which took a region of side X of the material into the ordered state. This would cost a free energy of the order of the volume of the region times the difference in free energy density between that at T and that at T_c, i.e.

$$X^s \varepsilon^{(2-\alpha)}. \tag{33}$$

But, the probability that such a fluctuation will occur is proportional to the exponential of the cost in free energy, divided by kT_c. Therefore, the maximum possible value of X, which is the coherence length is given by setting (33) to be of order kT_c or

$$X_{max} \sim \xi \sim \varepsilon^{-(2-\alpha)/s}. \tag{34}$$

Therefore, we find at once $\nu = (2 - \alpha)/s$.

Despite this simple derivation equations (14) are not unassailable. One could argue that, as $s \to \infty$, we know that the molecular field approximation is valid. This gives $\nu = 1/2$, $\alpha = 0$, and (14) fails. But, it may well be that as $s \to \infty$, the molecular field approximation becomes valid closer and closer to T_c; and for all finite s there remains a region within a very small neighborhood of T_c for which the molecular field approximation fails. Then (14) can still be true for all s, in the strict limit as $\varepsilon \to 0$.

Precisely this effect is expected to occur in the superconductor [8-10]. Here, the molecular field approximation (Landau-Ginzberg theory) is known to be valid until you get very close to T_c. This theory introduces a coherence length

$$\xi_{L.G.} = |\varepsilon|^{-1/2}\xi_0 \tag{35}$$

where ξ_0 is a very large number $\sim 10^4$ lattice constants for a pure superconductor. Over most of the temperature range $\xi_{L.G.}$ is larger than the coherence length defined by (34). Then the molecular field theory remains valid because fluctuations in the order parameter are not important. But very close to T_c, the coherence length of (34) will surpass that of (35), and then the molecular field theory will fail. In this tiny temperature region about T_c, equations (14) are expected to be true if the analysis of reference [10] is correct.

To check the seven relations among physical parameters we compare with known results for two and three dimensional Ising models. A variety of numerical results have been obtained in both two and three dimensions through evaluations of the singularities of power series. These are reviewed in references [2, 11-13]. In the two dimensional case there are also a variety of analytical results available [14-16] all based upon Onsager's solution [4]. For the two dimensional case, equations (28) would imply that the spin-spin correlation function is of the form

$$|r - r'|^{-1/4} H(\varepsilon|r - r'|)$$

when the magnetic field vanishes and $|r - r'| >> 1$. This result agrees with the conclusions of reference [16].

The second column of Table 1 lists the known values of the nine experimental parameters for the two dimensional Ising model. These all check exactly with Widom's relations.

TABLE 1

Parameter	Two dimensional case		Three dimensional case		
	Value	Reference	Previously calculated value	Reference	Fit to data
α	0	4	$0.0 \leqslant \alpha \lesssim 0.2$	2	0.085
α'	0	4	$0.0 \leqslant \alpha' \lesssim 0.06$	2	0.085
β	1/8	13	$0.303 \leqslant \beta \leqslant 0.318$	2	0.332
γ	7/4	2, 11, 16	1.250 ± 0.001	2	1.250
γ'	7/4	2, 16	$1.23 \leqslant \gamma' \leqslant 1.32$	2	1.250
δ	15	13	5.2 ± 0.15	13	4.78
ν	1	2	0.644 ± 0.003	2	0.638
ν'	1	16		2	0.638
η	1/4	14	0.060 ± 0.007	2	0.045

The first column listed under the three dimensional case gives information which has been obtained via Pade and other numerical methods. The last column, labeled "Fit to data", gives a set of values for these parameters which agrees with all seven of our relations including the more doubtful statement $\nu = \nu'$. This fit agrees with all the known numerical results to within about two and one-half standard deviations. These results indicate at least that the conclusions drawn from our model are not grossly inaccurate.

Acknowledgment

I would like to thank Dr. P.C. Martin and Dr. Brian Josephson for very helpful comments.

References

1. B. WIDOM, *J. Chem. Phys.* 43, 3898 (1965).

2. M. FISHER, *J. Math. Phys.* 5, 944 (1964).

3. B. WIDOM, *J. Chem. Phys.* 43, 3892 (1965).

4. For the two dimensional Ising model, the Onsager solution (L. ONSAGER, *Phys. Rev.* 65, 117, 1944) bears out equation (2).

5. M.J. BUCKINGHAM and W.M. FAIRBANK, in *Progress in Low Temperature Physics III*, (ed. C.J. Gorter), Ch. 3. North Holland, Amsterdam (1961).

6. J. SKALGO, Jr. and S.A. FRIEDBERG, *Phys. Rev. Letters* 13, 133 (1964).

7. J.W. ESSAM and M.E. FISHER, *J. Chem. Phys.* 39, 842 (1963).

8. A.B. PIPPARD, *Proc. Roy. Soc. Lond.* A216, 547 (1953).

9. D.M. GINSBERG and J.S. SHIER, to be published.

10. E.G. BATYEV, A.Z. PATASHINSKII and V.L. POKROVSKII, *Zh. Eksp. Teor. Fiz.* 46, 2093 (1964). English translation in *Sov. Phys.-JETP* 19, 1412 (1964).

11. C. DOMB, *Advanc. Phys.* 9, 34, 35 (1960).

12. M.E. FISHER, *J. Math. Phys.* 4, 278 (1963).

13. M.E. FISHER, in *Proceedings of the International Conference on Phenomena Near Phase Transitions*, to be published.

14. B. KAUFMAN, *Phys. Rev.* 76, 1232 (1949); B. KAUFMAN and L. ONSAGER, *Phys. Rev.* 76, 1244 (1949).

15. G.F. NEWELL and E.W. MONTROLL, *Rev. Mod. Phys.* 25, 353 (1953).

16. L. KADANOFF, to be published in *Nuovo Cimento*.

REVIEWS OF MODERN PHYSICS VOLUME 39, NUMBER 2 APRIL 1967

Static Phenomena Near Critical Points: Theory and Experiment

LEO P. KADANOFF,* WOLFGANG GÖTZE,† DAVID HAMBLEN, ROBERT HECHT, E. A. S. LEWIS
V. V. PALCIAUSKAS, MARTIN RAYL, J. SWIFT
Department of Physics and *Materials Research Laboratory,*‡ *University of Illinois, Urbana, Illinois*
DAVID ASPNES
Department of Physics, Brown University, Providence, Rhode Island
JOSEPH KANE
The Laboratory of Atomic and Solid-State Physics, Cornell University, Ithaca, New York

This paper compares theory and experiment for behavior very near critical points. The primary experimental results are the "critical indices" which describe singularities in various thermodynamic derivatives and correlation functions. These indices are tabulated and compared with theory. The basic theoretical ideas are introduced via the molecular field approach, which brings in the concept of an order parameter and suggests that there are close relations among different phase transition problems. Although this theory is qualitatively correct it is quantitatively wrong, it predicts the wrong values of the critical indices. Another theoretical approach, the "scaling law" concept, which predicts relations among these indices, is described. The experimental evidence for and against the scaling laws is assessed. It is suggested that the scaling laws provide a promising approach to understanding phenomena near the critical point, but that they are by no means proved or disproved by the existing experimental data.

CONTENTS

I. INTRODUCTION

In recent years, considerable attention has been drawn to the phenomena which occur very near critical points. Several recent conferences[1,2] have presented a wealth of new experimental data and theoretical ideas in this area. These conferences have broadcast the fact that there are quite marked similarities between apparently very different phase transitions. An antiferromagnet near its Néel point behaves quite similarly to a liquid near its critical point. The superconducting transition is not very different from several ferroelectric transitions. In all cases, there is an apparently rather simple behavior in the region right around the critical point.

This simplicity and similarity among phase transitions is not fully elucidated theoretically. Some of the qualitative features of this behavior are reasonably well understood; others remain a complete mystery.

In this paper we review the present status of theory and experiment in this area, concentrating on the time-independent properties of systems near T_c. Thus, we look at thermodynamic derivatives and time-independent correlations but ignore the very interesting work on transport coefficients and time-dependent correlations. The particular subject is what can be learned by comparing different phase transitions with each other and with the existing theories. How are different phase transitions alike? In what ways do they differ? Why should we expect these similarities and differences?

Because we are considering such a broad range of phenomena, we cannot expect our readers to be experts in any particular area we describe. Consequently, we attempt to provide explanations and discussions which will be comprehensible to the nonexpert. We are hopeful that our treatment will provide some picture of the interrelations within this broad field.

In the next section, the important theoretical ideas—the order parameter, the choice between different phases, long-range correlations, and fluctuations—are introduced via the molecular field approximation. These results are tested by comparing them with conclusions drawn from theoretical studies of the Ising model, which is a model of a ferromagnetic material that is particularly suitable for theoretical study. This comparison shows that the molecular field approximation gives a picture of the phase transition which is, at best, only qualitatively correct. The main quantitative predictions of the molecular field theory are a set of "critical indices" which turn out to have very different values from the indices found in the Ising model. In the third section we describe a recent theoretical proposal that there exist relations among the critical indices derived from the notion of "scaling laws".

In the next sections we turn successively to magnetic systems, critical points in classical liquids, the quantum liquid-gas phase transition, superfluids, and ferroelectrics. In each case, the experiments which determine the critical indices are tabulated and analyzed. These data are used to show many remarkable similarities between different phase transitions and a few equally remarkable differences. The relations among the indices predicted by the scaling laws are tested. Most often the theoretical ideas are supported by the experiments but the few discrepancies that exist allow one to leave open the question as to whether the scaling laws are in fact right.

II. MOLECULAR FIELD THEORIES

A. The Order Parameter

The most fundamental idea which helps elucidate the behavior near a critical point is the concept that this transition is describable by an order parameter.[3–5] This parameter, here indicated by the symbol $\langle p(\mathbf{r})\rangle$, is a numerical measure of the amount and kind of ordering which is built up in the neighborhood of the critical point. For example, in a single domain ferromagnetic crystal with an easy axis of magnetization along the z direction, a suitable order parameter is the statistically averaged z component of the magnetization at the point $\mathbf{r}$, $\langle p(\mathbf{r})\rangle=\langle M_z(\mathbf{r})\rangle$. Besides indicating how much spin ordering there is, the order parameter defined in this way has the following important properties:

(a) It may vanish above the critical point but is must be nonzero in the region just below T_c.

(b) It can approach zero continuously at $T\rightarrow T_c$ from below. (For example, at zero applied field the magnetization vanishes as the temperature is raised to the Curie temperature.) This condition ensures that the transition not be of first order.

(c) Below the phase transition, the order parameter is not fully determined by the external conditions, but may take on two or more different values under physically identical conditions. For example, at zero applied magnetic field, the magnetization may point in the plus or minus z direction with equal facility below T_c. Similarly, in the liquid-gas phase transition, the appropriate order parameter is the density minus the critical density, $\rho-\rho_c$. When liquid is in contact with vapor, this order parameter takes on two values: the positive value appropriate to the liquid phase and the negative value appropriate to the gaseous phase.

Table I lists a group of phase transitions which are of higher than first order—i.e., which can take place with zero latent heat—and gives the appropriate order parameter for each transition. Also included in this table are some indications of the amount of "free choice" which each $\langle p(\mathbf{r})\rangle$ may have below T_c.[4] For future reference we also include in Table I a list of the variables which are the thermodynamic conjugate to the order parameter in each case. These conjugate variables are indicated in general by the symbol h.

B. The Landau Theory of the Second-Order Phase Transition—Thermodynamics

A relatively simple view of phenomena near the critical point was provided by Landau.[5] Although this theory does not agree quantitatively with experimental observations very close to the critical point, it does provide a qualitatively correct view of the kinds of things which do happen. Landau's theory is most easily discussed in the case of a ferromagnet with a small magnetic field, B, and T near T_c. Then the magnetization $M_z(\mathbf{r})$ is necessarily small and it would seem reasonable to expand the free energy in a power series in this magnetization. This expansion takes the following form

$$G=\int d^3r g(\mathbf{r}) \qquad (2.1)$$

with

$$g(\mathbf{r})=g_0(T)-B_z(\mathbf{r})M_z(\mathbf{r})+a(T)[M_z(\mathbf{r})]^2 +b(T)[M_z(\mathbf{r})]^4+c(T)[\nabla M_z(\mathbf{r})\cdot\nabla M_z(\mathbf{r})]. \qquad (2.2)$$

The first term $g_0(T)$ represents the free energy per unit volume which would exist were there no magnetization. The second term B_zM_z represents the direct interaction between the applied magnetic field $\mathbf{B}$—assumed parallel to the easy axis z—and the magnetic moment of the spins within the sample. Direct spin-spin interactions produce the remaining terms in $g(\mathbf{r})$. Because these interactions do not change when we change the sign of M_z, these terms contain no odd powers of $M_z(\mathbf{r})$. This is the origin of the terms proportional to a and b in Eq. (2.2). The final term in this equation is inserted to make the free energy larger when $M_z(\mathbf{r})$ varies in space. Hence this term serves to damp out spatial variations in $M_z(\mathbf{r})$.

It is a general rule in statistical mechanics that the most probable value of any parameter is the one which minimizes the free energy. Thus, we would determine the most probable value of $M_z(\mathbf{r})$ by requiring that G

TABLE I. Partial list of transitions with critical points. In general, the symbol h denotes the conjugate to $\langle p\rangle$.

Transition	Meaning of $\langle p\rangle$	Free choice in $\langle p\rangle$	Thermodynamic conjugate of p
Liquid–gas	$\rho-\rho_c$	$p>0$=liquid $p<0$=vapor (2 choices)	μ
Ferromagnetic	magnetization $\langle\mathbf{M}\rangle$	if n equivalent "easy axes" $2n$ choices	applied magnetic field, H, along easy axes
Heisenberg model ferromagnet	magnetization $\langle\mathbf{M}\rangle$	direction of $\langle\mathbf{M}\rangle$ [can choose any value on surface of sphere.]	$\mathbf{H}$
Antiferromagnet	sublattice magnetization	if n "easy axes" $2n$ choices	not physical
Ising model	$\langle\sigma_r\rangle$	2 choices	h
Superconductors	Δ (complex gap parameter)	phase of Δ	not physical
Superfluid	$\langle\psi\rangle$ (condensate wave function)	phase of $\langle\psi\rangle$	not physical
Ferroelectric	lattice polarization	finite number of choices	electric field
Phase separation	concentration	2 choices	a difference of chemical potentials

be stationary under the infinitesimal change

$$M_z(\mathbf{r})\to M_z(\mathbf{r})+\delta M_z(\mathbf{r}). \tag{2.3}$$

The first-order change in G under this transformation is

$$\delta G=\int d^3r\,\delta M_z(\mathbf{r})\times\{-B_z(\mathbf{r})+2aM_z(\mathbf{r})+4b[M_z(\mathbf{r})]^3-2c\nabla^2M_z(\mathbf{r})\}. \tag{2.4}$$

Since this must vanish for all values of $\delta M_z(\mathbf{r})$, the brace must be zero. Hence the equation for the most probable value of $M_z(\mathbf{r})$ is

$$\{2a+4b[M_z(\mathbf{r})]^2-2c\nabla^2\}M_z(\mathbf{r})=B_z(\mathbf{r}). \tag{2.5}$$

Landau now makes the drastic assumption of neglecting all fluctuations in the magnetization. This is equivalent to the assertion that the most probable value of the magnetization, as defined by Eq. (2.5), is the only value and hence also the mean value.

The free energy is obtained by solving (2.5), and then substituting the solution into (2.2). When there is ambiguity arising from multiple solutions to (2.5), this is to be resolved by choosing the solution which actually minimizes G. For this procedure to work, b and c must both be greater than zero; otherwise there is no minimum for G. (If $b<0$, the transition is first order; if $c<0$ the magnetization is never uniform.)

Begin with the simplest case, that in which $B_z(\mathbf{r})$ is independent of $\mathbf{r}$. Then there is no reason for spatial variation of $M_z(\mathbf{r})$ and (2.5) reduces to

$$[2a+4bM_z^2]M_z=B_z. \tag{2.6}$$

When $B_z=0$ this has the solutions

$$M_z=0 \tag{2.7a}$$

$$M_z=\pm(-a/2b)^{1/2}. \tag{2.7b}$$

The first solution (2.7a) actually minimizes the free energy when $a>0$; the remaining solutions (2.7b) minimize G when $a<0$. But since we want the magnetization at $B_z=0$ to vanish above T_c and to be nonzero below T_c, we must have

$$a>0 \quad \text{for } T>T_c$$

$$a<0 \quad \text{for } T<T_c.$$

Landau chooses the simplest form of $a(T)$ which will accomplish this purpose: he takes

$$a(T)=a'(T-T_c) \tag{2.8}$$

near T_c. He also guesses that b and c approach constants as $T\to T_c$. This choice of $a(T)$ ensures that at the critical point ($B_z=0$, $T=T_c$) the system can produce a nonzero magnetization with a cost in free energy which is of fourth order in M_z. Hence near T_c the system can produce large-scale fluctuations in the order parameter with relatively little cost in free energy. According to the best present beliefs this unusual susceptibility to fluctuations is the cause of all the special phenomena which appear near the critical point.

TABLE II. Parameters describing phase transition.

Physical quantity		Range of variables $\epsilon=(T-T_c)/T_c$	h	Behavior of quantity	Parameter describing quantity
$\langle p\rangle$		>0	0	$\langle p\rangle=0$	
		<0	0	$\langle p\rangle\sim\pm\|\epsilon\|^{\beta}$	β
		0	$\neq0$	$\sim\pm\|h\|^{1/\delta}$	δ
$\chi=\partial\langle p\rangle/\partial h\|_{\epsilon}$		>0	0	$\sim\epsilon^{-\gamma}$	γ
		<0	0	$\sim\|\epsilon\|^{-\gamma'}$	γ'
$g(r,r')=\langle p_r p_{r'}\rangle-\langle p\rangle^2$		0	0	$\sim\|r-r'\|^{-d+2-\eta}$	η
ξ=range of $g(r,r')$		>0	0	$\sim\epsilon^{-\nu}$	ν
		<0	0	$\sim\epsilon^{-\nu'}$	ν'
C_h=specific heat at constant h		>0	0	$a\epsilon^{-\alpha}+b$	α
		<0	0	$a'\|\epsilon\|^{-\alpha'}+b'$	α'
	or	>0	0	$A\log\epsilon^{-1}+B$	$\alpha=0$
		<0	0	$A'\log\|\epsilon\|^{-1}+B'$	$\alpha'=0$

We now follow the thread of the Landau theory and derive the relation between critical phenomena and fluctuations from the free-energy formulation of Eqs. (2.2) and (2.5). The reader will recognize that there is a physical inconsistency in this whole calculation. Equation (2.5) was obtained by neglecting fluctuations —and we shall finally conclude that fluctuations are all-important! But the Landau theory, although it is essentially inconsistent near T_c, does give a good indication of the kind of behavior to be expected.

First, look at the temperature dependence of $\langle p(\mathbf{r})\rangle$ (the magnetization) at zero h (magnetic field). According to (2.7b) and (2.8) the magnetization just below T_c is proportional to $(T_c-T)^{1/2}$. Since we shall obtain and use several results of this type, it is convenient to collect them in tabular form. In the second line of Table II, we define an index β by the condition that the order parameter go to zero as $(T_c-T)^{\beta}$ when the thermodynamic conjugate to the order parameter h, is at its critical value.

We have just concluded that in the Landau theory $\beta=\frac{1}{2}$, as given in the first entry in Table III.

To continue this compilation, take Eq. (2.6) at $T-T_c=0$. Then $a=0$ so that Eq. (2.6) implies

$$M_z=\{B_z/4b\}^{1/\delta} \quad \text{with} \quad \delta=3. \tag{2.9}$$

This gives the next entries in Tables II and III.

The susceptibility, $\chi=(\partial M_z/\partial B)_T$ is calculated by differentiating Eq. (2.6) with respect to B. At zero magnetic field, the resulting susceptibility may be evaluated above and below T_c by employing respectively Eqs. (2.7a) and (2.7b). In the end the magnetic susceptibility is seen to diverge both above and below T_c as $|T-T_c|^{-1}$. Hence the values $\gamma=\gamma'=1$ are recorded in Table III.

The final thermodynamic property which we will obtain from the Landau theory is the specific heat at zero magnetic field, which is given by thermodynamics as

$$T^{-1}C_B=-\partial^2G/\partial T^2|_{B=0}. \tag{2.10}$$

But, at $B=0$, we can use (2.7) to find

$$G=\int d\mathbf{r}[g_0(T)] \quad \text{for } T>T_c$$

$$=\int d\mathbf{r}[g_0(T)-(a^2/4b)] \quad \text{for } T<T_c. \tag{2.11}$$

This extra term in the free energy below T_c produces, when it is twice differentiated, a constant term in the specific heat. Thus, there is a discontinuity in the specific heat at T_c predicted by the Landau theory.

C. Landau Theory—Fluctuations

Next consider fluctuations in the magnetization $M_z(\mathbf{r})$. These fluctuations are given by $[M_z(\mathbf{r})-\langle M_z(\mathbf{r})\rangle]$. The point to be studied is how the deviation of M_z from its average at one point in the material is tied to the similar fluctuations in neighboring regions. The mathematical description of this correction is given by the correlation function $g(\mathbf{r},\mathbf{r}')$ defined by

$$g(\mathbf{r},\mathbf{r}')=\langle[M_z(\mathbf{r})-\langle M_z(\mathbf{r})\rangle][M_z(\mathbf{r}')-\langle M_z(\mathbf{r}')\rangle]\rangle. \tag{2.12}$$

At first sight it is hard to see how g can be calculated within the context of the Landau theory. The difficulty arises because this theory begins by neglecting fluctuations, i.e., by saying that $M_z(\mathbf{r})-\langle M_z(\mathbf{r})\rangle$ is very small. Now we wish to use the theory in order to calculate a first correction to this initial statement.

Fortunately there is a well-defined procedure for calculating correlation functions starting from the free energy. This procedure arises from a quite general

TABLE III. Comparison of Landau theory and Ising models.[a]

Physical quantity	Parameter	Landau theory	Ising model 2-D	Ising model 3-D
$\langle p \rangle$	β	1/2	1/8	0.313 ± 0.004
	δ	3	15	5.2 ± 0.15
χ	γ	1	7/4	1.250 ± 0.001
	γ'	1	7/4	1.31 ± 0.05
$g(r, r')$	η	0	1/4	0.056 ± 0.008
ξ	ν	1/2	1	0.643 ± 0.0025
	ν'	1/2	1	?
C_h	α	0 discontinuity in C_h	0 log ∞ in C_h	$0.0\leq\alpha\leq0.25$
	α'	0	0	$0.066+0.16, -0.04$

[a] The values of the critical indices in this table are mostly taken from the reviews of Fisher (Refs. 1, 15–17). In addition, the numbers for β, γ', and α' are taken from a preprint of G. A. Baker and D. S. Gaunt. We wish to thank these authors for sending us their work prior to publication. The values for η and ν are taken from M. E. Fisher and R. J. Burford, Phys. Rev. (to be published).

theorem of *classical* statistical mechanics. This theorem[6] states that if the Hamiltonian contains the parameter $h(\mathbf{r})$ in the combination

$$-\int d\mathbf{r} h(\mathbf{r}) p(\mathbf{r}) \tag{2.13}$$

then if we allow $h(\mathbf{r})$ to change by adding to it the small increment $\delta h(\mathbf{r})$, this change induces a change in $\langle p(\mathbf{r})\rangle$, which is

$$\delta\langle p(\mathbf{r})\rangle = (kT)^{-1}\int d\mathbf{r}' \langle [p(\mathbf{r})) - \langle p(\mathbf{r})\rangle] \times [p(\mathbf{r}') - \langle p(\mathbf{r}')\rangle]\rangle \delta h(\mathbf{r}'). \tag{2.14}$$

Here k is the Boltzmann constant. In our case $h(\mathbf{r})$ corresponds to the magnetic field, $p(\mathbf{r})$ to the magnetization, and (2.14) reads

$$\delta\langle M_z(\mathbf{r})\rangle = (kT)^{-1}\int d\mathbf{r}' g(\mathbf{r}, \mathbf{r}')\delta B_z(\mathbf{r}'). \tag{2.15}$$

Our equation for the average magnetization in the Landau theory is (2.5). If we calculate the first-order change in that equation as $B_z(\mathbf{r})\rightarrow B_z(\mathbf{r})+\delta B_z(\mathbf{r})$ so that $M_z(\mathbf{r})\rightarrow M_z(\mathbf{r})+\delta M_z(\mathbf{r})$ we find that (2.5) implies

$$\{2a+12b\langle M_z(\mathbf{r})\rangle^2 - 2c\nabla^2\}\delta\langle M_z(\mathbf{r})\rangle = \delta B_z(\mathbf{r}). \tag{2.16}$$

Next, substitute (2.15) into (2.16) and move $\delta B_z(\mathbf{r})$ to the left-hand side of the equation to find

$$\int d\mathbf{r}'\{[2a+12b\langle M_z(\mathbf{r})\rangle^2 - 2c\nabla^2]g(\mathbf{r}, \mathbf{r}') - kT\delta(\mathbf{r}-\mathbf{r}')\} \times \delta B(\mathbf{r}') = 0.$$

Since $\delta B(\mathbf{r}')$ is arbitrary, the entire brace must vanish and the correlation function therefore obeys the equation

$$[2a+12b\langle M_z(\mathbf{r})\rangle^2 - 2c\nabla^2]g(\mathbf{r}, \mathbf{r}') = kT\delta(\mathbf{r}-\mathbf{r}'). \tag{2.17}$$

Equation (2.17) is easily solved when the magnetic field is zero. In this case, for $T>T_c$ the average magnetization vanishes and (2.17) reduces to

$$[2a'(T-T_c) - 2c\nabla^2]g(\mathbf{r}, \mathbf{r}') = kT\delta(\mathbf{r}-\mathbf{r}') \quad \text{for } T>T_c. \tag{2.18a}$$

For $T<T_c$ the squared magnetization is proportional to T_c-T according to (2.7b) so that (2.17) becomes

$$[4a'(T_c-T) - 2c\nabla^2]g(\mathbf{r}, \mathbf{r}') = kT\delta(\mathbf{r}-\mathbf{r}'). \tag{2.18b}$$

In three dimensions these equations have the solution

$$g(r, r') = \frac{\exp(-|\mathbf{r}-\mathbf{r}'|/\xi)}{|\mathbf{r}-\mathbf{r}'|}\frac{kT}{8\pi c} \tag{2.19}$$

with

$$\xi = (c/a')^{1/2}(T-T_c)^{-1/2} \quad \text{for } T>T_c$$
$$\xi = (c/2a')^{1/2}(T_c-T)^{-1/2} \quad \text{for } T<T_c. \tag{2.20}$$

Before discussing the consequences of (2.19), we must indicate why a g determined from the theorem of classical statistical mechanics [Eq. (2.15)] is correct for this quantum-mechanical problem. Equation (2.19) is not correct for all values of $(\mathbf{r}-\mathbf{r}')$, because quantum-mechanical fluctuations do, in fact, invalidate the derivation. But, when $|\mathbf{r}-\mathbf{r}'|$ is much greater than a lattice constant, we can replace the magnetization by its average over several cells of the lattice. Since this magnetization is now produced by many spins quantum fluctuations become insignificant. Hence the result (2.19) is a correct consequence of the Landau theory. whenever $|\mathbf{r}-\mathbf{r}'|$ is much bigger than a lattice constant.

But the important fact is that the correlation function (2.19) has such a very large spatial range near T_c. As T approaches T_c its characteristic range ξ

grows as $|T-T_c|^{-\nu}$, where $\nu=\frac{1}{2}$ in the Landau theory. This range measures the typical size of a region in which a coherent fluctuation in the magnetization occurs. As we get closer to the critical point, these fluctuations cover more space. Finally, at the critical point the spatial extend of the correlation functions becomes infinite and then g drops off very slowly in space. In three dimensions this drop-off occurs as $|\mathbf{r}-\mathbf{r}'|^{-1}$. In the general case of a d-dimensional lattice, the Landau theory predicts a drop-off as $|\mathbf{r}-\mathbf{r}'|^{-(d-2)}$.

All of the conclusions of this and the preceding section are consequences of the simple assumption that the free energy is expandable in power series in the order parameter and in $(T-T_c)$. A large number of theories of phase transitions make implicit use of this assumption. All of these approaches must necessarily lead to the same conclusions as Landau's about behavior near the critical point. In particular, they give the same values of the critical indices α, β, γ, δ, η, and ν. Examples of theories which are equivalent to Landau's are the van der Waals' equation for a liquid, the Weiss molecular field theory for a ferromagnet and its variants which have been applied to a variety of other systems, the Ornstein–Zernike equations for g, many varieties of the random phase approximation,[7] the Ginzburg–Landau equations for superconductors,[8] etc. Many of these approaches give correct information far from the critical point; but they all are expected to fail as the critical point is closely approached.

D. Connections Between Correlation Functions and Thermodynamics

All of these calculations do include some qualitatively correct statements. One is that large-scale fluctuations in the order parameter are the source of the singularities in thermodynamic derivatives near the critical point. To understand this point, refer back to Eq. (2.15) which gives an exact formula for the change in magnetization produced by a change in magnetic field in terms of a correlation function. If the change in magnetic field is independent of $\mathbf{r}$, Eq. (2.15) then gives the thermodynamic response of the magnetization to a change in B_z. This is precisely a definition of the magnetic susceptibility. Hence Eq. (2.15) implies that the susceptibility is

$$\chi \equiv \partial M_z/\partial B_z|_T = \delta M_z/\delta B_z$$
$$= (kT)^{-1}\int d\mathbf{r}' g(\mathbf{r},\mathbf{r}'). \qquad (2.21)$$

As $T\rightarrow T_c$ the range of the correlation function increases. Hence the $\mathbf{r}'$ integral covers a larger and larger region and the integral goes to infinity at T_c. Hence the divergence in the range of correlations is the precise cause of this and all the other singularities in thermodynamic derivatives.

It is possible to follow this point explicity through in the context of the Landau theory. Simply substitute the form (2.19) for g into (2.21) and perform the integral. Then, it follows that the susceptibility is, at $B=0$

$$\chi = (2c)^{-1}\xi^2. \qquad (2.22)$$

This result shows directly how a divergence in the range of correlations may produce a divergence in a thermodynamic derivative.

There are other useful exact correlation function expressions for thermodynamic derivatives. For example, if $E(\mathbf{r})$ is the energy density, the specific heat at fixed B is given by

$$C_B = \frac{1}{kT^2}\int d\mathbf{r}' \langle [E(\mathbf{r}) - \langle E(\mathbf{r})\rangle][E(\mathbf{r}') - \langle E(\mathbf{r}')\rangle]\rangle \qquad (2.23a)$$

and also

$$T[\partial\langle M_z(\mathbf{r})\rangle/\partial T]|_{B_z}$$
$$= \partial\langle E(\mathbf{r})\rangle/\partial B_z|_T$$
$$= (kT)^{-1}\int d\mathbf{r}' \langle [E(\mathbf{r}) - \langle E(\mathbf{r})\rangle]$$
$$\times [M_z(\mathbf{r}') - \langle M_z(\mathbf{r}')\rangle]\rangle. \qquad (2.23b)$$

E. The Ising Model

In order to investigate the correctness of the Landau theory we quote results from theoretical investigations of the Ising model. This is essentially a very anisotropic ferromagnet in which only the z components of the spins are coupled. The spins are denoted by $\sigma_\mathbf{r}$ where $\mathbf{r}$ gives the position on the lattice. Each $\sigma_\mathbf{r}$ can take on the values ± 1. Then the Hamiltonian is given by

$$H/kT = -h\sum_{\mathbf{r}}\sigma_\mathbf{r} - K\sum_{\mathbf{r},\mathbf{r}'}\sigma_\mathbf{r}\sigma_{\mathbf{r}'}. \qquad (2.24)$$

The first term in H represents an interaction with an external "magnetic field" h, which tends to make the spins point up if h is positive and down if h is nagative. The second term is an interaction among spins which, when K is positive, tends to line up neighboring spins with each other. To keep the interaction short-ranged, the second sum is carried out only over $\mathbf{r}$ and $\mathbf{r}'$ which are nearest neighbors.

To calculate the statistical mechanics of the Ising model it is necessary to compute the free energy, G, which is given by

$$\sum \exp(-H/kT) = \exp(-G/kT). \qquad (2.25)$$

Here, the sum covers all possible values of the $\sigma_\mathbf{r}$. That is, if there are N sites on the lattice, the sum has 2^N terms.

Some of the qualitative features of the resulting partition function are rather easy to guess on physical grounds. For example, consider the average "magnetization," $\langle\sigma_{\mathbf{r}}\rangle$, which is proportional to $(\partial G/\partial h)|_T$. At zero temperature and $h=0$, the aligning force represented by K wins out and the spins are perfectly aligned. Then (see Fig. 1), $\langle\sigma_{\mathbf{r}}\rangle$ is either plus or minus unity. As the temperature is increased, with h held equal to zero, the magnitude of $\langle\sigma_{\mathbf{r}}\rangle$ continually decreases until, at a critical temperature T_c, it disappears completely. Any applied field, h, will tend to increase the alignment, i.e., $|\langle\sigma_{\mathbf{r}}\rangle|$. This, too, is indicated in Fig. 1.

Onsager[9] calculated the sum (2.25) in two dimensions at $h=0$ and others have extended his work to get other physical properties.[10–12] The results have been described in review articles[13] and monographs.[3,14] This theory gives the values of all the critical indices defined in Table II except for δ. These theoretical values of critical indices are listed in Table III.

In three dimensions, there is no general solution. But there do exist numerical ways of attacking this problem. Various physical quantities can be expanded in power series about $T=0$ and about $T=\infty$. Then, there exist numerical tricks for guessing the radius of convergence of these power series and even their behavior near singularities. From this emerges numerical estimates for the parameters of the two-and three-dimensional Ising models. Table III records the values of the critical indices obtained from these numerical calculations. These are mostly taken from reviews of Fisher.[1,15–17]

The validity of these numerical methods may be checked by comparing the results with the exact two-dimensional calculations based upon the Onsager solution. It is found that the methods generally work well, except that occasionally the convergence is uncomfortably slow. The estimate of α' is, for example, not as accurate as we would like. On the other hand, γ is known to one part in a thousand for the sc, fcc, bcc, and diamond lattices.[15] According to these numerical estimates, the values of the parameters are, as far as we can tell, independent of the detailed nature of the lattice.

Table III reveals that the Ising model behaves very differently from the predictions of the Landau theory. Thus, for example, the specific heat is discontinuous in the Landau approach but has a logarithmic infinity in the two-dimensional Ising model and an infinity which is probably stronger than logarithmic in three dimensions. As another example, $\beta=\frac{1}{8}$ in the two-dimensional Ising model and is close to 0.31 in three dimensions; these results are clearly different from the $\beta=\frac{1}{2}$ predicted by the simple theory.

Because exact solutions exist for the two-dimensional Ising model at $h=0$ it is possible to calculate several of the relevant correlation functions in this case. For example, it is possible to show that the spin–spin correlation function has a relatively simple form in the case in which T is close to T_c and the spins are separated by a large distance. To express this result

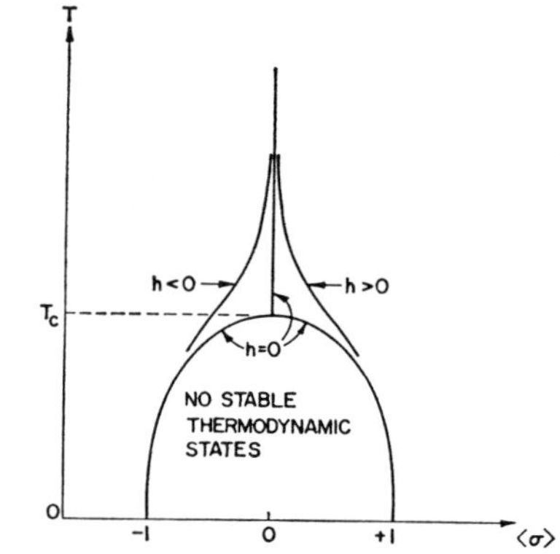

FIG. 1. Magnetization vs temperature (schematic) for various values of h. Below the $h=0$ curve, there are no stable thermodynamic states.

we use the variable

$$\epsilon=(T-T_c)/T_c \tag{2.26}$$

which is a dimensionless measure of the deviation from the critical temperature and

$$R=|\mathbf{r}-\mathbf{r}'|/(\text{lattice constant}). \tag{2.27}$$

When $\epsilon\ll 1$, $R\gg 1$, but ϵR arbitrary, the spin–spin correlation function has the structure[11]

$$g(\mathbf{r},\mathbf{r}')=|\epsilon|^{1/4}G(\epsilon R). \tag{2.28a}$$

The corresponding results for the energy density–energy density correlation function and the energy density–spin correlation function are, under the same restrictions for ϵ and R,[18]

$$\langle[E(\mathbf{r})-\langle E\rangle][E(\mathbf{r}')-\langle E\rangle]\rangle=|\epsilon|^2 G_{EE}(\epsilon R) \tag{2.28b}$$

$$\langle[\sigma(\mathbf{r})-\langle\sigma\rangle][E(\mathbf{r}')-\langle E\rangle]\rangle=|\epsilon|^{1+1/8}G_{\sigma E}(\epsilon R), \tag{2.28c}$$

where $G_{\sigma E}(\epsilon R)$ happens to vanish for all positive values of its argument, i.e., for all $T>T_c$.

Several interesting facts can be learned from these correlation functions. First, consider their values for large spatial separations between the points, i.e., for $\epsilon R\gg 1$. In this limit, each of the correlation functions decreases exponentially. The spin–spin correlation function behaves as [11]

$$g(\mathbf{r},\mathbf{r}')\sim[\epsilon^{1/4}/(\epsilon R)^{1/2}]\exp(-|\epsilon|R) \quad \text{for } T>T_c \tag{2.29a}$$

and

$$g(\mathbf{r},\mathbf{r}')\sim[|\epsilon|^{1/4}/(\epsilon R)^2]\exp(-2|\epsilon|R) \quad \text{for } T<T_c \tag{2.29b}$$

while the energy density–energy density correlation function behaves as[18]

$$[\epsilon^2/(\epsilon R)^2] \exp(-2|\epsilon| R) \qquad (2.29c)$$

and the energy density–spin correlation function has the behavior[18] for $\epsilon<0$

$$[|\epsilon|^{1+1/8}/(\epsilon R)^2] \exp(-2|\epsilon| R). \qquad (2.29d)$$

Hence in this limit all three correlation functions go exponentially to zero with the same characteristic range

$$\xi \sim |\epsilon|^{-1}. \qquad (2.30)$$

These correlation functions each reduce to a simple form when $\epsilon \to 0$. In this case even though R is very large ϵR is very small. Hence the functions in (2.28) are evaluated for small values of their arguments. Then the spin–spin correlation function takes the form[19]

$$g(\mathbf{r}, \mathbf{r}') \sim R^{-1/4} \qquad (2.31a)$$

the energy density–spin correlation function becomes proportional to

$$\langle\sigma\rangle/R \qquad (2.31b)$$

where the energy density–energy density correlation function becomes propertional to[18]

$$R^{-2}. \qquad (2.31c)$$

By using these results, we can see again how the divergences in thermodynamic derivatives are connected with the very long range of the correlation functions. For example, the specific heat at constant B according to (2.23) is proportional to the spatial intergral of the energy density–energy density correlation function. In the region

$$1 \ll R \ll (\xi/a_0),$$

where a_0 is the lattice constant Eq. (2.31c) gives the appropriate form for this correlation function. Hence this region contributes a part

$$\int_{\sim 1}^{\sim \xi/a_0} d^2R/R^2 \sim 2\pi \int_1^{\xi/a_0} \frac{dR}{R} \sim 2\pi \log(\xi/a_0) \qquad (2.32)$$

to the specific heat. Since the correlation function decreases very sharply as R becomes much bigger than ξ, Eq. (2.32) gives a correct estimate of the temperature dependence of the specific heat. From the fact that $\xi \sim |\epsilon|^{-1}$, we then infer the logarithmic divergence of the specific heat.

We expect that the long-ranged correlations are relatively insensitive to the details of the interactions between spins. If a correlation extends over a hundred or a million lattice constants, this correlation should be sensitive only to the grossest features of the interaction and should not be affected by a change from a bcc to a fcc lattice or to the introduction of some next-nearest-neighbor interaction. The Ising model solutions do, in fact, bear this out. The correlations in the two-dimensional case are basically the same for square and triangular lattices. In the three-dimensional case, so far as we can tell, $\gamma=1.25$ equally well for bcc, fcc, and simple cubic lattices.

F. Range of Validity of Landau Theory

Now that we have seen that the Landau theory does not work for Ising models, we ask when we expect this theory to work at all. Ginzburg[20] has proposed an argument which uses the Landau theory to predict its own range of validity. To do this, consider fluctuations in the order parameter—which we assumed small. Actual order parameters, e.g., the magnetization in a ferromagnet, fluctuate considerably. However, some of these fluctuations may be removed by averaging the order parameter over a suitable region of space. This averaging, however, cannot cover too large a region without ruining the local nature of the Landau theory of the second-order phase transition. The largest range over which we can average without ruining the theory is the coherence distance ξ.

Thus, for the Landau theory to be valid, fluctuations in the order parameter over distances comparable with ξ must be relatively small. In particular, they must be small in comparison with the order parameter itself. We must have then,

$$\langle [p(\mathbf{r})-\langle p\rangle][p(\mathbf{r}')-\langle p\rangle]\rangle\,|_{|\mathbf{r}-\mathbf{r}'|\sim\xi} \ll \langle p\rangle^2, \qquad (2.33)$$

where we are now using the symbol $\langle p(\mathbf{r})\rangle$ for the order parameter. We estimate the right- and left-hand sides of (2.33) below T_c at zero magnetic field from Eqs. (2.19) and (2.7b). The result is that this necessary condition for the validity of the Landau theory becomes

$$kT_c/4\pi ec\xi(T) \ll -a/b. \qquad (2.34)$$

We can rewrite this in terms of more physical quantities by using Eq. (2.20) for $\xi(T)$. We then extrapolate this expression to $T=0$, and define an extrapolated zero temperature coherence length

$$\lambda = (c/2a'T_c)^{1/2} \qquad (2.35)$$

so that the true correlation length of the Landau theory is, near T_c,

$$\xi(T) = \lambda |\epsilon|^{-1/2}. \qquad (2.36)$$

The jump in the heat capacity per unit volume (at $B=0$) predicted by the Landau theory is

$$\Delta C = [(a')^2/2b]T_c. \qquad (2.37)$$

In terms of these physical quantities (2.34) reads

$$[16\pi e(\Delta C/k)\lambda^3]^{-1} \ll |\epsilon|^{1/2}. \qquad (2.38)$$

Ginzburg[20] and Levanyuk[21] produced somewhat more careful versions of this argument. They came to the conclusion that the Landau theory could only be correct if $|\epsilon|$ were much greater than a critical value,

$$\epsilon_c = (1/32\pi^2)(k/\Delta C\lambda^3)^2. \qquad (2.39)$$

When $|\epsilon|\ll\epsilon_c$, fluctuations become important and the molecular field theories all fail. For $1\gg|\epsilon|\gg\epsilon_c$, we expect these theories to be valid. This calculation of ϵ_c is just an order of magnitude argument so one should, perhaps, not take the $32\pi^2$ in the denominator of Eq. (2.39) too seriously.

The factor λ^6 in the denominator of ϵ_c indicates that as the range of the forces, and hence the range of zero temperature correlations, gets longer and longer the Landau theory gets better and better. This point has been made explicitly in several recent papers[22–24] which show explicitly that the van der Waals theory (which is equivalent to Landau's near T_c) follows rigorously when you have an infinite range interaction.

Let us discuss ϵ_c for some examples:

Superconductors

For pure superconductors we expect λ to be of the same order as the zero temperature coherence length ξ_0. In tin,[25] $\xi_0=2300$ Å and $\Delta C=0.8\times10^4$ ergs/cm^{-3} deg^{-1}. Therefore, Eq. (2.39) would indicate that deviations from the molecular field theory (which is, in this case, the Ginzburg–Landau[8] theory) should not be expected until ϵ gets as small as 10^{-14}. This very small value of ϵ_c arises because of the very long range of coherence in the superconductor. Similar results for the width of the critical region have been obtained by several other authors, who have carried out more detailed calculations of the dependence of specific heat[26] and electrical resistance[27] upon ϵ.

The Superfluid

In the λ transition of liquid He^4, the situation is very different. The $T=0$ coherence length is quite comparable with the interatomic spacing instead of the huge number encountered in the superconducting transition. Consequently, there appears to be no region at all in which the molecular field theory is satisfactory.

Magnetic Transitions

In the case of iron ΔC is about 3×10^7 ergs cm^{-3} deg^{-1}. Here ξ can be measured directly by neutron scattering experiments. The total cross section σ for momentum $\hbar k$ is propertional to $(\xi^{-2}+k^2)^{-1}$. The experimental value for λ is about 2 Å[28] so that $\epsilon_c\sim10^{-2}$.

However, there is some experimental indication that the molecular field theory might be accurate for small values of ϵ, below T_c.[29,30] This might occur because for small ϵ, the weak magnetic dipole interactions between spins may become important. Since these interactions have a long range, the molecular field theory might perhaps again become correct in some range that includes small values of ϵ.

Since the magnetic fields produced by the spins in an antiferromagnet tend to cancel out, the magnetic interaction effectively has a much shorter range in an antiferromagnet. There is, then, no reason to expect molecular field theory to be correct for small ϵ in an antiferromagnet.

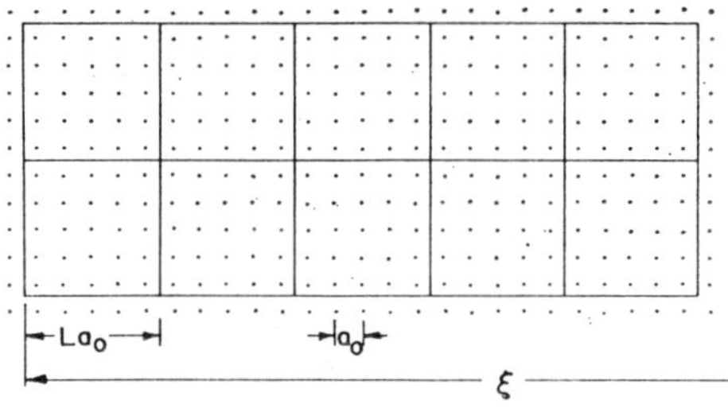

FIG. 2. Division of Ising model lattice into cells. $L\gg1$ but $La_0\ll\xi$.

Ferroelectrics

Ginzburg[20] has estimated λ for BaTiO$_3$ and uses Eq. (2.39) to find $\epsilon_c\sim10^{-4}$. The reason for this small value is that it is the smoothly varying coulomb force which is responsible for establishing the ferroelectric phase. Thus, in this case, the molecular field theory is expected to work for all $|\epsilon|\gg10^{-4}$. However, in the antiferroelectric, one should not expect a similar preservation of molecular field theory. The experimental data partially support this point of view since the only observation of γ different from the molecular field result ($\gamma=1$) is in the antiferroelectric phase of Rochelle salt.

III. SCALING LAWS

A. Formulation

Ideally, theory should predict values of the critical parameters, α, α', β, γ, γ', δ, ν, ν', and η. But, except for the case of the two-dimensional Ising model, no analytical theory exists.

A series of recent proposals[31–38] suggest that there exist relations among these indices, in particular, that there are not nine unknown quantities, but instead that the nine critical indices can all be expressed in terms of two fundamental unknown indices. There have been several suggested "derivations" of these relations among the critical indices. None really derive all the results at hand; they are plausibility arguments.

To see one of these consider an Ising model lattice. Imagine that we mentally divide the lattice into cells, each side of length La_0, where a_0 is the lattice constant. L is chosen to be very small compared to a coherence length, but is large enough so that the cell contains many spin sites (see Fig. 2).

$$1\ll L\ll\xi/a_0. \tag{3.1}$$

Let us assume for a moment that all the interactions between different cells were turned off. Then each cell would act independently of every other and there would be no correlations over distances larger than a cell size. But, these are precisely the correlations which produce the divergences in thermodynamic derivatives which are characteristic of the critical region. Hence correlations between different cells must be quite crucial—the interactions among cells must be all important.[39]

We have already noted that the correlations near the critical point should not depend upon the details of the interaction. Let us assume that the cell size La_0 is one of these irrelevant details. Then, in each cell there must be a variable $\mu_{\mathbf{a}}$, $\mathbf{a}$ being a label for the cell, which behaves in an essentially identical fashion to our original variable $\sigma_{\mathbf{r}}$. That is, interactions among different cells should produce correlations among the $\mu_{\mathbf{a}}$'s identical in structure to the correlations in the original Ising model problem.[40]

The only possible difference between the site problem and the cell problem can be in the values of the two parameters ϵ and h. Assume that the values of these parameters for the cell problem are $\tilde{\epsilon}$ and $\tilde{h}$. Since these, respectively, describe the crucial effects of the interactions among cells and the interaction between each cell and the external magnetic field, they can depend upon the size of the cell. As the magnetic field goes to zero, the effective field in the cell problem must also go to zero. Hence $\tilde{h}\sim h$. The constant of proportionality may depend on the cell size so we guess

$$\tilde{h}=L^x h, \tag{3.1a}$$

where x is unknown. Similarly guess

$$\tilde{\epsilon}=L^y \epsilon, \tag{3.1b}$$

so that the cell problem will become critical as the site problem becomes critical. It follows that all nine critical indices may be determined in terms of x and y.

B. Thermodynamic Results

First we calculate the change in the free energy as h changes by an infinitesimal amount. Assume the $h_{\mathbf{r}}$ varies so slowly with $\mathbf{r}$ that the field may effectively be considered constant in each cell. Then the change in the free energy is

$$\delta(G/kT)=-\sum_{\mathbf{r}}\langle\sigma_{\mathbf{r}}\rangle\delta h_{\mathbf{r}}=-\sum_{\mathbf{a}}\langle\mu_{\mathbf{a}}\rangle\delta\tilde{h}_{\mathbf{a}}. \tag{3.2}$$

This must be the same in the two problems because this change in the free energy is a physical quantity. But, since the variation in space is assumed to be very slow we can replace the sum over cells $\mathbf{r}$ by the number of sites per cell times a sum over cells. This number is just L^d, where d is the dimensionality of the system.

Thus, (3.2) implies

$$L^d\langle\sigma_{\mathbf{r}}\rangle\delta h_{\mathbf{r}}=\langle\mu_{\mathbf{a}}\rangle\delta\tilde{h}_{\mathbf{a}}. \tag{3.3}$$

From (3.1a) and (3.3) it follows that

$$\sigma_{\mathbf{r}}=L^{x-d}\mu_{\mathbf{a}}. \tag{3.4}$$

Next calculate $\langle\sigma_{\mathbf{r}}\rangle$ for the case in which the field $h_{\mathbf{r}}$ does not depend upon $\mathbf{r}$. This depends upon ϵ and h,

$$\langle\sigma_{\mathbf{r}}\rangle=F(\epsilon,h). \tag{3.5a}$$

Since $\mu_{\mathbf{a}}$ is supposed to describe a problem identical to the original one except that it has new values of ϵ and h, $\langle\mu\rangle$ is the *same* function of the *new* variables

$$\langle\mu\rangle=F(\tilde{\epsilon},\tilde{h}). \tag{3.5b}$$

Equations (3.2), (3.4), and (3.5) now combine as

$$\langle\sigma\rangle=F(\epsilon,h)=L^{x-d}F(L^y\epsilon,L^x h). \tag{3.6}$$

But the length L is only a mathematical construct. It must cancel out of the right-hand side of (3.6). This can only happen if $F(\epsilon,h)$ is of the form

$$\langle\sigma\rangle=F(\epsilon,h)=(h/|h|)\,|\epsilon|^{(d-x)/y}f(\epsilon/|h|^{y/x}). \tag{3.7}$$

The factor $h/|h|$ in (3.7) is inserted to ensure that $\langle\sigma\rangle$ changes sign as the sign of the magnetic field is changed.[41]

The functional form of the $f(z)$ defined by (3.7) is not known to us. Nevertheless, (3.7) permits us to compute critical indices in terms of x and y. For example, if $\langle\sigma\rangle\sim|\epsilon|^\beta$ for small negative ϵ, it must be true that $f(-\infty)=$const. Then (3.7) gives

$$\beta=(d-x)/y. \tag{3.8}$$

All other thermodynamic derivatives may be found by differentiating or integrating $\langle\sigma\rangle$ with respect to ϵ and h. Thus α, α', γ, γ', and δ may be expressed in terms of x and y. The result is that the six indices in question are expressed in terms of the two parameters, or alternatively, the six indices obey four relations. These are[42]

$$\gamma=\gamma', \tag{3.9a}$$

$$\alpha=\alpha', \tag{3.9b}$$

$$d/y=2-\alpha=\gamma+2\beta=\beta(\delta+1). \tag{3.9c}$$

The analysis which gives γ, γ', β, and δ is quite direct because it involves differentiating $\langle\sigma\rangle$. But the specific heat can only be found by integrating $\langle\sigma\rangle$ with respect to h. Even then there are no problems except when α is an integer. The interesting case is $\alpha=0$, which gives a singularity of the form

$$\begin{aligned} C_B(\epsilon,h=0)&=-A\log|\epsilon|+B \qquad \text{for } \epsilon>0\\ &=-A'\log|\epsilon|+B' \qquad \text{for } \epsilon<0. \end{aligned} \tag{3.10}$$

This analysis indicates that the same coefficient multiplies the logarithm when $\epsilon>0$ and $\epsilon<0$:

$$A=A' \qquad \text{for } \alpha=0. \tag{3.11}$$

Equation (3.11) is another conclusion drawn from the scaling laws which is subject to experimental verification.

TABLE IV. Check of scaling law equalities for Ising model. The scaling laws predict that, for a given transition, all the numbers listed in the table (except for Δ and y) should be the same.

-	$2-\alpha$	$2-\alpha'$	$d\nu$	$d\nu'$	$d\gamma/(2-\eta)$	$\gamma+2\beta$	$\gamma'+2\beta$	$\beta(\delta+1)$	y	$\Delta=2x/y$
Two-dimensional Ising model	2	2	2	2	2	2	2	2	1	3.75
Three-dimensional Ising model	1.87 ±0.12	1.93+0.04, −0.16	1.93 ±0.01	?	1.933 ±0.008	1.87 ±0.01	1.94 ±0.05	1.93 ±0.05	1.55 ±0.01	3.22 ±0.02

C. Correlation Function Results

The next object is the spin–spin correlation function

$$g(\mathbf{r},\mathbf{r}')=g(R,\epsilon,h)=\langle[\sigma_{\mathbf{r}}-\langle\sigma\rangle][\sigma_{\mathbf{r}}'-\langle\sigma\rangle]\rangle, \quad (3.12)$$

where R is the distance between the points in question, $R=|\mathbf{r}-\mathbf{r}'|/a_0$. We can equally well write a correlation function for the μ's defined by Eq. (3.4) which must be identical in structure to $g(R,\epsilon,h)$ except that the scales of length ϵ and h are changed

$$R\rightarrow R/L,$$
$$\epsilon\rightarrow\tilde{\epsilon}=\epsilon L^{y},$$
$$h\rightarrow\tilde{h}=hL^{x}. \quad (3.13)$$

It follows that

$$g(R,\epsilon,h)=L^{2(x-d)}\langle[\mu_{\mathbf{a}}-\langle\mu\rangle][\mu_{\mathbf{a}}'-\langle\mu\rangle]\rangle$$
$$=L^{2(x-d)}g(R/L,\epsilon L^{y},hL^{x})$$

so that $g(R,\epsilon,h)$ can turn out to be independent of L, only if it has the structure:

$$g(R,\epsilon,h)=|\epsilon|^{2(d-x)/y}\tilde{g}(R|\epsilon|^{1/y},\epsilon/|h|^{y/x}) \quad (3.14)$$

for $R\gg1$, $|\epsilon|\ll1$ and $h\ll1$.

Equation (3.14) gives the critical indices ν, ν', and η. It is immediately apparent that the $h=0$ coherence length must be proportional to $|\epsilon|^{-1/y}$. Hence[43]

$$y^{-1}=\nu=\nu'=(2-\alpha)/\mathrm{d}. \quad (3.15)$$

The last critical index is η, defined by

$$g(R,\epsilon=0,h=0)\sim(R^{d-2+\eta})^{-1}.$$

Then (3.9) and (3.13) give

$$d\gamma/(2-\eta)=2-\alpha. \quad (3.16)$$

To conclude, we notice that arguments almost identical to the ones which generated the form (3.14) for the spin correlation function may be invoked to describe correlations involving the variable $E_{\mathbf{r}}$, which is the energy per site in the neighborhood of $\mathbf{r}$. The results are that

$$\langle[E_{\mathbf{r}}-\langle E\rangle][E_{\mathbf{r}'}-\langle E\rangle]\rangle$$
$$=g_{EE}(\mathbf{r},\mathbf{r}')$$
$$=|\epsilon|^{2(d-y)/y}\tilde{g}_{EE}(R|\epsilon|^{1/y},\epsilon/|h|^{y/x}) \quad (3.17a)$$

and

$$\langle[E_{\mathbf{r}}-\langle E\rangle][\sigma_{\mathbf{r}'}-\langle\sigma\rangle]\rangle$$
$$=g_{\sigma E}(\mathbf{r},\mathbf{r}')$$
$$=|\epsilon|^{(2d-x-y)/y}\tilde{g}_{\sigma E}(R|\epsilon|^{1/y},\epsilon/|h|^{y/x}), \quad (3.17b)$$

where $\tilde{g}_{EE}$ and $\tilde{g}_{\sigma E}$ represent unknown functions.

D. Comparison of Ising Model Results with Scaling Law Conclusions

The simplest conclusion from the scaling law hypothesis is that the combinations of "experimental parameters" $2-\alpha$, $2-\alpha'$, $d\nu$, $d\nu'$, $d\gamma/(2-\eta)$, $\gamma+2\beta$, $\gamma'+2\beta$, and $\beta(\delta+1)$ are all just different ways of writing the same number, d/y. To check this against the known Ising model results, we have given in Table IV the values of all these supposedly equal quantities as given by the results of Table II. The errors quoted in these tabulations are numerical errors in the Ising model analysis. The close equality among these numbers in both the two-dimensional case and three-dimensional one serves to support the scaling law hypothesis.

In the last two columns of Table IV values of y and $\Delta=2x/y$ derived from the previous seven columns are quoted. [We use Δ as our basic variable because several other authors (e.g., Domb and Hunter) also use it.] The value of y is determined by using $\nu=1/y$. The value of Δ is derived from y from the scaling law relation

$$\Delta=2d\nu-2\beta \quad (3.18)$$

since ν and β are known quite accurately.

In the two-dimensional case, there are available several other checks of the scaling law results. First the specific heat singularity at $h=0$ is logarithmic in character ($\alpha=\alpha'=0$) so that one should expect the coefficients A and A' defined by (3.10) are equal. The Onsager solution[9] shows this to be true.

Also, the scaling laws predict the following forms for the correlation functions of the two-dimensional Ising model at $h=0$. From (3.13), (3.16), and (3.17) it follows that at $h=0$,

$$g(\mathbf{r},\mathbf{r}')=|\epsilon|^{1/4}G(\epsilon R), \quad (3.18a)$$

$$g_{\sigma E}(\mathbf{r},\mathbf{r}')=|\epsilon|^{1+1/8}G_{\sigma E}(\epsilon R), \quad (3.18b)$$

$$g_{EE}(\mathbf{r},\mathbf{r}')=|\epsilon|^{2}G_{EE}(\epsilon R). \quad (3.18c)$$

All three of these scaling law conclusions have been verified by actual calculations.[11,18]

In conclusion, there is excellent evidence for the correctness of the scaling laws for the two-dimensional Ising model. All the correlation functions and critical indices turn out to agree exactly with the scaling law hypotheses. In three dimensions, the critical indices agree quite closely with this hypothesis, but there are small discrepancies which appear to be outside of the statistical uncertainties in the indices. These discrepancies have caused Domb[32] to have some doubt about the exact validity of these scaling laws which are derived from considerations involving the correlation functions, namely the relations involving ν and η. Fisher[33] has expressed similar reservations and also has questioned whether the symmetry above and below T_c, e.g., $\gamma=\gamma'$, is exactly true.

E. Questions About Real Phase Transitions

Three questions about the scaling laws which the experimental evidence can answer, in principle, are considered in this section.

A. Are the scaling laws correct?

Experimental evaluation of the indices combined with a comparison like that of Table IV can disprove the scaling laws or it can be strong evidence in their favor.

B. Assuming that the scaling laws are correct, how many different phase transitions are there?

We started from the point of view that the details of the interaction do not matter near the critical point. But what is a detail and what is essential? The dimensionality certainly matters considerably. But, in the real, three-dimensional world would we say all ferromagnetic transitions are essentially similar? Are these in turn just like all liquid gas transitions? Measurements of critical indices and deductions of values of x and y can shed considerable light on this point.

C. What are the values of x and y? Are they simple numbers like 2.5 or $\pi/2$? If so, careful thinking might serve to predict them. Accurate experimental determinations of these numbers can serve as a useful guide to thinking about this difficult problem.

IV. MAGNETIC TRANSITIONS

A. Theoretical Models

In real magnetic materials the spins are coupled through the strong, short-range exchange interactions and the long-range magnetic dipolar interactions. For the low-temperature materials, such as dysprosium aluminum garnet (DAG) and some low-temperature salts, these dipole interactions are of the same size or larger than the exchange interactions.[44–46] In high-temperature materials the exchange interactions are much stronger; however, even a weak long-range interaction may affect the delicate correlations which occur quite close to the critical point.

No theories have been developed up to this point which can treat the complexity of a real magnetic material. Instead, theory deals with over-simplified models.

Molecular Field Theory

The simplest model is the molecular field theory or Landau theory. We discussed this extensively in Sec. II, where we mentioned that this model gave $\gamma=1$, $\beta=\frac{1}{2}$, $\delta=3$, and a finite discontinuity in C_H at the critical temperature. This model is appropriate whenever the force has a very long-range. In particular, if there are z equally interacting spins, then according to Brout[3] the molecular field theory works whenever $|\epsilon|=(T-T_c)/T_c|\gg z^{-1}$. The dipole interaction does not have a long enough range so that one can be sure that the Landau theory is valid for purely dipolar forces.

Ising Model

In the Ising model, one includes only one component of the spin. This model might be a suitable representation for highly anisotropic materials in which the coupling of one component of the spin is much stronger than that of the other two components.

Heisenberg Model

The Heisenberg model of the ferromagnet is, in some sense, an opposite limit to the Ising model. Here one assumes that the Hamiltonian contains terms of the form $J\mathbf{S}\cdot\mathbf{S}'$, where $\mathbf{S}$ and $\mathbf{S}'$ are the spin operators for neighboring particles.

This assumption of complete isotropy is stronger than just taking the crystal to contain three easy axes of magnetization—as in a cubic crystal. The Heisenberg model implies that all directions in space are equivalent. The order parameter is then the magnetization vector. Since this is qualitatively different from the scalar order parameter of the Ising model, the Heisenberg model may show a critical behavior which is qualitatively different from that of the Ising model.

Of the critical indices we are reviewing, the only numerical result for the three-dimensional Heisenberg model is for the critical index γ which describes the divergence of the susceptibility just above T_c. Numerical calculations give[47] $\gamma=1.43$ for spin $\frac{1}{2}$ and[48] $\gamma=1.33$ or[49] 1.36 for infinite spin. The three-dimensional Ising model gives $\gamma=1.250$. This difference between the two models appears to be larger than the numerical errors in the calculation.

Notice the apparent spin dependence of the Heisenberg model's γ. The possibility that a critical index depends upon a specific detail of the interaction such as the size of the spin is contrary to the assumptions of the scaling laws. Consequently, we must count this small spin dependence as an argument against the scaling laws.

Antiferromagnets

An antiferromagnetic ordering consists of two interpenetrating sublattices of equal and opposite magnetization. In this case the order parameter, $p(r)$, is the magnetization on each sublattice. The conjugate variable, h, is a magnetic field which is positive on one sublattice and negative on the other. There is no way of producing such a field macroscopically in the laboratory.

In the nearest-neighbor Ising model we convert a ferromagnetic ordering into an antiferromagnetic ordering by changing the sign of the interaction constant, K. Since this transformation may be undone by changing the signs of all the spin variables on one sublattice there is an exact equality between the partition functions of the ferromagnetic and antiferromagnetic Ising models. Therefore, all the zero field ($h=0$) conclusions of the Ising model apply equally well to the ferromagnet and antiferromagnet.[50]

An antiferromagnetic ordering may also be produced in the Heisenberg model by changing the sign of the interaction constant. However, in this situation, there is no exact correspondence between the problems for different signs of the interaction constant except at infinite spin. However, if the details of the interaction do not matter, and if the value of the spin is one such detail, then the Heisenberg model antiferromagnet would have the same critical indices as the Heisenberg model ferromagnet.

Because of the antiparallel sublattices, direct dipole–dipole interactions are of much shorter effective range than in ferromagnets. Therefore, these long-range forces should be expected to have a much smaller qualitative effect upon the transition in the antiferromagnetic case.

B. Problems in Interpreting Experimental Data

Concept of Critical Region and Determination of T_c

In interpreting the data we are not particularly interested in the absolute temperature, but in $\epsilon=\Delta T/T_c=(T-T_c)/T_c$ near a critical point. The temperature difference ΔT can be measured much more accurately than T itself.

Theoretically one assumes that the thermodynamic functions approach a simple behavior, namely, a power of ϵ, as $\epsilon\to 0$. Thus the critical region is defined as the region where this behavior dominates. Since molecular field theory fails for $\epsilon\lesssim 1/z$, where z is the number of nearest neighbors, we obtain a rough estimate that the critical region is for $\epsilon\lesssim 10^{-1}$. In fact, we shall show that for magnetic materials the experimental evidence indicates critical regions beginning when $\epsilon\lesssim 10^{-2}$ to 10^{-1}. [Since T_c varies from 0.1° to 1400°K in the materials studied, the critical region is most conveniently discussed in terms of ϵ rather than $(T-T_c)$ above.] In principle, one can make cross checks by comparing for example the regions in which the specific heat and the susceptibility seem to show critical behavior. In practice, there is rarely enough data on the same material to be confident that the various thermodynamic functions settle down to their critical behavior at the same ϵ.

A major experimental uncertainty is in the relative location of T_c itself. In specific heat measurements, a rounded peak is often observed, making the precise location of T_c uncertain (see Table IX). In resonance experiments one sometimes finds a small temperature region where both para- and antiferromagnetic lines exist (in external field) and overlap.[51,52] Measurements in applied field require extrapolation to zero field to fix T_c. Noakes, Tornberg, and Arrott[53] have taken some of this uncertainty into account by assuming a power-law temperature dependence and then fitting to the best value of T_c as well as the critical exponent. However, it would be most desirable to determine T_c independently of the above data, or at least through comparison of two critical quantities, such as specific heat and susceptibility.

An example illustrates how the choice of critical region and T_c are interdependent and affect the evaluation of critical indices: Heller[54] plots the cube of magnetization in MnF_2 vs the temperature. A linear relationship is found for $7\times 10^{-5}<|\epsilon|<8\times 10^{-2}$, and this straight line is extrapolated to zero to give $T_c=67.336\pm 0.003$°K. Using his published data, however, we plot the *square* of magnetization vs. temperature and find that a straight line also fits here for $|\epsilon|<10^{-3}$. Extrapolation gives $T_c=67.343\pm 0.004$°K. Consequently if we assume that the critical region is for $|\epsilon|\lesssim 10^{-1}$ we are forced to conclude that the cubic law is very nearly correct; if we assume that the critical region only includes $|\epsilon|\lesssim 10^{-3}$ then the data equally well permit the square and cube laws.

Domains

Ferromagnets and imperfect crystals of antiferromagnetic material have a tendency to break up into domains. The domain walls serve to break up the long-range correlations so essential to critical behavior. Hence, if one is to compare idealized theories (like the Heisenberg or Ising models) with experiments it is essential that the domains be larger than the theoretical coherence length. If this is not true there will be a rounding of the transition which will be very difficult to interpret.

Changes in Lattice Constants

The interaction constants which describe the coupling between spins depend upon the lattice parameters.

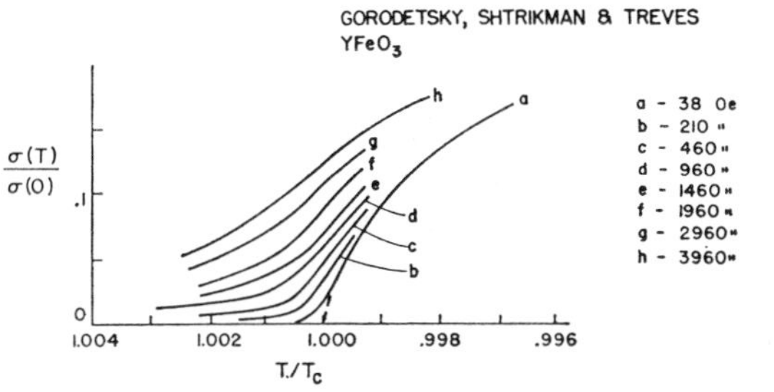

FIG. 3. Magnetization vs temperature of single-crystal $YFeO_3$ with the applied magnetic field as a parameter. The broken line is the magnetization extrapolated to $H=0$.

Since the internal energy of the lattice depends relatively strongly upon $T-T_c$ and T_c depends upon the lattice constants, it may be energetically favorable for the lattice constants to change near T_c. This effect should be included in the analysis of precise data.[54] In crude terms this is equivalent to say that $\epsilon=[T-T_c(T)]/T_c$, i.e., for each temperature in the critical region we need a different T_c in order to look for a power law to compare with a theory which has fixed lattice constants.

This connection between critical effects and lattice size and shape can in some cases make the transition first order.[55] Consequently, it is necessary, in every case, to be convinced either that the transition is indeed second order or that the latent heat in a first-order transition is too small to change the effects under study.

C. Experimental Results

Susceptibility

The experiments are of two kinds:

(1) An experimental mapping of the M-H-T surface, such as reported by Gorodetsky, Shtrikman, and Treves[30] for the weak ferromagnet $YFeO_3$. Their curves are shown in Fig. 3. From these data one can obtain the spontaneous magnetization, M vs H at T_c, and the susceptibility above and below T_c. To find T_c and the spontaneous magnetization one must extrapolate such data to $H=0$. To find the susceptibility below T_c one must subtract the spontaneous magnetization from the measured magnetization, which can introduce considerable uncertainty.

(2) Neutron scattering measures a time-dependent spin–spin correlation function from which one can deduce the correlation length and the susceptibility.

The experimental susceptibility data are summarized in Table V.[56–59] Except for $YFeO_3$ all the data are for $T>T_c$. The effective spin is listed in each case, but one cannot tell from these data whether the Heisenberg model behavior of larger γ for smaller spin is followed. The effects of single crystal versus polycrystal are more evident. Gadolinium is anisotropic, and Graham[60] has pointed out that for a single-crystal susceptibility measurements along the easy axis and along a hard axis show different behavior. Thus Develey's recent $\gamma\approx1.16$ for polycrystalline gadolinium is not surprising.[57] Possibly the $\gamma\approx1.21$ for polycrystalline cobalt has a similar explanation.[61] Iron and nickel each have three easy axes of magnetization and polycrystalline samples show the typical $\gamma\approx\frac{4}{3}$.

The recent experiment by Noakes, Tornberg, and Arrott shows that in iron the susceptibility and hence the long-range correlations are not altered by impurities.[53] The one value for $T<T_c$, namely $\gamma'\approx0.7$ in $YFeO_3$,[30] violates the scaling law prediction $\gamma=\gamma'$.

In experiments of Miedema, Van Kempen, and Huiskamp[45] on the low-temperature salt $CuK_2Cl_4\cdot2H_2O$, they also measured its specific heat (see Table IX). Our analysis indicates the specific heat settles down to its critical behavior for $\epsilon<10^{-1}$ whereas the susceptibility has settled down for $\epsilon<3\times10^{-1}$. Comparative data such as this should always be an objective of a critical phenomena experiment, for it helps establish where the critical region begins. In this case it appears that the critical region is approximately the same for these two thermodynamic functions.

Kouvel[62] has used M vs H and T data to construct a very direct check of the scaling laws. According to Eq. (3.7), for a given material the magnetization obeys the relation

$$M/\epsilon^{\beta}=f(H/\epsilon^{\beta\delta}), \tag{4.1}$$

where $f(H/\epsilon^{\beta\delta})$ is an unknown function. From (4.1) we conclude that if we knew M as a function of H for one value of ϵ and if we knew β then we could predict M vs H for any other value of ϵ. [Actually this is not quite true because $f(H/\epsilon^{\beta\delta})$ can have two different forms, one for $\epsilon>0$ and the other for $\epsilon<0$. But as long as we remain above T_c, the scaling laws predict that one measurement of M vs H would suffice to tell everything.]

Kouvel used the data of Weiss and Forrer[58] for Ni and his own data for CrO_2 [62] to construct a direct check of Eq. (4.1). He plotted M/ϵ^{β} as a function of $H/\epsilon^{\beta\delta}$ for different values of ϵ as indicated in Fig. 4. If Eq. (4.1) is correct this plot should give a simple curve, independent of the value of ϵ. The figure indicates that the points do fall on one curve. This then serves as a direct and convincing check of the scaling laws.

This analysis provides values of β and $\beta\delta$ which are different for the two materials. In particular, we have

$$\beta=0.42 \qquad \beta\delta=1.76 \qquad \text{for Ni}$$

$$\beta=0.33 \qquad \beta\delta=1.91 \qquad \text{for } CrO_2. \tag{4.2}$$

Equation (4.1) indicates that the critical index γ is given in terms of these as

$$\gamma=\beta(\delta-1). \tag{4.3}$$

TABLE V. Ferromagnetic susceptibility $T > T_c$.

Material	Experimenters	Ref.	Method	T_c (°K)	$\epsilon = \Delta T/T_c$ Range for fit	γ	Effective spin	Comments
Iron	Arajs, Colvin	56	Field gradient force technique	1044.0	10^{-3}–4×10^{-2}	1.33	1.1	
	Noakes, Tornberg, Arrott	53	Induction	1044.0	5×10^{-6}–1.5×10^{-2}	1.333±0.015		$\gamma = 1.32$–1.35 for alloys less than 2.4%
	Develey	57	Moving balance	1041.8	2×10^{-3}–3×10^{-2}	1.33±0.03		
Nickel	Weiss, Forrer	58	Magnetocaloric effect	627.2	4×10^{-3}–3×10^{-2}	1.35±0.02	0.3	
	Arajs	59	Field gradient force technique	626.5	2×10^{-3}–2×10^{-2}	1.29±0.03		
	Develey	57	Moving balance	631.0	5×10^{-3}–3×10^{-2}	1.32±0.02		
Gadolinium	Graham	60	Fluxmeter	292.5	3×10^{-3}–5×10^{-2}	1.33	7/2	Single crystal contains some Gd_2O_3
	Develey	57	Moving balance	292.85	2×10^{-3}–5×10^{-2}	1.16±0.02		Not single crystal
Cobalt	Colvin, Arajs	61	Field gradient force technique	1388.2	7×10^{-4}–10^{-2}	1.21±0.04	0.85	Not single crystal
$CuK_2Cl_4\cdot 2H_2O$ $Cu(NH_4)_2Cl_4\cdot 2H_2O$	Miedema, VanKempen, Huiskamp	45	ac bridge and ballistic galvanometer	0.88 0.70	3×10^{-2}–3×10^{-1} 4×10^{-2}–4×10^{-1}	1.36 1.37	1/2	
$YFeO_3$	Gorodetsky, Shtrikman, Treves	30	Vibrating sample magnetometer	643	2×10^{-4}–3×10^{-3}	1.33±0.04	1.1	$\gamma' = 0.7\pm0.1$
CrO_2	Kouvel, Rodbell	62		386.5	2×10^{-2}–2×10^{-1}	1.6±0.05	1	
Value used for scaling law analysis						1.33±0.03		
Molecular field theory						1.0		
3-dimensional Ising model						1.250±0.001		
3-dimensional Heisenberg model						1.33+0.05/spin (Ref. 63)		

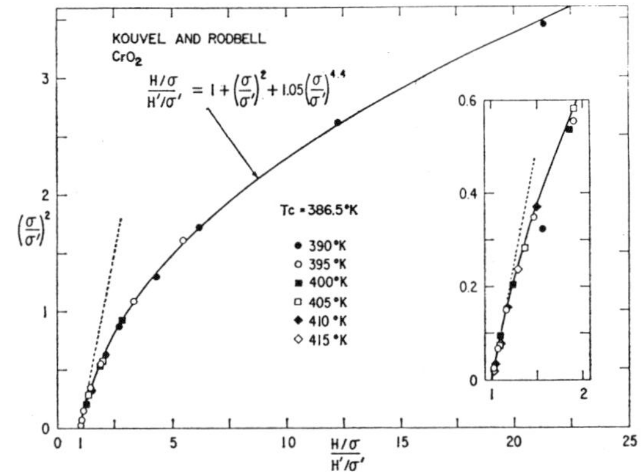

FIG. 4. Magnetization vs field for CrO_2. In our notation, $\sigma = M$, $\sigma' \sim \epsilon^\beta$, $H'/\sigma' \sim \epsilon^\gamma = \epsilon^{\beta(\delta-1)}$. Points for different ϵ fall on the same curve, which verifies the scaling law prediction, Eq. (4.1).

The resulting values of γ have been inserted in Table V. Also the index δ, defined by

$$M \sim H^{1/\delta} \quad \text{at} \quad T = T_c$$

can be identified with the quantity δ defined by Eq. (4.1). Therefore Kouvel's analysis gives the values of δ. These have been inserted in Table VIII.

If we assume the scaling law conclusion that (4.1) is equally correct for $\epsilon > 0$ and $\epsilon < 0$ with the same value of β and δ but a different f, then we find that

$$M \sim (-\epsilon)^\beta$$

for $\epsilon < 0$ and $H = 0$. This would permit us to identify the quantity, β, determined by Kouvel's analysis [see Eq. (4.2)] with the critical index, β, which defines the behavior of the magnetization for $T < T_c$. However, Kouvel provides no experimental evidence for (or against) this extrapolation into the region $T < T_c$.

Neutron Scattering

Magnetic systems have an anomalously large cross section for neutron scattering near their critical temperature. Van Hove[64] has shown that this is due to the large fluctuations in magnetization near T_c. An excellent review of this subject has been written by DeGennes.[65]

Passel, Blinowski, Brun, and Nielsen[66] have measured the neutron scattering cross section for iron. Their experiment can be interpreted in terms of the formula for the cross section

$$\frac{d}{dE}\frac{d\sigma}{d\Omega} \sim \int dt \exp(-i\omega t) \sum_{\mathrm{R}} \langle S_0 S_R(t) \rangle \exp(i\mathbf{q}\cdot\mathbf{R}). \tag{4.4}$$

Here E is the scattered neutron energy, q and ω are the momentum and energy transferred from the neutron to the magnetic material, R is a lattice site, and S is a spin operator. Van Hove[64] has described the time dependence of the correlation function by expressing the decay of a fluctuation in the magnetization by a diffusion equation with diffusion coefficient Λ.

This implies that the correlation function in Eq. (4.4) decays in time according to

$$\sum_R \exp(i\mathbf{q}\cdot\mathbf{R}) \langle S_0 S_{\mathrm{R}}(t) \rangle = \sum_R \exp(i\mathbf{q}\cdot\mathbf{R} - \Lambda q^2 t) \langle S_0 S_{\mathrm{R}} \rangle. \tag{4.5}$$

Thus, the cross section in (4.4) can be expressed in terms of the static correlation function $\langle S_0 S_R \rangle$. According to the scaling laws, at zero magnetic field,

$$\langle S_0 S_{\mathrm{R}} \rangle = G(R/\xi)/R^{1+\eta}, \tag{4.6}$$

where ξ is the coherence length. This is supposed to depend upon temperature as

$$\xi \sim \epsilon^{-\nu}.$$

The dependence of the correlation function upon $(T - T_c)$ is supposed to be hidden in the dependence of the unknown function, G, upon R/ξ.

In analyzing the experiments, it is convenient to guess[16] the form of G as

$$G(R/\xi) = A \exp(-R/\xi) \qquad \text{for } T > T_c, \tag{4.7}$$

where A depends only very weakly upon ϵ. [When $\eta = 0$, this gives the Ornstein–Zernike[67] result for the correlation function

$$\langle S_0 S_{\mathrm{R}} \rangle = A(e^{-R/\xi}/R).] \tag{4.8}$$

If we put together the results (4.5), (4.6), and (4.7) we find the differential cross section as

$$\frac{d\sigma}{d\Omega} \sim A(T) \int d\omega \frac{\Lambda_q^2}{(\Lambda^2 q^4 + \omega^2)(\xi^{-2} + q^2)^{1-1/2\eta}}. \tag{4.9}$$

Passel, Blinowski, Brun, and Nielsen obtained the parameters $A(T)$, Λ, and $\xi(T)$ by comparing their data with $d\sigma/d\Omega$ for various choices of η. For $\eta = 0$, the result is $\nu = 0.64 \pm 0.02$ in $7 \times 10^{-3} < \epsilon < 6 \times 10^{-2}$. For $\eta = 0.15$, the result is $\nu = 0.64 \pm 0.02$ in $4 \times 10^{-3} < \epsilon < 6 \times 10^{-2}$. So $\eta \neq 0$ fits the data better for smaller ϵ. However, experimental uncertainty for $\epsilon < 10^{-2}$ precludes drawing any conclusions about the value of η.

Above T_c, the susceptibility is given in terms of the spin–spin correlation function by the relation

$$\chi \sim \sum_{\mathrm{R}} \langle S_0 S_{\mathrm{R}} \rangle$$

From (4.6) and (4.7) it follows that

$$\chi \sim A(T) \xi^{2-\eta}. \tag{4.10}$$

If we express the coherence length $\xi \sim \epsilon^{-\nu}$, and if ϵ is small enough so that $A = $ constant, then $\chi \sim \epsilon^{-\gamma}$ implies $\gamma = (2 - \eta)\nu$. Unfortunately, this experiment did not extend to small enough values of ϵ for A to be a constant

TABLE VI. Correlation length $T > T_c$.

Material	Experimenters	Ref.	Method	T_c (°K)	$\epsilon = \Delta T / T_c$ Range for fit	ν	Comments
Iron	Passell, Blinowski, Brun, Nielsen	66	Neutron scattering	1044.0	5×10^{-3}–6×10^{-2}	0.64 ± 0.02	$\eta=0.07\pm0.07$
Cr_2O_3	Riste, Wanic	68	Neutron scattering	310.0	10^{-2}–5×10^{-2}	0.67 ± 0.03	Antiferro-magnets
α-Fe_2O_3				998.0	5×10^{-3}–7×10^{-2}	0.63 ± 0.04	
$KMnF_3$	Cooper, Nathans	69	Neutron scattering	88.06 ± 0.02	2×10^{-3}–10^{-1}	0.67 ± 0.04[a]	Antiferro-magnet
Value used for scaling law analysis						0.65 ± 0.03	$\eta=0.07\pm0.07$
Molecular field theory						0.5	$\eta=0$
3-dimensional Ising model						0.643 ± 0.003	$\eta=0.056\pm0.008$

[a] This paper reports the temperature dependence of the staggered susceptibility, which is simply related to the square of the correlation length, as discussed in our text for iron.

$[A\sim(r_1^2)^{-1}$ in Ref. 66$]$, so we cannot take $\gamma=(2-\eta)\nu$ here. Nevertheless one can plot $A\xi^{2-\eta}$ vs ϵ and obtain an approximate fit by $\epsilon^{-\gamma}$, to obtain $\gamma=1.30$. This agrees with γ for iron as listed in Table V.

In antiferromagnets, $\langle S_0 S_R\rangle$ is defined with respect to a sublattice, and again we write $\xi\sim\epsilon^{-\nu}$. Values of ν and η for three antiferromagnets, iron, and theoretical models are listed in Table VI.[66,68,69]

Magnetization

The most accurate measurements of magnetization, indeed the only ones possible for antiferromagnets, are measurements of internal fields, e.g., at nuclei through NMR or the Mössbauer effect. Can we be sure that the hyperfine field H_n is linearly related to the bulk or sublattice magnetization? The data we quote assume this, but it is not at all obvious in the study of hyperfine interactions.[70] Experimental evidence indicates nonlinearity between magnetization and hyperfine field in some cases. In iron, the Mössbauer measurement of Preston, Hanna, and Heberle[71] agrees with the NMR result (which is at low temperatures only) on relative H_n variation with temperature, but is in slight disagreement with the bulk magnetization measurement. Many authors find that *impurity* hyperfine fields are not directly proportional to bulk magnetization.[72] Callen, Hone, and Heeger[73] and more recently Hone, Callen, and Walker[74] deduce theoretically nonlinear relationships between impurity and host *magnetizations*. On the other hand, both experimental and theoretical results point to a linear relationship between H_n and M as $T\to T_c$, even for impurities, thus supporting the usefulness of hyperfine field measurement in the study of critical phenomena.

The experimental results are listed in Table VII. For the ferromagnetic materials they show $\beta\approx\frac{1}{3}$ for relatively high values of $-\epsilon$, i.e., $3\times10^{-3}\lesssim-\epsilon\lesssim10^{-1}$. However, for small values of ϵ there are two experiments, on Ni[29] and $YFeO_3$[30] which indicate higher values of β, namely $\beta\approx\frac{1}{2}$. This apparent change in index at $(-\epsilon)\sim3\times10^{-3}$ is perplexing. A theory due to Callen and Callen[75] predicts this kind of behavior, but one wonders what it would predict for other critical quantities, e.g., susceptibility and specific heat. Recent preliminary data of Arrott[75a] indicate $\beta\approx\frac{3}{8}$ for nickel in the same temperature region in which Ref. 29 reported $\beta=\frac{1}{2}$.

We suggest tentatively that this change in the value of β might be caused by the dipole forces in the ferromagnet. These dipole forces are weak in comparison with the exchange forces so that they should not be expected to modify the thermodynamic functions very much under usual circumstances. However, very near T_c, there are very delicate and long-range correlations which might be affected materially by these dipole fields. Since the dipole force is long-range, it might be expected to produce qualitatively different critical behavior from that produced by the short-range exchange forces. Hence, as the critical point, is approached, a change in β might occur at some value of ϵ for which the dipole forces just become important.

To estimate this critical value of $(-\epsilon)$, assume that, because large regions of spins fluctuate in their magnetization, there are fluctuations in the magnetization just about as big as the average magnetization. This would imply that there are fluctuations in the magnetic field, δB, which are of the order of the average magnetization:

$$\delta B\sim\langle M\rangle. \qquad (4.11)$$

Now let us try to estimate how much damage these fluctuations in B can do. When can they appreciably modify the results that we would get from a theory in which the magnetic forces were left out? To do this compare the spontaneous magnetization, which is given by

$$M/M_0\sim(-\epsilon)^\beta \quad \text{at } B=0 \qquad (4.12)$$

Table VII. Spontaneous magnetization.

Material	Experimenters	Ref.	Method	T_c (°K)	$\epsilon = \|\Delta T\|/T_c$ Range for fit	β	Comments
			Antiferromagnets				
MnF_2	Heller, Benedek	51	NMR on F^{19}	67.336±0.003	8×10^{-5}–2×10^{-2}	0.335±0.01	
$CuCl_2 \cdot 2\ H_2O$	Poulis, Hardeman	76	NMR, Protons	4.337±0.003	5×10^{-4}–10^{-2} 10^{-2}–10^{-1}	0.18±0.07 0.29±0.03	
$CoCl_2 \cdot 6\ H_2O$	Sawatzky, Bloom	52	NMR, Protons	2.275	10^{-2}–10^{-1}	0.15±0.05	
	Van der Lugt, Poulis	77	NMR, Protons	2.275	5×10^{-2}–2×10^{-1}	0.23±0.02	
$KMnF_3$	Cooper, Nathans	69	Neutron scattering	88.06±0.02	10^{-2}–10^{-1}	0.33	
			Ferromagnets				
Iron	Preston, Hanna, Heberle	71	Mössbauer Fe^{57}	1042.0±0.3	2×10^{-3}–10^{-1}	0.34±0.02	Only four data points in this region
	Potter	78	Magnetocaloric effect	1035.0±2.0	4×10^{-3}–2×10^{-1}	0.36±0.08	Large uncertainty possible in T_c
Nickel	Howard, Dunlap, Dash	29	Mössbauer Fe^{57}	629.4	5×10^{-4}–10^{-2} 10^{-2}–1.6×10^{-1}	0.51±0.04 0.33±0.03	
EuS	Heller, Benedek	79	NMR, Eu^{153}	16.50±0.03	10^{-2}–10^{-1}	0.33±0.015	
$YFeO_3$	Gorodetsky, Shtrikman, Treves	30	Vibrating sample magnetometer	643	2×10^{-4}–3×10^{-3}	0.55±0.04	
	Eibschutz, Shtrikman, Treves	80	Mössbauer Fe^{57}	640	10^{-2}–3×10^{-1}	0.354±0.005	β=0.348±0.005 is average for fourteen rare-earth orthoferrites
$CrBr_3$	Senturia, Benedek	81	NMR, Br^{79}, Br^{81}	32.56±0.015	7×10^{-3}–5×10^{-2}	0.365±0.015	
Value used for scaling law analysis						0.33±0.03	
Molecular field theory						0.5	
3-dimensional Ising model						0.313±0.004	

TABLE VIII. M vs H at T_c.

Material	Experimenters	Ref.	Method	T_c (°K)	δ	Comments
Nickel	Weiss, Forrer	58	Magnetocaloric effect	627.2	4.2 ± 0.1[a]	
Gadolinium	Graham	60	Fluxmeter	292.5	4.0 ± 0.1	
$YFeO_3$	Gorodetsky, Shtrikman, Treves	30	Vibrating sample magnetometer	643.0	2.8 ± 0.3	Weak ferromagnet
CrO_2	Kouvel, Rodbell	62		386.5	5.75	
Value used for scaling law analysis					4.1 ± 0.1	
Molecular field theory					3.0	
3-dimensional Ising model					5.2 ± 0.15	

[a] Recent preliminary data of Arrott indicate $\delta\approx4.7$ for nickel. This work was reported to the Twelfth Annual Conference on Magnetism and Magnetic Materials, Washington, D. C., November 1966.

with the induced magnetization

$$M/M_0\sim(\mu B/KT_c)^{1/\delta} \quad \text{at } \epsilon=0, \tag{4.13}$$

where M_0 is the zero temperature magnetization and μ_B is the Bohr magneton. From (4.11) and 4(.13), we would guess that the fluctuations in B would modify the magnetization by an amount of the order

$$(\delta M)\sim(\mu M/KT_c)^{1/\delta}M_0. \tag{4.14}$$

If this modification is negligible in comparison with the underlying magnetization given by (4.12), then we can neglect the long-range forces. This condition is

$$(\mu M/KT_c)^{1/\delta}\ll(-\epsilon)^{\beta}.$$

We substitute for M by using (4.12) and find

$$(-\epsilon)\gg(\mu M_0/KT_c)^{1/\beta(\delta-1)} \tag{4.15}$$

as our condition for neglecting the dipole forces.

To evaluate this condition we choose $\beta=\frac{1}{3}$, $\delta=4.2$, $M_0=\mu\times(\text{density})=\mu\times(10^{23}\text{ atoms/cm}^3)$, $T_c=600$°K and find that the condition (4.15) has become

$$(-\epsilon)\gg10^{-2} \quad \text{for Ni.}$$

From this point of view, it is not surprising that Ni should have a change in β at this point.

This calculation suggests that the $\beta\approx\frac{1}{2}$ observed in $YFeO_3$ and Ni and the $\gamma'\approx0.7$ in $YFeO_3$ might be characteristic of an effect of long-range dipole forces rather than the theoretically simpler short-range forces. If so, perhaps specific heat and susceptibility data below T_c should also approach molecular field values. The only applicable datum, however, is $\gamma'\approx0.7$ in $YFeO_3$.[30]

The experimental data for iron are not sufficiently clear cut to support or deny the calculation just made. Preston, Hanna, and Heberle's result for β is based on a very small number of data points[77a] but indicates $\beta\approx\frac{1}{3}$ for $-\epsilon>2\times10^{-3}$.[71] Potter's data may be interpreted either as fitting one power law with a large error or as gradually changing its β, with β increasing as $\epsilon\to0$.[78] There is clearly a need for further investigation in this area.

Of the antiferromagnets listed in Table VII, only MnF_2 has been carefully investigated for $\epsilon<10^{-2}$.[51] In several experiments NMR lines have been observed at the same temperature corresponding to the antiferromagnetic and paramagnetic state simultaneously.[51,52] This may be due to nonuniform T_c, or possibly to fluctuations in magnetization of sufficiently large space and time extent.

A clear-cut picture does not emerge from these measurements of β. However, on the basis of this rather scanty evidence we might propose that there are three separate behaviors here:

(a) When dipole forces are not important, $\beta\approx\frac{1}{3}$ in both ferromagnets and antiferromagnets. Hence, for our analysis of experiments which do not seem to be affected by these long-range forces we use $\beta=0.33\pm0.03$.

(b) In high-temperature ferromagnets very close to T_c, β seems to be about $\frac{1}{2}$. This might be caused by the dipole forces—which are unimportant far from T_c—but which might produce important corrections very near T_c.

(c) The experiments on $CuCl_2\cdot2H_2O$[76] and $CoCl_2\cdot6H_2O$ [52,77] yield still other values for β, suggesting that these materials are more complicated than can be accounted for by our simple picture. These are low-temperature antiferromagnets where dipole forces are comparable in magnitude with exchange forces.

M vs H at T_c

Molecular field theory predicts $M\sim H^{1/3}$ at T_c. More generally we write $M\sim H^{1/\delta}$. Table VIII lists four experimental values of δ. It is tempting to try to get values of δ for nickel,[59] iron,[56] and cobalt[61] from the data of Arajs and Colvin, since they also publish M vs H vs T, but in fact the magnetic fields they used are too small for this purpose. To find δ, one needs ϵ to be negligible. However, a comparison of Eqs. (4.12) and (4.13) indicates that ϵ can only be neglected if

$$|\epsilon|^{\beta}\ll(\mu H/KT_c)^{1/\delta}. \tag{4.16}$$

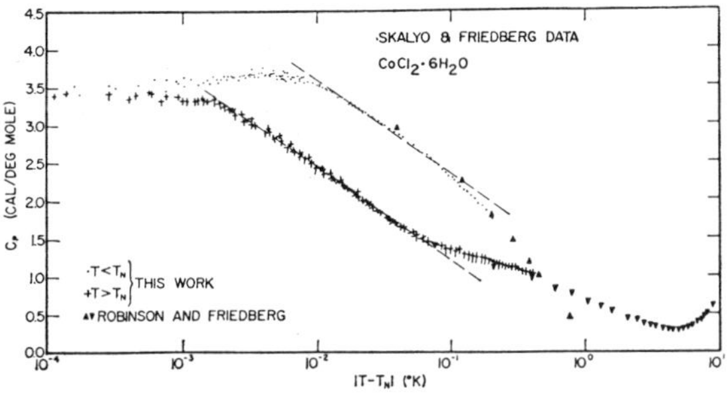

FIG. 5. Specific heat of $CoCl_2 \cdot 6\,H_2O$. The peaks round off for $\epsilon \approx 5 \times 10^{-3}$.

However, the uncertainty in T_c precludes satisfying Eq. (4.16). This uncertainty arises in the case of Arajs and Colvin, because they had to extrapolate from $H \neq 0$ to determine T_c. Thus their T_c for cobalt can easily be in error by $\pm 0.05°$K. Assuming $\delta \approx 4$ and $\beta \approx \frac{1}{3}$, the fields H_i (applied field corrected for demagnetizing effects) are not large enough to allow a determination of δ.

Specific Heat

The specific heat results are perhaps the most difficult critical data to categorize and understand In the two-dimensional Ising model, the specific heat behaves as

$$\begin{aligned} C_H &= -A \log |\epsilon| + B \qquad \text{for } T > T_c \\ &= -A' \log |\epsilon| + B' \qquad \text{for } T < T_c, \end{aligned} \tag{4.17}$$

with $A = A'$ and $B = B'$. In molecular field theory, there is a discontinuity in the specific heat. This corresponds to the result (4.17) in the special case $A = A' = 0$, $B \neq B'$. However, in the analysis of the three-dimensional Ising model, it has been suggested[16] that the specific heat diverges more strongly than indicated in (4.17), in fact that

$$\begin{aligned} C_H &\sim a\epsilon^{-\alpha} + b \qquad \text{for } T > T_c \\ &\sim a' |\epsilon|^{-\alpha'} + b' \qquad \text{for } T < T_c, \end{aligned} \tag{4.18}$$

with $\alpha \sim \alpha' \sim 0.1$. In writing Eq. (4.18), we have included terms in b and b' which are asymptotically much smaller than the divergent terms in (4.18). Hence they would be left out of the numerical analysis. However, there is no theoretical reason why these constant terms should not arise as slightly less divergent terms in the specific heat. Finally, there have been suggestions[82] that the specific heat might be less singular than indicated in (4.17) or (4.18). This behavior could arise, for example, if $b = b'$ in Eq. (4.18) and α and α' were both negative. Then C_H would show neither an infinity nor a discontinuity, but just a cusp equivalent to a third order phase transition.[83]

We would like to deduce the α's from the experimental data. This is difficult because what we wish to see is a weakly singular term above the background of b and b'. Given a bit of scatter in the experimental points, it becomes possible to make the three-parameter (a, b, α) fit indicated in Eq. (4.18) with a whole range of α's. A fit of the form (4.17) emerges as just a special case of the fit (4.18) since

$$\lim_{\alpha \to 0} (A/\alpha)[\epsilon^{-\alpha} - 1] + B = -A \log \epsilon + B. \tag{4.19}$$

It is hard enough to take data for, say $T > T_c$ and try to make a three-parameter fit as indicated in Eq. (4.18). But, actually the problem is much harder. One is supposed to reject all data for large values of $|\epsilon|$ since these data are "not in the critical region." This choice of which data are to be thrown out will influence considerably the range of α's which will fit the experiment. Also, there is often (see Fig. 5) some rounding of the peaks in the specific heat—which one might guess is due to imperfections in the crystals. It is tempting to say that this is the cause of the rounding and reject the data in the rounded region. But, if this is done, one more unknown parameter enters the problem: T_c. Clearly, if there is a broadening of the specific heat maximum, we cannot use the position of the peak to determine T_c with high precision.

Despite all this, it is possible to make some statements about α and α'. Consider, for example, the data of Skalyo and Friedberg[84] shown in Fig. 5. This plotting indicates that we can get a simple result if we reject all data for $|\epsilon| > 4 \times 10^{-2}$ as "not critical," and all data for $C_P > 3.3$ cal/(deg-mole) as "produced by crystalline imperfections." Then the figure indicates that the remaining data—which extends over only about one decade in ϵ—fit a logarithmic singularity. However, these remaining data also can be fitted by Eq. (4.18) with all values of α in the range

$$0 \leq \alpha' \lesssim 0.11$$

$$0 \leq \alpha \lesssim 0.19.$$

In Skalyo and Friedberg's data one is somewhat free to choose T_c because of the rounding off in the specific heat curve. Thus they chose $T_c = 2.289°$K because it gave $A'/A = 1$. We find $T_c = 2.288°$K gives an equally good logarithmic fit but yields $A'/A \approx 0.8$. Thus we see that A'/A is extremely sensitive to our choice of T_c.

Our range of α's becomes even larger if we consider data[85] like that in Fig. 6. If we say that the critical region is $|\epsilon| < 10^{-1}$, we are forced to fit the data with a large power of $(T - T_c)$. In particular, we find $\alpha \approx 0.6$ for $Co\,K_2(SO_4) \cdot 6\,H_2O$ and $\alpha \approx 0.7$ for $Co\,Cs_3\,Cl_5$. However, choosing the critical region to be $|\epsilon| < 2 \times 10^{-2}$ we can fit the data for α's between zero and roughly 0.6. In short, we cannot distinguish between $\alpha = 0$ and $\alpha = 0.6$ from these data.

Our experience is that at least two decades of data are desirable to establish the specific heat behavior. We also find that in order to fit data to a single logarithm or low power law we must often limit the critical region to $\epsilon \lesssim 10^{-2}$. This often leaves only one decade of critical data.

The rounding of the specific heat peak still may indicate that the specific heat is not a divergent function.

The experiment of Teaney[86] on the specific heat of MnF_2 best allows one to make a definite statement about α and α', for he obtained data over more than two decades and down to $\epsilon = 2\times10^{-4}$ with no rounding of the peak.

Figure 7 shows the specific heat of MnF_2 plotted on semilog paper. Both the logarithm and power law seem satisfactory over substantially the same temperature range. Figure 8 shows the effects of introducing various b' into the data. Sublattice magnetization data for MnF_2 [87] indicates that $\epsilon \lesssim 2\times10^{-2}$ specifies the critical region. If we use this cutoff to determine the range of α allowed in the three-parameter fit, then $0 \lesssim \alpha' \lesssim 0.18$. Similarly, for iron[88] the susceptibility data above T_c indicates a critical region of $\epsilon \lesssim 2\times10^{-2}$, leading to $0 \lesssim \alpha \lesssim 0.17$.

In Table IX are listed ranges for α and α' obtained from three parameter fits. In these cases critical data exist over a wide enough temperature interval to make such an analysis meaningful. There is a reasonable amount of data for which $\alpha \lesssim 0.16$. We use this value for the purposes of checking the scaling laws. But there are also considerable data which apparently do not fall into this category.

This table includes a column labeled A'/A which is found by taking the logarithmic fits to the data and finding the ratio of coefficients above and below T_c. In four of the five cases this ratio agrees with the scaling law conclusion $A'/A=1$. However, for the best experiment, MnF_2,[86] this result fails. Either the specific heat is not logarithmic for MnF_2 or this conclusion drawn from the scaling laws is wrong.

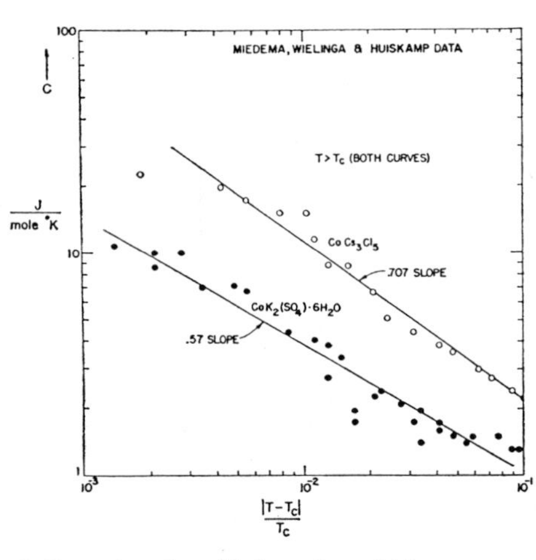

FIG. 6. Examples of specific heat data which are apparently fit by relatively large power laws.

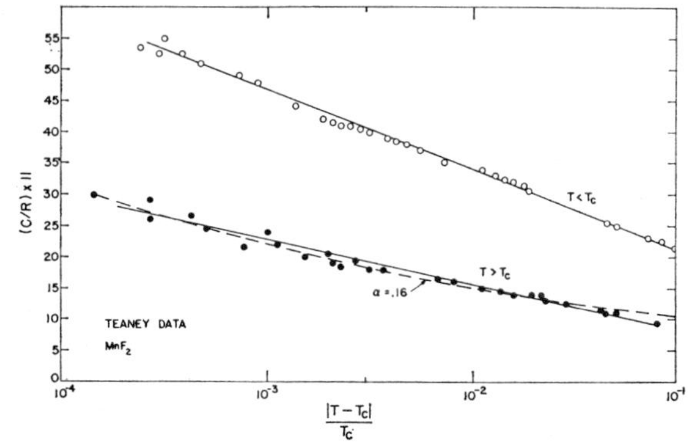

FIG. 7. Specific heat of MnF_2 as measured by Teaney. The solid lines represent logarithmic fits. The dashed line is the power law $C\sim\epsilon^{-0.16}$.

Specific Heat in External Fields

The second-order phase transition in an antiferromagnet is not destroyed upon application of an external field, as it is in a ferromagnet. According to a theory by Fisher[94] the critical temperature is lowered as $T_c(H) = T_c(0) - aH^2$. Recent results for dysprosium aluminum garnet[95] agree with this prediction close to T_c. Data for $MnCl_2\cdot 4\,H_2O$[96] indicate that the specific heat peak moves to lower T as H increases, but the peak also is broadened considerably, which makes the data difficult to interpret.

Comparison with Scaling Laws

Tables V–IX indicate the average values used in comparing the scaling law relations. Values which obviously contradict the chosen values are $\beta\approx\frac{1}{2}$ in nickel and $\beta\approx\frac{1}{2}$, $\gamma'\approx0.7$, $\delta\approx2.8$ in $YFeO_3$. This is a pattern of roughly molecular field theory behavior which might be explained by long-range forces which act in the presence of substantial spin alignment. For scaling law comparisons we neglect these data, but not because we consider them unreliable. The scaling laws are only applicable for short-range interactions, so that they might not apply to these data.

The comparisons are shown in Table X. Most of the data fit the pattern, although the column involving δ does not fit well. For antiferromagnets the scaling laws seem all right, upon very limited evidence, so definite conclusion cannot be drawn. For ferromagnets, the scaling laws relate a part of the data, but another part remains outside this framework. Further theoretical work on the effects of the long-range force might reconcile the apparent discrepancies.

TABLE IX. Specific heat.[a]

Material	Experimenters	Ref.	T_c (°K)	$\epsilon = \lvert \Delta T \rvert / T_c$ Range for fit	α	α'	A'/A	Comments
			Antiferromagnets					
MnF_2	Teaney	86	67.33 ± 0.01	2×10^{-4}–5×10^{-2}	$\lesssim0.16$	$\lesssim0.18$	2.0	
$CoCl_2\cdot6\,H_2O$	Skalyo, Friedberg	84	2.289 ± 0.002	10^{-3}–3×10^{-2}	$\lesssim0.11$			
				5×10^{-3}–4×10^{-2}		$\lesssim0.19$	1.0	Rounding of peak
$MnCl_2\cdot4\,H_2O$	Friedberg, Wasscher	89	1.622 ± 0.005	10^{-3}–10^{-1}		$\lesssim0.14$		$T>T_c$, Logarithm fits data for 4×10^{-3}–2×10^{-2}
$CuK_2(SO_4)_2\cdot6\,H_2O$	Miedema, Wielinga, Huiskamp	85	0.193 ± 0.001	10^{-3}–2×10^{-2}	$\lesssim0.6$			
$CoCs_3Cl_5$			0.52 ± 0.01	4×10^{-3}–2×10^{-2}	$\lesssim0.7$			
				4×10^{-3}–5×10^{-2}		$\lesssim0.25$		Rounding of peak
$RbMnF_3$	Teaney, Moruzzi, Argyle	90	0.83 ± 0.01	2×10^{-4}–5×10^{-2}	$\lesssim0.15$			
				2×10^{-3}–2×10^{-2}		$\lesssim0.15$	1.0	Rounding of peak
			Ferromagnets					
Iron	Kraftmakher, Romashina	91	1043.0 ± 1.0	2×10^{-3}–10^{-1}	$\lesssim0.17$			
				3×10^{-3}–7×10^{-2}		$\lesssim0.13$	1.0	
$CuK_2Cl_4\cdot2\,H_2O$	Miedema, Wielinga, Huiskamp	92	0.88 ± 0.01	10^{-3}–10^{-1}	$\lesssim0.10$	$\lesssim0.17$	1.0	
Nickel	Kraftmakher	93	627.0	5×10^{-3}–8×10^{-2}				Data have considerable scatter. Logarithms with $A'/A\approx1$ provide reasonable fit.
Value used for scaling law analysis					$\lesssim0.16$	$\lesssim0.16$		
Molecular field theory					0	0		Finite discontinuity
3-dimensional Ising model					$\lesssim0.2$	$\lesssim0.1$		

[a] Further information on magnetic specific heats may be found in Ref. 1, particularly the papers of Teaney and Yamamoto *et al.*

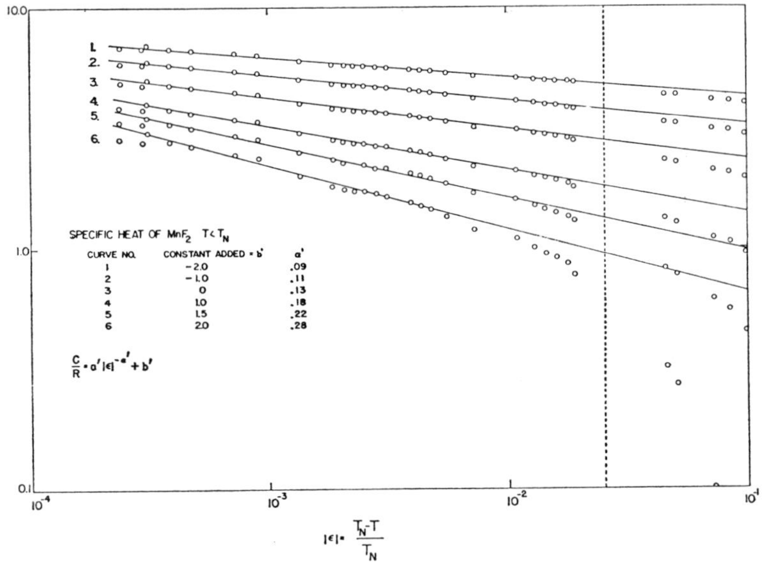

FIG. 8. Three-parameter fits for specific heat of MnF_2 for $T<T_N$. Dashed line near $|\epsilon|=2\times10^{-2}$ indicates extent of critical region.

V. CLASSICAL LIQUID–GAS TRANSITIONS

A. Theory

A simplified theory of critical behavior in the liquid–gas transition may be obtained from the Van der Waals equation of state:

$$[P+a(N/V)^2](V-bN)=NkT. \tag{5.1}$$

When this equation is used in conjunction with the Maxwell equal area construction,[97] results are obtained which are equivalent to those derived from the Landau theory of the second-order phase transition. In particular this approach implies that near the critical point:

(a) At the critical temperature, the density minus the critical density is given by

$$\rho-\rho_c\sim(P-P_c)^{1/3}, \tag{5.2a}$$

where $P-P_c$ is the deviation of the pressure from its critical value.

(b) On the coexistence curve (see Fig. 9), the densities of gas and liquid are given by

$$\rho_{\text{liquid}}-\rho_c=-(\rho_{\text{gas}}-\rho_c)\sim(T_c-T)^{1/2}. \tag{5.2b}$$

(c) When $T>T_c$ and $\rho=\rho_c$, the isothermal compressibility K_T diverges as

$$K_T\equiv\rho^{-1}(\partial\rho/\partial P)_T\sim(T-T_c)^{-1}. \tag{5.2c}$$

(d) There is a similar divergence for $T<T_c$ where

$$K_T\sim(T_c-T)^{-1} \tag{5.2d}$$

on the coexistence curve.

All these results look like the typical molecular field approximation answers when one identifies[16] the order parameter with $(\rho-\rho_c)$ and the conjugate variable (the analog of the magnetic field) with a quantity proportional to the pressure minus the critical pressure.

A further argument for the identification is provided by the lattice gas model for classical fluids.[98,17] In this

TABLE X. Scaling law comparison.

	$2-\alpha$	$2-\alpha'$	$d\nu$	$d\nu'$	$d\gamma/2-\eta$	$\gamma+2\beta$	$\gamma'+2\beta$	$\beta(\delta+1)$	$y=1/\nu$	Δ^b
Ferromagnets	1.92 ±0.08	1.92 ±0.08	1.95 ±0.09	···	2.08 ±0.12	1.99 ±0.09	···	1.7 ±0.2	1.54 ±0.07	3.2 ±0.2
Antiferromagnets	1.92 ±0.08	1.92 ±0.08	1.95 ±0.09	···	···	1.96 ±0.10[a]	···	···	1.54 ±0.07	3.2 ±0.2

[a] Here we use the approximation $\gamma\approx2\nu$.

[b] Calculated from $\Delta=2d\nu-2\beta$.

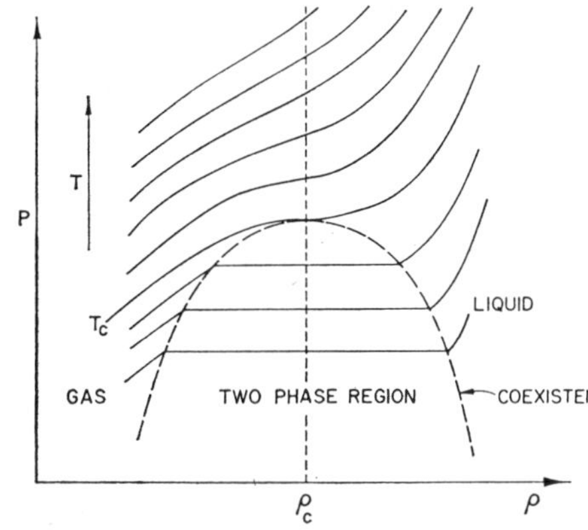

FIG. 9. Schematic pressure vs density diagram near the liquid–gas critical point.

model, the positions of the gas molecules are assumed to take on only discrete values, these values forming a specified lattice. To represent the fact that, in the real fluid, there is a short-range repulsion which keeps the molecules from getting too close to one another, it is assumed that each lattice site can be occupied by at most one molecule. To represent the fact that the real fluid exhibits an attraction between neighboring molecules, the lattice gas model includes an interaction potential which gives a negative contribution to the energy whenever nearest-neighbor sites are occupied.

The lattice gas model has been shown by Lee and Yang[98] to be mathematically identical to the Ising model of ferromagnetic phase transitions. This identity follows since for both models the essential features are a lattice, a nearest-neighbor interaction, and a number for each lattice site which can take on one of two values, i.e., for the Ising model, the z component of the spin

$$\sigma_{\mathbf{r}}=+1 \quad \text{if the spin is up at a site,}$$
$$=-1 \quad \text{if the spin is down at a site,} \qquad (5.3)$$

for the lattice gas the gas density at a lattice site,

$$\rho(\mathbf{r})=1 \quad \text{if the site is occupied,}$$
$$=0 \quad \text{if the site is unoccupied.} \qquad (5.4)$$

According to the above, to obtain a correspondence between the Ising model and the lattice gas model one should make the identification

$$\sigma_{\mathbf{r}}=2[\rho(\mathbf{r})-\tfrac{1}{2}]. \qquad (5.5)$$

For the lattice gas model the critical density turns out to be $\frac{1}{2}$, so the above statements say that the density minus the critical density plays the role of the magnetization in the lattice gas case. Since there is no difference between up and down magnetization for a ferromagnet we thus have the important quantitative conclusion that the lattice gas model will predict a perfect symmetry between behavior on the high- and low-density sides of the critical point. For example, if we define β_L and β_G by the requirement that

$$\rho_{\text{liquid}}-\rho_c\sim(-\epsilon)^{\beta_L}$$
$$\rho_{\text{gas}}-\rho_c\sim-(-\epsilon)^{\beta_G} \qquad (5.6)$$

on the coexistence curve, then we know that the lattice gas will give the result $\beta_L=\beta_G$.

A partial list of the corresponding quantities for the two models is given in Table XI. More detailed lists and a derivation of the correspondences are given in Refs. 98 and 17. Thus, a solution for the Ising model, discussed in Sec. IIE can be transcribed into a solution for the lattice gas model. If we then say that the lattice gas model is like the real fluid, we have determined the critical indices for the liquid–gas transition in terms of the indices α, β, γ, etc., for the Ising model. This proposed correspondence between thermodynamic derivatives for the fluid and critical indices for the Ising model is listed in Table XII.

In looking at the experimental data which follow, we focus on the following questions:

Are phase transitions in different fluids essentially similar?

Do they resemble the Ising model's critical behavior? Do they obey the scaling laws?

B. Experiments

In our discussion of experimental data, we concentrate upon the fluids Co_2, Xe, Ar, and O_2 because these are the ones which have been most extensively investigated in the critical region.[99–113] In these one-component systems, the critical point is defined as that temperature and pressure at which the densities of gas and liquid become identical. This point is the top of the coexistence curve of Fig. 9. Table XIII is a list of critical pressures, densities, and temperatures for these four fluids.

TABLE XI. Correspondences between lattice gas model and Ising model.

Ising model	Lattice gas model
Number of up spins	Number of molecules
Partition function	Grand partition function
Number of spins	Volume
Free energy minus magnetic field	Pressure
Magnetization	Density minus critical density
Specific heat C_m	Specific heat C_V

Some of the general difficulties inherent in these measurements may be mentioned. As with second order phase transitions in other materials in other effects, temperature control is critical. In some of the experiments involving the critical points in fluids, the data do not seem to settle down to their asymptotic critical behavior until $\epsilon=(T-T_c)/T_c$ gets smaller than 10^{-2}. A temperature control system must be able to maintain and reproduce temperatures to perhaps one part in 10^4 of T_c in order to provide meaningful data over a two-decade range in ϵ within this critical region. Since this control is most easily achieved near room temperature, the most complete data are available for the classical gases Xe ($T_c=289.6°K$) and CO_2 ($T_c=304.0°K$).

Given a good temperature control system, one must next contend with the extremely large compressibilities when critical conditions are approached. Due to the weight of the fluid, critical pressure is realized only over a very narrow vertical height range in a sample bomb (in theory, of course, only at a single horizontal plane), and what is measured in a *PVT* measurement is the average condition of the fluid. This may be quite different from the critical condition unless special precautions are taken, and can lead to a flat top in the coexistence curve[106] (liquid–gas density difference as a function of temperature). One of the most elegant methods of dealing with this was devised by Lorentzen,[105,106] who used a long vertical tube as his cell and measured the density of the fluid as a function of height near the critical region by the refraction of parallel light beams passing through the cell. All critical exponents except α and α' can be determined directly from an experiment of this type.

A further complication arises from the large heat capacity of a fluid near critical conditions.[101,107,108] Equilibration times become very long near the critical point as a result of this, necessitating waits of perhaps days before it is reasonably certain equilibrium conditions have been attained.

TABLE XII. Definitions of critical indices for liquid–gas transition. If this transition is described correctly by the lattice gas model, α, α', β, γ, γ', and δ all have the same values as in the Ising model.

ϵ	$\epsilon=(T-T_c)/T_c$	
α'	$C_v\sim(-\epsilon)^{-\alpha'}$	$\rho=\rho_c,\ \epsilon<0$
α	$C_v\sim\epsilon^{-\alpha}$	$\rho=\rho_c,\ \epsilon>0$
β	$(\rho_L-\rho_G)\sim(-\epsilon)^{\beta}$	$\epsilon<0$, coexistence curve.
γ'	$K_T\sim(-\epsilon)^{-\gamma'}$	$\epsilon<0$, coexistence curve.
γ	$K_T\sim\epsilon^{-\gamma}$	$\rho=\rho_c,\ \epsilon>0$
δ	$\|P-P_c\|\sim\|\rho-\rho_c\|^{\delta}$	$T=T_c$
μ	Surface tension $\sim(-\epsilon)^{\mu}$	$\rho=\rho_c,\ \epsilon<0$

TABLE XIII. Critical points of selected fluids.

Gas	P_c	ρ_c	T_c
Xe	57.636 ± 0.115 atm[a]	1.105 g[b]/cm^3	16.590 ± 0.001°C[b]
O_2	49.7 atm[c]	0.408 g[d]/cm^3	154.565°K[d]
Ar	48.34 atm[e]	0.5333 g[f]/cm^3	150.71°K[e]
CO_2	72.82 atm[g]	0.464 g[h]/cm^3	31.04°C[g]

[a] Reference 99.
[b] Reference 100.
[c] *Handbook of Chemistry and Physics* (Chemical Rubber Publ. Co., Cleveland, Ohio, 1961).
[d] Reference 101.
[e] Reference 103.
[f] Reference 102.
[g] Reference 104.
[h] Reference 105.

The most accurately determined parameter for the classical gases is β of the coexistence curve. This is determined from the slope of the log–log plot in Fig. 10. Plotted here are the results of four different experiments on CO_2 and Xe over almost a five decade range in ϵ. There is excellent agreement between the data and a line of slope 0.34 over the entire range and for both gases. The data taken by Lorentzen in 1965[106] was shifted vertically to coincide with the other sets; presumably there was a constant factor error in the density of this measurement, for the slope of the line is the same as the rest.

The results for β are summarized in Table XIV. All of this data is consistent with the statement that β is the same for all three fluids. We assume that this is true and take $\beta=0.346\pm0.01$ as the value to be used in further analysis. Notice that this is slightly higher than the lattice gas value, $\beta=0.31\pm0.01$.

The index δ can be calculated directly from the *PVT* data of Habgood and Schneider.[99] Our analysis is shown in Fig. 11 together with the results of Widom and Rice's[109] analysis of the data of Ref. 104. Since the Xe data cover a larger range, we use it to define the "best value" of δ listed in Table XV. There is a significant deviation from the lattice gas.

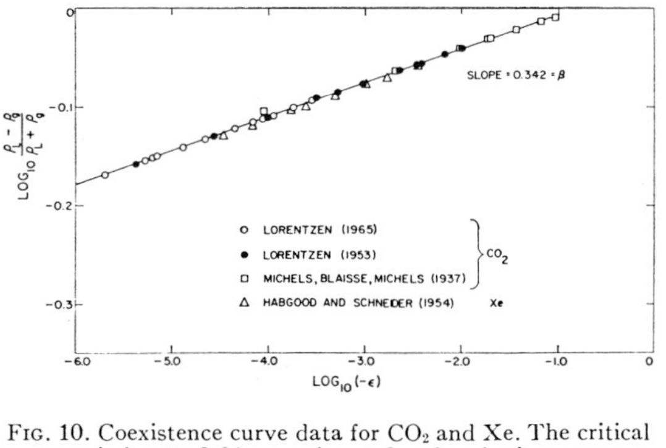

FIG. 10. Coexistence curve data for CO_2 and Xe. The critical index $\beta\approx0.34$ over almost five decades in ϵ.

TABLE XIV. Summary table for β.

Fluid	β	Reference	Range of ϵ
Xe	0.350 ± 0.015	99	$4\times10^{-5}<-\epsilon<4\times10^{-3}$
Ar	0.4 ± 0.2	103	$6\times10^{-5}<-\epsilon<8\times10^{-3}$
Ar	0.33 ± 0.05	103	$8\times10^{-3}<-\epsilon<10^{-1}$
CO_2	0.344 ± 0.01	105	$4\times10^{-6}<-\epsilon<10^{-1}$
"Best value"	0.346 ± 0.01		
Lattice gas	0.313 ± 0.004	Table III	

We employ three different modes of analysis to determine γ for Xe, CO_2, and Ar. For Xe, we employ the *PVT* data of Habgood and Schneider.[99] When these data are differentiated they give the K_T values plotted in Fig. 12. Then, γ is the slope of this curve. Unfortunately, this evaluation of γ is rather ambiguous because of the relatively small range of ϵ and the apparent change in slope at $\epsilon\approx10^{-3}$. However, the data points for the higher values of ϵ, $\epsilon\sim10^{-3}$, are believed to be more reliable than those for lower values, $\epsilon\sim10^{-4}$. If we weight the data in this way, we would conclude $\gamma=1.3\pm0.2$. Results for γ and γ' are listed in Tables XVI and XVII.

An application of the Ornstein–Zernike theory to the critical opalescence scattering of x rays in the critical region[16] has been used by Thomas and Schmidt[110] to calculate γ for argon. The theory is probably accurate only outside the critical region, but gives the result that reciprocal intensity plotted against momentum transfer at a constant temperature yields straight lines whose zero intercept with the ordinate is proportional to the reciprocal of the isothermal compressibility.[16] These intercepts as a function of temperature are plotted in Fig. 13.

For CO_2, γ may be obtained from the data of Michels, Blaisse, and Michels[104] and of Lorentzen.[105,106] Lorentzen measured the refraction R as a function of height x for light which traversed his tube. At the critical density, R as a function of x shows a point of inflection and a slope $\tan\alpha=dx/dR$ which is a function of temperature. Since R is proportional to the density,

$$(\tan\alpha)^{-1}\sim d\rho/dx=\partial\rho/\partial P\,|_T(dP/dx). \qquad (5.8)$$

At the critical density, $dP/dx=-\rho_c g$ is a constant and

$$\tan\alpha\sim K_T^{-1}\sim\epsilon^{\gamma}. \qquad (5.9)$$

Heller[111] has analyzed Lorentzen's 1953 data[105] in conjunction with the results of Michel *et al.*[104] He concludes in CO_2 that $\gamma=1.37\pm0.2$ for $10^{-5}<\epsilon<10^{-2}$ and $\gamma'=1.0\pm0.3$.

Lorentzen's more recent results[106] give a rather different answer. Figure 14 shows that $\tan\alpha$ is quite closely linear in ϵ. Thus, γ appears to be equal to 1.0 ± 0.1 in the range $10^{-6}<\epsilon<2\times10^{-5}$. These results tend to suggest that γ changes at $\epsilon\sim10^{-5}$ in CO_2. However, experimental difficulties are severe when ϵ is

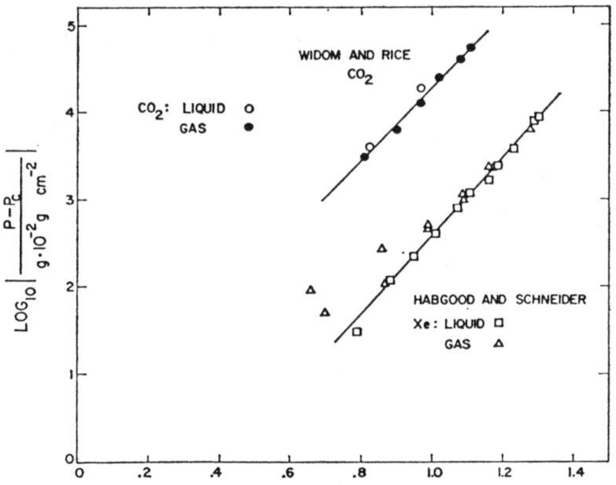

FIG. 11. Data from which the critical index δ is determined.

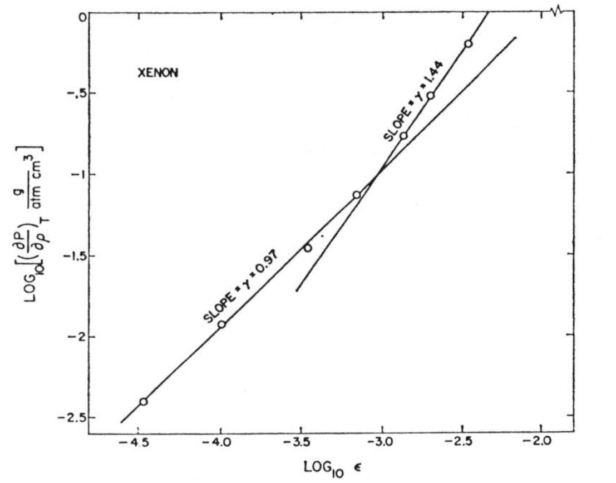

FIG. 12. PVT data for Xenon from which the critical index γ can be determined.

TABLE XV. Summary of results for δ.

Fluid	Value of δ	Reference	Range of variables
Xe	4.4 ± 0.4	99	$6\times10^{-2}<\|(\rho-\rho_c)/\rho_c\|<2\times10^{-1}$
CO_2	4.2	109	$1.4\times10^{-1}<\|(\rho-\rho_c)/\rho_c\|<1.8\times10^{-1}$
"Best value"	4.4 ± 0.4		
Lattice gas	5.2 ± 0.15	Table III	

as small as 10^{-5} and perhaps Lorentzen's later results should be discounted for this reason.

There is another reason for concern about using the data in Fig. 14 for finding γ. We want to know K_T for a homogeneous system. But, gravitational effects necessarily produce a pressure gradient of magnitude

$$dP/dx=\rho_c g. \qquad (5.10)$$

Can this gradient affect the measured value of K_T appreciably? Let us assume that the long-range correlations in the system basically average the pressure over one coherence length. Then we might expect that the result of the gradient (5.10) could be roughly equivalent to producing a deviation $P-P_c$ of the order

$$|P-P_c|\sim\rho_c g\xi. \qquad (5.11)$$

If this is large enough to push the system away from the critical point, we can evaluate K_T^{-1} by setting $\epsilon=0$. At $\epsilon=0$, the Widom and Rice[109] analysis of CO_2 data indicates

$$(\rho-\rho_c)/\rho_c\sim[(P-P_c)/P_c]^{1/\delta} \qquad \delta\approx4.2.$$

Then

$$\frac{\rho_c}{P_c}\frac{dP}{d\rho}\bigg|_T\sim\delta\left(\frac{P-P_c}{P_c}\right)^{1/\delta}\sim\left(\frac{\rho_c g\xi}{P_c}\right)^{1-1/\delta}. \qquad (5.12)$$

In the last step of (5.12) we have inserted the effective pressure deviation of Eq. (5.11). If this pressure deviation is small, we should expect that the contribution (5.12) to the inverse susceptibility is much smaller than the $P=P_c$ value which is roughly evaluable[111,105,104]

$$(\rho_c/P_c)(\partial P/\partial\rho)|_T\approx10^2\epsilon^\gamma \qquad \gamma\approx1.4. \qquad (5.13)$$

If we are to neglect the effects of gravity gradients, we should require that the contribution (5.12) to $dP/d\rho|_T$ be much less than the contribution (5.13). We evaluate this requirement by guessing ξ from the Ising model results as

$$\xi=(M/\rho_c)^{1/3}\epsilon^{-\nu} \qquad \nu\approx0.6,$$

where $(M/\rho_c)^{1/3}$ is the average distance between CO_2 molecules. Then our requirement for neglecting gravitational effects in CO_2 becomes

$$\epsilon\gg10^{-5}. \qquad (5.14)$$

The same procedure applied to Xe indicates that the gravitational pressure gradients are negligible for

$$\epsilon\gg3\times10^{-5}. \qquad (5.15)$$

Since Eq. (5.14) indicates that pressure gradients might be important in determining γ in Fig. 14, we omit the data of Ref. 106 from further analysis.

The variation of heat capacity with ϵ has been measured most accurately [102,110] for Ar and O_2. The data for Ar for $T<T_c$ are given in Fig. 15. In this log–log plot, we have subtracted constants from C_v/R so that we may fit C_v to a form

$$C_v/R=a\epsilon^{-\alpha}+b \qquad \text{for } T>T_c$$

$$=a'(-\epsilon)^{-\alpha'}+b' \qquad \text{for } T<T_c.$$

The values of b' are given on each curve together with the value of α' which best fits for that value of b'. A

TABLE XVI. Value of γ.

Fluid	Value of γ	Reference	Range of ϵ	Comments
Xe	1.3 ± 0.2	99	$3\times10^{-4}<\epsilon<3\times10^{-3}$	$\epsilon\lesssim10^{-4}$ neglected
Ar	0.6 ± 0.2	110	$3\times10^{-3}<\epsilon<6\times10^{-2}$	Is extrapolation procedure right?
CO_2	1.37 ± 0.2	105, 104, 111	$10^{-5}<\epsilon<10^{-2}$	
CO_2	1.0 ± 0.1	106	$10^{-6}<\epsilon<2\times10^{-5}$	Perhaps pressure gradients are important
"Best value"	1.37 ± 0.2			
Lattice gas	1.250 ± 0.001	Table III		

TABLE XVII. Values of γ'.

Fluid	Value of γ'	Reference	Range of ϵ	Comments
Ar	1.1 ± 0.2	110	$6\times10^{-4}<-\epsilon<6\times10^{-2}$	Is extrapolation procedure right?
CO_2	1.0 ± 0.3	105, 104, 111	$\epsilon\approx3\times10^{-2}$ $\epsilon\approx3\times10^{-5}$ $\epsilon\approx3\times10^{-6}$	If pressure gradients are important, this analysis gives no information.
"Best value"	1.0 ± 0.3			Very little information
Lattice gas	1.31 ± 0.05	Table III		

good fit cannot be obtained for α' as large as 0.37, but any value of α' in the region $0\leq\alpha'\leq.25$ can, with a suitable value of b', fit the data. Of course, $\alpha'=0$ corresponds to a logarithmic singularity in C_v.

For $T>T_c$, a similar plot indicates that any value of α in the range $0<\alpha<0.4$ fits the data for O_2. These results are summarized in Tables XVIII and XIX.

We consider one more critical index, μ, which is defined by the statement that in the two-phase region the surface tension is proportional to $(-\epsilon)^{\mu}$. Guggenheim[114] quotes results which tend to indicate $\mu\approx1.22$ but a recent analysis by Buff, Lovett, and Stillinger[115] of the data of Atack and Rice[116] on cyclohexone-aniline and of Stansfield[117] on argon and nitrogen gives $\mu\approx 1.27\pm0.02$. We use this value in our scaling law analysis.

C. Use of Inequalities

In this analysis, γ' is the quantity with the largest uncertainty. Fortunately there is an exact[118a,16] thermodynamic inequality which we can use to tie down γ'.

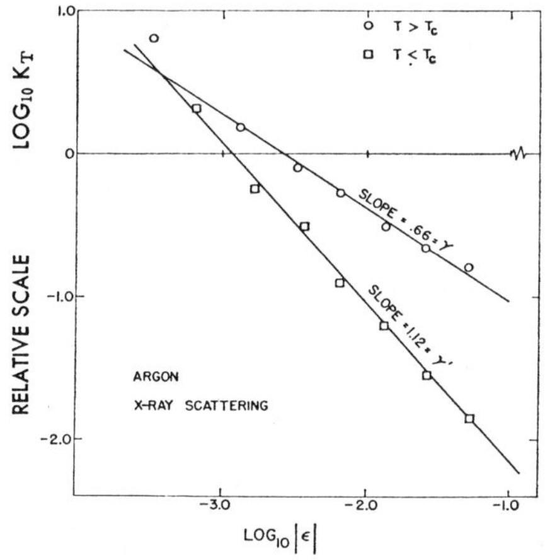

FIG. 13. Determination of γ and γ' from x-ray scattering data for Argon. However, the Ornstein–Zernike theory used in the data analysis may not be valid in the critical region.

This inequality is

$$2\beta+\gamma'\geq2-\alpha' \qquad (\text{for } \alpha'>0). \qquad (5.16)$$

When we substitute the known values of β and α', we see that

$$\gamma'\geq1.19\pm0.12.$$

Another exact inequality[118b]

$$\beta(\delta+1)\geq2-\alpha' \qquad (\text{for } \alpha'>0) \qquad (5.17)$$

implies

$$\delta\geq4.45\pm0.4.$$

This does not provide any useful limitations on δ but at least it indicates that the experimental results for δ are not wildly wrong.

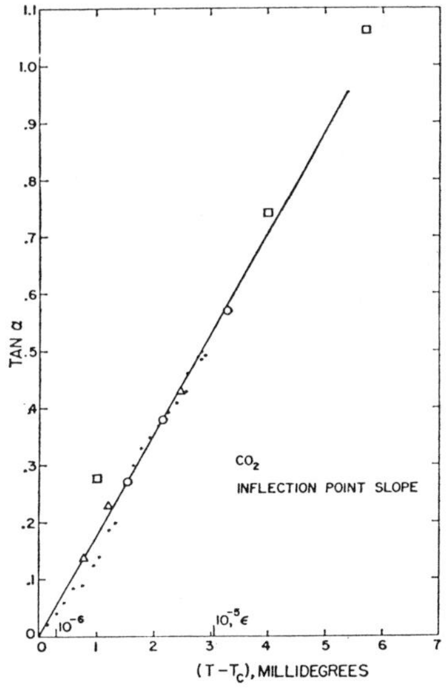

FIG. 14. Lorentzen's data near T_c for angle of refraction α of light passing through CO_2. Tan $\alpha\sim\epsilon^{\gamma}$.

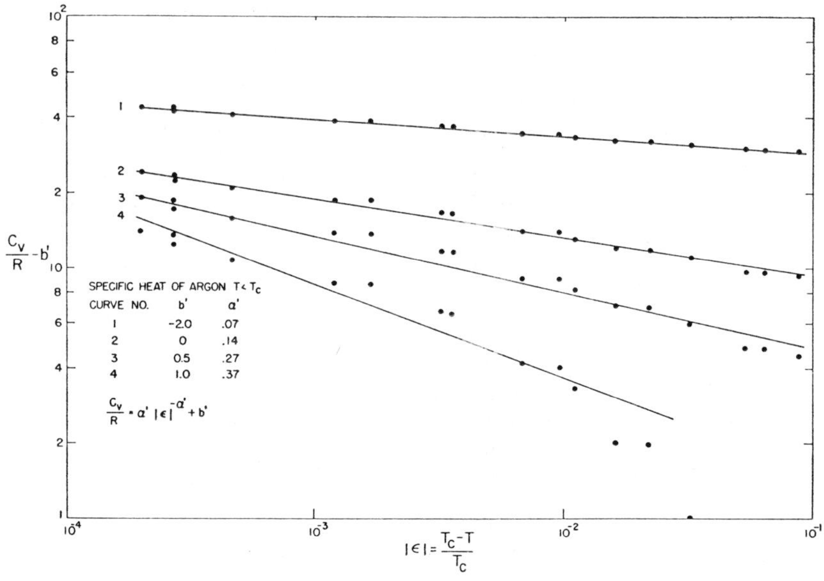

FIG. 15. Three-parameter fits to the specific heat of argon.

D. Comparison with Scaling Laws

According to the discussion in Sec. III, the scaling law approach would predict that the following combinations of indices were all precisely equal: $2-\alpha$, $2-\alpha'$, $\gamma+2\beta$, $\gamma'+2\beta$, $\beta(\delta+1)$. Widom, in his development[33] of the ideas we have called scaling laws, also proposed that the surface tension index, μ, should be equal to $d/(d-1)=\frac{2}{3}$ times these other combinations of indices. In Table XX are listed the quantities which the scaling laws would equate, and it is seen that they are roughly equal. In fact, the results presented here agree within experimental error with the conclusions of the scaling law theory.

From the value of μ in Table XX, we derive the basic scaling law quantity $y=\nu^{-1}=(d-1)/\mu$ as

$$y=1.57\pm0.03. \tag{5.18}$$

The other basic parameter of the scaling law analysis is $\Delta=2d/y-2\beta$, which turns out to be

$$\Delta=3.12\pm0.1. \tag{5.19}$$

These values of y and Δ are in agreement with the Ising model results (Table IV). However, the difference between the experimental value of $\beta=0.346\pm0.01$ and the Ising model value of $\beta=0.31\pm0.01$ indicates that real fluids and the Ising model show slightly different critical behavior.

VI. THE LIQUID GAS PHASE TRANSITION IN QUANTUM LIQUIDS

A. Theory

Some experimental evidence (see Sec. VIB) tends to indicate that the critical exponents characterizing the liquid–gas phase transition are markedly different for the light substances He, H_2, and D_2, when compared with the results for the heavier elements Xe, Kr, Ar, N_2, O_2, and Ne. If we assume for the moment, that the experiments are correct in indicating this difference, then we must look for an additional parameter which

TABLE XVIII. Values of α'. The experimental numbers are the maximum value of α' which appear consonant with our three parameter fit of the experimental data. Any smaller value of α', including $\alpha'=0$ (logarithmic singularity) will also fit the data.

Fluid	Value of α'	Reference	Range of ϵ
Xe	$<0.2\pm0.1$	112	$5\times10^{-3}<-\epsilon<3\times10^{-2}$
O_2	<0.25	101	$4\times10^{-4}<-\epsilon<1.4\times10^{-2}$
Ar	<0.25	102	$2\times10^{-4}<-\epsilon<10^{-1}$
CO_2	$<0.1\pm0.5$	113	$3\times10^{-3}<-\epsilon<2\times10^{-2}$
"Best value"	0.12 ± 0.12		
Lattice gas	$0.07^{+0.16}_{-0.04}$	Table III	

TABLE XIX. Values of α. All the data are consistent with α being as small as zero.

Fluid	Value of α	Reference	Range of ϵ
O_2	$\alpha<0.4$	101	$3\times10^{-5}<\epsilon<4\times10^{-2}$
Ar	$\alpha<0.4$	102	$10^{-4}<\epsilon<2\times10^{-2}$
"Best value"	0.2 ± 0.2		
Lattice gas	0.1 ± 0.1	Table III	

TABLE XX. Comparison with scaling laws. According to the analysis of Sec. III all of these numbers should be equal. The numbers used here are the "best values" of Tables XIV–XIX.

$2-\alpha$	$2-\alpha'$	$\gamma+2\beta$	$\gamma'+2\beta$	$\beta(\delta+1)$	$\frac{3}{2}\mu$
1.8	1.88	2.06	1.7	1.87	1.91
±0.2	±0.12	±0.2	±0.3	±0.14	±0.03

measures quantum behavior. It is quite reasonable to suggest that the different behavior of light gases is a manifestation of the quantum-mechanical dispersion[119] in the position of the molecules, which occurs when the kinetic energy of a molecule is comparable in magtitude to the effective potential energy of that molecule. To give this statement some quantitative substance we discuss the law of corresponding states as formulated by de Boer.[120] First we consider the classical problem.

Assume that the interaction between molecules is given by the potential energy $\rho(r)$. We can apply a law of corresponding states if, for different fluids

$$\rho(r)=\Delta f(r/\sigma), \tag{6.1}$$

where $f(x)$ has the same form for all fluids. Then Δ and σ represent the characteristic energy and length for the fluid in question. We may compare different fluids by using the reduced variables:

$$T^*=kT/\Delta, \qquad V^*=V/N\sigma^3, \qquad P^*=P\sigma^3/\Delta. \tag{6.2}$$

Since the potential energy function, (6.1), determines all of the statistical properties of a classical fluid, P^* should be the very same function of V^* and T^* for different fluids as long as quantum corrections are unimportant. Table XXI indicates that the critical reduced pressure, temperature, and specific volume are found to be nearly the same for the heavier fluids.

However, the kinetic energy is important in quantum statistical mechanics. If a particle is localized in a volume of order σ^3, it has a kinetic energy of the order

$$\hbar^2/m\sigma^2 \tag{6.3}$$

according to the uncertainly principle. The size of the quantum correction can be estimated by comparing (6.3) with the typical potential energy Δ. The ratio of these two energies

$$(\Lambda^*)^2\equiv\hbar^2/m\sigma^2\Delta \tag{6.4}$$

is a dimensionless measure of the importance of quantum effects in the liquid–gas transition. Table XXI shows that as Λ^* becomes larger than unity (in H_2, He^4, and He^3), the critical pressure, temperature, and volume deviate from their classical values.

The classical phase transition has been compared with the Ising model via the lattice gas (see Sec. V). Because, in the quantum case the Hamiltonian contains noncommuting terms, it is attractive to compare the quantum liquid–gas transition with a magnetic situation in which there are spins with noncommuting components. This has been done by Matsubara and Matsuda,[121] and Zilsel.[122] These models suggest that the quantum transition is like a ferromagnetic transition in an anisotropic situation, in which the zero field magnetization can only point in the z direction. Once more, one can identify $(\rho-\rho_c)$ with the z component of the magnetization. Since not very much is known about highly anisotropic ferromagnets near the critical point, it is difficult to judge the correctness of these models. However one point stands out clearly. Since the magnet shows complete symmetry between spin up and spin down, these quantum lattice gas models would predict a complete symmetry between liquid and vapor.

B. Experiment

The specific heats of both helium three and helium four have been measured by Moldover and Little.[124] We have fit their data to a formula:

$$C=a\epsilon^{-\alpha}+b \qquad \text{for } T>T_c$$
$$=a'(-\epsilon)^{-\alpha'}+b' \qquad \text{for } T<T_c \tag{6.5}$$

with a, a', b, and b' being adjustable parameters. The data for He^4 and some fits to it are shown in Fig. 16. We would conclude that α' lies between 0.0 (logarithmic singularity) and 0.2 and α between 0.0 and 0.3. However, this conclusion does not include a possible perturbation produced by gravitational effects. Accord-

TABLE XXI. Experimental values of P_{cr}^*, V_{cr}^*, T_{cr}^*, and Λ^* for substances having the potential energy function (6.1).

	Xe	Kr	Ar	N_2	Ne	H_2	He^4	He^3
P_{cr}^*	0.112	0.117	0.116	0.132	0.114	0.063	0.027	0.014[a]
V_{cr}^*	3.10	3.10	3.12	2.88	3.25	4.29	5.74	7.2[a]
T_{cr}^*	1.26	1.26	1.25	1.30	1.26	0.90	0.51	0.33[a]
Λ^*	0.064	0.102	0.187	0.225	0.591	1.73	2.64	3.05[a]

[a] Computed using present known values of P_{cr}, T_{cr}, and ρ_{cr}[132] and the values of Δ and σ appropriate for He^3 (the same as for He^4).[123] All other numbers are from Ref. 120.

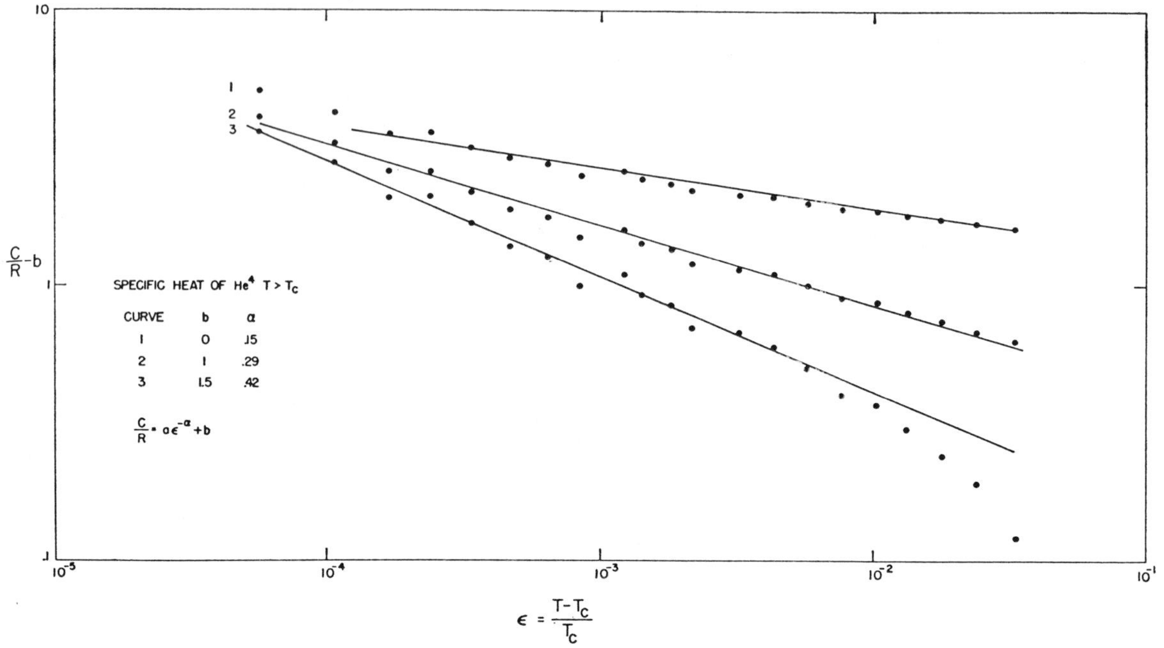

FIG. 16. Three-parameter fits to the specific heat of He^4 near the liquid–gas transition.

ing to a calculation based on the Landau theory,[125] using the data of Edwards and Woodbury,[126,127] a 1% density variation might exist in this experiment for $\epsilon < 5\times10^{-4}$. This could cause a rounding of the specific heat peak.

The order parameter for liquid–gas transitions is $\rho-\rho_c$, and therefore the analog of the magnetization in zero field is just the coexistence curve. For this the relevant critical exponent is β. Edwards and Woodbury[126] measured this for He^4 by using a Jamin interferometer. They measured the index of refraction of a "slice" of helium 1 mm thick, and from this determined the density using the Lorentz–Lorenz equation. The correction to this equation near the critical point has been estimated to be negligible using a theory of Larson, Mountain, and Zwanzig.[129] The results of this experiment have been fitted to many analytical expressions,[126–128] but for comparison with other data, we have considered only fits of the form

$$(\rho_L-\rho_G)/\rho_c \sim (-\epsilon)^{\beta}. \tag{6.6}$$

Unfortunately, this experiment does not permit an unambiguous conclusion about β. By varying T_c and the assumed size of the critical region within reasonable limits it appears possible to obtain fits for values of β which are in the range 0.40 to 0.50.

The results of recent measurements by Roach and Douglass[130] are that $\beta=0.35\pm0.01$ for $4\times10^{-4}<(-\epsilon)<2\times10^{-2}$. They measured the dielectric constant of helium between two capacitor plates, and obtained the density using the Clausius–Mosotti equation. Edwards[131] has also repeated his measurements, and his recent result is that $\beta=0.37\pm0.02$ for $2\times10^{-4}<(-\epsilon)<10^{-2}$. Thus it appears that β is about the same in He^4 as in the classical gases (see Table XXII).

The coexistence curve for He^3 has been determined by Sherman[132] by measuring the pressure of the vapor as a function of temperature using a constant volume bomb filled to 23 different densities. The pressure–temperature relations obtained in this way are nearly linear, and are extrapolated to the known vapor-pressure curve to give temperature–density data. This method has the advantage of giving the shape of the entire *PVT* surface near the critical point, and thus giving more information than just β. Furthermore, the effect of gravity and the infinite compressibility is avoided, since no part of the fluid is at the critical point. However, the extrapolation to the vapor curve is questionable, since it is not known definitely that the isochores (constant volume) continue to be linear in the immediate neighborhood of the critical point.[133]

We fit Sherman's data for $2.5\times10^{-2}<(-\epsilon)<2.5\times10^{-1}$ with $\beta=0.36\pm0.02$. Since his published data include only two data points for $(-\epsilon)<2.5\times10^{-2}$, we do not feel justified in estimating β in this region. However, the data suggest a possible increase in β towards 0.5.

TABLE XXII. Comparison of quantum and classical fluids.

	He^3	He^4	H_2	Classical gases[h]
α	$\lesssim 0.3$[a] $4\times10^{-4}<\epsilon<2\times10^{-2}$	$\lesssim 0.3$[a] $10^{-4}<\epsilon<3\times10^{-2}$		$\lesssim 0.4$
α'	$\lesssim 0.2$[a] $2\times10^{-4}<-\epsilon<3\times10^{-2}$	$\lesssim 0.2$[a] $2\times10^{-4}<-\epsilon<3\times10^{-2}$		$\lesssim 0.25$
β	0.36 ± 0.02[b] $2.5\times10^{-2}<(-\epsilon)<2.5\times10^{-1}$	0.352 ± 0.004[c] $4\times10^{-4}<(-\epsilon)<2\times10^{-2}$	0.36 ± 0.01[d] $10^{-2}<(-\epsilon)<?$	0.35 ± 0.01
γ	1.09 ± 0.05[b] $?<\epsilon<?$			1.37 ± 0.2
γ' (gas)	1.00 ± 0.05[b] $?<-\epsilon<?$			1.0 ± 0.3
γ' (liquid)	1.18 ± 0.10 $?<-\epsilon<?$			1.0 ± 0.3
δ	3.5 ± 0.1[e] $0.17<\|(\rho-\rho_c)/\rho_c\|<1.0$		4.2[f] $0.67<\|(\rho-\rho_c)/\rho_c\|<1.57$	4.4 ± 0.4
μ			1.25 ± 0.02[g] $4\times10^{-2}<(-\epsilon)<4\times10^{-1}$	1.27 ± 0.02

[a] Reference 124.
[b] Reference 132.
[c] Reference 130.
[d] Reference 135.
[e] Reference 136.
[f] References 109, 137.
[g] Reference 139.
[h] See Tables XI–XXII.

Fisher[134] proposed an explanation for the apparent ϵ dependence of β which was noted by Sherman. He suggested that the variation in β might reflect the behavior of a system in which quantum corrections are not very large. Then relatively far from T_c, for large ϵ, we should expect classical behavior—$\beta\approx\frac{1}{3}$. Nearer T_c, the more delicate quantum effects take over and β changes its value to one which is characteristic of a quantum system.

However this apparent ϵ dependence of β seems very questionable in view of the recent results for He^4, for as Roach and Douglass[130] point out, β seems to be independent of Λ^*, the quantum correction parameter, over the range $0.06<\Lambda^*<2.6$ (Xe to He^4), so that β would have to change very rapidly over the range $2.6<\Lambda^*<3.1$ (He^4 to He^3) to explain the apparent[132] $\beta\approx0.48$ in He^3.

The coexistence curve for hydrogen has been determined by the isochoric method,[135] and the data can be fit with $\beta=0.36\pm0.01$ for $\epsilon>10^{-2}$. There is an indication that β becomes larger (≈0.50) for $\epsilon<10^{-2}$, but this might be due to the effect of gravity.

Next consider the isothermal compressibility:

$$\chi=\partial\langle p\rangle/\partial h\mid_\epsilon=\partial(\rho-\rho_c)/\partial\mu\mid_\epsilon=\rho^2K_T\sim\epsilon^{-\gamma}. \qquad (6.7)$$

The only available compressibility data are those of Sherman,[132] who found $\gamma=1.09\pm0.05$. Below the phase separation γ' is evaluated along the two branches of the coexistence curve. This yields two different γ' values:

$$\gamma_G=1.00\pm0.05 \quad \text{(gas)}$$

$$\gamma_L=1.18\pm0.10 \quad \text{(liquid)}.$$

These values of γ are obtained from derivatives of PVT data, so they necessarily have large uncertainty. Also uncertain are the size of the critical region and the extrapolation procedure to the coexistence curve. In view of these difficulties one must conclude that a definitive measurement of γ probably requires data closer to the coexistence curve and for smaller values of ϵ.

The shape of the critical isotherm is characterized by the exponent δ.

$$\mid(P-P_c)/P_c\mid=A\mid(\rho-\rho_c)/\rho_c\mid^\delta \quad \text{at } \epsilon=0. \qquad (6.8)$$

By interpolating the data to the critical isotherm to obtain the density as a function of pressure, Sherman[132] finds $\delta=3.4\pm0.2$. Chase and Zimmerman[136] measured δ more directly by measuring the dielectric constant of He^3 as a function of pressure at $T=3.324$°K. Then using the Clausius–Mossotti relation to find the density, they determined $\delta=3.5\pm0.1$. Their data indicate that the coefficient A in Eq. (6.8), is twice as great for the high-density fluid as that for the low-density fluid. The exact ratio depends on the value of the critical density used in Eq. (6.8) but for reasonable values of ρ_c, the ratio is near 2. This difference between high and low density is also seen in Sherman's determination of γ^L and γ^G. This behavior contradicts one of the assumptions in the lattice gas model: that there is symmetry between vapor and liquid. If we were seeing here the true critical behavior this would be strong evidence against the applicability of the lattice gas model to these phase transitions. However, the sparse data for γ and the fact that measurements for

δ were carried out for relatively large values of $|(\rho-\rho_c)/\rho_c|$ prevent the formation of firm conclusions about these critical exponents.

The shape of the critical isotherm for H_2 has been calculated from the data of Johnson, Keller, and Friedman[137] by Widom and Rice,[109] who found $\delta=4.2$. However, these data required a fairly long extrapolation to obtain δ,[138] which may therefore be unreliable.

Finally, the critical index μ, which relates the surface tension of a liquid drop to $T-T_c$ as surface tension $\sim(-\epsilon)^\mu$, has been measured for H_2 by Blagoi and Pashkov.[139] Their result is $\mu=1.25$ in the range $4\times10^{-2}<-\epsilon<4\times10^{-1}$.

C. Summary and Comparison with Classical Fluids

The results are summarized in Table XXII. Because of the inherent difficulties in these experiments and because of the uncertainty in the size of the critical region all these results, except β for He^4 and the α's, should perhaps be considered to be provisional.

For comparison we list in Table XXII the "best values" for classical fluids as determined in Sec. V. There is no strong indication that the quantum fluids behave differently than classical fluids near the critical point. There are not yet sufficient data to make a scaling law comparison meaningful.

VII. SUPERFLUIDS

A. Order Parameter for this Transition

In the superfluid transition, a finite fraction of the entire number of helium atoms all fall into the very same quantum state.[140,141] The order parameter for this transition is then the wave function of this special state, which we write as $\langle\psi(r)\rangle$. Since this wave function is complex, we have here a two-component critical order parameter. Thus, the behavior of the superfluid might well be different from the liquid–gas transition, in which the order parameter has but a single component. In fact, the superfluid should be most like a hypothetical ferromagnet in which the magnetization can point with equal facility in an entire plane.[142] Unfortunately, we know of no ferromagnet with such an easy plane of magnetization.

Because the order parameter in the superfluid is so special, there is an especial difficulty in learning about this transition: we do not have experimental control over the thermodynamic conjugate to the order parameter. In fact, this conjugate is always zero. Hence, we cannot measure the derivative of the order parameter with respect to its conjugate—the analog to the magnetic susceptibility—in the superfluid case. Similarly, we have no experimental way of finding the (order-parameter)-(order-parameter) correlation function

$$g(r,r')=\langle\psi^+(r')\psi(r)\rangle \tag{7.1}$$

which is the one-particle density matrix.

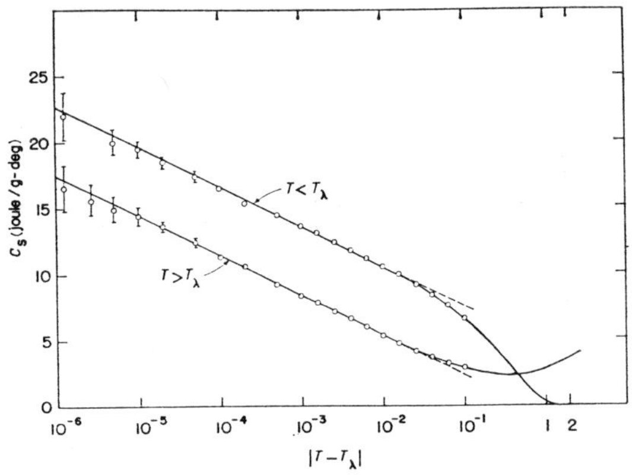

FIG. 17. Experimental specific heat of liquid He^4 near the superfluid (λ) transition.

B. Experimental Results

Since we cannot vary the variable we have been calling h, the conjugate to the order parameter, any specific heat measured in helium is C at constant h. Far and away the best measurement of C_h has been done by Buckingham, Fairbank, and Kellers,[143] who find C_h along the vapor pressure line for helium. Their results, shown in Fig. 17, indicate a logarithmic divergence in C_h over four decades of ϵ. It seems quite plausible that the divergence is truly logarithmic, i.e., $\alpha=\alpha'=0$, so that

$$C_h=-A\log|\epsilon|+B \qquad \text{for } \epsilon>0$$
$$=-A'\log|\epsilon|+B' \qquad \text{for } \epsilon<0.$$

Also, within experimental error $A=A'$. This is another piece of evidence in support of the scaling law idea, since the scaling laws predict that when there is a logarithmic singularity A must equal A'.

Another piece of useful information is provided by a measurement of the superfluid density ρ_s just below T_c. The definition of ρ_s is the statement that, when the superfluid is moving with velocity v_s, the flow of mass is given by a current

$$\mathbf{g}=\rho_s\mathbf{v}_s. \tag{7.2}$$

The most accurate critical data for ρ_s are provided by the experiment of Clow and Reppy[144] (Fig. 18) and that of Tyson and Douglass[145] which give

$$\rho_s\sim(-\epsilon)^\zeta \tag{7.3}$$

$$\zeta=0.666\pm0.006 \qquad \text{for } 3\times10^{-5}<-\epsilon\lesssim10^{-1}. \tag{7.4}$$

C. Scaling Law for ρ_s

In the relevant molecular field theory, the Landau–Ginzburg approach, ρ_s is simply proportional to the

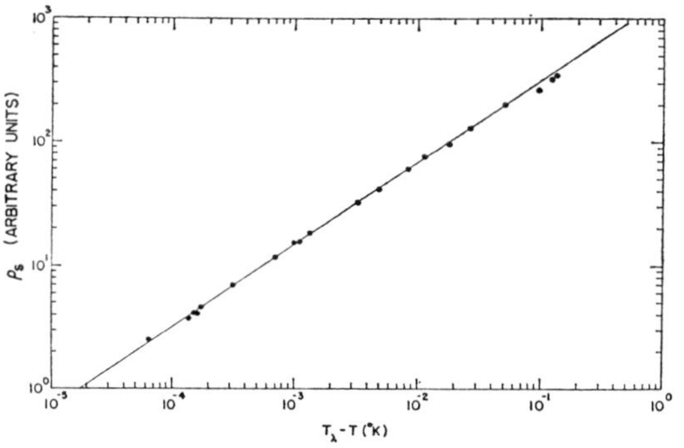

FIG. 18. Superfluid density of liquid He^4 just below T_λ.

order parameter squared

$$\rho_s \sim |\langle\psi\rangle|^2. \tag{7.5}$$

Then, if $\langle\psi\rangle$ goes to zero as $(-\epsilon)^\beta$, Eq. (7.4) would imply $\beta=\frac{1}{3}\pm0.003$, in agreement with the value for MnF_2.

However, Josephson[146] has argued that there is no particular reason for believing that the molecular field result, (7.5), should be correct near the λ point. [In particular he suggests that an extra factor proportional to $(-\epsilon)^{-\eta\gamma'}$ should appear on the right-hand side of Eq. (7.5). Then he uses the scaling laws to replace $2\beta-\eta\gamma'$ by $(d-2)(2-\alpha)/d$.] He concludes, in fact, that $\zeta=\frac{2}{3}$ is a direct consequence of the scaling law arguments and the fact that the specific heat singularity gives $\alpha=0$.

His argument is based upon the fact that superfluid flow arises from a gradient of the phase of $\langle\psi(r)\rangle$. In fact, v_s is derived from this phase via[141]

$$m\mathbf{v}_s(r)=i^{-1}\nabla\langle\psi(r)\rangle/\langle\psi(r)\rangle \tag{7.6}$$

for situations in which the magnitude of $\langle\psi(r)\rangle$ does not vary in space. However, if the superfluid is moving slowly, the free energy per unit volume has the form

$$g(\epsilon, v_s)=g(\epsilon,0)+\tfrac{1}{2}\rho_s v_s^2, \tag{7.7}$$

where the second term is just the kinetic energy of the superfluid in motion.

Equations (7.6) and (7.7) are definitions of ρ_s and v_s. We now use these definitions in conjunction with scaling law arguments to find ζ. To do this, the kinetic energy term is rewritten with the aid of (7.3) and (7.6) as

$$\tfrac{1}{2}\rho_s v_s^2 \sim (-\epsilon)^\zeta [\,|\nabla\langle\psi(r)\rangle|^2/|\langle\psi(r)\rangle|^2]. \tag{7.8}$$

However, under the scaling law transformations

$$\mathbf{r}\to\mathbf{r}/L$$

$$\epsilon\to\tilde{\epsilon}=\epsilon L^y \tag{7.9}$$

any singular term in the free energy density is scaled by a factor of L^d [cf. Eq. (3.3)]. In our case, the $(-\epsilon)^\zeta$ in (7.8) introduces a factor of $L^{\zeta y}$ and the gradients introduce two factors of L when the expression 7.8 is scaled. Thus we must have

$$L^d=L^{2+\zeta y}$$

or

$$\zeta=(d-2)/y. \tag{7.10}$$

But, according to Eq. (3.9c)

$$d/y=2-\alpha \tag{7.11}$$

so that (7.10) reduces to

$$\zeta=[(d-2)/d](2-\alpha). \tag{7.12}$$

Since we are working in a three-dimensional system, $d=3$ and from the specific heat experiments $\alpha=0$. Thus, the scaling laws give $\zeta=\frac{2}{3}$, in agreement with experiment.

For a two-dimensional system (7.12) reduces to the strange result $\zeta=0$. This would be rather difficult to justify were it not for the fact that there exist theoretical arguments[147] which tend to show that the superfluid transition does not take place or does not have a finite order parameter in a two-dimensional system.

D. Perturbation Theoretic Approach

Patashinskii and Pokrovskii[148] have discussed the nature of the critical singularity near the λ point of helium by attempting to sum the perturbation theory expansion for the one-particle Green's density matrix (7.1). They conclude that the specific heat and $(\partial n/\partial\mu)_T$ should diverge as $\log|\epsilon|$. This conclusion apparently agrees with experiment.[143] However, there have been several objections to this theory. At many points in their work, these authors have to guess the form of functions. They indicate that their guesses might be mutually consistent, but it is not clear that the guesses are correct. Furthermore, Abe[149] has extended their methods to the Ising model. He obtains results which are in clear disagreement with the known Ising model answers. For example, Abe finds $\eta=1$ in two dimensions and $\eta=\frac{1}{2}$ in three dimensions. The correct answers for these cases are, respectively, $\eta=\frac{1}{4}$ and $\eta=0.056\pm0.008$. (See Table III). This disagreement is one more argument against the theory of Patashinskii and Pokrovskii.

VIII. FERROELECTRIC TRANSITIONS

The ferroelectric transition is accompanied by a drastic change of the lattice structure and thus we always have a strong coupling between elastic, piezoelectric and ferroelectric phenomena of the crystal.[150] Stresses due to lattice imperfections, external forces and domain-clamping, influence the Landau theory parameters, thus producing an uncontrollable shift and smearing-out of the thermodynamic functions.

For these reasons one generally uses data with $\epsilon > 10^{-3}$ to determine the exponents α, β, and γ, obtaining agreement with the predictions of the phenomenological theory. Measurements of η and ν are not known.

Measurements of γ have been performed[151] for tri-glycine sulfate, KH_2PO_4 and the upper Curie point of Rochelle salt down to values for ϵ of 2×10^{-4}, 8×10^{-4}, and 4×10^{-4}, respectively. In all cases $\gamma=1$ within the experimental uncertainties of 2%. Hence the critical fluctuations have no influence on γ even for such small values of ϵ.

For temperatures below the lower Curie point in Rochelle salt, Craig[151] has found for the interval $4\times10^{-4} < |\epsilon| < 5\times10^{-2}$, $\gamma'=1.23\pm0.02$. The phase of the crystal is antiferroelectric there, and thus we must not expect to have the same critical region there as for temperatures near the upper transition point.

None of the other thermodynamic functions has been determined yet with the same accuracy as the values for χ. Thus we are not able to check the predictions of the scaling laws.

There are λ-shaped peaks in the specific heat of e.g., KH_2PO_4,[152] NH_4HSO_4[153] and Rochelle salt.[154] But the reported results do not allow a qualitative conclusion about the details of the singularity.

After having replotted Stephenson and Hooley's data, Grindlay[155] has pointed out that there are indications in favor of the existence of a logarithmic singularity in the specific heat of KH_2PO_4. For $0.6 < |T-T_c| < 5.3 (T_c=127.97°K)$ he obtained $A=1.2\pm0.3$ and $A'=14.2\pm1.4$. This statement cannot be accepted as a disproof of (3.11) for two reasons. Firstly, the discussed values for ϵ are possibly too large to get correct asymptotic data. One also would like to have more than 5 experimental points to determine the A's. Secondly, if one looks very close to the "critical point," one notices that the transition is actually first order.[151] So we have to expect a smeared out peak in the specific heat due to the latent heat of the transition. This peak has nothing to do with critical behavior. Furthermore the critical temperature T_c, seen in the normal phase, is lower than the critical temperature T_c', seen in the ordered state. T_c is the lowest temperature for an undercooling of the paraelectric state, while T_c' is the highest temperature for a superheating of the ferroelectric phase. T_c is[151] 121.062°K; T_c' is not known. Grindlay's fit depends sensitively on the value of T_c. The assumption that his critical temperature is T_c' can explain qualitatively why A is too small compared with A'.

REFERENCES

* A. P. Sloan Foundation Fellow.

† Stipendiat der Clemens Plassman-Stiftung. On leave from the Max Planck-Institut fur Physik und Astrophysik, Munich, Germany.

‡ Work supported in part by the Advanced Research Projects Agency under Contract ARPA SD-131 and the National Science Foundation Grant NSF GP-4937.

[1] *Proceedings of the International Conference on Phenomena near Critical Points*, Washington (1965) NBS Misc. Publication #273.

[2] J. Appl. Phys. **37**, No. 3 (March 1966).

[3] R. Brout, *Phase Transitions* (W. A. Benjamin, Inc., New York, 1965), discusses this concept of an order parameter in detail.

[4] P. W. Anderson, Rev. Mod. Phys. **38**, 298 (1966) discusses the meaning of the order parameter in the superfluid and superconducting transitions.

[5] L. D. Landau and E. M. Lifshitz, *Statistical Physics* (Pergamon Press, London, 1958), Chap. 14.

[6] This is readily derived by writing the Hamiltonian as $H=H_0-\int d\mathbf{r}p(\mathbf{r})h(\mathbf{r})$, using $\langle p(\mathbf{r})\rangle=\mathrm{Tr}\,p(\mathbf{r})e^{-\beta H}/\mathrm{Tr}\,e^{-\beta H}$, letting $h(\mathbf{r})\to h(\mathbf{r})+\delta h(\mathbf{r})$, and expanding to first order in $\delta h(\mathbf{r})$.

[7] These applications of ideas equivalent to the Landau theory are described in Brout's book (Ref. 3).

[8] The Ginzburg–Landau equations for a superconductor are described in E. A. Lynton, *Superconductivity* (John Wiley & Sons, Inc., New York, 1964), Chap. 5. The analogous equations for superfluids, the Gross–Pitaevskii equations, are discussed in E. P. Gross, J. Math. Phys. **4**, 195 (1963).

[9] L. Onsager, Phys. Rev. **65**, 117 (1944).

[10] C. N. Yang, Phys. Rev. **85**, 808 (1952); T. Schultz, D. Mattis, and E. Lieb, Rev. Mod. Phys. **36**, 856 (1964); G. V. Ryazanov, Zh. Eksperim. i Teor. Fiz. **49**, 1134 (1965) [English transl.: Soviet Phys.—JETP **22**, 789 (1966)].

[11] L. P. Kadanoff, Nuovo Cimento **44B**, 276 (1966).

[12] T. T. Wu, Phys. Rev. **149**, 380 (1966).

[13] C. Domb (Advan. in Phys.) Phil. Mag. Suppl. **9**, 151 (1960); G. F. Newell and E. W. Montroll, Rev. Mod. Phys. **25**, 253 (1953).

[14] D. Mattis, *The Theory of Magnetism* (Harper and Row, New York, 1965), Chap. 9; K. Huang, *Statistical Mechanics* (John Wiley & Sons, Inc., New York, 1965), Chap. 16.

[15] M. E. Fisher, J. Math. Phys. **4**, 278 (1963).

[16] M. E. Fisher, J. Math. Phys. **5**, 944 (1964).

[17] M. E. Fisher, in *Lectures in Theoretical Physics* (University of Colorado Press, Boulder, Colo. 1965), Vol. VII, Part C, p. 1.

[18] Unpublished calculations by one of the authors (R. Hecht). He finds that $G_{EE}(\epsilon R)\sim[K_1^2(|\epsilon|R)-K_0^2(|\epsilon|R)]$, where K_n is the modified Bessel function of the second kind (see, e.g., H. B. Dwight, *Tables of Integrals and Other Mathematical Data* (The Macmillan Co., New York, 1965) and

$$G_{\sigma E}(\epsilon R)\sim\int_{2\epsilon r}^{\infty}x^{-2}e^{-x}dx.$$

[19] This result is derivable from Eqs. (43), (71), and (72) of B. Kaufman and L. Onsager, Phys. Rev. **76**, 1244 (1949). See also Ref. 12.

[20] V. L. Ginzburg, Fiz. Tverd. Tela. **2**, 2031 (1960) [English transl.: Soviet Phys.—Solid State **2**, 1824 (1960)].

[21] A. P. Levanyuk, Zh. Eksperim. i Teor. Fiz. **36**, 810 (1959) [English transl.: Soviet Phys.—JETP **9**, 571 (1959)].

[22] M. Kac, G. E. Uhlenbeck, and P. C. Hemmer, J. Math. Phys. **4**, 216 (1963); G. E. Uhlenbeck, P. C. Hemmer, and M. Kac, *ibid.* **4**, 229 (1963); P. C. Hemmer, M. Kac, and G. E. Uhlenbeck, *ibid.* **5**, 60 (1964); P. C. Hemmer, *ibid.* **5**, 75 (1964).

[23] N. G. van Kempen, Phys. Rev. **135**, A362 (1964).

[24] J. L. Lebowitz and O. Penrose, J. Math. Phys. **7**, 98 (1966).

[25] See Lynton, Ref. 8.

[26] D. Thouless, Ann. Phys. **10**, 553 (1960); E. G. Batyev, A. Z. Patashinskii, and V. L. Pokrovskii, Zh. Eksperim. i Teor. Fiz. **46**, 2093 (1964) [English transl.: Soviet Phys.—JETP **19**, 1412 (1964)].

[27] J. S. Shier and D. M. Ginsberg, Phys. Rev. **147**, 384 (1966).

[28] H. A. Gersch, C. G. Shull, and M. K. Wilkinson, Phys. Rev. **103**, 525 (1956).

[29] D. G. Howard, B. D. Dunlap, and J. G. Dash, Phys. Rev. Letters **15**, 628 (1965).

[30] G. Gorodetsky, S. Shtrikman, and D. Treves, Solid State Commun. **4**, 147 (1966).

[31] L. P. Kadanoff, Physics **2**, 263 (1966).

[32] C. Domb and D. L. Hunter, Proc. Phys. Soc. (London) **86**, 1147 (1965); C. Domb, Ann. Acad. Sci. Fennicae, Ser. A, VI, Physica 210, Helsinki (1966).

[33] M. E. Fisher, J. Appl. Phys. **38,** 981 (1967).
[34] (a) B. Widom, J. Chem. Phys. **43,** 3892 (1965); (b) **43,** 3898 (1965).
[35] M. E. Fisher, Proceedings of the University of Kentucky Centennial Conference on Phase Transitions (March 1965).
[36] A. Z. Patashinskii and V. L. Pokrovskii, Zh. Eksperim. i Teor. Fiz. **50,** 439 (1966) [English transl.: Soviet Phys.—JETP **23,** 292 (1966)].
[37] J. W. Essam and M. E. Fisher, J. Chem. Phys. **39,** 842 (1963).
[38] G. E. Uhlenbeck and P. C. Hemmer, *Proceedings of the International Symposium on Statistical Mechanics and Thermodynamics, Aachen, Germany* (North-Holland Publ. Co., Amsterdam, 1965).
[39] The arguments given here are somewhat similar to arguments which have been given by Buckingham in Ref. 1.
[40] Reference 31 gives a particular description for the construction of μ_a as being proportional to the sum over the entire cell of σ_r. This particular description is not necessary for the conclusions drawn here so that we do not include it here. In other respects, our arguments follow quite closely those of Ref. 31.
[41] The cancellation of L in all physical quantities is a very stringent requirement upon the theory. In particular, it is hard to see how (3.13) could be generalized and still allow L to cancel out of $\langle\sigma_r\rangle$.
[42] A relation of this kind occurs in Ref. 37 which proposes $\alpha'+2\beta+\gamma'=2$. In Ref. 35, Fisher argues for Eq. (3.9) on the basis of Frenkel, Band, and Bijl's theory of condensation. In Ref. 34b, Widom derives all these relations from an assumption which is equivalent to the assertion that Eq. (3.7) is correct. Uhlenbeck and Hemmer, in Ref. 38, have given thermodynamic arguments in favor of $\gamma+2\beta=2$ when $\alpha=\alpha'=0$.
[43] The result $d\nu=2-\alpha$ was suggested by Widom in Ref. 34a. A result of this type follows from an argument given by A. B. Pippard, Proc. Roy. Soc. London **A216,** 547 (1953).
[44] W. P. Wolf and A. F. G. Wyatt, Phys. Rev. Letters **13,** 368 (1964).
[45] A. R. Miedema, H. van Kempen, and W. J. Huiskamp, Physica **29,** 1266 (1963).
[46] A. H. Cooke, D. T. Edmonds, C. B. P. Finn, and W. P. Wolf, J. Phys. Soc. Japan **17,** Suppl. B-1, 481 (1962).
[47] G. A. Baker, H. E. Gilbert, J. Eve, and G. S. Rushbrooke, Phys. Letters **20,** 146 (1966).
[48] C. Domb and M. F. Sykes, Phys. Rev. **128,** 168 (1962); J. Gammel, W. Marshall, and L. Morgan, Proc. Roy. Soc. (London) **275,** 257 (1963).
[49] P. J. Wood and G. S. Rushbrooke, Phys. Rev. Letters **17,** 307 (1966).
[50] See Ref. 17, p. 121.
[51] P. Heller and G. B. Benedek, Phys. Rev. Letters **8,** 428 (1962).
[52] E. Sawatzky and M. Bloom, Can. J. Phys. **42,** 657 (1964).
[53] J. E. Noakes, N. E. Tornberg, and A. Arrott, J. Appl. Phys. **37,** 1264 (1966).
[54] P. Heller, Phys. Rev. **146,** 403 (1966).
[55] O. K. Rice, J. Chem. Phys. **22,** 1535 (1954); O. K. Rice, J. Phys. Chem. **64,** 976 (1960); R. A. Farrell and P. H. E. Meijer, Physica **31,** 725 (1965); C. P. Bean and D. S. Rodbell, Phys. Rev. **126,** 104 (1962); D. C. Mattis and T. D. Schultz, *ibid.* **129,** 175 (1963).
[56] S. Arajs and R. V. Colvin, J. Appl. Phys. **35,** 2424 (1964).
[57] G. Develey, Compt. Rend. **260,** 4951 (1965).
[58] P. Weiss and R. Forrer, Ann. Phys. (Paris) **5,** 153 (1926). These data were analyzed near T_c by J. S. Kouvel and M. E. Fisher, Phys. Rev. **136,** A1626 (1964); and also in Ref. 62.
[59] S. Arajs, J. Appl. Phys. **36,** 1136 (1965).
[60] C. D. Graham, J. Appl. Phys. **36,** 1135 (1965).
[61] R. V. Colvin and S. Arajs, J. Phys. Chem. Solids **26,** 435 (1965).
[62] J. S. Kouvel and D. S. Rodbell, Phys. Rev. Letters **18,** 215 (1967).
[63] H. E. Stanley and T. A. Kaplan, J. Appl. Phys. **38,** 975, 977 (1967).
[64] L. Van Hove, Phys. Rev. **35,** 1374 (1954).
[65] P. G. DeGennes, in *Magnetism*, G. T. Rado and H. Suhl, Eds. (Academic Press Inc., New York, 1963), Vol. III.
[66] L. Passell, K. Blinowski, T. Brun, and P. Nielsen, Phys. Rev. **139,** A1866 (1965).
[67] See Ref. 3, Chap. 2.
[68] T. Riste and A. Wanic, Phys. Letters **16,** 231 (1965).
[69] M. J. Cooper and R. Nathans, J. Appl. Phys. **37,** 1041 (1966).
[70] G. T. Freeman and R. E. Watson, in *Magnetism*, G. T. Rado and H. Suhl, Eds. (Academic Press Inc., New York, 1965), Vol. IIA.
[71] R. S. Preston, S. S. Hanna, and J. Heberle, Phys. Rev. **128,** 2207 (1962).
[72] For example, see J. G. Dash, B. D. Dunlap, and D. G. Howard, Phys. Rev. **141,** 376 (1966).
[73] H. B. Callen, D. Hone, and A. Heeger, Phys. Letters **17,** 233 (1965).
[74] D. Hone, H. Callen, and L. R. Walker, Phys. Rev. **144,** 283 (1966).
[75] E. Callen and H. Callen, J. Appl. Phys. **36,** 1140 (1965).
[75a] A. Arrott *et al.*, J. Appl. Phys. **38,** 969 (1967).
[76] N. J. Poulis and G. E. G. Hardeman, Physica **19,** 391 (1953).
[77] W. Van der Lugt and N. J. Poulis, Physica **26,** 917 (1960).
[77a] *Note added in proof.* More recent work by Preston has filled in more data points for iron with $-\epsilon\gtrsim10^{-3}$. These results seem consistent with $\beta\approx1/3$. We wish to thank Dr. Preston for sending us his work prior to publication.
[78] H. H. Potter, Proc. Roy. Soc. (London) **146,** 362 (1934).
[79] P. Heller and G. B. Benedek, Phys. Rev. Letters **14,** 71 (1965).
[80] M. Eibschutz, S. Shtrikman, and D. Treves, Solid State Commun. **4,** 141 (1966).
[81] S. D. Senturia and G. B. Benedek, Phys. Rev. Letters **17,** 475 (1966).
[82] For example, if we use the scaling law relation $2\beta+\gamma=2-\alpha$ and Kouvel and Rodbell's data [Ref. 62 and our Eqs. (4.2)] for Ni ($\beta=0.42$, $\gamma=1.35$) and CrO_2 ($\beta=0.33$, $\gamma=1.6$) we find $\alpha=-0.19$ for Ni and $\alpha=-0.26$ for CrO_2.
[83] A. B. Pippard, *The Elements of Classical Thermodynamics* (Cambridge University Press, Cambridge, England, 1964), Chap. 9.
[84] J. Skalyo, Jr., and S. A. Friedberg, Phys. Rev. Letters **13,** 133 (1964).
[85] A. R. Miedema, R. F. Wielinga, and W. J. Huiskamp, Phys. Letters **17,** 87 (1965).
[86] D. T. Teaney, Phys. Rev. Letters **14,** 898 (1965).
[87] See Table VII.
[88] See Table V.
[89] S. A. Friedberg and J. D. Wasscher, Physica **19,** 1072 (1953).
[90] D. T. Teaney, V. L. Moruzzi, and B. E. Argyle, J. Appl. Phys. **37,** 1122 (1966).
[91] Ya. A. Kraftmakher and T. Yu. Romashina, Fiz. Tverd. Tela **7,** 3532 (1965) [English transl.: Soviet Phys.—Solid State **7,** 2040 (1966)].
[92] A. R. Miedema, R. F. Wielinga, and W. J. Huiskamp, Physica **31,** 1234 (1965).
[93] Ya. A. Kraftmakher, Fiz. Tverd. Tela **8,** 1306 (1966) [English transl.: Soviet Phys.—Solid State **8,** 1048 (1966)].
[94] M. E. Fisher, Proc. Roy. Soc. (London) **A254,** 66 (1960).
[95] B. E. Keen, D. Landau, B. Schneider, and W. P. Wolf, J. Appl. Phys. **37,** 1120 (1966).
[96] W. H. M. Voorhoeve and Z. Dokopuil, Physica **27,** 777 (1961).
[97] See Ref. 3, Chap. 3.
[98] T. D. Lee and C. N. Yang, Phys. Rev. **87,** 410 (1952); see also Ref. 14, K. Huang.
[99] H. W. Habgood and W. G. Schneider, Can. J. Chem. **32,** 98 (1954).
[100] M. A. Weinberger and W. G. Schneider, Can. J. Chem. **30,** 422 (1952).
[101] A. V. Voronel', Yu. R. Chashkin, V. A. Popov, and V. G. Simkin, Zh. Eksperim. i Teor. Fiz. **45,** 828 (1963) [English transl.: Soviet Phys.—JETP **18,** 568 (1964)].
[102] A. V. Voronel', V. G. Snigirev, and Yu. R. Chashkin, Zh. Eksperim. i Teor. Fiz. **48,** 981 (1964) [English transl.: Soviet Phys.—JETP **21,** 653 (1965)].
[103] A. Michels, J. M. Levelt, and W. de Graaf, Physica **24,** 659 (1958); A. Michels, J. M. Levelt, and G. J. Walkers, *ibid.* 24, 769 (1958).
[104] A. Michels, B. Blaisse, and C. Michels, Proc. Roy. Soc. (London) **160A,** 358 (1937).
[105] H. L. Lorentzen, Acta Chem. Scand. **7,** 1335 (1953).
[106] H. L. Lorentzen, *Statistical Mechanics of Equilibrium and Nonequilibrium* (North-Holland Publ. Co., Amsterdam, 1965), p. 262.

[107] M. I. Bagatskii, A. V. Voronel', and V. G. Gusak, Zh. Eksperim. i Teor. Fiz. **43**, 728 (1962) [English transl.: Soviet Phys.—JETP **16**, 517 (1963)].
[108] Kh. I. Amirkhonov and I. G. Gurvich, Dokl. Akad. Nauk. SSSR **91**, 221 (1953); Chemical Abstracts **48**, 13363g (1954).
[109] B. Widom and O. K. Rice, J. Chem. Phys. **23**, 1250 (1955).
[110] J. E. Thomas and P. W. Schmidt, J. Chem. Phys. **39**, 2506 (1963).
[111] P. Heller, to be published in a review for Rept. Progr. Phys. We wish to thank Dr. Heller for telling us of his results prior to publication.
[112] A. V. Voronel', Zh. Fiz. Khimii **35**, 958 (1961).
[113] A. Michels and J. Strijland, Physica **18**, 613 (1952).
[114] E. A. Guggenheim, J. Chem. Phys. **13**, 253 (1945).
[115] F. P. Buff and R. A. Lovett, *Simple Dense Fluids: Data and Theory*, H. L. Frisch and Z. W. Salsburg, Eds. (Academic Press Inc., New York, to be published); F. P. Buff, R. A. Lovett, and F. H. Stillinger, Jr., Phys. Rev. Letters **15**, 621 (1965), footnote 7.
[116] D. Atack and O. K. Rice, Discussions Faraday Soc. **15**, 210 (1953).
[117] D. Stansfield, Proc. Phys. Soc. (London) **72**, 854 (1958).
[118] (a) G. S. Rushbrooke, J. Chem. Phys. **39**, 842 (1963); (b) R. B. Griffiths, *ibid.* **43**, 1958 (1965).
[119] C. N. Yang and C. P. Yang, Phys. Rev. Letters **13**, 303 (1964).
[120] J. deBoer, Physica **14**, 139 (1948).
[121] T. Matsubara and H. Matsuda, Progr. Theoret. Phys. (Kyoto) **16**, 569 (1956).
[122] P. R. Zilsel, Phys. Rev. Letters **15**, 476 (1965).
[123] J. deBoer and R. J. Lunbeck, Physica **14**, 318 and 510 (1948).
[124] M. R. Moldover and W. A. Little, Phys. Rev. Letters **15**, 54 (1965).
[125] A. V. Voronel' and M. Sh. Giterman, Zh. Eksperim. i Teor. Fiz. **39**, 1162 (1960) [English transl.: Soviet Phys.—JETP **12**, 809 (1961)].
[126] M. H. Edwards and P. W. C. Woodbury, Phys. Rev. **129**, 1911 (1963).
[127] L. Tisza and C. E. Chase, Phys. Rev. Letters **15**, 4 (1965).
[128] M. H. Edwards, Phys. Rev. Letters **15**, 349 (1965); L. Mistura and D. Sette, *ibid.* **16**, 268 (1966).
[129] S. Y. Larsen, R. D. Mountain, and R. Zwanzig, J. Chem. Phys. **42**, 2187 (1965).
[130] P. R. Roach and D. H. Douglass, Jr., Phys. Rev. Letters **17**, 1083 (1966). We wish to thank Dr. Roach and Dr. Douglass for sending us a preprint of their work.
[131] M. H. Edwards, *Proceedings of the Tenth International Conference on Low Temperature Physics, Moscow*, 1966 (to be published). We wish to thank Dr. Edwards for sending us a preprint of his work.
[132] R. H. Sherman, Phys. Rev. Letters **15**, 141 (1965).
[133] B. Widom (private communication).
[134] M. E. Fisher, Phys. Rev. Letters **16**, 11 (1966).
[135] H. M. Roder, D. E. Diller, L. A. Weber, and R. D. Goodwin, Cryogenics **3**, 16 (1965).
[136] C. E. Chase and G. O. Zimmerman, Phys. Rev. Letters **15**, 483 (1965).
[137] H. S. Johnson, W. E. Keller, and A. Friedman, J. Am. Chem. Soc. **76**, 1482 (1954).
[138] R. H. Sherman and E. F. Hammel, Phys. Rev. Letters **15**, 9 (1965).
[139] Yu. P. Blagoi and V. V. Poshkov, Zh. Eksperim. i Teor. Fiz. **49**, 1453 (1965) [English transl.: Soviet Phys.—JETP **22**, 999 (1966)].
[140] O. Penrose and L. Onsager, Phys. Rev. **104**, 576 (1956).
[141] D. Pines, *Theory of Quantum Liquids* (W. A. Benjamin, Inc., New York, to be published), Vol. 2.
[142] V. G. Vaks and A. I. Larkin, Zh. Eksperim. i Teor. Fiz. **49**, 975 (1965) [English transl.: Soviet Phys.—JETP **22**, 678 (1966)].
[143] C. F. Kellers, thesis, Duke University (1960). M. J. Buckingham and W. M. Fairbank, *Progress in Low Temperature Physics* (North-Holland Publ. Co., Amsterdam, 1961), Vol. III.
[144] J. R. Clow and J. D. Reppy, Phys. Rev. Letters **16**, 887 (1966).
[145] J. A. Tyson and D. H. Douglass, Jr., Phys. Rev. Letters **17**, 472 (1966); and erratum, Phys. Rev. Letters **17**, 622 (1966).
[146] B. D. Josephson, Phys. Letters **21**, 608 (1966).
[147] T. M. Rice, Phys. Rev. **140**, A1889 (1965). J. W. Kane, thesis, Univ. of Illinois (1966).
[148] A. Z. Patashinskii and V. L. Pokrovskii, Zh. Eksperim. i Teor. Fiz. **46**, 994 (1964) [English transl.: Soviet Phys.—JETP **19**, 677 (1964)].
[149] R. Abe, Progr. Theoret. Phys. (Kyoto) **33**, 600 (1965).
[150] P. W. Forsbergh, Jr., in *Handbuch der Physik*, S. Flügge, Ed. (Springer-Verlag, Berlin, 1956), Vol. 17; W. Känzig, in *Solid State Physics*, F. Seitz and D. Turnbull, Eds. (Academic Press Inc., New York, 1957), Vol. 4, p. 1; F. Jona and A. Shirane, *Ferroelectric Crystals* (The Macmillan Co., New York, 1962).
[151] P. P. Craig, Phys. Letters **20**, 140 (1966).
[152] C. C. Stephenson and J. A. Hooley, J. Am. Chem. Soc. **66**, 1397 (1944).
[153] B. A. Strukov and M. N. Danilychewa, Fiz. Tverd. Tela. **5**, 1724 (1963) [English transl.: Soviet Phys.—Solid State **5**, 1253 (1963)].
[154] A. A. Rusterhold, Helv. Phys. Acta **8**, 39 (1935).
[155] J. Grindlay, Phys. Rev. **139**, A1603 (1965).

Reprinted from THE PHYSICAL REVIEW, Vol. 165, No. 1, 310–322, 5 January 1968
Printed in U. S. A.

Transport Coefficients near the Critical Point: A Master-Equation Approach

LEO P. KADANOFF AND JACK SWIFT
Department of Physics and Materials Research Laboratory, University of Illinois, Urbana, Ill.*
(Received 18 July 1967)

Kawasaki has shown how to construct a nonequilibrium theory which relaxes to an equilibrium described by the standard Ising model. The main significance of Kawasaki's work is his proof that transport coefficients do not diverge near the critical point in his model. In this paper, his approach is generalized. Two master-equation models of transport are examined: one gives spin diffusion and thermal diffusion but no sound waves; the other gives thermal diffusion and sound waves. The first model involves a Hermitian transition matrix in the master equation. The Hermiticity enables one to prove a variational theorem which requires the transport coefficients to be finite. However, sound waves appear as complex eigenvalues of the relaxation time. Hence, they must come from a non-Hermitian master equation. A model is constructed which includes sound waves. In this case, the proof of the finiteness of transport coefficients fails. Aside from formal questions, the main physical point of this paper is the speculation that infinities in transport coefficients might be tied to the existence of oscillatory transport modes (like sound waves) coupled into the dynamics of the phase transition.

I. INTRODUCTION

IN a very interesting series of papers,[1,2,3] Kawasaki has investigated the behavior of transport coefficients near the critical point. Basically, his model is an Ising model in which spins on neighboring sites are interchanged at a given rate. The details of the interchange are arranged so that the system relaxes to the usual Ising model equilibrium state. This model then describes spin diffusion in an insulator or particle diffusion in a binary alloy like β brass.

Kawasaki formulates his model in terms of a master equation. Glauber[4] and Heims[5] have also applied master equations to Ising models. However, a special feature of Kawasaki's work is a variational theorem which enables him to calculate an upper bound to the transport coefficients in his model. This upper bound is finite so that the exact transport coefficients derived from this model cannot possibly diverge near the critical point.

This model is a rough description of the situation in materials[6] such as β brass. Indeed, experiment indicates that the particle self-diffusion coefficient does not appear to diverge near the critical point in these materials.

However, other critical points are accompanied by divergent transport coefficients. In the liquid gas phase transition the thermal conductivity λ_T appears to diverge as the critical point is approached.[7,8] In iron,[9] the spin diffusivity does not appear to go very strongly to zero, so that the spin diffusion transport coefficient is probably diverging rather strongly. There are theoretical arguments[10–12] and experimental evidence[13] that the shear viscosity η diverges at the critical point of fluid mixtures. The thermal conductivity apparently diverges as the superfluid transition of He^4 is approached from above.[14,15]

Therefore, Kawasaki's arguments seem to work well for the specific situation in which the physics is closest to the conditions of his model. However, in other situations, Kawasaki's basic conclusion—that the transport coefficients are finite—fails.

At first sight this failure appears to be a great mystery. The arguments for finite transport coefficients seem very simple and general: The existence of a master equation, the proof of a variational theorem, and finally the explicit calculation of a finite upper limit. However, there turns out to be one feature of the argument which is not universally applicable. The proof of the variational theorem depends upon the Hermitian nature of

* Work supported in part by the Advanced Research Projects Agency under Contract No. SD-131.
[1] K. Kawasaki, Phys. Rev. **145**, 224 (1966).
[2] K. Kawasaki, Phys. Rev. **148**, 375 (1966).
[3] K. Kawasaki, Phys. Rev. **150**, 285 (1966).
[4] R. J. Glauber, J. Math. Phys. **4**, 294 (1963).
[5] S. P. Heims, J. Chem. Phys. **45**, 370 (1966).
[6] J. D. Noble and M. Bloom, Phys. Rev. Letters **14**, 250 (1965); A. B. Kuper, D. Lazarus, J. R. Manning, and C. T. Tomizuka, Phys. Rev. **104**, 1536 (1956).
[7] L. A. Guildner, Proc. Natl. Acad. Sci. **44**, 1149 (1958) and J. Res. Natl. Bur. Std. **66A**, 333 (1962); **66A**, 341 (1962); A. Michels, J. V. Sengers, and P. S. van der Gulik, Physica **28**, 1201, 1216 (1962); A. Michels and J. V. Sengers, *ibid.* **28**, 1238 (1962).
[8] N. C. Ford and G. B. Benedek, Phys. Rev. Letters **15**, 649 (1965); H. Z. Cummins and H. L. Swinney (to be published). We wish to thank both of these groups for informative discussions.
[9] B. Jacrot, J. Kostantinovic, G. Perette, and D. Cribier, Symposium on Inelastic Scattering of Neutrons in Solids and Liquids, Chalk River, 1962 (unpublished); L. Passell, K. Blinowski, T. Brun, and P. Nielsen, J. Appl. Phys. **35**, 933 (1964); Phys. Rev. **139A**, 1866 (1965).
[10] M. Fixman, J. Chem. Phys. **36**, 310 (1962); W. Botch and M. Fixman, *ibid.* **36**, 3100 (1962); M. Fixman in *Advances in Chemical Physics*, edited by I. Prigogine (Interscience Publishers, N. Y., 1964), Vol. VI.
[11] K. Kawasaki, Phys. Rev. **150**, 291 (1966).
[12] J. M. Deutch and R. Zwanzig, J. Chem. Phys. **46**, 1612 (1967).
[13] See the review of the experimental situation by J. V. Sengers, in *Proceedings of the Conference on the Phenomena near the Critical Point, Washington, D. C., 1965*, edited by M. S. Green and J. V. Sengers (Natl. Bur. Std. Misc. Publ. No. 273, 1965).
[14] R. A. Ferrel *et al.*, Phys. Rev. Letters **18**, 891 (1967).
[15] J. F. Kerrisk and W. E. Keller, Bull. Am. Phys. Soc. Ser. II **12**, 550 (1967); J. A. Lipa *et al.*, Phys. Rev. **155**, 75 (1967).

the transition matrix in the master equation. The transition matrix then has real eigenstates and these eigenvalues have the physical significance of being the relaxation times of the different modes in the system. Of course, diffusion processes involve real relaxation times. For example, thermal diffusion produces an inverse relaxation time for a disturbance with wave vector q which is $\tau^{-1}=(\lambda_T/\rho C_p)q^2$. But, sound waves and spin waves are oscillatory phenomena. Insofar as they are not damped, they are represented by pure imaginary relaxation times. When damping is included, these eigenvalues of the transition matrix become complex. Hence, as soon as we admit sound waves, we must abandon the Hermitian transition matrix in the master equation. Then the proof of the finiteness of transport coefficients fails because the variational principle no longer exists. It seems possible, at least, that infinities in transport coefficients are tied to the existence of oscillatory modes coupled into the dynamics of the phase transition and their consequent non-Hermitian representation in the transition matrix.

In this paper, these basic ideas are examined through the consideration of master-equation models of transport phenomena. In the next section, the formalism of master equations is described with a view to seeing how equilibrium behavior and conservation laws are fixed into the structure of these equations. Section III describes a model for transport which relaxes to the standard Ising model in equilibrium. The spin-exchange processes are constructed in such a manner as to conserve the Ising model energy and the total spin. Therefore, the master equation implies two diffusive transport modes: spin diffusion and thermal diffusion. For $T>T_c$ and zero magnetic field, we prove that the two transport coefficients involved are finite at T_c. (As a consequence, the spin diffusivity goes to zero as the inverse susceptibility and the thermal diffusivity vanishes as the inverse specific heat).

Section IV describes a model of dynamical behavior that includes sound waves. This model is then appropriate for fluids. To bring in sound waves, a momentum variable is introduced in addition to the "spin" variable of the standard lattice gas.[14,15] The model is arranged so that, in equilibrium, the momentum may be eliminated from the partition function in a trivial way. This same elimination occurs in classical statistical mechanics. Then, the remaining partition function is chosen to represent the lattice gas or Ising model.

In this model, the momentum is tied into the dynamics in a nontrivial manner. Particles tend to move in the direction of their momentum. Coupled conservation laws for momentum, number, and energy are shown to reduce to standard linearized hydrodynamic equations. The model then includes both thermal diffusion and sound waves.

However, the transition matrix of this model is non-Hermitian. The proof that the transport coefficients are finite at the critical point fails. Therefore, the divergence or nondivergence of these transport coefficients in the fluids is left as an open question in this analysis.

II. MASTER EQUATIONS

In this section, we review some of the formalism of master equations and construct a notational system which will be useful for the rest of the paper.

A. Notation

We describe a system like the classical Ising model. There is a lattice and at each lattice site, **r**, there is a spin variable $\sigma_{\mathbf{r}}$ which can take on two values, ± 1. The complete state of the system is given by specifying all the values of all these spin variables. We find it convenient to associate a state vector with every spin configuration. This state vector is written as $|\alpha\rangle$, where α is an index which defines all the values of all the different spins.

The time dependence of the model is given by specifying the equation of motion obeyed by $p_\alpha(t)$, the properly normalized probability of finding the system in the state α. We take this to be a first-order equation

$$\frac{d}{dt}p_\alpha(t)=\sum_\beta T_{\alpha\beta}p_\beta(t)\,, \tag{2.1}$$

so that a determination of all the $p_\alpha(t)$ at any time will define the values of the probabilities at all later times. If $\beta\neq\alpha$, $T_{\alpha\beta}$ is the probability per unit time that the system will hop from state β to state α while $-T_{\alpha\alpha}$ is the rate of hopping out of the state α.

It is convenient to write the purely classical master equation, Eq. (2.1), in a matrix form. To do this, let us assume that the basis states $|\alpha\rangle$ are orthonormal so that

$$\langle\alpha|\beta\rangle=\delta_{\alpha\beta}. \tag{2.2}$$

This equation means that two basic states $|\alpha\rangle$ and $|\beta\rangle$ are orthogonal if the value of the spin at any site is different in state $|\alpha\rangle$ from that in state $|\beta\rangle$. The statistical state of the system at time t is given by

$$|t\rangle=\sum_\alpha p_\alpha(t)|\alpha\rangle\,,$$

so that $\langle\alpha|t\rangle$ is the probability of finding the system in the state $|\alpha\rangle$. If we also define a transition matrix

$$\mathbf{T}=\sum_{\alpha\beta}|\alpha\rangle T_{\alpha\beta}\langle\beta|\,, \tag{2.3}$$

then Eq. (2.1) can be written in the simple form

$$(d/dt)|t\rangle=\mathbf{T}|t\rangle. \tag{2.4}$$

An important auxiliary quantity in the analysis is the special-state vector

$$|\;\rangle=\sum_\alpha|\alpha\rangle. \tag{2.5}$$

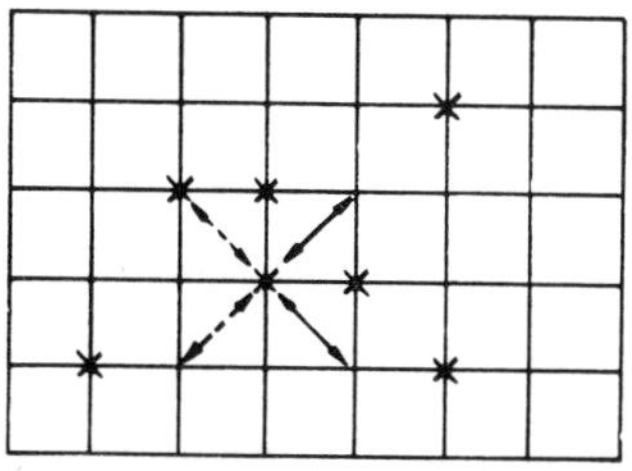

FIG. 1. Some interchanges described by the T^A matrix. The x's represent $\sigma_r=+1$ at the lattice site while the unmarked intersections represent $\sigma_r=-1$ at that site. The solid lines indicate allowed interchanges between spin "up" and spin "down". The broken lines indicate forbidden interchanges. The forbidden interchanges shown above involve either a change in energy or the interchange of two like spins. There are other allowed and nonallowed interchanges for the situation shown above which are not shown for the sake of clarity.

For example, we can use this quantity to give a compact formulation of the concept of conservation of probability. This is the statement

$$\frac{\partial}{\partial t}\langle\,|\mathbf{t}\rangle=\frac{\partial}{\partial t}\sum_\alpha p_\alpha(\mathbf{t})=\frac{\partial}{\partial t}1=0.$$

From Eq. (2.5) we then see that the statement of conservation of probability or detailed balance is

$$\langle\,|\mathbf{T}=0 \qquad \text{detailed balance.} \tag{2.6}$$

Given any physical quantity, X which has values X_α in the state α we define a diagonal operator representation of that quantity as

$$\mathbf{X}=\sum_\alpha |\alpha\rangle\langle\alpha|\,X_\alpha. \tag{2.7}$$

For example, the spin at site $\mathbf{r}$ is given by taking $X_\alpha=(\sigma_r)_\alpha$ to have the values ±1 depending upon the sign of σ_r in the state α. The statistical average of X at time t is given by

$$\langle X\rangle_t=\langle\,|\mathbf{X}|t\rangle. \tag{2.8}$$

From Eq. (2.5) this obeys the equation of matrix

$$(d/dt)\langle X\rangle_t=\langle\,|\mathbf{XT}|t\rangle, \tag{2.9}$$

when we apply the detailed balancing condition (2.6) we find that this equation of motion can also be written as

$$(d/dt)\langle X\rangle_t=\langle\,|[\mathbf{X},\mathbf{T}]|t\rangle. \tag{2.10}$$

Transport properties are intimately linked with conservation laws. We denote the n conserved quantities in the system by the symbols C_i, $i=1, 2, \cdots n$. For example, in the analysis of the usual Ising-model ferromagnet, we would like to have the total magnetization

$$\mathbf{M}=\sum_r \sigma_r=\mathbf{C}_1, \tag{2.11a}$$

and the Hamiltonian

$$\mathbf{H}=-\sum_{\langle r,r'\rangle}\sigma_r\sigma_{r'}=\mathbf{C}_2 \tag{2.11b}$$

be conserved. [In Eq. (2.11b), $\langle \mathbf{rr}'\rangle$ indicates that the sum is to be taken over all pairs of nearest neighbors.] From Eq. (2.10) we can see that the conservation laws

$$(d/dt)\langle C_i\rangle_t=0$$

can be equivalently stated as commutation relations

$$[\mathbf{C}_i,\mathbf{T}]=0 \qquad \text{conservation laws} \tag{2.12}$$

B. A Specific Example

It is easy to construct transition matrices, $T_{\alpha\beta}$, which satisfy the requirements stated above. For example, consider a model in which Ising spins are distributed upon a square or cubic lattice. Let there be a probability per unit time w^A that any pair of next-nearest-neighbor spins will interchange their values of σ_r. This interchange is only allowed if the process does not change the total energy H. Examples of possible and impossible interchanges are shown in Fig. 1.[16]

Mathematically, this model is represented by

$$\mathbf{T}=\mathbf{T}^A=\tfrac{1}{2}\sum_{r,r'}\mathbf{T}_{r,r'}{}^A, \tag{2.13}$$

with

$$\mathbf{T}_{r,r'}{}^A=w^A(\mathbf{I}_{r,r'}-1)\Delta_{r,r'}{}^A. \tag{2.14}$$

Here $\mathbf{T}_{r,r'}{}^A$ is the operator which describes the transition probability resulting from interchanges of spins at the nearest neighbor sites $\mathbf{r}$ and $\mathbf{r}'$. The off diagonal operator $\mathbf{I}_{r,r'}$ represents the actual interchange process

$$\begin{aligned}&\mathbf{I}_{r,r'}\sigma_r=\sigma_{r'}\mathbf{I}_{r,r'},\\&\mathbf{I}_{r,r'}\sigma_{r'}=\sigma_r\mathbf{I}_{r,r'}.\\&[\mathbf{I}_{r,r'},\sigma_{r_1}]=0 \text{ for } \mathbf{r}_1 \text{ different from } \mathbf{r} \text{ and } \mathbf{r}'.\end{aligned} \tag{2.15}$$

Of course, $\mathbf{I}_{r,r'}$ obeys

$$\begin{aligned}&\mathbf{I}_{r,r'}=\mathbf{I}_{r',r},\\&(\mathbf{I}_{r,r'})^2=1.\end{aligned} \tag{2.16}$$

Note that $\mathbf{I}_{r,r'}$ is not a matrix in $\mathbf{r}$ and $\mathbf{r}'$; instead these coordinates are labels for $\mathbf{I}_{r,r'}$. The matrix elements of $\mathbf{I}_{r,r'}$ are between states with specified values of the spin variables, i.e.,

$$\langle\bar\sigma_r\bar\sigma_{r'}|\,\mathbf{I}_{r,r'}|\,\sigma_r\sigma_{r'}\rangle=\delta(\bar\sigma_r,\sigma_{r'})\delta(\bar\sigma_{r'},\sigma_r).$$

The delta symbol, $\Delta_{r,r'}{}^A$, delimits the allowed interchanges. It is defined by

$$\Delta_{r,r'}{}^A=\begin{cases}1 \text{ if } \begin{cases}\mathbf{I}_{r,r'}\mathbf{H}\mathbf{I}_{r,r'}=\mathbf{H}\\ \sigma_r=-\sigma_{r'}\\ \mathbf{r} \text{ and } \mathbf{r}' \text{ are next-nearest neighbors}\end{cases}\\ 0 \text{ otherwise}\end{cases} \tag{2.17}$$

[16] This choice of the transition matrix makes the magnetization on two different sublattices separately conserved. However, we may add nearest neighbor or next-next-nearest neighbor interchanges to the transition matrix to destroy this conservation law without changing the results in any essential way.

The first condition in Eq. (2.17) is energy conservation; the next is inserted because exchange of equivalent spins has no effect; the third condition limits the interchanges to next-nearest-neighbor sites. The condition of energy conservation can be restated with the aid of Eq. (2.11b) as

$$u_{\mathbf{r}}=u_{\mathbf{r}'}. \tag{2.18}$$

Here $u_{\mathbf{r}}$ is the "potential energy" for a spin at site $\mathbf{r}$,

$$u_{\mathbf{r}}=\sum_{\mathbf{r}'\supset\mathbf{r}}\sigma_{\mathbf{r}'}. \tag{2.19}$$

The notation $\mathbf{r}'\supset\mathbf{r}$ indicates that the sum covers those values of $\mathbf{r}'$ which are nearest neighbors to $\mathbf{r}$. Thus the "energy conserving" operator $\Delta_{\mathbf{r},\mathbf{r}'}{}^{A}$ can be written as

$$\Delta_{\mathbf{r},\mathbf{r}'}{}^{A}=\delta(u_{\mathbf{r}},u_{\mathbf{r}'})\tfrac{1}{2}(1-\sigma_{\mathbf{r}}\sigma_{\mathbf{r}'}), \tag{2.20}$$

when $\mathbf{r}$ and $\mathbf{r}'$ are next-nearest neighbors.

Notice that the transition matrix (2.14) contains two terms. These terms are best understood if we return to the $p_{\alpha's}$. Equation (2.14) indicates a contribution to dp_α/dt which is

$$\sum_{\beta} w^{A}(\mathbf{I}_{\mathbf{r},\mathbf{r}'})_{\alpha\beta}(\Delta_{\mathbf{r},\mathbf{r}'}{}^{A})_{\beta}p_{\beta}(t)-w^{A}(\Delta_{\mathbf{r},\mathbf{r}'}{}^{A})_{\alpha}p_{\alpha}(t).$$

The first term, the "scattering-in term" indicates an increase in $p_\alpha(t)$ resulting from the scattering from all other states, β, into the state α. The second term, the "scattering-out term" shows the decrease in $p_\alpha(t)$ coming from the scattering out of that state. The projection operators $\Delta_{\mathbf{r},\mathbf{r}'}$ simply delimit the states in which the scattering is permitted.

The model we have just defined satisfies all the formal requirements described in Sec. (2A). Since $\langle\,|$ contains all states equally,

$$\langle\,|\mathbf{I}_{\mathbf{r},\mathbf{r}'}=\langle\,|. \tag{2.21}$$

Hence the definition (2.14) automatically guarantees the detailed balancing condition (2.6) since

$$\langle\,|\mathbf{T}_{\mathbf{r},\mathbf{r}'}{}^{A}=\langle\,|(\mathbf{I}_{\mathbf{r},\mathbf{r}'}-1)\Delta_{\mathbf{r},\mathbf{r}'}{}^{A}w^{A}=0.$$

The conservation laws (2.12) for energy and magnetization are also automatically satisfied in this model. Since $\mathbf{I}_{\mathbf{r},\mathbf{r}'}$ only interchanges spins, it commutes with the total magnetization, $\sum_{\mathbf{r}}\sigma_{\mathbf{r}}$. Also, the projection operation which insists that $u_{\mathbf{r}}=u_{\mathbf{r}'}$ for all interchanges then guarantees that no interchange will change the value of the "Hamiltonian," $\mathbf{H}$.

C. Other Formal Requirements

We would like to demand that the equilibrium solution of the master equation be precisely the standard Ising-model density matrices

$$p_\alpha=\exp(-\beta H_\alpha+hM_\alpha)/Z, \tag{2.22a}$$

with

$$Z(\beta,h)=\sum_{\alpha}\exp(-\beta H_\alpha+hM_\alpha), \tag{2.22b}$$

where β is a dimensionless inverse temperature and h is a dimensionless magnetic field. This is the same as the requirement that the state

$$|eq\rangle=(1/Z(\beta,h))\exp(-\beta\mathbf{H}+h\mathbf{M})|\,\rangle \tag{2.23}$$

be a time-independent solution of the master equation. Since the master reads

$$(\partial/\partial t)|t\rangle=\mathbf{T}|t\rangle,$$

we must require

$$\mathbf{T}\exp(-\beta\mathbf{H}+h\mathbf{M})|\,\rangle=0.$$

But, the transition matrix $\mathbf{T}$ commutes with $\mathbf{H}$ and $\mathbf{M}$. The condition that Eq. (2.23) be an equilibrium solution is then the requirement

$$\mathbf{T}|\,\rangle=0. \tag{2.24}$$

Equation (2.23) is automatically satisfied in this simple model. For $\mathbf{I}_{\mathbf{r},\mathbf{r}'}$ commutes with $\Delta_{\mathbf{r},\mathbf{r}'}$ so that

$$\mathbf{T}_{\mathbf{r},\mathbf{r}'}{}^{A}=\Delta_{\mathbf{r},\mathbf{r}'}{}^{A}(\mathbf{I}_{\mathbf{r},\mathbf{r}'}-1)w^{A}.$$

Then, $\mathbf{I}_{\mathbf{r},\mathbf{r}'}|\,\rangle=|\,\rangle$ implies

$$\mathbf{T}_{\mathbf{r},\mathbf{r}'}{}^{A}|\,\rangle=0.$$

Below, we shall consider another model in which $\mathbf{I}_{\mathbf{r},\mathbf{r}'}$ does not commute with $\Delta_{\mathbf{r},\mathbf{r}'}$. In this situation, there will be real difficulty in constructing a model with the correct equilibrium properties.

Notice the structure of the matrix $\mathbf{T}$. It has the form

$$T_{\alpha\beta}=(\mathbf{T}^{\text{in}})_{\alpha\beta}-\delta_{\alpha\beta}T_\alpha{}^{\text{out}}. \tag{2.25}$$

The first term in $\mathbf{T}$ comes from the "scattering in" term in each of the $\mathbf{T}_{\mathbf{r},\mathbf{r}'}$. Since these terms are all proportional to $\mathbf{I}_{\mathbf{r},\mathbf{r}'}\frac{1}{2}(1-\sigma_{\mathbf{r}}\sigma_{\mathbf{r}'})$, they are completely off diagonal. Furthermore, they are positive semidefinite, i.e.,

$$T_{\alpha\beta}{}^{\text{in}}\geqslant 0,$$
$$T_{\alpha\alpha}{}^{\text{in}}=0. \tag{2.26}$$

Also, the detailed balancing condition requires $\langle\,|\mathbf{T}=0$, or

$$T_\beta{}^{\text{out}}=\sum_{\alpha}T_{\alpha\beta}{}^{\text{in}}, \tag{2.27a}$$

while the statement that we have a correct equilibrium solution $\mathbf{T}|\,\rangle=0$ insures

$$T_\alpha{}^{\text{out}}=\sum_{\beta}T_{\alpha\beta}{}^{\text{in}}. \tag{2.27b}$$

In the model we have just described Eqs. (2.27a) and (2.27b) are essentially identical, because $T_{\alpha\beta}$ is a symmetrical matrix. However, below we shall consider a T matrix which is not symmetrical but nonetheless satisfies both (2.27a) and (2.27b).

D. Eigenstates of T

Let the matrix $\mathbf{T}$ have the eigenstate $|\nu\rangle$ such that

$$\mathbf{T}|\nu\rangle = -s_\nu|\nu\rangle, \tag{2.28}$$

so that s_ν is an eigenvalue of $-\mathbf{T}$. If a state $|t\rangle$ is a superposition of such eigenstates

$$|t\rangle = \sum_\nu a_\nu(t)|\nu\rangle,$$

then the equation of motion

$$(\partial/\partial t)|t\rangle = \mathbf{T}|t\rangle$$

implies that

$$a_\nu(t) = e^{-s_\nu t}a_\nu(0). \tag{2.29}$$

The stability of the system demands that exponentials in Eq. (2.29) never become infinite. This is equivalent to the statement

$$\mathrm{Re}s_\nu \geqslant 0 \tag{2.30a}$$

or

$$\mathrm{Re}\langle\nu|\mathbf{T}|\nu\rangle \leqslant 0. \tag{2.30b}$$

In our specific example, $\mathbf{T}^A$ is a Hermitian operator since $\mathbf{I}_{\mathbf{r},\mathbf{r}'}$ commutes with $\Delta_{\mathbf{rr}'}$. This Hermiticity guarantees that all the eigenvalues s_ν are real. Later on we shall consider a non-Hermitian $\mathbf{T}$. However, Eqs. (2.30) must be equally true whether or not $\mathbf{T}$ is Hermitian.

To see this point, write the eigenstate $|\nu\rangle$ as

$$|\nu\rangle = \sum_\alpha a_\alpha|\alpha\rangle,$$

$$\langle\nu| = \sum_\alpha \langle\alpha|a_\alpha^*. \tag{2.31}$$

If $|\nu\rangle$ is properly normalized

$$\begin{aligned}2\,\mathrm{Re}s_\nu &= -2\,\mathrm{Re}\langle\nu|\mathbf{T}|\nu\rangle \\ &= -2\,\mathrm{Re}\sum_{\alpha\beta} T_{\alpha\beta}a_\alpha^* a_\beta \\ &= -\sum_{\alpha\beta} T_{\alpha\beta}(a_\alpha^* a_\beta + a_\alpha a_\beta^*).\end{aligned}$$

The last line follows because $T_{\alpha\beta}$ is real. Equations (2.25) and (2.27) now give

$$2\,\mathrm{Re}s_\nu = \sum_{\alpha\beta} T_{\alpha\beta}^{\mathrm{in}}|a_\alpha - a_\beta^*|^2. \tag{2.32}$$

The desired result, Eq. (2.30a), follows immediately from the fact that $T_{\alpha\beta}^{\mathrm{in}} \geqslant 0$.

III. ANALYSIS OF HERMITIAN TRANSPORT MODEL

In this section, we analyze the transport properties of the model introduced in Sec. 2.2.

A. Conservation Laws and Currents

The two conserved operators, $\mathbf{M}$ and $\mathbf{H}$, are sums over all sites of a magnetization density operator and an energy density

$$\begin{aligned}m(\mathbf{r}) &= \sigma_\mathbf{r} - \langle\langle\sigma_\mathbf{r}\rangle\rangle, \\ \epsilon(\mathbf{r}) &= -\tfrac{1}{2}u_\mathbf{r}\sigma_\mathbf{r} + \langle\langle\tfrac{1}{2}u_\mathbf{r}\sigma_\mathbf{r}\rangle\rangle.\end{aligned} \tag{3.1}$$

The double bracket indicates an equilibrium average in the Ising model and

$$u(\mathbf{r}) = \sum_{\mathbf{r}'\supset\mathbf{r}} \sigma_{\mathbf{r}'}. \tag{3.2}$$

These two densities obey local conservation laws

$$\begin{aligned}(\partial/\partial t)\langle\,|m(\mathbf{r})|t\rangle + \nabla\cdot\langle\,|\mathbf{j}^m(\mathbf{r})|t\rangle &= 0, \\ (\partial/\partial t)\langle\,|\epsilon(\mathbf{r})|t\rangle + \nabla\cdot\langle\,|\mathbf{j}^\epsilon(\mathbf{r})|t\rangle &= 0.\end{aligned} \tag{3.3}$$

To find the currents we make use of the master equation. For example,

$$\nabla\cdot\langle\,|\mathbf{j}^m(\mathbf{r})|t\rangle = -\langle\,|m(\mathbf{r})\mathbf{T}|t\rangle, \tag{3.4}$$

so that

$$\begin{aligned}\nabla\cdot\langle\,|\mathbf{j}^m(\mathbf{r}) &= \langle\,|[\mathbf{T},\sigma_\mathbf{r}] \\ &= \tfrac{1}{2}w^A\sum_{\mathbf{r}'\mathbf{r}''}\langle\,|[(\mathbf{I}_{\mathbf{r}',\mathbf{r}''}-1),\sigma_\mathbf{r}]\Delta_{\mathbf{r}',\mathbf{r}''}{}^A \\ &= w^A\sum_{\mathbf{r}'}\langle\,|[\mathbf{I}_{\mathbf{r},\mathbf{r}'},\sigma_\mathbf{r}]\Delta_{\mathbf{r}',\mathbf{r}}{}^A \\ &= w^A\sum_{\mathbf{r}'}\langle\,|(\sigma_\mathbf{r} - \mathbf{I}_{\mathbf{r},\mathbf{r}'}\sigma_\mathbf{r}\mathbf{I}_{\mathbf{r},\mathbf{r}'})\Delta_{\mathbf{r},\mathbf{r}'}{}^A.\end{aligned}$$

The line of argument follows because $\langle\,|$ is an eigenstate of $\mathbf{I}_{\mathbf{r},\mathbf{r}'}$ with eigenvalue unity and also because $\mathbf{I}_{\mathbf{r},\mathbf{r}'}$ commutes with $\Delta_{\mathbf{r},\mathbf{r}'}{}^A$. Since $\mathbf{I}_{\mathbf{r},\mathbf{r}'}$ converts $\sigma_\mathbf{r}$ into $\sigma_{\mathbf{r}'}$ we find

$$\nabla\cdot\mathbf{j}^m(r) = w^A\sum_{\mathbf{r}'}\Delta_{\mathbf{r},\mathbf{r}'}{}^A(\sigma_\mathbf{r} - \sigma_{\mathbf{r}'}). \tag{3.5}$$

To eliminate the divergence in Eq. (3.5) multiply by $e^{-i\mathbf{q}\cdot\mathbf{r}}$ and sum over all $\mathbf{r}$.

Then,

$$\sum_\mathbf{r} i\mathbf{q}\cdot\mathbf{j}^m(r)e^{-i\mathbf{q}\cdot\mathbf{r}} = \tfrac{1}{2}\sum_{\mathbf{r},\mathbf{r}'}(e^{-i\mathbf{q}\cdot\mathbf{r}} - e^{-i\mathbf{q}\cdot\mathbf{r}'})w^A\Delta_{\mathbf{r},\mathbf{r}'}{}^A(\sigma_\mathbf{r} - \sigma_{\mathbf{r}'}).$$

For small q, the difference of exponentials reduces to $i\mathbf{q}\cdot(\mathbf{r}'-\mathbf{r})e^{-i\mathbf{q}\cdot\mathbf{r}}$. An inversion of the Fourier transform then gives

$$\mathbf{j}^m(\mathbf{r}) = w^A\sum_{\mathbf{r}'}(\mathbf{r}'-\mathbf{r})\Delta_{\mathbf{r},\mathbf{r}'}{}^A\tfrac{1}{2}(\sigma_\mathbf{r} - \sigma_{\mathbf{r}'}). \tag{3.6}$$

A similar but slightly more complex calculation gives a form for the energy current.

B. Local Equilibrium Solution

Equations (3.3) are exact local conservation laws. However, these conservation laws cannot be used without a solution for the nonequilibrium state of the system. We follow Kawasaki[1] in employing a local equilibrium approximation in which the nonequilibrium state is approximated by

$$|t\rangle = -\sum_{\mathbf{r}} [\delta\beta(\mathbf{r},t)\epsilon(\mathbf{r}) - \delta h(\mathbf{r},t)m(\mathbf{r})]|eq\rangle. \quad (3.7)$$

In Eq. (3.7), $\delta\beta(\mathbf{r},t)$ and $\delta h(\mathbf{r},t)$ have the physica significance of being the local deviation of inverse temperature and magnetic field from their equilibrium values.

We employ (3.7) as an approximate eigenstate of $\mathbf{T}$. The low-lying eigenstates of $\mathbf{T}$ determine the slow relaxation toward equilibrium which is characteristic of transport processes. In particular, the two smallest eigenvalues of $-\mathbf{T}$ for small q are

$$s_1 = D_s q^2,$$
$$s_2 = D_T q^2, \quad (3.8)$$

where D_s and D_T are the spin and thermal diffusivities.

To find approximate eigenvalues of $\mathbf{T}$, we assume that the disturbance has wave vector $\mathbf{q}$ and eigenvalue s. The n, $|t\rangle$ can be written as

$$|t\rangle = p_{eq} e^{-st}[\,|m,\mathbf{q}\rangle\delta h - |\epsilon,\mathbf{q}\rangle\delta\beta].$$

Here we have taken (3.9)

$$\delta h(\mathbf{r},t) = (1/\sqrt{N})e^{i\mathbf{q}\cdot\mathbf{r}-st}\delta h,$$

and written

$$|m,\mathbf{q}\rangle = (1/\sqrt{N})\sum_{\mathbf{r}} e^{i\mathbf{q}\cdot\mathbf{r}} m(\mathbf{r})|\,\rangle,$$

$$|\epsilon,\mathbf{q}\rangle = (1/\sqrt{N})\sum_{\mathbf{r}} e^{i\mathbf{q}\cdot\mathbf{r}}\epsilon(\mathbf{r})|\,\rangle. \quad (3.10)$$

N is the number of sites and p_{eq} is the equilibrium density matrix. The master equation now reads

$$-sp_{eq}[\,|m,\mathbf{q}\rangle\delta h - |\epsilon,\mathbf{q}\rangle\delta\beta] = \mathbf{T}p_{eq}[\,|m,\mathbf{q}\rangle\delta h - |\epsilon,\mathbf{q}\rangle\delta\beta]. \quad (3.11)$$

To form the local spin conservation law, we multiply this equation on the left by $\langle m,\mathbf{q}|$; to form the energy conservation law, we multiply by $\langle\epsilon,\mathbf{q}|$. The results of this multiplication are the pair of equations

$$-s[\chi_{mm}\delta h - \chi_{m\epsilon}\delta\beta] = -q^2[\lambda_{mm}\delta h - \lambda_{\epsilon m}\delta\beta]$$
$$-s[-\chi_{\epsilon m}\delta h + \chi_{\epsilon\epsilon}\delta\beta] = -q^2[-\lambda_{m\epsilon}\delta h + \lambda_{\epsilon\epsilon}\delta\beta], \quad (3.12)$$

where

$$\chi_{mm} = \langle m,\mathbf{q}|p_{eq}|m,\mathbf{q}\rangle,$$
$$-q^2\lambda_{mm} = \langle m,\mathbf{q}|\mathbf{T}p_{eq}|m,\mathbf{q}\rangle. \quad (3.13)$$

Equations (3.12) have a direct physical significance. The left-hand side of the first of these equations gives the time derivative of the magnetization produced by a magnetic field variation δh and an inverse temperature variation $\delta\beta$. The χ's are all thermodynamic derivatives. For example,

$$\chi_{mm} = (1/N)\sum_{\mathbf{r},\mathbf{r}'} e^{i\mathbf{q}\cdot(\mathbf{r}-\mathbf{r}')}\langle\langle(\sigma_{\mathbf{r}} - \langle\langle\sigma\rangle\rangle)(\sigma_{\mathbf{r}'} - \langle\langle\sigma\rangle\rangle)\rangle\rangle$$

is just a dimensionless version of the ordinary spin susceptibility as calculated from the equilibrium behavior of the ordinary Ising model. Similarly, $\chi_{\epsilon\epsilon}$ is the specific heat and $\chi_{m\epsilon}$ the derivative of the magnetization with respect to temperature. All these thermodynamic derivatives diverge near the critical point.

The right-hand side of Eq. (3.12) gives the divergence of the currents in the presence of the gradients of magnetic field and temperature. Hence, the λ's are the transport coefficients of the model as determined by our local equilibrium approximation. For example, $\lambda_{\epsilon\epsilon}$ is the thermal conductivity, λ_{mm} is the spin-diffusion transport coefficient.

It is relatively easy to calculate the transport coefficients in this local equilibrium approximation. For example, the spin-diffusion coefficient is given by

$$-q^2\lambda_{mm} = (w^A/2N)\sum_{\mathbf{r}_1,\mathbf{r}_2}\sum_{\mathbf{r},\mathbf{r}'} e^{-i\mathbf{q}\cdot(\mathbf{r}_1-\mathbf{r}_2)} \times\langle\,|\sigma_{\mathbf{r}_1}\Delta_{\mathbf{r},\mathbf{r}'}{}^A(\mathbf{I}_{\mathbf{r},\mathbf{r}'} - \mathbf{1})\sigma_{\mathbf{r}_2}|eq\rangle.$$

In the sums over $\mathbf{r}_1$ and $\mathbf{r}_2$, the only nonvanishing terms are those with

$$\mathbf{r}_1 = \mathbf{r} \text{ or } \mathbf{r}',$$
$$\mathbf{r}_2 = \mathbf{r} \text{ or } \mathbf{r}'.$$

Since $\sigma_{\mathbf{r}} = -\sigma_{\mathbf{r}'}$, the expression simplifies to

$$-q^2\lambda_{mm} = (w^A/N)\sum_{\mathbf{r},\mathbf{r}'} 2[1 - \cos\mathbf{q}\cdot(\mathbf{r}-\mathbf{r}')]\langle\,|\Delta_{\mathbf{r},\mathbf{r}'}{}^A|eq\rangle.$$

For small q we may expand the cosine and find

$$\lambda_{mm} = (w^A/N)\sum_{\mathbf{r},\mathbf{r}'}(x-x')^2\langle\langle\Delta_{\mathbf{r},\mathbf{r}'}{}^A\rangle\rangle.$$

In three dimensions we find

$$\lambda_{mm} = 8w^A\langle\langle\tfrac{1}{2}(1-\sigma_{\mathbf{r}}\sigma_{\mathbf{r}'})\delta(u_{\mathbf{r}},u_{\mathbf{r}'})\rangle\rangle, \quad (3.14a)$$

where $\mathbf{r}$ and $\mathbf{r}'$ are any pair of next-nearest neighbors. Similar calculations give

$$\lambda_{m\epsilon} = \lambda_{\epsilon m} = 4(-\tfrac{1}{2})w^A \times\langle\langle(u_{\mathbf{r}} + 2v_{\mathbf{r}})\tfrac{1}{2}(1-\sigma_{\mathbf{r}}\sigma_{\mathbf{r}'})\delta(u_{\mathbf{r}},u_{\mathbf{r}'})\rangle\rangle, \quad (3.14b)$$

where $u_{\mathbf{r}}$ is defined in Eq. (2.19) and $v_{\mathbf{r}}$ is the sum of all nearest-neighbor spins of $\mathbf{r}$ which are not also nearest-neighbor spins of $\mathbf{r}'$ and

$$\lambda_{\epsilon\epsilon} = 4(\tfrac{1}{2})^2 w^A\langle\langle[(u_{\mathbf{r}} + 2v_{\mathbf{r}})^2 + (\sigma_{\mathbf{r}-\hat{e}_x} - \sigma_{\mathbf{r}'+\hat{e}_y})^2] \times\tfrac{1}{2}(1-\sigma_{\mathbf{r}}\sigma_{\mathbf{r}'})\delta(u_{\mathbf{r}},u_{\mathbf{r}'})\rangle\rangle, \quad (3.14c)$$

where $\hat{e}_x$ and $\hat{e}_y$ are unit vectors in the x and y directions respectively. Equations (3.14b) and (3.14c) are given for two dimensions since the expressions are slightly more complicated in three dimensions due to the greater number of nonequivalent nearest neighbors.

The significant feature of the results (3.14) is that the transport coefficients, λ, are all finite even at the critical point. This follows because $\Delta_{\mathbf{r},\mathbf{r}'}$ and the σ's are all bounded operators. Therefore, this local equilibrium approximation gives all transport coefficients finite values.

To determine the relaxation rate spectrum, one would have to find the eigenvalues, s_1 and s_2, of the Eqs. (3.12). There is one particularly simple case. If the magnetic field is zero and the temperature is above the critical temperature, there is complete symmetry between spin up and spin down. Then $\chi_{m\epsilon}$ and $\lambda_{m\epsilon}$ both vanish so that Eqs. (3.12) become

$$-s\chi_{mm}\delta h = -q^2\lambda_{mm}\delta h\,,$$
$$-s\chi_{\epsilon\epsilon}\delta\beta = -q^2\lambda_{\epsilon\epsilon}\delta\beta\,. \tag{3.15}$$

Equations (3.15) indicate a diffusive type relaxation with diffusivities

$$D_s = \lambda_{mm}/\chi_{mm}\,,$$
$$D_T = \lambda_{\epsilon\epsilon}/\chi_{\epsilon\epsilon}\,. \tag{3.16}$$

Therefore, this local equilibrium approximation implies that the spin diffusivity D_s goes to zero as the inverse spin susceptibility near the critical point, while the thermal diffusivity, D_T, vanishes as the inverse specific heat.

C. Variational Statements

As Kawasaki[2] pointed out, the approximate evaluations of the transport coefficients as given by Eq. (3.13) are particularly significant, because there exist variational statements which indicate that these results are in some cases upper bounds on the coefficients.

Consider, in particular, the case $T>T_c$ and $h=0$. In that case, we have determined two approximate eigenvalues of $-\mathbf{T}$,

$$s_1^{\text{approx}} = \lambda_{mm}q^2/\chi_{mm} \tag{3.17a}$$

and

$$s_2^{\text{approx}} = \lambda_{\epsilon\epsilon}q^2/\chi_{\epsilon\epsilon} \tag{3.17b}$$

for small q. Here λ_{mm} and $\lambda_{\epsilon\epsilon}$ are given by Eqs. (3.14).

We claim that the exact transport coefficients will be smaller than these approximate coefficients. The exact coefficients, λ_T = thermal conductivity and λ_s = spin-diffusion transport coefficient, may be defined in terms of the smallest eigenvalues of $-\mathbf{T}$ for wave vector q. These eigenvalues are

$$s_1 = \lambda_s q^2/\chi_{mm}\,,$$
$$s_2 = \lambda_T q^2/\chi_{\epsilon\epsilon}\,, \tag{3.18}$$

where the χ's are the spin-susceptibility and the specific heat.

Equations (3.18) describe a slow, diffusive decay of long wavelength excitations. This slow decay occurs because $\mathbf{T}$ has a pair of eigenvalues which go to zero as q goes to zero. In fact, there are many such eigenvalues corresponding to different values of the equilibrium temperature and magnetic field. However, in any situation of small deviations from equilibrium at a given temperature and field, only two eigenvalues will count—all the eigenvalues corresponding to different equilibrium parameters will have zero weight.[17]

The two exact eigenvalues correspond to eigenstates $|1,\mathbf{q}\rangle$ and $|2,\mathbf{q}\rangle$ which obey

$$s_\nu p_{eq}|\nu,q\rangle = -\mathbf{T}p_{eq}|\nu,q\rangle \quad (\nu=1,2). \tag{3.19}$$

The eigenstates are normalized so that

$$\langle\nu,\mathbf{q}|p_{eq}|\nu,\mathbf{q}\rangle = 1\,. \tag{3.20}$$

The requirement that you cannot go too far away from the equilibrium state for the given value of β and h can be stated as the demand that $|\nu,\mathbf{q}\rangle$ be a sum of quasi-local operators times $|\ \rangle$.

Since $-\mathbf{T}$ is Hermitian and $|1,\mathbf{q}\rangle$ is the lowest eigenvalue with wave vector $\mathbf{q}$,

$$s_1 \leqslant \langle x,\mathbf{q}|(-\mathbf{T})p_{eq}|x,\mathbf{q}\rangle/\langle x,\mathbf{q}|p_{eq}|x,\mathbf{q}\rangle\,, \tag{3.21}$$

where $|x,\mathbf{q}\rangle$ is any state with wave vector $\mathbf{q}$. In our calculation, we used as an approximate eigenstate

$$|m,\mathbf{q}\rangle = \frac{1}{(\chi_{mm}N)^{1/2}}\sum_{\mathbf{r}} e^{i\mathbf{q}\cdot\mathbf{r}}m(\mathbf{r})|\ \rangle\,.$$

With this approximate state, the right-hand side of Eq. (3.21) becomes $\lambda_{mm}q^2/\chi_{mm}$. Therefore, it follows that

$$s_1 \leqslant \lambda_{mm}q^2/\chi_{mm}\,.$$

Equation (3.18) then indicates that the exact spin-diffusion transport coefficient obeys

$$\lambda_s \leqslant \lambda_{mm}\,. \tag{3.22}$$

Hence this transport coefficient cannot diverge.

The exact spin-diffusion state has, for $h=0$ and temperatures above T_c, odd parity under an operation in which all spins flip sign. The thermal conduction eigenstate has even parity under this operation. Since $\mathbf{T}$ commutes with this spin-parity operation, there exist separate variational principles for the lowest eigenstates of even and odd parity. The lowest even-parity state of $-\mathbf{T}$ has eigenvalue $D_T^{\text{exact}}q^2$. The variational principle for the lowest exact even-parity state gives the fact that

$$\lambda_T \leqslant \lambda_{\epsilon\epsilon}\,, \tag{3.23}$$

where λ_T is the exact thermal conductivity and $\lambda_{\epsilon\epsilon}$ is our approximate conductivity.

The statement (3.22) holds for all temperatures and magnetic fields. However, Eq. (3.23) requires the spin-parity symmetry which only holds for $h=0$ and tem-

[17] R. Haag, N. M. Hugenholtz, and M. Winnink in an unpublished report have discussed how the apparent temperature dependence of eigenvalues may arise in many a particle system.

peratures above the critical temperature. Notice that both variational statements are only valid because $-\mathbf{T}$ is Hermitian.

IV. ANALYSIS OF A MODEL WHICH INCLUDES SOUND WAVES

The last chapter discussed a model with a Hermitian $\mathbf{T}$. Now, we turn to a model rich enough to include sound waves. In this case, we must have eigenvalues of $-\mathbf{T}$ given by $s_\nu(q)=\pm icq$ where c is the sound velocity and i is $\sqrt{-1}$. Clearly, we must abandon any hope of a Hermitian $\mathbf{T}$ at this point. Hermitian $\mathbf{T}$'s cannot give the imaginary eigenvalues which are the hallmark of sound waves.

A. A Model Which Includes Sound Waves

In this section we describe a model which has sound waves among its nonequilibrium modes and has as its equilibrium solution an Ising model.

Again in this model we consider a three-dimensional cubic lattice with N sites. There is again a spin variable, $\sigma_\mathbf{r}=\pm 1$, for each site. In addition, there is a vector variable, a "momentum," for each site. For a two-dimensional lattice, $\mathbf{g}_\mathbf{r}$ has the values

$$\mathbf{g}_\mathbf{r}=0 \quad \text{if} \quad \sigma_\mathbf{r}=-1\,,$$
$$\mathbf{g}_\mathbf{r}=(1,1),\ (1,-1),\ (-1,1) \text{ or } (-1,-1) \quad \text{if} \quad \sigma_\mathbf{r}=+1\,, \tag{4.1a}$$

while for a three dimensional lattice

$$\mathbf{g}_\mathbf{r}=0 \quad \text{if} \quad \sigma_\mathbf{r}=-1\,,$$
$$\mathbf{g}_\mathbf{r}=(1,1,0),\ (1,-1,0),\ (-1,1,0),\ (-1,-1,0),$$
$$(1,0,1),\ (1,0,-1),\ (-1,0,1),\ (-1,0,-1),$$
$$(0,1,1),\ (0,-1,1),\ (0,1,-1), \text{ or } (0,-1,-1) \quad \text{for} \quad \sigma_\mathbf{r}=+1\,,$$

where $\mathbf{g}_\mathbf{r}$ in the "momentum" of a particle at the site $\mathbf{r}$.

A basis "state" of the system, $|\alpha\rangle$, will then be given by specifying all the spin variables and all the momentum variables.

It will be demanded that in this model the following quantities by conserved quantities:

$$\mathbf{N}=\sum_\mathbf{r} \tfrac{1}{2}(1+\sigma_\mathbf{r}) \tag{4.2a}$$

$$\mathbf{H}=-\sum_{\langle \mathbf{r},\mathbf{r}'\rangle} \sigma_\mathbf{r}\sigma_{\mathbf{r}'}\,, \tag{4.2b}$$

$$\mathbf{G}=\sum_\mathbf{r} \mathbf{g}_\mathbf{r}\,. \tag{4.2c}$$

$\mathbf{N}$ is called the number operator, $\mathbf{G}$ the total "momentum" operator, and $\mathbf{H}$ is the Hamiltonian. In view of the well-known connection between the lattice gas model of a fluid and the Ising model,[18,19] $\frac{1}{2}(1+\sigma_\mathbf{r})$ can be thought of as the density of the lattice gas at the site $\mathbf{r}$ so that $\sum_r \sigma_\mathbf{r}=2\times$ (number of gas atoms)+constant—hence the terminology employed above. If there is no particle present $\sigma_\mathbf{r}=-1$. In that case, we have chosen to say that there is no momentum at the site. The momentum variable is chosen to be discrete in order to simplify the calculations. This discreteness may be a serious defect of the model.

Since $\mathbf{N}$, $\mathbf{H}$, and $\mathbf{G}$ are all conserved an appropriate equilibrium state will be

$$|eq.\rangle=\exp(-\beta\mathbf{H}+\mu\mathbf{N}+\mathbf{v}\cdot\mathbf{G})|\ \rangle/Z(\beta,\mu,\mathbf{v})\,, \tag{4.3}$$

with

$$Z(\beta,\mu,\mathbf{v})=\langle\ |\exp(-\beta\mathbf{H}+\mu\mathbf{N}+\mathbf{v}\cdot\mathbf{G})|\ \rangle.$$

Here β is a dimensionless inverse temperature, μ a dimensionless chemical potential, and $\mathbf{v}$ a dimensionless velocity. At $\mathbf{v}=\mathbf{0}$, in two dimensions the state $\sigma_\mathbf{r}=+1$ is four times as likely as the state $\sigma_r=-1$. Hence the critical value of the chemical potential is defined by

$$4e^{\mu_c}=1 \tag{4.4}$$

for the two dimensional case. At $\mathbf{v}=\mathbf{0}$, Z differs from the Ising-model partition function by only a multiplicative factor. The Ising-model variable h, which is the dimensionless magnetic field, is related to μ by $h=\mu-\mu_c$.

We once again make use of a master equation of the form

$$(d/dt)|t\rangle=\mathbf{T}|t\rangle. \tag{4.5}$$

Here the state $|t\rangle$ is of the form

$$|t\rangle=\sum_\alpha p_\alpha(t)|\alpha\rangle,$$

where the index α defines the values of both spin and momentum at each site.

$\mathbf{T}$ must satisfy the general requirements given in Sec. II of this paper and commute with $\mathbf{N}$, $\mathbf{H}$, and $\mathbf{G}$. A $\mathbf{T}$ meeting all these requirements is

$$\mathbf{T}=\mathbf{T}^A+\mathbf{T}^B+\mathbf{T}^C. \tag{4.6}$$

Here $\mathbf{T}^A$ is the matrix described in the previous section. It describes processes in which particles fall into holes at next-nearest-neighbor sites. The next term, $\mathbf{T}^B$, has a very similar structure to $\mathbf{T}^A$:

$$\mathbf{T}^A=\tfrac{1}{2}w^A\sum_{\mathbf{r},\mathbf{r}'}(\mathbf{I}_{\mathbf{r},\mathbf{r}'}-\mathbf{1})\Delta_{\mathbf{r},\mathbf{r}'}{}^A\,, \tag{4.7a}$$

$$\mathbf{T}^B=w^B\sum_{\mathbf{r},\mathbf{r}'}(\mathbf{I}_{\mathbf{r},\mathbf{r}'}-\mathbf{1})\Delta_{\mathbf{r},\mathbf{r}'}{}^B. \tag{4.7b}$$

Now $\mathbf{I}_{\mathbf{r},\mathbf{r}'}$ is an interchange matrix which interchanges both the spin and the momentum at $\mathbf{r}$ and $\mathbf{r}'$. In

[18] T. D. Lee and C. N. Yang, Phys. Rev. **87**, 410 (1952).
[19] K. Huang, *Statistical Mechanics* (John Wiley & Sons, Inc., N. Y., 1963).

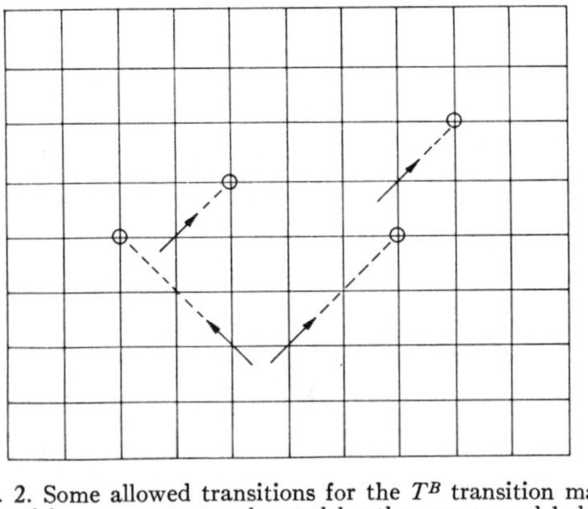

FIG. 2. Some allowed transitions for the T^B transition matrix. The particles momenta are denoted by the arrows and holes by unmarked intersections. The particles interchange with the circled holes lying on the line determined by particles momenta; the circled holes being the first holes for which an exchange is energetically allowed.

particular

$$\mathbf{I}_{\mathbf{r},\mathbf{r}'}\sigma_{\mathbf{r}}=\sigma_{\mathbf{r}'}\mathbf{I}_{\mathbf{r},\mathbf{r}'},$$
$$\mathbf{I}_{\mathbf{r},\mathbf{r}'}\mathbf{g}_{\mathbf{r}}=\mathbf{g}_{\mathbf{r}'}\mathbf{I}_{\mathbf{r},\mathbf{r}'},$$
$$[\mathbf{I}_{\mathbf{r},\mathbf{r}'},\sigma_{\mathbf{r}''}]=0 \quad \text{for} \quad \mathbf{r}''\neq\mathbf{r} \text{ or } \mathbf{r}',$$
$$[\mathbf{I}_{\mathbf{r},\mathbf{r}'},\mathbf{g}_{\mathbf{r}''}]=0 \quad \text{for} \quad \mathbf{r}''\neq\mathbf{r} \text{ or } \mathbf{r}',$$
$$\mathbf{I}_{\mathbf{r},\mathbf{r}'}=\mathbf{I}_{\mathbf{r}',\mathbf{r}},$$
$$(\mathbf{I}_{\mathbf{r},\mathbf{r}'})^2=1. \tag{4.8}$$

In Eq. (4.7a) $\Delta_{\mathbf{r},\mathbf{r}'}{}^A$ is the same diagonal matrix as defined in Eq. (2.17). Therefore, the term $\mathbf{T}^A$ describes once again the diffusion of particles onto neighboring unoccupied sites.

Notice that this diffusion does not depend upon the momentum of the particle. The transition matrix, $\mathbf{T}^B$, introduces the momentum. Physically, we want a particle to move in the direction of its momentum and land in the first unoccupied site which is permissible from the point of view of energy conservation. The projection operator $\Delta_{\mathbf{r},\mathbf{r}'}{}^B$ in Eq. (4.7b) is designed to produce a proper operator for describing the motion of a particle from $\mathbf{r}$ to $\mathbf{r}'$. We write

$$\Delta_{\mathbf{r},\mathbf{r}'}{}^B=1 \quad \text{if}$$

(a) $\sigma_{\mathbf{r}}=+1 \quad \sigma_{\mathbf{r}'}=-1 \quad u_{\mathbf{r}}=u_{\mathbf{r}'}$,

and (b) there is a positive integer n such that

$$\mathbf{r}+n\mathbf{g}_{\mathbf{r}}=\mathbf{r}',$$

and (c) there is no site $\mathbf{r}'$ which satisfies conditions a and b for a smaller positive value of n.

Otherwise, $\Delta_{\mathbf{r},\mathbf{r}'}{}^B=0$. In symbols

$$\Delta_{\mathbf{r},\mathbf{r}'}{}^B=\tfrac{1}{2}(1+\sigma_{\mathbf{r}})\tfrac{1}{2}(1-\sigma_{\mathbf{r}'})\delta(u_{\mathbf{r}},u_{\mathbf{r}'}) \times\sum_{n=1}^{\infty}\delta(\mathbf{r}+n\mathbf{g}_{\mathbf{r}},\mathbf{r}')\theta_{\mathbf{r},\mathbf{r}'}. \tag{4.9}$$

Here $\theta_{\mathbf{r},\mathbf{r}'}$ is the projection operator which enforces condition (c). It has the structure

$$\theta_{\mathbf{r},\mathbf{r}'}=\prod_{n''=1}^{n-1}(1-\delta(u_{\mathbf{r}},u_{\mathbf{r}+n''\mathbf{g}_{\mathbf{r}}}))\tfrac{1}{2}(1-\sigma_{\mathbf{r}+n''\mathbf{g}_{\mathbf{r}}}), \tag{4.10}$$

where $\mathbf{g}$ is a vector of length $\sqrt{2}$ pointing from $\mathbf{r}$ to $\mathbf{r}'$.

Some allowed and forbidden exchanges are shown in Fig. 2.

The whole point of this rather complex construction of $\Delta_{\mathbf{r},\mathbf{r}'}{}^B$ is to make sure that every particle has one and only one place to go. In symbols

$$\sum_{\mathbf{r}'}\Delta_{\mathbf{r},\mathbf{r}'}{}^B=\tfrac{1}{2}(1+\sigma_{\mathbf{r}}). \tag{4.11}$$

Equation (4.11) underlies the condition which allows the complicated transition matrix $\mathbf{T}^B$ to have an Ising-model equilibrium solution.

$\mathbf{T}^A$ was discussed in Sec. II of this paper so we know that it commutes with $\mathbf{N}$, $\mathbf{H}$, and $\mathbf{G}$, conserves probability and has the proper Ising-model equilibrium solution.

$\mathbf{T}^B$ is a more complicated object since $\mathbf{I}_{\mathbf{r},\mathbf{r}'}$ does not commute with $\Delta_{\mathbf{r},\mathbf{r}'}{}^B$. In fact, if we define

$$\mathbf{I}_{\mathbf{r},\mathbf{r}'}\Delta_{\mathbf{r},\mathbf{r}'}{}^B\mathbf{I}_{\mathbf{r},\mathbf{r}'}=\tilde{\Delta}_{\mathbf{r}',\mathbf{r}}{}^B, \tag{4.12}$$

we discover that

$$\tilde{\Delta}_{\mathbf{r},\mathbf{r}'}{}^B=\tfrac{1}{2}(1+\sigma_{\mathbf{r}})\tfrac{1}{2}(1-\sigma_{\mathbf{r}'})\delta(u_{\mathbf{r}},u_{\mathbf{r}'}) \times\sum_{n=1}^{\infty}\delta(\mathbf{r}-n\mathbf{g}_{\mathbf{r}},\mathbf{r}')\theta_{i,\mathbf{r}'}. \tag{4.13}$$

Notice the minus sign in the term involving $\mathbf{g}_{\mathbf{r}}$ in Eq. (4.13). The same term is Eq. (4.9) has a plus sign. Hence $\tilde{\Delta}_{\mathbf{r},\mathbf{r}'}{}^B$ is different from $\Delta_{\mathbf{r},\mathbf{r}'}{}^B$. This difference means that $\mathbf{T}^B$ is not a Hermitian matrix—it is real but not symmetrical. In fact, not only is $\mathbf{T}^B$ not Hermitian but $\mathbf{T}^B$ is also not normal. That is, $\mathbf{T}^B$ does not commute with its adjoint which implies that $\mathbf{T}^B$ does not possess a complete set of orthonormal eigenvectors.

Physically, the difference between Δ and $\tilde{\Delta}$ arises because these functions answer different questions. When $\Delta_{\mathbf{r},\mathbf{r}'}{}^B=1$, a particle at $\mathbf{r}$ can go to $\mathbf{r}'$. Thus $\mathbf{g}_{\mathbf{r}}$ points from $\mathbf{r}$ to $\mathbf{r}'$. When $\Delta_{\mathbf{r},\mathbf{r}'}{}^B=1$, a particle at $\mathbf{r}$ *can have come from* $\mathbf{r}'$. Then $\mathbf{g}_{\mathbf{r}}$ must point from $\mathbf{r}'$ to $\mathbf{r}$.

Notice that every particle can come from one and only one place. This means that

$$\sum_{\mathbf{r}'}\tilde{\Delta}_{\mathbf{r},\mathbf{r}'}{}^B=\tfrac{1}{2}(1+\sigma_{\mathbf{r}}). \tag{4.14}$$

Notice that similarity between Eqs. (4.11) and (4.14). As a result of this symmetry, $\mathbf{T}^B$ has one important left-right symmetry property. According to Eqs. (4.7b) and (4.12)

$$\mathbf{T}^B = w^B \sum_{\mathbf{r},\mathbf{r}'} (\mathbf{I}_{\mathbf{r},\mathbf{r}'} - \mathbf{1}) \Delta_{\mathbf{r},\mathbf{r}'}{}^B$$

$$= w^B \sum_{\mathbf{r},\mathbf{r}'} [\tilde{\Delta}_{\mathbf{r}',\mathbf{r}}{}^B \mathbf{I}_{\mathbf{r},\mathbf{r}'} - \Delta_{\mathbf{r},\mathbf{r}'}{}^B]. \qquad (4.15)$$

But, according to Eqs. (4.14) and (4.11) the last Δ^B in Eq. (4.15) can be replaced by $\tilde{\Delta}^B$. Hence,

$$\mathbf{T}^B = w^B \sum_{\mathbf{r},\mathbf{r}'} \tilde{\Delta}_{\mathbf{r},\mathbf{r}'}{}^B (\mathbf{I}_{\mathbf{r},\mathbf{r}'} - \mathbf{1}). \qquad (4.16)$$

Equation (4.16) has the same structure as Eq. (4.7b)—only now the $\mathbf{I}$ appears on the right.

This structual similarity is quite important indeed. The form (4.7b) for $\mathbf{T}^B$ clearly indicates that

$$\langle \, | \mathbf{T}^B = 0,$$

so that $\mathbf{T}^B$ satisfies the detailed balancing condition. On the other hand, the form (4.16) is necessary in order to see that

$$\mathbf{T}^B | \, \rangle = 0.$$

This latter condition ensures that $\mathbf{T}^B$ has the correct Ising-model equilibrium solution.

Notice that $\mathbf{T}^A$ and $\mathbf{T}^B$ each conserve the total number of particles with a given value of $\mathbf{g}_\mathbf{r}$. These terms alone give too many "momentum" conservation laws. In order to mix different values of $\mathbf{g}_\mathbf{r}$ we introduce a scattering term

$$\mathbf{T}^C = \tfrac{1}{2} w^C \sum_{\mathbf{r},\mathbf{r}'} (\mathbf{J}_{\mathbf{r},\mathbf{r}'} - \mathbf{1} j) \Delta_{\mathbf{r},\mathbf{r}'}{}^C. \qquad (4.17)$$

The term (4.17) is designed to allow isotropic scattering of nearest-neighbor particles if the total momentum of these particles is zero as depicted in Fig. 3. Hence

$$\Delta_{\mathbf{r},\mathbf{r}'}{}^C = \tfrac{1}{2}(1+\sigma_\mathbf{r})\tfrac{1}{2}(1-\sigma_{\mathbf{r}'})\delta(\mathbf{g}_\mathbf{r} + \mathbf{g}_{\mathbf{r}'}, \mathbf{0}), \qquad (4.18)$$

when $\mathbf{r}$ and $\mathbf{r}'$ are nearest neighbors. Otherwise $\Delta_{\mathbf{r},\mathbf{r}'}{}^C$ is zero. The operator $\mathbf{J}_{\mathbf{r},\mathbf{r}'}$ changes the values of $\mathbf{g}_\mathbf{r}$ and $\mathbf{g}_{\mathbf{r}'}$. In matrix notation

$$\langle \bar{\mathbf{g}}_\mathbf{r}, \bar{\mathbf{g}}_{\mathbf{r}'} | \mathbf{J}_{\mathbf{r},\mathbf{r}'} | \mathbf{g}_\mathbf{r} \mathbf{g}_{\mathbf{r}'} \rangle = \delta(\bar{\mathbf{g}}_\mathbf{r} + \bar{\mathbf{g}}_{\mathbf{r}'}, \mathbf{0}) \delta(\mathbf{g}_\mathbf{r} + \mathbf{g}_{\mathbf{r}'}, \mathbf{0}). \qquad (4.18')$$

To ensure

$$\mathbf{T}^C | \, \rangle = \langle \, | \mathbf{T}^C = 0,$$

it is only necessary to choose

$$j = 4 \quad \text{in two dimensions},$$
$$j = 12 \quad \text{in three dimensions}. \qquad (4.19)$$

The model of a liquid-gas phase transition we have just described has one very serious drawback: It has cubic (or square) symmetry rather than a complete rotational invariance. On the other hand, it has some very important virtues. Energy, momentum, and particle number are exactly conserved. These quantities can flow from one point to a neighboring point but they are never increased or decreased. Since the transport

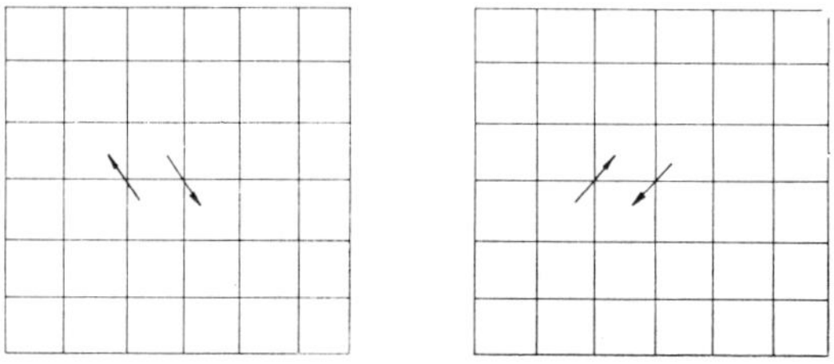

(a) (b)

FIG. 3. A process represented by the transition matrix T^C. Two particles, whose momenta, represented by the arrows add up to zero, scatter from the configuration shown in Figs. (3a) to that shown in (3b) where the final momenta again add up to zero. These particles can scatter with equal probability into any momentum values which add up to zero total momentum.

modes are crucially connected with these conservation laws, it is probably very important that they be included exactly rather than in an averaged sense.[1,4] The model used here has the further virtues of keeping the density matrix always diagonal—as it should be in a classical system and introducing the oscillatory modes via the memory inherent in a momentum variable—as actually occurs in a real fluid.

C. Local Equilibrium Solution

In this section, we describe the transport properties of the model defined in the last section. We will show that standard transport behavior of sound waves and heat waves will emerge from the local equilibrium approximation. Hence this model has some very satisfactory qualitative features. However, one should recognize that because $\mathbf{T}$ is non-Hermitian, we have not been able to prove that the local equilibrium approximation gives the correct temperature dependence for the transport coefficients very close to the critical point. In fact, we are hopeful that the local equilibrium approximation is very inaccurate near the critical point. Real fluid phase transitions show infinities in transport coefficients at the critical point; the local equilibrium approximation does not. Preliminary calculations indicate that this model gives an infinite thermal conductivity at the critical point when one employs a more accurate calculational scheme that the local equilibrium approximation.

The local equilibrium approximation uses the approximate eigenstate

$$|t\rangle = e^{-st} p_{eq} [\, |n,\mathbf{q}\rangle \delta\mu - |\epsilon,\mathbf{q}\rangle \delta\beta + |\mathbf{g},\mathbf{q}\rangle \cdot \delta\mathbf{v}], \qquad (4.20)$$

where

$$|n,\mathbf{q}\rangle = (1/\sqrt{N}) \sum_\mathbf{r} e^{i\mathbf{q}\cdot\mathbf{r}} m(\mathbf{r}) | \, \rangle,$$

$$|\epsilon,\mathbf{q}\rangle = (1/\sqrt{N}) \sum_\mathbf{r} e^{i\mathbf{q}\cdot\mathbf{r}} \epsilon(\mathbf{r}) | \, \rangle,$$

$$|\mathbf{g},\mathbf{q}\rangle = (1/\sqrt{N}) \sum_\mathbf{r} e^{i\mathbf{q}\cdot\mathbf{r}} \mathbf{g}_\mathbf{r} | \, \rangle. \qquad (4.21)$$

In Eq. (4.20), $\delta\mu$, $\delta\beta$, and $\delta\mathbf{v}$ represent variations in the local chemical potential, inverse temperature, and velocity. The equilibrium density matrix is taken at $\mathbf{v}=\mathbf{0}$.

In order to eliminate the simplest case first, let $\delta\mathbf{v}$ point in the x direction and $\mathbf{q}$ point in the y direction. This transverse momentum flow does not involve a $\delta\mu$ or $\delta\beta$. Then the approximate master equation becomes

$$-s\rho_{eq}|\mathbf{g},\mathbf{q}\rangle\cdot\delta\mathbf{v}=\rho_{eq}\mathbf{T}|\mathbf{g},\mathbf{q}\rangle\cdot\delta\mathbf{v}.$$

We multiply this equation on the left by $\langle g_x,\mathbf{q}|$ and find

$$-s\langle g_x,\mathbf{q}|\rho_{eq}|g_x,\mathbf{q}\rangle\delta v=\langle g_x,\mathbf{q}|\rho_{eq}\mathbf{T}|g_x,\mathbf{q}\rangle\delta v. \quad (4.22)$$

The left-hand side of Eq. (4.22) is simply the time derivative of the average momentum which appears in response to the velocity disturbance. Since g's at different sites are uncorrelated in equilibrium, the left-hand side is easily evaluable. In two dimensions

$$-s\tfrac{1}{2}(1+\langle\langle\sigma_{\mathbf{r}}\rangle\rangle)\delta v=-s\langle g_x,\mathbf{q}|\rho_{eq}|g_x,\mathbf{q}\rangle\delta v. \quad (4.23)$$

In three dimensions, there is an extra factor of $\frac{2}{3}$ on the left-hand side of Eq. (4.23). The right-hand side of Eq. (4.22) has the meaning of $\nabla^2 v$ times the viscosity η. Hence the approximate eigenvalue of $-\mathbf{T}$ is for three dimensions

$$s=3q^2\eta/(1+\langle\langle\sigma_{\mathbf{r}}\rangle\rangle), \quad (4.24)$$

while

$$\eta=-(1/q^2)\langle g_x,\mathbf{q}|\rho_{eq}\mathbf{T}|g_x,\mathbf{q}\rangle. \quad (4.25)$$

Because $\mathbf{T}^B$ drives particles in the direction of their momentum, it makes no contribution to Eq. (4.25). The contributions from $\mathbf{T}^A$ and $\mathbf{T}^C$ are

$$\eta^A=-\frac{1}{q^2}\sum_{\mathbf{r},\mathbf{r}'}\frac{w^A}{2N}\sum_{\mathbf{r}_1\mathbf{r}_2}e^{-iq(y_1-y_2)}$$
$$\times\langle\,|(\mathbf{g}_{\mathbf{r}_1})_x(\mathbf{I}_{\mathbf{r},\mathbf{r}'}-1)\Delta_{\mathbf{r},\mathbf{r}'}{}^A(\mathbf{g}_{\mathbf{r}_2})_x|eq\rangle$$
$$=-\frac{1}{q^2}\sum_{\mathbf{r},\mathbf{r}'}\frac{w^A}{4N}[1-\cos q(y-y')]$$
$$\times\langle\,|[(\mathbf{g}_{\mathbf{r}})_x-(\mathbf{g}_{\mathbf{r}'})_x]\Delta_{\mathbf{r},\mathbf{r}'}{}^A[(\mathbf{g}_{\mathbf{r}})_x-(\mathbf{g}_{\mathbf{r}'})_x]|eq\rangle.$$

For small q in two dimensions this gives

$$\eta^A=\sum_{\mathbf{r},\mathbf{r}'}\langle\langle\Delta_{\mathbf{r},\mathbf{r}'}{}^A\rangle\rangle\frac{w^A}{4N}$$
$$=w^A\langle\langle\tfrac{1}{2}(1-\sigma_{\mathbf{r}}\sigma_{\mathbf{r}'})\delta(u_{\mathbf{r}},u_{\mathbf{r}'})\rangle\rangle. \quad (4.26a)$$

The last form of writing holds when $\mathbf{r}$ and $\mathbf{r}'$ are any pair of next-nearest neighbors. In writing Eq. (4.26a), we have taken the lattice constant to be unity.

A roughly similar calculation gives

$$\eta^C=w^C\langle\langle\tfrac{1}{2}(1+\sigma_{\mathbf{r}})\tfrac{1}{2}(1+\sigma_{\mathbf{r}'})\rangle\rangle \quad (4.26b)$$

in two dimensions. Here $\mathbf{r}$ and $\mathbf{r}'$ are any pair of nearest neighboring sites. Hence the local equilibrium approximation correctly describes the diffusion of the transverse component of the momentum and predicts a finite shear viscosity.

Now, return to the more complex structure in Eq. (4.20) and take $\mathbf{v}$ and $\mathbf{q}$ to be parallel to the x direction. The master equation reads

$$-s|t\rangle=\mathbf{T}|t\rangle.$$

If we take matrix elements of the master equation by multiplying on the left by $\langle n,\mathbf{q}|$, $\langle\epsilon,\mathbf{q}|$, and $\langle g_x,\mathbf{q}|$, we obtain the following set of equations for small q:

$$-s\left[\left(\frac{\partial n}{\partial\mu}\right)_\beta\delta\mu-\left(\frac{\partial n}{\partial\beta}\right)_\mu\delta\beta\right]=-q^2[\lambda_{nn}\delta\mu-\lambda_{n\epsilon}\delta\beta]$$
$$-i\mathbf{q}\cdot\rho\delta\mathbf{v}, \quad (4.27\text{a})$$

$$-s\left[-\left(\frac{\partial n}{\partial\beta}\right)_\mu\delta\mu+\left(\frac{\partial\epsilon}{\partial\beta}\right)_\mu\delta\beta\right]=-q^2[-\lambda_{\epsilon n}\delta\mu+\lambda_{\epsilon\epsilon}\delta\beta]$$
$$-i\mathbf{q}\cdot\rho^\epsilon\delta\mathbf{v}. \quad (4.27\text{b})$$

Finally, in three dimensions, the time derivative of the momentum density is

$$-s\left[\frac{2}{3}\frac{(1+\langle\langle\sigma_{\mathbf{r}}\rangle\rangle)}{2}\right]\delta\mathbf{v}=-i\mathbf{q}[\rho\delta\mu+\rho^\epsilon\delta\beta]$$
$$-q^2[\zeta+\tfrac{4}{3}\eta]\delta\mathbf{v}. \quad (4.27\text{c})$$

In two dimensions, the factor $\frac{2}{3}$ on the left-hand side of Eq. (4.27c) is replaced by unity.

These equations look quite complex, but they have a simple physical interpretation. The left-hand sides are the time derivatives of the number density, energy density, and momentum density evaluated for local equilibrium. The thermodynamic derivatives are evaluated in the lattice gas. The only difference from the usual case is that our μ is the usual μ/kT. The derivative

$$\left(\frac{\partial n}{\partial\mu}\right)_\beta=\sum_{\mathbf{r},\mathbf{r}'}\frac{1}{N}\langle\langle[\sigma_{\mathbf{r}}-\langle\langle\sigma_{\mathbf{r}}\rangle\rangle][\sigma_{\mathbf{r}'}-\langle\langle\sigma_{\mathbf{r}'}\rangle\rangle]\rangle\rangle \quad (4.28)$$

describes the infinite compressibility of the usual lattice gas at the critical point. Also $(\partial\epsilon/\partial\beta)_\mu$ is proportional to C_v. The terms involving λ's are the transport coefficients. For example,

$$\lambda_{\epsilon\epsilon}=-(1/q^2)\langle\epsilon,\mathbf{q}|\rho_{eq}\mathbf{T}|\epsilon,\mathbf{q}\rangle \quad (4.29)$$

is essentially the thermal conductivity. Because the translation invariance has been lost there is a particle diffusion coefficient λ_{nn} and a thermal diffusion coefficient $\lambda_{n\epsilon}$ as in a two-component gas. The approximation also includes an Onsager relation $\lambda_{n\epsilon}=\lambda_{\epsilon n}$.

The standard longitudinal viscosity which appears in Eq. (4.27c) is

$$\zeta+\tfrac{4}{3}\eta=-(1/q^2)\langle g_x,\mathbf{q}|\rho_{eq}\mathbf{T}|g_x,\mathbf{q}\rangle, \quad (4.30)$$

where q points in the x direction. Finally Eq. (4.27c) contains an analog of the gradient of the pressure as the first square bracket on its right-hand side.

Sound waves arise from Eq. (4.27) because of the drift currents $\rho\delta\mathbf{v}$ and $\rho^\epsilon\delta\mathbf{v}$ in Eqs. (4.27a) and (4.27b) and because of the "pressure gradient"

$$-i\mathbf{q}[\rho\delta\mu+\rho^\epsilon\delta\beta]$$

in Eq. (4.27c). These terms appear as

$$-i\mathbf{q}\rho=\langle \mathbf{g},\mathbf{q}|\mathbf{T}p_{eq}|n,\mathbf{q}\rangle =\langle n,\mathbf{q}|\mathbf{T}p_{eq}|\mathbf{g},\mathbf{q}\rangle, \quad (4.31a)$$

$$-i\mathbf{q}\rho^\epsilon=-\langle \mathbf{g},\mathbf{q}|\mathbf{T}p_{eq}|\epsilon,\mathbf{q}\rangle =-\langle \epsilon,\mathbf{q}|\mathbf{T}p_{eq}|\mathbf{g},\mathbf{q}\rangle. \quad (4.31b)$$

Notice that ρ appears in two places: The drift current is proportional to $\rho\mathbf{v}$ and the "pressure gradient" contains $\rho\delta\mu$. These two ρ's are identical because there is a kind of "Onsager relation" between the off-diagonal matrix elements of $\mathbf{T}$. In the usual transport theory these relations appear as the statement that the particle current is $n\mathbf{v}$ while the pressure obeys $(\partial p/\partial\mu)_T=n$.

We can evaluate ρ most conveniently by employing the second of Eqs. (4.31). The only part of $\mathbf{T}$ which contributes to ρ is the non-Hermitian term $\mathbf{T}^B$. We find when q points in the x direction

$$-iq\rho=\langle n,\mathbf{q}|\mathbf{T}^Bp_{eq}|g_x,\mathbf{q}\rangle$$
$$=\sum_{\mathbf{r},\mathbf{r}'}\frac{w^B}{N}\sum_{\mathbf{r},\mathbf{r}'}e^{-i\mathbf{q}\cdot(\mathbf{r}_1-\mathbf{r}_2)}\langle\,|\sigma_{\mathbf{r}_1}(\mathbf{I}_{\mathbf{r},\mathbf{r}'}-\mathbf{1})\Delta_{\mathbf{r},\mathbf{r}'}{}^B(\mathbf{g}_{r2})\,|eq\rangle$$
$$=-\frac{2w^B}{N}\sum_{,\mathbf{r}_2\mathbf{r}_1\mathbf{r}_2}[e^{-i\mathbf{q}\cdot(\mathbf{r}-\mathbf{r}_2)}-e^{-i\mathbf{q}\cdot(\mathbf{r}'-\mathbf{r}_2)}]\times\langle\,|\Delta_{\mathbf{r},\mathbf{r}'}{}^B(\mathbf{g}_{\mathbf{r}_2})_x|eq\rangle.$$

Since $\Delta_{\mathbf{r},\mathbf{r}'}{}^B$ only depends upon the value of $\mathbf{g}_{\mathbf{r}_1}$ at $\mathbf{r}_2=\mathbf{r}$, only the term with $\mathbf{r}_2=\mathbf{r}$ contributes to the sum. Thus, for small q,

$$-iq\rho=-\frac{2w^B}{N}\sum_{\mathbf{r},\mathbf{r}'}i\mathbf{q}\cdot(\mathbf{r}'-\mathbf{r})\langle\,|\Delta_{\mathbf{r},\mathbf{r}'}{}^B(\mathbf{g}_\mathbf{r})_x|eq\rangle$$

or

$$\rho=\frac{2w^B}{N}\sum_{\mathbf{r},\mathbf{r}'}(x'-x)\langle\,|\Delta_{\mathbf{r},\mathbf{r}'}{}^B(\mathbf{g}_\mathbf{r})_x|eq\rangle. \quad (4.32)$$

Notice that $\Delta_{\mathbf{r},\mathbf{r}'}{}^B$ sends a particle from $\mathbf{r}$ to $\mathbf{r}'$ in the direction of $\mathbf{g}_\mathbf{r}$. Then, inside the sum

$$(x'-x)=|x'-x|(\mathbf{g}_\mathbf{r})_x.$$

Consequently,

$$\rho=\frac{2w^B}{N}\sum_{\mathbf{r},\mathbf{r}'}|x'-x|\langle\,|\Delta_{\mathbf{r},\mathbf{r}'}{}^B[(\mathbf{g}_\mathbf{r})_x]^2|eq\rangle. \quad (4.33)$$

Equation (4.33) indicates that ρ is positive and finite.

Direct calculations of matrix elements show that all quantities in Eqs. (4.27) are finite everywhere save the thermodynamic derivatives $(\partial n/\partial\beta)_\mu$, $(\partial n/\partial\mu)_\beta$, $(\partial\epsilon/\partial\beta)_\mu$ which may diverge at the critical point. As an example, a calculation of λ_{nn} is given in an appendix.

Without further ado, we may conclude that all transport coefficients are finite in this local equilibrium approximation. However, the approximation is not variational because $\mathbf{T}$ is not Hermitian. Hence we really know nothing about the exact transport coefficients.

The approximate eigenvalues of s in the local equilibrium approximation are determined by the condition that the 3×3 matrix of Eqs. (4.27) have zero determinant. For small q this condition gives: first, a diffusion mode with

$$s=\frac{\lambda_{nn}(\rho^\epsilon)^2+2\lambda_{n\epsilon}\rho\rho^\epsilon+\lambda_{\epsilon\epsilon}\rho^2}{(\partial n/\partial\mu)_\beta(\rho^\epsilon)^2+2(\partial n/\partial\beta)_\mu\rho^\epsilon\rho+(\partial\epsilon/\partial\beta)_\mu\rho^2}q^2. \quad (4.34)$$

Near the critical point, the term in $(\partial n/\partial\mu)_\beta$ dominates the denominator. Hence according to this model the thermal diffusivity goes to zero as the inverse compressibility.

The other modes are sound waves with

$$s^2=-c^2q^2 \quad (4.35)$$

and sound velocity squared

$$c^2=\frac{\rho}{\langle\langle[(g_r)_x]^2\rangle\rangle}[\rho(\partial\mu/\partial n)_s+\rho^\epsilon(\partial\beta/\partial n)_s], \quad (4.36)$$

where fixed s means

$$\rho dn=\rho^\epsilon d\epsilon$$

or

$$\rho[(\partial n/\partial\mu)_\beta d\mu-(\partial n/\partial\beta)_\mu d\beta] =\rho^\epsilon[-(\partial n/\partial\beta)_\mu d\mu+(\partial\epsilon/\partial\beta)_\mu d\beta]. \quad (4.37)$$

Thus c^2 goes to zero as $T\rightarrow T_c$.

The modes are then qualitatively correct, but further work is necessary to establish whether or not the transport coefficients are actually finite.

APPENDIX: CALCULATION OF λ_{nn} IN TWO DIMENSIONS

The expression for λ_{nn} is

$$\lambda_{nn}=-(1/q^2)\langle n,\mathbf{q}|p_{eq}\mathbf{T}|n,\mathbf{q}\rangle. \quad (A1)$$

Since the process described by the transition matrix $\mathbf{T}^c$ involves no interchange of particles, $\mathbf{T}^c$ does not contribute to λ_{nn}. The contribution to λ_{nn} from the $\mathbf{T}^A$ term is λ_{mm} which is given in Eq. (3.14a) and is obviously finite. Consequently, we concentrate on the contribution due to $\mathbf{T}^B$

$$\lambda_{nn}{}^B=-(1/q^2)\langle n,\mathbf{q}|\mathbf{T}^Bp_{eq}|n,\mathbf{q}\rangle$$
$$=-\frac{1}{q^2}\frac{w^B}{N}\sum_{\mathbf{r},\mathbf{r}',\mathbf{r}_1,\mathbf{r}_2}e^{-i\mathbf{q}\cdot(\mathbf{r}_1-\mathbf{r}_2)}\times\langle\,|\sigma_{\mathbf{r}_1}(\mathbf{I}_{\mathbf{r},\mathbf{r}'}-\mathbf{1})\Delta_{\mathbf{r},\mathbf{r}'}{}^B\sigma_{\mathbf{r}_2}|eq\rangle. \quad (A2)$$

If the equilibrium state has $\mathbf{v}=\mathbf{0}$, we can easily average over the directions of all the momenta. We represent the result of this averaging by the replacement

$$\Delta_{\mathbf{r},\mathbf{r}'}{}^{B} \rightarrow \tfrac{1}{2}(1+\sigma_{\mathbf{r}})\tfrac{1}{2}(1-\sigma_{\mathbf{r}'})\bar{\Delta}_{\mathbf{r},\mathbf{r}'}{}^{B}/[2d(d-1)], \quad \text{(A3)}$$

where d is the dimensionality of the lattice. The factor $2d(d-1)$ gives the number of different $\mathbf{g}_{\mathbf{r}}$'s which can contribute. The factor $\bar{\Delta}_{\mathbf{r},\mathbf{r}'}{}^{B}$ is

$$\bar{\Delta}_{\mathbf{r},\mathbf{r}'}{}^{B} = \delta(u_{\mathbf{r}},u_{\mathbf{r}'})\tfrac{1}{2}(1-\sigma_{\mathbf{r}}\sigma_{\mathbf{r}'})\theta_{\mathbf{r},\mathbf{r}'},$$

when $\mathbf{r}-\mathbf{r}'$ is parallel to a nearest-neighbor direction and is zero otherwise. Therefore, the matrix element in Eq. (A2) is

$$\mathbf{M}_{\mathbf{r},\mathbf{r}'} = \langle\, |\sigma_{\mathbf{r}_1}(\mathbf{I}_{\mathbf{r},\mathbf{r}'}-1)\tfrac{1}{2}(1+\sigma_{\mathbf{r}})\tfrac{1}{2}(1+\sigma_{\mathbf{r}'})\bar{\Delta}_{\mathbf{r},\mathbf{r}'}{}^{B}\sigma_{\mathbf{r}_2}|\, eq\rangle \times 1/[2d(d-1)]. \quad \text{(A4)}$$

Because of the sum over $\mathbf{r}$ and $\mathbf{r}'$ we can replace $\mathbf{M}_{\mathbf{r},\mathbf{r}'}$ inside the sum by

$$\mathbf{M}_{\mathbf{r},\mathbf{r}'} \rightarrow \tfrac{1}{2}(\mathbf{M}_{\mathbf{r},\mathbf{r}'}+\mathbf{M}_{\mathbf{r}',\mathbf{r}}) = \langle\, |\sigma_{\mathbf{r}_1}(\mathbf{I}_{\mathbf{r},\mathbf{r}'}-1)\bar{\Delta}_{\mathbf{r},\mathbf{r}'}{}^{B}\sigma_{\mathbf{r}_2}|\, eq\rangle/[4d(d-1)], \quad \text{(A5)}$$

but $\mathbf{I}_{\mathbf{r},\mathbf{r}'}$ commutes with $\bar{\Delta}_{\mathbf{r},\mathbf{r}'}{}^{B}$. Therefore we can insert a factor $\frac{1}{2}(1-\mathbf{I}_{\mathbf{r},\mathbf{r}'})$ just to the right of $\bar{\Delta}_{\mathbf{r},\mathbf{r}'}{}^{B}$ without changing the matrix element at all, since this factor is just a projection operator. When this replacement is made and $\mathbf{M}_{\mathbf{r},\mathbf{r}'}$ is reinserted in Eq. (A2), this equation becomes

$$\lambda_{nn}{}^{B} = \frac{1}{q^2}\frac{w^B}{4d(d-1)N}\sum_{\mathbf{r}_1,\mathbf{r}_2}\sum_{\mathbf{r},\mathbf{r}'} e^{-i\mathbf{q}\cdot(\mathbf{r}_1-\mathbf{r}_2)} \times\langle\, |\sigma_{\mathbf{r}_1}\tfrac{1}{2}(1-\mathbf{I}_{\mathbf{r},\mathbf{r}'})\bar{\Delta}_{\mathbf{r},\mathbf{r}'}{}^{B}\tfrac{1}{2}(1-\mathbf{I}_{\mathbf{r},\mathbf{r}'})\sigma_{\mathbf{r}_2}|\, eq\rangle. \quad \text{(A6)}$$

Now, the only terms which can possibly contribute to the sum over $\mathbf{r}_1$ and $\mathbf{r}_2$ are those with $\mathbf{r}_1=\mathbf{r}$ and $\mathbf{r}'$ and also $\mathbf{r}_2=\mathbf{r}$ and $\mathbf{r}'$. Otherwise the projection operators give zero. With this restriction, Eq. (A6) becomes

$$\lambda_{nn}{}^{B} = \frac{w^B}{4d(d-1)N}\sum_{\mathbf{r},\mathbf{r}'}\frac{2-2\cos\mathbf{q}\cdot(\mathbf{r}-\mathbf{r}')}{q^2}\langle\langle\bar{\Delta}_{\mathbf{r},\mathbf{r}'}{}^{B}\rangle\rangle. \quad \text{(A7)}$$

For small q this reduces to

$$\lambda_{nn}{}^{B} = \frac{w^B}{4d^2(d-1)N}\sum_{\mathbf{r},\mathbf{r}'}(\mathbf{r}-\mathbf{r}')^2\langle\langle\bar{\Delta}_{\mathbf{r},\mathbf{r}'}{}^{B}\rangle\rangle. \quad \text{(A8)}$$

Since $\mathbf{r}-\mathbf{r}'$ must be parallel to a next-nearest-neighbor axis

$$\lambda_{nn}{}^{B} = \frac{w^B}{d}\sum_{n=1}^{\infty} n^2\langle\langle\bar{\Delta}_{\mathbf{r},\mathbf{r}+\mathbf{e}n}{}^{B}\rangle\rangle, \quad \text{(A9)}$$

where $\mathbf{e}$ is any vector which takes you from a particle to its next-nearest neighbor. More explicitly

$$\lambda_{nn}{}^{B} = \frac{w^B}{d}\sum_{n=1}^{\infty} n^2 \times\langle\langle\delta(u_{\mathbf{r}},u_{\mathbf{r}+n\mathbf{e}})\tfrac{1}{2}(1-\sigma_{\mathbf{r}}\sigma_{\mathbf{r}+n\mathbf{e}})\theta_{\mathbf{r}',\mathbf{r}+n\mathbf{e}}\rangle\rangle. \quad \text{(A10)}$$

Since θ requires that there be no holes which lie between $\mathbf{r}$ and $\mathbf{r}+n\mathbf{e}$, the sum over n gains exponentially small contributions from large n. Hence the local equilibrium approximation transport coefficient is finite.

Reprinted from THE PHYSICAL REVIEW, Vol. 166, No. 1, 89–101, 5 February 1968
Printed in U. S. A.

Transport Coefficients near the Liquid-Gas Critical Point*

LEO P. KADANOFF AND JACK SWIFT
Department of Physics, University of Illinois, Urbana, Illinois
(Received 23 August 1967; revised manuscript received 23 October 1967)

A perturbation theory for the determination of transport coefficients near the critical point is presented. This perturbation theory is based upon processes in which one transport mode decays into several low-wave-number modes. Scaling-law concepts are used to calculate the order of magnitude of the matrix elements and frequency denominators which appear in this theory. This permits the estimation of the order of magnitude of the transport coefficients near the critical point. In particular, this approach indicates that the thermal conductivity should diverge roughly as $(T-T_c)^{-2/3}$ on the critical isochore and coexistence curve, while the viscosity η should be either weakly divergent or strongly cusped at the critical point. On the other hand, the bulk viscosity ζ should diverge roughly as $(T-T_c)^{-2}$ for low frequencies, and as $(T-T_c)^{-2/3}$ for higher frequencies on the critical isochore near the critical point. Specific predictions are made for these quantities in terms of critical indices, and the connection between these relations and the scaling of frequencies is discussed.

I. INTRODUCTION

IN several recent papers,[1–6] correlation function or equivalent response function techniques have been applied to the problem of predicting and explaining the apparent divergences in transport coefficients near the critical point.[7] The present paper is devoted to an extension of these methods and their application to the liquid-gas phase transition. The methods of analysis are purely classical, so that the work is directly relevant to classical fluids. However, most of this work deals with long-wavelength limits in which the quantum corrections are quite small. For this reason, it is hoped that the analysis here can be appropriate for either classical or quantum fluids.

Our work is very closely related to Kawasaki's[1] formulation of Fixman's[3–5] theoretical approach. There are two main differences between our methods and Kawasaki's. First, we estimate correlation functions with the aid of the "scaling-law" idea,[8–10] which has proved very successful in describing the correlations in the two-dimensional Ising model[11,12] and moderately successful in describing the three-dimensional Ising model[13] and real three-dimensional phase transitions.[14] Kawasaki and Fixman estimated correlation functions with the aid of ideas drawn from the Ornstein-Zernike[15] theory of critical correlations, and their estimates are probably less accurate than estimates drawn from the scaling-law ideas.

The second difference between this work and Kawasaki's is a matter of formalism. Kawasaki evaluates correlation functions involving currents of conserved quantities by expanding these currents in the densities of the conserved quantities. At first sight, this expansion appears to be no better justified than the analogous expansion of the energy density to second order in the order parameter[16,17] which predicts an incorrect, $(T-T_c)^{-1/2}$, divergence in the specific heat. In this paper, we construct a formal perturbation theory for transport coefficients, which turns out to be equivalent to the expansion procedure of Kawasaki. In this way, we produce a partial justification for the basic ideas used by him and by Fixman.

The next section of this paper is devoted to the development of formal techniques; the following section applies these techniques to the estimation of transport coefficients; the final section lists the conclusions of this analysis.

II. FORMULATION

A. Liouville Equation

Any nonequilibrium problem in classical statistical mechanics can be stated in terms of the Liouville equation. We employ a state notation to describe this equation. The state vector $|t\rangle$ describes the statistical

* This research was supported in part by the National Science Foundation under Grant Nos. NSF GP 4937 and 776S.

[1] K. Kawasaki, Phys. Rev. **150**, 291 (1966).
[2] K. Kawasaki, J. Phys. Chem. Solids **28**, 1277 (1967).
[3] M. Fixman, J. Chem. Phys. **36**, 310 (1961); W. Botch and M. Fixman, *ibid.* **36**, 3100 (1962).
[4] M. Fixman, in *Advances in Chemical Physics*, edited by I. Prigogine (Interscience Publishers, Inc., New York, 1963), Vol. VI.
[5] M. Fixman, J. Phys. **47**, 2808 (1967).
[6] J. M. Deutch and R. Zwanzig, J. Chem. Phys. **46**, 1612 (1967).
[7] See the review by J. V. Sengers, in *Proceedings of the Conference on Phenomena in the Neighborhood of Critical Points, Washington, D. C., 1965*, edited by M. S. Green and J. V. Sengers (Natl. Bur. Std., Washington, D. C., Misc. Publ., 1965), No. 273.
[8] L. P. Kadanoff, Physics **2**, 263 (1966).
[9] B. Widom, J. Chem. Phys. **43**, 3892 (1965); **43**, 3898 (1965).
[10] M. E. Fisher, in Proceedings of the University of Kentucky Centennial Conference on Phase Transitions, March 1965 (unpublished).
[11] L. P. Kadanoff, Nuovo Cimento **44B**, 276 (1966).
[12] R. Hecht, thesis, University of Illinois, 1967 (unpublished); Phys. Rev. **158**, 557 (1967).
[13] M. E. Fisher and R. J. Burford, Phys. Rev. **156**, 583 (1967). There is a 3% discrepancy between scaling-law predictions and critical indices apparently uncovered by their work.
[14] L. P. Kadanoff *et al.*, Rev. Mod. Phys. **39**, 395 (1967).
[15] L. S. Ornstein and F. Zernike, Proc. Acad. Sci. Amsterdam **17**, 793 (1914); **19**, 1312 (1917); Physik Z. **27**, 761 (1926).
[16] R. Brout, *Phase Transitions* (W. A. Benjamin, Inc., New York, 1965).
[17] M. Fixman, J. Chem. Phys. **36**, 1957 (1962); W. Botch and M. Fixman, *ibid.* **42**, 196 (1965).

state of the system at time t. This state is defined so that its components

$$\langle \mathbf{p}_1,\mathbf{p}_2\cdots\mathbf{p}_N,\mathbf{r}_1,\mathbf{r}_2\cdots\mathbf{r}_N|t\rangle=\langle p,r,N|t\rangle \equiv f_N(p,r,t) \quad (2.1)$$

are the probabilities for finding N particles in the system with one particle havingco ordinate $\mathbf{r}_1$ and momentum $\mathbf{p}_1$, another with coordinate $\mathbf{r}_2$ and momentum $\mathbf{p}_2$, etc. The time development of the system is described by the Liouville equation

$$\left(\frac{\partial}{\partial t}+L\right)|t\rangle=0, \quad (2.2)$$

with L having the matrix element

$$\langle p',r',N'|L|p,r,N\rangle=\sum_{\alpha=1}^{N}\left[\frac{\partial\mathcal{H}}{\partial\mathbf{p}_\alpha}\frac{\partial}{\partial\mathbf{r}_\alpha}-\frac{\partial\mathcal{H}}{\partial\mathbf{r}_\alpha}\frac{\partial}{\partial\mathbf{p}_\alpha}\right]\times\langle p',r',N'|p,r,N\rangle \quad (2.3a)$$

and

$$\langle p',r',N'|p,r,N\rangle=\delta_{N,N'}\prod_{\alpha=1}\delta(\mathbf{p}_\alpha-\mathbf{p}_\alpha')\delta(\mathbf{r}_\alpha-\mathbf{r}_\alpha'). \quad (2.3b)$$

There are two basic states which will prove to be particularly useful in our analysis. The first of these is the equilibrium state $|\rangle$ defined by

$$\langle p,r,N|\rangle=\exp\{-\beta[\mathcal{H}(p,r)-\mu N]\}/[h^{3N}N!Z(\beta,\mu)], \quad (2.4)$$

which gives the grand canonical ensemble equilibrium distribution with chemical potential μ and inverse temperature β. The second is the "summational state"

$$\langle|=\sum_N\int\left(\prod_{\alpha=1}^{N}d\mathbf{p}_\alpha d\mathbf{r}_\alpha\right)\langle p,r,N|. \quad (2.5)$$

Notice that $|\rangle$ is not the conjugate vector to $\langle|$. However, they do have a conjugate significance relative to L since

$$L|\rangle=0 \quad\text{and}\quad \langle|L=0. \quad (2.6)$$

The importance of $\langle|$ lies in its usefulness for determining averages. For example,

$$\langle|t\rangle=1$$

expresses the proper normalization of the state $|t\rangle$. The average of the physical quantity X in the state $|t\rangle$ is given by

$$\langle X\rangle_t=\langle|X_{\mathrm{op}}|t\rangle, \quad (2.7a)$$

while the average of X in the grand canonical ensemble is

$$\langle X\rangle=\langle|X_{\mathrm{op}}|\rangle. \quad (2.7b)$$

In Eq. (2.7), X_{op} is a diagonal matrix in the p's and r's. (In fact, the only off-diagonal matrix in our presentation is L.)

B. Operators and States

For our purposes, the important operators in the theory are the densities and currents of conserved quantities. We write[18] the number density operator as $n_{\mathrm{op}}(\mathbf{r})$, the momentum density as $\mathbf{g}_{\mathrm{op}}(\mathbf{r})$, and the energy density as $\epsilon_{\mathrm{op}}(\mathbf{r})$. For example, the matrix element of the momentum density is

$$\langle p',r',N'|\mathbf{g}_{\mathrm{op}}(\mathbf{r})|p,r,N\rangle=\sum_{\alpha=1}^{N}\mathbf{p}_\alpha\,\delta(\mathbf{r}-\mathbf{r}_\alpha)\langle p',r',N'|p,r,N\rangle.$$

The current corresponding to these densities are $\mathbf{j}(\mathbf{r})$ (number current), $\mathbf{j}^\epsilon(\mathbf{r})$ (energy current), and $\tau_{ij}(\mathbf{r})$ (stress tensor). (In cases in which it does not appear to cause confusion, we shall drop the subscripts "op" on operators.) These currents are defined by

$$\begin{aligned}-\nabla\cdot\mathbf{j}_{\mathrm{op}}(\mathbf{r})&=[L,n_{\mathrm{op}}(\mathbf{r})],\\ -\nabla\cdot\mathbf{j}^\epsilon{}_{\mathrm{op}}(\mathbf{r})&=[L,\epsilon_{\mathrm{op}}(\mathbf{r})],\\ -\nabla\cdot\tau(\mathbf{r})&=[L,\mathbf{g}(\mathbf{r})].\end{aligned} \quad (2.8)$$

Before we begin our analysis, it is useful to recall several important properties of these currents. The first is that the diagonal element of the stress tensor $\tau_{xx}(\mathbf{r})$ is, for a system at rest, the pressure operator $p(\mathbf{r})$. The second is that products of the currents have very simple momentum averages. If $\langle\;\rangle_{\mathrm{q}}$ represents a momentum average in the grand canonical ensemble,

$$\beta\langle g_i(\mathbf{r})j_k(\mathbf{r}')\rangle_{\mathrm{q}}=\delta(\mathbf{r}-\mathbf{r}')n_{\mathrm{op}}(\mathbf{r})\delta_{ik}, \quad (2.9a)$$

$$\beta\langle g_i(\mathbf{r})j_k{}^\epsilon(\mathbf{r}')\rangle_{\mathrm{q}}=\delta(\mathbf{r}-\mathbf{r}')[\epsilon_{\mathrm{op}}(\mathbf{r})+p_{\mathrm{op}}(\mathbf{r})]\delta_{ik}, \quad (2.9b)$$

$$\beta\langle g_i(\mathbf{r})\tau_{kl}(\mathbf{r}')\rangle_{\mathrm{q}}=0. \quad (2.9c)$$

Because Eqs. (2.9) are so important to us, we define a linear combination of conserved quantities with a current which has a particularly simple correlation with $\mathbf{g}(\mathbf{r})$; namely,

$$s_{\mathrm{op}}(\mathbf{r})=\frac{1}{T}\left[\epsilon_{\mathrm{op}}(\mathbf{r})-\frac{\langle\epsilon+p\rangle}{\langle n\rangle}n_{\mathrm{op}}(\mathbf{r})\right] \quad (2.10)$$

as the symbol indicates, $s_{\mathrm{op}}(\mathbf{r})$ plays a role of an entropy density. In particular, the entropy current

$$\mathbf{j}_{\mathrm{op}}{}^s(\mathbf{r})=\frac{1}{T}\left[\mathbf{j}_{\mathrm{op}}{}^\epsilon(\mathbf{r})-\frac{\langle\epsilon\rangle+\langle p\rangle}{\langle n\rangle}\mathbf{j}_{\mathrm{op}}(\mathbf{r})\right] \quad (2.11)$$

obeys

$$\beta\langle g_i(\mathbf{r})j_k{}^s(\mathbf{r}')\rangle_{\mathrm{q}}=\frac{1}{T}\delta(\mathbf{r}-\mathbf{r}')\left[\epsilon_{\mathrm{op}}(\mathbf{r})+p_{\mathrm{op}}(\mathbf{r})-\frac{\langle\epsilon\rangle+\langle p\rangle}{\langle n\rangle}n_{\mathrm{op}}(\mathbf{r})\right]\delta_{ik}. \quad (2.12)$$

The full thermodynamic average of the right-hand side of Eq. (2.12) vanishes.

[18] In this and several other regards we follow the notation of L. P. Kadanoff and P. C. Martin, Ann. Phys. (N. Y.) **24**, 419 (1963).

In the discussion of hydrodynamic phenomena, the most important states are local equilibrium states which describe situations in which the equilibrium parameters (temperature, chemical potential, and velocity) are varying slowly from point to point. To form these states, we begin with linear combinations of the densities of conserved operators: $a_i(\mathbf{r})$ with $i=1, 2 \cdots 5$. These densities are used in the form of their Fourier transforms

$$a_i(\mathbf{q})=\int d^3r\, e^{-i\mathbf{q}\cdot\mathbf{r}}a_i(\mathbf{r}).$$

The linear combinations are set up so that the "local equilibrium states"

$$|i,\mathbf{q}\rangle=a_i(\mathbf{q})|\rangle, \quad \langle i,\mathbf{q}|=\langle|a_i(-\mathbf{q}), \tag{2.13}$$

are properly orthonormal

$$\langle i,\mathbf{q}|j,\mathbf{q}'\rangle=\delta_{i,j}(2\pi)^3\delta(\mathbf{q}-\mathbf{q}'). \tag{2.14}$$

To be specific: We choose the states by writing

$$a_1(\mathbf{q})=\frac{s_{\text{op}}(\mathbf{q})}{[k_B\rho C_p(\mathbf{q})]^{1/2}}. \tag{2.15a}$$

The state orthogonal to $|1,\mathbf{q}\rangle$ and properly normalized is $|2,\mathbf{q}\rangle=a_2(\mathbf{q})|\rangle$ with

$$a_2(\mathbf{q})=\frac{(\rho\beta)^{1/2}}{\langle n\rangle}c(\mathbf{q})n_{\text{op}}(\mathbf{q}) +\left(\frac{1}{k_B\rho}\left[\frac{1}{C_V(\mathbf{q})}-\frac{1}{C_p(\mathbf{q})}\right]\right)^{1/2}s_{\text{op}}(\mathbf{q}). \tag{2.15b}$$

Finally

$$a_3(\mathbf{q})=g_x(\mathbf{q})(\beta/\rho)^{1/2}, \tag{2.15c}$$

$$a_4(\mathbf{q})=g_y(\mathbf{q})(\beta/\rho)^{1/2}, \tag{2.15d}$$

$$a_5(\mathbf{q})=g_z(\mathbf{q})(\beta/\rho)^{1/2}. \tag{2.15e}$$

The q-dependent "thermodynamic quantities" in Eqs. (2.15a) and (2.15b) reduce to the standard thermodynamic values at $q=0$. These are defined to have the correct value so that $a_1(\mathbf{q})$ and $a_2(\mathbf{q})$ are orthogonal and properly normalized for all q.

In particular,

$$C_p(\mathbf{q})=\langle|s_{\text{op}}(-\mathbf{q})s_{\text{op}}(\mathbf{q})|\rangle/k_B\rho \tag{2.16a}$$

reduces to the specific heat at constant pressure at $q=0$, while the quantities $C_V(\mathbf{q})$ and $c(\mathbf{q})$ defined by

$$C_V(\mathbf{q})[c(\mathbf{q})]^2=\frac{T\langle n\rangle}{m\rho}\frac{\langle|s_{\text{op}}(-\mathbf{q})s_{\text{op}}(\mathbf{q})|\rangle}{\langle|n_{\text{op}}(-\mathbf{q})n_{\text{op}}(\mathbf{q})|\rangle} \tag{2.16b}$$

and

$$\frac{\beta}{\langle n\rangle}c(\mathbf{q})=\frac{\left(k_B\beta\left\{\dfrac{[C_p(\mathbf{q})]^2}{C_V(\mathbf{q})}-C_p(\mathbf{q})\right\}\right)^{1/2}}{\langle|n_{\text{op}}(-\mathbf{q})s_{\text{op}}(\mathbf{q})|\rangle} \tag{2.16c}$$

reduce to the specific heat at constant volume and the adiabatic sound velocity at $q=0$.

In our further analysis, it will be important to make a contrast between $a_1(\mathbf{q})$ and $a_2(\mathbf{q})$. Notice that $a_1(\mathbf{q})$ as defined by Eq. (2.15a) has a normalization factor $1/\sqrt{C_p}$ while $a_2(\mathbf{q})$ has normalization factors $c(\mathbf{q})$ and $(1/C_V-1/C_p)^{1/2}$. Near the critical point, C_p diverges very strongly. For example, on the critical isochore it diverges as $(T-T_c)^{-\gamma}$, with $\gamma\approx\frac{4}{3}$. On the other hand, $1/c^2$ and C_V diverge much more weakly near the critical point, with a roughly logarithmic dependence upon $T-T_c$. Therefore, we must conclude that $a_1(\mathbf{q})$, which is proportional to the entropy density that appears in $a_1(\mathbf{q})$, has much stronger fluctuations near the critical point than the combination of operators in $a_2(\mathbf{q})$.

The point is further borne out by the thermodynamic role of the two operators within a correlation function. If we have some operator $X_{\text{op}}(\mathbf{q})$ which does not depend explicitly upon the thermodynamic parameters, then

$$\lim_{q\to 0}\langle|s(-\mathbf{q})X_{\text{op}}(\mathbf{q})|\rangle=\beta^{-1}\left[\frac{\partial}{\partial T}\langle X\rangle\right]_p.$$

Since the condition of fixed p does not preclude any of the wild variations which occur near the critical point, these derivatives can be very large indeed. On the other hand,

$$\lim_{q\to 0}\langle|a_2(-\mathbf{q})X(\mathbf{q})|\rangle\frac{\langle n\rangle}{\sqrt{(\rho\beta)}}c(\mathbf{q})=\left.\frac{\partial\langle X\rangle}{\partial\mu}\right|_{S/N}.$$

The condition of fixed S/N precisely holds constant the strongest variations which occur near the critical point. Therefore, the thermodynamic derivative at fixed S/N tends to be, at worst, weakly divergent.

The idea that there is a characteristic size to the operators a_2 and a_1 can be carried even further, to say that the addition of an extra factor of $s_{\text{op}}(\mathbf{q}=\mathbf{0})$ to an expectation value, which is already undergoing critical fluctuations, multiplies this correlation function by a characteristic factor. To compute this characteristic factor notice that on the coexistence curve

$$\lim_{q\to 0}\langle s_{\text{op}}(-\mathbf{q})n_{\text{op}}(\mathbf{q})\rangle\sim(-\epsilon)^{-\gamma};\quad \epsilon=(T-T_c)/T_c,$$

while the difference between $\langle n\rangle$ and its critical value is given by

$$|\langle n\rangle-\langle n\rangle_c|\sim(-\epsilon)^{\beta}.$$

Consequently, in this case the extra factor of s_{op} has had the effect of multiplying the thermodynamic quantity by a factor $\epsilon^{-(\beta+\gamma)}$, which diverges roughly as $(T-T_c)^{-5/3}$. In general, we expect that, for small q,

$$s_{\text{op}}(\mathbf{q})\sim\epsilon^{-(\beta+\gamma)}. \tag{2.17a}$$

A similar analysis may be applied to $a_2(\mathbf{q})$. Except for the prefactor of $c(\mathbf{q})\sim 1/[C_V(\mathbf{q})]^{1/2}$, a_2 acts as a temperature derivative inside a correlation function. For this reason, at small q, a_2 is of the order of

$$a_2(\mathbf{q})\sim c\epsilon^{-1}, \tag{2.17b}$$

when it appears as a factor in an already fluctuating correlation function. On the basis of this logic, one would, for example, estimate that

$$\langle|a_2(\mathbf{q})s(\mathbf{q}')s(-\mathbf{q}-\mathbf{q}')|\rangle\sim c\frac{\partial}{\partial T}\langle|s(\mathbf{q}')s(-\mathbf{q}')|\rangle\sim c\frac{\partial}{\partial T}\epsilon^{-\gamma}$$
$$\sim c\epsilon^{-\gamma-1}$$

on the critical isochore if q and q' are very small.

This hypothesis that there is a natural size to quantities near the critical point can be extended to give an estimate of the q dependence of different correlation functions. According to the scaling hypothesis, all lengths near the critical point should be referred to a characteristic range of correlations, ξ. On the critical isochore $\xi\sim\epsilon^{-\nu}$, with $\nu\approx\frac{2}{3}$. Then, near the critical point, correlation functions like $\langle|s_{\text{op}}(\mathbf{q})s_{\text{op}}(-\mathbf{q})|\rangle$ and $\langle|n_{\text{op}}(\mathbf{q})n_{\text{op}}(-\mathbf{q})|\rangle$ should depend upon q only in the combination $q\xi$.

There is one more scaling-law result which we shall need in our arguments—a result which is much more questionable than the ones we have stated so far. The results stated above seem to give at least roughly correct estimates of divergences in the critical region; the hypothesis we are about to state has never been checked except in the two-dimensional Ising model. This extra idea is that there are essentially only two different fluctuating quantities near the critical point, e.g., $a_1(\mathbf{r})$ and $a_2(\mathbf{r})$, and that all other critically fluctuating quantities can be considered to be linear combinations of these two. As a consequence of this assumption, in its leading or most singular behavior $n_{\text{op}}(\mathbf{q})$ is proportional to $s_{\text{op}}(\mathbf{q})$. Hence, the ratio of correlation functions $\langle|n_{\text{op}}(\mathbf{q})n_{\text{op}}(-\mathbf{q})|\rangle/\langle|s_{\text{op}}(\mathbf{q})s_{\text{op}}(-\mathbf{q})|\rangle$ should be essentially a ratio of identical quantities so that it is very weakly dependent on either ϵ or $q\xi$ in the critical region. The product $C_V(\mathbf{q})[c(\mathbf{q})]^2$ defined by Eq. (2.16b) will then be almost a constant independent of $q\xi$ or ϵ in the critical region. This result will have important implications for sound-wave damping. However, we should point out once more that this conclusion is much less reliable than the other conclusions we have drawn from the scaling hypothesis, because this result requires the very strongest and most dubious form of that hypothesis.

C. Transport Processes

The transport modes of the system appear as slowly relaxing solutions to the Liouville equation. An eigenstate of L with eigenvalue s will relax in time as e^{-st}. Consequently, the slowly relaxing modes have a small real part to the eigenvalue s. Since L is translationally invariant, its eigenvalues may be classified according to the value of the wave number q. The transport modes appear as states whose relaxation time goes to infinity as $q\rightarrow 0$.

To find the eigenvalues of L we start with the equation for the νth right eigenstate of L corresponding to the eigenvalue s_ν:

$$s_\nu|\nu,\mathbf{q}\rangle_R=L|\nu,\mathbf{q}\rangle_R. \tag{2.18}$$

If this is a transport eigenstate $|\nu,\mathbf{q}\rangle_R$ is mostly composed of the states $|i,\mathbf{q}\rangle$. (Notice that $\nu=1,2\cdots\infty$ labels eigenvalues of L while $i=1,2\cdots 5$ labels the local equilibrium states.) Then we apply $\langle i,\mathbf{q}|$ to Eq. (2.18) and find[19]

$$s_\nu\langle i,\mathbf{q}|\nu,\mathbf{q}\rangle_R=\langle i,\mathbf{q}|L|\nu,\mathbf{q}\rangle_R=\sum_j\langle i,\mathbf{q}|L|j,\mathbf{q}\rangle$$
$$\times\langle j,\mathbf{q}|\nu,q\rangle_R+\langle i,\mathbf{q}|LP|\nu,\mathbf{q}\rangle_R, \tag{2.19}$$

where P is the projection operator which rejects the states $|i,\mathbf{q}\rangle$

$$P=1-\sum_{j=1}^{5}|j,\mathbf{q}\rangle\langle j,\mathbf{q}|. \tag{2.20}$$

According to Eq. (2.18),

$$s_\nu P|\nu,\mathbf{q}\rangle_R=PL|\nu,\mathbf{q}\rangle_R$$
$$=PLP|\nu,\mathbf{q}\rangle_R+\sum_j PL|j,\mathbf{q}\rangle\langle j,\mathbf{q}|\nu,\mathbf{q}\rangle_R,$$

so that

$$P|\nu,\mathbf{q}\rangle_R=\frac{1}{s_\nu-PLP}\sum_j PL|j,\mathbf{q}\rangle\langle j,\mathbf{q}|\nu,\mathbf{q}\rangle_R.$$

Thus, Eq. (2.19) can be written as

$$\sum_j[s_\nu\delta_{ij}-L_{ij}(\mathbf{q})-U_{ij}(\mathbf{q},s_\nu)]\langle j,\mathbf{q}|\nu,\mathbf{q}\rangle_R=0, \tag{2.21}$$

with

$$L_{ij}(\mathbf{q})=\langle i,\mathbf{q}|L|j,\mathbf{q}\rangle, \tag{2.22}$$

$$U_{ij}(\mathbf{q},s)=\langle i,\mathbf{q}|LP\frac{1}{s-PLP}PL|j,\mathbf{q}\rangle. \tag{2.23}$$

The eigenvalues of L are, of course, determined by the condition that the matrix $s\delta_{ij}-L_{ij}-U_{ij}$ have zero determinant.

The standard transport theory emerges from Eqs. (2.21), (2.22), and (2.23), if we identify L_{ij} with the set of thermodynamic derivatives which appear in the nondissipative part of the theory and U_{ij} with the matrix of transport coefficients appearing in the dissipative part of the theory. Once these identifications are made, we can see that Eq. (2.19) represents the usual linearized hydrodynamic equations.

To evaluate L_{ij} notice that

$$L_{ij}=\langle i,\mathbf{q}|L|j,\mathbf{q}\rangle$$
$$=\langle|a_i(-\mathbf{q})La_j(\mathbf{q})|\rangle$$
$$=\langle|[a_i(-\mathbf{q}),L]a_j(\mathbf{q})|\rangle$$
$$=i\mathbf{q}\cdot\langle|\mathbf{j}_i(-\mathbf{q})a_j(\mathbf{q})|\rangle, \tag{2.24}$$

[19] In Eq. (2.19) and below, we have taken the volume of the system to be unity.

TABLE I. The matrix $(s\delta_{ij}-L_{ij}-U_{ij})$.

	Heat flow	Sound waves		Viscous flow	
Heat flow	$s-\frac{\lambda q^2}{\rho C_p}$	$-\frac{\lambda q^2}{\rho}\left(\frac{1}{C_V C_p}-\frac{1}{C_p^2}\right)^{1/2}$	0	0	0
Sound waves	$-\frac{\lambda q^2}{\rho}\left(\frac{1}{C_V C_p}-\frac{1}{C_p^2}\right)^{1/2}$	$s-\frac{\lambda q^2}{\rho}\left(\frac{1}{C_V}-\frac{1}{C_p}\right)$	$-icq_x$	0	0
	0	$-icq_x$	$s-\frac{(\zeta+\frac{4}{3}\eta)q^2}{\rho}$	0	0
Viscous flow	0	0	0	$s-\frac{\eta q^2}{\rho}$	0
	0	0	0	0	$s-\frac{\eta q^2}{\rho}$

where $\mathbf{j}_i$ is the current corresponding to the ith conserved quantity. If L_{ij} is to be nonzero, a_i must be a vector with a component parallel to $\mathbf{q}$, and a_j must be a scalar or vice versa. If $\mathbf{q}$ points in the x direction, the only possible nonvanishing elements of L_{ij} couple the scalars $(i=1,2)$ with $g_x(j=3)$ and vice versa. Furthermore, we picked the form of $a_1(q)\sim s_{\rm op}(q)$ with the idea in mind of making

$$\langle|\,\mathbf{j}_1(-\mathbf{q})g_x(\mathbf{q})\,|\rangle\sim\langle|\,\mathbf{j}^s(-\mathbf{q})g_x(\mathbf{q})\,|\rangle=0.$$

[See Eq. (2.12).] Therefore, L_{13} and L_{31} vanish.

The only remaining terms in L_{ij} are L_{32} and L_{23}. These terms are

$$\begin{aligned}L_{32}(\mathbf{q})&=L_{23}(\mathbf{q})\\&=i\mathbf{q}\cdot\langle|\,\mathbf{j}_2(-q)a_3(\mathbf{q})\,|\rangle\\&=iq_xc(\mathbf{q})\frac{\beta}{\langle n\rangle}\langle|\,j_x(-\mathbf{q})g_x(\mathbf{q})\,|\rangle.\end{aligned}\quad(2.25)$$

In writing the last line of Eq. (2.25), we have made use of expressions (2.15b) and (2.15c) for a_2 and a_3, and also of the fact that the entropy current $j^s(-\mathbf{q})$ can generate no contribution to the average (2.25). Equation (2.9a) now enables us to evaluate L_{23} and L_{32} and find

$$L_{23}=L_{32}=iq_xc(\mathbf{q}).\quad(2.26)$$

If the U_{ij} in Eq. (2.23) were set equal to zero, then this hydrodynamic equation would contain two nonzero "relaxation times"

$$s_\pm=\pm icq_x,\quad(2.27)$$

and the other three eigenvalues would be zero. These pure imaginary "relaxation times" reflect the oscillatory behavior of undamped sound waves. The vanishing of the remaining relaxation times indicates that all the diffusive processes must arise from the neglected term U_{ij}.

In fact, all the transport coefficients arise from U_{ij}. To see this, we rewrite Eq. (2.23) as

$$U_{ik}(\mathbf{q},s)=-\langle|\,\mathbf{q}\cdot\mathbf{j}_i(-\mathbf{q})P\frac{1}{s-PLP}P\mathbf{q}\cdot\mathbf{j}_k(\mathbf{q})\,|\rangle.\quad(2.28)$$

This result has essentially the same structure as the Kubo[20] formulas for transport coefficients. It is a correlation function involving a product of currents. In place of the usual time integral, Eq. (2.28) contains a denominator with differences of relaxation times.

There are 25 terms in U_{ik}; but there are only three independent transport coefficients λ, η, ζ.

Our next task must be the elimination of redundant terms. Notice that the current for particle flow $\mathbf{j}$ is proportional to the momentum density. The projection operators eliminate this current. Hence, of the terms U_{11}, U_{21}, U_{12}, and U_{22}, there is only one independent combination: that arising from the energy current $\mathbf{j}^\epsilon$. We have

$$q^2\lambda(\mathbf{q},s)=-\langle|\,s_{\rm op}(-\mathbf{q})LP\frac{1}{PLP-s}PLs_{\rm op}(\mathbf{q})\,|\rangle/k_B,\quad(2.29)$$

with the U's being given by

$$U_{11}(\mathbf{q},s)=q^2\lambda(\mathbf{q},s)/\rho C_p(\mathbf{q}),\quad(2.30)$$

$$U_{22}(\mathbf{q},s)=q^2\lambda(\mathbf{q},s)\left[\frac{1}{\rho C_V(\mathbf{q})}-\frac{1}{\rho C_p(\mathbf{q})}\right],\quad(2.31)$$

$$U_{12}(\mathbf{q},s)=U_{21}(\mathbf{q},s)=\frac{q^2\lambda(\mathbf{q},s)}{\rho}\times\left[\frac{1}{C_p(\mathbf{q})}\left(\frac{1}{C_V(\mathbf{q})}-\frac{1}{C_p(\mathbf{q})}\right)\right]^{1/2}.\quad(2.32)$$

[20] See the article by R. Kubo in *Lectures in Theoretical Physics* (Interscience Publishers, Inc., New York, 1959), Vol. I, Chap. 4.

The terms in U involving both vector currents and tensor currents like U_{23} are higher order in q and are probably negligible in describing the transport.

In summary, we list the significant terms in the matrix $(\delta_{ij}s-L_{ij}-U_{ij})$ in Table I. This matrix is almost diagonal. The last two rows and columns describe the diffusion of the transverse component of the momentum. The coupling term between the first row and column and the second is small and may be neglected. Then, the first row describes the heat-flow process, and the second and third rows describe the sound wave. When the sound-wave damping is small compared to its rate of oscillation, the sound waves obey a dispersion relation

$$s=\pm iq_x c(\mathbf{q})+\tfrac{1}{2}(q^2)D_s(\mathbf{q},s)\,, \tag{2.33}$$

with

$$D_s(\mathbf{q},s)=\frac{\frac{4}{3}\eta(\mathbf{q},s)+\zeta(\mathbf{q},s)}{\rho}+\frac{\lambda(\mathbf{q},s)}{\rho}\left(\frac{1}{C_V(\mathbf{q})}-\frac{1}{C_p(\mathbf{q})}\right). \tag{2.34}$$

The eigenstates for these modes are

$$|\pm,\mathbf{q}\rangle=a_\pm(\mathbf{q})|\rangle=\frac{a_2(\mathbf{q})\pm a_3(\mathbf{q})}{\sqrt{2}}|\rangle. \tag{2.35}$$

We call the solution to Eq. (2.34)$s_\pm(q)$. Similarly, the heat-flow mode has a relaxation time which is a solution of

$$s=\lambda(\mathbf{q},s)q^2/\rho C_p(\mathbf{q})\,, \tag{2.36}$$

and we call this solution $s_T(\mathbf{q})$. Finally the viscous-flow mode has

$$s=\frac{\eta(\mathbf{q},s)q^2}{\rho}\,, \tag{2.37}$$

with a solution we call $s_\eta(\mathbf{q})$.

III. PERTURBATION THEORY FOR THE TRANSPORT COEFFICIENTS

A. Intermediate States

In the expressions (2.23), (2.28), and (2.29) for the transport coefficients, there appear structures of the form

$$X=1/(PLP-s)\,, \tag{3.1}$$

which play the role of frequency denominators in the Kubo formula. To gain a convenient representation for X, we employ a representation of L in terms of its right eigenstates $|\nu,\mathbf{q}\rangle_R$, its left eigenstates ${}_L\langle\nu,\mathbf{q}|$, and its eigenvalues $s_\nu(\mathbf{q})$ by writing

$$L=\sum_{\nu'}\int\frac{d^3q'}{(2\pi)^3}|\nu',\mathbf{q}'\rangle_R s_{\nu'}(\mathbf{q}')\,{}_L\langle\nu',\mathbf{q}'|\,. \tag{3.2}$$

Here the eigenstates are normalized so that

$${}_L\langle\nu,\mathbf{q}|\nu',\mathbf{q}'\rangle_R=\delta_{\nu,\nu'}\delta(\mathbf{q}-\mathbf{q}')(2\pi)^3. \tag{3.3}$$

Because the projection operator P in X discriminates against local equilibrium states with wave vector $\mathbf{q}$, this projection operator almost entirely removes the lowest eigenstates of L (the transport states) and leaves the remaining states almost untouched. For this reason, we write the part of X with wave vector q as

$$X_{\mathbf{q}}=\sum_{\nu'=6}^{\infty}\frac{|\nu',\mathbf{q}\rangle_R\,{}_L\langle\nu',\mathbf{q}|}{s_{\nu'}(\mathbf{q})-s}. \tag{3.4}$$

We are interested in divergences in the transport coefficient $\eta(\mathbf{q},s)$ near the critical point. These divergences can be expected to arise from states ν' which give small values of $s_{\nu'}(q)$, that is slowly decaying intermediate states. There is one set of intermediate states which is particularly attractive for this consideration: those states involving multiple transport processes with long wavelengths. For example, we can consider a state which involves two independent transport processes with wave vectors $\mathbf{q}'$ and $\mathbf{q}-\mathbf{q}'$. These states would be of the structure

$$\sum_{\nu'}|\nu',\mathbf{q}\rangle_R\,{}_L\langle\nu',\mathbf{q}|=\tfrac{1}{2}\sum_{\nu_1=1}^{5}\sum_{\nu_2=1}^{5}\int\frac{d^3q'}{(2\pi)^3}\times a_{\nu_1}(\mathbf{q}')a_{\nu_2}(\mathbf{q}-\mathbf{q}')|\rangle\langle|a_{\nu_1}(-\mathbf{q})a_{\nu_1}(\mathbf{q}'-\mathbf{q})\,, \tag{3.5}$$

where the a_ν's are the linear combinations of the a_j's which generate specific transport processes. The eigenvalues for the states (3.5) are

$$s_{\nu'}(\mathbf{q})=s_{\nu_1}(\mathbf{q}')+s_{\nu_2}(\mathbf{q}'-\mathbf{q})\,, \tag{3.6}$$

since two noninteracting disturbances have an inverse relaxation time which is the sum of the inverse relaxation times for the individual disturbances.

With this logic, $X_{\mathbf{q}}$ gains a representation:

$$X_{\mathbf{q}}=\frac{1}{2!}\sum_{\nu\nu'}\int\frac{d^3q'}{(2\pi)^3}\frac{a_\nu(\mathbf{q}')a_{\nu'}(\mathbf{q}-\mathbf{q}')|\rangle\langle|a_\nu(-\mathbf{q}')a_{\nu'}(\mathbf{q}'-\mathbf{q})}{s_\nu(\mathbf{q}')+s_{\nu'}(\mathbf{q}-\mathbf{q}')-s}+\frac{1}{3!}\sum_{\nu\nu'\nu''}\int\frac{d^3q'}{(2\pi)^3}\frac{d^3q''}{(2\pi)^3}$$
$$\times\frac{a_{\nu'}(\mathbf{q}')a_{\nu''}(\mathbf{q}'')a_\nu(\mathbf{q}-\mathbf{q}'-\mathbf{q}'')|\rangle\langle|a_{\nu'}(-\mathbf{q}')a_{\nu''}(-\mathbf{q}'')a_\nu(\mathbf{q}'+\mathbf{q}''-\mathbf{q})}{s_\nu(\mathbf{q}-\mathbf{q}'-\mathbf{q}'')+s_{\nu'}(\mathbf{q}')+s_{\nu''}(\mathbf{q}'')-s}+\cdots. \tag{3.7}$$

B. Viscous Flow → Heat Modes

To show the utility of the representation (3.7), we consider the formula for the viscosity

$$-q^2\eta(\mathbf{q},s)=\langle|g_y(-\mathbf{q})LPXPLg_y(\mathbf{q})|\rangle\beta. \quad (3.8)$$

In Eq. (3.8), the projection operators P may both be replaced by unity since $P-1$ makes no contribution to the matrix element. Consider the contributions to the right-hand side of (3.8) from intermediate states which involve two heat-flow modes. These give a contribution to $\eta(\mathbf{q},s)$ which is

$$-q^2\eta_{TT}(\mathbf{q},s)=\tfrac{1}{2}\beta\int\frac{d^3q'}{(2\pi)^3}\frac{\langle|g_y(-\mathbf{q})La_1(\mathbf{q}')a_1(\mathbf{q}-\mathbf{q}')|\rangle}{s_T(\mathbf{q}')+s_T(\mathbf{q}-\mathbf{q}')-s}\times\langle|a_1(\mathbf{q}'-\mathbf{q})a_1(-\mathbf{q}')Lg_y(\mathbf{q})|\rangle.$$

Since

$$\langle|a_1(-\mathbf{q}')a_1(\mathbf{q}'-\mathbf{q})Lg_y(\mathbf{q})|\rangle=-\langle|g_y(-\mathbf{q})La_1(\mathbf{q}')a_1(\mathbf{q}-\mathbf{q}')|\rangle^*,$$

this result may be rewritten as

$$q^2\eta_{TT}(\mathbf{q},s)=\frac{\beta}{2k_B{}^2}\int\frac{d^3q'}{(2\pi)^3}\times\frac{|M_{\mathbf{q},\mathbf{q}'}|^2}{[\rho C_p(\mathbf{q}')\rho C_p(\mathbf{q}-\mathbf{q}')][s_T(\mathbf{q}')+s_T(\mathbf{q}-\mathbf{q}')-s]}, \quad (3.9)$$

with

$$M_{\mathbf{q},\mathbf{q}'}=\langle|g_y(-\mathbf{q})Ls_{\rm op}(\mathbf{q}')s_{\rm op}(\mathbf{q}-\mathbf{q}')|\rangle. \quad (3.10)$$

To evaluate this matrix element, we successively commute L to the right and use $L|\rangle=0$ to write

$$M_{\mathbf{q},\mathbf{q}'}=\langle|g_y(-\mathbf{q})(-i\mathbf{q}')\cdot\mathbf{j}_{\rm op}{}^s(\mathbf{q}')s_{\rm op}(\mathbf{q}-\mathbf{q}')|\rangle+\langle|g_y(-\mathbf{q})[i(\mathbf{q}-\mathbf{q}')]\cdot\mathbf{j}_{\rm op}{}^s(\mathbf{q}-\mathbf{q}')s_{\rm op}(\mathbf{q}')|\rangle. \quad (3.11)$$

In this expression $\mathbf{j}^s$ is the heat-flow current which is

$$\mathbf{j}^s(\mathbf{r})=\left[\mathbf{j}^\epsilon(\mathbf{r})-\frac{\langle\epsilon+p\rangle}{\langle n\rangle}\mathbf{j}(r)\right]\Big/T. \quad (3.12)$$

This current has short-range correlations with the momentum density such that, when one averages over the momenta of all the particles, in the system, one finds

$$\beta\langle j_x{}^s(\mathbf{r})g_x(\mathbf{r}')\rangle_q=\frac{1}{T}\delta(\mathbf{r}-\mathbf{r}')\left[\epsilon(\mathbf{r})+p(\mathbf{r})-\frac{\langle\epsilon+p\rangle}{\langle n\rangle}n(\mathbf{r})\right]=\delta(\mathbf{r}-\mathbf{r}')[s_{\rm op}(\mathbf{r})+p_{\rm op}(\mathbf{r})/T], \quad (3.13)$$

where $p_{\rm op}$ is the pressure operator. Thence, Eq. (3.11) reduces to

$$M_{\mathbf{q},\mathbf{q}'}=(-i\mathbf{q}_y')\beta^{-1}\langle|[s_{\rm op}(\mathbf{q}'-\mathbf{q})+(1/T)p_{\rm op}(\mathbf{q}'-\mathbf{q})]\times s_{\rm op}(\mathbf{q}-\mathbf{q}')|\rangle+i(\mathbf{q}_y'-\mathbf{q}_y)\beta^{-1}\times\langle|[s_{\rm op}(-\mathbf{q}')+(1/T)p_{\rm op}(-\mathbf{q}')]s_{\rm op}(\mathbf{q}')|\rangle. \quad (3.14)$$

The entropy operator s has been defined so that its correlations with the pressure vanishes, and so that its autocorrelations are related to $\rho C_p(\mathbf{q})$. Consequently,[21]

$$M_{\mathbf{q}\mathbf{q}'}=[i\mathbf{q}_y'C_p(\mathbf{q}-\mathbf{q}')+i(\mathbf{q}_y'-\mathbf{q}_y)C_p(\mathbf{q}')]\rho k_B{}^2T. \quad (3.15)$$

Since $q_y=0$, Eq. (3.9) reduces to

$$q^2\eta_{TT}(\mathbf{q},s)=\frac{1}{2\beta}\int\frac{d^3q'}{(2\pi)^3}(\mathbf{q}_y')^2\frac{[C_p(-\mathbf{q}+\mathbf{q}')-C_p(\mathbf{q}')]^2}{C_p(-\mathbf{q}'+\mathbf{q})C_p(\mathbf{q}')}\times\frac{1}{s_T(\mathbf{q}')+s_T(\mathbf{q}-\mathbf{q}')-s}. \quad (3.16)$$

In the static, long-wavelength limit $q\to 0$, $s\to 0$, Eq. (3.16) gives

$$\eta_{TT}(0,0)=\frac{1}{4}\int\frac{d^3q'}{(2\pi)^3}(q_y')^2\frac{\left[\dfrac{\partial}{\partial q_x'}C_p(\mathbf{q}')\right]^2}{s_T(\mathbf{q}')[C_p(\mathbf{q}')]^2}. \quad (3.17)$$

Because the thermal relaxation rate $s_T(\mathbf{q}')$ becomes very small for long wavelengths, this integral may contain large contributions from small values of q'. According to the scaling-law hypothesis, there is an inverse length, ξ^{-1}, which measures the characteristic wave vector for all phenomena near the critical point. From this hypothesis, one would conclude that main contributions to the q' integral come from $q'\lesssim\xi^{-1}$. Furthermore, in this region

$$\frac{\partial}{\partial q_x'}C_p(\mathbf{q}')\sim\frac{q_x'}{(q')^2}C_p(\mathbf{q}')$$

and

$$s_T(\mathbf{q}')|_{q'=\xi^{-1}}\sim s_T{}^*=(\lambda^*/\rho C_p)\xi^{-2}, \quad (3.18)$$

where $s_T{}^*$ stands for the inverse thermal relaxation time at $q'\sim\xi^{-1}$, and λ^* stands for the thermal conductivity at this wave vector and characteristic frequency. With these estimates in hand, we can estimate the right-hand side of Eq. (3.17) as

$$\eta_{TT}(\mathbf{q},s)\sim\frac{1}{\beta}\frac{\rho C_p}{\lambda^*}\xi^{-1} \quad (3.19)$$

for $q\lesssim\xi^{-1}$ and $s\lesssim s_T{}^*$.

Notice the restrictions on Eq. (3.19). For $q\gg\xi^{-1}$ or $s\gg s_T{}^*$, the frequency denominator in Eq. (3.16) is necessarily considerably increased. Therefore, if the restrictions in Eq. (3.19) are not satisfied, this contribution to $\eta(\mathbf{q},s)$ must necessarily be considerably reduced.

---

[21] Kawasaki evaluated matrix elements like (3.10) by a slightly different technique. He moved L to the left in (3.10) and obtained a matrix element with a factor of $q_x\tau_{xy}(-\mathbf{q})$, where τ_{xy} is the momentum current. He then argued that, at small q, $\tau_{xy}(\mathbf{q})$ reduced to $\beta^{-1}q_y'(\partial/\partial q_x')$. One can see that our result supports this conclusion.

Equation (3.19) is our first sight of a necessary divergence in a transport coefficient. If $\rho=\rho_c$ and $T>T_c$, according to the conventional notation C_p and ξ diverge as

$$C_p\sim\epsilon^{-\gamma},\quad \xi\sim\epsilon^{-\nu},$$

where $\epsilon=(T-T_c)/T_c$. Therefore, the product of the transport coefficients is

$$\eta_{TT}(0,0)\lambda^*\sim\epsilon^{-\gamma+\nu}. \tag{3.20}$$

Since γ is greater than ν, this result tends to indicate that one or both of the transport coefficients should diverge.

The same type of analysis may be employed to discuss the decay of a viscous mode into three or more heat modes. This analysis yields

$$q^2\eta_{TTT}(\mathbf{q},s)=\frac{\beta}{6k_BT}\int\frac{d^3q'}{(2\pi)^3}\frac{d^3q''}{(2\pi)^3}\times\frac{|M_{\mathbf{q},\mathbf{q}',\mathbf{q}''}|^2}{\rho^3C_p(\mathbf{q}')C_p(\mathbf{q}'')C_p(\mathbf{q}'+\mathbf{q}''-\mathbf{q})}\times\frac{1}{s_T(\mathbf{q}')+s_T(\mathbf{q}'')+s_T(\mathbf{q}-\mathbf{q}'-\mathbf{q}'')-s}. \tag{3.21}$$

A typical term in M is

$$M_{\mathbf{q},\mathbf{q}',\mathbf{q}''}=\beta^{-1}q_xq_y'\frac{\partial}{\partial q_y'}\times\langle s_{\mathrm{op}}(\mathbf{q}')s_{\mathrm{op}}(\mathbf{q}'')s_{\mathrm{op}}(-\mathbf{q}''-\mathbf{q}')\rangle. \tag{3.22}$$

We now employ the scaling idea to argue that the contribution (3.21) to η is of the same order of magnitude as the two heat-mode contribution, (3.9), which we have already evaluated. Consider how (3.21) differs from (3.9). There is an extra factor of C_p^{-1} in the denominator of (3.21) and an extra wave-number integral. If the wave-number integral contributes over $q''<\xi^{-1}$, this integral gives us an extra factor of ξ^{-3}. Finally, the matrix element (3.22) contains one more factor of $s_{\mathrm{op}}(\mathbf{q}'')$ than (3.9). According to the scaling idea this extra factor enhances the matrix element by a factor of $|\epsilon|^{-\gamma-\beta}$ on the critical isochore. Of course, the frequency denominator is of the same order of magnitude, s_T^*, in both cases.

Hence the extra factors which appear in (3.21) and over and above the factors in Eq. (3.9) are $(\xi^{-3}/C_p)(\epsilon^{-\gamma-\beta})^2$ or $\epsilon^{3\nu-\gamma-2\beta}$. But according to the scaling idea, $3\nu=\gamma+2\beta$. (See Refs. 8 and 14.) Therefore, the contribution (3.21) is of the same order of magnitude as the contribution (3.9). An extension of the above argument indicates that all contributions from intermediate states with any number of heat modes in them are of the same order of magnitude.

C. Two Heat-Mode Contributions to $\zeta+\frac{4}{3}\eta$

The same kind of calculation works for $\zeta+\frac{4}{3}\eta$, with the only difference being that the predominant contributions come from the terms involving the local equilibrium part of the projection operator P.[22] In particular, the two heat-mode contribution to $\zeta+\frac{4}{3}\eta$ is

$$q^2[\zeta_{TT}(\mathbf{q},s)+\tfrac{4}{3}\eta_{TT}(\mathbf{q},s)]=\tfrac{1}{2}\beta\int\frac{d^3q'}{(2\pi)^3}\frac{|L_{\mathbf{q},\mathbf{q}'}|^2}{s_T(\mathbf{q}')+s_T(\mathbf{q}-\mathbf{q}')-s}. \tag{3.23}$$

Here

$$L_{\mathbf{q},\mathbf{q}'}=\langle|g_z(-\mathbf{q})LPa_1(\mathbf{q}')a_1(\mathbf{q}-\mathbf{q}')|\rangle\approx-\langle|g_z(-\mathbf{q})La_2(\mathbf{q})|\rangle\langle|a_2(-\mathbf{q})a_1(\mathbf{q}')a_1(\mathbf{q}-\mathbf{q}')|\rangle.$$

For small q and q' the matrix element is readily estimated as

$$L_{q,q'}\sim-iqc\frac{(k_B)^{1/2}}{(\beta C_V)^{1/2}}T\left(\frac{\partial}{\partial T}\right)_{S/N}$$

so that the whole contribution to the viscosity is obtained by replacing the integrals in Eq. (3.23) in the same way as above. In this way we obtain

$$\zeta_{TT}(\mathbf{q},s)+\tfrac{4}{3}\eta_{TT}(\mathbf{q},s)\sim\frac{\rho C_p\xi^{-1}}{\lambda^*}\frac{k_Bc^2}{C_V}\left(\frac{T(\partial/\partial T)C_p|_{S/N}}{C_p}\right)^2 \tag{3.24}$$

for $q\lesssim\xi^{-1}$ and $s\lesssim s_T^*$. Or, if we use the scaling-law equality $3\nu=2-\alpha$ in the dimensional form

$$\xi^{-3}\left[\frac{1}{C_p}\left[\frac{\partial C_p}{\partial T}\right|_{S/N}\right]^2\sim\beta^2k_B\rho C_V$$

on the critical isochore we have finally that

$$\zeta_{TT}(\mathbf{q},s)+\tfrac{4}{3}\eta_{TT}(\mathbf{q},s)\sim\frac{\rho^2c^2C_p\xi^2}{\lambda^*}$$

for $q\lesssim\xi^{-1}$ and $s\lesssim s_T^*$ on the critical isochore.

D. Heat → Heat+Viscous Flow

We use Eq. (2.29) to find $\lambda(\mathbf{q},s)$. The situation in which the intermediate state contains one heat-flow mode and one viscous-flow mode gives a contribution to $\lambda(\mathbf{q},s)$ which is

$$q^2\lambda_{\eta T}(\mathbf{q},s)=\frac{\beta}{k_B^2}\int\frac{d^3q'}{(2\pi)^3}\times\frac{|N_{\mathbf{q},\mathbf{q}'}|^2}{\rho^2C_p(\mathbf{q}')[s_T(\mathbf{q}')+s_\eta(\mathbf{q}-\mathbf{q}')-s]}. \tag{3.25}$$

[22] K. Kawasaki and M. Tanaka [Proc. Phys. Soc. (London) **90**, 791 (1967)] have calculated the contribution to $\zeta+\frac{4}{3}\eta$ from the local equilibrium part of the projection operator for a two-component liquid mixture.

TABLE II. Contributions to transport coefficients.

	Region I	Region II	Region III
	→ Increasing s →		
	$s \sim s^*_T \sim \frac{\lambda^* \xi^{-2}}{\rho C_p} \sim \epsilon^2$	$s \sim s^*_\eta \sim \frac{\eta^* \xi^{-2}}{\rho} \sim \epsilon^{4/3}$	$s \sim c\xi^{-1} \sim \epsilon^{2/3}$
Contributions to λ:			
From viscous flow plus heat modes	$\lambda \sim \frac{\rho C_p \xi^{-1}}{\beta \eta^*} \sim \epsilon^{-2/3}$		
From sound waves plus heat modes	$\lambda \sim \frac{\xi^{-2} C_p}{c\beta} \sim \epsilon^0$		
Contributions to η:			
From heat modes	$\eta \sim \frac{\rho C_p \xi^{-1}}{\beta \lambda^*} \sim \epsilon^0$		
From sound waves plus heat modes		$\eta \sim \frac{C_p \xi^{-3}}{c\beta C_V} \sim \epsilon^0$	
Contributions to ζ:			
From heat modes	$\zeta \sim \frac{\rho^2 c^2 C_p \xi^2}{\lambda^*} \sim \epsilon^{-2}$		
From sound waves plus heat modes		$\zeta \sim \rho c \xi \sim \epsilon^{-\nu+\alpha/2} \sim \epsilon^{-2/3}$	
From high q processes		$\eta \sim \zeta \sim \lambda \sim$ constant	

If the direction of the momentum in the intermediate state is described by the unit vector $\hat{n}$, then

$$N_{\mathbf{q},\mathbf{q}'} = \langle | s_{\mathrm{op}}(-\mathbf{q}) L P \hat{n} \cdot \mathbf{g}(\mathbf{q}-\mathbf{q}') s_{\mathrm{op}}(\mathbf{q}') | \rangle .$$

In the thermal conductivity matrix elements, the contribution of $(P-1)$ is once again negligible. Consequently,

$$\begin{aligned} N_{\mathbf{q},\mathbf{q}'} &= i\mathbf{q} \cdot \langle | \mathbf{j}_{\mathrm{op}}{}^{s}(-\mathbf{q}) \hat{n} \cdot \mathbf{g}(\mathbf{q}-\mathbf{q}') s_{\mathrm{op}}(\mathbf{q}') | \rangle \\ &= i\mathbf{q} \cdot \hat{n} \beta^{-1} \langle | [s_{\mathrm{op}}(-\mathbf{q}') + p_{\mathrm{op}}(-\mathbf{q}')/T] s_{\mathrm{op}}(\mathbf{q}') | \rangle \\ &= i\mathbf{q} \cdot \hat{n} \beta^{-1} k_B \rho C_p(\mathbf{q}') . \end{aligned}$$

After averaging over directions of $\mathbf{q}'$ and summing over polarization vectors, $\hat{n}$, perpendicular to $\mathbf{q}-\mathbf{q}'$, we find

$$\lambda_{\eta T}(q,s) = \tfrac{1}{3} \beta^{-1} \int \frac{d^3 q'}{(2\pi)^3} \frac{C_p(\mathbf{q}')}{s_T(\mathbf{q}') + s_\eta(\mathbf{q}-\mathbf{q}') - s} . \quad (3.26)$$

The factor $C_p(\mathbf{q}')$ in the numerator of Eq. (3.26) allows this contribution to λ to be large. The integral contributes for $q' \lesssim \xi^{-1}$. Since the thermal diffusion rate is very slow near the critical point, the viscous relaxation rate $s_\eta(\mathbf{q}')$ dominates the thermal relaxation rate in the denominator. As a result

$$\lambda_{\eta T}(\mathbf{q},s) \sim \frac{\rho C_p \xi^{-1}}{\eta^* \beta}, \quad \text{for} \quad q \lesssim \xi^{-1}, \quad s \lesssim s_\eta^* . \quad (3.27)$$

Here s_η^* is the viscous relaxation rate at the wave number ξ^{-1}, i.e.,

$$s_\eta(q') |_{q'=\xi^{-1}} \sim s_\eta^* = (\eta^*/\rho) \xi^{-2} . \quad (3.28)$$

The limitations on Eq. (3.27) indicate that for large s and large q the denominator in Eq. (3.26) becomes large enough to reduce the size of the contribution to the thermal conductivity quite appreciably.

From scaling-law arguments we conclude that contributions with two or more heat-flow modes in the intermediate state together with a viscous-flow mode give a contribution to λ of the same order as (3.27).

If there are no further contributions to η and λ, we would now have enough information to compute both η and λ. To do this calculation, it is necessary to recognize that the large factor C_p in the denominator of s_T guarantees that, as indicated in Table II, $s_T^* \ll s_\eta^*$.

Therefore, η^* is the viscosity evaluated at "frequencies" much higher than s_T^*. The contributions (3.19) to η cut off at the characteristic thermal relaxation rate s_T. Hence, η^* gains nothing from the processes involving heat-flow modes in intermediate states.

If these be the only processes contributing to the analomous viscosity, there cannot be any critical fluctuation term in η^*. Then η^* will be finite at the critical point and Eq. (3.27) will predict

$$\lambda(\mathbf{0},0)\sim\epsilon^{-\gamma+\nu}, \tag{3.29}$$

which diverges as roughly the $-\frac{2}{3}$ power of $T-T_c$ on the critical isochore. Furthermore, the contribution (3.29) will persist for all $s\lesssim s_\eta^*$. Hence, it will be quite relevant up to and beyond the thermal cut off s_T^*. This means that we can use (3.29) to evaluate λ^* in Eq. (3.19). We then find

$$\eta_{TT}(\mathbf{q},s)\sim\epsilon^0,\quad\text{for}\quad q<\xi^{-1}\quad\text{and}\quad s<s_T^*. \tag{3.30}$$

Since η^* is, by our hypothesis of leaving out other contributions, quite finite, Eq. (3.30) predicts $\eta_{TT}(\mathbf{0},0)$ does not diverge as a power of $T-T_c$. This result does not preclude a logarithmic behavior[23] or a very strong cusp in the low-frequency viscosity. In fact, this analysis suggests that one of these two types of singularities might well hold for η.

Our conclusions about the low-frequency behavior of λ and η will depend quite crucially upon the high-frequency form of η. High-frequency processes which can contribute to η include those in which sound waves are produced. The characteristic frequency for sound waves is $c\xi^{-1}$, which is much higher than s_η^* or s_T^*. Therefore, as indicated sound-wave processes are good candidates for producing contributions to λ, η, and ζ, which will not cut off until high frequencies.

E. Sound-Wave Intermediate States for λ

Intermediate states with two sound waves do not produce an appreciable contribution to $\lambda(\mathbf{q},s)$. However, three-sound-wave intermediate states do produce a contribution, which can be computed from

$$q^2\lambda_{ppp}(\mathbf{q},s)=\frac{1}{3!k_B}\int\frac{d^3q'}{(2\pi)^3}\frac{d^3q''}{(2\pi)^3}\sum_{\sigma,\sigma',\sigma''=\pm1}\frac{|\langle|\mathbf{q}\cdot\mathbf{j}_{\mathrm{op}}^{s}(-\mathbf{q})a_{\sigma'}(\mathbf{q}')a_{\sigma''}(\mathbf{q}'')a_\sigma(\mathbf{q}-\mathbf{q}'-\mathbf{q}'')|\rangle|^2}{s_{\sigma'}(\mathbf{q}')+s_{\sigma'}(\mathbf{q}'')+s_\sigma(\mathbf{q}-\mathbf{q}'-\mathbf{q}'')-s}, \tag{3.31}$$

with

$$a_{\sigma'}(\mathbf{q})=\frac{1}{\sqrt{2}}\left[a_2(\mathbf{q})+\sigma'\frac{\mathbf{q}\cdot\mathbf{g}(\mathbf{q})}{|\mathbf{q}|}\left(\frac{\beta}{\rho}\right)^{1/2}\right], \tag{3.32}$$

$$s_{\sigma'}(\mathbf{q})=\sigma' ic(\mathbf{q})|\mathbf{q}|+\tfrac{1}{2}D_s(\mathbf{q},s_{\sigma'}(\mathbf{q}))q^2. \tag{3.33}$$

The significant terms in the product of three a's are the ones which involve a product of two a_2's with one momentum. After the momentum average is performed, we find that the matrix element in Eq. (3.31) is

$$\frac{1}{\sqrt{(\rho\beta)}}\frac{1}{(\sqrt{2})^3}\langle|\left\{\frac{\mathbf{q}\cdot\mathbf{q}'}{|\mathbf{q}'|}\sigma'[s_{\mathrm{op}}(-\mathbf{q}+\mathbf{q}')+p_{\mathrm{op}}(-\mathbf{q}+\mathbf{q}')/T]a_2(\mathbf{q}'')a_2(\mathbf{q}-\mathbf{q}'-\mathbf{q}'')\right.$$
$$+\frac{\mathbf{q}\cdot\mathbf{q}''}{|\mathbf{q}''|}\sigma''[s_{\mathrm{op}}(-\mathbf{q}+\mathbf{q}'')+p_{\mathrm{op}}(-\mathbf{q}+\mathbf{q}'')/T]a_2(\mathbf{q}')a_2(\mathbf{q}-\mathbf{q}'-\mathbf{q}'')$$
$$\left.+\frac{\mathbf{q}\cdot(\mathbf{q}-\mathbf{q}'-\mathbf{q}'')}{|\mathbf{q}-\mathbf{q}'-\mathbf{q}''|}\sigma[s_{\mathrm{op}}(-\mathbf{q}'-\mathbf{q}'')+p_{\mathrm{op}}(-\mathbf{q}'-\mathbf{q}'')/T]a_2(\mathbf{q}')a_2(\mathbf{q}'')\right\}|\rangle.$$

If q, q', and q'' are each $\lesssim\xi^{-1}$, this matrix element may be estimated to be of the order of

$$\frac{1}{\sqrt{(\rho\beta^3)}}q\frac{1}{C_V}\left(\frac{\partial}{\partial T}C_V\right)_p,$$

since $s_{\mathrm{op}}(-q)$ generates $\beta^{-1}(\partial/\partial T)$ as $q\to0$ and $\langle|a_2(-\mathbf{q})a_2(\mathbf{q})|\rangle$ is of order unity.

The damping terms, i.e., the real part, of the frequency denominator of Eq. (3.31) are smaller or of the same order as the imaginary part of this denominator. Therefore, to get an order of magnitude estimate we can replace the denominator by the δ function

$$\delta(\sigma c(\mathbf{q}')q'+\sigma''c(\mathbf{q}'')q''+\sigma c(\mathbf{q}-\mathbf{q}'-\mathbf{q}'')|\mathbf{q}-\mathbf{q}'-\mathbf{q}''|-\mathrm{Im}s),$$

which generates a contribution

$$\sim\frac{1}{c\xi^{-1}},\quad\text{for}\quad|s|\lesssim c\xi^{-1}.$$

Since each q intergral covers $q\lesssim\xi^{-1}$, the resulting estimate of λ is

$$\lambda_{ppp}(\mathbf{q},s)\sim\frac{\xi^{-5}}{c}\frac{1}{C_V^2}\left[\left(\frac{\partial C_V}{\partial T}\right)_p\right]^2\frac{1}{\rho\beta^3k_B}. \tag{3.34}$$

The scaling-law arguments inform us that the two $(\partial/\partial T)_p$ each produce a factor $\epsilon^{-\beta-\gamma}$ multiplying the

[23] The scaling-law analysis is based upon exponents and it cannot distinguish a finite but discontinuous result from a logarithmic infinity.

singular part of C_V while $\xi^{-3}\sim\epsilon^{3\nu}=\epsilon^{2\beta+\gamma}$. If we put this argument in dimensional form, we find that

$$\beta^{-2}\xi^{-3}(\partial/\partial T)_p^2\sim k_B C_p \rho$$

so that Eq. (3.34) can be written

$$\lambda_{ppp}(\mathbf{q},s)\sim\frac{\xi^{-2}}{c}\frac{C_p}{C_V^2}\frac{1}{\beta}\{[C_V]_{\text{sing}}\}^2,$$

$$\text{for } q\lesssim\xi^{-1},\quad |s|\lesssim c\xi^{-1}. \quad (3.35)$$

Here the subscript "sing" reminds us to take the singular part of C_V in our estimates and leave out any constant term which might appear. If C_V diverges as $\epsilon^{-\alpha}$ Eq. (3.35) gives an estimate of a singular contribution to λ, which is

$$\lambda_{ppp}(0,0)\sim\epsilon^{-\gamma+2\nu-\alpha/2} \quad (3.36)$$

on the critical isochore. Since $\gamma-2\nu\approx0$, Eq. (3.36) describes a weakly divergent or strongly cusped contribution to the high-frequency sound-wave damping constant.

If the intermediate state includes in addition to sound waves viscous flow and heat modes, the scaling-law arguments imply that the estimate (3.35) still gives the correct order of magnitude for the singular parts of $\lambda(\mathbf{q},s)$ at high frequencies.

F. Sound-Wave Intermediate States for η

The contribution to η from a three-sound-wave intermediate state is given by

$$q^2\eta_{ppp}(q,s)\sim\beta\int d^3q'd^3q''|M_{\mathbf{q},\mathbf{q}',\mathbf{q}''}|^2$$

$$\times\sum_{\sigma\sigma'\sigma''=\pm1}\frac{1}{s_{\sigma'}(\mathbf{q}')+s_{\sigma''}(\mathbf{q}'')+s_\sigma(\mathbf{q}-\mathbf{q}'-\mathbf{q}'')-s}, \quad (3.37)$$

where a typical term in M is

$$M_{\mathbf{q},\mathbf{q}',\mathbf{q}''}=\langle|g_y(-\mathbf{q})[L,a_2(\mathbf{q}')]\times a_2(\mathbf{q}'')a_2(\mathbf{q}-\mathbf{q}'-\mathbf{q}'')|\rangle. \quad (3.38)$$

Equation (2.15b) implies that

$$[L,a_2(\mathbf{q}')]=\frac{(\rho\beta)^{1/2}}{\langle n\rangle}c(\mathbf{q}')(-i\mathbf{q}')\cdot\mathbf{j}_{\text{op}}(\mathbf{q}')+\left\{\frac{1}{k_B\rho}\left[\frac{1}{C_V(\mathbf{q}')}-\frac{1}{C_p(\mathbf{q}')}\right]\right\}^{1/2}(-i\mathbf{q}')\cdot\mathbf{j}_{\text{op}}^{\,e}(\mathbf{q}'),$$

so that a momentum average gives

$$\beta\langle g_y(-\mathbf{q})[L,a_2(\mathbf{q}')]\rangle_{\mathbf{q}}=\frac{(\rho\beta)^{1/2}}{\langle n\rangle}c(\mathbf{q}')(-iq_y')n_{\text{op}}(\mathbf{q}'-\mathbf{q})+\left\{\frac{1}{k_B\rho}\left[\frac{1}{C_V(\mathbf{q}')}-\frac{1}{C_p(\mathbf{q}')}\right]\right\}^{1/2}(-iq_y')\times\left[s_{\text{op}}(\mathbf{q}'-\mathbf{q})+\frac{p_{\text{op}}(\mathbf{q}'-\mathbf{q})}{T}\right]. \quad (3.39)$$

At this point, a serious question arises. What is the behavior of the collection of operators in Eq. (3.39) for $q\sim\xi^{-1}$, $q'\sim\xi^{-1}$? At $q=0$, the answer is clear. Since p_{op} has only weak fluctuations and may be neglected, expression (3.39) is $-iq_y'a_2(q')$ at $q=0$. But for $q\neq0$ each of the two terms on the right-hand side of (3.39) is of order $c(\mathbf{q}')a_1(\mathbf{q}'-\mathbf{q})$, which is much more singular than $a_2(\mathbf{q}'-\mathbf{q})$. The question then boils down to: Do the most singular parts of the two terms in Eq. (3.39) cancel against one another?

There is an alternative way of stating this problem. If the product $[C_V(\mathbf{q}')]^{1/2}c(\mathbf{q}')$ is independent of q', then (3.39) is proportional to $a_2(\mathbf{q}'-\mathbf{q})$. If, however,

$$\frac{\partial}{\partial q'}[C_V(\mathbf{q}')]^{1/2}c(\mathbf{q}')\sim\xi, \quad (3.40)$$

then expression (3.39) is of the order of $q'ca_1(\mathbf{q}')$. This idea was discussed at the end of Sec. IIC. We stated there the point that the product $[C_V(\mathbf{q}')]^{1/2}c(\mathbf{q}')$ would be essentially independent of $q'\xi$ if the strongest way of stating the scaling-law idea were right. However, we believe this to be probably an overextension of the scaling-law idea. In any case, we here use the estimate (3.40) and concomitant estimate of the matrix element defined in Eq. (3.38) as

$$M_{\mathbf{q},\mathbf{q}'\mathbf{q}''}\sim\frac{q}{\beta}\left(\frac{1}{k_B\rho C_V}\right)^{1/2}\times\langle|s_{\text{op}}(\mathbf{q}'-\mathbf{q})a_2(\mathbf{q}'')a_2(\mathbf{q}-\mathbf{q}'-\mathbf{q}'')|\rangle.$$

From this point on we can use the same analysis as for λ_{ppp}, and we find

$$\eta_{ppp}(\mathbf{q},s)\sim\lambda_{ppp}(\mathbf{q},s)/C_V. \quad (3.41)$$

G. Sound-Wave Intermediate States for $\zeta+\frac{4}{3}\eta$

Two-sound-wave processes contribute a term to the longitudinal viscosity

$$q^2[\zeta_{pp}(\mathbf{q},s)+\tfrac{4}{3}\eta_{pp}(\mathbf{q},s)]=\tfrac{1}{2}\beta\sum_{\sigma,\sigma'=\pm1}\int\frac{dq'}{(2\pi)^3}\frac{|H_{\mathbf{q},\mathbf{q}'}|^2}{s\sigma(q')+s\sigma'(q-q')-s},$$

with

$$H_{\mathbf{q},\mathbf{q}'}=\langle|g_x(-\mathbf{q})LPa_2(\mathbf{q}')a_2(\mathbf{q}-\mathbf{q}')|\rangle$$
$$\approx-\langle|g_x(-\mathbf{q})La_2(\mathbf{q})|\rangle$$
$$\times\langle|a_2(-\mathbf{q})a_2(\mathbf{q}')a_2(\mathbf{q}-\mathbf{q}')|\rangle. \quad (3.42)$$

For small q and q', H reduces to

$$H_{\mathbf{q},\mathbf{q}'}\sim -iq\frac{(k_B)^{1/2}}{(\beta C_V)^{1/2}}T\frac{\partial}{\partial T}c|_{S/N}, \quad (3.43)$$

so that the contribution to $\zeta+\frac{4}{3}\eta$ may be estimated as

$$\zeta_{pp}(\mathbf{q},s)+\tfrac{4}{3}\eta_{pp}(\mathbf{q},s)\sim\frac{1}{C_Vc}\xi^{-2}k_B\left[T\frac{\partial c}{\partial T}\bigg|_{S/N}\right]^2 \quad (3.44)$$

for $q\lesssim\xi$, $|s|\lesssim c\xi^{-1}$. Or if we use the scaling-law result $3\nu=2-\alpha$ in the dimensional form

$$\xi^{-3}\left[\frac{1}{c}\frac{\partial c}{\partial T}\bigg|_{S/N}\right]^2\sim\beta^2k_B\rho C_V$$

on the critical isochore we find

$$\zeta_{pp}(\mathbf{q},\mathbf{s})\sim\rho c\xi$$

on the critical isochore.

IV. SUMMARY

In this section we summarize our results. (See Table II.) Note that the rough temperature dependences of quantities are also given in Table II taking ξ to diverge roughly as the $-\frac{2}{3}$ power of ϵ, C_p to diverge roughly as the $-\frac{4}{3}$ power of ϵ, and C_V to be roughly logarithmically divergent.

There are basically three different frequency regions: the low-frequency domain

$$s\lesssim s_T^*=(\lambda/\rho C_p)\xi^{-2}, \quad \text{(region I)} \quad (4.1a)$$

the intermediate region

$$s_T\ll s\lesssim s_\eta^*=\eta^*\xi^{-2}/\rho, \quad \text{(region II)} \quad (4.1b)$$

and the high-frequency region

$$s_\eta^*\ll s\lesssim c\xi^{-1}. \quad \text{(region III)} \quad (4.1c)$$

To state our results, we start from the high-frequency region and work down. In all our statements we assume $q\lesssim\xi^{-1}$.

In region III, $\lambda(q)$ has a singular part given by Eq. (3.35). If C_V is divergent at the critical point, this singular part of λ is given by

$$\lambda\sim\frac{\xi^{-2}}{c}\frac{C_p}{\beta}. \quad \text{(region III)} \quad (4.2a)$$

On the critical isochore, this gives

$$\lambda\sim\epsilon^{2\nu-\gamma-\alpha/2}. \quad (4.2b)$$

The exponent here probably lies within one- or two-tenths of zero. In addition to this strong cusp or weak infinity, there is also a constant term in λ coming from high wave numbers. In this region there are probably also weakly divergent or strongly cusped terms in η which we estimate as

$$\eta\sim\frac{\xi^{-2}}{c\beta}\frac{C_p}{C_V}, \quad \text{(region III and region II)} \quad (4.3a)$$

which behave like

$$\eta\sim\epsilon^{2\nu-\gamma+\alpha/2} \quad (4.3b)$$

on the critical isochore. On the other hand, ζ is much more strongly infinite than η. From Eq. (3.44)

$$\zeta\sim\frac{\xi^{-2}}{cC_V}k_B\left[T\frac{\partial c}{\partial T}\bigg|_{S/N}\right]^2 \quad \text{(region III and region II)} \quad (4.4a)$$
$$\sim\xi c\rho,$$

so that on the critical isochore

$$\zeta\sim\epsilon^{2\nu-2+\alpha/2}, \quad (4.4b)$$

which is roughly a $(-\frac{2}{3})$-power divergence.

The terms in Eq. (4.3) also seem to be the most important singular terms in η in region II. If these terms go to infinity, they dominate the behavior of η so that Eq. (3.27) gives

$$\lambda\sim\rho C_Vc\xi, \quad (4.5a)$$

and on the critical isochore we have

$$\lambda\sim\epsilon^{-\alpha/2-\nu}, \quad (4.5b)$$

if C_V diverges as $\epsilon^{-\alpha/2}$. Since α is small and $\nu\approx0.6$, this result indicates a rather strong divergence in the thermal conductivity. If, on the other hand, the singularities in Eq. (4.3) lead to a cusp rather than an infinity in η, then the constant term in η will dominate near the critical point. If we call this constant η_∞, we have from Eq. (3.27)

$$\lambda\sim\frac{\rho C_p\xi^{-1}}{\beta\eta_\infty} \quad \text{(regions I and II)} \quad (4.6a)$$

and on the critical isochore

$$\lambda\sim\epsilon^{-\gamma+\nu}, \quad (4.6b)$$

which is again roughly a $\frac{2}{3}$ power-law divergence. If the scaling laws were so fully correct as to prevent the high-frequency singularities in η, then Eqs. (4.6) would be correct rather than Eqs. (4.5).

Whatever form the singularities in λ might take, our approach predicts that λ will vary smoothly as one goes from region II into region I. That is to say that both the coefficient in front of the singularity and the critical exponent will remain constant as s passes through s_T^*. However, as one passes into region I, new processes become possible which make for new contributions to η. If η turns out to be divergent in region II, our approach predicts that it will have the same critical exponent in region I, but the coefficient preceding the divergent term might well increase markedly as one passed into region I. If η turns out to be nondivergent in region II, the integrals which defines it appears to diverge as ϵ^0 in region I. This should be read to mean that η can diverge logarithmically in region I if there is no divergence in regions II and III. On the other hand, ζ has a very strong divergence in region I. According to Eq. (3.24)

$$\zeta \sim \frac{\rho C_p \xi^{-1}}{\lambda^*} \frac{k_B c^2}{C_V} \left[\frac{T(\partial C_p/\partial T)_{S/N}}{C_p} \right]^2$$

$$\sim (C_p/C_V)\xi c\rho, \quad \text{(region I)} \tag{4.7a}$$

so that on the critical isochore

$$\zeta \sim \epsilon^{-\gamma-\nu+(3/2)\alpha}, \tag{4.7b}$$

which diverges roughly as ϵ^{-2}.

Finally, we notice one characteristic feature of the sound-wave damping constant:

$$D_s = \frac{\lambda}{\rho C_V} + \frac{\zeta + \frac{4}{3}\eta}{\rho}.$$

According to Eqs. (4.5) and (4.4), in regions II and III

$$D_s = A\xi c, \tag{4.8}$$

where A is a constant of order unity. Then in this region the sound-wave dispersion relation will read

$$s = \pm icq + \tfrac{1}{2}Ac(q\xi)q$$

or

$$s = cq[\pm i + \tfrac{1}{2}Aq\xi]. \tag{4.9}$$

Notice the dependence of this expression upon the characteristic parameter $q\xi$. Several recent authors[24,25] have used the assumption that frequencies of modes near the critical point depend upon $q\xi$ to relate apparently different transport phenomena near the critical point. In particular, they have estimated the order of magnitude of damping terms by assuming that the complex frequency of the oscillations were functions of the form

$$s = q^x f(q\xi). \tag{4.10}$$

Equation (4.9) is precisely of the form of (4.10). Hence, our arguments have provided one case in which this scaling assumption about frequencies can be derived from microscopic considerations. Notice, however, that in our case Eq. (4.4) only holds for the relatively high frequencies of regions II and III. However, in region I, ζ is considerably enhanced in size. Hence, in this region, the assumptions of Ferrel *et al.*,[24] and Halperin and Hohenberg[25] do not serve to predict the sound-wave damping constant.

The divergence of λ predicted by this work seems to agree with the experimental results of Cummins and Swinney[26] for CO_2, while it disagrees, for $T > T_c$, with the experiment of Ford and Benedek[27] on SF_6.

[24] R. A. Ferrell, N. Menyhárd, H. Schmidt, F. Schwabl, and P. Szepfalusy, Phys. Rev. Letters **18**, 891 (1967).

[25] B. I. Halperin and P. C. Hohenberg, Phys. Rev. Letters **19**, 700 (1967).

[26] H. Z. Cummins and H. L. Swinney (to be published).

[27] N. C. Ford and G. B. Benedek, Phys. Rev. Letters **15**, 649 (1965).

Reprinted From
Critical Phenomena
© 1971, LI Corso
Academic Press Inc. - New York

The Droplet Model and Scaling.

L. P. KADANOFF

Department of Physics, Brown University - Providence, R. I.

Consider a magnetic system near its critical point. The entire volume of the system will be filled by « droplets » or micro-domains which represent local fluctuations in the order parameter.

We describe a particular droplet by specifying a typical dimension, R. The system will contain in profusion droplets of all sizes up to a characteristic size ξ. The coherence length ξ is a function of

$$\varepsilon = \frac{T - T_c}{T_c} \tag{1a}$$

and of the dimensionless magnetic field,

$$h = \frac{\mu H}{kT_c}\,. \tag{1b}$$

For $R \gg \xi$, there are no droplets; however for any $R \ll \xi$, there are so many droplets with size of order R that these droplets essentially fill the entire volume. We shall assume one more property of these droplets: a droplet of size R will produce a fluctuation in the energy density and order parameter with an order of magnitude

$$\begin{cases} \delta\varepsilon_R = \dfrac{1}{R^{x_{\varepsilon_r}}}\,, \\[2ex] \delta\sigma_R = \dfrac{1}{R^{x_\sigma}}\,. \end{cases} \tag{2}$$

Now consider a correlation function formed by n-spin variables:

$$\langle \sigma_{r_1} \sigma_{r_2} \dots \sigma_{r_n} \rangle\,.$$

The completely correlated part of this function is the part which cannot be factorized into a product of several lower order correlation functions. We represent this by a subscript c. This part of the correlation function involves simultaneous correlations among all the different spin sites. Consequently, it must be produced by a droplet which spans all the sites, a droplet with

$$R \geqslant \max |r_i - r_j| . \tag{3}$$

Call this maximum r. Then the correlation function is the product of:

a) the probability that you are in a droplet with $R \geqslant r$, and

b) the fluctuation in each spin variable inside the droplet.

The first term is of order unity if $r \ll \xi$ and very small if $r \gg \xi$. The second term is according to eq. (2):

$$(\delta\sigma_R)^n|_{R\sim r} .$$

You put both terms together and find:

$$\langle \sigma_{r_1} \sigma_{r_2} \dots \sigma_{r_n} \rangle_c \sim \frac{1}{r^{n x_\sigma}} f_n(r/\xi) .$$

In a more precise statement, we would like the correlation functions to depend upon the distance variables, $|r_{ij}|$ divided by ξ. In this way, we find:

$$\langle \sigma_{r_1} \sigma_{r_2} \dots \sigma_{r_n} \rangle_c = \frac{1}{r^{n x_\sigma}} f_n(\{r_{ij}/\xi\}) , \tag{4a}$$

with $r_{ij} = |\mathbf{r}_i - \mathbf{r}_j|$.

Similarly we conclude that

$$\langle \sigma_{r_1} \sigma_{r_2} \dots \sigma_{r_n} \varepsilon_{r_{n+1}} \varepsilon_{r_{n+2}} \dots \varepsilon_{r_{n+m}} \rangle_c = \frac{1}{r^{n x_\sigma + m x_{\varepsilon}}} f_{n,m}(\{r_{ij}/\xi\}) , \tag{4b}$$

where ε_r represents the deviation of the energy density from its critical value.

Notice that our discussion so far has not given the dependence of the correlation function upon ε and h. We only know the dependence upon ξ. Therefore, let us for now set $h = 0$, $T > T_c$. Here,

$$\xi = \frac{1}{\varepsilon^\nu} .$$

Later on, we can see the behaviour for all h and ε.

Now we can calculate the n-th derivative of the free energy with respect to the magnetic field by making use of the formula

$$\frac{\partial^n G}{\partial n^n} = \int \mathrm{d}r_1 \dots \mathrm{d}r_n \langle \sigma_{r_1} \dots \sigma_{r_n} \rangle_c \,,$$

according to eq. (40), this integral may be written as:

$$\frac{\partial^n G}{\partial n^n} = \int \mathrm{d}r_1 \, \mathrm{d}(r_2 - r_1)\, \mathrm{d}(r_3 - r_1) \dots \mathrm{d}(r_n - r_1) \frac{1}{r^n \chi_\sigma} f_n(\{r_{ij}/\xi\}) \,.$$

The integral over $\boldsymbol{r}_1$ gives the volume of the system. By using the variables of integration,

$$\boldsymbol{q}_i = (\boldsymbol{r}_i - \boldsymbol{r}_1)/\xi \qquad (i = 2, 3, \dots, n) \,,$$

we can rewrite the integral as:

$$\frac{\partial^n G}{\partial n^n} = V \frac{\xi^{d(n-1)}}{\xi^{n\chi_\sigma}} \int \mathrm{d}q_2 \dots \mathrm{d}q_n f_n(\{q_{ij}\}) \,,$$

where d is the dimensionality of the system. The integral is now just a constant factor independent of ε. We call this factor f_n and we have:

$$\frac{\partial^n G}{\partial h^n} = V(\xi^{-d}/\xi^{-n(d-\chi_\sigma)}) f_n \,. \tag{5}$$

We can view eq. (5) as a statement about the power series expansion of σ in h. This power series would hold if

$$G(\varepsilon, h) = V\xi^{-d} f(h/\xi^{-n(d-\chi_\sigma)}) + \text{const} = V\varepsilon^{2-\alpha} f(h/\varepsilon^{\Delta}) + \text{const} \,, \tag{6}$$

with

$$\Delta = (d - \chi_\sigma)\nu \,; \qquad 2 - \alpha = d\nu \,. \tag{7}$$

Equation (6) now can be used to find all the thermodynamic derivatives on the critical region. For example, the heat capacity is:

$$C_h = \frac{\partial^2}{\partial \varepsilon^2} \sigma(\varepsilon, h) \,.$$

At $h = 0$, we get a contribution to the heat capacity

$$C_h = V \frac{\partial^2}{\partial \varepsilon^2} \varepsilon^{2-\alpha} f(0) \sim \varepsilon^{-\alpha} \,.$$

Consequently, α is the usual specific heat index.
The spontaneous magnetization is:

$$\frac{\partial G(\varepsilon, h)}{\partial h}\bigg|_{\substack{h=0\\ \varepsilon>0}} = V\varepsilon^{2-\alpha}\left[\frac{\mathrm{d}}{\mathrm{d}x} f(x)\right]_{x=h/\varepsilon^{\Delta}} \cdot \varepsilon^{-\Delta} = V\varepsilon^{2-\alpha-\Delta} f'(0) .$$

But the spontaneous magnetization is also:

$$\langle\sigma\rangle \sim (-\varepsilon)^{\beta} .$$

Therefore

(8a) $$\beta = 2 - \alpha - \Delta .$$

Similarly, the susceptibility is:

$$\chi \sim \frac{\partial^2 G}{\partial h^2}\bigg|_{h=0} = V\varepsilon^{2-\alpha-2\Delta} .$$

If γ is the susceptibility index, defined by:

$$\chi \sim \varepsilon^{-\gamma} \qquad \text{for } h = 0 ,$$

we find

(8b) $$\gamma = 2\Delta - (2 - \alpha) .$$

From (8a) and (8b) follows the relation among critical indices

(9) $$\gamma + 2\beta + \alpha = 2 .$$

Many more relations analogous to (7), (8) and (9) are easily derivable from eq. (6).

Let me list a few more consequences of this droplet idea. From eq. (4b), we find:

$$\frac{\partial^m}{\partial\varepsilon^m}\frac{\partial^n G}{\partial h^n}\bigg|_{h=0} = \int \mathrm{d}r_1 \ldots \mathrm{d}r_{n+m} \langle \sigma_1 \ldots \sigma_{r_n} \varepsilon_{r_{n+1}} \ldots \varepsilon_{r_{n+m}} \rangle_c \sim \left(\frac{\xi^d}{\xi^{x_{\varepsilon_r}}}\right)^m \frac{V\xi^{d(n-1)}}{\xi^{n x_\sigma}} .$$

Consequently

$$\frac{\partial^m}{\partial\varepsilon^m}\varepsilon^{2-\alpha-n\Delta} \sim \left(\frac{\xi^d}{\xi^{x_{\varepsilon_2}}}\right)^m \varepsilon^{2-\alpha-n\Delta} ,$$

as a result

$$\varepsilon^{-1} \sim \xi^d / \xi^{x_{\varepsilon_r}} ,$$

or

(10) $$(d-\chi_{\varepsilon_r})\nu = 1 .$$

All critical indices are evaluable from the two independent indices, χ_σ and χ_{ε_r}.

Finally, let us return to the h-dependence of the correlation functions. From statistical mechanics, we know:

$$\frac{\partial^q}{\partial h^q}\langle\sigma_{r_1}\ldots\sigma_{r_n}\varepsilon_{r_{n+1}}\ldots\varepsilon_{r_{n+m}}\rangle_c = \int d\tilde{r}_1\ldots d\tilde{r}_q\langle\sigma_{\tilde{r}_1}\ldots\sigma_{\tilde{r}_q}\sigma_{r_1}\ldots\sigma_{r_n}\varepsilon_{r_{n+1}}\ldots\varepsilon_{r_{n+m}}\rangle_c .$$

From eq. (4*b*), the last line may be rewritten as:

$$\int d\tilde{r}_1\ldots d\tilde{r}_q(1/r^{(n+q)\chi_\sigma+m\chi_{\varepsilon_r}})f_{n+q,m}(\{r_{ij}/\xi\}) .$$

By the same analysis as before, this integral is of the order:

$$\frac{1}{\varepsilon^{q\Delta}}\frac{1}{r^{n\chi_\sigma+m\chi_{\varepsilon_2}}}f_{n+q,m}(\{r_{ij}/\xi\}) .$$

Since the derivatives with respect to h bring down a factor $\varepsilon^{-\Delta}$, the result (4*b*) can be generalized to $h\neq 0$ as:

(11) $$\langle\sigma_{r_1}\ldots\sigma_{r_n}\varepsilon_{r_{n+1}}\ldots\varepsilon_{r_{n+m}}\rangle_c = \frac{1}{r^{n\chi_\sigma+m\chi_{\varepsilon_r}}}f_{n,m}(\{r_{ij}/\xi\},\{h/\varepsilon^\Delta\}) .$$

Equation (11) looks rather complex. But in one sense, it is really a very simple statement. Consider a correlation function formed from $2n$ spins. Let all of the distances be of the same order:

$$|r_{ij}|\sim r .$$

The eq. (11) states:

$$\langle\sigma_{r_1}\ldots\sigma_{r_{2n}}\rangle_c\sim\frac{1}{r^{n\chi_\sigma}}f_n(r/\xi)\sim[\langle\sigma_{r_1},\sigma_{r_2}\rangle]^n .$$

Consequently, all we have said is that when the r_{ij}'s are of the same order of magnitude, the completely correlated function is of the same order as the correlation function itself.

The scaling discussion of the n-spin correlation function was first carried out by A. PATASHINSKII and V. POKROWSKII (*Sov. Phys. JETP*, **23**, 292 (1966)).

Reprinted From
Critical Phenomena
© 1971, LI Corso
Academic Press Inc. - New York

Critical Behavior. Universality and Scaling.

L. P. KADANOFF
Department of Physics, Brown University - Providence, R. I.

1. – Introduction.

All phase transition problems can be described in precisely similar terms. The key to this description is the concept of an order parameter, which takes on different values in coexisting phases and hence jumps discontinuously in the course of a phase transition. The magnitude of the jump describes the amount of difference between the coexisting phases. If the jump is finite, the transition is first order; at a critical point, the jump goes to zero.

For example, in the liquid gas phase transition, the order parameter is the number of particles in a given volume. At low temperatures, the density difference between the gas and the saturated vapor is large. As the temperature increases (Fig. 1), the discontinuity in density decreases until—at the critical temperature—it goes to zero. At all temperatures higher than T_c, the phase transition disappears. A single-axis ferromagnet behaves in a precisely analogous fashion. The order parameter is the magnetization, which changes its sign in the course of the transition between the two possible directions of the magnetization. As the temperature increases, the zero-field magnetization has a decreasing magnitude, and it goes to zero at the critical or Curie temperature. Above T_c, the zero-field magnetization remains zero.

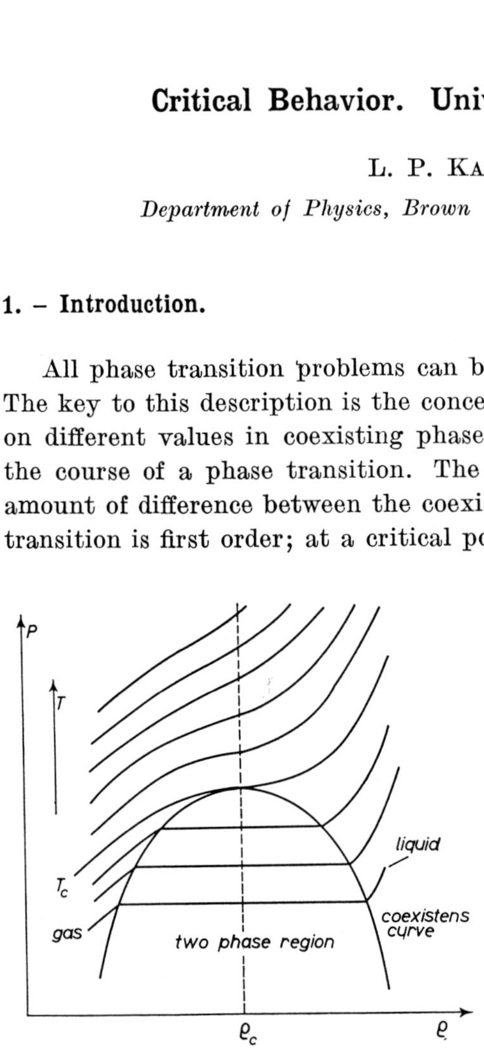

Fig. 1. – Schematic pressure *vs.* density diagram near the liquid-gas critical point.

In each case, a full description of the phase transition requires the use of a free energy and of thermodynamic « field » variables. The fields are intensive thermodynamic quantities which vary continuously across the phase transition. Typical field variables include temperature, chemical potential, magnetic field, and pressure. Usually, two field variables are employed. One field—for example, the magnetic field, h, in the ferromagnetic case—can be used to drive the system from one coexisting phase to the other. The other field—for example,

$$\varepsilon = \frac{T - T_c}{T_c} \tag{1}$$

is intended to drive the « orthogonal » change, that is, toward or away from the critical point.

The free energy is then written in terms of these field variables: $F = F(h, \varepsilon)$. The differential of the free energy

$$\mathrm{d}F = M\,dh + \langle \mathscr{H} \rangle \mathrm{d}\varepsilon\,, \tag{2}$$

defines pairs of thermodynamically conjugate variables (M, h) and $(\langle \mathscr{H} \rangle, \varepsilon)$, defined so that the conjugate to the first field is the order parameter. In most cases, the conjugate to the second field has the interpretation of an energy.

Near the critical point, the qualitative similarity among the different phase transition problems becomes even more apparent. Consider, for example the second derivatives of the free energy:

$$\left\{\begin{aligned} &\frac{\partial^2 F}{\partial h^2} \sim \langle (\delta M)^2 \rangle = \text{strongly singular}\,, \\ &\frac{\partial^2 F}{\partial h\, \partial \varepsilon} \sim \langle \delta M\, \delta \mathscr{H} \rangle = \text{intermediate strength singularity}\,, \\ &\frac{\partial^2 F}{\partial \varepsilon^2} \sim \langle \delta \mathscr{H} \partial \mathscr{H} \rangle = \text{weakly singular}\,. \end{aligned}\right. \tag{3}$$

In the magnetic case, the first of these is the magnetic susceptibility, χ, the second the temperature derivative of the magnetization, and the constant field specific heat. The second derivatives can always be written in terms of correlations in the fluctuations in the order parameter, M, and the energy, H. In all phase transition problems, there are strong fluctuations in the order parameter which make the first expression in eq. (3) highly singular at the critical point. The weaker fluctuations in the energy produce a far weaker singularity in the specific heat.

To cover this qualitative statement into quantitative terms, one notices that at the critical value of the first field, $h = 0$, the experimental results for these derivatives can be written in terms of power-law singularities in $\varepsilon = (T - T_c)/T_c$. For example, the most singular derivative is usually written:

$$\langle (\delta M)^2 \rangle \sim \begin{cases} -\varepsilon^{-\gamma} & \text{for } T > T_c, \\ (-\varepsilon)^{\gamma'} & \text{for } T < T_c. \end{cases} \tag{4}$$

For $T < T_c$, the jump in the order parameter goes to zero as

$$\text{jump} \sim \langle M \rangle \sim (-\varepsilon)^{\beta}, \tag{5a}$$

so that the second line in eq. (3) becomes proportional to the derivative of the jump with respect to ε or

$$\langle \delta M \delta \mathscr{H} \rangle - (-\varepsilon)^{-(1-\beta)}. \tag{5b}$$

Finally, the specific heat is usually represented as

$$\begin{cases} \langle (\delta \mathscr{H})^2 \rangle - \dfrac{A}{\alpha} [\varepsilon^{-\alpha} - 1] + B, & \text{for } T > T_c, \\ \dfrac{A'}{\alpha'} [(-\varepsilon)^{-\alpha'} - 1] + B', & \text{for } T < T_c, \end{cases} \tag{5c}$$

In the special case $\alpha = 0$, the singularity in eq. (5c) becomes logarithmic in character.

TABLE I. – *Singularities in various thermodynamic derivatives.*

Phase transition	Strong singularity γ	Intermediate singularity $1-\beta$	Weak singularity α
Mean field theory	1	$\frac{1}{2}$	0
$d = 2$ ISING	$\frac{7}{4}$	$\frac{7}{8}$	0
CO_2, Xe, ^{4}He	1.25	0.65	~ 0.1
$d = 3$ ISING	1.25	0.685	0.125

Table I shows a selection of data on the critical indices. These data clearly confirm the segregation of singularities into strong, intermediate and weak with

$$\gamma > (1 - \beta) > \alpha. \tag{6}$$

Furthermore, this Table indicates a remarkable regularity in the values of α, β and γ. Several apparently different systems show quite similar values of these indices. This empirical evidence has led several different authors to suggest that the details of the system undergoing the phase transition might be irrelevant and that many different phase transition problems might be essentially identical (*). This idea can be stated more precisely in the following *hypothesis of universality.*

All phase transition problems can be divided into a small number of different classes depending upon the dimensionality of the system and the symmetries of the order state. Within each class, all phase transitions have identical behaviour in the critical region, only the names of the variables are changed.

To understand the physical source of the universality concept, we should look into the source of the critical singularities. Consider, for example, the magnetic susceptibility χ which can be written in terms of the fluctuations in the order parameter as:

$$\chi = \langle (\delta M)^2 \rangle .$$

If the magnetization is rewritten as an integral over all space of the magnetization density

$$M = \int \mathrm{d}\boldsymbol{r}\, m(\boldsymbol{r}) ,$$

then the susceptibility is

$$\chi = \int \mathrm{d}\boldsymbol{r}\, \mathrm{d}\boldsymbol{r}' \langle \delta m(\boldsymbol{r})\, \delta m(\boldsymbol{r}') \rangle ,$$

so that the suceptibility per unit volume becomes

$$\frac{\chi}{V} = \int \mathrm{d}\boldsymbol{r} \langle \delta m(\boldsymbol{r})\, \delta m(0) \rangle . \tag{7}$$

The susceptibility is known to diverge as the critical point is approached. How can this occur? The magnetization density in eq. (7) is known to be bounded so that the integrand can never become infinite. The only possible source of the observed singularity is that, as the critical point is approached, the range of the integrand becomes larger and larger. Physically, we are seeing

(*) The ideas stated up to this point are very old ones in the theory of phase transitions. They were, for example, discussed by LANDAU and LIFSHITZ (*Statistical Physics*, Chapt. 5 (London, 1958)) in the context of mean field theory. Recently, GRIFFITHS (*Phys. Rev. Lett.*, **24**, 1479 (1970)) and KADANOFF (*Newport Beach Conference on Phase Transitions and Invited Paper at Washington APS Meeting* (1970)) hawe stressed these ideas once more.

correlations in fluctuations involving larger and larger special regions. Our conclusion is then that critical phenomena are produced by fluctuations in the order parameter within domains which may become larger and larger as criticality is approached.

The range over which appreciable correlations exists is called the coherence length, ξ. As the critical point is approached, the thermodynamic derivatives go to infinity because ξ goes to infinity. The source of the infinity in ξ can be understood by looking at fluctuation phenomena in the liquid-gas phase transition.

Just below the critical temperature, the system has a free choice to make between being a liquid at high density or a vapor at low density. Both choices are equally good. But since liquid wants to be in contact with liquid and vapor with vapor, the system must decide.

At any temperature and pressure near the critical values, this decision-making process is going on throughout the volume of the fluid. A natural fluctuation produces a droplet of the wrong phase. This droplet secretes other material of the same density, and it grows larger and larger. It stops growing when the cost in available—*i.e.* free-energy for making the droplet of wrong phase becomes comparable with kT. Since near the critical point it cost very little free energy per unit volume to make the wrong phase, the droplet can grow very large.

The coherence length ξ is a size of a typical droplet. As the critical point is approached, the free energy cost of making a droplet goes to zero, and the size of a typical droplet ξ, goes to infinity. Notice how this droplet making can be viewed as the oscillations in the decision making while the system tries to decide what phase it is to enter.

Incidentally, this effect is literally easily visible. Far from the critical point, the fluctuation droplets exist, but they are too small to scatter an appreciable amount of light. (Remember the light scattering for a small droplet is proportional to the radius to the sixth power.) But, as the fluid approaches the critical point, it begins to contain fluctuation droplets which are large enough to scatter light appreciably. Then, the liquid takes on a cloudy appearance. This effect is called critical opalescence.

Now let us return to the universality hypothesis. The physical origin of this hypothesis is the observation that critical phenomena are dominated by fluctuations in M and H which appear in a special scale which is very long compared to the force range. Then these fluctuations cannot probe all the details of the interatomic potential. Rather they only see certain gross features of the potential: for example, the amount of breaking of an exact symmetry (as measured by h_z) or the distance from the critical point (as measured by ε). Furthermore, the nature of the fluctuations is very much determined by the symmetries of the order parameter and the dimensionality of the system.

Universality describes the relationship among different phase transition problems. The theorist can discuss this relation in the following way: He imagines that yet another field is inserted into the free energy. Call that other field λ and the operator which is its thermodynamic conjugate, U. Here λ represents a parameter in the Hamiltonian. Continuous variation from $\lambda = 0$ to $\lambda = 1$ might represent the change in the Hamiltonian which takes us from the Ising model to the Heisenberg model, or from Ni to Fe or from a nearest neighbor interaction to a next nearest neighbor interaction. Therefore, the discussion of λ and its thermodynamic conjugate U is in effect the discussion of the relationship among different phase transition problems.

This mathematical redefinition of the problem is expressed by writing the free energy as a function of h, ε and λ

$$F = F(\varepsilon, h, \lambda)\,,$$

with a differential

$$\mathrm{d}F = \langle M\rangle \mathrm{d}h + \langle \mathscr{H}\rangle \mathrm{d}\varepsilon + \langle U\rangle \mathrm{d}\lambda\,. \tag{8}$$

TABLE II. – *Variables for phase transition problems.*

Phase transition	First pair		Next pair		Third pair	
	Order parameter	Conjugate			Conjugate	λ
ferro-magnetic	M_z	H_z	H	ε	$H_{n.n}-H_{n.n.n.}$	ratio of n.n. to n.n.n. coupling
anti ferro-magnetic	sublattice magneti-zation	alternating magnetic field	H	ε	M_z	H_z
liquid-gas	$N-N_c$	$\mu-\mu_c(T)$	$H-\mu N$	ε	kinetic energy-potential energy	De Boer's parameter
super-fluid	$\langle\psi\rangle$	un-physical	H	ε	N	μ

We hope that, in some sense, λ is irrelevant. Table II gives examples of this new conjugate pair of variables (λ, U).

The experimental consequences of universality can be seen most easily in the simplest case: a magnetic transition in a single-axis ferromagnet. We choose our basic variables ε and h so that the phase transition remains at $\varepsilon = 0$ and $h = 0$ and imagine that λ is some internal parameter—like the ratio of nearest neighbor coupling constant to the next nearest neighbor coupling constant—which leaves unchanged the basic symmetry about $h = 0$. Then universality implies that the basic thermodynamic functions and correlation functions only depend on λ via a trivial change of variables. The functional form is the same as at $\lambda = 0$. However, the variables in these functions are changed in that h, ε, and M are each multiplied by parameters a, b, c, d which depend upon λ. If this conclusion is right, then the critical indices—which determine the power laws appearing in various singularities—as well as the form of the correlation and thermodynamic functions should be independent of λ (see Table III).

TABLE III. – *Example of universality hypothesis.* Here λ is a ratio of nearest-neighbor to next-nearest-neighbor coupling. GRIFFITHS and KADANOFF (see ref. [1]) have each independently come to this formulation of universality.

Variables

$$\varepsilon = (T - T_c(\lambda)(/T_c(\lambda) , \qquad h = \mu_\beta H_z/KT_c , \qquad \lambda = J_{\text{n.n.n.}}/J_{\text{n.n.}} .$$

Write

$$\langle M \rangle = m_0(h, \varepsilon) , \qquad \text{at } \lambda = 0 ,$$
$$\langle \sigma_z(0)\, \sigma_z(r) \rangle = g_0(h, \varepsilon, r) .$$

Universality implies for $\lambda \neq 0$:

$$\langle M \rangle = am_0(\bar{h}, \bar{\varepsilon}) , \qquad \langle \sigma_z(0)\, \sigma_z(r) \rangle = (ab/d^3)\, g_0(\bar{h}, \bar{\varepsilon}, \bar{r}) ,$$
$$\bar{h} = bh , \qquad \bar{\varepsilon} = c\varepsilon , \qquad \bar{r} = dr .$$

One set of data on this point is contained in the equation-of-state work for CO_2, Xe and He as done by VICENTINI-MISSONI, LEVELT-SENGERS, GREEN and by SCHOFIELD, LISTER and HO. Each group concluded that these diverse gases could—near the critical point,—be described by an equation of state of universal form. Table IV shows some critical indices and ratios of susceptibilities and specific heats above and below T_c.

TABLE IV. – *Critical data for testing universality.*

E.g.	CO_2	Xe	^{4}He
β	0.350	0.350	0.359
γ	1.26	1.26	1.24
$C_0/C^\pm$	4.4	4.1	3.6
$A^\pm/A_0$	1.27	1.27	1.34

P. SCHOFIELD, J. D. LITSTER and J. T. HO: *Phys. Rev. Lett.*, **23**, 1098 (1969).
M. VICENTINI-MISSONI, J. M. H. LEVEL SENGER and M. S. GREEN: *Phys. Rev. Lett.*, **22**, 389 (1969).

If universality is correct, each of these parameters should be materially independent. As we can see, even for these diverse materials, the parameters are approximately constant. Although it is difficult to be sure, the difference may well be within experimental error.

TABLE V. – $-\beta\varkappa \sum_{\langle I^0 I^{0\prime}\rangle} [\boldsymbol{S}_{I^0}\cdot\boldsymbol{S}_{I^{0\prime}} + \lambda S^z_{I^0} S^z_{I^{0\prime}}]$.

Region	Spin symmetry of ordered state	Value of λ	Value of γ	Value of 2ν
$-2<\lambda<0$	two-dimensional rotation	– –	1.31	1.33
		-1	1.32	1.34
		-0.4	1.32	1.34
		-0.2	1.28 (?)	1.31 (?)
$\lambda=0$	three-dimensional rotation	0	1.38	1.40
$\lambda>0$	single easy axis	0.25	1.19	1.20
		0.83	1.24	1.25
		4	1.23	1.25
		—	1.23	1.25

D. JASNOW and M. WORTIS: *Phys. Rev.*, **176**, 739 (1968).

Another case in point is the calculation of Jasnow and Wortis related to an anisotropic spin-infinity Heisenberg model. The Hamiltonian included a parameter λ which measures the couplings—the z components of the spin relative to the x and y component. The symmetries of the ordered state fall into three classes depending upon the value of λ. Apparently, as the symmetry changes so does the value of the critical indices, but within each symmetry group, the indices seem to remain constant, as required by universality.

Therefore, the conjugate to λ, *i.e.* U, must have much weaker fluctuations than either M or $\mathscr{H}$. Furthermore, this must be true for any U conjugate to λ at fixed $\bar{\varepsilon}$ and $\bar{h}$. The conditions of fixed $\bar{\varepsilon}$ and $\bar{h}$ essentially guarantee that U do not contain any terms which fluctuate as M or $\mathscr{H}$. In effect, our arrangement of the free energy has subtracted from a general operator, U, any terms which might be proportional to M or H. According to universality, what is left is very weakly fluctuating.

This is a very strong result. It implies, for example, that we need not concern ourselves about a too careful definition of the order parameter or energies. For example, if M and M' are two possible candidate for the order parameter, then we can always write:

$$M' = aM + b\mathscr{H} + U ,$$

where a and b are constants and $b\mathscr{H}+U$ will always be much more weakly fluctuating than M. If $a\neq 0$, M' is as good an order parameter as M.

Later on, we shall have to look in more detail at this result of small fluctuations in U.

2. – Scaling derived from universality.

The next set of ideas which I wish to derive from the universality hypothesis go under the general name of scaling. Scaling theory basically gives a set of relations among critical exponents by assuming that universality includes the case in which the «irrelevant» parameter λ is the scale of length in the problem. To see how scaling arises, consider the particular example of the Ising model in which the Hamiltonian is given by

$$-\frac{\mathscr{H}}{k_B T} = K \sum_{\langle I^0 I' \rangle} \sigma_{I^0}\sigma_{I^{0'}} + h \sum_{I^0} \sigma_{I^0} , \tag{12}$$

while the free energy is

$$\exp[-F(h, \varepsilon)/k_B T] = \sum_{\{\sigma_r=\pm 1\}} \exp[-\mathscr{H}/R_B T] . \tag{13}$$

Basically, we have a problem with a group of spins on sites, $\boldsymbol{r}$, each spin being capable of pointing up ($\sigma_r = +1$) or down ($\sigma_r = -1$). The Hamiltonian gives an energetic preference for lining up nearest neighboring spins with each other and for having spins aligned with the external magnetic field, h. Two of the key descriptions of critical behaviour for this problem are the order parameter

$$\langle\sigma\rangle = n_0(h, \varepsilon) \tag{14a}$$

and the spin-spin correlation function

$$\langle\sigma_0\sigma_r\rangle = g_0(h, \varepsilon, \boldsymbol{r}) \tag{14b}$$

Now imagine that we change our point of view about this problem. We mentally divide the lattice into cells, each side of length La_0, where a_0 is the lattice constant. L is chosen to be very small compared to a coherence length, but is large enough so that the cell contains many spin sites (see Fig. 2).

$$1 \ll L \ll \xi/a .$$

Let us assume for a moment that all the interactions between different cells were turned off. Then each cell would act independently of every other and there would be no correlations over distances larger than a cell size. But, these are precisely the correlations which produce the divergences in thermodynamic derivatives which are characteristic of the critical region. Hence, correlation between different cells must be quite crucial—the interactions among cells must be all important.

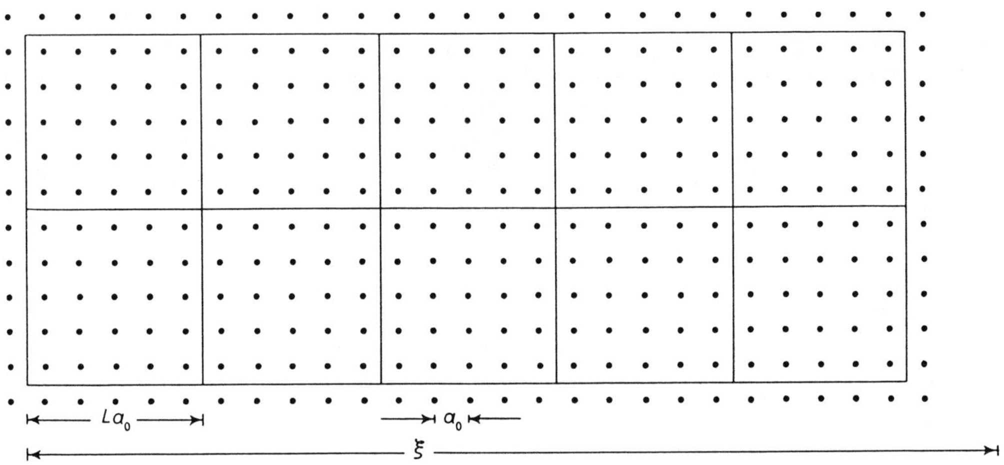

Fig. 2. – Division of Ising model lattice into cells $L \gg 1$ but $La_0 \ll 1$.

We have already noted that the correlations near the critical point should not depend upon the details of the interaction. Let us assume that the cell size La_0 is one of these irrelevant details. Then, in each cell there must be a variable μ_α, α being a label for the cell, which behaves in an essentially identical fashion to our original variable σ_r. That is, interactions among different cells should produce correlations among the μ_α's identical in structure to the correlations in the original Ising model problem.

The only possible difference between the site problem and the cell problem can be in the values of the two parameters ε and h. Assume that the values of these parameters for the cell problem are $\tilde{\varepsilon}$ and $\tilde{h}$. Since these, respectively, describe the crucial effect of the interaction between each cell and the external magnetic field, they can depend upon the size of the cell. As the magnetic field goes to zero, the effective field in the cell problem must also go to zero. Hence, $\tilde{h} \sim h$. The constant of proportionality may depend on the cell size so we guess

$$\tilde{h} = bh , \tag{15a}$$

where b is a function of cell size. Similarly, $\tilde{\varepsilon}$ must go to zero as ε goes to zero since the cell problem must become critical as the site problem does. For this reason, we write

$$\tilde{\varepsilon} = c\varepsilon \, . \tag{15b}$$

Finally, the statement that site and cell problems are identical implies that the new variables have the same averages as the old, *e.g.*

$$\langle \mu_\alpha \rangle \;\; = m_0(\tilde{h}, \tilde{\varepsilon}) \, , \tag{16a}$$

$$\langle \mu_\alpha \mu_0 \rangle = g_0(\tilde{h}, \tilde{\varepsilon}, \alpha) \, . \tag{16b}$$

Relations (**16**) require that the 0 and α be separated by many cells since universality only applies to long-ranged correlations.

How should we define the local order parameter, μ_α, in the cell problem? According to universality, the precise choice of order parameter is not too important because any variable with very large fluctuations will behave like M plus a more weakly fluctuating quantity. Therefore, we can choose μ_α as proportional to the total magnetization in a cell and conclude from universality that our choice is appropriate. We write:

$$\mu_\alpha = a \sum_{r\alpha} \sigma_r / L^d \, , \tag{17}$$

with a being a function of L and d the dimensionality of the system. Now put everything together. Insert Eq. (17) into (16a). Since there are L^d sites per cell, we find

$$a \langle \sigma_r \rangle = m_0(\tilde{h}, \tilde{\varepsilon}) \, ,$$

so that from eq. (14a)

$$a m_0(h, \varepsilon) = m_0(\tilde{h}, \tilde{\varepsilon}) \, .$$

Finally, eqs. (15) imply

$$m_0(h, \varepsilon) = \frac{1}{a} m_0(bh, c\varepsilon) \, . \tag{18a}$$

A similar analysis applied to g_0 gives:

$$g_0(h, \varepsilon, r) = \frac{1}{a^2} g_0(bh, c\varepsilon, r/L) \, . \tag{18b}$$

The solutions of eqs. (18) are very restricted. The only physically reasonable solutions have a, b, and c of L:

(19)
$$\begin{cases} a = L^{x_\sigma} = L^{\beta/\nu}\,, \\ b = L^{x_h} = L^{\Delta/\nu}\,, \\ c = L^{x_\varepsilon} = L^{1/\nu}\,. \end{cases}$$

Here the symbols β, Δ and ν are the conventional critical indices for this problem. The solution of eqs. (18) are then

(20a)
$$m_0(h, \varepsilon) \;\;= \varepsilon^\beta m^*(h/\varepsilon^\Delta)\,,$$

(20b)
$$g_0(h, \varepsilon, r) = \varepsilon^{2\beta} g^*(h/\varepsilon^\Delta, \varepsilon^2 r)\,.$$

Apparently, this scaling analysis shows that all the critical indices depend upon three indices, β, Δ, and ν. However, there is one more relationship among those which implies that there are only two independet indices. One can calculate the magnetic susceptibility either as

$$\chi = \frac{\partial m_\sigma(h, \varepsilon)}{\partial h}\,,$$

or as

$$\chi = \int \mathrm{d}^2 r\, g_0(h, \varepsilon, r)\,.$$

The identity of these two results implies at once that

(21)
$$x_\sigma + x_n = d\,,$$

or

(22)
$$\beta + \Delta = d\nu\,,$$

so that all critical indices can be determined from β and Δ alone.

3. – Dimensional analysis.

The argument I have just gone through is rather involved. It would be helpful to have a simple set of arguments which would enable us to derive in a fairly automatic and quick fashion the main results of the scaling analysis. This kind of useful «engineering» result will emerge if we restate scaling as a kind of dimensional analysis. To do this, notice that scaling describes the result od making the replacement

$$\mathbf{r} \to \mathbf{r}/L\,, \tag{23a}$$

as equivalent to making replacements in our other variables. As another example consider the correlation function:

$$\langle \sigma_0 \sigma_{r_1} \sigma_{r_2} \sigma_{r_3} \rangle\,, \tag{23b}$$

where the four points form a square of side R. We know that this correlation function behaves as $(1/R^{\chi})$ at the critical point. What is x? From eqs. (23) the correlation function scales as $L^{4\chi_\sigma}$. Hence, $\chi = 4\chi_\sigma$.

There exists one more important quantity, the energy density ε_r. We add to eqs. (23) the statement that the critical part of ε_r scales as

$$\varepsilon_r \to L^{\chi_{\varepsilon_r}} \varepsilon_r\,. \tag{24a}$$

Notice that

$$\mathscr{H} = \int \mathrm{d}r\, \varepsilon_r$$

and ε are conjugate variables.

$$\sigma_r \to a\sigma_r = L^{\chi_\sigma} \sigma_r\,, \tag{24b}$$

$$h \to bh = L^{\chi_h} h\,, \tag{24c}$$

$$\varepsilon \to c\varepsilon = L^{\chi_\varepsilon} \varepsilon\,. \tag{24d}$$

All critical relations are supposed to be invariant under the simultaneous transforms (23). For example, consider the relationship

$$\left.\frac{\partial \sigma}{\partial h}\right|_{\varepsilon, h=0} \sim \varepsilon^{-\gamma}\,.$$

The replacement (23) applied to this relation gives

$$L^{x_\sigma - x_h} \frac{\partial \sigma}{\partial h} \sim \varepsilon^{-\alpha} L^{-\gamma x_\varepsilon}\,,$$

or

$$\gamma = \frac{x_\sigma - x_h}{x\varepsilon}\,.$$

From eq. (19), this reduces to the well-known result

$$\gamma = \Delta - \beta .$$

For our other pair of conjugate variables (σ, h), we have the relation

$$x_\sigma + x_h = d . \tag{25a}$$

A careful analysis shows that a similar relation is satisfied by this pair

$$x_\varepsilon + x_{\varepsilon_r} = d . \tag{25b}$$

Then, the specific heat

$$\left.\frac{\partial \varepsilon_r}{\partial \varepsilon}\right|_{h=0} \sim \varepsilon^{-\alpha}$$

scales as $L^{x_{\varepsilon_r} - x_\varepsilon}$ from the left-hand side and as $L^{-\alpha x_\varepsilon}$ from the right-hand side. We conclude that

$$\alpha = -\frac{\chi_{\varepsilon_r}}{\chi_\varepsilon} + 1 = -\frac{d - \chi_\varepsilon}{\chi_\varepsilon} + 1 = 2 - \alpha\nu .$$

4. – Back to universality.

Notice that the variables with lower powers of x are the ones with the larger fluctuations. Thus,

$$x_\sigma < x_{\varepsilon_r} .$$

We could have either $x_{\varepsilon_r} < x_\varepsilon$ or $x_{\varepsilon_r} > x_\varepsilon$ depending upon whether α is greater or less than zero. However, when $\alpha < 0$, it is convenient to redefine the free energy by subtracting a term ε, ε_r from it. Then ε_r and ε interchange roles. Therefore we can always choose

$$0 < x_\sigma < x_{\varepsilon_r} < x_\varepsilon < x_h < d .$$

Now let is try to fit U and λ into this scheme. If U has less strong fluctuations than λ or M, we should expect that the density, $u(r)$ defined by:

$$U = \int \mathrm{d}r\, u(r) ,$$

might scale as

$$u(r) \to L^{x_u} u(r) , \tag{26a}$$

with

$$x_u > x_{\varepsilon_r}$$

Then, the conjugate variable λ scales as

$$\lambda \to L \lambda^{\alpha - x_u} = L^{x_\lambda} \lambda . \tag{26b}$$

Large values of x_u correspond to weakly fluctuating operators. How large need x_u be before we can neglect the critical effects of λ? The answer which arises from this scaling analysis is quite simple: $\partial/\partial\lambda$ is negligible if

$$d > x_u . \tag{27}$$

To see this simply calculate $(\partial/\partial\lambda)\langle\sigma_r\rangle$. Assume that $\langle\sigma_r\rangle \sim \tilde{\varepsilon}^\beta$ and $(\partial/\partial\lambda)\langle\sigma\rangle|_{\widehat{\varepsilon, h}} \sim \sim \tilde{\varepsilon}^{\beta'}$, If $\beta' > \beta$, the change in λ will produce only a small correction to the previous result for the magnetization since

$$\langle\sigma_r\rangle_{\lambda=1} - \langle\sigma_r\rangle_{\lambda=0} \sim \tilde{\varepsilon}^{\beta'} \ll \tilde{\varepsilon}^{\beta} .$$

But, according to scaling, the condition $\beta' > \beta$ is a consequence of eq. (27).

By this argument, the only local fluctuating variables which we need consider obey

$$0 < x_u < d .$$

If a variable falls in this range, we must include it in the scaling analysis. As a result, we find that the thermodynamic functions are now dependent upon three variables:

$h =$ conjugate to order parameter,

$$\varepsilon = \frac{T - T_c(\lambda)}{T_c(\lambda)} ,$$

$$\lambda - \lambda_a = \delta\lambda = \text{new variable} .$$

The magnetization, for example, appears in the form:

$$\begin{cases} m(\varepsilon, h, \delta\lambda) = \varepsilon^\beta m^*\left(\dfrac{h}{\varepsilon^{\Delta'}}, \quad \dfrac{h}{\delta\lambda^{\Delta'}}\right), \\ \Delta' = (d - \chi)/(d - \chi) . \end{cases} \tag{28}$$

The form (28) describes a generalization of scaling to include the possibility of more than two «field» variables. The concepts of scaling and universality remain unchanged by this trivial generalization.

A new situation arises, however, when $x_u = d$, so that $\Delta' = \infty$. In this case, dimensional analysis is not sufficient to determine the behaviour of thermodynamic functions. A more careful analysis shows that the critical indices can vary with λ in this case. Hence, universality can fail when $x_u = d$.

To sum up: If $0 < x_u < d$, a new variable must be included in the scaling functions but scaling holds; if $x_u = d$, scaling may fail; if $x_u > d$; the fluctuating variable, U, is irrelevant to critical behaviour.

5. – Operator algebras.

The foregoing analysis indicates that critical phenomena are described in terms of a finite number of local fluctuating operators, $O_\gamma(\boldsymbol{r})$. Here γ is a label to tell us which operator we are talking about. For example, $O_0(\boldsymbol{r})$ might be the unit operator, $O_1(\boldsymbol{r})$ the local order parameter, $O_2(\boldsymbol{r})$ the energy density, $O_3(\boldsymbol{r})$ the local value of ε, etc. Each of these operators scales with a characteristic power of L, *i.e.*

$$O_\gamma(\boldsymbol{r}) \to O^\gamma(r) L^{x_\gamma} . \tag{29}$$

Here, $x_0 = 1$, $x_1 = x_\sigma$, etc. This scaling analysis implies that at the critical point,

$$\langle O_\alpha(\boldsymbol{r}) O_\beta(0)\rangle = \frac{\alpha_{\alpha,\beta,0}}{r^{x_\alpha + x_\beta}} , \tag{30}$$

where the a's are numbers independent of $\boldsymbol{r}$.

Now consider the operator formed from a product of the basic operators, *e.g.*

$$X(\boldsymbol{R}) = O_\alpha\left(\boldsymbol{R} + \frac{\boldsymbol{r}}{2}\right) O_\beta\left(\boldsymbol{R} - \frac{\boldsymbol{r}}{2}\right). \tag{31}$$

When r becomes much smaller than the coherence length, this product is itself a local operator, localized about the point, $\boldsymbol{R}$, and then its critical behaviour must be expressible in terms of the known set of operators O_γ. In particular, it must be a linear combination of the known operators

$$X(\boldsymbol{R}) = \sum_\gamma A_\gamma O_\gamma(\boldsymbol{R}) . \tag{32}$$

Equation (32) means that $X(\boldsymbol{R})$ has critical fluctuations which behave in just the same way as the right hand side of eq. (32). If $Y(R')$ is an operator localized about R' with

$$|R'-R| \gg r\,,$$

then eq. (32) means

$$\langle X(\boldsymbol{R})\, Y(\boldsymbol{R}')\rangle = \sum_{\gamma} A_{\gamma} \langle O_{\gamma}(\boldsymbol{R})\, Y(\boldsymbol{R}')\rangle\,,$$

with precisely the same values of A for all different possible Y's.

The coefficients A_{γ} depend upon α and β and also upon the separation vector r. For this reason, we write:

$$A_{\gamma} \equiv A^{\alpha,\beta,\gamma}(\boldsymbol{r})\,,$$

so that eq. (32) becomes:

$$O_{\alpha}\left(\boldsymbol{R}+\frac{\boldsymbol{r}}{2}\right) O_{\beta}\left(\boldsymbol{R}-\frac{\boldsymbol{r}}{2}\right) = \sum_{\gamma} A_{\alpha,\beta,\gamma}(\boldsymbol{r})\, O_{\gamma}(\boldsymbol{R})\,. \tag{33}$$

Furthermore if $r \gg a_0$, where a_0 is a lattice constant, the dimensional analysis concepts apply to eq. (33). As a result, we can find the r-dependence of A as

$$A_{\alpha,\beta,\gamma}(\boldsymbol{r}) = \frac{a_{\alpha,\beta,\gamma}(\boldsymbol{r}/|\boldsymbol{r}|)}{r_{x_{\alpha}+x_{\beta}-x_{\gamma}}}\,, \qquad \text{for } r \gg a_0\,. \tag{34}$$

It is interesting to speculate about whether a knowledge of the structure of the operator algebra—that is whose coefficients vanish—might be sufficient to determine the values of the critical indices ν_{α}.

There is one case in which we can check out this idea: the two-dimensional Ising model. A direct calculation enables us to examine a structure of correlations which gives the reduction relations. These reduction relations include some unexpected symmetry properties in addition to the expected ones. The expected symmetries include flipping the sign of a spin, special rotation, and the interchange of the sign of $T-T_c$. This last symmetry was initially predicted by KRAMERS and WANNIER. But there is also an unexpected symmetry which involves a kind of « quantum number » with addition properties like those of angular momenta. To exploit this quasi-angular momentum symmetry, it is convenient to pick an axis of quantization—say the x-axis—and allow all the operators being multiplied to be on this axis. When this representation is used, the algebra simplifies considerably because many of the coefficients A vanish.

An identification of two of the elements of the algebra as σ_r and $\partial\sigma_r/\partial y$ permits the imposition of consisting conditions on the algebra. This conditions, together with rotational invariance, result in the determination of all the critical indices, ν_γ.

Consequently, the symmetry properties of the theory can indeed be used to construct all the critical indices in this case of the two-dimensional Ising model. Perhaps the algebraic idea will also permit the symmetries to be exploited in finding the critical indices for other cases. I do not know. One difficulty is that the symmetry properties used in describing the two-dimensional Ising model were very far from obvious. It is probably that hidden symmetries underlie the three-dimensional cases. Right now, I am trying to find them. But I have no successes to report so far.

Reprinted from *Phase Transitions and Critical Phenomena*,
Vol. 5A, ed. C. Domb and M. S. Green, 1976, Academic Press.

1. Scaling, Universality and Operator Algebras

Leo P. Kadanoff

*Barus and Holley Building, Brown University,
Providence, Rhode Island 02912, USA*

Introductory Note

There are three identifiable periods in the development of the theory of critical point behaviour. The first and longest period begins with Van der Waal's development of a mean field theory approach to the liquid–gas phase transition. This approach proved to be tremendously rich and extendable to the description of a whole series of different phase transitions.[1] A beautiful extension and culmination of this approach is found in the BCS theory of superconductivity.[2] However, in the decade preceding 1965 both numerical calculations[3] and experiments[4, 5] showed with increasing clarity that the mean field idea was quantitatively incorrect in the neighbourhood of thermodynamic critical point.[6]

The next era in the development of the field started around 1965 with the incorporation of these facts into more or less phenomenological theories of critical behaviour.[7] These phenomenological approaches can all be understood in terms of the concept of *scaling invariance*,[8] which was brought forward at roughly the same time. The scaling concept is one of the ideas described in detail in this paper.

In this same period, another key idea in critical phenomena theory, *universality*, was extensively developed. The idea of universality is that apparently dissimilar systems show considerable similarities near their critical points. It is an old idea dating back to the nineteenth century,[9] but its mathematical formulation[10] has led to a considerably better understanding of the underlying nature of critical phenomena and also to a much more useful analysis of experiments.[11] The recent developments in the concepts of universality form another main section of this paper.

The most recent period in the theory was opened with a brilliant series of papers by K. Wilson and co-workers.[12–15] Wilson essentially took the semi-phenomenological concepts of scaling and universality and converted these ideas into real calculations of critical point behaviour. Literally hundreds of other calculations have followed upon this breakthrough. In addition, F. Wegner[15] has shown how the concepts of universality and scaling naturally follow from Wilson's calculational method.

In this paper, I shall utilize the hindsight offered by Wegner's formulation of Wilson's approach to explain and develop the concepts included in the theory of critical phenomena.

I. Critical Points and Thermodynamic Singularities

In this section of the paper, the basic concepts of critical phenomena theory are introduced, using the notation and example of the Ising model.

A. The Ising model

1. *Definition*

Given a lattice of points $\vec{r}$ at each point there is a spin $\sigma(\vec{r})$ which can take on two values:

$$\sigma(\vec{r}) = \pm 1. \tag{1.1}$$

There is an attractive interaction between spins at neighbouring sites which gives an energy $-J$ if the spins point in the same direction and $+J$ if they point in opposite directions. Also, there is an external magnetic field H which gives a negative energy $-H$ if the spin is lined up with the field and $+H$ if it is not. In total, then, the energy of the system is:

$$\mathscr{H} = -J \sum_{\substack{\text{nearest}\\ \text{neighbours}}} \sigma(r)\sigma(r') - H\sum_{r} \sigma(r) \tag{1.2}$$

All thermodynamic properties are defined by the free-energy function, $\mathscr{F}(\sigma)$ defined by

$$\exp[+\mathscr{F}(\sigma)] = \exp[-\mathscr{H}/kT] \tag{1.3}$$

where k is the Boltzman constant and T is the absolute temperature. Here, the free energy function $\mathscr{F}(\sigma)$ depends upon two extensive operators

$$S_1 = \sum_{r} \sigma(\mathbf{r}) \tag{1.4}$$

$$S_2 = \sum_{\substack{\text{nearest}\\ \text{neighbours}}} \sigma(\mathbf{r})\sigma(\mathbf{r}') \tag{1.5}$$

as

$$\mathscr{F}\{\sigma\} = hS_1 + KS_2 \tag{1.6}$$

The quantities S_1 and S_2 are termed extensive operators because they are sums over the entire lattice. In contrast $\sigma(\mathbf{r})$ and

$$\mathscr{E}(\mathbf{r}_1) = \sum_{\substack{\mathbf{r}_2\ \text{nearest}\\ \text{neighbour to}\ \mathbf{r}_1}} \sigma(\mathbf{r}_1)\sigma(\mathbf{r}_2) \tag{1.7}$$

are called local operators. These local operators are, of course, the summands in the definitions (1.4) and (1.5) of the extensive operators.

The two parameters h and K appearing in the definition (1.6) of $\mathscr{F}(\sigma)$ are given by

$$h = H/kT \tag{1.8}$$

$$K = J/kT. \tag{1.9}$$

In general, we shall utilize the notation S_α for the αth extensive operator in the theory. The corresponding local operators will be written as $s_\alpha(r)$, while the parameters in the free energy will be denoted by K_α. In particular, $K_1 = h$ and $K_2 = K$.

In terms of these quantities, we can define a free energy density $f(\mathbf{K})$, as

$$Z(\mathbf{K}) = \exp\left[Nf(\mathbf{K})\right] = \sum_{\{\sigma\}\pm 1} e^{\mathscr{F}\{\sigma\}} \tag{1.10}$$

Here N is the number of spin sites and $\mathbf{K}$ is a vector with components $(K_1, K_2) = (h, k)$.

2. *Thermodynamic averages*

The average of any function of the spins $\Theta\{\sigma\}$ is given by

$$\langle\Theta\{\sigma\}\rangle = \frac{1}{Z}\sum_{\{\sigma\}} \Theta\{\sigma\} \exp\left(\mathscr{F}\{\sigma\}\right) \tag{1.11}$$

The most important averages are the expectation value of the local operators,

$$m_\alpha(K) = \langle s_\alpha(r)\rangle = \frac{\partial f(\mathbf{K})}{\partial K_\alpha}, \tag{1.12}$$

the correlation functions,

$$g_{\alpha\beta}(r, \mathbf{K}) = \langle s_\alpha(0) s_\beta(r)\rangle \tag{1.13}$$

and the generalized susceptibility

$$\chi_{\alpha\beta}(\mathbf{K}) = \langle s_\alpha(r) S_\beta\rangle - \langle s_\alpha(r)\rangle \langle S_\beta\rangle. \tag{1.14}$$

The expression (1.12) which relates the average of the local densities with derivations of the free energy density follows directly from eqn (1.10) and the definition (1.11) of an average since

$$S_\alpha = \sum_r s_\alpha(r). \tag{1.15}$$

In an analogous fashion, the susceptibility (1.14) can also be related to derivatives of the free energy density as

$$\chi_{\alpha\beta}(\mathbf{K}) = \frac{\partial^2 \mathrm{f}(\mathbf{K})}{\partial \mathrm{K}_\alpha \partial \mathrm{K}_\beta} \tag{1.16}$$

Moreover, χ can be written in terms of the connected part of the correlation function

$$g^c_{\alpha\beta}(\mathbf{r}, \mathbf{K}) = g_{\alpha\beta}(\mathbf{r}, \mathbf{K}) - \langle s_\alpha(\mathrm{o})\rangle \langle s_\beta(\mathbf{r})\rangle \tag{1.17}$$

$$\chi_{\alpha\beta}(\mathbf{K}) = \sum_r g^c_{\alpha\beta}(\mathbf{r}, \mathbf{K}). \tag{1.18}$$

The most important averages and correlations are those involving the spin-variable. We use a special notation for these, i.e.,

$$\langle \sigma(r) \rangle = m(\mathbf{K}) \tag{1.19}$$

$$\langle \sigma(\mathbf{o})\sigma(r) \rangle = g(\mathbf{r}, \mathbf{K}) \tag{1.20}$$

$$\langle \sigma(\mathbf{o})\sigma(r) \rangle - [m(\mathbf{K})]^2 = g^c(\mathbf{r}, \mathbf{K}). \tag{1.21}$$

Finally, the heat capacity, C_h, is defined as

$$C_h = \frac{\partial \langle s_2(r) \rangle}{\partial K} = \frac{\partial^2}{\partial K^2} f(\mathbf{K}). \tag{1.22}$$

3. *Qualitative description of behaviour*

For large $T(K \to 0)$ and $h = 0$, $\langle \sigma \rangle = 0$. For small T, there is a tendency for spins to line up. If the dimensionality of the lattice is greater than 1, then for $T < T_c(K > K_c)$,

$$\langle \sigma \rangle = \pm\, m(0+, K) \tag{1.23}$$

since the spins can line up in either the positive or negative direction.

The system can jump from the state with

$$\langle \sigma \rangle = +\, m(0+, K) \tag{1.24a}$$

to the state with

$$\langle \sigma \rangle = -\, m(0+, K). \tag{1.24b}$$

This jump is called a first order phase transition. Thus at $h = 0$, the state of the system is *not* uniquely defined as a function of K.

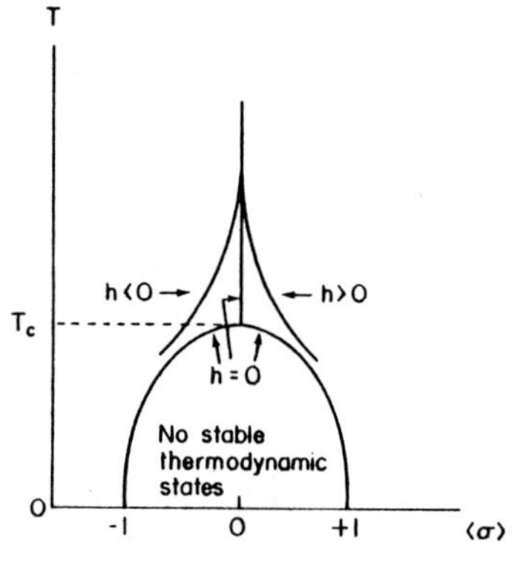

FIG. 1.1. Magnetization vs temperature (schematic) for various values of h. Below the $h = 0$ curve, there are no stable thermodynamic states.

If $h \neq 0$, the state is uniquely defined. For $h > 0$, $\langle\sigma\rangle > 0$; for $h < 0$, $\langle\sigma\rangle < 0$. These results are shown in Fig. 1.1.

If we look at behaviour as a function of T and h, we see, in Fig. 1.2, that the first order phase transition produces a line of singularities which culminate in the critical point.

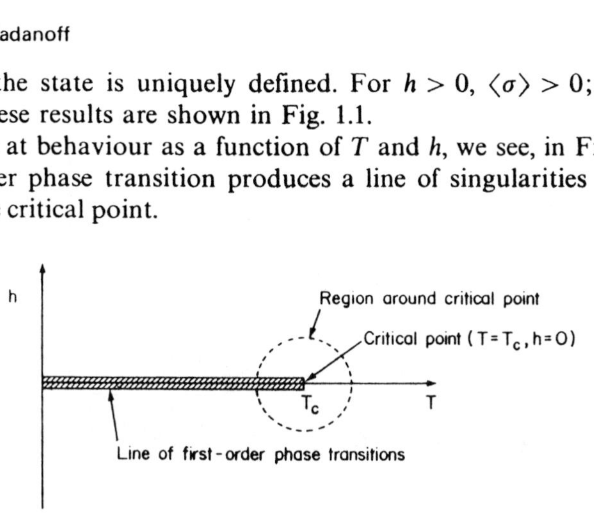

FIG. 1.2. The phase diagram of the Ising model.

The critical point is the point at which the first order phase transition disappears. Since

$$\sigma(h = 0+, K) = 0 \text{ for } K < K_c$$

and

$$\sigma(h = 0+, K) \neq 0 \text{ for } K > K_c$$

the critical point is necessarily a point of non-analyticity for the system. In this paper we are interested in the behaviour in the neighbourhood of this critical point.

B. Other problems

The most natural parameters to use in the description of the critical behaviour of the Ising model are the magnetic field variable h and

$$t = (T - T_c)/T_c.$$

In the free energy function $\mathscr{F}\{\sigma\}$, h is paired with the magnetization density $\sigma(r)$ and t (or K) with the energy density $\mathscr{E}(r)$. As Griffiths and Wheeler have pointed out,[10a] these parameters play a very natural role in a phase diagram like Fig. 1.2. Changes in the parameter h drive the system across the phase boundary. Changes in t produce motion parallel to the phase boundary. Hence, it is these two variables which produce the significant changes in the state of the system in the thermodynamic region. Griffiths introduced the term, *fields*, to describe parameters like these which appear in the free energy and which vary continuously across a first order phase transition.

TABLE 1.1. Fields and operators used in the description of different phase transition problems.

Phase transition problem	First field = h	First operator order parameter	Second field = t	Second operator	Third field	Third operator
Ising model dimensionality $\geqslant 2$	h	$\sigma(r)$	$(T - T_c)/T_c$	Energy density	Next nearest neighbour coupling	Next neighbour energy density
Ising model dimensionality = 1	h	$\sigma(r)$	e^{-k}	Energy density		
Ferromagnetic transition	h[a]	Magnetization density[a]	$(T - T_c)/T_c$	Energy density	Staggered field	Staggered magnetization density
Antiferromagnetic transition	Staggered magnetic field[b]	Staggered magnetization density[b]	$(T - T_c)/T_c$	Energy density	Magnetic field	Magnetization density
Liquid–gas	Chemical potential	Particle density minus initial density	$(T - T_c)/T_c$[c]	Energy density[c]	Parameter measuring size of quantum effects	—
Superconducting	Not physical	$\Delta(r)$[d]	$(T - T_c)/T_c$	Energy density	Pressure	Density
Superfluid	Not physical	$\psi(r)$[e]	$(T - T_c)/T_c$	Energy density	Pressure	Density

[a] These quantities are scalars when there is an easy axis of magnetization. When there is an easy plane, they are two dimensional vectors; when (as in the Heisenberg model) the zero field magnetization may point in any direction in space, then they are vectors.

[b] The antiferromagnetic lattice splits into two sublattices. Staggered quantities point in opposite directions on the sublattices. The quantities involved may be scalars or vectors as in note (a).

[c] A linear combination of this field or operator must be taken with the first one to provide changes orthogonal to the phase boundary.

[d] An annihilation operator for Cooper pairs.

[e] The wave field annihilation operator.

Other phase transition problems can equally well be defined in terms of these two fields:

(a) A field h, which drives the system across the phase boundary and vanishes at criticality;

(b) A field t, which moves the system along the phase boundary and also vanishes at the critical point. Table 1.1 lists the fields and conjugate operator densities which we shall use in the description of different phase transition problems. For later reference, we have included another set of fields and operators in the last two columns of Table 1.1. The reader is asked to pass these over for the time being.

C. Critical singularities

The theory must explain a very rich collection of singularities. These singularities are generally of the form of power laws. For example, at $h = 0$, in the variable t we see singularities of the form $t^{-\lambda}$ or logarithms, which are limits of these power law singularities, i.e.

$$\ln t^{-1} = \lim_{\lambda \to 0} \frac{t^{-\lambda} - 1}{\lambda}.$$

Exponents like λ, which appear in critical singularities are termed *critical indices.* Table 1.2 gives a definition of the critical indices used in this paper.

TABLE 1.2. Definition of critical indices

Critical index	Definition	Condition	
β	$m \simeq \pm(-t)^{\beta}$	$t < 0$	$h = 0$
γ	$\chi \simeq t^{-\gamma}$	$t > 0$	$h = 0$
γ'	$\chi \simeq (-t)^{-\gamma'}$	$t < 0$	$h = 0$
δ	$m \simeq h^{1/\delta}$	$t = 0$	
α	$C_h \simeq t^{-\alpha}$	$t > 0$	$h = 0$
ν	$g^c(r) \simeq e^{-r/\xi}$	$r \to \infty$	
	$\xi \simeq t^{-\nu}$	$t > 0$	$h = 0$
x_σ	$g(r) \simeq (1/r^{2x_\sigma})$	$t = h = 0, r \to \infty$	
x_ε	$g_{22}(r) \simeq (1/r^{2x_\varepsilon})$	$t = h = 0, r \to \infty$	

Table 1.3 show some of the values of the indices observed in experiments and numerical calculations. Here d represents the spatial dimensionality of the lattice. The last column (MFT) refers to mean field theory. We shall describe this approximation in the next section.

The important points to notice from Table 1.3 are the following:

1. All indices seem to vary smoothly with dimensionality;

TABLE 1.3. Values of critical indices.

Critical index	Ising model $d = 1$	Ising model $d = 2$	Ising model $d = 3$	Fluids $d = 3$	Ferromagnetic[d] models $d = 3$: Easy plane of magnetization	Ferromagnetic[d] models $d = 3$: Magnetization can point in any direction	Ising model $d = 4$[a]	Ising model $d > 4$	MFT
β	—	$\frac{1}{8}$	0·31	0.34	0·33	0.34	$\frac{1}{2}$	$\frac{1}{2}$	$\frac{1}{2}$
γ	2	$\frac{7}{4}$	1.250	1.22 ± 0.05	1.32	1.38	1	1	1
γ'	—	$\frac{7}{4}$	1.30 ± 0.05	—	—	—	1	1	1
δ	∞	15	5.0 ± 0.2	4.4 ± 0.2	5	5	3	3	3
α	—	0[c]	0.12	0.1	0.02	-0.1	0	0	0[b]
ν	2	1	0.640 ± 0.003	0.65 ± 0.05	0.675	0.7	$\frac{1}{2}$	$\frac{1}{2}$	$\frac{1}{2}$
x_σ	0	$\frac{1}{8}$	0.55	0.5	0.51	0.51	$\frac{1}{2}$	$\frac{1}{2}$	$\frac{1}{2}$
x_ε	—	1	$\simeq\frac{3}{2}$	—	—	—	2	2	2

[a] Extra logorithms can appear at $d = 4$.
[b] A discontinuity at $t = 0$.
[c] A logarithmic singularity.
[d] Data adapted from H. E. Stanley, "*Introduction to Phase Transitions*" (Oxford University Press, London and New York, 1971), p. 127.

2. The real fluids show indices close to but not exactly equal to the indices of the three-dimensional Ising model;

3. The ferromagnetic models show a dependence of the critical indices upon whether there is any easy axis of magnetization, or an easy plane, or the magnetization can point in any direction;

4. There is a very rich set of data to be explained by the theory.

D. The importance of correlation functions

One of the most dramatic features of critical behaviour is the infinity in the magnetic susceptibility at criticality. Notice that this susceptibility is the sum over sites of a correlation function which is strictly bounded to be < 1, since

$$\chi = \sum_r g^c(r, h, t) = \sum_r [\langle \sigma_0 \sigma_r \rangle - \langle \sigma \rangle^2].$$

The infinity at criticality is achieved because at criticality, $g(r)$ is not integrable. Instead, it is a power of the separation distance

$$g^c(r, 0, 0) \sim \frac{1}{r^2 x^\sigma}$$

for large r. We conclude that the characteristic infinities of thermodynamic functions at the critical point are reflections of correlations which extend over infinite distances in space.

At any finite distance from criticality, correlations tend to fall off with exponential rapidity as $r \to \infty$, i.e.

$$\lim_{r \to \infty} g^c(r, h, t) \sim e^{-r/\xi} \times \text{weaker function of } r. \tag{1.25}$$

Here ξ, has the physical significance of being a range of correlations within the system. This correlation length tends to diverge at the critical point. In particular, at $h = 0$,

$$\xi \sim t^{-\nu} \tag{1.26}$$

The divergence, (1.26), in the correlation length is the fundamental source of all thermodynamic singularities at the critical point.

E. Droplet picture of correlation behaviour

The physical source of this correlation length divergence may be seen by considering fluctuations for $t < 0$ and $h = 0$. Begin at low temperatures. Assume that the system has a net magnetization pointing in the $+z$ direction. A fluctuation may drive a region of the material into a state in which the magnetization points in the "wrong" direction. This droplet of the "wrong" phase has the same energy per unit volume as the material with the "right" magnetization. However, there is an extra free energy added to the system proportional to the area of the region and to a cost per unit area for forming the boundary

$$\text{free energy cost} \sim \text{area} \times \left(\frac{\text{energy}}{\text{unit area}}\right). \tag{1.27}$$

The formation of such a drop will be very unlikely if the cost in free energy is much greater than kT. For this reason, we find

$$\text{area} \leqslant \frac{kT}{\left(\dfrac{\text{energy}}{\text{unit area}}\right)}. \tag{1.28}$$

However, as criticality is approached, the difference in magnetization between the two different phases gets smaller and smaller. Hence the energetic cost per unit area of producing a region of the wrong phase approaches zero. For this reason, the area of a droplet and its radius both can get very, very large. Critical phenomena are connected with large-scale but weak fluctuations in the magnetization.

So far, our picture of critical fluctuations is like that in Fig. 1.3. Droplets

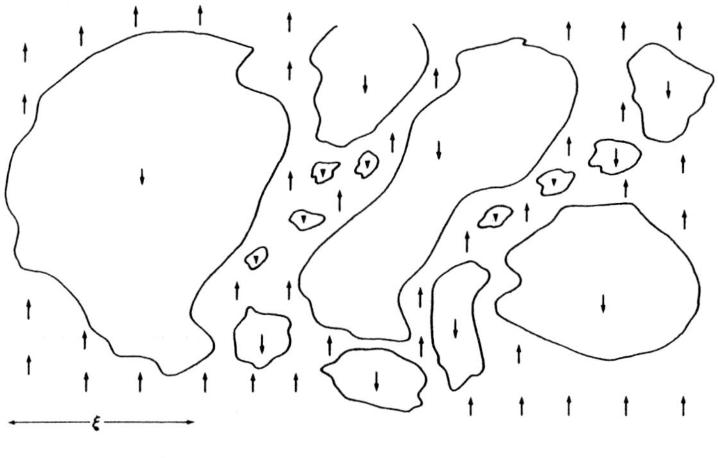

FIG. 1.3. Droplet picture of the critical region.

with spin down of all sizes up to a maximum size ξ appear near the critical point. The size, ξ, is determined by condition (1.28).

However, this picture is incomplete. Each fluctuating region is also a nearly-critical system.

Fluctuations appear within the droplets. And within these fluctuations yet more appear. Hence, each droplet in Fig. 1.3 has a characteristic appearance like that shown in Fig. 1.4. This clustering of droplets within droplets appears until a purely microscopic scale of distances is reached.

FIG. 1.4. Droplets inside of droplets inside droplets . . .

From this picture we conclude that critical phenomena are connected with fluctuations over all length scales between ξ and the microscopic distance between particles.

II. Mean Field Theory

A. Results

A first qualitative picture of near-critical behaviour can be obtained by neglecting fluctuation phenomena. In one sense, fluctuations are at the heart of critical phenomena. However, in another sense, one can discuss phase transitions by considering the important effect to be the lining up of spins to form an order in the system. Then, one can take into account the average order through $\langle\sigma_r\rangle$. Consider a case in which the magnetic field, h_r, varies with r. Then,

$$\exp\left[\mathscr{F}\langle\sigma\rangle\right] = \exp\left[K \sum_{\langle rr'\rangle} \sigma(r)\sigma(r') + \sum_r h_r\sigma(r)\right]. \tag{2.1}$$

Now focus your attention upon one spin, the one at r. Replace every other spin by its average. Then, the free energy function (2.1) becomes

$$\exp\left[\mathscr{F}\{\sigma\}\right] \rightarrow \text{const} \times \exp\left[\sigma(r)h_r^{\text{eff}}\right] \tag{2.2}$$

where the effective or mean field is

$$h_r^{\text{eff}} = h_r + \sum_{r' \text{ neighbours of } r} K\langle\sigma(r')\rangle \tag{2.3}$$

From eqn (2.2) the average of $\sigma(r)$ is

$$\langle\sigma(r)\rangle = \tanh h_r^{\text{eff}}. \tag{2.4}$$

If $\sigma(r)$ varies slowly in space, eqn (2.3) may be replaced by

$$h_r^{\text{eff}} = h_r + zK\langle\sigma(\vec{r})\rangle + c\nabla^2\langle\sigma(\vec{r})\rangle \tag{2.5}$$

where z is the number of nearest neighbours and

$$c = K\sum_{r'}\frac{(r'-r)^2}{d} \tag{2.6}$$

with d being the dimensionality. The sum in 2.6 extends over all nearest neighbours to r. If eqn (2.4) is expanded to first order in h_r and c and to third in $K\langle\sigma\rangle$ we find

$$h_r = (1 - zK)\langle\sigma(r)\rangle - \tfrac{1}{3}\langle\sigma(r)\rangle^3(zK)^3 - c\nabla^2\langle\sigma(r)\rangle. \tag{2.7}$$

The condition for criticality in eqn (2.7) is that $h \to 0$ and

$$t = 1 - zK$$

be zero. Hence we can replace (2.7) by

$$[t + \tfrac{1}{3}\langle\sigma(r)\rangle^2 - c\nabla^2]\langle\sigma(r)\rangle = h_r. \tag{2.8}$$

Equation (2.8) is our basic mean field theory approximation.

If h is independent of r, we derive the equation of state

$$[t + \tfrac{1}{3}\langle\sigma\rangle^2]\langle\sigma\rangle = h. \tag{2.9}$$

From this result, we can derive all the thermodynamic properties of this mean field approximation. For example, if $h = 0$ and $t > 0$, we find $\langle\sigma\rangle = 0$. But if $t < 0$, i.e. we are below the critical temperature, there exists the solution

$$\langle\sigma\rangle = \pm(-3t)^{1/2} \tag{2.10}$$

so that $\beta = \frac{1}{2}$. The indices γ, δ, and γ' are derived in similar fashion. Since $\langle\sigma\rangle = (\partial f/\partial h)_t$, we can derive an expression for f by writing

$$f'(\langle\sigma\rangle, t) = f(\sigma, t) - \langle\sigma\rangle h \tag{2.11}$$

so that

$$\left.\frac{\partial f'}{\partial\langle\sigma\rangle}\right|_t = -h = -[t\langle\sigma\rangle + \tfrac{1}{3}\langle\sigma\rangle^3]. \tag{2.12}$$

The last equality follows from eqn (2.9). Equation (2.12) may be integrated to yield

$$f' = -\left[\frac{t\langle\sigma\rangle^2}{2} + \tfrac{1}{12}\langle\sigma\rangle^4\right]. \tag{2.13}$$

Then, the specific heat is

$$\begin{aligned} C_h &= \left.\frac{\partial^2}{\partial t^2}\right|_h f(h,t) \\ &= \left.\frac{\partial^2}{\partial t^2}\right|_h [f'(\langle\sigma\rangle, t) + \langle\sigma\rangle h] \\ &= \left.\frac{\partial}{\partial t}\right|_h \left(\left.\frac{\partial f'}{\partial t}\right|_{\langle\sigma\rangle} + \left.\frac{\partial\langle\sigma\rangle}{\partial t}\right|_h h\right) \\ &= +h\left.\frac{\partial^2\langle\sigma\rangle}{\partial t^2}\right|_h - \tfrac{1}{2}\left.\frac{\partial\langle\sigma\rangle^2}{\partial t}\right|_h \end{aligned}$$

at $h = 0$, this result reduces to

$$\begin{aligned} C_h &= 0 \qquad \text{for } t > 0 \\ &= +\tfrac{2}{3} \qquad \text{for } t < 0. \end{aligned} \tag{2.14}$$

Thus the specific heat shows a discontinuity at $t = 0$ in this mean field approximation.

Finally, the spin–spin correlation function obeys

$$\begin{aligned} \langle\sigma(0)\sigma(r)\rangle &= \langle\sigma\rangle^2 + \frac{\delta\langle\sigma(r)\rangle}{\delta h(0)} \\ &= \langle\sigma\rangle^2 + g^c(r,h,t) \end{aligned} \tag{2.15}$$

From eqn (2.8), the connected part of the correlation function obeys

$$[t + \tfrac{1}{2}\langle\sigma(r)\rangle^2 - cV^2]g^c(r,h,t) = \delta_{r0}.$$

At $h = 0$, at $t > 0$ this gives a Fourier transform

$$(t + cq^2)\hat{g}(q,h,t) = 1.$$

Then

$$\left.\hat{g}(r,h,t)\right|_{h=0} = \int \frac{d^dq}{(2\pi)^d} \exp i\mathbf{q}\cdot\mathbf{r} \frac{1}{t + cq^2}$$

In three dimensions we then find

$$\hat{g}(r,h,t)\Big|_{h=0} = \frac{1}{4\pi rc} e^{-r/\xi}; \qquad \xi = \sqrt{c/t}. \tag{2.16}$$

Hence $x_\sigma = \frac{1}{2}$ and $\nu = \frac{1}{2}$. More generally, to an order of magnitude

$$\hat{g}(r,h,t) \sim \frac{1}{cr^{d-2}} e^{-r/\xi}. \tag{2.17}$$

so that $x_\sigma = \frac{1}{2}(d-2)$ in d-dimensions.

B. Generalizations

For most phase transitions, one can define an order parameter analogous to $\langle\sigma(r)\rangle$. This order parameter has the following properties:

(1). It jumps discontinuously across the first order phase transition;
(2). As criticality is approached, this jump gets smaller and smaller;
(3). Large values of the order parameter imply that one is far from criticality.

One can also usually define a free energy density $f(h_r, t)$ such that

$$\frac{\delta}{\delta h_r}\sum_{r'} f(h_{r'}, t) = \langle\sigma(r)\rangle \tag{2.18}$$

where $\langle\sigma(r)\rangle$ is the order parameter. Then it is also possible to define

$$f'(\langle\sigma\rangle, t) = f(h_r, t) - \langle\sigma(r)\rangle\, h_r \tag{2.19}$$

so that

$$\sum_{r'} \frac{\delta}{\delta\langle\sigma(r)\rangle} f'(\langle\sigma(r')\rangle, t) = -h_r. \tag{2.20}$$

The mean field theory is the assumption that one can expand the free energy function f' in a power series in $\langle\sigma(r)\rangle$ as

$$f'(\langle\sigma(r)\rangle, t) = -\frac{t}{2}\langle\sigma(r)\rangle^2 + b\langle\sigma(r)\rangle^4 + c[\nabla\langle\sigma(r)\rangle]^2 \tag{2.21}$$

All the consequences of mean field theory follow directly from this expansion assumption.

C. Failure of mean field theory

In one sense, mean field theory is a great success. It does give first order phase transitions and also a qualitative indication of the types of singularities which might be expected in the second order phase transition.

In a quantitative sense, however, it is a complete failure in the critical region. All the critical indices derived from mean field theory are wrong in two and three dimensions. The source of this failure is clear. Critical phenomena involve fluctuations. Mean field theory is based upon the assump-assumption of small fluctuations.

To see this quantitative failure, we shall use mean field theory to predict its own short-comings. For $t < 0$ and $h = 0$, it is exactly true that

$$C_h \sim -\frac{\partial}{\partial t}\langle \sigma_r \sigma_{r+a} \rangle \tag{2.22}$$

where a is the distance between two nearest neighbours. Our free-energy calculation approximates the expression (2.22) by

$$C_h = -\frac{\partial}{\partial t}\langle \sigma_r \rangle^2 = +3.$$

However, a more careful calculation based upon mean field theory would give

$$\begin{aligned} C_h &= -\frac{\partial}{\partial t}[\langle \sigma_r \rangle^2 + g(r = a, h = 0, t)] \\ &= \tfrac{3}{2} - \tfrac{1}{2}\frac{\partial}{\partial t}\int \frac{d^d q}{(2\pi)^d}\frac{\exp i\mathbf{q}.\mathbf{a}}{t + q^c c} \\ &= \tfrac{3}{2} + \tfrac{1}{2}\int \frac{d^d q}{(2\pi)^d}\frac{\exp i\mathbf{q}.\mathbf{a}}{(q^2 c + t)^2}. \end{aligned} \tag{2.23}$$

Therefore

$$\begin{aligned} C_h - \tfrac{3}{2} &\sim \left.\frac{q^d}{(q^2 c)^2}\right|_{q \sim (t)^{1/2}/c^{1/2} = \xi^{-1}} \\ &\sim \frac{1}{c^{d/2}}(t)^{-(4-d)/2} \end{aligned}$$

If the mean field theory is to be correct, the right-hand side must be small. But this term diverges as $t \to 0$ whenever

$$d < 4. \tag{2.24}$$

Hence we know that mean field theory must fail for all dimensionalities less than 4.

III. The Roots of The Theory: Universality

A. Fields and operators

Since critical phenomena arise from long-ranged correlations, it is reasonable to expect that some of the details of the interatomic potential might be quite irrelevant to the behaviour in the critical region. Thus, for example, it is usually asserted that the values of the critical indices are independent of lattice structure. Similarly we expect that if the Hamiltonian is a mixture of nearest-neighbour and next nearest neighbour interactions the critical behaviour is independent of the exact mixing ratio.

To convert these qualitative statements into quantitative form, we write the basic free energy function for the system as

$$\exp[\mathscr{F}\{\sigma\}] = \exp[\mathscr{F}^*\{\sigma\}] + \sum h_\alpha S_\alpha\{\sigma\} \tag{3.1}$$

Here $\mathscr{F}^*\{\sigma\}$ is a free energy function which produces some kind of critical point behaviour. The fields h_α represent some kinds of deviation from this critical point. For example, h_1 might be the magnetic field; h_2 might be t—the temperature deviation from criticality—and so forth. The S_α's are then operators (i.e. functions of the σ's) which are thermodynamically conjugate to the h_α's. Thus S_1 might be the total magnetization, S_2 the energy, etc. Each S_α is translationally invariant and may be written as a sum over all lattice sites of a local density $s_\alpha(r)$, i.e.

$$S_\alpha = \sum_r s_\alpha(\vec{r}). \tag{3.2}$$

Thus for example, $s_1(r)$ might be $\sigma(r)$, $s_2(r)$ might be

$$\sum_{\vec{a}} \sigma(\vec{r})\sigma(\vec{r} + \vec{a})$$

where the sum over $\vec{a}$ is a sum over vectors to nearest neighbours, etc.

The most important descriptors of critical behaviour are the free energy per site, $f(\vec{h})$, the average of the operators, and their correlations. If N is the number of sites, we define

$$\begin{aligned} f(\vec{h}) &= \frac{1}{N} \ln \sum_{\{\sigma\}} \exp \mathscr{F}\{\sigma\} \\ m_\alpha(h) &= \langle s_\alpha(r) \rangle \\ g_{\alpha\beta}(\vec{r}, \vec{h}) &= \langle s_\alpha(0) s_\beta(r) \rangle \end{aligned} \tag{3.3}$$

and notice that we can define a generalized susceptibility $\chi_{\alpha\beta}(\vec{h})$ in three different ways:

$$\chi_{\alpha\beta} = \frac{\partial^2 f}{\partial h_\alpha \partial h_\beta} = \frac{\partial m_\alpha}{\partial h_\beta} = \sum_{\vec{r}} [g_{\alpha\beta}(\vec{r}, \vec{h}) - \langle s_\alpha \rangle \langle s_\beta \rangle]. \tag{3.4}$$

B. Statement of universality hypothesis

In its simplest terms, the universality hypothesis is the statement that all critical problems may be divided into classes differentiated by:

(a) The dimensionality of the system;

(b) The symmetry group of the order parameter; and
(c) Perhaps other criteria.

Within each class, the critical properties are supposed to be identical or, at worst, to be a continuous function of a very few parameters.

To convert this physical statement into mathematical form, we consider a reference problem characterized by fields $h^{\rm I}_\alpha$, local operators $s_\alpha(\mathbf{r})$, a free-energy per site $f^{\rm I}(\mathbf{h}^{\rm I})$, an expectation value of the fields $m^{\rm I}_\alpha(\mathbf{h}^{\rm I})$ and a set of correlation functions $g^{\rm I}_{\alpha\beta}(r, \vec{h}^{\rm I})$. A comparison problem is defined by fields $h^{\rm II}_\alpha$, operators $s^{\rm II}_\alpha(\mathbf{r})$, free-energy per site $f^{\rm II}(\mathbf{h}^{\rm II})$, an expectation value $m^{\rm II}_\alpha(\mathbf{h}^{\rm II})$ and correlation functions $g^{\rm II}_{\alpha\beta}(r, h^{\rm II})$. Then, the two problems are said to be in the same universality class if there exists a transformation upon problem II which will convert its solution into a form identical with the solution of problem I. In particular, the transformation is hypothesized to take the form

$$\begin{aligned} h^{\rm II}_\alpha &\to h'_\alpha = b_{\beta\alpha} h^{\rm II}_\beta \\ s^{\rm II}_\alpha(r) &\to s'_\alpha(r) = s^{\rm II}_\beta(r) c_{\beta\alpha} \\ r &\to r' = r/l \end{aligned} \tag{3.5}$$

with $b_{\alpha\beta}$, $c_{\alpha\beta}$, and l being analytic functions of $h^{\rm II}_\alpha$ for small values of $h^{\rm II}_\alpha$. Here and below we use a summation convention for repeated Greek indices.

The exact meaning of the identity between the two problems is that the free-energies per site are related by

$$f^{\rm II}(\mathbf{h}^{\rm II}) = b_0 f^{\rm I}(\mathbf{h}') + \text{a non-singular function of } h^{\rm II} \tag{3.6}$$

where b_0 is analytic in $h^{\rm II}$, and that the operator averages are given by

$$\begin{aligned} \langle s'_\alpha(\mathbf{r})\rangle &\equiv m^{\rm II}_\beta(\mathbf{h}^{\rm II}) c_{\beta\alpha} \\ &= m^{\rm I}_\alpha(\mathbf{h}') + \text{a non-singular function of } h^{\rm II}\text{'s}. \end{aligned} \tag{3.7}$$

Moreover, the correlation functions are related by

$$\begin{aligned} g'_{\alpha\beta}(\vec{r}, \vec{h}^{\rm II}) &\equiv g^{\rm II}_{\mu\nu}(r, \vec{h}^{\rm II}) c_{\mu\alpha} c_{\nu\beta} \\ &= g^{\rm I}_{\alpha\beta}(\vec{r}/l, \vec{h}') \end{aligned} \tag{3.8}$$

for $r \gg a$ lattice constant.[16, 17]

C. A simple example

In the Ising model, there are two relevant fields, h and t. Let our reference problem be the ordinary nearest neighbour Ising model. Our comparison problem could be the Ising model of the form

$$\{\sigma\} = \sum_r h^{\rm II}\sigma_r + K_1 \sum_{\substack{\text{nearest} \\ \text{neighbours}}} \sigma_r \sigma_{r'} + K_2 \sum_{\substack{\text{next} \\ \text{nearest} \\ \text{neighbours}}} \sigma_r \sigma_{r''}. \tag{3.9}$$

Then, let this problem have a critical point at $h^{\mathrm{II}} = 0$, $K_1 = U(K_2)$. Fix the value of K_2 and define

$$t^{\mathrm{II}} = -K_1 + U(K_2).$$

Because of the symmetry of the problem, the transformation coefficients in eqns (3.5) are diagonal in α and β. Hence the statement of the universality hypothesis is that the free-energy per site for problem (3.9) is

$$f^{\mathrm{II}}(h^{\mathrm{II}}, K_1, K_2) = b_0(K_2) f^{\mathrm{I}}(b_h(K_2)h^{\mathrm{II}}, b_t(K_2)t^{\mathrm{II}}) + \text{ non-singular terms} \tag{3.10}$$

where $f^{\mathrm{I}}(h, t)$ is the free-energy per site for the ordinary Ising model. For another example, the spin–spin correlation function defined by the solution to the problem (3.9) is

$$\langle \sigma(0)\sigma(\mathbf{r})\rangle = g(r, h^{\mathrm{II}}, K_1, K_2).$$

According to universality, this can be expected to take the form

$$g(r, h^{\mathrm{II}}, K_1, K_2) = [c_h(K_2)]^{-2} g^{\mathrm{I}}\left(\frac{r}{l(K_2)}, b_h(K_2)h^{\mathrm{II}}, b_t(K_2)t^{\mathrm{II}}\right) \tag{3.11}$$

for r much bigger than a lattice constant. Here $g^{\mathrm{I}}(r, h, t)$ is the spin–spin correlation function for the pure nearest neighbour case.

Clearly if the statements we have just made are correct, they are very useful in simplifying any discussion of critical phenomena. In particular, they imply that the critical indices are independent of K_2.

D. Additional universality statements

The transformation among problems just defined involves the following unknown parameters: $c_{\alpha\beta}$, $b_{\alpha\beta}$, b_0, l. However, one can argue that these parameters are not all independent. In particular, one can use the identities (3.4) to obtain relations among these coefficients. To obtain these assume that

$$\chi^{\mathrm{II}}_{\alpha\beta} = \frac{\partial^2 f^{\mathrm{II}}}{\partial h^{\mathrm{II}}_\alpha \partial h^{\mathrm{II}}_\beta} \tag{3.12}$$

is singular and that the leading singularities in this function can be obtained from the singular terms in f^{II}, m^{II} and $g^{\mathrm{II}}_{\alpha\beta}$. From (3.4) and (3.6), we find

$$\chi^{\mathrm{II}}_{\alpha\beta}(\mathbf{h}^{\mathrm{II}}) = b_0 \chi^{\mathrm{I}}_{\mu\nu}(\mathbf{h}') b_{\alpha\mu} b_{\beta\nu} \tag{3.13a}$$

if we can neglect the derivatives of the b's with respect to the h's. From eqns (3.7) and (3.8) we can obtain alternative evaluations of $\chi_{\alpha\beta}$. The results are

$$\chi^{\mathrm{II}}_{\mu\nu}(\mathbf{h}^{\mathrm{II}})C_{\mu\alpha}C_{\nu\beta} = \sum_r [g^{\mathrm{I}}_{\alpha\beta}(r/l, \mathbf{h}') - \langle s_\alpha\rangle_{h'}\langle s_\beta\rangle_{h'}]$$

$$= l^d\chi^{\mathrm{I}}_{\alpha\beta}(\mathbf{h}') \tag{3.13b}$$

where d is the dimensionality of the system and

$$\chi^{\mathrm{II}}_{\beta\gamma}(\mathbf{h}^{\mathrm{II}})C_{\gamma\alpha} = \gamma^{\mathrm{I}}_{\gamma\alpha}(\mathbf{h}')b_{\beta r}. \tag{3.13c}$$

If we further assume that $c_{\alpha\beta}$ has an inverse, then the mutual consistency of eqns (3.13) demands

$$b_{\gamma\alpha}C_{\gamma\beta} = \delta_{\alpha\beta}l^d$$

$$b_0 = l^{-d}$$

In the previous section, we considered a case in which b and c were diagonal matrices. In general we shall use Latin indices, $i, j, k \ldots$, to describe results in this diagonal representation. In this case

$$b_{ij} = \delta_{ij}b_i; \qquad C_{ij} = \delta_{ij}C_i \tag{3.16}$$

with no sum on i intended. Then, eqn (3.14) reduces to

$$b_iC_i = l^d. \tag{3.17}$$

IV. The Roots of The Theory: Scaling

A. The irrelevance of the length scale

The physical picture that we have of critical phenomena involves the notion that at the critical point fluctuations at all length scales contribute to the various thermodynamic functions. As a result, one might expect that critical phenomena will show an invariance under the change of length scales

$$\vec{r} \to \vec{r}' = \vec{r}/l. \tag{4.1}$$

This invariance can be defined by the statement that the transformation (4.1) takes us to a new description of the critical problem which is identical to the old description in the sense that they be within the same universality class.

B. A formal description of the length transformation

Consider a general Ising model problem on some simple lattice, for example, the square lattice of Fig. 4.1a. The most general translationally invariant description of this problem would be to write

$$\exp - [Nf(k)] = \sum_{\{\sigma\}} \exp[-\mathscr{F}\{\sigma\}]$$

$$\mathcal{F}\{\sigma\} = \sum_{\alpha} S_{\alpha}\{\sigma\}K_{\alpha}. \tag{4.2}$$

Here the S_α's are all the translationally invariant operators which can be formed from the spins in the problem. Thus, for example S_1 might be the

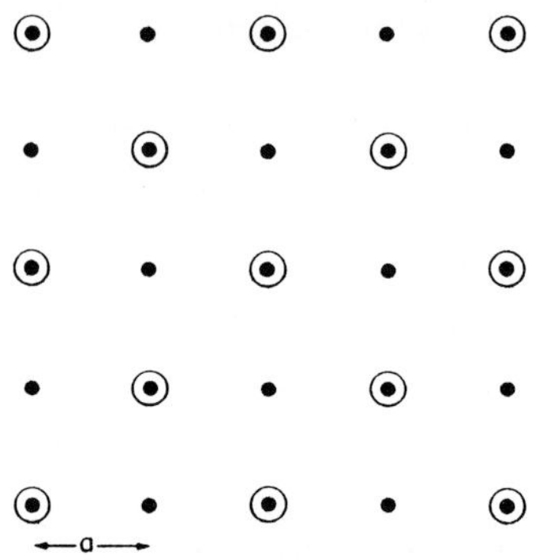

FIG. 4.1a. A two-dimensional lattice. The points inside the circles will be kept in the new lattice of Fig. 4.1b.

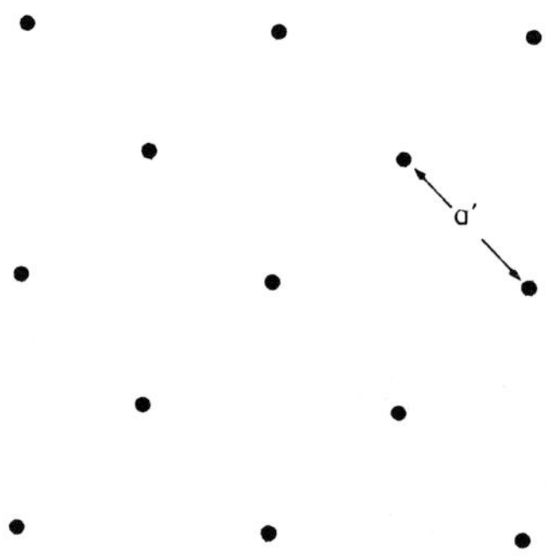

FIG. 4.1b. The new lattice.

total magnetization, S_2 the sum over all products of nearest neighbour spins, etc. Each of these operators is of the form

$$S_\alpha = \sum_r s_\alpha(r).$$

To obtain a complete set of operators, we must include the constant term:

$$S_0 = \sum_r 1 = N$$

where N is the number of sites on the lattice.

Now consider a transformation which reduces the number of degrees of freedom in the problem. In formal terms, we can write this transformation as

$$\exp[\mathcal{F}'\{\mu\}] = \sum_{\{\sigma\}} T\{\mu, \sigma\} \exp[\mathcal{F}\{\sigma\}]. \tag{4.3}$$

We demand that this transformation leave the free energy invariant, i.e. that

$$\sum_{\{\mu\}} \exp[\mathcal{F}'\{\mu\}] = \sum_{\{\sigma\}} \exp[\mathcal{F}\{\sigma\}] \tag{4.4}$$

The condition (4.4) will be true if

$$\sum_{\{\mu\}} T\{\mu, \sigma\} = 1. \tag{4.5}$$

Let us exhibit one transformation which meets these conditions. The lattice in Fig. 4.1a contains the points,

$$\vec{r} = (n, m)a$$

where n and m are integers. The lattice in Fig. 4.1b contains only the subset of these points

$$\vec{r}\,' = (n, m)a \quad \text{for} \quad n + m \text{ even.}$$

If there are N spins in the original lattice, there are $N/2$ spins in the new lattice. In fact, the new lattice is exactly the same as the old except that the lattice constant has been increased by the amount

$$\begin{aligned} a \to a' &= la \\ l &= \sqrt{2}. \end{aligned} \tag{4.6}$$

Therefore, two points separated by n lattice constants in the original lattice are separated by n/l lattice constants in the new lattice.

If we define new spin variables

$$\mu_{r'} = \pm 1$$

at all the points of the new lattice, we have a new Ising model on a transformed lattice with a new length scale, defined as in eqn (4.1).

In particular define

$$T_1\{\mu, \sigma\} = \prod_{r'} (1 + \mu_{r'}\sigma_{r'})/2. \tag{4.7}$$

Then if T is T_1, the transformation (4.3) is a summation over the spins at the eliminated sites of the original lattice. For all the sites which lie on the new lattice $\mu_{r'}$, $\sigma_{r'}$. Notice that the transformation function (4.7) manifestly obeys eqn (4.4).

More generally, we can write

$$T\{\mu, \sigma\} = \sum_{\{\mu'\}\pm 1} T_2\{\mu, \mu'\} T_1\{\mu', \sigma\} \tag{4.8}$$

Here $\mu'_{r'}$ and $\mu_{r'}$ are spins on the same lattice, i.e. that of Fig. 4.1b. $T_2\{\mu, \mu'\}$ is any function of μ's and μ''s, which obeys

$$\sum_{\{\mu\}=\pm 1} T_2\{\mu, \mu'\} = 1.$$

The combined transformation (4.8) represents a very general transformation which includes a change in the length scale.

Just as the original free energy function $\mathscr{F}\{\sigma\}$ can be described by a set of coupling constants K_α, in exactly the same way the new function $\mathscr{F}'\{\mu\}$ can be written as

$$\mathscr{F}'(\mu) = \sum_\alpha K'_\alpha S_\alpha\{\mu\} \tag{4.9}$$

Here the S_α's are exactly the same function of the spins as defined previously. The arguments of these functions have been changed. If, we recognize that

$$S_\alpha\{\sigma\} = \sum_r s_\alpha(r)$$

then we notice that $S_\alpha\{\mu\}$ contains only half as many terms as $S_\alpha\{\sigma\}$.

In general, we can regard the transformation (4.3) as a change in the values of the coupling constants

$$K_\alpha \to K'_\alpha(\vec{K}), \tag{4.10}$$

where as indicated the new coupling constants are functions of the old.

C. Fixed points

The basic concept in the Wilson formulation of critical phenomena theory is the idea that for a suitable choice of T the transformation (4.10) may have a "fixed point". This fixed point is simply a value of the vector of coupling constants $K = K^*$ such that the transformation (3.35) leads to exactly the same value once again.

$$K'_\alpha(\vec{K}^*) = \vec{K}^*_\alpha \quad \text{for all } \alpha \neq 0 \tag{4.11}$$

At this fixed point, the length transformation does not change the problem at all. Hence we identify the fixed point with the critical point.

Notice that eqn (4.11) does not apply to the constant term in the free energy

$$K_0 S_0 = K_0 \sum_r s_r \tag{4.12}$$

Because this constant term does not change any expectation values, it is allowed to change under the transformation (4.10). In our later work, we shall see that the transformation properties of this term are intimately related to the evaluation of the free energy. For now, we need only note that the "trivial" constant K_0 does not enter the evaluation of the K'_α for $\alpha \neq 0$.

Notice also that the numerical value of the vector K^*_α depends upon our choice of $T\{\mu, \sigma\}$.

D. Deviations from the fixed point

To get physically interesting results from the fixed point theory, we must consider deviations from the fixed point. Write

$$K_\alpha = h_\alpha + K^*_\alpha \quad \text{for } \alpha \geqslant 1 \tag{4.13}$$

where h_α measures the deviations. We think of h_α as small. The transformation (4.10) can then be written as

$$K_\alpha \to K'_\alpha = K'_\alpha(\vec{K})$$

or

$$K^*_\alpha + h'_\alpha = K'_\alpha(\vec{K}^* + \vec{h}) \tag{4.14}$$

when $\vec{h} = 0$, $h'_\alpha = 0$. Therefore, we can consider h'_α to be at least proportional to h_α, i.e.

$$h'_\alpha = b_{\alpha\beta} h_\beta \tag{4.15}$$

where $b_{\alpha\beta}$ is a function of the h's which reduces to a constant as the h's approach zero. We assume that $b_{\alpha\beta}$ is analytic in $\vec{h}$ for small $\vec{h}$.

A reasonably clear physical picture of what is going on may be obtained if we assume that the transformation (4.15) may be diagonalised by picking a suitable linear combination of the h_α's. In diagonal form (4.15) reads

$$h'_i = b_i h_i = h_i l^{y_i}. \tag{4.16}$$

The indices y_i defined in eqn (4.16) are the principal determinants of critical behaviour.

We wish to distinguish three cases:

$$\begin{aligned} &y_i > 0;\ h_i \text{ is termed a relevant field} \\ &y_i < 0;\ h_i \text{ is termed irrelevant} \\ &y_i = 0;\ \text{the borderline case.} \end{aligned} \tag{4.17}$$

Imagine that we make a succession of scaling transformations like (4.16). After n such transformations

$$\begin{aligned} r &\to r' = r/L \\ h_i &\to h'_i = h_i L^{y_i} \end{aligned} \qquad L = l^n$$

Since the critical behaviour is determined by long-range effects, one can probe the particularly critical effects by looking at the problem after such a transformation. The results of this transformation are that

$h'_i \to 0$ for irrelevant variables

h'_i grows quite large for relevant variables

$h'_i = h_i$ for the borderline variables

Hence the irrelevant variables drop out of the critical behaviour; the relevant variables grow; the border-line variables remain fixed.

Thus we can identify the irrelevant variables with those which were described as irrelevant in our previous discussion of universality.

The relevant variables represent relevant deviations from criticality, e.g. h and t. As we look at longer and longer length scales these variables become more and more important.

The borderline variables play a more subtle role in the theory which we shall discuss below.

E. Scaling results for the free energy

For the time being, set both the borderline and the irrelevant fields equal to zero. We are left with only the relevant fields. We wish to calculate the free energy as a function of these relevant fields. On one hand, we can calculate the free energy as

$$\exp[Nf(h)] = \sum_{\{\sigma\}} \exp[\mathscr{F}^*\{\sigma\} + \sum_i h_i S_i\{\sigma\}]. \tag{4.18}$$

On the other hand, we can calculate the free energy as

$$\begin{aligned} \exp[Nf(h)] &= \sum_{\{\mu\}} T\{\mu, \sigma\} \exp[\mathscr{F}^*\{\sigma\} + \sum_i h_i S_i\{\sigma\}] \\ &= \sum_{\{\mu\}} \exp[\mathscr{F}^*\{\mu\} + h'_i S'_i\{\mu\} + f_0(h)N'] \end{aligned} \tag{4.19}$$

where N' is the number of sites in the new lattice. The h'_i are the transformed field variables. The term in $f_0(h)$ is a constant term in the free energy, independent of all the variables $\mu(r')$. Because the transformation only involves summations over small regions of the original lattice, we argue that $f_0(h)$ is an analytic function of the h_i's near $h_i = 0$. A further evaluation of the sum on the right-hand side of eqn (4.19) gives

$$\exp[Nf(h)] = \exp\{N'[f_0(h) + f(h')]\} \tag{4.20}$$

since the sum over μ's in (4.19) is of exactly the same form as the sum over

σ's in (4.18). Since $N' = N/l^d$, we see that

$$\begin{aligned} f(h) &= \frac{1}{l^d}[f_0(h) + f(h')] \\ &= \frac{1}{l^d}[f_0(h) + f(b_i h_i)]. \end{aligned} \tag{4.21}$$

Now take a particular case. Assume that there are two thermodynamically significant variables h and t. Write $b_h = l^{y_h}$, $b_t = l^{y_t}$. Equation (4.21) reads

$$f(h, t) = \frac{1}{l^d}[f_0(h, t) + f(l^{y_h}h, l^{y_t}t)].$$

A solution to this equation is

$$f(h, t) = t^{d/y_t}\hat{f}(h/t^{y_h/y_t}) + \text{regular function of } h \text{ and } t. \tag{4.22}$$

Using this form of the solution we can identify the variables α, β, γ, γ', δ in terms of the pair of variables d/y_t and y_h/y_t as

$$\begin{aligned} 2 - \alpha &= d/y_t \\ \beta &= \frac{d}{y_t} - y_h/y_t \\ \gamma = \gamma' &= d/y_t - 2y_h/y_t \\ 1 + \delta^{-1} &= d/y_h \end{aligned} \tag{4.23}$$

so that we find the relationships

$$2 - \alpha = \gamma + 2\beta = \gamma' + 2\beta = \beta(\delta + 1). \tag{4.24}$$

The data in Table 1.1 support these assertions.

F. A simple example of the theory

As the simplest example of the theory consider the nearest neighbour, one-dimensional Ising model[18]:

$$\exp(\mathscr{F}\{\sigma\}) = \prod_i \exp(K\sigma_i\sigma_{i+1} + h\sigma_i) \tag{4.25}$$

This model has a kind of critical point at $h = 0$ and zero temperature ($K = \infty$). We describe the solution in terms of the two variables h and

$$t = \exp -K. \tag{4.26}$$

In terms of these variables, the solution to this problem in the limit $t \to 0$, $h \to 0$ is given by

$$f(h, t) = -\ln t + \sqrt{(t^4 + h^2)} \tag{4.27}$$

$$\langle\sigma\rangle = \frac{h}{\sqrt{(h^2 + t^4)}} \tag{4.28}$$

$$\langle\sigma_i\sigma_{i+r}\rangle = \langle\sigma\rangle^2 + [1 - \langle\sigma\rangle^2] \exp -r/\xi \tag{4.29}$$

where

$$\xi^{-1} = 2\sqrt{(t^4 + h^2)}. \tag{4.30}$$

Now obtain the results of the renormalization method for this problem. Choose

$$\mu_i = \sigma_{2j}$$

and sum expression (4.25) over all σ_i for odd values of i. This particular choice of renormalization procedure corresponds to picking

$$l = 2 \tag{4.31}$$

and T equal to the T_1 defined by eqn (4.7). Then we have

$$\exp(\mathscr{F}\{\mu\}) = \sum_{\substack{\sigma_{2j+1}=\pm 1 \\ \sigma_{2j}=\mu_i}} \prod_j \exp[\sigma_{2j+1}(h + K\sigma_{2j} + \sigma_{2j+2})] \times \exp[h(\sigma_{2j} + \sigma_{2j+2})/2]$$

$$= \prod_j \exp[h(\mu_i + \mu_{j+1})/2] \times 2\cosh[h + K(\mu_j + \mu_{j+1})].$$

Notice that this result can exactly be written in the form

$$\exp(\mathscr{F}\{\mu\}) = \prod_j \exp[f_0 + K'\mu_j\mu_{j+1} + (h'/2)(\mu_j + \mu_{j+1})]$$

with

$$(h'/2)(\mu_j + \mu_{j+1}) + f_0 + K'\mu_j\mu_{j+1} = (h/2)(\mu_j + \mu_{j+1}) + \ln 2\cosh[h + K(\mu_j + \mu_{j+1})]$$

so that

$$h' = h + \tfrac{1}{2}\ln\frac{\cosh(2K + h)}{\cosh(2K - h)}$$

$$f_0 = \ln 2 + \tfrac{1}{4}\ln\cosh(2K + h)\cosh(2K - h) + \tfrac{1}{2}\ln\cosh h$$

$$K' = \tfrac{1}{4}\ln\cosh(2K + h)\cosh(2K - h) - \tfrac{1}{2}\ln\cosh h.$$

In the limit as $K \to \infty + h \to 0$ these results reduce to

$$h' = 2h$$

$$f_0 = K + \frac{\ln 2}{2} = \ln t + \frac{\ln 2}{2} \tag{4.32}$$

$$t' = \sqrt{2}t$$

As a result, eqn (4.21) reads

$$f(h,t) = \tfrac{1}{2}\left[-1 \ln t + \frac{\ln 2}{2} \right] + \frac{1}{l} f(l^{y_h}h, l^{y_t}t) \tag{4.33}$$

with $y_h = 1$, $y_t = \frac{1}{2}$. The solution of (4.33) is

$$f(h,t) = -\ln t + t^2 \hat{f}(h/t^2) \tag{4.34}$$

which is exactly of the form (4.27). We identify the first term in (4.34) as the analogue of the "regular" term in critical phenomena theory and the second term as the analogue of the "scaling" part.

V. Scale Transformations for Correlation Functions

A. General discussion

The scaling transformation, as described in the previous section, can be described in the following fashion. A length transformation

$$r \to r' = r/l \tag{5.1}$$

converts the critical phenomena problem into another, essentially similar problem in which the basic field transform according to

$$h_\alpha \to h'_\alpha = b_{\alpha\beta} h_\beta \tag{5.2}$$

where $b_{\alpha\beta}$ are power series expandable in $\vec{h}$ for small h. Compare this statement to the definitions of the universality in Section III. B. It seems reasonable to believe that the transformation defining the change in length scale is a particular example of a transformation which takes us from one problem to another in the same universality class. If this be true then the scaling theory should yield restrictions upon the correlation functions quite analogous to the restrictions implied by the universality arguments.

Another way of making the same point is to say that the scale transformation idea is based upon the scale invariance of the fluctuations in the problem. These fluctuations are best studied in the correlation functions. Hence the scale-transformation theory should be most fully reflected in the behaviour of correlation functions.

B. Formal theory

The most direct way of studying this behaviour is to consider the operators $S_\alpha\{\sigma\}$ to be sums of local operators

$$S_\alpha\{\sigma\} = \sum_r s_\alpha(\{\sigma\}, r). \tag{5.3}$$

Each $s_\alpha(\{\sigma\}, r)$ is expected to depend upon only the $\sigma(r')$ for r' in the immediate neighbourhood of r. Now consider a generalization of the transformation (4.3)

$$\exp(\mathscr{F}\{\mu\} = \exp[\mathscr{F}^*\{\mu\} + \sum_{\alpha=0}^{\infty} h'_\alpha S_\alpha\{\mu\}]$$

$$= \sum_{\{\sigma\}} T\{\mu, \sigma\} \exp[\mathscr{F}^*\{\sigma\} + \sum_{\alpha=0}^{\infty} h_\alpha S_\alpha\{\sigma\}]$$

to the case in which h_α depends upon r

$$\exp[\mathscr{F}^*\{\mu\} + \sum_{r'} \sum_{\alpha=0}^{\infty} h'_\alpha(r') s_\alpha(\{\mu\}, r')]$$

$$= \sum_{\{\sigma\}} T\{\mu, \sigma\} \exp[\mathscr{F}^*\{\sigma\} + \sum_{r} \sum_{\alpha=0}^{\infty} h_\alpha(r) s_\alpha(\{\sigma\}, r)] \quad (5.4)$$

$$= \sum_{\{\sigma\}} T\{\mu, \sigma\} \exp(\mathscr{F}\{\sigma\}).$$

Here $s_\alpha(\{\mu\}, r')$ is exactly the same function of the neighbouring μ's as $s_\alpha(\{\sigma\}, r)$ is one of the neighbouring σ's.

We shall apply eqn (5.4) to the case in which $h_\alpha(r)$ varies slowly in r. When h_α was independent of r and small, we asserted that

$$h'_\alpha = b_{\alpha\beta} h_\beta \quad (5.5)$$

Now make the analogous assertion for the use of slow spacial variation.

$$h'_\alpha(r') = b_{\alpha\beta} h_\beta(r) \quad \text{for } r' \approx \vec{r}/l$$

Or, if we wish to be more careful we can write

$$h'_\alpha(r') = \sum_r b_{\alpha\beta}(r', r) h_\beta(r) \quad (5.6)$$

with

$$b_{\alpha\beta}(r', r) \approx 0 \text{ unless } (\vec{r}' - \vec{r}/l) \leqslant \text{a few lattice constants} \quad (5.7)$$

By considering the case $h_\beta(r)$ independent of r and comparing (5.6) with (5.5), we find

$$\sum_r b_{\alpha\beta}(r', r) = b_{\alpha\beta} \quad (5.8a)$$

On the other hand, sum eqn (5.6) over r', when $h_\beta(r)$ is independent of r'. The equation becomes:

$$N' h'_\alpha = \sum_r [\sum_{r'} b_{\alpha\beta}(r', r)]\, h_\beta = N(\sum_{r'} b_{\alpha\beta}(r', r) h_\beta)$$

or

$$l^{-d}h'_\alpha = \sum_{r'} b_{\alpha\beta}(r', r)h_\beta$$

consequently

$$\sum_{r'} b_{\alpha\beta}(r', r) = l^{-d}b_{\alpha\beta}. \tag{5.8b}$$

The difference between (5.8a) and (5.8b) arises precisely because there are more terms in a sum over r than in a sum over r'.

Now we are ready to calculate averages and correlation functions.

$$\langle s_\alpha(r)\rangle_h = \frac{[\delta/\delta h_\alpha(r)]\ \Sigma_{\{\sigma\}} \exp(\mathscr{F}\{\sigma\})}{\Sigma_{\{\sigma\}} \exp(\mathscr{F}\{\sigma\})}$$

$$\langle s_\alpha(r_1)s_\beta(r_2)\rangle_h = \frac{[\delta/\delta h_\alpha(r_1)]\ [\delta/\delta h_\beta(r_2)]\ \Sigma_{\{\sigma\}} \exp(\mathscr{F}\{\sigma\})}{\Sigma_{\{\sigma\}} \exp(\mathscr{F}\{\sigma\})}$$

but,

$$\sum_{\{\sigma\}} \exp(\mathscr{F}\{\sigma\}) = \sum_{\{\mu\}} \exp(\mathscr{F}'\{\mu\})$$

consequently,

$$\langle s_\alpha(r)\rangle_h = \frac{[\delta/\delta h_\alpha(r)]\ \Sigma_{\{\mu\}} \exp(\mathscr{F}'\{\mu\})}{\Sigma_{\{\sigma\}} \exp(\mathscr{F}'\{\mu\})} \tag{5.9}$$

If we take $b_{\alpha\beta}(r', r)$ in eqn (5.6) to be independent of h_β, as will be true for sufficiently small h_β, then eqn (5.9) becomes

$$\langle s_\alpha(r)\rangle_h = \frac{\Sigma_{r'}\ \Sigma_{\{\mu\}}\ s_\beta(\{\mu\}, r')b_{\beta\alpha}(r', r) \exp(\mathscr{F}'\{\mu\})}{\Sigma_{\{\mu\}} \exp(\mathscr{F}'\{\mu\})}$$

The left-hand side of this equation is $M_\alpha(h)$. The right-hand side is

$$\sum_{r'} M_\beta(h')b_{\beta\alpha}(r', r).$$

With the aid of (5.8b), we find

$$M_\alpha(h) = M_\beta(h')l^{-d}b_{\beta\alpha}. \tag{5.10}$$

An analogous derivation applied to the correlation function yields

$$G_{\alpha\beta}(\vec{r}_1 - \vec{r}_2, h) = \langle s_\alpha(r_1)s_\beta(r_2)\rangle_h =$$

$$\sum_{r_1 r_2} \langle s_\mu(r'_1)s_\nu(r'_2)\rangle_{h'} b_{\mu\alpha}(r'_1, r_1)b_{\nu\beta}(r'_2, r_2) + \sum_{r_2'} M_\nu(h') \times \frac{\delta}{\delta h_\alpha(r)} b_{\nu\beta}(r'_2, r_2) \tag{5.11}$$

Equation (5.11) simplifies greatly if

$$|\vec{r}_1 - \vec{r}_2| \gg \text{a lattice constant.}$$

In this case the correlation function is sufficiently slowly varying so that it can be taken outside the first sum. Also, the coefficient $b_{\nu\beta}(r_2', r_2)$ should only depend upon $h_\alpha(r)$ in its immediate neighbourhood. Consequently the second term can be dropped. As a result, (5.11) simplifies to

$$G_{\alpha\beta}(r_1 - r_2, h) = G_{\mu\nu}\left(\frac{r_1 - r_2}{l}, h'\right) b_{\mu\alpha} b_{\nu\beta} l^{-2d}$$
$$\text{for } |r_1 - r_2| \gg \text{a lattice constant.} \tag{5.12}$$

Equations (5.10) and (5.12) are the objects of our derivations. They are a direct result of the extension of the transformation theory to slowly varying $h_\alpha(r)$, with the assumption that the transformations are quasi-local in $\vec{r}'$.

C. Connections with universality

To make a connection with the universality concept define a matrix $c_{\alpha\beta}$ by

$$b_{\alpha\gamma} c_{\gamma\beta} l^{-d} = 1; \tag{5.13}$$

of course this definition will fail if b is not invertible. Then define

$$s_\alpha'(r') = s_\beta(\{\sigma\}, r) c_{\beta\alpha} \tag{5.14}$$

for all $\vec{r}' = \vec{r}/l$ for which r' lies on the new lattice. According to eqns (5.10) and (5.12), all the critical point equations will remain true if we make the transformations

$$\begin{aligned} h &\to h_\alpha' = b_{\alpha\beta} h_\beta \\ r &\to r' = r/l \\ s_\alpha(r) &\to s_\alpha'(r') = s_\beta(r) c_{\beta\alpha}. \end{aligned} \tag{5.15}$$

Equations (5.15) are exactly of the same form as our universality transformation. Hence one can describe scaling theory as equivalent to the statement that the universality hypothesis applies to scale transformations.

D. Consequences

To obtain the physical consequences of the scaling theory consider a representation in which b and c are diagonal

$$\begin{aligned} b_{ij} &= \delta_{ij} l^{y_i} \\ c_{ij} &= \delta_{ij} l^{x_i} \\ x_i + y_i &= d. \end{aligned} \tag{5.16}$$

The last relation follows in virtue of eqn (5.13).

Then eqn (5.10) imples that

$$M_i(h_1, h_2, \ldots) = l^{-x_i} M_i(l^{y_1}h_1, l^{y_2}h_2, \ldots)$$

and consequently

$$M_i(h_1, h_2, \ldots) = h_1^{x_i/y_1} M_i\left(\frac{h_2}{h_1^{y_2/y_1}}, \frac{h_3}{h_1^{y_3/y_1}}, \ldots\right)$$

In the special case in which h_1 is h, h_2 is t and there are no other variables in question, we find

$$\begin{aligned} M_1(h, t) &= \langle\sigma\rangle \\ &= h^{x_y/y_t} M_1\left(\frac{t}{h^{y_t/y_r}}\right) \end{aligned}$$

or equivalently

$$\langle\sigma\rangle = t^{x_t/y_r} M'\left(\frac{h}{t^{y_r/y_t}}\right). \tag{5.17}$$

If $y_i + x_i = d$, this equation contains no new information beyond that in eqn (4.22).

However, the correlation function information is new. From eqn (5.12)

$$G_{ij}(r, h, t) = \frac{G_{ij}(r/l, l^{y_h}h, l^{y_t}t)}{l^{x_i + x_j}}.$$

Hence

$$G_{ij}(r, h, t) = \frac{1}{r^{x_i + x_i}} g\left(rt^{1/y_t}, \frac{h}{t^{y_h/y_t}}\right). \tag{5.18}$$

Equation (5.18) enables us to identify the value of ξ at $h = 0$ as proportional to t^{-1/y_t}. Hence ν is given by

$$\nu = 1/y_t. \tag{5.19}$$

The additional information about critical indices contained in eqn (5.19) is derived by comparing the first of eqns (4.23) with (5.19). We then see that

$$d\nu = 2 - \alpha. \tag{5.20}$$

Furthermore, eqn (5.18) implies that x_h and x_t are directly observable since

$$g(r, h = 0, t = 0)\frac{1}{r^{2x_h}} \quad \text{for } r \gg a \tag{5.21}$$

$$\langle \varepsilon(0)\varepsilon(r)\rangle \underset{\substack{h=0\\ t=0}}{\sim} \frac{1}{r^{2x_t}}.$$

These observable quantities are derivable from thermodynamic data and the last of eqns (5.16), i.e.

$$x_h + y_h = d$$

$$x_t + y_t = d$$

hence,

$$\begin{aligned} x_t &= d - 1/\nu \\ &= d\left(1 - \frac{1}{2-\alpha}\right) = d\frac{1-\alpha}{2-\alpha} \end{aligned} \tag{5.22}$$

while from (4.23)

$$x_h = d\left(1 - \frac{\delta}{\delta+1}\right) = \frac{d}{\delta+1}. \tag{5.23}$$

Equations (5.20), (5.22), and (5.23) are exactly satisfied for the two-dimensional Ising model. However, the available evidence suggests that (5.20) is wrong for the three-dimensional Ising model.

References

1. A summary of the Mean Field Theory method can be found in "Phase Transitions" R. Brout, W. A. Benjamin, Inc., New York; 1965.
2. J. Bardeen, L. Cooper, J. R. Schrieffer, *Phys. Rev.* **108**, 1175 (1957).
3. Reviews of Numerical Calculations by M. E. Fisher can be found in *J. math. Phys.* **4**, 278 (1963), *J. math. Phys.* **5**, 944 (1964) and Lectures in Theoretical Physics, University of Colorado, 1965, Vol VII, Part C, p. 1.
4. Reviews of experimental data can be found in P. Heller, *Rep. Prog. Phys.* **30**, 731–826 and in Ref. 5.
5. L. P. Kadanoff, *et al.*, *Rev. mod. Phys.* **39**, 395 (1967).
6. A more recent review is H. E. Stanley "Phase Transitions and Critical Phenomena." Oxford University Press, London and New York, 1971.
7. B. Widom, *J. chem. Phys.* **43**, 3892 (1965) and *ibid* page 3898. A. Z. Patashinskii and V. L. Pokrovskii, *Zh. Eksperim. i Theo. Fiz.* **50**, 439 (1966) (*English* Translation *Soviet Phys. JETP* **23**, 292 (1966). C. Domb and D. L. Hunter, *Proc. phys. Soc. Lond.*, **86**, 1147 (1965). C. Domb, *Ann. Acad. Sci. Fennicae* A, VI No. 210, p. 167, Helsinki (1966). M. E. Fisher, *Physics* **3**, 255 (1967). The ideas in this latter paper were first presented in March, 1965.
8. L. P. Kadanoff, *Physics* **2**, 263 (1966).
9. For a relatively "recent" application see e.g. the seminal paper: E. A. Guggenheim, *J. chem. Phys.* **13**, 253 (1945).

10. The initial concepts in this area were developed independently by two groups of papers:
(a) R. B. Griffiths, *Phys. Rev. Letts.* **24**, 1479 (1970). R. B. Griffiths and J. C. Wheeler, *Phys. Rev.* **A2**, 1047 (1970).
(b) L. Kadanoff, Newport Beach Conference 1970, (unpublished) and "Proceedings of the Enrico Fermi Summer School of Physics, Varenna 1970," (M. S. Green ed.) Academic Press, London and New York, 1971.
11. See, for example, Ref. 5 and J. M. H. Levelt-Sengers in "Proceedings of the Enrico Fermi Summer School of Physics, Varenna, 1970" (M. S. Green ed.), Academic Press, London and New York, 1971.
12. K. G. Wilson, *Phys. Rev.* **B4**, 3174 (1971); *ibid*, page 3184.
13. K. G. Wilson and M. E. Fisher, *Phys. Rev. Letts.* **28**, 240 (1972).
14. K. G. Wilson and J. Kogut: *Phys. Reports* (in press); also Princeton University lecture notes, July, 1972.
15. F. Wegner, *Phys. Rev.* **135**, 1429–36 (1972).
16. The two scale factor universality has been verified by some series calculation. See e.g.:
(a) M. Ferer and M. Wortis, *Phys. Rev.* **136**, B426 (1972), also
(b) D. Stauffer, M. Ferer and M. Wortis, *Phys. Rev. Lett.* **29**, 345 (1972).
(c) M. Ferer, M. A. Moore and M. Wortis, *Phys. Rev.* **B8**, 5205 (1973).
17. For a discussion of 2 scale factor universality with lattice type see—
D. D. Betts, A. J. Guttman, and G. S. Joyce, *J. Phys.* C **4**, 1944.
P. G. Watson, *J. Phys.* C **2**, 1883 (1969).
S. Fisk and B. Widom, *J. chem. Phys.* **50**, 3219 (1969).
18. Some renormalization group transformations have been constructed for one-dimensional Ising models by D. R. Nelson and M. E. Fisher, *Annals of Physics*, (to be published).

PHYSICAL REVIEW B VOLUME 3, NUMBER 11 1 JUNE 1971

Determination of an Operator Algebra for the Two-Dimensional Ising Model

Leo P. Kadanoff and Horacio Ceva
Department of Physics, Brown University, Providence, Rhode Island 02912
(Received 18 November 1970)

A previous publication showed how the critical indices for the two-dimensional Ising model could be derived from an assumed form of an operator algebra which describes how the product of two fluctuating variables may be reduced to a linear combination of the basic fluctuating variables. In this paper, the previously used algebra is derived from the Onsager solution of the two-dimensional Ising model. The calculation makes use of a "disorder" variable which is mathematically the result of applying the Kramers-Wannier transformation to the Ising-model spin variable. The average of products of spin and disorder variables are evaluated at the critical point for the special case in which all the variables lie on a single straight line. The ordering of these variables on the line determines a "quantum number" Γ such that the average is nonzero only for $\Gamma = 0$. Composition rules for this quantum number are derived and used to develop an algebra for the multiplication of complex variables at the critical point. Arguments are given to suggest the identifications of elements of the algebra as the spin, the energy density, the Kaufman spinors, and a stress density. The result of this calculation is the operator algebra which formed the starting point of the previous paper.

I. INTRODUCTION

Critical-point fluctuations can be described in terms of correlations among a set of local fluctuation variables, $O_\gamma(\vec{r})$. This set will include, for example, a local energy density and a local order parameter, as well as several other quantities which have fluctuations on a large spatial scale. The concept of reducibility[1-4] starts from the suggestion that the set $O_\gamma(\vec{r})$ might include only a finite number of fluctuation variables with really different large-scale fluctuations. If this is true, then any product

$$X = \prod_{i=1}^{n} O_{\gamma_i}(\vec{r}_i)$$

of variables within some small neighborhood must be reducible to a linear combination of the basic operators

$$X = \textstyle\sum_\gamma A_\gamma O_\gamma(\vec{R}) \quad ,$$

$$\vec{R} = (1/n) \textstyle\sum_i \vec{r}_i \quad .$$

There the A_γ's are a set of numbers which can depend upon the γ_i and the values of the spacial differences $\vec{r}_i - \vec{r}_j$. In particular, the reducibility hypothesis suggests that a product of two nearby basic operators is reducible as

$$O_\alpha(\vec{r}_1)\, O_\beta(\vec{r}_2) = \textstyle\sum_\gamma A_{\alpha\beta,\gamma}(\vec{r})\, O_\gamma(\vec{R}) \quad ,$$

$$\vec{r} = \vec{r}_1 - \vec{r}_2 \quad , \quad R = \tfrac{1}{2}(\vec{r}_1 + \vec{r}_2). \qquad (1.1)$$

The coefficients A in the reducibility relations describe a kind of algebra for critical fluctuations. We believe that an understanding of the structure of this algebra is a powerful tool which perhaps might be used for predicting critical indices. This belief is based upon the result of a previous calculation,[2] in which all critical indices for the two-dimensional Ising model were derived from: (a) a listing of the O_γ, (b) the scaling concept, (c) symmetry properties, and (d) a knowledge of which of the coefficients $A_{\alpha\beta,\gamma}$ were nonvanishing.

The present paper is devoted to defining the reduction algebra for the two-dimensional Ising model so as to exhibit the justification for the assumptions used in Ref. 2. We begin by introducing, in Sec. II, a new fluctuating variable $\mu_{\vec{r}}$, which roughly represents the amount of disorder in the neighborhood of the point $\vec{r}$. More precisely, this new variable is the transform of the standard local magnetization variable $\sigma_{\vec{r}}$ under the Kramers-Wannier[5] transformation. This transformation describes an important symmetry of the Ising model which is very useful in determining symmetries in the coefficients A.

Section III lists the important fluctuation variables O_γ. In addition to $\sigma_{\vec{r}}$ and $\mu_{\vec{r}}$, this list includes the energy density and the components of a stress tensor t_{ij}. Kawasaki[6] has used this tensor for discussions of the liquid-gas transition, but its importance in the Ising model has not been emphasized in the past. The remaining variables on this list are the two-component spinor or "fermion" variables used by Kaufman,[7] by Schultz, Mattis, and Lieb,[8] and by one of the present authors.[9]

In Sec. IV, we consider correlations among the basic variables for the special case in which the variables all lie on a single straight line. Previous calculations of spin correlations[1] have shown that special simplicities ensue for this case. Even more remarkable symmetry relations appear when we consider correlations of products of σ's and μ's along a line. The ordering of the operators determines a "quantum number" Γ, where 2Γ is an integer. The value of Γ, which is in appearance similar to an angular momentum label, determines whether or not the correlation vanishes. By using operators D_γ each of which has a well-defined value of Γ, we can make the reduction relation (1.1) take a particularly simple form. Furthermore, the critical-point correlations of any product of D_γ's on the line is found and written explicitly.

Section V lists the coefficients in the reduction formulas (1.1).

All the calculations of the present paper are, of course, derived from the Onsager solution[10] of the Ising model. However, the main results are sets of symmetry relations for $A_{\alpha\beta,\gamma}$ which might just have been guessed without the Onsager solution. Hence, similar symmetries might be found for other critical fluctuation problems. If so, then the methods of Ref. 2 could perhaps provide a technique for finding critical indices from first principles.

II. DISORDER VARIABLES

A. Magnetic Dislocations

At zero magnetic field, the Ising model has a partition function which can be expressed as a sum over all spins according to

$$Z\{K\} = \sum_{\{\sigma_{jk}\}=\pm 1} e^{G\{K,\sigma\}} \quad , \tag{2.1}$$

$$G\{K,\sigma\} = \sum_{j,k} \sigma_{j,k}\left[K_x(j+\tfrac{1}{2},k)\,\sigma_{j+1,k} + K_y(j,k+\tfrac{1}{2})\,\sigma_{j,k+1}\right].$$

Here we have a lattice in which j increases in the x direction, k increases in the y direction, and we have allowed all the coupling constants to be unequal. A basic cell in this lattice is depicted in Fig. 1(a).

Eventually, we shall wish to set all the coupling constants equal to one another. However, before doing this, we introduce a magnetic dislocation into the lattice by letting[11]

$$K \to -K \tag{2.2}$$

for all the coupling constants along the path indicated in Fig. 1(b). Then, imagine that all the coupling constants indicated by light bars, are equal and positive, whereas those with heavy bars are equal to minus the others. Since there is an energetic advantage for spins connected by light bars to be equal and spins connected by heavy bars to be opposite, this reversal of coupling constants tends to introduce a Bloch wall into the spin system. If $\{K\}$ is the set of coupling constants before the reversal and $\{\tilde{K}\}$ the set after, we define a correlation function

$$\langle \mu_{r_1}\mu_{r_2}\rangle\,\{K,\Gamma\} = Z\{\tilde{K}\}/Z\{K\} \quad . \tag{2.3}$$

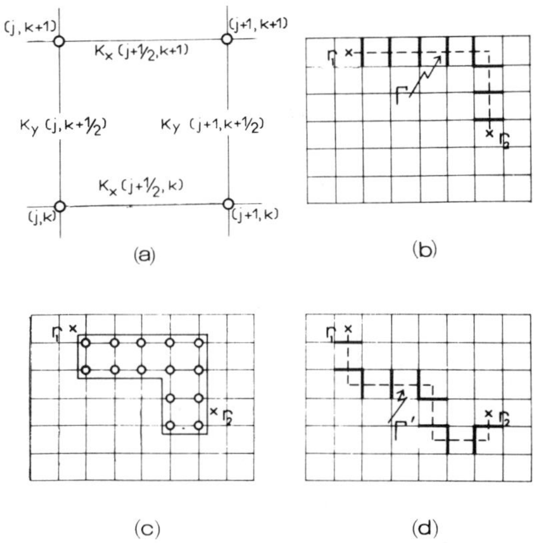

FIG. 1. (a) A basic cell of the model used in the present work. All the coupling constants K_x and K_y are different. (b)–(d) The μ variables are denoted symbolically by a cross (×). Heavy bars indicate the bonds subject to the transformation $K\to -K$. The correlation function involving two μ variables is path independent. We pass from the path Γ in (b) to the path Γ' in (d) simply by changing the sign of the spins inside the region outlined in (c). The partition function is invariant under such a change, whereby the path independence follows.

The right-hand side of (2.3) is physically an exponential of minus the cost in free energy for introducing a magnetic dislocation between $\vec{r}_1$ and $\vec{r}_2$ along the path Γ.

The result is independent of the path Γ. To see this path independence, consider the effect upon $Z\{\tilde{K}\}$ of changing the sign of a group of the dummy summation variables $\sigma_{j,k}$ in Eq. (2.1). Let us flip the signs of all the σ's at the positions circled in Fig. 1(c). Since the K's appear in the form $K\sigma\sigma'$, the effect of changing the sign of these summation variables is to flip the sign of all coupling constants connecting the region outlined with the remainder of the lattice. Under this transform

$$Z\{\tilde{K}\} \to Z\{\tilde{K}'\} \quad ,$$

where $\tilde{K}'$ no longer has minus signs on the path Γ but now has them on the path Γ' shown in Fig. 1(d). Since the change of a summation variable does not change the result of the sum,

$$Z\{\tilde{K}\} \equiv Z\{\tilde{K}'\} \quad .$$

Consequently, the correlation function defined in Eq. (2.3) has equal values for all possible paths connecting the points $\vec{r}_1$ and $\vec{r}_2$.

In a similar way, we can define correlations of any even number of μ_r's, as depicted, for example,

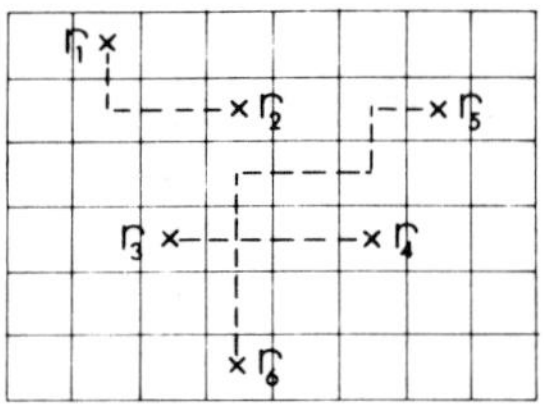

FIG. 2. The correlation function of several μ's is defined by a straightforward generalization of the case of two μ variables. The paths shown are arbitrary, because of the path independence.

in Fig. 2. These correlation functions are defined with the aid of paths through the lattice. But the result is path independent.

In just the same way as σ_r describes the order in the lattice, μ_r describes the disorder. At infinite temperatures, all K's vanish so that $Z\{\tilde{K}\} = Z\{K\}$. Hence, in the completely disordered system, $\langle \mu_r \mu_{r'} \rangle = 1$, for all r, r'.

Introducing $\langle \mu \rangle^2$ as

$$\langle \mu \rangle^2 = \lim_{|r-r'| \to \infty} \langle \mu_r \mu_{r'} \rangle \quad ,$$

we have

$$\langle \mu \rangle^2 = 1 \; , \quad T = \infty \quad . \tag{2.4a}$$

For $T < T_c$, a long spin dislocation costs a "surface" free energy equal to the "surface" tension t times the length of the path.[11] Hence, for $T < T_c$,

$$\lim_{|r-r'| \to \infty} \ln \langle \mu_r \mu_{r'} \rangle \sim -|r - r'| t \quad , \quad T < T_c \; . \tag{2.4b}$$

Therefore when there is long-range order, for $T < T_c$, the long-range disorder vanishes. If μ_r is directly analogous to σ_r,[12] we might expect that $\langle \mu \rangle^2$ decreases and goes to zero as T is reduced from infinity to T_c. To see this property, we must calculate correlations of the μ_r. The easiest way of doing this calculation is to employ the Kramers-Wannier transformation to relate μ correlations to σ correlations.

B. Path Formulation of Spin Correlations

To make this relation, one must introduce a path formulation for spin correlations. Consider a path Γ on the lattice as shown in Fig. 3(a). Take a new set of coupling constants $\{K'\}$ such that

$$K' = \begin{cases} K & \text{off the path} \\ K + i\,\tfrac{1}{2}\pi & \text{on the path} \end{cases} \quad . \tag{2.5}$$

Notice that if K lies on the path, the structure $e^{K\sigma\sigma'}$, which appears in $Z\{K\}$, transforms according to

$$e^{K\sigma\sigma'} \to e^{K'\sigma\sigma'} = e^{K\sigma\sigma'}\, i\sigma\sigma' \quad .$$

Therefore, in an n-step path,

$$Z\{K'\} = \sum_{\{\sigma\}} e^{G\{K',\sigma\}} = \sum_{\{\sigma\}} e^{G\{K,\sigma\}} \prod_{j=1}^{n} (i\,\sigma_j\,\sigma_{j+1}) \quad ,$$

where σ_j is the spin at the beginning of the jth step and σ_{j+1} is the spin at the end of it. Since $\sigma_j^2 = 1$, and since all spins, save the first and last, appear twice,

$$\frac{Z\{K'\}}{Z\{K\}} = \sum_{\{\sigma\}} e^{G\{K,\sigma\}}\, i^n \sigma_{r_1}\sigma_{r_2} \Big/ \sum_{\{\sigma\}} e^{G\{K,\sigma\}} = i^n \langle \sigma_{r_1}\sigma_{r_2} \rangle \quad .$$

As a result, the spin-spin correlation function is

$$\langle \sigma_{r_1}\sigma_{r_2} \rangle \{K\} = (-i)^n\, Z\{K'\}/Z\{K\} \; . \tag{2.6}$$

The results for the two kinds of correlations can be written in an even more symmetrical fashion in terms of a symmetric partition function $Y\{K\}$ defined by

$$Y\{K\} = Z\{K\}\; 2^{-\mathfrak{N}/2} \prod_{j,k} [\cosh 2K_x (j + \tfrac{1}{2}, k) \times \cosh 2K_y (j, k + \tfrac{1}{2})]^{-1/2} \quad , \tag{2.7}$$

where $\mathfrak{N}$ is the number of spins in the lattice. Since $\cosh 2K$ is invariant under $K \to -K$ and changes sign under $K \to K + \frac{1}{2}\pi i$, the results (2.3) and (2.6) may be expressed as

$$\langle \mu_{r_1}\mu_{r_2} \rangle \{K\} = Y\{\tilde{K}\} / Y\{K\} \quad , \tag{2.8a}$$

$$\langle \sigma_{r_1}\sigma_{r_2} \rangle \{K\} = Y\{K'\}/Y\{K\} \quad . \tag{2.8b}$$

Here, as before, $\tilde{K}$ has an opposite sign to K along Γ and is identical to K elsewhere; K' is $K + \frac{1}{2}\pi i$ along Γ' and equals K everywhere else.

C. Kramers-Wannier (K-W) Transform

The introduction of μ_r was motivated by the existance of an exact symmetry relating the $T > T_c$ and the $T < T_c$ regions in the two-dimensional Ising model. To describe this symmetry, define a function of K by

$$K^*(K) = \tfrac{1}{2} \operatorname{arcsinh}\left(\frac{1}{\sinh 2K}\right) \tag{2.9a}$$

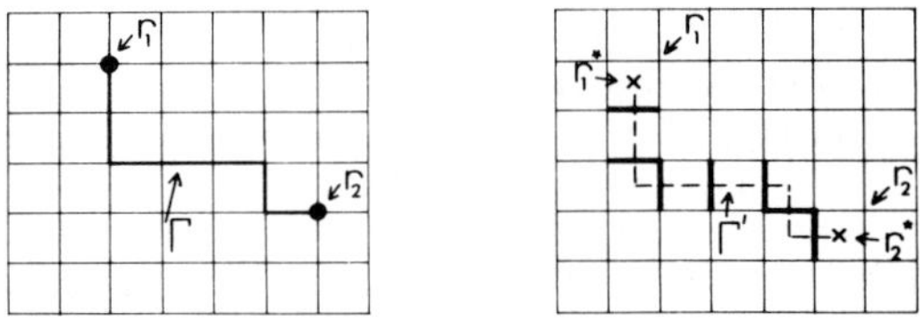

(a) (b)

FIG. 3 Spins are denoted by a dot (•); μ's by a cross (×). Under the K-W transform the path Γ in (a) becomes the path Γ' shown in (b). After setting all the coupling constants equal, we find

$$\langle \sigma_{r_1}\sigma_{r_2} \rangle_{\Gamma} (\epsilon, u) = \langle u_{r_1^*} u_{r_2^*} \rangle_{\Gamma'} (-\epsilon, u).$$

or

$$\sinh 2K^*(K) = \frac{1}{\sinh 2K} \quad . \tag{2.9b}$$

Let us define a new set of coupling constants $\{K^*\{K\}\}$ by

$$K_x^*(j+\tfrac{1}{2},k) = K^*[K_y(j+1,k+\tfrac{1}{2})] \quad ,$$
$$K_y^*(j,k+\tfrac{1}{2}) = K^*[K_x(j+\tfrac{1}{2},k+1)] \quad . \tag{2.10}$$

The K-W transform corresponds to $K \to K^*(K)$ in $Y\{K\}$ and in correlation functions. Notice that at low temperatures the original coupling constants K are large but the new ones K^* are small. Conversely, at high temperatures where the K's are small, the K^*'s are large. Hence, the transformation (2.9) and (2.10) interchanges high and low temperatures.

The statement of the basic symmetry between high and low temperatures is very simple: For all possible $\{K\}$,

$$Y\{K^*\{K\}\} \equiv Y\{K\} \quad . \tag{2.11}$$

This result is proved in Appendix A.

To see one consequence of the symmetry (2.11), consider the homogeneous case in which all the K_x's are equal to each other and all the K_y's are the same also, but K_x is not necessarily equal to K_y. Then, $Y\{K\}$ can be written as a function of $\sinh 2K_x$ and $\sinh 2K_y$:

$$Y\{K\} = Y(\sinh 2K_x, \ \sinh 2K_y) \quad .$$

Then, the symmetry (2.11) implies that for large systems

$$Y(\sinh 2K_x, \ \sinh 2K_y) = Y\left(\frac{1}{\sinh 2K_y}, \ \frac{1}{\sinh 2K_x}\right)$$

or

$$Y(a,b) = Y(b^{-1}, \ a^{-1}) \quad .$$

Moreover, since there is no distinction between the x and y directions, $Y(a,b) = Y(b,a)$. As a result of these symmetries, Y can be considered to be a function of ϵ^2 and $\frac{1}{2}(u+u^{-1})$, where u is the asymmetry parameter

$$u^4 = a/b = \sinh 2K_x/\sinh 2K_y \quad , \tag{2.12}$$

such that $u - u^{-1}$ measures the asymmetry between the x and y directions, and ϵ is

$$-\epsilon = \tfrac{1}{4}(a+b-a^{-1}-b^{-1})$$
$$= \tfrac{1}{4}[\sinh 2K_x + \sinh 2K_y - (\sinh 2K_x)^{-1} - (\sinh 2K_y)^{-1}] \quad . \tag{2.13}$$

Near T_c, ϵ is proportional to $(T-T_c)/T_c$, so that ϵ measures the deviation from T_c.

Under the K-W transform, $\epsilon \to -\epsilon$ and $u \to u$. Under the transform which interchanges x and y, $u \to u^{-1}$ and $\epsilon \to \epsilon$. The result of the K-W transform is that, except for a trivial multiplicative term, the partition function is an even function of ϵ. Hence, the singular term in the energy $\sim \partial \ln Z/\partial\epsilon$ is odd in ϵ, while the singular term in the specific heat is even in ϵ.

Now apply the K-W transform to the spin-spin correlation function shown in Fig. 3(a). Let K^* be the transform of the set of coupling constants K, and K'^* be the transform of the set K' on the path Γ'. The transformed path is shown in Fig. 3(b). From Eqs. (2.8b) and (2.11) we find

$$\langle \sigma_{r_1}\sigma_{r_2}\rangle\{K\} = \frac{Y\{K'\}}{Y\{K\}} \equiv \frac{Y\{K'^*\}}{Y\{K^*\}} \quad . \tag{2.14}$$

But, the transformation operations act differently upon K and K^*. According to Eq. (2.10),

$$K \to K + \tfrac{1}{2}\pi i$$

implies

$$K^*(K) \to -K^*(K) \ , \tag{2.15a}$$

whereas

$$K \to -K$$

implies

$$K^*(K) \to K^*(K) + \tfrac{1}{2}\pi i \quad . \tag{2.15b}$$

Hence, we evaluate the ratio on the right-hand side of Eq. (2.14) by employing either one of the paths shown in Figs. 3(a) or 3(b). On the path Γ', K'^* is $-K^*$. Therefore,

$$Y\{K'^*\}/Y\{K^*\} = \langle u_{r_1^*} u_{r_2^*}\rangle\{K^*\} \quad , \tag{2.16}$$

where

$$r_1^* = (j_1^*, k_1^*) = (j_1 - \tfrac{1}{2}, \ k_1 - \tfrac{1}{2}) \quad . \tag{2.17}$$

When we set all the coupling constants equal, we then find from (2.14) and (2.16) that

$$\langle \sigma_{r_1}\sigma_{r_2}\rangle(\epsilon,u) \equiv \langle u_{r_1^*} u_{r_2^*}\rangle(-\epsilon,u) \quad . \tag{2.18}$$

When r_1 and r_2 are far separated, we find

$$\langle \sigma_r\rangle(\epsilon,u) = \pm\langle \mu_r\rangle(-\epsilon,u) \quad , \tag{2.19}$$

so that as the temperature is reduced from ∞, the $\langle\mu_r\rangle$ decreases and approaches zero as $(T-T_c)^{1/8}$ in the neighborhood of the critical point.

The K-W transformation can be reduced to a set of simple rules. All correlation functions are invariant under the simultaneous replacement: $\epsilon \to -\epsilon$, $u \to u$, $\sigma \to \mu$, $\mu \to \sigma$, and $r = (j,k) \to r^* = (j - \frac{1}{2}, k - \frac{1}{2})$.

D. Correlation Functions Involving both μ's and σ's

The definition of correlation functions involving both μ's and σ's is very straightforward. Consider the product of two μ's and two σ's shown in Fig.

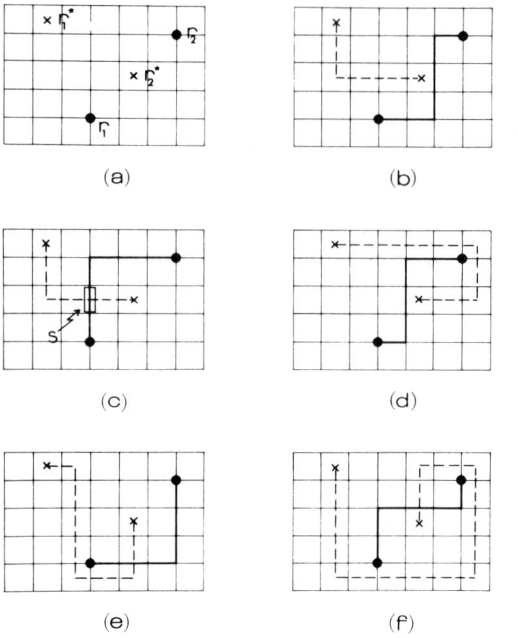

FIG. 4. In the cases involving both σ's and μ's only the *absolute value* of the correlation function is path independent, because a μ path crossing a σ variable produces an over-all change of sign. In the text it is proved that the correlations calculated with the paths shown in (b), (c), and (f) are the same, and have opposite sign to the ones in (d) and (e). The crossing S shown in (c) illustrates a case where there is a basic indeterminacy of sign produced by the noncommutativity of the operations defining σ and μ paths.

4(a). To define the correlation function, employ the paths shown in Fig. 4(b) and write

$$\tilde{K}=\begin{cases} K+\frac{1}{2}\pi i & \text{on } \_\_\_\_\_\_ \\ -K & \text{on } ------ \\ K & \text{elsewhere ,} \end{cases} \quad (2.20a)$$

so that

$$\langle\mu_{r_1}\mu_{r_2}\sigma_{r_1'}\sigma_{r_2'}\rangle = Y\{\tilde{K}\}/Y\{K\} \quad . \quad (2.20b)$$

There is only one further difficulty. Notice that the paths shown in Fig. 4(b) do not intersect. This is important because the operations

$$K \rightarrow K+\tfrac{1}{2}\pi i \quad , \quad (2.21a)$$

$$K \rightarrow -K \quad (2.21b)$$

do not commute. If (2.21a) comes before (2.21b), then

$$e^{K\sigma\sigma'} \rightarrow e^{-K\sigma\sigma'} i\sigma\sigma' \quad , \quad (2.22a)$$

whereas if the order is reversed,

$$e^{K\sigma\sigma'} \rightarrow e^{-K\sigma\sigma'}(-i\sigma\sigma') \quad . \quad (2.22b)$$

For this reason, we always imagine that all operations (2.21a) are performed before all operations (2.21b). This assumption defines the meaning of intersecting paths like those shown in Fig. 4(c).

Nonetheless, the resulting correlation functions are not quite path independent. No possible deformations of σ paths can cause minus signs. Therefore, it is quite irrelevant whether we draw these paths, as in Fig. 4(b), or leave them out. However, the results are not quite independent of the μ path. The proof of μ-path independence fails if the correlation function includes a σ_r. Consider that one of the circled points in Fig. 1(c) contained a σ variable. Then as the path was changed from Γ to Γ', the change in sign of the spin variable would produce an over-all sign change in the correlation function.

This result can be reduced to a rule: As a μ path is deformed through a spin variable, the correlation function changes sign. There is no other path dependence. According to this rule, the correlation function defined by Fig. 4(d) or 4(e) is opposite in sign to the correlation function in Fig. 4(b) or 4(c) whereas the correlation function in Fig. 4(f) is the same as Fig. 4(b) or 4(c).

III. OTHER BASIC VARIABLES

A. General Considerations

So far, we have discussed in detail two important fluctuation variables, the local order variable σ_r and the disorder variable μ_r. We call any function of these variables an operator. There are two important kinds of operators: extensive and intensive. An *intensive operator* $x(r)$ depends only upon spin and disorder variables within a few lattice constants of the point $\vec{r}$; an *extensive operator* X is the sum over the entire lattice of an intensive operator:

$$X=\textstyle\sum_r x(r) \quad . \quad (3.1)$$

Typical extensive operators are the Hamiltonian $\mathcal{H}$ and the total magnetization; their intensive partners are the energy density and the local magnetization σ_r.

Fluctuations in extensive operators are closely connected with thermodynamic derivatives. To see this, consider the effect of adding to the Hamiltonian at the critical point a set of terms of the form $x_\alpha X_\alpha$, where x_α are parameters and the X_α are operators. In symbols,

$$-\mathcal{H}/kT \rightarrow -\mathcal{H}/kT+\textstyle\sum_\alpha x_\alpha X_\alpha \quad .$$

An added term of the form

$$x_\alpha X_\alpha=(\delta T/T_c)\mathcal{H}/kT_c$$

would effectively change the temperature from the critical value T_c to $T=T_c+\delta T$, whereas a term

$$x_\alpha X_\alpha = (\mu H/kT_c)\, M$$

would represent the addition of a magnetic field H. Then the derivative of any physical quantity $\langle Y\rangle$ with respect to one of the parameters would have the form

$$\frac{\partial}{\partial x_\alpha}\langle Y\rangle = \frac{\partial}{\partial x_\alpha} \sum_{\{\sigma_r\}=\pm1} e^{-H/kT}\, Y \Big/ \sum_{\{\sigma_r\}=\pm1} e^{-H/kT}$$

$$= \langle Y\,(X - \langle X\rangle)\rangle \quad . \tag{3.2}$$

If the fluctuations in X are large enough so that the right-hand side of Eq. (3.2) might be more strongly divergent at the critical point than $\langle Y\rangle$, then we say that the operator X is *thermodynamically significant*. Since derivatives with respect to temperature and magnetic field increase the rate of divergence of various physical quantities, the Hamiltonian and total magnetization are, in these terms, thermodynamically significant.

The scaling concept permits a more precise definition of the size of fluctuations in various operators[4] and hence a more precise statement of the concept of thermodynamic significance. According to scaling, when the length scale changes according to

$$\vec{r} \to \vec{r}/L \quad , \tag{3.3a}$$

basic intensive operators scale according to

$$x_\alpha(\vec{r}) \to L^{\nu_\alpha} x_\alpha(\vec{r}/L) \quad , \tag{3.3b}$$

whereas the extensive operators obey

$$X_\alpha = \int d\vec{r}\; x_\alpha(r) \to L^{-(d-\nu_\alpha)} X_\alpha \quad . \tag{3.3c}$$

Here d is the dimensionality of the system and the factor L^{-d} comes from the transformation of $d\vec{r}$. In particular, if $x_\alpha(\vec{r})$ scales as ν_α and $x_\beta(\vec{r})$ scales as ν_β, then their correlation function at the critical point must have an r dependence of the form

$$\langle x_\alpha(\vec{r})\, x_\beta(\vec{r}\,')\rangle \approx \frac{1}{|\vec{r} - \vec{r}\,'|^{\nu_\alpha+\nu_\beta}} A_{\alpha\beta} \quad , \tag{3.4}$$

in order that the invariance property under the transformation (3.3a) and (3.3b) may hold. [In this equation $A_{\alpha\beta}$ may depend upon the angular orientation of $(\vec{r} - \vec{r}\,')$ if x_α and x_β are spinor, vector, or tensor variables.]

The general rule that arises from this form of scaling is that operators which scale as low powers of L can have large fluctuations whereas higher powers of L indicate smaller fluctuations. In particular, negative powers of L on the right-hand side of (3.3b) or (3.3c) indicate the possibility for infinite fluctuations in question whereas positive powers denote bounded fluctuations. Since both σ_r and μ_r are bounded, no intensive operator has infinite fluctuations, hence $\nu_\alpha > 0$. On the other hand, if infinite fluctuations in X_α are to be possible,

$$d - \nu_\alpha \geq 0 \quad . \tag{3.5}$$

This last statement is then the condition for thermodynamic significance.

By definition, thermodynamically significant operators produce infinities at $T = T_c$. Physical arguments are then available to discuss these infinities. For this reason, we shall confine our attention almost completely to thermodynamically significant operators.

B. Order and Disorder Variables

The variables μ_r and σ_r are each thermodynamically significant. Since at the critical point μ and σ each have the same autocorrelation function, their scaling indexes are identical. Their scaling index, which we shall call $\nu_{1/2}$, is

$$\nu_\sigma = \nu_\mu \equiv \nu_{1/2} = \tfrac{1}{8} \quad . \tag{3.6}$$

But the σ's and μ's also give other thermodynamically significant operators, since these and the other operators of interest have the property that their correlation functions for large spacial separations are slowly varying in space. Hence, if $\vec{l}$ is any vector which goes from one lattice site to another nearby site, we know that a correlation function

$$\langle[\sigma(\vec{r} + \vec{l}) - \sigma(\vec{r})]\, x(\vec{r}\,')\rangle$$

is approximately equal to a dot product of the form $\vec{l}\cdot\vec{V}$ when $|\vec{r} - \vec{r}\,'| \gg l$. The coefficient $\vec{V}$ is identified with a new correlation function of the form $\langle\vec{x}(\vec{r})\, x(\vec{r}\,')\rangle$. It is natural to give to $\vec{x}(\vec{r})$ the name $\vec{\nabla}\sigma(\vec{r})$. From the identification of the gradient operation it follows at once that $\vec{\nabla} x_\alpha(\vec{r})$ has the scaling index $1 + \nu_\alpha$. In particular the variables $\nabla\mu_r$ and $\nabla\sigma_r$ have $\nu = 1 + \frac{1}{8}$ and are also thermodynamically significant.

C. Energy Density

The intensive variable conjugate to $\mathcal{E} \sim (T - T_c)/T_c$ is an energy density. We choose as our definition of energy density

$$\begin{aligned}\mathcal{E}(j+\tfrac{1}{2}, k+\tfrac{1}{2}) = {} & [\sigma_{jk}\sigma_{j+1,k} - \tanh 2K_x(j+\tfrac{1}{2}, k)] \\ & \times \tanh 2K_x(j+\tfrac{1}{2}, k) \\ & + [\sigma_{jk}\sigma_{j,k+1} - \tanh 2K_y(j, k+\tfrac{1}{2})] \\ & \times \tanh 2K_y(j, k+\tfrac{1}{2}) \quad . \end{aligned} \tag{3.7}$$

This particular form is taken to produce a simple transformation law for $\mathcal{E}$ under the K-W transform. Since

$$\langle\sigma_{jk}\sigma_{j+1,k}\rangle\{K\} = \frac{\partial}{\partial K_x(j+\frac{1}{2}, k)} \ln Z\{K\} \quad ,$$

it follows from Eq. (3.7) that

$$\langle \mathcal{E}(j+\tfrac{1}{2}, k+\tfrac{1}{2})\rangle \{K\}$$

$$= \left(\tanh 2K_x(j+\tfrac{1}{2}, k)\frac{\partial}{\partial K_x(j+\tfrac{1}{2}, k)} + \tanh 2K_y(j, k+\tfrac{1}{2})\frac{\partial}{\partial K_y(j, k+\tfrac{1}{2})}\right)\ln Y\{K\} \; .$$

Application of the transform then indicates that

$$\langle \mathcal{E}(r)\rangle \{K\} \equiv -\langle \mathcal{E}(r^*)\rangle \{K^*\} \quad ,$$

so that under the K-W transformation

$$\mathcal{E}(r) \rightarrow -\mathcal{E}(r^*) \quad .$$

From the known[13] energy-energy correlation function and Eq. (3.4) it follows that $\mathcal{E}$ scales as $1/r$ and also that the gradient of $\mathcal{E}$ is also thermodynamically significant.

D. Stress Tensor

From the definition of $\mathcal{E}(\vec{r})$, the integral over all space of the energy density is an operator which has, as its main effect, a derivative with respect to temperature, or more properly a derivative with respect to $\mathcal{E}$ at fixed u. But, we also need an intensive variable whose extensive partner varies u, the asymmetry parameter.

Near the critical point, all correlation functions depend upon a distance variable with a kind of elliptical symmetry[14]:

$$\mathcal{R}^2 = x^2/u^2 + u^2y^2 \tag{3.8}$$

in the limit of $\mathcal{R}$ much larger than a lattice constant. Therefore, a derivative with respect to u is equivalent to straining the system according to

$$u\frac{\partial}{\partial u} = -x\frac{\partial}{\partial x} + y\frac{\partial}{\partial y} \quad . \tag{3.9}$$

There is an extensive operator which quite precisely performs the differentiation operation indicated in Eq. (2.29). Its intensive partner is

$$t_1(j+\tfrac{1}{2}, k+\tfrac{1}{2})$$

$$= \tfrac{1}{2}\tanh 2K_x(j+\tfrac{1}{2}, k)\left[\sigma_{jk}\sigma_{j+1,k} - \tanh 2K_x(j+\tfrac{1}{2}, k)\right]$$

$$-\tfrac{1}{2}\tanh 2K_y(j, k+\tfrac{1}{2})\left[\sigma_{jk}\sigma_{j,k+1} - \tanh 2K_y(j, k+\tfrac{1}{2})\right] \quad , \tag{3.10}$$

so that

$$\langle t_1(j+\tfrac{1}{2}, k+\tfrac{1}{2})\rangle \{K\}$$

$$= \tfrac{1}{2}\left[\tanh 2K_x(j+\tfrac{1}{2}, k)\frac{\partial}{\partial K_x(j+\tfrac{1}{2}, k)} - \tanh 2K_y(j, k+\tfrac{1}{2})\frac{\partial}{\partial K_y(j, k+\tfrac{1}{2})}\right]\ln Y\{K\} \quad . \tag{3.11}$$

Notice that t_1 is even under the K-W transform and odd under the transform which replaces x by y. Note that the expression for t_1 is the same as the expression for $\mathcal{E}$ except for a sign change which produces the opposite parity under the K-W transform.

The effect of the extensive operator

$$T_1 = \sum_{j,k} t_1(j+\tfrac{1}{2}, k+\tfrac{1}{2}) \tag{3.12}$$

is given by Eq. (3.9). In particular, whenever the distance between r_1 and r_2 is large,

$$\langle T_1\sigma_{r_1}\sigma_{r_2}\rangle = \left(-j_1\frac{\partial}{\partial j_1} + k_1\frac{\partial}{\partial k_1} - j_2\frac{\partial}{\partial j_2} + k_2\frac{\partial}{\partial k_2}\right)\langle\sigma_{r_1}\sigma_{r_2}\rangle$$

$$= \left(\frac{-2(j_1-j_2)^2}{u^2} + 2(k_1-k_2)^2u^2\right)\frac{\partial}{\partial \mathcal{R}_{12}^2}\langle\sigma_{r_1}\sigma_{r_2}\rangle \quad ,$$

where

$$\mathcal{R}_{12}^2 = (j_1-j_2)^2/u^2 + (k_1-k_2)^2u^2$$

is the distance variable which appears in the correlation functions.

According to Eq. (3.9), the effect of T_1 is that of a second-rank tensor. Since the two-dimensional Ising model has, near its critical point, an underlying rotational symmetry, the other component of this second-rank tensor

$$T_2 \rightarrow -\left(x\frac{\partial}{\partial y} + y\frac{\partial}{\partial x}\right) \tag{3.13}$$

must also exist together with an associated density $t_2(j, k)$.

A plausible but so far unchecked expression for $\langle t_2\rangle$ is

$$\langle t_2(j+\tfrac{1}{2}, k)\rangle = -\tfrac{1}{2}\left(\tanh 2K_x(j+\tfrac{1}{2}, k)\tanh 2K_y(j+1, k+\tfrac{1}{2})\frac{\partial}{\partial K_x(j+\tfrac{1}{2}, k)}\frac{\partial}{\partial K_y(j+1, k+\tfrac{1}{2})}\right.$$

$$\left. - \tanh 2K_x(j+\tfrac{1}{2}, k)\tanh 2K_y(j+1, k-\tfrac{1}{2})\frac{\partial}{\partial K_x(j+\tfrac{1}{2}, k)}\frac{\partial}{\partial K_y(j+1, k-\tfrac{1}{2})}\right)\ln Y\{K\} \quad . \tag{3.14}$$

This identification agrees with Eq. (3.13), in the sense that t_2 is even under the interchange of x and y. The main characteristic of Eq. (3.14) is the presence of the correlations $\langle \sigma_{jk}\sigma_{j+1,k+1}\rangle$ and $\langle \sigma_{jk}\sigma_{j+1,k-1}\rangle$.

In the present work we will not need to use Eq. (3.14) but rather the symmetry properties of t_2, which are independent of its explicit form. Both t_1 and t_2 must be even under the K-W transformation. The intensive variables t_1 and t_2 form a second-order traceless tensor:

$$t_{ij}=\begin{pmatrix} t_1 & t_2 \\ t_2 & -t_1 \end{pmatrix} \quad . \tag{3.15}$$

The scaling properties of t_{ij} are very simple. According to Eqs. (3.9) and (3.13), T_1 and T_2 are invariant under length transforms. Therefore,

$$d-\nu_t=0 \quad ,$$

so that the index for t_{ij} is $\nu_t=2$. Since t_{ij} is itself at the very edge of thermodynamic significance, ∇t_{ij} is not significant.

E. Spinor Variables

The remaining fundamental fluctuating variables needed in the discussion of the two-dimensional Ising model are the variables employed by Onsager, by Kaufman, and later by many other authors. The original Onsager variables are composed of a product of a spin and an adjacent disorder variable as in Fig. 5(a). We call these spinor variables because near T_c correlation functions formed from products of these variables have the rotational properties associated with spinor correlations.

According to the reduction hypothesis, any product of nearby local variables can be expressed as a linear combination of fundamental variables. Consequently, we might expect the product of a μ and a nearby σ to be a linear combination of two fundamental spinor components. The relative placement of the μ and σ will determine the coefficients in the expansion. When $K_x=K_y$, we choose to interpret a product of σ and μ to be analogous to the spinor which has angular momentum directed along the line pointing from σ to μ. Hence, the products shown in Figs. 5(a) are proportional to spinors pointing in directions differing by $\pm 45°$ from the y direction. If a_+ and a_- represent, respectively, spinors with angular momentum in the $+y$ and $-y$ directions, then the two Onsager variables are related to a_+ and a_- by

$$b_+=a_+\cos\tfrac{1}{8}\pi+a_-\sin\tfrac{1}{8}\pi \quad ,$$

$$b_-=i\,(a_+\sin\tfrac{1}{8}\pi+a_-\cos\tfrac{1}{8}\pi) \quad .$$

The two spinor components a_+ and a_-, which we take to be fundamental, are defined by correlations with orientations as shown in Fig. 5(b).

To check that a_+ and a_- have some of the rotational properties of spinors, we study the correlation function composed of one a_+ or a_-, a μ, and a σ as shown in Fig. 6(a). Now rotate the orientation of the a_+ and the a_- clockwise through 180°. If we never let the paths cross in the course of the rotation, the result of the rotation is shown in Fig. 6(b). To bring the paths back to the form shown in Fig. 6(a), we must let the μ paths on the left-hand figure cross through the σ, thereby picking up a minus sign. We find then that a 180° rotation produces $a_+\to -a_-$ and $a_-\to a_+$. Two 180° rotations would then have the effect $a_+\to -a_+$ and $a_-\to -a_-$. These are precisely the correct transformation properties for spinors. The spinor variables have a correlation function which was called g in Ref. 9. Hecht has evaluated the asymptotic form of this correlation at $T\to T_c$ and finds an inverse first-power dependence upon the separation distance [see Hecht[15] Eq. (4.50)]. Then from Eq. (3.4) it follows that the spinor variables scale as $r^{-1/2}$. Therefore, their gradients are thermodynamically significant.

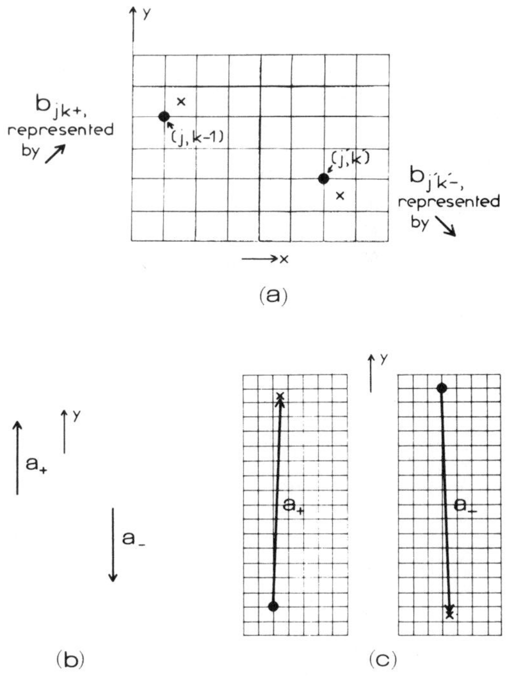

FIG. 5. In the case $K_x=K_y$ the spinors b_+ and b_- admit a graphic representation indicated in (a). As before, a cross (×) represents a μ variable, a dot (•) a σ variable. The spinors a_+ and a_- are shown in (b). This figure needs to be interpreted as the limiting case of (c) when the y distance between the σ and μ goes to infinity. Alternatively, they can be interpreted as the limiting case in which the interaction along the y axis is vanishingly small.

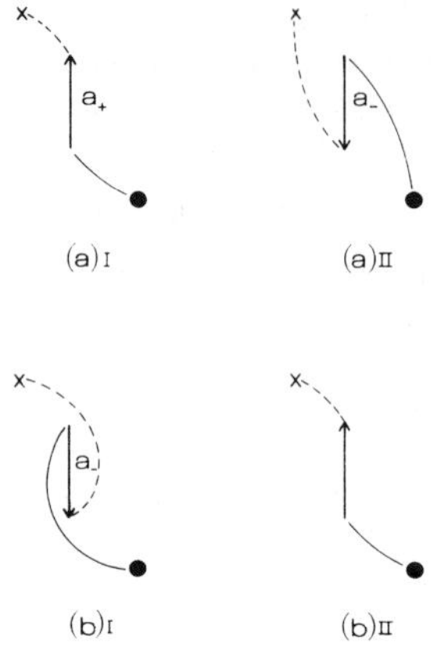

FIG. 6. We draw a spinor variable, paths, and a σ and μ. The lattice has been omitted and the paths are drawn as curves. The operations indicated sustain the interpretation of a_+ and a_- as having spinor character. (i) Rotating a_+ by 180° clockwise, we go from (a)I to (b)I. To go from (b)I to (a)II the μ path needs to cross the "tail of the arrow," i.e., the σ contained in a_-. Hence $a_+ \rightarrow -a_-$ under the clockwise rotation. (ii) Rotating a_- by 180° clockwise in (a)II we end up in (b)II, which is identical to (a)I. Therefore $a_- \rightarrow a_+$ under this rotation. By repeating this process twice we find that a 360° rotation produces $a_+ \rightarrow -a_+$; $a_- \rightarrow -a_-$, the well known rotation property of spinors.

IV. CORRELATIONS ALONG A LINE

A. Basic Result

In Appendix B, the Onsager theory is employed to evaluate a $T = T_c$ correlation function composed of μ's and σ's being on a single line and separated by many lattice constants. The type of correlation function is shown in Fig. 7. Notice that the μ's lie slightly to the right of the y axis, and their paths all lie to the right of the σ's.[16] To write this result, we define

$$D_{1/2}(r) = \sigma_r\,, \quad D_{-1/2}(r) = \mu_r\,, \tag{4.1}$$

and consider an average of a product of operators $D_{\gamma_i}(r_i)$

$$\left\langle \prod_{i=1}^{N} D_{\gamma_i}(r_i) \right\rangle = \langle X \rangle\,, \tag{4.2}$$

arranged on the y axis so that $y_i < y_{i+1}$. This average is evaluated in Appendix B for the special case $T = T_c$. The key part of this result is the statement that there is a number Γ describing the product in (4.2). This "quantum number" is defined recursively as

$$\Gamma_1 = \gamma_1\,, \quad \Gamma_{i+1} = \Gamma_i + (-1)^{2\Gamma_i}\gamma_{i+1}\,, \tag{4.3}$$

and $\Gamma \equiv \Gamma_N$. Our basic result can be stated as

$$\langle X \rangle = 0 \quad \text{for all } \Gamma \neq 0\,. \tag{4.4}$$

It is important to understand why the Γ defined in these terms determines the vanishing or nonvanishing of the correlation.

B. "Quantum Number" Γ

It is easy to understand why correlations vanish for half-integral values of Γ, for if Γ is half-integral, then the correlation function is necessarily a product of an odd number of terms. Therefore it must contain either an odd number of μ's[17] or an odd number of σ's. But every critical-point correlation function must be even under the spin-flip operation

$$\sigma_r \rightarrow -\sigma_r \quad \text{for all } r \tag{4.5}$$

and by the K-W symmetry also even under the μ flip

$$\mu_r \rightarrow -\mu_r \quad \text{for all } r\,. \tag{4.6}$$

But all correlation functions with odd N are odd under the simultaneous application of (4.5) and (4.6). Since they must be both odd and even, they are zero.

The vanishing of the correlations when Γ is an odd integer is also easy to understand. In this case, the product contains an odd number of both σ's and μ's, so that either the symmetry (4.5) or the symmetry (4.6) is sufficient to ensure a vanishing average.

But, the case in which Γ is an even integer is much harder. Why do the correlation functions shown in Figs. 7(d) and 7(e) behave so differently? They contain the same operators, just ordered differently upon the line. The first of these has $\Gamma = 0$, and is certainly nonvanishing; the second has $\Gamma = 2$ and vanishes. Why?

A partial understanding of this behavior can be obtained if we remember that when r_1 and r_2 are close together and $y_1 < y_2$, then

$$\sigma_{r_1}\mu_{r_2} \sim a_+\,, \tag{4.7a}$$

$$\mu_{r_1}\sigma_{r_2} \sim a_-\,. \tag{4.7b}$$

If we were working in three-dimensional space, we would agree that a product of operators of the form (4.7a) adds $\frac{1}{2}$ unit of J_y whereas the combination (4.7b) adds $-\frac{1}{2}$ unit of J_y, where J_y is the angular momentum in the y direction. Then the correlation function in Fig. 7(e) would have $J_y = 1$ whereas the correlation function in Fig. 7(d) would have $J_y = 0$. Therefore, in three-dimensional space, the correlation function Fig. 7(d) would be nonvanishing and the one in Fig. 7(e) would vanish.

This angular momentum argument gives the right answer for all even values of N. To find J_y, we pair adjacent operators in the product. We ascribe

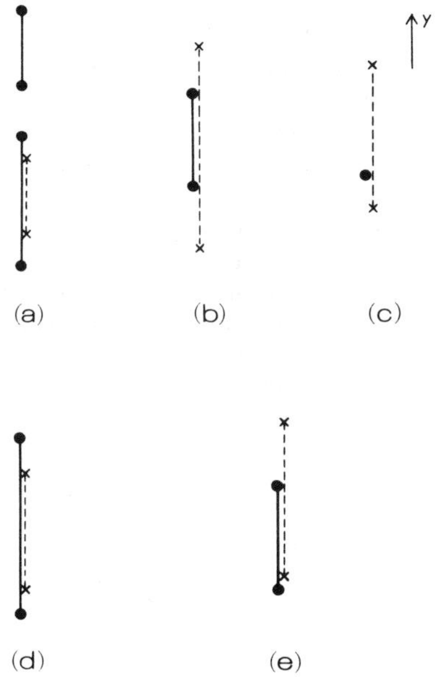

FIG. 7. Different cases of correlation functions along a line. The μ paths lie slightly to the right of the σ paths. The γ values for the different cases shown are $\gamma_a = \gamma_b = \gamma_d = 0$, $\gamma_c = -\frac{3}{2}$, $\gamma_e = 2$. The correlation functions of the operators in (c) and (e) vanish. Note that (e) has the same operators as in (b) and (d) but in a different arrangement.

$J_y = +\frac{1}{2}$ to all products of the type (4.7a), $J_y = -\frac{1}{2}$ to all products of the type (4.7b), and $J_y = 0$ to products $\mu_{r_1}\mu_{r_2}$ and $\sigma_{r_1}\sigma_{r_2}$. Then the total J_y is the same as $\frac{1}{2}\Gamma$. The average is nonvanishing if and only if J_y is zero.

Since we are not working in three dimensions, this angular momentum analogy is far from perfect. Nonetheless, it is clear that the quantum number Γ, which describes the relative ordering of σ's and μ's, describes an absolutely essential quality of the operators on a line and that this quality bears some relation to angular momentum.

From Eq. (4.3) we can work out a composition rule: Given two operators $X_\alpha(r_1)$ and $X_\beta(r_2)$ with quantum numbers α and β, such that all spins and μ's in the X_α lie below, i.e., have smaller y coordinate, all spins and μ's in X_β, then the product of these two is another operator with well-defined Γ,

$$X_\alpha(r_1)X_\beta(r_2) = X_\gamma, \qquad \gamma = \alpha + (-1)^{2\alpha}\beta. \tag{4.8}$$

Consequently, the Γ values add if the first operator has integral γ, and subtract if the first has half-integer γ.

The Γ quantum number will play an essential role in our further discussions. Its important properties are: (a) the composition rules (4.3) and (4.8); (b) the fact that nonvanishing operators with different Γ's behave differently near the critical point; (c) only operator products with total Γ equal to zero have nonvanishing averages at $T = T_c$.

C. Higher-Order Operators

The other key result for our analysis is the actual value of the average (4.2)

$$\left\langle \prod_{i=1}^{N} D_{\gamma_i}(r_i)\right\rangle = \begin{cases} 0 & \text{if } \Gamma \neq 0 \\ \prod_{1\le i<j\le N} [f(i,j)]^{\gamma_i\gamma_j p_i p_j} & \text{if } \Gamma = 0\,. \end{cases} \tag{4.9}$$

Here

$$p_i = (-1)^{2\Gamma_i},$$
$$f(i,j) = \langle \sigma(r_i)\sigma(r_j)\rangle^{-4} \approx |r_i - r_j|\, c, \tag{4.10}$$

where c is a known constant. The results (4.9) and (4.10) hold at $T = T_c$ whenever $|r_i - r_j|$ is much greater than a lattice constant and all the r_i lie on a single straight line. (We neglect the fact that the μ's are not exactly on the same line than the σ's because this is unimportant here. But see Sec. IV D.)

The result (4.9) describes correlations of operators D_γ with γ value equal to $\pm\frac{1}{2}$. The reduction idea permits us to use (4.9) to define and determine the properties of a much larger set of operators: the set $D_\gamma(r)$ with γ being any positive or negative integer or half-integer. The basic idea is that the higher-order operator is formed from a set of σ's and μ's closely spaced upon the line. For convenience we allow the σ's and μ's to alternate and the spacing between the neighboring operators to be the small distance a, as in Fig. 8. We then define, for positive γ,

$$X_\gamma(\vec{R}) = \sigma(\vec{r}_1)\times\mu(\vec{r}_2)\times\sigma(\vec{r}_3)\times\mu(\vec{r}_4)\cdots \quad (2\gamma \text{ terms}), \tag{4.11a}$$

with

$$y_{i+1} = y_i + a, \quad x_i = 0,$$
$$\vec{R} = \frac{1}{2\gamma}\sum_{i=1}^{2\gamma}\vec{r}_i\,.$$

The same definition holds for negative γ, except that each σ is replaced by a μ, and vice versa,

$$X_\gamma(\vec{R}) = \mu(\vec{r}_1)\times\sigma(\vec{r}_2)\times\mu(\vec{r}_3)\cdots \quad (2|\gamma| \text{ terms}). \tag{4.11b}$$

For completeness, we also define

$$X_0(\vec{R}) = 1\,. \tag{4.11c}$$

From the reduction idea, we might expect that as $a \to 0$, the $X_\gamma(\vec{R})$ becomes a function of a times an operator independent of a. In symbols

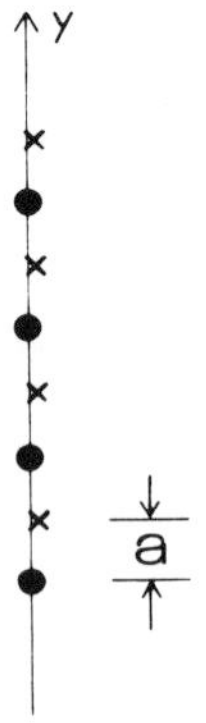

FIG. 8. Operators with $|\gamma| > \frac{1}{2}$ are constructed with the use of the two types of fundamental entities, σ's and μ's spaced along the line. In this figure, $\gamma = 4$.

$$X_\gamma(\vec{R}) \to A_\gamma(a) D_\gamma(\vec{R}) , \tag{4.12}$$

where A_γ is a set of coefficients dependent upon a and γ, and $D_\gamma(\vec{R})$ is an operator.

Equation (4.9) can then be extended to include correlations among the higher-order D_γ. Let $r_1, r_2, \ldots, r_{2\gamma}$ be spaced as in Eq. (4.11a) and let

$$|r_k - r_{2|\gamma|}| \gg a \quad \text{for } 2|\gamma| < k \leq N .$$

We require also that for $i \leq 2|\gamma|$ the operators alternate as in Fig. 8. Then from Eq. (4.9) the resulting correlation function takes the form

$$\left\langle X_\gamma(\vec{R}) \prod_{k=2|\gamma|+1}^{N} D_{\gamma_k}(r_k) \right\rangle = \begin{cases} 0 & \text{if } \Gamma \neq 0 \\ ABC & \text{if } \Gamma = 0 \end{cases} ,$$

where

$$A = \prod_{1 \leq i < j \leq 2|\gamma|} [f(i,j)]^{1/4} ,$$

$$B = \prod_{1 \leq i \leq 2|\gamma|} \prod_{2|\gamma| < k \leq N} [f(i,k)]^{\gamma_i \gamma_k p_i p_k} , \tag{4.13}$$

$$C = \prod_{2|\gamma| < k < m \leq N} [f(k,m)]^{\gamma_k \gamma_m p_k p_m} .$$

Notice that C does not depend upon a. As $a \to 0$ all the r_i in B approach one another and B is a product of $2|\gamma|$ identical terms. Hence,

$$B \to \prod_{2|\gamma| < k \leq N} [f(R - r_k)]^{\gamma \gamma_k p p_k} \left[1 + O\left(\frac{a^2}{(R - r_k)^2} \right) \right] . \tag{4.14}$$

The correction term becomes negligible as $a \to 0$. In Eq. (4.14) the p is equal to the p value of the operator with the smallest y coordinate in X_γ as defined in Eq. (4.9). So, in this case it is $p = -1$.

Finally, the A defined by Eq. (4.13) is a number which depends only upon γ and a but not upon the values of r_k and γ_k for $k \geq 2|\gamma|$. Hence, it is possible to identify the A in Eq. (4.13) with the quantity $A_\gamma(a)$ defined by Eq. (4.12). Therefore, when Eq. (4.12) is substituted in Eq. (4.13), the A's cancel out and we find that

$$\left\langle D_\gamma(R) \prod_{k=2|\gamma|+1}^{N} D_{\gamma_k}(r_k) \right\rangle = \begin{cases} 0 & \text{for } \Gamma \neq 0 \\ BC & \text{for } \Gamma = 0 \end{cases} . \tag{4.15}$$

Furthermore, a term by term comparison of Eq. (4.15) with Eq. (4.9) indicates that these equations have precisely the same structure, except that the first γ has been extended to have possible values $\pm 1, \pm \frac{3}{2}, \pm 2, \ldots$, also. Therefore, successive applications of the same logic enable us to extend our arguments and prove that Eq. (4.9) and the associated Eq. (4.3) hold for operators D_{γ_k} with all possible values of γ_k.

The only case not discussed so far is $\gamma_k = 0$. But if we define

$$D_0(r) = 1 , \tag{4.16}$$

then this case is covered also.

D. Symmetry Properties

The operators D_γ have particularly simple symmetry properties under the symmetry transformations of the Ising model. For example, under $y \to -y$, the X_γ go into themselves for γ half-integral and into $X_{-\gamma}$ for γ integral (see Fig. 9). Hence, the D_γ's obey

$$y \to -y ,$$

which implies

$$D_\gamma(y) \to D_{\gamma'}(-y), \quad \gamma' = -(-1)^{2\gamma}\gamma . \tag{4.17a}$$

The transformation $x \to -x$ is a bit more subtle. At first sight, it would appear that since all operators are defined by the placement of operators on the y axis the transformation $x \to -x$ cannot possibly make any difference, i.e., under $x \to -x$, $D_\gamma \to D_\gamma$. But this statement neglects the fact that the D_γ are defined with the μ_r appearing infinites-

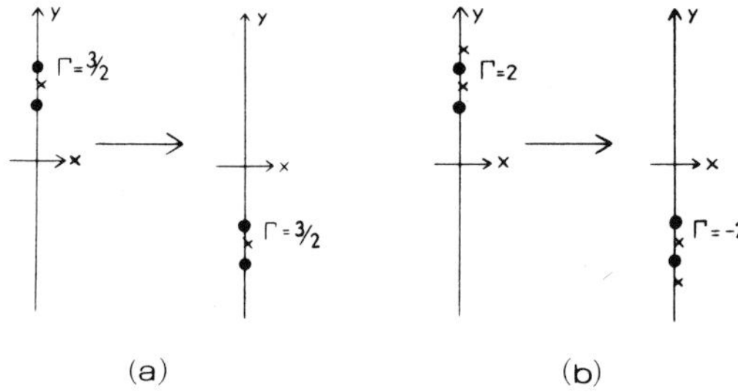

FIG. 9. Upon the transformation $y \to -y$ the operators X_γ, with γ being a half-integral, go into themselves (a), and the ones with γ integral into $X_{-\gamma}$ (b).

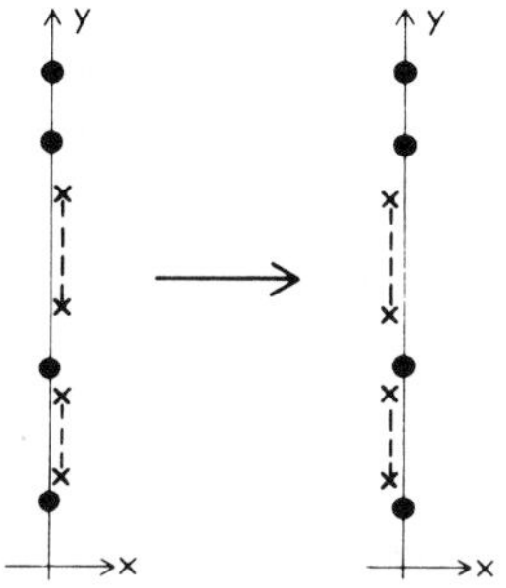

FIG. 10. Due to the fact that the μ's are not exactly on the y axis, the transformation $x \to -x$ is not the identity. The restoration of the operators to their "canonical" arrangement can introduce a minus sign in X_γ, related with the number of σ variables crossed by the μ paths in this process.

imally to the right, say, of the y axis. Hence, under $x \to -x$ the μ paths are displaced to the left of the y axis (see Fig. 10). Extra minus signs can appear as the paths are returned to their "canonical" position, just to the right of the y axis. Hence, we conclude that

$$x \to -x$$

implies

$$D_\gamma(r) \to D_\gamma(r)\Lambda_\gamma(r) \ , \qquad (4.17b)$$

where $\Lambda_\gamma(r)$ is ± 1, depending upon whether or not the path displacement gives a net change of sign.

A similar problem complicates the discussion of the K-W transformation. If path complications are neglected, the K-W transform simply replaces all μ_r by σ_r, and vice versa. Therefore, in the absence of these complications the transform of D_γ is $D_{-\gamma}$. But as the σ's and μ's are interchanged, the μ's end up to the left of the y axis. The displacement of the μ's to their canonical position once again gives minus signs, so that K-W transform implies

$$D_\gamma(r) \to D_{-\gamma}(r)\Lambda_\gamma(r) \ . \qquad (4.18)$$

Since the "parity" factors $\Lambda_\gamma(r)$ are the same in (4.17b) and (4.18), both transformations together give a result which does not contain these factors. We call this combined transformation CP and note that

$$CP = (\text{K-W transform}) \times (x \to -x)$$

implies

$$D_\gamma(r) \to D_{-\gamma}(r) \ . \qquad (4.19)$$

Equation (4.19) evades the question of the parity of the operators D_γ under $x \to -x$. But the value of $\Lambda_\gamma(r)$ is important to our physical identification of the operators D_γ in terms of the already mentioned $a_\pm$, σ, μ, ϵ, and t_{ij}. Therefore, we must write down the values of $\Lambda_\gamma(r)$. By definition $\Lambda_\gamma(r) = \pm 1$, depending upon whether the translation of the μ paths across the y axis requires the μ paths to cut through the σ's in D_γ an even or odd number of times. The evaluation of $\Lambda_\gamma(r)$ requires only a careful enumeration of the possible cases.

For $\gamma = 0$ or $-\frac{1}{2}$ there are no σ's to be considered, and Λ is 1. More generally, for integral n if

$$\gamma = 4n$$

or

$$\gamma = 4n - \tfrac{1}{2} \ ,$$

then

$$\Lambda_\gamma = 1 \ . \qquad (4.20a)$$

Consider for instance the case $\gamma = 3\frac{1}{2}$. This corresponds to an operator with four σ's and three μ's ordered as in Eq. (4.11a). There are two possible forms of drawing the μ paths, depending on the relative location of D_γ on the line [see Figs. 11(a) and 11(b)]. The first μ in D_γ (i.e., the one with smallest y coordinate) is either connected to another μ variable "below" it, with the second and third interconnected by a path, or the first one is connected to the second, in which case the last need to be connected to another μ not belonging to D_γ, with bigger y. In both cases, there are two μ paths, each of which is bound to cut through a σ variable. As each cut gives a minus sign, the net result is $\Lambda_\gamma(r) = 1$, as given in Eq. (4.20a).

On the other hand, σ_r and $D_1 \sim \sigma_{r1}\mu_{r2}$ each flip sign if there is a μ path going below the position of these operators and remain unchanged if there is no such path. Therefore, if

$$\gamma = 4n + \tfrac{1}{2}$$

or

$$\gamma = 4n + 1,$$

then

$$\Lambda_\gamma = \Lambda_\sigma \ , \qquad (4.20b)$$

$$\Lambda_\sigma(r) = \begin{cases} -1 & \text{if there is a } \mu \text{ path going below } r \\ 1 & \text{otherwise} \ . \end{cases}$$

This factor $\Lambda_\sigma(r)$ appears in all Λ_γ in which γ describes an operator with odd numbers of σ's in it. Thus, if

$$\gamma = 4n - 1$$

or

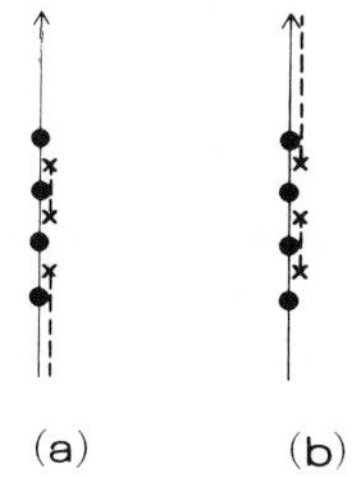

FIG. 11. The figure illustrates the discussion about the determination of the parity Λ_γ. With $\gamma = 3\frac{1}{2}$ there are two possible forms of drawing the μ paths, depending upon the relative placement of $X_{3/2}$ along the line. In both cases the parity is $\Lambda_{3/2} = +1$.

$$\gamma = 4n - \tfrac{3}{2} ,$$

then

$$\Lambda_\gamma(r) = -\Lambda_\sigma(r) , \qquad (4.20c)$$

and if

$$\gamma = 4n + 2$$

or

$$\gamma = 4n + \tfrac{3}{2} ,$$

then

$$\Lambda_\gamma(r) = -1 . \qquad (4.20d)$$

We can summarize our results by saying that the total parity $\Lambda_\gamma(r)$ is a product of a kind of "intrinsic parity" Λ_γ^i and an "orbital parity" $\Lambda_\gamma^0(r)$ such that

$$\Lambda_\gamma(r) = \Lambda_\gamma^i \Lambda_\gamma^0(r) . \qquad (4.21)$$

The orbital parity is just the parity of an individual σ raised to the power of the number of σ's in D_γ, while the intrinsic parity is

$$\Lambda_\gamma^i = \begin{cases} 1 & \text{for } \gamma = 4n - \tfrac{1}{2},\ 4n,\ 4n + \tfrac{1}{2},\ 4n+1 \\ -1 & \text{for } \gamma = 4n - 2,\ 4n - \tfrac{3}{2},\ 4n - 1,\ 4n + \tfrac{3}{2} . \end{cases} \qquad (4.22)$$

E. Physical Identifications

We have already agreed that $D_0 = 1$, $D_{1/2} = \sigma$, $D_{-1/2} = \mu$, $D_1 = a_+$, and $D_{-1} = a_-$. In this section we try to give plausible physical arguments for the identifications[16]

$$D_{3/2} = +\lambda\, \partial\mu/\partial x, \quad D_{-3/2} = -\lambda\, \partial\sigma/\partial x , \qquad (4.23)$$

$$D_{\pm 2} = \pm v t_{12} + w \frac{\partial \mathcal{E}}{\partial x} . \qquad (4.24)$$

First let us consider $D_{\pm 2}$. These operators must be even under each of the separate operations

$$\mu_r \to -\mu_r, \quad \sigma_r \to -\sigma_r .$$

Hence, if $D_{\pm 2}$ is to be identified in terms of our previously defined basic operators, it must be proportional to $\mathcal{E}$, t_{ij}, and/or spacial derivatives of $\mathcal{E}$ and t_{ij}. According to Eq. (4.20d), these D's must be odd under $x \to -x$. A proportionality to $\mathcal{E}$, $\partial\mathcal{E}/\partial y$ or t_{11}, is ruled out since each of these operators is even under $x \to -x$. The simplest possible odd operators are t_{12} and $\partial\mathcal{E}/\partial x$. We then try a linear combination

$$D_{\pm 2} = v_\pm t_{12} + w_\pm \frac{\partial \mathcal{E}}{\partial x} ,$$

where $v_\pm$ and $w_\pm$ are coefficients to be determined. Under $y \to -y$, $D_2 \to D_{-2}$ from Eq. (4.16). Under this operation, $\partial\mathcal{E}/\partial x$ is even and t_{12} is odd. As a result

$$v_\pm = \pm v, \quad w_\pm = w ,$$

so that our symmetry principles naturally lend to Eq. (4.24).

As a check, let us apply the *CP* transform. Under this transform t_{12} changes sign because it is even under K-W but odd by $x \to -x$. On the other hand, $\partial\mathcal{E}/\partial x$ is odd under both parts of the transform and, hence, even in total. As a result, under *CP*

$$D_2 = v t_{12} + w \frac{\partial \mathcal{E}}{\partial x} \to -v t_{12} + w \frac{\partial \mathcal{E}}{\partial x} \equiv D_{-2} .$$

This result checks with Eq. (4.19).

As a further check, notice that Eq. (4.9) implies

$$\langle D_2(r_1) D_{-2}(r_2)\rangle = [f(r_1 - r_2)]^{-4} \sim |r_1 - r_2|^{-4} .$$

But $\nabla\mathcal{E}$ and t_2 are each known to scale as r^{-2}. Hence the correlation of two such operators should be expected to scale as r^{-4}.

Next, consider $D_{\pm 3/2}$. We know that $D_{3/2}$ changes sign when μ changes sign but not when σ changes sign since $X_{3/2} = \sigma(r_1)\mu(r_2)\sigma(r_3)$. Hence, $D_{3/2}$ may be expected to be proportional to μ or gradients thereof, whereas $D_{-3/2}$ should be proportional to σ and gradients of σ. According to Eq. (4.20), $D_{\pm 3/2}$ behaves the same way as $(\partial/\partial x) D_{\mp 1/2}$ under $x \to -x$. Hence, it is natural to guess

$$D_{\pm 3/2} = \lambda_\pm \frac{\partial}{\partial x} D_{\mp 1/2} .$$

Then *CP* invariance implies $\lambda_+ = -\lambda_-$, so that Eq. (4.23) follows. To check this result, notice that $D_{\pm 3/2}$ are each even under $y \to -y$, as required by Eq. (4.16). Note also that

$$\langle \sigma_{r_1} \sigma_{r_2} \rangle \sim |r_1 - r_2|^{-1/4}$$

implies via Eq. (4.23) that

$$\langle D_{-3/2}(r_1) D_{-3/2}(r_2)\rangle \sim |r_1 - r_2|^{-9/4} .$$

But Eq. (4.9) implies

$$\langle D_{-3/2}(r_1) D_{-3/2}(r_2)\rangle = [f(r_1 - r_2)]^{-9/4} ,$$

so that the correlation function has its expected value. This check further confirms our hypothesis (4.23).

V. REDUCTION FORMULAS

In this section we make use of the symmetry properties of the A's to suggest a set of reduction formulas for products of two nearly D_γ on the line $x = 0$. The results thereby obtained are then checked against the known results for multiple correlations of D_γ.

A. Hypothesis

To write down the reduction formulas we consider products of the form

$$D_\alpha(\vec{r}_1) D_\beta(\vec{r}_2) = P , \qquad (5.1)$$

with r_1 and r_2 lying on the y axis and $y_1 < y_2$. Since the product (5.1) has "quantum number"

$$\gamma = \alpha + (-1)^{2\alpha}\beta\ , \tag{5.2}$$

it is reasonable to guess that

$$P = A_{\alpha,\beta}(r) D_\gamma(\vec{\bar{R}})\ , \tag{5.3}$$

where $A_{\alpha,\beta}$ is a number, $r = |\vec{r}_1 - \vec{r}_2|$, and $\vec{\bar{R}}$ is a coordinate on the y axis in the neighborhood of $\vec{r}_1$ and $\vec{r}_2$. If we use the reference point

$$\vec{R} = \tfrac{1}{2}(\vec{r}_1 + \vec{r}_2)\ , \tag{5.4}$$

we can assume that $\vec{\bar{R}}$ differs from $\vec{R}$ by only a small amount and rewrite the hypothesis (5.3) as

$$D_\alpha(\vec{r}_1) D_\beta(\vec{r}_2) \approx A_{\alpha,\beta}(r) D_\gamma(\vec{R}) + B_{\alpha,\beta}(r) \frac{\partial}{\partial Y} D_\gamma(\vec{R}) \tag{5.5}$$

for $\gamma = \alpha + (-1)^{2\alpha}\beta \neq 0$.

Equation (5.5) is a reasonable consequence of the hypothesis that $D_\gamma(R)$ is the only fluctuating operator with quantum number equal to γ. This hypothesis appears reasonable for $\gamma \neq 0$. However, it is possible to identify several operators with $\gamma = 0$. Explicit calculations show that for r_1 and r_2 close together on the y axis,

$$D_{1/2}(\vec{r}_1) D_{1/2}(\vec{r}_2) = \sigma_{r_1}\sigma_{r_2}$$
$$= A_{1/2,1/2}(r) + \bar{B}_{1/2,1/2}(r)\mathcal{E}(\vec{R})$$
$$+ C_{1/2,1/2}(r)\, t_{11}(\vec{R})\ .$$

Hence $D_0(R) = 1$, $\mathcal{E}(\vec{R})$, and $t_{11}(\vec{R})$ all have $\gamma = 0$. Therefore, when the γ defined by Eq. (5.2) vanishes, we replace Eq. (5.5) by

$$D_\alpha(\vec{r}_1) D_\beta(\vec{r}_2) = A_{\alpha,\beta}(r) + \bar{B}_{\alpha,\beta}(r)\mathcal{E}(\vec{R}) + C_{\alpha,\beta}(\vec{R}) t_{11}(\vec{R}) \tag{5.6}$$

when $\gamma = \alpha + (-1)^{2\alpha}\beta = 0$.

B. Symmetry Properties

The coefficients in Eqs. (5.5) and (5.6) are limited by the symmetry properties listed in Sec. IV D. For example, the *CP* transformation which takes

$$\vec{r}_1 \to \vec{r}_1,\quad \vec{r}_2 \to \vec{r}_2,\quad \vec{R} \to \vec{R}\ ,$$
$$D_\gamma \to D_{-\gamma},\quad \mathcal{E}(R) \to -\mathcal{E}(R),\quad t_{11}(R) \to t_{11}(R) \tag{5.7}$$

is an exact symmetry of the Ising model. The application of this symmetry to Eq. (5.6) indicates that

$$D_{-\alpha}(r_1) D_{-\beta}(r_2) = A_{\alpha,\beta}(r) - \bar{B}_{\alpha,\beta}(r)\mathcal{E}(R) + C_{\alpha,\beta}(r) t_{11}(R) \tag{5.8}$$

for $\gamma = \alpha + (-1)^{2\alpha}\beta = 0$. But Eq. (5.6) indicates directly that the right-hand side of (5.8) is

$$A_{-\alpha,-\beta}(r) + \bar{B}_{-\alpha,-\beta}(r)\mathcal{E}(\vec{R}) + C_{-\alpha,-\beta}(r) t_{11}(\vec{R})\ .$$

Therefore, C is even under the change in sign of both subscripts and $\bar{B}$ is odd under such an interchange. The general conclusion drawn from this symmetry and Eqs. (5.5) and (5.6) is

$$A_{-\alpha,-\beta} = A_{\alpha,\beta},\qquad B_{-\alpha,-\beta} = B_{\alpha,\beta}\ ,$$
$$\bar{B}_{-\alpha,-\beta} = -\bar{B}_{\alpha,\beta}\ ,\quad C_{-\alpha,-\beta} = C_{\alpha,\beta}\ . \tag{5.9}$$

A precisely similar logic may be applied to the transformation $y \to -y$, which takes

$$\vec{r}_1 \to -\vec{r}_1\ ,\quad \vec{r}_2 \to -\vec{r}_2\ ,\quad \vec{R} \to -\vec{R}\ ,$$
$$\alpha \to \alpha' = -(-1)^{2\alpha}\alpha\ ,\quad \beta \to \beta' = -(-1)^{2\beta}\beta\ ,$$
$$\gamma \to \gamma' = -(-1)^{2\gamma}\gamma\ .$$

After this transformation, Eq. (5.5) reads

$$D_{\beta'}(-r_2) D_{\alpha'}(-r_1) = A_{\alpha,\beta}(r) D_{\gamma'}(-R)$$
$$+ B_{\alpha,\beta}(r)\frac{\partial}{\partial Y} D_{\gamma'}(-R)\ . \tag{5.10}$$

A direct application of Eq. (5.5) indicates that the right-hand side of (5.10) is also

$$A_{\beta',\alpha'}(r)\, D_{\gamma'}(-R) + B_{\beta',\alpha'}(r)\left(-\frac{\partial}{\partial Y}\right) D_{\gamma'}(-R)\ .$$

By this logic we find

$$A_{\beta',\alpha'} = A_{\alpha,\beta}\ ,\qquad B_{\beta',\alpha'} = -B_{\alpha,\beta}\ , \tag{5.11}$$

with

$$\alpha' = -(-1)^{2\alpha}\alpha\ ,\quad \beta' = -(-1)^{2\beta}\beta\ .$$

The application of this logic to Eq. (5.6) indicates that (5.6) satisfies this symmetry identically.

C. Check of Reduction Formulas

From Eq. (4.9),

$$\left\langle \prod_{i=1}^{N+1} D_{\gamma_i}(r_i) \right\rangle = \prod_{1 \le i < j \le N+1} [f(|r_i - r_j|)]^{\gamma_i \gamma_j p_i p_j}\ . \tag{5.12}$$

Now allow r_N and r_{N+1} to approach each other. Write $\gamma_N = \alpha$, $\gamma_{N+1} = \beta$; $\gamma = \alpha + (-1)^{2\alpha}\beta$. Equation (5.12) has a right-hand side which is a product of three terms:

$$A = f(|r_N - r_{N+1}|)^{\alpha\beta(-1)^{2\alpha}}\ ,$$

$$U = \prod_{1 \le i \le N-1} [f(|r_i - r_N|)^\alpha f(|r_i - r_{N+1}|)^{\beta(-1)^{2\alpha}}]^{\gamma_i p_i p_N}\ , \tag{5.13}$$

$$V = \prod_{1 \le i \le j \le (N-1)} [f(|r_i - r_j|)]^{\gamma_i p_i \gamma_j p_j}\ .$$

When $\gamma \neq 0$, U may be simplified to

$$U=\prod_{1\leq i\leq N-1}[f(|r_i-\overline{R}|)]^{\gamma\gamma_i p_i p_N}\left[1+O\left(\frac{r^2}{|r_i-r_N|^2}\right)\right], \tag{5.14}$$

with

$$\vec{\overline{\mathrm{R}}}=[\alpha\vec{\mathrm{r}}_N+(-1)^{2\alpha}\beta r_{N+1}]/\gamma\,,$$
$$r=r_{N+1}-r_N\,. \tag{5.15}$$

When this simplification is made, the right-hand side of (5.12) is precisely of the form

$$A\times\left\langle\prod_{i=1}^{N-1}D_{\gamma_i}(r_i)D_\gamma(\overline{R})\right\rangle.$$

Consequently, we find that for $\gamma\neq 0$,

$$D_\alpha(r_1)D_\beta(r_2)=A\,D_\gamma(\overline{R})\,, \tag{5.16}$$

as asserted in Eq. (5.3). In this way, we verify the basic correctness of our reduction hypothesis. To calculate A and B, note

$$\vec{\overline{\mathrm{R}}}=\vec{\mathrm{R}}+\frac{\alpha-(-1)^{2\alpha}\beta}{2[\alpha+(-1)^{2\alpha}\beta]}(\vec{\mathrm{r}}_1-\vec{\mathrm{r}}_2)\,,$$

so that a comparison of Eqs. (5.13) and (5.5) indicates that $A=A_{\alpha,\beta}$ is

$$A_{\alpha,\beta}(r)=[f(r)]^{\alpha\beta(-1)^{2\alpha}}$$
$$=[\langle\sigma_r\sigma_0\rangle]^{-4\alpha\beta(-1)^{2\alpha}}\,, \tag{5.17a}$$

whereas

$$B_{\alpha,\beta}(r)=-A_{\alpha,\beta}(r)\,\frac{\alpha-(-1)^{2\alpha}\beta}{2[\alpha+(-1)^{2\alpha}\beta]}\,r\,. \tag{5.17b}$$

When $\gamma=0$, this approach does not work since the denominator of B diverges. Instead, we note that in this case, as $r=r_{N+1}-r_N$ goes to zero,

$$U=1-\alpha r p_N\sum_{i=1}^{N-1}\frac{\gamma_i p_i}{|r_i-R|}$$
$$+\tfrac{1}{2}\alpha^2r^2\sum_{i=1}^{N-1}\sum_{j=1}^{N-1}\frac{\gamma_i\gamma_i p_j p_j}{|r_i-R|\,|r_j-R|}+O\left(\frac{r^3}{|r_i-R|^3}\right). \tag{5.18}$$

The first term in U is identified with correlations of $D_0(R)=1$, the second term with correlations of $\mathcal{E}(R)$, and the third with correlations of $t_{11}(R)$. We then find for $\gamma=0$, $A_{\alpha,\beta}$ is still given by Eq. (5.17a), whereas

$$\overline{B}_{\alpha,\beta}(r)=-\alpha rb\,A_{\alpha,\beta}(r)\,, \tag{5.19a}$$

$$C_{\alpha,\beta}(r)=\tfrac{1}{2}\alpha^2r^2d\,A_{\alpha,\beta}(r)\,, \tag{5.19b}$$

where b and d are constants. With this identification, the correlations of $\mathcal{E}(R)$ and $t_{11}(R)$ with a group of D_γ's may be computed by writing

$$X=\prod_{i=1}^{N}D_{\gamma_i}(r_i)\,,$$

$$\langle X\mathcal{E}(R)\rangle=\frac{\langle X\rangle}{b}\sum_{i=1}^{N}\frac{\gamma_i p_i q_R}{(r_i-R)}\,, \tag{5.20a}$$

$$\langle Xt_{11}(R)\rangle=\frac{\langle X\rangle}{d}\sum_{i,j=1}^{N}\frac{\gamma_i\gamma_j p_i p_j}{|r_i-R|\,|r_j-R|}\,, \tag{5.20b}$$

where $q_R=(-1)^{\Gamma_{\rm inf}}$, and $\Gamma_{\rm inf}$ is the Γ value for the D operator being just below the position of $\mathcal{E}(R)$, i.e., q_R is equal to the p defined in (4.9) if the corresponding operator has $\gamma=\pm\frac{1}{2}$. Equations (5.20) describe the correlations of X with $\mathcal{E}(R)$ and $t_{11}(R)$ for the special case in which all points in question r_i as well as R lie well separated upon the y axis.

D. Correlation Functions with Several $\mathcal{E}$'s and t's

The next step is to use the same reduction technique to evaluate correlation functions containing several $\mathcal{E}$'s or t_{11}'s.

As an illustration, consider $\langle X\mathcal{E}(R_1)\mathcal{E}(R_2)\rangle$. We start with Eq. (5.20a) and let two operators D_{γ_j} and $D_{\gamma_{j+1}}$ approach each other. We also require that

$$\gamma=\gamma_j+(-1)^{2\gamma_j}\gamma_{j+1}=0.$$

Hence, by using the same argument as in (5.18), we obtain

$$\langle X\mathcal{E}(R_1)\mathcal{E}(R_2)\rangle=\frac{\langle X\mathcal{E}(R_1)\rangle\langle X\mathcal{E}(R_2)\rangle}{\langle X\rangle}+\frac{\langle X\rangle}{b^2}\frac{1}{(R_2-R_1)^2}\,. \tag{5.21}$$

The last term in the right-hand side of (5.21) originates in the terms with γ_j and γ_{j+1} in the sum in Eq. (5.20a).

We notice that setting $X=1$ in (5.21) then gives (note that at $T=T_c$, $\langle\mathcal{E}\rangle=0$)

$$\langle\mathcal{E}(R_1)\mathcal{E}(R_2)\rangle=\frac{1}{b^2}\frac{1}{(R_2-R_1)^2}\,,$$

which coincides with previous calculations.[13] This can be considered as another indication of the consistency and the power of the method. Exactly the same type of calculation gives

$$\langle X\mathcal{E}(R_1)t_1(R_2)\rangle=\frac{\langle X\mathcal{E}(R_1)\rangle\langle Xt_{11}(R_2)\rangle}{\langle X\rangle}$$
$$+\frac{2}{d}\frac{q_{R_1}q_{R_2}}{(R_2-R_1)^2}\langle X\mathcal{E}(R_2)\rangle\,. \tag{5.22}$$

APPENDIX A

We will prove here[18] Eq. (2.11), i.e.,

$$Y\{K\}=Y\{K^*(K)\}\,, \tag{2.11}$$

where

$$Y\{K\}=Z\{K\}P\{K\}\equiv Z\{K\}2^{-\mathfrak{N}/2}\prod_{j,k}[\cosh 2K_x(j+\tfrac{1}{2},k)\cosh 2K_y(j,k+\tfrac{1}{2})]^{-1/2}$$

and $\mathfrak{N}$ is the number of spins in the lattice. By using $\tanh 2K^*(K)=(\cosh 2K)^{-1}$ in P, we obtain

$$P\{K\}=2^{-\mathfrak{N}/2}\prod_{j,k}\left(\frac{\cosh 2K_y^*(j,k-\frac{1}{2})}{\sinh 2K_y^*(j,k-\frac{1}{2})}\,\frac{\cosh 2K_x^*(j-\frac{1}{2},k)}{\sinh 2K_x^*(j-\frac{1}{2},k)}\right)^{-1/2}$$
$$=P\{K^*\}\,2^{(\mathfrak{N}^*-\mathfrak{N})/2}\prod_{j,k}[\sinh 2K_y^*(j,k-\tfrac{1}{2})\sinh 2K_x^*(j-\tfrac{1}{2},k)]^{1/2}, \qquad \text{(A1)}$$

because $\Pi_{j,k}$ covers the whole lattice. Here $\mathfrak{N}^*$ is the number of sites in the dual lattice.

Next, we proceed to write $Z\{K\}$ in terms of K_x and K_y. Use

$$e^{K\sigma\sigma'}=\cosh K+\sigma\sigma'\sinh K, \qquad \cosh K=(\tfrac{1}{2})^{1/2}(\cosh 2K+1)^{1/2}, \qquad \tanh K=e^{-2K^*(K)}$$

to get

$$Z\{K\}=\sum_{\{\sigma_{j,k}\}=\pm 1}\prod_{j,k}\{\tfrac{1}{2}(e^{K_y^*(j,k-1/2)}+\sigma_{jk}\sigma_{j+1,k}e^{-K_y^*(j,k-1/2)})(e^{K_x^*(j-1/2,k)}+\sigma_{jk}\sigma_{j,k+1}e^{-K_x^*(j-1/2,k)})$$
$$\times[\sinh 2K_y^*(j,k-\tfrac{1}{2})\sinh 2K_x^*(j-\tfrac{1}{2},k)]^{-1/2}\}.$$

The last square bracket in the right-hand side of this expression is not a function of σ_{jk}, and can be taken out of the summation. It can be seen to cancel the product appearing in Eq. (A1).

Consequently,

$$Y\{K\}=P\{K\}Z\{K\}$$
$$=2^{(\mathfrak{N}^*-\mathfrak{N})/2}P\{K^*\}\sum_{\{\sigma_{jk}\}=\pm 1}\left\{\prod_{jk}(\tfrac{1}{2})(e^{K_y^*(j,k-1/2)}+\sigma_{jk}\sigma_{j+1,k}e^{-K_y^*(j,k-1/2)})(e^{K_x^*(j-1/2,k)}+\sigma_{jk}\sigma_{j,k+1}e^{-K_x^*(j-1/2,k)})\right\}$$
$$\equiv 2^{(\mathfrak{N}^*-\mathfrak{N})/2}P\{K^*\}T\{K^*\}. \qquad \text{(A2)}$$

The last line of Eq. (A2) serves as definition of $T\{K^*\}$.

But $Y\{K^*\}=P\{K^*\}Z\{K^*\}$; hence we must compare $2^{(\mathfrak{N}^*-\mathfrak{N})/2}T\{K^*\}$ with $Z\{K^*\}$, where

$$Z\{K^*\}=\sum_{\{\sigma_{jk}\}\pm 1}\prod_{j,k}e^{K_x^*(j+1/2,k)\sigma_{jk}\sigma_{j+1,k}}$$
$$\times e^{K_y^*(j,k+1/2)\sigma_{jk}\sigma_{j,k+1}}.$$

We now argue in a manner similar to the case in which all the coupling constants are equal. By expanding the product in $T\{K^*\}$ we notice that it is possible to classify the resulting terms, considering the number of products $(\sigma\sigma')$ included in each one.

So, there will be one term without any $(\sigma\sigma')$, i.e., the one with all the exponentials with positive sign, plus terms with just one pair $(\sigma\sigma')$, ... up to a term with all the pairs $(\sigma\sigma')$'s, i.e., with all the exponentials with negative sign.

Since each one of these terms is summed over all the $\sigma_{jk}=\pm 1$, any term with a σ elevated to an odd power will cancel out.

The ones left actually are not a function of $\sigma_{jk}(\sigma_{jk}^{2n}=1)$. Hence each one of them can be taken out of the sum. All will have a common factor (including the factor $2^{(\mathfrak{N}^*-\mathfrak{N})/2}$)

$$2^{(\mathfrak{N}^*-\mathfrak{N})/2}\left(\prod_{j,k}(\tfrac{1}{2})\right)\sum_{\sigma_{jk}=\pm 1}1.$$

The summation gives $2^{\mathfrak{N}}$.

To calculate the product, remember that it comes from the transformation of $Z\{K\}$, each factor $(\frac{1}{2})^{1/2}$ being related to one of the coupling constants, K_x or K_y. Thus, we have

$$\prod_{j,k}(\tfrac{1}{2})=(\tfrac{1}{2})^{s/2},$$

with s being the number of coupling constants.

But it has been indicated[19] that the K-W transform is an *exact* symmetry *only* in the case that the lattice can be extended on a simply connected surface. Moreover, to be completely defined, in the sense that *all* the transformed coupling constants really represent an interaction, the surface needs to be spheroidal. In this case, the following relation holds: $\mathfrak{N}+\mathfrak{N}^*=s+2$.

Hence the factor which multiplies each of the terms in $T\{K\}$ is

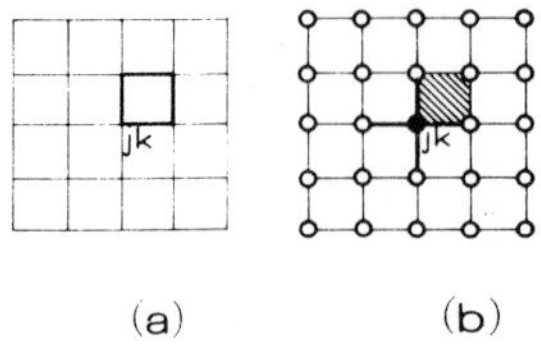

FIG. 12. (a) Simplest of the cells or "closed loops" contributing to $Y\{K\}$. (b) Under the K-W transform the *boundary* of the original cell (cross-hatched here) goes to the arms of the cross centered at (j, k). In this figure O denotes positive spins, • negative spins, and heavy bars indicate the coupling constants that will appear with a minus sign in the partition function. Configuration shown illustrates the term-by-term correspondence discussed in the text, between the lattices (a) and (b). Notice that the configuration in (b) is degenerate with the one obtained by flipping over all spins.

$$2^{(\mathfrak{N}^*-\mathfrak{N})/2}\left(\tfrac{1}{2}\right)^{(\mathfrak{N}+\mathfrak{N}^*-2)/2}2^{\mathfrak{N}}=2\ .$$

Notice that all the terms obtained will be different from each other, because all the K^*'s are different. Moreover, it is easy to see that all of them correspond to "closed loops" of bonds.

Let us compare these terms with the corresponding terms of $Z\{K^*\}$:

(i) The first term in $T\{K^*\}$, with all exponentials with plus signs, is present in $Z\{K^*\}$ *twice*, once for the configuration with all spins $\sigma_{jk}=+1$, and also for the case $\sigma_{jk}=-1$. (This is also the case for the term with all the exponentials with minus signs.)

(ii) To get the first term with some negative exponents in T, we need four products of $(\sigma\sigma')$, whose generic form is [see Fig. 12(a)]

$$(\sigma_{jk}\sigma_{j+1,k})(\sigma_{j+1,k}\sigma_{j+1,k+1})(\sigma_{j+1,k+1}\sigma_{j,k+1})(\sigma_{j,k+1}\sigma_{jk})\ .$$

This term is

$$e^{-K_y^*(j,k-1/2)}\,e^{-K_x^*(j+1/2,k)}\,e^{-K_y^*(j,k+1/2)}\,e^{-K_x^*(j-1/2,k)}\,M\ ,$$

where M is the product of all the other exponentials with plus signs.

In $Z\{K^*\}$ we obtain the same term if the following conditions are fulfilled:

$$-1=\sigma_{j,k-1}\sigma_{jk}=\sigma_{jk}\sigma_{j,k+1}=\sigma_{jk}\sigma_{j+1,k}=\sigma_{j-1,k}\sigma_{jk}$$

and all the other products $\sigma\sigma'=+1$. In Fig. 12(b) we show that this is precisely one allowed configuration in $Z\{K^*\}$. Once more there is a factor of 2, because the configuration with all spins flipped over is also allowed.

We stress that the fundamental topological property here is that the original configuration, a "square", goes to a "cross," where the arms of the cross represent the K-W transform of the sides of the square (this is of great importance, because all the K's are assumed to be different).

By looking at the derivation of $T\{K^*\}$, we can see that the K-W transform is *uniquely* determined, save for unimportant translations of the lattice as a whole.

(iii) Consider the slightly more complicated case shown in Fig. 13(a). In $T\{K^*\}$ it corresponds to the product

$$(\sigma_{jk}\sigma_{j+1,k})(\sigma_{j+1,k}\sigma_{j+2,k})(\sigma_{j+2,k}\sigma_{j+2,k+1})(\sigma_{j+2,k+1}\sigma_{j+1,k+1})$$
$$\times(\sigma_{j+1,k+1}\sigma_{j+1,k+2})(\sigma_{j+1,k+2}\sigma_{j,k+2})(\sigma_{j,k+2}\sigma_{j,k+1})(\sigma_{j,k+1}\sigma_{jk})\ .$$

The corresponding conditions to be fulfilled in $Z\{K^*\}$ are

$$-1=\sigma_{j,k-1}\sigma_{jk}=\sigma_{j+1,k-1}\sigma_{j+1,k}=\sigma_{j+1,k}\sigma_{j+1,k+1}=\sigma_{j,k+1}\sigma_{j,k+2}$$
$$=\sigma_{j-1,k}\sigma_{jk}=\sigma_{j-1,k+1}\sigma_{j,k+1}=\sigma_{j,k+1}\sigma_{j+1,k+1}=\sigma_{j+1,k}\sigma_{j+2,k}$$

and all the other products $\sigma\sigma'=+1$. We illustrate this configuration in Fig. 13(b). Again, it is doubly degenerate.

It should be clear by now that any closed loop in $T\{K^*\}$ can be "translated" to a corresponding configuration in $Z\{K^*\}$ by inserting "crosses" in the lower-left corner of each one of the elementary "squares" of the original graph, and considering "double bonds" as positive. In all the cases, there is a double degeneracy. Since the reciprocal is also true, we get

$$2^{(\mathfrak{N}^*-\mathfrak{N})/2}\,T\{K^*\}=Z\{K^*\}\ ,$$

which completes the proof of Eq. (2.11).

APPENDIX B

We choose as our starting point Eqs. (2.20a) and (2.20b) extended[17] to include N operators $D_{\gamma_i}(i)$ on the y axis, at the points $1=(0,k1)$, $2=(0,k_2),\ldots$, and $N=(0,k_N)$, such that $k_1\le k_2\le\cdots\le k_N$. In this form we have

$$\left\langle\prod_{i=1} D_{\gamma_i}(i)\right\rangle=\langle X\rangle=\frac{Z\{K'\}}{Z\{K\}}(-i)^{N_\sigma}\ . \qquad \text{(B1)}$$

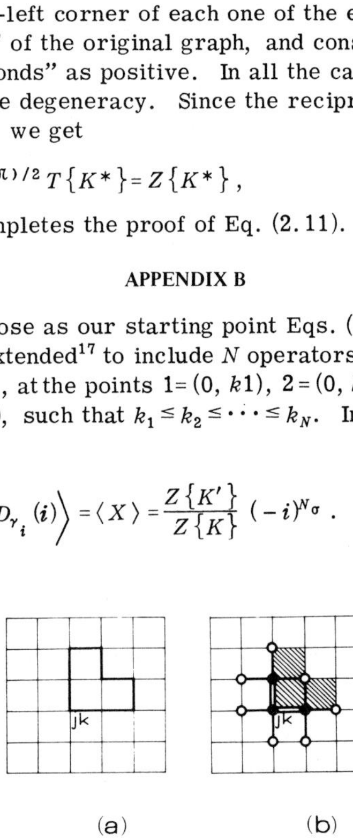

FIG. 13. A more complicated closed loop (a) is seen to be transformed into the pattern shown in (b). Bonds with double line have positive sign. Again, the configuration (b) is twofold degenerate.

Here N_σ is the number of coupling constants on the σ path.

To calculate $Z\{K'\}/Z\{K\}$ we use the Onsager solution in the particular form developed in Refs. 9 and 1, conveniently generalized. Equations from these papers will be preceded by the numerals I and II, respectively. Hence, the "equation of motion" for the spinor variables is replaced by [see Eq. (I 2.12)]

$$P(j)\,bP^{-1}(j)=\tilde{Q}b\ , \tag{B2}$$

where $P(j)$ is the "transfer matrix" for the case in which the coupling constants are functions of position with j indicating the jth column of spins, and the matrix Q generalizes Eq. (I 2.26) for the same case

$$\tilde{Q}=\tilde{Q}_1\tilde{Q}_2\ ,$$

$$\tilde{Q}_1=e^{-iP_y(1+\tau_3/2)}\,e^{2K_y^*\tau_2}e^{iP_y(1+\tau_3/2)}\ , \tag{B3}$$

$$\tilde{Q}_2=e^{-iP_y/2}\,e^{-2K_y\tau_2}\,e^{iP_y/2}\ .$$

In Eq. (B3), $e^{\pm iP_y}$ is a translation operator in the y direction, such that[20]

$$e^{\pm iP_y}\,|\,j,k\rangle=|\,j,k\mp1\rangle\ .$$

We interpret K_y and K_y^* to be matrices, diagonal in the (j,k) and (τ) spaces, i.e.,

$$K_y|\,j,k,\tau\rangle=K_y(j,k)\,|\,j,k,\tau\rangle\ ,$$

and similarly for K_y^*. Here $K_y(j,k)$ is a c number.

We also find that Eq. (I 2.25) transforms simply to read

$$e^{\,iP_x}g-\tilde{Q}g=\tilde{Q}\ , \tag{B4}$$

with $e^{\,iP_x}$ being a translation operator in the x direction, defined similarly to $e^{\,iP_y}$ and g given by Eq. (I 3.21).[21] As we will show below, only the matrix elements of g with $j=j'=0$ will be needed; their form is [see Eq. (I 3.21)]

$$2g(0k,0k')=\delta_{k,k'}-\int_0^{2\pi}\frac{dp_y}{2\pi}\,e^{ip_y(k-k')}[\Phi(p_y)]^{-\tau_3}\tau_2\ . \tag{B5}$$

We will come back shortly to discuss the function $\Phi(p_y)$.

With only these modifications, the scheme of calculation developed in I, Sec. 2, is now applicable. with the result that

$$\ln Z\{K\}=\tfrac{1}{2}\sum_{j,k}\ln[2\sinh 2K_x(j+\tfrac{1}{2},k)]$$

$$+\tfrac{1}{2}\mathrm{Tr}\ln(\tilde{Q}-e^{iP_x})+\text{const}\ . \tag{B6}$$

Before writing $Z\{K'\}$, we pause to change slightly the notation for future convenience. We replace the notation

$$k_1,k_2,\ldots,k_{N-1},k_N$$

used in Eq. (B1) for the y coordinate of the operators, by

$$k_1,k_{1'},\ \ldots,k_n,k_{n'},\quad 2n=N\ .$$

To get $Z\{K'\}$, we simply "follow the instructions" given in Eq. (2.20a).

Hence, consider $\tilde{Q}_1$. Assume there is a μ variable at k_i, and another at $k_{i'}$.

All the coupling constants crossed by the μ path will change sign, $K_x\to-K_x$; in turn, this implies that the corresponding K_y^* transform as given in Eq. (2.15b), i.e.,

$$K_y^*\to K_y^*+\tfrac{1}{2}\pi i\ .$$

Consequently, the function

$$e^{2K_y^*\tau_2}\to e^{i\pi\tau_2}e^{2K_y^*\tau_2}=(-1)e^{2K_y^*\tau_2}\ .$$

This leads us to introduce a projection operator η_μ to take care of this factor (-1), in complete analogy with I, Sec. 3:

$$\tilde{Q}_1\to\tilde{Q}_1'=e^{-iP_y(1+\tau_3/2)}(1-2\eta_\mu)e^{2K_y^*\tau_2}\,e^{iP_y(1+\tau_3/2)}$$

$$=(1-2\tilde{\eta}_\mu)\tilde{Q}_1\ , \tag{B7}$$

with

$$(1-2\tilde{\eta}_\mu)=e^{-iP_y(1+\tau_3/2)}(1-2\eta_\mu)\,e^{iP_y(1+\tau_3/2)}\ .$$

In Eq. (B7), $\tilde{\eta}_\mu$ is a matrix diagonal in (j,k) and (τ) spaces, such that its matrix elements are

$$\langle jk\tau|\tilde{\eta}_\mu|jk\tau\rangle\equiv\tilde{\eta}_\mu(jk\tau)=\delta_{j,0}$$

if

$$k_i+\tfrac{1}{2}(1+\tau_3)\le k<k_{i'}+\tfrac{1}{2}(1+\tau_3)\ , \tag{B8}$$

and zero otherwise. Here and later on, the notation τ_3 is also used to represent the eigenvalues of the vector $|\tau\rangle$, i.e., $\tau_3=\pm1$. This double use of τ_3 should not produce any confusion.

It is interesting to point out that the same result can be obtained by the use of the commutation relations of the operator μ with $b_\pm$, as in I.

The case of $\tilde{Q}_2$ can be handled in an identical fashion, by making $K_y\to K_y+\frac{1}{2}\pi i$ on the σ path.

If there are two spins at k_i and $k_{i'}$, then

$$\tilde{Q}_2\to\tilde{Q}_2'=\tilde{Q}_2(1-2\tilde{\eta}_\sigma), \tag{B9}$$

with $\tilde{\eta}_\sigma$ diagonal and

$$\langle jk\tau|\tilde{\eta}_\sigma|jk\tau\rangle\equiv\tilde{\eta}_\sigma(j,k,\tau)=\begin{cases}\delta_{j,0} & \text{if } k_i<k\le k_{i'}\\ 0 & \text{otherwise}\end{cases}\ . \tag{B10}$$

This result is identical with Eq. (II 2.4). The extension to the case of several μ's and σ's is obtained, roughly speaking, by having $\tilde{\eta}_\mu(j,k,\tau)$ and $\tilde{\eta}_\sigma(j,k,\tau)$ equal to 1 on the μ and σ paths, respectively, and zero otherwise. The exact handling of the "end effects" is as in Eqs. (B8) and (B10).

The new system, i.e., the original system in which $K \to K'$ as indicated by Eq. (2.20a), is described by the partition function $Z\{K'\}$, whose form is, therefore

$$\ln Z\{K'\} = \tfrac{1}{2}\sum_{j,k} \ln[2\sinh 2K_x'(j+\tfrac{1}{2}, k)] + \tfrac{1}{2}\mathrm{Tr}\ln(\tilde{Q}' - e^{iPx}) + \text{const},$$

$$\tilde{Q}' = \tilde{Q}_1'\tilde{Q}_2' = (1-2\tilde{\eta}_\mu)\tilde{Q}_1\tilde{Q}_2(1-2\tilde{\eta}_\sigma). \quad (B6')$$

By performing a similarity transformation inside the trace in Eq. (B6') it is possible to redefine $\tilde{Q}'$ as

$$\tilde{Q}' = \tilde{Q}(1-2\tilde{\eta}), \quad (B11a)$$

without changing Eq. (B6').

The operator $\tilde{\eta}$ is given by

$$(1-2\tilde{\eta}) = (1-2\tilde{\eta}_\sigma)(1-2\tilde{\eta}_\mu),$$

that is, (B11b)

$$\tilde{\eta} = \tilde{\eta}_\sigma(1-\tilde{\eta}_\mu) + \tilde{\eta}_\mu(1-\tilde{\eta}_\sigma).$$

The projection operator $\tilde{\eta}$ is also represented by a diagonal matrix.

With one operator $D_{\gamma_i}(i)$ at k_i, and another at $k_{i'}$, $D_{\gamma_{i'}}(i')$, the diagonal matrix elements of $\tilde{\eta}$ are

$$\langle jk\tau|\tilde{\eta}|jk\tau\rangle \equiv \tilde{\eta}(j,k,\tau) = \delta_{j,0}$$

if

$$[k_i - \tfrac{1}{2}(1-\tau_3)\tfrac{1}{2}(1-2\gamma_i)] < k \le [k_{i'} - \tfrac{1}{2}(1-\tau_3)\tfrac{1}{2}(1-2\gamma_{i'})], \quad (B12)$$

and zero otherwise. Here γ_i and $\gamma_{i'}$ are $+\frac{1}{2}$ if $D_\gamma = \sigma$, and $-\frac{1}{2}$ if $D_\gamma = \mu$. This expression is valid for all pairs $D_{\gamma_i}(i)$ and $D_{\gamma_{i'}}(i')$, with $i = 1, 2, \ldots, n$. We will denote the values of k inside the i region as given in Eq. (B12) by $\{i\}$. Whenever we want to specify the value of τ_3, we will write $\{i_+\}$ and $\{i_-\}$ to represent the subregions of $\{i\}$ with $\tau_3 = \pm 1$. For future reference, we also introduce the notation N_+^i and N_-^i, for the number of terms in $\{i_+\}$ and $\{i_-\}$, respectively. Finally, the set of all possible values of k for which $\tilde{\eta}(j,k,\tau) \ne 0$ will be indicated by $\{\tilde{\eta}\}$ or $\{\tilde{\eta}_+\}$ and $\{\tilde{\eta}_-\}$.

As a result [see also Eqs. (I 3.12) and (I 3.14)] we have

$$\ln\langle X\rangle = \tfrac{1}{2}\ln\det(1-2\tilde{\eta}g\tilde{\eta}) + \tfrac{1}{2}(\pm i\pi N_\mu - i\pi N_\sigma)$$
$$= \tfrac{1}{2}\ln\det_{\tilde{\eta}}(h) + \tfrac{1}{2}\ln(-1)^{\pm N_\mu - N_\sigma}, \quad (B13)$$

with

$$h = [\Phi(p_y)]^{-\tau_3}\tau_2 .$$

In the last line of Eq. (B13), $\det_{\tilde{\eta}}$ covers the region $\{\tilde{\eta}\}$, and N_μ is the total number of coupling constants crossed by the μ path. The term in N_μ is due to the fact that $\sinh(-\alpha) = -\sinh\alpha$. As mentioned before, only the $j=0$ matrix elements of g (or h) are needed. Equation (B13) can also be written as

$$\langle X\rangle^4 = [\det_{\tilde{\eta}}(h)]^2 , \quad (B14)$$

whereby we eliminate a spurious factor of (-1).[22] This equation can be seen to be the generalization of Eq. (I 3.13).

We carry on the calculation of Eq. (B14) in the case $T = T_c$, $A = \infty$ as in II, i.e., a critical correlation function for infinitesimal coupling strength along the direction in which the operators D_{γ_i} are placed. Then, in Eq. (I 3.18) or (I 3.19) set $B = 1$, i.e., $T = T_c$, and $A = \infty$; we obtain for $\Phi(p_y)$,[23]

$$\Phi(p_y) = -ie^{(i/2)p_y}, \quad 0 \le p_y < 2\pi .$$

To write down the matrix elements of $\tilde{\eta}h\tilde{\eta}$ in a convenient form, let us introduce a *composite variable* z, associated with the vector $|0,k,\tau\rangle$ by $z = (k,\tau)$; also $z_+ = (k,+)$ and $z_- = (k,-)$.

The introduction of z is motivated by the fact that the solution of Eq. (B14), written below, calls for an ordering of the elements of the matrix $\tilde{\eta}h\tilde{\eta}$.

The matrix elements of $\tilde{\eta}h\tilde{\eta}$ are

$$h(z,z') = \langle 0,k,\tau|\tilde{\eta}h\tilde{\eta}|0,k',\tau'\rangle = -\frac{i}{\pi}\frac{1}{k-k'-\frac{1}{2}\tau}$$

if

$$z \in \{\tilde{\eta}_+\}, \quad z' \in \{\tilde{\eta}_-\} \quad (B15)$$

or

$$z \in \{\tilde{\eta}_-\}, \quad z' \in \{\tilde{\eta}_+\}.$$

Otherwise, they vanish. Then, the elements diagonal in τ space vanish. We can rewrite $h(z,z')$ as

$$h(z,z') = \lim_{C\to\infty}\frac{1}{a(z)+b(z')}, \quad (B16a)$$

with

$$a(z) = a(k,\tau) = (\pi/i)(k-\tfrac{1}{2}\tau) + \tfrac{1}{2}C\tau,$$
$$b(z) = b(k,\tau) = -(\pi/i)k + \tfrac{1}{2}C\tau. \quad (B16b)$$

The introduction of the terms in C, which up until here is just a simple device to make the terms diagonal in $|\tau\rangle$ vanish, leads immediately to one of our basic results, Eq. (4.2), i.e., $\Gamma = 0$.

The determinant of $h(z,z')$ is given by[24]

$$\det_{\tilde{\eta}} h(z,z') = \lim_{C\to\infty} \prod_{z'<z}[a(z)-a(z')][b(z)-b(z')] \Big/ \prod_{z,z'}[a(z)+b(z')] \quad (B17)$$

and $z, z' \in \{\tilde{\eta}\}$.

The rows and columns of $h(z,z')$ are arranged in order of increasing regions $\{i\}$; inside one region, the elements belonging to the $\tau = 1$ subregion precede the elements of the $\tau=-1$ subregion. Finally, inside one subregion the elements are set in order of increasing k. This ordering defines the condition $z'<z$.

Because *a posteriori* we will take the lim $C\to\infty$, the products in Eq. (B17) can be expanded by considering C much bigger than any k, k'. Then, for instance

$$a(z)-a(z')\simeq C \quad \text{if } z\in\{\tilde{\eta}_+\} \text{ and } z'\in\{\tilde{\eta}_-\}.$$

To make further progress, let us group the terms of Eq. (B17) into C terms and no-C terms, i.e., terms which do or do not contain the factor C.

So, for instance, the C terms of the numerator are the ones for which

$$z'<z,\quad z'\in\{\tilde{\eta}_+\},\ \text{and } z\in\{\tilde{\eta}_-\},$$

or

$$z'<z,\quad z'\in\{\tilde{\eta}_-\},\ \text{and } z\in\{\tilde{\eta}_+\}.$$

Next, inside each one of these two types of products, we proceed to group terms with z and z' belonging to the same region $\{j\}$, or to two different regions, $\{i\}$ and $\{j\}$.

Let us call S_j and S_{ij} the products of C terms with z and z' in the region $\{j\}$ or the regions $\{i\}$ and $\{j\}$, respectively. Analogously, let us call T_j and T_{ij} the similar products for the no-C terms.

The result of this process is that

$$\det_{\tilde{\eta}} h(z,z')=\lim_{C\to\infty}\Big(\prod_j S_j \prod_{i<j} S_{ij}\Big)\Big(\prod_j T_j \prod_{i<j} T_{ij}\Big). \quad \text{(B18)}$$

Consider the C part first:

$$\prod_j S_j \prod_{i<j} S_{ij}=\prod_j (-1)^{(N_-^j)^2} C^{-(N_+^j-N_-^j)^2}\prod_{i<j} C^{-2(N_+^i-N_-^i)(N_+^j-N_-^j)}$$

$$=(-1)^{\Sigma_j N_-^j} C^{-[\Sigma_i (N_+^i-N_-^i)]^2}. \quad \text{(B19)}$$

But the exponent of C must vanish if the determinant is different from zero, because we need to consider the case $C\to\infty$.

Hence, we require

$$\textstyle\sum_i (N_+^i-N_-^i)=0$$

and, consequently, have

$$\prod_j S_j \prod_{i<j} S_{ij}=(-1)^{N_-},$$

where $N_-=\sum_j N_-^j$.

Equation (4.2) follows, with the identification

$$\textstyle\sum_i (N_+^i-N_-^i)\equiv\Gamma\,. \quad \text{(B20)}$$

All the properties described in Sec. IV B can be derived with the help of Eq. (B20). We note that the difference of $(N_+^i-N_-^i)$ vanishes if the operators at k_i and $k_{i'}$ are two σ's or two μ's, whereas the cases $(\sigma_i\mu_{i'})$ and $(\mu_i\sigma_{i'})$ give $(+1)$ and (-1), respectively. We now sketch the derivation of Eq. (4.9), the other basic result. Going back to Eq. (B18), with the condition $\Gamma=0$,

$$\langle X\rangle=[\det_{\tilde{\eta}} h]^{1/2}=\prod_j T_j^{1/2}\prod_{i<j} T_{ij}^{1/2}, \quad \text{(B21)}$$

where

$$T_j^{1/2}\equiv F_{jj'}=\prod_{\substack{z,z'\in\{j+\}\\ z'<z}}[\pi(k-k')]\prod_{\substack{z,z'\in\{j-\}\\ z'<z}}[\pi(k-k')]\Big/\prod_{\substack{z\in\{j+\}\\ z'\in\{j-\}}}[\pi(k-k'-\tfrac{1}{2})], \quad \text{(B22a)}$$

$$T_{ij}^{1/2}\equiv F_{ii'jj'}=\prod_{k'\in\{i+\};\,k\in\{j+\}}[\pi(k-k')]\Big/\prod_{\substack{k\in\{i+\}\\ k'\in\{j-\}}}\pi(k-k'-\tfrac{1}{2})\Big(\prod_{k'\in\{i-\};\,k\in\{j-\}}[\pi(k-k')]\Big/\prod_{\substack{k\in\{j+\}\\ k'\in\{i-\}}}[\pi(k-k'-\tfrac{1}{2})]\Big).$$

(B22b)

As stated in Ref. 21, we know that our $\langle X\rangle$ is never negative. Because of this we need not keep factors of (-1) in either Eq. (B22) or (B23).

The expressions for $F_{ii'}$ and $F_{ii'jj'}$ look very impressive. However they are actually very simple in structure. For $F_{ii'}$ we obtain [see Eq. (II 2.10)]

$$F_{\sigma_i\sigma_{i'}}=F_{\mu_i\mu_{i'}}=\langle\sigma_{k_i}\sigma_{k_{i'}}\rangle\ , \quad \text{(B23a)}$$

$$F^2_{\sigma_i\mu_{i'}}=\pi|r_\mu-r_\sigma|\langle\sigma_{k_i}\sigma_{k_{i'}}\rangle\langle\sigma_{k_i}\sigma_{k_{i'}-1}\rangle\ , \quad \text{(B23b)}$$

$$F^2_{\mu_i\sigma_{i'}}=\pi|r_\sigma-r_\mu|\langle\sigma_{k_i}\sigma_{k_{i'}}\rangle\langle\sigma_{k_i-1}\sigma_{k_{i'}}\rangle\ , \quad \text{(B23c)}$$

so that for $|r_i-r_{i'}|$ much larger than a lattice constant,

$$F_{\sigma_i\mu_{i'}}=F_{\mu_i\sigma_{i'}}=\pi^{1/2}|r_\sigma-r_\mu|^{1/2}\langle\sigma_{k_i}\sigma_{k_{i'}}\rangle\ . \quad \text{(B23d)}$$

The terms $F_{ii'jj'}$ are best handled by a generalization of the "contraction" method of II, Eqs. (II 2.15)–(II 2.18).

Then we write

$$\langle D_{\gamma_1}(1)\,D_{\gamma_{1'}}(1')\cdots D_{\gamma_n}(n)\,D_{\gamma_{n'}}(n')\rangle$$

in terms of $F_{ii'}$ and $F_{ii'jj'}$. By setting $(n-1)'=n$, one obtains the correlation function

$$\langle D_{\gamma_1}(1)\,D_{\gamma_{1'}}(1')\cdots D_{\gamma_{n-1}}(n-1)\,D_{\gamma_{n'}}(n')\rangle\ .$$

But, this last correlation function can also be written directly. By equating these two expressions, we get

$$F_{ii'jj'} = F_{ij'}F_{i'j}/F_{ij}F_{i'j'} \quad . \tag{B24}$$

Notice that Eq. (II 2.18) turns out to be just a special case of our Eq. (B24).

This form for $F_{ii'jj'}$ allow us to write $\langle X\rangle$ as

$$\langle X\rangle = \left(\prod_i F_{ii'}\right)\left(\prod_j \prod_{i<j} \frac{F_{ij'}F_{i'j}}{F_{ij}F_{i'j'}}\right) \quad . \tag{B25a}$$

At this point we return to the original notation, i.e., form

(1), (1'),...,(n), (n')

back to

1, 2,...,N = 2n ,

and introduce $p_i = (-1)^{2\Gamma_i}$, which is (+1) or (−1) if i is even or odd, respectively.

The correlation function takes the form

$$\langle X\rangle = \prod_{1\le i<j\le N} (F_{ij})^{-p_i p_j} \quad . \tag{B25b}$$

Finally, by considering the case, $|r_i - r_j| \gg$ lattice const,

$$F_{ij} = \begin{cases} \langle\sigma_i\sigma_j\rangle & \text{if } \gamma_i = \gamma_j \\ \pi|r_i - r_j|^{1/2}\langle\sigma_i\sigma_j\rangle = \dfrac{\text{const}}{\langle\sigma_i\sigma_j\rangle} & \text{if } \gamma_i = -\gamma_j \ , \end{cases} \tag{B26}$$

Eq. (B25b) can be seen to be Eq. (4.9).

---

*Work supported in part by the National Science Foundation under Grant No. 287-910-4647-4.

[1]L. Kadanoff, Phys. Rev. 188, 859 (1969). Referred to as II in Appendix B.

[2]L. Kadanoff, Phys. Rev. Letters 23, 1430 (1969).

[3]A. M. Polyakov, Zh. Eksperim. i Teor. Fiz. [Sov. Phys. JETP (to be published)].

[4]K. Wilson, Phys. Rev. 179, 1499 (1969).

[5]H. A. Kramers and G. H. Wannier, Phys. Rev. 60, 252 (1941).

[6]K. Kawasaki, Phys. Rev. 150, 291 (1966).

[7]B. Kaufman, Phys. Rev. 76, 1232 (1949).

[8]T. D. Schultz, D. C. Mattis, and E. H. Lieb, Rev. Mod. Phys. 36, 856 (1964).

[9]L. Kadanoff, Nuovo Cimento 40, 276 (1966). Referred to as I in Appendix B.

[10]L. Onsager, Phys. Rev. 65, 117 (1944).

[11]M. E. Fisher and A. E. Ferdinand [Phys. Rev. Letters 19, 169 (1967)] have used a technique like this one to determine "surface tensions" for the two-dimensional Ising model. See also Ref. 10.

[12]Note that μ_r is not local in the same sense as σ_r since μ_r appears to depend upon σ's infinitely far away from the point r. In particular $\langle\mu_r\mu_{r'}\rangle$ is proportional to an average of a product of terms like $e^{-2K\sigma\sigma'}$ over any path between r and r'. But this average is local in the sense that $\langle\mu_r\mu_{r'}\rangle$ depends only upon the values of the coupling constants in the neighborhood of r and r' (i.e., within one coherence length) and not upon the values of the K's throughout the entire lattice.

[13]R. Hecht, Phys. Rev. 158, 577 (1967); J. Stephenson, J. Math. Phys. 7, 1123 (1966).

[14]The symbol $\mathcal{R}$ has been used in this definition of the distance variable to distinguish it from the distance variable $R = [x^2 + u^4y^2]^{1/2}$ used in Ref. 9. We also employ a slightly different definition of ϵ in this work from that used in Ref. 9 in that ϵ (for this work) $= u^{-2}\epsilon$ (Ref. 9). These redefinitions provide variables which have more convenient symmetry properties when u is varied.

[15]R. Hecht, thesis (University of Illinois, 1967) (unpublished).

[16]In Ref. 2 this formalism was presented for all the operators on the x axis. Here we prefer the y axis instead, because in this form the present scheme of calculation reduces to the one in Ref. 9 for the case K_x, K_y const. Of course, the calculation is identical in any of the two axes, and the final results differ only in notation.

[17]A correlation function with an odd number of μ's (σ's) can be calculated, in general, with the use of a path joining the μ (σ) with biggest y to a μ (σ) variable at infinity.

[18]The material in this Appendix is not original. It is designed to make explicit the ideas contained in implicit form in the literature (see, e.g., Ref. 19).

[19]G. H. Wannier, Rev. Mod. Phys. 17, 50 (1945); see also C. Domb, Advan. Phys. 9, 1 (1960).

[20]This definition is identical to the one given in Ref. 9

[21]Equation (I 3.21), as well as (I 3.14) and (I 3.19) have some misprints and incorrect signs. Instead of these, we used in this work the following equations:

$$q(p_y) = \tfrac12\{1 - [\Phi(p_y)]^{-\tau_3}\tau_2\}e^{\gamma(p_y)} + \tfrac12\{1 + [\Phi(p_y)]^{-\tau_3}\tau_2\}e^{-\gamma(p_y)}, \tag{I 3.14}$$

$$\Phi(p_y) = \left(\frac{Be^{ip_y}-1}{B-e^{ip_y}}\ \frac{Ae^{ip_y}-1}{A-e^{ip_y}}\right)^{1/2} \ , \tag{I 3.19}$$

$$g(j, k; j', k') = \int_0^{2\pi}\frac{dp_y}{2\pi}\ \exp[-\gamma(p_y)|j-j'| + ip_y(k-k')]$$

$$\times\begin{cases}\tfrac12\{1-[\Phi(p_y)]^{-\tau_3}\tau_2\} & \text{for } j\le j' \\ \tfrac12\{1+[\Phi(p_y)]^{-\tau_3}\tau_2\} & \text{for } j>j' \ .\end{cases} \tag{I 3.21}$$

[22]The present calculation determines the correlation function $\langle X\rangle$, save for a multiplicative factor of a power of (−1). However, we are able to prove that *for operators on a line* and $\Gamma = 0$, $\langle X\rangle$ is never negative, by borrowing results already obtained by B. Kaufman and L. Onsager [Phys. Rev. 76, 1244 (1949)]. In their paper they calculated the average values $\langle iP_mQ_e\rangle$, $\langle Q_mQ_e\rangle$, and $\langle P_mQ_e\rangle$. The first one is nothing but

$-\langle\sigma_m\mu_m\mu_{e+1}\sigma_e\rangle$ if $m > e$;

or

$+\langle\mu_{e+1}\sigma_e\sigma_m\mu_m\rangle$ if $e > m$.

In both cases, $\langle X\rangle \ge 0$ for all temperatures [see their

Eq. (13)]. More complicated correlations can be handled by taking recourse to the factorization property of averages of b's, as explained in Hecht (Ref. 16). The other two averages $\langle Q_m Q_e \rangle$ and $\langle P_m P_e \rangle$ provide us with a cross-check of our result that correlation functions with $\Gamma \neq 0$ vanish for infinitesimal coupling strength along the line.

[23]See the discussion in Ref. 1, immediately below Eq. (II 2.6). Incidentally, this equation is misprinted; it should read

$$\Phi(p_y) = [-(Ae^{ip_y} - 1)/(A - e^{ip_y})]^{1/2}.$$

Also, the operators σ of this reference are located on the y axis, as indicated by their coordinates $(0, k)$ and not on the x axis as stated in the text.

[24]See, for instance, N. I. Achieser, *Theory of Approximation* (Ungar, New York, 1956), p. 19.

Teaching the renormalization group[a)]

Humphrey J. Maris and Leo P. Kadanoff
Department of Physics, Brown University, Providence, Rhode Island 02912
(Received 24 August 1977; accepted 29 October 1977)

The renormalization group theory of second-order phase transitions is described in a form suitable for presentation as part of an undergraduate statistical physics course.

INTRODUCTION

The problem of communicating exciting new developments in physics to undergraduate students is of great importance. There is a natural tendency for students to believe that they are a long way from the frontiers of science where discoveries are still being made. The presentation to undergraduates of important new discoveries can show them that they are nearer the frontier than they had imagined. The effect on students' interest and enthusiasm is especially marked when they find that the "great new discovery" that they had heard mentioned so many times without actually knowing what it was turns out to be rather simple, and something they might have thought of themselves given a little time and a better background. Students tend to think that the nature and manner of physics research changed abruptly at some time (possibly about 40 years ago). Before this time, physicists thought deeply and had sudden inspirations that enabled them to proceed by giant strides. Later on, however, physics became very technical and detailed. In this later period, which includes the present time, so students believe, progress could only be made after enormous study and effort because all of the simple things had been done. The realization that the new developments in physics are really not essentially different from the old dispels these ideas about the degeneration of our subject.

In this paper, we want to describe how an important recent development in theoretical physics, namely, the renormalization group, can be successfully incorporated into a junior-level course in statistical physics. We assume that the material is presented at the end of a one-semester course using a text such as Reif's *Statistical and Thermal Physics*.[1] A student thus already knows that:

(1) Substances have various phases which can be represented on a phase diagram in the pressure-temperature plane. Going from one phase to another is accompanied by discontinuous changes in entropy and volume.

(2) All thermodynamic quantities can be calculated from the partition function Z, defined by

$$Z = \sum_n e^{-E_n/k_B T}, \tag{1}$$

where E_n is the nth energy state of the system, k_B is Boltzmann's constant, and T is the temperature.

(3) Ordinary phase transitions (i.e., first-order transitions) can be understood in a relatively simple way be treating the various phases as separate entities each with its own Gibbs free energy.[2]

SECOND-ORDER PHASE TRANSITIONS

By way of introduction to the topic of second-order phase transitions some examples are useful to the student. Suitable choices include the following:

(a) Order-disorder transitions. Certain materials undergo a phase transition in which the specific heat C has a singularity of the general form shown in Fig. 1(a). Close to the transition temperature T_c, C is proportional to

$$|T - T_c|^{-\alpha}, \tag{2}$$

where α is typically around 0.1. There is no latent heat. The standard example of this type of transition is β-brass. This is an alloy of copper and zinc, in equal amounts, with a body-centered-cubic crystal structure. Above the critical temperature T_c of 460 °C a copper atom has, on the average, as many copper nearest neighbors as zinc nearest neighbors. As the temperature is lowered below T_c, however, there develops an increasing probability for a copper atom to have more zinc nearest neighbors than would result from a completely random arrangement of copper and zinc atoms. This probability increases so that at temperatures much less than T_c each copper atom has only zinc neighbors and each zinc atom has only copper neighbors. Thus one says that there is a transition from complete disorder above T_c to a phase which has continuously increasing order as the temperature is lowered below T_c.

(b) Paramagnetic-ferromagnetic transition. Above T_c the magnetic dipoles in a solid are randomly oriented, and so in the absence of an applied magnetic field there is no net magnetic moment. Below T_c there is a partial alignment of the dipoles and a so-called "spontaneous magnetization" [Fig. 1(b)]. This magnetization **M** is a vector whose direction is not unique. Thus upon heating a magnet above T_c and recooling the same magnitude of magnetization will occur but the direction may be different. It is found experimentally that for T close to T_c

$$|\mathbf{M}| \propto (T_c - T)^\beta \tag{3}$$

for $T < T_c$. The critical exponent lies in the range 0.3–0.4. The specific heat also has a singularity proportional to

$$|T - T_c|^{-\tilde{\alpha}} \tag{4}$$

with $\tilde{\alpha}$ in most cases being close to zero.

These two examples show clearly that there are in nature more subtle transitions than those of the simple first-order variety. To understand how this can occur it is useful to consider a very simplified model, the Ising model, which nevertheless manages to retain enough of the essential features to be useful. One can choose to think of this model as a model of the magnetic transition, but it is also just as good a representation of the order-disorder transition. The model involves N magnetic dipoles which we will call spins for short. The simplifications made are the following:

(1) Each spin can point only up or down. These directions are chosen to be the positive and negative z directions. Thus

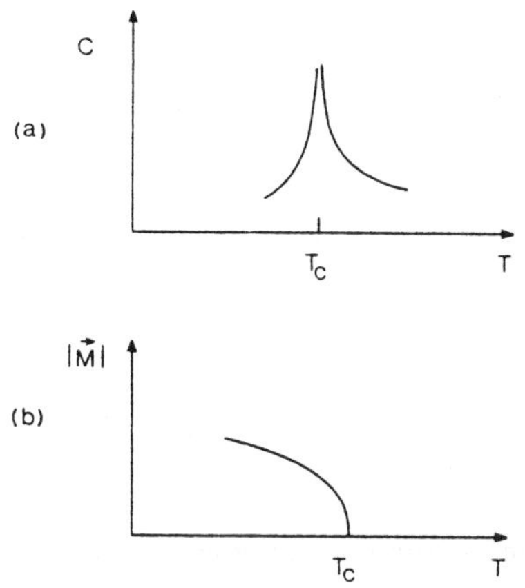

Fig. 1. Examples of singularities in thermodynamic functions at second-order phase transitions. (a) Specific heat C at an order-disorder transition. (b) Magnitude $|\mathbf{M}|$ of the spontaneous magnetization at a magnetic phase transition.

one can describe the state of any spin by a number σ which is $+1$ for up and -1 for down, and the state of the whole system is specified by giving the value of σ for each of the N spins.

(2) The spins are arranged in a simple-cubic lattice. Thus there is a spin at each of the positions

$$\mathbf{X}_{lmn} = l\hat{i} + m\hat{j} + n\hat{k}, \tag{5}$$

where $\hat{i}$, $\hat{j}$, and $\hat{k}$ are unit vectors parallel to the Cartesian axes, and l, m, and n are integers. To be more precise this is the arrangement of spins in the three-dimensional cubic Ising model. In the square two-dimensional model the spins are located in a square array in the x-y plane,[3] and in the one-dimensional model they are at equally spaced intervals along the x axis.

(3) There is an interaction energy between spins that tends to align them. Each pair of nearest-neighbor spins that are parallel makes a contribution $-J$ to the total energy of the system, and each nearest-neighbor pair that are antiparallel gives a contribution $+J$. J is a positive constant. Thus the total energy can be written

$$E = -J \sum_{pq} \sigma_p \sigma_q, \tag{6}$$

where the sum is restricted to be over nearest-neighbor pairs pq of spins.

It is easy to see the model exhibits some of the qualitative characteristics of real magnetic systems. There are two lowest-energy states which we can call + and −. In these states all spins are either up (+ state) or down (− state). In either case the energy of the state is

$$\begin{aligned} E_0 &= -J \times (\# \text{ of nearest-neighbor bonds}) \\ &= -J \times (\# \text{ of atoms}) \times (\# \text{ of bonds per spin}). \end{aligned} \tag{7}$$

In three dimensions, for example, each spin has 6 nearest neighbors but one must divide by 2 to avoid counting bonds twice. So

$$E_0 = -3NJ, \tag{8}$$

where N is the number of spins. In both of these states the spins are all aligned and so there is a large magnetization. Since at zero temperature there is a probability of 0.5 of finding the system in either of these two lowest-energy states we can see that the Ising model does indeed exhibit a spontaneous magnetization. Higher-energy states, which will be important at nonzero temperatures, are produced by reversing the direction of some of the spins. It is clear that these states will have a smaller spontaneous magnetization, and so we expect that the magnitude of the spontaneous magnetization will decrease as the temperature goes up, again in qualitative agreement with what happens for a real magnetic system.

To "solve" the Ising model means to evaluate the sum for Z [Eq. (1)] using the expression for the energy levels [Eq. (6)]. This can be done with not too much difficulty for the 1D model with the result[4]

$$Z = [2\cosh(J/k_BT)]^N. \tag{9}$$

For the 2D model the problem is much harder, and the solution was first given in a celebrated paper by Onsager.[5] Although his method has been simplified somewhat[6] there is still no easy derivation of the result. He found that

$$Z = [2\cosh(2J/k_BT)e^I]^N, \tag{10}$$

where

$$I = (2\pi)^{-1} \int_0^\pi d\phi \ln\{(1/2)[1 + (1 - \kappa^2 \sin^2\phi)^{1/2}]\}, \tag{11}$$

$$\kappa = 2\sinh(2J/k_BT)/\cosh^2(2J/k_BT). \tag{12}$$

In 3D no analytic solution has yet been found, and there are only numerical calculations of the thermodynamic quantities. From Z one can calculate the internal energy U and the specific heat C using the formulas

$$U = k_BT^2\, d\ln Z/dT, \tag{13}$$

$$C = dU/dT. \tag{14}$$

For the 1D model one finds there is no phase transition in the sense that U and C are both smoothly varying functions of T. The increase in the alignment of the spins as the temperature decreases occurs in a very gradual way. In 2D the specific heat is as shown in Fig. 2. There is a singularity at the temperature T_c such that

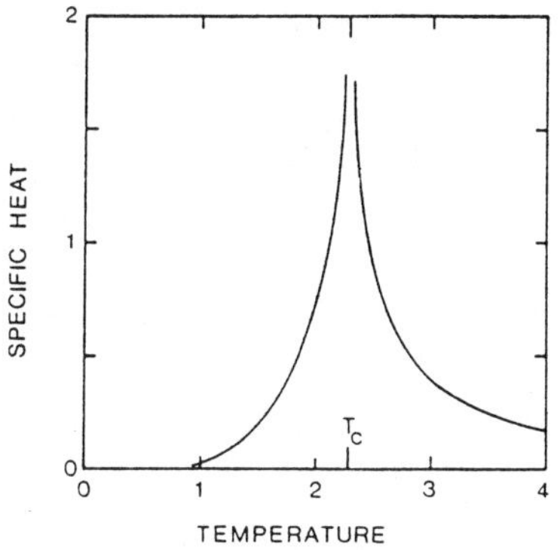

Fig. 2. Specific heat of the 2D Ising model as a function of temperature. The specific heat is plotted in units of Nk_B, and the temperature in units of J/k_B. In these units the critical temperature is 2.269.

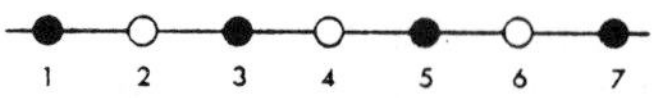

Fig. 3. Section of a 1D Ising model. The sums over the spin variables σ_2, σ_4, σ_6, etc. (spins denoted by open circles) are performed first. This leaves a new sum that involves only the spin variables σ_1, σ_3, σ_5, etc. (solid circles).

$$\sinh(2J/k_BT_c) = 1. \tag{15}$$

Near this temperature C has a singular contribution which varies with temperature as

$$(8k_B/\pi)(J/k_BT_c)^2 \ln|1/(T-T_c)| \tag{16}$$

In 3D there is also a phase transition and the singular term in the specific heat is proportional to

$$|T - T_c|^{-\alpha}, \tag{17}$$

where α is close to 0.125.

RENORMALIZATION GROUP

The aim of the renormalization group theory is to calculate directly the critical exponents such as α and β occurring in Eqs. (2), (3), (4), and (17). If the theory also happens to give the partition function at all temperatures so much the better, but the primary emphasis is on the determination of the partition function in the critical region, i.e., for T close to T_c. The renormalization group arose from Wilson's brilliant translation[7] of the conceptual picture of phase transitions which arose in the 1960s into a calculational tool. The history of these developments is complex and involves the appreciation of such concepts as scaling, universality, and correlations in the critical region.[8] The approach we want to present here is deliberately directed away from these concepts. In this way we certainly lose both history and what many current workers in the field consider to be the deep physics of the renormalization group (RG). What we gain is the possibility of explaining the RG to a wider audience. We will simply take the view that the RG approach is a practical but approximate way of calculating the sums involved in Z.

Consider first the 1D Ising model.[9] We show a section of the chain of spins in Fig. 3. The problem is to evaluate the sum

$$Z = \sum \exp[J(\cdots \sigma_1\sigma_2 + \sigma_2\sigma_3 + \sigma_3\sigma_4 + \sigma_4\sigma_5 + \cdots)/k_BT]. \tag{18}$$

The sum is over all possible values of σ_1, σ_2, etc. For convenience let us introduce a quantity K, called the coupling constant, defined by

$$K \equiv J/k_BT. \tag{19}$$

We partition Z into the form

$$Z = \sum \cdots e^{K(\sigma_1\sigma_2+\sigma_2\sigma_3)} e^{K(\sigma_3\sigma_4+\sigma_4\sigma_5)} \cdots. \tag{20}$$

The only place σ_2 appears in this equation is in the first exponential. We therefore carry out the sum over σ_2 with the result

$$Z = \sum \cdots [e^{K(\sigma_1+\sigma_3)} + e^{-K(\sigma_1+\sigma_3)}] \times e^{K(\sigma_3\sigma_4+\sigma_4\sigma_5)} \cdots. \tag{21}$$

We continue by summing over σ_4, and σ_6, etc. This gives

$$Z = \sum \cdots [e^{K(\sigma_1+\sigma_3)} + e^{-K(\sigma_1+\sigma_3)}] \times [e^{K(\sigma_3+\sigma_5)} + e^{-K(\sigma_3+\sigma_5)}] \cdots. \tag{22}$$

The sum is now over the possible values of the remaining spin variables, i.e., σ_1, σ_3, σ_5, etc. The next idea is to try to find a value of K' and a function f such that

$$e^{K(\sigma_1+\sigma_3)} + e^{-K(\sigma_1+\sigma_3)} \equiv f(K)e^{K'\sigma_1\sigma_3} \tag{23}$$

for all possible values of σ_1 and σ_3. The function f must not depend on σ_1 or σ_3. The solution is (see Appendix)

$$K' = (1/2)\ln\cosh(2K), \tag{24}$$

$$f(K) = 2\cosh^{1/2}(2K). \tag{25}$$

Thus we can write

$$Z = \sum \cdots f(K)e^{K'\sigma_1\sigma_3} f(K)e^{K'\sigma_3\sigma_5} \cdots = f(K)^{N/2} \sum e^{K'(\cdots \sigma_1\sigma_3+\sigma_3\sigma_5+\cdots)}. \tag{26}$$

We still have not done the sum, but we notice that if we had been trying to solve a problem where the coupling constant had the value K' and there were only $N/2$ spins exactly this sum would appear. Thus we have shown that the partition function $Z(N,K)$ for N spins interacting with coupling constant K is related to the partition function $Z(N/2,K')$ for $N/2$ spins and coupling constant K' by the equation

$$Z(N,K) = f(K)^{N/2} Z(N/2,K'). \tag{27}$$

For a large system we know how Z depends on N. Since we believe that the free energy F is proportional to the size of the system[10] it must be true that

$$\ln Z = N\zeta, \tag{28}$$

where ζ depends on K but is independent of the system size. Then from Eq. (27)

$$\zeta(K) = (1/2)\ln f(K) + (1/2)\zeta(K') \tag{29}$$

$$\therefore \zeta(K') = 2\zeta(K) - \ln[2\cosh^{1/2}(2K)]. \tag{30}$$

Equations (24) and (30) are the essential results of the renormalization group[11] analysis. If the partition function is known for one value of the coupling constant K, or equivalently the temperature T, these equations provide a recursion relation that can be used to calculate Z for other values. Using Eqs. (24) and (30) this recursion, or "renormalization," is always towards lower values of K, i.e., K' is always less than K. One can find recursion relations that work in the opposite direction by solving Eq. (24) for K. These recursion relations are

$$K = (1/2)\cosh^{-1}(e^{2K'}), \tag{31}$$

$$\zeta(K) = (1/2)\ln 2 + (1/2)K' + (1/2)\zeta(K'). \tag{32}$$

These results can be used to find the partition function $\zeta(K)$ in the following way. For K' very small, e.g., 0.01, the interaction between the spins is negligible. For free spins the partition function is just the number of ways of arranging them. So for $K' = 0.01$

$$Z \approx 2^N \tag{33}$$

$$\therefore \zeta(0.01) \approx \ln 2. \tag{34}$$

We now calculate K from Eq. (31) and obtain the result 0.100 334. The value of ζ for this value of K is found from Eq. (32) to be 0.698 147. The procedure is then repeated with K' equal to 0.100 334, and by continuing in this way one obtains the results shown in Table I. This table includes

Table I. Values of ζ for the 1D Ising model calculated from the recursion formulas Eqs. (31) and (32) of the renormalization group and from the exact formula derived from Eq. (9). ζ is related to the partition function Z through Eq. (28).

K	$\zeta(K)$ Renormalization group	Exact
0.01	ln 2	0.693 197
0.100 334	0.698 147	0.698 172
0.327 447	0.745 814	0.745 827
0.636 247	0.883 204	0.883 210
0.972 710	1.106 299	1.106 302
1.316 710	1.386 078	1.386 080
1.662 637	1.697 968	1.697 968
2.009 049	2.026 876	2.026 877
2.355 582	2.364 536	2.364 537
2.702 146	2.706 633	2.706 634

the exact values of ζ calculated from Ising's formula Eq. (9). The agreement is excellent. A remarkable feature of this method is that small errors in the first value of ζ actually lead to smaller and smaller errors as the calculation proceeds. If one attempts the same calculation starting from a large coupling constant, on the other hand, one obtains progressively larger errors.

This process may be represented by a "flow diagram," as shown in Fig. 4, which shows how K moves under successive recursion. Of significance are points where recursion does not change K. For the 1D Ising model the only such "fixed points" are at $K = 0$ and $K = \infty$.

To use the renormalization group to study phase transitions we must apply it to the 2D Ising model, as shown in Fig. 5. In analogy with the 1D problem the first step is to sum over the spin variables of half of the spins. A choice of which spins to sum over is shown in Fig. 5. After the summation the expression for Z becomes

$$Z = \sum \cdots [e^{K(\sigma_1+\sigma_2+\sigma_3+\sigma_4)} + e^{-K(\sigma_1+\sigma_2+\sigma_3+\sigma_4)}] \cdots. \quad (35)$$

The sum is over all possible values of the remaining spin variables. There is a square bracket term for every large square, such as 1234, in the lattice of remaining spins. To follow the same method as in the 1D problem we would now like to do something so that Z involves exactly the same form of difficult summation that was present in the original expression. This is not possible, however. In fact, it is straightforward to show that Eq. (35) is equivalent to the result (see Appendix)

$$Z = f(K)^{N/2} \sum \exp\Big(K_1 \sum_{nn} \sigma_p\sigma_q + K_2 \sum_{nnn} \sigma_p\sigma_q + K_3 \sum_{sq} \sigma_p\sigma_q\sigma_r\sigma_s\Big), \quad (36)$$

where

$$f(K) = 2\cosh^{1/2}(2K)\cosh^{1/8}(4K), \quad (37)$$

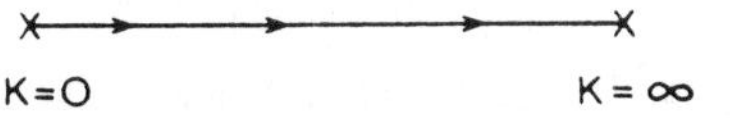

Fig. 4. Flow diagram for the 1D Ising model. The only fixed points are at $K = 0$ and ∞ and are marked ×. The arrows show the direction of flow when the recursion formula Eq. (31) is used. Equation (24) gives a flow in the opposite direction.

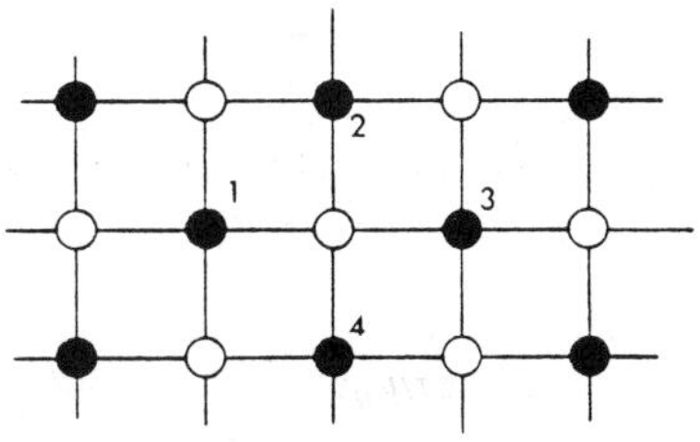

Fig. 5. Section of the 2D Ising model. The summed over spins are denoted by open circles, and the spins that remain by solid circles. Notice that the lattice formed by the remaining spins is still a simple cubic lattice, but is rotated by 45° relative to the original lattice.

and the sums in the exponential are over all nearest-neighbor pairs, next-nearest-neighbor pairs, and sets of four spins $pqrs$ around all squares. Explicit expressions for K_1, K_2, and K_3 are

$$K_1 = (1/4)\ln\cosh(4K), \quad (38)$$

$$K_2 = (1/8)\ln\cosh(4K), \quad (39)$$

$$K_3 = (1/8)\ln\cosh(4K) - (1/2)\ln\cosh(2K). \quad (40)$$

At this point an essential difference between the renormalization group in the 1D and 2D models appears. The expression for Z in the 2D case [Eq. (36)] does not just involve the same sort of sum that appeared in the original expression for Z. Hence, unless we can somehow get rid of K_2 and K_3 the method fails. To eliminate K_2 and K_3 it is necessary to make some sort of approximation, and this requires considerable intuition. For example, two possibilities are the following.

(1) Simply ignore K_2 and K_3. This gives recursion relations

$$K' = (1/4)\ln\cosh(4K), \quad (41)$$

$$\zeta(K') = 2\zeta(K) - \ln[2\cosh^{1/2}(2K)\cosh^{1/8}(4K)]. \quad (42)$$

These recursion relations have the same sort of flow pattern as in the 1D model (Fig. 4). They do not predict a phase transition, i.e., ζ turns out to be an analytic function of K.

(2) A better result is obtained by correcting the theory in an approximate way for the presence of the term in K_2. K_2 is the coupling between next-nearest spins, such as 1 and 3. Both K_1 and K_2 are positive, and hence both have the effect of increasing the alignment of spins. Hence, a possible procedure is to drop K_2 but simultaneously increase K_1 to a new value K' so that the net "aligning tendency" remains the same. A crude way of determining K' is as follows. Consider the energy of the lattice when all the spins are aligned. In a lattice of $N/2$ spins there are N nearest-neighbor bonds and N next-nearest bonds. Thus if we retain K_1 and K_2 the total energy is

$$-Nk_BTK_1 - Nk_BTK_2.$$

We choose K' so that this gives by itself the same energy. Thus

$$K' = K_1 + K_2 = (3/8)\ln\cosh(4K). \quad (43)$$

The partition function is still determined by Eq. (42). The recursion relations now have a remarkable new feature. The

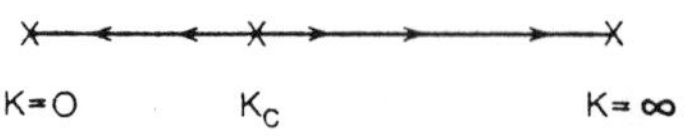

Fig. 6. Flow diagram for the 2D Ising model based upon the recursion formula (43).

flow diagram, as shown in Fig. 6, splits into two separate parts separated by a fixed point at $K_c = 0.506\ 98$. If one starts with a K just to the right of this point recursion increases K eventually to ∞. Starting just to the left decreases K to zero. Detailed study of the partition function using these recursion relations shows that there is a singularity in ζ and Z at K_c, and so this fixed point is to be associated with the phase transition. The value of K_c is surprisingly near the exact value (much nearer than it should be considering the naivete of the approximation). The exact solution of Onsager [Eq. (15)] shows that the correct value of K_c is

$$J/k_B T_c = (1/2)\sinh^{-1}(1) = 0.440\ 69. \tag{44}$$

Another test is to calculate the specific heat. By expanding ζ and Z around K_c it is a simple exercise (see Appendix) to show that near the transition

$$C \propto |1 - T/T_c|^{-\alpha}, \tag{45}$$

where

$$\alpha = 2 - \ln 2/\ln (dK'/dK|_{K=K_c}), \tag{46}$$

α comes out to be 0.131. This is to be compared with Onsager's exact result, which has a logarithmic singularity in the specific heat, and hence a critical exponent α of zero.

To obtain a more accurate value of the critical exponent is not trivial. In fact, the second example given above is deceptive in the sense that seemingly logical improvements can easily lead to worse results! For example, if K' is calculated in the same way as above but the energy contribution from K_3 in the aligned state is included a poorer value of α is obtained. For a discussion of better approximation schemes see Ref. 11.

Our experience has been that this material can be covered at a fairly leisurely pace in two lectures, and successfully gives students an idea of what the renormalization group is, and how it can be used in the theory of second-order phase transitions.

APPENDIX

We give here some of the intermediate steps in the derivations. Equations (24) and (25) are obtained as follows. Equation (23) must be true for all values of σ_1 and σ_3. For $\sigma_1 = 1$ and $\sigma_3 = 1$, Eq. (23) becomes

$$e^{2K} + e^{-2K} = f e^{K'}. \tag{A1}$$

The same result is obtained with $\sigma_1 = -1$ and $\sigma_3 = -1$. For $\sigma_1 = 1$ and $\sigma_3 = -1$ or $\sigma_1 = -1$ and $\sigma_3 = 1$ we obtain

$$2 = f e^{-K'}. \tag{A2}$$

Solution of Eqs. (A1) and (A2) leads to the expressions for K' and f given in the text.

The derivation of Eqs. (36)–(40) proceeds in a similar way. We first try to find K_1, K_2, K_3, and f such that

$$e^{K(\sigma_1+\sigma_2+\sigma_3+\sigma_4)} + e^{-K(\sigma_1+\sigma_2+\sigma_3+\sigma_4)} = f \exp[(1/2)K_1(\sigma_1\sigma_2 + \sigma_2\sigma_3 + \sigma_3\sigma_4 + \sigma_4\sigma_1) + K_2(\sigma_1\sigma_3 + \sigma_2\sigma_4) + K_3\sigma_1\sigma_2\sigma_3\sigma_4] \tag{A3}$$

for all possible values of σ_1, σ_2, σ_3, σ_4. For example, when all the σ's are equal to $+1$, or all equal to -1, we obtain the condition

$$e^{4K} + e^{-4K} = f e^{2K_1+2K_2+K_3}. \tag{A4}$$

Investigation of all of the other possible values of the σ's gives the conditions

$$2 = f e^{-2K_1+2K_2+K_3}, \tag{A5}$$

$$e^{2K} + e^{-2K} = f e^{-K_3}, \tag{A6}$$

$$2 = f e^{-2K_2+K_3}. \tag{A7}$$

Solution of these equations gives the expressions for K_1, K_2, K_3, and f given in the text [Eqs. (38)–(40)]. The result for Z now becomes

$$Z = \sum \cdots [f \exp((1/2)K_1(\sigma_1\sigma_2 + \sigma_2\sigma_3 + \sigma_3\sigma_4 + \sigma_4\sigma_1) + K_2(\sigma_1\sigma_3 + \sigma_2\sigma_4) + K_3(\sigma_1\sigma_2\sigma_3\sigma_4))] \cdots. \tag{A8}$$

There is one square bracket for each large square of the type 1234. Since there are $N/2$ such squares Z contains a factor

$$f^{N/2}.$$

A nearest-neighbor term such as

$$e^{(1/2)K_1\sigma_1\sigma_2}$$

will appear both in the square bracket that has been written out explicitly and in one other square bracket. Thus the *total* term coming from the neighbors 1 and 2 is

$$e^{K_1\sigma_1\sigma_2}.$$

Hence there appears in Z a factor

$$\exp\left(K_1 \sum_{nn} \sigma_p\sigma_q\right).$$

The understanding of the terms in K_2 and K_3 that appear in Eq. (36) is straightforward.

To derive Eq. (46) assume that there is a nonanalytic term in $\zeta(K)$ which is

$$a|K - K_c|^{2-\alpha}$$

where a is a constant. Near to K_c we have to first order in $(K - K_c)$

$$K' = K_c + (K - K_c)\left.\frac{dK'}{dK}\right|_{K=K_c} \tag{A9}$$

Now expand Eq. (42) around $K = K_c$. On the right-hand side the only term which is a $(2 - \alpha)$ power is

$$2a|K - K_c|^{2-\alpha}. \tag{A10}$$

On the left-hand side the corresponding term is

$$a|K' - K_c|^{2-\alpha} = a\left|(K - K_c)\left.\frac{dK'}{dK}\right|_{K=K_c}\right|^{2-\alpha} \tag{A11}$$

using Eq. (A9). Comparison of Eqs. (A10) and (A11) gives

$$2a|K - K_c|^{2-\alpha} = a|K - K_c|^{2-\alpha} \left.\frac{dK'}{dK}\right|_{K=K_c}^{2-\alpha}, \quad (A12)$$

and Eq. (46) follows.

[a]Work supported in part by the National Science Foundation through the Materials Research Laboratory at Brown University and through Grant Nos. DMR 75-14761 and DMR 73-04886.

[1]F. Reif, *Statistical and Thermal Physics* (McGraw-Hill, N.Y., 1965).
[2]See, for example, Chap. 8 of Ref. 1.
[3]Choice of the x-y plane is completely arbitrary since there is no interaction between the spin direction and the plane in which the spins are arranged.
[4]E. Ising, Z. Phys. **31**, 253 (1925).
[5]L. Onsager, Phys. Rev. **65**, 117 (1944).
[6]T. D. Schultz, D. C. Mattis, and E. H. Lieb, Rev. Mod. Phys. **36**, 856 (1964).
[7]K. Wilson, Phys. Rev. B **4**, 3174, 3184(1971). Reviews appear in M. Fisher, Rev. Mod. Phys. **46**, 597 (1974), and K. Wilson and J. Kogut, Phys. Rep. **12C**, 75(1974).
[8]Some review papers from this period are the following: P. Heller, Rep. Prog. Phys. **30**, 731(1967); M. Fisher, *ibid.* p. 615; L. Kadanoff *et al.*, Rev.Mod. Phys. **39**, 395 (1967); L. Kadanoff, in *Phase Transitions and Critical Phenomena*, edited by C. Domb and M. S. Green, (Academic, New York, 1976), Vol. 5A; S. K. Ma, *Modern Theory of Critical Phenomena* (Benjamin, New York, 1976).
[9]D. Nelson and M. Fisher, Ann. Phys. (N.Y.) **91**, 226 (1975); A. Houghton and L. Kadanoff, in *Proceedings of 1973 Temple University Conference on Critical Phenomena and Quantum Field Theory*, edited by J. D. Gunton and M. S. Green (Department of Physics, Temple University, 1973); M. Nauenberg, J. Math. Phys. **16**, 703 (1975).
[10]An explicit proof of this point for some spin systems has been given by R. B. Griffiths, J. Math. Phys. **5**, 1215 (1964).
[11]This particular type of application is called the "real-space" renormalization method. For a review see Th. Niemeijer and J. M. J. Van Leeuwen in Vol. 6 of the Domb-Green series mentioned in Ref. 8.

Physica A 163 (1990) 1–14
North-Holland

SCALING AND UNIVERSALITY IN STATISTICAL PHYSICS

Leo P. KADANOFF
The Research Institutes, The University of Chicago, Chicago, IL 60637, USA

The twin concepts of Scaling and Universality have played an important role in the description of statistical systems. Hydrodynamics contains many applications of scaling including descriptions of the behavior of boundary layers (Prandtl, Blasius) and of the fluctuating velocity in turbulent flow (Kolmogorov, Heisenberg, Onsager).

Phenomenological theories of behavior near critical points of phase transitions made extensive use of both scaling, to define the size of various fluctuations, and universality to say that changes in the model would not change the answers. These two ideas were combined via the statement that elimination of degrees of freedom and a concomitant scale transformation left the answers quite unchanged. In Wilson's hands, this mode of thinking led to the renormalization group approach to critical phenomena.

Subsequently, Feigenbaum showed how scaling, universality, and renormalization group ideas could be applied to dynamical systems. Specifically, this approach enabled us to see how chaos first arises in those systems in which but a few degrees of freedom are excited. In parallel Libchaber developed experiments aimed at understanding the onset of chaos, the results of which were subsequently used to show that Feigenbaum's universal behavior was in fact realized in honest-to-goodness hydrodynamical systems. More recently, Gemunu Gunaratne, Mogens Jensen, and Itamar Procaccia have indicated that they believe that a different (and weaker) universality might hold for the fully chaotic behavior of low dimensional dynamical systems.

Dynamically generated situations often seem to show kinds of scaling and universality quite different from that seen in critical phenomena. A technical difference which seems to arise in these intrinsically dynamical processes is that instead of having a denumerable list of different critical quantities, each with their critical index, instead there is continuum of critical indices. This so-called multifractal behavior may nonetheless show some kinds of universality. And indeed this might be the kind of scaling and universality shown by those hydrodynamical systems in which many degrees of freedom are excited.

1. Scaling in hydrodynamics: early developments

Concepts of scaling, which are an extension of Fourier's [1] idea of dimensional analysis, reached a very high degree of refinement in the study of hydrodynamics. So I start my story from one of the simplest set of equations which are used to describe fluid flow [2], the Navier–Stokes equations

$$\begin{aligned} &\boldsymbol{u}_t + \boldsymbol{u}\cdot\nabla\boldsymbol{u} = -(\nabla p)/\rho + \nu\nabla^2\boldsymbol{u}\,, \\ &\nabla\cdot\boldsymbol{u} = 0\,. \end{aligned} \tag{1.1}$$

These equations assume an incompressible fluid in which $\boldsymbol{u}$ represents the local fluid velocity, p and ρ are respectively its pressure and density, and ν is its kinematic viscosity.

There are two kinds of scaling solutions to eqs. (1.1). One kind has as its prototype the Blasius description [3] of a two dimensional time-independent flow past a semi-infinite flat plate. The velocity at infinity is parallel to the plate and has a magnitude U. The coordinate along the plate, x, is used to set the scale for the problem. Then suitable simplifying approximations are made in which small terms are eliminated from the Navier–Stokes equation. Specifically, one says that the variation in x is much slower than that in y and that the pressure term may be neglected in the x-component of the Navier–Stokes equation. Finally, a scaling or similarity solution is worked out in which each variable is expressed as an appropriately chosen dimensional constant times a suitable function of the dimensionless variables left in the problem. In this example, we write

$$u_x = Ug^*(y/Y(x)) ,$$
$$u_y = U(Y(x)/x)h^*(y/Y(x)) , \tag{1.2}$$
$$Y(x) = (\nu x/U)^{1/2} .$$

Here $Y(x)$ sets the scale for y while g^* and h^* are unknown functions which are determined by eqs. (1.1).

Kolmogorov [4], Heisenberg [5] and Onsager [6] analyzed turbulent flow by applying a more sophisticated kind of scaling to eq. (1.1). Turbulence involves a complicated flow pattern in which there are velocity fluctuations over a wide range of length scales. Let b denote a particular length scale and u_b describe the typical variation in velocity in that scale. In a turbulence problem, the kinematic viscosity is often a very small quantity. To express this smallness, form the dimensionless number

$$\mathrm{Re} = UL/\nu , \tag{1.3}$$

called the Reynolds number from a typical length scale of the enclosure, L, and a typical velocity of large-scale motion, U. It is quite often true that Re is very large indeed. (For an airplane flying at jet speeds Re is 10^9 or so.) Nonetheless the viscosity cannot be ignored because it provides the sole mechanism by which the kinetic energy which enters the fluid may be dissipated. Turbulence is then described by using a scaling argument, now in a statistical form, that says that kinetic energy is cascaded downward from longer wavelengths to shorter ones. Notice that a typical contribution to the kinetic energy density from terms at scale b is u_b^2 while the current or flux of kinetic energy at this scale is u_b^3. Of course, the divergence of a current (here of order (u_b^3/b) is a rate of change of the corresponding density produced by the downward cascade. Since the process is steady, the same amount is carried downward at all scales and hence

$$(u_b^3)/b = \varepsilon \tag{1.4}$$

with ε being a constant independent of the scale, b. Eq. (1.4) can then be converted into an estimate of u_b namely

$$u_b = (b/L)^{1/3} U . \tag{1.5}$$

This statement has proven to give a roughly correct indication of the typical velocities encountered in turbulent flow. It works in the socalled inertial range, i.e. for scales below L and above the Kolmogorov length, $L\ \mathrm{Re}^{-3/4}$, at which the viscosity terminates the kinetic energy cascade.

Notice that the two different scaling arguments described above have quite a different character. In the Blasius argument, for each x there is one characteristic distance in the y-direction (set by $Y(x)$) and the structure of the flow is describable by that scale alone. In the Kolmogorov argument, there are a continuum of characteristic distances, described by the variable b. The same basic structure reappears at all length scales and, in fact, one expects that structures are nested inside of one another.

One can also describe the results of both arguments by kind of universality idea. Universality simply means that the solution to a given problem is, in an appropriate sense, independent of the details of the problem set-up. In the Blasius case, we neglected some terms in the hydrodynamic equations, because the answer was insensitive to these terms. In the Kolmogorov case, the reasoning is once again deeper. One argues that for each volume of side b, with b in the appropriate scaling range, one can produce essentially the same kind of theory as one uses for the entire turbulent flow. That's scaling. But furthermore one argues that the only important characteristics of that volume are its average velocity and its dissipation rate, ε. That's universality. (See refs. 22 below.) We still do not know whether the Kolmogorov theory is essentially correct.

2. Scaling and universality at critical points

The same ideas have been applied quite successfully to the study of behavior near phase transitions. The essential scaling ideas were stated by B. Widom

[7]. Imagine that we have a system, like for example the Ising model, and that this system has a local fluctuating quantity $O(\boldsymbol{r})$. Here O might be, for example, the magnetization or the energy density. The main point is that at criticality fluctuations appear at all length scales. Let $O_b(\boldsymbol{r})$ represent the result of averaging $O(\boldsymbol{r})$ over a volume of order b^d – with d being, of course, the dimensionality of the space. One can expect that this quantity might scale as some power of b/a_0, a_0 being a microscopic distance, say as $(b/a_0)^{-x}$. The critical index, x, will of course be different for different fluctuating quantities, O. There are two aspects of the theory which should be mentioned here. One is that any quantity O can be expanded in a set of basic fluctuating quantities (or operators) which each have simple scaling properties. The sum converges rapidly for $(b/a_0) \gg 1$ because there are only a few operators with small values of x. Thus there is a natural classification and ordering of operators in the theory. The second aspect arises specifically because we are in a system described in equilibrium statistical mechanics. For each fluctuating quantity $O(r)$ there is a corresponding change in the quantity – variously called the free energy, Hamiltonian, or action – which generates the Gibbsian probability distribution. In units in which $kT = 1$, this change is

$$\Delta F = \sum_{r} O(\boldsymbol{r}) h \,. \tag{2.1}$$

Here h is the thermodynamic quantity conjugate to O. In this context, it is generally called a 'field'. For example if $O(\boldsymbol{r})$ is the magnetization density then h is the magnetic field, while if O is the energy density, h is the temperature change. Since ΔF is a logarithm of a probability, it is unchanged under scaling transformations. Therefore, if the index of the operator O is x, then the corresponding scaling index of the field h is [8]

$$y = d - x \,. \tag{2.2}$$

The few fields with y greater than zero play a special role in the theory. These fields are termed 'relevant'. A perturbation including relevant terms will drive one away from the critical point. Conversely, perturbations including only fields with y less than zero will have no effect on the critical behavior. Systems which have Hamiltonians which are different only in terms proportional to these irrelevant operators are said to lie in the same 'Universality Class'. In this way, all critical behavior is classified and Hamiltonian arranged in equivalence classes. (There is an exceptional intermediate case in which there is a field with $y = 0$. Changes in the Hamiltonian proportional to such a marginal operator can lead to continuous variations in the universality class [9]. But most critical systems do not contain such a marginal operator.)

Near the critical point, these operators and indices can be used to derive scaling statements similar in character to those in eq. (1.2). For example, in a critical system of size L, a quantity O_b with critical index x would have fluctuations which obey

$$\langle O_b^2 \rangle = (b/a_0)^{-2x} C^*(b/L) . \tag{2.3}$$

Here $C^*(\eta)$ is a correlation function expressed in a coordinate, $\eta = b/L$, appropriate for this kind of 'finite sized scaling'. By using scaling relations like these one can estimate the sizes and length dependence of all kinds of quantities in the critical region.

A large variety of different workers contributed to the formulation of this scaling and universality picture during the period centered about 1965. I point with pride to a review paper of 1966 [10] in which a group of us at the University of Illinois surveyed *all* (!) the literature which dealt with experiment and theory near critical points and concluded that the results could very well be correlated by this picture [11].

A few years later, K.G. Wilson capped off this line of development by converting the essentially phenomenological considerations of the earlier period into a calculational tool via the method of the renormalization group [12]. The basic idea was to consider the Hamiltonian H which generates the statistical weight. This Hamiltonian depends upon some basic set of variables. For the Ising model this is the local magnetization $\sigma(\boldsymbol{r})$. So the weight is $\exp - H\{\sigma\}$. Wilson then looked at what happens when one eliminates degrees of freedom, for example by eliminating short wavelength components from $\sigma(\boldsymbol{r})$. This change should result in a new statistical weight with a new form of the spin–spin coupling. By construction, new and old weights have the same consequences at the critical point. Wilson's key insight was that universality and scaling would be achieved if after many transformations the coupling structure of the weight approached a limit, a 'fixed point'. He showed how this fixed point could be calculated, and from it all aspects of critical behavior followed [13].

Wilson's work was performed in the context of field theory. In fact, the name 'renormalization group' had long been applied to related calculations in electrodynamics. Other work showed that this problem of critical phenomena was somehow very deeply connected to the Lagrangian field theories studied in particle physics. For example, Schultz, Mattis and Lieb [14] worked out the Onsager [15] solution of the two-dimensional Ising model as an example of a fermion field theory. The work of Kadanoff and Ceva [16] showed how scaling ideas implied that stress tensors and energy densities were natural parts of two-dimensional Ising model behavior. Polyakov [17] saw that the existence of the stress tensor in a rotationally symmetric situation implied that the critical field theory had a far richer symmetry than scaling. It is invariant under a larger group of transformations called the conformal group. In two dimensions this conformal symmetry is particularly rich and, in fact [18], it almost fully defines the possible critical behaviors.

The end result is that we understand critical phenomena. Each critical problem defines its own little world, described mostly by the dimensionality and symmetry of the underlying problem. As a result of universality the basic laws within each of these worlds are fixed and unchangeable. But because of fluctuations in the operators, things do happen. Each near-critical system supports elementary excitations which can be interpreted as massive elementary particles. These 'particles' scatter off of one another and behave as lively inhabitants of the universe in question. The operator classification ensures that the things that happen can be well described in terms of a known list of possibilities and measurements. Because of the field classification, there are a finite list of different significant ways of disturbing the world. In fact, the disturbances can be fully described by giving position-dependent values of the relevant fields. These fields play the same role as the velocity and dissipation rate in the Kolmogorov theory of turbulence. Because of scaling, the resulting geometrical structures are particularly simple. Basically the very same spatial structures are repeated again and again, at different places, in different magnifications, one inside the other. Thus, we find a beautiful toy world, almost fully understood, and correctly describing critical phenomena. Of course, the interesting question is whether other problems can be fitted into the same framework.

3. Chaos: scaling and universality

Mitchell Feigenbaum was responsible for a particularly elegant and important application and extension of the ideas described above. He was looking at dynamical systems, examining the patterns of x-values produced by the successive applications of mapping function, G. That is he was concerned with the character of the sequence x_j, $j = 1, 2, \ldots,$ generated by the successive applications of the function, G, via

$$x_{j+1} = G(x_j)\,. \tag{3.1}$$

In one example, he examined the so-called logistic map in which $G(x) = rx(1 - x)$, where r is a parameter which could be varied to change the character of the x-sequence. He noted, following earlier authors [19], that as r is increased from one the long term behavior of the sequence changes from a fixed point (i.e. an approach to a fixed value, $x^* = 1 - 1/r$), to a cycle of length two (x approaching different values on even and odd values of j), to a cycle of length four, . . . to a cycle of length 2^n. Then a specific value of r, called r_∞, results in a cycle of length 2^∞. For higher r yet, chaotic behavior ensues in which the sequence essentially never repeats itself. Feigenbaum observed two kinds of scaling: one that the length 2^n cycle first appears at an r-value, r_n, which obeys:

$$r_n = r_\infty - (\text{const})\delta^{-n} \tag{3.2}$$

in the limit of large n. The other scaling was a special behavior which occurred near the x-value for which the map is extremal ($x = \frac{1}{2}$ in the logistic map). If you started out at this value for x_0 then for large n

$$x_{2^n} = x_0 - (\text{const})\alpha^{-n} . \tag{3.3}$$

Here α and δ are two scaling indices which turn out to be 'universal'. In this case, universality means that these very same values of the critical indices arise from infinite period-doubling in a wide variety of different mapping problems. Building upon the analogy to critical phenomena, Feigenbaum [20] used the universality and scaling ideas to develop a renormalization theory for the period doubling. In this case, the theory was based upon the idea that the high order cycles (say of order 2^n) could be realized as fixed points of the function, G, composed with itself 2^n times. Further, Feigenbaum noted that, at r_∞, the large-n result of this multiple functional composition (plus a scale change in x) was a fixed point function, G^*. Under iterations of this renormalization process, small deviations from the fixed point function grow with a discrete spectrum of eigenvalues. Once again, there are only a few relevant (growing) eigenvalues. Thus the renormalization group analysis was carried over to the new problem with the function (and its compositions) playing the role of the Hamiltonian while the x_j sequence plays the role of the observable quantities.

There is a technical difference between the critical phenomena case and the one of period doubling: In the former, there is a duality between operators and changes in the free energy; in the latter, the connection between observables and G has been broken. To get observables, imagine getting 2^n successive x_j values and then forming the small differences:

$$\Delta_j = |X_{j+2^{n-1}} - X_j| . \tag{3.4}$$

It is observed that these differences have a whole range of critical indices.

Specifically Δ_j achieves an index-value [21] α_j by having

$$\Delta_j = (2^{-n})^{\alpha_j} . \tag{3.5a}$$

Define a weight for the critical index α by saying that $W(\alpha)\,\mathrm{d}\alpha$ is the number of times that α_j lies within an interval $\mathrm{d}\alpha$ about a given value α. After a normalization $W(\alpha)$ is a kind of weight function for the critical index. In the limit as n goes to finity, we can write this weight as

$$W(\alpha) = (2^n)^{f(\alpha)} . \tag{3.5b}$$

Evidently, $f(\alpha)$ serves as a kind of entropy function for the occurrence of the continuously varying critical index, α. For many problems, including this period doubling, $f(\alpha)$ turns out to be independent of n for large n. When this happens over an interval of α, we say that we have a multifractal behavior [22].

Like the critical indices of eqs. (3.2) and (3.3), the function $f(\alpha)$ is universal. My experimental colleagues at the University of Chicago constructed a hydrodynamic situation in which a fluid heated from below was pushed to the onset of chaos. Then, the $f(\alpha)$ curves derived from two of the theoretically described 'routes to chaos' [23] were compared with the corresponding curves derived from the experiments [24]. The excellent agreement between the two clearly showed that the universality concept extended to the real experimental system.

In many ways, Feigenbaum's work was a very big surprise. In particular, the mathematicians who had studied chaos earlier noticed, quite correctly, that these problems were far too rich to be characterized within an ordered list of universality classes. Indeed, typically, these problems seem to be classifiable only with the aid of infinities of continuously varying parameters. But by focusing upon the onset of chaos, Feigenbaum had picked out a group of problems which could, once again, be divided into sets, universality classes and described via scaling and renormalization.

Despite this further success of the scaling/universality/renormalization approach, there are some hints that the critical phenomenon example will not be infinitely extendable. We know about dynamical systems problems which give scaling but no universality. One such example is the construction of Julia sets [25–27], which are mathematical structures somewhat analogous to chaotic attractors.

Nonetheless it is worthwhile to look for some remnant of universality in low-dimensional chaos, carried beyond onset. The analysis of Gunaratne, Jensen, and Procaccia [28] shows there is a kind of topological universality visible in the chaotic regime, but no real metric universality. More specifically, they look at mapping problems and focus upon the cycles, calculating their lengths and seeing the structure of very long cycles. They ask questions about which cycle is close to another, without asking 'how close?'. And they see that, if they look at the right kinds of problems, the set of all cycles is stable under appropriately chosen small changes in the mapping problem. They then hope to use this observation to obtain a classification of chaotic problems by working about structurally stable points in the phase space. Thus while they aim for universality, they have given up all elements of scaling. On the contrary, they conclude that in really chaotic problems there is no universality in distance measurements. This conclusion agrees with our experience of Julia sets. For myself, I worry whether there can be deep structure to a universality theory which leaves out all measurements of distance.

4. Dynamical systems with many degrees of freedom

4.1. Dynamic critical phenomena

The next step is to look at dynamical systems with many degrees of freedom to see how much scaling and universality they show. The simplest case is dynamical critical phenomena in which the static correlations of near-critical behavior combine with conservation laws to engender interesting correlations in space and time [29]. There are well developed renormalization theories [30] based upon Lagrangians with extra fluctuating operators beyond those needed in the equilibrium theory [31]. Of course the extra operators depend upon both space and time. The resulting theory indicates universal behavior, with the universality depending upon both the equilibrium critical behavior and also the types of conservation laws. Thus, we have one more successful extension of the standard synthesis which I have been expounding.

4.2. Self-organized criticality

But there do exist dynamical processes which produce scaling results without having any obvious underlying equilibrium critical behavior. When one gets to dynamical systems with many degrees of freedom, we no longer know the nature of the fundamental theory and so, for example, we cannot know whether or how renormalization group concepts apply. I discuss two examples here.

The first example is called DLA for diffusion limited aggregation. It is a model invented by Witten and Sander [32], in which a fractal [33] aggregate is grown by a step by step process. The aggregate sits on a lattice. A walker, added at infinity, undergoes a random walk process until it comes to a neighboring site to one already occupied. Then the walker stops and its final site is added to the aggregate. This whole process starts once again with another walker added at infinity. The aggregate thus produced is extremely tenuous and contains long branching arms with considerable space between the branches. Question: can this object be understood via concepts of scaling and universality? Is there somehow an underlying renormalization group description? Despite considerable study, it is fair to say that we simply do not know. It does appear that the aggregate itself is properly described as a fractal, but the measurement of its fractal dimension has been peculiarly difficult. Alternatively, one can describe the growth process by asking what is the probability that a new particle will be added at a given surface site. This probability varies over a wide range and is best described by a range of critical indices, in a multifractal formalism [34].

In DLA, the dynamics produces its own critical ordering. This kind of situation is termed by Bak, Tang and Wiesenfeld [35] to be one of "self-organized criticality". There are several other interesting examples in which the dynamical process produces an object which is marginally stable, and hence shows very long-ranged correlations [36]. One situation studied by Bak et al. and by others [37], is one in which model sand is added grain by grain to a model sandpile, built upon a regular d-dimensional lattice. In between additions, there are cascades of events in which sand falls downhill in response to a too-large local slope of the pile. These 'avalanches' can be small or they can cover the entire system many times over. Once again, it is interesting to ask about whether there is some scaling or universality. For example, one can study the nature of the probability distributions $\rho(X, L)$ for the probability that an event of size X will occur in a system with spatial extent L. As before, Widom scaling is the statement

$$\rho(X, L) = L^{-\beta}\rho^*(X/L^{\nu}) \tag{4.1a}$$

while multifractal behavior is one in which

$$(\ln \rho(X, L))/\ln L = f(\alpha) \qquad \text{with } \alpha = (\ln X)/(\ln L) \tag{4.1b}$$

for X and L much bigger than one. Numerical work suggests that the multifractal behavior seems preferred in one dimension, while for two dimensions the question remains open. For both dimensionalities, the numerics indicates several universality classes, but universality within each class [38].

Up to now, we have had neither a completely convincing formalism nor a fully credible phenomenology for describing these essentially dynamical objects and processes. It is interesting to notice that they tend to be multifractal. Since none of the most conventional field theory/statistical mechanics systems seem to have a continuum of critical indices, we might wish to question whether these dynamical systems are really described by a simple field theory. This question is an important one in that many of our methods of setting up and thinking about such problems are closely based in Lagrangian field theory. But, perhaps these problems do have a Lagrangian someplace after all. One does know problems (e.g. percolation and random resistor networks) in which the behavior is multifractal but the system is understandable as a limiting case of one with a Lagrangian description. Alternatively, they could be describable by a Lagrangian which has no symmetry between space and time. However, notice that in both DLA and sand slides, the step between elementary addition events involves complex processes: either an entire random walk or a whole avalanche. Thus, the elementary processes in this proposed Lagrangian are far from simple. For this reason, the Lagrangian formulation may fail. My own hope is that sand slides and DLA and many other dynamics problems are essentially new and have some – as yet unknown – formation, combining some elements of scaling, universality, and renormalization group.

4.3. The binomial distribution: a simple multifractal example

However, one can set up renormalization examples which do have a multifractal character. I do this here. Since the work of Billingsley [39], it has been known that the binomial distribution provides an example of a probability which has a rich asymptotic structure and is, in modern terms, multifractal. Consider Q objects which are distributed randomly between 2 bins. The total number of ways that this can be done is $L = 2^Q$. (If we visualize adding the objects one at a time then Q can be considered to be an analog of a temporal variable.) The probability that P objects will show up in the first bin is, of course,

$$\rho(P, Q) = Q!/(P!(Q-P)!2^Q) . \tag{4.2}$$

Let $P, Q \gg 1$. From the Stirling approximation for factorials we then find that $\rho(P, Q)$ has a multifractal form with

$$f(\alpha) = -\ln 2 - \alpha \ln \alpha - (1-\alpha)\ln(1-\alpha) \tag{4.3}$$

and $\alpha = P/Q$. Near the peak (at $\alpha = \frac{1}{2}$) the distribution is Gaussian, but the result (4.3) also works far into the wings of the distribution.

To derive a renormalization method for computing $f(\alpha)$ recall Pascal's Triangle for binomial coefficients which is the statement

$$\rho(P, Q) = [\rho(P-1, Q-1) + \rho(P, Q-1)]/2 . \tag{4.4}$$

Now let us make the change of variables:

$$f(\alpha, Q) = \frac{\ln \rho(P, Q)}{Q} ,$$

substitute into eq. (4.4), and expand in $1/P$ and $1/Q$ to find the simple renormalization equation

$$Q \frac{\partial f}{\partial Q} = \ln\left\{\exp\left[-f - (1-\alpha)\frac{\partial f}{\partial \alpha}\right] + \exp\left[-f + \alpha \frac{\partial f}{\partial \alpha}\right]\right\} - \ln 2 . \tag{4.5}$$

The fixed point solution makes the right hand side of (4.5) vanish. The $f(\alpha)$ of eq. (4.3) is a correct solution to this equation in the limit as q goes to infinity.

Can one attack the sandslide problem in this fashion? I cannot yet see how to achieve this but I would dearly love to do so. But, notice that the calculation I have just done is an essentially traditional application of real space renormalization methods. If, as I hope, there is something really new in DLA and sand slides then no progress is likely without a new idea, as yet unforeseen.

The systems with novel scaling states based upon self-organized criticality are going to be an interesting subject of study during the next few years. Many workers have the tantalizing feeling that solutions to one or another of these problems is just within reach, but so far they seem to have eluded us.

4.4. Back to turbulence

Fully turbulent systems, which are even more interesting than the ones with chaos or those with self-organized criticality, have proven very hard. One hopes, nevertheless, that some important simplification will arise because of the very large number of excited degrees of freedom. Experiment shows some suggestion of multifractal behavior [40], and other suggestions that simple scaling laws might hold [41]. Apparently similar spatial structure reappear in a wide variety of problems [42]. In that weak sense, there is some universality in turbulence. But, we cannot be really sure about how many parameters (fields) we need to define what is going on in a given turbulent region. Can we get by with a finite number? Is Kolmogorov's four (u_b and the dissipation rate ε) the right number? Are there structures inside of structures? Which are the really instructive situations: a randomly stirred system [43], or one in which turbulence is decaying in time [44] or one in which turbulence is spreading in space, or one in which there is in continual non-random forcing? How does one begin to attack this kind of problem?

If one asks how much universality and scaling really hold for a turbulent system, the only possible answer is 'nobody knows'.

Acknowledgements

My research has been very generously supported by the University of Illinois, Brown University, and the University of Chicago as well as such agencies as the National Science Foundation and the Office of Naval Research. It would not have occurred without the collaboration and exchange of views with many colleagues. I owe my largest debt to these fellow scientists who were at once audience, inspiration, and teachers.

References

[1] M. Fourier, Théorie Analytique de la Chaleur (Didot, Paris, 1822) p. 154. He introduces the term 'exponent of dimension', closely analogous to modern scaling indices.
[2] One of the themes of this talk is that hydrodynamic systems and other dynamical problems, especially those described by partial differential equations, are likely to provide a fruitful area of study for people trained in statistical physics.
[3] See G.K. Batchelor, An Introduction to Fluid Mechanics (Cambridge Univ. Press, London, 1983), p. 308.
[4] A.N. Kolmogorov, Dokl. Akad. Nauk SSSR 26 (1941) 115.
[5] W. Heisenberg, Proc. Roy. Soc. (London) A 195 (1948) 402.
[6] L. Onsager, Nuovo Cimento, Supplement (9) 6 (1949) 279.
[7] B. Widom, J. Chem. Phys. 43 (1965) 3892; 43 (1965) 3898.
[8] An expression of the character of (2.2) was called a 'hyperscaling' relation to suggest that this was more scaling than really occurred. However, by now, it is rather generally believed that this kind of relation between the scaling of operators and fields is quite generally characteristic of critical phenomena.
[9] L. Kadanoff and F. Wegner, Phys. Rev. B 4 (1971) 2909.
[10] L. Kadanoff, W. Gotze, D. Hamblen, R. Hecht, E.A.S. Lewis, V.V. Palciauskas, M. Rayl, J. Swift, D. Aspnes and J.W. Kane, Rev. Mod. Phys. 39 (1967) 395.
[11] In this review paper, we tried to list the people who had been mainly responsible for the scaling ideas. In our list we included C. Domb and D.L. Hunter, M.E. Fisher, B. Widom, A.Z. Patashinskii and V.L. Pokrovshii, G.E. Uhlenbeck and P.C. Hemmer and M. Buckingham. Ideas of universality played a fundamental role in this review paper. In critical phenomena, universality has many parents, the first being L. Landau and then A.B. Pippard, and afterwards Leo Kadanoff (in Critical Phenomena, Proc. Int. School of Physics, "Enrico Fermi", Course LI, M.S. Green, ed. (Academic Press, New York, 1971), p. 101); R.B. Griffiths (Phys. Rev. Lett. 24 (1970) 1479); D. Jasnow and Michael Wortis (Phys. Rev. 176 (1968) 739). See also the work of the King's College School, especially C. Domb and Michael Fisher.
[12] An excellent review can be found in K.G. Wilson and J.B. Kogut, Phys. Rep. 12C (1974) 75.
[13] F. Wegner in Phase Transition and Critical Phenomena, C. Domb and M.S. Green, eds. (Academic Press, London, 1976) Vol. 6, p. 8, expressed the consequences of the fixed point idea in a particularly powerful and elegant form.
[14] T.D. Schultz, D.C. Mattis and E.H. Lieb, Rev. Mod. Phys. 36 (1964) 856.
[15] L. Onsager, Phys. Rev. 76 (1949) 1232.
[16] L. Kadanoff and H. Ceva, Phys. Rev. B 3 (1971) 3918.
[17] A.M. Polyakov, JETP Lett. 12 (1970) 10.
[18] D. Friedan, Z. Qiu and S.H. Shenker, Phys. Rev. Lett. 52 (1984) 1575. This paper is reprinted in Conformal Invariance and Applications to Statistical Mechanics, C. Itzykson, H.

Saleur and J.-B. Zuber, eds. World Scientific, Singapore, 1988). This is an excellent collection of reprints.
[19] R.B. May, Nature 261 (1976) 459.
[20] M. Feigenbaum, Los Alamos Science 1, 4 (1980).
[21] This meaning of α is different from that in the scaling index of eq. (3.3). This multiple use of the same symbol is a fault for which I am partially responsible.
[22] B. Mandelbrot, J. Fluid Mech. 62 (1974) 331. U. Frisch and G. Parisi in Turbulence and Predictability in Geophysical Fluid Dynamics and Climate Dynamics, M. Ghil, R. Benzi and G. Parisi, eds. (North-Holland, Amsterdam, 1985) p. 84.
[23] D. Bensimon, M.H. Jensen and Leo Kadanoff, Phys. Rev. A 33 (1986) 3622. M. Feigenbaum, Leo Kadanoff and Scott J. Shenker, Physica D 5 (1982) 370.
[24] M.H. Jensen, L.P. Kadanoff, A. Libchaber, I. Procaccia and J. Stavans, Phys. Rev. Lett. 55 (1985) 2798. J.A. Glazier, M.H. Jensen, A. Libchaber and J. Stavans, Phys. Rev. A 34 (1986) 1621.
[25] D. Ruelle, Ergodic Th. & Dynam. Sys. 2 (1982) 99. R. Bowen, Equilibrium States and the Ergodic Theory of Anosov Diffeomorphisms (Lect. Notes in Math. 470, Springer, 1975). Ya. Sinai, Gibbs Measures in Ergodic Theory, Russ. Math. Surveys 166 (1969) 21.
[26] H.-O. Peitgen and P.H. Richter, The Beauty of Fractals (Springer, Berlin, 1986).
[27] M.H. Jensen, Leo Kadanoff and I. Procaccia, Phys. Rev. A 36 (1987) 1409.
[28] Gemunu Gunaratne, Mogens Jensen and Itamar Procaccia, Nonlinearity 1 (1988) 157. Gemunu Gunaratne, Paul Linsay and Michael Vinson, Univ. Chicago preprint, 1989.
[29] R. Ferrell et al., Ann. Phys. (N.Y.) 47 (1968) 565. B.I. Halperin and P.C. Hohenberg, Phys. Rev. 177 (1969) 952. M. Fixman, J. Chem. Phys. 36 (1962) 310. K. Kawasaki in Phase Transition and Critical Phenomena, C. Domb and M.S. Green, eds. (Acad. Press, London, 1976), Vol. 5a, p. 165. J. Swift and L. Kadanoff, Phys. Rev. 166 (1968) 89.
[30] B.I. Halperin, P.C. Hohenberg and S. Ma, Phys. Rev. B 10 (1974) 139; S. Ma and G. Mazenko, Phys. Rev. B 11 (1975) 4077.
[31] P.C. Martin, H.A. Rose and E. Siggia, Phys. Rev. A 8 (1973) 423.
[32] T.A. Witten and L.M. Sander, Phys. Rev. Lett. 47 (1981) 1400.
[33] B. Mandelbrot emphasized the geometric side of scale invariance during the period in which this side of the problem was partly neglected by the critical phenomena people. See B. Mandelbrot, The Fractal Geometry of Nature (Freeman, New York, 1983) but also M.E. Fisher, Proc. Univ. Kentucky Centennial Conf. on Phase Transitions (1965).
[34] T. Halsey, P. Meakin and I. Procaccia, Phys. Rev. Lett. 56 (1986) 854.
[35] Per Bak, Chao Tang and Kurt Weisenfeld, Phys. Rev. Lett. 59 (1987) 381.
[36] Model earthquakes, as in J.M. Carlson and J.S. Langer, Phys. Rev. Lett. 62 (1989) 2632, are an interesting application of this idea.
[37] Deepak Dhar and Ramakrishna Ramaswamy, Jawaharial Nehru Univ. preprint, 1989. T. Hua and M. Kardar, Phys. Rev. Lett. 62 (1989) 1813.
[38] Leo Kadanoff, Sid Nagel, Lei Wu and Su-Min Zhou, Phys. Rev. A 39 (1989) 6524.
[39] P. Billingsley, Ergodic Theory and Information (Wiley, New York, 1965), p. 139.
[40] D. Schertzer and S. Lovejoy in Turbulence and Chaotic Phenomena in Fluids, S. Tatsumi, ed. (North-Holland, Amsterdam, 1983).
C. Meneveau and K.R. Sreenivasan, J. Fluid Mech. 173 (1986) 357.
F. Anselmet, Y. Gagne, E.J. Hopfinger and R.A. Antonia, J. Fluid Mech. 140 (1984) 63.
[41] B. Castaing, G. Gunaratne, F. Heslot, L. Kadanoff, A. Libchaber, S. Thomae, Xiao-zhong Wu, S. Zaleski and G. Zanetti, J. Fluid Mechanics (June, 1989).
[42] M. Van Dyke, An Album of Fluid Motion (Parabolic Press, Stanford, CA, 1982).
[43] D. Forester, D. Nelson and M.J. Stephen, Phys. Rev. A 16 (1977) 732.
[44] Ya.G. Sinai and V. Yakhot, Princeton Univ. preprint, 1989.

Section C

Simulations, Urban Studies, and Social Systems

Models and Arguments

Science includes the critical application of ideas to real world situations. In the 1950s, 60s, and 70s I found many applied problems very interesting. While I was still in college, I worked for the Guided Missile Division of Republic Aviation Corporation, helping with the proposal stage of the design of robot aircraft. In graduate school and thereafter, I worked for the AVCO corporation doing heat transfer calculations related to the design of guided missiles and space probes. I found the latter job extremely instructive. It enabled me to interact with some scientists whose paths I would not normally have crossed, including Hans Bethe, Jim Keck, and Arthur Kantrowitz. I was very impressed by the intellectual vitality of the work at AVCO, particularly the branch in Everett, Massachusetts.

But after a while, I became convinced that this kind of work was socially unproductive. During the Eisenhower years I had been quite uncritical about most of our (U.S.) foreign policy. However, while I was a postdoc in Copenhagen, our government did things I felt that I could not explain to the 'foreigners' around me. The Bay of Pigs invasion of Cuba was the example that I best remember. Having no explanation for others, I became less convinced that our role was always a good one. One defining event for me was sitting in the cafeteria at lunch time at AVCO's Missile Division in Wilmington, Massachusetts discussing the dangers to the world of nuclear weapons. I even wondered whether it was prudent to pick up stakes and go someplace like Brazil. This discussion was taking place during the Cuban Missile Crisis. On the way to work that day, I had driven by Boston's Logan airport which had a small section covered with Strategic Air Command bombers, all probably loaded with nuclear weapons. At that time, I could not see why the US could not tolerate the same kind of proximity of enemies that the Soviets had to endure. So I began to lose sympathy with US Cold War policy. This progression in my thinking continued through the period of the beginning of the Vietnam War. Naturally, I could no longer work in the missile industry.[23]

So, for a while, my applied interests had no good outlet. But then, starting in about 1967 I began work on the use of scientific techniques to describe and control urban social and economic phenomena.

I got started in this work though the efforts of Dale Compton who was then head of the Coordinated Sciences Laboratory at Urbana. He introduced me to Jerrold Voss, an urban studies specialist. I worked on their project for studying tendencies in urban real estate prices and land use in the little town of Kankakee, Illinois. I do not know what I or they expected to gain from my involvement in this work. I brought essentially no knowledge whatsoever to the project. I do not know any economics or sociology. I got involved in the computer side of the project, supervising the computer specialist in the Urbana laboratory. In this computer work, we absorbed great piles of historical data about real estate prices in Kankakee and tried to display them in some way which would enable us to understand what

[23] To complete my story: The lunch-table discussion ended when one of the executives suggested, with some annoyance, that we should all get back to work. We did.

was going on. Little by little I gained some rudiments of information about the urban scene, and some small knowledge of computers. (The actual computer at the Coordinated Sciences Laboratory[24] was large and far too complex for me to program.) I did not contribute much but I was proud, and somewhat flattered to be included in their work, and its writeup, C1 in this section.

Then, I moved to Providence, Rhode Island, and Brown University. At that time there was a great national push toward understanding the dynamics of urban development. Jay Forrester of MIT had developed a computer model of urban change, which took a very simplified view of a urban society, produced a quantification of that view in terms of a computer model, and then used the output of that model to prescribe social policy. I did not like the policy prescribed. He basically suggested that by removing housing for poor people the city could free up land which could be used for industrial expansion, to — in Forrester's view — the long-range benefit of all. In my view, this policy made the poorer section of society pay for the economic development which would occur in the city, and tried to sanctify that policy by giving it the blessing of apparently-objective computer output. Consequently, I set out to use the modeling tools that Forrester had developed to reach conclusions which were more to my liking. Brown University was a good environment for this because they had a computer system which was set up so that anyone could use it. I went to work reproducing Forrester's model in a more flexible computer environment.

Brown University contained a group of people interested in urban public policy. Eventually, the group of urban research coworkers at Brown would include Benjamin Chinitz, Graham Crampton, Bennett Harrison, Susan Jacobs, John Tucker, and Herbert Weinblatt. We took as one of our tasks the incisive criticism of Forrester's work. The parameters and structure of his model had never been verified against real data. We could have tried to disprove the model by looking at the world. However, that would be difficult or impossible since many of the social concepts in the model are very hard to quantify. Instead, we chose another route. (See C2 and C3). Our group changed the focus of the model from a one-city application to a national application. The nation was viewed as being completely made up of cities. We insisted on keeping track of the poor people. This social group would be tend to be squeezed out of the city by a local application of the housing-removal strategy, and would be just squeezed by a national application of that policy. We also tried to develop a more objective (or perhaps just different) set of criteria for the success of the policy. Our first result was that while not changing the model at all, we could reach opposite conclusions from that of the Forrester group. By our criteria, the poor population was highly squeezed with little overall gain to any population segment in the urban society.

This first work can be construed in two ways. One truth was that you got out of such models just about what you put into them. They were mostly a way of recording your preconceptions. The second truth was that such preconceptions could be fluently incorporated into models of this type and one could, in fact, reach any conclusions that one

[24]This Coordinated Sciences Laboratory (CSL) work eventually replaced my work in the missile industry. Life does have its ironies. This laboratory was almost fully supported on Department of Defense contracts. As a result of complaints about the Vietnam war by people like myself, after a while it became impossible for labs like CSL to use military funds to contribute to civilian research.

wanted. Hence, at one level, this work could be construed as a criticism of all model-building in which one's preconceptions were not tested against data.

I am not sure that we listened to our own criticism fully. We went on to build other models which more accurately recorded our own prejudices (see paper C4) and points of view. But, after a while the point we had made began to sink in. If these models really represented little more than we could say in words, why not leave out the computer? The construction of this sort of computer model seemed to be a rather pointless endeavor. For this reason and others, I moved away from urban studies.

Nonetheless the experience had taught me much. My collaborators had valuable insights into the functionings of a city and I was pleased to absorb some of these. And I had learned a new technique: the use of the computer to model experience. The computer would prove crucial in the next stage of my career, in which I would work on hydrodynamics and chaos.

My experience in urban work extended beyond research. I worked (unpaid) for the Rhode Island State Planning Department as chair of a committee which constructed criticisms of all proposals for federal funding of public programs. The point of such a committee is to see that everyone talks to everyone so that there is no unnecessary unhappiness engendered by such proposals. Since I enjoy talking, this was a fun job for me.

At the same time, I taught courses for undergraduate majors in urban studies at Brown. We discussed modeling, and the spatial structure of cities, and social policy and lots of other things. Such teaching was a rewarding experience for me, particularly because I could no longer do the kind of teaching which dominated my early career. At Urbana, I spent a lot of time teaching graduate students either in large classes or one-on-one as their graduate thesis adviser. But by 1971 or so, this kind of teaching became much less attractive. Within a few years, a large fraction of the jobs which had been available to physics graduate students dried up. Our students were not finding appropriate work. Consequently, graduate-school teachers like myself felt their work to be unnecessary.

The climate for employment of advanced-degree students in physics had basically changed because of the Vietnam War. Many physicists and other scientists were critical of the policy of the government. Military contractors looked for potential workers who would not criticize governmental aims and policies. Gradually engineers replaced physicists in the 'defense' industries. Fewer physicists hired meant fewer graduates needed, meant fewer teachers needed to train the graduate students, means fewer physicists hired .... In any case, there was little satisfaction in training large numbers of graduate students in this period.

However, the job of training urban studies specialists is not very satisfactory either. As pointed out by Alan Altshuler,[25] one problem of the urban specialist is that there exists no large base of real knowledge, information, or technique which would enable the trained person to better prescribe urban planning policy than the ordinary intelligent citizen. The specialist has nothing special to add. As you can imagine this situation is rather discouraging to the teacher, who then has nothing special to teach.

So my urban public policy phase gradually withered away. When I accepted a job at the University of Chicago in 1978, I knew that the City of Chicago and its University were both far too professional to allow an easy outlet for my urban interests. After I came to Chicago,

[25] Alan Altshuler, *The Urban Planning Process* (Cornell University Press, Ithaca, 1965).

I devoted myself single-mindedly to teaching and research in the physical and mathematical sciences. But some residue of my social interests remained. Within the framework provided by an occasional column in *Physics Today* I tried to comment upon the social and economic context in which physicists work.[26] Three more columns are included at the end of this Section on the urban scene and public policy. These columns are indeed about the relation between the physics community and public policy. I enjoy writing the columns, and I even think that many people enjoy reading them. But Altshuler's criticism of the urban planning specialist can equally well be extended to most columnists. The question to ask is: "By what process have you become qualified to offer us advice?".

[26]The reader has probably noted that I have also taken advantage of these introductions to pontificate in this direction.

Computer Display and Analysis of Urban Information Through Time and Space*

LEO P. KADANOFF

JERROLD R. VOSS

and

WENDELL J. BOUKNIGHT

Introduction

The physical environment of our cities is shaped by a complex interplay between the private sector—the real estate market—and the public sector. Private investors, in a myriad of individual decisions, attempt to maximize their economic return or simply try to find a good place to live. Ideally, public policy attempts both to respond to the needs of this private sector and also to guide it in a manner which will improve the life of all. However, public policymakers have important but limited tools at their command: the placement and nature of transportation facilities; water and sewers; the creative use of zoning and taxing authority; etc. These tools are supposed to aid and influence the private decisions and thereby improve the workings and quality of the city.

Unfortunately neither public policymakers, nor private investors, nor even students of urban life understand very well the urban market place. As a result, many private real estate decisions are incorrect and wasteful. Moreover, great public programs, such as urban renewal or the interstate highway program, have had very serious and unexpected negative side effects.

To avoid errors in future public and private decisions, one needs both a better

JERROLD R. VOSS is an architect and a city planner who, for a number of years, has been researching the factors influencing urban growth and development. Mr. Voss was an Associate Professor of City Planning at the University of Illinois until September of 1969 when he joined the faculty of Harvard University as an Associate Professor of City and Regional Planning and was appointed as an IBM Fellow.

LEO P. KADANOFF is by training, a physicist whose research specialty is the theory of solids. During the last two years, Professor Kadanoff has participated in the urban studies project at CSL and has divided his research and teaching efforts equally between physics and urban studies. Mr. Kadanoff was Professor of Physics at the University of Illinois until the fall of 1969 when he joined the faculty of Brown University where he is currently Professor of Physics and University Professor.

WENDELL J. BOUKNIGHT is an electrical engineer and since joining the CSL as a Research Engineer in 1966 has been concerned with the application of computers in many fields, especially that of graphics. His work (for the urban studies project at CSL) in developing and writing the graphic display programs for the Centralia display and the Ford Motor Company movie provided the groundwork for the design of the graphics program currently being developed for the Kankakee project.

*This work was first reported in an article entitled "A City Grows Before Your Eyes," *Computer Decisions*, I December 1969), 16–23.

The Project was supported in major part by the Ford Motor Company through a grant to the Coordinated Science Laboratory, the University of Illinois. Auxiliary support for computer services was furnished by the Joint Services Electronics Program (U.S. Army Research Office, Office of Naval Research and the Air Force Office of Scientific Research under Contract No. DAAB-07-67-C-0199). In addition partial support for personnel was provided from a grant by the National Science Foundation (NSF GR-60).

Technological Forecasting and Social Change **2** (1970), 77–103

theoretical understanding of urban growth and a means of presenting this knowledge in a form which can be meaningful to the citizen or policymaker.

In recent years, considerable effort has been directed toward developing an understanding of the various public and private determinants of urban structure. Mathematical theories of the urban real estate market have been constructed by location theorists such as William Alonso or Lowdon Wingo. These theories have been simplified and converted into "urban growth models" which have been used to predict development in particular cities. Moreover, these mathematical models have been used extensively in the process of making decisions about the nature and placement of urban transportation facilities.

Yet, all this model-making has not been fully successful. First, it has been virtually impossible to focus on individual decision units such as households as this would greatly increase the complexity and difficulty of the analysis. Second, most of the models have concentrated on urban extension and expansion rather than renewal and change in already developed areas. Third, the analyses have tended to be cross-sectional rather than dynamic because of the difficulty of obtaining reliable time-series information and consequently they are unable to capture, reproduce, or simulate the way in which an urban area actually evolves. Fourth, the models are built to serve a definite purpose in a particular city and in many cases have not been adequately tested and evaluated in the city for which the model was prepared. And, finally, the other major reason for our lack of progress has been our inability to analyze, examine, and arrange urban data in an interactive and dynamic fashion.

The complexity of a city is great and requires the most innovative use of our data-manipulation technology. It is no wonder that our response to this problem has been more artistic than scientific. The techniques that are currently in use have done little to expand the synthesis capability of the researcher and the policymaker. There is now a ubiquitous use of statistics, and especially correlation and factor analysis, and though using these techniques is an improvement, they fail to capture either the complexity or dynamics of urban development. Indeed, these techniques are especially suited to cross-sectional analyses and in the absence of good time-series data there has been little need to apply spectral analysis in order to examine lags and hysteresis effects. On the other hand we have progressed somewhat in our graphic representation of the city, but unfortunately, these developments have limited dynamic qualities which in turn limit the opportunities for analysis. Although there are other reasons why we have made such a modest impact on coping with our urban problems, those mentioned seem to be the most important, and therefore have provided the basis for the development of this study.

The ultimate goal of our research is the construction and assessment of an urban growth model which can be used to understand alternative urban growth patterns in such a way that they may be evaluated by policymakers. In order to do this it is necessary to move forward along two paths: one which involves urban growth theory; and, the other which attempts to make optimum use of our data handling and display capabilities. In the former case the residential development of a city of manageable size (50,000) population) was selected for an in-depth study of all real estate transactions (land and total property value) and the construction and use histories of each structure. The time period examined extends from 1854 (the date of the first real estate transaction) to 1969. The object is to construct a complete history of all of the changes in these three areas so that the dynamic aspects of urban change can be analized. In addition and through time, data will be assembled on a wide variety of public and private variables which might

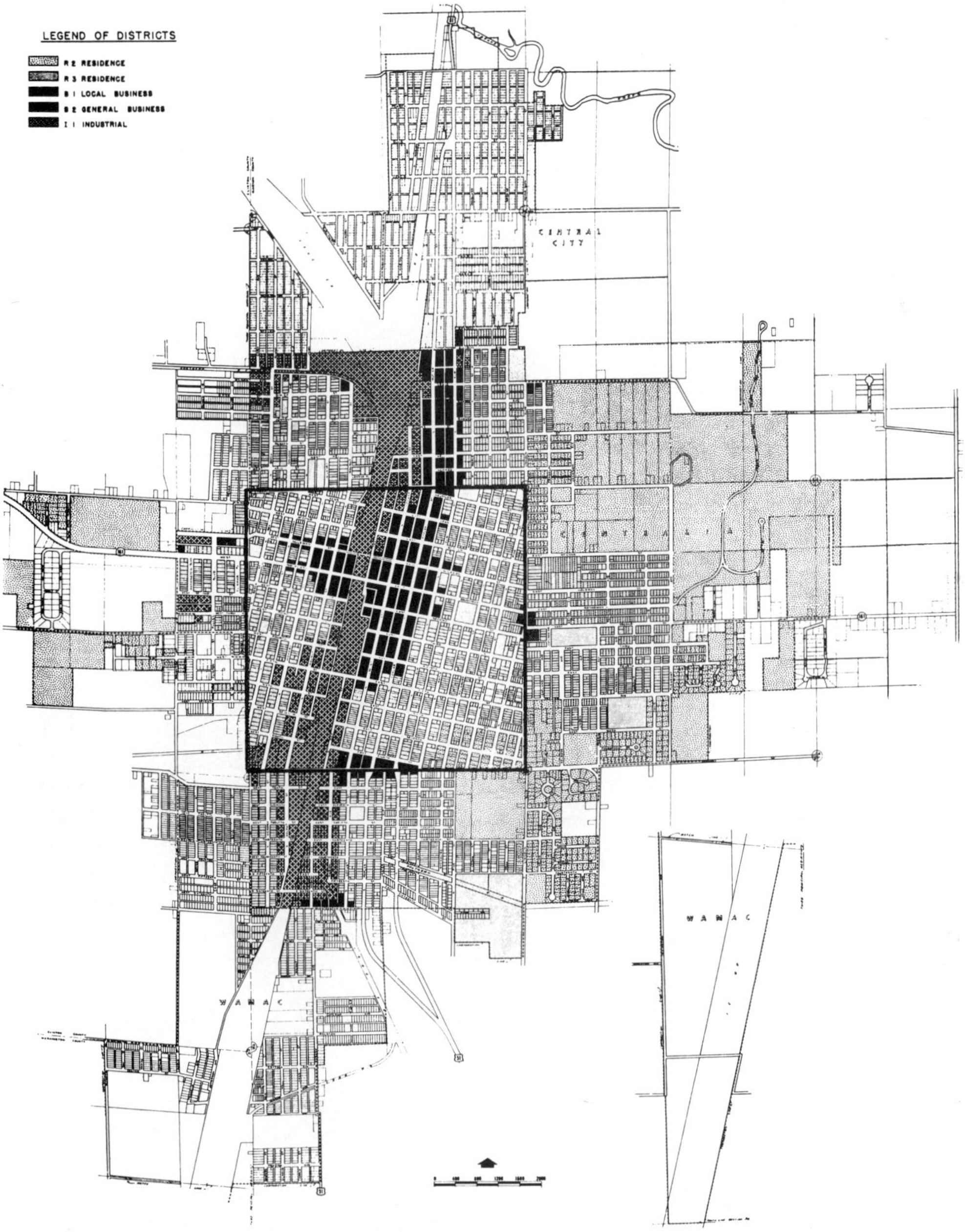

Fig. 1. Centralia, Illinois and vicinity in 1964 showing the outlines of the original town. Source: Metropolitan Planners, Inc., *Comprehensive Development Plan for Centralia*, Illinois, 1964.

have influenced the changes in these three areas. Substantively then, an effort is being made to model the physical evolution of a city with the emphasis on the residential environment.

In order to execute this analysis and exploit the data resources our other major effort is directed at improving our synthesis techniques by developing ways in which a researcher and/or policymaker may work with a computer in an interactive mode. Therefore, our intent is to design computer programs which will enable an analyst to interact with a computer using a light pen and a CRT (cathode ray tube) or using a typewriter. As a result we will be able to demonstrate and operationalize the use of computers and their calculational and display capabilities in expanding greatly one's ability to make informed choices cheaply and efficiently. The remainder of this article will describe the progress we have made in developing computer-generated display and presentation techniques of urban data.

A First Cut—The Centralia Study

Our first attempt to develop new methods of data presentation utilized information about land sales by the Illinois Central Railroad in Centralia, Illinois (see Fig. 1). The original portion of this small city (outlined area) was subdivided and developed by the railroad. One of us (JRV) had collected data about the price and land use for the first sale of all land within this original square-mile area. This very small data base—comprising about 1100 sales of raw land—could permit experimentation and innovation without excessive use of computer time.

To study these data, the analyst, sitting at the console, is first confronted by a display on the CRT which asks which display options he wishes to employ (see Figs. 2 and 3). Does he wish to study the data on a month by month or a year by year basis? He chooses

Fig. 2. The CSL display unit with operator examining display options.

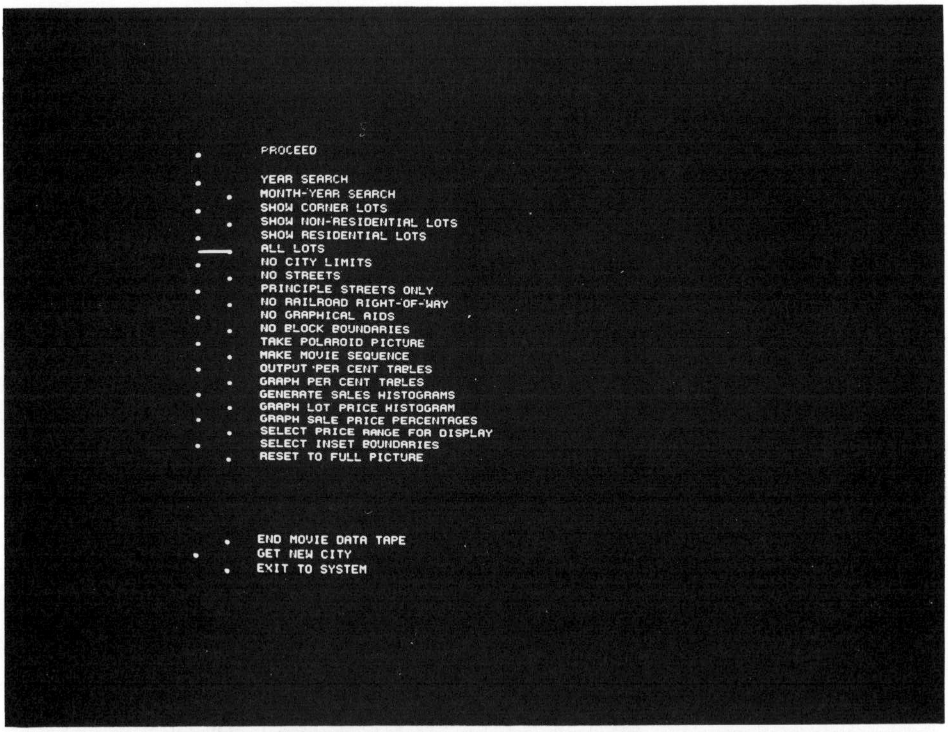

Fig. 3. The display options for the Centralia study as they appear on the face of the CRT (cathode ray tube).

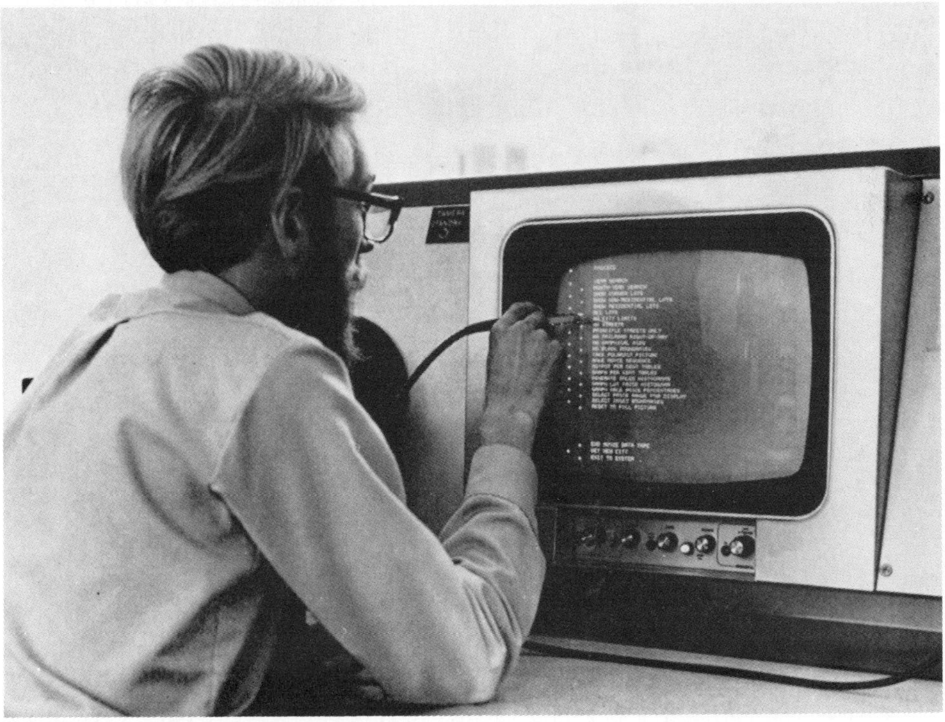

Fig. 4. Operator selecting the option of viewing "all lots."

the mode he wishes by touching the CRT at the desired mode with a light pen (see Fig. 4). Another possible choice is the type of property to be seen, residential or commercial, all lots or just corner lots. Further he can study only properties within a particular price range by touching the price-range option with the light pen and then utilizing the typewriter at the display facility to insert a minimum and a maximum price per square foot. The desired price range appears on the upper edge of the display and is shown in Figs. 5, 6, 7, and 8. An inset routine was developed in order to study a limited portion of the data. After picking "inset" on the original display, a map of the community and a square appear on the CRT (see Fig. 5). Using the light pen the operator can move the square to any location and by touching "larger" or "smaller" can increase or decrease the size of the area to be examined. When "done" is touched the area within the square is presented on an expanded scale so that it fills the full scope-face (see Fig. 6).

Now, the computer begins to present data. First appears a map of the part of the community chosen for study. This map depicts the street plan laid out by the railroad at the founding of Centralia. It is held in a buffer unit in the core of the computer memory. The

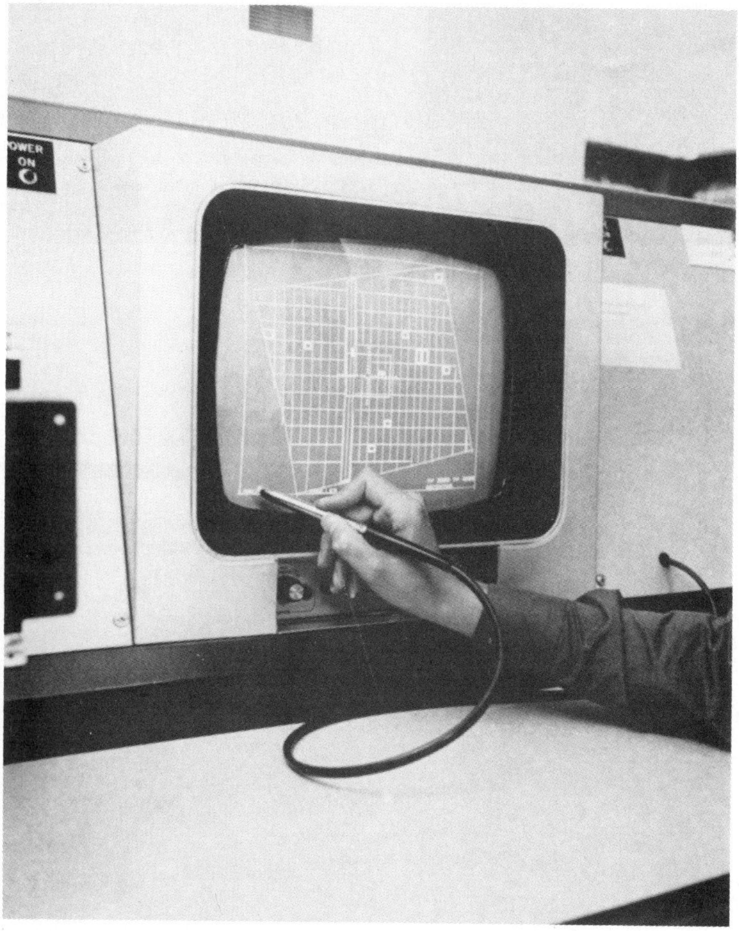

Fig. 5. Operator increasing the size of the inset by touching the face of the CRT with the light pen opposite "larger."

Fig. 6. Operator advancing the date within the previously determined inset area by touching the face of the CRT with the light pen opposite 1858.

screen shows the date, 1858, since year search was selected, the price range of 0–10.00 cents per square foot, and the category of "all lots." When the light pen contacts the screen the date advances and all sales in the chosen category of price and land use which occurred in the intervening period are presented as spots of light. These spots are placed in the center of the lots sold. As the light pen hits the screen again and again, the time counter advances and the additional sales are added to the picture (see Figs. 6, 7, and 8). At any point the operator can pause to record what he has learned. The photographs in Fig. 8 were made at separate points in time with a polaroid camera. Alternatively the whole sequence can be captured in a motion picture (Figs. 20 through 24).

The computer produces this time-sequence plotting by holding in core all the data records including sale price, date of sale, type of property, and x, y coordinates of property. As the data register is advanced by the light pen, the computer looks through its memory for transactions within the time interval which satisfy the criteria set by the operator when he initiated the display. All sales which meet the requirements have their x, y coordinates transferred to a buffer unit in memory. Thereafter these sales remain in the buffer and on the scope until the sequence is completed.

The displays produced in this manner can serve as a powerful tool in visualizing and analyzing patterns of urban growth. For example, Figs. 7 and 8 show the expansion in time of a diamond-shaped pattern set about a town center. The basic diamond is produced because people wanted to live at a minimum possible over-the-street distance from the center of the town. The viewer can also see an anomaly such as the arrested pattern of growth in the lower left-hand corner of the picture.

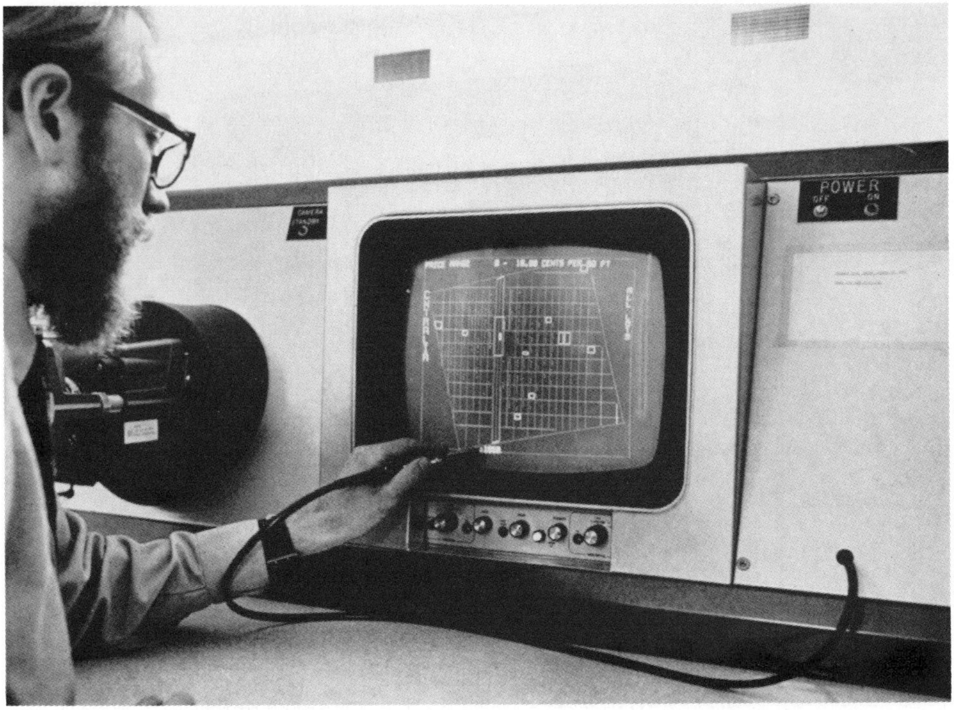

Fig. 7. Operator advancing the date for sales of all lots in Centralia by touching the surface of the CRT with the light pen opposite the displayed date. Notice the diamond shape that land sales have taken up to 1858.

The Centralia study demonstrated the feasibility of our methods and the usefulness of sequential displays for time-series studies. However, improvements in our techniques would be necessary if we were to move from the Centralia work. First of all, we sought to study a larger community with a more complex history of development. Second, we wished to include more kinds of data including all raw and developed land sales, the precise land use at each sale, and the existence of various factors which might have influenced development such as water and sewer service, transportation facilities, schools, employment, and topography. Since all these data could not be fitted into core, we needed new data-handling procedures which could transfer data from tape to core in the course of the display. Furthermore, our Centralia display program, which could add spots of light but not remove them, would clearly be inadequate for the next stage of display. Now we wished to show the whole history of an urban area and we were required to develop methods for adding and subtracting and aggregating data on the displays. A new beginning was clearly necessary.

Data Gathering for Kankakee

For the next stage, we chose Kankakee, Illinois (see Fig. 9) which is the center of a relatively small metropolitan area of 50,000, so that all relevant data could be gathered without exorbitant costs and so that we and our students would have a reasonable chance of understanding its history. However, there were other factors also influencing our selection of Kankakee. Specifically, it has, up to now, been almost entirely self-contained

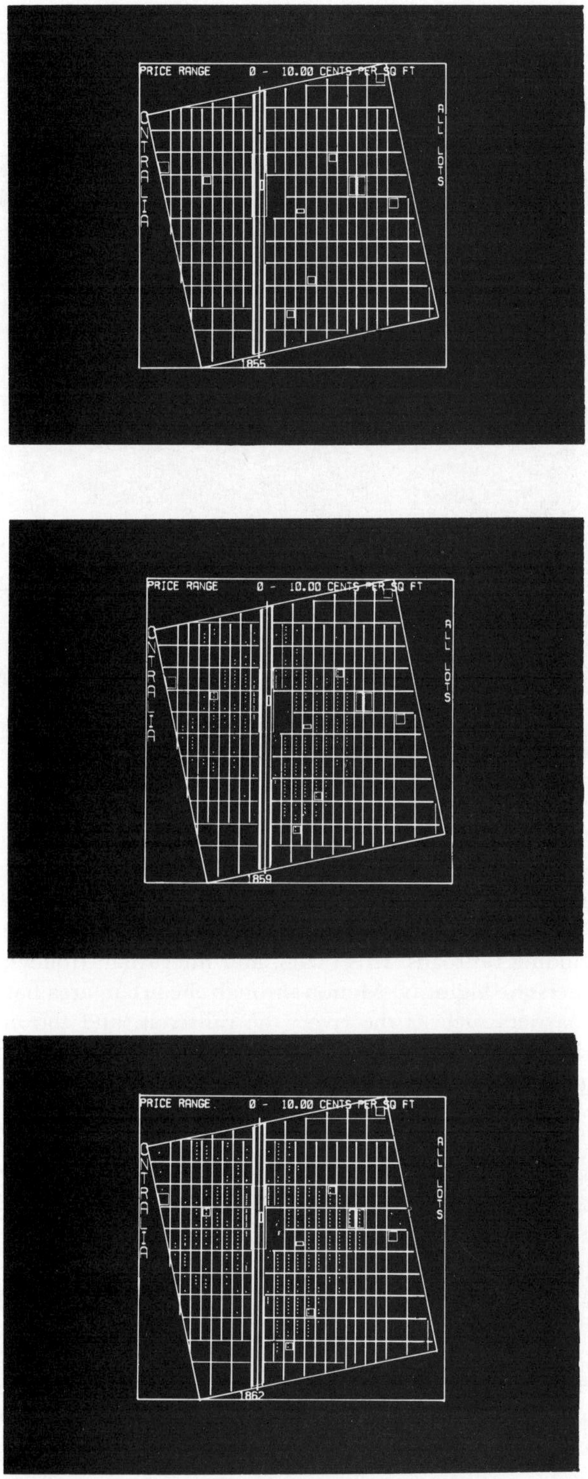

Fig. 8. A three-photo sequence showing land sales in Centralia from its origin in 1855 through 1862. The diamond pattern suggests that accessibility to the town center was an important factor influencing the time at which parcels sold.

Fig. 9. An aerial view of Kankakee in 1965, looking Northwest.

and not strongly influenced by any nearby community. It is an old town for Illinois situated on an interesting site, and has had a long and rich urban life and today possesses a diversity of income, ethnic, and racial groups. The transportation history is also quite diverse including railroads, street cars, and interurban trolleys, a complex street system, and an interstate highway. Motion through the urban area has been modified by the presence of barriers such as the river, the railroads, and the interstate highway. Finally, the new interstate highway (I-57) on the east may be expected to produce significant changes in the city in the near future because of the greatly increased accessibility to Chicago.

The job then, was to gather as much information as possible about Kankakee. We attempted to obtain sales prices for every real estate transaction since the founding of the city in 1853 for every other block in the city. These data were collected from the records in the Kankakee County Title and Trust Company and the County Recorder's office by examining sale prices and tax stamps affixed to the deeds. The history of the street and transportation system was obtained from the recorded plats of the original subdivisions (see Fig. 10) from the city's engineer, and from old maps and aerial views. These last sources are also useful in helping to suggest when homes were built or torn down and whether or not changes in land use took place. In addition we have assembled data through time on the amount and distribution of employment, family size, and the occupational status of the population. At this time we are gathering information on

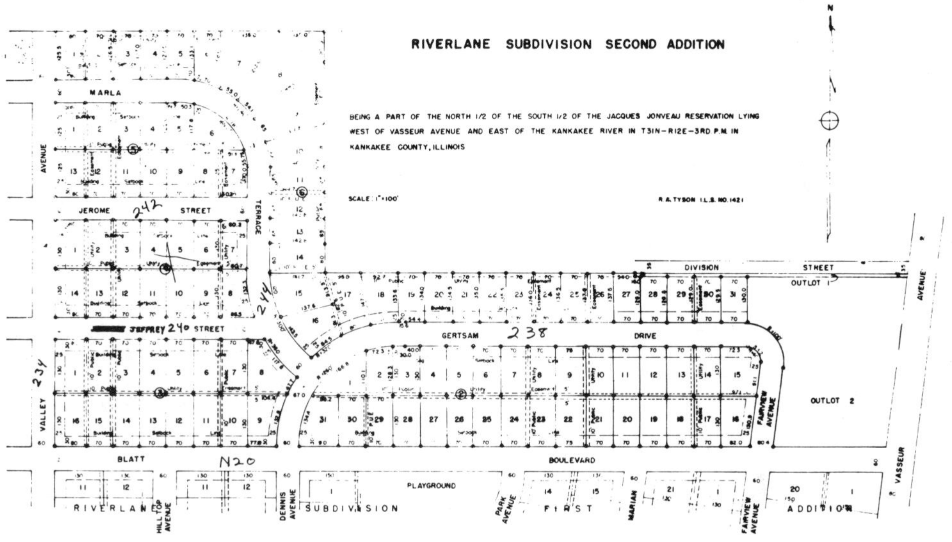

Fig. 10. A typical subdivision plat as recorded in the Kankakee County Recorder's Office.

water, sewer, and educational services which should be in machine-readable form during the fall of 1970.

At the center of the entire data base is price information (sales price or mortgage amount) on approximately 120,000 transactions. Even the data on a single block are quite formidable. Figure 11 shows a computer printout which contains price, deed type, and date of transaction for a single block! In Fig. 12 a portion of this record is shown in readable form.

To use these data, we must determine what is being sold. By consulting maps such as Sanborn insurance maps illustrated in Fig. 13 and aerial views, it is possible to discover the history of each residential structure in the study area. Each structure is given a number (see Fig. 14) and the dates of construction and possible demolition are recorded. The number of the home sold is added to the printout or if the land is vacant, a V is inserted. These notations can be seen in Fig. 12. Finally, these data are added to the original data cards producing an inventory through time of the number of residential structures on each block, and an indication of the price history of all vacant land transactions is prepared and recorded. And finally this stage also includes the construction of an inventory which records how much land on each block is devoted to resdiential, commercial, industrial, and public uses.

The assemblage of this information will permit a detailed examination of how various factors influence dwelling unit and vacant land prices and the rate of change of dwelling unit and vacant land prices.

Presentation of Kankakee Data

In a parallel effort, a dynamic map of the town has been constructed for CRT display which presents railroads, streets, subdivision boundaries, city limits, and the river. Each of these elements is added to the map at the appropriate point in time and removed when the depicted feature changes or disappears. Figure 15 presents the growth of Kankakee in subdivision increments from June 1854 to January 1969.

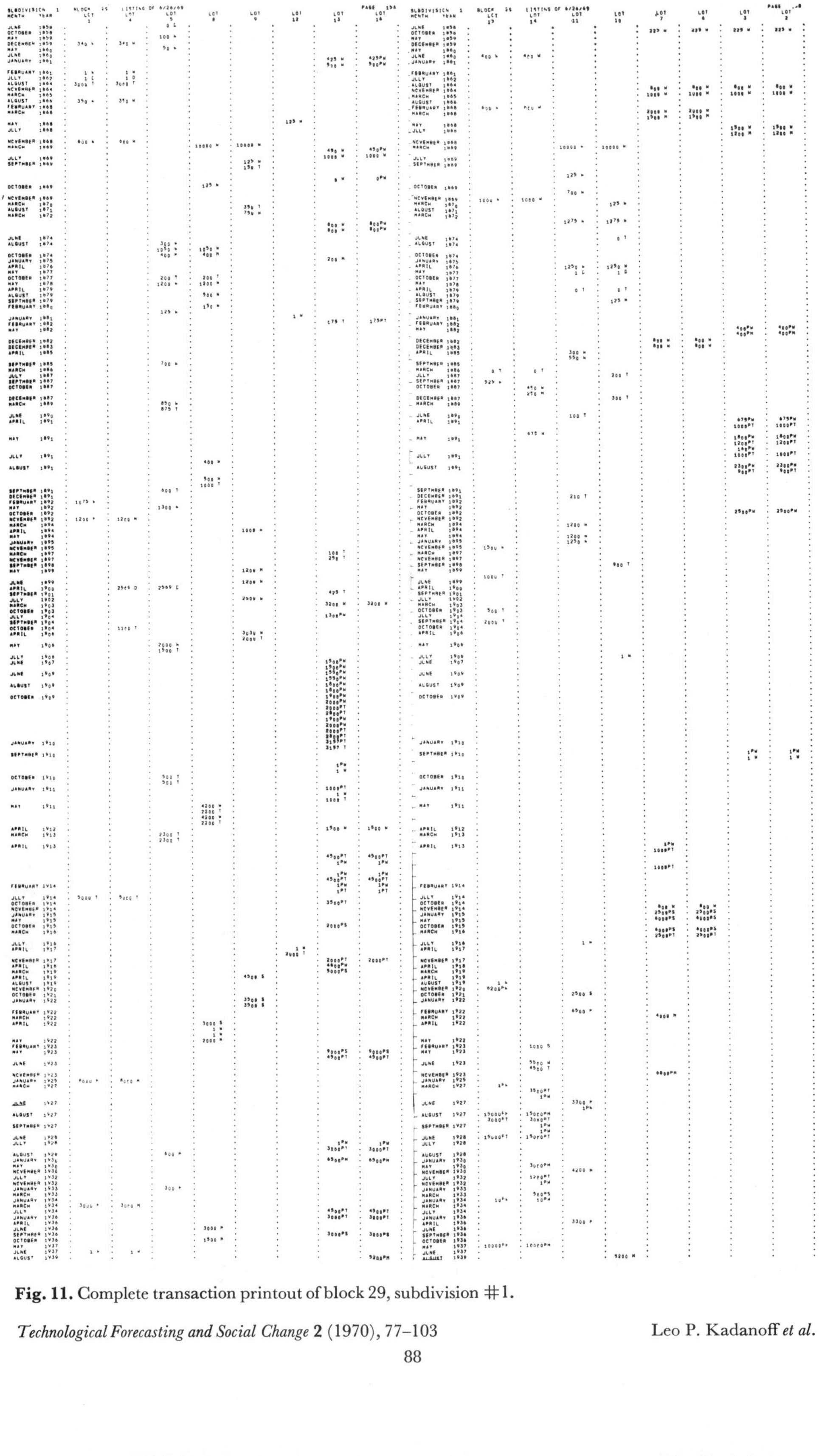

Fig. 11. Complete transaction printout of block 29, subdivision #1.

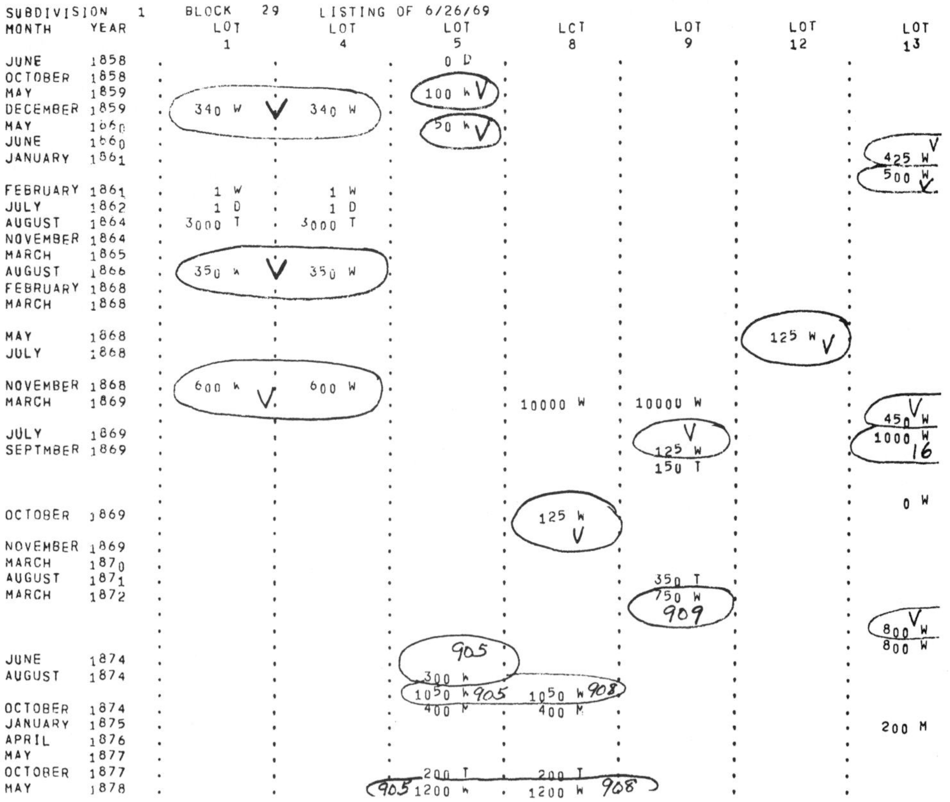

SUBDIVISION 1 BLOCK 29 LISTING OF 6/26/69

MONTH	YEAR	LOT 1	LOT 4	LOT 5	LOT 8	LOT 9	LOT 12	LOT 13
JUNE	1858			0 D				
OCTOBER	1858							
MAY	1859			100 W				
DECEMBER	1859	340 W	340 W					
MAY	1860			50 W				
JUNE	1860							
JANUARY	1861							425 W
								500 W
FEBRUARY	1861	1 W	1 W					
JULY	1862	1 D	1 D					
AUGUST	1864	3000 T	3000 T					
NOVEMBER	1864							
MARCH	1865							
AUGUST	1866	350 W	350 W					
FEBRUARY	1868							
MARCH	1868							
MAY	1868						125 W	
JULY	1868							
NOVEMBER	1868	600 W	600 W					
MARCH	1869				10000 W	10000 W		
								450 W
JULY	1869							1000 W
SEPTMBER	1869					125 W		
						150 T		
								0 W
OCTOBER	1869				125 W			
NOVEMBER	1869							
MARCH	1870							
AUGUST	1871					350 T		
MARCH	1872					750 W		
								800 W
								800 W
JUNE	1874							
AUGUST	1874			300 W				
				1050 W	1050 W			
OCTOBER	1874			400 W	400 W			
JANUARY	1875							200 M
APRIL	1876							
MAY	1877							
OCTOBER	1877			200 T	200 T			
MAY	1878			1200 W	1200 W			

Fig. 12. A portion of the transaction printout of block 29, subdivision #1 showing typical analysis notations.

Unfortunately, we have only a small computer (see Fig. 16), a CDC 1604 which has 32K word memory that is not large enough to contain all the mapping information. To conserve core space, we hold most of the mapping data on magnetic tape arranged according to the date the feature is added. Each piece of mapping data is added to the core as it is needed in the time-sequence presentation and arranged in core according to the time it is to be removed from the map. Then when the feature disappears from the city and the data are no longer needed on the computer map, it is also automatically withdrawn from core.

As in the Centralia display, portions of the city may be examined with the aid of an inset device (see Fig. 17). In addition, particular subdivisions can be selected for special study. When the inset feature is not used, the map may be set to expand or contract automatically to fill the entire screen (see Figs. 18 and 19).

To depict the economic history, spots of light are placed within blocks which contain a specific economic feature. One can show a spot for every block which contains, for example: (a) rapid building of new homes; (b) commercial land; (c) residential overcrowding; (d) rapid deterioration of home values; or (e) sales in a given price range; as well as many other possibilities. From viewing all these displays, we expect to gain a feeling for the kinds of development which occurred in the community and as we become more experienced, our ideas can be checked against the visual image. For example, did land values go up or down in the neighborhood of the new shopping centers? Did

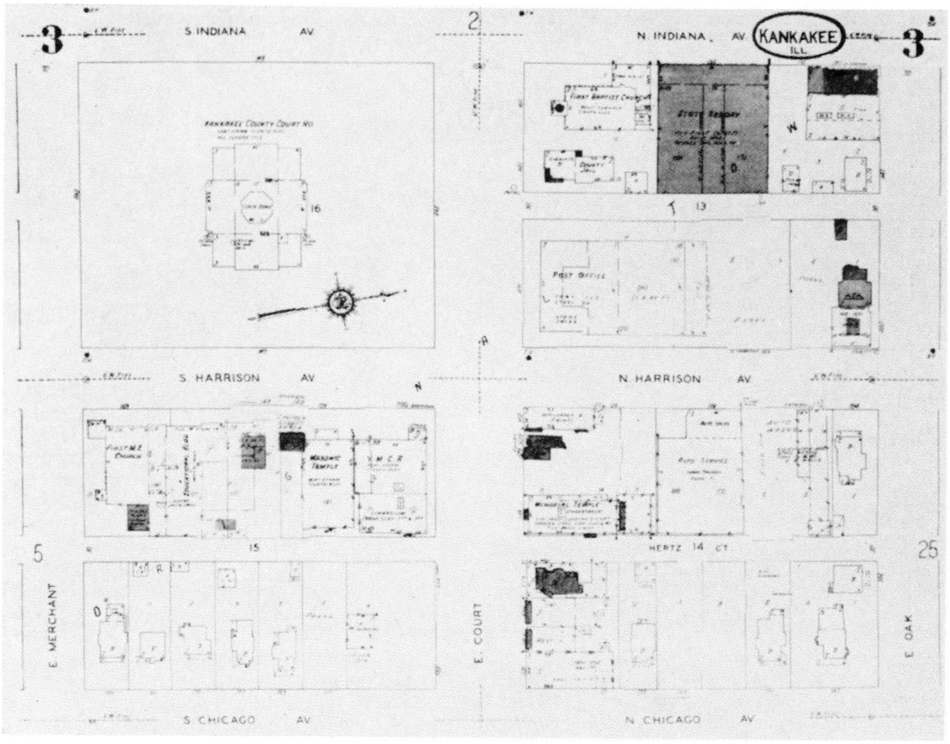

Fig. 13. A typical page from a Sanborn Map. Source: Sanborn Map Company, Pelham, New York.

development in the eastern part of the city precede or follow the extension of sewers? Some ideas may require new programming: for example, to show a spot of light for every block which has both deteriorating home values and new commercial land uses in the neighborhood. In this way, our computer program and our understanding of the community will grow together.

There are three problems which make the data-handling and presentation task for Kankakee very different from the one for Centralia. First, we have very many different kinds of data for Kankakee; land use, structure inventory, price information, streets, schools, etc. Second, we have much more data. It will obviously not fit into memory. Third, we cannot predict before we begin to know what kinds of relationships we shall see and wish to explore.

To overcome the core-storage problem, we must hold data on magnetic tape, feed it into core as it is needed in presentation, and free the storage space as it is no longer needed. As with the map, the best way to arrange for this reading, use, and elimination is to arrange our data sequentially in time on the tape and in the sequence they are to be eliminated within core.

In order to study the large variety of relationships which will arise in our problem, it is quite necessary that the tape records be both flexible and compact. For different types of studies, tapes are required which contain various combinations of the different types of data. It is undesirable to have blank spaces and unnecessary data on the tapes because they will slow down the presentation and take too long to read. Therefore, we have

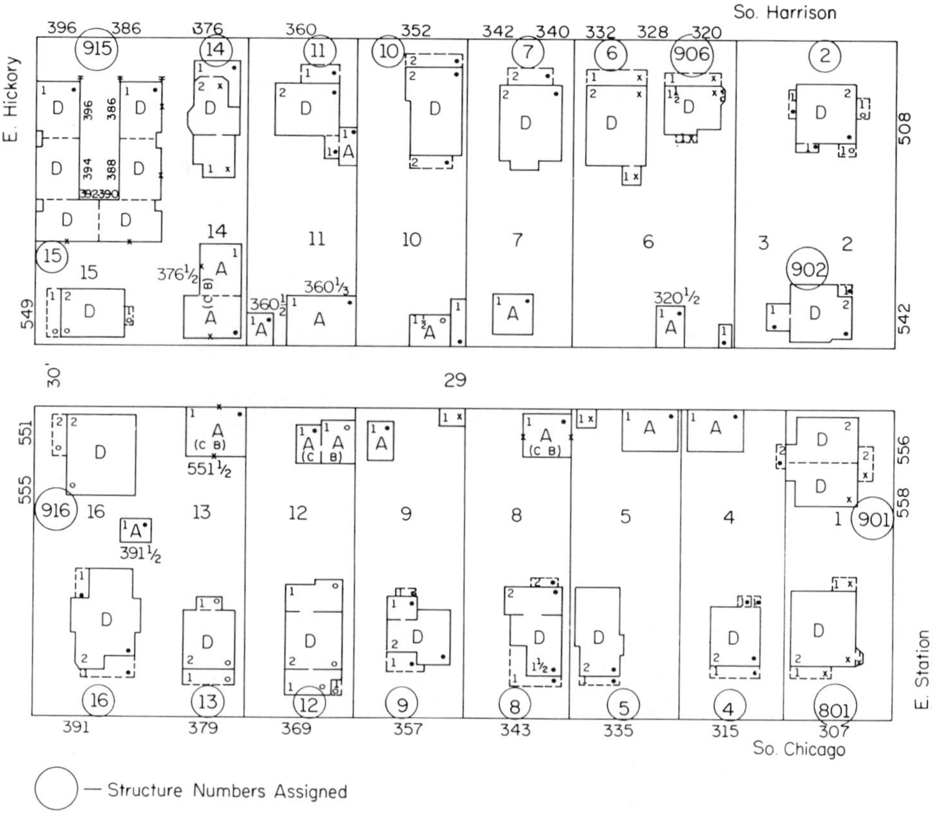

Fig. 14. A typical block in Kankakee with structures numbered for analysis.

constructed programs to permit a variable data format. By using relatively simple instructions, a bit analogous to Fortran, the programmer can select, recombine, read, and write data records which contain different data of variable length and composition. Thus, after a few minutes of programming time and a few more minutes on the computer, the programmer could make up a tape containing, for instance, only the price per square foot for sales of vacant land with date and location, together with the number of residences per acre or block at the time of sale, and a notation to indicate whether there were industrial land uses on the block. Then we could see for example whether industries on overcrowded blocks increase or decrease the price of vacant land. This production of a variable data format and its associated read, write, sort, and merge routines is a complex programming problem. However, a result of the effort expended in this direction is a flexible and quick data-handling facility.

Using these data-handling methods we have the potential for flexible and fast visual presentation of the Kankakee data. The expected result of this visual study is an appreciation of the variables which determine and limit urban growth. Following this stage of our analysis we will then attempt to develop mathematical relationships or "models" which might provide a numerical explanation of that which we have seen. For example, the predicted quantities might be prices for homes and vacant land as well as the extent of residential development. These variables or their rates of change in time might be

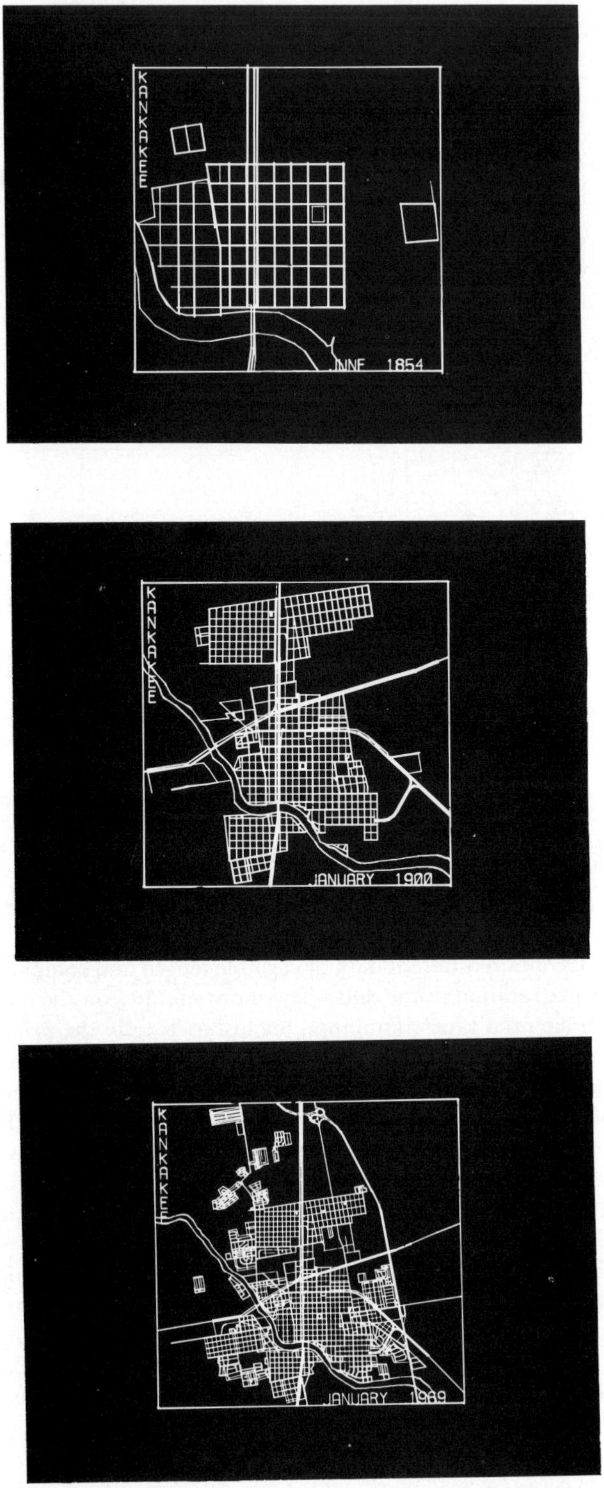

Fig. 15. A photo sequence made with the polaroid camera depicting the growth of Kankakee in subdivision increments from the original platting of the town in January of 1854 through January of 1969. Notice how each display is automatically scaled to fill the screen of the CRT.

Fig. 16. The computer room of the Coordinated Science Laboratory showing the display unit on the left, tape banks, console, printer, and computer in the far corner.

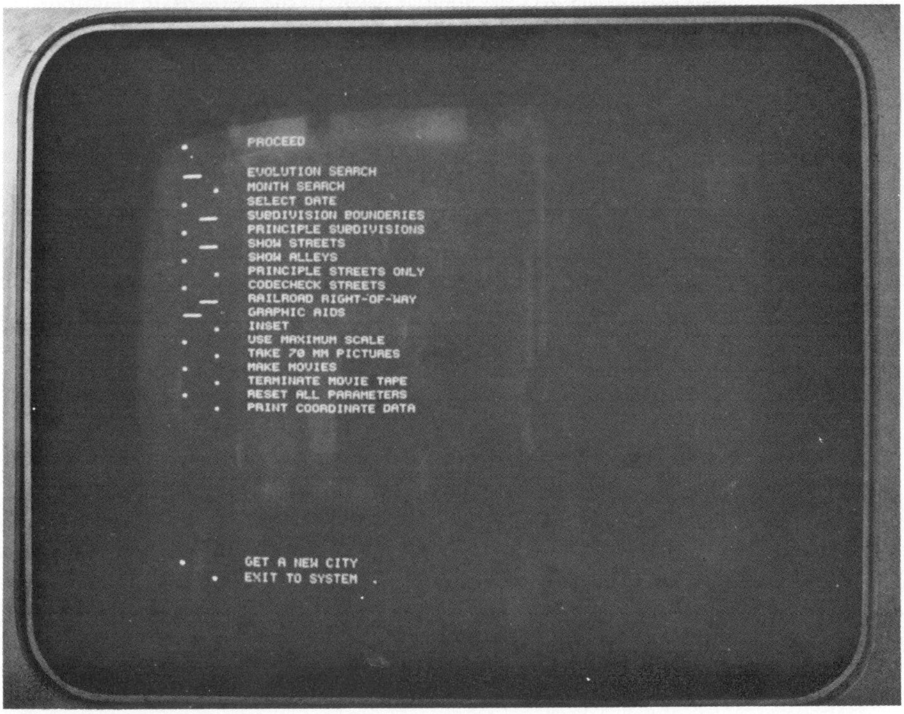

Fig. 17. Kankakee options as they appear on the screen of the CRT.

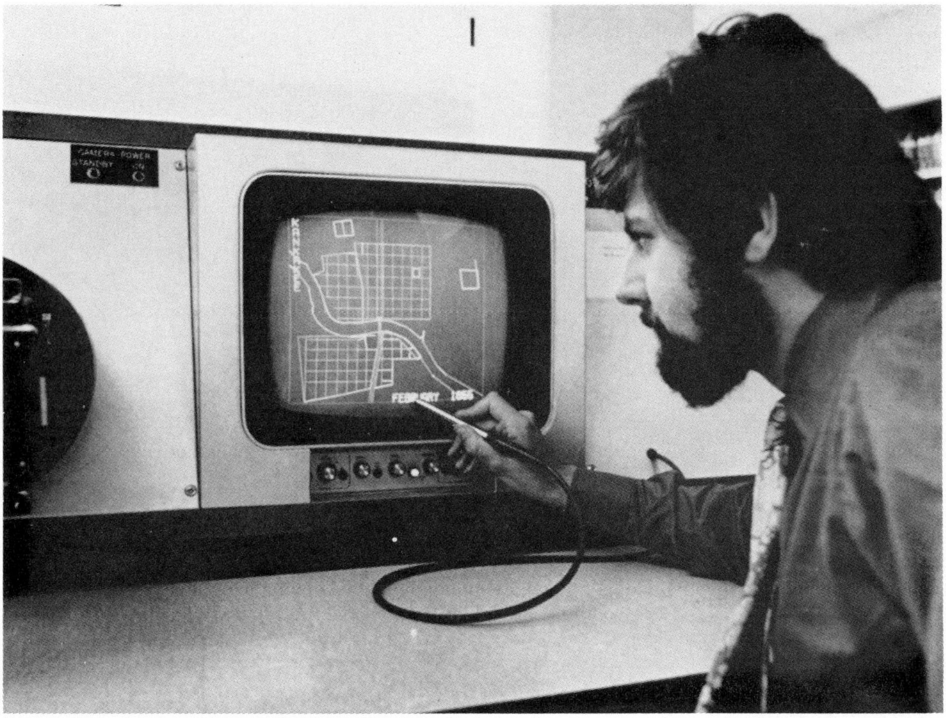

Fig. 18. An operator examining early subdivision development in Kankakee.

predictable in terms of causative variables such as prices in the immediate neighborhood, time-to-travel to the center of town, quality of schools, presence of sewers, topographic features. This effort then seeks to predict the cities' growth.

While the model is being constructed and tested, the display facility would still be in use. Using a spot of light the computer could show a particular sale or block which disagreed with the model's predictions. With such a visual display of errors, we would have a guide to the correction and improvement of our mathematical relations.

In the end, we would have a model which would have a known reliability in the context of Kankakee. We would also have a considerable understanding of which variables proved significant in determining the growth through time of this community. At that point, it would be appropriate to see whether the knowledge gained in the Kankakee study could be extended and applied to other communities. Hopefully, our deeper understanding of the significance of the different variables which enter the problem could make our analysis of another city more speedy and less complex.

A Hypothetical Example: The Location of a Regional Expressway

As mentioned at the beginning of this article, the ultimate objective of our study is to provide a decisionmaker with the means, both substantive and methodological, to improve his ability to make a choice among alternatives. An example of what such an improvement might be like was developed in a movie prepared by the Coordinated Science Laboratory in conjunction with the Ford Motor Company which depicts the regional development consequences of alternative locations for an expressway. The movie was

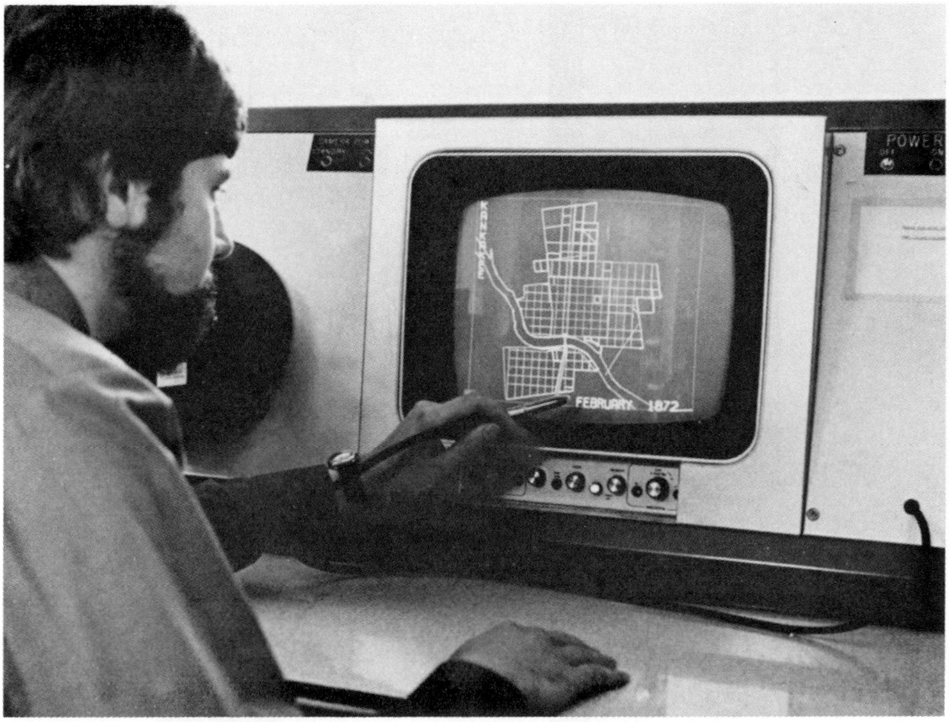

Fig. 19. An operator examining subdivision growth of Kankakee up to February, 1872. The display always fills the screen as new subdivisions are added or older ones subtracted when time is advanced by touching the light pen to the face of the CRT opposite the date.

designed to portray how a policymaker might solve a problem using on-line computer graphics in an interactive mode. Consequently, the demonstration simulates both a regional growth model and the way in which an analyst would interact with the computer using a light pen or a typewriter.

Our example assumed two major conditions:

(a) that there is a reliable regional growth model which will spatially predict and distribute population as a function of changes in accessibility; and

(b) that computer graphics techniques exist which will permit an analyst to input information into the model by drawing with a light pen on the face of the CRT.

Further we constructed the following hypothetical situation:

(a) a region with a relatively level surface containing a distribution of varying sized urban places linked together by state and country roads and an interregional railroad passing through the major urban centers;

(b) some arbitrarily located city boundaries;

(c) that the state highway department intends to construct an interstate highway through the region, and that there are two possible locations for the facility; and, finally,

(d) that the ultimate choice of the highway planners depends exclusively on the criterion of how the new expressway will influence the spatial distribution of the population in the region.

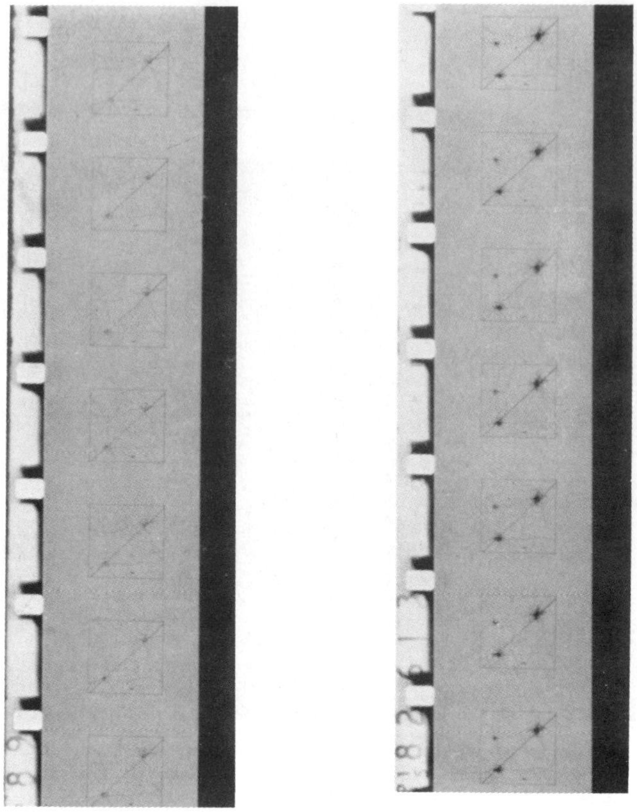

Fig. 20. A film strip from the Ford movie showing the region and its growth assuming that an expressway is not constructed.

The problem, then, is where should the facility be located. The existing structure and development of the region is shown in Fig. 20, with each spot of light corresponding to a quantity of population. To this the computer adds and distributes spots of light (population (corresponding to the amount of growth that takes place each year. The growth period extends from 1950 to 2001. Our assumed regional growth model distributes spots of light or new growth using a random number of generator with development density decreasing exponentially with increases in distance from urban centers. From time to time, new industries have been arbitrarily added to and distributed in the region, and are represented by the symbol I. In addition, city boundaries have been enlarged to encompass new growth around urban centers. These changes were also arbitrarily decided upon. Taken together, our hypothetical model is obviously and at best a very crude simulation of an actual regional development pattern. Nevertheless, it is adequate enough to permit the computer to display on the CRT what a regional growth pattern might look like at different times in the future.

Figure 20 shows changes and growth in the region assuming that an expressway is not built. The operator, then, in Figs. 21 and 22 simulates drawing on the face of the CRT the location of the two alternative expressways and Figs. 23 and 24 present the development

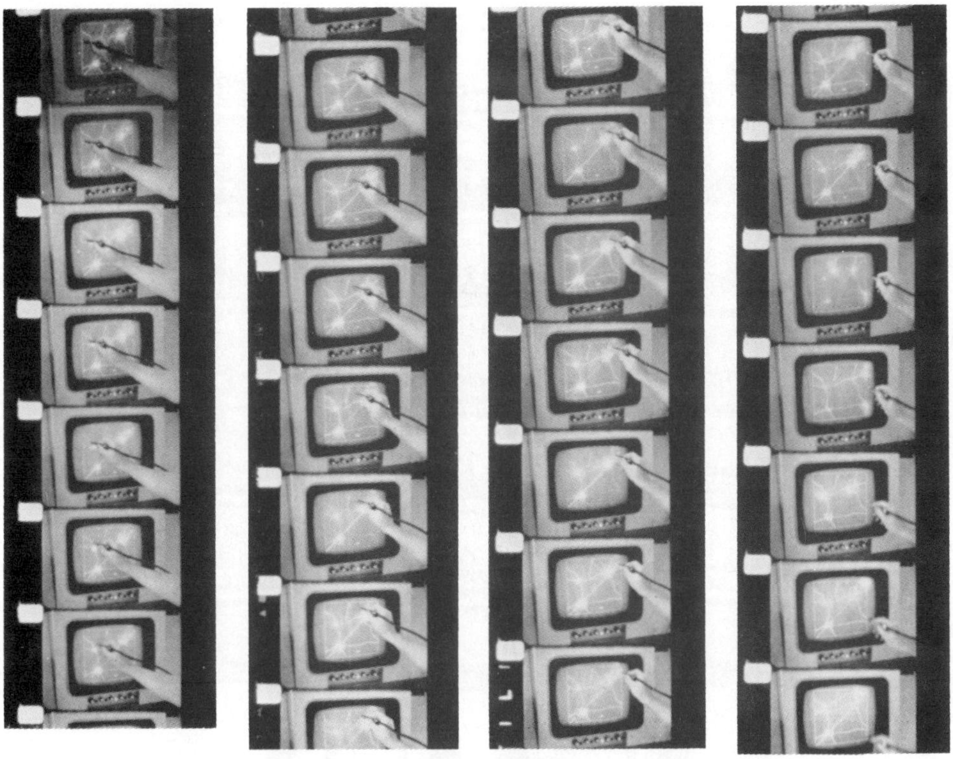

Fig. 21. A film strip from the Ford movie showing an operator simulating the drawing of the expressway through the northern portion of the region. The operator simulates drawing the location of the expressway from left to right and the bottom frames show the computer drive display of the located road.

consequences associated with each possibility. Figures 25 and 26 are enlargements of selected years from the film sequences. The operator is now able to view and analyze the various development patterns and presumably reach a decision as to which is most satisfactory.

This hypothetical example illustrates how an analyst or a planner could interact with a computer to explore a number of locational alternatives. Although a region was used for the simulation, the principles could be applied to an urban area. Consider a planner in Kankakee trying to decide simultaneously the consequences of locating a new school, extending sewer lines, improving some city streets, building a bypass expressway to the west, renewal of a deteriorated area, and building a new shopping center on the east. These elements as well as others could be input into the existing structure of the city using a light pen or a typewriter and then using a calibrated growth model the computer could quickly present what the city would look like as it evolved through time. Obviously, the nature of the problem could be, and probably would be, much more complex if one introduced such things as changes in interest rates or sequential capital improvement programs. Nevertheless, if a powerful theory is available, the analyst could examine a wide range of differently structured alternatives and with this knowledge our chances of designing an improved living environment would be greatly increased.

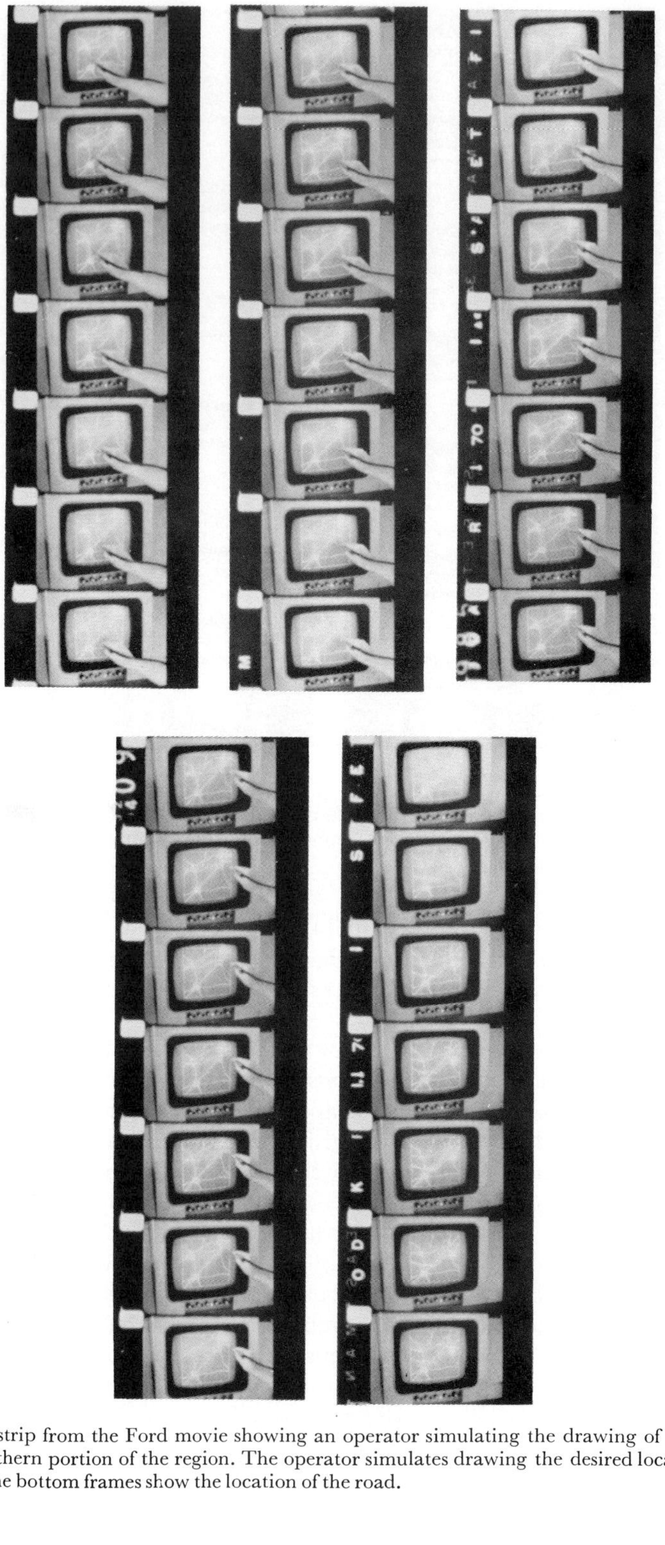

Fig. 22. A film strip from the Ford movie showing an operator simulating the drawing of the expressway through the southern portion of the region. The operator simulates drawing the desired location of the expressway, and the bottom frames show the location of the road.

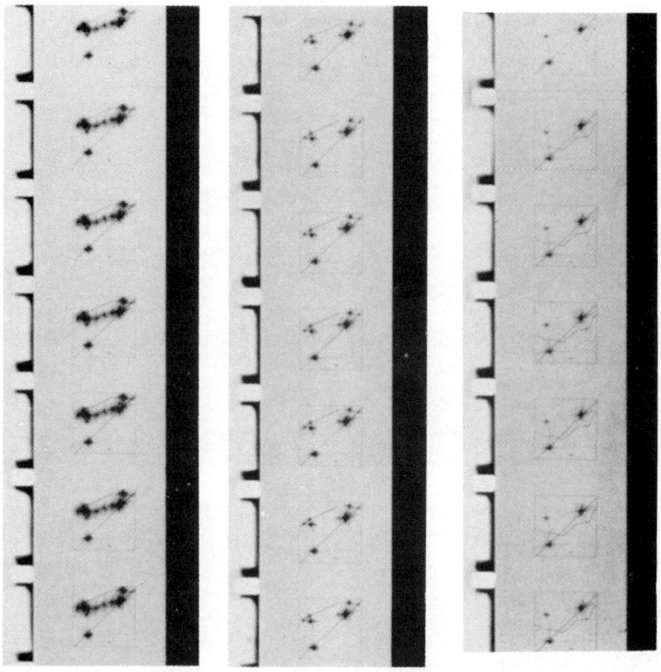

Fig. 23. A film strip from the Ford movie depicting the anticipated population distribution assuming that the northern alternative is constructed.

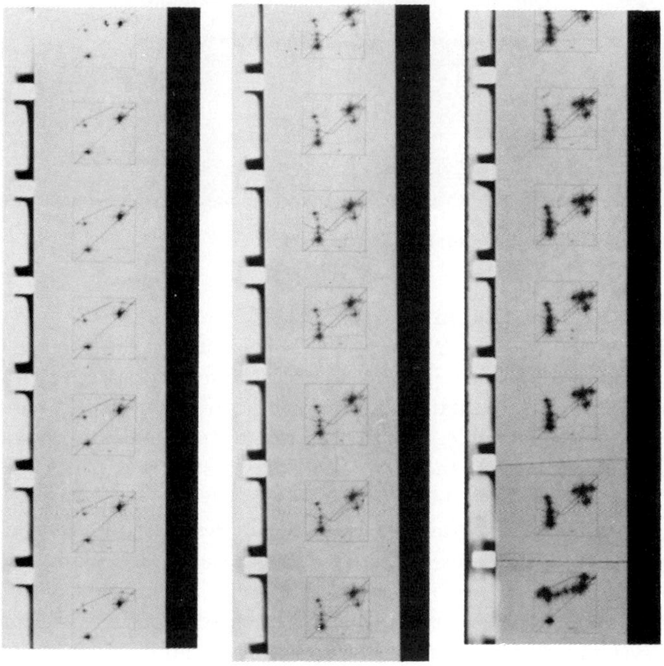

Fig. 24. A film strip from the Ford movie depicting the anticipated population distribution assuming that the southern alternative is constructed.

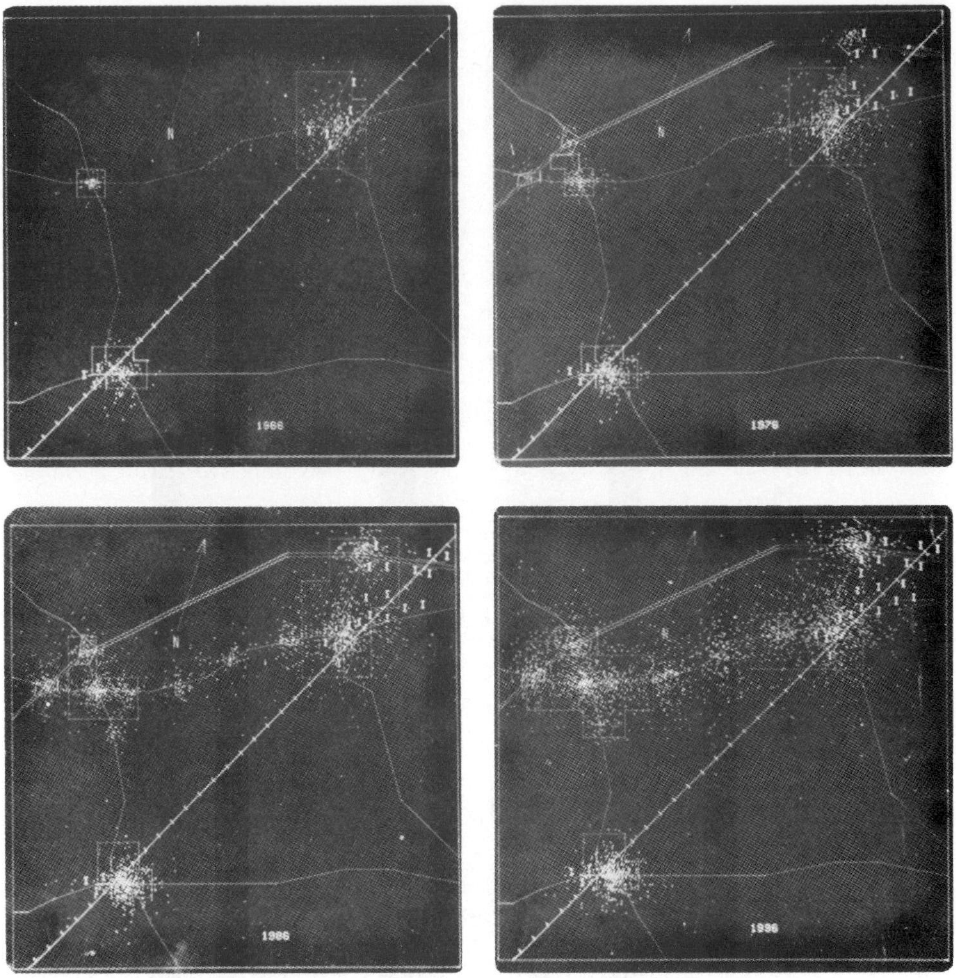

Fig. 25. Enlargement of selected years for Northern alternative.

A Technical Note on the Display System

The display system (see Fig. 2) used in this project was designed and built at the Coordinated Science Laboratory in order to explore the graphic representation of a wide variety of computer-generated data. The major objectives guiding its construction were that the system should be flexible, easy to program and operate, and require a minimum of computer memory space. At the time the display system was constructed it possessed a number of unique qualities which distinguished it from other contemporary display facilities either commercially available or in use in research installations. Since 1967, when this display system became operational, there have been many improvements in graphic output facilities. However, it is worth emphasizing those aspects of our system that are unique or of unusually high quality. First, and perhaps the most significant feature of the system is that it possesses the capability of an electronic resolution of 4096 positions along both the x and y axes for twenty-four bit words with twelve bits for x and

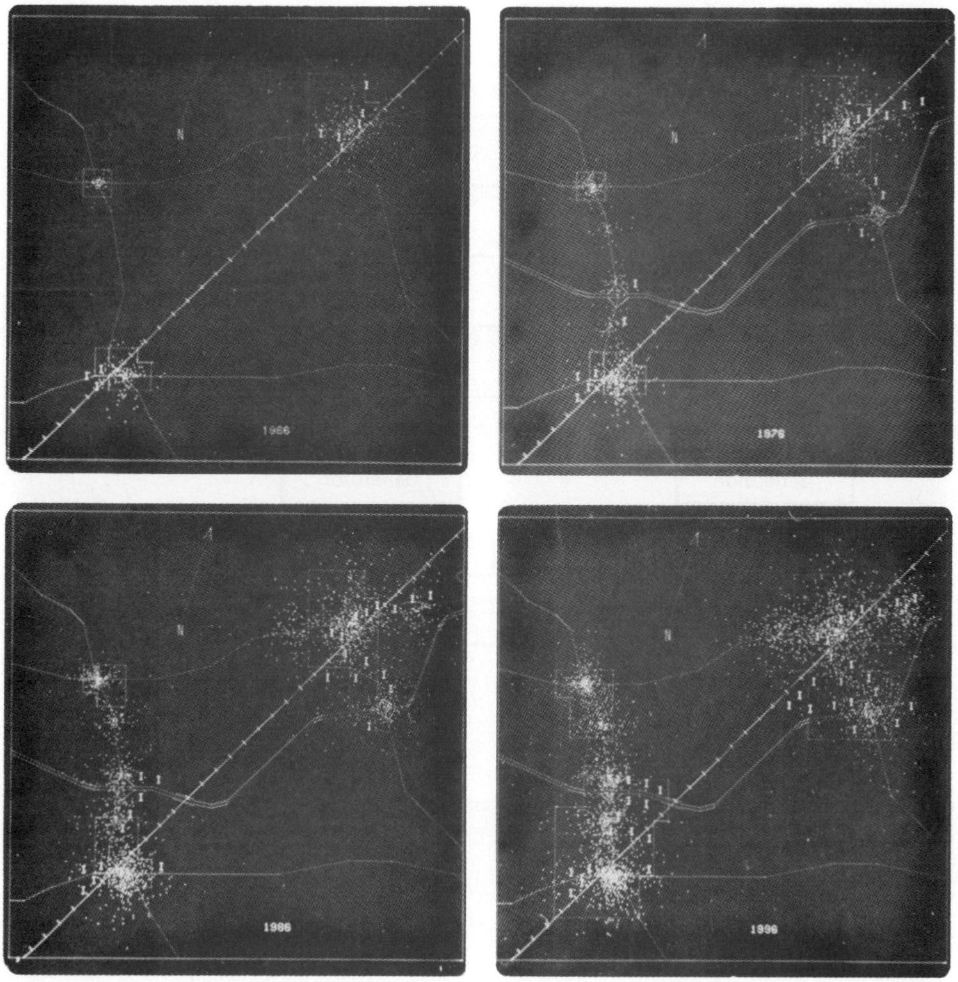

Fig. 26. Enlargement of selected years for Southern alternative.

twelve bits for y. Unfortunately, the CRT is not of sufficient quality to exploit this capability, and we actually operate with 2084 resolvable points in the center of the screen. Nevertheless, our operational capability, let alone our electronic potential, compares quite favorably with other systems which usually operate with a resolution of 1024×1024 positions. Second, we have developed a very flexible reproductive system which can accommodate a variety of camera units including: a Mitchell 16-mm animation camera; a 70-mm roll film camera; a 35-mm camera; and, a polaroid camera. The photographs included in this article were made using all but the 35-mm camera. Third, the user of the system has available eight operating modes, with the potential of expanding to sixteen. The modes being used at this time include: point plotting; graph plotting using x as the independent variable; graph plotting using y as the independent variable; a matrix mode which may be used to construct special symbols; line segments; continuous line generating; character typing; and a TV raster for shading. Fourth, by changing the interface unit the

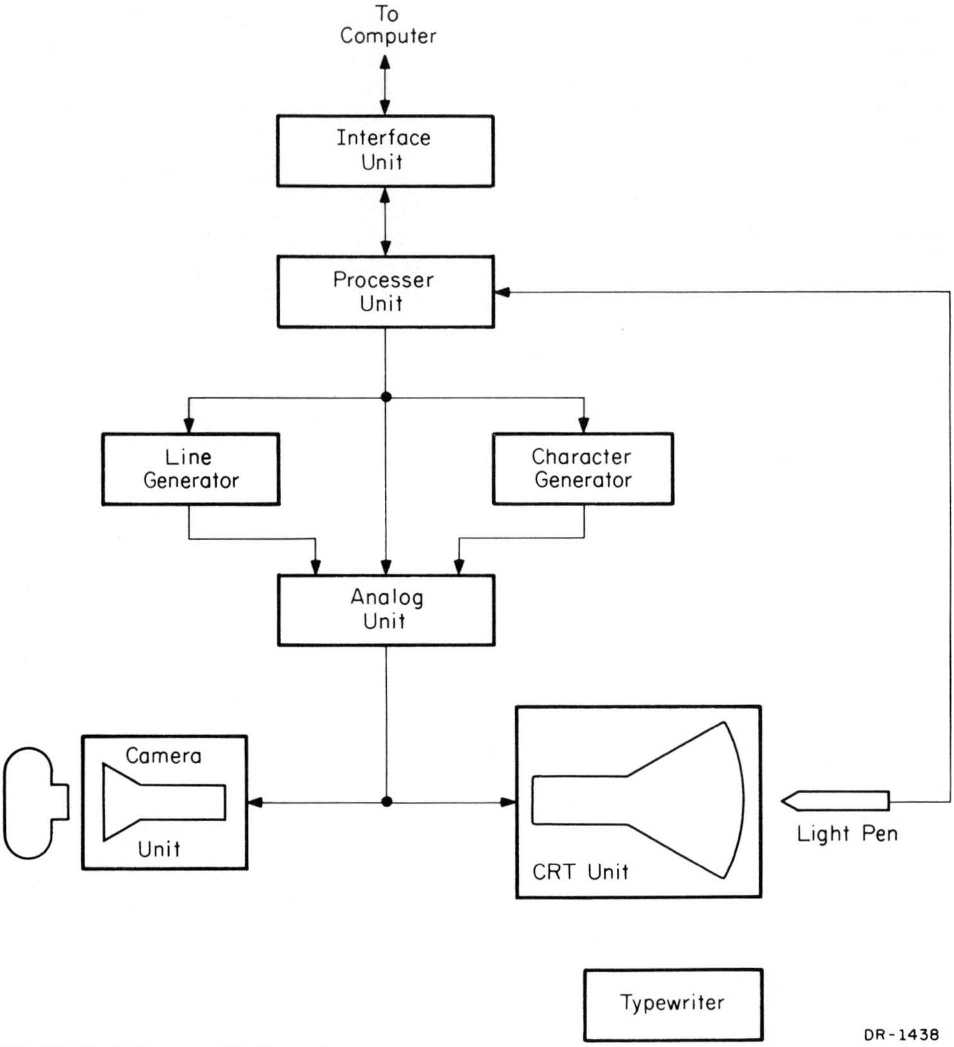

Fig. 27. Block diagram of display unit.

system can be adapted to most computers. Since this is the only portion of the display that is computer dependent, the interface unit has been designed to operate with our CDC 1604 computer. Fifth, the CRT unit is a Fairchild Type 737A Large Screen Indicator containing an electrostatically deflected 17-inch rectangular cathode ray tube and x and y deflection amplifiers with a full output band width of 1 megahertz. Slaved to this unit is a magnetically deflected, electrostatically focused 5-inch CRT for high resolution (1000 lines per inch) photographic recording. Both moving pictures and stills may be made with this camera unit by either manually or automatically operating the shutter unit by the computer program. Sixth, the system is designed to operate at speeds of 333,000 points per second, 50,000 characters per second, and 16,000 lines per second. Seventh, and finally, the system employs two means of communicating with the computer—by typewriter or by light pen. The display typewriter can be used as a substitute for the 1604 typewriter and can function either as a keyboard input or as an output device for producing printed copy. The light pen interacts with the computer program directly from the face of the CRT by detecting light emanating from a point on the CRT and, after assembling the word, mode, and control bits in the 24-bit light pen register, passes this information to the computer.

In general the display system consists of eight functional units (see Fig. 27) and includes the following: an interface unit which controls the exchange of information between the display system and the computer; a processor unit which performs all arithmetic and logical operations on the data being displayed; an analog unit which converts data from digital to analogue form for the display; a line generator; a character generator; a CRT unit which provides the visual display viewed by the operator; a camera unit for the photographic recording of the data; and the input-output devices for the operator control.

The engineering and programming details of the display system are contained in a report published by the Coordinated Science Laboratory and referenced at the end of this article.

Some Selected References

1 Alonso, William *Location and Land Use*. Cambridge, Mass.: Harvard Univ. Press, 1964.

2 Bouknight, Wendell J. *Preliminary Users Manual, CSL6 and CSL7*. Urbana, Ill.: Coordinated Science Laboratory, Univ. of Illinois, December, 1967.

3 Brigham, E. F., *A Model of Residential Land Values*, Memorandum, RM-4043-RC. Santa Monica, California: The RAND Corporation, August, 1964.

4 Chapin, F. Stuart, and Shirley F. Weiss, *Factors Influencing Land Development*. Chapel Hill, N.C.: Institute of Research in Social Science, Univ. of North Carolina, August, 1962.

5 Delaware Valley Regional Planning Commission, *The Construction of an Urban Growth Model*, DVRPC Plan Report # 1, Technical Supplement, Volume A.

6 Donnelly, Thomas G., F. Stuart Chapin, Jr., and Shirley F. Weiss, A Probabilistic Model for Residential Growth. Chapel Hill, N.C.: Center for Urban and Regional Studies, Univ. of North Carolina, May, 1964.

7 Highway Research Board, *Urban Development Models*. Special Report 97, Proceedings of a Conference Held June 26–30, 1967, Publication 1628. Washington, D.C.: National Academy of Sciences, 1968.

8 Hoyt, Homer, *The Structure and Growth of Residential Neighborhoods in American Cities*, U.S. Federal Housing Administration. Washington, D.C.: U.S. Government Printing Office, 1939

9 Kain, John, *The Journey to Work as a Determinant of Residential Location*. California: The RAND Corporation, 1700 Main Street, Santa Monica, 1961.

10 Kilbridge, Maurice D., Robert P. O'Block, and Paul V. Teplitz, A Conceptual Framework for Urban Planning Models, *Management Science*, **15,** (February 1969), B-246–B-266.

11 Knowlton, Kenneth C., Computer-Produced Movies, *Science*, **150** (November 26, 1965), 1116–1120.

12 Muth, Richard F., *Cities and Housing, The Spatial Pattern of Urban Residential Land Use*. Chicago, Ill.: Univ. of Chicago Press, 1969.

13 Schlager, Kenneth J., *A Mathematical Approach to Urban Design—A Progress Report on a Land Use Plan Design Model and a Land Use Simulation Model, Technical Report No. 3*. Waukesha, Wis.: Northeastern Wisconsin Regional Planning Commission, January, 1966.

14 Stifle, Jack, *A Cathode Ray Tube Display*, Report R-357 Urbana, Ill.: Coordinated Science Laboratory, Univ. of Illinois, June, 1967.

15 Sutherland, Ivan E., Sketchpad: *A Man-Machine Graphical Communication System*, Lincoln Laboratory Technical Report No. 296. Lexington, Mass.: MIT, 1963.

16 Thornhill, D. E., *et al.*, *An Integrated Hardware-Software System for Computer Graphics in Time Sharing*, Electronic Systems Laboratory Report No. ESL-R-356. Cambridge, Mass.: MIT, 1968.

17 Traffic Research Corporation, *Reliability Test Report: Empiric Land Use Forecasting Model*. (Prepared for Boston Regional Planning Project, 219 East 44th Street, New York.) January, 1969.

18 *Visual Information Display Systems: A Survey*. NASA SP-5049. Philadelphia, Pa.: Auerbach Corp, 1968.

19 Warner, Sam B., *Streetcar Suburbs, The Process of Growth in Boston*, **1810–1900**, Cambridge, Mass.: Harvard Univ. Press, 1962.

20 Wingo, Lowdon, Jr., *Transportation and Urban Land*. Washington, D.C.: Resources for the Future, 1961.

April 10, 1970

From simulation model to public policy:

An examination of Forrester's "Urban Dynamics"

by

LEO P. KADANOFF
Brown University
Providence, Rhode Island

LEO P. KADANOFF was educated as a physicist, but since 1965, his interests have been turning more and more toward problems of city growth. He was educated at Harvard, where he got his Bachelor's, Master's, and PhD degrees in theoretical physics. He then studied for a year and a half at the Bohr Institute for Theoretical Physics in Copenhagen. During his subsequent tenure at the University of Illinois (1962-1969), his research interests included the theory of solids, particularly behavior near phase transitions, and the use of urban growth models. He was also involved with computer display of urban growth patterns at the Coordinated Sciences Laboratories at the University of Illinois. Since 1969 he has been a professor at Brown University and has maintained in approximately equal measure an interest in solids and cities.

ABSTRACT

Forrester's model is described and critically analyzed with a view to understanding the relationship between his conclusions and his normative scheme. He claims that his model produces "counterintuitive" results; it is argued here that the main results follow directly from his goals. An alternative method of evaluating the results of the model is proposed, based upon the model's calculation of the city's attractiveness for various groups. This alternative normative scheme leads to quite different conclusions from those reached by Forrester.

INTRODUCTION

In the book *Urban Dynamics*[1] Jay W. Forrester constructs a simulation model of urban growth and then utilizes this model to assess and evaluate a variety of possible strategies for public policy. This work is provocative in several respects. First, computations based upon the model are used to reject as harmful or of mixed value a variety of the traditional "liberal" schemes for city improvement, including provision of jobs for unskilled workers, job training to increase the skills of the unskilled, financial aid for the city, construction of low-cost housing, and income maintenance schemes. Concurrently, Forrester presses for some policies which have considerably less appeal for the liberal thinker, including discouragement of the construction of housing for workers, destruction of low-cost housing, and encouragement of industrial growth to the further detriment of housing.

Even more provocative than these specific conclusions, however, is his explanation of why he reached the "result that past programs designed to solve urban problems may well be making matters worse...while policy changes in exactly the opposite direction from present trends are needed if the decaying inner city is to be revived."[2] According to Forrester, he has gotten better results than his predecessor because planners and public policy makers have applied intuitive reasoning to the complex system that is a city. For this reason, their proposed policies turn out to be palliatives rather than cures. "With a high degree of confidence we can say that the intuitive solutions to the problems of complex social systems will be wrong most of the time. Here lies much of the explanation for...troubles of urban area."[3]

To reach a more effective treatment of city problems, Forrester proposes that we analyze them with the aid of the diagnostic techniques provided by simulation models. Public policies can be mathematically tested by working out the models and seeing all of the policies' effects, intended and unintended. In this way, one can reach conclusions, unhampered by the defects and perils of intuitive thought.

But, as Moynihan has pointed out,* this point of view raises perplexing difficulties for planners, public policy makers, and ordinary citizens. Must we all be experts in systems analyses before we can make intelligent conclusions about public programs and policies? Must we train all planners** and policy makers in computer programming so that they can avoid the necessary errors of "intuitive thinking?" Clearly these questions bear very seriously upon the educational experiences we propose for our future experts and our ordinary citizens.

To study this point, we shall examine Forrester's work in some detail and draw upon some earlier criticism.[5,6,7] The main conclusions of this paper are that:

(a) The main policy recommendations of *Urban Dynamics* are in no sense counterintuitive; they follow directly from Forrester's implied normative scheme and intuitive thought.

(b) Some of the apparently counterintuitive features of this book result from questionable representations of urban dynamics and incorrect representations of proposed public policies.

Despite these criticisms of Forrester's conclusions, I would argue that his model-making is so brilliant and beautiful that his ideas are certainly worthy of examination and further development. I would reject the conclusions, but accept the model as an appropriate basis for further work.

THE MODEL: FROM BASIC VARIABLES TO POLICY CONCLUSIONS

Forrester reaches his conclusions via a five-step process. First, he isolates a few basic variables which describe the social and economic composition of the city. Second, he writes down equations which describe the "natural" city development. These equations tell how the values of the variables at a given point in time determine their values at a later time. Third, public policies are introduced as modifications in these equations. Therefore, Forrester can as the fourth step find the composition of the city which results from each of the proposed policies or no policy at all. In the fifth and final step, he compares the resulting composition with his conception of a desirable city and thereby chooses the policies he would like to recommend.

The basic variables are chosen with an eye to city problems: the existence of large slum areas, the unemployment caused by industrial flight from the city, and insufficient tax revenue for city needs. The variables chosen are:

(1) The numbers of people in various socioeconomic groups, namely:

  (a) Management and professional workers
  (b) Skilled workers (called Labor in the model)
  (c) Unskilled workers (called Underemployed in the model)

(2) The number of acres of housing devoted to each of the above groups.

(3) The number of acres devoted to business and industrial uses. Maximum economic activity occurs in the newer areas, called "New Enterprise." As the enterprises age to "Mature Business" and then to "Declining Industry," the economic activity per acre declines.

(4) Taxes. Here there are two important variables, the taxes needed and the actual taxes collected. The taxes the city needs to collect are assumed to be proportional to the number of people in the various social and economic groups, with the management and professional people requiring the least tax expenditure and the underemployed, the most. The actual tax collected usually lies below the taxes needed because the city can only respond imperfectly to an increase in its tax needs.

(5) Land. The city is assumed to have a fixed area of 100,000 acres. Each unit of enterprise and housing subtracts from the pool of land available for further development.

Forrester's city is then a fixed land area, like an island, containing people, housing, and enterprises. It has a uniform tax rate. All the potential jobs and workers lie within this fixed area. Of course, this island-city is a poor representation[9] of either our central cities (which do have a fixed area but include many jobs filled by suburban workers) or of our metropolitan areas (which are continually growing). For this reason, this model does not include the effects of city-suburb interactions and in particular leaves out the influence of suburban growth upon the central city.[10]

In the model, the only interaction between the city and the outside world arises through the migration of people into and out of the city. Of course, the model includes the fact that a city which is more attractive for a given type of worker will have more immigration and less emigration of that group. Thus, the model includes the idea that--all other things being equal--a city which is more attractive for unskilled workers will tend to have more unskilled workers.

This point is important in understanding Forrester's conclusions, because his normative scheme seems to be one in which a "healthy" city contains relatively few unskilled workers. Forrester does not devote much attention to his goals, apparently because he does not consider them to be very controversial. Instead, he focuses attention upon the model's predictive methods. "The approach presented in this book is suggested as a method for evaluating urban policies *once the proposed dynamic model or a modification of it has been accepted as adequate.*"[11]

However, a careful reading does indicate the goals implicit in Forrester's work. These include the "minimization of the average per capita tax rate"[12] and "to diminish the population share of the underemployed."[13]

Given this point of view, the trend of Forrester's policy conclusions becomes obvious. Any policies which will make the city more attractive to unskilled workers will be classified under *Failures in urban programs,* because of this normative framework. Under this category we find the provision of jobs for the unskilled, the provision of housing for them, a tax subsidy, and also job training to increase their skills. All these programs draw the unskilled to the city and hence "fail" in Forrester's terms. On the other hand, he applauds policies designed to force out the unskilled. His most preferred scheme is to destroy their housing and to limit the construction of new housing for skilled workers, thus preventing filtering down. The resulting reduction in unskilled worker population and in tax rates is described as "Urban Revival."

In short, Forrester's conclusions follow from his goals without any counterintuitive steps.

* From speech at Hendrix College, April 6, 1970.

** On this point see Ernest Erher, Editor[4], especially the articles by Britton Harris, George M. Raymond, and Lawrence Mann.

DYNAMIC PROCESSES

Nonetheless, it is instructive to study the detailed logic which leads to these conclusions. The model focuses upon rates of change. Each of the important variables change because of the flows which occur within the city and between the city and its external environment. For example, one of the key equations of the model calculates the number of "Underemployed"--that is, unskilled workers--as

(Number of unskilled workers this year)
= (Number last year)
+ (Net flow into this group during year). (1)

Then, each of the rates is further broken down into its component parts.

For example:

(Net flow into unskilled group during year)
= (Migration of Underemployed into the city per year)
+ (People added to this category via births)
+ (People added via downward mobility from the skilled workers category during year)
- (People who have moved upward into skilled workers category during year) (2)

The determination of the various levels then depends upon an accurate evaluation of the various component flow processes like those listed in Equation 2. All of the flows in the model have the same basic form:

Flow per year = (Rate constant) + (Some level) (3)

Equation 4 looks technical, but several examples should serve to illustrate its meaning. For example:

(People added to Underemployed category per year via births)

= (Birth rate) x (Number of unskilled workers) (4a)

The flow is the expression on the left, which is a rate constant (the birth rate) times a level (in this case the number of Underemployed). As another example:

(Workers added to the unskilled group via downward mobility from the skilled worker category during year) = (Rate of downward mobility) x (Number of skilled workers). (4b)

So far, we have seen how the job of determining levels--like the number of unskilled workers--can be reduced to a problem of determining rates of flow. Then, the rates of flow are written in terms of the known levels and rate constants, as in Equation 4. To finish the story, we need to know the rate constants. Once the rate constants are known, the model is completely determined.

Some of the rate constants are rather easy to know. For example, the "birth rate" of Equation 4--which is actually a birth rate minus a death rate--can be determined from tables of vital statistics once the age of distribution of the unskilled workers is estimated. Others are harder. The "rate of downward mobility" of Equation 4 is not known. But Forrester makes a plausible guess by saying that this rate depends upon the ratio of workers to available jobs. He writes this guess as a graph (see Figure 1) which is incorporated into the model. This graph says that for small values of worker unemployment, the downward mobility rate is very small; while for larger values of unemployment, the ratio grows roughly in proportion to the amount of unemployment.

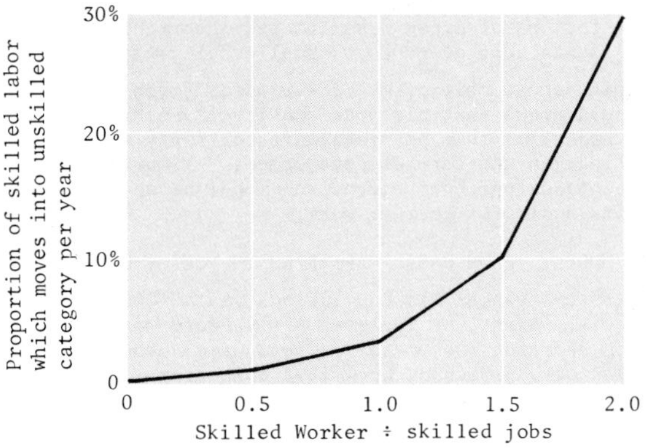

Figure 1

If the aim of *Urban dynamics* were an accurate prediction of the transition rate from labor to underemployed, the use of guesswork like that in Figure 1 would be unacceptable. However, the purpose is rather the comparison of different public policy alternatives. The model need only predict the kinds of changes caused by the different programs. In this case, it may be sufficient to obtain a qualitatively right form for the transition rate,[14] for the relative effects of different public policies may well be quite insensitive to variations of curves like the one in Figure 1.

The most important flow rates are those due to the in-migration and out-migration of underemployed people. In this case, the flow rates are determined by rate constants which can be respectively interpreted as:

(a) The attractiveness of the city as perceived by unskilled potential immigrants

(b) The unattractiveness of the city for the unskilled. The relevant rate equations are:

(Rate of in-migration) = (Perceived attractiveness for unskilled) x (Number of skilled and unskilled workers) (4c)

and

(Rate of out-migration) = (Unattractiveness for unskilled) x (Number of unskilled workers) (4d)

"Unattractiveness for the unskilled" appears in Equation 4d as a rate constant which determines the rate of out-migration from the city. Double the unattractiveness, while holding the number of unemployed fixed, and the rate of out-migration will double. Similarly, from Equation 4c, if you halve the perceived attractiveness, the rate of in-migration of the underemployed will go down by a factor of two.

Forrester gives a precise numerical meaning to "attractiveness" by guessing the strength of the various forces which draw the unskilled to the city. Attractiveness for unskilled workers grows as their economic opportunity grows, as the density in their housing diminishes, as their unemployment rate diminishes, as the public expenditure per capita increases, and also as the underemployed housing program produces superior housing units. Mathematically, the attractiveness is a product of separate factors describing each of these separate components of attractiveness. For example, Figure 2 gives the dependence of the attractiveness upon residential density in the housing for the underemployed. This figure indicates that as the density rises from 120 people per acre to 180 people per acre, the attractiveness of the city diminishes by a factor of five.

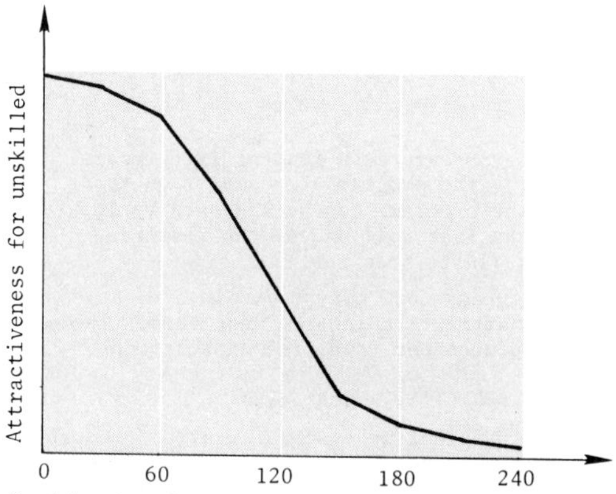

Figure 2

By using guesses like that in Figure 2, Forrester can give a numerical value to attractiveness. The unattractiveness which governs emigration is then given as the inverse of attractiveness

$$(\text{Unattractiveness}) = (\text{Attractiveness})^{-1} \qquad (5)$$

The perceived attractiveness which governs immigration of unskilled workers is assumed to differ from the actual attractiveness because people outside the city do not immediately find out about changed conditions within the city. The model assumes a twenty year time lag so that this perceived attractiveness at a given moment is approximately equal to the actual attractiveness twenty years before.

In this way, Forrester gives concrete mathematical expression to his ideas about the flow of people into and out of the city, and achieves a model in which an increased attractiveness for the unskilled will draw more unskilled into the city.

Another key part of the model is the mechanism for producing new jobs via the creation of new enterprise. The rate constant for this kind of new job production is very sensitive to the amount of land still unutilized in the city. For the conditions most characteristic of the mature cities studied in the model, a 1% drop in the land occupied by housing will cause a 5% increase in this kind of new job formation.

Once the rate constants are specified, Forrester's model is complete. He can then set the city at some initial point, year zero, and let his rate equations calculate the changes in all the variables between year zero and year one. Successive applications of this procedure give the year-by-year growth of the city. At each point, the model calculates the values of all the level-variables. Eventually, the city begins to fill most of its available land area with housing and industry. Thereafter, the city begins to settle down to an equilibrium in which all the levels remain roughly constant. We then have a description of the mature city.

THE MATURE CITY: STAGNATION AND REMEDIES

As Forrester looks at the mature city of his model, he finds that it contains many of the defects of our real cities. The story is summarized in Table 1. Notice the high unemployment rate among the unskilled, the lack of skilled labor, and the high land fraction (31%) occupied by the unskilled and their families. Their area is identified as slums even though the residential density is rather low. As Forrester points out, these slums are harmful to the city because they occupy land which could be utilized by industry which could provide more jobs. This condition is then termed "urban stagnation."

	Population group			
	Unskilled workers ="Underemployed"	Skilled workers = "workers"	Management + Professionals	Total
Number of potential workers	377,300	392,600	71,100	841,000
Jobs for this category	208,300	403,500	51,800	663,600
Unemployed	169,000	- 10,900	19,300	177,400
Potential workers plus families	3,000,000	2,370,000	350,600	5,720,600
Land area occupied (acres)	31,000 "slums"	33,500	11,100	75,600
Density (people/acre)	98	71	32	76

Table 1 - Forrester's "stagnant" city. The unemployment is calculated as the first row in the table minus the second.

Next, Forrester examines a possible set of alternative strategies for city improvement; these strategies are inserted as changes in the model. For example, the underemployed job program simply provides jobs for 10% of the unskilled workers over and above those jobs naturally provided by the business sector. A job training program moves 5% of the underemployed into the skilled worker group without changing any of the other flow processes. The tax subsidy program makes $100.00 per capita per year available to the city from outside sources. In the model, this permits extra tax expenditure which then has the effect of increasing the upward mobility of the underemployed.

All of these "liberal programs" are directly designed to reduce unemployment among the poor. To evaluate how well they work, the model is run for 50 years. After this time, a new equilibrium is reached as shown in Table 2. From the data in the first three rows, all three programs seem to have failed. The training program seems to have had no

Program	"Natural" development	Some "liberal" programs: Job program	Tax subsidy	Low-cost housing	Training program	New enterprise construction	Some programs for city revival — Demolition of slum housing: Alone	+Discouragement of worker construction	+Encouragement of new enterprise construction
Number of unskilled workers = underemployed (in thousands)	377	417 (+ 10%)	407 (+ 8%)	372 (- 1%)	382 (+ 1%)	452 (+ 20%)	317 (- 16%)	290 (- 23%)	336 (- 11%)
Unemployment in this group (in thousands)	169	173 (+ 3%)	201 (+ 19%)	214 (+ 27%)	168 (+ 0%)	153 (- 9%)	57 (- 66%)	4 (- 97%)	22 (- 87%)
Land area occupied by this group = slums (thousands of acres)	31.0	32.6 (+ 8%)	31.9 (+ 3%)	45.1 (+ 46%)	30.7 (- 1%)	32.0 (+ 3%)	17.4 (- 44%)	14.2 (- 54%)	17.5 (- 43%)
Net upward social mobility (net flow underemployed to labor - thousands of workers per year)	5.5	6.8 (+ 24%)	7.6 (+ 38%)	3.8 (- 31%)	16.8 (+ 206%)	13.0 (+ 136%)	5.6 (+ 2%)	7.1 (+ 29%)	9.2 (+ 68%)
Attractiveness for Underemployed	1	1.03	1.05	1.01	1.25	1.16	.89	.91	.95
Attractiveness for Labor	1	.90	.90	.91	.68	.78	1.08	1.03	.98
Attractiveness for Management and Professionals	1	.98	.91	.94	.94	.95	.98	.98	.97

Table 2 - Effects of programs in the Forrester model. The programs are each run for fifty years. The numbers in parentheses are the changes produced by the programs in comparison to the results of no programs at all. The effects of the programs can be assessed by looking at the net upward social mobility (fourth from the last row) and at the resulting attractiveness of the city for various social groups (last three rows).

effect. The job program and tax subsidy have increased the number of unskilled, the amount of unemployment in this group, and the amount of land devoted to slums.

If Forrester is right, the job program and the tax subsidy are harmful to the city. Those who have proposed them are the victims of "intuitive thinking" applied to a situation too complex for any simple method of thought.

The final liberal program in Table 2, low-cost housing, is inserted into the model as a simple addition to the stock of housing for the underemployed. Land available for industry decreases, jobs decline, and disastrous unemployment results.

To replace these "unsuccessful" liberal programs, Forrester proposes a set of programs aimed at city revival. The most obvious, the direct encouragement of additional new enterprise formation, is inserted into the model as an increase in the rate constant for new enterprise construction. However, this program is rejected both because its effects are too small and also because Forrester sees no way of directly effecting this encouragement of new enterprise.

An indirect method is proposed: a program of slum housing demolition which removes 5% of the slum housing each year. This is most effective when it is coupled with diminished worker housing construction or with increased new enterprise construction. From Table 2, these programs seem quite successful in alleviating unemployment among the unskilled and reducing the size of slums.

Of course, the results of these calculations are, in no sense, counterintuitive. Each of the liberal programs increases the city's attractiveness for the unskilled (see row 5 of Table 2) and draws them to it. The programs for "city revival" are, as expected, unattractive for the unskilled. The model merely reproduces our intuitive expectations.

ANALYSIS OF BENEFITS: LOCAL VIEW

Each of the programs under consideration produces both gains and losses. For example, the demolition of slum housing combined with the discouragement of worker housing construction does decrease unemployment and diminish the area devoted to "slums." (The slum area goes down both because houses are pulled down and also because the "filtering down" of housing from the labor group to the underemployed group is inhibited). However, the density in the "slums" increases from 98 people per acre to 160 people per acre, while the density in the "labor" sections increases by 9%. How are we to balance the benefits of increased job opportunity against the disadvantage of more crowded housing?

Forrester does not exactly perform this balancing process. Rather, he has in mind an improved version of the city with fewer low-skilled workers, fewer slums, fewer people out of jobs, and more industrial growth. If a program produces results which approach these ends, he judges it to be successful. In essence, Forrester is working toward a goal which is to improve a given area of land--a city.

To see this reasoning at work, consider Forrester's evaluation of the training program. This program has the advantage that it increases the net flow of people from the "Underemployed" category to skilled "Labor." (This flow is recorded as the fourth row in Table 2). However, Forrester focuses upon the losses to the "city."

> The training program has created a flow through the area with a much increased

> underemployed-arrival rate and a much increased labor-departure rate. People come to the area because of the training program and leave when they find there is no use for the skills they have acquired. *As a service to society*, the program might be considered successful. But as a service to the city, its value is far less clear. The area is more crowded, the land fraction occupied has risen slightly, housing conditions are more crowded, the total of underemployed has risen very slightly, and the ratio of labor to jobs is higher, indicating a higher degree of unemployment.[15]

Forrester's assumption that there is an object called "the city" to which we can assign benefits or debits is, I think, incorrect. We should only assign benefits and hurts to people, since the goals of our policies should be to enable people to live more satisfactory lives.

Is there anything in the model which would permit the estimation of benefits to the people involved? There is. The attractiveness functions give numerical estimates of the worth of the city as perceived by various groups. The model computes the changes in attractiveness resulting from each of the proposed programs. These attractiveness numbers are listed in the last three rows of Table 2. The numbers in each row are divided by a constant factor so that the attractiveness is unity in the absence of any public program. These attractiveness numbers provide a numerical way of estimating the worth of any public program for the various groups involved.

The benefits and losses to the unskilled have already been discussed. Each program which is attractive to the unskilled reduces the city's attractiveness for the skilled group. This effect occurs because the attractiveness for the skilled includes a "social attractiveness" which decreases as the city draws in a larger proportion of unskilled workers. Conversely, the reduction in the proportion of unskilled workers produced by the destruction of their housing increases the city's attraction for the skilled group. However, when this program is coupled with new enterprise construction, the increased attractiveness for the skilled group is cancelled because not enough land is available for housing.

From the point of view of the attractiveness concept, Forrester's favored program of slum-housing demolition plus discouragement of labor-housing construction does not look very good. The attractiveness for Labor only increases 3%, while Underemployed and Management are respectively 9% and 2% worse off than before. On this basis, we should probably reject this program. Furthermore, the last program in Table 2 is favored by Forrester even though it seems bad for everyone!

However, one might argue that this result is not really fair to Forrester. After all, there is more industry in the city. Is this not a gain? It is true that, in general, industry is good for a city by providing jobs and paying taxes. But the attractiveness measures already include these benefits. We cannot count them again. All the other beneficial effects of increased industry are harder to evaluate because they are largely benefits which accrue to the entire nation rather than to the city in question. However, it is possible that the appearance of this industry in this special city prevented the construction of competitive industry elsewhere. Perhaps the other location would have been better for the nation. We cannot know.

But there is one benefit which is possibly undervalued in the attractiveness measures. It is possible that the underemployed do not have a sufficiently long view to perceive the real value to themselves and their children of upward economic mobility. In the construction of city programs, we might consider the upward mobility from the low-skilled underemployed group to the higher-skilled labor group to be valuable in itself. In fact, one might argue that the main role of cities in American history has been to foster this upward mobility. Then, in evaluating the different proposed public policies in the context of this model, we should also consider--as Forrester does--the total number of potential workers who have been raised from the low-skilled group to the high. In the model, the traffic goes both ways: from "Underemployed" to "Labor" as well as vice versa. The key number is the net flow from "Underemployed" to "Labor". Table 2 shows this number for the models under consideration. From this point of view, the "favorable" program has a much less favorable impact than the direct training program. Both seem preferable to the "stagnation" result.

It is true, nonetheless, that the programs favored by Forrester do increase the net upward mobility. Or at least the model says that they do so. This mobility increase is supposed to occur because the programs result in increasing industry, and the increased jobs help the upward mobility. Furthermore, the improved "social atmosphere" caused by an increased ratio of skilled labor to unskilled is also supposed to increase upward mobility.

It is, however, extremely dangerous to base any public policy decisions upon the upward mobility predictions of this model. As Banfield has emphasized,[16] we know very little about the conditions which help upward mobility. Moynihan [17] has suggested that this mobility might be tightly interwoven with family structure considerations which are certainly not in the model. The mobility predictions of this model cannot be used to justify any public policy because they are completely unreliable.

A NATIONAL VIEW

The most striking fact about the changes in attractiveness listed in Table 2 is that they are very small. If you momentarily increase the attractiveness for any group, more of that group will enter the city, consume jobs and housing, and thereby reduce the attractiveness. To see this result in operation, consider the attractiveness effects of the job training program as indicated in Table 3.

		Before	After 10 years	After 50 years
Relative Attractiveness for	Underemployed	1	1.68	1.25
	Labor	1	.74	.68
	Management & Professionals	1	1.09	.95

Table 3 - Attractiveness Changes Produced by Training Program

According to the following table, after the first ten years, the training program produces a very large favorable effect for the unskilled in that the attractiveness for this group increases 68%. On the other hand, there is an immediate harmful effect to the skilled laboring population produced by the increased job and housing competition felt by this group. This effect is initially smaller than the benefit to the unskilled, being only a 26% decrease in attractiveness.

Notice that after 50 years a large fraction of the benefits of this job training program for the unskilled group disappears, while the losses for the fully-skilled labor group increase. This kind of dissipation of benefits occurs because the increased attractiveness of the city draws more unskilled into the city so that a larger group of people must compete for a roughly fixed number of jobs. Hence, everyone is worse off at year 50 compared to their state at year 10.

Forrester points out that this dissipation of benefits produced by increased in-migration is a general effect of all programs designed to give direct aid to any group of people. However, it is important to notice that this analysis only applies to a program which is applied only to the single city in question. If the program were applied nationwide, the attractiveness of all areas for the unskilled would increase equally. As a result, there would be no increase in the migration into any city. The long-term deterioration shown in the last two columns of Table 2 would then be replaced by a long-term improvement.

This discussion then leads us to the following conclusion: *Programs for improving the lot of the unskilled should be applied nationally rather than locally in order to prevent the partial neutralization of these policies as a result of the concentration of the unskilled in the program areas.* Forrester's model automatically assumes that all his programs are locally employed; hence, his work is simply inapplicable to the analysis of the long-range effects of any policy applied nationwide.

To see the striking effects of policies applied nationwide, imagine that nothing at all were changed within the city under study, but the rest of the nation improved its conditions suddenly to make its attractiveness for the unskilled group a factor of two better than before. Then immediately this group's in-migration decreases by a factor of two. Even though nothing has changed within the city itself, the results of this nationwide change would be quite substantial, at least for the unskilled group. These changes are summarized in Table 4. The conditions of this group have bettered very substantially, without anyone else in the city being the worse off.

Naturally, the course we have just described is not a realistic policy alternative. Forrester's published analysis does not permit us to study and evaluate the results of realistic policies applied nationwide.

	Before nation-wide change	After 10 years	After 50 years
Number of unskilled potential workers = "Underemployed" (in thousands)	377	296 (- 20%)	322 (- 14%)
Unemployed in this category	166	86 (- 41%)	102 (- 39%)
Land occupied by "Underemployed" = "slums" (in thousands of acres)	31.0	29.5 (- 5%)	28.7 (- 8%)
ATTRACTIVENESS FOR Unskilled workers ("Underemployed")	1	2.58	1.88
ATTRACTIVENESS FOR Skilled workers ("Labor")	1	1,08	1.02
ATTRACTIVENESS FOR Management	1	1.02	1.00
Underemployed to Labor net in thousands	5.5	5.9 (+ 8%)	5.6 (+ 2%)

Table 4 - Effects of an increase in the national level of the attraction for unskilled workers upon a city which itself is not changed in any structural sense. Numbers in parentheses refer to the percentage changes in this city caused by the change in the environment.

REFERENCES

1 FORRESTER J W
Urban dynamics
MIT Press Cambridge 1969

2 FORRESTER J W
Urban dynamics
MIT Press Cambridge 1969 p 109

3 FORRESTER J W
Urban dynamics
MIT Press Cambridge 1969 p 110

5 KAIN JOHN F
A computer version of how a city works
Fortune November 1969

6 GARN HARVEY A
An urban systems model: A critique of urban dynamics
The Urban Institute Washington D C Working Paper 113-25 unpublished

7 INGRAM GREGORY K
book review *AIP Journal* May 1970

9 INGRAM GREGORY K
book review *AIP Journal* May 1970 p 207

10 GARN HARVEY A
An urban systems model: A critique of urban dynamics
The Urban Institute Washington D C Working Paper 113-25 unpublished p 5

11 FORRESTER J W
Urban dynamics
MIT Press Cambridge 1969 p 2 italics added

12 KAIN JOHN F
A computer version of how a city works
Fortune November 1969

13 INGRAM GREGORY K
book review *AIP Journal* May 1970 p 206
Garn also reaches similar conclusions about Forrester's goals

14 FORRESTER J W
Urban dynamics
MIT Press Cambridge 1969 appendix B

15 FORRESTER J W
Urban dynamics
MIT Press Cambridge 1969 p 59 italics added

16 BANFIELD EDWARD C
The heavenly city
Little, Brown and Company 1968

17 MOYNIHAN R
as reported in Rainwater and Yancey
The Moynihan report and the politics of controversy
MIT Press 1967

The SCi Editorial Board procedure requiring a minimum of three reviews for technical articles was relaxed in this case to expedite publication of this timely material.

IEEE TRANSACTIONS ON SYSTEMS, MAN, AND CYBERNETICS, VOL. SMC-2, NO. 2, APRIL 1972 159

Public Policy Conclusions from Urban Growth Models

LEO P. KADANOFF AND HERBERT WEINBLATT

***Abstract*—This paper is intended to be a case study in the use of simulation models to test public policy alternatives. Suggested programs for dealing with urban poverty are tested with the aid of different models which describe the linked growth of housing, population, and industry in an urbanized area. The tested programs are: 1) training, to provide the unskilled with job skills; 2) job provision, to make extra jobs for skilled workers; 3) clearance, to eliminate "excess" housing and thereby free land which may be used by industry. The models used are: the original Forrester [1] model, which treats a single city as a unit in an unchanging national environment; an extension of this model to include all the central cities of the nation and describe the migration between these areas; and finally a complete revision of the Forrester model to obtain a simulation of the national economy including both cities and suburbs.**

The three different models give very different results. The original model, which focuses upon applying programs to a single city, strongly indicates that clearance is the only one of these programs which is effective in eliminating urban poverty. The second model indicates that when applied throughout the nation, both the clearance and training programs are effective, but job provision is ineffective. However, the third model uses its more complete picture of the national economy to conclude that job provision can indeed be effective in reducing poverty. In fact, the third model suggests that a combination of training, clearance, and the provision of more jobs will reduce poverty with a minimum of undesirable side effects. This analysis is intended to show that the choice of focus for the model will strongly affect the policy conclusions reached.

I. MODELS

THIS PAPER compares the public policy implications of three different models of urban growth. All three are based upon *Urban Dynamics* [1]. This simulation model is a set of equations for the development of an urbanized area over a period of time. It attempts to catch the essential features of the city without including any unnecessary detail.

In all three models, the nation is divided into districts of fixed land areas. The basic variables are the amounts of population, housing, and industry within each of the different regions. The population is divided into three groups: a) unskilled workers (U); b) skilled workers (L); and c) management–professional workers (MP).

Housing is divided into categories suitable for each of these groups. Industry provides jobs for these people and competes with housing for the available land. The models include real estate taxes charged against housing and industry.

These models each consist of a set of linked first-order difference equations which describe the construction, decline, and demolition of industry and housing, as well as changes in the size of each of the population groups. For example, the equation describing the rate of change of the number of skilled workers is

$$L(\text{this year}) - L(\text{last year}) = \text{LA} - \text{LD} + \text{SMN} + \text{LB} \quad (1)$$

where LA represents the rate of arrivals of skilled workers into the area, LD is their departure rate, SMN represents the net flow into this group which occurs via social mobility from the MP and U groups, and LB is the change via births and deaths.

The most important flow rates in the models are those which describe the arrivals and departures of workers. In particular, the departure rate for skilled workers is

$$\text{LD} = \text{constant} \times L/\text{ATTL}. \quad (2)$$

This equation states that the number of departures per year is proportional to the number of people in the skill group divided by a number ATTL, which represents the attractiveness of the city for the L group. The higher the attractiveness number, the lower the departures. In addition,

$$\text{LA} = (\text{another constant}) \times L \times \text{ATTLP}. \quad (3)$$

Here the arrival rate is proportional to the size of the city, as measured by the total number of skilled workers (1), and to the attractiveness of the city as perceived by outsiders ATTLP. Similar equations describe the arrivals and departures of the other two social groups.

To complete the specification of these rates, all three models follow Forrester in giving a numerical meaning to the "attractiveness' by guessing the strengths of the various forces which draw workers to the city. For example, attractiveness for unskilled workers grows as their economic opportunity grows, as more housing becomes available to them, as their unemployment rate diminishes, and as the public expenditures per capita increase. Mathematically, the attractiveness is a product of separate factors describing each of these separate components of attractiveness. The models assume a time lag in the diffusion of information about the city so that the perceived attractiveness is essentially equal to the actual attractiveness at a previous time.

We can put these equations together to get a picture of the migration from one district to another. Let district 1 have a skilled labor population L_1 and an attractiveness ATTL_2 for these workers. District 2 has a perceived attractiveness ATTPL_2 and a proportion of the national L group population given by L_2 divided by the total national skilled labor population. If we assume that the migration from 1 to 2 is governed by the attractiveness in the same

Manuscript received October 5, 1971. This work was supported in part by the Urban Mass Tránsit Administration, the NSF, and the Rhode Island Urban Observatory.

The authors are with Brown University, Providence, R.I. 02912.

fashion as in (2) and (3) and that the probability that a worker will migrate to district 2 is proportional to the number of workers in that district relative to the number in the nation, then the migration rate from 1 to 2 is

$$\text{constant} \times \frac{(L_1/\text{ATTL}_1) \times (L_2 \times \text{ATTPL}_2)}{\text{total number of } L \text{ in the nation}} \quad (4)$$

when the two regions lie in different metropolitan areas. A similar equation describes migration within the metropolitan area except that the multiplicative constant is different and the denominator contains the L group population of the metropolitan area.

All three models use (4) to describe the migration patterns. However, they use very different frameworks to set up the environment for their cities. These frameworks are illustrated in Fig. 1. The original Forrester model (Fig. 1(a)) puts a single city into an unchanging national environment. Population and industry migrate into and out of the city in response to changes in its attractiveness relative to national standards which the city is too small to affect. In the second model, we set up a nation composed of many cities, each described by Forrester's equations. The only difference between this many-city Forrester model and the original model is that the different cities are linked together by migration equations (see Fig. 1(b)). This many-city Forrester model was programmed by one of the authors in the APL language for Brown University's IBM 360/67 computer. This new model also includes migration to and from the suburbs and rural areas. However, the dynamics of these regions are not really included in this many-city model. The suburbs are given a fixed attractiveness for each population group and a fixed proportion of the metropolitan area population. The rural regions are also given a fixed attractiveness and a fixed proportion of the national population. These different frameworks are depicted in Figs. 1(a) and (b). The primary difference between the original Forrester model and the many-city Forrester model is that, in the latter, the area under study is a much larger proportion of the total nation. We may use this many-city model to study the effects of nationally applied urban programs and be sure that we have at least taken into account the purely urban part of the national environment.

Our third model is a major revision of the Forrester model to include an explicit picture of the suburbs. It is also written in the APL language and runs on Brown University's IBM 360/67 computer. This model attempts to combine Forrester's picture of the social and economic nature of an urban area with the distributional ideas contained in the Lowry model [2] and its successors [3]. In this national metropolitan model, any metropolitan area of the nation may be split into a number of districts, each with its own characteristic transportation connections with the other parts of the metropolitan area and of the nation. For simplicity, in this paper, we limit ourselves to situations in which all metropolitan areas are identically modeled as consisting of a central city and a suburban ring. Both districts are modeled with equal care. However, the rural sector of the nation is represented only by a fixed attractiveness. Fig. 1(c) depicts the structure of the nation represented by this application of the national metropolitan model.

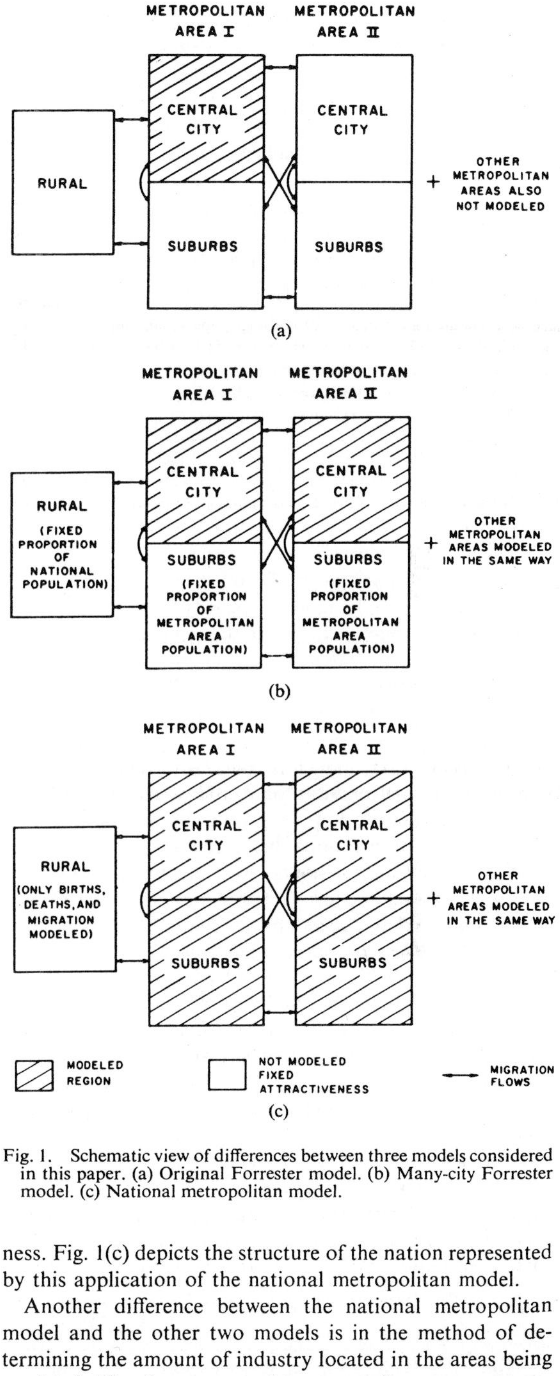

Fig. 1. Schematic view of differences between three models considered in this paper. (a) Original Forrester model. (b) Many-city Forrester model. (c) National metropolitan model.

Another difference between the national metropolitan model and the other two models is in the method of determining the amount of industry located in the areas being modeled. The first two models essentially assume an unlimited demand for the goods and services produced by these areas. The rate of construction for new industrial enterprise is

$$\text{constant} \times (\text{amount of industry already present}) \times \text{EM}. \quad (5)$$

Here, the quantity EM is an enterprise multiplier which measures how attractive each area is for the location of new industry. This EM increases as the proportion of vacant land increases, as more managers and skilled workers become available, and as the tax rate decreases. Consequently, both the original Forrester model and its extended version include no effects upon industrial location arising from the demand side of the urban economy. The quantity EM is constructed purely from proxies for the prices of inputs and is hence restricted to supply-side considerations.

The national metropolitan model includes a very simple model of the location-specific relationship between supply and demand. The model estimates the total demand for goods and services generated by the industry and workers within each metropolitan district. This demand is then split into sectors which will be satisfied on the local market, on a metropolitan-wide market, and on a national market. The demands thereby generated are distributed among the different districts of the nation in proportion to an enterprise multiplier, to the size of the districts (as measured by the total amount of industry already present), and to the quality of the assumed transportation connections between the district where the demand is generated and the district where it is to be supplied. When the generated and distributed demand exceeds the output of a given district, more industry is constructed.

There are other differences between the first two models and the national metropolitan model. For example, the national metropolitan model permits workers to live in one part of the metropolitan area and work in another, while the other two do not. However, for the purpose of the analysis to be presented here, the main differences among the models are the kinds of regions under study (as described in Fig. 1) and the fact that the national metropolitan model includes demand-side limitations upon the amounts of new industry constructed.

II. Applications of Public Programs

Each model consists of a set of equations which will determine the state of the system in one year on the basis of the state in the previous year. Additional iterations can be used to simulate development over an extended period of time.

The basic procedure for evaluating public policies is to: a) start with the city in some reasonable initial state; b) insert the public programs as alternative changes in the model equations; c) run the models with and without the various programs in operation; d) compare the results obtained. The original Forrester model and its many-city version both have a very simple and natural original state to start from. If these models are run starting from any state at all, they will eventually reach a state of dynamic equilibrium in which none of the variables change. In this dynamic equilibrium, most of the available land is filled up. In any time interval, equal amounts of housing are constructed and destroyed. This is also true of industry. Net migration is just right to maintain the number of people at a constant value. This equilibrium city shows many of the defects of our real cities. There is a high rate of unemployment among the unskilled (45 percent) and a very high real-estate tax rate to pay for the needs of the unskilled and their families.

It is considerably harder to choose a natural starting point for the national metropolitan model. In this model, the total population of the metropolitan area must grow continually as a response to the natural increase in national population. If we try to fix the central city population at a constant value, then the suburban population must grow. Since the central city and suburb interact with one another, the suburban growth will inevitably change the central city. In order to see the effects of urban policies in the clearest possible fashion, we have chosen an initial state for the typical metropolitan area in this national metropolitan model which makes for rather slow changes in the course of the natural development; i.e., the development without any policies being applied. Table I lists some of the important parameters of this typical metropolitan area both at the initial time and after 30 years. For clarity in presentation, the attractiveness numbers in Table I and succeeding figures have been normalized by dividing each group's attractiveness by a different number. For the original and many-city Forrester models, these numbers are chosen to make the attractiveness for each group equal to one in the central city when no programs are applied. For the national metropolitan model, the normalization constants are chosen to make the average attractiveness in the metropolitan area equal to one just before the programs are applied. It also

TABLE I
Parameters of Models when No Programs Are Applied

Situation	Area	Parameters								
		Workers in thousands			Unemployment in per cent			Attractiveness (normalized)		
		U	L	MP	U	L	MP	U	L	MP
National metropolitan model — starting point	Central city	37.4	50.0	3.6	29	4.1	7.6	1.070	.860	.575
	Suburbs	18.3	71.2	34.7	37	6.6	3.5	.855	1.035	1.048
National metropolitan model — after 30 years	Central city	42.0	47.9	4.2	27	5.1	6.8	1.142	.775	.600
	Suburbs	37.2	122.6	50.3	34	7.0	3.6	.878	.961	.961
Original Forrester Model and Many-City Forrester Model	Central city	38.0	39.5	5.3	45.2	-2.4	27.8	1	1	1

gives the corresponding parameters for the original Forrester model and its many-city version. The extension has been set up in such a way that the many-city Forrester model has the same equilibrium solution as the original model.

The size of the cities is set to be comparable with that of Providence, R.I. In the national metropolitan model, the proportion of MP is highest in the suburbs. The central city is more attractive for the *U* group because of the presumed existence of public transportation in the central city. Notice the relatively small changes which have occurred in unemployment rates and attractivenesses in the course of 30 years of "natural" development of the national model.

The public programs are inserted as changes in the model equations. These changes are the following.

1) The training program represents an effort to provide more job skills for unskilled workers. The models represent this program by moving 5 percent of *U*-group workers into the *L* group during each year;

2) The job program gives more jobs for skilled workers. It is represented by increasing the number of skilled jobs to a value 10 percent higher than that provided by the industry in the model.

3) The clearance program represents the attempt to clear housing from central-city land which may then be filled by industry. To do this, the models destroy 5 percent per year of the *U*-group housing. This is in addition to the housing which would naturally be torn down. Also, there is a discouragement of the construction of *L*-group housing to hold that construction to half of the value which would result from the natural working of the model.

III. Results

Fig. 2 shows the changes in the *U*-group population and its unemployment rate produced by the application of each of the three programs to a single city. These results of the original Forrester model seem to show that the clearance program is very effective in reducing the number of unskilled workers in the city and decreasing their unemployment rate. In the runs, we have made a minor change in the Forrester model. The constants in the migration equations (2), (3) have been changed to make them compatible with the results of the many-city model. This revision causes no qualitative change in the results presented. Besides, one of the main defects of the Forrester model is that the equations are completely uncalibrated [4]–[9]. Consequently, our changes in these constants do not necessarily make the model less realistic. This program primarily works in two ways. As the *U* group's housing is destroyed, the city becomes less attractive for them. Therefore, the number of unskilled workers decreases, and their unemployment rate goes down. After a while, we find that the city contains vacant land suitable for industry. Then industry begins to move into the city, increasing the job opportunities for the unskilled. They gain job skills and move up to the skilled ranks, thereby producing a further reduction in the number of unskilled workers.

Fig. 2 also seems to indicate that both the training program and the job program are rather ineffective in reducing

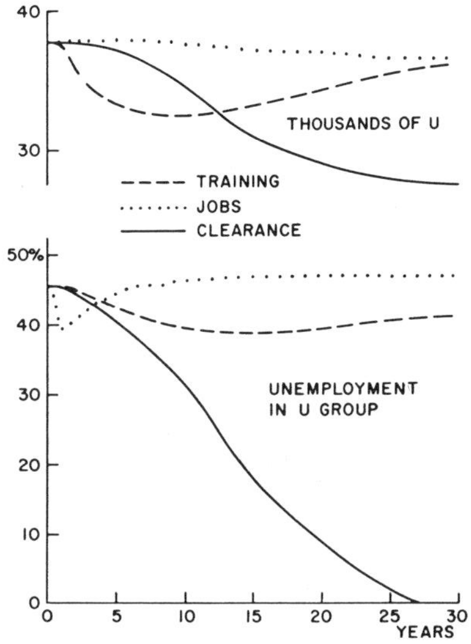

Fig. 2. Changes in numbers of unskilled workers and their unemployment rate. Changes in this figure are produced by public programs as applied to single city in context of original Forrester model. All programs start at year zero. If no programs were applied, city would remain the same.

the amount of poverty in the city. Initially, the training program reduced the number of *U*-group workers. But, it also makes the city far more attractive for the unskilled. They move into the city in larger numbers, effectively canceling out the reduction in their ranks produced by the increased upward mobility. At first, the job program reduces unemployment in the *U* group because the extra *L*-group jobs tend to filter down to the unskilled. But, after a time, more skilled workers enter the city to fill the extra jobs provided. When this has occurred, the *U*-group unemployment rate rises once again.

The results, however, are very different if we apply the same programs on a nationwide basis. Fig. 3 includes data produced by the many-city Forrester model for nationally applied clearance, training, and skilled-job programs. Once again, the job program appears to be almost completely ineffective. However, both the training program and the clearance program now seem to work quite well in reducing urban poverty. The training program appeared to work badly in the one-city application because, as *U*-group people were trained, they were replaced by new unskilled workers drawn into the city by the very high attractiveness produced by the program. However, when the programs are applied nationally, the attractiveness of each and every city for the unskilled increases. This nationwide increase cannot change intermetropolitan migration patterns. Hence the replenishment of the poor which occurred in the single-city training program is much smaller in the nationwide program.

The data shown in Fig. 3 seems to rule out the job program. However, it does not enable us to choose between the clearance and the training programs. Training seems to be somewhat more effective in reducing the numbers of

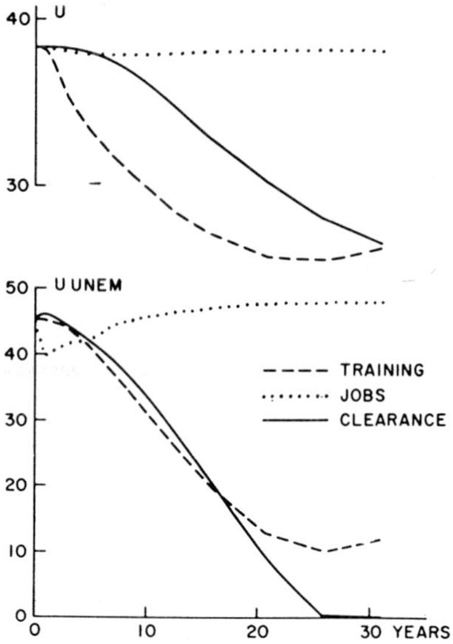

Fig. 3. Changes in number of unskilled workers and their unemployment rate as described by many-city Forrester model. All programs are applied on nationwide basis.

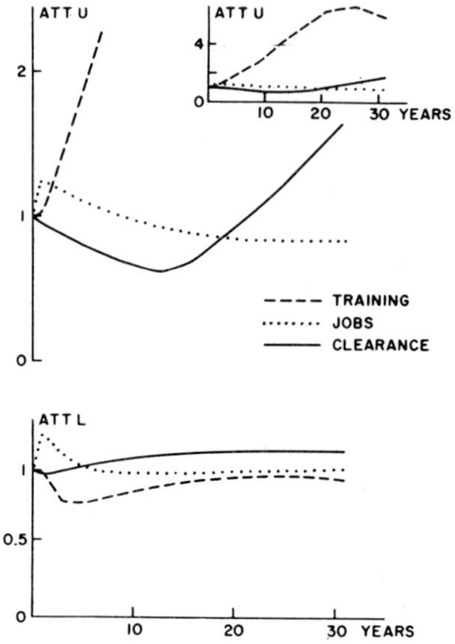

Fig. 4. Changes in attractiveness for U and L groups produced by nationwide programs in many-city Forrester model.

unskilled workers; clearance seems to be more effective in reducing their unemployment rate. Fig. 3, however, does not tell the full story about possible side effects of the programs. The clearance program has the unfavorable side effect of reducing the amount of housing available for both the L and the U. The training program hurts the L group by producing extra competitors for skilled jobs.

The model provides a method for balancing these unfavorable side effects against the benefits produced by the programs. The attractiveness numbers, which are used in determining migration, also describe the quality of city life for the people concerned. Higher attractiveness means that the people feel that their life is better so that they are less inclined to leave the city. For that reason, it has been suggested [6], [9] that these numbers should be used as indices of merit for the various programs.

Fig. 4 shows the attractiveness for the two main population groups as it is modified by the programs in question. If these programs were not applied, the attractiveness would remain equal to one. As expected, the training program is very good for the U group, but bad for the L group because of increased job competition. However, the rise in attractiveness for the U group is many times greater than the corresponding decline for the L group. This indicates that, on balance, the U-group gains more than the L-group loses. On the other hand, the clearance program produces a modest gain for the L group (for whom the attractiveness increases by a maximum of about 15 percent) while, at first, causing a larger loss (40 percent at maximum) for the U group. In the short run, therefore, this program looks less favorable than training. However, in the long run, unskilled workers also seem to obtain a net benefit from the extra jobs produced by this program. Hence, if we focus on a time scale of about 15 years, we should favor training; if we take a very long view, we might favor clearance.

Notice also that the provision of additional skilled jobs seems, in the long run, to be disadvantageous to the U group. This result occurs because these jobs draw into the city additional skilled workers, who then compete with the U group for the stock of urban goods, including the additional jobs and land for housing. This competition hurts the U group in the long run, at least within the context of this model.

These harmful effects of the job program, however, might well be an artifact produced by the particular assumptions of the many-city Forrester model. In reality, as skilled workers leave their suburban jobs to fill the extra jobs provided in the central city, the jobs they have vacated become available for other workers. Furthermore, the provision of extra jobs will increase national demand and thereby stimulate the entire national economy. Since the many-city model does not really include the suburban portion of the economy, this model cannot provide a full description of these beneficial effects of the job program.

To obtain a better evaluation, we turn to the national metropolitan model, which includes both a description of the suburbs and a rough picture of the national economy. Fig. 5 plots the total number of unskilled workers in a metropolitan area and their unemployment rates for four runs of the national metropolitan model, including a run in which no programs are applied. In this run the new model gives a continually increasing U-group population and a roughly fixed unemployment rate.

The programs considered are essentially identical to those considered earlier. The training and the provision of skilled jobs are performed through the entire metropolitan area,

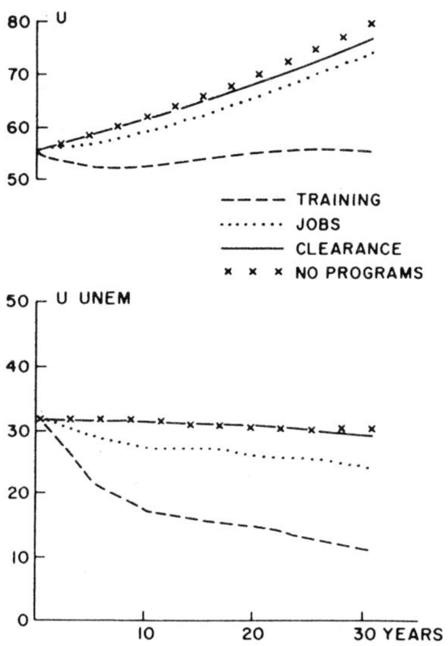

Fig. 5. Changes in numbers of unskilled workers and their unemployment rate resulting from runs of programs in national metropolitan model. Data plotted are total number of unskilled workers in metropolitan area, including suburbs, and average unemployment rate. Note that largest growth in unskilled workers occurs when no programs are applied.

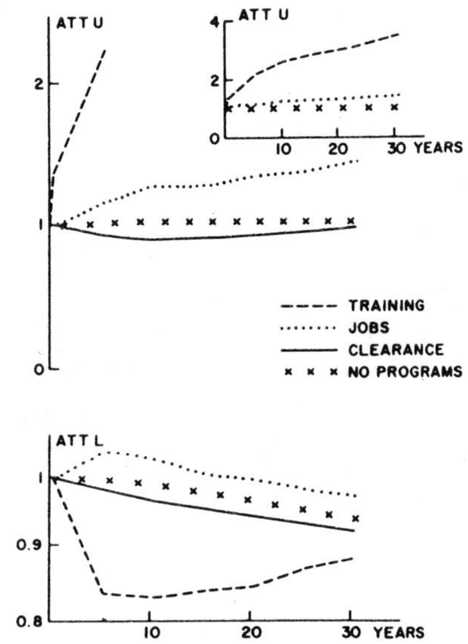

Fig. 6. Attractiveness plots for programs run with national metropolitan model.

while the clearance program is applied only in the central city. This last limitation is introduced because there is no shortage of land in the suburbs, so that clearance is unnecessary there.

From Fig. 5, we see that the skilled-labor-job program has been moderately successful in reducing the number of unskilled workers and their unemployment rate. The program both increases the upward mobility of unskilled workers and causes some of the excess skilled jobs to filter down to the unskilled. As this filtering occurs, the unskilled experience a modest decrease in unemployment. Furthermore, the gain in employment for both the U and the L tends to increase national demand and thereby stimulate the national economy. This stimulation produces more jobs for all, which produces more stimulation, and so on. In the end, the provision of skilled jobs has served to markedly reduce the number of unskilled and their unemployment rate.

As expected, the training program is quite effective in holding down the number of unskilled workers and decreasing their unemployment rate. However, Fig. 5 indicates that, in the context of this national model, the clearance program is almost totally ineffective. This result is in striking contrast to those of the other two models.

The reason for this change is simple. In all three models the major effect of the clearance program is an increase in the construction of new industry in the central city. In the first two models only the dynamics of the central cities are represented, and so these effects appear as increases in the total amounts of industry within the areas modeled. These increases, in turn, create more jobs and improve the general economic health of these areas. In the national metropolitan model, however, the suburbs are explicitly represented. The total amount of industry in the nation is determined by demand for goods and services and is essentially unaffected by a clearance program; only the distribution of industry is affected. The increased industry in the central city is produced at the expense of a commensurate decrease in the growth rate of suburban industry. Many of the jobs in the city are filled by people who would otherwise have filled suburban jobs. The resulting improvement in the overall economy is minimal.

The failure of the clearance program is shown quite clearly in the attractiveness plots of Fig. 6. These plots give the attractiveness averaged over both suburbs and central city as they are affected by the programs. The clearance program produces a net decline in attractiveness for both population groups. This decline arises from increased crowding in their housing which is not compensated for by any substantial change in employment opportunities. So much for clearance.

The job program produces benefits for both the L group and the U group. Once again, the training program produces a substantial gain for the U but a large loss for the L. This loss arises because their unemployment rate has risen from 5.6 to 9.1 percent. For this reason, perhaps, the training program should be considered to be unacceptable.

The increase in the L-group unemployment rate caused by the training program could be eliminated if such a program were combined with a job program for skilled labor. A combination of these two programs could be expected to improve conditions for both the U and the L and considerably reduce the percentage of unskilled workers in the general population. For this reason, a program to eliminate some U-group housing might well be added to the other

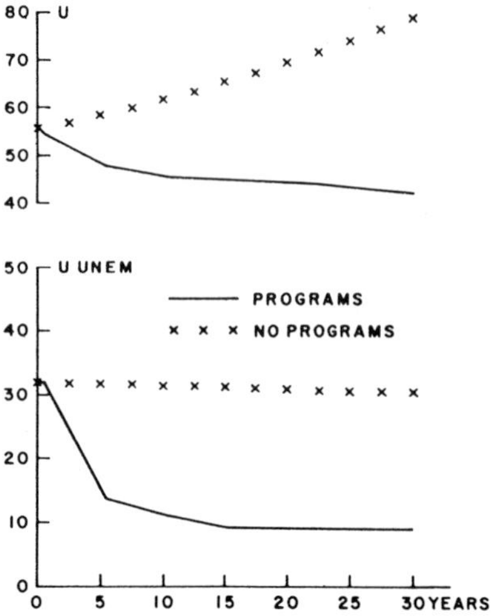

Fig. 7. Effect of combined programs of training, job provision, and slum clearance upon reducing poverty as predicted by the national metropolitan model.

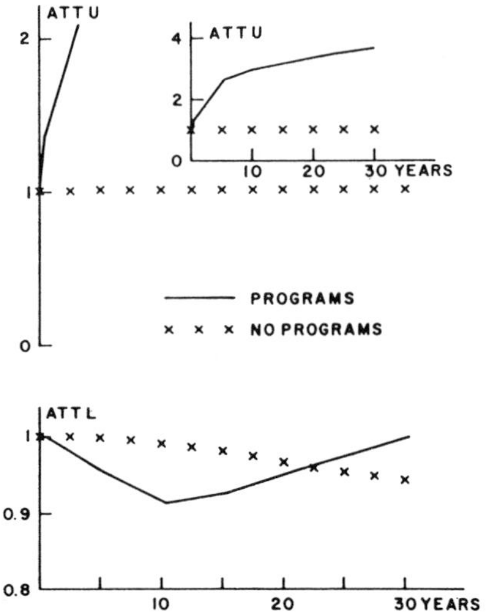

Fig. 8. Effects of combined programs upon attractiveness indices.

two programs. The effects of one possible combination of these three programs are shown in Figs. 7 and 8. This combination consists of training and job programs which are identical to those previously modeled, along with a clearance program which demolishes *U*-group housing at the annual rate of 10 percent in the central city and 5 percent in the suburbs. The effects of this mixed program are shown in Figs. 7 and 8. The numbers of unskilled workers and their unemployment rate have decreased quite dramatically. The *L* feel no extra unemployment. In fact, in the long run the average attractiveness for the *L* increases. Attractiveness for the *U* soars. The combination of programs appears to be quite successful.

The outcome of the foregoing analysis might suggest that we advocate the enactment of massive programs for the production of public service jobs, combined with a national effort to train the unemployed, and further combined with an urban renewal program to reuse the land freed from the slums. In fact, we do advocate such a combination of programs. However, one should recognize that the national metropolitan model does not necessarily provide a compelling case for this program mix. Many important features of the cities and of the nation are left out of the model. For example, it does not include any information about what proportion of the unskilled can be trained for skilled jobs. Nor does the model include any of the costs of the proposed programs which would appear as increased taxes and/or chronic budgetary deficits, with the consequent effects upon the national economy.

In fact, we do not believe that incomplete and uncalibrated models like the national metropolitan model can provide any definitive answers whatsoever. Rather these models provide novel ways of phrasing qualitative arguments. What we have done here is essentially to put our subjective beliefs into numerical form. One might be more inclined to believe a model which had been carefully calibrated and checked against experience. For this reason, a group at Brown University is working to extend the model described here and also to calibrate it against the real urban environment. We are hopeful that such a calibrated model might provide a very helpful tool in discussing the advantages and disadvantages of alternative urban strategies.

Acknowledgment

We would like to thank G. Crampton and J. Tucker for their helpful discussions and criticism of this paper and B. Chinitz, G. Crampton, and P. Symonds for their advice and consultation in the development of the national metropolitan model.

References

[1] J. W. Forrester, *Urban Dynamics*. Cambridge, Mass.: M.I.T. Press, 1969.
[2] I. Lowry, "A model of metropolis," Rand Corp., Santa Monica, Calif., 1964.
[3] M. Echenique, "Urban systems: toward an explorative model," Cent. Environmental Studies, Cambridge, England, 1969.
[4] H. A. Garn, "An urban systems model: a critique of *Urban Dynamics*," The Urban Institute, Washington, D.C., Working Paper 113-25, unpublished.
H. A. Garn and R. H. Wilson, "A look at *Urban Dynamics*: the Forrester model and public policy," this issue, pp. 150–155.
[5] H. Weinblatt, "*Urban Dynamics*: a critical examination," *Policy Sci.*, vol. 1, no. 3, pp. 377–383, Fall 1970.
[6] A. Yezer, "Migration and models of urban development," M.I.T., Cambridge, Mass, Rep., 1971.
[7] D. L. Babcock, "Analysis and improvement of an urban model," Ph.D. dissertation, University of California, Los Angeles, 1970.
——, "Assumptions in Forrester's *Urban Dynamics* model and their implications," this issue, pp. 144–149.
[8] J. A. Hester, "Dispersal, segregation and technological change," Ph.D. dissertation, M.I.T., Cambridge, Mass., 1970.
[9] L. P. Kadanoff, "From simulation model to public policy: an examination of Forrester's *Urban Dynamics*," *Simulation*, vol. 16, pp. 261–268, June 1971.

A simulation model of urban labour markets and development policy

Leo Kadanoff
Bennett Harrison
Benjamin Chinitz

1 Introduction

Among the many social problems in the United States which came into sharp focus in the early 1960's was the problem of individual poverty and inequality in the distribution of income. Awareness of poverty and policies to ameliorate it had been a major preoccupation of the New Deal in the 1930's. But widespread poverty had persisted despite two decades of rapid economic growth and sharp rises in per capita income along with expanded income maintenance programs for large segments of the poverty population. The persistence of poverty in the midst of great affluence sparked new interest and commitment and the exploration of alternative approaches to its reduction.

Poverty had also become more visible as a consequence of the rapid urbanisation which brought large numbers of rural poor to the big cities, and the rapid suburbanisation of middle and upper income groups which created large concentrations of poverty in the cores of our largest metropolitan areas. Two other features of the new poverty were particularly disturbing. One was the substitution of the large, high density city ghetto for the rural South as a setting for poor blacks. The other was the prevalence of poverty among actual and potential members of the labour force because of low wages as well as high unemployment. It was during the 1960's that America discovered the existence of 'the working poor'.

Nor surprisingly, therefore, the new policy thrust in the 1960's, carrying the self-styled designation of a 'War Against Poverty', took on a new dimension absent in the New Deal era. Now the emphasis was on creating a greater ubiquity of economic opportunity, thus providing the conditions which would enable the poor to move out of poverty and participate more widely in contributing to, as well as sharing in, the fruits of economic progress. The new strategy called for a great variety of new programs to bring the poor into the economic system and equip them to command higher earnings and levels of public services: community action, Head Start, manpower training, anti-discrimination laws, job creation and black capitalism (actual job creation did not become an important part of the 'mix' until 1972). Nobody knew for sure which lever, if pulled or pushed, would achieve the greatest impact, so all levers received some attention.

Reprinted from *Urban Development Models*, ed. R. Baxter, 1975, The Construction Press.

Today, after more than ten years of 'combat', the uncertainty about cause and effect, and considerable disillusionment and disappointment with the results achieved so far, are widely prevalent. Considerable intellectual support has developed for a retreat from the 'opportunity' strategy in favour of a cash strategy in the form of a beefed-up universal income maintenance program. The latter is slow in coming and would in any case, in the opinion of many, fall far short of eliminating poverty while exacerbating social tensions between those who receive income from work and those who receive income without work despite the absence of any physical infirmities.

Thus, there is still considerable interest in exploiting the labour market as an anti-poverty device. But the diagnosis of labour market failure, especially in the core cities of our major metropolitan areas, remains at issue, and there is, therefore, sharp disagreement on the choice of policy tools.

Urban chronic under- and unemployment have been attributed to at least four different causes. One relates to the mismatch between the location of the workers in the city and the location of jobs in the suburbs. The mismatch becomes operational on the assumption that low income city dwellers are unable to commute outward to the suburbs in search of jobs in the short run. The validity of this 'explanation' of urban unemployment has been hotly debated in the literature (Kain and Persky, 1969; Harrison, 1974; Danziger and Weinstein, 1974). But, if this is the nature of the problem, the policy implications are clear. We must either create new jobs which are accessible to the unemployed or improve access to existing jobs. The latter, in turn, calls for subsidies to commuters and for the creation of low income housing opportunities near the relevant existing centres of employment. These objectives, in turn, can be satisfied in alternative ways. We can give vouchers for the purchase of existing transport and/or housing services, or we can operate on the supply side by expanding transit services and/or building subsidised low income housing in the suburbs.

A second explanation emphasises the mismatch of supply and demand in the more conventional sense of skills required and skills available, abstracting from geographic access, in the relevant labour market. In this view, the issue is not city and suburb but, rather, low marginal productivity relative to minimum wage levels, which may be set either by legislation or by the 'competition' of welfare programmes (Feldstein, 1973; Thurow, 1972; Harrison, 1974). This diagnosis suggests the need for reducing the effective cost of labour to firms. That can be accomplished in many different ways. We can try to induce private employers to take on these workers by offering the former a wage or training subsidy. Alternatively, we can pay for training programmes and hope that, after they are trained, the workers will find jobs in the customary fashion. In any case, the augmentation of underemployed workers' 'human capital' constitutes the basic approach.

Many observers find neither explanation convincing and insist that racial discrimination is the critical factor, even if access and labour quality are also aspects of the problem. This view generates a variety of policy prescriptions, reflecting the fundamental schism in values and expectations which has evolved in recent years. The integrationists would emphasise laws and measures to reduce discrimination in housing, education and employment. The separatists would prefer to get resources committed to the black population and speak of black capitalism and community control over the local public sector. Some observers call for a mix of both types of policy (Harrison, 1972).

This latter perspective has more recently become nested in a more comprehensive view of the labour market which has come to be known as the 'dual labour market' theory. In this view, which provides the fourth explanation, there are strong barriers to upward mobility for a large segment of the labour force comprised of blacks, females, and even white male unskilled workers; and the excess demand for the labour of this group, confined to a significant extent to a 'secondary labour market', is insufficient to generate satisfactory levels of incomes and/or stability of employment (Doeringer and Piore, 1971; Gordon, 1972; Harrison, 1972). Those who believe in the existence of strongly segmented labour markets call for policies designed to selectively expand the primary sector at the expense of the secondary, and to promote greater intersectoral mobility through restructuring of private hiring procedures. Such policies are usefully described as 'economic development plus institutional change'.

Since these theories of urban under- and unemployment emphasise radically different (although not mutually exclusive) policy prescriptions, it is of the utmost importance to provide some basis for testing and choosing among them and/or assigning weights since it is, of course, possible that all four account in some measure for the observed phenomena. We are currently engaged in the development and testing of just such a tool of analysis. Our purpose in this project is to shed light on one basic question: what kinds of intervention by public agencies or private agencies with government funds are most likely to be effective in creating more advantageous labour market conditions for the urban poor?

2 A model of labour market segmentation

The most general theory of labour market structure to emerge from twenty years of research is the dual - or more generally the segmented - labour model. Segments may exist (by race, class, sex, or human capital), but affirmative action, human capital, or 'ghetto dispersal' policies may be sufficient to permit intersegment mobility, thereby weakening the segmentation. On the other hand, such policies may prove insufficient to guarantee mobility across segments. The very rules of transition may have to be changed. Dual labour market theory - focuses on the operation of labour markets, especially those in the large cities. It has important antecedents in the work of Robert Averitt and John Kenneth Galbraith, who discovered and measured the 'uneven development' of industrial capitalist economies into a highly concentrated 'core' of powerful firms operating within a largely planned, non-market framework, and a competitive 'periphery' of smaller, low wage, low profit firms largely devoid of economic or political power; and in the earlier work of Philip Cairnes and Clark Kerr on 'non-competing groups' within the economy. The scholars most frequently associated with the 'dualist' paradigm are Peter Doeringer at Boston University, Michael Piore and Bennett Harrison at MIT, Barry Bluestone at Boston College, David Gordon and Thomas Vietorisz at The New School, and Howard Wachtel at The American University.

As indicated, dual labour market analysts believe the economy to be stratified into a 'core' and a 'periphery'. The division is functional and not simply semantic: workers, employers, and even the underlying technologies in the two strata behave very differently in important qualitative ways. The central institution of the core is the 'primary labour market'. In the primary labour market, the attributes of jobs and the behavioural traits of workers interact to produce a structure characterised by relatively high productivity, non-poverty wages, and employment stability. The high productivity of primary labour is a function not only of the knowledge and skills of the workers but also, and perhaps more fundamentally, of the capital equipment with which they work and the degree of training which their employers invest in them. High productivity permits the payment of high wages, as does the economic power of these employers, who are able to pass part of any wage increases along to consumers in the form of higher prices. High wages make these jobs relatively attractive, so that workers value stability. Because of the specificity of many of the tasks involved and because of the extent to which they have invested in their employees' training, employers in the primary labour market also value stability. The longer it takes a worker to reach maximum efficiency at a particular job, the greater the cost to the employer of losing that worker and of having to replace him with one who is unfamiliar with the tasks to be performed.

Primary labour markets tend to be highly structured, with intra-firm wages being determined by technical studies or institutionalised by custom, and fixed for a period of time for an entire class of jobs, quite regardless of which individual workers fill those jobs (civil service systems are excellent examples of such 'internal labour markets'). There is little or no competitive, i.e. continuous wage/productivity adjusting or wage 'bargaining' between individual workers and employers in this sector.

The most prominent part of the periphery of the economy is called, by contrast, the 'secondary labour market'. Jobs here involve simple tasks, performed within the context of relatively modest technologies. Workers display low productivity and high instability, and receive low wages. Since tasks are simple, workers do not require much training and reach maximum efficiency rather quickly; thus, employers have little incentive to induce stability among their workers. In fact, stability may be quite undesirable from the point of view of a secondary employer since workers who remain on the job for long periods of time usually require raises in pay and benefits and are more likely to join unions (a major empirical distinction between the primary and secondary labour markets seems to be the degree of union membership in each group). Secondary employers have little or no economic power: they behave much more like the textbook model of the perfect competitor. Consequently, they cannot afford to pass salary raises or other cost increases along and are, therefore, far more reluctant to grant such increases.

For all of these reasons and related ones having to do with the low status and often physically debilitating aspects of the work, employees also place little value on stability. Indeed, this aspect of the secondary labour market - the extraordinary turnover rates in such jobs and among such classes of workers - has caught the attention of many economists who are otherwise unsympathetic to the theory.

Thus, where primary employers and employees interact in an institutional setting characterised by high productivity, non-poverty wages, and high stability, the firms and work forces in the secondary labour market tend to organise themselves into production systems displaying low productivity, poverty level wages, and low stability (i.e. high turnover). How large are the respective sectors? Definitive numbers await the development of more precise, quantitatively verified definitions of the boundaries between the primary and the secondary labour markets. (Gordon, Vietorisz, Harrison, Robert Cohen, Paul Osterman, and Les Boden are among those economists currently attempting to measure the size of the labour market segments under various definitions or topologies. All of this empirical work moves beyond the simplified two-sector model described above.) By the simplistic criterion

Job control?	Internal labour market?	Appren-ticing?		
yes	no	no	Upper tier	Primary labour market
no	yes	yes	Lower tier	
no	no	no		Secondary labour market

Figure 1 Three-sector labour market model

Wage rate (dollars per hour)	Job type		
	Blue collar	White collar	Management-Professional
More than $5.00	Lower tier - Primary labour market		Upper tier - Primary labour market
$3.50-$5.00			
$2.00-$3.50	Secondary labour market		
Less than $2.00			

Figure 2 Segmentation of the labour market in the labour mobility model

of wages alone, using the hourly equivalent of the Bureau of Labor Statistics' 1970 lower level standard of income adequacy for a four-person urban family ($3.50 an hour), 40% or more of the jobs in America are in the secondary labour market. Using the new 'subemployment rate' as an index, Spring et al. (1972) estimate that anywhere from a third to 60% of the inner-city labour force may be so confined. In an ingenious (if still tentative) quantitative analysis, David Gordon (1971) has developed a dichotomisation of a large list of jobs according to stability, commitment of the worker towards his present job, job search techniques, and so forth. In two samples - non-ghetto Detroit and the Harlem ghetto areas in the City of New York - Gordon found an almost 50/50 split in the labour market. Moreover, this stratification appeared to be extraordinarily stable over time.

Doeringer and Piore have studied the structure of manpower and personnel management within the concentrated firms in the core of the economy. In the process, three important new variables have been isolated: (1) 'job control', the extent to which the worker controls his own tasks, in design or implementation; (2) 'the internal labour market', whether the firm manages its internal labour through the use of structured norms, work rules, and job lattices; and (3) 'apprenticing', the extent to which new workers receive training on the job through formal or informal relationships with older, more experienced workers. Using these categories, Piore (1972) has suggested a three-sector modification of the 'dual' labour market model, according to the scheme in Figure 1. Thus, in effect, he isolates as a 'lower tier of the primary labour market' those high laying, but often alienating, blue-collar production jobs (for example, in the automobile industry) which have become the object of attention by sociologists studying 'job satisfaction'.

Piore's three-sector model provides the most direct basis for the current version of our own model, shown in Figure 2. (Another insight of Piore's and Paul Osterman's, which our project will attempt to model, is the suggestion that many or even <u>most</u> young workers enter the 'world of work' through the secondary labour market - in the 'Macdonald's jobs' - so that the crucial question becomes: why do some pass through and 'up' while others remain behind indefinitely?). The purpose of this model is to provide us with a kind of 'wind-tunnel' for evaluating the implications of manpower and economic development policies, given the belief that labour markets are indeed segmented. This is the unique potential contribution of our model and defines its place in the field.

3 Probabilistic mobility models

The earliest published work in the labour economics literature to apply probabilistic transition analysis to the study of job mobility was completed at Cornell University in the mid-1950's. In their pioneering monograph, Isadore Blumen (1955) and his colleagues defined 'jobs' by major industry groups. All subsequent work (until Gordon defined 'jobs' as specific industry-occupation combinations) has been in terms of the standard occupational classification created by the federal government.

Robert Hodge (1966) modelled both inter- and intra-generational mobility. He found a high degree of similarity between empirically estimated transition matrices and matrices generated by a first-order Markov process.

Several studies at the Rand Corporation have applied Markov and 'semi-Markov' (e.g. 'mover-stayer') techniques to the study of labour markets. E.P. Durbin (1968) modelled the passage through manpower programs as a Markov process. Each program is characterised by four states: (1) no service, (2) counselling only, (3) training, and (4) 'out-employed'. (His model should have included a fifth state, 'out-unemployed'; many quantitative evaluations of the federal manpower programs indicate this to be the status of large numbers of graduates.)

Durbin's Rand colleague, J.J. McCall (1968), has published several different Markov and semi-Markov models of the process by which low income people move from year to year across one or another official 'poverty line'. McCall has also applied multiple regression analysis to explain the annual variation in transition probabilities in his sample (the Continuous Work History Sample of the Social Security Administration, useful for studying income flows but not job changes). The regression results indicate that macroeconomic growth (measured by changes in real Gross National Product over time) is indeed positively correlated with the probability of transition from poverty to non-poverty status for those who did, in fact, experience both states during the period 1957-1966, but they indicate that there exists a large subclass of individuals who remain in poverty throughout the period, regardless of changes in GNP. Terrence Kelly (1970) has applied a similar analysis to the one-step 1966 to 1967 transitions, using the Office of Economic Opportunity's Survey of Economic Opportunity.

More recently, Herbert Parnes (1969) and his colleagues at the Ohio State University Research Center for Human Resources Research have generated data with which to study interoccupational mobility. These 'Parnes data' have been used by Andrisani (1973) as the basis for a regression analysis of the determinants of youth mobility between primary and secondary jobs, and as the object of an ongoing factor analysis designed by David Gordon to identify job segments. At the Johns Hopkins University, James Coleman and his colleagues have studied occupational mobility in a sample of young workers generated by the staff of their Center for the Study of the Social Organization of Schools (Coleman et al., 1970; Ornstein, 1971).

Finally, researchers at MIT and The New School for Social Research are currently engaged in the analysis of job mobility and the transition between work and welfare. In this Research Center for Economic Planning project, transition probabilities are being estimated as dummy dependent variables in multiple regressions, in order to adjust the 1970 Census data for the heterogeneous life histories of those in the sample, and to identify the determinants of successful transition. The researchers are trying to identify sets of jobs with the 'dualistic' property that mobility is significantly greater within, than among, sets. Spatial relocation (migration) and aging are of particular interest as explanatory variables.

4 Specification of the labour mobility model

We have constructed a computer model of the segmented labour market which it is hoped will be relevant to the understanding of how the poor move from job to job. In conception, we have a kind of accounting scheme which keeps track of the numbers of workers, people, and jobs in various categories in the different metropolitan areas of the nation. The nation is divided into functionally different types of metropolitan areas. For example, one type might represent a declining area like Newark, another an economically growing SMSA (Standard Metropolitan Statistical Area) like Dallas, a third a set of areas to which a particular type of public programme is relevant.

The model describes the time development of a metropolitan labour market by an iterative process in which the values of the basic variables at a given time determine the rates of change of these variables and hence their values at the next iteration. There are different modes in which the model may be operated. In the 'many metropolitan area mode', the nation is divided into a few different types of metropolitan areas. Each type of area may be 'replicated' several times within the model. These types of areas may differ in their size or internal characteristics, or they may differ in the presence or absence of public programs. The other mode of operation is the 'one-city' mode. In this case, all metropolitan areas save one are given characteristics which are independent of time. Then, the single metropolitan area under study develops in an unchanging external environment.

The basic variables in the model include:

1 The numbers of people in the metropolitan areas, divided into different groups by their race, age, and educational attainment (see Figure 3).

2 The numbers of jobs, divided into three job types (blue-collar, white collar, and management-and-professional jobs) and four wage ranges (see Figure 2).

3 The most important variable in the model is the number of people in the work force, described by age, education, sex, and race as above. In our preliminary work, we have not included the division of the population into males and females. The reason for this omission is that our data source refers to 'family heads', who are overwhelmingly chosen to be male by the surveyors.

A. By residence 1. Rural areas 2. Different metropolitan areas
B. By race 1. White 2. Blacks and other 'non-Whites' including Spanish surnamed people
C. By age 1. Children or people still in school 2. Out of school but less than 65 3. Over 65 years old
D. By education 1. Family head with less than a high-school diploma 2. With a high-school diploma but no college education 3. With at least some college training
E. By sex 1. Male 2. Female

Figure 3 Population disaggregations in the model

The model includes a description of how jobs are distributed among different people. It is in fact a reasonably elaborate model of the supply of workers. However, we take the demand side to be perfectly rigid. That is, the demand for workers at a given job-type and wage level is fixed exogenously. Of course, this assumption of demand rigidity is much more characteristic of institutional theories of the labour market than of any of the neoclassical approaches. Thus, from the first, this model includes the presuppositions of the model's constructors in a very explicit fashion. Later, in the policy simulations, we will vary the inter-job distribution of the demand for labour instrumentally, in order to examine the consequences of changes in demand.

The model includes a simple demographic section which describes births, aging, and deaths within the population. Children are given educations. The educational level they attain is probabilistically determined by their parents' educational level and income.

The main component of entry into the job market is that of new workers who have just finished their education. In addition, more mature workers may also enter, partially in response to the availability of jobs which have been opened through turnover, and partially in response to optimistic expectations (as measured, for example, by reductions in the aggregate unemployment rate). Workers start out by entering the ranks of the unemployed. The types of jobs they seek are determined by exogenously defined branching ratios and by the numbers of workers and turnover rates for their racial group in the different job categories. The branching ratios depend upon the educational background of the workers. College-trained workers may enter jobs which pay between \$2.00 and \$5.00 per hour (in 1970 dollars), but entrants with a lower educational attainment enter only at levels below \$3.50 per hour. Figure 4 shows the entrance scheme. Obviously these institutional 'rules' of entry can be modified to study the model's sensitivity to such assumptions as these.

Work force entrants may migrate from one metropolitan area to another or to and from rural areas. These entrants have a migration rate which is dependent upon unemployment rates in both the area they are leaving and the area they are entering. A similar, but smaller, componentof migration represents the motion of mature workers seeking better job opportunities. Finally, the higher income workers may migrate to obtain a particular job in another metropolitan area. The migrants in the first two categories are added to the unemployed group in their destination, while the workers in the final category enter into the ranks of the employed.

People leave the work force either when they retire or during their normal working years. In the latter situation, the rate of leaving is large for low-wage jobs and for unemployed workers and is much smaller for employed workers and for higher-wage jobs. Thus, one possible response for an unemployed worker having experienced an unsuccessful job search is to leave the work force. Another possible response is to become 'discouraged' and lower his reservation wage, thereby becoming eligible for employment in a 'lower level' job in Figure 2.

Several of the processes mentioned above, including retirement, migration, and job-leaving, produce job openings. The most important part of the model is the filling of these available jobs.

We visualise the following type of job distribution process. Employers know how many available jobs they have to offer per year. (Our use of annual time periods is dictated by the data to be described below.) Call this rate at which jobs become available JA(jc), in which jc is an index which stands for the different types of jobs to be offered. Each employer makes a list of people who apply or can be induced to apply for the job. From among these applicants, the employer makes a list of workers who will be judged suitable for the job. The number of such listed workers, who have worker characteristics wc for all jobs with job characteristics jc, is defined to be WL(jc,wc).

Then, the available jobs are fairly distributed to a proportion of the listed workers until all available jobs are given out. The number of jobs of type jc distributed to workers of type wc per unit time is

$$JA(jc) \cdot WL(jc,wc)/\sum_{wc} WL(jc,wc)$$

Thus, all is determined if we but know how many workers of a particular type get listed for a particular type of job. In the model, the probability that a given type of worker will apply for and then be listed for a given type of job is described by an exogenously defined matrix called a job distribution function, JDF(jc,wc). The most important worker characteristic is the type of job he or she presently holds (or is applying for). Given that datum, the job distribution function defines the possible types of transitions and promotions which are permitted within the model.

Of course, every employer hires mostly people who have had similar types of jobs in the past. Thus, the job distribution function is constructed so that the probability of assignment to any job is largest for those who hold or have just held jobs in the particular job category being considered. However, in addition, some transitions will involve promotions. The only permitted promotions involve an increase of one step in wage rate. These promotions are depicted in Figure 5. The model also permits some changes in job type without changes in income level. In particular, higher wage blue-collar and white-collar workers can move into management jobs. Also, secondary work-force members may change their job type. These lateral changes are indicated in Figure 6. Figures 5 and 6 thus show all the possible transitions caused by hirings within the model. (As mentioned above, unemployed workers may become discouraged and move downward to a lower category in Figure 5.)

In general, employers prefer workers with more educational experience. Hence, the job distribution functions are constructed so that the probabilities of employment or promotion are higher for more educated workers. This educational preference is stronger in the management-professional jobs, intermediate within white-collar jobs, and weakest in blue-collar jobs. Similarly, workers who are out of a job are more likely to seek and accept a 'new' job than are fully employed workers. For this reason, the probability of entry into a job is made to be higher for unemployed than for employed workers (because of the postulated labour supply behaviour).

An additional important component of the job distribution process is racial discrimination. We represent this discrimination as a kind of conservatism in the job distribution process which discourages a very rapid change in the racial composition of any occupation or industry. Following Barbara Bergmann (1971), we visualise some jobs or job categories as labelled 'mostly for whites' or 'mostly for blacks'. When such expectations exist, employers, unions, and employees all act in such a manner as to slow down any possible changes in racial composition.

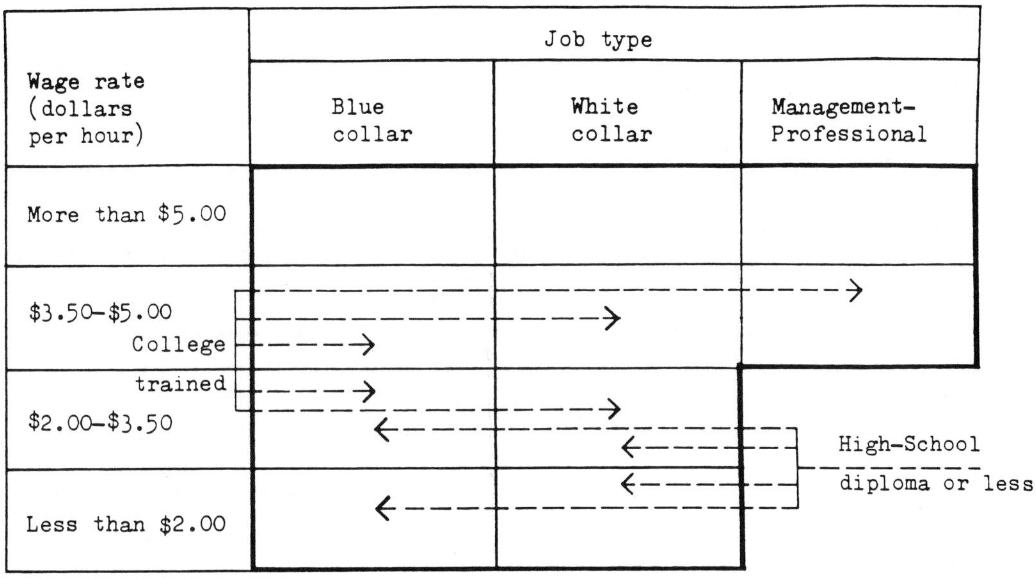

Figure 4 Allowed types of entry into the work force

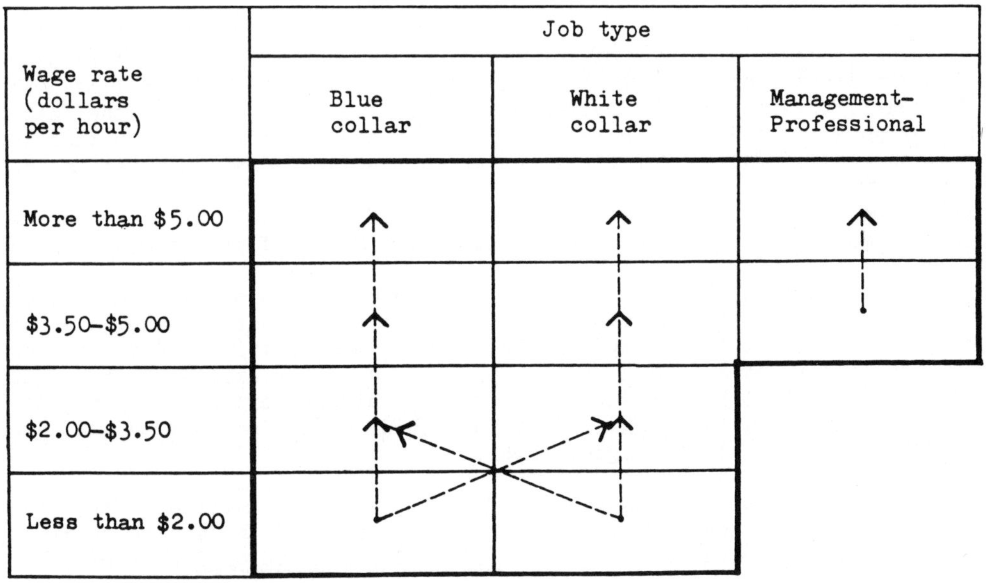

Figure 5 Possible promotions in the model

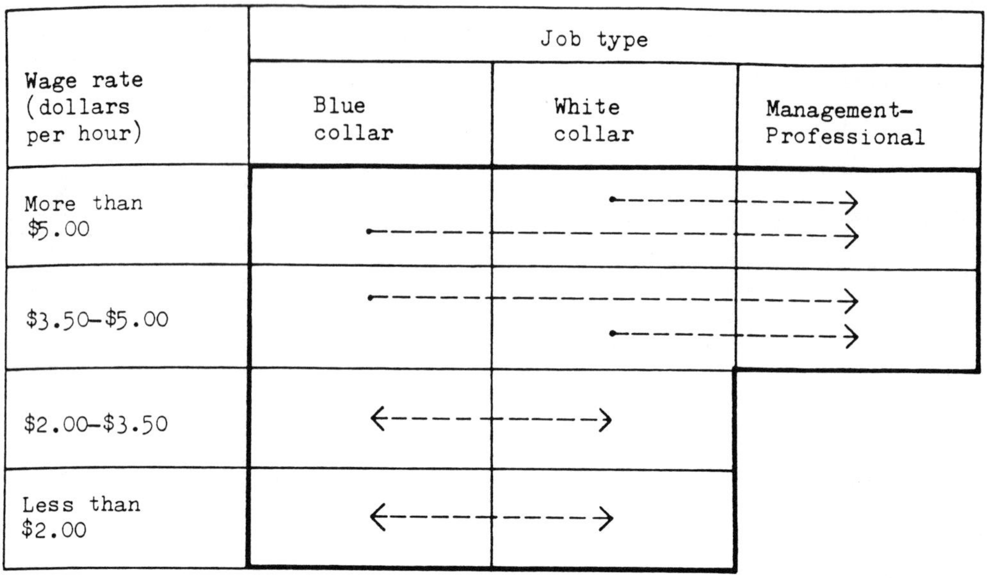

Job type
Wage rate (dollars per hour)
Blue collar
White collar
Management-Professional
More than $5.00
$3.50–$5.00
$2.00–$3.50
Less than $2.00

Figure 6 Possible lateral job changes

To represent this variant of statistical discrimination, we imagine that each available job is labelled with the race of its last holder. In the job offering process, a worker of the same race is much more likely than a worker of a different race to apply for the job and be accepted on the list. Within the model, this concept is represented by including the race of the last holder as a job characteristic and by making the job distribution function take on smaller values for processes which involve racial turnover in the job.

Since the promotion process is the key part of our model, we must consider it in particular detail. The model includes two kinds of promotions:

1. An institutionalised mechanism. In some occupations (e.g. the police), higher wage jobs are given almost exclusively to workers who have 'served their time' at lower wage levels. In the model, this kind of process is represented by moving a certain proportion of the employed workers during each year into higher wage categories. This proportion is chosen to be independent of the tightness of the labour market so long as sufficient jobs are open at the higher level.

2. A demand-activated mechanism. If any jobs remain unfilled after the 'institutionalised' promotion processes described above, the remaining jobs are distributed among both employed and unemployed workers.

To put this distinction in a more specific context, in most universities the hiring of assistant professors is demand-actuated, but the promotion from one tenured position to another is usually institutionalised.

Wage rate (dollars per hour)	Total employed work force			Black employed work force			Total employed work force		
	Blue collar	White collar	Management/ Professional	Blue collar	White collar	Management/ Professional	Blue collar	White collar	Management/ Professional
More than $5.00	23.7	10.0	44.3	2.3	0.9	1.0	10.1	7.9	5.2
$3.50–$5.00	30.4	13.8	14.1	5.1	1.2	0.9	6.7	5.6	4.2
$2.00–$3.50	31.3	29.1		8.0	3.5		8.3	6.6	
Less than $2.00	14.8	10.2		6.9	2.2		16.8	11.5	

Figure 7a Employed work force (in thousands of people) for the job types and wages in the model for a hypothetical metropolitan area of 600,000 people

Figure 7b Micro-unemployment indices (in per cent) for the job types and wages in the model

5 Initialisation and calibration

The model works as a set of differential equations which describe the time-dependence of the state variables. These variables are:

1. The population in different age, race, and educational categories, and

2. the numbers of jobs filled in the various categories by different types of workers.

To initialise the model, we must pick starting values of these variables. The basic data sources for the initialisation are the University of Michigan Survey Research Center's Panel Study of Income Dynamics and the 1970 Census of Population. The Panel Study describes employment and changes in employment patterns for heads of households for the five years between 1968 and 1972.

We concern ourselves here only with those heads of metropolitan households whc are less than 65 years old and have some previous history in the work force. Their job-type and wage category is determined by their present job or, if they are unemployed, by their last job. In this way, we construct a data table which lists the number of employed workers by their personal characteristics and by their job characteristics. These starting data come from the 1972 data in the Panel Study. This table is multiplied by an appropriate constant so that the employment figures are appropriate for a typical metropolitan area of about 600,000 people.

To show the meaning of these data in a clearer fashion various marginals are given in Figure 7 and Table 1. Of course, the blacks are relatively concentrated in low-wage and blue-collar jobs. The white population has a higher proportion of high-wage and highly educated workers. By our initial definitions, the ratio of primary to secondary jobs is 2:1 for whites and 1:2 for blacks.

The Michigan data also enable us to calculate unemployment indices for each cell as given in Figure 7. A worker's fractional unemployment is given by

fractional unemployment = 1 − weeks' fraction worked x hours' fraction worked

The fractional number of weeks he or she works is the actual weeks worked divided by (52 less the number of weeks of vacation). Similarly, the hours' fraction worked is the average number of hours worked per week divided by 35 and cut off at a maximum of one (thus, 'overtime' implies 'overemployment' - although not necessarily an escape from poverty). Therefore, both a short work-week and also the weeks out of work can contribute to our definition of unemployment. However, the great majority of the observed unemployment comes from workers who work a full week but spend several weeks during the year out of work.

Census data are used to define the number of blacks and whites in the different age categories. These are further disaggregated by the educational attainments of their household heads. This division is done in proportion to the numbers of family heads in the relevant categories as set in the Michigan file.

Categories of workers	Employed workers (in thousands)	
	White	Black
TOTAL EMPLOYMENT	190.3	32.0
JOB TYPE		
Primary (wage more than $3.50 per hour)		
Management-Professional	56.5	1.9
White collar	21.8	2.1
Blue collar	46.9	7.4
Secondary (wage less than $3.50 per hour)	65.0	20.6
EDUCATIONAL ATTAINMENT		
Some college	80.2	5.3
High-school diploma	64.1	9.8
Less than high-school diploma	46.0	16.9
WAGE LEVEL (in dollars per hour)		
More than $5.00	74.0	4.2
$3.50 - $5.00	51.3	7.2
$2.00 - $3.50	48.9	11.5
Less than $2.00	16.1	9.1

Table 1 Numbers of employed workers (in thousands) in various categories of workers - from Michigan file, 1972 data. The numbers apply to a typical metropolitan area of some 600,000 people. All numbers refer only to employed heads of families. Wage rates are in 1970 dollars.

The most important data for the calibration of the model are the transition rates from job to job. These data are also taken from the Michigan file. The estimation involves asking two sets of questions:

1. Where will workers who now have a particular job-type end up after n years? How many will leave the work force? How many will work for new employers? How many will work at the different wages and job types?

2. Alternatively, one may start with a given number of workers with a particular wage and job type. One may look back n-years and ask: How many of these workers were out of the work force n-years ago? How many have changed employers? How many were in the different wage and job-type categories?

Tables 2 and 3 show the answers to these questions for a particular group - the lowest-wage blue-collar category. (Because of the limited number of cases in the Michigan file, there is no control on age or race in constructing these transition tables.) These kinds of data form the basic raw material of our model. Note that the mobility patterns revealed by the empirically estimated transitions shown in Tables 2 and 3 are strongly supportive of dual labour market theory since over the five years, low-wage blue-collar workers tend to 'recirculate' within the secondary labour market.

1000
workers start from lowest wage
blue-collar jobs. After one year

132
leave work
force

661
stay with old
employers. These
are divided as:

New wage	BC	WC	MP
>$5.00/hr			
$3.50–$5.00	1		3
$2.00–$3.50	178	9	
<$2.00	349	121	

207
work for new
employers. These
are divided as:

BC	WC	MP
2		
11		
63	15	8
94	14	

Table 2a One–year transitions for workers in lowest wage blue–collar category

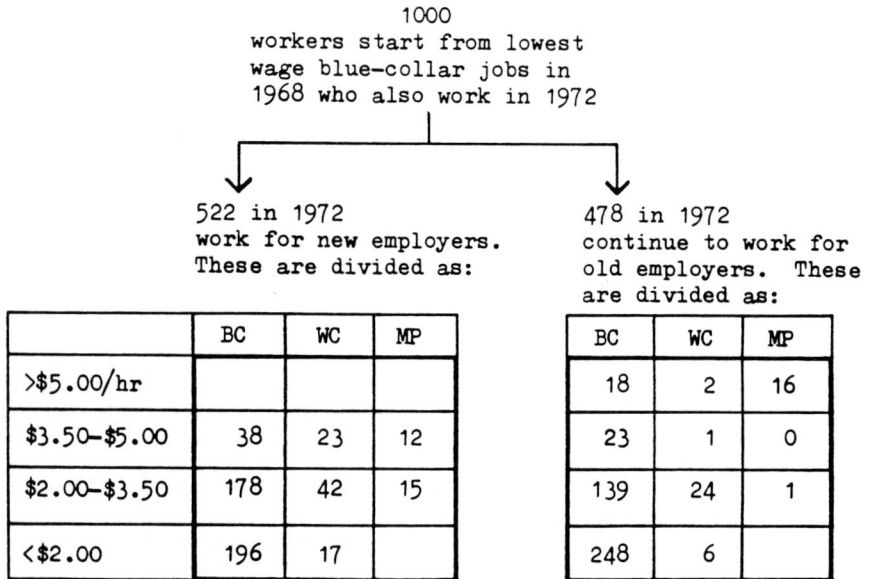

1000
workers start from lowest
wage blue–collar jobs in
1968 who also work in 1972

522 in 1972
work for new employers.
These are divided as:

	BC	WC	MP
>$5.00/hr			
$3.50–$5.00	38	23	12
$2.00–$3.50	178	42	15
<$2.00	196	17	

478 in 1972
continue to work for
old employers. These
are divided as:

BC	WC	MP
18	2	16
23	1	0
139	24	1
248	6	

Table 2b Four–year transitions for workers in lowest wage blue–collar category

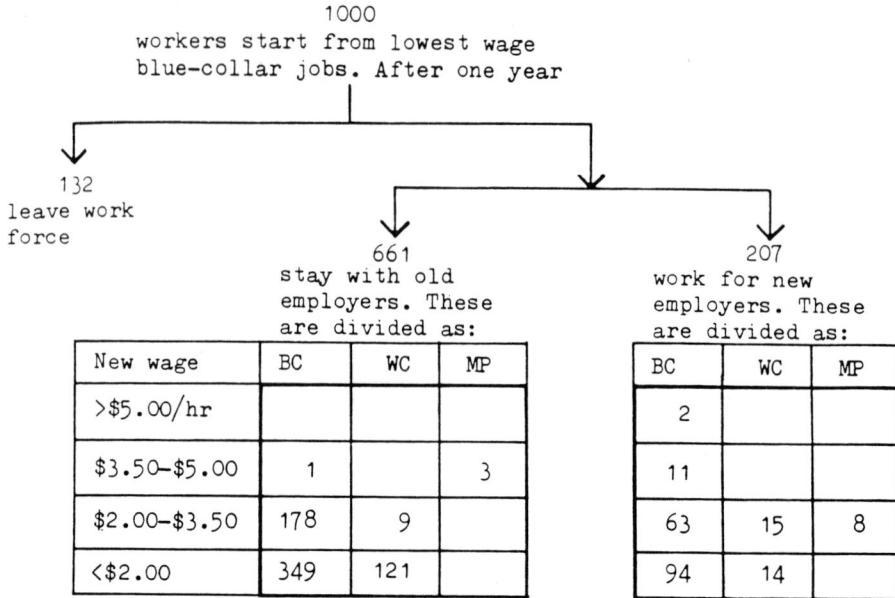

1000
workers have the lower wage blue-collar job in 1972 who also worked in 1968. These came from

496
worked for same employer in 1968, divided as:

	BC	WC	MP
>$5.00	72		
$3.50–$5.00	2	17	
$2.00–$3.50	104		
<$2.00	300	1	

504
worked for another employer in 1968. These were divided as:

BC	WC	MP
1	18	
52		
127	13	
239	54	

Table 3 Sources of workers for low-wage blue-collar jobs

To fully define the model, one must set the values of some 200 or so parameters which appear in the differential equations. Some of these parameter values, for example, birth rates and death rates, are known with reasonable accuracy. Others, such as the transition rates between different job types, have been estimated from the Michigan file. Some others – for example the elasticities of migration with respect to intercity unemployment rates – have been studied by other workers. Yet others, for example the parameters describing the effects of racial discrimination in the hiring process, cannot be easily measured.

At this point, only a very few of our parameters have been set in a satisfactory way. In fact, the great majority of the parameters have been fixed only by guesswork. As a partial definition of parameters, we have required that the parameter values be consistent with measured rates of change of the state variables. Roughly 100 conditions on the parameter values have been set in this manner. However, as yet we still have more parameters than equations to determine them.

For this reason, we must interpret the quantitative output of the model with a great deal of caution. At most, we can say that the apparently quantitative statements of the models are, at this point, only qualitative arguments.

6 Model output

A first indication of the behaviour of the model can be given by instrumentally improving the economic climate in a particular metropolitan area. To do this, we compare two hypothetical metropolitan areas. In the first, the base case, we set up the rate of increase of jobs so as to just about equal the rate of population increase. Then, in this 'status quo' base case, employment simply grows at a constant rate, which is chosen to be the same for each wage and job category. The parameters in the model are arranged so that unemployment rates remain essentially constant in this case.

For comparison, consider a 'program' metropolitan area in which the job climate is improved by the following process. For 2.5 years, extra jobs are added to the area. They are added to the different categories in proportion to the number of jobs which already exist in each job category. In total, 2,500 jobs are added. In each category, there is an increase in the total number of jobs by one per cent. Then, for the next 2.5 years, the metropolitan area keeps these additional jobs, but no extra jobs are added in comparison to the base case.

Then, a comparison between the base case and the program area serves to tell us about the differential effects of economic growth upon different classes of the population. The changes in unemployment rates are summarised in Table 4. The two periods mentioned are 2.5 years, when the job creation program terminates, and 5 years, which includes an additional period of 'natural' development in the presence of the new jobs. The data in this table are changes in unemployment rates. For example, the first entry in the table is obtained by comparing the total unemployment rate among whites in the program case, which is 5.49%, with the rate in the base case, 6.13%. The difference is the reduction in the white unemployment rate created by the program, i.e. 0.64%.

Since 1% more jobs are added to each category, if there were no change in the number of workers in each category, this program would reduce the unemployment rate in every category by 1.0%. However, the flows of workers into the metropolitan area by migration and from category to category change this figure considerably. Extra net migration into the area reduces the net effect on unemployment by about one-third after 2.5 years and makes the effect only 40% of the ideal after 5.0 years.

Furthermore, the effects of the improvement in the climate are unevenly distributed among the different segments. In general, the most disadvantaged workers feel the biggest and longest lasting effect. For example, after 5.0 years, the reduction in the number of unemployed among workers in the lowest-wage jobs is ten times that in any other wage category. The extra job openings in higher-wage categories tend to grab up low-wage workers and draw workers from other metropolitan areas.

The management-professional category sees a particularly small long-term effect upon unemployment because in-migration works particularly effectively to fill vacancies in this category.

In short, the model has the result that more disadvantaged workers tend to experience greater decreases in unemployment as a result of proportional economic growth. Conversely, a decline in economic health is felt most strongly by the most disadvantaged workers. To see this, we ran a case in

Disaggregation	Class	Decrease in unemployment rates after	
		2.5 years	5.0 years
Race	White	.64	.36
	Black	.83	.61
Education	Less than high-school diploma	.96	.65
	High-school diploma	.62	.38
	Some college	.45	.19
Job segment	Primary		
	Management-Professional	.25	.01
	White collar	.48	.12
	Blue collar	.54	.24
	Secondary	1.04	.77
Wage level (dollars per hour)	Less than $2.00	2.10	1.70
	$2.00 - $3.50	.26	.16
	$3.50 - $5.00	.32	.07
	More than $5.00	.47	.16

Table 4 Decrease in unemployment rates (the rates expressed in per cents) caused by across-the-board economic improvement. The most disadvantaged groups benefit most because in addition to the direct benefit they see from the expansion of low-wage jobs, they also can, to some extent, compete for the added higher-wage jobs.

Disaggregation	Class	Increase in unemployment rates after	
		2.5 years	5.0 years
Race	White	2.17	3.51
	Black	3.17	5.57
Education	Less than high-school diploma	3.25	5.70
	High-school diploma	2.41	3.72
	Some college	1.65	2.36
Job segment	Primary		
	Management-Professional	.92	1.05
	White collar	1.57	2.06
	Blue collar	1.83	2.65
	Secondary	3.62	6.41
Wage level (dollars per hour)	Less than $2.00	7.63	13.35
	$2.00 - $3.50	0.87	1.35
	$3.50 - $5.00	1.17	1.47
	More than $5.00	1.65	2.24

Table 5 Increase in unemployment rates (the rates expressed in per cents) caused by no-growth situation

which over a period of 5.0 years there was zero net growth in all the job categories. Since the population is growing at about 1.4% per year, this cessation of growth is a disaster for the metropolitan area. In Table 5, we show increases in unemployment rates caused by this cessation of growth. The differences shown are obtained by subtracting the unemployment rates in the base case (normal growth) from those in the no-growth situation. Note the huge increase in unemployment at the bottom of the wage scale and for the lowest educational category. The large increase in unemployment at the bottom has two causes:

1. The workers in the lowest-wage category are least geographically mobile, and so they tend to stay in the area even though the economic climate is very bad, and

2. Unemployed workers from higher-wage categories 'filter down' and compete with the original group of low wage workers.

An essential feature of the 'segmented' view of the labour market is the idea that an improvement in the economic climate in one segment has at best a weak influence on the other segments. To test this, we consider the effect of selectively adding blue-collar jobs at the highest-wage rates. As before, we add these jobs over a period of 2.5 years and compare with a base case in which the number of jobs in the metropolitan area remains fixed.

The basic effect of this job addition is shown in Figure 8. This figure shows a very substantial reduction in unemployment. 2,500 jobs have been added during the first 2.5 years. At the end of this time, there are 2,040 fewer unemployed blue-collar workers in the sector with wages above $3.50 per hour.

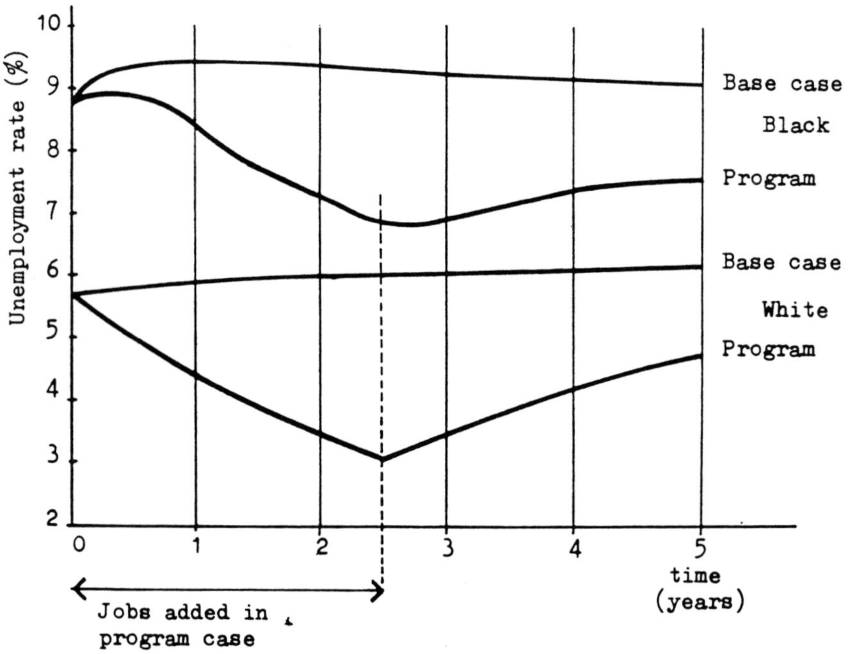

Figure 8 Effects of job addition

Category to which jobs added			Changes in unemployment					
			Total changes after 2.5 years		Changes in segments			
					Primary			Secondary
Segment	Job Type	Wage (dollars per hour)	White	Black	MP	WC	BC	
Primary	Management-Professional	> $5.00	0.36	0.18	0.82	0.51	0.03	0.10
		$3.50–5.00	0.29	0.30	0.20	0.50	0.05	0.31
	White collar	> $5.00	0.54	0.54	0.02	2.90	0.01	0.51
		$3.50–5.00	0.44	0.58	-0.02	1.04	0.02	0.78
	Blue collar	> $5.00	0.89	0.53	0.0	0.0	2.66	0.40
		$3.50–5.00	0.63	0.94	0.0	0.03	0.94	1.40
Secondary	White collar	$2.00–3.50	0.53	0.92	0.01	0.03	0.05	1.60
		< $2.00	0.65	1.15	0.0	0.02	0.03	1.77
	Blue collar	$2.00–3.50	0.60	0.98	0.0	0.04	0.00	1.63
		< $2.00	0.63	1.27	0.0	0.02	0.02	1.82
All	All	All	0.64	0.88	0.25	0.48	0.54	1.02

Table 6 Changes in unemployment rates caused by alternative job creation programs. Over a period of 2.5 years, 2500 jobs are added in each case.

Naturally, the reduction in unemployment in all other sectors must be very small. Table 6 shows this reduction by giving the difference between the unemployment rates in the program and the base case for the different segments of the job market. The data in this table represent, we believe, roughly what the segmented labour market theorists would expect: selective expansion of high-wage blue-collar jobs has a very small effect upon reducing unemployment outside the primary blue-collar segment. Even in the long run (2.5 years after the last jobs are added), the change in unemployment rate remains twice as large in the primary blue-collar segment as in the secondary segment (see Table 7). Thus, the effect of adding these jobs to the primary blue-collar segment remains mainly limited to improving conditions in that segment.

However, this result is not an inescapable conclusion from the data available to us and the model structure. On the contrary, we can reach quite the opposite conclusion by guessing parameters differently. The particular parameters which influence this result most strongly reflect the mechanism for promotion within and into the primary blue-collar category. Remember that we have two promotion mechanisms:

1. An institutional mechanism in which the promotion rate is, in good times, independent of the demand for new workers.

2. A demand activated mechanism, in which the promotion rate is proportional to the availability of jobs at the upper-wage level.

The distinction between these mechanisms is essentially a difference in the elasticity of promotion with respect to job availability. For the first mechanism, the elasticity is zero; for the second, it is one.

In order to clarify this distinction with a specific numerical example, we use data from the Michigan study. Table 8 gives a breakdown of the origin of 1,000 blue-collar workers who earned more than $5.00/hr. (1970 dollars) in 1972. This group was selected to include only people who also worked in 1968. The breakdown shows how many worked for the same or another employer and also how many did or did not pass over the $5.00/hr. wage level.

Imagine that 100 new jobs were added. A pure shift-share analysis would be based upon the idea that these new jobs would be parcelled out to new workers in proportion to the numbers who changed employer or wage level previously. In essence, this shift-share analysis is a case of pure demand activated promotion. The result is shown in Table 9a.

However, it is also reasonable to believe that promotions within the same firm depend upon a 'ladder' and do not change very much with the economic climate. If we then apply the shift-share analysis only to new hirings, we get the result shown in Table 9b. This table results from the assumption that within-firm promotions are so institutionalised that they do not increase when the economic climate improves. Notice the net result of the institutionalised promotion mechanism is to decrease the amount of upward mobility caused by the new jobs. In fact, 77 workers move up in the pure 'demand activation' case of Table 9a, but only 48 workers move up in the second case.

The same thing happens in the model. In constructing the equations of the model, we have made assumptions about promotions similar to the modified shift-share analysis of Table 9b. Table 10 shows the result of replacing

Time (in years)	Reduction in unemployment rate				
	All workers	Primary segment			Secondary segment
		MP	WC	BC	
1	.42	0	0	1.58	.07
2	.75	0	0	2.61	.27
3	.83	0	0	2.51	.53
4	.69	0	0	1.83	.65
5	.59	0	0	1.36	.68

Table 7 Reduction in white unemployment rates (the rates expressed in per cents) caused by added high-wage blue collar jobs

	Same employer 1968 and 1972	Another employer	Total
Same wage level > 5.00/hr.	500	114	614
Lower wage level < 5.00/hr.	280	106	386
Total	780	220	1,000

Table 8 A breakdown showing the origin (in 1968) of workers who had high wage (> 5.00/hr.) blue collar jobs in 1972. Data from ISR survey.

	Same employer	Another employer	Total
Same wage level	X	23	23
Below	56	21	77
Total	56	44	100

Table 9a New hirings caused by adding 100 new jobs. Pure shift-share analysis of the result of adding new high-wage blue collar jobs. These jobs are apportioned. proportionally to the number of workers who change their employer or wage level as shown in Table 8.

	Same employer	Another employer	Total
Same wage level	X	52	52
Below	X	48	48
Total	0	100	100

Table 9b New hirings caused by adding 100 new high-wage blue collar jobs. In this case, we compute the new hirings by assuming that all the extra hirings come from outside the firm. These hirings are apportioned proportionally to the number of workers who change employers and obtain high-wage blue collar jobs as shown in Table 8.

Time (in years)	Reduction in unemployment rate		
	All workers	Primary blue collar segment	Secondary segment
1	0.36	1.10	0.24
2	0.65	1.67	0.62
3	0.71	1.44	0.93
4	0.60	0.97	0.94
5	0.51	0.71	0.88

Table 10 Reduction in white unemployment rates (the rates being expressed in per cents) caused by added high-wage blue collar jobs. The main difference between this case and the case reflected in Table 7 is the substitution of pure 'demand activated' promotion rates.

Parameters chosen for promotion	Reduction in unemployed workers	
	Primary blue collar segment	Secondary segment
Pure demand activation	670	810
40% demand activation 60% 'Institutional' mechanism	1270	460

Table 11 Effect of promotion mechanism upon reduction in number of unemployed workers in two segments. The reduction shown is the difference between the number of unemployed in the program cases and those in the base cases.

this assumption by a pure 'demand activated' process like that of Table 9a. In other words, the distinction between intra- and inter-firm job changes is relaxed.

It would be hard to say that Table 10 represents an especially good support for segmented labour market theory. After five years, the effect on the secondary market has surpassed the effect upon the primary market segment. To make this point rather sharply, take the actual reduction in the number of unemployed workers three years after the beginning of the program. These numbers are shown in Table 11. The pure-demand theory makes for a much stronger effect of the primary jobs upon the secondary market; the theory with partially institutionalised promotions does just the opposite.

It should be no surprise that an institutionalised picture of promotion processes gives support to a segmented picture of the labour market, while the more free-market (human capital, neoclassical) type of picture leads to relatively little segmentation. So in one sense, the answer to the question of whether our work supports a segmented view of the labour market is: we can have it either way. If we choose some parameters to fit an institutionalist view, we get a segmented picture. If we make the opposite choice of parameters, we get the opposite answer.

In the next sections, we continue to employ parameter values reflecting institutionalised barriers to mobility. However, the little example given here should warn the reader that the conclusions we reach are closely related to our initial presuppositions. Later econometric work will focus as sharply as possible on 'positive' estimation of these crucial parameters. We have not been working entirely in a vacuum, however. The pairs of matrices in Tables 2a, 2b and 3 clearly show that intra-firm transition probabilities do differ significantly from inter-firm transition probabilities in the Michigan Panel Survey. In his research at M.I.T., Bader is discovering the same thing in the Ohio State ('Parnes') Longitudinal Survey, a completely independent data set covering a comparable period of time.

We now continue the description of the modelled effects of adding new jobs in various wage and job-type categories. Once again, we compare all output to a base case in which SMSA employment grows at the natural (i.e. population) rate. We then consider a series of program cases in which 2,500 additional jobs are provided during the first 2.5 years.

Table 6 gives the reduction in unemployment rates produced by several different programs. In the first ten cases, jobs are added at the top of the respective segments of the job market. The last case shown is one in which jobs are added to all job categories. The predominant effect of adding the jobs at the highest wage level is to reduce unemployment in the segment in question. When jobs are added in the \$3.50-\$5.00 range, there is a spillover of benefits into the secondary segment. This spillover is largest from adding jobs in the blue-collar category and smallest when the jobs are added to the management-professional category. In the latter case, a major spillover occurs into the white-collar primary segment. In general, induced immigration substantially reduces the unemployment rate gains for the addition of management-professional jobs. Jobs added to the secondary segment hardly affect unemployment rates in other segments.

The data in Table 6 suggest that, if one desires to aid secondary workers and blacks (in terms of lower unemployment rates, if not necessarily higher incomes),

Category to which jobs added			Changes in unemployment					
Segment	Job type	Wage (dollars per hour)	Total changes after 5 years		Changes in segments			
					Primary			Secondary
			White	Black	MP	WC	BC	
Primary	Management Professional	> $5.00	.10	.09	.06	.17	.01	.06
		$3.50–5.00	.12	.22	- .05	.0	.04	.24
	White collar	> $5.00	.27	.42	- .03	.72	.01	.51
		$3.50–5.00	.23	.42	- .02	.17	.03	.53
	Blue collar	> $5.00	.59	.73	.00	.00	1.30	.70
		$3.50–5.00	.36	.70	.00	.02	.20	.92
Secondary	White collar	$2.00–3.50	.36	.64	.01	.01	.05	.99
		< $2.00	.44	.90	.0	.01	.03	1.30
	Blue collar	$2.00–3.50	.34	.67	.0	.02	- .03	1.04
		< $2.00	.43	.97	.00	.01	.03	1.32
All	All	All	.36	.61	.01	.12	.24	.79

Table 12 Changes in unemployment rate caused by job creation programs. Eleven alternate programs are shown. In each case, 2500 jobs are added in the first 2.5 years; then more jobs are added and natural growth continues for 2.5 years

the best places to add jobs are in the secondary categories and the lower-wage part of the primary blue-collar category. The upper-wage blue-collar jobs have the most beneficial effect upon white unemployment rates and a moderate effect upon overall black unemployment rates. The reason that these jobs have a relatively small effect upon blacks is that the model includes a particularly large amount of racial discrimination in this wage and job-type category. We, in turn, derive this picture from the observed fact that, within this category, black unemployment is much higher than white.

That there can be no such thing as a perfectly general 'classless' development policy - a major tenet of radical political economics, which argues that conflict is more prevalent than harmony in the American economy - seems clear from Table 6. If one aims at improving all-over unemployment rates in a given metropolitan area, top-wage blue-collar jobs are best. If one is concerned with unemployment rates among blacks, the lowest-wage jobs are best, followed by the lower-wage primary blue-collar jobs, and jobs in the $2.00-$3.50 per hour wage range. If one is concerned with unemployment among secondary workers, secondary jobs are best and the lower-wage primary blue-collar jobs are next best. From this point of view, the upper-wage blue-collar jobs are a poor third. Moreover, if the elimination of 'working poverty' is the objective, none of these policies helps very much (note, however, that this policy analysis assumed given institutional rules of transition).

One can carry this same analysis considerably further. The job creation program discussed in Table 6 terminated 2.5 years after its beginning. Then we observed the program areas as they developed, with no additional jobs added, during a period of another 2.5 years. During the latter period, some of the benefits shown in Table 6 are dissipated. Additional workers move into the program area. Some nonparticipants residing in the program area enter the labour force. Thus, the reductions in unemployment detailed in Table 6 will become less marked after five years.

Table 12 shows the reductions in unemployment rates of the program areas five years later, compared to the base case for the different programs indicated. Most of the general features shown in Table 6 are continued at the later time shown by Table 12. The management-professional job programs remain very ineffective in reducing unemployment. The white-collar programs are relatively ineffective in reaching the secondary group in comparison with the blue-collar programs. However, one marked difference is shown in the comparison of the two tables. Over the long term, there is a greater amount of filtering down of benefits from the top-wage white-collar and blue-collar jobs into the secondary group. In fact, the benefit felt in secondary unemployment as caused by additional top-wage blue-collar jobs increases over the later 2.5 years. Creation of new jobs in this segment is still not as beneficial to secondary workers as the $3.50-$5.00 blue-collar jobs or secondary jobs themselves. But the high-wage jobs serve some function in reducing unemployment among low-wage workers.

We now consider an educational program in which a predetermined number of children during each year are given extra education. There are two kinds of benefits given: either raising high-school dropouts to the level of high-school graduates or raising high-school graduates to the level attained by the people with some college training.

Advocates of human capital-oriented anti-poverty policy anticipate two different impacts on the job market when educational levels are increased. First, workers gain 'human capital', which enables them to win better jobs and,

PROGRAM: In every year, 600 people are given extra education			RESULTING CHANGES: In different wage ranges, change in number of employed workers				
Racial group	Level educated from	Level educated to	Effect upon racial group	Effect upon wage range (dollars per hour) Less than $2.00	$2.00–$3.50	$3.50–$5.00	More than $5.00
White	Less than high-school diploma	High-school diploma	White	- 88	- 3	4	0
			Black	78	- 12	- 5	0
	High-school diploma	Some college	White	- 99	-104	87	17
			Black	84	85	- 74	- 15
Black	Less than high-school diploma	High-school diploma	White	160	-106	- 26	- 2
			Black	-172	109	27	2
	High-school diploma	Some college	White	373	351	-594	-120
			Black	-403	-388	635	124

Table 13 Changes in employment resulting from education in which 600 people per year are given a higher level of education. Results after five years.

presumably, to perform better in any job. Secondly, the increase in the time in school reduces the pressure of new work force entrants upon the market. There is also some feedback through the macroeconomy, in orthodox human capital-neoclassical theory. Additional human capital formation raises productivity, which in turn increases the national output, the derived demand for labour, and the wherewithal to pay the additional labour. In fact, if wages are paid in proportion to marginal productivity, and if the 'aggregate production function' is linear homogeneous in labour, physical capital, and human capital, then a ceteris paribus increase in the latter will produce just enough extra GNP to finance just the additional amount of labour needed to 'equilibrate' the labour market.

We wish to study the first effect on the job market in its purest form. For this reason, we assume that the extra education is given without increasing the total time the children spend in school (perhaps the extra education is all obtained in summer school). In any case, we consider the effects of providing higher educational attainment without changing the rate of work force entry. We assume that 600 people per year are given extra education, and we test the differential effect of applying this education to the two racial groups.

Table 13 shows the results of five years of such educational programs. Four alternative programs are considered and compared with a base case in which no extra education is provided. In each case, a total of 3,000 extra people are educated. For example, the last two rows of the table show changes produced by educating extra blacks from the level of a high-school diploma to a situation in which they have had some college training. As a result, more blacks can compete for high-wage jobs. Blacks gain 635 more jobs in the range of \$3.50–\$5.00 per hour and 124 more jobs at the highest-wage level. Correspondingly, fewer blacks are in low-wage jobs. Of course, blacks have gained these jobs at the expense of whites. There is a rough equality within each wage group between the increase in black employment and the decrease in white employment.

This picture, in which added education for a particular individual (or group) enables him (or it) to compete better for the more favourable jobs, is fully compatible with at least a portion of the 'human capital' view of the job market. However, we should emphasise that, in our model, what one individual or group gains from increased education, another individual or group will lose.

The result of giving 3,000 blacks higher education is, in our model, quite large. On the other hand, we see blacks reaping a much more modest gain from high-school diplomas. In this case, only 29 more blacks get jobs with wages above \$3.50 per hour and only 139 more get jobs at above \$3.00 per hour. Thus, in our model, the higher education has more apparent marginal value to the group than the lower. This result is essentially built into the structure of the model. College training opens doors for young people, enabling them to enter higher-wage occupations as they get out of school. A high-school diploma only gives preferential treatment on hiring but does not essentially change the segment which one enters. Therefore, the impact of increasing the incidence of black high-school graduates takes longer to become apparent.

The group effects of educating more whites are much smaller than the effects of black education. The reasons for this are very simple. The main effect of educating any group is to enable it to compete more effectively with the remaining people. When 3,000 more whites are educated, they gain and, in

Going to:	Job-type	Blue collar				White collar				Management-Professional		
	wage	1	2	3	4	1	2	3	4	2	3	4
Starting from: Job-type	Wage											
BC	1	44.4	31.6	6.1	1.7	2.3	6.5	2.3	-	1.5	1.1	1.6
	2	9.9	43.6	20.5	9.9	3.2	5.2	3.3	-	2.1	1.1	-
	3	1.9	18.9	40.7	19.2	1.8	4.2	4.1	-	3.2	-	1.6
	4	2.6	5.8	23.1	45.7	3.3	4.3	1.9	4.9	1.2	1.5	5.1
WC	1	7.9	4.6	7.8	-	35.5	22.0	2.4	8.4	-	7.2	1.4
	2	1.1	14.0	6.6	4.2	6.4	31.1	15.8	6.1	6.5	9.7	4.6
	3	1.2	3.5	5.0	1.3	1.2	7.2	38.4	12.0	7.7	4.6	14.3
	4	2.3	-	4.4	-	1.9	12.3	10.1	46.9	2.5	-	29.7
MP	2	-	4.1	-	-	-	10.9	-	3.4	22.8	32.6	13.7
	3	-	2.4	4.1	-	2.4	4.5	8.3	6.0	6.7	27.5	41.8
	4	-	-	1.7	-	-	3.5	2.8	3.1	4.9	6.6	75.0

Table 14 Four-year transition table. Numbers of workers (in per cents) starting from a given job-type and wage who finish in a given job category. The table is broken down by final job-type and wage. All transitions with less than 1% of the workers are omitted. Data on metropolitan area family heads drawn from Panel Study of Income Dynamics. Starting date is 1968, finishing 1972. The wage levels are: 1 is < \$2.00/hr.; 2 is between \$2.00/hr. and \$3.50/hr.; 3 is between \$3.50/hr. and \$5.00/hr.; and 4 is above \$5.00/hr. All wages are in 1970 dollars. Some rows do not add up to 100% because of rounding errors.

almost equal measure, other people in the metropolitan area lose. The fact that blacks form a minority group means that they feel only a minority of the loss.

There is another effect of education which should be mentioned even though it is numerically small. Whenever any group gets extra education, the net result is to decrease the number of workers in that group. This effect occurs because the metropolitan area in question gains highly qualified workers without gaining job opportunities, so that it will tend to lose workers to other areas. In fact, this loss is apparently the main result of educating extra whites to the level of a high-school diploma.

7 Summary and future directions

Quantitative model building is an exercise in logic. Most people who espouse particular public policies or programs have some picture of the way in which they expect the programs to operate. As a part of this they have a picture of the underlying workings of the 'social system' which they wish to affect. A model builder should try to catch the logic of these putative program builders and express this logic in the most precise possible form. In this step the model builder can be of value to the program builders in forcing them to express clearly and face up to their underlying assumptions. In favourable situations the model builder can be of further value in adding some quantitative facts to the qualitative pictures. The running of the model then represents no more than the working out of the logic used by the model's builders in the context of known parameter values. The numerical output of the model is the model makers' presuppositions written clearly.

Hopefully, this entire process - when exposed to view in the appropriate scholarly and policy-oriented publications - can help people think about the underlying logic upon which programs are based. In many cases model output is only a demonstration of relatively obvious connections between initial presuppositions and derived conclusions. In some cases, the numerical data can help illustrate the magnitude of various policy effects. Sometimes, enough data are available to show that a particular program approach is unlikely to be fruitful.

Given the point of view expressed above, the following criterion for a good model emerges:

> A good model is one which illustrates in a clear and understandable fashion the relationship between initial assumptions and conclusions about program effectiveness. A very good model may even clarify the numerical magnitude of certain relationships.
>
> A bad or unsuccessful model is one which is so complex or so poorly explained that the relationships between initial assumptions and final conclusions are only obscured by the modelling process.

Operating according to this criterion, our future work on the labour mobility model must first begin to incorporate quantitative estimates of as many important parameters as possible. We have already begun to estimate first-fourth order job transition probabilities from the Michigen data (see Table 14). Charles Holt and his associates at the Urban Institute have estimated monthly transition probabilities for the aggregate states (labour force,

	Employment to unemployment E → U	Employment to out of labour force E → N	Unemployment to employment U → E	Unemployment to out of labour force U → N	Out of labour force to labour force entrant N → L	Labour force entrant to employment NE/NL
Teens (16–19)						
White males	.033	.121	.338	.332	.203	.707
Black males	.065	.151	.242	.365	.199	.533
White females	.026	.142	.302	.406	.146	.683
Black females	.056	.188	.167	.438	.133	.432
Adults (20+)						
White males	.009	.013	.385	.124	.072	.754
Black males	.018	.021	.350	.157	.099	.701
White females	.009	.058	.316	.336	.043	.771
Black females	.014	.064	.241	.377	.075	.661

Table 15 Average of monthly transition probabilities, July 1967–June 1972. All figures are probabilities per month except the right column which is a probability. Supplied by Charles Holt, based on current population survey data.

employed, unemployed) from current population survey data (see Table 15). Holt's work involves the econometric estimation of all parameters in a complex group of non-linear equations. In our work, however, we expect that some parameters will not be satisfactorily estimated by econometric means. For these parameters, ranges of values will be selected, based upon professional judgment and repeated simulation runs to gauge the sensitivity of results to parametric variations.

As our modelling effort proceeds, we shall examine the effects of additional disaggregations of the basic variables. Three disaggregations of particular interest are central city/suburban location of jobs and residences (requiring the modelling of a commuting function), major industry (especially public versus private sector), and voluntary part-time (especially youth) workers. We shall also attempt to respecify the race/sex preferences of employers to permit examination of the Bergmann hypothesis (1972) that, by trying to bid already fully employed white adult males away from their current jobs, employers force wages upward and then pass these increased costs on to consumers in the form of higher prices. This alleged 'inflation-discrimination tradeoff' has received far too little attention in the literature or in policy discussions in Washington.

The basic question to be asked of this model is: What types of public intervention in the job market are likely to produce worthwhile results? We have several variables at our command to serve as social indicators: proportion of well-paying jobs, the unemployment rates of various groups, and the degree of mobility within the work force. In our future work, several different types of intervention will be studied within this context:

1. The direct provision of new jobs of various different kinds. Here, the question to be asked is which job-types can produce the maximum worthwhile effect for a given governmental expenditure level. For example, given the existing barriers to upward mobility in the labour market, how does a policy of selectively increasing the demand for (say) \$7,000+ white-collar workers (i.e. increasing job vacancies in that 'job slot') trade-off against applying a given budget to the expansion of vacancies in the \$4-7,000 white-collar category, in terms of reducing unemployment among the urban poor, or in terms of changing the distribution of incomes (e.g. as more work experience in the lower-wage jobs produces greater long-run upward mobility)? How crucial is the 'Forrester problem', i.e. that provision of opportunities in one city (in our case, in the labour rather than in the housing market) will attract large numbers of migrants who overload any redevelopment efforts? Do the data on migration show the existence of particular jobs to which migrants seem relatively less attracted?

2. Government stimulation of industrial - rather than job - growth in particular sectors. Various Census publications provide listings of numbers and types of employees by industry for selected urban areas. With the aid of these job disaggregations by industry, we can ask which types of industry are likely to provide the best conditions for improving the job status of the poor. Of course, a truly systematic impact analysis would require linking such industry-occupation matrices to an input-output table of the region under study, as has been suggested by Alterman and Isard. The scope of such a task is immense. Still, it is worth contemplating the long-term potential for policy analysis of a linked series of models: input-output → industry-occupation → labour market simulator. With such a system, the consequences of selective government stimulation of particular industries (or conversely,

the consequences of the decline in growth of certain industries, such as automobiles or aircraft), as opposed to unselective stimulation via conventional Keynesian fiscal and monetary policy, could be described in terms of changes in the mix of occupational demands, and then 'played through' the simulator to predict changes in the distributions of income and unemployment by demographic group.

3. The government as an employer. We can consider the effects of local, state, and federal governmental employment. To what extent will the experience that workers gain in temporary governmental employment aid their future job prospects? If the government really seeks to create 'full employment' through public service job creation, how many jobs will it have to offer and of what kinds? To model this process, we will assign (or estimate) different parameter values to (for) the entry, promotion, and job search equations for the public sector labour market, and compare these public and private structures. What must be true of the public structure in order for public service job creation to be a more cost-effective anti-poverty policy than government stimulation of the private sector? Are these requisite structural characteristics in fact realised in the U.S., and for which public employers (state or local or federal)? The most difficult task would be to model transitions from the subsidised public to the unsubsidised private sector (a major goal of the Public Employment Program and the new Comprehensive Employment and Training Act), and the extent to which fully employed private sector workers are attracted to the subsidised public jobs, out-competing the disadvantaged for the limited supply of new job vacancies (the 'Forrester problem' all over again!).

4. Structural changes in the job market. Holt has already written extensively on the utility of labour market simulation models for 'experimenting' with policies for reducing the duration of unfilled job vacancies, an obvious source of social inefficiency and thought by many to be contributory to inflationary pressure. The Department of Labor has been especially concerned with the role of education and experience requirements as factors contributing to inefficient job matching: highly qualified workers are often over-educated for their jobs, with attendant effects on their productivity and job attachment, while disadvantaged workers lacking credentials but possessed of the requisite skills are 'locked out' by such employer practices. Our model will permit us to examine how the distributions of unemployment and income respond to changes in these institutional barriers to mobility. In other words, while the previous policy simulations involve instrumental changes in the stocks of vacancies, with flow parameters held constant, these 'institutional change' simulations will involve experimentation with the flow parameters themselves.

5. Again with given flow parameters - given institutional practices - we may examine the impact of more or less conventional human capital programs on the various performance criteria. With given employer attitudes (manifested in the hiring and firing parameters), what kinds of training - general or specific, vocational or on-job training - seem most effective?

REFERENCE FRAME

THE BIG, THE BAD AND THE BEAUTIFUL

Leo P. Kadanoff

This should be a good time for physics. Our President, several governors and many industrial leaders have eulogized science as significant both for its own sake and also in helping our nation maintain its long-term competitive position. But despite all this public goodwill, US physics continues to decay.

The first and most serious decay is on the industrial side. For example, Ed David estimates that over the last three years oil industry support of basic science and engineering has declined by about one-third. The breakup of AT&T has threatened the support base of the world's best industrial laboratory. Other labs have deteriorated, probably as part of the usual "life history" cycle of industrial laboratories. Because we have not seen any large offsetting growth of new private-sector centers in the physical sciences, the net result is decline.

Government-based science and technology have also done poorly. This result is most obvious in the NASA-based programs in planetary sciences, astronomy, atmospheric sciences and so on. As an apparent result of program misdirection, management incompetence and pursuit of publicity, during a period of several years we totally lost the capability of launching anything—shuttles or Atlases or balloons. So our once excellent planetary and space programs are in trouble. Still NASA continues to put publicity above science and technology. Corresponding losses have occurred in other areas of physics. Basic research in plasma physics has been too tightly coupled with fusion R&D, so that the recent cuts in the fusion program have hurt the basic work. "Environmental" science has disappeared precipitously from the national laboratories.

The universities have not done very well either. Undergraduate training suffers because high-school teaching in mathematics and physics is so very poor. An adequate supply of really good American graduate students is unavailable, so that graduate education (and the supply of teaching assistants needed to teach undergraduates) must rely upon the availability of students from abroad.

Another university difficulty is the spottiness of governmental support. There has been oversupport for some chosen areas (mostly based on big science) like supercomputer-based numerical computation or high power laser experiments. However, over the last seven years, we have seen attrition in support for small-group efforts. For example, condensed matter experiment, which is strongly based in the small-group mode, has had considerable difficulty in keeping sufficient support to remain competitive with analogous efforts in Europe or in industry. Additional cutbacks in Federal funding of condensed matter experiment are in progress.

The decline in the support for small science is particularly painful because the small-science mode is particularly appropriate for graduate student training. Students can plan small experiments, build the equipment, take the data and assess the connection of the output with theory. The time constants and the scale of large science are much less appropriate for training new scientists.

Governmental action to improve this situation seems to be largely misdirected. Instead of investing in small science our leaders seem to be entranced by less valuable but more arresting big projects: Star Wars before mine sweepers; billions for Superconducting Super Colliders, but inadequate funds to even run the neutron sources that might give insights into the new superconductors; money poured into supercomputer centers, but no support for the more economical use of local or individual computers on individual campuses; mathematics institutes funded, but a steep decline in support for faculty doing mathematics research; and so forth.

These tendencies can be seen in the NSF-sponsored Science and Technology Centers, which are intended to provide a meeting place for university and industrial scientists to work on common problems. The nation would indeed be helped by some carefully chosen additional possibilities for innovative research of the highest quality. Perhaps some of this would be helped by an interdisciplinary context in which new groupings of investigators would appear. But it looks as if these S&T centers are going to be a replacement for basic-science projects rather than a supplement to them.

These centers are also likely to focus upon the immediate needs of industry, rather than on basic science. In my view, this direction of development arises from a complete misunderstanding of what industry needs from the university world. By and large, big industrial concerns have sufficient resources to attack any technical problem they face. In some measure they need the universities to help in areas of sudden and unexpected need. For example, when oil firms suddenly realized that they could use tomography to "see into" rocks in a three-dimensional manner, they hired the knowledge of outside experts. But industry's most important need is for really new, "far out" ideas. Universities can serve as an important intellectual source. Big, project-directed science will be too focused and too goal oriented to provide the breadth and the innovation sorely needed by industry.

In this context, it is salutary to remember the disappointing results of the NSF Presidential Young Investigators Program. This program was intended to fund a group of promising young engineers and scientists and guide them into types of research that could be of long-term use to the US. The NSF was to give some money, and some of that money was to be matched

Leo P. Kadanoff, a condensed matter theorist, is John D. MacArthur Professor of Physics at the University of Chicago.

REFERENCE FRAME

by industry, producing an incentive to joint exploration of new areas of interest. For many of the investigators, the match didn't work out. Often industry did not wish to invest in areas in which there would be only a long-term payoff—while promising young scientists did not wish to channel their research into areas with only short-term interest.

Much of the government sponsorship of research is forcing us to choose between the immediately fashionable and what will be interesting in the long term. The fashionable work is being structured into big projects and large teams, at considerable potential cost to the quality of US science and technology. Even in superconductivity, where almost all previous advances have come from very small groups, there is considerable governmental pressure for the construction of team-based "initiatives."

But I do not wish to suggest that the trouble lies entirely outside our physics community. In some measure, our own people are responsible. We scientists have chosen to follow fashion and have not continued research in less fashionable areas. We have thoughtlessly pushed our own subjects of research without concern for whether these areas have a value commensurate with their cost. And we have indeed pursued big science. While giving lip service to statements that a balanced and solid physics program requires a major investment in small science, physicists have pushed for each of the big projects. It is easier to argue for another synchrotron light source or for a supercomputer or supercollider than for the broad programs that will provide the real intellectual base for understanding our natural world. The big projects provide jobs. They have sufficient "sex appeal" and economic impact to command attention. However, in the long run an overinvestment in such efforts detracts from good science.

The true value of science is in the development of beautiful and powerful ideas. Overinvestment in big science detracts from what is really worthwhile. I do not think that the nation's or the government's budget for research or for R&D is too small. It is, however, increasingly misdirected toward grandiose projects. We physicists have a responsibility to understand what is truly valuable in science and use this understanding to help the nation develop and express its priorities. In doing this, we should push for the kinds of work that are of enduring intellectual and technical value, rather than projects that are "in the right field" or are simply easy to sell. ■

"I'LL BE WORKING ON THE LARGEST AND SMALLEST OBJECTS IN THE UNIVERSE—SUPERCLUSTERS AND NEUTRINOS. I'D LIKE YOU TO HANDLE EVERYTHING IN BETWEEN."

REFERENCE FRAME

HARD TIMES

Leo P. Kadanoff

In recent years, physics has seen much of its support base disappear. In Eastern Europe the decline of governments and the changes in the economy have moved interest away from all activities with long-range payoffs. For different reasons, a short-term focus has begun to dominate American life also. Here, all the props for science have begun to weaken. Government has become unpopular. The military has started to shrink. Corporations are concerned with tomorrow's stock value and the next quarterly income statement, and have lost interest in promoting applied research. Antiscientific threads have become evident in many parts of popular thinking. Congress enjoys exposing the misbehavior of some of the leading figures in the biological sciences. Both the animal rights movement and the environmental movement have considerable antiscientific components. Science is in low regard.

These fundamental difficulties arose first. Now more specific symptoms are beginning to appear. All branches of activity in US physics have insufficient support. It looks as if experimental particle physics will have to accept a postponement of its latest accelerator project, the SSC; a substantial cut in the support given to the science in the field; or both. Plasma physics has put all its eggs in a fusion-engineering basket, and fusion does not appear to be economical in any reasonable time frame. Condensed matter cannot support 50% of its researchers. Space research must contend with a NASA that has lost its way. The National Science Foundation has begun to move away from science.

As our support becomes weaker, our demands and claims become more strident. Very often the expected benefits of any given scientific advance are puffed to many times their real value. Think of cold fusion (or hot), of the statements coming from the Texas Congressional delegation about the SSC, or of the more recent history of high-temperature superconductivity. We are fast approaching a situation in which nobody will believe anything we say in any matter that touches upon our self-interest.

Nothing we do is likely to arrest our decline in numbers, support or social value. Too much of our real base depended on events that are now becoming ancient history: nuclear weapons and radar during World War II, silicon and laser technology thereafter, American optimism and industrial hegemony, socialist belief in rationality as a way of improving the world. Now China and much of Eastern Europe have moved away from science. America's problems cannot be given a quick fix by science or technology. For the world's corporations, "high tech" now tends to mean not the applications of the physical sciences but instead of computer sciences or biology or of the innovative use of financial instruments. Most important is science's loss of reputation. Einstein and Sakharov caught the popular imagination, for excellent reasons. But today when the public thinks of the products of science it is likely to think about environmental problems, an unproductive armament industry, careless or dishonest "scientific" reports, Livermore cheers for "nukes forever" and a huge amount of self-serving noise on every subject from global warming to "the face of God." At this point, we must ask ourselves how we should proceed in a time of declining financial and societal support for all of science and especially for physics. We must face up to two types of difficulties:

First, individual difficulties. Many excellent students now in the pipeline will not be able to find careers in the mainstream of science. Some older scientists will be forced into early retirement or other forms of enforced unproductivity. Other physicists will continue to follow our profession but with greatly reduced possibilities for productivity or appreciation from society. We can only do a little to help our colleagues in trouble. We can try to make all of them feel that they remain part of our professional world. We all can realize that individual success or failure is in part a product of market forces and might not be an indicator of individual worth.

In the past, a career in science has served as a mechanism for social mobility. Now, this mechanism is likely to work less well. Many trained scientists will have to look for new kinds of employment in fields of activity that reward socialization better than technical or intellectual accomplishments. My own field of condensed matter will have to downsize itself considerably. Industry's interests have turned elsewhere, and support continues to shrink across the board. In the US we shall have to help people from Eastern Europe and China, as well as the native crop, in relocating their interests.

Second, problems in the social and economic fabric of our science. There is no doubt that the world of science and scientists will change as resources become scarcer. But we can help by asking what portion of what we hold in common is really worth preserving, and what strategy will best work to save the valuable elements for future generations. Let me try to define what is truly valuable:

▷ First and foremost is a rational view of the world. Things happen because of laws, and are in that sense knowable and predictable. One should oppose this view to that of Dan Quayle's grandmother, who said one could do anything one truly wanted. We should and must communicate to our students and to the world at large a belief that rational study can delineate between what is feasible and

Leo Kadanoff, a condensed matter theorist, is John D. and Catherine T. MacArthur Professor of Physics and Mathematics at the University of Chicago.

REFERENCE FRAME

unfeasible and provide methods of accomplishing the former.

▷ Second, our reputation as reliable observers. The public has demanded that our politicans say things that are untrue, and then feels shocked that it is not well served by its leaders. Science should be better. If society is to entrust to us the testing of advanced technological systems, we should say when they do not work. When did we ever report that any activity in which we were involved had failed? With supercomputers? (Most of us have noticed that personal computers and workstations are more economical for most tasks at hand.) Or high-temperature superconductivity theory? (Its contact with experiment is often problematical or worse.) Who points out to the public that in particle physics, the imaginations of experimentalists and of theorists have been focused upon almost entirely different topics?

▷ Third, we have a set of techniques that should be passed on to future generations. As physics contracts, we should try to preserve in reduced form the main elements of our culture. These include the capability for careful and thoughtful design of experiments, an insightful and intelligent use of mathematics, and a creative use of gadgets and technology. The last two are not in terrible danger. However, small-scale experiments must be explicitly protected lest the siren calls for big science, beautiful technology or "deep" mathematical developments absorb all attention and resources.

We shall see many wrenching dislocations in our field. But we can do some things to help. Those of us who are in relatively secure positions should, in our own self-interest, broaden the reach of our society. Every person trained in physics and everyone who wants to be associated with us should be made to feel a part of the world of physics. APS meetings should include more activities for teachers, students and just interested individuals. Universities and governments should sponsor more outreach programs for informing citizens of all ages about science and engineering.

Also, we should communicate a realistic view of the future to people trying to join our profession. We can try to be helpful by finding and using all the possible outlets for talent and training, for example, by supporting teaching and by taking a very broad view of technology. In a period of decreasing support of science throughout the world, no nation can expect to be a leader in all fields of research. To help each nation maintain its technological base, we should encourage governments to provide additional sponsorship for undergraduate, graduate and postdoctoral study abroad. For similar reasons, there should be government sponsorship for individuals at mid-career to pursue a year of training or study at home or abroad in industrial, university or government laboratories. But we should not expect too much from these efforts.

Looking for truth may be expected to be lonely and outwardly unrewarding. In recent decades, science has had high rewards and has been at the center of social interest and concern. We should not be surprised if this anomaly disappears. We will all be disappointed and hurt by this likely development. But if we can back away and look at the situation with some perspective, all of us in science can say that we have been lucky to be part of a worthwhile enterprise. ■

"THE ROYAL ACADEMY OF SCIENCE IS WILLING TO PAY YOU FOR THIS APPLE TREE, IF YOU'LL SHARE WITH US ANY IDEAS YOU GET FROM IT."

REFERENCE FRAME

GREATS

Leo P. Kadanoff

I had a tiny idea that I thought might improve physics education a bit. It is neither original nor profound but I thought it might be helpful. I suggest that all of us, when asking for continued financial support for existing programs, state what became of the people previously supported on similar grants or contracts. We would name them and describe the first jobs, and, where appropriate, the second or even third.

I brought the idea to the dinner table and pointed out that it might help the Federal agencies evaluate proposals better. Some schools have claimed they had exceptionally good records in job placement. Maybe someone could look at those claims and help put the money for training graduate students, postdocs and undergraduates into the places that had proved most effective. My wife wondered whether this suggestion would be good for me. The change might, she suggested, be really bad for my own grant renewals. After all, some of my young coworkers got pretty unconventional jobs: in banking or in a Japanese government lab or in a hospital or (going back a little in time) in the atomic energy program of a nonaligned nation or as an administrator in a Soviet university, or . . . . In fact, a very large majority of the people whom I might have described under this proposal were doing something quite distant from the topic of their supported research. How should I evaluate my own accomplishments as a mentor? What kind of outcome should I count as a success and what kind as a failure for an undergraduate physics program?

My next occasion for thinking about this topic arose on a visit to Haverford College, an excellent institution with serious students and a particularly good undergraduate research program.

Leo Kadanoff, a condensed matter theorist, is John D. and Catherine T. MacArthur Professor of Physics and Mathematics at the University of Chicago.

Jerry Gollub, my host, showed me a list of professions pursued by the school's physics-major alumni. Doctor, lawyer, captain of industry . . . every profession under the sun, with only a small minority in physics and allied fields. Should this outcome be counted as a failure for an undergraduate physics program?

I do have a strong opinion about the purposes of undergraduate education in physics: I am in favor of it. (Surprise, surprise!) I think that we physicists have a lot to offer to many different kinds of people who wish to become educated. Specifically, an undergraduate education in physics potentially offers the following benefits to students.

Physics teaches that many problems can be well enough formulated so that they have answers. In our field, there is a distinction between correct and incorrect. A main goal of any education should be to teach that truth is not relative. (A wise student might recognize that a distinction between right and wrong applies also to other aspects of life.)

Physics teaches some concepts that should be part of the knowledge of every educated human. In studying physics we learn that events in the natural world occur through the working out of laws of nature and that many aspects of these laws are accessible to human intelligence. One sees that the development of the universe or the Earth or a piece of granite is the result of natural law. As a corollary, one begins to understand that humans are also part of the natural world and subject to nature's laws. The existence and ubiquity of law is the primary lesson, and the exact subset of laws learned less significant.

But even more important than any set of facts is the basic fabric of science itself, with its major components: curiosity about the world. The belief that problems can often be isolated and understood. The importance of successful predictions. The damning importance of unsuccessful ones. Problem isolation and prediction form the keystone of all science, indeed of most thoughtful human endeavor, and they appear in their purest form in physics. (Correspondingly, the weakness of physics as an educational tool is that our problems are too pure, too clean, and do not partake of the messiness of many real world conundrums.)

In addition the student of physics might learn technique and method. On the most basic level, physics method consists of problem isolation, the building of (often mathematical) models, the deduction of results from the models and the generalization of these results. A student may also learn the importance of more specific tools: laboratory technique, computer technique, mathematical analysis.

Finally the student will begin to see and understand a connected body of knowledge. Physics is a beautiful and complex intellectual creation, with an intricate interconnection among its parts.

So what kind of education do we wish to design for our students? What do we wish them to carry away? To where? So far, I have focused most on undergraduate education; next I turn to graduate education.

A PhD student should learn enough so that he or she will be able to find a first job that makes use of some portion of the skills and knowledge gained in grad school. Most students will directly use what they have learned about problem isolation, modeling, generalization and maybe some technique (perhaps use of computers) rather than any specific information close to their PhD theses. Nonetheless the thesis, if done right, will be the crucial learning experience in which they bring their technique and knowledge to bear on a real problem and advance, incrementally, some field of human knowledge. As the new PhDs enter the job market, they should be deeply and broadly enough trained so that they can make the expected switches from one career area to another. We uni-

REFERENCE FRAME

versity people should be pleased if our students have learned enough to start out in materials development, switch to the analysis of financial markets, and perhaps switch again to management or science writing. Any student should expect and be prepared for one or two major switches in the course of a career. Teachers should be pleased to see any previous student who is using his or her intellect fully and enthusiastically on work with some real value. Conversely we should be disappointed with a student who doggedly followed a straight-line career path in the direction of diminishing utility and diminishing rewards.

Many students can go in a straight line; few students can be flexible. Our graduate teaching should encourage the latter and discourage the former. Every student should have an opportunity to see the full range of a problem, from problem isolation to generalization. Through it all, the student and the teachers should keep their eyes on the two significant goals: Learn the right things to get a productive first job, and learn to learn so that one can switch fields as needed. (The professors may find they will need that skill too!)

Thus in graduate education we should give up narrow professional training and aim for flexibility and breadth. Correspondingly, on the undergraduate side, we should mostly give up thinking about the education of professional physicists. Instead we should try to develop a curriculum that exposes people who have a reasonably wide range of abilities and interests to the genuinely worthwhile, culturally valuable parts of our physics tradition. We should not aim to teach all of physics but merely representative portions of it, so that the student can see how it, and we, work. The particular portion taught might depend upon the resources of a particular institution. These portions should include considerable material from "applied" fields—astrophysics, geophysics, materials science—so that the student may see ideas interact with the real world. Broadly useful techniques should be stressed: more spreadsheets and small computer methods, less fancy mathematics; more model-building, less quantum mechanics.

In undergraduate education our goal should be to produce people who appreciate the capabilities of science, who might understand a discussion about global warming or the likely value of astrology. Our goals should not be narrowly professional, and we should certainly not judge the outcome on the number of people who go on to graduate school in physics. We might wish to ask something about what fraction of the students we have trained go on to productive lives independent of their field of activity.

So after having thought it over, I am reasonably happy that our former students chose such diverse careers. It would appear that most of the students have learned, and branched out, and it seems that their minds have found worthy tasks.

But, the reader might ask, what does the title of this piece have to do with all of this? During the high period of the British Empire, its public servants were largely given a classical education (Greek and Latin). "Greats" was the name for one portion of this funny education, which seemed in part to work. We need to inquire into the appropriate education for our time. ■

"DR. GROMMET IS FUNDED BY A MAJOR HOLLYWOOD FILM STUDIO. HE'S BEEN ASKED TO COME UP WITH AN ANTI-GRAVITY DEVICE AND AN INVISIBLE-RAY GUN."

Section D

Turbulence and Chaos

Questions Without Answers

In the last dozen or so years, my interests have turned to yet another field of science: the description of complexity. My own view of this field starts with a major intellectual problem: We know that the laws of physics are rather simple in structure. Newton's laws or the Schrodinger equation or even string theory is described by a rather simple system of equations. One's expectation might be that such simplicity in formulation should lead to simplicity in outcome. However, all our experience in life contradicts any expectation of simple outcomes. The world is wondrously complicated and bewilderingly diverse. How can it be that from simple beginnings one gets complex endings.

This problem is not illuminated by much of the traditional practice of physics. Most systems picked for study by physicists are picked precisely because they have simple outcomes. Kepler's orbits or the quantum harmonic oscillator or the simple pendulum have been extensively studied and are used as examples just because they have simple predictable outcomes. Even the many-body systems studied in critical phenomena have simple predictable outcomes. But the world tends not to be predictable in the same way. Once again our life experience suggests opposite outcomes from those of the exemplary physical systems. A pot of boiling water, or a double pendulum, or any person will not behave in expectable or predictable ways. Why should our professional experience with physics belie our civilian experience as human observers of the world.

But the conundrum is worse yet. Most of the systems traditionally studied by physicists remain equally complicated throughout their history. A microwave cavity, any pendulum, or any electrical circuit with passive linear elements does not gain extra complexity as time goes on. However, if you take a hot mass of rock of an appropriate size and throw it into orbit around the sun, you might observe it to gain all kinds of organizing features. Oceans and continents will develop and move around. Mountain ranges will grow and decay. Great deserts will spread out and then be covered with ice. Little cells of organization will emerge, become more complex, and form into trees and us. Is all this development of complexity and organization outside the laws of physics and of science?

Our prejudice as scientists is to say *no*! *Physics encompasses all.* But our job as scientists is to see and understand. How is it that simple laws can give complex outcomes? How can organizations develop from blind law and naked chance? A major thread of development in the modern physical, mathematical, and biological sciences has been devoted to answering this class of questions. My own work on this area has been modest in outcome, but quite exciting and meaningful to me. These two last sections of the volume are devoted to work on the borders of this great subject.

For me, the story starts in about 1982. I am working at the University of Chicago, doing phase transitions and critical phenomena, and being somewhat dissatisfied with my work. In the great days of the 1960s I was all excited with this subject and its creative possibilities. Later on, I got considerable joy in seeing what Fisher, Wilson, Wegner, Nelson, Polyakov, and many others had created based upon the earlier work. But, by this time the freshest joy of discovery, either directly or though the work of others, has begun to go out of the field. Then, at a crucial moment, Bob Gomer, a colleague in Chemistry, asks why I am devoting so much work to a particular model system. His implication is that the model is not so real as to be of practical interest, and perhaps not so deep as to have real intellectual interest. My reply seems superficially convincing, but I know that his implied criticism is right. So, I resolve to learn something new.

The new subject I find is *dynamical systems theory* the study of the time development of relatively simple systems. There are portions of the subject which are absolutely beautiful closed topics, not as deep as critical phenomena but roughly similar in structure. Two of these beautiful topics are *iterated one-dimensional maps* and the *onset of chaos.* The first is about problems in which a system is described by a single variable, say its degree of excitation, and jumps from one value of the excitation to another via discrete time steps. The second is about a kind of continuous transition which might occur when a system first shows chaotic behavior. In 1982 or so, my students Albert Zisook, Michael Widom, Scott C. Shenker and I set out to learn about dynamical systems theory. Our work was filled with great thrills of invention and discovery and then the partial disappointments of finding that our creations had been preempted by previous workers. Some of the structure of this subject in reviewed in papers D1 and D2.

Eventually, we found our own area which could serve as an entry to this research field. We built upon the work of Mitchell Feigenbaum, who had first understood the onset of chaos via a renormalization group calculation. His work was built upon the tools I had previously known: scaling, universality, and renormalization. When we began, there were two kinds of onsets which had been studied both of which involved a period doubling route to chaos. We looked to other routes and learned about them through the development of renormalization calculations. Some treatment of this subject can be found in paper D3.

Thus, our strategy was to enter this new field by considering systems which produces time-sequences of data — like annual populations, or economic outputs, or daily stock prices — and to see how such a sequence may develop chaotic attributes. These systems were picked to be just complicated enough so that we could fully encompass them with the computers we then had available. (Here was the point at which I would see the computer knowledge gained by studying urban problems pay off for me in genuine, hard science.) Papers D3 and D4 describe the results of this kind of analysis. All the examples studied are just simple mathematical toy systems picked to be analyzed with computer and then with techniques like the renormalization group. In this way, we began to understand the development of complexity in toy examples.

However, these examples also proved to be unsatisfactory in the long run. They were both too simple and too complicated. At onset of chaos they were simple. In their chaotic regimes they could show a behavior which, when studied in detail, was bewildering diverse and did not seem to lead to useful and interesting general principles. We were stuck once again.

The best source of theoretical inspiration is the real world, and the best window to that world is provided by experiments. The systems studied by experimentalists have a much better chance of organizing themselves than computer toys. The real world has much, much more time to organize than our toy examples. A computer might analyze 10^8 or so very simple events per second. (In looking at our models, we looked at much fewer events, perhaps a few thousand time steps in every second of computer analysis.)

However, a real system — say a cubic centimeter of gas — will have 10^{32} or so collision events per second. Naturally, the real system will have a greater opportunity to 'get organized'. Conversely, our computer models will have to be picked with very great care if they are to give any inkling of true natural phenomena.

However, with luck or skill, one can relate model systems to real-world outcomes. One very pleasing example of how to do so is given in paper D5, authored by Jensen, Libchaber, Procaccia, Stavans and myself. This work is devoted to showing how a real system, a pot of liquid mercury heated from below, first develops chaos. The onset in this real system follows one of the universal routes to chaos, the so called quasiperiodic route, which was extensively studied by the Chicago theoretical group. This paper itself is mostly experimental in content. The system studied is the brainchild of Albert Libchaber, its actualization is due in large part to Joel Stavans. Libchaber understood how to adjust the system to push it into the domain investigated by the theorists. And pleasingly enough, after adjustment the system behaved in exactly the fashion predicted by theory.

It is important to recognize that our understanding of the physical world is due in very large measure to the experimental scientist. Pure thought gets us only so far. The worlds which the theorist may construct are only powerful and rich because they are modeled on the real worlds that nature produces. And our only access to such real worlds is through the studies of the experimental scientist. Without experimental check, the universality of the onset of chaos is a theorists' possibility. The experiment shows that it is a reality.

In addition to illustrating the craft of the experimentalist, this paper also has a pleasing theoretical content. According to the renormalization theory, near onset dynamical systems produce a characteristic structure in which very delicate and small structures are produced in the plot excitation versus time. The incipient chaos is, in fact, built into structural details which persist to arbitrary small scales. The word for such a delicate and scale invariant geometrical structure is a *fractal.* This word was coined by Benoit Mandelbrot, who brilliantly argued that such structures were rather pervasive in nature, and then introduced many clever and incisive techniques for producing and analyzing them. Our description of the experiment was based upon a study of the properties of the fractal produced at onset in this hydrodynamic system. This in turn pointed back to a previous paper on the nature of fractals of this type, D6 and to my concern about the proper use of the fractal concept (see D7). (See D8 for a summary of some research on multifractals.)

Paper D6 is in some ways an embarrassment to me. It is an excellent expository work which in fact helped make a particular form of analysis of fractal behavior very fashionable. The form of analysis is called multifractal or multiscaling analysis and follows from the work of Mandelbrot,[27] Hentschel and Procaccia[28] and others. The analysis closest to the one in our paper was done by Georgio Parisi and Uriel Frisch, published in a slightly obscure place,[29] but explained to me in some detail by Parisi. Unfortunately I either forgot or did

[27]B. Mandelbrot, *J. Fluid Mech.* **62**, 331 (1974).
[28]H. G. E. Hentschel and I. Procaccia, *Physica* **8D**, 435 (1983).
[29]Georgio Parisi and Uriel Frisch, "On Turbulence and Predictability", in *Geophysical Fluid Dynamics and Climate Dynamics* eds. M. Ghil, R. Benzi, and G. Parisi (North-Holland, Amsterdam, 1985).

not understand his explanation, so that we did not appropriately cite this earlier work in ours. Fortunately, an erratum[30] and a bit of publicity[31] set the record straight.

I was involved in a few more research projects in dynamical systems theory, one with Charles Amick, Emily Ching and Vered Rom-Kedar, another with Mahito Kohmoto and Chao Tang (see #91), and still another with Oreste Piro and Mario Feingold (see #127). This work was pleasing, but we were never again able to use dynamical systems theory to achieve the generality or experimental relevance of the work on the mercury cell. We were stuck once more.

But we did follow another theoretically motivated line of research, based upon simple mathematical models of extended physical systems. In 1981, Thomas Witten and Leonard Sander[32] developed a model of the dynamical aggregation of clusters called DLA. This resulting clusters exhibited a kind of universality along with scaling and fractal behavior. At Chicago, we followed up on these ideas and applied them to other model systems.[33]

But, then we turned back to experimental systems. With David Bensimon, who was my student, Boris Shraiman, a postdoc, and Albert Libchaber I started looking at fluid flow, which can produce a genuine and rich complexity in both space and time. The example they suggested is the flow of two liquids confined in a very narrow space between two glass plates. This setup is called a Hele-Shaw cell. It is capable of producing both a very simple behavior and also a rich complexity. A portion of the theory of such systems is outlined in the review paper D9 in this section. This paper was followed by many others[34] as we once again constructed theory in response to the experiments of the Libchaber laboratory.

Another closely related experiment is discussed in paper D10.

Paper D11 describes how concepts drawn from dynamical systems and other chaotic situations can become part of undergraduate teaching. I have spent a very considerable period of time designing courses which use computers to teach about chaos. This teaching is a fitting and satisfactory outcome for the research directed at chaos and dynamical systems.

[30] T. C. Halsey, M. H. Jensen, L. P. Kadanoff, I. Procaccia, and B. Shraiman, *Phys. Rev.* **A34**, 1601E (1986).

[31] Barbara G. Levi, *Physics Today* (April 1986), p. 17.

[32] Thomas Witten and Leonard Sander, *Phys. Rev. Lett.* **47,** 1400 (1981).

[33] Some of the original work can be found in the publication list. See #90 (Alexander, Domany, LPK, and Bensimon) and #103 (LPK and Liang) as well as #99 (Bensimon, Shraiman, and LPK).

[34] This work eventually grew into a very big activity, finally involving extensive interactions with people in the Math Department at Chicago. So far, I had the pleasure of collaborating with Bensimon, Shraiman, Libchaber, Giovanni Zocchi, Bruce Shaw, Wei-shen Dai, Su-min Zhou, Peter Constantin, Todd Dupont, Ray Goldstein, Michael Shelley, and Andrea Bertozzi in papers on the Hele-Shaw cell.

Roads to chaos

Simple mathematical systems exhibit complex patterns of behavior that can serve as models for chaotic behavior, including perhaps turbulent flow in real hydrodynamic systems.

Leo P. Kadanoff

Hydrodynamic systems often show an extremely complicated and apparently erratic flow pattern of the sort shown in figure 1. These turbulent flows are so highly time-dependent that local measurements of any quantity that describes the flow—one component of the velocity, say—would show a very chaotic behavior. However, there is also an underlying regularity in which the motion can be analyzed (see figure 1 again) as a series of large swirls containing smaller swirls, and so forth. One approach to understanding this turbulence is to ask how it arises. If one puts a body in a stream of a fluid—for example, a piece of a bridge sitting in the stream of a river—then for very low speeds (figure 2a) the fluid flows in a regular and time-independent fashion, what is called laminar flow.[1] As the speed is increased (figure 2b), the motion gains swirls but remains time-independent. Then, as the velocity increases still further, the swirls may break away and start moving downstream. This induces a time-dependent flow pattern—as viewed from the bridge. The velocity measured at a point downstream from the bridge gains a periodic time-dependence like that shown in figure 2c. The parameter that characterizes these changes in the flow pattern is the dimensionless Reynolds number $\mathcal{R}$, which is the product of the velocity and density times a characteristic length (the size of the bridge pier, for example) divided by the viscosity. As $\mathcal{R}$ is increased still further, the swirls begin to induce irregular internal swirls as in the flow pattern of figure 2d. In this case, there is a partially periodic and partially irregular velocity history (see the second column of figure 2d). Raise $\mathcal{R}$ still further and a very complex velocity field is induced, and the v(t) looks completely chaotic as in figure 2e. The flow shown in figure 1 has this character.

These different flow patterns can also be characterized by looking at the power spectrum of the flow. The power spectrum $P(\omega)$ is the square of the Fourier transform of the velocity field:

$$\hat{V}(\omega) = \frac{1}{\sqrt{T}} \int_0^T \mathrm{d}t\, \mathrm{e}^{2\pi i \omega t} v(t)$$

$$P(\omega) = |V(\omega)|^2$$

The fourth column of figure 2 shows the power spectra for the flow patterns we've discussed. For the time-independent flows, figures 2a and 2b, $P(\omega)$ shows a spike at zero frequency. In the periodic region (figure 2c), additional spikes appear at the frequency of the oscillation and at its harmonics, that is, at integer multiples of this frequency. As the motion becomes partially chao-

Leo Kadanoff is professor of physics at the University of Chicago and a member of the James Franck and Enrico Fermi Institutes there.

tic, as in figure 2d, a broad, slowly varying background appears behind the spectral lines. Finally, the fully chaotic flow has a power spectrum that is apparently continuous.

We would, of course, like to understand this transition to turbulence in hydrodynamical systems. Unfortunately, after many years of study by scientists in many different disciplines, we still do not have a fully satisfactory approach to this problem. In this article, I would like to describe an extremely simplified model that shows a kind of transition to chaos and to discuss how the features of this model can be, and have been, observed in hydrodynamic systems.

The spirit of this approach is similar to the one used in the theory of critical phenomena in condensed-matter physics: To understand a complicated phase transition—that is, a change in behavior of a many-particle system—choose a very simple system that shows a qualitatively similar change. Study this simple system in detail. Abstract those features of the behavior of the simple system that are "universal"—that is, appear to be independent of the details of the system's makeup. Apply these universal features to the more complex problem.

Our simple problem is so simple that one might, at first glance, imagine that it contains nothing of interest. But it has an amazingly intricate and regular structure. Consider a dynamical system characterized by one variable, x. At time zero, the value of this variable is x_0; at discrete later times $t = j\tau$ it has the value x_j. The major assumption is that the value of the variable at one step x_j determines the value at the next. Mathematically we write

$$x_{j+1} = R(x_j) \qquad (1)$$

where $R(x)$ is a function that describes the dynamics. Our job is to find time histories of the system.[2] That is, we start from x_0, find $x_1 = R(x_0)$, $x_2 = R(x_1)$ and so forth; we then look for patterns in the sequence $x_0, x_1, x_2, \ldots$. One simple visualizable model for such a system is an island containing an insect population which breeds in the summer and leaves eggs that hatch the next summer; our variable is the population each summer. Specifically, x_j is the ratio of the actual population in the summer of the jth year to some reference population. To make our model explicit, we assume that the population next summer x_{j+1} is determined by the population this summer via the relation

$$x_{j+1} = rx_j - sx_j^{\,2}$$

Here there are two terms. The first term rx_j represents the natural growth rate of the population; the term $sx_j^{\,2}$ represents a reduction of this natural growth caused by overcrowding of the insects. When r is greater than 1, the first term simply expresses an increase in the population by a factor r in each year. If this were the only term, the population would grow exponentially. The second term represents the reduction in the population growth caused by, for example, competition for resources (or perhaps shyness) of the insects when the population is large. By rescaling, that is, by letting x_j be replaced by $(r/s)x_j$, one can convert this equation into a standard form:

$$x_{j+1} = rx_j(1 - x_j) \qquad (2)$$

We wish to examine the long-term behavior of the population, or of x_j, based upon equation 2. In particular, we are interested in how this behavior depends upon the growth rate r. We can think of r as being akin to the Reynolds number in the hydrodynamic example. To keep the insect population ratio in the interval between 0 and 1 we limit our examination to values of r between 0 and 4.

First we study the behavior for small r. If r is less than 1, the insects are living in such an inhospitable environment that their population will diminish each year. Their population pattern is like that shown in figure 3a. If, for example, $r = 1/2$ and one starts from $x_0 = 1/2$, then x_1 is $1/8$ and each succeeding x_j is less than $2^{-(j+2)}$. The population simply dies away to zero, for all starting values. This result is summarized in figure 4 in which we plot eventual population values as a function of r. For values of r below 1, the eventual population is zero. Roughly speaking, we might think of this behavior as akin to the laminar (smooth) flow in figure 2a.

The region of r between 1 and 3 shows another kind of simple behavior, perhaps akin to the time-independent swirls of figure 2b. If we start with any x_0 between zero and one, the population approaches a constant but non-zero value. This constant population, x^* can be found by replacing both x_j and x_{j+1} in equation 2 by x^*:

$$x^* = rx^*(1 - x^*)$$

which has the two solutions $x^* = 0$ and

$$x^* = 1 - 1/r \qquad (3)$$

Such a self-generating value of x is called a fixed point. The zero-population solution is unstable. If we start with a very low population (see figure 3b), the population will increase year by year until it settles down to the value given in equation 3. The final populations $1 - 1/r$ are also plotted in figure 4. This behavior might be considered to be compatible with the time-dependent flow of figure 2b.

Thus the region in which r is below 3 is easily understood. No chaos has arisen so far. Now jump to $r = 4$. Figure 3c shows the values of x induced at this value of r, starting from $x_0 = 0.707$. Apparently the population x_j assumes all values of x in the range between zero and one starting from this point. Although x_{j+n} is uniquely determined by x_j, for large n the pattern of determination looks chaotic rather than—as it is—deterministic. For small n, one can see patterns (for example, that small x_j produces small x_{j+1}) but these correlations become invisible as $n \rightarrow \infty$. What we see is apparent chaos.

For $r = 4$ (only!) one can solve equation 2 by the simple change of variables

$$x_j = (1 - \cos 2\pi\theta_j)/2$$

Then equation 2 can be converted into the statement

$$\begin{aligned} &1/2(1 - \cos 2\pi\theta_{j+1}) \\ &\quad = 4[1/2(1 - \cos 2\pi\theta_j)] \\ &\qquad \times [1/2(1 + \cos 2\pi\theta_j)] \\ &\quad = 1/2(1 - \cos 4\pi\theta_j) \end{aligned}$$

which has as one solution $\theta_{j+1} = 2\theta_j$ or

$$\theta_j = 2^j\theta_0 \qquad (4)$$

One can see the chaos in the solution quite directly. Since x_j is related to $\cos 2\pi\theta_j$, adding an integer to θ_j (or changing its sign) leads to the very

Turbulent flow patterns as drawn by Leonardo da Vinci. Note how the large swirls break into smaller ones, and these again break up. (From the Royal Library, Windsor Castle; reproduced with permission.) Figure 1

same x_j. Hence if one writes θ_j in ordinary base-10 notation as, say, $\theta_j = 11.2693\ldots$ one can simply throw away the 11. Better yet, if one writes θ_0 as a "decimal" base 2, as for example

$$\theta_0 = 1/2 + 1/8 + 1/16 + 1/64 + \ldots$$
$$= 0.101101\ldots$$

then the multiplication by 2 is simply a shift in the "decimal" point, so that

$$\theta_1 = .01101\ldots$$
$$\theta_2 = .1101\ldots$$
$$\theta_3 = .101\ldots$$
$$\theta_4 = .01\ldots$$

Thus, if we start out with any θ_0, the θ_j produced will depend on the jth and higher digits in θ_0. This gives us one possible definition of chaos: For large j the dynamical variable x_j has a value which is extremely sensitive to the exact value of x_0. In our case, suppose we have two starting values x_0 and x_0' that differ by a small number ϵ and generate two sequences of populations x_j and x_j' based, respectively, upon x_0 and x_0', then after j steps, the difference grows to the value $2^j\epsilon$. (See also Joseph Ford's article, "How random is a coin toss?" April, page 40.)

In fact, the calculation represented in the picture of Figure 3c is, in some sense, incorrect. It was calculated on a computer with 16 decimal digits. After about 50 steps, an initial error of 10^{-16} grows to be an inaccuracy of order 1. Consequently, all the points after step 50 are wrong, representing some random effect in the computer, and not a selected initial value.

This might be an appropriate point to mention that one of the major sources of modern stochastic theory is the work of the meteorologist Edward Lorenz.[3] As an analog to weather forecasting, he studied systems like our $r = 4$ system, in which the final state is an extremely sensitive function of the initial state. For this kind of system, as the prediction period grows longer, both the initial data needs and the computational power required grow exponentially. True long-range detailed weather prediction is, in practical terms, impossible.

The system at $r = 4$ is chaotic in another sense. For almost any randomly chosen x_0—or θ_0—the set of resulting θ_j will be uniformly distributed between 0 and 1. Correspondingly, we will get a set of x_j's in which the probability $p(y)$ that x_j will have the value between y and $y + dy$ is proportional to $(y(1-y))^{-1/2}$ for almost any starting point. Thus the time average for this chaotic system is the same for almost every starting point.

We have inserted on the right-hand side of figure 4 this distribution for this value of r by showing all x between 0

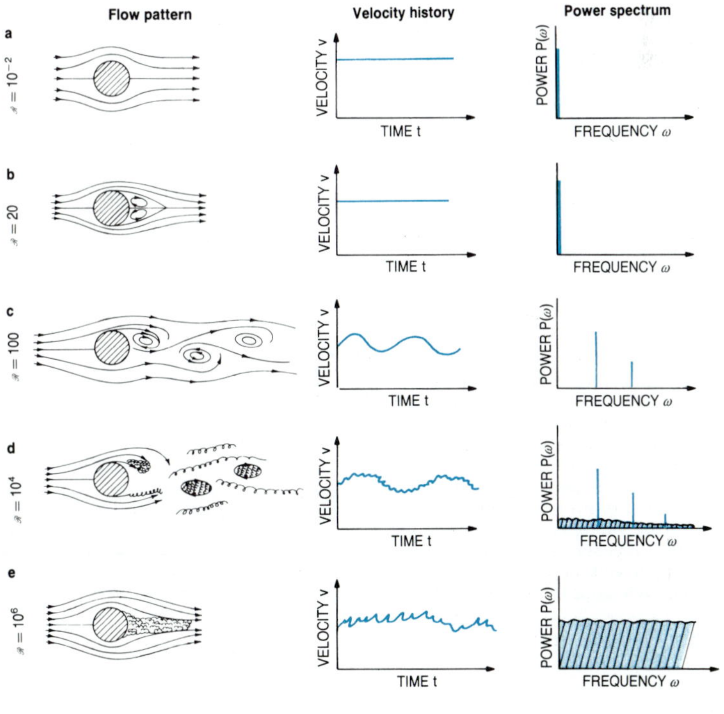

Patterns of hydrodynamic flow for various Reynolds numbers $\mathscr{R}$. At small values of $\mathscr{R}$ the flow is laminar (**a**); as $\mathscr{R}$ is increased, the flow becomes first periodically undulating (**c**) and finally turbulent (**e**). In the graphs for each Reynolds number, we plot the time variation of one component of the velocity as measured, for example, at the indicated point in the sketches; we also show the power spectrum $P(\omega)$ for these time variations of the velocity. Figure 2 and 1 as possible long-term values.

A final definition of chaos is that the power spectrum

$$P(\omega) = \frac{1}{N}\left|\sum_{j=1}^{N} x_j e^{2\pi i\omega_j}\right|^2 \qquad (5)$$

has broad spectral features. For $r = 4$ and large N, the power spectrum can be calculated exactly. For almost any x_0 it is perfectly flat.

There are some special values for x_0 that generate exceptional patterns of x_j. We can see these patterns more readily if we first note that one can always choose θ_j to be in the interval between 0 and 1/2, because θ_j can be flipped in sign and shifted by an integer without changing x_j. Thus, the recursion relation for θ_j can be written as

$$\theta_{j+1} = \begin{cases} 2\theta_j & \text{for } 0 \leqslant \theta_j \leqslant 1/4 \\ 1 - 2\theta_j & \text{for } 1/2 \leqslant \theta_j \leqslant 1/2 \end{cases} \qquad (6)$$

Any rational number that is chosen for θ_0 will lead to a recurring pattern of θ_j and x_j. For example, if we take $\theta_0 = 1/3$ then all subsequent θ_j are also 1/3, so this is a fixed point. If one starts with $\theta_0 = 1/5$ then the subsequent θ_j are 2/5, 1/5, 2/5, 1/5 and so forth. Thus we have a cyclic behavior with a period of length 2. One solution that has period of 3 is 1/2, 2/9, 4/9. Equation 6 has periodic solutions with all possible periods.[4]

Now we have long-term solutions for equation 3 that are time-independent for r between 0 and 3 and one that is quite chaotic for $r = 4$. Next, increase r from 3 and observe the first hints of chaos which arise. As r increases just above 3 the fixed point at x^* near 2/3 becomes unstable; instead, the population keeps flipping back and forth between high and low values. The insects start from a low population. They reproduce avidly, leaving a large number of eggs. But the resulting population next year is too crowded (or shy), so the population the year after will be low. Thus, odd years will have high populations, even years low populations. The Bible records[5] that Joseph predicted such a periodic behavior with a basic time step of seven years. In our model the exact values of x for the two-cycles are

$$x = 1/2(1 + 1/r) \pm 1/2[(1 + 1/r)(1 - 3/r)]^{1/2}$$

These $q = 2$ cycles remain stable over the range of r between 3 and $1 + \sqrt{6} = 3.4495$. I will call this upper value of r at which the two-cycle becomes unstable r_2. For r slightly larger than r_2, the stable behavior is a four-cycle, as shown in figure 3e. The basic period of the behavior has doubled once more. This behavior is also only stable up to a limiting value, r_4. Above r_4 an eight-cycle appears and remains stable between r_4 and r_8, whereupon a sixteen-cycle appears. These stable beha-

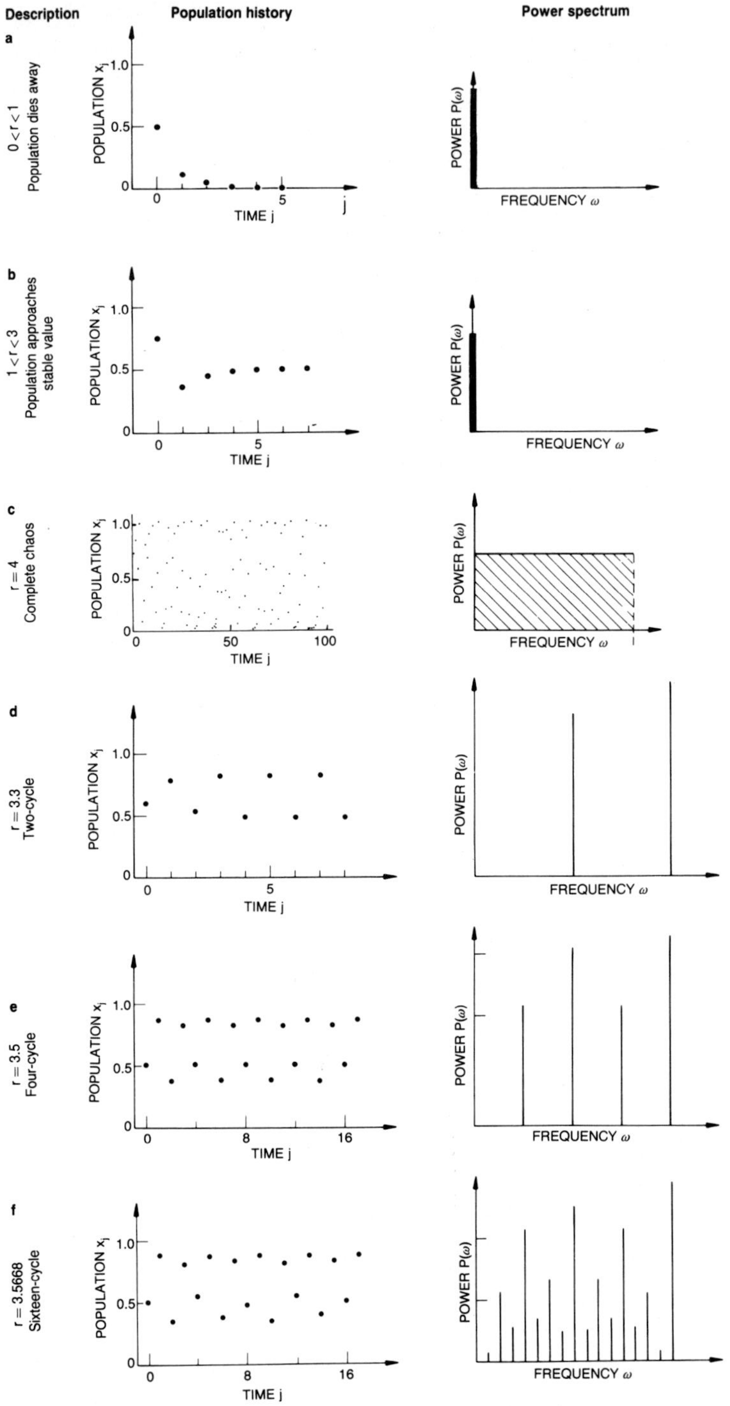

Behavior of a population that obeys a simple nonlinear reproduction equation (see the text). We show both the time evolution and the corresponding power spectrum of the population for various values of the growth parameter r. For growth parameters above 3.0, no stable population is reached; in the cases shown here the population oscillates over a varying number of cycles. The oscillations show up as peaks in the power spectrum, with each higher-order cycle contributing less than the earlier cycles. Figure 3 viors are shown on figure 4 as doublings in the number of x-values which the system assumes as j becomes infinite. Successive doublings continue until at $r_c = 3.5699\ldots$, when a cycle of infinite length appears.

To see the beginnings of the onset of chaos in this model, look at the power spectrum defined by equation 5. When the fixed point is the stable behavior, the power spectrum is a spike at zero frequency. When the two-cycle appears, another frequency, $\omega = 1/2$, appears in the spectrum, as figure 3d shows. This frequency is, of course, equal to the inverse period of the motion. Between r_2 and r_4, $\omega = 1/4$ and $\omega = 3/4$ also enter, as in figure 3e. As the period increases, more and more spectral lines enter until at r_c there are infinitely many lines.

Of course, a spectrum with an infinite number of discrete lines is not the same as a broad-band, continuous spectrum. Even at $r = r_c$ there is no fully developed chaos of the type that occurs at $r = 4$. However, in this model, the development of an infinite number of lines through successive period doublings is a major step toward the production of chaos.

To see the remaining steps, turn to figure 5. This picture is, like figure 4, a depiction of the populations that arise after a huge number of iterations of an initial x_0. Our job is to understand how this picture changes between r_c and 4. We shall discuss the predominant features of this development, leaving out for the moment small regions—which appear as white stripes in the more or less uniformly gray area—in which stable cycles once again dominate the picture.

We start at $r = 4$ and decrease r. As we saw, an initial x_0 generates points that move erratically through an entire band of permitted x-values. As r decreases, this band narrows slightly to be between $r(1 - r/4)$ and $r/4$, but the behavior is otherwise qualitatively unchanged. The spectrum contains no sharp peaks. However, below the value labeled r_2' the behavior changes. The band splits into two. Between r_2' and r_4', the population is somewhere in the lower band on even steps; on odd steps it lies somewhere in the upper. The spectrum then has a broad background produced by the erratic values the population assumes in each band and a sharp peak at $\omega = 1/2$ produced by the regular way it jumps from band to band. Then at r_4' the behavior changes once more. There are four bands, which we can number from bottom to top as 1, 2, 3, 4, and the motion goes from band 1 to 3 to 2 to 4—which is, not accidentally, exactly the same ordering as the motion in the four-cycle. As r decreases beyond r_8' there are eight bands, then beyond r_{16}' sixteen, and so forth. When there are 2^n bands, the population returns to a given band after 2^n steps but the exact point at which it returns is chaotic in exactly the same sense as there is chaos at $r = 4$. In this region of 2^n bands there are sharp spectral lines at 2^{-n} times an integer together with a broad background produced by the erratic behavior in the band. To the naked eye this erratic behavior looks very much the same as the $r = 4$ chaotic motion. The only differences are ones of scale. In this 2^n-band case, the erratic motion is confined to a set of narrow regions, inside each band. Furthermore, the motion only returns to the band every 2^n steps, so that as we change values of r, there is a change in the time-scale as well. (A reader who is acquainted with renormalization and scaling might wonder whether this change of scale might be used to build a renormalization-group analysis of the period-doubling process. It can, and has.[6])

This process of successive band splittings enables the system to interpolate smoothly between the full chaos at $r = 4$ and the 2^∞ cycle at r_c. As r approaches r_c from above, we get more and more bands until at r_c there are 2^∞ bands, each of them infinitesimally narrow, and merging into the infinite number of lines just below r_c.

All of this so far has applied to our simple model, given by equation 2. However, it is important to notice that the general nature of the processes of period doubling and band splitting are independent of the details of the model and will occur with any mapping of the form

$$x_{j+1} = rf(x_j)$$

with $f(0) = f(1)$ and f being a smooth function with a single maximum in the interval 0 to 1.

Furthermore, Mitchell Feigenbaum has looked[6] at several maps and demonstrated that some of the quantitative properties of the behavior of band splitting and period doubling near r_c—that is, for large numbers of periods—apply equally well to almost all maps with a smooth maximum in which $f''(x)$ is negative at the maximum. In particular recall that r_q and r_q' are, respectively, the values of r at which a $q = 2^n$ cycle first appears or $q = 2^n$ bands merge. (Again, refer to figures 4 and 5 for a depiction of these bands and cycles.) As n becomes infinite, these limiting values both approach the same limiting value r_c. Feigenbaum showed

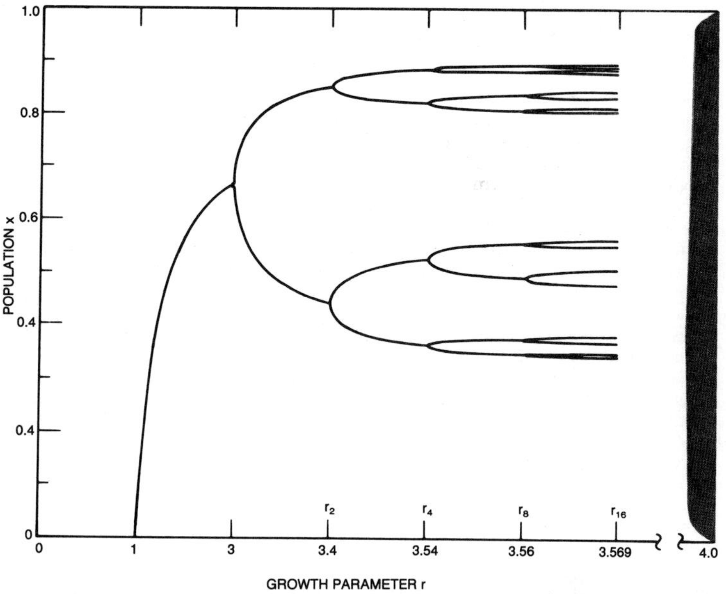

Stable values for the population as a function of the growth parameter r. The solid lines show the values of the population x_j that recur as j becomes infinite. For growth parameters r below 1, the population decays to 0. For r between 1 and 3.4, there is no single stable value—instead the population oscillates between the two values in the upper and lower branches of the curve. Above 3.54, the population oscillates among the four branches, and so forth. These oscillations give rise to the peaks in the power spectrum shown in figure 2. Note the highly nonlinear scale. Note also the break at 3.569; the subsequent behavior for growth parameters between 3.569 and 4 is shown in figure 5. Here we indicate only the continuum of values allowed as the population approaches the truly chaotic behavior at $r = 4$. Figure 4 that for large n they approached r_c in a very simple manner, namely:

$$r_q - r_c = A\delta^{-n}$$

$$r_q' - r_c = A'\delta^{-n}$$

Here A and A', naturally, depend in detail upon the mapping function $f(x)$. However, the exciting and surprising result of his work was that δ is universal: It does not change as we change the mapping function f. A similar result is obtained for the splitting of the x_j values. When one chooses in an appropriate fashion two neighboring values for x_j that lie in a 2^n-cycle, one finds that the separation between these values decreases with n as $B\alpha^{-n}$, where α is universally equal to 2.503 .... Feigenbaum has also given a renormalization group treatment that verifies this universality and generates the universal numbers δ and α.

If these quantitative results for large n apply to all mathematical maps, might they not also apply to real dynamical systems? In particular, if we use x_t to refer to some dynamical property at time t and let $f(x_t)$ describe how the dynamical property at time t determined its value at time $t+\tau$, will we be able to see the period-doubling behavior our maps show?

Nonlinear electrical circuits have been shown[7] to give a very similar behavior to the one described above. As a control parameter, r, in the circuit is varied, one of the circuit variables, such as a voltage, traces patterns in which successive period doublings occur. Moreover, the observed values of α and δ are the same as the values mentioned above.

It is not surprising that simple circuits, which can be described in terms of a few variables, show identical behavior to the mapping problems. But, what of real hydrodynamical systems? They are much more complex. Can they show this behavior also? One class of studies that could make contact with this theory of dynamical behavior is the experimental work on the Rayleigh–Bénard instability in fluid systems. When an enclosed fluid is heated from below, at low heating rates, no flow occurs. At higher rates, time-independent convection is set up. At higher rates yet, a periodic time dependence appears. At still higher rates, the time dependence looks very chaotic and has a broad band spectrum. In a small system containing helium at low temperatures, Albert Libchaber and Jean Maurer observed[8] a series of successive period doublings. When they adjusted the heating rate very carefully, they could see the power spectrum shown in figure 6. Notice the qualitative similarity between this spectrum for a real hydrodynamic system and the spectrum shown in figure 4f. This connection is, however, more than qualitative. The relative heights of the weaker spectral lines are predicted to be universally determined by the properties of the high-n band splittings. In particular, one can determine the quantity α from these heights. There is a quite satisfactory agreement between this experimental value of α and the one calculated by Feigenbaum's renormalization-group analysis. Hence, one route to chaos in one real system may be said to be largely understood.

However, this is only the beginning of the story—not the end. Libchaber and Maurer's cells are rectangular in cross section. Günter Ahlers and Robert Behringer have done[8] a series of parallel experiments on cylindrical cells. They observe a different route to chaos. In fact, the period-doubling route appears to be rather rare. Are there other relatively universal routes to chaos observable in real systems? Can they also be analyzed in terms of very simple models? We do not know, but there are a large number of workers trying to find out.

Because the experiments show additional roads to chaos, it is sensible to look back at the general mapping problems described by equation 1 and see whether they, too, have additional paths to interesting behavior. In fact, the recursion relation we have been studying, equation 2, does show one more class of chaotic transitions. Notice the vertical white stripes in the broad bands of figure 5, in particular the widest one, labeled C_3. At the right-hand margin of C_3 the motion becomes disordered in the way I have described. But at the left-hand boundary of this region the long-term stable motion is the three-cycle. As r increases above this boundary value, r_3, there are periods of disordered motion followed by long periods (which have a length of order $|r_3 - r|^{-1/2}$) of very orderly motion in which the system looks very much like it is undergoing cyclical motion of period three. As this almost cyclical motion progresses, there is a gradual drift away from the period-three cycle elements. Finally the population elements get far enough away so that once again there is a period of apparently random motion. That is, there are long orderly periods mixed with bursts of disorder (see figure 7). This kind of behavior is called intermittency and has been studied in some detail.[9] It also has been observed in experimental systems, but to date the detailed correspondence between the model systems and the real ones has not been fully worked out.

In the examples I have mentioned so far, the fact that the models all exhibit chaotic behavior is related to the fact that the mapping function $f(x)$ has a

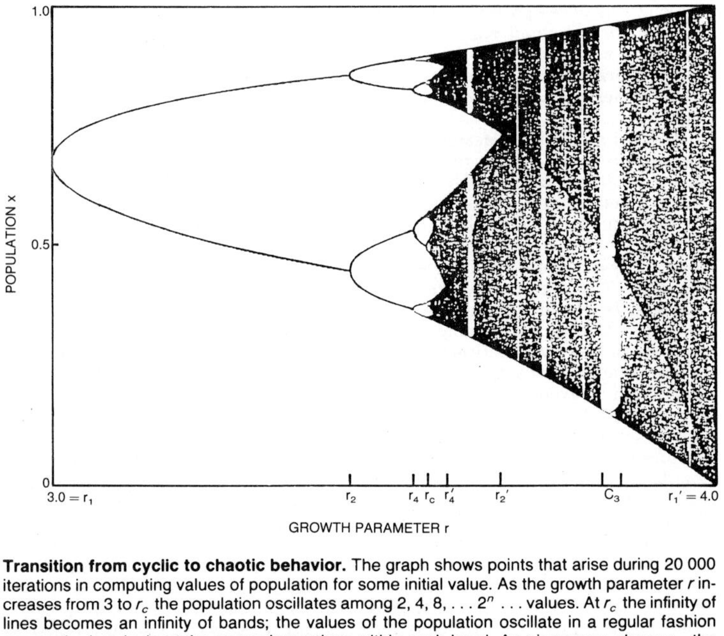

Transition from cyclic to chaotic behavior. The graph shows points that arise during 20 000 iterations in computing values of population for some initial value. As the growth parameter r increases from 3 to r_c the population oscillates among 2, 4, 8, ... 2^n ... values. At r_c the infinity of lines becomes an infinity of bands; the values of the population oscillate in a regular fashion among the bands, but take on random values within each band. As r increases above r_c, the bands merge, until for values of r above r_2' there is only a single band of values that the population assumes chaotically. The thin white stripes, such as the region labeled C_3 represent periods in which the population assumes regular values for much of the time and is only intermittently chaotic. While the scale in figure 4 is highly nonlinear, the scale here is linear. Figure 5

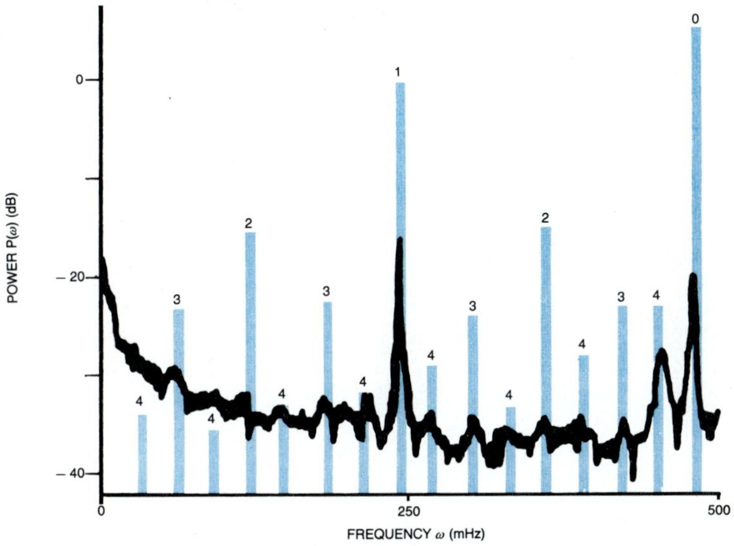

Power spectrum for convective flow in a small cell containing liquid helium, just above the Rayleigh–Bénard instability. The cell is heated from below and cooled from above. Under the conditions of this experiment, the regular convection has given way to a chaotic behavior, but with some regular components—like that shown in figure 2d—similar to the behavior of the population for a growth parameter somewhat above r_c. The experimental curve is in black; the colored vertical lines indicate the theoretical positions of the peaks and reflect their predicted heights; the numbers on the theoretical lines give their order, that is the number of period doublings involved in the line. There is a fundamental frequency of oscillation (near 500 mHz); the other oscillations occur at multiples of $f/2^n$, with n the order of the oscillations. (Adapted from reference 8) Figure 6

maximum at some value of x. Maps that do not have a maximum—such as

$$x_{j+1} = x_j + \Omega - (k/2\pi)\sin 2\pi x_j \qquad (7)$$

for $|k|$ below 1—cannot show any chaotic structure. These are called "no-passing" systems for the following reason. Imagine that you start with two points x_0 and x_0' and go through j steps to obtain x_j and x_j', respectively. In these systems it turns out that if $x_0 < x_0'$ then it is always true that $x_j < x_j'$, that is, the sequence x_j never passes the sequence x_j'. For the map above, with $|k| < 1$ there are two kinds of stable motions, both being smooth and unchaotic. For some values of Ω, the system falls into a cycle of length q in which x advances by p units in the q steps. In this motion $x_{j+q} = x_j + p$ and the average rate of advance of x, w, which in this case is p/q, is a rational number. This motion may be described as commensurate in the sense that the cycles are commensurate with the period of $\sin 2\pi x$ in equation 7. On the other hand, Ω may also be chosen so that the average rate of advance per step is irrational. In this incommensurate motion, if one starts from $x_0 = 0$ then the subsequent motion will be of the form

$$x_j = jw + \phi(jw) \qquad (8)$$

where $\phi(t)$ is a periodic function of t with period 1.

As k passes through 1, the cyclical or commensurate motion persists. Near the Ω-values which produced commensurate motion for $k < 1$, there is also orderly cyclical motion for $k > 1$. However, infinitesimally close to each value of Ω that produces an incommensurate motion for k just below 1, there is for k just above 1 a domain of Ω in which the motion is chaotic. That is, the incommensurate motion becomes unstable to chaos at $k = 1$.

This instability has not yet been analyzed in detail. However, the incommensurate motion at $k = 1$ has been analyzed by two groups,[10] in particular for the case in which the average speed is $w = (\sqrt{5} - 1)/2$. (Other irrational values of w will probably show qualitatively similar but quantitatively different behavior.) They conclude that equation 8 still describes the motion, but that the continuous function $\phi(t)$ is very bumpy indeed at $k = 1$, while for k below 1, it is quite smooth. By quite bumpy I mean something rather specific and rather specifically awful. Consider the derivative of ϕ, $\phi'(t)$, in some small region of t. Pick the interval to be as small as you like. Furthermore pick some big number (say 10^{50}) and a small one (say 10^{-50}). Now let k approach closer to one, but

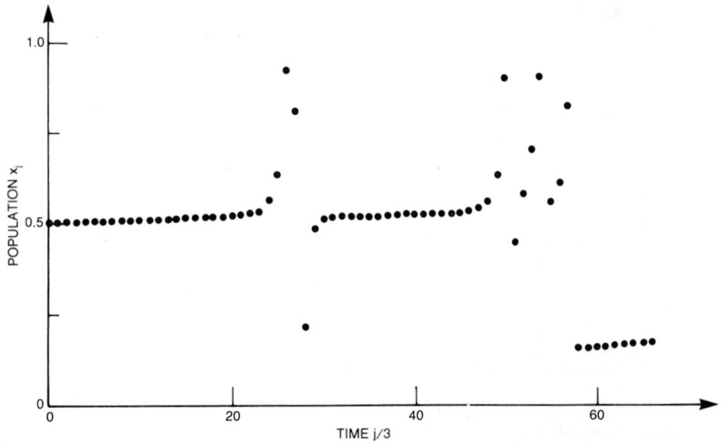

Intermittency: a brief stop at nearly regular behavior on the road to chaos. The graph shows values of the population for a growth parameter just above the lower boundary of the region marked C_3 in figure 4. At that slightly smaller value, the behavior is periodic, with three stable values for x: 0.5, 0.96 and 0.16. Here we have plotted only every third value of x. The population behaves in an orderly, period-three manner for long stretches of time, but intermittently, and at irregular intervals, behaves chaotically. Figure 7 always from below. Just so long as k is below 1, $\phi'(t)$ is smooth and is always greater than zero but not infinite. However, although ϕ' is smooth, it is very steeply varying. Thus, for example, I can always find some value of k, close to one, in which $\phi'(t)$ takes on both the value of your big number and also that of your small one in the specific interval you have chosen. If you choose more extreme numbers, I just have to go closer to $k=1$. Clearly, we have reached a situation in which $\phi(t)$ exists and describes a more or less physical problem, but the function in question is, at $k=1$, not differentiable anywhere.

This strange mathematical behavior can be seen experimentally. It results in a power spectrum which contains an infinite number of discrete lines which are bunched together and pile up toward $\omega=0$.

Experimentalists will, no doubt, be looking for power spectra of this character to observe the onset of chaos in the theoretically predicted manners. Also, theorists will, of course, be looking in their models and at experimental data hoping to see new forms of the onset of chaos.

* * *

This paper was written while I was in Israel enjoying the hospitality and support of the Israel Academy of Sciences and Humanities of Tel-Aviv University, and of the Weizmann Institute. It has been my pleasure to learn about dynamical systems from M. Feigenbaum and my students David Bensimon, Scott J. Shenker, Michael Widom, and Albert Zisook. Bensimon has helped in the production of some of the figures shown here.

References

1. R. P. Feynman, R. B. Leighton, M. Sands, *The Feynman Lectures on Physics* volume II, chapter 41, Addison-Wesley, Reading, Mass. (1964).
2. R. B. May, Nature **261**, 459 (1976) is a review of this subject.
3. E. N. Lorenz, J. Atmos. Sci. **20**, 130 (1963).
4. See, for example, P. Collet, J.-P. Eckmann, *Iterated Maps on the Interval as Dynamical systems*, Birkhauser, Boston (1980).
5. Genesis: 41.
6. M. J. Feigenbaum, J. Stat. Phys. **19**, 25 (1978); **21**, 669 (1979).
7. See, for example, P. S. Lindsay, Phys. Rev. Lett. **47**, 1349 (1981); J. Testa, J. Perez, C. Jeffries, Phys. Rev. Lett. **48**, 715 (1982).
8. A. Libchaber, J. Maurer, J. Phys. (Paris) **41** C3, 51 (1980).
9. G. Ahlers, R. P. Behringer, Prog. Theor. Phys. Suppl. **64**, 186 (1978).
10. P. Manneville, Y. Pomeau, Phys. Lett. **75A**, 1 (1979).
11. M. J. Feigenbaum, L. P. Kadanoff, S. J. Shenker, Physica **5D**, 370 (1982). D. Rand, S. Ostlund, J. Setna, E. D. Siggia, Phys. Rev. Lett. **49**, 132 (1982). □

Chaos: A View of Complexity in the Physical Sciences

Leo P. Kadanoff

Leo Kadanoff is a theoretical physicist who has contributed widely to research in the properties of matter and upon the fringes of elementary particle physics. Most recently he has been involved in the understanding of the onset of chaos in simple mechanical and fluid systems.

He was born and received his early education in New York City. He did his undergraduate and graduate work at Harvard University, and after some post-doctoral work at the Niels Bohr Institute in Copenhagen he joined the staff at the University of Illinois in 1962. In 1966 and 1967 he did research on the organization of matter in "phase transitions," which led to a substantial modification of physicists' ways of looking at these changes in the state of matter. For this work he received the Buckley Prize of the American Physical Society (1977) and the Wolf Foundation Prize (1980).

He went to the University of Chicago in 1978 and became John D. MacArthur Distinguished Service Professor of Physics there in 1982.

Professor Kadanoff is a member of the National Academy of Sciences, a Fellow of the American Physical Society, and a member of the American Academy of Arts and Sciences.

Reprinted from *The Great Ideas Today*, 1986.

Introduction

Definition of chaos: complexity and order

The word *chaos* has suddenly come to be popular in the physical sciences. It is used to describe situations in which we can see a very complex behavior in space and time. Because the term is used imprecisely, it is best explained by example (*see* illustrations on the following two pages). Figure 1 shows a kind of atmospheric disturbance which, over the course of many years, has been observed on the surface of the planet Jupiter. This close-up shows a quite intricate pattern of atmospheric swirling or turbulence. Observers of the TV news will recognize that somewhat similar swirling patterns also exist in the Earth's atmosphere. On both planets, the turbulence takes on fantastic forms in which we can nonetheless see some underlying regularity and order. One kind of regularity is that the storm, according to what we believe, has continued to exist on the surface of Jupiter for millions of years. Another kind of regularity is that the storm contains large, rather uniform regions.

The predominant impression that one gets from weather maps on either planet is nevertheless one of considerable complexity. Chaotic patterns are characteristically quite varied in their details, but they may have quite regular general features. For example, clouds are sufficiently orderly so that one can give a meaningful classification of their general types, but each type exhibits endless variations in its detailed shapes.

Look at another example. Figure 2 shows a dried-up lake in which mud has hardened itself into a complex pattern. We can see that the pattern is almost the same in different places, but it repeats itself with apparently unpredictable variations and is hence "chaotic." Additional familiar examples of chaotic behavior are provided by the fantastically rich patterns of snowflakes or of the frost which can appear on the inside of a window in winter. Figure 3 shows the result of the solidification of water on a cool, flat surface. New ice forms in contact with the old. Because a piece of the ice surface which sticks out can more effectively move forward, projections upon the ice surface grow into longer and longer branches. But then, if a branch has a little bump on

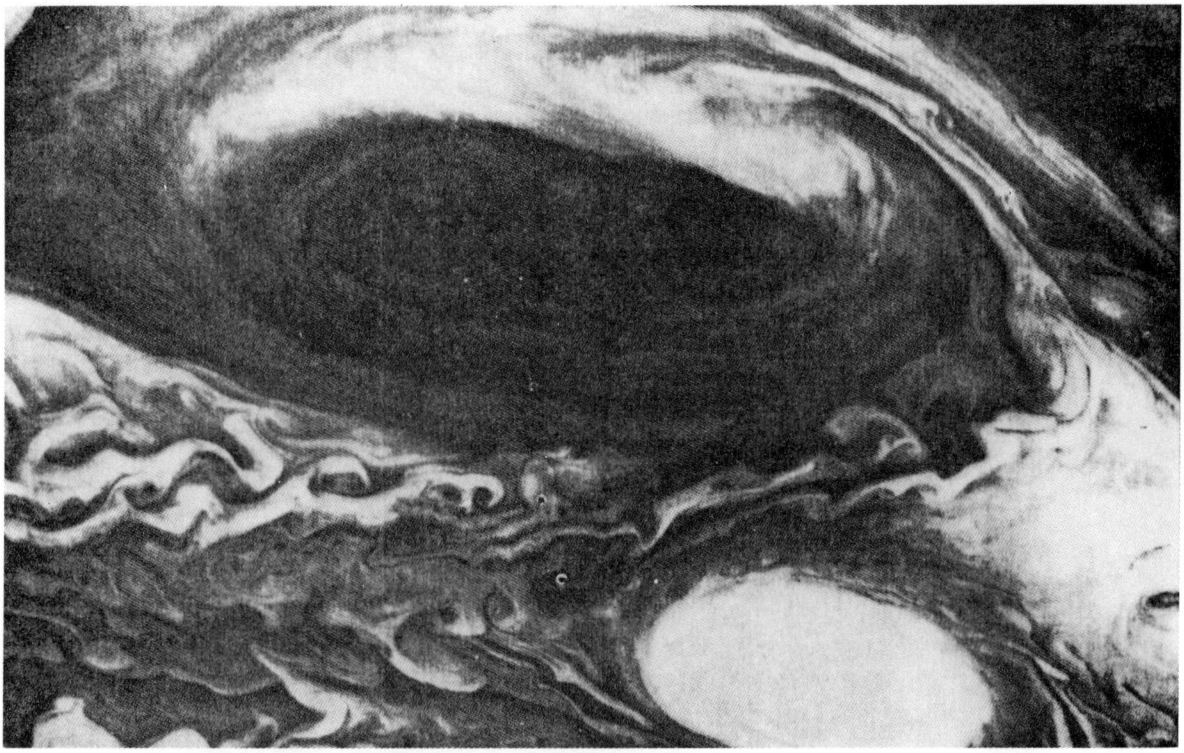

Figure 1. A "storm" in the atmosphere of Jupiter.

it, that bump will also tend to grow and become a branch itself. And, by the same logic, branches will grow on branches, and so on indefinitely until a beautiful treelike shape arises.

All these examples of chaos have several striking features in common. One, which I wish to emphasize now, is the outcome's sensitivity to the conditions under which the pattern is formed. A degree's change in temperature today, and next week's weather map would change totally. If you blow upon the windowpane, you can completely change the details of a branching pattern like that of Figure 3.

Chapter I: Simple laws, complex outcomes

The physical sciences are divided into many disciplinary subfields, among them meteorology, astronomy, aerodynamics, and physics. Except for physics, all these subfields try to gain a deep and solid understanding of particular areas of nature. Thus, if an astronomer looks at a galaxy and sees it to be chaotic, his or her natural reaction is likely to be a desire to understand that particular problem. Concentration on a complex behavior is natural to fields of activity like astronomy.

Physicists, on the other hand, consider themselves to be looking for the fundamental laws of nature. They seek basic principles, ideas, and mathematical formulations on which all further understanding can be

Figures 2 and 3. "*Chaotic patterns are characteristically quite varied in their details, but they may have quite regular general features.*" The pattern of the dried up lake, in Figure 2 (top), and frost on a flat surface, in Figure 3, are examples.

built. To look at complexity is to some extent a new endeavor for physicists. It runs counter to the idea of physics as the science that seeks to understand nature in simplest terms. Newton gave us three simply stated laws to describe all the motions of the heavenly bodies—and many aspects of earthly motion as well. The laws of general relativity or of quantum mechanics are also simple to state. Moreover, such laws tend to result from a study of their "simplest," most elementary realizations. For Newton's gravity this realization is found in Kepler's rules for the motion of two gravitating bodies; for general relativity it is found in black holes; for quantum mechanics it lies in hydrogen atoms.

For the student of physics, or the practitioner, the science is interesting and beautiful precisely because it summarizes the complexity of the world in a few simple laws and then describes the consequences of these laws by almost equally simple examples. However, many students and even some practitioners suspect that something is lost in the process. When physics concentrates on three laws, or five, or seven, when those laws are mostly applied only to the very simplest examples, we have lost something of the real world. We have chosen to ignore the wonderful diversity and exquisite complication that really characterize our world. This choice has led to wonderful descriptions of nature in our theories of quantum mechanics, relativity, cosmology, and so forth. But, these theories are so focused upon the simple and "basic" that they run the danger of providing a peculiar caricature of nature. Focus upon simplicity and you leave out Jupiter's storms, the diversity of galaxies, the intricacies of organic chemistry, and indeed life itself.

In recent years there has been some change in the attitude of many physicists toward complexity. Indeed, the very existence of this article reflects the change. Physicists have begun to realize that complex systems might have their own laws, and that these laws might be as simple, as fundamental, and as beautiful as any other laws of nature. Hence, more and more the attention of physicists has turned toward nature's more complex and "chaotic" manifestations, and to the attempt to construct laws for this chaos.

In some sense, this change in attention has resulted from a natural attempt to understand interesting situations such as the ones I have shown in the figures. In another sense, the concentration upon chaos has been a part of a change in our understanding of what it means for a law to be "fundamental" or "basic." Physical scientists have sometimes been tempted to take a reductionist view of nature. In this view, there are fundamental laws and everything else follows directly and immediately from them. Following this line of thought, one would construct a hierarchy of scientific problems. The "deepest" problems would be those connected with the most fundamental things, perhaps the largest issues of cosmology, or the hardest problems of mathematical logic, or maybe the physics of the very smallest observable units in the universe. To the reductionist the important problem is to understand these deepest matters and to build from them, in a step-by-step way, explanations of all other observable phenomena.

Here I wish to argue against the reductionist prejudice. It seems to me that considerable experience has been developed to show that there are levels of aggregation that represent the natural subject areas of different groups of scientists. Thus, one group may study quarks (a variety of subnuclear particle), another, atomic nuclei, another, atoms, another, molecular biology, and another, genetics. In this list, each succeeding part is made up of objects from the preceding level. Each level might be considered to be less fundamental than the one preceding it in the list. But at each level there are new and exciting valid generalizations which could not in any very natural way have been deduced from any more "basic" sciences. Starting from the "least fundamental" and going backward on the list, we can enumerate, in succession, representative and important conclusions from each of these sciences, as Mendelian inheritance, the double helix, quantum mechanics, and nuclear fission. Which is the most fundamental, the most basic? Which was derived from which? From this example, it seems rather foolish to think about a hierarchy of scientific knowledge. Rather, it would appear that grand ideas appear at any level of generalization.

With exactly this realization in mind, one might look at the rich variety of chaotic systems and wonder whether there are broad and general principles which can be derived from them. In fact, I have already mentioned one such "law": *chaotic systems show a detailed behavior which is extremely sensitive to the conditions under which they are formed.* The consequences of this sensitivity are further examined in the next section.

Practical predictability

Many of the modern concepts of chaos were formed by Henri Poincaré, a nineteeth-century French astronomer and mathematician. He recognized very clearly that there was a qualitative difference between the motion of two gravitating bodies (Earth-Sun, for example) and that of three (Moon-Earth-Sun). In the former case, when we have two bodies each moving under the gravitational influence of the other, the orbits are simple and easily predictable. They are Kepler's ellipses, and these orbits are certainly not chaotic. The latter situation, the famous "three-body problem," is chaotic. Three bodies develop complex orbit structures in which the positions of the objects in the distant future are extremely sensitive to their positions now. And this sensitivity and complexity is not just theoretical nonsense. It has practical consequences. To predict the future, one needs information about the present, and the longer the forecast, the better the information required. In the chaotic problem, the accuracy required of the input data must be very sharply improved as the forecasting period becomes longer and longer.

For astronomical systems, the data initially needed are the position and velocities of the gravitating bodies. Imagine that we are looking ahead and trying to forecast the positions of the planets, perhaps with a view to predicting the time of eclipses. Imagine further that there is a certain error in our present knowledge of planetary positions, perhaps by only a few feet. In both the nonchaotic and the chaotic cases our forecasting uncertainty will get larger as we look further forward into the future. The difference is in the type of growth. In the nonchaotic case, for each further year of forecast, the uncertainty grows by the addition of an increment proportional to the original uncertainty. In the chaotic case, for each additional year of forecast, the uncertainty grows by an increment proportional to the uncertainty in *that year's* forecast. The latter type of growth, so-called exponential growth, is akin to compound interest and is very rapid compared with the growth in the nonchaotic case, which is akin to simple interest. In the long run, the uncertainties in the "compound interest" case are far, far larger than in the corresponding case of "simple interest."

Thus our ability to forecast, for example, eclipses, is far, far worse in the chaotic case than in the more orderly example of the motion of two gravitating bodies.

This line of thought was picked up by a meteorologist from the Massachusetts Institute of Technology, Edward Lorenz. He was interested in the implications of the idea that Earth's atmosphere might exhibit a sensitivity to initial conditions similar to the one which had been thought about in the gravitational case. His work, published in 1963,* in some sense marked the beginning of our "modern era" in the study of chaotic systems. He looked at convection, that is, flows in which a heated fluid rises because it is less dense than its surroundings. He set up a simple mathematical model for convection, solved it on a computer, and showed that even in this oversimplified case the system's behavior was wonderfully rich and complex. In addition, he showed that its long-term behavior exhibited the kind of sensitivity to initial conditions that was described above for chaotic planetary systems. He made the point that if actual weather prediction were like the model he studied, it would be terribly hard to predict very far ahead.

Naturally, this practical unpredictability has very important implications for all kinds of engineering arts involving chaotic situations—not only weather prediction, but also airplane wing design, the flow of fluids through chemical plants, and many other cases. It has also inspired

*E. N. Lorenz, "Deterministic Non-Periodic Flows," *Journal of the Atmospheric Sciences* 20 (1963): 130.

some rethinking about such familiar philosophical questions as free will and determinism. For several different reasons, then, we may wish to understand this result in somewhat more detail.

In the next chapter I shall further discuss the nature of chaos by showing how it arises. In the final chapter I shall describe how chaos reflects itself in beautiful and complicated geometrical structures and illustrate this with an example from Lorenz's work.

Chapter II: Routes to chaos

One way of understanding the nature of chaos is to ask how it arises. One can start from a very orderly situation, gradually change the situation, and then see chaos set in little by little. Consider, for example, the flow of water that might occur in a river as it flows past an obstacle. Imagine that we are standing on a bridge, looking down at the water as it flows past a buttress of the bridge sitting in the water. If the water is flowing slowly, it flows in smooth and unswirly paths like those in Figure 4a. The rate of flow is listed in the different parts of Figure 4 by giving the value of a "parameter," R, which is proportional to the rate or speed at which the river is flowing. (The word *parameter* is often used in the sciences to mean a numerical value which defines a natural situation; for example, the birthrate is an important parameter for determining the future quality of life on our planet.)

Successive rows in Figure 4 show the situation for successively higher values of the flow rate. As this parameter is increased, the flow gets successively more complex. This increase in complexity is depicted in two ways. The first column shows spatial patterns by depicting the flow path of typical particles in the water—or of debris on the surface—as the particles (or the debris) move around the buttress. In the second column we plot the speed of the water at a particular spot, the one marked with an x in Figure 4a. For small speeds, as in 4a, the flow pattern would be completely time-independent and totally lacking in swirls. If the river were running a bit faster, as in Figure 4b, there would be a few swirls or vortices fixed in place near the bridge, but because these are fixed, the pattern would remain time-independent. Increase the speed still more, as in Figure 4c, and the swirls come loose and start moving slowly downstream, away from the bridge. New swirls are produced near the buttress at a regular rate, and these too move downstream in a regular progression. In this case, our instrument, which measures the speed at point x, will show a repetitive time dependence in which the speed goes through a maximum as the swirl passes by.

Such periodic behavior simply repeats itself again and again as time goes by and is certainly not chaotic. But a further increase in the speed does produce chaos. Figure 4d indicates that at this higher value of R

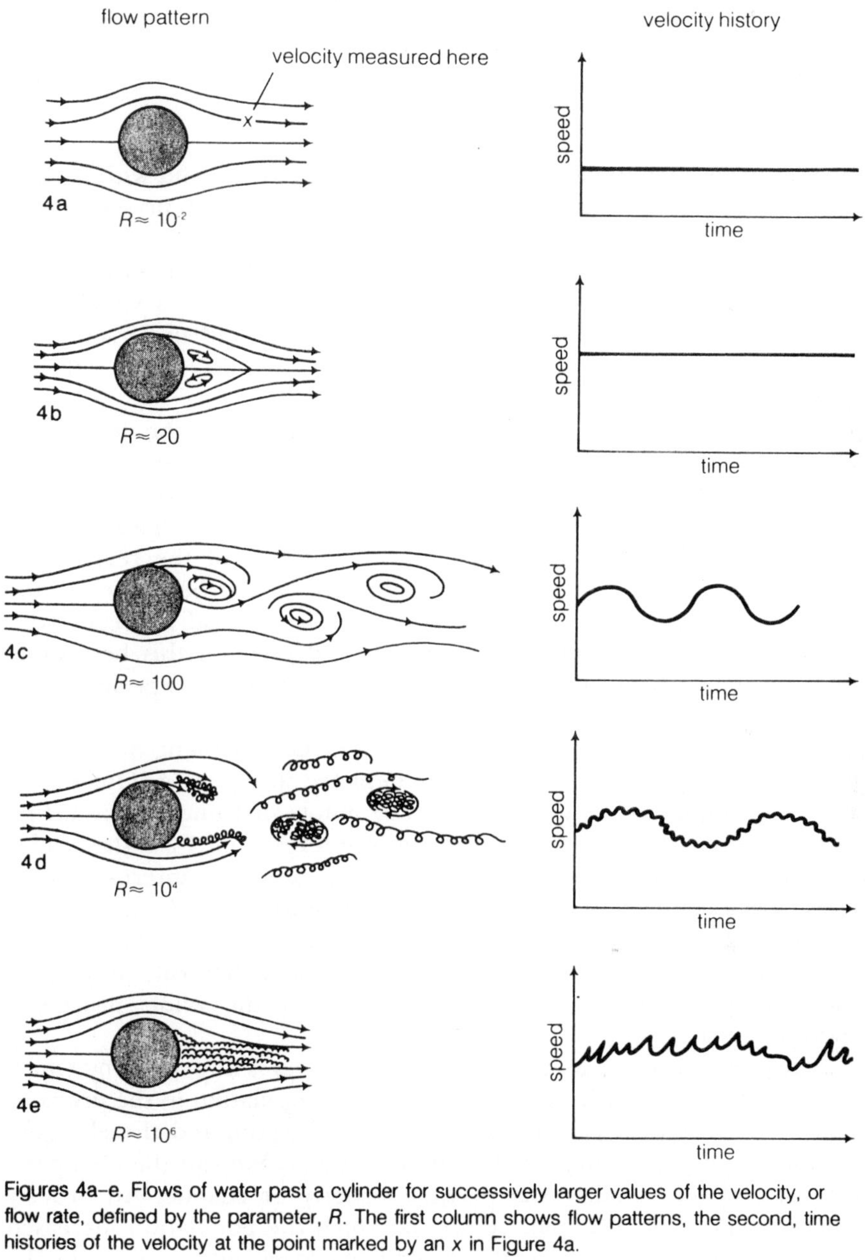

Figures 4a–e. Flows of water past a cylinder for successively larger values of the velocity, or flow rate, defined by the parameter, R. The first column shows flow patterns, the second, time histories of the velocity at the point marked by an x in Figure 4a.

the individual swirls have begun to look a bit ragged and chaotic. The time dependence shows a basically periodic pattern, similar to that of Figure 4c, but there is a small amount of chaotic jiggling superposed upon the regular motion. Finally, if the river is moving very fast indeed, as in Figure 4e, the turbulent region moves out and fills the entire wake behind the bridge. Then the time dependence seems totally unpredictable and chaotic.

A model of chaos

Next, I would like to explain in somewhat more detail and with greater precision exactly how the chaos arose in the hydrodynamic system just described. I would like to, but I cannot. Nobody has a real understanding of chaos in any fluid dynamical context. So, instead of that, I shall turn my attention to a simpler problem, one with a very simple mathematical structure which we can encompass and understand. (A simpler problem used to illuminate a more complex one is called a "model.")

This chapter and the next are largely concerned with the description of several mathematical models of chaos. These models are sets of equations which are easier to understand and study than the realistic cases which are our actual concern. However, if deftly chosen, such a model might just capture some important feature of the real system and exhibit it in a transparent form. In fact, in the best of cases the model will capture the essential nature of the physical process under study and will leave out only insignificant details. In this best situation, the model can be used to predict the results of experiments in the real system.

Our present interest is in the onset and development of chaos in fluid mechanical systems like the one depicted in Figure 4. Our model system for understanding this onset is so simple that one might, at first glance, assume that it contains nothing of interest. But I ask the reader to suspend disbelief, at least for a time. The model is interesting and does have a connection to hydrodynamic systems.

Consider, therefore, an island with a population of insects. In every year, during one month, the insects are hatched, they eat, they mate, they lay eggs, and they die. In the next year the whole process is repeated over again.

A mathematical model for this kind of process is a formula by which we can infer each year's population from the population in the previous year. We can repeat this inference again and again, and thereby generate a list of populations in the different years. We can then examine the list and see whether the result is orderly or chaotic. An orderly pattern might, for example, be one in which the population increased year by year but, after a while, started to "level out" so that, in the long run, the population approached closer and closer to some final value. A chaotic pattern could be one in which the population went up and down in an apparently disorderly way, like a stock market average. A given island might behave in either fashion, depending upon the formula used to generate one year's population from the last.

Now imagine a whole group of islands, each of which provides a different kind of environment for our insects. The different islands are distinguished by a growth-rate parameter, called r, which is a qualitative indication of how well the particular island supports the insect population. We must visualize two basic processes going on. One is the natural increase in population year by year in a manner rather similar to compound interest. The other is the effect of overcrowding. If the population gets too large, destructive competition ensues and the next year's population is considerably smaller than it would otherwise have been.

Given these two processes, there are several outcomes we may imagine. I will describe these outcomes verbally here and mathematically later on:

Case a. (The lowest values of r.) A poor environment provides a negative natural increase for the insects; year by year the population decreases until, finally, no insects remain. This pattern is totally orderly and not at all chaotic. The time-evolution of the population for this case is depicted in Figure 5a.

Case b. (Very high population increase. Large r.) In this case, a quite disorderly pattern may ensue. For example, imagine that in the first year the population is small. With a large growth factor, it could well be true that the next year's population will be very large. Then the unfavorable effects of crowding could cause a precipitous drop in the population for the third year. Over the next few years the population could grow again, and then once again collapse due to overcrowding. We could thus have a situation in which the population increased and decreased in a disorderly and apparently chaotic fashion. This kind of behavior is depicted in Figure 5b for two separate islands.

Case c. (Intermediate values for the population increase.) Imagine a natural growth rate just large enough to sustain a slow population growth in the absence of any overcrowding effect. In this kind of island, we would see a population that increases year by year until overcrowding limits its growth. In the end, the population would settle down to a steady value in which crowding and natural growth balance each other. Such a population pattern is quite orderly. The pattern is shown in Figure 5c.

We have said, in sum, that different islands might be described by the same kind of model, but that each island would have to be distinguished by different values of the population growth parameter, r. Depending on the value of this parameter, a given island might show either orderly behavior (for low rates of population increase) or chaos (for high rates).

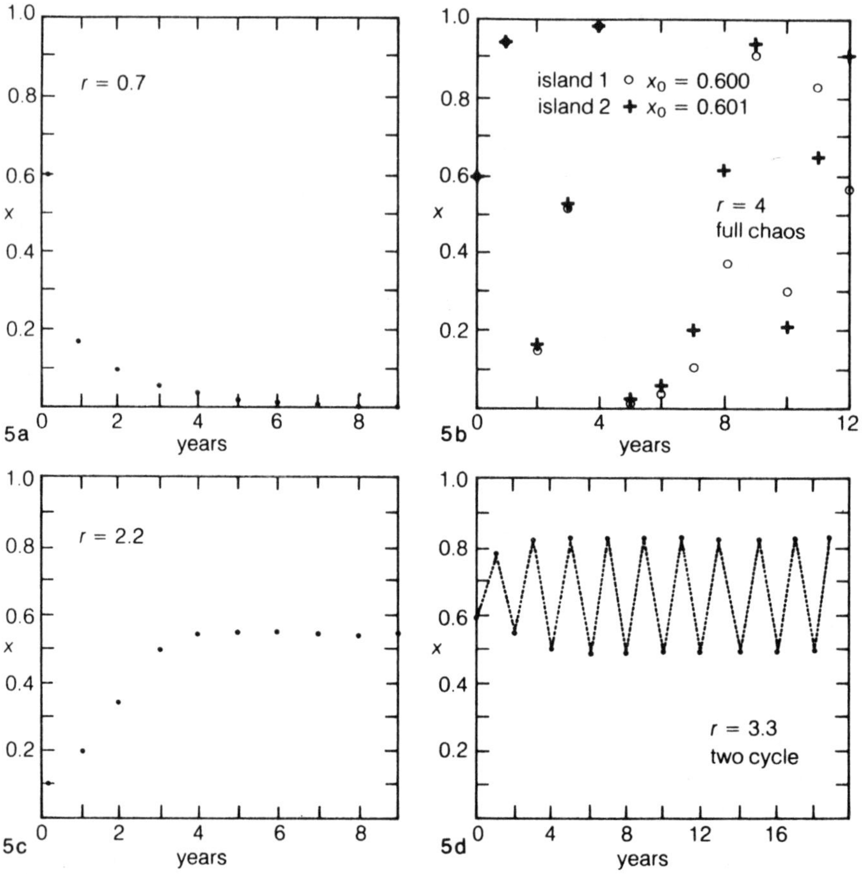

Figures 5a–d. Time histories of insect populations for four different values of the parameter, *r*. Figure 5b shows two histories with slightly different starting populations. In the course of time this difference becomes larger and larger. For the other *r*-values shown, the effect of such a small mutual difference would never become noticeable. The case shown in 5b is chaotic, the other cases depicted are not. Dotted lines in Figure 5d highlight the fact that the diagram is one oscillating insect population, not two.

Now we wish to convert the verbal argument into a mathematical one by giving a well-defined formula to determine one year's population from that of the year before.

The mathematical description of this situation will enable us to give a far more detailed account of the possible outcomes than I could give in the verbal descriptions above. There is a further advantage however. We can imagine a succession of different islands, each with a slightly larger value of the growth-rate parameter. In our minds, we can examine the history of each of these islands. Those with small values of *r* will be orderly; those with the largest will be chaotic. By studying the intermediate values of r we can ask ourselves, "just how does the chaos first arise?"

The model in mathematical form

One can describe an island in terms of a variable, p, which tells you the population of insects in a given year. The mathematical model is an equation that gives the population next year, p_{next}, in terms of the population this year, p. For example, the simplest such model might predict that the population will increase by 10% during each year. This process would be represented by a simple equation for next year's population, namely,

$$p_{next} = rp. \tag{1}$$

To represent the 10% per year increase in population, we will choose the growth-rate parameter, r, to be 1.1. To look for chaos, we will take the equation that predicts each year's population from the last and then use it to generate a list of populations in different years,

$$p_0, p_1, p_2, \ldots, p_j, \ldots. \tag{2}$$

Here the subscripts 0, 1, and 2 are used to describe the populations in years 0, one, and two, . . . ,while p_j represents the population in the jth year. The game then is to use an equation like equation (1) to calculate each year's population in terms of the last and thereby generate, in year-by-year fashion, a list like that in expression (2). (The problem, and its solution, is exactly the same as the one for compound interest.) We then look for patterns in the list and ask whether the pattern is orderly or chaotic, and why. We especially ask whether the different islands, which are represented by different values of r, show different types of behavior. The answer is yes. There are three different categories of behavior corresponding to three qualitatively different types of environments for the insects and consisting of three different ranges of r. These are:

First case: A poor environment. Here the growth-rate parameter, r, lies in the range between zero and one. For these islands the population is smaller each year until eventually it becomes invisible. The resulting population pattern is orderly but dull.

Second case: An equally orderly and dull result will ensue in an island described by equation (1) with $r = 1$. In this balanced environment the population would simply remain unchanged year by year.

Third case: In a favorable environment, r would be greater than one. An island with this environment would have a population that increases year by year. The population would grow without limit.

This last case is unrealistic, of course. In the long run, something must limit the insect population. Thus, the simple model of equation (1) is unsatisfactory as a natural prediction. Furthermore, it shows no chaos. We must go on to develop a slightly more complex model.

The next simplest model could show that when the insect population gets large the reproductive process is inhibited and next year's population is diminished. This reduction might occur because individuals would compete for food or for nesting space. Or maybe the insects are simply shy and do not reproduce well when they are crowded. In any case, the model we need is one that reduces the population predicted in equation (1) by an amount proportional to the number of possible interactions among the different individuals in the population. Since the number of possible interactions is proportional to the population squared, we might try a model of the form:

$$p_{next} = rp - sp^2, \tag{3}$$

where s is another parameter which measures the effectiveness of the various interactions in a diminishing population.

Note that equation (3) only makes sense if the population is smaller than r/s. If the population is larger than this value, equation (3) gives the non-sense result of a population in the next year as negative. For this reason, we limit our attention to situations in which the population is a positive number but smaller than r/s.

One final step is required to convert this model into a form suitable for further study. Instead of using a variable, p, for population, we will use instead a variable x and say that $x = (s/r)p$. By this we mean that x measures the ratio of the actual population of the island to its maximum possible one. Thus, x varies between zero and one, for the population cannot be less than zero, nor greater than the maximum population the island can sustain. According to equation (3), the population ratio next year is determined from the population ratio this year as follows:

$$\begin{aligned} p_{next} &= rp\,(1 - sp^2/rp) \\ p_{next} &= rp\,[1 - (s/r)\,p] \\ (r/s)\,x_{next} &= r\,(r/s)\,x\,(1 - x). \qquad [\text{since } p = (r/s)\,x] \end{aligned}$$

We can now cancel out the common factor, r/s, and find

$$x_{next} = rx\,(1 - x). \tag{4}$$

Once again, r has the significance of a growth factor. When the population is small, i.e., x is close to zero, then $(1 - x)$ is close to one, and the population will be multiplied only by a factor of r during each year. Here then is our model. Next we can look at its consequences.

Order and chaos

For r (the growth-rate parameter) less than one, our first model, equation (1), gave a uniformly diminishing population. Since the modification that led to equations (3) and (4) will further decrease the population through decreased reproductive ability, it is reasonable to expect that this decline will also occur in our new model. To see how it works, consider, for example, the case in which $r = 0.7$. Choose some initial value of the population ratio x, for example $x_0 = 0.6$. Then it is very easy to use equation (4) to calculate the next year's population ratio to be $x_1 = rx_0 (1 - x_0) = .7 \times .6 \times .4 = 0.168$. The next year's calculation gives $x_2 = 0.0978$, showing that the population has diminished. It continues to diminish, as shown in Figure 5a.

In contrast, consider an island in which the growth-rate parameter, r, has the value 4. Then, by equation (4), if the population starts out low, it will in the next year quadruple. Hence, it cannot stay low long. However, if the ratio of the actual population to the maximum ever gets close to its highest possible value, 1, then in the next year the population will become very small (from the combined obstacles to reproduction). The resulting population pattern can be seen with the data points shown as circles in Figure 5b. It goes up and down in an apparently unpredictable manner. To see this unpredictability in even more detail, compare the circle-points to the data points shown as plus signs. The only difference between the population patterns in the two cases is the starting value of the population. The circle-points represent an island in which the ratio of the initial population to the maximum is given by $x_0 = 0.600$. The plus points represent another island which has a very slightly different starting value, $x_0 = 0.601$. At the beginning and for the first few years, we cannot tell the difference between the two islands. The plus points lie on top of the circle ones. However, after a few years the difference between the two cases becomes noticeable. By the end of the twelve-year period shown in Figure 5b, there seems to be no correlation between the populations of the two islands. The diverging behavior of the population patterns, which were at first so similar, is a demonstration of the sensitivity to initial conditions that I described in the first chapter as a sign of chaos.

Now we have two situations which we can describe in words. For a growth factor, r, between zero and one the pattern is very orderly: the population simply dies down to zero. At $r = 4$, the behavior is highly chaotic; the population pattern keeps jumping around, and for most starting values of the population ratio, x, it never settles down to any orderly behavior. So this system exhibits both order and chaos. What lies in between, or how do we get from one to the other?

Period doubling and the onset of chaos

To repeat, for a growth-rate parameter, r, less than one, eventually the population dies away and x settles down to a specific value, namely, 0, by equation (4). Try to visualize what happens for r just greater than one. Imagine that for this case, too, a settling down occurs. Use the

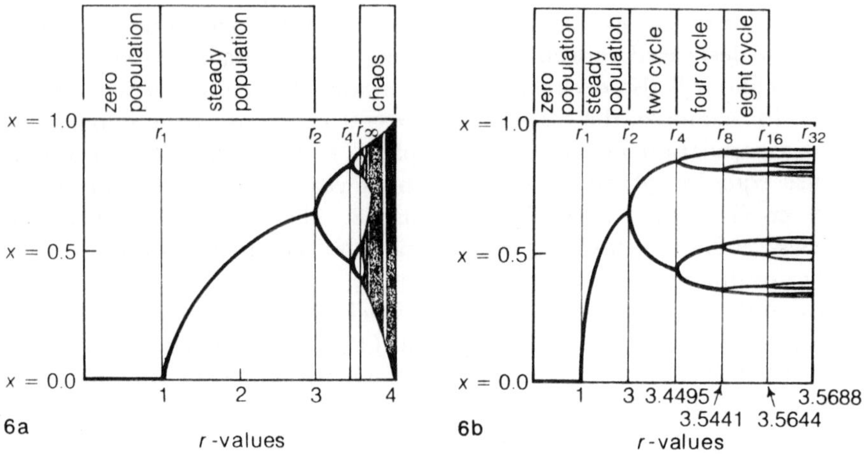

Figure 6. A summary of possible long-run behavior of insect populations for islands with different values of the growth-rate parameter, r. Two views of the same basic plot are shown. The left view has an ordinary linear scale for r in which the distance between $r = 1$ and $r = 2$ is the same as that between $r = 3$ and $r = 4$. At right, the scale of r is distorted to emphasize the region in which the higher order period doublings occur.

symbol x^* to denote the long-run value of x, i.e., the value into which the population settles after many years. Then, if the value of x this year is x^*, the value next year will also be x^*. What is the possible value of x^* itself? Look back at equation (4), substitute x^* for both x and x_{next} and we find that x^* must obey

$$x^* = rx^* (1 - x^*). \tag{5}$$

There are two possible solutions,

$$x^* = 0 \quad \text{and} \quad x^* = 1 - 1/r. \tag{6}$$

We observe that for r greater than one, the first solution is unstable in the sense given by Malthus: Even if the initial population is small, since r is greater than one, the population will grow until it is limited by overcrowding. We can see this behavior by looking at Figure 5c. This plots the population change starting from the very small value $x_0 = 0.1$ for the case in which $r = 2.2$. If we substitute this value for r in equation (6), we find x tending to 0.5454 . . . after a long period of time. From Figure 5c, we see that this value for the population ratio, x, has essentially been achieved after ten years.

In Figure 6 I will summarize the knowledge we have gained so far. This plot shows what x-values are obtained in the long run for different values of the growth-rate parameter, r. We know that at $r = 4$, all x-values between zero and one show up (fig. 5b), while for r between zero and one, only $x^* = 0$ is possible, i.e., in the long run the population will decline to nothing (fig. 5a). For r between one and three, the only possible long-term value of the insect population is the x^* given by the second part of equation (6). These three regions of behavior—$r = 4$, $0 < r < 1$, and $1 < r < 3$—correspond, respectively, to the chaotic behavior shown for the fluid in Figure 4e, and to the time-independent fluid behaviors shown in Figures 4a and 4b. In Figure 6, these three regions are marked as "chaos," "zero population," and "steady population." However, in the actual fluid, when the flow rate, R, was increased beyond the value shown in Figure 4b, then the motion became time dependent.

The analogy between the model example and the real system continues to work. As r is further increased, beyond the value 3 the steady behavior represented by equation (6) disappears. As in the fluid example of Figure 4c, a time dependence suddenly appears. One can see this time dependence by looking at Figure 5d, where the growth rate, r, is given as 3.3. Notice that after a few years the insect population settles down into a regular pattern. But the pattern is time dependent. There are *two* eventual population values: one high, one low. In good years the insect population is low and they reproduce avidly. The next year is a bad one in the sense that there is much overcrowding. Hence, reproduction is impaired, so that the following year's population is low. This alternation continues forever. Such a situation, in which the behavior repeats after two steps, is called a two cycle or a cycle of length two. The cycle is depicted in Figure 6 by showing two values of x for each r in this region.

But time dependence is not chaos. Chaos will not arise until the x-values cease to settle into a regular pattern. To see how that happens, imagine increasing r still further. As r increases above 3.4, a cycle of four years dominates the long-run behavior of the population. For almost all starting values of x in this region, the insect population will, after many years, fall into a pattern in which the population cycles through four different values before it repeats. This pattern is also shown in Figure 6 in the region marked four cycle. Increase r a bit more and you get an eight cycle, a tiny bit more and the period doubles yet again, until at $r = 3.59946\ldots$ an infinite number of period doublings have occurred and we reach a situation which might fairly be described as chaos. In this system, chaos first appears as a result of many successive doublings of the period of the cyclic population pattern. Hence, we call what we have just described the period-doubling route to chaos.

The successive values of r at which cycles of length 1, 2, 4, 8, . . . first appear are denoted in Figure 6 by $r_1, r_2, r_4, r_8, \ldots$, while the r-value for which the cycle of infinite length appears is denoted by r_∞. At r_∞ the insect population never repeats itself, never settles down to a fixed value or values. We say then that we have reached the onset of chaos. Likewise, for most *higher* values of the growth-rate parameter, r, those between r_∞ and 4 (i.e., the region marked chaos in fig. 6a), the typical behavior of the insect system is one in which it does not repeat itself but shows a chaotic behavior. The crucial value of r is thus r_∞, since it is at this value that chaos first appears. We call this point on the x versus r curve a Feigenbaum point, for Mitchell J. Feigenbaum, who first elucidated its properties and thus enabled us to understand the period-doubling route to chaos.

Universality and contact with experiment

It may seem that we have lost contact with the hydrodynamic systems that served as our starting point. The last section's argument about the onset of chaos seems very specific to population problems, or at most to problems that involve a single variable, x, and a dynamics in which x is determined again and again in a step-by-step fashion. However, the work on our simple model offered a hint that the results obtained might be more generally applicable. For there are some aspects of the answers which seem quite independent of the exact form of the problem under study.

Recall that we studied a problem in which the population was determined by the equation:

$$x_{next} = rx\,(1 - x).$$

But, we could have used a slightly different equation, for example,

$$x_{next} = rx\,(1 - x^2).$$

If it were true that the answers obtained were equally valid for both types of equations, and for many others like them, we might guess that these answers would have the potential for being much more general than the particular starting point we used would suggest. They might even be applicable to real systems, in the laboratory or in nature. When some result is much more general than its starting point, the situation is described by mathematicians as one of ***structural stability***. Physicists describe a similar situation by saying that it exhibits ***universality***.

Where can we find universality in the period-doubling route to chaos? There are two places. First, the overall structure of Figure 6, with its doublings and chaotic regions, is insensitive to the choice of the exact equation that will determine x_{next}. Changing the equation will distort the picture somewhat, as if it were drawn upon a piece of rubber and stretched, but will leave its essential features quite unchanged. Since this picture describes a variety of different mathematical problems that lead to an infinite period doubling, perhaps it also describes some real physical cases that show infinite period doubling.

In this way, we can hope to make contact between the "insect system" and real experimental systems. For example, we can set up electrical circuits that are unstable and "go chaotic" as some control parameter, roughly analogous to r, is changed. The most familiar example is an audio system, where r would describe the position of the microphone relative to the speaker. As these are brought closer together, the system may become unstable, that is, a hum may develop.

Analogous purely electrical circuits have been constructed to test the theory described above. For example, Testa, Perez, and Jeffries* performed an experiment in which an electrical circuit containing a transistor was controlled with a voltage, v_c, which played a role analogous to our parameter r. They noticed that their circuit had a natural oscillation that changed character as v_c varied. As v_c was increased, the period of the oscillation doubled, and doubled, and doubled again. By observing peak voltages, v_p, at one point in the circuit, they were able to trace out a v_p versus v_c picture that looked very much like Figure 6. Thus the electrical circuit showed a behavior very much like the one we have described. We can say, therefore, that we understand the period-doubling route to chaos in the real electrical system because we understand it in the insect model, and the two are very much the same.

Feigenbaum pointed to another way in which universality would manifest itself. As r approaches r_∞, he said, some aspects of the time pattern would remain the same even if insect behavior was different. For example, he looked at r-values for which cycles of length 1, 2, 4, 8, 16, . . . , ∞ first appeared. In Figure 6 these are denoted by r_1, r_2, r_4, r_8, r_{16}, . . . , r_∞. There is nothing universal or general about the appearance of the first few cycles. Hence there is nothing very useful to say about r_1 or r_2 or r_4. But the r-values at which very long cycles would appear turned out to be much more predictable. As the cycles get longer and longer, the spacing between successive r-values gets smaller and smaller (*see* the numbers at the bottom of fig. 6b). Indeed, for long cycles, the spacing forms a geometrical series in which the successive terms are divided by a constant factor called δ. It is surprising but true that *this constant has a value which is universal,* i.e., independent of insect behavior specifically. The spacing ratio, δ, takes the value $\delta = 4.8296$ . . . for *all* growth of the type indicated here, the type, that is, which follows a pattern of period doubling.

At first, other workers in the field were resistant to Feigenbaum's work, and particularly to the proposition that a number like δ could be universal. Feigenbaum derived an elaborate theory of this universality, based upon the "renormalization group" theory that Kenneth Wilson† had invented for quite another area of physics. The argument

*J. Testa, J. Perez, and C. Jeffries, "Evidence for Universal Chaotic Behavior of a Driven Non-Linear Oscillator," *Physical Review Letters* 48 (1982): 714.

†Kenneth G. Wilson, "Problems in Physics with Many Scales of Length," *Scientific American* 241 (August 1979): 158.

was settled by two developments: (1) a mathematical proof that in an appropriate sense the result *was* universal, and (2) experimental verifications that Feigenbaum's predictions about the quantitative aspects of successive period doubling held in other examples far removed from simple population growth models. In fact, δ-values are obtained for experiments involving instabilities in electrical circuits and also fluid systems. Within the limited accuracy of the experiments, Feigenbaum's predictions were fully verified.

The end result is a remarkable intellectual achievement. We can say with some truth that we understand how chaos arises in the simple model system described by equation (4). This is a satisfying achievement, and even though the system is simple, it is impressive. But we also have evidence, partially based upon theory and partially upon experiment, that exactly the same route to chaos is obtained in other much more complex systems. We believe that if we took one of these complex systems, measured the period doublings, and thereby found the value of δ, that number would be exactly and precisely the same number as the corresponding δ-value obtained from the simple model of equation (4). Thus the pattern of the simpler system is exactly duplicated in the more complex ones, and we can see that in understanding one case, we understand many.

In the years since Feigenbaum's work, several other scenarios for the onset of chaos have been explored both experimentally and theoretically. Each of these "routes to chaos" is universal in the sense that many different systems will exhibit the same pattern. We can therefore say that we are beginning to understand how chaos arises.

In the next chapter we will look at fully developed chaos and ask how well *that* is understood.

Chapter III: The geometry of chaos

In the last chapter the onset of chaos was considered. A full understanding of chaos, beyond its onset, still eludes us. However, we have built up a few substantial ideas about its geometrical structure. The purpose of this chapter is to discuss these geometrical ideas.

Well-developed turbulence

Chaos has been defined here as a physical situation in which the basic patterns never quite repeat themselves. One vivid example of such a nonrepetitive pattern is in a flow pattern depicted by Leonardo, shown in Figure 7. Notice how within this big swirl there are smaller ones and within them smaller ones yet. This kind of flow within flow within flow is called well-developed turbulence. The nonrepetitive nature of the pattern arises precisely because it is a Chinese box in which structures

Figure 7. Turbulent flow patterns as drawn by Leonardo da Vinci. Note how the large swirls break into smaller ones, and these again break up.

appear within structures. This point was later made into a little poem by L. F. Richardson, who wrote:

Big whorls have little whorls,
Which feed on their velocity;
And little whorls have lesser whorls,
And so on to viscosity
(in the molecular sense).*

The Soviet mathematician A. N. Kolmogorov picked up on this picture and developed a useful theory of such behavior by writing down the mathematical consequences of the idea that similar structures reappear again and again inside of one another. We do not believe that Kolmogorov's theory is entirely right, but we don't yet have a replacement for it.

The Chinese box example has to contend with two major complications. One is that any well-developed chaos is hard to understand. The other is that this particular chaos of flow within flow occurs in space.

*This poem is quoted in Benoit B. Mandelbrot, *The Fractal Geometry of Nature* (New York: W. H. Freeman and Company, 1983), p. 402.

To describe it fully, we would have to specify the velocity at every single point in the entire system. Since there are an infinite number of points, we would have to have an infinity of different numbers just to specify the chaotic situation at one time! But we have already seen a chaotic situation that could be specified by giving only one number, x, at each time. (A number like x, which can change in time, is called a "variable.") We were able to understand this example reasonably well and to see how chaos arose in it. Of course this situation was intentionally constructed to be simple. When we go out and look at the real world, we can find situations which are intermediate in complexity between the one-variable case described in the last chapter and the cases with an infinite number of variables depicted by Leonardo. These cases can often be described by specifying the time dependence of just a few variables, perhaps two, x and y, or three, x, y, and z. In the next sections, we will try to describe the kinds of behavior that could arise in these few-variable systems.

Attractors, strange and otherwise

To describe these relatively simple systems, we will direct our attention to the mathematical world in which they live. In the case of our insect island, we can fully specify its future behavior by giving one number, which describes this year's population in relation to the maximum possible one. This ratio, x, must be a number between zero and one. Indeed, for the mathematician studying our example, the relevant world is not the hypothetical island upon which the insects live, it is the mathematical world which is the set of all numbers between zero and one. That kind of world is called the "phase space" for the problem, so as to distinguish it from the physical space (the island) in which the events occur. In this example, we can depict the phase space by drawing a line segment, as in Figure 8, *case a*, and imagining that a specific value of x is depicted by putting a point upon that segment. Notice that this segment is drawn as a line that goes up and down, instead of the more conventional drawing that would go from left to right. I ignore the conventions so as to have my pictures look like the ones drawn in the previous chapter, that is, Figures 5 and 6.

Now let us return to the kind of thinking that we used in constructing these earlier figures. Consider some fixed value of the growth-rate parameter, r, say the $r = 2.2$ of Figure 5c. Then, as shown in that figure, year by year x, the ratio of the existing population to the maximum one, approaches a specific value, namely, 0.5454 . . . . For almost any value of the initial population, the long-term result is precisely the same. As the years go by, the population ratio will get closer and closer to that particular value of x. This is graphically described by drawing a point at $x = 0.5454$ within the phase space of Figure 8, as in *case c*. We then say that this point is the "attractor" for

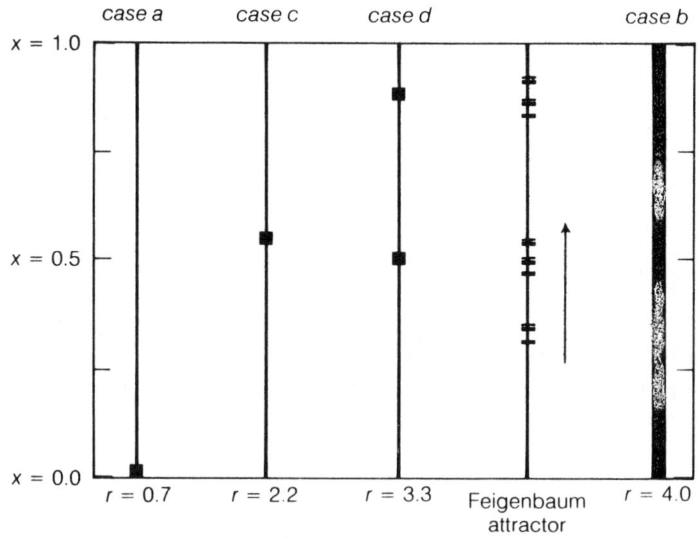

Figure 8. A description of some attractors for different r-values in our insect system. The vertical axis on the far left describes the "space" in which the attractor will fit, namely, the line between $x = 0.0$ and $x = 1.0$. The line labeled *case a* describes what happens when $r = 0.7$. It has a point at $x = 0.0$, showing that for this r-value the insect population ratio goes to zero. The filled-in regions for the other r-values similarly depict the possible population ratio values for each one of these situations.

the insect system at $r = 2.2$. By this we mean that in the year-by-year development of the insect systems, their population ratios approach or "are attracted to" this point.

For a higher growth-rate parameter, the attractor might be more complicated. For example, as we already know from Figure 5d, at $r = 3.3$ the long-term behavior of the population is a two-cycle one. Hence, the motion is attracted to a pair of points, as shown in Figure 8, *case d*. In this way, for each value of r, we can plot out the attractor, that is, the value to which the population ratio converges. In fact, Figure 6 is simply a plot that shows the attractors for all values of r.

For a still higher growth-rate parameter, the population graph exhibits chaotic behavior. The attractors in the chaotic regions of this parameter, r, are not collections of points, as in the last example, but instead *regions*. For example, in the full chaos of $r = 4.0$, as shown in Figure 8, *case b*, all values of the population ratio, x, between zero and one arise within the course of a typical pattern of time development. Hence, for this value of r, the attractor is the entire interval between zero and one. Whenever there is chaos, the attractor is an interval, or perhaps a collection of different intervals. This behavior can also be seen in the right-hand portion of Figure 6a.

In the cases described so far, the attractors are relatively simple and straightforward: a point, a few points, an interval, or a few intervals.

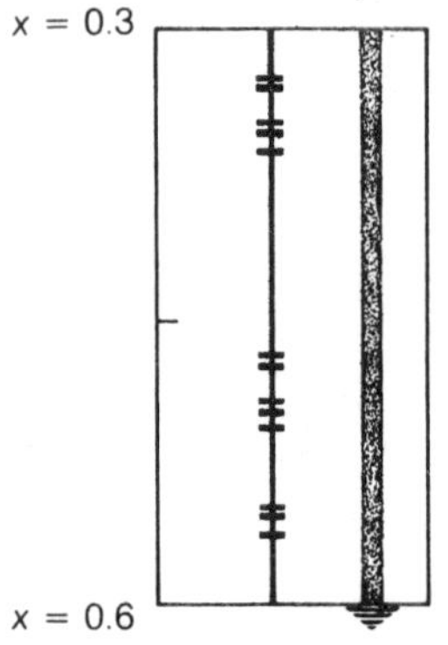

Figure 9. A portion of the Feigenbaum attractor blown up and replotted. This portion looks essentially identical to the entire attractor (*see* fig. 8).

However, when we follow Feigenbaum and look at the point for which a cycle of infinite length first appears, we see a much stranger and richer behavior. What we must do is draw the attractor for a two cycle, as in Figure 8, *case d,* then for a four cycle, next for an eight cycle, next sixteen. Then, in the limit, one gets the picture shown in the part of Figure 8 labeled "Feigenbaum attractor." This configuration contains an infinite number of points arranged in a rather interesting pattern. Such an attractor, in which there is structure inside of structure inside of structure, is called a "strange" attractor. The reader will notice that I have just defined *strange* as a technical word. In doing this, I am following the standard terminology in the field. The term *strange attractor* is due to David Ruelle of the Institut pour Haut Étude Scientifique in Paris.

Before going further, I should compare the structure within structure of Figure 7, Leonardo's drawing, with that of the Feigenbaum attractor in Figure 8. They are both "strange." However, the pictures show two different types of worlds. Leonardo draws a picture of one time in our real three-dimensional world. Figure 8, on the other hand, is drawn in phase space and is a superposition of infinitely many pictures at different times.

To see the strange character of the Feigenbaum attractor, we take the portion of it indicated by the arrow in Figure 8, blow it up, turn it over, and plot it again. The result is shown in Figure 9. Notice that, except for the values of the x-coordinates, the picture looks essentially identical to the one in Figure 8. This identity strongly suggests that there is a succession of almost identical structures nested within the Feigenbaum attractor, in the same way that Russian dolls are nested within one another.

Such nested behavior is very common in physical systems. Look back at the solidification patterns shown in Figure 3. Notice once again how

Figure 10. An example of a fractal object. Notice how the fine treelike object is represented again and again at different sizes.

the ice consists of a group of arms, upon which lie smaller arms and upon them smaller ones yet. This kind of behavior was first described by the nineteenth-century mathematician Georg Cantor, and later on by Felix Hausdorff and others. In more recent work, such nested behavior is often described by the terms *scale-invariant* or *fractal*. Figure 10 shows such a pattern. The term *scale-invariant* merely says that when you blow up the picture in Figure 9 (i.e., change its scale) and look at a portion of the result, you get much the same thing as before. Thus it is unchanged or invariant. The word *fractal* was introduced by Benoit Mandelbrot of IBM, who has discovered and publicized many examples of scale-invariant behavior. The term is intended to remind us of another property of these strange objects. They can be described by using a variant of the concept of a dimension. It is commonplace to say that a point has no dimension, a line is one-dimensional, an area two-dimensional, and a volume three-dimensional. For strange objects, we extend the meaning of *dimension* to include possibilities in which a dimension is not just an integer (1, 2, or 3) but instead any positive number (say 0.41). There is then a technical definition that enables us to calculate this fractional or "fractal" dimension from the picture of the object, as, for example, Figure 9. (This particular attractor has dimension 0.538 . . . which, as one might expect, is larger than the value for a point and smaller than the value for a line.)

Incidentally, I should note that the motion upon the Feigenbaum attractor is really rather orderly and cannot be described in any sense as chaotic. For example, imagine starting off with a population ratio of $x_0 = 0.5$, which is indeed a point lying on the attractor. We can look at the points that arise after 1, 2, 4, 8, 16, or 32, successive iterations of equation (4), using the value of the growth-rate parameter,

r, appropriate to the Feigenbaum attractor. The placement of these x-values is very orderly. The points $x_1, x_2, x_4, x_8, x_{16}, \ldots$ approach $x_0 = 0.5$, with alternate members of the list lying above and below x_0. Thus the Feigenbaum attractor may be "strange," but the motion on it is certainly not chaotic.

*Chaos on strange attractors**

In our insect example, chaos and strange attractors tend to occur for different values of r. However, in slightly more complicated systems, chaotic behavior almost always produces a strange attractor. So far, we have worked mostly with an example in which the present and future behavior of the system could be defined by giving the value of one number, x, representing a population ratio. In the next, more complicated, example the future state of the system is defined by two numbers, called x and y. For example, these might be the populations of two different age-groups in a given year. A model system could be defined by saying how the values of x and y in the next year depended upon the values this year.

Another example of such a dependence is given by the following equations:

$$x_{next} = rx\,(1 - x) + y, \text{ and} \qquad (7)$$
$$y_{next} = xb.$$

We might arrive at such equations by visualizing a case in which there is once again an insect population and where x is the population in a given year. (For simplicity's sake we will here talk of x as if it were the population, rather than a population ratio, as previously.) But we could introduce one difference. We could suppose that a proportion, b, of the insects would live for two summers. Call the number of "old" insects y. In that case, the same analysis as before would lead us to a result like equation (7).

The point here is that, to define the situation fully, we now need to specify two numbers: the existing population, x, and the previous year's population, y. This can be rendered geometrically by drawing a common kind of graph with x and y axes and specifying some situation by a point on the graph, as shown in Figure 11. An attractor for this situation can be constructed by starting out with some initially chosen value of x and y, constructing successively the next values via equation (7), and then imagining that after some large number of steps the values of the pair (x,y) have moved in toward the attractor.

This model, or rather one equivalent to it, has been constructed by

*For a slightly more technical presentation of similar material about Hénon's and Feigenbaum's work, *see* Douglas R. Hofstadter, "Metamagical Themas," *Scientific American* (November 1981): 22.

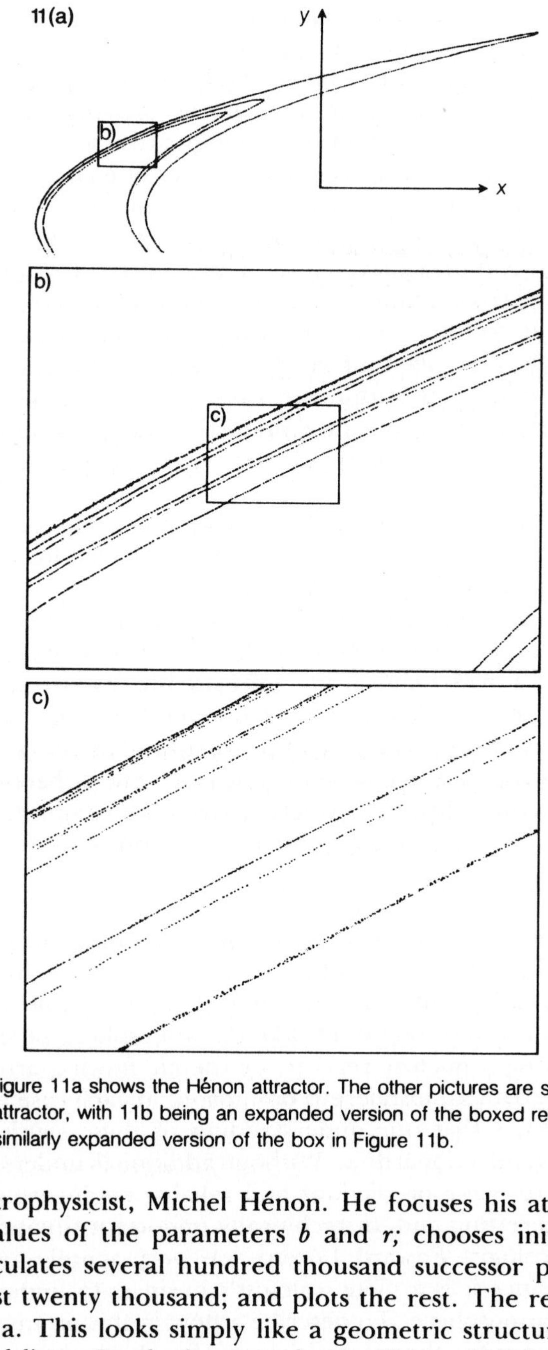

Figures 11a–c. Figure 11a shows the Hénon attractor. The other pictures are successive blowups of this attractor, with 11b being an expanded version of the boxed region in 11a, and 11c a similarly expanded version of the box in Figure 11b.

a French astrophysicist, Michel Hénon. He focuses his attention upon particular values of the parameters *b* and *r;* chooses initial values of *x* and *y;* calculates several hundred thousand successor points; throws away the first twenty thousand; and plots the rest. The result is shown in Figure 11a. This looks simply like a geometric structure containing a few parallel lines. But look more closely. Figure 11b is an expanded view of the box shown in Figure 11a. This expanded view also contains lines, but when one blows up a box within that figure, obtaining as a result 11c, one sees a familiar looking picture with a few lines within it. In this way, Hénon demonstrated that the attractor from his map could be scale-invariant and fractal. And, in contrast to the previous example, we do not have to do any careful adjustment of r to find a strange attractor. On the contrary, strange attractors will pop up for many reasonable and arbitrarily chosen values of b and r.

In contrast to the motion on the Feigenbaum attractor, the motion on the Hénon attractor is chaotic. In the first case, it is easy to predict the x-value that will be achieved after a large number of iterations. For example, if the large number is a high power of two, e.g., 2^{99}, the achieved x-value will be almost identical to the starting x-value. In the Hénon example, if we start off on the attractor, we know that the (x,y) point will continue to lie within the attractor, but there is no similarly simple rule that permits accurate prediction of just where the points will lie after many steps. Moreover, the result is extremely sensitive to initial conditions.

Hence, in Hénon's model chaos and strange attractors exist together. In fact, they are believed to have a causative relaxation: the attractors are strange exactly because they are chaotic.

Hénon is an astrophysicist. He is interested in motion in the solar system and galaxies. His work on the model described above is not just the construction of a mathematical toy, unrelated to his astronomical interests. On the contrary, all kinds of astronomical systems—for example, our own solar system—can be usefully thought of as being described by a few variables that in the course of time trace out a chaotic motion on a strange attractor. The real attractors are more complicated than the one in Figure 11, but probably in many ways not essentially different.

Tracing chaos through time

Return to Figure 7, and the complicated swirls of Leonardo's picture of chaos in a fluid. From a practical point of view, it is distressing that we do not have a decent understanding of these turbulent flows. The flow of energy through real fluids like the atmosphere of the Earth, or the water cooling a nuclear reactor, or the air flowing around a body entering the Earth's atmosphere is dominated in each case by turbulent swirls. The fact is that our understanding of these swirls has hardly progressed beyond Leonardo's. Without additional understanding, we lack the tools to make predictions and reliable engineering designs in all kinds of interesting and/or technically important situations.

The meteorologist Edward Lorenz was very acutely aware of this imperfection, since it is a meteorologist's business to understand flows in the Earth's atmosphere. To describe a flow in this or any other fluid, we write equations for the rate of change of such properties as the fluid velocity, the temperature, and the pressure at each point in the fluid. As I have already mentioned, since there are an infinite number of

The Lorenz Equations

A solution of these equations is depicted in Figure 12. In this set of equations $\mathring{x}$, $\mathring{y}$, and $\mathring{z}$ stand for the rate of change of x, y, and z with respect to time:

$$\mathring{x} = 10\,(y - x)$$
$$\mathring{y} = 28x - y - xz$$
$$\mathring{z} = 8z/3$$

"points" in every geometrical body, we must solve an infinite number of equations. Lorenz sought a ruthless simplification of the problem. Instead of describing his fluid by giving the values of an infinite number of quantities, he assumed that the fluid could be described by three, which he called, naturally enough, x, y, and z. His phase space could then be described in terms of the three coordinates. I show the exact form of his equations in the box on this page.

This detailed form is irrelevant to all the arguments that follow. The main idea, however, is not at all irrelevant. Lorenz's goal was to describe the particular kind of swirling motion called "convection." In this flow, the lower layer of a fluid heated from below rises because it is lighter (less dense) than the material above it. As the air above one portion of the Earth rises, air in another region flows downward. The net result is a complicated swirling flow. Lorenz's equations were an attempt to catch the essence of a swirling region in the very simplest fashion.

The major point about these equations is that if you give numerical values for x, y, and z at a particular time, the system will determine the values of these quantities at subsequent times. Hence, we can picture the system at a given time by drawing a point on a standard x, y, z coordinate system of the kind shown in Figure 12a. The subsequent motion of the system is shown by giving the x, y, z coordinates at later times (fig. 12b) and then connecting up these points with arrows that show the direction of increasing time along the trajectory. After an initial time to settle down, the motion approaches an orbit that covers only a small portion of the x, y, z space (*see* fig. 12c). This orbit is, of course, a strange attractor. The path traced out by the time development of the system is an object of both impressive simplicity and imposing complexity.

First, the simplicity. Two basic kinds of motions are shown in Figure 12c. There are loops tilted leftward and loops tilted rightward. These

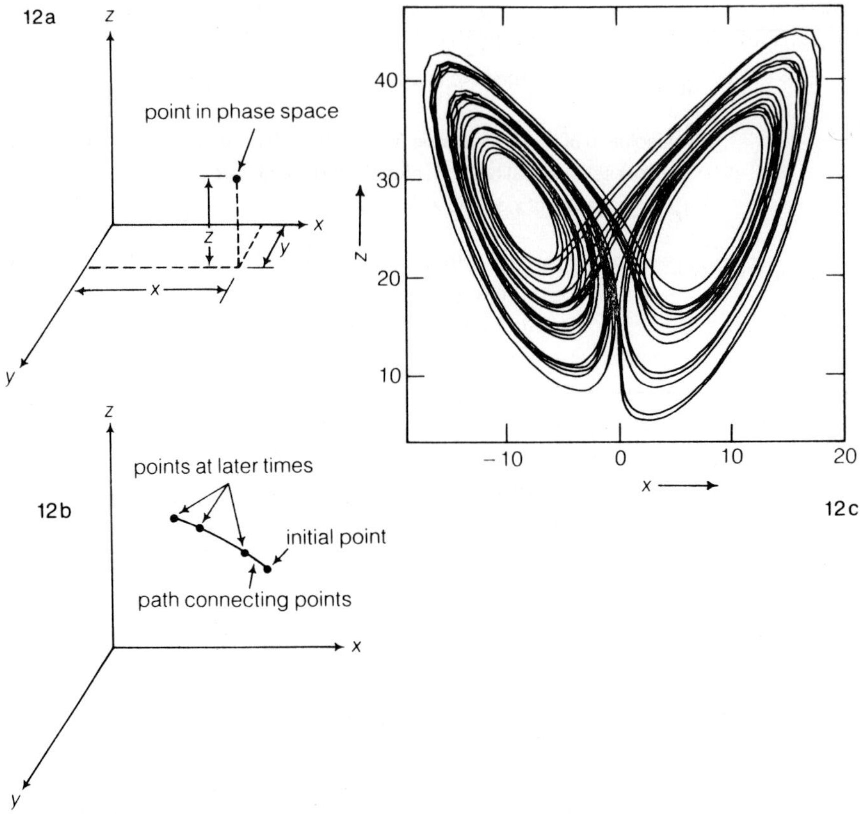

Figures 12a–c. Solution of the Lorenz equations. 12a shows the x, y, z space in which the equations are solved, 12b shows a fragment of the solution, while 12c is a projection upon the x, z plane of the solution over a long period of time.

come together in the region near the bottom of the diagram. As the system develops, it goes through each of these two kinds of loops in turn. To describe the sequence of events, we may list, in order, the loops that are traversed. For example, to describe the orbit in Figure 12c, which covers first a rightward loop and then two leftward ones, we may write "right-left-left." Over a very wide range of starting points, that is starting values of x, y, and z, the system will go through this loop-type behavior.

Now, the complexity: Depending upon the exact starting values of x, y, and z, the subsequent motion will be different. For one set of starting values, the motion might be

right-left-left-right-right-left-right-right-left-left-right- . . . .

Change the starting values just a little and the initial looping will change hardly at all. But the later stages may change quite considerably. Thus, with a small change in starting point we might have

right-left-left-right-right-left-right-right-left-left-**left-** . . . .

(For clarity, the changed values are shown in bold face.) A larger change in starting point will lead to an early change in the looping, for example,

right-left-left-right-right-left-**left-left-left-left-right-** . . . .

The looping structure is fully predictable in the sense that for any initial values of the x, y, z coordinates we can know and calculate the subsequent order of loops. But the structure is very sensitive to the initial conditions in that a small change in the beginning will cause a complete reshuffling of the loops at later times.

Calculating the motion in the Lorenz model is a quite nontrivial undertaking. We must carry through all the work of solving a set of differential equations. That requires a larger computer and considerable skill in its use. But the real fluids and the real world must be described with many many more variables than the three used by Lorenz. Nobody knows whether the more complicated "realistic" situations will show the same kind of complicated algebraic and geometric structure as the simplified models described here. I suspect and hope that many of the features presented here will reappear in the "real world." But now I have reached about as far as our present knowledge of the subject runs.

Physica Scripta. Vol. T9, 5–10, 1985

From Periodic Motion to Unbounded Chaos: Investigations of the Simple Pendulum*

Leo P. Kadanoff

The James Franck and Enrico Fermi Institutes, The University of Chicago, Chicago, IL 60637, USA

Received June 11, 1984; accepted June 16, 1984

Abstract

The simplest example of the onset of chaos in a Hamiltonian system is provided by the "standard" or Chirikov–Taylor model. As a nonlinearity parameter, k, is increased the long term behavior of the momentum, p, is examined. At $k = 0, p$ is conserved. For $k < k_c$, for all starting points, p is of bounded variation. For some starting points its behavior is periodic, for others quasi-periodic, for others chaotic. At some critical value of k, unbounded chaotic variation first appears. A scaling analysis to describe this onset is described.

The problem we will discuss in this lecture has a long history. The basic work in the problem was done by Kolmogorov, Arnold and Moser [1–3]; more recent work has been done by J. Greene, B. V. Chirikov, D. Escande, F. Doveil, and R. MacKay. I have worked on this problem in collaboration with S. J. Shenker; parts of this and related work were done in collaboration with M. Widom, A. Zisook, M. Feigenbaum, D. Bensimon and Subir Sarkar.

The phenomenology of Hamiltonian systems is quite different from that of dissipative systems. In this lecture we shall analyze in detail the breakdown of a KAM curve and the onset of unbounded chaotic motion in a particular map. First, let us give three physical systems to motivate the study of this map.

First consider a pendulum (Fig. 1)

$$ml\, 2\pi\ddot{\theta} = -mg \sin(2\pi\theta) \tag{1}$$

in which we choose units of θ so that $\theta = 1$ corresponds to 360°. Let us act upon this system periodically by modulating g, the force due to gravity (say, by wiggling the support of the pendulum)

$$g = g_0 + g_1 \sin(\omega t) \tag{2}$$

We get equations

$$\begin{aligned} \dot{r} &= -k(t)\sin(2\pi\theta) \\ \dot{\theta} &= r \end{aligned} \tag{3}$$

Here $k(t)$ is a periodic function of line with frequency ω and r is the velocity of the pendulum. We can now use a trick due to Poincaré to transform this differential equation into a map. Observe the pendulum once each period of the force; let $r_j = r(t_j)$ and $\theta_j = \theta(t_j)$ where $t_j = (2\pi/\omega)_j$. Since the phase of the external force at time t_j is independent of j, one can integrate the equations of motion (3) over this period, expressing the new state of the pendulum in terms of its state one period earlier:

$$\begin{aligned} r_{j+1} &= F(r_j, \theta_j) \\ \theta_{j+1} &= G(r_j, \theta_j) \end{aligned} \tag{4}$$

In general, F and G will be some nonlinear functions periodic in θ. The simplest model which seems to capture the physics of this system is the standard map

$$\begin{aligned} r_{j+1} &= r_j - (k/2\pi)\sin(2\pi\theta_j) \\ \theta_{j+1} &= \theta_j + r_{j+1} \end{aligned} \tag{5}$$

also known as the Chirikov–Taylor model [4]. (Fradkin and Huberman [5] have studied this periodically modulated pendulum and have indicated how eq. (4) can be converted to eq. (5) in several limiting situations.)

The second system we will use to motivate this map is an accelerator model (Fig. 2). Envision a particle moving around a circular track, accelerated each time it enters a small box; the acceleration is provided by an a.c. field in the box at the time the particle enters; the eqs. (5) describe the state of the particle as it enters the box for the $j+1$st time in terms of its state as it entered the time before. One can reexpress eq. (5) in the form

$$\theta_{j+1} - 2\theta_j + \theta_{j-1} = -(k/2\pi)\sin(2\pi\theta_j) \tag{6}$$

which as a discrete version of eq. (1) perhaps makes the connection to the pendulum problem more clear.

Finally, consider a solid state model of a one-dimensional array of atoms adsorbed on a periodic substrate (Fig. 3). The jth atom feels a force from the springs connecting it to its two neighbors, and from the gradient of the potential energy at its position on the periodic substrate. If we choose $\theta_j = x_j/a$ to be the position of the jth atom x_j divided by the substrate lattice constant a, then

$$k_{\text{spring}}[(\theta_j - \theta_{j+1}) + (\theta_j - \theta_{j-1})] = k_{\text{substrate}} \sin(2\pi\theta_j) \tag{7}$$

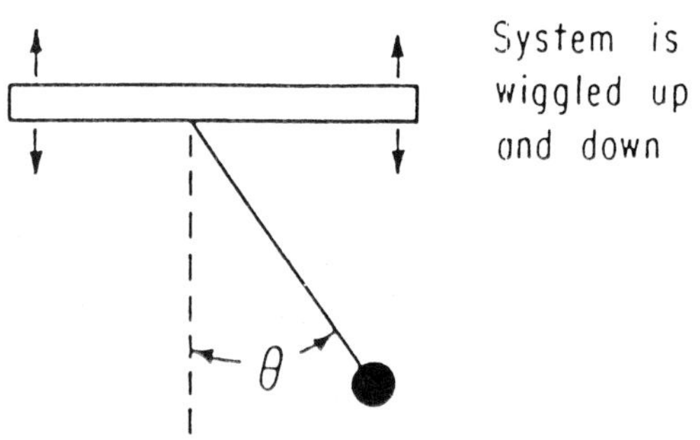

Fig. 1. Our first model. A pendulum is accelerated up and down.

* From lectures originally delivered at Erice in 1983. The writeup comes from notes by R. de la Llave and J. Sethna.

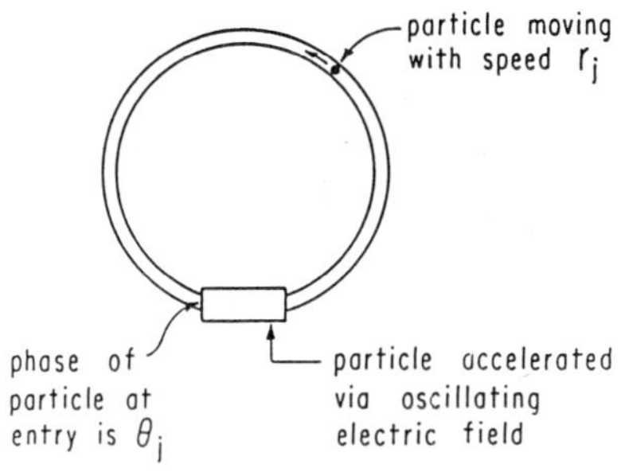

Fig. 2. The second model. A simplified accelerator.

describes a static configuration of atoms. This is of the form (6). The breakdown of the KAM surface has a physical meaning here as a pinning of the density wave of adsorbed atoms; this problem has been studied by Coppersmith et al. [6], Aubry [7], and others.

The standard map eq. (5) has no conserved energy; as noted above, it can represent an externally forced pendulum which exchanges energy with its environment. It does, however, obey Liouville's theorem

$$\frac{\partial(r_{j+1}, \theta_{j+1})}{\partial(r_j, \theta_j)} = 1. \tag{8}$$

The map possesses at least three qualitatively different kinds of orbits: periodic, chaotic, and KAM curves. To describe these orbits I shall draw two kinds of pictures: Figure 4 plots θ_j vs j to show the "average velocity" and Figure 5 plots (r_j, θ_j) in phase space to show the nature of the typical orbits.

The top orbit is periodic; after three iterations θ_j has increased by two while r_j has returned to its initial value; since θ and $\theta - 1$ are identified, this is a return to the initial state. The orbit progresses $q = 2$ units in its period $p = 3$, so its average speed W is $p/q = 2/3$.

The bottom orbit is chaotic. Chaotic orbits appear to fill areas which we shall argue are bounded by the KAM curves. The middle curve is an example of such a KAM curve. All the points on the orbit fall on a smooth periodically climbing curve; it has a well-defined average speed of $W = (\sqrt{5} - 1)/2$.

For large values of k, the chaotic region becomes unbounded in the vertical direction along r (Fig. 6). For many (but not all) initial conditions the motion in this regime appears diffusive, with a diffusion constant

$$D = \lim_{n \to \infty} \frac{2\overline{(r_{j+n} - r_j)^2}}{n} \tag{9}$$

which depends on k (Fig. 6). Numerical studies due to Chirikov indicate that D grows from zero at a critical value (now known to be given by $k_c = 0.971\ldots$) like $(k - k_c)^{2.56}$. On the other hand, for large values of k, $D \sim k^2$. The latter fact can be understood; write

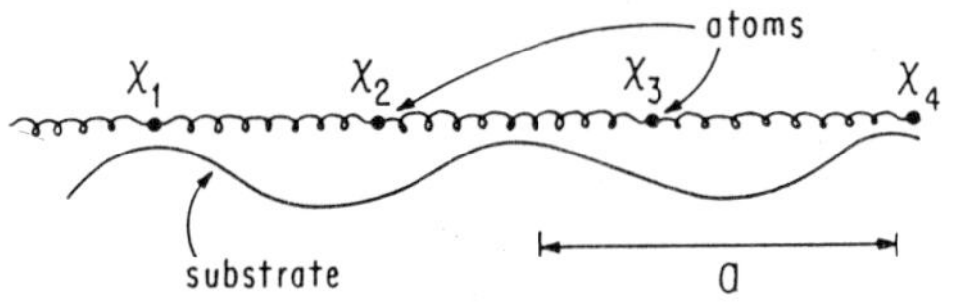

Fig. 3. A solid state example. Particles on a wavy surface.

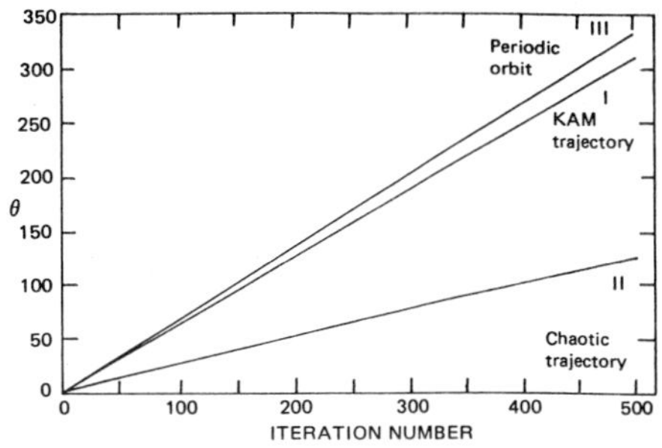

Fig. 4. Value of θ_j versus j for the three kinds of orbits. On this scale, θ seems to simply increase linearly.

$$r_{j+n} - r_j = \sum_{l=j}^{j+n-1} \left(\frac{k}{2\pi}\right) \sin(2\pi\theta_l) \tag{10}$$

$$\theta_{j+1} = \theta_j + r_{j+1}$$

For large k, the successive θ_j's can be considered virtually independent random numbers mod one, since the r_j change by a number of order k after each iteration. Thus

$$\begin{aligned} \overline{(r_{j+n} - r_j)^2} &= (k/2\pi)^2 \overline{(\Sigma \sin 2\pi\theta_l)^2} \\ &\approx (k/2\pi)^2 \, \Sigma \overline{(\sin 2\pi\, \theta_l)^2} \\ &\approx (k/2\pi)^2 n/2 \\ &= D/2n \end{aligned} \tag{11}$$

and $D = (k/2\pi)^2$.

Why is the chaos bounded for $k \leqslant k_c$? Let's start at $k = 0$, where the orbits of eq. (5) lie on the straight horizontal lines $r = W$. For rational W, the orbits along $r = W$ are periodic; for irrational W any orbit fills the entire line densely. Some of these latter orbits will form the KAM trajectories as we increase k; the rational orbits will destroy nearby irrational orbits to form chaotic regions.

For $k > 0$ but small, there are lots of KAM trajectories; in between any two, lies a chaotic region (in between any two irrationals lies a rational). We assert that the chaotic regions are confined by the persistence of horizontal KAM trajectories (as in Fig. 5). Consider the chaotic region, for example, containing the fixed point $r = 0$, $\theta = 0$. The union of the images of a small neighborhood of the origin (Fig. 7) under successive iterations of the map will in general form a very contorted open set. However, it must always include the origin, and cannot

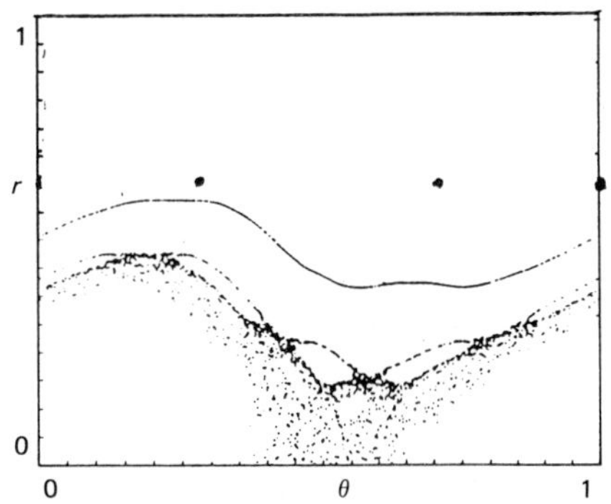

Fig. 5. Three kinds of orbits in the $r - \theta$ plane.

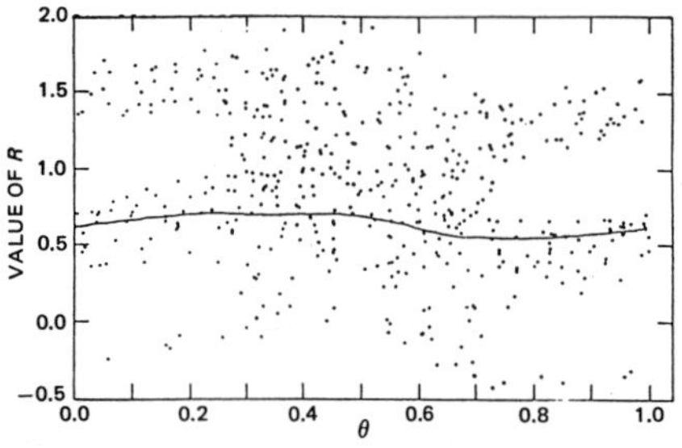

Fig. 6. Unbounded Chaos for $k > k_c$. Here 500 iterations of a single starting point are plotted when $k = 1.8$. The central curve is the best KAM trajectory to break up and is shown here to display the diffusion across it.

intersect a KAM surface (since points on a KAM surface are images only of other points on the surface). Thus no point in the neighborhood can ever leave the region bounded by horizontal KAM surfaces.

As we continue to increase k, more and more KAM curves break up, and the chaotic regions merge (Fig. 8) until at $k = k_c$ the last remaining horizontal KAM curves disappear and unbounded vertical chaotic diffusion ensues. Empirically, the last surfaces to go have $W = (\sqrt{5} - 1)/2$, the inverse of the Greek's golden mean [8–10] (up to an integer). That is, at k_c (Fig. 9) only isolated horizontal curves are left amid the chaos, and these have become very crinkled. For $k > k_c$ these curves also break up; gaps form in them, changing the curve into a Cantor set (Fig. 10).

The remainder of this lecture is devoted to describing the detailed mechanism for the break up according to the lines of the research carried on by Scott J. Shenker and myself [11].

How can we study this transition in detail? Greene has observed that these last KAM curves can be thought of as a limit of the particular periodic cycles with $W_n = p_n/q_n$ and $q_n = p_{n+1} = F_n$, the Fibonacci numbers. (F_n satisfies $F_0 = F_1 = 1, F_{n+1} = F_n + F_{n-1}$). In Fig. 11, for example, we see two periodic cycles with $W = 2/3$ above the KAM surface, and two below with $W = 3/5$. These periodic cycles alternately converge to the KAM surface with $W = (\sqrt{5} - 1)/2$ from above and below. Indeed, the solid line in Fig. 11 actually is composed of two cycles with $q = 4181 = F_{18}$, as we can see in the expanded scale of Fig. 12. Here note how magnificantly smooth the KAM surface appears on this length scale.

Why do we choose precisely this sequence of periodic orbits to look at $W = (\sqrt{5} - 1)/2$? Perhaps there are better answers, but roughly this sequence provides an orderly progression, converging rapidly and uniformly in its properties to those of the surface. One can think of the $n + 1$st cycle as "almost"

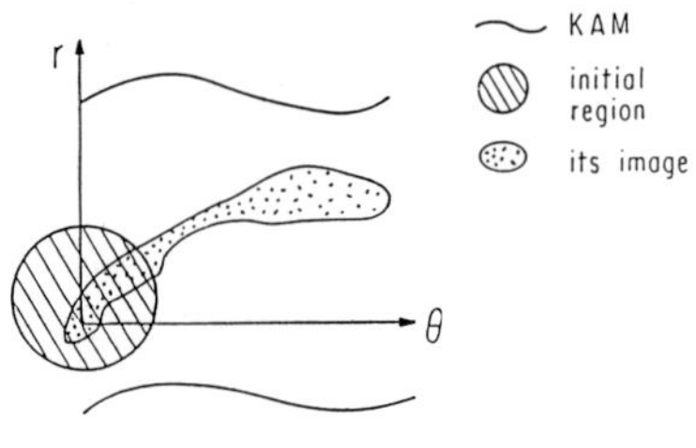

Fig. 7. Image of initial region including origin. Note that the image can never cross a KAM curve.

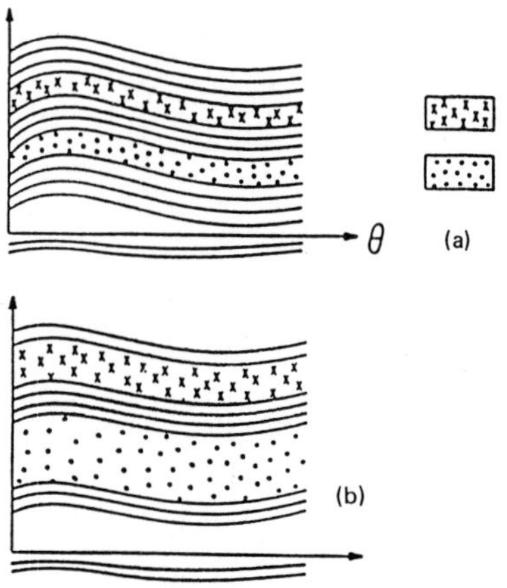

Fig. 8. Separation of chaotic regions by KAM trajectories. The X's and dots represent two different chaotic regions, separated by many KAM trajectories. In (b) $|k|$ is larger than in (a). As $|k|$ increases, KAM trajectories disappear and chaotic regions merge.

being the nth cycle followed by the $n - 1$st cycle. Since the convergence is alternating a proper weighting of the previous two cycles is a natural choice for the next; for $W = (\sqrt{5} - 1)/2$ the proper weighting is equal.

At this point it will be useful to introduce the Moser representation. Remember that W is an average speed for $\theta : \theta_j = \theta_0 + Wj + O(1)$ on a periodic orbit or a KAM surface. If we define a time $t_j = t_0 + W_j$. Moser has shown for small k that $\theta_j = \theta(t_j)$ where $\theta(t)$ can be written

$$\theta(t) = t + u(t) \tag{12}$$

with $u(t + 1) = u(t)$. On the KAM curve for $k < k_c$, $u(t)$ is a smooth function of period one, and we may define a Fourier transform

$$u(t) = \sum_{\omega=1}^{\infty} A_\omega \sin(2\pi\omega t). \tag{13}$$

For cycles $W = p/q$, $u(t)$ is defined only on a discrete set of points. Nonetheless, one may define a discrete Fourier transform

$$u(t) = \sum_{\omega=1}^{q} A_\omega(p/q) \sin(2\pi\omega t) \tag{14}$$

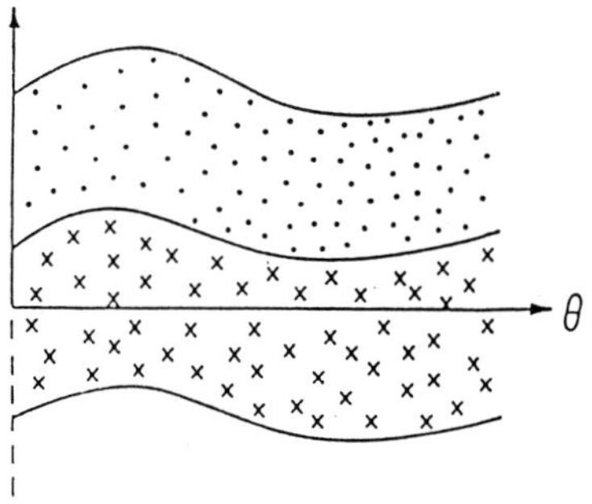

Fig. 9. At k_c, only a few well-separated KAM trajectories are left so that there are large chaotic regions.

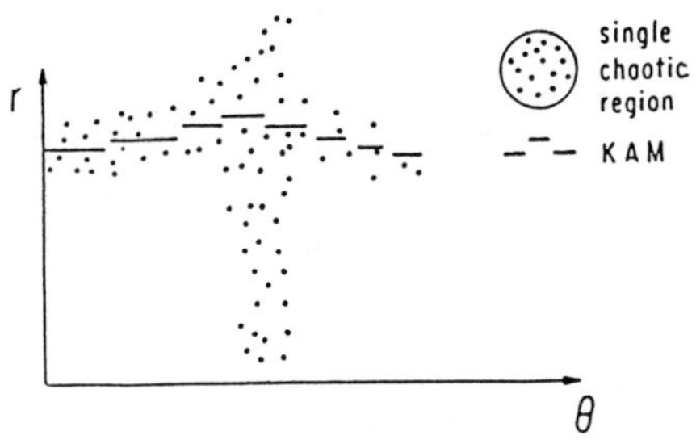

Fig. 10. As k is increased beyond k_c, the KAM trajectory breaks up into pieces so that the chaotic orbit can spread out over an infinite range of r-values.

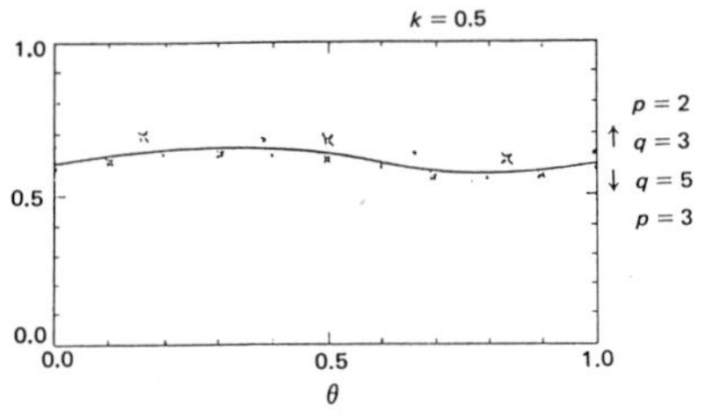

Fig. 11.

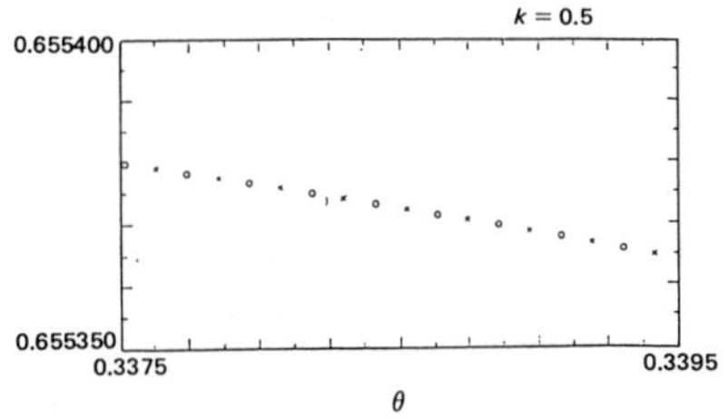

Fig. 12.

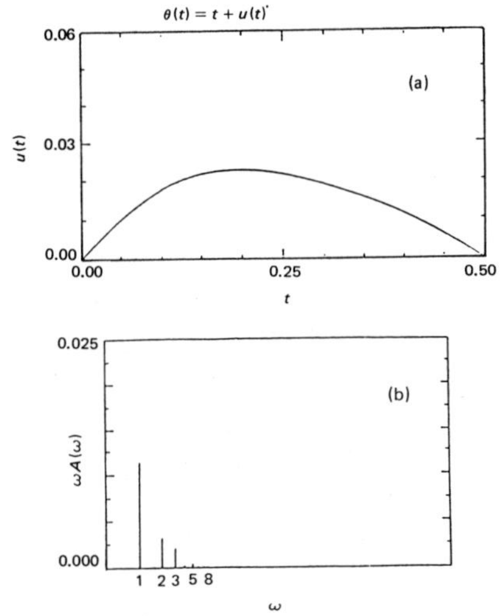

Fig. 13.

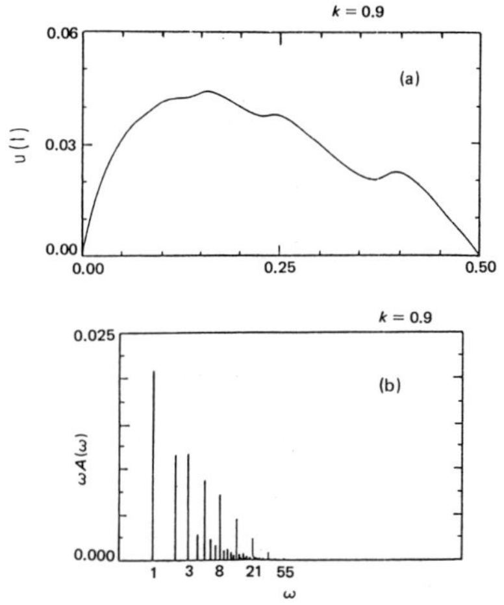

Fig. 14.

Clearly, $A_\omega(W_n)$ with $W_n = F_{n-1}/F_n$ vanishes if $\omega \geqslant F_n$. However, (Greene asserts) the values $A_\omega(W_n) \to A_\omega(W)$ as $n \to \infty$.

In Fig. 13 we see $u(t)$ and $\omega A_\omega(W)$ for $W = \sqrt{5} - 1/2$, $k = 0.5$. (We plot $u(t)$ only to $t = 0.5$; u is odd about this point $u(1/2 + t) = u(1/2 - t)$.) The obvious smoothness of u is reflected on $A(\omega)$, which is exponentially small as ω gets large. Note that already A_5 is larger than A_4; 4 is not a Fibonacci number.

In Fig. 14, we see u and ωA_ω for $k = 0.9$. New bumps have appeared in u, and the Fourier transform reflects this with prominent structure at the low Fibonacci numbers. (Notice that these peaks do not reflect numerical effects from our use of Fibonacci length orbits in approximating the KAM surface. These frequencies are naturally arising in the spectrum of the surface.)

Finally, in Fig. 15 we look at $k = k_c$. Figure 15(a(1)) is a graph of $u(t)$ and a small part of the phase trajectory is shown expanded in Fig. 15(a(2)). In striking contrast to Fig. 12, it is very bumpy even on this small length scale.

A longer orbit than in Fig. 15(a(2)) would fill the gaps between these dots with more bumps like these. Figure 15(b) shows the spectrum $\omega \cdot A_\omega$ at k_c. One should ignore the low frequencies (long wavelengths) as they depend upon the details of the map and are not universal. One must also ignore the very high frequencies (short wavelengths) as the finite length of the orbit numerically introduces errors. The scale region of the spectrum is expanded in Fig. 15(c). The big lines at the Fibonacci numbers have settled down: $|q_n A(q_n)|$ is clearly constant as $n \to \infty$. Also the smaller peaks are settling down, giving a beautiful self-similarity.

Let us look in more detail at $\Theta(t)$ near $t = 0$ (Fig. 16). ($t = 0$ and $t = 1/2$ are special symmetry points of our map (3). The function $\theta(t)$ at k_c is self-similar, scaling [12] and like [13]

$$\Theta(t) = |t|^{x_0} \operatorname{sign}(t)\, \Theta(\tau) \qquad (15)$$
$$\tau = \ln |t| / \ln W$$

(LPK [13]) where $\Theta(\tau)$ is a universal function. Overall, the function develops an infinite slope at zero, with

$$\Theta(1/F_n) = \text{constant} \cdot (1/F_n)^{x_0} \qquad (16)$$

The exponent x_0 is characteristic of the behavior of Θ in this transition. A similar exponent y_0 can be developed governing the behavior of $r(1/F_n) - r(0)$. Near $t = 1/2$, two more expo-

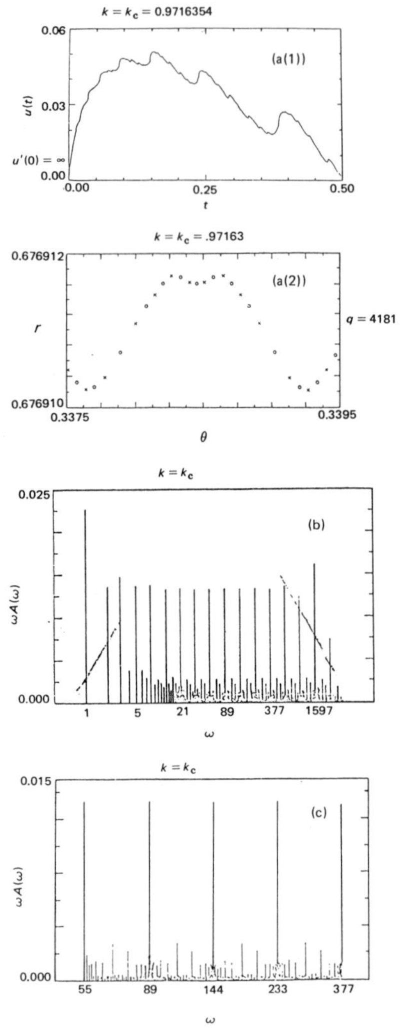

Fig. 15

nents x_1 and y_1 governing θ and r can be defined. Finally, two more exponents R_s and R_u can be defined as follows. To each Fibonacci ratio approximating the golden mean there are two periodic orbits – one stable (elliptic) and one unstable (hyperbolic). (In the figures the circles denote elliptic, the crosses hyperbolic orbits.) The derivative of the q-times iterated map for a p/q cycle has two eigenvalues e^{λ} and $e^{-\lambda}$ (with λ pure imaginary for elliptic orbits). The residue is defined to be $1/4\,(2 - e^{\lambda} - e^{-\lambda})$; its behavior for $W = F_{n-1}/F_n$ as $n \to \infty$ at $(k_c - k) \ll 1$ is described by the exponents R_s and R_u.

These exponents are tabulated in Table I. (These are old numbers – now each is known to at least one more decimal point.) The first column shows the numerical values of these exponents for the breakdown of the last KAM curve in the standard map (5). The last column shows the breakdown of the last curve in a different area preserving map of the plane; the agreement in the exponents shown has persisted using the better accuracy available today. The second column shows the breakdown of another KAM curve, whose winding number W has continued fraction which ends in [1, 1, 1 . . .] [but begins with the digits of π]. Again, the exponents agree.

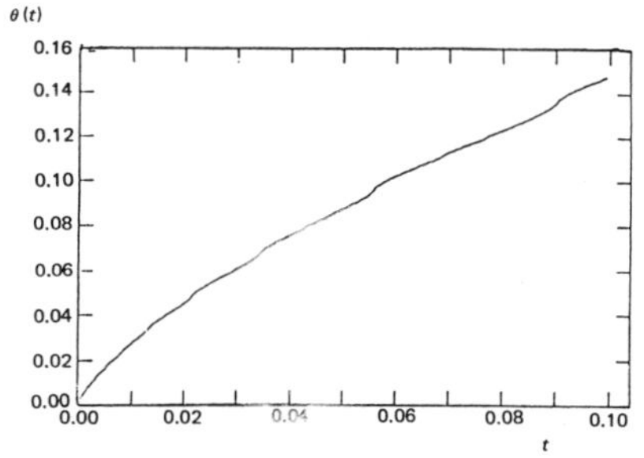

Fig. 16. Singular behavior at $k = k_c$. Each bump in this plot of $\theta(t)$ vs t marks an infinity in the slope.

Table I. *Potentially universal observables*

	W = [1, 1, 1 . . .]	[3, 1, 4, 1 . . .]	[2, 2, 2 . . .]	[1, 1, 1 . . .]
x_0	0.7211	0.72	0.72	0.72
x_1	1.093	1.09	1.10	1.09
y_0	2.329	2.33	2.34	2.33
y_1	1.957	1.96	1.96	1.95
R_s	0.250	0.25	0.23	0.25
R_u	−0.255	−0.25	−0.23	−0.26

However, choosing $W = \sqrt{2} - 1 = [2, 2, 2 \ldots]$ and varying k until this KAM curve breaks down, gives distinctly different values for R_s and R_u. The greater precision available today indicates that this difference is real, and extends to the other exponents as well. It is thought that the curve with winding number ending in ones separates chaotic regions at its breakdown, while (e, e, . . .) is surrounded by surviving curves (e.g. [2, 2, 2 . . . , 2, 1, 1, 1, . . .]) at its breakdown – its environment is quite different as it goes.

Thus we see universal behavior; the singularities at k_c are independent of the exact map, or (within limits) the exact W. We see scale invariant behavior (15) with structures which recur on all t-scales as $t \to 0$, 1/2, and which occurs again and again as the frequency W goes through Fibonacci numbers. We see behavior which is connected to Fibonacci numbers. We need an explanation for the scaling and universality behavior we have seen. This can be obtained from a renormalization group treatment [14–16] which explains some but not all of the features noted here.

References

1. Kolmogorov, A. N., Dokl. Akad. SSSR **98**, 527 (1954) (Russian).
2. Arnol'd, V. I., Izv. Akad. Nauk **25**, 21 (1961).
3. Moser, J., Nachr. Akad. Wiss. Gottingen Math. Phys. Kl. **IIa**, 1 (1962).
4. Chirikov, B., Phys. Rev. **52**, 265 (1979).
5. Fradkin, E. and Huberman, B., Urbana preprint.
6. Coopersmith, S. N. et al., Phys. Rev. Lett. **46**, 459 (1981).
7. Aubry, S., in Solitons in Condensed Matter Physics (eds. A. R. Bishop and T. Schnieder), Springer Verlag, New York (1978).
8. Greene, J. M., J. Math. Phys. **9**, 760 (1968).
9. Greene, J. M., J. Math. Phys. **20**, 1183 (1979).
10. Greene, J. M. and Percival, I., Physica **3D** (1981).
11. Shenker, Scott J. and Kadanoff, Leo P., J. Stat. Phys. **27**, 631 (1982).
12. Widom, B., J. Chem. Phys. **43**, 3989 (1983).
13. Kadanoff, Leo P., Phys. Rev. Lett. **47**, 1641 (1981).
14. Escande, D. and Doveil, F., J. Stat. Phys. **26**, 257 (1981).
15. MacKay, R., Princeton Thesis (1983).
16. Kadanoff, Leo P., in Melting, Localization and Chaos (eds. R. K. Kalia and P. Vashista), p. 202. North Holland, New York (1982).

Proc. Natl. Acad. Sci. USA
Vol. 81, pp. 1276–1279, February 1984
Physics

Escape from strange repellers

(dynamical system/mapping/cycles/derivative matrix)

LEO P. KADANOFF AND CHAO TANG

The James Franck Institute, The University of Chicago, 5640 Ellis Avenue, Chicago, IL 60637

Contributed by Leo P. Kadanoff, October 17, 1983

ABSTRACT **In a dynamical system described by a map, it may be that a "strange" sets of points is left invariant under the mapping. The set is a repeller if points placed in its neighborhood move away. An escape rate is defined to describe this motion. An alternative method of evaluating the escape rate, based on the consideration of repulsive cycles, is proposed. In the several cases examined numerically and analytically, the escape rate is shown to agree with the proposed formula.**

1. Introduction

The description of dynamical systems often involves the consideration of sets of points that are left unchanged by the flow. When these invariant sets have a complex topological structure, they are termed "strange." Both strange attractors and strange repellers play a major role in our description of dynamical systems. Attractors are important because, as the system advances, the motion can approach the attractor more and more closely.

Conversely, of course, the motion of the system tends to move away from repellers. Nonetheless a repeller might be important because, for example, it might describe a seperatrix that serves to divide two different attractors or two different types of motion. Alternatively, the motion might be one in which almost all initial points lead to an orbit that escapes to infinity. The remaining nonescaping points will then be a repeller, which might be sufficiently complex to term "strange."

One can introduce a wide variety of numbers that characterize these strange sets and the motion on them. Many of these characterizers have the nature of one kind or another of fractal dimension (1–6). In this paper, we describe the motion in the immediate vicinity of the set by an "escape rate" that states quantitatively how fast the repulsion occurs.

To define the escape rate, imagine that we have a mapping f that maps a point $\mathbf{r}$ in a manifold M into another point in the manifold. Consider some finite-sized region R within the manifold. Unlike the repeller, R is a set that is supposed to be very simple and not strange in any way whatsoever. Start with a set of N_0 initial points that are distributed uniformly (with Lebesque measure) within R. Let the mapping f operate n times and find out how many of the initial points lie in R after n iterations. Call the number N_n. As N_0 goes to infinity and n remains fixed, the staying ratio $\Gamma_n = N_n/N_0$ will approach a limit (6, 7). The escape rate α is then defined via

$$\alpha = -\lim_{n\to\infty} \frac{\ln \Gamma_n}{n}, \qquad [1.1]$$

so that exponential decrease of the number of points in R implies nonzero α. If R contains an attractor, α will be zero; if it contains a repeller but no attractor, α may be a finite positive number; if R contains neither repeller nor attractor, one may expect α to be infinite.

The quantity Γ_n is then of considerable physical interest, but it is hard to calculate, especially if M has a high dimensionality. There is an alternative approach based on the set of repulsive cycles of f, which gives a related quantity that is much easier to calculate. Let $\mathbf{r}$ be an element of the set Fix f^n, if $\mathbf{r} = f^n(\mathbf{r})$ and if this fixed point of f^n is repulsive. Then define

$$A_n = \sum_{\mathbf{r}\in \mathrm{Fix}\, f^n} \frac{1}{|\det[1 - \mathrm{D}f^n(\mathbf{r})]|}. \qquad [1.2]$$

Here, if the manifold is of dimension d, 1 is a d by d unit matrix and $\mathrm{D}f^n$ is the derivative matrix. The basic idea we propose is that, for large n, A_n and Γ_n are proportional to one another. In particular we define an exponential decay rate for A_n in analogy to Eq. **1.1** as

$$\delta = -\lim_{n\to\infty} \frac{\ln A_n}{n}. \qquad [1.3]$$

We then assert the basic identity that, for a wide variety of maps,

$$\delta = \alpha. \qquad [1.4]$$

It is our hope that this assertion can be proven for some wide class of maps, perhaps all maps that have a hyperbolic repelling set. However, we do not know any general proof of statement **1.4**.

Lacking such a theorem, we must examine some more fragmentary evidence. In the next section, we reformulate condition **1.4** and list a variety of simple cases in which this condition is known to be satisfied. We then notice that all strange repellers known to satisfy Eq. **1.4** arise in a situation in which there is a hyperbolic, but fully repulsive, set. For this reason, we examine in *Section 3* a map with both expansion and contraction (7) and show that the basic result **1.4** equally holds in this case.

2. Formulation

The definition of Γ_n in terms of Lebesque measures can be converted into an integral statement, namely

$$\Gamma_n = \int_R d\mathbf{r}' \int_R d\mathbf{r}\,\delta[\mathbf{r}' - f^n(\mathbf{r})] \Big/ \int_R d\mathbf{r}''. \qquad [2.1a]$$

An analogous expression for A_n is obtained from Eq. **1.2**. If all the unstable cycles of f^n lie in R, we have

$$A_n = \int_R d\mathbf{r}\,\delta[\mathbf{r} - f^n(\mathbf{r})]. \qquad [2.1b]$$

To understand the asymptotic relation between these two

The publication costs of this article were defrayed in part by page charge payment. This article must therefore be hereby marked "*advertisement*" in accordance with 18 U.S.C. §1734 solely to indicate this fact.

expressions, introduce the Frobenius–Perron operator P, defined as operating to the left on states $\langle\phi|$. If the state $\langle\phi|$ corresponds to a function $\phi(\mathbf{r})$, then the state $\langle\phi|P$ has

$$\langle\phi|P = \langle\psi| \quad [2.2]$$

if

$$\psi(\mathbf{r}) = \phi[f(\mathbf{r})].$$

Now let us assume that the set R has the very special property that if $\mathbf{r}_0$ lies within R but $\mathbf{r}_1 = f(\mathbf{r}_0)$ does not, then $\mathbf{r}_n = f^n(\mathbf{r}_0)$ will not be elements of R for each value $n = 2, 3, \ldots$. If all this is true, we can comfortably define an inner product

$$\langle\phi_1|\phi_2\rangle = \int_R d\mathbf{r}\phi_1(\mathbf{r})\phi_2(\mathbf{r}) \quad [2.3]$$

and then write expressions **2.1** as

$$\Gamma_n = \langle 0|P^n|0\rangle/\langle 0|0\rangle \quad [2.4a]$$

$$A_n = \text{trace } P^n. \quad [2.4b]$$

Here $\langle 0|$ is a state with $\phi_0(\mathbf{r}) = 1$ and P^n is an operator with a matrix realization $\langle\mathbf{r}'|P^n|\mathbf{r}\rangle = \delta[\mathbf{r}' - f^n(\mathbf{r})]$, as is obtained by multiplying P n times.

If P were a function of a hermitian operator, all the rest would be easy. Then P would have a set of eigenvalues $\exp(-\varepsilon_\mu)$, $\mu = 0, 1, 2, \ldots$. The eigenvalue (say the one with $\mu = 0$) having the smallest real part of ε_μ would dominate for large n. In the large n limit, Eqs. **2.4** would reduce to

$$\Gamma_n \rightarrow \gamma\, e^{-n\varepsilon_0}$$

$$A_n \rightarrow I_0\, e^{-n\varepsilon_0}. \quad [2.5]$$

Here I_0 is the multiplicity of the hypothetical lowest lying state. Then Eq. **2.5** would guarantee the correctness of our basic result (**1.4**), with α being given by the lowest eigenvalue ε_0.

But this last paragraph is a pipe dream because P is certainly not known to be a function of a hermitian operator. But, we can get closer to the results (**2.5** and **1.4**) if the mapping f is hyperbolic and has a Markov decomposition. In that case, one can use one-dimensional statistical mechanics to show (8) that A_n does indeed have a kind of property analogous to a spectral decomposition—namely, that one can write

$$A_n = \sum_\mu I_\mu\, e^{-n\varepsilon_\mu}, \quad [2.6]$$

where the I_μs are integers but are not necessarily positive. When the eigenvalues are widely spaced, representation **2.6** is a great help in obtaining accurate estimates of ε_0 based on values of A_n for relatively low n.

One case in which Eq. **1.4** is certainly valid is the one in which the region R is a small neighborhood of a hyperbolic fixed point, $\mathbf{r}_0$. Let $\Lambda = \mathrm{Df}(\mathbf{r}_0)$ be the derivative matrix at the fixed point. Then, according to Eq. **1.2**,

$$A_n = |\det(1 - \Lambda^n)|^{-1}. \quad [2.7]$$

For large n, we need consider only the eigenvalues of the matrix Λ that lie outside the unit circle. If these eigenvalues are of the form $\Lambda' = e^{\delta_1}$, Eqs. **2.7** and **1.3** give

$$\delta = \sum_1 \delta_1. \quad [2.8]$$

Here δ_1s are the logarithms of the eigenvalues of the Floquet matrix, Λ, and the sum is constrained by the condition Re $\delta_1 > 0$. Thus only the expanding directions enter.

To see that Eq. **2.8** is right for the fixed point, visualize a case with one expanding direction, "x," and one contracting direction, "y," and let R be a rectangle with sides parallel to x and y (Fig. 1). The image of this rectangle is the region R'. One can calculate Γ_1 (which is $e^{-\alpha}$) as the ratio of the area of overlap between R and R' to the total area of R'. This ratio gives the staying probability. But notice that the ratio does depend on the expansion rate but is independent of the contraction rate. This argument then establishes the result $\alpha = \delta$ for this kind of fixed point.

The case in which $\mathbf{r}$ is a number so that f is a mapping on the real line was considered in a previous publication (16). The situation in which $f(x) = x^2 + p$ with x and p real and $p < -2$ was investigated in detail. In this case, the strange repeller is a Cantor set. The quantities Γ_n could be accurately evaluated by using the inverse images of $x = 0$, while A_n could be found from the cycles. The speculation $\alpha = \delta$ was substantiated by this calculation, at least for the example in question.

This same publication also considered the complex version of this mapping problem, $f(z) = z^2 + p$. Now the repeller is a Julia set (9). Numerical and analytical arguments were presented that strongly suggested that $\alpha = \delta$ in this case. The analytic nature of the map leads to a simpler form of Eq. **1.2**—namely,

$$A_n = \sum_{z\in \mathrm{Fix}\, f^n} \frac{1}{\left|1 - \frac{df^n}{dz}(z)\right|^2}. \quad [2.9]$$

In these two examples of escape from strange sets, all the cycles have only repulsive directions. There is no attraction. To increase our range of experience with Eq. **1.4**, we consider in the next section a case in which there is both an expanding and a contracting direction.

3. Escape Rates on a Two-Dimensional Manifold

In an earlier paper (7), a localization problem led the consideration of the escape rate for a mapping in a three-dimensional space in which $\mathbf{r} = (x, y, z)$ and

$$f((x, y, z)) = (2xy - z, x, y). \quad [3.1]$$

It turns out that the mapping has a simple "time-reversal" symmetry. If you have an orbit $\mathbf{r}_{j+1} = f(\mathbf{r}_j)$ that has x-values $\ldots x_j, x_{j+1}, x_{j+2} \ldots$, then one has an equally good solution with x-values $\ldots x_{j+2}, x_{j+1}, x_j \ldots$. As a result, for every cycle,

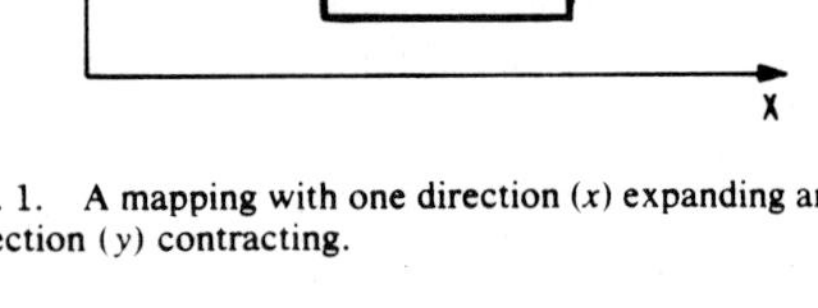

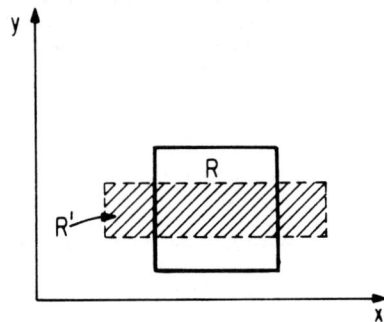

FIG. 1. A mapping with one direction (x) expanding and the other direction (y) contracting.

Table 1. Comparison of α and δ for the mapping **3.1**

n	0	0.5	1	10	50
2	0		0.183919		
4	0		0.286357		
5	0		0.306945		
6	0		0.267650		
7	0		0.301200		
8	0		0.300088		
9	0		0.311232		
10	0	0.13794706	0.300361	1.3986	2.226
11	0	0.13796868	0.300457	1.4140	2.293
12	0	0.14021587	0.298761	1.4026	2.235
13	0	0.13795662	0.300434	1.4107	2.281
14	0	0.13795567	0.300426	1.4048	2.242
15	0	0.13821219	0.301026	1.4092	2.273
16	0	0.13795594	0.30042795	1.4059	2.246
17	0	0.13795606	0.30042868	1.4983	—
∞		0.13795600 ± 0.00000006	0.3004283 ± 0.0000003	1.4071 ± 0.0012	2.26 ± 0.014
α		0.13796 ± 0.00005	0.3003 ± 0.0005	1.42 ± 0.02	2.29 ± 0.1
1/2 ln 2λ				1.50	2.30

Values of $\delta_n = -\ln A_n/n$ as a function of λ and n are presented. The line labeled infinity gives the extrapolated values of δ_∞, with a quoted error that is the half difference between the last two δ_n shown, which may not be conservative because δ_n has larger variations when $n = 0$ (mod 3) than for other values of n. The α-values given are a direct calculation of escape rate, with a statistical quoted error. Note the satisfactory agreement between δ_∞ and α. For completeness, the values of the asymptotic estimate $\delta \sim (1/2) \ln 2\lambda$ $(\lambda \to \infty)$, which was derived in Eq. **A.1**, are also given.

if there is a Floquet multiplier outside the unit circle, there is also one inside.

There is a conserved quantity associated with the mapping (**3.1**). Form the combination:

$$\lambda^2(\mathbf{r}) = \lambda^2((x, y, z)) = x^2 + y^2 + z^2 - 1 - 2xyz. \quad [\mathbf{3.2}]$$

A brief calculation shows that

$$\lambda^2[\mathrm{f}(\mathbf{r})] = \lambda^2(\mathbf{r})$$

so that combination is unchanged in the course of the mapping.

For this reason, we do not consider an escape problem in the entire three-dimensional euclidean space R^3 but instead focus our attention on manifolds in which the quantity **3.2** is fixed. We focus on the case in which the fixed value of the right-hand side of Eq. **3.2** is a number greater than or equal to zero, so that we can define our manifold by giving a real value of λ between 0 and ∞.

Notice that the manifold in question is certainly not compact. Topologically, it is similar to the surface of a sphere with four arms coming out of the sphere and moving out toward infinity. The manifold contains points with x, y, and z all very large but with the requirement that the product xyz be positive.

In addition to these regions at infinity, the manifold contains a central region where x, y, and z are all of order unity—assuming that λ itself is of order unity. If a point is placed "at random" within the central region, it is very likely that after a few iterations its coordinates will start to grow with greater than exponential rapidity. When this happens we say that a point "has escaped."

To make this definition more precise notice that the recursion relation $\mathbf{r}_{j+1} = f(\mathbf{r}_j)$ may be combined with Eq. **3.1** to give the relation

$$\mathbf{r}_j = (x_j, x_{j-1}, x_{j-2}) \quad [\mathbf{3.3}]$$

and the statement

$$x_{j+2} + x_{j-1} = 2x_j x_{j+1} \quad [\mathbf{3.4}]$$

From Eq. **3.4**, one can prove (10) that escape to infinity will occur whenever

$$|x_j| > 1$$
$$|x_{j-1}| > 1$$
$$|x_{j-1}|\,|x_j| > |x_{j-2}|. \quad [\mathbf{3.5}]$$

We then use conditions **3.5** as our requirement for asserting that $\mathbf{r}_j$ "has escaped."

In ref. 7, we chose a set of initial points with x and y each uniformly distributed between -1 and 1. We then calculated the escape rate α via Eq. **1.1** (Table 1).

The other approach involves finding cycles of the mapping. A very straightforward way for doing this is presented in ref. 10.

To work out A_n in the two-dimensional constant λ manifold, one may first calculate the derivative matrix $\mathrm{Df}^n(\mathbf{r})|_{\mathbf{r}=\mathrm{f}^n(\mathbf{r})}$ of Eq. **3.1** in three-dimensional euclidean space by a simple matrix multiplication. The eigenvalues of the matrix are of the form 1, η, $(-1)^n/\eta$. The last two of them are also the eigenvalues of the derivative matrix tangent to the λ manifold. Once we know A_n, Eq. **1.3** gives the value of δ. In Table 1 are shown the results for different ns and λs. Notice that for large n, δ and α agree within error.

Appendix

Here we discuss the derivation of the formula

$$\delta = \frac{1}{2} \ln 2\lambda \qquad (\lambda \to \infty). \quad [\mathbf{A.1}]$$

Ref. 10 describes a simple way to calculate all the cycles for $\lambda \geq 0$. Any n-length cycle can be expressed symbolically by writing a string containing n symbols from the set L, S, and $\bar{L}$. For large λ, these stand respectively for a value of x of order λ, of order unity, and of order $-\lambda$. If B stands for L or $\bar{L}$, the permitted strings include all possibilities save ones in which two Bs are adjacent, or there are three Ss in a row, or in which the combinations $LSSL$ or $\bar{L}SS\bar{L}$ appear. Each permitted string of length n that does not repeat itself corresponds to one and only one cycle of length n.

We can represent points on the constant-λ manifold by giving the value of x_j, x_{j-1} and an additional quantity $\varepsilon_{j-1/2}$—with values plus or minus one.

In terms of the auxiliary quantity

$$Z_{j-1/2} = \sqrt{\lambda^2 + (1 - x_j^2)(1 - x_{j-1}^2)}\,,$$

the mapping can then be written as

$$(u', v', \varepsilon') = T(u, v, \varepsilon) \qquad \text{[A.2]}$$

with

$$\begin{aligned} u' &= uv - \varepsilon z(u, v) \\ v' &= u \\ \varepsilon' &= \mathrm{sign}(v - u''). \end{aligned} \qquad \text{[A.3]}$$

The derivative matrix has the simple form

$$M_j = \frac{\partial(u_{j+1}, v_{j+1})}{\partial(u_j, v_j)} = \begin{bmatrix} \dfrac{u_j - u_{j+1}v_j}{\varepsilon_{j-1/2}z_{j-1/2}} & \dfrac{\varepsilon_{j+1/2}z_{j+1/2}}{\varepsilon_{j-1/2}z_{j-1/2}} \\ 1 & 0 \end{bmatrix} \qquad \text{[A.4]}$$

and for an n-length cycle

$$\frac{\partial(u_n, v_n)}{\partial(u_0, v_0)} = M_{n-1} \cdot M_{n-2} \cdot \ldots \cdot M_0 \equiv M^n. \qquad \text{[A.5]}$$

Let η, $(-1)^n/\eta$ be the eigenvalues of M^n. In the limit $\lambda \to \infty$, Eqs. **A.2**, **A.4**, and **A.5** may be analyzed to give

$$|\eta| \propto \lambda^m, \qquad \text{[A.6]}$$

where m is the number of Ss in the cycle. Particularly, for an even-length cycle of the type $BSBS \ldots BS \ldots BS$ (alternate B and S, $m = n/2$), we can show that

$$|\eta| = 2^n \lambda^{\frac{n}{2}} \qquad (\lambda \to \infty). \qquad \text{[A.7]}$$

Since $|\eta| \gg 1$,

$$A_n = \sum_{r \in \mathrm{Fix} f^n} \frac{1}{|\eta|}. \qquad \text{[A.8]}$$

For large λ, the contributions of the cycles with smallest m (the type described above) dominate. A little counting work shows that the total number of such cycle elements is $2^{\frac{n}{2}+1}$. Therefore,

$$A_n = \frac{2^{\frac{n}{2}+1}}{2^n \lambda^{\frac{n}{2}}} = 2(2\lambda)^{-\frac{n}{2}} \qquad (\lambda \to \infty) \qquad \text{[A.9]}$$

and

$$\delta = \frac{1}{2} \ln 2\lambda$$

where Eq. **1.3** has been used.

We would like to thank John Guckenheimer, Oscar Lanford, Scott J. Shenker, and Michael Widom for helpful discussions. This research was supported in part by National Science Foundation Grants MRL DMR 82-16892 and 80-20609.

1. Mandelbrot, B. (1977) *Fractals: Form, Chance, and Dimension* (Freeman, San Francisco).
2. Farmer, J. D., Ott, E. & Yorke, J. A. (1983) *Physica D* **7**, 153–180.
3. Abraham, N. B., Gollub, J. P. & Swinney, H. L. (1983) *Physica D*, in press.
4. Guckenheimer, J. (1982) *Nature (London)* **298**, 358–361.
5. Grassberger, P. & Procaccia, I. (1983) *Phys. Rev. Lett.* **50**, 346–349.
6. Widom, M., Bensimon, D., Kadanoff, L. P. & Shenker, S. J. (1983) *J. Stat. Phys.* **32**, 443–454.
7. Kohmoto, M., Kadanoff, L. P. & Tang, C. (1983) *Phys. Rev. Lett.* **50**, 1870–1872.
8. Bowen, R. (1975) *Lect. Notes Math.* **470**, 1–89.
9. Brolin, H. (1965) *Ark. Mat.* **6**, 103–144.
10. Kadanoff, L. P. (1983) Preprint (Univ. of Chicago, Chicago, IL).

VOLUME 55, NUMBER 25 PHYSICAL REVIEW LETTERS 16 DECEMBER 1985

Global Universality at the Onset of Chaos: Results of a Forced Rayleigh-Bénard Experiment

Mogens H. Jensen, Leo P. Kadanoff, and Albert Libchaber
The James Franck Institute, The University of Chicago, Chicago, Illinois 60637

Itamar Procaccia
Department of Chemical Physics, The Weizmann Institute of Science, Rehovot 76100, Israel

and

Joel Stavans
The James Franck Institute, The University of Chicago, Chicago, Illinois 60637
(Received 15 October 1985)

We study an experimental orbit on a two-torus with a golden-mean winding number obtained from a forced Rayleigh-Bénard system at the onset of chaos. This experimental orbit is compared with the orbit generated by a simple theoretical model, the circle map, at its golden-mean winding number at the onset of chaos. The "spectrum of singularities" of the two orbits are compared. Within error, these are identical. Since the spectrum characterizes the metric properties of the entire orbit, this result confirms theoretical speculations that these orbits, taken as a whole, enjoy a kind of universality.

PACS numbers: 47.20.+m, 05.45.+b, 47.25.−c

In the study of the transition to chaos most theoretical attention has been paid to the behavior near special points in phase space. Thus Feigenbaum[1] concentrated upon the region of the maximum of the period-doubling map, Shenker[2] looked near the inflection point of the circle map, etc. In experimental situations, such distinguished points in phase space are not readily discernible. If one is to look experimentally for universality, one would do well to seek more global,[3] but still universal, features of the phase-space orbits.

In this Letter we report experimental results which, together with theoretical analysis, show that critical orbits in phase space at the onset of chaos exhibit global universal properties. The example discussed here is the cycle with golden-mean winding number at the point of breakdown of a 2-torus. The experiment is a periodically forced Rayleigh-Bénard system with mercury as a fluid. Recent measurements on this system revealed two scaling indices[4]: the index for the ratio between two successive Fibonacci resonances[2] and the dimension of the structure of mode locking.[5] Both were in agreement with the indices found for circle maps. We therefore compare the experimentally observed critical orbit with the corresponding orbit in the circle map.

In order to examine global scaling properties it is not sufficient to measure the dimension of the attracting set; the set certainly contains more topological information than can be characterized by a single number.[1b] To achieve a characterization that more fully describes those properties of such sets which remain unchanged under smooth changes of coordinates, it has been proposed to use a continuous spectrum of scaling indices.[6] These spectra display the range of scaling indices and their density in the set. To clarify what we mean, consider the experimental cycle displayed in Fig. 1. One sees with bare eyes that the time series is concentrated with various intensities in different regions. The spectrum that we use quantifies this variation in density on the attractor, and allows us to show the similarity of

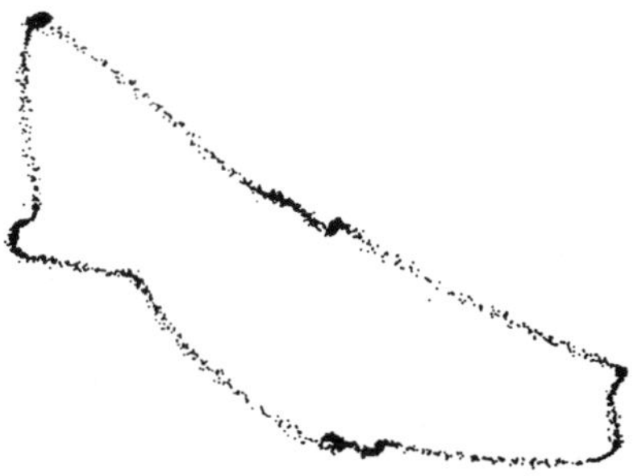

FIG. 1. The experimental attractor in two dimensions. 2500 points are plotted. Note the variation in the density of points on the attractor. Part of this variation is, however, due to the projection of the attractor onto the plane. The attractor is nonintersecting in three dimensions, in which it was embedded for the numerical analysis. In the absence of experimental noise the points should fall on a single curve. The smearing of the observed data set is mostly due to the slow drift in the experimental system during the run over about 2 h. Our method of analyzing the data to secure f vs α (see Fig. 2) is intended to minimize the effect of the slow drift. This is realized by estimating the recurrence times which experimentally are matters of minutes rather than hours.

this cycle to sets produced by model equations that describe the onset of chaos via quasiperiodicity.[2,7,8] In fact, the approach proposed here constitutes a rare opportunity for an extensive quantitative comparison of experiments with universal results obtained from theoretical models.

The experiment which yielded the critical golden-mean trajectory has been described previously.[4] The experiment studies a small-aspect-ratio Rayleigh-Bénard system of size $0.7\times0.7\times1.4$ cm^3 with two convective rolls present. For a low-Prandl-number fluid like mercury, as the heat flux increases beyond the convection threshold R_c, the system undergoes a Hopf bifurcation, called the oscillatory instability, into a time-dependent periodic mode. This mode is characterized by an ac vertical vorticity otherwise absent in the static roll pattern. This oscillation is one of our two oscillators (frequency ≈ 230 mHz). The second oscillator is introduced electromagnetically, mercury being an electrical conductor. An ac current sheet is passed through the mercury and the system is immersed in an horizontal magnetic field ($H\approx 200$ G) parallel to the rolls' axes. The geometry of electrode and field is such that the Lorentz force on the fluid produce ac vertical vorticity. In this way the oscillators are dynamically coupled. During the experiment the Rayleigh number is kept fixed at $R=4.09R_c$ giving a large amplitude to the first oscillator.

The nonlinear interaction between the oscillators is controlled by the amplitude of the injected ac current. A signal is obtained from the experiment by means of a thermal probe located in the bottom plate of the cell. The winding number, which is the ratio between the two frequencies, is kept close to the golden mean, i.e., within 10^{-4}. Time series are obtained by observation of the temperature signal at discrete times separated by the period of the forcing.

The theoretical work is aimed at estimating in a quantitative fashion how "bunched" the density on the orbit might be. In technical terms this bunching is a description of singularities in the probabilities of the orbit points.[9] Less technically, one can view a particular orbit point x_i, in a phase space like that of Fig. 1, and ask what is the probability for other points falling within the small distance, l, of this one. Call this probability $p_i(l)$. One can describe this probability by defining an index $\alpha_i(l)$ via

$$p_i(l)=l^{\alpha_i(l)}. \tag{1}$$

In typical sets the scaling index α_i takes, for small l, a range of values between $\alpha_{\min}$ and $\alpha_{\max}$. We refer to this situation as a spectrum of singularities.

To analyze the experimental time series, we make use of a key theoretical idea that fractal sets in general, and critical orbits in particular, can be described as interwoven sets of singularities.[9] The density of singularities of type α, $\alpha_{\min}<\alpha<\alpha_{\max}$, is determined by an index f that can be interpreted as the dimension of the set of singularities of this type. In other words, if the system is divided into pieces of size l, then the number of times, $n(\alpha,l)$, that α takes on a value between α and $\alpha+d\alpha$ is of the form

$$n(\alpha,l)=d\alpha\rho(\alpha)l^{-f(\alpha)}, \tag{2}$$

where $\rho(\alpha)$ is nonsingular with respect to l. The intuitive meaning of $\alpha_{\max}$ is that it is associated with the most rarefied regions of the measure, whereas $\alpha_{\min}$ with the most concentrated. Typically, $f(\alpha_{\max})=f(\alpha_{\min})=0$. Other types of singularities between $\alpha_{\max}$ and $\alpha_{\min}$ live on subsets of dimension f, $0<f<D_0$ (D_0 being the dimension of the set). The functions $f(\alpha)$ are universal functions for critical cycles like the trajectory with golden-mean winding number at the onset of chaos via quasiperiodicity.[2] Another key point is that these functions are smooth, in contrast to the universal scaling functions of the type suggested by Feigenbaum,[1] which are nowhere differentiable (see for example Fig. 10 of Ref. 1b). The reason for this important difference is that the latter functions are constructed by following the *local* changes in scaling everywhere,[1] whereas the former are based on finding the *global density* of scaling indices of each type.

The $\alpha_i(l)$ of Eq. (1) are estimated in a very simple fashion: Start from the point x_i on the trajectory. Count the number of steps along the time series required before a point returns to within l of the starting point. We call the number the recurrence time and denote it m_i. We shall now make use of the fact that the orbit is conjugate to a pure rotation[2,7,8] with an irrational winding number, and is therefore ergodic. Thus, we simply estimate $p_i(l)$ as the inverse recurrence time, $(m_i)^{-1}$, and find

$$\alpha_i(l)=-\ln m_i/\ln l. \tag{3}$$

In principle, the remainder of the analysis is very simple. One estimates how many $\alpha_i(l)$ values live in a given range, substitutes that estimate into Eq. (2), and then chooses some very small value of l to find $f(\alpha)$. In practice, given only a moderate amount of data, one cannot obtain a good estimate of $f(\alpha)$ by this direct method. Instead, we employ[6] an indirect method which smooths the data and gives an efficient calculation of $f(\alpha)$. To obtain this smoothness, we use the data to calculate the auxilary quantity

$$\Gamma(q,l)=\langle p_i(l)^{q-1}\rangle=\langle m_i^{1-q}\rangle, \tag{4}$$

where the brackets represent an average over all the trajectory elements i. The whole point of using the "partition function" Eq. (4) is that it is a smooth function of l and q and, for $l \ll 1$, is given by a power of l,

$$\Gamma(q,l) \sim l^{\tau(q)}. \tag{5}$$

This $\tau(q)$ is related to the generalized dimensions, D_q, of Hentschel and Procaccia[10] by $D_q = (q-1)^{-1}\tau(q)$.

From the point of view of this paper $\tau(q)$ is not important in itself. Instead it is a kind of generating function which can be used to determine the function $f(\alpha)$ via the pair of formulas (derived in Ref. 6),

$$\alpha(q) = d\tau(q)/dq, \qquad f(q) = \tau(q) - q\, d\tau/dq. \tag{6}$$

This is essentially a Legendre transformation, as used in statistical thermodynamics. Once q is eliminated from the pair of Eqs. (6), we discover that we have $f(\alpha)$ defined in a range of α values $\alpha_{\min} < \alpha < \alpha_{\max}$.

An example of a theoretical $f(\alpha)$ curve calculated in this manner is shown as the curve in Fig. 2. The attractor is the critical cycle of the circle map[2,7,8] $\theta_{n+1} = \theta_n + \Omega - (K/2\pi)\sin 2\pi\theta_n$ at the critical value $K=1$ and $\Omega = \Omega_{\rm gm}$, where the orbit has a golden-mean winding number. Here, $\tau(q)$ was evaluated by calculation of the average Eq. (4) for several l values and then finding τ as the slope of a straight line fit of a plot of $\ln\Gamma$ vs $\ln l$. The value of $\alpha_{\max}$, which is also $D_{-\infty}$ agrees with the theoretical expectation[6]

$$\alpha_{\max} = \ln\omega^*/\ln\alpha_{\rm gm}^{-1} = 1.8980\ldots, \tag{7}$$

where ω^* is the golden mean $\omega^* = (\sqrt{5}-1)/2$ and $\alpha_{\rm gm}$ is the universal local scale factor[2] in the vicinity of the critical point $\theta = 0$, $\alpha_{\rm gm} = 1.2885\ldots$. This is the most rarefied region in the trajectory. This region is mapped onto the most concentrated region in the set[6] which is characterized by the α value $\alpha_{\min} = \ln\omega^*/\ln\alpha_{\rm gm}^{-3} = 0.6326\ldots$. The curve turns around at the value of f which is $f = D_0 = 1$. This is also to be expected since the support of the measure is the circle, which is one dimensional.

The experimental data were similarly analyzed. On the basis of a time series of 2500 points embedded in a three-dimensional space we first calculated Γ as required by Eq. (4). Plotting again $\ln\Gamma$ vs $\ln l$, we typically fitted the τ's with over fifty different values of l ranging over two decades. The $f(\alpha)$ values were computed via Eqs. (6) with the result shown as the dots in Fig. 2. For small q ($|q| < 1$) the scaling was in general best for the largest values of l. As $|q|$ was increased, the best scaling regime gradually moved towards lower values of l. This is expected since high $|q|$ values correspond to isolated regimes on the attractor. The accuracy of the fits was always very good for positive q's (corresponding to the leftmost branch of the curve), and we estimate the error bar on the point $(D_\infty, 0)$ to be a few percent. The accuracy was less for negative values of q (corresponding to the rightmost branch) and the error bar on the point $(D_{-\infty}, 0)$ is around (10–12)%. The accuracy of the maximum point of the curve (i.e., D_0) is indicated by a vertical error bar. Theory and experiment agree. This agreement supports our conjecture that this Rayleigh-Bénard system at the onset of chaos and the critical circle map belong to the same universality class. We note in passing that from the value of $\alpha_{\max}$ (and also of $\alpha_{\min}$) one can read immediately $\alpha_{\rm gm}$, cf. Eq. (7). To the best of our knowledge this is the first direct measurement of this universal scaling number.

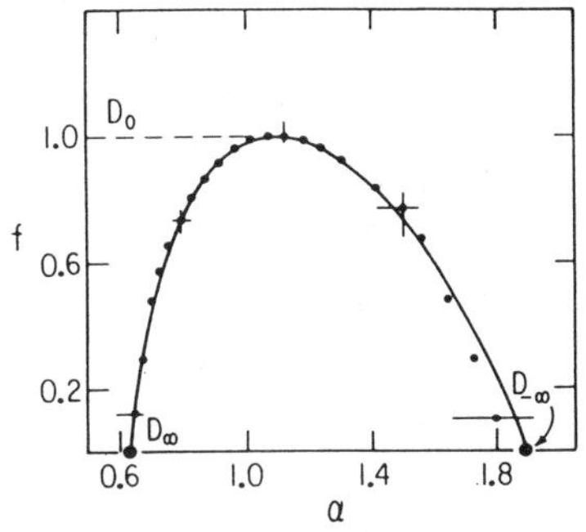

FIG. 2. The $f(\alpha)$ spectrum calculated for a critical circle map with golden-mean winding number is shown by the curve (Ref. 6). The curve ends in the points $(D_\infty, 0)$ and $(D_{-\infty}, 0)$, which are shown by the two large dots. The $f(\alpha)$ estimates for the experimental time series are marked by the smaller dots. The error bars are estimated by varying the range of l used to fit the data.

To conclude we note that the raw experimental orbit in its reconstructed phase space looks nothing like the orbit of the circle map. To the eye, it does not appear to lie on a circle. It is twisted and contorted in a complicated way. Our results demonstrate, however, that from the metric point of view these two sets *are the same* within experimental accuracy. To date we are not aware of any other approach that can lead to such a strong conclusion.

We thank Thomas C. Halsey for many stimulating discussions. This work has been partially supported by the Materials Research Laboratory of the University of Chicago, the National Science Foundation through Grants No. DMR 83-16626 and No. DMR 83-16204, the Office of Naval Research, and the Minerva Foundation, Munich, Germany.

[1a]M. J. Feigenbaum, J. Stat. Phys. **19**, 25 (1978), and J. Stat. Phys. **21**, 669 (1979).

[1b]M. J. Feigenbaum, Los Alamos Sci. **1**, 4 (1980).

[2]Scott J. Shenker, Physica (Amsterdam) **5D**, 405 (1982).

[3]In parameter space however, global universal features have been proposed [M. H. Jensen, P. Bak, and T. Bohr,

Phys. Rev. Lett. **50**, 1637 (1983), and Phys. Rev. A **30**, 1960 (1984); P. Cvitanović, M. H. Jensen, L. P. Kadanoff, and I. Procaccia, Phys. Rev. Lett. 55, 343 (1985)] and measured experimentally [J. Stavans, F. Heslot, and A. Libchaber, Phys. Rev. Lett. 55, 596 (1985); see also A. Libchaber, C. Laroche, and S. Fauve, Physica (Amsterdam) **7D**, 73 (1983), and J. Phys. (Paris), Lett. **43**, L211 (1982)] for the structure of modelocking.

[4]Stavans, Heslot, and Libchaber, and Libchaber, Laroche, and Fauve, Ref. 3.

[5]Jensen, Bak, and Bohr, and Cvitanović *et al.*, Ref. 3.

[6]T. C. Halsey, M. H. Jensen, L. P. Kadanoff, I. Procaccia, and B. I. Shraiman, Phys. Rev. A (to be published).

[7]M. J. Feigenbaum, L. P. Kadanoff, and Scott J. Shenker, Physica (Amsterdam) **5D**, 370 (1982).

[8]S. Ostlund, D. Rand, J. P. Sethna, and E. D. Siggia, Phys. Rev. Lett. **49**, 132 (1982), and Physica (Amsterdam) **8D**, 303 (1983).

[9]T. C. Halsey, P. Meakin, and I. Procaccia, to be published.

[10]H. G. E. Hentschel and I. Procaccia, Physica (Amsterdam) **8D**, 435 (1983); I. Procaccia, Phys. Scr. **T9**, 40 (1985).

PHYSICAL REVIEW A VOLUME 33, NUMBER 2 FEBRUARY 1986

Fractal measures and their singularities: The characterization of strange sets

Thomas C. Halsey, Mogens H. Jensen, Leo P. Kadanoff, Itamar Procaccia,* and Boris I. Shraiman†
The James Franck Institute, The Enrico Fermi Institute for Nuclear Studies, and Department of Chemistry, The University of Chicago, 5640 South Ellis Avenue, Chicago, Illinois 60637
(Received 26 August 1985)

We propose a description of normalized distributions (measures) lying upon possibly fractal sets; for example those arising in dynamical systems theory. We focus upon the scaling properties of such measures, by considering their singularities, which are characterized by two indices: α, which determines the strength of their singularities; and f, which describes how densely they are distributed. The spectrum of singularities is described by giving the possible range of α values and the function $f(\alpha)$. We apply this formalism to the 2^∞ cycle of period doubling, to the devil's staircase of mode locking, and to trajectories on 2-tori with golden-mean winding numbers. In all cases the new formalism allows an introduction of smooth functions to characterize the measures. We believe that this formalism is readily applicable to experiments and should result in new tests of global universality.

I. INTRODUCTION

Nonlinear physics presents us with a perplexing variety of complicated fractal objects and strange sets. Notable examples include strange attractors for chaotic dynamical systems,[1,2] configurations of Ising spins at critical points,[3] the region of high vorticity in fully developed turbulence,[4,5] percolating clusters and their backbones,[6] and diffusion-limited aggregates.[7,8] Naturally one wishes to characterize the objects and describe the events occurring on them. For example, in dynamical systems theory one is often interested in a strange attractor (the object) and how often a given region of the attractor is visited (the event). In diffusion-limited aggregation, one is interested in the probability of a random walker landing next to a given site on the aggregate.[8] In percolation, one may be interested in the distribution of voltages across the different elements in a random-resistor network.[6]

In general, one can describe such events by dividing the object into pieces labeled by an index i which runs from 1 up to N. The size of the ith piece is l_i and the event occurring upon it is described by a number M_i. For example, in critical phenomena, we can let M_i be the magnetization of the region labeled by i. Such a picture is natural in the droplet theory of the Ising model, where one argues that if the region i has a size of order l, the magnetization has a value of the order of

$$M_i \sim l^y , \tag{1.1}$$

where y (or y_σ) is one of the standard critical indices.[9] Since these droplets are imagined to fill the entire space, the density of such droplets is simply

$$\rho(l) \sim \frac{1}{l^d} , \tag{1.2}$$

where d is the Euclidean dimension of space. In fact, in critical phenomena we define a whole sequence of y_q's by saying that the typical values of $(M_i)^q$ vary with q and have the form[10]

$$(M_i)^q \sim l^{y_q}, \quad q=1,2,3,\ldots . \tag{1.3}$$

Typically, our attention focuses upon the values of y_q that are greater than zero and *we have only a few distinct values of these.*[11]

In this paper we are interested not in critical phenomena but instead in a broad class of strange objects. However, we specialize our treatment to the case in which M_i has the meaning of a probability that some event will occur upon the ith piece. For example, in experiments on chaotic systems one measures a time series $\{\mathbf{x}_i\}_{i=1}^{N}$. These points belong to a trajectory in some d-dimensional phase space. Typically, the trajectory does not fill the d-dimensional space even when $N \to \infty$, because the trajectory lies on a strange attractor of dimension D, $D<d$. One can ask now how many times, N_i, the time series visits the ith box. Defining $p_i = \lim_{N\to\infty}(N_i/N)$, we generate the measure on the attractor $d\mu(\mathbf{x})$, because

$$p_i = \int_{i\text{th box}} d\mu(\mathbf{x}) .$$

In many nonlinear problems, the possible scaling behavior is richer and more complex than is the case in critical phenomena. If a scaling exponent α is defined by saying that

$$p_i^q \sim l_i^{\alpha q} , \tag{1.4}$$

then α [roughly equivalent to y_q/q in Eq. (1.3)] can take on a range of values, corresponding to different regions of the measure. In particular, if the system is divided into pieces of size l, we suggest that the number of times α takes on a value between α' and $\alpha'+d\alpha'$ will be of the form

$$d\alpha' \rho(\alpha') l^{-f(\alpha')} , \tag{1.5}$$

where $f(\alpha')$ is a continuous function. The exponent $f(\alpha')$ reflects the differing dimensions of the sets upon which the singularities of strength α' may lie. This expression is roughly equivalent to Eq. (1.2), except that now, instead of the dimension d, we have a fractal dimension $f(\alpha)$

which varies with α. Thus, we model fractal measures by interwoven sets of singularities of strength α, each characterized by its own dimension $f(\alpha)$. The rest of our formalism attempts to unravel this complexity in a workable fashion.

The concept of a singularity strength α was stressed in the context of diffusion-limited aggregation in independent work of Turkevich and Scher[12] and of Halsey, Meakin, and Procaccia.[8] The latter group pointed out the significance of the density of singularities and expressed it in terms of f.

In order to determine the function $f(\alpha)$ for a given measure, we must relate it to observable properties of the measure. We relate $f(\alpha)$ to a set of dimensions which have been introduced by Hentschel and Procaccia, the set D_q defined by[13]

$$D_q = \lim_{l \to 0} \left[\frac{1}{q-1} \frac{\ln \chi(q)}{\ln l} \right], \tag{1.6}$$

where

$$\chi(q) = \sum_i p_i^q . \tag{1.7}$$

D_0 is just the fractal dimension of the support of the measure, while D_1 is the information dimension and D_2 is the correlation dimension.[14]

As q is varied in Eq. (1.7), different subsets, which are associated with different scaling indices, become dominant. Substituting Eqs. (1.4) and (1.5) into Eq. (1.7), we obtain

$$\chi(q) = \int d\alpha' \rho(\alpha') l^{-f(\alpha')} l^{q\alpha'} . \tag{1.8}$$

Since l is very small, the integral in Eq. (1.8) will be dominated by the value of α' which makes $q\alpha' - f(\alpha')$ smallest, provided that $\rho(\alpha')$ is nonzero. Thus, we replace α' by $\alpha(q)$, which is defined by the extremal condition

$$\frac{d}{d\alpha'}[q\alpha' - f(\alpha')] \bigg|_{\alpha' = \alpha(q)} = 0 .$$

We also have

$$\frac{d^2}{d(\alpha')^2}[q\alpha' - f(\alpha')] \bigg|_{\alpha' = \alpha(q)} > 0 ,$$

so that

$$f'(\alpha(q)) = q , \tag{1.9a}$$

$$f''(\alpha(q)) < 0 . \tag{1.9b}$$

It then follows from Eq. (1.6) that[8]

$$D_q = \frac{1}{q-1}[q\alpha(q) - f(\alpha(q))] . \tag{1.10}$$

Thus, if we know $f(\alpha)$, and the spectrum of α values, we can find D_q. Alternatively, given D_q, we can find $\alpha(q)$ since

$$\alpha(q) = \frac{d}{dq}[(q-1)D_q] , \tag{1.11}$$

and, knowing $\alpha(q)$, $f(q)$ can be obtained from Eq. (1.10).

Equations (1.9)—(1.11) are the main formal results used in this paper. In the next section we develop the formalism outlined here in somewhat more detail and apply it to systems with strong self-similarity properties. In Sec. III we apply the formalism to some important examples of measures arising in dynamical systems. We examine the 2^∞ cycle of period doubling,[15] the devil's staircase of mode locking in circle maps,[16,17] and the elements of the critical cycle at the onset of chaos in circle maps with golden-mean winding number.[18–20] Although all of these cases have been examined previously, we are able to find a *smooth* function with which to characterize them. Furthermore, these characterizations are universal. Other attempts to study these measures have led to nowhere smooth scaling functions.[15,21] Since the characterizations are functions rather than numbers, they offer much more information than fractal dimensions. Unlike power spectra, these functions possess an immediate connection to the metric properties of the measures involved, and do not call for cumbersome interpretation. Therefore, we believe that experimental measurements of D_q, and thus of $f(\alpha)$, should replace more common tests of universality in the transition to chaos. We give many examples of the procedures employed, and we hope to encourage experiments to follow these lines.

II. EXACTLY SOLUBLE STRANGE SETS

A. Preliminaries

We begin by introducing a more general definition of the dimensions D_q. Consider a strange set S embedded in a finite portion of d-dimensional Euclidean space. Imagine partitioning the set into some number of disjoint pieces, $S_1, S_2, \ldots, S_N$, in which each piece has a measure p_i and lies within a ball of radius l_i, where each l_i is restricted by $l_i < l$. Then define a partition function

$$\Gamma(q,\tau,\{S_i\},l) = \sum_{i=1}^{N} \frac{p_i^q}{l_i^\tau} . \tag{2.1}$$

Eventually we shall argue that, for large N, this partition function is of the order unity only when

$$\tau = (q-1)D_q . \tag{2.2}$$

To make this argument, consider now two regions:

$$\text{region } A\text{: } q \geq 1, \ \tau \geq 0 , \tag{2.3a}$$

$$\text{region } B\text{: } q \leq 1, \ \tau \leq 0 . \tag{2.3b}$$

In region A, adjust the partition $\{S_i\}$ so as to maximize Γ. In region B, adjust it so that Γ is as small as possible. Then define

$$\Gamma(q,\tau,l) = \text{Sup } \Gamma(q,\tau,\{S_i\},l) \ \text{(region } A\text{)} , \tag{2.4a}$$

$$\Gamma(q,\tau,l) = \text{Inf } \Gamma(q,\tau,\{S_i\},l) \ \text{(region } B\text{)} . \tag{2.4b}$$

The supremum in region A will exist as long as there are constants $a > 0$ and $\alpha_0 > 0$, so that for any possible subset of S, $\{S_i\}$, we have

$$p_i \leq a(l_i)^{\alpha_0} . \tag{2.5}$$

Then $\Gamma(q,\tau,l)$ will exist and be less than infinity whenever

$$\alpha_0(q-1)>\tau \, . \tag{2.6}$$

Next define

$$\Gamma(q,\tau)=\lim_{l\to 0}[\Gamma(q,\tau,l)] \, . \tag{2.7}$$

Notice that $\Gamma(q,\tau)$ is a monotone nondecreasing function of τ and a monotone nonincreasing function of q. One can argue that there is a unique function $\tau(q)$ such that

$$\Gamma(q,\tau)=\begin{cases}\infty & \text{for } \tau>\tau(q), \\ 0 & \text{for } \tau<\tau(q) \, .\end{cases} \tag{2.8}$$

Equation (2.8) permits us to define D_q as

$$(q-1)D_q=\tau(q) \, . \tag{2.9}$$

Once D_q is known, Eqs. (1.10) and (1.11) will then give $\alpha(q)$ and $f(q)$. Notice that our definition of D_q is precisely the one which makes D_0 the Hausdorff dimension.

B. Connection to previously defined D_q

Hentschel and Procaccia[13] also defined a D_q, which we now denote as D_q^{HP}. To relate the two quantities, recall that the authors of Ref. 13 defined a partition in which all the diameters l_i had the same value l. We know that

$$\Gamma(q,\tau,l)\begin{cases} >l^{-\tau}\sum_{i=1}^{N} p_i^q & \text{(region } A) \\ <l^{-\tau}\sum_{i=1}^{N} p_i^q & \text{(region } B) \, . \end{cases} \tag{2.10}$$

If τ is chosen correctly, i.e., $\tau=\tau(q)$, the left-hand side of Eq. (2.10) will neither go to zero nor diverge very strongly as $l\to 0$. In particular, we guess that $\Gamma(l)$ is no worse than logarithmically dependent upon these quantities. Then

$$\lim_{l\to 0}[\ln\Gamma(q,\tau(q),l)/\ln l]\to 0 \, .$$

We have now

$$\frac{\tau}{q-1}\le\lim_{l\to 0}\left[\frac{\ln\left[\sum_{i=1}^{N} p_i^q\right]}{(\ln l)(q-1)}\right] . \tag{2.11}$$

The right-hand side of (2.11) is D_q^{HP}. We thus find

$$D_q\le D_q^{\text{HP}} \, . \tag{2.12}$$

Since we believe that Eq. (2.10) will often be an order of magnitude equality when $\tau=\tau(q)$, we think that Eq. (2.12) will be an equality in most cases of interest.

At this point we turn to some simple examples to illustrate the quantities $\tau(q)$. These examples will enable us to gain intuition about the quantities $\alpha(q)$ and $f(\alpha)$.

C. Exactly soluble examples

1. Power-law singularity

One of the simplest possible applications of this formalism is to a probability measure with only one power-law singularity. Imagine a probability density $\rho(x)=\tilde{\alpha}x^{\tilde{\alpha}-1}$ on $x\in[0,1]$, where $0<\tilde{\alpha}<1$. Let us partition the interval into N segments $[x_i, x_i+\Delta x]$, with $\Delta x=N^{-1}$. The total probability measure on all of these intervals except for that adjoining zero is well approximated by $\rho(x_i)\Delta x$. The probability upon the segment adjoining zero possesses a probability $\rho_0=(\Delta x)^{\tilde{\alpha}}$. The partition function is therefore

$$\Gamma(q,\tau,\Delta x)\approx\frac{(\Delta x)^{\tilde{\alpha}q}}{(\Delta x)^{\tau}}+\sum_{i\neq 0}\frac{\tilde{\alpha}x_i^{\tilde{\alpha}-1}(\Delta x)^q}{(\Delta x)^{\tau}} \, . \tag{2.13}$$

There are $(\Delta x)^{-1}$ terms in the sum, so that

$$\Gamma(q,\tau,\Delta x)\sim(\Delta x)^{\tilde{\alpha}q-\tau}+(\Delta x)^{q-1-\tau} \, . \tag{2.14}$$

Thus, since we require that Γ neither go to zero nor infinity, we have that

$$\tau=\min\{q-1,\tilde{\alpha}q\} \, , \tag{2.15a}$$

or

$$D_q=\frac{1}{q-1}\min\{q-1,\tilde{\alpha}q\} \, . \tag{2.15b}$$

Thus for $q>q^*=1/(1-\tilde{\alpha})$, the dimensions correspond to a value of $\alpha=\tilde{\alpha}$ and of $f=0$, while for $q<q^*$ the dimensions correspond to $\alpha=1$ and $f=1$. Thus, in this example the f-α spectrum consists of two points, corresponding to the two types of behavior in the measure.

2. Cantor sets and generators

If a measure possesses an exact recursive structure, one can find its D_q. Suppose that the measure can be generated by the following process. Start with the original region which has measure 1 and size 1. Divide the region into pieces S_i, $i=1,2,\ldots,N$, with measure p_i and size l_i. Suppose that the maximum of l_i is given by l. Then at the first stage we can construct a partition function,

$$\Gamma(q,\tau,l)=\sum_i\frac{p_i^q}{l_i^{\tau}} \, . \tag{2.16}$$

Continue the Cantor construction. At the next stage each piece of the set is further divided into N pieces, each with a measure reduced by a factor p_j and size by a factor l_j. At this level the partition function will be

$$\Gamma(q,\tau,l^2)=[\Gamma(q,\tau,l)]^2 \, . \tag{2.17}$$

We see at once that, for this kind of measure, the first partition function $\Gamma(l)$ will generate all the others, and that $\tau(q)$ is defined by

$$\Gamma(q,\tau(q),l)=1 \, . \tag{2.18}$$

If a partition with finite N yields a Γ which obeys (2.17), that partition is called a generator.[22]

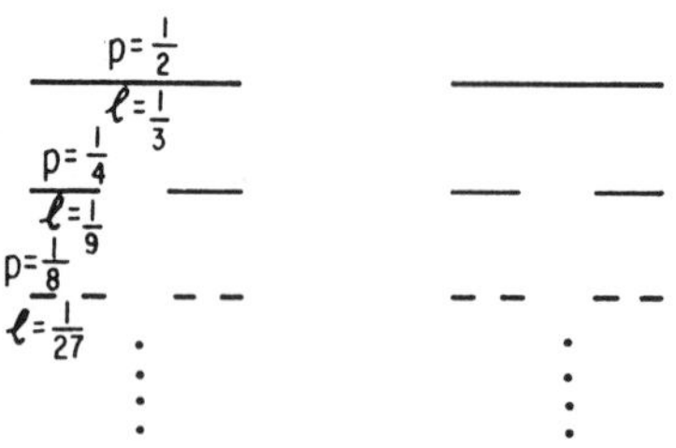

FIG. 1. The construction of the uniform Cantor set. At each stage of the construction the central third of each segment is removed from the set. Each segment has measure $p_0=(\frac{1}{2})^n$ and scale $l_0=(\frac{1}{3})^n$, where n is the number of generations.

3. Uniform Cantor set

A simple example is the classical Cantor set obtained by dividing the interval [0,1] as shown in Fig. 1. We initially replace the unit interval with two intervals, each of length $l=\frac{1}{3}$. Each of these intervals receives the same measure $p=\frac{1}{2}$. At the next stage of the construction of the measure this same process is repeated on each of these two intervals. Thus, for this measure we require

$$2\left[\frac{(\frac{1}{2})^q}{(\frac{1}{3})^\tau}\right]=1\ , \tag{2.19}$$

which yields

$$\tau=(q-1)[\ln(2)/\ln(3)]\ \ [\text{or } D_q=\ln(2)/\ln(3)]\ . \tag{2.20}$$

If l_0 is the length scale of the intervals at a particular level of the partitioning, and p_0 is the measure for such an interval, then

$$p_0=l_0^{\ln(2)/\ln(3)}\ . \tag{2.21}$$

Calling the index of the singularity α, i.e., $p_0\sim l_0^\alpha$, we have here $\alpha=\ln(2)/\ln(3)$. If we further ask what is the density of these singularities, we find immediately that it is simply the density of the set,

$$\rho(l_0)=\frac{1}{l_0^{\ln(2)/\ln(3)}}\ , \tag{2.22}$$

and Eq. (1.5) leads to $f=\ln(2)/\ln(3)$. Thus in this example, $\alpha=f$, and also

$$\tau(q)=q\alpha-f\ . \tag{2.23}$$

Although Eq. (2.23) is trivial here, we shall see that its analog, Eq. (1.10), also holds in the most general cases.

4. Two-scale Cantor set

A somewhat less trivial example is obtained by constructing a Cantor set as in Fig. 2. Here we use two rescaling parameters l_1 and l_2 and two measures p_1 and p_2, and then continue to subdivide self-similarly. We assume that $l_2>l_1$. It is apparent that this example also has a generator, since the condition

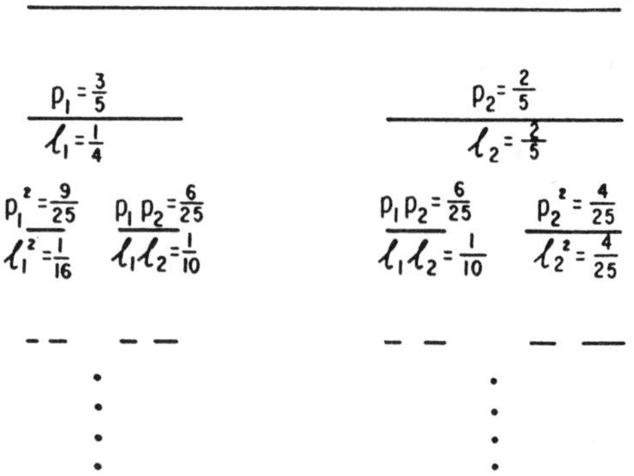

FIG. 2. A Cantor-set construction with two rescalings $l_1=0.25$ and $l_2=0.4$ and respective measure rescalings $p_1=0.6$ and $p_2=0.4$. The division of the set continues self-similarly.

$$\Gamma(q,\tau,l_2^n)=\left[\frac{p_1^q}{l_1^\tau}+\frac{p_2^q}{l_2^\tau}\right]^n=1 \tag{2.24}$$

results in a τ that does not depend on n. The value of τ depends, however, on q. In Fig. 3 we show $D_q=\tau(q)/(q-1)$ as a function of q, as obtained numerically by solving Eq. (2.24). To further understand this curve, we can examine the quantity $\Gamma(l_2^n)$ for this case explicitly:

$$\Gamma(q,\tau,l_2^n)=\sum_m \begin{bmatrix} n\\ m\end{bmatrix} p_1^{mq}p_2^{(n-m)q}(l_1^m l_2^{n-m})^{-\tau}=1\ . \tag{2.25}$$

We expect that in the limit $n\to\infty$ the largest term in this sum should dominate. To find the largest term we compute

$$\frac{\partial \ln\Gamma(l_2^n)}{\partial m}=0\ . \tag{2.26}$$

Using the Stirling approximation, we find that Eq. (2.26) is equivalent to

$$\tau=\frac{\ln(n/m-1)+q\ln(p_1/p_2)}{\ln(l_1/l_2)}\ . \tag{2.27}$$

Since we expect that the maximal term dominates the sum, we have a second equation,

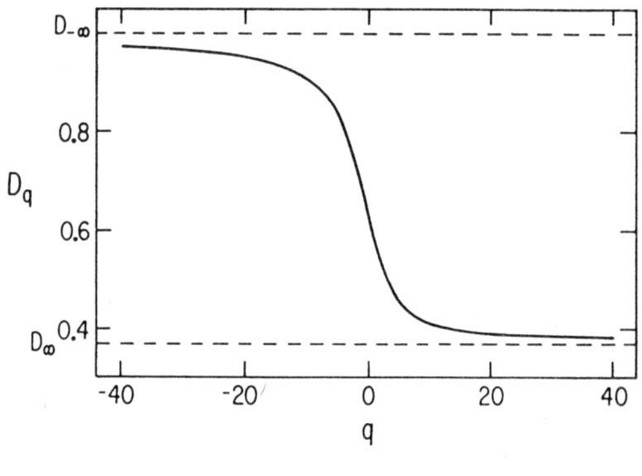

FIG. 3. D_q plotted vs q for the two-scale Cantor set of Fig. 2.

$$\begin{bmatrix} n \\ m \end{bmatrix} p_1^{mq} p_2^{(n-m)} \left[\frac{1}{l_1^m l_2^{(n-m)}} \right]^{\tau} = 1 . \quad (2.28)$$

Inserting Eq. (2.27) into Eq. (2.28) leads to an equation for n/m. After some algebraic manipulation, one finds

$$\ln(n/m)\ln(l_1/l_2) - \ln(n/m-1)\ln l_1 = q(\ln p_1 \ln l_2 - \ln p_2 \ln l_1) . \quad (2.29)$$

We thus see that for any given q there will be a value of n/m which solves Eq. (2.29) and, in turn, determines τ from Eq. (2.27). This maximal term which determines τ actually comes from a set of $\binom{n}{m}$ segments, all of which have the same size $l_1^m l_2^{(n-m)}$. Their density exponent f is determined by

$$\begin{bmatrix} n \\ m \end{bmatrix} (l_1^m l_2^{(n-m)})^f = 1 , \quad (2.30)$$

or

$$f = \frac{(n/m-1)\ln(n/m-1) - (n/m)\ln(n/m)}{\ln l_1 + (n/m-1)\ln l_2} . \quad (2.31)$$

The exponent determining the singularity in the measure, α, is determined by

$$p_1^m p_2^{(n-m)} = (l_1^m l_2^{(n-m)})^{\alpha} , \quad (2.32)$$

or

$$\alpha = \frac{\ln p_1 + (n/m-1)\ln p_2}{\ln l_1 + (n/m-1)\ln l_2} . \quad (2.33)$$

Thus, for any chosen q, the measure scales as $\alpha(q)$ on a set of segments which converge to a set of dimension $f(q)$. As q is varied, different regions of the set determine D_q. It can be shown that Eqs. (2.27), (2.29), (2.31), and (2.33) again lead to

$$\tau = (q-1)D_q = q\alpha(q) - f(q) . \quad (2.34)$$

We can also understand the spectrum of scaling indices α by considering the "kneading sequences" for the segments. In the first level of the construction there are two segments of sizes l_1 and l_2 and measures p_1 and p_2 which we can label L (left) and R (right). At the next level we have four segments, which we can reach by going left or right: LL, LR, RL, and RR. Thus the measure and the size of any segment are determined by its address, the kneading sequence of L's and R's. For example, the size of a segment is $l_1^m l_2^{(n-m)}$, where m and $n-m$ are, respectively, the numbers of L's and R's in the kneading sequence. Clearly, the sequence $LLL\ldots LLL\ldots$ is associated with $\alpha = \ln(p_1)/\ln(l_1) = D_{\infty}$, which lies on the edge of the spectrum, while the sequence $RRR\ldots RRR\ldots$ is associated with the singularity lying on the other edge of the spectrum. Other, less trivial kneading sequences lead to values of α between these two extremes. We note, however, that it is only the infinite "tail" of the sequence that determines the asymptotic scaling behavior. The number of sequences leading to the same singularity α may be simply found, and leads via Eq. (2.30) to exactly the same results for $f(\alpha)$ as the partition-function analysis above.

Finally, in Fig. 4 we display the curve $f(\alpha)$. The curve has been obtained for $l_1 = 0.25$, $l_2 = 0.4$ and $p_1 = 0.6$, $p_2 = 0.4$. The leftmost point on the curve is $f = 0, \alpha = \ln(0.6)/\ln(0.25)$. This is the value that in Eqs. (2.31) and (2.33) obtains for $n = m$. At any level of the construction there is exactly one such segment ($f = 0$) and the singularity is

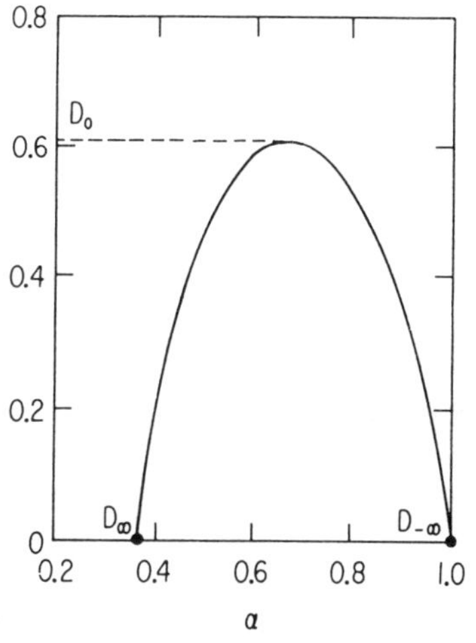

FIG. 4. The plot of f vs α for the set in Fig. 2. Note that $f=0$ corresponds to α values $D_{-\infty} = \ln(0.4)/\ln(0.4) = 1.0$ and $D_{\infty} = \ln(0.6)/\ln(0.25) = 0.3684$.

$$\ln p_1 / \ln l_1 = \text{Inf}\{\ln p_1/\ln l_1, \ln p_2/\ln_2\} .$$

This value of α is also D_{∞}. The rightmost point on the graph again corresponds to $f = 0$, but now

$$\alpha = \ln p_2 / \ln l_2 = \text{Sup}\{\ln p_1/\ln l_1, \ln p_2/\ln l_2\} .$$

This is also $D_{-\infty}$. Whereas D_{∞} corresponds to the region in the set where the measure is most concentrated, $D_{-\infty}$ corresponds to that where the measure is most rarefied. For $q = 0$ we simply obtain $f = D_0$, where D_0 is the Hausdorff dimension of the set. This is the maximum of the graph $f(\alpha)$.

Certain features of this curve are quite general, and follow from Eqs. (1.9)–(1.12). From Eq. (1.9) we find immediately that

$$\frac{\partial f}{\partial \alpha} = q , \quad (2.35a)$$

$$\frac{\partial^2 f}{\partial \alpha^2} < 0 . \quad (2.35b)$$

Thus, for any measure the curve $f(\alpha)$ will be convex, with a single maximum at $q = 0$, and with infinite slope at $q = \pm\infty$. Also from Eq. (1.10) with $q = 1$, we find that $\alpha(1) = f(1)$. The slope $\partial f/\partial \alpha$ there is unity. This general behavior of the curve $f(\alpha)$ will be seen in all cases where the measure possesses a continuous spectrum.

Although this example is rather simple, it contains many of the properties of the richer sets considered in Sec. III. In particular, we will not lose this intuitive view of the meaning of α and f.

5. *Other types of spectra*

We can obtain more insight into the meaning of the f-α spectrum for a measure by considering two examples of measures on continuous supports. Many of the most interesting measures encountered in applications lie on continuous supports, including the growth measure for diffusion-limited aggregates and strange attractors for systems of ordinary differential equations.

The first example is a simple generalization of the two-scale Cantor set defined by (2.24). A unit interval is subdivided into three segments, two of length l_2 and one of length l_1. The two former intervals each receive a proportion of the total measure given by p_2, and the latter interval receives a proportion given by p_1. We imagine that $l_1+2l_2=1$ and that $p_1+2p_2=1$. We also imagine, for the sake of the argument below, that $p_2/l_2>p_1/l_1$ and that $l_2>l_1$. Each of these three intervals is then subdivided in the same manner, and so forth. Although the measure on the line segment is rearranged at each step of the recursive process, the support for the measure remains at each step the original line segment. Thus we expect that D_0 for this measure will be 1. Furthermore, the densest intervals on the line segment contract not to one point (as was the case in the two-scale Cantor set), but to a set of points of finite dimension. Thus, we expect the lowest value of α, and hence the value of D_∞, to correspond to a nonzero value of f. Note that there is always only one segment at the lowest value of the density, so that we still expect $D_{-\infty}$ to correspond to a value of $f=0$. The condition (2.18) above on Γ requires that

$$\Gamma(q,\tau,l_2)=\frac{p_1^q}{l_1^\tau}+2\frac{p_2^q}{l_2^\tau}=1\ . \tag{2.36}$$

The solution is simple and is displayed in Fig. 5. As predicted above, $f(q\to\infty)\neq 0$, so that the leftmost part of the f-α curve resembles a hook.

The second example is a set generated according to a different rule than the Cantor sets. The method is displayed in Fig. 6. At each stage, only the regions which

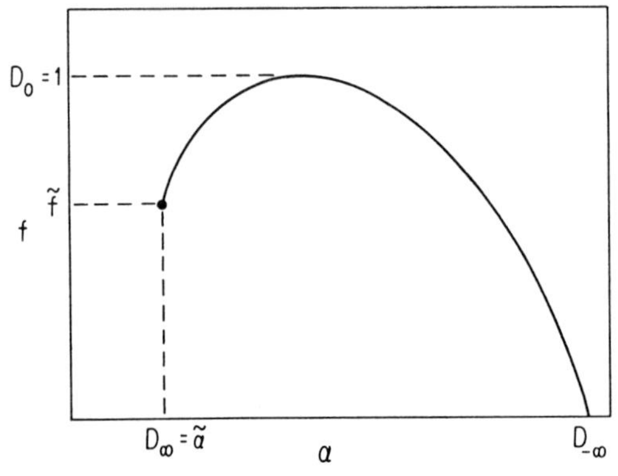

FIG. 5. The function $f(\alpha)$ for the measure defined by Eq. (2.36). Note that D_∞ corresponds to a nonzero value of f. Also, $D_0=1$.

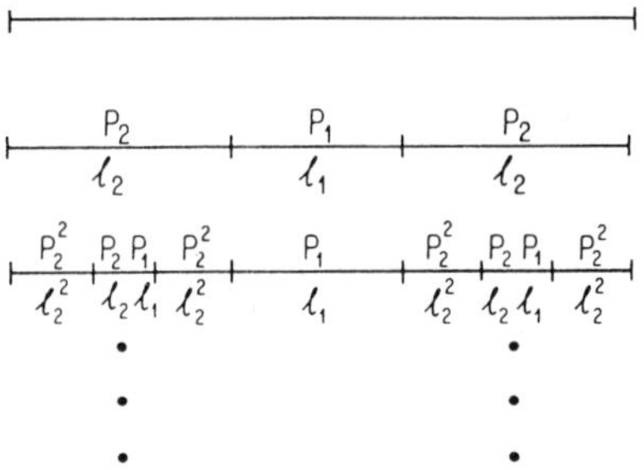

FIG. 6. The partitioning process for the measure yielding the partition function (2.37). Only those segments receiving a measure multiplied by p_2 at any stage of the construction are further subdivided. This measure is far less self-similar than those generated by the Cantor process.

have had their measure multiplied by a factor p_2 in the preceding stage are subdivided further, while the regions which have had their measure multiplied by a factor p_1 are not subdivided further. Thus the expression for the measure density of any region, at any stage of the iterative construction, will have, at most, one factor of p_1. The measure generated by this construction is much less self-similar than that considered in Sec. III C 4. For this measure the partition function is given for large n by

$$\Gamma(q,\tau,l_2^n)=(p_1^q/l_1^\tau)\Gamma^U(l_2^{n-1})+2(p_2^q/l_2^\tau)\Gamma(l_2^{n-1})\ , \tag{2.37}$$

where Γ^U is the partition function for a uniform measure on a line segment. It is easy to show that

$$\tau(q)=\min\{q-1,q\tilde{\alpha}-\tilde{f}\}\ , \tag{2.38}$$

with $\tilde{\alpha}=\ln(p_2)/\ln(l_2)$, and $\tilde{f}=\ln(\frac{1}{2})/\ln(l_2)$. This example corresponds to a discrete, rather than a continuous, f-α curve, consisting of a point at $(\tilde{\alpha},\tilde{f})$ and a point at (1,1). This result should not surprise us, as this measure is properly described as a nonsingular background interrupted by singularities upon a Cantor set of dimension $\tilde{f}$.

III. EXAMPLES FROM DYNAMICAL SYSTEMS

In this section we examine the implications of the formalism of Sec. II for three examples: (i) the 2^∞ cycle at the accumulation point of period doubling, (ii) the set of irrational winding numbers at the onset of chaos via quasiperiodicity, and (iii) the critical cycle elements at the golden-mean winding number for the same problem. In all cases we calculate numerically the D_q, and use Eqs. (1.10) and (1.11) to extract $\alpha(q)$, $f(q)$, and a plot of $f(\alpha)$. In all three cases we can find theoretically $D_\infty,D_{-\infty}$, and thus $\alpha(q=\pm\infty)$.

A. The 2^∞ cycle of period doubling

Dynamical systems that period double on their way to chaos can be represented by one-parameter families of maps $M_\lambda(\mathbf{x})$, where M_λ: $R^F\to R^F$, and F is the number of degrees of freedom. At values of $\lambda=\lambda_n$ the system

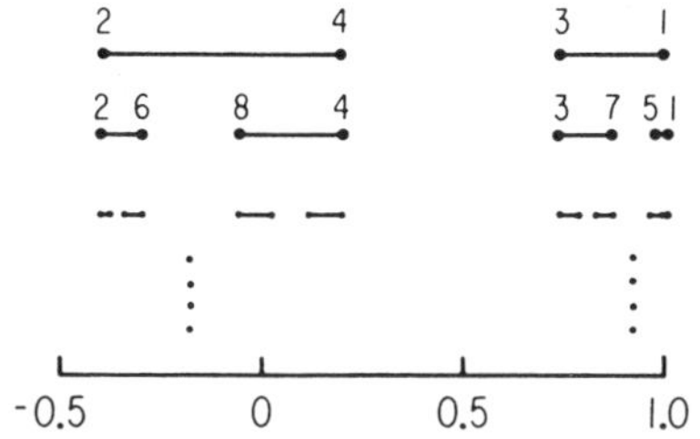

FIG. 7. The construction of the period-doubling attractor; the indices refer to the number of the iterate of $x=0$. The lines represent the scales l_i. Note the similarity with Fig. 2.

gains a stable 2^n-periodic orbit. This period-doubling cascade accumulates at λ_∞, where the system possesses a 2^∞ orbit. We generated numerically the set of elements of this orbit for the map $x'=\lambda(1-2x^2)$, with $\lambda_\infty \approx 0.837\,005\,134\ldots$.[15] The points making up the cycle are displayed in Fig. 7. The iterates of $x=0$ form a Cantor set, with half the iterates falling between $f(0)$ and $f^3(0)$ and the other half between $f^2(0)$ and $f^4(0)$. The most natural partition, $\{S_i\}$, for this case simply follows the natural construction of the Cantor set as shown in Fig. 7. At each level of the construction of this set, each l_i is the distance between a point and the iterate which is closest to it. The measures p_i of these intervals are all equal.

With 2^{11}-cycle elements we solved numerically $\Gamma=1$, thereby generating the D_q-versus-q curve shown in Fig. 8. From these results we calculated $\alpha(q)$ from Eq. (1.11) and $f(\alpha)$ from (1.10). The curve $f(\alpha)$ is displayed in Fig. 9.

To understand the shape of the curve in Fig. 9 we first consider the end points of the curve (for which $f=0$). As with the example solved in Sec. II C 4, we expect these two points to be determined by the most rarefied and the most concentrated intervals in the set. As has been shown by Feigenbaum,[15] these have scales $l_{-\infty}\sim\alpha_{\text{PD}}^{-n}$ and $l_{+\infty}\sim\alpha_{\text{PD}}^{-2n}$, respectively, where $\alpha_{\text{PD}}=2.502\,907\,875\ldots$ is the universal scaling factor.[15] Since the measures there

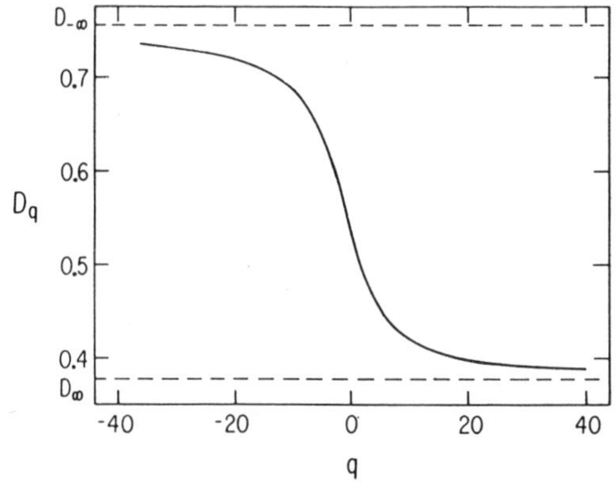

FIG. 8. D_q vs q calculated for the period-doubling attractor of Fig. 7.

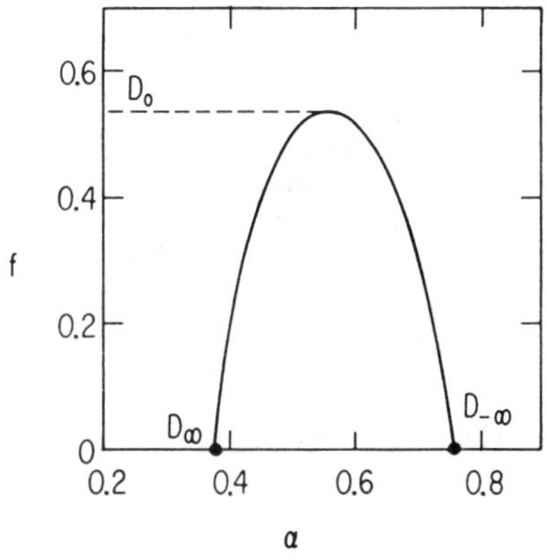

FIG. 9. The function $f(\alpha)$ for the period-doubling attractor of Fig. 7.

are simply $p_{-\infty}\sim 2^{-n}$, we expect these end points to be $\ln p_{-\infty}/\ln l_{-\infty}$ and $\ln p_\infty/\ln l_\infty$, respectively. These values are also $D_{-\infty}$ and D_∞, so that we find

$$D_{-\infty}=\frac{\ln 2}{\ln\alpha_{\text{PD}}}=0.755\,51\ldots, \tag{3.1a}$$

$$D_{\infty}=\frac{\ln 2}{\ln\alpha_{\text{PD}}^2}=0.377\,75\ldots. \tag{3.1b}$$

These values are in extremely good agreement with the numerically determined endpoints of the graph. The curve $f(\alpha)$ is perfectly smooth. The maximum is at $D_0=0.537\ldots$, in agreement with previous calculations of the Hausdorff dimension for this set. Since the slope of the curve $f(\alpha)$ is q, $\alpha(q)$ will be very close to $D_{\pm\infty}$ even for $|q|\sim 10$. However, Fig. 8 indicates that D_q is far from converged to $D_{\pm\infty}$ even for $q\sim\pm 40$. Thus, the transformations (1.10) and (1.11) lead more easily to good estimations of $D_{\pm\infty}$ than do direct calculations of the D_q's.

B. Mode-locking structure

Dynamical systems possessing a natural frequency ω_1 display very rich behavior when driven by an external frequency ω_2. When the "bare" winding number $\Omega=\omega_1/\omega_2$ is close to a rational number, the system tends to mode lock. The resulting "dressed" winding number, i.e., the ratio of the response frequency to the driving frequency, is constant and rational for a small range of the parameter Ω. At the onset of chaos the set of irrational dressed winding numbers is a set of measure zero, which is a strange set of the type discussed above. The structure of the mode locking is best understood in terms of the "devil's staircase" representing the dressed winding number as a function of the bare one. Such a staircase is shown in Fig. 10 as obtained for the map[16,17]

$$\theta_{n+1}=\theta_n+\Omega-\frac{K}{2\pi}\sin(2\pi\theta_n), \tag{3.2}$$

with $K=1$, which is the onset value above which chaotic orbits exist.

To calculate $D_{\pm\infty}$ analytically we make use of previous findings that the most extremal behaviors of this staircase are found at the golden-mean sequence of dressed winding numbers

$$F_n/F_{n+1} \to w^* = (\sqrt{5}-1)/2 \approx 0.6108\ldots ,$$

where F_n are the Fibonacci numbers, ($F_0=0$, $F_1=1$, and $F_n=F_{n-1}+F_{n-2}$ for $n \geq 2$) and at the harmonic sequence $1/Q \to 0$.[16,17] The most rarefied region of the staircase is located around the golden mean. Shenker found that the length scales l_i vary in that neighborhood as $l_{-\infty} \sim F_n^{-\delta} \sim (w^*)^{n\delta}$, where $\delta = 2.1644\ldots$ is a universal number.[18] The corresponding changes in dressed winding number are

$$p_{-\infty} \sim F_n/F_{n+1} - F_{n+1}/F_{n+2} \sim (w^*)^{2n} .$$

We thus conclude that

$$D_{-\infty} = \frac{\ln p_{-\infty}}{\ln l_{-\infty}} = \frac{2}{\delta} = 0.9240\ldots . \tag{3.3a}$$

For the $1/Q$ series it has been shown that changes in dressed winding number go as the square root of changes in bare winding number, i.e., that $p_i \sim l_i^{1/2}$.[16] This series determines the most concentrated portion of the staircase (Fig. 10), which means that $p_\infty \sim l_\infty^{1/2}$, leading to

$$D_\infty = \ln p_\infty / \ln l_\infty = \tfrac{1}{2} . \tag{3.3b}$$

To construct the curve $f(\alpha)$ we generated 1024 mode-locked intervals following the Farey construction, which also defines the partition $\{S_i\}$.[17] For each two neighboring intervals (see Fig. 10) we measured the change both in bare and in dressed winding numbers. The changes in bare winding numbers determined the scales l_i of the partition $\{S_i\}$, whereas the changes in dressed winding numbers were defined to be the measures p_i. Solving then the equation $\Gamma=1$ we generated D_q as shown in Fig. 11 (for $q>0$ we accelerated the convergence as will be described shortly). Figure 12 shows $f(\alpha)$ for this case. Again the curve is smooth, in contrast to scaling functions found for

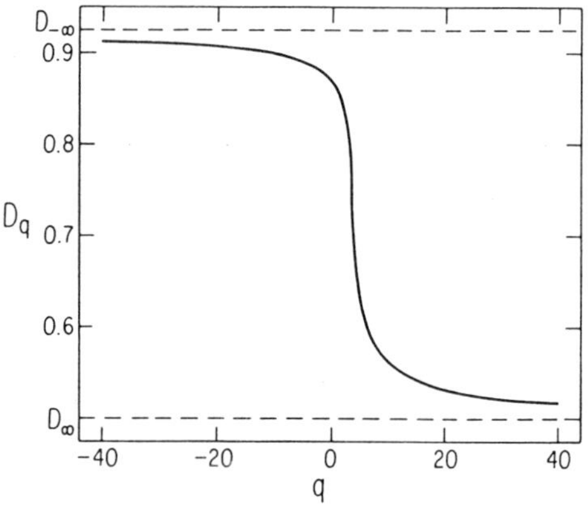

FIG. 11. D_q vs q for the staircase of Fig. 10.

the same problem by other authors.[15,21] Note that the maximum on Fig. 12 gives the fractal dimension D_0 of the mode-locking structure as $D_0 \approx 0.87\ldots$, in agreement with the predictions of Refs. 16 and 17. The rightmost branch of the curve $f(\alpha)$ in Fig. 12 (i.e., for $q<0$) converges vary rapidly within the Farey partition. This is, however, not the case for the leftmost branch (i.e., for $q>0$). To improve the convergence of this portion of the curve substantially, we made use of the following trick. In general, the partition function (2.1) will be of the form

$$\Gamma(l) = a e^{\gamma \ln l} , \tag{3.4}$$

where a and γ are constants. The convergence is often slowed down by the prefactor a and by the logarithmic dependence on l. However, by considering instead the ratio

$$\frac{\Gamma(l)}{\Gamma(2l)} = e^{-\gamma \ln 2} , \tag{3.5}$$

we find that a and l do not appear in the equation. We thus determine $\tau(q)$ by requiring that

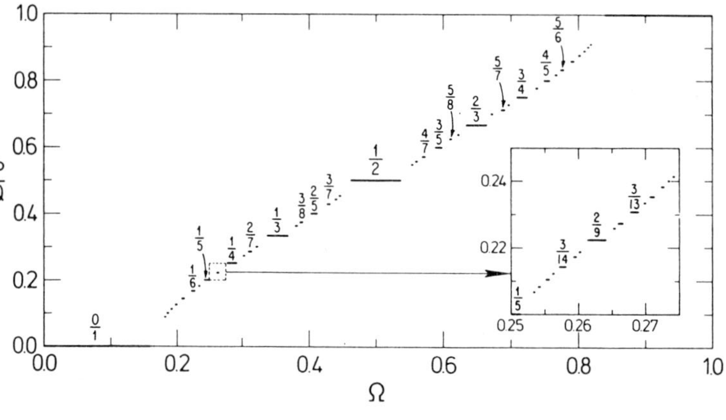

FIG. 10. The "devil's staircase" for the critical circle map of Eq. (3.2). The "dressed" winding number is plotted vs the "bare" winding number (Ref. 16).

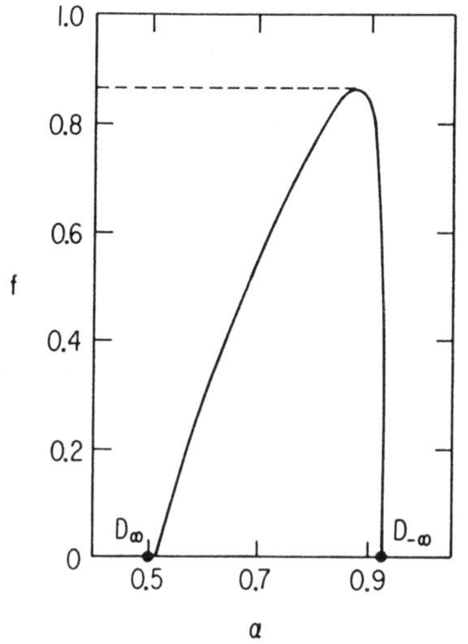

FIG. 12. A plot of f vs α for the mode-locking structure of the circle map. The left portion of the curve is found by accelerated convergence as described in Sec. III B.

$$\frac{\Gamma(l)}{\Gamma(2l)}(\tau,q)=1\ .$$

In general, the denominator can be chosen to be of the form $\Gamma(bl)$, where b is a constant. The leftmost portion of the curve was generated with this method by calculating $\Gamma(l(1452))/\Gamma(l(886))=1$ [where $l(1452)$ and $l(886)$ are the maximal scales for partitions with 1452 and 886 intervals, respectively], and we observe that it passes through the point $(D_\infty,0)$. We found empirically that this method usually did not give reliable results for large values of $|q|$. Still, this method did successfully generate the entire curve in Fig. 12. We emphasize the ease of this measurement. The rightmost branch of the f-α curve of Fig. 12 converges very rapidly, even when only 8–16 mode-locked intervals are available.

C. Quasiperiodic trajectories for circle maps

Circle maps of the type (3.1) exhibit a transition to chaos via quasiperiodicity. A well-studied transition takes place at $K=1$ with dressed winding number equal to the golden mean, w^*.[18–20] We have at this point studied the structure of the trajectory $\theta_1,\theta_2,\ldots,\theta_i,\ldots$. To perform the numerical calculation we chose $\theta_1=f(0)$ and truncated the series θ_i at $i=2584=F_{17}$. The distances $l_i=\theta_{i+F_{16}}-\theta_i$ (calculated mod 1) define natural scales for the partition with measures $p_i=1/2584$ attributed to each scale. Figure 13 shows D_q versus q calculated for this set and Fig. 14 shows the corresponding function $f(\alpha)$. Again the curve is smooth. Shenker found for this problem that the distances around $\theta\sim 0$ scale down by a universal factor $\alpha_{\rm GM}=1.288\,5\ldots$ when the trajectory θ_i is truncated at two consecutive Fibonacci numbers,

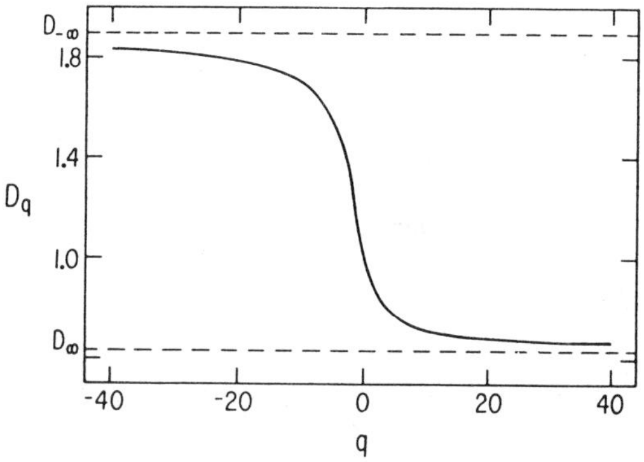

FIG. 13. D_q plotted vs q for the critical trajectory of a circle map with golden-mean winding number.

F_n,F_{n+1}.[18] This corresponds to the most rarefied region so that $l_{-\infty}\sim\alpha_{\rm GM}^{-n}$. The corresponding measure scales as $p_{-\infty}\sim 1/F_n\sim(\omega^*)^n$, leading to

$$D_{-\infty}=\frac{\ln w^*}{\ln\alpha_{\rm GM}^{-1}}=1.898\,0\ldots\ . \tag{3.6a}$$

The map (3.2) for $K=1$ has at $\theta=0$ a zero slope with a cubic inflection and is otherwise monotonic. The neighborhood around $\theta=0$, which is the most rarefied region of the set, will therefore be mapped onto the most concentrated region of the set. As the neighborhood around $\theta=0$ scales as $\alpha_{\rm GM}$ when the Fibonacci index is varied, the most concentrated regime will scale as $\alpha_{\rm GM}^3$ due to the cubic inflection. This means that $l_\infty=\alpha_{\rm GM}^{-3n}$ and $p_\infty=(w^*)^n$, so that we obtain

$$D_\infty=\frac{\ln w^*}{\ln\alpha_{\rm GM}^{-3}}=0.632\,6\ldots\ . \tag{3.6b}$$

Figure 14 shows that the curve passes very close to the points $(D_\infty,0)$ and $(D_{-\infty},0)$. Again, however, we find

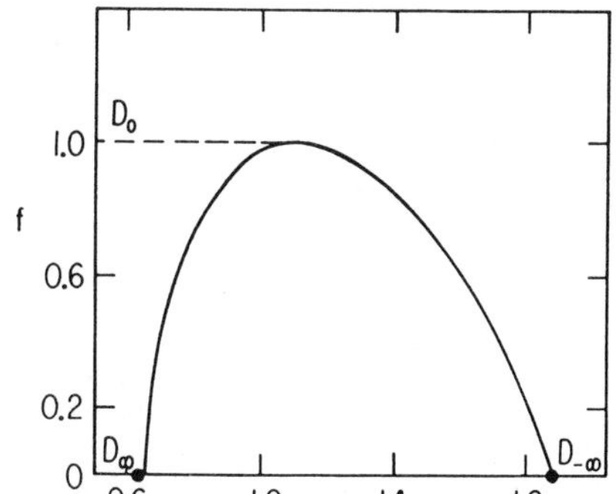

FIG. 14. A plot of f vs α for the golden-mean trajectory for the circle map.

that the dimensions D_q are far from $D_{\pm\infty}$ even for $q\sim\pm40$.

To check for universality it is important to investigate $f(\alpha)$ for a higher-dimensional version of a circle map. We chose the dissipative standard map,

$$\theta_{n+1}=\theta_n+\Omega+br_n-\frac{K}{2\pi}\sin(2\pi\theta_n)\ ,$$
$$r_{n+1}=br_n-\frac{K}{2\pi}\sin(2\pi\theta_n)\ , \tag{3.7}$$

and studied the critical cycle for $b=0.5$, again truncated at $i=2584=F_{17}$. We defined the scales by the Euclidean distances

$$l_i=[(\theta_{i+F_{16}}-\theta_i)^2+(r_{i+F_{16}}-r_i)^2]^{1/2}\ . \tag{3.8}$$

We found that the convergence for the two-dimensional (2D) case was slightly slower than for the one-dimensional (1D) case. This is, however, to be expected since it was found by Feigenbaum, Kadanoff, and Shenker that the convergence of the scaling number $\alpha_{\rm GM}$ is slower for the 2D case than for the 1D case.[19] To improve the convergence we again made use of the ratio trick as embodied in Eqs. (3.4) and (3.5). For this case we calculated the partition function for two consecutive Fibonacci numbers, $F_{16}=1597$ and $F_{17}=2584$, and found τ from the requirement

$$\frac{\Gamma(l(F_{17}))}{\Gamma(l(F_{16}))}(\tau,q)=1$$

[$l(F_i)$ are the maximal scales for the partitions]. This improves the convergence significantly, and the f-α curve for this 2D case coincides almost completely with the curve found for the 1D case and displayed in Fig. 14.

IV. CONCLUSION

Most previous characterizations of strange sets arising in physics have followed the example of critical phenomena in relying upon a few universal numbers to characterize the physical systems generating these sets. Thus, strange attractors are characterized by their Hausdorff dimensions, or by the scaling exponents of particularly divergent regions of their measure. However, these numbers reflect only a small part of the universal scaling structure of these systems. Feigenbaum introduced scaling functions in order to describe the complex scaling properties of attractors at the onset of chaos.[15] These scaling functions contain all of the geometric information about the attractor, in contrast to the partial information furnished by local scaling exponents. These functions are, however, nowhere differentiable, and are thus very difficult to use. The full complexity of this scaling structure is more conveniently reflected by the continuous spectrum of exponents α and their densities $f(\alpha)$, of which previously investigated scaling exponents and Hausdorff dimensions comprise only a part.

Not only does this spectrum enrich our conceptual vocabulary, it should enrich our experimental vocabulary as well. The numerical studies of Sec. III were straightforward and did not require large investments of computer time in order to obtain extremely accurate results. Furthermore, this spectrum can be measured, and has been measured, in experiments upon physical realizations of dynamical systems.[23] The measurement of this spectrum should result in new tests of scaling theories of nonlinear systems.

ACKNOWLEDGMENTS

We would like to thank P. Jones and J. Rudnick for stimulating discussions. This work was supported by the U.S. Office of Naval Research, by the National Science Foundation through Grant No. DMR-83-16626, and through the Materials Research Laboratory of the University of Chicago. One of us (I.P.) wishes to thank Professor S. Berry for his warm hospitality and also acknowledge the support of the Minerva Foundation (Munich, West Germany).

*Permanent address: Department of Chemical Physics, The Weizmann Institute of Science, 76100 Rehovot, Israel.

†Present address: AT&T Bell Laboratories, 600 Mountain Avenue, Murray Hill, NJ 07974.

[1]R. Bowen, *Equilibrium States and the Ergodic Theory of Anisov Diffeomorphisms*, Vol. 470 of *Lectures Notes in Mathematics* (Springer, Berlin, 1975); M. Widom, D. Bensimon, L. P. Kadanoff, and Scott J. Shenker, J. Stat. Phys. **32**, 443 (1983).

[2]I. Procaccia, in Proceedings of the Nobel Symposium on Chaos and Related Problems [Phys. Scr. T 9, 40 (1985)].

[3]K. G. Wilson, Sci. Am. **241**, (2) 158 (1979).

[4]B. B. Mandelbrot, Ann. Isr. Phys. Soc. **225** (1977); H. Aref and E. D. Siggia, J. Fluid Mech. **109**, 435 (1981).

[5]I. Procaccia, J. Stat. Phys. **36**, 649 (1984).

[6]L. de Arcangelis, S. Redner, and A. Coniglio, Phys. Rev. B **31**, 4725 (1985).

[7]T. A. Witten, Jr. and L. M. Sander, Phys. Rev. Lett. **47**, 1400 (1981).

[8]T. C. Halsey, P. Meakin, and I. Procaccia (unpublished).

[9]B. Widom, J. Chem. Phys. **43**, 3892 (1965); M. E. Fisher, in Proceedings of the University of Kentucky Centennial Conference on Phase Transitions (March 1965) (unpublished); A. Z. Patashinskii and V. L. Prokovskii, Zh. Eksp. Teor. Fiz. **50**, 439 (1966) [Sov. Phys.—JETP **23**, 292 (1966)]; L. P. Kadanoff, W. Gotze, D. Hamblen, R. Hecht, E. A. S. Lewis, V. V. Palciauskas, M. Rayl, J. Swift, D. Aspnes, and J. Kane, Rev. Mod. Phys. **39**, 395 (1967).

[10]L. P. Kadanoff, Physics 2, 263 (1966).

[11]In fact, for many two-dimensional critical phenomena we know [see, for instance, D. Friedan, Z. Qiu, and Stephen Shenker, Phys. Rev. Lett. **52**, 1575 (1984)] that there is only a finite set of dist t y's.

[12]L. Turkevich and H. Scher, Phys. Rev. Lett. **55**, 1024 (1985).

[13]H. G. E. Hentschel and I. Procaccia, Physica **8D**, 435 (1983).

[14]P. Grassberger and I. Procaccia, Physica **13D**, 34 (1984).

[15]M. J. Feigenbaum, J. Stat. Phys. **19**, 25 (1978); **21**, 669 (1979).

See also the two reprint compilations: P. Cvitanović, *Universality in Chaos* (Hilger, Bristol, 1984); Hao Bai-lin, *Chaos* (World Scientific, Singapore, 1984).

[16]M. H. Jensen, P. Bak, and T. Bohr, Phys. Rev. Lett. **50**, 1637 (1983); Phys. Rev. A **30**, 1960 (1984); **30**, 1970 (1984).

[17]P. Cvitanović, M. H. Jensen, L. P. Kadanoff, and I. Procaccia, Phys. Rev. Lett. **55**, 343 (1985).

[18]Scott J. Shenker, Physica **5D**, 405 (1982).

[19]M. J. Feigenbaum, L. P. Kadanoff, and Scott J. Shenker, Physica **5D**, 370 (1982).

[20]S. Ostlund, D. Rand, J. P. Sethna, and E. D. Siggia, Phys. Rev. Lett. **49**, 132 (1982); Physica **8D**, 303 (1983).

[21]P. Cvitanović, B. Shraiman, and B. Soderberg (unpublished).

[22]A. Cohen and I. Procaccia, Phys. Rev. A **31**, 1872 (1985), and references therein.

[23]M. H. Jensen, L. P. Kadanoff, A. Libchaber, I. Procaccia, and J. Stavans, Phys. Rev. Lett. **55**, 2798 (1985).

reference frame

Fractals: Where's the physics?

Leo P. Kadanoff

Why all the fuss about fractals? *Physical Review Letters* complains that every third submission seems to concern fractals in some way or another. Corporate research labs such as Exxon's and IBM's expend perceptible fractions of their entire basic-research budgets on the study of fractal systems. Perhaps a half-dozen conferences during the past year were devoted to the subject. Why?

But first what. What are fractals? Different people use the word *fractal* in different ways, but all agree that fractal objects contain structures nested within one another like Chinese boxes or Russian dolls as, for example, in figure 1. This Sierpinski gasket consists of triangles within triangles and so on to the finest level. The littlest triangles, in color, are hard to draw but easy to describe: They are little copies of the entire object. Some fractal objects are more random in structure. The cover shows a "tree" produced by a computer algorithm called DLA, for diffusion-limited aggregation. The basic DLA algorithm was introduced by Thomas A. Witten and Leonard M. Sander in *Physical Review Letters* **47**, 1400 (1981). In DLA, the tree grows unit by unit by the following process: A particle is inserted above the tree with a randomly chosen x-coordinate. The particle then undergoes a random walk in which each step is to a neighboring lattice site; the choice of neighbor is determined by a "throw of dice." The walk continues until the particle reaches a site neighboring the tree. It then stops walking, the tree grows one unit by the addition of that particle and the entire process continues with the insertion of a new particle. Figure 2 shows another example, this time experimental. The basic units are little gold balls, and the process is essentially the same, except the real-world object is three-dimensional while the computer object on the cover is two-dimensional. One final example: Figure 3 shows a top view of a block of Lucite 2 cm thick; monoenergetic MeV electrons have been shot into it and stopped within a thin layer perhaps 1 cm below the surface. After the injection is completed, a nail is inserted at the side and the electrons shoot out, leaving a trail shaped like a river system behind them.

A fractal object. For many other examples of such objects see Benoit Mandelbrot's book *The Fractal Geometry of Nature* (Freeman, New York, 1983). Figure 1

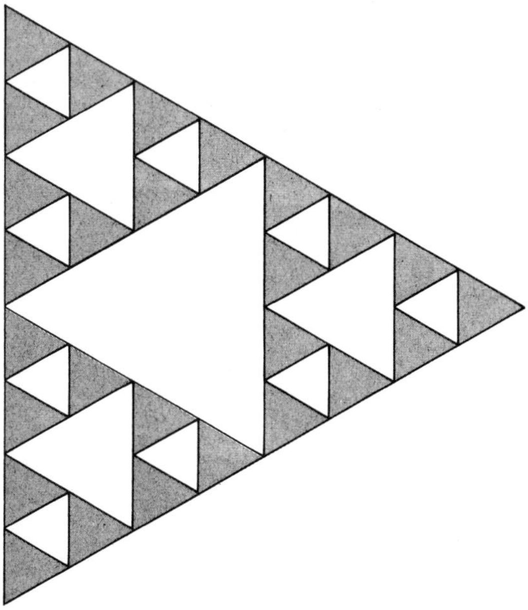

One reason for our interest in fractal objects is their practical importance. Materials scientists want to produce entirely new structures with entirely new properties, like the sponge of figure 2. When this kind of object is grown to be very large it has huge empty spaces and a density that decreases roughly as $\rho = \rho_0 (R/a)^{-1/2}$, where R/a is the ratio of the radius of the object to that of the balls. Dielectric breakdown, shown in figure 3, is of considerable technical importance. As a final example, consider petroleum-bearing rock layers. These typically contain fluid-filled pores of many sizes, which might be effectively understood as a fractal object.

The technical interest of fractals is matched by their intellectual interest. Two of the fundamental symmetries of nature are dilation ($\mathbf{r} \to a\mathbf{r}$) and translation ($\mathbf{r} \to \mathbf{r} + \mathbf{b}$). We can represent them verbally by talking about a change in our unit of length or in the origin of our coordinate system. Fractal objects are highly nontrivial representations of these symmetries. Thus, for example, an expanded piece of the Sierpinski gasket of figure 1 can be moved in such a way as to make it coincide with the entire gasket—and this operation can be performed in an infinite number of ways. The more random fractals of figures 2 and 3 probably obey similar combined translation–dilation symmetries.

Another example of dilation-symmetric behavior that has been studied in great detail involves the critical phenomena that arise near "second-order" phase transitions. The phenomenology of these transitions has been

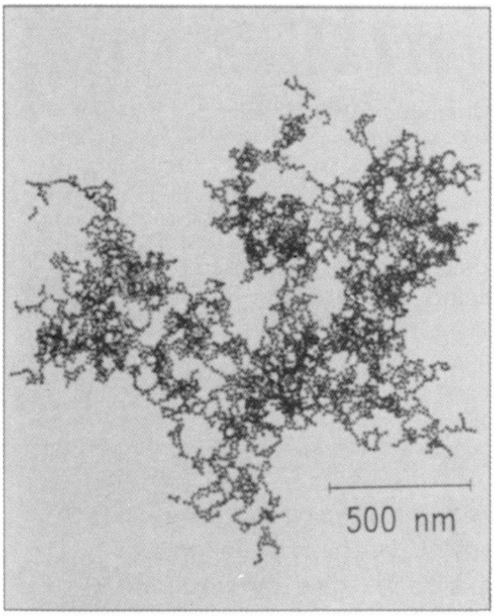

A laboratory-grown fractal produced by David A. Weitz and M. Oliveria (*Physical Review Letters* **52**, 1433 1984). (Photo courtesy of Exxon.) Figure 2

A "beam tree" produced by dielectric breakdown. Object courtesy of SLAC and Sidney Drell. (Photo by Oscar Kapp.) Figure 3

elaborated in considerable detail with the aid of the concept of *universality*, that is, the notion that disparate phase transitions may exhibit quantitatively identical behavior. This concept might be extended to processes that produce fractals. For example, might the fine details of dielectric-breakdown patterns formed in different contexts be identical? Moreover, might real breakdowns in two dimensions (see figure 3) be identical with theoretical models of step-by-step growth, perhaps even with the model simulated on the cover?

One way of answering such questions is to perform appropriate measurements on the objects in question. One quantity commonly measured is the fractal dimension. (This dimension is a variant of a concept due to the mathematician Felix Hausdorff.) The fractal dimension is defined as $d \ln M(R)/d \ln R$, where $M(R)$ is the mass contained within a distance R from a typical point in the object. If two objects are the same they must at least have the same fractal dimension.

Unfortunately, although this single, rather primitive measurement enables us to distinguish among objects, it never enables us to give a convincing case for their essential identity. Some progress has been made in identifying other qualities, beyond the fractal dimension, that might be universal. However, further progress in this field depends upon establishing a more substantial theoretical base in which geometrical form is deduced from the mechanisms that produce it. Lacking such a base, one cannot define very sharply what *types of questions* might have interesting answers. One might hope, and even expect, that eventually a theoretical underpinning—like that of Kenneth Wilson's renormalization approach—will be developed to anchor this subject.

Without that underpinning much of the work on fractals seems somewhat superficial and even slightly pointless. It is easy, too easy, to perform computer simulations upon all kinds of models and to compare the results with each other and with real-world outcomes. But without organizing principles, the field tends to decay into a zoology of interesting specimens and facile classifications. Despite the beauty and elegance of the phenomenological observations upon which the field is based, the physics of fractals is, in many ways, a subject waiting to be born. □

Nonlinear Phenomena in Fluids, Solids
and Other Complex Systems
P. Cordero, B. Nachtergaele (Editors)
Elsevier Science Publishers B.V., 1991

SCALING AND MULTISCALING (FRACTALS AND MULTIFRACTALS)

Leo P KADANOFF

The Research Institutes, The University of Chicago, 5640 S. Ellis Avenue, Chicago, Il, 60637

Many different physical situations can be described by multifractal distributions. A general framework is presented. Several specific examples are discussed.

1. INTRODUCTION: SIMPLE SCALING

For many years, we have been used to describing physical situations in terms of a scaling analysis. Scaling is a variant of dimensional analysis in which we argue that there is a characteristic size or scale by which we can set the order of magnitude of things of interest. Hydrodynamics gives us many simple examples such scaling analysis.

A. Blasius profile

Consider for example the Blasius analysis of flow past a flat plate. In this situation, see Figure 1.1, a stream of fluid is flowing past a semi-infinite flat plate. The basic equations to describe this situation are the Navier Stokes equations:

$$\mathbf{u}_t + \mathbf{u}\cdot\nabla\, \mathbf{u} = -(\nabla\, p)/\rho\; + \nu\, \nabla^2\, \mathbf{u} \qquad (1.1)$$

Here $\mathbf{u}$ is the fluid velocity, p the pressure, ρ the density and ν the kinematic viscosity. This statement is $\mathbf{F}=m\mathbf{a}$, applied to the fluid mass with forces from pressure and viscous drag. The second of our equations is the condition for the conservation of mass in this fluid, which assumes that the fluid is incompressible. It is

$$\nabla\cdot\mathbf{u}\; = 0 \qquad (1.2)$$

We employ the usual boundary conditions, namely that $\mathbf{u}=0$ on solid boundary. In our particular problem we that the flow be steady and the $\mathbf{U}$ goes to a constant, (U,0) as y goes to ± infinity. We consider the flow to be two dimensional and completely neglect the third direction.

To see the structure of the result take the curl of Eq. (1.1) to find

$$(\mathbf{u}\cdot\nabla)\ \nabla\times\mathbf{u} = \nu\ \nabla^2\ \nabla\times\mathbf{u} \tag{1.3}$$

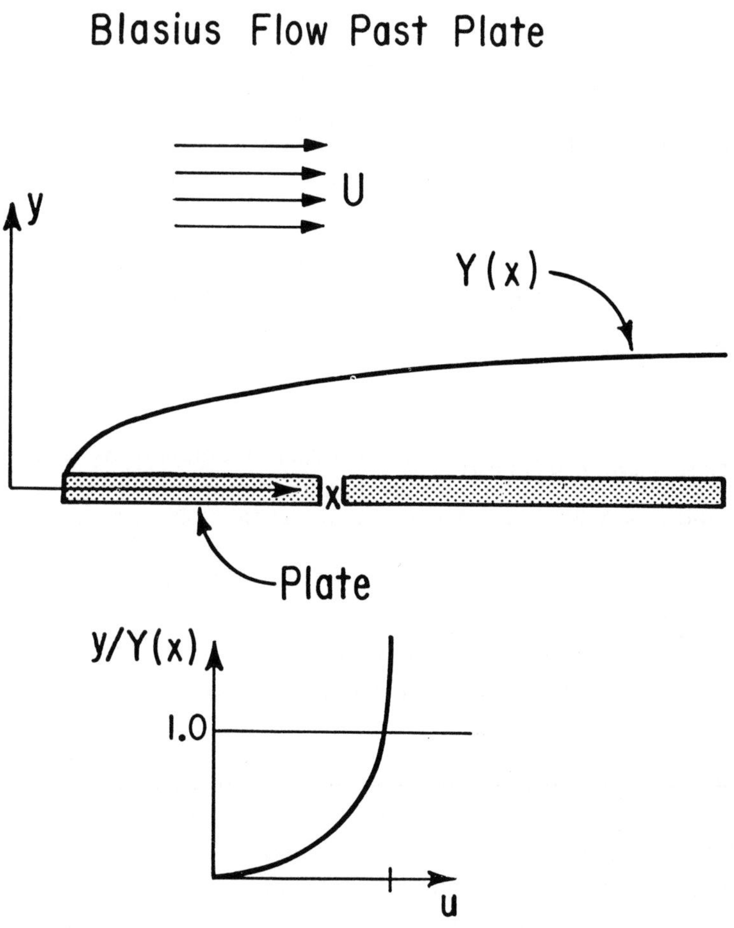

FIGURE 1.1.
Blasius Figure. This figure shows first the geometry of the situation described by Blasius and then the form of the velocity profile achieved.

The Blasius solution is based upon the idea that there is a characteristic scale of the y coordinate, which varies with x, the distance along the plate measured from its beginning. Therefore, every function of y will depend upon y in the form of functions of y/Y(x) where Y(x) is the characteristic scale of y. The scale of x is simply the distance from the leading edge of the plate. We look far downstream, where the scale of Y is much smaller than the scale of x. In this limit, one can neglect the $(\partial/\partial x)^2$ term in ∇^2 in comparison with the $(\partial/\partial y)^2$ term. Thence (1.3) reduces to

$$(\mathbf{u}\cdot\nabla)\ \nabla\times\mathbf{u} = \nu \frac{\partial^2}{\partial y^2} \nabla\times\mathbf{u} \tag{1.4}$$

To an order of magnitude one can estimate the size of a y-derivative as being proportional top $Y(x)^{-1}$, while an x derivative is of the order of the inverse x-scale x^{-1}. A typical value of u_x is just U. Thus, if the two terms in Eq (1.4) are to balance out, we must have that, to an order of magnitude:

$$U/x \sim \nu / Y(x)^2 \tag{1.5}$$

so that the y scale is

$$Y(x) = (\nu\, x / U)^{1/2} \tag{1.6}$$

Using this idea, one can in fact, construct a solution of (1.4) assuming that the x-component of the velocity has the form

$$u_x = U\, \Psi'(\ y/Y(x))$$

Then, the continuity equation, Eq (1.2) provides a solution for the y-component of the velocity, namely

$$u_y = -U\, Y'(x)\ [\Psi(\eta) - \eta\, \Psi'(\eta)]\ ,\ \eta = y/Y(x)$$

One then uses Eq. (1.4) to write down an ordinary differential equation for Ψ,

$$-\frac{\partial}{\partial\eta}(\Psi\, \Psi'') = 2 \frac{\partial^4}{\partial^4\eta} \Psi$$

which can be solved numerically.

The physical idea upon which this is all based is that the flow is the same for all x except for the change of scale, represented by the Y(x). Notice that the answer can be represented by simple power laws, for example that Y(x) is proportional to $x^{1/2}$. We see here that the simple power laws are an outcome of the idea that as one goes to larger x nothing changes except the scale of y.

B. Widom scaling[1]

A second example is provided by the Widom analysis of simple scaling behavior in critical phenomena. Consider a magnet near its critical point. There are three quantities of interest to us. Two of these measure the deviations from the critical point: a dimensionless magnetic field $h=\mu H/kT$ and a dimensionless measure of the deviation of the temperature, T, from its critical value, T_c. This second dimensionless quantity is then $t=\frac{T-T_c}{T_c}$. Near the critical point both of these are much smaller than unity. The statement of scaling is that the powers of the temperature deviation provide a characteristic scale of all physical quantities. For example, the magnetic field always appears in the theory in the combination h/t^{Δ}, where Δ is the critical index for the magnetic field. Correspondingly, the magnetization appears in the combination m/t^{β}, with β being another critical index. The content of this statement is then that the magnetization appears in the scaling form:

$$m(h,t) = t^{\beta} \, m^{*}\left(\frac{h}{t^{\Delta}}\right) \tag{1.7}$$

If h is of the same order of magnitude as t^{Δ} then m^* is of order one and m is of order t^{β}. One can check a statement like (1.3) experimentally. One takes data on m as a function of h and t. One then guesses values for β and Δ. Then one plots $m(h,t) / t^{\beta}$ on one axis and h/t^{Δ} on the other. If β and Δ have been correctly chosen, then all the data for different values of t should fall on a single curve. One adjusts the parameters to make this happen as closely as possible. Figure (1.2) shows the results of one such fitting. Notice that it really does work.

---

[1] See the description in Radu Balescu, Equilibrium and Nonequilibrium Statistical Mechanics, Wiley (1975).

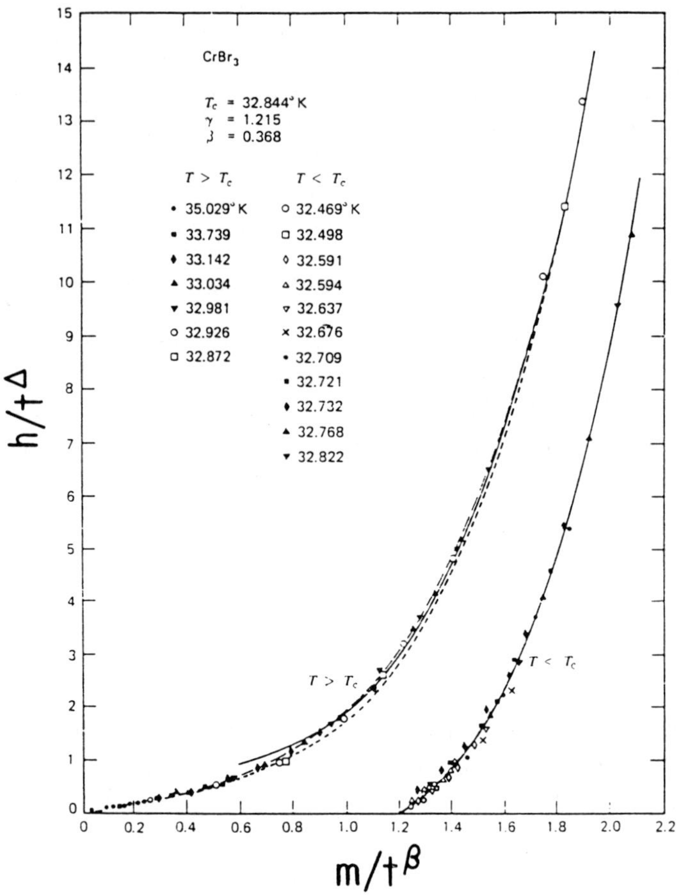

FIGURE 1.2

Experimental Data for scaled magnetic field versus scaled magnetization. The experiment was done by T.J. Ho and J.D. Litster, Phys. Rev. Letts **22**, 603 (1969). The fact that all the data fits on two curves (one for $T>T_c$, the other for $T<T_c$) shows that this systems obeys the Widom scaling hypothesis. (The figure is drawn from Balescu, ref 1).

C. Kolmogorov theory of turbulence[2]

Our third example is drawn from the Kolmogorov theory of turbulence. Imagine that one measures the velocity of of fluid in a

---

[2]See A.S. Monin and A.M. Yaglom, Statistical Fluid Mechanics, The MIT Press.

sensor moving rapidly through a turbulent fluid. The fluid is described by a Reynolds number $R= U\ L\ /\upsilon$ which is large. In this Reynolds number U is a typical relative velocity of the flow, L is a typical length scale and υ is the kinematic velocity of the fluid. Now consider the Fourier transform of the velocity signal

$$V(\omega) = \int_0^T dt\ v(t)\ e^{i\omega t} \qquad (1.8)$$

and form $P(\omega)$ the squared magnitude of Fourier Transform

$$P(\omega) \sim |\ V(\omega)|^2 \qquad (1.9)$$

This $P(\omega)$ is called the power spectrum. According to the Kolmogorov theory there is a characteristic dissipative frequency $\omega_D \sim (U/L)\ R^x$ which scales as a power of the Reynolds number and all other frequencies are to be compared with that one. The result is that there is a scaling form for the power spectrum. Kolmogorov also predicted the values of the scaling indices and ended up with a answer in the form

$$P(\omega) \sim \omega^{-5/3} \times p^*(\omega/\ \omega_D) \qquad (1.10)$$

with the x in ω_D being 1/4. In Figure 1.3, this theoretical result is compared with experiment. Once again the fit is good.

The argument for the Kolmogorov theory arises from the analysis of the Navier Stokes equation. One looks at the density of kinetic energy $|\mathbf{u}(\mathbf{r},t)|^2/2$ and notices that it obeys an equation of the form

$$\frac{\partial}{\partial t} |\mathbf{u}(\mathbf{r},t)|^2/2 + F = -D \qquad (1.11)$$

here the dissipation term is

$$D= \qquad \nu\ |\nabla\ \mathbf{u}|^2$$

and represents a loss of kinetic energy to heat. The flux term is

$$F= \mathbf{u}\cdot(u.\nabla)\ \mathbf{u}\ -\nu\nabla^2\ |u|^2 \qquad (1.12)$$

and represents the effect of moving energy up and back between different wave number components. In fact, one visualizes putting in energy at long wavelengths and seeing it cascade down into shorted and shorter wavelengths. Let u_k be the typical velocity which the system has in wave vectors (inverse wavelengths) of order k. Then, at the longer wavelengths (for which the dissipations is very small) the flux on scale k is of the order

$$F\sim k\ u_k^3 = \varepsilon_k \qquad (1.13)$$

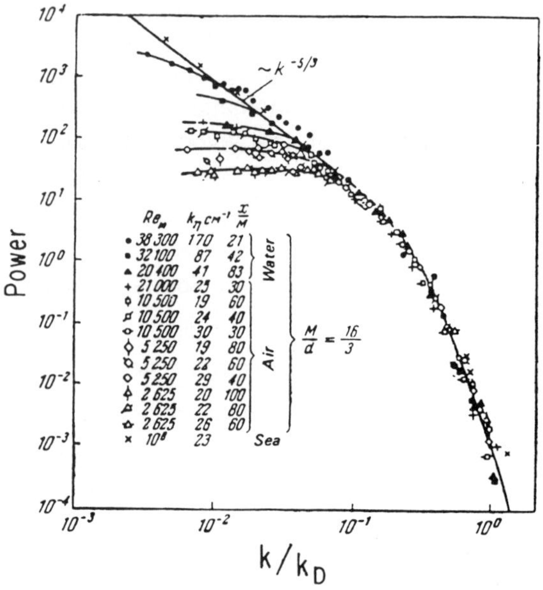

FIGURE 1.3

Power spectrum, Kolmogorov theory compared with experiment. Here $\eta=(k_D)^{-1}$. The figure is drawn from Monin and Yaglom (reference 2) Volume 2, Figure 75. The theoretical prediction is that all the data fall on a single curve, which has on the left-hand end a straight line portion with slope -5/3. The fit is excellent. The data falls off the curve on the left-hand side because the fluctuations are limited by the finite size of the region containing the turbulence.

Kolmogorov argument is that ε_k must be independent of k in this longer wavelength region since the flux moves energy from scale to scale and it cannot be lost. Thus, the flux is independent of k. The final estimate here is then that ε_k has the same value for all of the longer wavelengths. At the largest spatial scale, L, we assume a typical velocity U and find

$$u_k = (\varepsilon\, k)^{-1/3} = U\, (kL\,)^{-1/3} \qquad (1.14)$$

This result holds until the dissipation is of the same order as the flux. which then arises when

$$\varepsilon \sim \nu\, (k\, u_k)^2 \qquad (1.15)$$

Which then occurs when

$k=k_D=L^{-1}(Re)^{3/4}$ (1.16)

Here Re is the Reynolds number

$Re = U L/\nu$ (1.17)

Now, we can apply the scaling result to the power spectrum. Imagine that we move rapidly through the fluid and measure the velocity as a function of time, u(t). Our speed V is much greater than any turbulent velocities. Then a fluctuation with wavelength k will appear us to be a fluctuation with frequency ω=k V. The power we measure is the size of fluctuations with this frequency $u_{\omega/V}$ squared times the typical time interval over which these fluctuations will be visible, ω^{-1}. Thus, we estimate

$P(\omega) \sim \omega^{-1}(u_{\omega/V})^2$ (1.18)

From equation (1.14), we get

$P(\omega) \sim \omega^{-1} U^2 (\omega L/V)^{-2/3}$ (1.19)

Equation (1.19) will apply until the frequencies get high enough so that dissipative effect will be come important. This will arises when k is or the order of the k_D estimated above. Then, to take care of this region of wave vectors we will need another factor $p^*(k/k_D)$ to represent the loss of power via dissipation. When this factor is inserted into Eq. (1.19), we get the scaling estimate

$P(\omega) \sim \omega^{-1} U^2 (\omega L/V)^{-2/3} p^*(\omega L(Re)^{-3/4}/V)$ (1.19)

This is the estimate which was used for comparison with experiment.

D. Central limit theorem

I give one final example of simple scaling. Let X be a sum of N weakly correlated random variables $\mathbf{x}_j$:

$$\mathbf{X} = \sum_{j=1}^{N} \mathbf{x}_j$$

According to the central limit theorem, under very general conditions, there is a simple form for the probability of observing **X** with the value X in the limit $N\to\infty$. The result is

$$\rho(X,N) = \text{const} \times N^{-1/2} \times \exp[-a\,[X-<X>]^2/N]$$

Notice that the result is *universal* in the sense that it always has the same form independent of the probability distribution for the individual $\mathbf{x}_j$. In fact, universality is characteristic of many scaling results. In the critical phenomena example, the function m* is experimentally known to be the same for many different forms of the interactions among the basic entities. It varies almost only when the symmetry of the underlying phenomenon changes.

2. SOME THEORY

A. Binomial distribution

Q objects are distributed randomly between 2 bins. The probability that P will be in bin 1 is

$$\rho(P|Q) = \frac{Q!}{P!\,(Q-P)!\,2^Q} \qquad (2.1)$$

For large Q, the most likely result is that about half the objects will appear in bin 1. In fact, if we ask what is the probability that P will differ from Q/2 by an amount which is not too large, we once again get the central limit theorem result

$$\rho(P|Q) = (2\,\pi\,Q)^{-1/2} \times \exp[-\,[P-Q/2]^2/2\,Q] \qquad (2.2)$$

However, we can also ask about large deviation of P from Q/2. These large deviations are very unlikely, but one can estimate their probability by using the Sterling approximation for the factorials

$$N! \sim \frac{N^N\,e^{-N}}{(2\pi\,N)^{1/2}} \qquad (2.3)$$

which applies for large N. Then for large P and Q, the distribution (2.1) becomes

$$\rho(P|Q) = \text{const} \times Q^{-1/2} \times \exp[-Q\,f(P/Q)\,] \qquad (2.4)$$

The major point is that the large parameter Q appears in the exponential multiplying a function of a quantity $\alpha = P/Q$ which is of order unity. In this case

$$f(\alpha) = -\ln 2 - \alpha \ln \alpha - (1-\alpha) \ln (1-\alpha) \qquad (2.5)$$

Near the peak (at $\alpha = 1/2$) the distribution is Gaussian, but this result also works far into the wings.

We have here a very general result: Unlikely events are given by exponential distributions in which large parameters appear in the exponents multiplying functions which are of order unity.

B. A little bit of theory

I would now like to derive a general form of distributions which look like the probability we got for the binomial case. The result in the binomial case can be written as

$$\rho(P|Q) \sim \left(e^{-Q}\right)^{f(\alpha)} \tag{2.6}$$

Here we are saying that exp(-Q) is the small parameter in the problem and that we are raising the small parameter to a power. The power, $f(\alpha)$, is of course a scaling index. The characteristic and special feature of this result is that in this case the power is not a constant but instead a continuously varying function of α. In some sense this is a problem in which there are an infinite number of critical indices. Since there are many critical indices this way of thinking is called a multi-scaling or (multifractal) approach

Let us generalize this approach. Consider some conditional probability $\rho(X \mid L)$ where X, L are large numbers. (Here L is like our previous e^P and x is like our previous e^{Q}.) In a simple scaling approach one might say that ρ scales as the $-\nu$ power of L and X scales as the μ power of L. Then one would get a Widom-like form:

$$\rho(X \mid L) = L^{-\nu} \rho^*(X/L^{\mu}) \tag{2.7}$$

But now we generalize this formula to the case in which there are many critical indices labeled by subscripts i. Then this formula will become

$$\rho(X \mid L) = \sum_i L^{-\nu_i} \rho_i^*(X/L^{\mu_i}) \tag{2.8}$$

Now we make some assumptions. Let us order the indices in such a fashion that μ_i , ν_i both increase with i and assume that the scaling functions have the order of magnitude behavior

$$\rho_i^*(x) \sim \begin{cases} 1 & \text{for } x << 1 \\ 1 & \text{for } x \sim 1 \\ 0 & \text{for } x >> 1 \end{cases} \tag{2.9}$$

Now look at the structure of the sum. To get an order of magnitude estimate take $X \sim L^{\mu_j}$. Then by our assumptions the term with i=j sum dominates the sum and to an order of magnitude

$$\rho(X| L) \sim L^{-\nu_i} \tag{2.10}$$

Finally assume that there are in infinite number of terms in the sum. Then the sum can be replaced by an integral. Write instead of μ_i, α. Use α as the integration variable. Since ν_i depends upon i it is a function of α. Write this function as $f(\alpha)$. Now one can write instead of (2.8), the multiscaling expression

$$\rho(X| L) = \int d\alpha \, L^{-f(\alpha)} \rho_\alpha^*(X/L^{\alpha}) \tag{2.11}$$

One can now obtain a result analogous to (2.10) in this new formalism. Assume that to an order of magnitude $X \sim L^{\alpha}$. Then just the same argument which led to (2.10) now gives

$$\rho(X|\,L) \sim L^{-f(\alpha)} \qquad \text{with } \alpha = \ln X / \ln L \tag{2.12}$$

We can convert this result into a recipe. Take a system which we suspect is multifractal described by some large (or small) parameter like reduced temperature or Reynolds number. Call this parameter L. Let us measure a quantity which varies over a considerable range called X. To see if Eq (2.12) is right plot log probability divided by log L against log X divided by log L. If all the results for different L's fall on the same curve, one has a multifractal spectrum of critical indices.

Now we have all our formal apparatus developed[3]. We can turn to specific examples.

3 SPECIFIC EXAMPLES

A. Sand slides

Bak Tang and Wiesenfeld[4] invented a cute dynamical model which shows how richly complicated events can arise in a relatively simple dynamical system. Let model sand[5] be added grain by grain to a model sandpile, built upon a regular d-dimensional lattice. A one-dimensional version of the model is shown in Figure 3.1. To start the cascade, one grain of sand is added to one of the columns, picked at random. In between additions, there are cascades of events in which sand falls downhill in response to a too-large local slope of the pile. In the particular model depicted in Figure 3.1, the sand falls over whenever the column in question stand more than two above its right-hand neighbor. In that case, two grains of sand 'fall over'

---

[3] This 'multifractal' analysis was well known in the mathematical literature as a method for studying low-probability events. In more physical contexts, it was introduced by Mandelbrot J. Fluid Mech. **62**, 331 (1974). for the study of turbulence and then further applied by U. Frisch and G. Parisi in *Turbulence and Predictability in Geophysical Fluid Dynamics and Climate Dynamics,* edited by M. Ghil, R. Benzi and G. Parisi (North-Holland, Amsterdam, 1985) p. 84. See T. Halsey, M.,H. Jensen, L. Kadanoff, I. Procaccia, and B. Shraiman Phys. Rev. A33 1141 (1986) for an exposition of this thinking.

[4] P. Bak, C. Tang and K. Wiesenfeld, Phys. Rev. Lett. 59, 381 (1987); P. Bak, C. Tang and K. Wiesenfeld, Phys. Rev. A 38, 364 (1988)

[5] Real sand does not behave in precisely the same manner as model sand. See H.M.Jaeger, Chu-heng Liu and Sidney R. Nagel, Phys. Rev. Lett. **62**, 40 (1989) but also G.A.Held, D.H. Solina II, D.T. Keane, W.J..Haag, P.M. Horn and G. Grinstein IBM Yorktown Heights preprint (1990).

and land on the two columns to the immediate right of the unstable site. If a grain of sand reaches the right-hand end of the system, it falls off and disappears from view. At any given time, all columns that can fall do so, simultaneously. Then, if any columns remain unstable, they fall and so on. Thus the system can sustain 'avalanches'. The cascade of events continues until no more columns are unstable. Then another grain of sand is added at random and the entire process begins once more. The algorithm for the model is shown in the box below:

An Avalanche Model

Square Sand stacked up in a region of size L

A. Add a Grain at a Random Site

(**Avalanche begins**)

B. If the Slope (Height Difference) is greater than 2, two grains from stack fall over.

At right hand end, grains fall off

Continue until no more stacks are unstable.

(**Avalanche Ends**)

C. Return to A.

These avalanches in this system can be small or they can cover the entire system many times over. In our work[6] , we studied the nature of the probability distributions $\rho(X|L)$ for the probability that an event of size X will occur in a system with spatial extent L. We looked at two different quantities in some detail:

a. The drop number D. In this case X=D is the number of grains which fall of the end between two addition events.

[6] Leo Kadanoff, Sidney Nagel, Lei Wu and Su-min Zhou, Phys Rev. **A39** 6524 (1989)

b. The flip number F. In this case X is the number of falling events which occur between two additions. and ask whether there is some scaling or universality. In low dimensions, $\rho(X,L)$ is most likely multifractal.

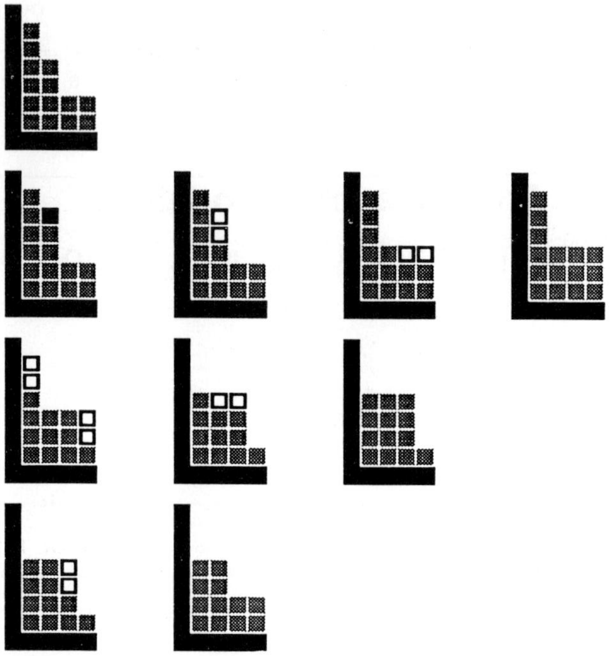

FIGURE 3.1
The Sand Model This picture shows a short avalanche in progress in a small model sandpile.

Like most of the examples discussed below, sands slides can be considered to be an example of self-organized criticality. That is, the system organizes itself in such a way that it is marginally stable against the occurrence of events within it. In this marginal stability, the system is just at the edge of stability. An event, once started, has a finite probability for growing larger at every stage of its early existence. It also has some probability for growing smaller and dying. These two probabilities balance out so that events of all sizes occur. In the sand pile. However, by the nature of the process $\rho(D|L)$ must have some weight for large D. In a large sandpile if the new grain falls far from the edge, the most likely events are sufficiently small so that the edge will never be reached and no grains will drop off. Hence, for D=0, ρ(D|L) is close to unity. For small D falling events will most likely occur only when the first grain falls near the edge. Hence they have a likelihood of order 1/L

$$\rho(D|L) = c(D)/L \quad \text{for } D=1,2, \ldots \tag{3.1}$$

However, since the pile is in a kind of steady state, on the average one grain must fall for each grain added. In symbols

$$\sum_D \rho(D,L)\, D = \langle D \rangle = 1 \tag{3.2}$$

Eqs. (3.1) and (3.2) can only coexist if ρ(D|L) has some weight for D of order L and hence non-trivial scaling structure. The same argument can be extended to show that the other probability functions, e.g. ρ(F|L), also have non-trivial scaling structures. This is an indication that events which involve large values of D and F must play an important role in determining the steady state dynamics of avalanches.

We want to know the answer to two questions:

a. What kind of scaling occurs? Is it the simple scaling or is it a multifractal distribution? Perhaps it is something else altogether.

b. How universal is the result. Do different sets of rules give the same answers?

To answer these questions, we turn to simulations of the flow in sandpiles.

Consider a variant of the model indicated in Figure 3.1. We work with a one dimensional array of stacks of sand. (See Figure 3.2) The cascade occurs when the height difference between neighboring stacks is higher than NF. For all stacks which satisfy this condition, then NF grains on sand will fall onto the right-hand neighboring stack. This process continues until the cascade ends.

The case with NF=1 is trivial. The sandpile gets into a state where the slope is one everywhere. See Figure 3.3. If a grain is added it just goes step by step to the right until it falls off the end. Thus, the probability distribution is trivial

$$\rho(D|L) = \delta_{D,1} \tag{3.3}$$

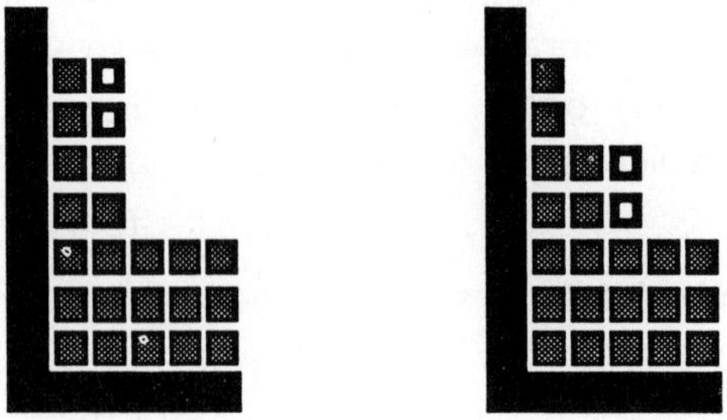

LIMITED LOCAL NF=2

FIGURE 3.2.
Another sand model. Here all the sand falls onto the nearest neighbor stack.

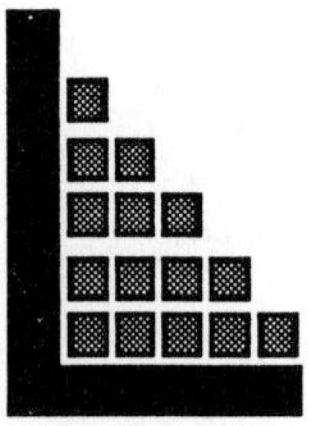

NF =1

FIGURE 3.3 .
The case of NF=1. Here the sand falls into a stack of constant slope and stays in this configuration.

However, as soon as NF=2, the situation becomes much more complex and interesting. Figure 3.4a shows the probability distribution obtained from computer simulations for cases in which L varies from 32 to 2048.

This figure is drawn for simulations in which NF=2. Notice the very wide distribution of probabilities. One can try a simple scaling analysis of the form

$$\rho(D|L) = L^{-p}\, G(D/L^{q}) \tag{3.3}$$

Figure 3.4b shows the best fit of this form which we were able to obtain. Notice that although the data is well-fit in the central region, the fit is not wonderful for either small D or large. On the other hand we might try a multifractal fit in which logs of physical quantities are divided by logs of the large parameter in the problem. For this case the appropriate functional fit is to try

$$\frac{\log \rho(D|L)}{\log L} = f(\log D/\log L) \tag{3.4}$$

A slight improvement in the fit can be obtained by taking the measured quantities and multiplying by a constant and so using the fit

$$\frac{\log \rho(D|L)}{\log A\ L} = f(\log (B\ D)\ /\log A\ L) \tag{3.5}$$

Here we choose A=1.5 and B=0.5. The fit is considered to be good if a plot of log ρ(D|L)/log A L against log (B D) /log A L, gives the same curve for each L. Figure 3.4c shows a fit of this form for the data under consideration. The fit is excellent. We conclude that a multiscaling or multifractal analysis is the right one.

Notice incidentally that there is no way that the curves in figure 3.4 indicate that there is really simple scaling. If there were, the pictures would show a large straight line region representing power law behavior of ρ(D|L). There is really no readily apparent single slope, or single power law.

Now we are in a position to ask another question. What is the range of applicability of the fit (3.5)? Is there some 'universal' behavior of large cascades In particular, if we change the model somewhat will the answer (*viz* the form of the function f) change. Figure (3.4c) shows the form of the function f for a different model, or rather for the same model with NF=20. There is no trivial sense in which these models are identical. Nonetheless the forms of f(α) for the two models ,are, as far as one can tell, exactly the same. (See especially the inset in the figure in which the two f's are superposed.)

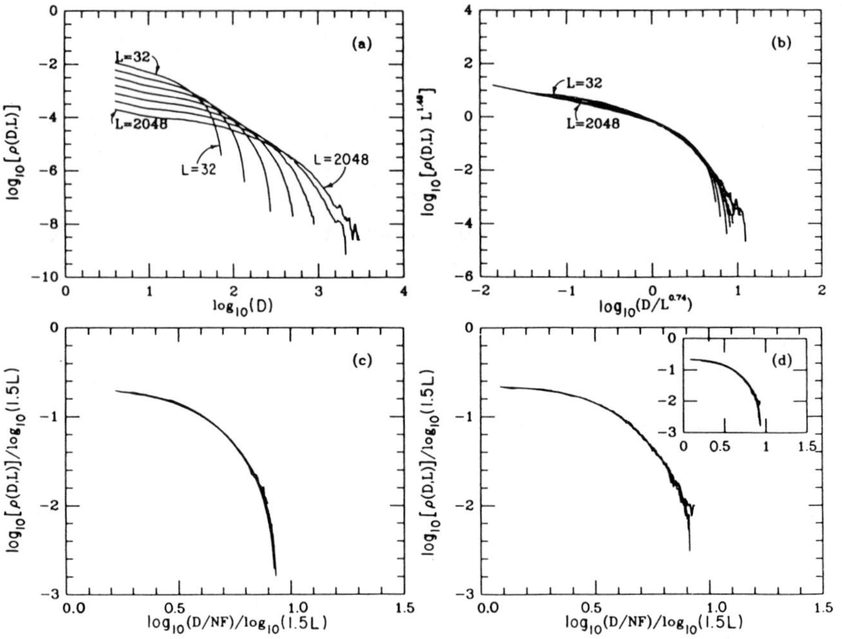

FIGURE 3.4 (from Kadanoff, et. al (see reference 6. Figure 1.) Distribution of drop number in a one dimensional model. Part a shows the raw data for NF=2. The system size, L, ranges from 32 to 2048. Part b shows the best scaling fit to these data. This fit is much worse than the multiscaling fit shown in part c. In part d, the same fit is shown for NF=20. The inset compares NF=20 top NF=2, showing that the f(α) is the same for both cases.

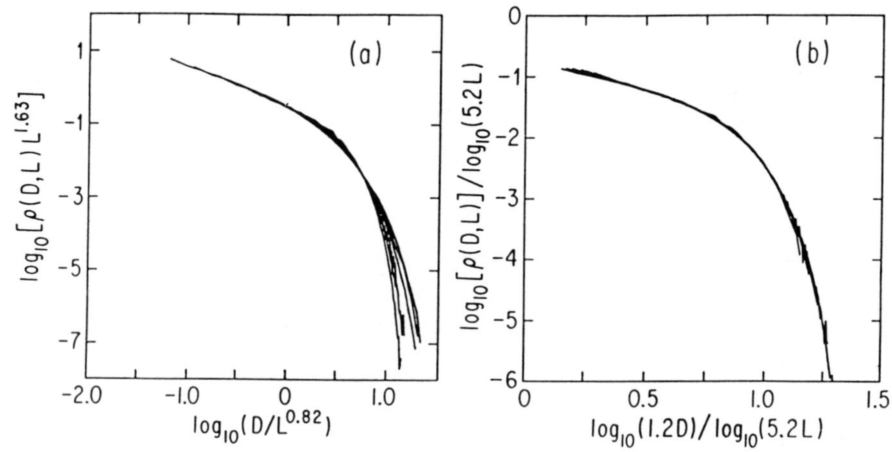

FIGURE 3.5 (Taken from O'Brien et al (see reference **7** Figure 2.) The same as the previous figure but for the three-dimensional model. The left-hand plot shows the best simple scaling fit that the authors could obtain. The right hand plot shows the multifractal fit. The latter is much better.

In a more recent work[7], O'Brien et al have consider what happens in higher dimensions. Perhaps, one might say, the multiscaling behavior which was apparent in one dimension does not apply for higher dimensions.

---

[7] Kevin O'Brien, Lei Wu, and Sidney R. Nagel, Avalanches in Three and Four Dimensions, University of Chicago preprint,1990.

Figure 3.5 shows the analysis of three dimensional arrays of stacks for a model which is a simple generalization of the one described above. The net result is that once again the multiscaling analysis gives a better fit. There is in three dimensions, a somewhat longer region of power law behavior, which can be fit by either the simple scaling or the multifractal approach. However, at the right hand end of the curve, the multifractal analysis clearly wins the day. The right hand side of the curve represents large events in which many grains of sand do fall off the edge. These drop events reduce the number of grains of sand in the pile and eventually dissipate the avalanche. Apparently, this dissipation is well-represented by a multi-fractal analysis.

B. Convective turbulence

Albert Libchaber has described his experiments on convective flow in helium gas elsewhere in this lecture volume.[8] We turn our attention to the part of the experiment which measures the temperature as a function of time in the center of the cell.[9] In this region, we are seeing quite well-developed turbulence. In fact, since the strongest shears are toward the side walls of the container, perhaps it is appropriate to say that what is looking at is the decay in space and time of well-developed turbulence. The quantity under examination is the power spectrum, the squared magnitude of the fourier transform of temperature ass a function of time.

The control parameter which describes the size of the forcing in the cell is called the Rayleigh number. It is given by

$$Ra = \frac{g\alpha\Delta L^3}{\kappa\upsilon} \tag{3.6}$$

Here, L is the height of the cell, g is the acceleration of gravity, α is the volume thermal expansion coefficient, Δ is the temperature difference between the bottom and the top of the cell and κ and ν are respective the thermal diffusivity and the kinematic viscosity. The frequencies are measured in units of the characteristic viscous diffusion time for the cell, υ/L^2.

---

[8] A. Libchaber, this volume.

[9] A longer description of the experiment and of the scaling analysis of it appears in B. Castaing, G. Gunaratne, F. Heslot, L. Kadanoff, A. Libchaber, S. Thomae, Xiao-zhong Wu, S. Zaleski and G. Zanetti, J. Fluid Mechanics, **204**, 1 (1989).

The basic data[10] for one cell is shown in Figure 3.6. In the spirit of the usual scaling analysis we try to fit the data by considering $P(\omega)$ to be of the form:

$$\ln P(\omega)/P_h(Ra) = F(\ln \omega/\omega_h(Ra)) \qquad (3.7)$$

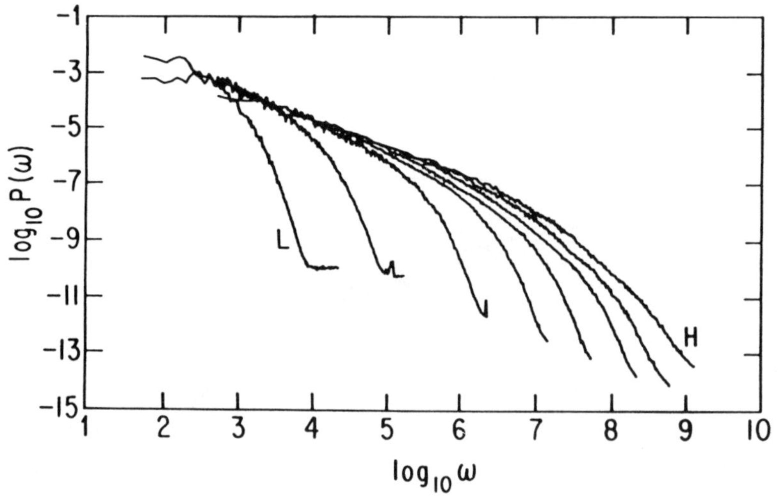

FIGURE 3.6

The experimental data. Power, P, is plotted as a function of angular frequency, ω. Taken from reference 10. The symbol H marks the highest frequency, L the lowest.

Here we take P_h and ω_h to be adjustable constants, which we try to fit for each data set (value of Ra) so that all the data in Figure 3.6 falls upon a single curve. We show such a scaling fit in Figure 3.7. This figure includes all the samples for lower values of the Rayleigh number, $10^7 < Ra < 7\times10^{10}$. As one can see, in the central region of frequency this scaling

---

[10] This work is described in Xiao-Zhong Wu, Leo Kadanoff, Albert Libchaber, and Masaki Sano, Phys. Rev. Letts **64** 2140 (1990).

fit is extremely good. Hence, we do have a region in which simple scaling works. The failure of the fit at high frequencies is an experimental artifact due to the finite resolution of the thermometer. The low frequency failure is a different story. There is a characteristic frequency, ω_p, which represents a typical frequency for the overall flow around the cell. It is in fact, a typical transit time for a revolution of the liquid around the cell. The physics of this large-scale effect does not fit into the scaling picture. Hence the failure of the scaling fit a low frequencies. As Ra gets larger and larger, in this region, the fit works over a large region of dimensionless frequency.

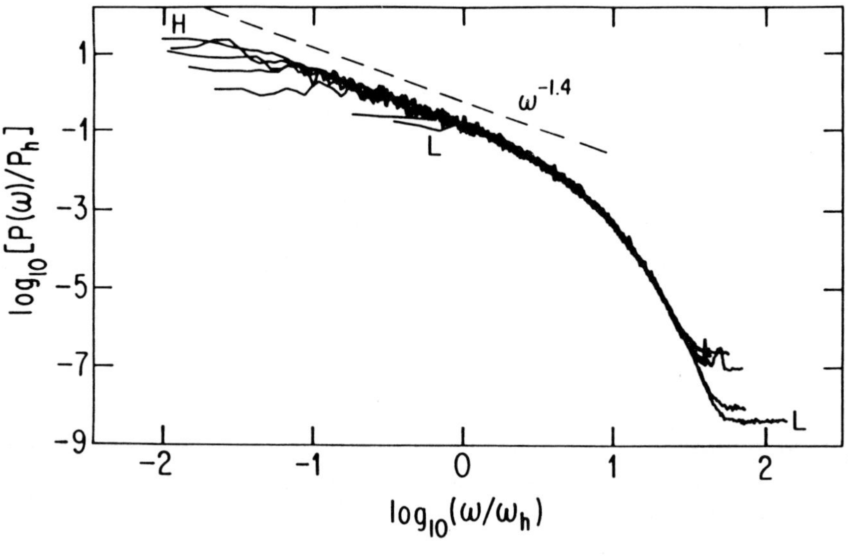

FIGURE 3.7

The experimental data normalized so that it fits upon a single curve.. Taken from reference 10. The normalizing factors, P_h and ω_h depend upon Rayleigh number. The data shown here includes all cases where Ra<7× 10^{10}. The symbol H marks the highest frequency, L the lowest.

Notice the straight line region of the plot. This is a result of the region in which there is a Kolmogorov style cascade, and in which the dissipative effects have not yet become important. (They only become important at higher frequencies.) In this region we have that the power goes as ω raised to the -1.4 power[11]. Thus our general fit is of the form

$$P(\omega) \sim \omega^{-7/5} \times p^*(\omega/\omega_h(Ra)) \tag{3.8}$$

Equation (3.8) fails for higher values of Ra. Hence, we try a multifractal fit, one of the form

$$\frac{\ln P(\omega)/P^*}{SF(Ra)} = \frac{F(\ln \omega/\omega^*)}{SF(Ra)} \tag{3.9}$$

In the spirit for multifractal analysis, we have taken the log log representation of our data and compressed the data by using a scale factor which depends upon Ra. In doing this, we have the other normalizing factors (now called P^* and ω^*) and taken them to be independent of Ra. The scaling factor, SF, is taken to be of the form

$$SF(Ra) = A + B \ln Ra \tag{3.10}$$

Figure (3.8) shows a plot of the appropriate range of the data, $7\times10^{10} < Ra < 7\times10^{14}$, using the data correlation of Equation (3.9). If this equation is right, all the data should fall on a single curve. Once again, in the central regions of frequency, the fit works in an excellent fashion. The failures for higher and lower frequency are explained as before.

Equation (3.8) suggests that there is a characteristic critical index ($\approx$-1.4) associated with the power spectrum. If the plot in Figure 3.8 had straight-line regions, each of those regions would have a slope which is a critical index for the data. However, there is no substantially large straight-line region except perhaps the one at the lower end of the fit frequencies. Therefore, one says once more that there is a whole spectrum of critical indices which describe the data.

---

[11] This is the exponent expected in this case of convective flow. The convective theory was worked out initially by A.M. Obuhkov *Dokl Akad Nauk USSR* **121** 1246 (1959) and R. Bolgiano, J. Geophy Res. **64**, 2226 (1959), and has more recently been examined by Shraiman and Siggia, preprint (1990).

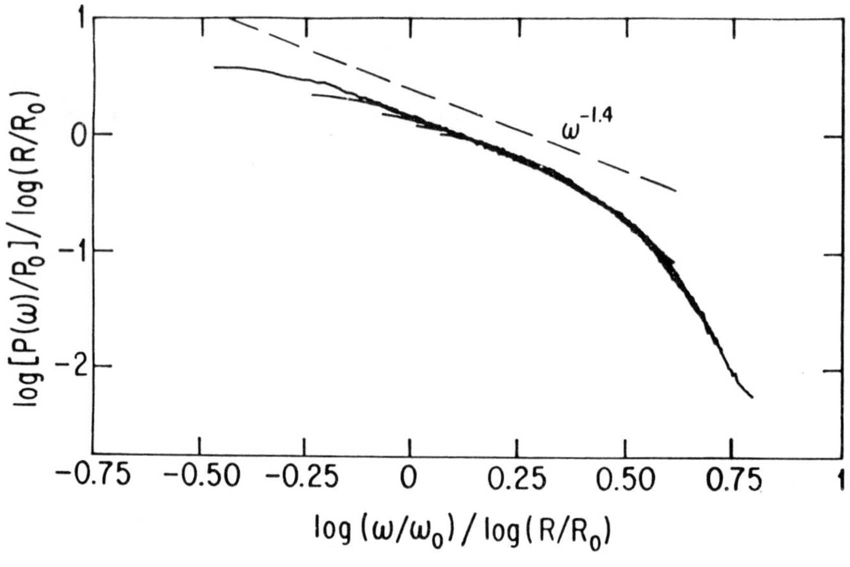

FIGURE 3.8
The high Rayleigh number experimental data with the logarithmic scales normalized so that the data fits upon a single curve..Taken from reference 10. The normalizing factors, P_0 and ω_0 and R_0 do not depend upon Rayleigh number. The data shown here includes all cases where Ra > 7× 10^{10}. The symbol H marks the highest frequency, L the lowest.

Because the data set for Ra=7×10^{10} falls in both the scaling and the multiscaling plots, one sing curve with one value of F will fit all the data. The general fir is of the form

$$\frac{\ln P(\omega)/P_h}{SF} = \frac{F(\ln \omega/\omega_h)}{SF} \tag{3.11}$$

Here, for Ra < 7×10^{10}, the scaled factor, SF, is independent of Ra while for Ra > 7×10^{10} , P_h and ω_h are independent of Ra. The very same F describes all data.

C. Glasses

The work of Dixon, Wu, Nagel, Williams, and Carini[12] has led to the surprising result that dielectric relaxation in glasses can also be described by a variant of a multifractal analysis. Glasses relax slowly to equilibrium. The lower the temperature, the slower the relaxation. Of course, this slowness is the reason why we can use a liquid, window glass, in our buildings and have the windows remain in place for many centuries. After a while, they will flow downward and form a puddle.

This discussion assume that as the temperature is lowered, but the material remains glassy, there is never a true phase transition to a qualitatively different state of matter. If the phase transition occurs at some temperature or another, below this temperature there will emerge a qualitatively different form of the behavior of the material. Hence by examining the form of the dielectric relaxation in the results of Dixon et. al., one can see whether a phase transition does indeed occur. As we shall see, there is never an apparent change in behavior. Thus, within the experimental range, the phase transition appears to be ruled out.

This slowness of the relaxation can be studies by looking at the frequency dependent of the dielectric 'constant', $\varepsilon(\nu)$. Figure (3.9) shows the imaginary part of the dielectric function plotted as a function of frequency, here called ν. The characteristic microscopic frequency is 10^{13} or so Hertz. The typical frequencies seen in this glass, salol, range from 10^9 Hertz to 10^{-2} Hertz for the temperature range shown here. The solid curves are fits to the data obtained from a stretched exponential form of fit

$$\varepsilon''(\nu) = -\ \mathrm{Im}\left[\text{fourier transform of } \left(\frac{d}{dt}\exp(-\,|\,\nu_p t|^{\beta}\,)\right]\right. \tag{3.12}$$

Here β and ν_p are taken to be adjustable function of temperature. As the data is presented in figure (3.14), the stretched exponential seems to fit just fine. However, if one looks in the tail of the data the fit of eq (3.12) fails by up to a factor of thirty or so. However, one can make use of the structural form of Eq (3.12) to get an excellent fit to the glassy relaxation data. Eq (3.12) implies that

$$\frac{\log\left(\varepsilon''(\nu)\,\nu_p/(\nu\,\Delta\varepsilon)\right)}{w} = F\left((1+w^{-1})\,\frac{\log \nu/\nu_p}{w}\right) \tag{3.13}$$

[12] Paul Dixon, Lei Wu, Sidney Nagel, Bruce Williams, and John Carini, Scaling in the Relaxation of Supercooled Liquids, University of Chicago preprint,1990.

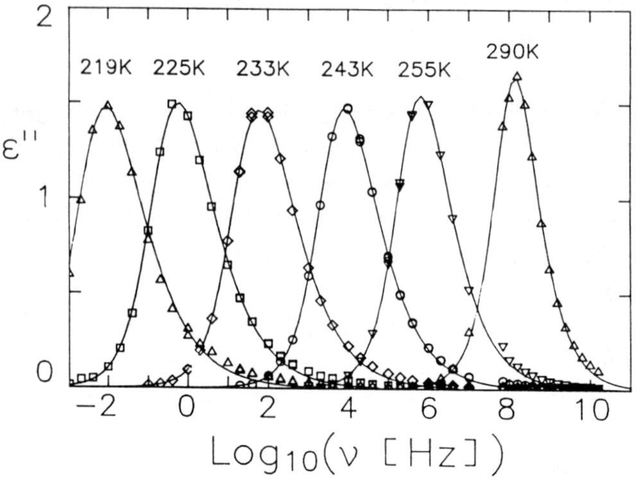

FIGURE 3.9

Figure 2b of Dixon et al (reference 12) The imaginary part of the dielectric constant, ε'', plotted against frequency, ν, for various temperatures. The data seems to be well fit by a stretched exponential form as shown by the solid lines. However, this form does not work well in the tails.

Here $\Delta\varepsilon$ is a fitted normalization factor. The other fitting parameter, w^{-1}, is of the order of β. It is taken to be the half-width of the ε'' curves. There is a particular form of $F(\alpha)$ which goes with the stretched exponential. Notice that Eq (3.13) is, except for one small difference, exactly of the same form which we have used heretofore. In a log log representation, physical quantities are divided by scaling parameters, here w, which depends upon the key control parameter, here temperature.

The only difference is that here there is an extra factor $(1+w^{-1})$ which has no analog in our previous fits.

Despite this 'imperfection', the authors of this paper used Eq (3.12) because it gave an excellent way of representing their data. For each material and each temperature, they fitted a value of w and of $\Delta\varepsilon$ and then plotted the data in the form suggested by Eq (3.13). The result for salol (and all the data shown in Figure 3.9) is given in Figure 3.10a. Each data point fits beautifully on one smooth curve. Figure 3.10b shows the same plot with many different glasses superposed. Apparently, in this representation, the glassy data is universal. Furthermore, it cannot be fit by a stretched exponential. Figure 3.10c shows the best stretched exponential fit to the data in glycerol. The simple fit clearly fails for the higher frequencies. [13] Furthermore, the stretched exponential does not work at lower frequencies either since for $\nu \sim \nu_p$, the Fourier transform of the stretched exponential agrees with neither Eq. (3.13) nor the experimental data.

[13] Equation (3.13) fits the high frequency data (that for $\nu > \nu_p$) quite well. However, at lower frequencies, it appears that some additional physics enters, and the universal multifractal fit fails.

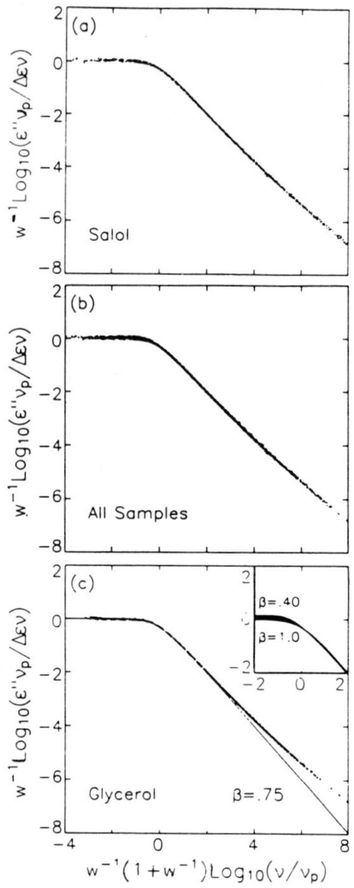

FIGURE 3.10.

Drawn from figure 3 of Dixon et al (reference 12) The first part shows the same data as in Figure 3.9, but now in a log log plot. The result plotted in this way is temperature independent. Part b shows the data for the seven different glasses studied in this reference. Note that they all fall onto the same curve. The third plot shows the data for the glassy material, glycerol. The line is the best stretched exponential fit. Notice that this stretched exponential fit is not too good. The inset shows the range of fits possible using streached exponentials with different values of β. On the other hand, since the data of plot c fits beautifully onto the curve of plot b, the multifractal fit gives an excellent result.

Once again the straight line of the simple fit describes a single critical exponent, here β. Once again, the single exponent fails to fit the data but the multiscaling fit, which uses many exponents, does work. Clearly there is some magic in these fits. But what does it all mean?

ACKNOWLEDGEMENT

This research was supported by the University of Chicago MRL. Many of the successes (and failures) reported here would have been impossible without the interactive and collaborative atmosphere provided by that Laboratory and by the Research Institutes of the University of Chicago.

Reprinted from *Directions in Condensed Matter Physics*,
ed. G. Grinstein and G. Mazenko, 1986, World Scientific.

COMPLEX ANALYTIC METHODS FOR VISCOUS FLOWS IN TWO DIMENSIONS

D. Bensimon, L. P. Kadanoff, S. Liang, B. I. Shraiman & C. Tang
The James Franck Institute
The University of Chicago
5640 S. Ellis Avenue, Chicago, IL 60637
USA

This paper is an expository treatment of recent work on using complex analytical methods for understanding the stability of hydrodynamic flow patterns in two-dimensional or almost two-dimensional geometries. We want to know the instabilities which might arise when a more viscous fluid is displaced by a less viscous one and also how surface-tension effects can restore the stability of non-trivial flow patterns.

In the first section, we describe the physical situation, restate the description in terms of partial differential equations, and summarize our state of knowledge about the solutions to the equations and the physical phenomena that arise. In two-dimensional problems, one can often make considerable progress by using calculational methods based upon analytic functions of complex variables. Section 2 describes how these methods can be used to obtain exact solutions for zero surface tension, while Sec. 3 sets up the interface equations for nonzero surface tension. Finally, the fourth section uses complex-variable methods to describe the stabilization of finger-like flow patterns.

1. The Saffman-Taylor Problem. Where Do We Stand?

1.1. Introduction: phenomenology and the basic equations

The formation and evolution of dynamical structures is one of the most exciting areas of nonlinear phenomenology. Such pattern formation problems are common in hydrodynamic systems. Perhaps the best studied ones involve the patterns formed by the interface between two phases: a solid and a fluid or two fluids. In turn, one of the simplest problems of this class is the Saffman-Taylor (1958) problem in which two fluids move in the narrow space between two plates. This geometry is called a Hele-Shaw (Hele-Shaw, 1898) cell, see Fig. 1.1. When the plate separation, b, is very small the problem is effectively two-dimensional. If we call the coordinates perpendicular to the plates z, and the other two x and y, we can specify the problem by the two components of the velocity, v_x and v_y, the pressure, $P(x,y)$ and a two-component vector $\gamma(s)$ which sweeps out the position of the interface as s is varied.

The basic equations involved are very simple indeed. In each fluid, the average velocity parallel to the plates is proportional to a local force (Saffman and Taylor, 1958)

$$\mathbf{v}(x,y) = -K_i\left[\nabla P(x,y) - \rho_i \mathbf{g}\right] . \qquad (1.1)$$

Here $i = 1, 2$ labels the different fluids, ρ_i is the density, and the constant K_i is given in terms of the fluid viscosity, μ_i, and the plate-spacing, b, as

$$K_i = \frac{b^2}{12\mu_i} , \qquad (1.2)$$

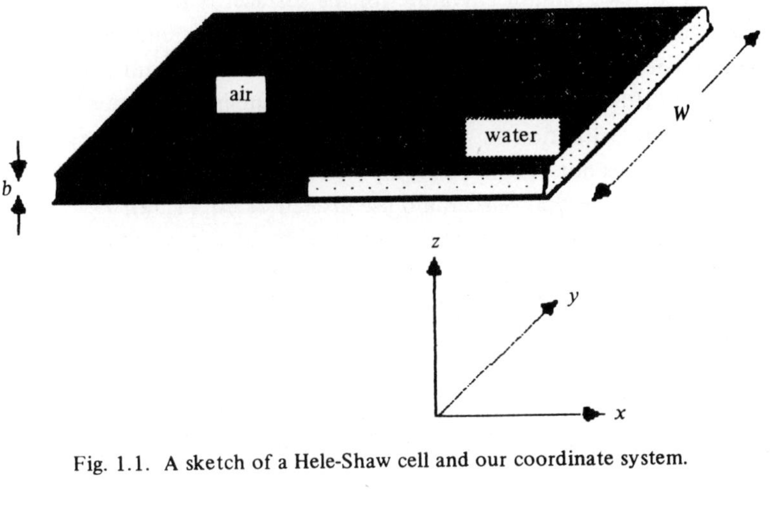

Fig. 1.1. A sketch of a Hele-Shaw cell and our coordinate system.

while **g** is the component of the gravitational acceleration parallel to the plates. Equations (1.1) and (1.2) constitute the Darcy approximation. They are derived in a trivial way from the Navier-Stokes equation by considering a parabolic flow profile parallel to the plates with a velocity which vanishes at both plates. Then, **v** is the average over the perpendicular direction of the actual velocity.

The remaining equations are easy to write down. Assume that the fluids are incompressible so that the divergence of the velocity vanishes. Then, in each fluid

$$\nabla^2 P = 0 \; . \tag{1.3}$$

Continuity also implies a boundary condition that, at the interface, the normal components of the velocity are equal to each other and to the speed of the interface:

$$\mathbf{v}_n = -K_1(\nabla P_1)_n = -K_2(\nabla P_2)_n \; , \tag{1.4}$$

where the gradients are evaluated at the points $\gamma(s)$.

One more boundary condition is needed to give the jump in pressure across the interface. Theorists working on this problem often choose to take the pressure jump to be the surface tension, T, times the curvature, κ, observed in the x-y plane, i.e.,

$$\Delta P = T\kappa \; . \tag{1.5a}$$

This formula would follow were the classical Gibbs-Thomson equations for the pressure jump really applicable. This, in turn, would be true were the Hele-Shaw cell really two-dimensional. But, the cell lives in a three-dimensional world in which there are two radii of curvature for the surface. The larger one, R, has the smaller effect upon the pressure drop, while the smaller one (which is roughly $b/2$) dominates. Hence Eq. (1.5a) makes very little experimental sense. Park and Homsy (1984) suggested an alternative boundary condition which might better describe a situation in which a fluid which wets the plates is displaced by one which does not. From an asymptotic analysis they derived an expression for the pressure jump that one should use instead of Eq. (1.5a),

$$\Delta P = \frac{T}{b/2}\left[1 + 3.80\left(\frac{\mu v_n}{T}\right)^{2/3}\right] + \frac{\pi}{4}T\kappa \; . \tag{1.5b}$$

The first term in Eq. (1.5b), $2T/b$, is independent of x and y so it does not really affect the motion. The other two terms act together, producing forces which tend to flatten out the interface.

We shall look at the simplest possible situation. Let the plates be very long rectangles with width W (Fig. 1.1). The second fluid is, say, air so that we may take it to have a negligible density and viscosity. Let the first fluid be, say, water and let the air be pushing it so that it moves with an average velocity, U. The plates are horizontal so that gravity does not enter.

The sidewalls are rigid. This is represented by using a slip boundary condition so that at the sidewalls the normal component of the velocity vanishes

$$(\mathbf{v})_n = 0 , \tag{1.6}$$

This free slip boundary condition may not be realistic for the true experimental situation, but it may be an essentially correct approximation for small b, where there is a boundary layer of width b near the surface over which $\mathbf{v}$ may vary quite rapidly.

Now, both in experiments and simulations, one observes three different types of motion:

Case A: Small U or negative U. (The latter implies that the water is pushing on the air.) An initial interface which perhaps has a few bumps in it eventually flattens out and forms a straight boundary between the two fluids.

Case B: Intermediate U. Any initially present bump grows and forms a stable finger (Fig. 1.2). The width of the finger is a multiple, λ, of the channel width W and varies with velocity. Under the stated conditions, in which we can neglect the second fluid, there is only one dimensionless parameter[1] entering these equations, namely

$$d_0 = \frac{\pi^2}{3} \frac{b^2}{W^2} \frac{T}{\mu U} , \tag{1.7}$$

which, therefore, acts as a control parameter. Here U is the fluid velocity in the region far downstream from the finger. We call d_0 the "surface tension parameter." The dependence of the finger width on d_0 is an interesting quantity to predict theoretically. Below, we discuss our results for this dependence and those of others.

Case C: Large U. This corresponds to the parameter d_0 being very small. In this domain, several types of time-dependent behavior may be observed. For the

[1] Since there is no agreement on the exact form of the surface-tension parameter we will, for the convenience of the reader, relate our parameter d_0 to the one used by other workers in the field. Thus the parameter κ used by McLean and Saffman (1981) is $\kappa = d_0 \lambda/(1-\lambda)^2$. The parameters B, τ introduced respectively by Tryggvason and Aref (1983) and DeGregoria and Schwartz (1985) are $\tau = B = d_0/(2\pi)^2$.

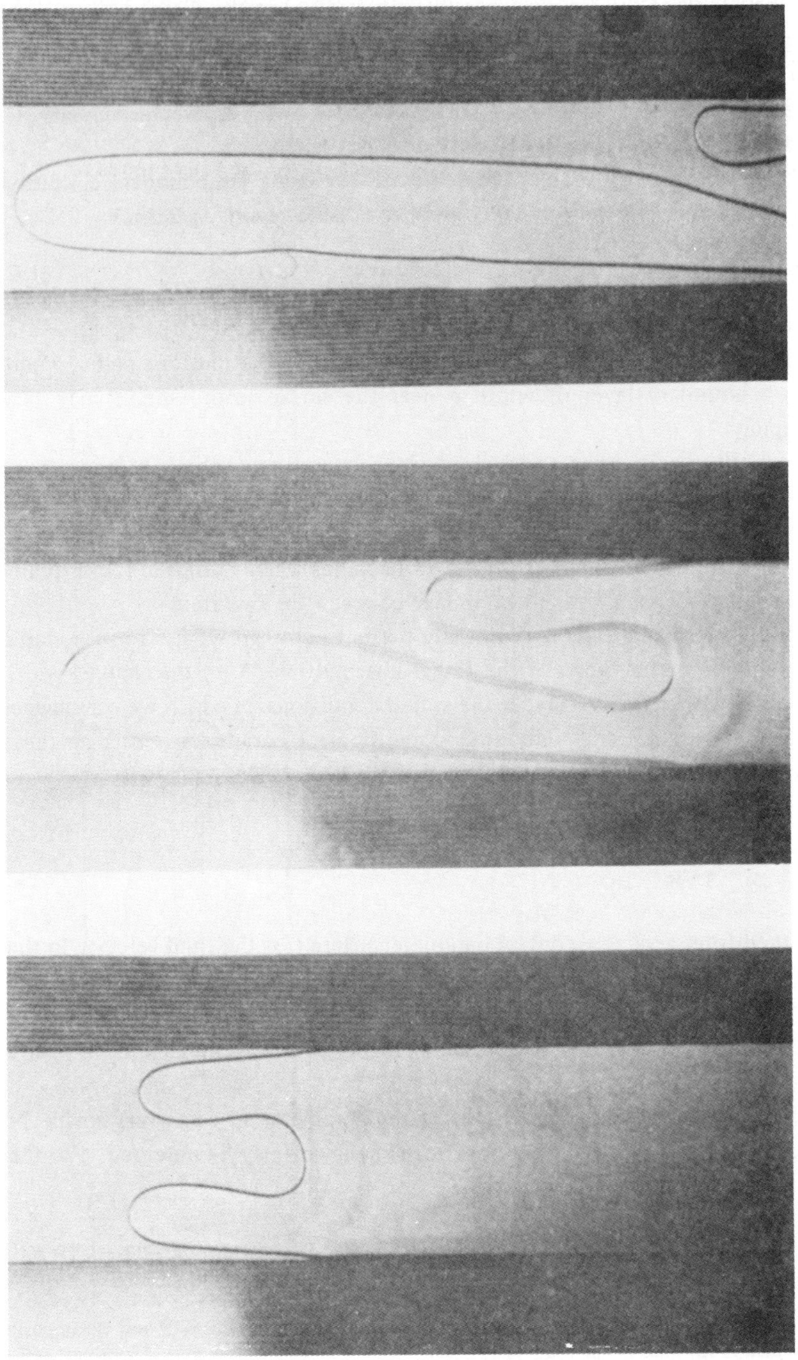

Fig. 1.2. Competition between two bumps leading to the emergence of a single propagating finger; courtesy of Tabeling and Libchaber (1985).

very largest values of U a kind of chaotic behavior is observed in which several fingers are formed which may branch and split (Fig. 1.3). There is a tendency for the tallest fingers to get ahead and leave the smaller ones well behind. Thus there is essentially a cascade into large length scales, which saturates when the fingers become of width comparable to that of the cell. Even if there is only one finger in the channel, for these large values of d_0 the finger tends to wiggle up and down, partially split, and in general show quite an unstable behavior.

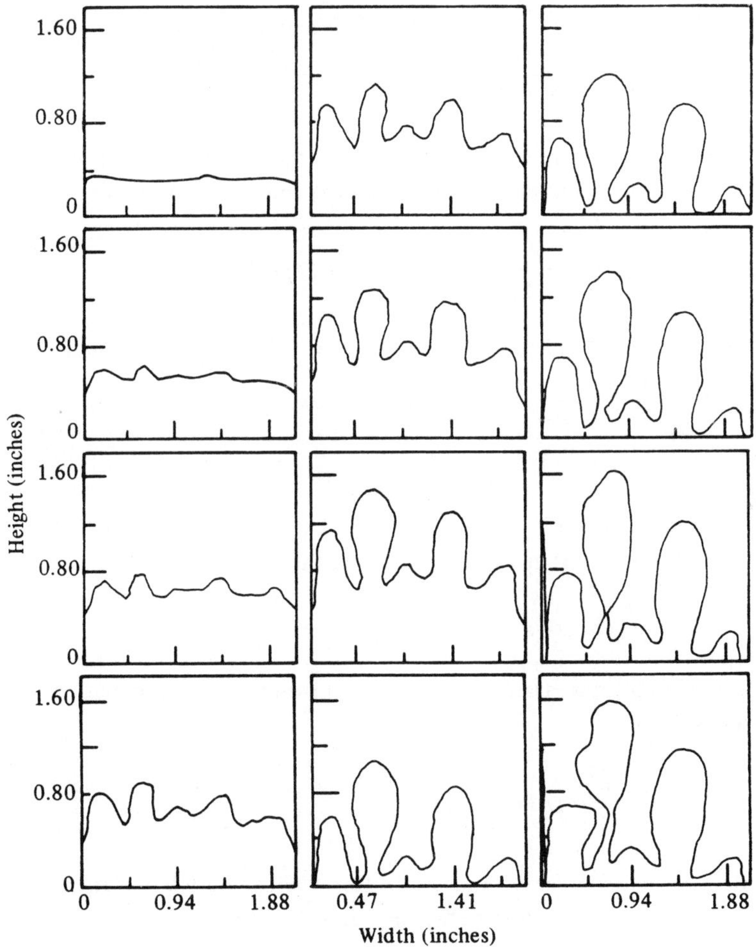

Fig. 1.3. Chaotic behavior in a Hele-Shaw cell; courtesy of Maher (1984).

1.2. Stability analysis (Chuoke, et al., 1959)

The first step is to look at the stability of an almost-flat interface. Let the flat interface be at a position $x(y) = Ut$, which moves with velocity U relative to the walls. A small deviation from flatness may be represented by writing

$$x(y) = Ut + A(t)\cos qy \ , \tag{1.8a}$$

where $A(t)$ is considered to be small. If A vanished, the velocity U would be produced by a pressure gradient $-U/(b^2/12\mu)$. If we add to this zero-order term a term produced by the deviation from flatness we find a result like (1.8a), namely

$$P(x,y) = P_0 - \frac{U}{(b^2/12\mu)}(x - Ut) + B(x,t)\cos qy \ .$$

For $P(x,y)$ to obey Laplace's equation $B(x,t)$ must vary as e^{qx} or e^{-qx}. The former is impossible if the pressure is to remain finite as $x \to \infty$. Thus, we find

$$P(x,y) = P_0 - \frac{U}{b^2/12\mu}(x - Ut) + B(t)e^{-qx}\cos qy \ , \tag{1.8b}$$

where P_0 is constant. The boundary condition at the sidewalls then requires the wavevector q to be

$$q = \frac{2\pi n}{W} \ , \tag{1.9}$$

where n is a positive integer.

If A is small, Eq. (1.8a) gives the velocity of the interface to be

$$U_n = U + \dot{A}(t)\cos qy \ .$$

On the other hand, the Darcy equation (1.1) and pressure equation (1.8b) together imply

$$U_n = U + \frac{b^2}{12\mu} qB(t)\cos qy \ ,$$

if we neglect terms of order A^2 or AB. In this way, we derive two equations for U_n and thus get one relationship between A and B, namely

$$\dot{A}(t) = B(t)\frac{b^2}{12\mu}q \ . \tag{1.10}$$

The final relationship is derived by calculating the terms in the pressure jump which are proportional to $\cos qy$, using the fact that pressure in the air is constant. On one hand from Eqs. (1.8a) and (1.8b), this part of the pressure jump is

$$\Delta P(y) = \left(\frac{U}{b^2/12\mu} A - B\right)\cos qy \ . \tag{1.11}$$

On the other hand Eq. (1.5a) gives the pressure jump as

$$\Delta P(y) = -T\kappa \approx -T\frac{d^2}{dy^2}x(y) \ . \tag{1.12}$$

Putting this result together with equations (1.10) and (1.11) one finds that A satisfies

$$\dot{A} = A\left(\frac{U}{b^2/12\mu} - Tq^2\right)\frac{b^2}{12\mu}q \ . \tag{1.13}$$

This result is easily interpreted. First of all, notice that when the quantity in parentheses is positive the flat interface is unstable, when it is negative the interface is stable against a disturbance of the given wavenumber. Since, according to Eq. (1.9) the minimum value of q is $2\pi/W$, the flat interface will be unstable against some perturbation whenever the d_0 defined by Eq. (1.7) obeys $d_0 < 1$. Alternatively expressed, a very long interface will be unstable against a perturbation of wavelength $l = 2\pi/q$ whenever $l > W\sqrt{d_0}$. Hence for very small surface tension, the system will be unstable against even very short-wavelength perturbations.

The physical source of this instability lies in the geometry of the moving interface. Imagine a situation in which the pressure difference along the length of the channel is fixed. Then, since the pressure in the air is constant, the larger gradients in the pressure appear at the end of the largest fingers of air. Hence these fingers move faster than the rest. Hence they get further ahead. The entire system, is in this way, destabilized by the motion.

Conversely, the surface tension tends to stabilize and smooth out the smallest fingers, those with a radius of curvature less than $W\sqrt{d_0}$. These smallest fingers have at their ends a large pressure drop from across the air-water interface. Water flows in toward these low pressure regions, pushing the smallest fingers backward. Hence they are smoothed out by the surface tension.

Something very peculiar happens when the surface tension goes to zero. Then the most unstable wavelength becomes shorter and shorter. In the limit, $T \to 0$, the shortest wavelengths are the most unstable. One suspects that, in this limiting case the entire physical problem may well be poorly defined.

1.3. The small surface tension puzzle

Let's for a moment ignore the problem of the short wavelength instability in the absence of surface tension and ask about the steady states only. In their classic paper, Saffman and Taylor (1958) found a one parameter family of finger shaped steady state solutions. These solutions correspond to different values of λ, the ratio of the finger width to the width of the cell. The finger shape is described by the formula to be given in Sec. 2,

$$\frac{x}{W} = \frac{(1-\lambda)}{\pi} \ln \cos\left(\frac{\pi y}{\lambda W}\right) . \tag{1.14}$$

These shapes seemed to be quite similar to those that were observed experimentally. However, there were two serious problems. First, in the experiment, a finger of a well defined width was observed at each given velocity. The zero surface tension theory could not predict that, since by varying λ one could obtain fingers of any width at all. Second, from our analysis above, one should expect the $T = 0$ solutions to be completely unstable, while the fingers that were observed were quite stable. Both problems were related to the singular nature of the zero surface tension limit. Another point made by Saffman and Taylor in their paper was that no fingers with λ less than $\frac{1}{2}$ were seen in the experiment at all, and they asserted that $\frac{1}{2}$ was the asymptotic width of the finger in the $d_0 \to 0$ limit.

The question of "velocity," or finger width λ, selection was again taken up by McLean and Saffman (1981) who looked for the steady state solutions in the presence of a small but finite surface tension. Numerically solving the integral equation for the interface, they found a unique solution for a given value of the surface-tension parameter, rather than a one parameter family. Their work was further extended by Vanden-Broeck (1983) who found not just one, but a discrete set of solutions. However, for all of his solutions, as d_0 goes to zero, λ goes to one half.

Thus, the "degeneracy" of the steady states is lifted by the effects of the surface tension, which is a *singular* perturbation in this problem (see Bender and Orsag (1978) for a discussion of singular perturbations). This phenomenon, common to a large class of nonlinear problems arising in physics, was studied by Barenblatt and Zel'dovich (1972) (see also Barenblatt, 1977) in the general context of similarity solutions to partial differential equations. (The propagating solution, such as the Saffman-Taylor finger, may be thought of as a kind of similarity solution as well.) Barenblatt and Zel'dovich point out that in cases where singular perturbations are involved, the search for the similarity solutions leads to nonlinear eigenvalue problems. These eigenvalues then determine the scaling, or in case of propagation, the velocity, of the similarity solution. The existence of a continuous family of solutions would then correspond to a continuous spectrum. More commonly, a discrete spectrum is found. Thus, the work of McLean and Saffman and Vanden-Broeck fit nicely into this general[2] framework.

The results of McLean and Saffman (1981) and Vanden-Broeck (1983) produced a theoretical prediction for the finger-width dependence on the control parameter. Alas, the stability problem remained unresolved since the analysis performed by McLean and Saffman found that the fingers remained unstable even in the presence of surface tension! A result contradictory to the experiment and numerical simulations. Hence, the point about stability remained open.

All of these difficulties arise from the subtlety of the zero surface tension limit, which is singular indeed. In fact the short wavelength instability leads to the appearance of finite time singularities in the dynamical equations for a large class of initial conditions as was shown by Shraiman and Bensimon (1984) and Sarkar (1984) (see also the work of Meyer (1982) and Howison (1985a)). These singularities correspond to $\frac{2}{3}$ power cusps in the interface. After the appearance of the cusp the calculations (and probably the solutions) break down. Some of the time dependent solutions evolving into such cusps can be found explicitly (Meyer, 1982; Shraiman and Bensimon, 1984 and Howison, 1985a). While many initial conditions lead to cusps, there are also some special initial conditions which give instead $\lambda = \frac{1}{2}$ steady fingers. These $T = 0$ results can be derived using the conformal mapping method which we shall describe in Sec. 3.

In the last year, as a result of the experimental studies of Tabeling and Libchaber (1985) and theoretical work of Kessler and Levine (1985), DeGregoria and Schwartz (1985) and Bensimon (1985) a new understanding of the stability problem began to emerge (see also the earlier experiments of Aribert (1970), as well as the more recent work of Maher (1985)). First of all, experimentally the fingers at high velocity (small d_0) *are* unstable (a fact that was observed, but for some reason ignored in Saffman and Taylor (1958)). Naively, one can try to explain this by noting that for small d_0 the unstable wavelength $l = W\sqrt{d_0}$ is much shorter than the characteristic curvature and width of the finger. Thus

[2] There is reason to believe that some of the other puzzling "selection" problems will be resolved along the same lines: for example, recently Pomeau and Pelce (1985) derived a nonlinear eigenvalue equation governing the shape and velocity of a dendrite (in the low Peclet number limit of the "two-sided" model). The work of Barenblatt and Zel'dovich also largely anticipated the "microscopic solvability" principle put forward by Kessler, Koplik and Levine (1984) and Ben-Jacob and coworkers (1984) to explain the growth velocities in their models of solidification.

on the length scale l the finger appears to be essentially flat and therefore should be unstable. If this were indeed the case one would expect the finger to become unstable at $d_0 \approx 1$, that is, shortly after the "primary" instability which lead to the appearance of the finger in the first place. Instead, the instability is observed at $d_0 = 10^{-2}$ and a different scenario is required. Kessler and Levine (1985) suggested that the interaction of the finger with the rigid walls makes the finger stable with respect to infinitesimal perturbations (contrary to the result of McLean and Saffman (1981)). This was corroborated by the observation by DeGregoria and Schwartz (1985) that in the numerical simulations the disturbances generated at the tip decay in amplitude as they are subvected along the side of the finger. DeGregoria and Schwartz (1985) and Bensimon (1985) then proposed that the experimentally observed behavior is due to a finite amplitude instability. Furthermore, from numerical stability analysis and simulations, Bensimon found that the noise amplitude required for destabilization decreases rapidly with d_0, and is consistent with the expression

$$\ln(noise) \sim -d_0^{-1/2} \,, \tag{1.15}$$

(see Sec. 4.3 below). He also found the most unstable modes, which are in excellent agreement with the experiments of Tabeling and Libchaber (1985).

The physical mechanism that is involved here appears to be very similar to the one proposed earlier by Zel'dovich and coworkers (1980) in connection with the stability of cellular flames. The growth rate of a disturbance is proportional to the normal velocity of the interface, so that it is large at the tip and approaches zero toward the side of the fingers. As the finger moves forward, the disturbance moves more slowly than the tip, so that it gradually moves toward less unstable regions. When the instability becomes weak, even a little surface tension is sufficient to damp out the disturbance. The dependence of the finger width λ on the control parameter d_0 is shown in Fig. 1.4. While the dependence is similar to that observed in the experiment of Saffman and Taylor (1958), the direct comparison is not quite satisfactory. The reason is that the theorists have simplified the problem by assuming that the pressure jump on the interface is velocity independent, Eq. (1.5a), rather than a more appropriate condition given by Eq. (1.5b). Patrick Tabeling has argued that the latter, more complicated boundary condition can be heuristically incorporated as an effective surface tension parameter. In the physically important region it differs by about a factor of two from the actual surface tension of the oil used in the experiment. With this correction the theoretical predicted finger width agrees with his experimental data to within 25%, which is reasonably good in view of all the complications of the actual flow: the three-dimensional effects at the meniscus, the wetting film that the finger leaves behind, etc. Figure 1.5 shows that the theoretical obtained finger shape does agree reasonably well with the experimental data.

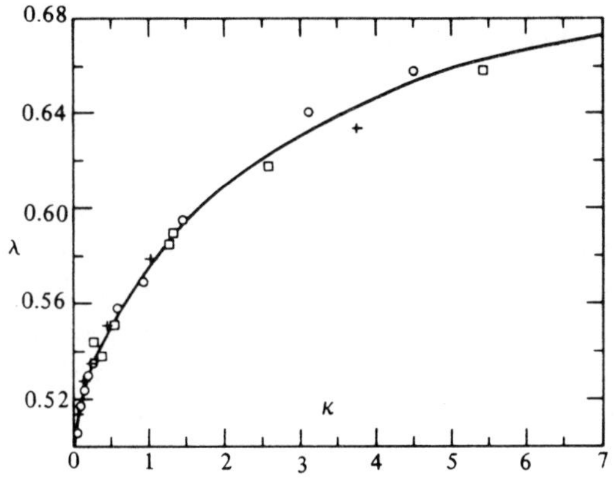

Fig. 1.4. The finger width λ as a function of the control parameter $\kappa(\kappa = \lambda d_0(1-\lambda)^{-2})$. Solid line is the McLean-Saffman (1981) numerical result. Circles are Bensimon's (1985) simulations. Crosses are from the simulations of DeGregoria and Schwartz (1985). Boxes are from Liang's (1985) random walk simulations.

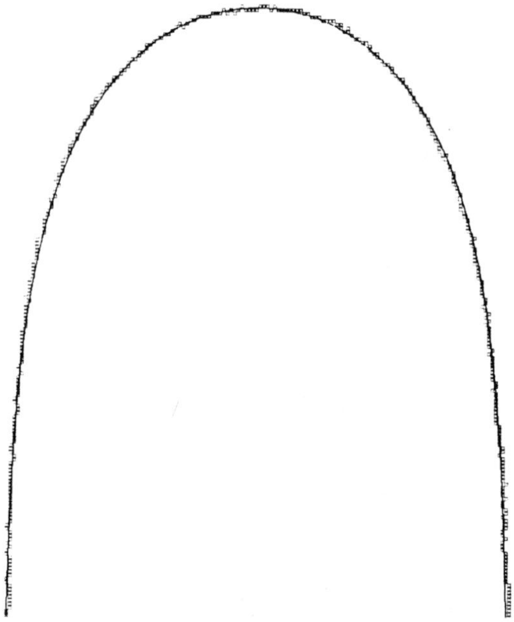

Fig. 1.5. Comparison of the finger shape obtained from simulations and experiment. The boxes are from a random walk simulation, Liang (1985), the solid line is Pitts' (1980) phenomenological scaling hypothesis which agrees with the experimental data of Saffman and Taylor (1958).

2. The Hodograph Method

2.1. *Complex analytical methods*

In this section we summarize what is known about the Saffman-Taylor problem at zero surface tension. We shall describe a method, involving analytic functions of complex variables which enables one to study the dynamics of the interface evolution in the absence of surface tension.

In the first section the Saffman-Taylor problem was formulated mathematically in terms of a Laplace equation for the pressure (or velocity potential) with two boundary conditions. As is usual in two-dimensional problems involving the Laplace equation, one can be helped considerably by using complex variable techniques (Birkhoff, 1952; Carrier, Krook, Pearson, 1966). The basic idea is to think of the velocity potential, $\phi(x,y)$ as the real part of a complex field

$$\Phi(x,y) = \phi(x,y) + i\psi(x,y) . \tag{2.1}$$

The demand that $\nabla^2\phi = 0$ can be automatically satisfied by simply requiring that Φ be an analytic function of the complex variable

$$z = x + iy . \tag{2.2}$$

For reasons which will become more obvious later, it is better to invert the functional dependence and describe not how Φ is determined by z but rather that the potential Φ defines where we are in space. In symbols

$$z = f_t(\Phi) . \tag{2.3}$$

The subscript t indicates that the dependence changes in time. In fact we shall use the boundary conditions (1.4) and (1.5) to define the time-dependence of f_t.

Equation (2.3) is the central equation in the hodographic method, widely used in fluid mechanics. For the Saffman-Taylor problem the hodograph technique has been used to obtain a partial solution in the absence of surface tension (see Secs. 2.3 and 2.4 below), to do simulations (as discussed in Sec. 3 below), and to discuss the stability of small surface tension solutions (see Sec. 4 below).

2.2. *Basic equations*

The real velocity potential $\phi(x,y)$ is defined so that its gradient is the velocity vector. The corresponding statement for the complex case is that

$$v_x(x,y) - iv_y(x,y) = \frac{d}{dz}\Phi(z) . \tag{2.4}$$

Equation (2.4) can then be re-expressed to give several useful boundary conditions. Since v_y vanishes on the side walls, $y = \pm W/2$, the imaginary part of Φ must be constant on each wall. Denote the velocity downstream by U. Since v_y vanishes far downstream, as $\mathrm{Re}(z) \to \infty$, $\Phi \to Uz + \mathrm{const}$. Take the constant to be real and then notice that $\mathrm{Im}\,\Phi = \pm UW/2$ on the two walls. What we have just said can be converted into boundary conditions upon the unknown function f_t of Eq. (2.3), i.e.,

$$\text{if} \quad \mathrm{Re}\,\Phi \to \infty\,, \quad \text{then} \quad f_t(\Phi) \to \frac{\Phi}{U} + C_0 \quad \text{with} \quad \mathrm{Im}\,C_0 = 0\,, \tag{2.5a}$$

$$\text{if} \quad \mathrm{Im}\,\Phi = \pm UW/2\,, \quad \text{then} \quad \mathrm{Im}\,f_t(\Phi) = \pm W/2\,. \tag{2.5b}$$

Finally, at zero surface tension, the pressure and hence the velocity potential is constant on the interface between the two fluids. Thence, if we take this constant to be zero, we can choose Φ to be of the form is for s real, on the interface.

To define the interface precisely we must determine a curve for each value of the time t. One way of doing this is to define a complex function of two real variables, s and t, i.e., $\gamma(s, t)$. For each value of t, as s sweeps over its entire range, $\gamma(s, t)$ sweeps over a set of points $z_t(s)$. These points are the complex variables which give the values of $x + iy$ for all x and y on the interface. This function, $\gamma(s, t)$, is thus the solution to our problem. The third boundary condition is that, in the case of zero surface tension,

$$\gamma(s, t) = f_t(is)\,, \qquad s \in \left(\frac{-UW}{2}, \frac{UW}{2}\right). \tag{2.5c}$$

Notice that the three boundary conditions (Eqs. 2.5) have defined the physical region of Φ to be the strip:

$$\Phi = \phi + i\psi\,, \qquad -\frac{UW}{2} \leqslant \psi \leqslant \frac{UW}{2}\,, \qquad 0 \leqslant \phi \leqslant \infty\,. \tag{2.6}$$

Within this strip, f_t must be analytic and its derivative must be nonzero. These two conditions together ensure that $\nabla^2\phi = 0$ in the region filled by water.

One more condition must be fulfilled: the interface must move with the same velocity as the fluid. Consider the time-dependence of $\gamma(s, t)$. For each value of s, $\gamma(s, t)$ specifies the value of $x + iy$ for some point on the interface at time t. An infinitesimal time interval later, this same piece of fluid will have moved forward by an amount $(v_x + iv_y)dt$. In this way, we find one term in the time derivative of γ:

$$\frac{d}{dt}\gamma(s,t) = v_x + iv_y \ . \tag{2.7a}$$

Using Eq. (2.4), we can rewrite (2.7a) as

$$\frac{d}{dt}\gamma(s,t) = \left(\frac{d\Phi}{dz}\right)^* = \frac{1}{(dz/d\Phi)^*} = \frac{-i}{\partial_s\gamma(s,t)^*} \ . \tag{2.7b}$$

The total derivative on the left-hand side of (2.7b) represents the possibility that, as the front advances, the value of the parameter s labeling a particular piece of fluid might change. If it does at a rate ds/dt then we must add to Eq. (2.7b) a term reflecting this change to obtain

$$\partial_t\gamma(s,t) = -i\,\frac{\partial_s\gamma(s,t)}{|\partial_s\gamma(s,t)|^2} - \frac{ds}{dt}(s,t)\,\partial_s\gamma(s,t) \ . \tag{2.8}$$

Equation (2.8) will determine the motion of the interface. At first sight, it does not look as if the derivation of Eq. (2.8) is a huge amount of progress. For one unknown function, $\gamma(s,t)$ we have traded another, ds/dt. However, in fact, a considerable advance has been made. We know that $\gamma(s,t)$ is analytic and has a nonzero derivative for all values of s in the strip (2.5c). This analyticity essentially determines the parameterization of $\gamma(s,t)$. In turn the analyticity, plus the condition that ds/dt is real fully determines the solution to Eq. (2.8).

There are several different methods for obtaining solutions to Eq. (2.8). The easiest is to eliminate ds/dt by multiplying by the complex conjugate of $\partial_s\gamma$ and then taking the imaginary part of the result, to find

$$\partial_t\gamma(s,t)\,\partial_s\gamma(s,t)^* - \partial_t\gamma(s,t)^*\,\partial_s\gamma(s,t) = -2i \ . \tag{2.9}$$

Equation (2.9) plus the statement that $\partial_s\gamma(s,t)$ is analytic and nonzero whenever s lies in the strip (2.5c), yields the evolution of the interface.

2.3. An example

To show what all this means, we develop an example analogous to the stability analysis of Sec. 1.2, but now specific to the zero surface tension case. We shall now get not just an expansion for small amplitudes but an exact solution for the interface. To derive this, replace the guesses, (1.8), about the form of the pressure and the interface by a corresponding guess for $f_t(\Phi)$,

$$f_t(\Phi) = C_0(t) + \Phi/U + C_1(t)\,e^{-q\Phi/U} \ . \tag{2.10}$$

Here the first two terms give a flat interface while the third represents a "correction" with wave vector q. The actual interface is given by the curve traced out by γ as a function of s, where

$$\gamma(s,t) = C_0(t) + is/U + C_1(t)e^{-isq/U} \quad . \tag{2.11}$$

Notice that the nontrivial dependence upon Φ and s is given in terms of

$$\omega = e^{-q\Phi/U} = e^{-isq/U} \quad . \tag{2.12}$$

Our boundary conditions insist that C_0 is real. Choose C_1 to be real also.

To see that (2.11) is an exact solution, simply differentiate and substitute into Eq. (2.9). The result is

$$(\dot{C}_0 + \omega\dot{C}_1)(1 - C_1 q\omega^{-1}) + (\dot{C}_0 + \omega^{-1}\dot{C}_1)(1 - C_1 q\omega) = 2U \; . \tag{2.13}$$

The expression (2.11) will be an exact solution if we can insure that (2.13) is satisfied for all s. This will, in turn, be true if we can make the coefficients of ω^j for $j = 0, \pm 1$ each vanish. The resulting differential equations for C_0 and C_1 are

$$\frac{d}{dt}(2C_0 - C_1^2 q) = 2U \; , \tag{2.14a}$$

$$\frac{d}{dt}\ln C_1 = q\frac{dC_0}{dt} \; . \tag{2.14b}$$

If $C_1 = 0$ at time zero, it remains zero. Then C_0 increase linearly in time, $C_0 = U(t - t_0)$ and correspondingly the interface moves forward with speed U. If, however, C_1 starts out positive but small it will continually grow larger. The solution will remain acceptable until

$$\frac{\partial}{\partial s}\gamma(s,t) = \frac{i}{U}[1 - qC_1(t)\omega] \; , \tag{2.14c}$$

vanishes at some s. This will happen at some finite time when C_1 becomes equal to $\pm q^{-1}$. At this time the interface acquires a cusp and after that the evolution is not defined.

2.4. *Finger solution*

There exist a few special solutions which do not go to cusps. One kind is the family of finger solutions found by Saffman and Taylor (1958). Here if the finger has a width λW and the fluid moves with speed U at $x = +\infty$, then the speed at which the interface advances is U/λ. Thence as $\mathrm{Re}\,\Phi \to \infty$, we can expect a solution of the form

$$f_t(\Phi) = Ut/\lambda + \Phi/U + g(\Phi) , \tag{2.15a}$$

where g vanishes rapidly as $\mathrm{Re}\,\Phi \to +\infty$. In order to construct a finger, we need a singularity as Φ goes to the corners of the strip, which lie at $\Phi = \pm iUW/2$. One guess is that there is a logarithmic singularity at these points, i.e., that

$$g(\Phi) = \alpha \ln(1 + e^{-2\pi\Phi/UW}) . \tag{2.15b}$$

Given this guess, one finds that

$$\frac{\partial}{\partial s}\gamma(s,t) = \frac{i}{U} - \frac{2\pi i\alpha}{WU}\frac{1}{1 + e^{2\pi is/(WU)}} . \tag{2.16}$$

Then, a brief calculation based upon Eq. (2.9) shows that (2.15a) is indeed a solution if

$$\alpha = \frac{W}{\pi}(1-\lambda) . \tag{2.17}$$

This solution gives the profile described in Eq. (1.14), i.e., a single finger with width λ.

There are indeed other solutions of the form

$$f_t(\Phi) = Ut/\lambda + \Phi/U + \sum_{j=1}^{m} a_j \ln(e^{iq_j} + e^{-2\pi\Phi/UW}) , \tag{2.18}$$

with a_j real and positive and q_j real. These solutions have m "channels" going off to $x \to -\infty$, and are thus a generalization of the original Saffman-Taylor solution (2.15). These solutions are singular in the sense that the interface extends to infinity, the mapping has logarithmic singularities on the unit circle. The form of the steady-state solution given in Eq. (2.15) suggest the following ansatz for the time dependent solutions:

$$f_t(\Phi) = \Phi/U + C_0(t) + \sum_{j=1}^{m} a_j \log(\omega - p_j(t)) , \tag{2.19}$$

where ω is determined as a function of the complex potential Φ

$$\omega = e^{-2\pi\Phi/UW} . \tag{2.20}$$

The singularities of this map, $p_j(t)$ move with time but are confined outside the physical domain: $|p_j(t)| > 1$. When one of them hits the circle a channel going to $x = -\infty$ is formed. We can write down the evolution equation for these singularities. It turns out, however, to be more convenient to track the zeroes of $\partial_\Phi f$ instead. They must also lie outside the disk — otherwise the conformality requirement is not satisfied. We have

$$\partial_\Phi f = \alpha_0 \frac{\prod_{j=1}^{m} [\omega - \alpha_j(t)]}{\prod_{j=1}^{m} [\omega - p_j(t)]} . \tag{2.21}$$

Proceeding along the lines developed by Shraiman and Bensimon (1984), after some algebraic manipulations one can obtain the "pole dynamics" equations: a system of m ordinary differential equations governing the motion of the critical points of the map, $\alpha_j(t)$, of the form:

$$\partial_t \alpha_j = F_j(\alpha, p) , \tag{2.22a}$$

$$\partial_t p_j = G_j(\alpha, p) . \tag{2.22b}$$

Note, that while the p_j's completely determine the α_j's and vice versa (from Eq. (2.19–2.21) since the relation involves high order algebraic equations), it is more convenient to track the evolution of zeroes and poles by differential equations.

The simplest example is the case in which there is only one term in the sum in Eq. (2.19)

$$f_t(\Phi) = a_0(t) + \Phi/U + \frac{W}{2\pi} \ln(\omega - a_1(t)) . \tag{2.23}$$

Equations (2.22) then have the solution

$$a_0(t) = 2Ut ,$$

$$a_1^2(t) = 1 + [a_1^2(0) - 1]\, e^{-2\pi t U/W} . \tag{2.24}$$

In this case a cusp does not appear and instead the solution asymptotically approaches the Saffman-Taylor $\lambda = \frac{1}{2}$ steady-state solution.

The existence of a "pole" decomposition is somewhat surprising, since its existence is more commonly associated with integrable systems such as the

Burgers equation (Calogero, 1975; Chudnovski, 1977), and the KdV equation, (Kruskal, 1974; Moser, 1975), although it has been discovered in few other systems as well, see Lee and Chen (1982) and Thual, Frish and·Henon (1985).

The differential equations (Eqs. 2.22) can be solved explicitly in some cases, otherwise they can be studied numerically. It can be shown (Sarkar, 1984, Howison, 1985a), that most initial conditions of the form (2.21), lead to the appearance of $\frac{2}{3}$ power cusps on the interface which appear when one of the zeroes of $\partial_\Phi f$ hits the unit disk. Figure 2.1 shows an example of the appearance of such a cusp in the evolution of the initial interface given by

$$f(\Phi) = \Phi/U + \sum_{1}^{4} \log(\omega - p_j(t)) + C_0(t) , \tag{2.25}$$

with $p = (3,\ 10i,\ -9i,\ -6+4i)$. For most initial conditions, cusps form and after they form the equations seem to stop having solutions.

Physically, one expects the surface tension to prevent the singularities from appearing. Its effect should become important when the curvature near the cusp becomes of the order of the capillary length $W\sqrt{d_0}$. It may be possible to understand this problem using the method of matched asymptotic expansions (see Bender and Orzag, 1978), however, this has not yet been done.

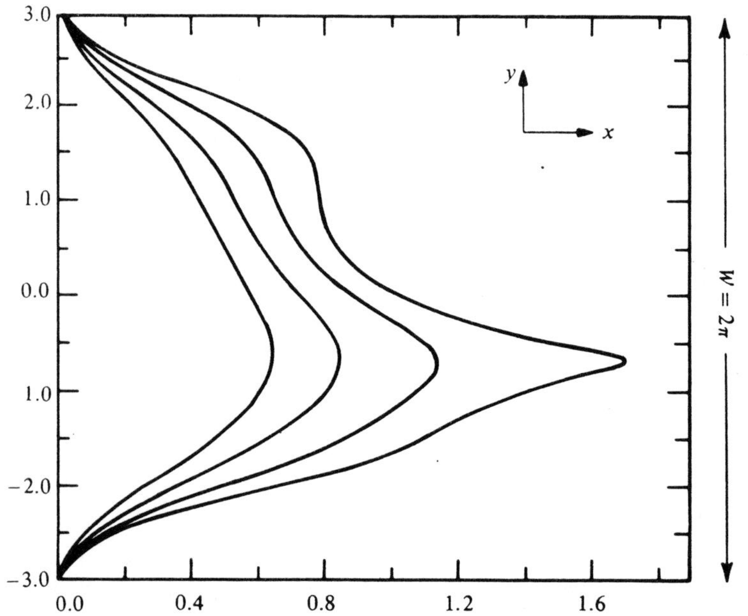

Fig. 2.1. Generation of a finite time cusp in the evolution of an arbitrary initial interface in the absence of surface tension.

3. The Conformal Mapping Algorithm

3.1. Introduction

There are several different ways of studying the fluid flow which results when the surface tension is not zero. Some of them have already been mentioned. Tryggvason and Aref (1983 and 1985) have applied a method in which the interface is described as a vortex sheet and thereby constructed a set of very appealing simulations of time development. A slightly different boundary integral method was employed by DeGregoria and Schwartz (1985) in their simulations of interface evolution. Both groups find that apparently stable fingers are generated, at least for the not-too-small values of d_0 for which their calculations are accurate. Their fingers, in turn, look very much like the fingers obtained from a direct solution of the steady-state problem as set up by McLean and Saffman (1981).

In this section we shall describe yet another method. This one is based upon the conformal mapping approach of Shraiman and Bensimon (1984) and used in the simulations carried out by Bensimon (1985). Similar ideas were used by Menikoff and Zemach (1983) in simulating motion of interfaces. Our motivation for this particular focus is that this method seems particularly well suited to the study of the limit of small d_0 where the behavior remains a bit of a puzzle.

3.2. Conformal method for the problem with surface tension

In Sec. 2, the hodograph technique was used to derive the equation of motion of the interface. In this section we will rederive this equation in the presence of surface tension ($d_0 \neq 0$) and from a slightly different approach.

The velocity potential ϕ obeys

$$\nabla^2 \phi = 0 \, , \tag{3.1a}$$

and the boundary conditions on the interface

$$\phi = \frac{Tb^2}{12\mu} \kappa \, , \tag{3.1b}$$

$$\hat{\mathbf{n}} \cdot \nabla \phi = \hat{\mathbf{n}} \cdot \frac{\partial \gamma}{\partial t} \, , \tag{3.1c}$$

where γ is the interface between the two fluids, κ its curvature and $\hat{\mathbf{n}}$ indicates the direction normal to the interface. Instead of including boundary conditions at the side walls, we assume here a periodicity under $y \to y + W$. This corresponds physically to a cylindrical Hele-Shaw cell such as the one used by Aribert (1970).

Equations (3.1) determine the evolution of the interface: the Laplace equation with the Dirichlet boundary condition on γ, Eq. (3.1b), completely determines the flow field, then, the value of the normal velocity (the normal gradient of ϕ) at the boundary determines the velocity of the interface, Eq. (3.1c). We have to solve a Stefan, or, moving boundary value problem. The two-dimensionality of the problem greatly simplifies the task by allowing the use of the conformal mapping technique.

The idea which is standard in all textbooks on complex variables, e.g., Carrier, Krook and Pearson (1966) is based on the Riemann mapping theorem. This theorem ensures the existence of a conformal map from the complicated but simply connected domain enclosed by the interface γ into a standard domain, the interior of the unit disk. Within the disk the Dirichlet problem for the potential ϕ, Eq. (3.1a, b) can be readily solved. That solution then enables us to rewrite Eq. (3.1c) as an evolution equation for the mapping. As in Sec. 3, we introduce the complex potential $\Phi(z) = \phi(x,y) + \psi(x,y)$ (with $z = x + iy$). We then conformally map the domain of interest, i.e., the space occupied by the driven fluid, into the unit disk ($|\omega| \leqslant 1$: $z = f_t(\omega)$) (see Fig. 3.1). Since the interface γ between the two fluids in the image of the unit circle ($|\omega| = 1$) under the map $f_t(\omega)$:

$$\gamma(t,s) = f_t(e^{is}) , \tag{3.2}$$

specifying the mapping, $f_t(\omega)$, at a given time t, is identical to specifying the interface, $\gamma(t,s)$ together with its parametrization, s.

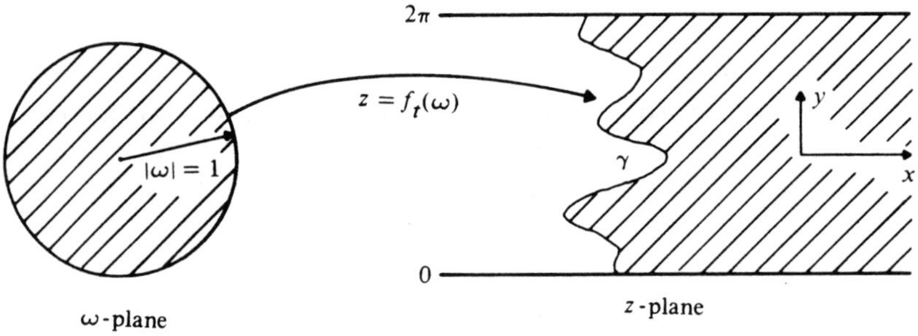

Fig. 3.1. The conformal map from the space occupied by the driven fluid to the unit disk.

In the unit disk the solution of the Dirichlet problem is standard. One has to find the function analytic inside the unit disk, $\Phi(\omega)$, real part which on the boundary, $\omega = e^{is}$, is specified, $\phi(s) = (Tb^2/12\mu)\kappa(s)$, where $\kappa(s)$, the local curvature of $\gamma(t, s)$ is

$$\kappa(s) = -\operatorname{Im} \frac{\partial_s^2 f / \partial_s f}{|\partial_s f|} \quad . \tag{3.3}$$

The solution is known to be given by the Poisson integral formula. That formula states that the function analytic for $|\omega| < 1, \tilde{g}(\omega) = g(\omega) + ih(\omega)$, for which the real part on the unit disk $g(s)$ can be written as

$$g(s) = a_0 + \sum_{n=0}^{\infty} (a_n e^{ins} + a_n^* e^{-ins}) \tag{3.4a}$$

must be

$$\tilde{g}(\omega) \equiv A\{g\}(\omega) = a_0 + 2 \sum_{n=0}^{\infty} a_n \omega^n \ . \tag{3.4b}$$

Here we interpret $A\{g\}$ as an integral operator applied to a real-valued function defined on the unit circle, giving a new complex-valued function defined and analytic within that circle.

This approach can be directly applied to the determination of the potential. First we do the trivial case in which we have a flat interface and the curvature vanishes. In this case the potential is $\Phi = Uz$, while the mapping that takes one from the unit disk $|\omega| \leqslant 1$ onto the strip $x > 0; -W/2 \leqslant y \leqslant W/2$ is $z = -(W/2\pi)\ln\omega$. Hence in this example

$$\Phi(\omega) = -\frac{UW}{2\pi}\ln\omega \ . \tag{3.5}$$

The singularity of Φ at $\omega = 0$ is simply a reflection of the $x \to \infty$ behavior of the problem. This behavior and this singularity will persist even the presence of a nontrivial interface. Thus we can say that $\Phi(\omega) + (UW/2\pi)\ln\omega$ must be analytic for $|\omega| \leqslant 1$, and must have the value given by (3.1b) on the circle. We use Eq. (3.4) to conclude that

$$\Phi(\omega) = -\frac{UW}{2\pi}\ln\omega + \frac{Tb^2}{12\mu}A\{\kappa\}(\omega) \ .$$

Following the notation of the theorem, we use the notation $\tilde{\kappa}(\omega)$ for $A\{\kappa\}(\omega)$ and hence get an expression for the complex potential

$$\Phi(\omega) = -\frac{UW}{2\pi}\ln\omega + \frac{Tb^2}{12\mu}\tilde{\kappa}(\omega)\ . \tag{3.6}$$

A comparison between Eq. (2.20) and (3.6) will show that the ω used here will reduce to the ω of Sec. 2 as $T \to 0$. However the s's of the two sections are different. As $T \to 0$, the s of this section is $-2\pi/UW$ times the s of the previous section. Now, the normal velocity of the interface, given by Eq. (3.1c), is $(\hat{\mathrm{n}} \cdot \nabla)\phi$ where $\hat{\mathrm{n}}$ is a unit vector normal to the interface. This vector can be rewritten in complex notation as

$$n = n_x + in_y = i\frac{\partial_s f}{|\partial_s f|} = -\frac{\omega\partial_\omega f}{|\omega\partial_\omega f|}\ . \tag{3.7a}$$

Here and in the rest of this section, ω is specified to lie on the unit circle. Given n one can calculate the normal component of the gradient as

$$\begin{aligned}(\hat{\mathrm{n}}\cdot\nabla)\phi &= \partial_x\phi n_x + \partial_y\phi n_y = \partial_x\phi n_x - \partial_x\phi n_y \\ &= \mathrm{Re}\,(n\partial_z\Phi) = \mathrm{Re}\left(n\frac{\partial_\omega\Phi}{\partial_\omega z}\right)\ . \\ &= -\frac{\mathrm{Re}\,(\omega\partial_\omega\Phi)}{|\omega\partial_\omega f|}\ .\end{aligned} \tag{3.7b}$$

Notice that Eq. (3.1c) only specifies the normal velocity of the interface. Of course there is no physical significance to a tangential velocity which simply corresponds to a reparametrization of the interface.

However the analyticity of the mapping function $f(\omega)$ fixes a particular "analytic," parameterization "gauge." This parametrization has to be maintained for all t. For that purpose, it is sufficient to make the time derivative of the map, $\partial_t f$, analytic inside the unit disc. To achieve this as in Eq. (2.8), we add to the right-hand side of Eq. (3.6) an appropriate tangential velocity component.

$$\begin{aligned}\partial_t f &= n(\hat{\mathrm{n}}\cdot\nabla)\phi + inC' \\ &= \omega\partial_\omega f\left\{\frac{\mathrm{Re}\,(\omega\partial\omega\Phi)}{|\omega\partial\omega f|^2} + iC\right\}\ .\end{aligned} \tag{3.8}$$

Here C and C' are real functions of ω. To make the right-hand side of Eq. (3.9) analytic, the function C has to be the harmonic conjugate of the first term in the brackets $\mathrm{Re}(\omega\partial_\omega\Phi)/|\omega\partial_\omega f|^2$. In other words the terms in the brackets have to represent the function analytic in $|\omega| \leqslant 1$ and which real part on $|\omega| = 1$ is specified. We have seen previously that this is achieved by the Poisson integral formula as expressed in Eq. (3.4). Therefore using Eq. (3.4) upon (3.8) and then substituting from Eq. (3.6) yields the desired evolution equation for the interface.

$$\frac{\partial f}{\partial t} = -\omega\partial_\omega f A\left\{\frac{1-(d_0 W/2\pi)\,\mathrm{Re}(\omega\partial_\omega\tilde{\kappa}(\omega))}{|\omega\partial_\omega f|^2}\right\}\frac{UW}{2\pi}\,. \tag{3.9}$$

3.3. *Numerical simulations*

If we know $f_t(\omega)$ for a given value of t we can obtain the entire right-hand side of Eq. (3.9) and thus find $\partial_t f$. This enables one to set up a numerical algorithm for simulating the evolution of $\gamma(s, t)$. This algorithm is rather efficient. One measure of the quality is the number of operations needed per time step. In this case, if one fits the interface at N points and thus retains N coefficients in a Fourier series like (3.4), then the computer code requires $O(N \log N)$ operations per time step. (Most of the time is spent in computing Fourier series, using a Fast Fourier Transform algorithm.)

The algorithm was checked against known results in the asymptotic regime $(t \to \infty)$. A typical outcome of such a simulation fits well the finger shape obtained by the phenomenological scaling hypothesis of Pitts (1980), see Figs. 1.5 and 3.2b. The dependence of the finger width on the McLean-Saffman surface tension parameter κ is shown in Fig. 1.4 and agrees with their numerical results for the steady-state interface. In the absence of surface tension, the time evolution of the interface was in complete agreement with the exact time dependent solution. It developed finite time singularities, see Figs. 3.1 and 3.2a. In its presence one observes two regimes. One is at low velocities ($d_0 > 10^{-2}$) for which an initial arbitrary interface evolves into the corresponding McLean-Saffman steady-state propagating finger, Fig. 3.2b. The other is at high velocities ($d_0 < 10^{-2}$) for which the finger is unstable, and shows wobbling and tip splitting, Fig. 3.2c. This is in qualitative agreement with recent numerical and experimental work, Tabeling and Libchaber (1985), Park and Holmsy (1985), Liang (1985), DeGregoria and Schwartz (1985).

In Sec. 4 we will use this algorithm to study the linear and nonlinear stability of the propagating finger.

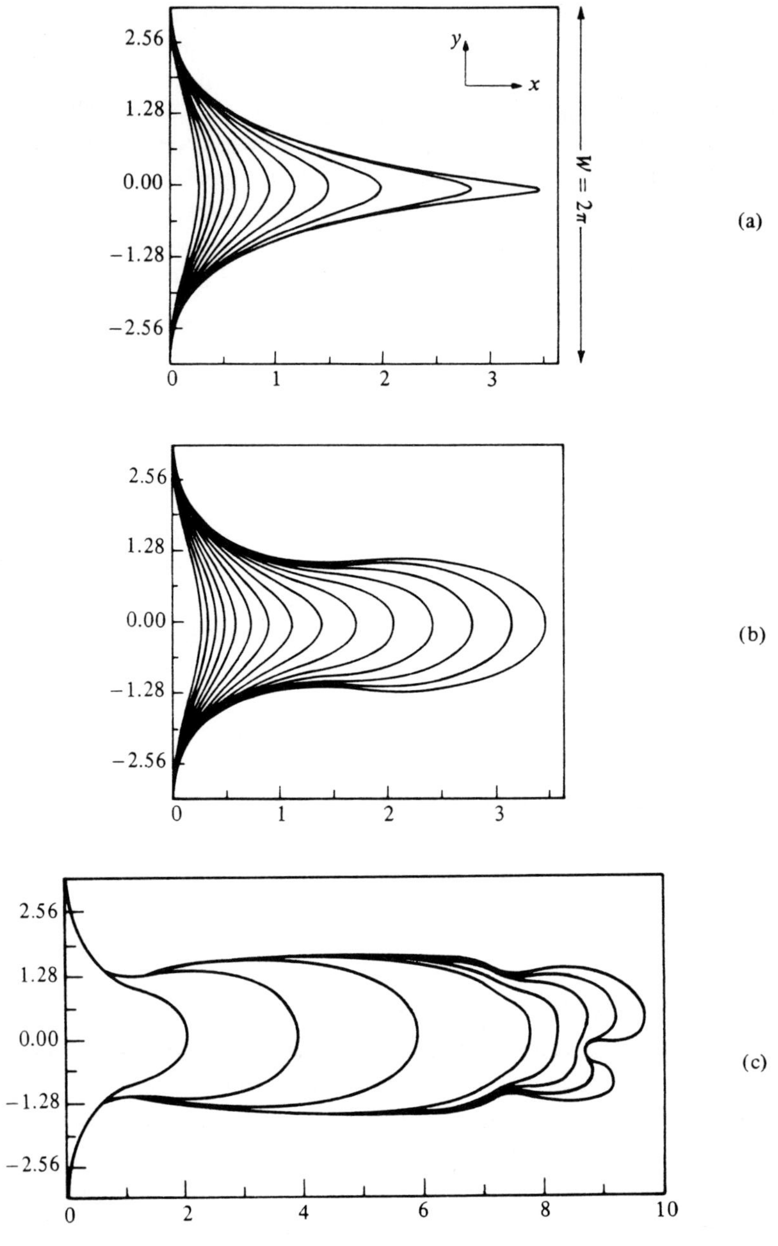

Fig. 3.2. Time evolution of an interface, from Bensimon (1985). Here the cell width W is set equal to 2π. (a) Evolution of an arbitrary initial interface without surface tension. (b) Evolution of the same initial interface as in (a), but in the presence of surface tension ($d_0 = .01$). (c) Tip splitting in the evolution of an interface at low surface tension ($d_0 = .01$).

4. Stability of the Fingers

4.1. Anomalous stability

In Sec. 1.3 we discussed the difficulties associated with the stability analysis of the finger solutions. A naive argument given there suggested that since the capillary length, $W\sqrt{d_0}$ is — for small d_0 — small compared to the characteristic curvature of the finger, one might expect the finger to be unstable just like the flat surface. This is, of course, at odds with experimental observations as well as numerical simulations. On the other hand, the direct linear stability analysis of Kessler and Levine (1985) and Bensimon (1985), carried out numerically, have shown the fingers to be stable to infinitesimal perturbation for all values of d_0. Kessler and Levine (1985) suggested that the stability is due to the interaction with the walls of the cell.

A particularly illuminating observation was made by P. Tabeling and A. Libchaber (1985), who studied the behavior of the perturbed fingers in the real Hele-Shaw cell, and by DeGregoria and Schwartz (1985), who simulated the evolution of the fingers numerically. This observation was that the localized disturbance of the interface drifts along the finger onto its side, whereupon it slowly disappears. The stabilization is thus due to the fact that the disturbance is expelled from the region of instability and does not have "enough time" to grow! This provides a valuable physical insight.

As pointed to us by P. Pelce, the physical mechanism involved is exactly the one proposed by Zel'dovich and coworkers (1980) to explain the stabilization of cellular flames. In Bensimon *et al.* (1986) we apply their ideas to the Saffman-Taylor problem. To verify the results obtained by the heuristic (and approximate) argument we present the results of a more direct study of stability carried out by Bensimon (1985) using the complex analytic method outlined in Sec. 3.

4.2. Stability analysis: the complex analytic method

The complex method can be directly used to study the stability of a steadily propagating finger. The method is directly analogous to the one used in Sec. 1.2. Assume that the conformal map describing the moving interface is

$$f_t(\omega) = f^0(\omega) + A_t(\omega) \ . \qquad (4.1)$$

Here $f^0(\omega)$ is the steady state finger solution, which then depends upon d_0, and $A_t(\omega)$ is a small time-dependent deviation from this solution. Since $f_t(\omega)$ is analytic inside the unit disk we may assume

$$f^0(\omega) = \sum_{n=0}^{\infty} f_n \omega^n ,$$

$$A_t(\omega) = \sum_{n=0}^{\infty} A_n(t)\,\omega^n . \tag{4.2}$$

The stability analysis is done by expanding Eq. (3.9) and keeping terms of first order in A. Since f^0 is a steady-state solution, the result is an equation of the form

$$\dot{A}_k(t) = \sum_{n=1}^{\infty} M_{kn}[f^0]\,A_n(t) + M'_{kn}[f^0]\,A_n^*(t) . \tag{4.3}$$

Notice that the matrices M and M' depend upon the presumed steady-state solution f^0. The $k=0$ term in Eq. (4.3) gives $\partial_t A_0 = 0$, thus yielding a marginal mode that corresponds to the translation of the finger. The fact that it decouples from the rest is quite convenient and is an advantage of the method.

The solution to the linear stability problem can now be clearly seen. Consider M and M' in Eq. (4.3) to be matrices. Form the supermatrix

$$M = \begin{bmatrix} M & M' \\ M'^* & M^* \end{bmatrix} . \tag{4.4}$$

If that matrix has an eigenvalue E, then $A_n(t)$ is exponential in t, e^{Et}. Stability then requires that the real part of the eigenvalue be negative so that any deviation from the finger solution would vanish as $t \to \infty$.

Once the the matrix elements are written down, the eigenvalues can be calculated numerically. Bensimon (1985) expanded f^0 in a power series in d_0 and then used that expansion to calculate M and M' to first order in d_0. The eigen-spectrum for $d_0 = 0.05$ is shown in Fig. 4.1. Notice the continuum of of symmetric modes (A_n real) and antisymmetric modes (A_n imaginary) with a negative real eigenvalue preceded by a discrete set of asymmetric modes with complex eigenvalues (their number increases as $d_0 \to 0$). The eigenvalues all have negative real part for values of d_0 down to 10^{-3}, so that the interface appears to be linearly stable. This is in agreement with the heuristic argument described in Bensimon *et al.* This stability result, as well as the eigenvalue spectrum, is also in agreement with previous results of Kessler and Levine (1985). It disagrees with the results of McLean and Saffman (1981), where instability was predicted. However Sarkar has argued that the instability prediction might be wrong because of neglected terms of order d_0 while Levine suggested that

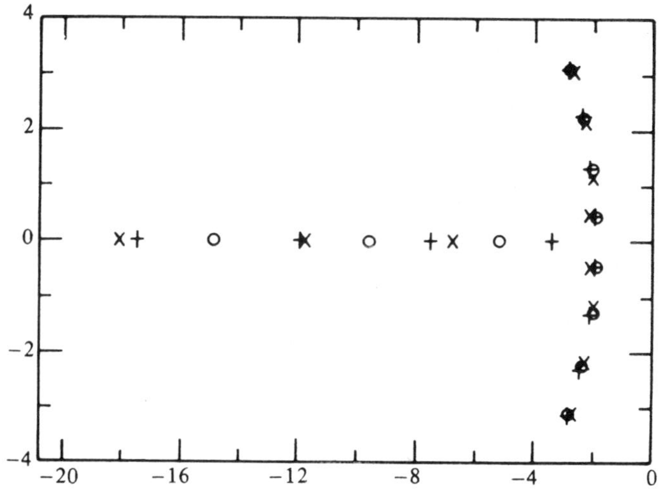

Fig. 4.1. The eigenspectrum at $d_0 = .05$, for various truncations: $N = 130$ (+); $N = 80$ (×); $N = 100$ (○). Notice the discrete spectrum of asymmetric modes (complex eigenvalues) and the continuum of symmetric and antisymmetric modes (negative real eigenvalues), from Bensimon (1985).

the number of mesh points was not sufficient to ensure accuracy. It appears to us that the results of Kessler and Levine (1985) and Bensimon (1985) are more reliable.

4.3. Structural stability and nonlinear instability

We have previously seen that the finger is destabilized due to the existence of a finite amplitude instability. Its proximity shows up as a sensitivity of the eigenspectrum of the linearized problem. Bensimon (1985) argued, that there is a relation between the structural stability of the linearized problem and the nonlinear instability of the full problem.

One may then study the dependence of the critical amplitude for destabilization on d_0 by looking at the appearance of unstable modes in response to a random distortion of the interface (letting the f_m in Eq. (4.2) contain a random term), an instability arising at a typical perturbation strength v_c which depends upon d_0. For example, we can obtain a fit with a form

$$v_c \sim d_0^{1/2} \exp(-\beta d_0^{-1/2}) \,, \tag{4.5}$$

with $\beta \approx 1.3$. This particular form for the fit was suggested by the argument of

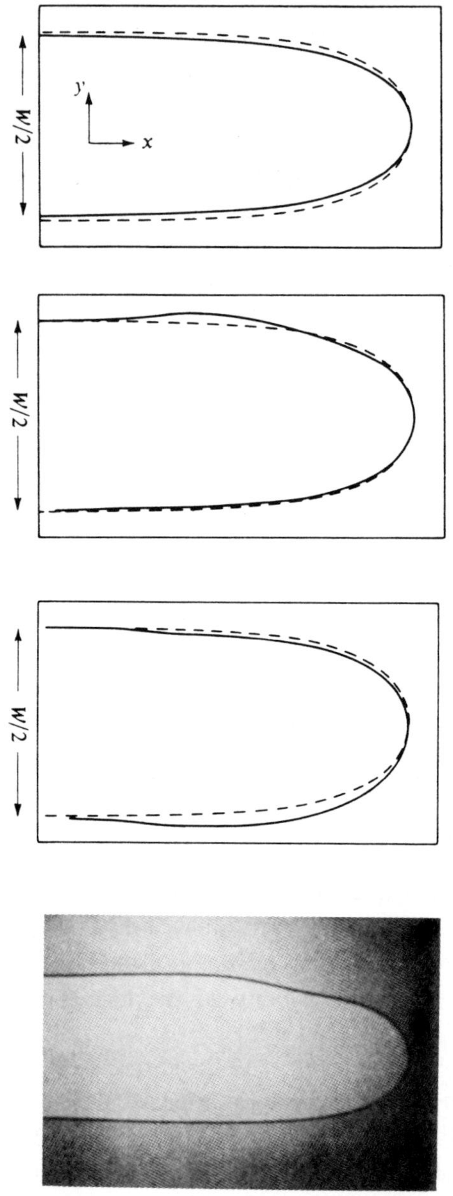

(a)

(b)

(c)

(d)

Fig. 4.2 The three most unstable modes (dashed line: unperturbed finger). (a) The symmetric non-oscillatory mode, which may be responsible for the experimentally and numerically observed fingers of width $\lambda < 1/2$. (b) The asymmetric "hump" mode. (c) The asymmetric "tip wobbling" mode. (d) The asymmetric "hump" mode observed in an experiment (courtesy of P. Tabeling and A. Libchaber (1985)).

Bensimon *et al.* Other fits are possible, for example $v_c \approx \exp(-\gamma d_0^{-\beta})$, with $\beta \approx .61$ and $\gamma \approx .72$. They all exhibit a singular behavior as $d_0 \to 0$.

When the noise in f^0 is larger than v_c unstable modes arise. The most unstable modes are shown in Fig. 4.2: there are two asymmetric oscillatory modes, and a symmetric non-oscillatory one which corresponds to a change in the width λ of the finger. These modes have apparently been observed in numerical simulations (Liang, 1985), and in experiments (Tabeling, 1985).

The conjecture that the structural stability of the spectral problem is related to a finite amplitude instability was verified numerically with the help of the algorithm based on the conformal mapping technique described in Sec. 3. This was done by studying the threshold of the instability at a given value of d_0 as a function of the amplitude of an initial random analytic perturbation of the interface. This comparison confirms the existence of a finite amplitude instability with threshold depends singularly on d_0, possible of the form predicted by Eq. (4.5).

Acknowledgments

The research reported here was supported by the DOE, ONR, NSF, and the University of Chicago's MRL. Every step of the work was aided by a lively interchange with the experimental group of Tabeling, Libchaber, and Heslot. At several points S. Sarkar joined us in our work. In addition we have been helped by a frank and timely interchange of views and data with many of the other workers in the field, including E. Ben-Jacob, N. Goldenfeld, J. Langer, A. DeGregoria, L. Schwartz, S. Howison, D. Kessler, H. Levine, J. Koplik, R. Lenormand, P. Saffman, J. Nittman, Y. Pomeau, P. Pelce, J. Kertesz, U. Frisch, G. Tryggvason, H. Aref, T. Vicsek, F. Family, and T. Witten. Work in this area has been made much more pleasant by the vigorous exchanges mentioned above.

References

Aribert, J.-M. (1970) PhD thesis, Toulouse.

Barenblatt, G. I. (1979) *Similarity, Self-similarity and Intermediate Asymptotics* (Consultants Bureau, New York).

Barenblatt, G. I., Zel'dovich, Ya. B. (1972) *Ann. Rev. Fluid Mech.* **4**, 285.

Bender C. M., Orsag, S. A. (1978) *Advanced Mathematical Methods for Scientists and Engineers* (McGraw-Hill, New York).

Ben-Jacob, E., Goldenfeld, N., Kotliar, G., Langer, J. (1984) *Phys. Rev. Lett.* **53**, 2110.

Bensimon, D. (1986) *Phys. Rev.* **A33**, 1302.

Bensimon, D., Kadanoff, L. P., Liang, S., Shraiman, B. I. and Tang, C. (1986) submitted to *Rev. Mod. Phys.*

Birkhoff, G., Zarantonello, E. (1957) *Jets, Wakes and Cavities*, New York.

Calogero, F. (1975) *Lett. Nuovo Cimento* **13**, 411.

Carrier, G. G., Krook, M. and Pearson, C. E. (1966) *Functions of Complex Variable* (McGraw-Hill, New York).

Chudnovsky, D. V. and Chudnovsky, G. V. (1977) *Nuovo Cimento* **40B**, 339.

Chuoke, R. L., van Meurs, P. and van der Pol, C. (1959) *Trans. AIME* **216**, 188.

DeGregoria A. J. and Schwartz, L. W. (1985) Exxon preprints.

DiBenedetto, E. and Friedman, A. (1984) *T. Am. Math. Soc.* **282**, 183.

Elliot, C. M. and Janovsky, V. (1981) *Proc. Roy. Soc.* Edin. **88A**.

Hele-Shaw, H. J. S. (1898) *Nature* **58**, 34.

Howison, S. D. (1985a) *J. Appl. Math.*, to appear in *SIAM*.

Howison, S. D., Ockendon, T. R. and Lacey, A. A. (1985b) to appear in *QJMAM* 267.

Kessler, D. A. and Levine, H. (1985) preprint.

Kessler, D. A., Koplik, J. and Levine, H. (1984) *Phys. Rev.* **A30**, 3161 and *Proceedings of the Electrochemical Soc. of Amer.*, Toronto (1985).

Kruskal, M. D. (1974) in *Nonlinear Wave Motion*, Newell, A. ed., (Am. Math. Soc., Prov. R. I.) p. 61.

Langer, J. S. (1985) ITP preprint.

Lee, Y. C. and Chen, H. H. (1982) *Phys. Scripta* **T2**, 41.

Liang, S. (1986) *Phys. Rev.* **A33**, 2663.

Maher, J. (1985) *Phys. Rev. Lett.* **54**, 1498.

McLean, J. W. and Saffman, P. G. (1981) *J. Fluid Mech.* **102**, 455, and McLean, J. W. (1980) PhD thesis, Caltech.

Menikoff, R., Zemach, C. (1983) *J. Comp. Phys.* **51**, 28.

Meyer, G. H. (1982) in *Numerical Treatment of Free Boundary Value Problem*, J. Albercht, ed., (Birkhauser, Basel).

Moser, J. (1975) *Adv. Math.*, **16**, 197.

Nittman, J., Daccord, G. and Stanley, H. E. (1985) *Nature* **314**, 141.

Park, C.-W. and Homsy, G. M. (1984) *J. Fluid Mech.* **139**, 291.

Park, C.-W. and Homsy, G. M. (1985) *Phys. Fluids* **28**, 1583, 1621.

Pietronero, L. and Wiesmann, H. J. (1984) *J. Stat. Phys.* **36**, 909.

Pitts, E. (1980) *J. Fluid Mech.* **97**, 53.

Pomeau, Y. and Pelce, P. (1985) preprint.

Richardson, S. (1972) *J. Fluid Mech.* **56**, 609.

Saffman, P. G. and Taylor, G. I. (1958) *Proc. R. Soc.* Lond. **A245**, 312.

Sarkar, S. (1984) *Phys. Rev.* **A31**, 3468.

Shraiman, B. I. and Bensimon, D. (1984) *Phys. Rev.* **A30**, 2840.

Tabeling, P. and Libchaber, A. (1986) *Phys. Rev.* **A33**, 794.

Thual, O., Frish, U. and Henon, M. (1985) Nice Observatory preprint.

Tryggvason, G. and Aref, H. (1983) *J. Fluid Mech.* **136**, 1, and to be published, 1985, *J. Fluid Mech.*

Vanden-Broeck, J.-M. (1983) *Phys. Fluids* **26**, 2033.

Zel'dovich, Ya. B., Istratov, A. G., Kidin, N. I. and Librovitch, V. B. (1980) *Comb. Sci. and Tech.*, **24**, 1.

Note added in proof:

The reader may also be interested in more recent papers:

Combescot, R., Dombre, T., Hakim, V., Pomeau, Y. and Pumir, A. (1986) *Phys. Rev. Lett.* **56**, 2036.

Hong, D. C. and Langer, J. S. (1986) *Phys. Rev. Lett.* **56**, 2032.

Shraiman, B. I. (1986) *Phys. Rev. Lett.* **56**, 2028.

reference frame

On complexity

Leo P. Kadanoff

The physicist is largely concerned with abstracting simple things from a complex world. Newton found simple laws that could predict planetary motion. Results from atomic physics have demonstrated that the Schrödinger equation is a correct description of almost all observed atomic behavior. In fact, the hydrogen atom is used as a metaphor to describe the search for simple and illuminating examples in all areas of the physical sciences.

In some ways, this search for the simple makes our description of the world seem like a caricature rather than a portrait. Most elementary physics textbooks describe a world that seems filled with very simple, regular and symmetrical systems. A student might get the impression that atomic physics is the hydrogen atom, that electromagnetic phenomena appear most often in the world in guises like the parallel plate capacitor or dipole radiation, and that regular crystalline solids are "typical" materials. (Hydrodynamics is the one subject in the standard physics curriculum that deals most fully with richly complicated outcomes of physical laws. It is now only rarely taught.) If our hypothetical students look at the world with an unbiased eye, they will see a richness quite unlike anything described in physics texts. In the real world, simplicity is a rare exception. Waves break upon reaching a beach, producing swirls, foam and spray. Every planet is different from the others, some with moons (again all different), some with beautifully structured rings. Atomic and molecular spectra show many different characteristic energy or frequency differences and a richness of structure not seen in the hydrogen atom. Even "elementary" particle physics includes a complexity that belies its name. Almost every decade of energy from millielectron volts to hundreds of GeV is described by its own individual phenomenology. Although complexity is the most obvious feature of the world about us, the textbook description of physics ignores this diversity or treats it as an aberration to be circumvented by choosing simple situations that might give us "characteristic" examples.

But I paint too extreme a picture. Physics *is* concerned with understanding complexity, particularly in those parts of the science aimed at more applied examples. Here I intend "ap-

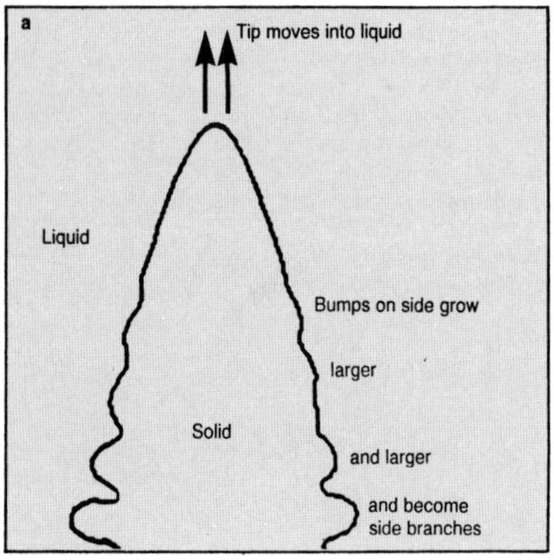

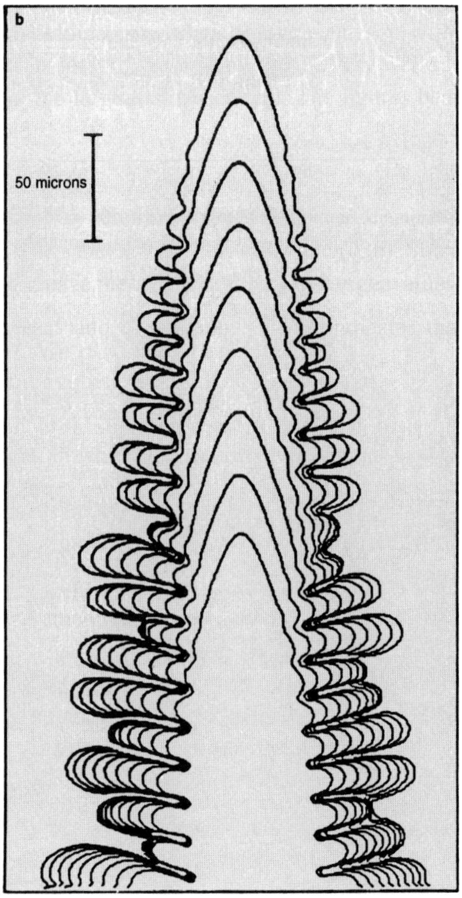

Dendritic growth, a process in which a needle-shaped crystal grows into a liquid via the forward motion of the needle's point and the continual growth of bumps on its sides. **a:** A schematic view of the region near the point at a given time. **b:** The output of an experiment in which the growth of an ammonium bromide dendrite is observed at 20-second intervals and the resulting contours are superposed. (Reproduced from a preprint of A. Dougherty, P. D. Kaplan and J. P. Gollub.) Figure 1

reference frame

plied" not to refer to commercial usefulness, but rather to mean focused upon the particular types of systems that really arise in the physical world. The astrophysicist, who must understand the distribution of matter in the universe; the biophysicist, asking perhaps how life arose; the plasma physicist, working with the intertwined structure of flux lines in a swirling ionized gas; the solid-state scientist, looking at the crystallization of a piece of steel—all these scientists must deal with complexity as an everyday issue. Until recently, many physicists have dismissed examples such as these as "dirt physics" or "squalid-state physics"—perhaps intending to suggest that these examples somehow contain less intellectual content than, say, a simple and easily interpreted spectrum. Here, I wish to suggest the possibility of the opposite view: that the observed complexity in the world around us raises questions that are absolutely fundamental to our understanding of the nature of physical law. Three such questions are:

▶ How do very simple laws give rise to richly intricate structures?

▶ Why are such structures so ubiquitous in the observed world?

▶ Why is it that these structures often embody their own kind of simple physical laws?

There is a discipline in and around physics that does study complexity as something interesting in itself, both in general and more especially in the most accessible examples in hydrodynamic flow, astrophysics, solid-state physics, lasers and so on. This subject of study has been given various names, but recently the most popular terms have been "pattern formation" and "chaos." Let me describe one of the central ideas that have emerged from these studies.

A major reason that complexity emerges in the working out of physical laws is that in many situations one has what is called sensitive dependence upon initial conditions. Sensitivity can be seen very directly in a process called "dendritic growth," which occurs when a crystalline solid freezes. Under appropriate conditions the solid moves forward into the liquid in the form of a needle with bumpy sides. (See figure 1.) Each bump is produced by an instability in the solidification process. The bumps are initially formed near the tip of the needle. At that point, they are very, very small. As time goes on, the tip moves forward and the instability magnifies each bump. Thus, eventually, the bumps form a complex pattern of very many side branches pushing out of the needle's sides. The heights and placements of the side branches vary because each branch is produced by the amplification of an initially microscopic perturbation on the interface. Hence the variation of the observed structures is produced by the *repeated occurrence* of the sensitive dependence of outcomes upon the growth conditions. This repeated sensitivity is a major source of natural complexity.

Leo Kadanoff, a condensed matter theorist, is John D. MacArthur Professor of Physics at the University of Chicago.

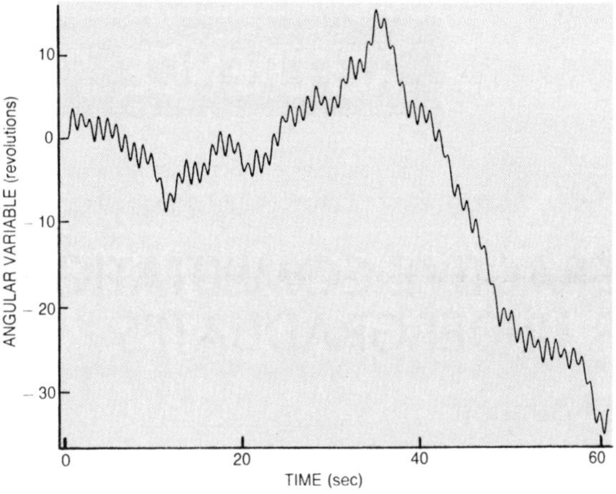

Displacement angle θ for a damped forced pendulum as a function of time. Note that θ is measured in revolutions; the plot shows that the pendulum goes through full circles in a complex and never repeating pattern. Figure 2

Dendrites provide an example of chaotic irreversible growth. (The growth of snowflakes is more familiar and probably roughly analogous.) Another example arises when motion occurs in a kind of closed environment so that the variables describing the system cannot ever get out of a bounded domain. In some sense the system must do the same thing again and again. However, if this kind of system exhibits repeated sensitivity to conditions, it too will be chaotic. It will move through its bounded domain, with each go-around being different from all that came before. The system will never precisely retrace the same path.

A prototypical example of this chaotic behavior is one in which a damped pendulum is pushed by an external periodic force. The differential equation involved is terribly simple. If the pendulum's angular position is θ, and it has mass m and length l, the equation is

$$ml^2\frac{d^2\theta}{dt^2} + \gamma\frac{d\theta}{dt} = -mgl\sin\theta + F\cos\Omega t$$

For a correct choice of damping constant γ, forcing strength F and forcing frequency Ω, the motion never exactly repeats itself. (See figure 2.) Here, a very intricate motion is produced from the simplest of ingredients.

Some progress has been made in understanding motions such as that plotted in figure 2. We do have a partial phenomenological description in terms of the concepts of fractals and strange attractors. Moreover, there is a good theoretical and experimental description of how, as the parameters in the equation of motion are varied, systems like these go from an orderly to a chaotic motion.

However, our present understanding is limited to systems in which we can describe the motion in terms of a few variables or a few excited modes. Hydrodynamic systems can exhibit much richer behavior in which there are swirls within swirls within swirls and simultaneous sensitive dependences upon conditions at many different length scales. Despite a few tantalizing hints, understanding of this fully turbulent behavior has not yet been been achieved.

So what do we say to the student who asks why the laws of physics are so simple, but the world so complicated? Clearly the right answer is "Come into physics and help us find out."

* * *

My series of columns in PHYSICS TODAY has been improved by the helpful criticism of my colleagues Ruth Ditzian Kadanoff, Albert Libchaber, Sidney Nagel, Stephen H. Shenker, Robert Wald and others, including many members of the staff of PHYSICS TODAY. I especially mention this help here because Shenker guided me away from a real blooper in an early draft of this column.

Further reading

• P. Berge, Y. Pomeau, C. Vidal, *Order Within Chaos*, Wiley, New York (1986). □

REFERENCE FRAME

INTERACTIVE COMPUTATION FOR UNDERGRADUATES

Leo P. Kadanoff

The practice of physics, and indeed of all the sciences, has changed greatly because of the existence and ready availability of computers. Naturally, practice has changed more rapidly than education. Nonetheless, in the last few years the computer revolution has resulted in some revision in the style and content of undergraduate physics instruction.

The tedium of lab data analysis has been mitigated by the use of graphing programs and other similar software for data manipulation and display. (For example the spreadsheet, which is a business oriented display and analysis program, can very easily take a hundred estimates for V and I, convert them into a hundred estimates for R and display the entire result.) A few undergraduate and first-year graduate courses are now aimed at teaching future physicists how medium and large-scale computations are performed.[1]

In this column, however, I would like to concentrate upon another style of computation, in which the student or the scientist uses the computer interactively to answer questions. In the best realization of this style, a student works at a graphics terminal or workstation, enters instructions to the computer in a "higher level" language like Basic, Fortran, Pascal, C or APL, and immediately sees the result of these instructions in the form of numbers and pictures. The goal is to answer questions, such as *Given a particle trapped in central force potential that goes like* $1/r^4$, *is a typical orbit closed?* (Almost never) or *What can we say about the frequency-dependence of the response of a forced pendulum?* (It is very complex and most often chaotic).

Questions like these can be asked and answered by the typical undergraduate physicist or engineer who has mastered a year of calculus. In my view, they should be posed to students working in the setting of a computational lab, with perhaps one lab partner. The two would share a workstation or terminal, have available some device to produce hard copy, and would have a lab assistant or professor available to answer or pose questions. This computational lab setup would have a somewhat lower cost than the usual undergraduate experimental lab setup. Some texts related to interactive computer use are available.[2]

Classical mechanics can provide much material for such a course. The US university typically teaches classical mechanics to potential physicists at least thrice: freshman physics (Halliday and Resnick), sophomore–junior mechanics (Marion) and graduate level (Goldstein). (I list common texts in parentheses.) None of these books treats any modern topic in mechanics. Thus there are probably many young physicists who do not know that classical mechanics is a subject of intense research activity.[3]

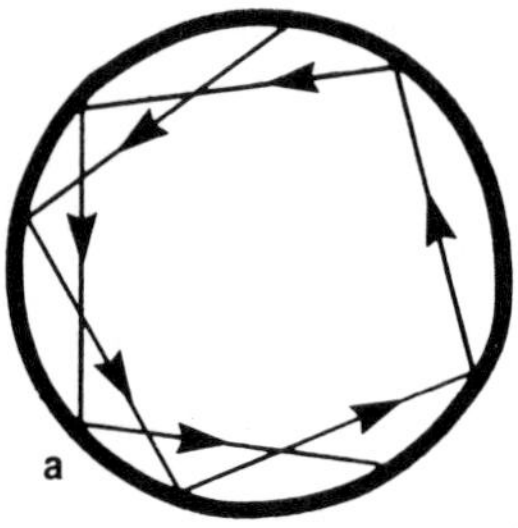

a

Motion inside a circle.

The modern period starts with Poincaré, who was interested in the qualitative properties of motion in classical mechanics and how long it might take for the motion to repeat itself. This question translates in part into whether the motion is orderly or chaotic. On the mathematical side these studies led to the work of A. M. Liapunov, Andrei Kolmogorov, Stephen Smale, Vladmir Arnol'd, Jascha Sinai and others. On the physical side the question was translated into a line of excellent computer experiments conducted, for example, by Enrico Fermi, John Pasta and Stanislaw Ulam, by Norman J. Zabusky and Martin D. Kruskal, and most recently by Gerald Sussman and Jack Wisdom of MIT,[4] who showed that the orbit of Pluto (and hence of the entire solar system) is chaotic. Such questions have consequences for experiment and engineering. One recent example: John Hoffnagle and his collaborators[5] showed that the orbit of two barium atoms caught in an ion trap makes a transition from regular to chaotic behavior as the depth of the well is increased. In this case, the experimental study was backed up by a simulation showing that the presumed transition did indeed occur.

The recent progress in dissipative systems[6] has in a remarkable degree combined experimental, theoretical and computational components. On the computational side, most of the

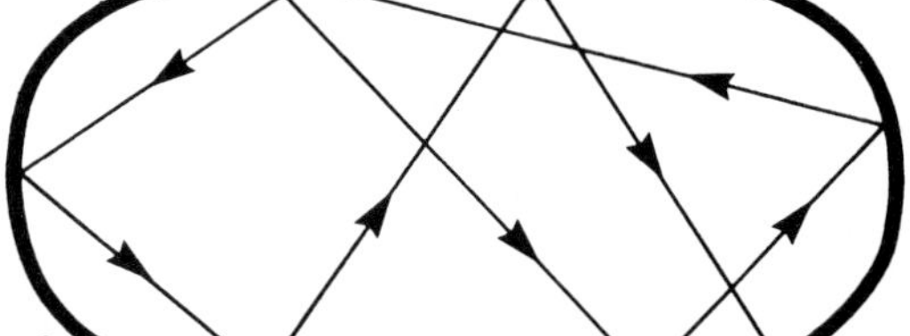

b

Motion inside a stadium.

Leo P. Kadanoff, a condensed matter theorist, is John D. MacArthur Professor of Physics at the University of Chicago.

REFERENCE FRAME

progress has been achieved via interactive methods. Students can themselves follow the actual path taken by this kind of research, and thus see for themselves how the period-doubling route to chaos was discovered, or how people first understood the nonlinear interactions of solitons.[7] These hands-on studies are well-suited to help students develop their own physical intuitions.

There are certainly many different ways of designing such a course. In our University of Chicago course,[2] Michael Vinson and I include treatment of phase-plane portraits, the Lorenz equations, closure of orbits, order and chaos in dissipative maps, stability analysis, Liapunov indices, period doubling, fractals, random walks and solitons. These topics are quite elegant and have mathematical depth. In addition, they are closely related to some of the most fundamental questions of present-day physics.

To see the flavor I want, consider the following "trivial" question: Start with the circular and "stadium" shapes shown respectively in parts a and b of the figure on page 9 and imagine that each container includes a particle that bounces elastically off its walls. Given typical values of starting position and direction of the particle, what is the long-run probability distribution for finding the particle at point r in the container? Of course the student will need a few hints. But any well-trained physicist can see at once that one of these problems has a trivial answer.... But which one?

References

1. S. E. Koonin, *Computational Physics*, Benjamin/Cummins (1986).
2. D. A. Niguidula, A. van Dam, *Pascal on the Macintosh*, Addison-Wesley (1988) (for first-year computer majors); H. Gould, J. Tobochnik, *An Introduction to Computer Simulation Methods*, Addison-Wesley (1988); L. Kadanoff, M. Vinson preprint (1988) (for junior and senior physics majors).
3. There are a few texts that do take a more modern point of view. See for example I. Percival, D. Richards, *Introduction to Dynamics*, Cambridge U. P. (1986), for a modern but not-too-mathematical treatment at the senior/first-year graduate level.
4. G. Sussman, J. Wisdom, Science **241**, 433 (1988).
5. J. Hoffnagle, R. G. DeVoe, L. Reyna, R. G. Brewer, Phys. Rev. Lett. **61**, 255 (1988). See also the report by R. Pool in Science **241**, 787 (1988).
6. See the text by P. Berge, Y. Pomeau, C. Vidal: *Order within Chaos*, Wiley, New York (1986).
7. N. J. Zabusky, M. D. Kruskal, Phys. Rev. Lett. **15**, 240 (1965). ■

Section E

Complex Patterns

From Correlation to Complexity

In the previous section, I discussed how apparently complex patterns could be produced by dynamical systems. My own study of complexity has been going through a kind of transition in recent years. I have begun to realize that dynamical systems often show only a very limited and partial complexity. Often these systems produce complex time patterns in one or a few variables. However, the true richness of natural systems go well beyond few-variable chaos. This richness has been emphasized by the people at the Santa Fe Institute, and most particular Murry Gell-Mann[35] and Stuart Kauffman.[36] Specifically, Stuart argued that a sufficiently complex dynamical system would naturally develop a rapidly increasing complexity. Structures would emerge of a fantastic sort. Details of history would determine exactly which structures were produced, but — in this view — the production of structures was an inevitable outcome of a system with a sufficient number of variables and interconnections.

There is another possible aspect to this structure generation. Most truly interesting dynamical systems (you and me for example) involve dynamics in both space and time. The spatial dimension permits a structured complexity. First there is translational invariance. This invariance permits structures to repeat again and again at a different point in space. But in a chaotic system, nothing quite repeats — instead it is reproduced with different variations each time. For example, all the people in the world are in one sense similar structures, but they have interesting and important individuality that in the end, makes life interesting. Another aspect of this structured complexity is that it exists on all length scales. There are people, and organ systems, and cell tissues, and individual cells, and the working parts of cells, and individual molecules, and each of these different scales shows an organized behavior — different on each scale. Each scale's behavior is repetitive and complex and history dependent.

Recently, condensed matter physics (following biology and many other disciplines) has moved into the study of such multi-structured complexity.

As part of a kind of study of this subject, I have been working on turbulence. In my whole career, and particularly at Chicago, I have been blessed by many coworkers who have led and immeasurably enriched various aspects of my research. Theoretical visitors, especially Procaccia, and a whole rich collection of graduate students and postdocs have made my research what it is. But, probably, I learned most from my collaboration with Albert Libchaber.

For several years, Libchaber and I worked as a team in studying complexity. He led the way in visualizing and realizing complexity, I tried to apply theoretical tools to understanding what he had done. We studied three systems together. I have already mentioned the onset of chaos in the mercury cell and the Hele-Shaw system. But, the richest case which we studied together was convective turbulence. Turbulence is chaos in both space and time.

[35]M. Gell-Mann, *The Quark and the Jaguar* (W. H. Freeman, 1994).
[36]S. Kauffman, *At Home in the Universe.*

When a relatively large volume of fluid is heated from below, the fluid on the bottom expands and tends to rise. Thus the fluid is set into motion. At high heating rates the motion becomes unstable and turbulence results. This kind of system is called a Rayleigh-Bénard cell. Our work is summarized in E1, E2 and E3. My work in this area has been the product of collaboration with many people including Bernard Casting, G. Gunaratne, F. Heslot, S. Thomae, Xiao-Zhong Wu, S. Zaleski, and S. Zanetti, Anton Kast, and Masaki Sano. An additional study was done by Dan Rothman and myself as paper E4. This paper involved using a very simple kind of computer model, called a cellular automaton, to produce the kinds of swirling motion seen in the Rayleigh-Bénard cell. It was a very pleasant kind of *tour de force* and produced lovely pictures and patterns.

This turbulence work is and was a dream shared by Libchaber and myself. We hoped we could and would really understand how a hydrodynamic system developed a very rich complexity. Something about this dream came to reality. Our work did expose many aspects of the behavior in the cell, including many aspects of the complicated geometry produced by the motion. (The cover of this book shows four different pictures of this geometrical structure.) The experimental work cast into doubt the general applicability of the Kolmogorov theory to systems of this kind. This theory involves a cascade of energy from larger scales to smaller ones. The cascade produces some kind of fractal structure. But then, as always, the work reached a barrier. We had gotten all the knowledge we could from the experiments on convective turbulence, and it was time to move on. Albert and I turned to different approaches to complexity, he in biological systems, and me through the mathematical sciences.

In papers E5, E6, and E7, I turned away from real turbulence and instead studied a simplified model of the cascade process built into the Kolmogorov theory. This work involved a system simpler than fluids, but one which nevertheless showed a rich structure on a wide range of length scales. Of course, because this model does not begin to show the richest kind of complexity and structure, it is in the end slightly unsatisfactory. (Please see #168, #170, and #171 for more detailed write-ups of my collaborations on this subject with Roberto Benzi, Detlef Lohse, Jane Wang, and Norbert Schrghofer.)

One kind of structure repeated again and again in hydrodynamic systems arise from the propensity of these systems to produce mathematical singularities. Papers E7, E8, and E9 describe the universal structures which arise from this kind of singularity mechanism. But once again one gets nothing like the complexity shown by the natural world, so the analysis feels incomplete and somewhat unsatisfactory. At least this is somewhat the feeling I get from re-reading the papers. When the chase is on, investigations like these are always great fun and wonderfully satisfying. There is a neatness and completeness in the classification of the universal properties of relatively simple similarity solutions, which stands in contrast to the complexity and messiness of the real world.)

Systems in Statistical Equilibrium can only show a small amount of complexity. This limitation on complexity is built into the equilibrium statistical mechanics of Boltzmann, Gibbs, and Maxwell. Their statistical mechanics formalism is one in which the probability of a given configuration is proportional to an exponential of the Hamiltonian divided by minus the temperature times the Boltzmann constant. Such a weight prevents the most complex configurations from dominating the system. However, if a system is not in statistical equilibrium, this exponential weight no longer holds. Something much more complex may emerge. The next series of papers, is based upon particles which cannot go to true equilibrium because their interparticle collisions do not conserve energy. Among the physical systems which show these properties are glass balls or ordinary sand. We use the phrase 'granular material' to describe these systems. Paper E10 is aimed at elucidating the properties of these theoretical models of 'sand' in the simplest possible geometries. In fact, we observe that the granular material does have a far richer and more structured behavior than the usual equilibrium systems. Their main characteristic is that portions of them may freeze or slow down into a glassy state. This slowing or freezing then serves as a memory of the history of the system. Sand castles are examples of such memory.

Memory is the subject of the next paper, E11, which aims at explaining how non-equilibrium phenomena can serve as part of a solid state memory device. None of these works actually reach the levels of complexity familiar from everyday life, but they are all essays in the right direction. This volume closes with one more paper on complexity. In E12, G. McNamara, Gianluigi Zanetti and I use a cellular automaton model to simulate the very simplest situation in hydrodynamics, flow down a pipe — and find a rich renormalization of the viscosity caused by the development of large-scale swirls in the flow.

Scientific work on turbulence and complexity is far from completed. Science cannot yet answer, or even properly formulate, questions related to how complexity does actually arise in the world. I started saying that computers can do 10^8 or so calculations per second while a bit of gas sees 10^{32} events/second. But biological systems are even more richly complex. In the course of evolution there might have been 10^{55} or so collisions among atoms in biological systems. This provides lots and lots of steps in which different levels of complexity might have developed. We certainly do not understand the development of complexity. In fact, we do not even understand how the patterned formed by the clumping of matter as it is pushed around by the motion of a fluid.[37] But, unanswered questions and partially formulated problems are the bread and butter of science. I am pleased to see that my table is still full.

[37]This has been a subject of study for me in recent years with my principal collaborators, being Mario Feingold, Oreste Piro, and Peter Constantin, Itamar Procaccia, and Emily Ching.

Scaling and Structures in the Hard Turbulence Region of Rayleigh Benard Convection

reported by

Leo P Kadanoff

The Research Institutes
The University of Chicago
5640 S. Ellis Avenue
Chicago Illinois 60637 USA

Abstract

Experimental and theoretical studies of Rayleigh-Bénard convection at high Rayleigh numbers ($10^8 < Ra < 10^{13}$) were performed. by Bernard Castaing, Gemunu Gunaratne, François Heslot, Leo Kadanoff, Albert Libchaber, Stefan Thomae, Xiao-Zhong Wu, Stéphane Zaleski, and Gianluigi Zanetti (J. Fluid Mech. (1989)). The results of these studies are further examined in the light of visualization in Rayleigh-Bénard flow in water (Steve Gross, Giovanni Zocchi, and Albert Libchaber, C.R. Acad Sci. Paris t. 307, Série 2, 447 (1988)). The previously developed theory is shown to be incomplete in leaving out many of the structures in the flow. We take special note of the coherent flow throughout the entire water tank. Despite the many omissions of geometrical structures from the scaling analysis, most of the order of magnitude estimates seem right.

Reprinted from *On Turbulence: Proc. of the 5th EPS Liquid State Conf.*, 1989.

A recent paper[1], describing convective flow in Helium gas at low temperatures, about 5 degrees Kelvin, included an overall theoretical description of the flow. We considered that the cell could be divided into regions (see the cartoon of Figure 1) which have different characteristic scaling behaviors. As in the classical descriptions of the system[2], the regions include a boundary layer, where the viscosity and thermal conductivity dominate and a central region region, in which convective effects and buoyancy forces are important. In addition these is a mixing zone in which we imagined structures composed of hot fluid (see the cartoon of Figure 2) coming up and merging into the central region flow.

The scaling behavior can be specified in terms of the Rayleigh number

$$Ra = \frac{g\alpha\Delta L^3}{\kappa\upsilon} \quad (1)$$

Here, g is the acceleration of gravity, α is the volume thermal expansion coefficient, Δ is the temperature difference between the bottom and the top of the cell, κ and ν are respective the thermal diffusivity and the kinematic viscosity, and L is a characteristic size of the cell. In the theory described in reference (1), the thickness of the boundary layer and of the mixing zone are respectively given by

$$\lambda = L\, Ra^{-2/7} \quad (2a)$$

$$d_m = L\, Ra^{-1/7} \quad (2b)$$

In addition, the Nusselt number is

---

1 Bernard Castaing, Gemunu Gunaratne, François Heslot, Leo Kadanoff, Albert Libchaber, Stefan Thomae, Xiao-Zhong Wu, Stéphane Zaleski, and Gianluigi Zanetti (J. Fluid Mech. (1989).

2 Malkus W. V. R. : Discrete transitions in turbulent convection, Proc. Roy. Soc. **A 225** 185-195 (1954), The heat transport and spectrum of thermal turbulence Proc. Roy. Soc. **A 225** 196-212 (1954).; Spiegel, E. A. : On the Malkus theory of turbulence, Mécanique de la turbulence (Colloque International du CNRS a Marseille), Paris,'Ed CNRS 181-201, (1962); Kraichnan R. H. : Turbulent thermal convection at arbitrary Prandtl number, Phys. Fluids. **5**, 1374, (1962).

$$Nu = Ra^{2/7} \tag{2c}$$

a typical fluctuation in the central region is given by

$$\Delta_c = \Delta\ Ra^{-1/7} \tag{2d}$$

while the velocity in the central region has a characteristic size

$$u_c = \frac{\kappa}{L}\ Ra^{3/7} \tag{2e}$$

The Helium experiment gave direct support for relations (2c) and (2d). The observed fact that temperature differences across the boundary layer were of order $\Delta/2$ and expression (2c) implies (2a). Other experiments[3] give support for (2e). However, there was at the time that paper was written (January, 1988) no evidence for the existence of the mixing zone. In addition, the helium experiment measured a characteristic frequency of oscillations in the cell, given by

$$\omega_p = \frac{\kappa}{L^2}\ Ra^{1/2} \tag{2f}$$

The theory's prediction of this frequency seems at bit *ad hoc*.

Robert Kraichnan has pointed out that 'the wonderful thing about scaling it that you can get everything right without understanding anything'[4]. I think he means that one can estimate the orders of magnitudes involved without really having a good picture of the geometrical structures. So next, we should look at the geometry of the flow.

It is very hard to see anything in the helium system. However, after the helium paper of reference 1 was written, the experimental group constructed an experiment in which they could see the temperature variations in a water cell at a similar Rayleigh number[5]. It is

---

[3] H. Tanaka and H. Miyata, Int. J. Heat and Mass Transfer 23, 1273 (1980); A.M. Garron and R.J. Goldstein, Int. J. Heat and Mass Transfer 23, 738 (1980)

[4] private communication.

[5] Steve Gross, Giovanni Zocchi, and Albert Libchaber, C.R. Acad Sci. Paris t. 307, Série 2, 447 (1988).

interesting to compare a picture they took (see Figure 3) with the result of the theory. The picture was obtained by shining light through the cell and observing the shadows cast by the variations in the water's index of refraction as in figure 1B of reference 5.

The picture clearly shows the three different regions, that had been anticipated by the theory. The dark shadows at the very bottom and top are caused by the gross variations of temperature in the boundary layers. In the center, there is a region of relatively slow variation in temperature corresponding to the central region of Figure 1. Furthermore, the mixing zone hypothesized by the theory is seen in the picture as a relatively active region near the top and bottom walls. So far so good.

However, notice also the heightened activity near the right and left-hand walls in comparison with the central region. By looking at the flow, one sees that there is a rapid and coherent directed upward motion of fluid on the right hand wall and a downward motion on the left. These motions are concentrated in the jet-like regions which are seen in Figure 3 as the areas of heightened activity. This coherent flow is closed by left to right motion in the bottom half of the cell and corresponding right to left motion of the top. We call those motions 'wind'. These portions of the vortical flow are apparently rather more spread out than the jets near the walls. These jets were certainly left out of the original scaling picture. But, they were seen in a followup study of the Helium system[6] in which average velocities of the fluid were measured. The authors of reference 6 pointed out that these jets provide a quantitative explanation of the frequency ω_p which they interpreted as simply the result of convecting hot blobs round and round the cell. A consequence of this picture and of the observed value of ω_p is that in addition to the velocity scale (2b) the overall coherent motion of the system would have to have the velocity $\omega_p L$, which would then scale roughly as $Ra^{1/2}$. This scaling behavior was not mentioned in reference 1 and in fact it is different from the $Ra^{3/7}$ behavior of Eq. (2e).. I find it disconcerting to have two nearby scaling indices for two closely related quantities[7].

---

6 Masaki Sano, Xiao Zhong Wu, and Albert Libchaber 'Study of Turbulence in Helium Gas Convection'

7 Sano, Wu, and Libchaber (reference 6) point out that there is a loophole in that the unknown Prandtl number dependence of the various measured

Next notice that the boundary layers look bumpy. Reference 5 contains some pictures which show a wave-like disturbance being advected along the bottom of the container in the direction of the wind. Apparently, these waves are propagating deviations in the thickness of the boundary layer. Moreover, one sees in the mixing zone some concentrated disturbances which perhaps have a character like that in the cartoon of Figure 2. A more characteristic shape is that of a mushroom which is apparently rooted in the boundary layer and then is advected by the flow. Neither wave nor mushroom was a portion of the original scaling theory. The picture indicates that some thermal have come loose from their moorings and are moving through the central region, but these free plumes are few and far between. Many features of this picture have been seen before. Van Dyke[8] shows pictures of mushroom-like thermals while Chu and Goldstein[9] say "As the Rayleigh number is increased, the thermals show a persistent horizontal movement near the bounding surface and the majority of the thermals are dissipated without reaching the opposite surface; the central region of the layer becomes an isothermal core".

I have drawn in the waves and thermals in the cartoon of figure 4. This particular drawing shows the thermals as rising from the crest of the wave. It is not implausible that there is a secondary three-dimensional instability that breaks up the two-dimensional wave-structure. However it is also plausible that thermals might arise independently of the waves[10]. Because the pictures are two-dimensional projections, it is not yet known just how the waves and plumes might be associated.

All in all, the hard-turbulence cell appears to be a very complicated object with many working parts. Nonetheless a simple scaling description close to that given in Equs. (2) seems to hold.

---

quantities might perhaps produce changes in the observed critical index values.

8 M. Van Dyke, *An Album of Fluid Motion*, The Parabolic Press, Stanford, California, (1982).p. 63.

9 Chu T.Y. and R.J. Goldstein: Turbulent convection in a horizontal layer of water, J. Fluid Mech., **60**,141-159 (1973).

10 Andrew J. Majda, private communication.

Acknowledgements

This description of experimental observations in this report is a report upon knowledge I have gained from many hours of talking with Albert Libchaber and with his students and colleagues. The research reported here was supported by the University of Chicago MRL.

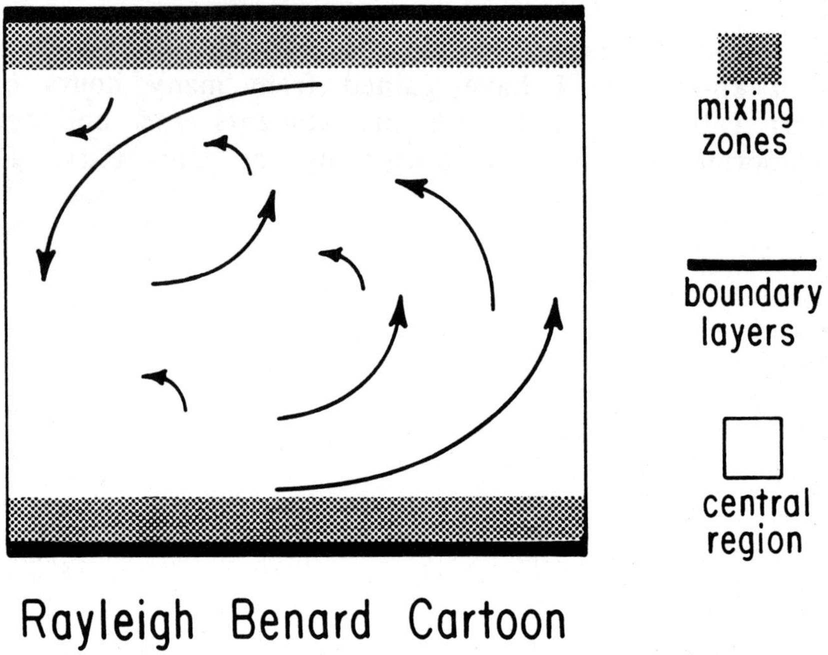

Figure 1.

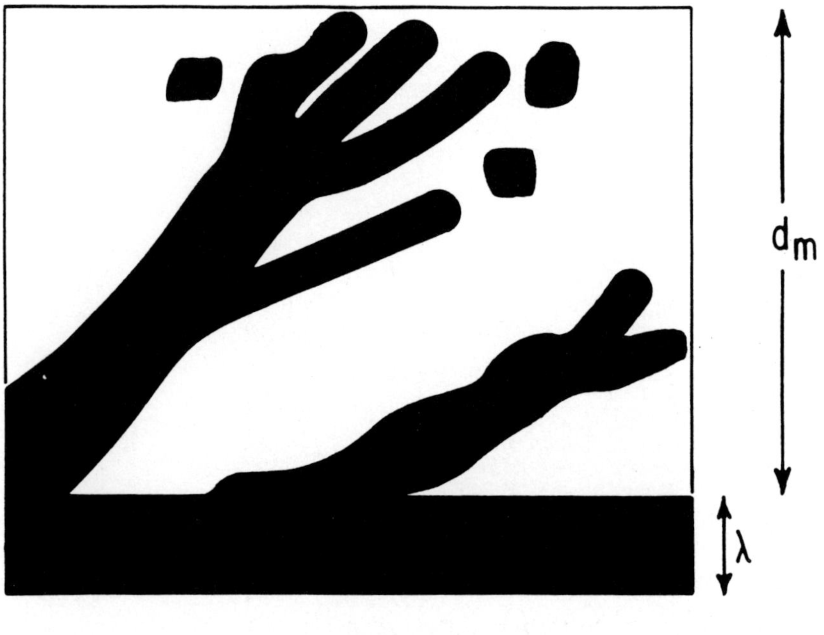

Figure 2.

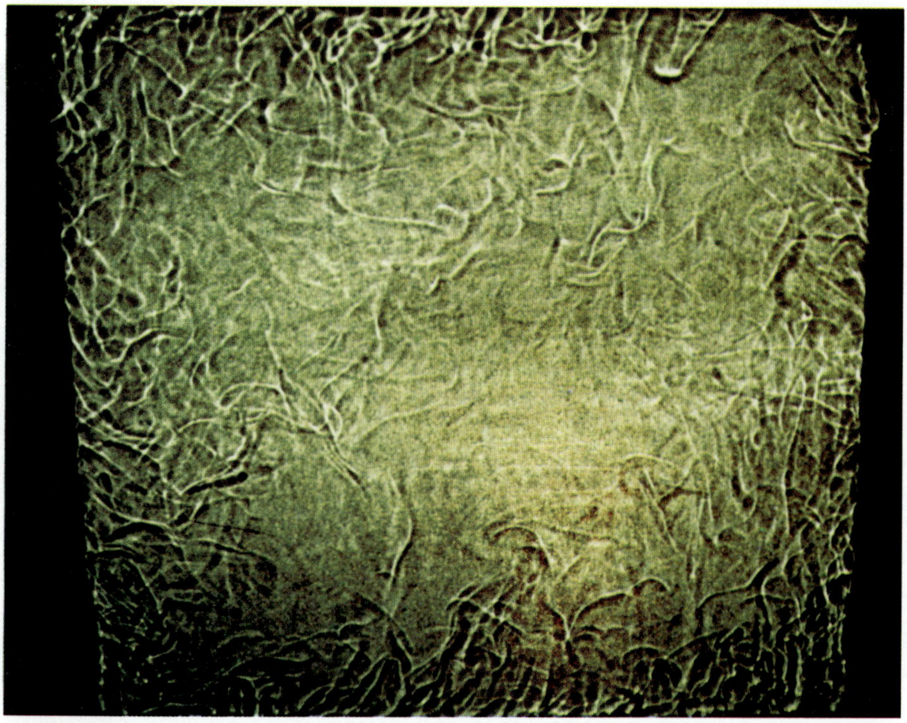

Figure 3. An experimental picture of the flow in a water tank from reference 5.

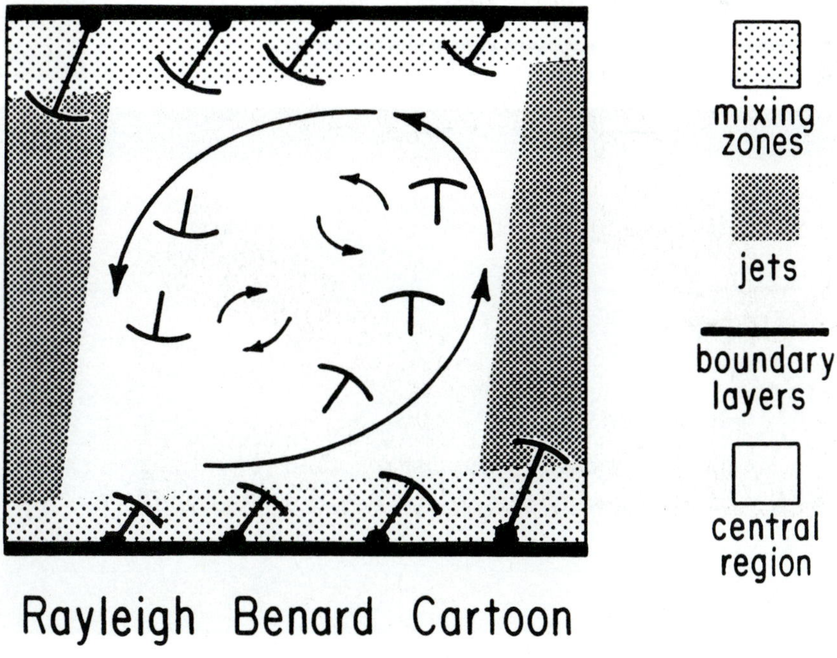

Figure 4.

TURBULENCE DANS UNE BOÎTE

L'UNIVERSITÉ DE CHICAGO ABRITE L'UNE DES EXPÉRIENCES LES PLUS AMBITIEUSES JAMAIS TENTÉES POUR COMPRENDRE LE DÉSORDRE DES FLUIDES TURBULENTS. SIMPLIFIÉE À L'EXTRÊME, ELLE CONSISTE À FORCER UN COURANT DE CHALEUR À TRAVERS DE L'HÉLIUM GAZEUX À -268 °C. POURTANT, LA COMPLEXITÉ DES STRUCTURES OBSERVÉES EST GRANDE. COUCHES EN MOUVEMENT, JETS DE GAZ, « PLUMES THERMIQUES » NAISSENT, TOURBILLONNENT, MEURENT ET FLUCTUENT D'UNE MANIÈRE QUI SEMBLE TOTALEMENT ALÉATOIRE ; MAIS L.P. KADANOFF, A. LIBCHABER, E. MOSES, G. ZOCCHI ET LEURS NOMBREUX COLLABORATEURS ONT RÉUSSI À ÉTABLIR DES LOIS SIMPLES QUI DÉCRIVENT BIEN CE DÉSORDRE. COMME L'EXPLIQUE S. ZALESKI EN ENCADRÉ, CES RÉSULTATS BOUSCULENT TRENTE-CINQ ANS D'IDÉES ANCIENNES ET SUSCITENT DES DÉBATS INTENSES. IL SEMBLE QU'ON S'APPROCHE ENFIN D'UNE COMPRÉHENSION DES ORIGINES PHYSIQUES DE LA TURBULENCE.

LEO P. KADANOFF, ALBERT LIBCHABER, ELISHA MOSES ET GIOVANNI ZOCCHI

■

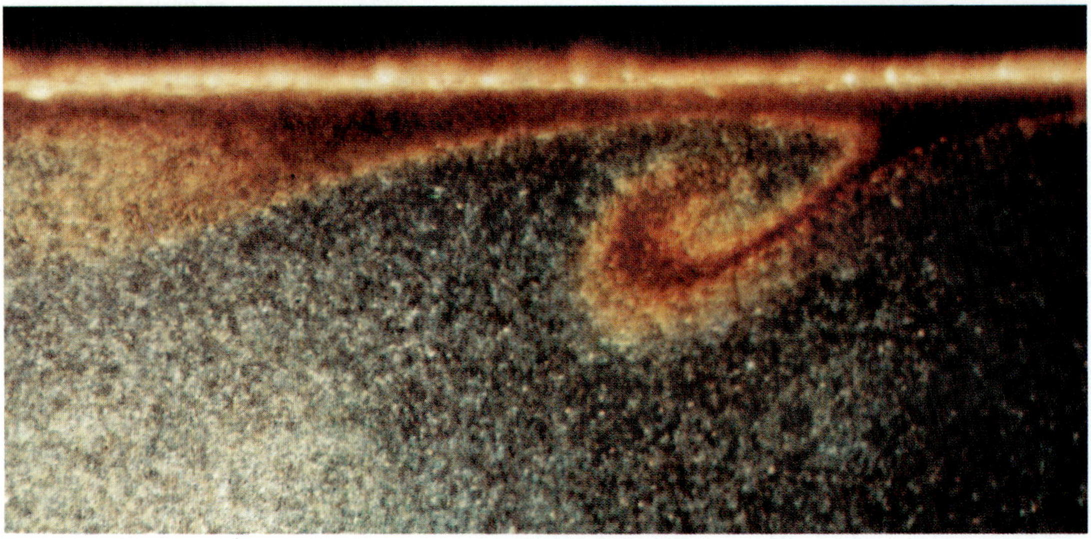

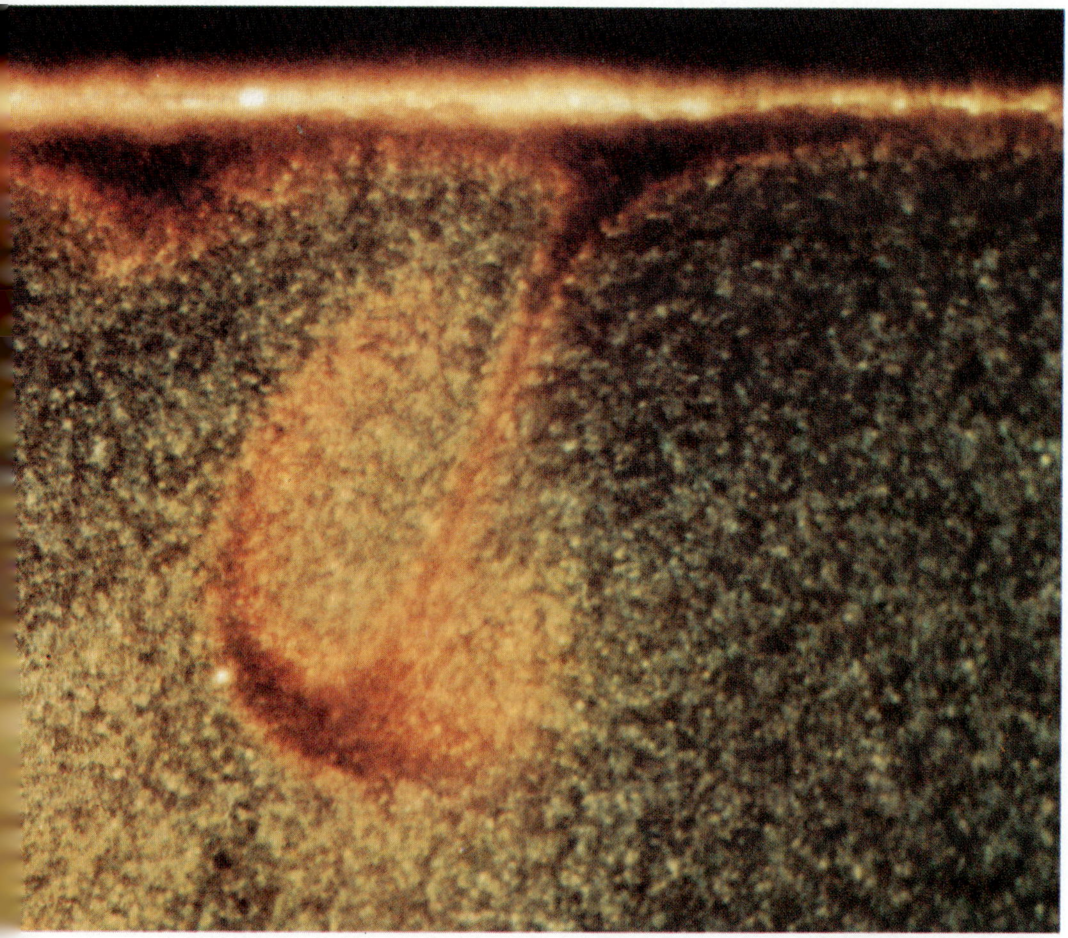

Vous êtes en train de lire *La Recherche*. Dans la pièce où vous vous trouvez, tout semble calme. Pourtant, autour de vous, l'air est animé de mouvements en tous sens que les physiciens qualifient de « turbulents ». Si vous mesuriez sa vitesse en un point donné, à intervalles de temps réguliers, vous trouveriez des résultats aléatoires, comme si vous les aviez tirés au sort. Vous pourriez aussi comparer la vitesse de l'air en différents endroits de la pièce. Et vous auriez la même surprise : l'air bouge différemment partout, apparemment au hasard. Or c'est bien ainsi qu'en physique on définit la turbulence : un écoulement est dit turbulent si la vitesse du fluide semble varier de façon aléatoire, dans le temps comme dans l'espace. De fait, la plupart des écoulements que nous rencontrons dans la vie courante sont turbulents. Par exemple, l'écoulement de l'air autour d'un marcheur, ou celui de l'eau autour d'un nageur, *a fortiori* autour d'un bateau, d'une voiture, etc.

La turbulence est, bien sûr, l'un des grands problèmes de la technologie moderne. Mais qu'ils conçoivent des engins, des machines ou des véhicules, les ingénieurs ne savent en tenir compte que de façon empirique. Si donc les physiciens s'intéressent tant à la turbulence, aujourd'hui, c'est tout d'abord parce qu'ils n'en possèdent toujours qu'une compréhension très incomplète ; de plus, ils ont l'impression que pour trouver une description rigoureuse, satisfaisante, de la turbulence, les concepts actuels de la physique ne vont pas suffire, il va leur falloir en inventer d'autres.

Situation intéressante ? Certes, mais pas si nouvelle. Au milieu du siècle dernier, les lois de la mécanique étaient bien connues. On savait donc décrire la trajectoire d'un corps soumis à une force donnée. Pourtant, on ne savait pas en déduire les propriétés d'un gaz. C'est plusieurs décennies plus tard que la mécanique statistique y parvint,

Figure 1. Soumis à un courant de chaleur, un fluide se met généralement en mouvement. Cette « convection thermique » peut devenir désordonnée si le chauffage est suffisamment intense. Ce désordre résulte de l'existence de différentes structures dans l'écoulement du fluide, qui apparaissent, disparaissent, se propagent ou simplement fluctuent d'une façon qui semble aléatoire. Pour observer la répartition et le mouvement des régions froides ou chaudes du fluide, on peut injecter des microsphères de cristaux liquides cholestériques, dont la couleur change avec la température. Ici, les régions froides apparaissent en brun ou jaune, les régions chaudes en vert. On voit deux exemples de « plumes thermiques », près du bord supérieur d'un récipient contenant de l'eau chauffée par le bas. Structures typiques en turbulence convective, ces plumes thermiques sont des bouffées de liquide froid ou chaud qui naissent à partir de fluctuations d'épaisseur d'une autre structure typique, la « couche limite » de fluide qui réside dans le voisinage immédiat des parois. (Clichés auteurs)

L.P. KADANOFF et **A. LIBCHABER** sont professeurs à l'université de Chicago. **E. MOSES** effectue des recherches post-doctorales dans l'équipe que dirige A. Libchaber. **G. ZOCCHI,** après avoir passé son doctorat à Chicago, a été accueilli au département de physique de l'Ecole normale supérieure de Paris dans le cadre d'un séjour post-doctoral.

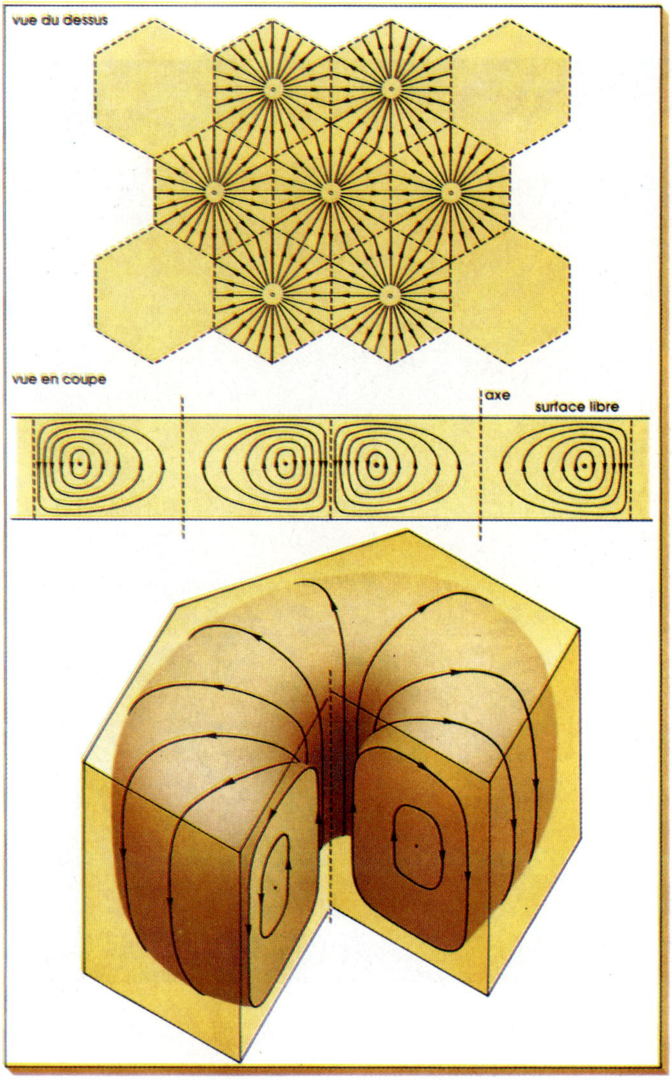

Figure 2. Ces dessins représentent l'expérience historique que réalisa Bénard en 1900. Il découvrit qu'une couche fluide, chauffée par le bas, s'organise spontanément en cellules, hexagonales ici. Dans chacune de ces cellules, environ deux fois plus larges que hautes, du fluide chaud monte au centre et redescend froid sur les bords. Dans le cas particulier étudié par Bénard, la surface supérieure du liquide était en contact libre avec l'air, ce qui faisait jouer à la capillarité un rôle important. Dans les expériences que décrit l'article, le fluide est enfermé entre deux plaques rigides, ce qui simplifie les phénomènes observés, mais le chauffage est suffisamment intense pour que l'écoulement semble totalement désordonné.

après que de nombreux concepts nouveaux, « énergie libre », « entropie », « température absolue », etc. eurent été développés.

Aujourd'hui, on sait que toute description de la turbulence devra, elle aussi, être statistique : prédire la vitesse exacte d'un fluide turbulent en un point donné est impossible. Tout au plus peut-on espérer prédire la probabilité pour que la vitesse ait telle ou telle valeur. Or même cela semble difficile : on connaît les équations de la dynamique des fluides, on sait que les vitesses fluctuent dans un écoulement turbulent, mais on ne sait pas vraiment selon quelle loi de probabilité.

La première question qu'il faut se poser est « qu'y a-t-il de commun entre différents écoulements turbulents ? » Existe-t-il des éléments génériques qui soient les mêmes pour tous ? D'une situation à l'autre, les différences observables ne sont-elles dues qu'à des questions de forme générale ? Personne ne sait vraiment répondre à ces questions non plus. Deux stratégies s'offrent cependant si l'on veut s'attaquer au problème.

LA STRATÉGIE À ADOPTER POUR COMPRENDRE LA TURBULENCE : SIMPLIFIER L'EXPÉRIENCE À L'EXTRÊME

■

La première consiste à comparer de nombreux exemples de turbulence, disons les écoulements autour d'une hélice, dans un jet d'eau et dans une rivière, et à essayer de trouver des ressemblances. En fait, on ne sait pas très bien comment s'y prendre. Si l'on compare la vitesse en un point de la rivière à la vitesse en un point près de l'hélice, on trouve qu'il s'agit de deux quantités aux valeurs aléatoires n'ayant aucun rapport entre elles. On peut aussi comparer leurs propriétés statistiques : dans chaque système, la vitesse fluctue autour d'une certaine valeur moyenne, et l'on peut comparer la probabilité d'observer un certain écart à cette valeur moyenne. Mais, même si l'on trouvait le même résultat dans chaque système, pourrait-on en conclure que la turbulence est la même ? Ce n'est pas clair. Prenons un exemple, plus familier mais différent, celui de la gravitation. A la surface de la Terre, tout corps est soumis à la même accélération (la pesanteur). Cependant, si l'on étudie ce phénomène de la façon la plus simple qui vienne à l'esprit, en regardant différents corps tomber, on s'aperçoit qu'une plume tombe plus lentement qu'une pierre. C'est bien sûr à cause de la friction de l'air qui, elle, est différente sur chaque corps et c'est donc seulement en prenant certaines précautions qu'on peut trouver le même mouvement engendré par la même gravité : en observant la chute des corps dans le vide.

Cet exemple nous amène à la deuxième stratégie que l'on peut adopter. C'est celle que nous avons choisie. Elle consiste à étudier le système le plus simple possible, à éliminer tout ce qui ne semble pas essentiel, comme l'air

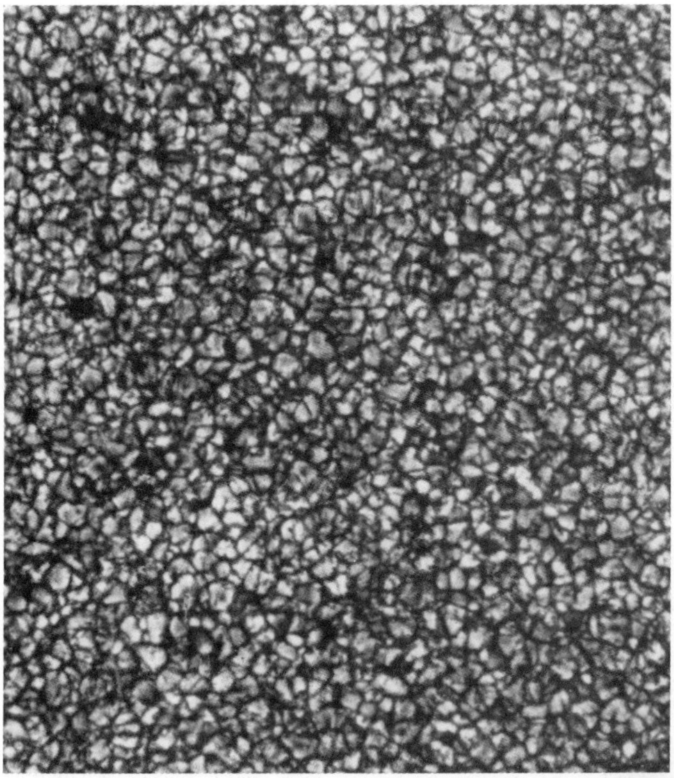

Figure 3. Le Soleil est une boule d'hydrogène et d'hélium gazeux ionisés. La chaleur, produite par des réactions nucléaires, va du centre vers la surface, et l'on peut observer des cellules de convection analogues à celles qu'observa Bénard au début de ce siècle. Cette photographie a été prise le 9 juillet 1978, à l'observatoire du Pic du Midi. Chaque grain est une cellule dont la dimension est de l'ordre de mille kilomètres. Dans les régions claires, les gaz montent vers la surface, et dans les régions sombres, ils s'enfoncent vers le centre. (Cliché Observatoire du Pic du Midi)

Figure 4. La « plume thermique » est l'une des structures d'écoulement que l'on rencontre dans un fluide mis en mouvement par un courant de chaleur. Ces photographies, prises toutes les trois secondes, montrent la croissance d'une telle plume, dans l'eau, à partir d'une résistance chauffante située en bas. Les régions sombres correspondent ici à du liquide chaud qui monte à partir de la source de chaleur avant de s'élargir en se refroidissant pour former le chapeau supérieur. Les champignons atomiques, tristement célèbres depuis la Seconde Guerre mondiale, sont des exemples de plumes thermiques turbulentes. (Clichés auteurs) ▼

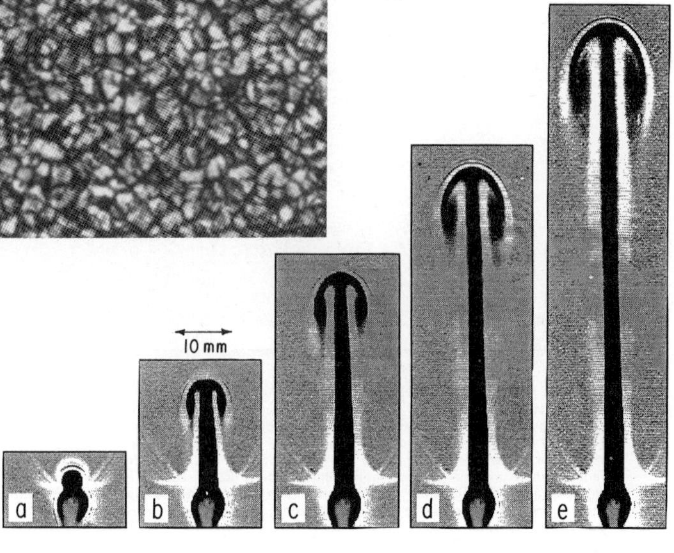

dans le problème de la chute des corps, pour arriver au cœur du problème. Nous avons donc étudié une boîte de forme simple (un cylindre) remplie d'un fluide. Cette boîte est fermée, parce que nous pensions qu'en isolant notre système, au moins en contrôlant ses interactions avec son environnement, nous serions mieux à même de mesurer les quantités pertinentes qui définissent son état, ses « variables d'état », et de trouver leurs relations mutuelles. On peut reprendre l'exemple de la thermodynamique des gaz : si l'on mesure la température T et la pression P de l'air, dehors, on ne trouve aucune corrélation entre ces deux quantités ; mais si l'on mesure P et T dans une boîte de volume V, on peut obtenir l'« équation d'état » du gaz contenu dans la boîte (le produit PV est approximativement proportionnel à T).

Toutefois, prendre une boîte fermée ne suffit pas. Il faut aussi produire un écoulement. On pourrait, par exemple, agiter le fluide avec une hélice. Mais on introduirait des perturbations, une sorte de bruit difficilement contrôlable associé aux vibrations des différentes pièces mécaniques. De plus, il ne serait pas facile de mesurer quelle quantité d'énergie est exactement transmise par l'hélice au fluide dans la boîte. Heureusement, on peut entretenir un écoulement dans une boîte de nombreuses manières, sans nécessairement utiliser de pièces mécaniques mobiles. La plus simple consiste à produire une convection thermique avec un courant de chaleur : on impose une différence de température dans la boîte, en chauffant le bas et en refroidissant le haut. Les couches du bas sont plus chaudes, donc plus légères, que celles du haut, puisque, presque toujours, un fluide se dilate lorsqu'on le chauffe. On observe alors que, de même que la force d'Archimède fait remonter à la surface une bouée de sauvetage qu'on plonge sous l'eau, de même ici, le fluide chaud du bas a tendance à monter et le fluide froid du haut à descendre. A condition de chauffer suffisamment, un mouvement de convection du fluide s'établit donc entre le haut et le bas de la boîte ; un courant de chaleur suffit ainsi à agiter le fluide (fig. 1 et 2). Autre avantage, il est maintenant facile de mesurer l'énergie injectée : c'est le courant de

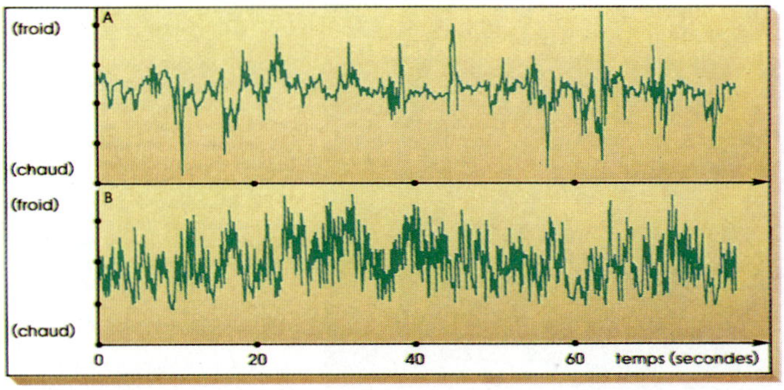

Figure 5. On a ici deux exemples de fluctuations de la température dans l'expérience, telle que la mesurent deux thermomètres situés l'un au centre du fluide en mouvement (A), l'autre dans la région appelée « couche limite », au voisinage immédiat du fond (B). Au centre, les petites bouffées successives sont dues aux plumes qui passent ; chaque division correspond à un vingtième de la différence de température totale Δ entre le haut et le bas du fluide. Dans l'enregistrement du bas, les fluctuations sont plus intenses (chaque division correspond maintenant à un cinquième de la différence Δ).

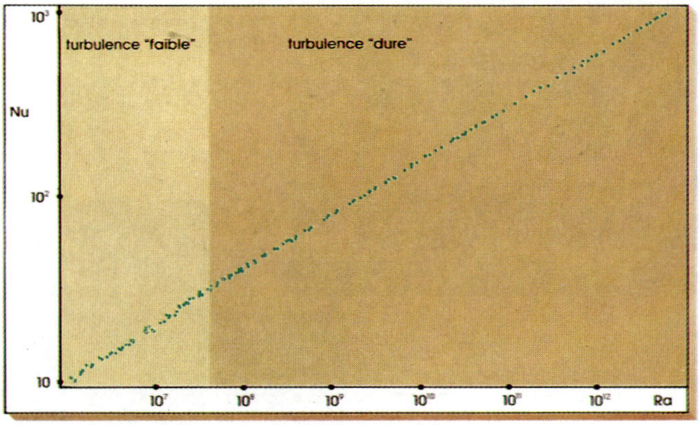

Figure 6. Cette courbe représente la réponse du fluide lorsqu'il est chauffé par le bas. Les échelles sont logarithmiques. En ordonnée, le nombre de Nusselt, Nu, est égal au courant de chaleur qui passe effectivement à travers le fluide divisé par le courant qui passerait en l'absence de convection. En abscisse, le nombre de Rayleigh Ra est, lui, proportionnel à la différence de température imposée entre le haut et le bas. La pente de la courbe s'infléchit, légèrement, entre le régime de « turbulence faible » ($Ra \approx 10^7$) et le régime de « turbulence dure » ($Ra > 10^8$). Dans un très grand domaine de variation du nombre de Rayleigh (cinq ordres de grandeur), la réponse du fluide obéit à une loi simple et remarquable : $Nu \simeq (Ra)^{2/7}$.

(1) H. Bénard, *Revue générale des Sciences pures et appliquées*, 2161, 1900 ; F.H. Busse, *Rep. Prog. Phys.* 41, 1929, 1978.
(2) D.J. Tritton, *Physical Fluid Dynamics*,, Van Nostrand Reinhold, 1984.

chaleur qui entre dans la boîte. La puissance du chauffage est même le seul paramètre qui contrôle le degré de turbulence du fluide. Enfin, comme on s'intéresse aux propriétés statistiques de la turbulence, on peut mesurer les fluctuations de la température en différents points de la boîte avec de petits thermomètres, ce qui est plus facile, compte tenu des techniques actuelles, que de mesurer des vitesses.

C'est en 1900 à Paris que H. Bénard effectua les premières études scientifiques de la convection[1]. Il observa qu'un fluide, chauffé par le bas, se met effectivement en mouvement ; il s'organise en cellules telles que le fluide chaud monte au centre et le fluide froid descend sur les bords (fig. 2). Ces « cellules de Bénard » sont un exemple, parmi d'autres, de ces structures hydrodynamiques que l'on rencontre dans les fluides en mouvement. D'une manière très générale, ces cellules tendent à être deux fois plus larges que hautes. Dans une couche mince chauffée par le bas, de nombreuses cellules apparaissent donc, chacune avec une région centrale où le mouvement est vers le haut et des bords où il est vers le bas. Et ce phénomène est très répandu. La surface du soleil, par exemple, est couverte de cellules de Bénard (fig. 3).

De minces régions actives, où la vitesse d'écoulement varie rapidement, comme les bordures de ces cellules solaires, constituent ce qu'en hydrodynamique on appelle des « couches limites ». Il s'agit, après les cellules, d'un deuxième exemple de structures que l'on rencontre en convection, et il en existe un troisième, les « plumes »[2]. Ces plumes thermiques ressemblent à des champignons et sont tristement célèbres depuis la Seconde Guerre mondiale (fig. 4). Elles résultent généralement de l'existence d'une source de chaleur ponctuelle, au fond d'un réservoir de fluide[3]. Le fluide chaud monte dans le pied du champignon, et, quand il atteint le chapeau, s'étend avant de redescendre pour remplir une région plus large. L'inverse existe aussi, avec, du fluide froid descendant dans le pied d'un champignon à l'envers (fig. 1).

Tout suggère actuellement que ces trois types de structures apparaissent dans des contextes extrêmement divers. Nous venons de parler des champignons atomiques. On observe des plumes semblables se détachant de la surface du Soleil. On pense aussi que c'est la montée du magma terrestre sous forme de plumes qui est responsable de l'essentiel des volcans localisés et de l'activité sismique. Les « zones de turbulence » qu'on traverse parfois en avion, sont elles aussi généralement des plumes thermiques, qui se sont détachées de la surface de la Terre et se propagent ensuite librement dans l'atmosphère. Enfin, à une échelle nettement plus réduite, on peut aussi voir de telles plumes se former à la surface d'une route goudronnée, si le temps est suffisamment chaud et ensoleillé. L'aspect miroitant de la lumière, au dessus de la route, provient des plumes qui sont ancrées au sol et déforment les images en infléchissant les rayons de la lumière.

Dans notre boîte à fluide, sorte de bouilloire scientifique, l'écoulement est constitué d'un mélange toujours brassé de cellules, de couches limites et de plumes. Ce que nous voulons comprendre, c'est pourquoi leur structure et leurs interactions mutuelles conduisent à des lois simples, qui, comme nous allons le voir, décrivent la turbulence dont notre boîte est le siège.

Il faut remarquer que, dans tous les exemples que nous venons de mentionner, un chauffage continu, c'est-à-dire stable dans le temps, produit un résultat qui, au contraire, varie en fonction du temps. Les plumes montent et remontent à partir de la route, chaque fois un peu différentes. Il n'est pas facile d'étudier ce genre de manifestations naturelles de la variabilité de la convection. Les phénomènes de convection naturelle sont soit trop grands, soit trop lointains, soit trop difficiles à isoler d'un environnement qui produit des effets incontrôlables.

ORDRE, CHAOS ET TURBULENCE SE SUCCÈDENT DANS UN CYLINDRE D'HÉLIUM GAZEUX QUE L'ON CHAUFFE PAR LE BAS

■

Née à Paris il y a une dizaine d'années, puis développée à Chicago, notre expérience a été conçue, au contraire, pour étudier une convection simple et bien contrôlée(4). Notre fluide est de l'hélium gazeux, à basse température, environ 5 K soit −268 °C. L'hélium est contenu dans un cylindre vertical de 20 centimètres de diamètre, chauffé par le bas et refroidi par le haut. Pourquoi de telles températures ? Principalement parce qu'elles permettent de minimiser l'effet des fluctuations thermiques environnantes et de réaliser des mesures extrêmement sensibles. Pourquoi l'hélium gazeux? Parce que sa viscosité est faible et qu'il permet d'atteindre plus facilement de grandes vitesses d'écoulement, donc un haut degré de turbulence.

On a l'habitude de décrire une expérience de convection grâce à un nombre, issu des travaux en 1916 d'un autre pionnier de l'hydrodynamique, Lord Rayleigh(5). Ce « nombre de Rayleigh », noté Ra, mesure la contrainte appliquée au fluide. Il est proportionnel à la différence de température imposée, mais dépend aussi des caractéristiques du fluide (épaisseur, coefficient de dilatation thermique, viscosité, conductivité thermique) et de la gravité. Grâce au choix de l'hélium gazeux, il est possible d'étudier une gamme remarquablement large de degrés de turbulence, correspondant à des nombres de Rayleigh variant de 10^3 à 10^{15}. Les températures sont mesurées, et contrôlées, grâce à des petits thermomètres semiconducteurs, morceaux de silicium dopé à l'arsenic d'environ 0,2 millimètre.

Cette expérience d'apparence simple donne, à elle seule, une idée des différents types d'ordre et de désordre qui se manifestent en hydrodynamique. Lorsque le nombre Ra est faible, la chaleur diffuse par conduction à travers le système, sans que le fluide se mette en mouvement ; un gradient de température constant s'établit entre le haut et le bas, et les thermomètres indiquent chacun une température fixe dans le temps. L'ordre est stable. Pour des nombres de Rayleigh d'environ 2 000, la convection s'installe, et une cellule de Bénard se forme, régulière, dans la boîte. Si l'on augmente encore le chauffage, donc pour de plus grandes valeurs de Ra, la vitesse du fluide commence à osciller avec une période bien définie. C'est seulement à encore plus fort chauffage que ce mouvement d'oscillation périodique de la cellule de Bénard devient désordonné et que les thermomètres indiquent des températures qui semblent varier de façon aléatoire. On dit, habituellement, que le système est entré dans un régime « chaotique ». Lorsqu'on étudie la température qu'indiquent des thermomètres placés en différents endroits de la cellule, on s'aperçoit que tout le fluide se déplace en masse, mais d'une façon irrégulière dans le temps.

Pourtant, ce qui nous intéresse ici se passe à des valeurs de Ra supérieures à 10^8. Dans ce dernier régime, la température moyenne ne varie pas de façon homogène entre le haut et le bas, mais essentiellement dans deux régions très minces, situées l'une près du fond de la boîte cylindrique, l'autre près de son sommet. C'est là que la présence d'une paroi rigide force la vitesse de l'hélium à s'annuler, un phénomène caractéristique de l'existence d'une viscosité non-nulle. On pourrait penser, par analogie, à un circuit électrique, où la tension appliquée serait l'analogue de la différence de température (ou du nombre de Rayleigh) et où le courant électrique serait l'analogue de notre courant de chaleur. Les couches limites du haut et du bas de notre boîte seraient comme deux grosses résistances, et tout l'intérieur du fluide, qui bouge à grande vitesse, comme un court circuit entre les deux résistances. De même que la tension électrique chuterait essentiellement aux bornes des deux résistances, de même, dans la convection turbulente que nous avons observée, toute la chute de température est localisée dans les deux couches limites situées en haut et en bas du cylindre de fluide.

A partir d'un tel niveau de turbulence, le comportement du fluide est tellement erratique que chacune de ses différentes portions se déplace de façon à peu près indépendante du reste. Si l'on regarde les températures affichées par deux thermomètres suffisamment espacés, on ne leur trouve qu'un degré de corrélation très faible. C'est ce comportement que nous avons appelé la « turbulence dure ». Ici, nous mesurons des températures qui montent et descendent de façon complètement désordonnée (fig. 5). A première vue, il semble difficile de parvenir à une compréhension satisfaisante d'un comportement aussi compliqué. La tur-

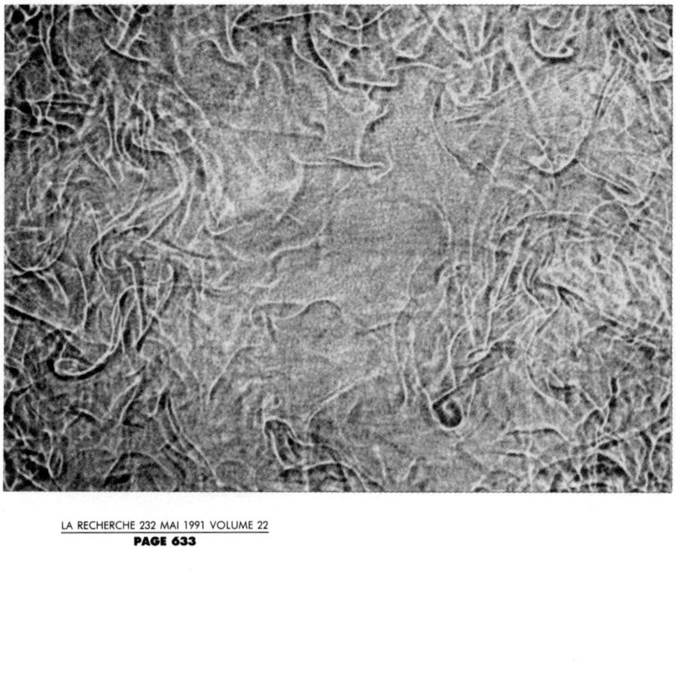

Figure 7. Cette photographie donne une idée de l'aspect typique d'un fluide en régime de convection turbulente dure. Ce sont les variations locales de la température qui, en induisant des variations de densité donc d'indice optique, infléchissent les rayons lumineux et permettent cette visualisation. La structure désordonnée est produite par les plumes thermiques émises par les couches du haut et du bas. La plupart des plumes se déplacent au voisinage des parois. Au cœur de la boîte, elles sont moins nombreuses car entraînées sur les bords par l'écoulement. (Cliché auteurs)

(3) J. von Neumann, *The point source solution*, Los Alamos Report LA-2000, août 1947 ; G. I. Taylor, *Proc. Roy. Soc. A 201*, 159, 1950 ; E. Moses *et al.*, *Europhysics Letters 14*, 55, 1991.
(4) B. Castaing *et al.*, *J. Fluid Mech. 204*, 1, 1989 ; D.C. Threlfall, *J. Fluid Mech. 67*, 17, 1975.
(5) Lord Rayleigh, *Phil. Mag.*, 32, 529, 1916. S. Chandrasekhar, *Hydrodynamic and hydromagnetic stability*, Dover, 1981.

bulence dure est l'état de désordre maximal jamais réalisé dans une expérience de convection thermique contrôlée.

Pour comprendre ce type de turbulence, on peut emprunter deux approches. La première consiste à regarder des propriétés moyennées dans le temps ou dans l'espace. En effectuant de telles moyennes, on élimine les fluctuations erratiques dont nous venons de parler, et on peut espérer retrouver des phénomènes suffisamment simples pour donner lieu à des lois, simples elles aussi. L'autre approche consiste à étudier les structures présentes dans le fluide turbulent, plumes, couches limites et cellules.

Commençons par les propriétés moyennes. On a l'habitude d'utiliser un nouveau nombre, le « nombre de Nusselt » cette fois, pour décrire une quantité particulièrement intéressante dans notre problème, qui est proportionnelle au flux de chaleur moyen traversant le fluide du bas vers le haut. Le nombre de Rayleigh était l'analogue, en hydrodynamique, d'une tension électrique et le nombre de Nusselt est l'analogue d'un courant. Ce nouveau nombre, noté Nu, a été introduit en 1915 par le physicien Ernst Kraft Wilhelm Nusselt, qui fut par la suite titulaire de la chaire de mécanique théorique à la Technische Hochschule de Munich. Le nombre Nu est égal au rapport de la chaleur effectivement transportée à travers le système turbulent à la chaleur (plus faible) qui serait transportée s'il n'y avait que de la conduction, pas de mouvement convectif du fluide. Lorsque le chauffage est faible, c'est-à-dire aux faibles valeurs de Ra, le nombre de Nusselt vaut un, mais en régime turbulent, Nu peut être beaucoup plus grand que un : la chaleur est essentiellement transportée par convection (fig. 6). A ce propos, si, dans une bouilloire, l'eau ne chauffait que par conduction, elle prendrait une dizaine d'heures à bouillir. Grâce à la convection, cela ne prend que quelques minutes.

Dans notre expérience de turbulence convective, nous avons étudié comment Nu varie en fonction de Ra, et nous nous sommes aperçus que les résultats des mesures sont bien décrits par une loi simple[4] : Nu est proportionnel à la puissance p-ième de Ra, où p est un nombre voisin de 2/7. Cette « loi de puissance » relie donc simplement le flux de chaleur moyen à travers le fluide à la différence de température appliquée. On sent ici à quel point la turbulence est un phénomène non-linéaire, contrairement à l'électricité ordinaire, où la réponse (le courant) est proportionnelle à la force appliquée (la tension).

D'autres quantités moyennes suivent des lois de puissance analogues, toujours en fonction du nombre de Rayleigh. Physiciens ou hydrodynamiciens, les théoriciens en ont déduit que, dans tout le domaine où les mêmes lois de puissance décrivent les mesures, les mêmes phénomènes doivent avoir lieu, qu'ils sont simplement plus intenses à fort chauffage.

Mais de quels phénomènes s'agit-il ? Il n'est pas très facile de regarder à travers une grande boîte d'hélium gazeux aux températures nécessaires à notre expérience. Aussi avons-nous construit une expérience semblable, cette fois avec une sorte d'aquarium de 20 centimètres de haut, rempli d'eau, chauffé à nouveau par le bas et refroidi par le haut[6]. Il est moins facile d'y faire des mesures précises, mais beaucoup plus simple de faire des observations qualitatives. On peut éclairer le récipient avec un laser et regarder ce qui se passe quand on chauffe (fig. 7). En effet, l'indice optique de l'eau dépend de sa densité, donc de sa température, et les rayons lumineux sont infléchis dans les régions où la température varie, comme dans l'air au dessus des routes au Soleil. On obtient donc une image du fluide en mouvement. A première vue, c'est remarquablement complexe, mais un peu de patience permet de voir émerger quelques phénomènes simples.

L'eau, sur nos photos, tourne dans le sens inverse des aiguilles d'une montre. Un léger écoulement a lieu de gauche à droite, près du fond, là où l'eau est chauffée. Elle y est aussi accélérée avant de monter rapidement vers le haut sous forme d'un jet, à droite. Ensuite, l'eau traverse la partie supérieure et redescend sous forme d'un autre jet rapide, à gauche cette fois. Ainsi, le mouvement global de l'eau est organisé sous forme d'une cellule qui ressemble à celles qu'étudiait Bénard au début du siècle. En outre, près du fond ou du sommet, on peut voir une mince couche de liquide relativement calme. Comme nous l'avons déjà remarqué, c'est l'effet conjugué de la viscosité de l'eau et de l'existence d'une paroi rigide : les parois empêchent l'eau, fluide visqueux, de s'écouler sur une faible épaisseur. Le liquide, qui est ainsi maintenu près du fond, est chauffé de façon très efficace par la plaque inférieure de l'aquarium. Il existe donc, à cet endroit, une « couche limite » beaucoup plus chaude que le fluide se trouvant juste au dessus. Cette structure est particulièrement instable, et émet, à la moindre perturbation, une plume thermique (fig. 1 et 4).

On sera peut-être surpris de constater que certaines de ces plumes n'ont pas l'air perturbées par les mouvements désordonnés du fluide qui les entoure, elles sont presque aussi régulières que celles que l'on peut produire au sein d'un fluide calme (fig. 4). Mais c'est l'un des paradoxes de la turbulence que l'on a déjà souligné : le mouvement de chaque portion d'un fluide turbulent est pratiquement indépendant de celui des autres. D'autres structures, de type plume, apparaissent ailleurs, dues au mouvement relatif des différentes couches du fluide (fig. 7). En étudiant soigneusement ces photographies, on peut voir les plumes s'étirer ou glisser en travers puis, toujours enracinées, atteindre les jets des parois latérales. Elles sont alors emportées par le courant. Beaucoup en meurent, quelques-unes réussissent à en ressortir pour arriver dans des régions plus calmes. Finalement, on remarque que les plumes sont nombreuses près de chacune des quatre parois et que la région centrale est, en somme, moins active. Il y a toutes les raisons de penser que ce qu'on observe dans l'eau se passe de la même manière dans l'hélium gazeux de notre boîte.

Résumons la manière dont les deux couches limites, en bas et en haut, échangent de la chaleur *via* une circulation générale du fluide, se « parlent » entre elles, en quelque sorte. Une

suite page 637

Interprétations concurrentes

L'expérience de Chicago a bousculé le consensus que trente-cinq ans d'expériences et de théories avaient à peu près réussi à établir sur la théorie de Malkus : elle a montré qu'à suffisamment fort chauffage, le nombre de Nusselt varie comme la puissance 2/7 (et non pas 1/3) du nombre de Rayleigh. Mais l'interprétation de cette nouvelle loi n'échappe pas pour autant à un certain nombre de controverses que Libchaber, Kadanoff, Moses et Zocchi n'auraient pu mentionner sans alourdir leur exposé. Par exemple, l'existence du régime de « turbulence dure » dans l'eau a été contestée à Minneapolis par R.J. Goldstein[10] mais confirmée en Pennsylvanie par T.H. Solomon et J.P. Gollub[11]. Un débat subsiste aussi autour des transitions entre les différents types de turbulence possibles. Sur la base d'une théorie différente reprenant les idées de Bob Kraishnan[12], Boris Shraiman et Eric Siggia ont suggéré à Cornell (Etats-Unis) que si la couche limite devient turbulente, on devrait observer une puissance 1/2[13]. Poursuivant dans la même voie, j'ai moi-même prédit que la transition entre turbulence faible (puissance 1/3) et turbulence dure (puissance 2/7) doit dépendre de la forme de la boîte et même de la nature du fluide (conduction thermique et viscosité)[14]. Ainsi, l'extrapolation à l'hélium des observations sur l'eau n'est pas admise par tous. Une certaine turbulence agite donc aussi l'actuel débat scientifique sur ces questions.

Stéphane Zaleski
Ecole normale supérieure (Paris)

suite de la page 634

plume chaude monte et atteint la couche limite froide ; là, elle crée une perturbation dans la distribution locale des températures, ce qui émet une sorte d'onde le long de la couche ; à son tour, l'onde émet une plume, froide cette fois, qui descend, etc. Les plumes chaudes ascendantes et les plumes froides descendantes sont couplées entre elles par ces ondes de surface. Tels sont, schématiquement, les phénomènes qui animent le mouvement de la cellule.

Il nous reste maintenant à expliquer pourquoi, aussi complexes que soient les fluctuations qui agitent notre cellule, la turbulence y est décrite par une loi simple : $Nu \approx (Ra)^{2/7}$. Le raisonnement qui aboutit à ce résultat ne nécessite pas de calcul véritable et mérite d'être, presque intégralement, reproduit ici. Il procède en effet d'une méthode simple et exemplaire en physique, introduite dès 1822 par J. Fourier, l'« analyse dimensionnelle »[(7)]. Cette méthode est elle-même à la base des actuelles théories d'échelles (*scaling*) dont l'usage est fructueux dans l'étude des changements d'état de la matière.

Les équations qui décrivent notre écoulement contiennent des quantités (vitesse, température, viscosité, densité, accélération de la gravité, etc.) qui possèdent certaines dimensions. Par exemple, une vitesse a la dimension d'une longueur divisée par un temps. Mais la nature des phénomènes ne peut dépendre du fait que nous exprimons une vitesse en mètres par seconde plutôt qu'en yards par minute, c'est-à-dire du fait que nous choisissons tel ou tel système d'unités. On doit donc pouvoir rapporter chaque quantité physique considérée à une grandeur typique de notre expérience ; par exemple, au lieu de considérer une longueur quelconque x dans les équations, considérer le rapport x/L où L est la dimension de notre boîte. On doit pouvoir faire de même pour toutes les autres quantités physiques nécessaires. Finalement on doit ainsi obtenir des équations qui ne font intervenir que des combinaisons de quantités devenues « sans dimension » et où n'interviennent que des nombres comme le nombre de Rayleigh Ra et le nombre de Nusselt Nu.

Quel est l'intérêt d'une telle démarche ? Le nombre de Rayleigh s'obtient, entre autres, à partir du rapport Δ/ν de la différence de température Δ entre le haut et le bas de la boîte à la viscosité ν du fluide. Comme nos équations ne dépendent que de la valeur de Ra, on doit obtenir la même turbulence si le nombre Ra est le même, par exemple en prenant une différence de température dix fois plus petite mais aussi un fluide de viscosité dix fois plus faible (toutes choses égales par ailleurs). Ce sont d'ailleurs ces raisonnements dimensionnels qui autorisent les ingénieurs, comme le firent D'Alembert, Condorcet et Bossut dès 1777[(8)], à étudier l'écoulement autour de maquettes de bateaux à échelle réduite. Il suffit de changer la vitesse ou la viscosité du fluide afin de garder la même valeur au nombre sans dimension qui contrôle la turbulence du système (ici, un troisième nombre, plus simple et plus connu que les deux précédents, le nombre de Reynolds).

COMMENT UNE THÉORIE D'ÉCHELLE MÈNE À UNE LOI SIMPLE

■

Revenons maintenant à notre expérience, et considérons l'épaisseur δ de la couche limite où la vitesse du fluide est faible, en haut et en bas de notre boîte de longueur L. Le rapport δ/L est une quantité sans dimensions qui ne doit dépendre que du nombre de Rayleigh. Pour différentes raisons qu'il est inutile de développer ici, il est vraisemblable que δ est de la forme $L(Ra)^{\alpha}$, où l'exposant α est inconnu. De même, on peut considérer les fluctuations typiques de la température au centre de la cellule, leur attribuer une amplitude Δ_c et, en se souvenant que la différence de température appliquée aux bornes de la boîte est Δ, supposer que $\Delta_c \approx \Delta(Ra)^{\beta}$. Continuons en vous demandant d'admettre que la diffusivité thermique $\varkappa$ (une quantité proportionnelle à la conductivité thermique du fluide) est homogène à une vitesse que multiplie une longueur. On peut donc comparer la vitesse typique v_c du fluide au centre de la boîte au rapport $\varkappa/L$, et introduire un troisième exposant en écrivant $v_c \approx \varkappa/L\ (Ra)^{\gamma}$. Enfin, le nombre de Nusselt lui-même doit aussi s'écrire sous la forme $(Ra)^{\varepsilon}$. Il s'agit maintenant de déterminer cette série d'exposants en trouvant quelles relations les lient les uns aux autres. Il faut utiliser des lois de conservation générales et estimer des ordres de grandeurs, ce qui n'a rien d'évident. Mais commençons par reconsidérer le nombre de Nusselt. C'est le rapport du flux de chaleur qui passe effectivement à travers la boîte à ce qui passerait s'il n'y avait pas de convection, uniquement de la conduction. Chacun de ces flux est inversement proportionnel à une distance, δ dans le premier cas, L dans le second. Nu est donc proportionnel à L/δ , et l'on trouve une relation simple entre deux de nos exposants :

(6) G. Zocchi *et al.*, *Physica A 166*, 387, 1990.
(7) J. Fourier, *Théorie analytique de la chaleur*, Didot, 1822, p. 154-158.
(8) D'Alembert, Condorcet et Bossut, *Nouvelles expériences sur la résistance des fluides*, Paris, Jombert, 1777.
(9) W.V.R. Malkus, *Proc. Roy. Soc. London, Ser. A 225*, 196, 1954 ; L.N. Howard in H. Gortier (ed.), *Proc. of the 11th Int. Congress of Appl. Mech.*, Springer-Verlag, 1966.
(10) R.J. Goldstein *et al.*, *J. Fluid Mech.*, *213*, 11, 1990.
(11) T.H. Salomon et J.P. Gollub, *Phys. Rev. Lett.*, *64*, 2382, 1990.
(12) R.J. Kraishnan, *Phys. Fluids*, *5*, 1374, 1962.
(13) B.I. Shraiman et E.D. Siggia, *Phys. Tev. A 42*, 3650, 1990.
(14) S. Zaleski, *C.R. Acad. Sc. Paris*, 1991.
(15) A.N. Kolmogoroff, *C.R. Acad. Sci., URSS*, *30*, 299, 1941.
(16) B. Castaing, *J. Physique*, *50*, 147, 1989.

$\varepsilon = -\alpha$. Peut-être êtes-vous déçus par l'apparente simplicité d'un tel argument. Le problème, dans ce genre de raisonnement, c'est de choisir les bonnes quantités, et ce n'est pas facile, comme va le prouver la suite.

On pourrait penser que l'épaisseur δ de la couche limite est si faible qu'elle ne peut dépendre de L. Cela signifierait qu'il n'y a pas d'échange entre les deux couches limites, qu'elles ne s'influencent pas l'une l'autre. Si l'on se souvient que le nombre de Rayleigh est proportionnel au cube de la longueur L, on s'aperçoit que si l'exposant ε valait 1/3, on obtiendrait bien une épaisseur δ indépendante de L. Bien que ce résultat ne s'applique pas à notre situation de turbulence « dure », précisément parce que les deux couches limites ne sont pas indépendantes, c'est un résultat intéressant. Il fut obtenu en 1954, au Massachussetts Institute of Technology, par W.V.R. Malkus et le raisonnement que nous venons de développer apparaît dans les travaux ultérieurs de L.N.Howard(9). La loi $Nu \simeq (Ra)^{1/3}$ s'applique en fait à un régime de turbulence que nous avons appelé turbulence « faible » et qui est le précurseur de la turbulence « dure », celui qui a lieu pour des nombres de Rayleigh intermédiaires (jusqu'à 10^8).

Au-delà de 10^8, les lois sont légèrement différentes. Il faut écrire que le flux de chaleur qui passe à travers l'épaisseur δ de la couche limite est égal à celui qui traverse ensuite tout le reste de la boîte, où des fluctuations de température d'amplitude Δ_c se propagent à la vitesse v_c. Cela suffit à obtenir la relation $\beta + \gamma = -\alpha$. Bien qu'il soit impossible de poursuivre ce raisonnement jusqu'au bout, on comprend qu'en ajoutant deux autres relations de ce type on peut déterminer les valeurs de tous les exposants. On trouve $\alpha = -2/7$, $\beta = -1/7$, $\gamma = 3/7$.

Cette démarche s'avère évidemment fructueuse dans la mesure où elle aboutit à une loi simple qui décrit bien les observations expérimentales. C'est un nouvel exemple de la puissance des « théories d'échelles », qui prouvent aujourd'hui leur utilité dans un autre domaine, l'hydrodynamique. On peut remarquer que pour obtenir une loi qui décrit bien la turbulence dure de notre boîte, nous n'avons eu besoin de considérer aucune des propriétés spécifiques des structures que nous observons, des plumes par exemple. Il nous a suffi de supposer qu'il existait une interaction entre les couches limites, et que leur épaisseur dépendait donc de la longueur de la boîte, et cela illustre à la fois la puissance et la faiblesse de notre raisonnement. Sans nécessairement comprendre le détail des processus physiques en jeu, nous avons obtenu la bonne loi. Mais l'écoulement convectif est beaucoup plus riche que cela ; si l'on reste à ce niveau de généralité, on ignore toutes sortes de phénomènes intéressants, tels que la manière dont les couches limites produisent des plumes, l'interaction entre plumes, etc.

Il est bon, avant de conclure, de dire encore un mot de ces lois d'échelle. En matière de turbulence, la plus célèbre de ces lois fut introduite dès 1941 par le physicien soviétique Kolmogoroff(15). Au lieu de considérer seulement deux ou trois longueurs comme nous venons de le faire (la longueur de la boîte, l'épaisseur des couches limites et éventuellement la dimension caractéristique des plumes), Kolmogoroff supposait l'existence de structures de toutes tailles, d'une infinité d'échelles de longueur. Il aboutissait à des lois différentes qui s'appliquent peut-être à un régime que notre boîte pourrait atteindre si on forçait sa convection encore davantage, à des nombres de Rayleigh encore plus grands. Il semble en effet que de plus en plus de structures doivent apparaître, avec des dimensions caractéristiques de plus en plus variées, qu'on tende donc vers ce qu'on entend communément par une « turbulence développée », sorte d'état de désordre ultime. Contrairement à la nôtre, la théorie de Kolmogoroff ne tient pas compte de l'existence de parois autour de l'écoulement, de « conditions aux limites ». C'est une théorie encore plus générale qui ne dépend même pas de la manière dont la turbulence est engendrée, seulement de bilans énergétiques entre les mouvements de différentes tailles. Cette ancienne théorie est donc différente, bien qu'elle procède du même état d'esprit, et rapproche, comme la nôtre, la turbulence de la thermodynamique. Récemment, à Grenoble, B. Castaing a combiné une théorie « à la Kolmogoroff » avec des méthodes de thermodynamique pour rendre compte, au-delà des propriétés du flux de chaleur que nous venons de décrire, de la statistique des fluctuations de notre écoulement(16).

Au point où nous en sommes de cette étude, une question au moins reste évidemment à poser : que se passe-t-il au-delà du régime que nous avons appelé la « turbulence dure » ? Atteint-on vraiment un régime de turbulence développée encore plus complexe ? D'un point de vue expérimental, il est malheureusement difficile d'atteindre des nombres de Rayleigh beaucoup plus grands que nos 10^{15}. Bien sûr, on pourrait concevoir une boîte cinq fois plus grande dans chaque dimension, et atteindre 10^{17}. L'expérience devrait changer de nature. Les problèmes techniques, concernant l'isolation et le contrôle thermiques par exemple, deviendraient tels qu'il s'agirait d'un projet lourd coûtant plusieurs millions de dollars. De plus, il n'est sûrement pas inutile de préciser que nous sommes encore très loin d'entrevoir les applications pratiques de nos travaux. Nous essayons de comprendre les principes fondamentaux de la turbulence, et nous sommes loin de fournir une justification des connaissances semi-empiriques de la turbulence qu'utilisent les ingénieurs. Si l'on nous demandait par exemple, à partir de ce que nous sommes en train d'apprendre, de construire un avion, nous n'aurions aucune chance de parvenir à un résultat comparable à ce que la technologie moderne sait faire actuellement. Mais, en même temps, c'est justement ce type de constat qui stimule notre désir de poursuivre. Nous sommes persuadés que la turbulence recèle de nombreux phénomènes à comprendre, bien davantage que ceux que nous avons déjà découverts. L'aventure d'une expérience sur plusieurs mètres cubes d'hélium nous tente donc.

Quel que soit l'avenir exact de nos études, il nous semble qu'on peut tirer des résultats déjà acquis une leçon de portée générale. Nous avons étudié une petite boîte, remplie de fluide, de forme simple et chauffée par le bas. C'est une sorte de monde minimal dont nous avons compris les lois fondamentales. Or, dans le détail, nous nous sommes rendu compte qu'il était le siège de phénomènes d'une grande complexité. Il est divisé en régions, dans lesquelles naissent et meurent différentes sortes de structures mouvantes qui donnent à l'ensemble l'apparence d'un désordre total. Il n'est pas question, bien sûr, d'extrapoler ce que nous avons appris de la turbulence pour l'appliquer de façon simpliste à une explication du monde, de sa complexité physique ou biologique. Au moins avons-nous réussi à comprendre et à faire sentir comment, même dans des situations très simples, des phénomènes d'une grande complexité peuvent apparaître. Là réside une bonne partie de ce qui nous fascine dans l'étude de la turbulence. ■

POUR EN SAVOIR PLUS

■ L. Landau et E. Lifchitz, *Mécanique des fluides*, Editions MIR, Moscou, 1989.
■ J. S. Turner, *Buoyancy effects in fluids*, Cambridge University Press, 1973.
■ V. Lighthill, *An informal introduction to theoretical fluid mechanics*, Oxford University Press, 1986.
■ D. J. Tritton, *Physical fluid dynamics*, Van Nostrand Reinhold, 1977.
■ R. P. Behringer, « Rayleigh-Bénard convection and turbulence in liquid helium », *Rev. Mod. Phys.*, *57*, 657, 1985.
■ Pour se procurer une bibliographie plus complète voir page 626.

REFERENCE FRAME

COMPLEX STRUCTURES FROM SIMPLE SYSTEMS

Leo P. Kadanoff

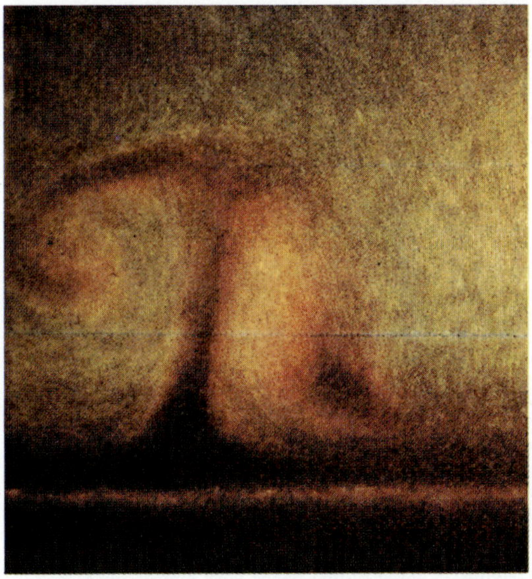

A plume observed at the edge of a tank of water undergoing convective turbulence. (See G. Zocchi, E. Moses, A. Libchaber, *Physica A* **166**, 387, 1990.) The colors are obtained by floating plastic-coated spheres of liquid crystal material in the water. The color of the liquid crystals serves as a sensitive thermometer. (Cognoscenti might wish to note that the picture is upside-down.)

A simple physical system might be defined as one that obeys simple laws. But simplicity of the rules of the game does not necessarily imply triviality of outcome. On the contrary, the action of elementary laws on many particles over long periods of time will often give rise to interesting structures and events. (We ourselves are probably results of this principle.) Thus, all the richness of structure observed in the natural world is not a consequence of the complexity of physical law, but instead arises from the many-times repeated application of quite simple laws.

Over the last few years, physicists have studied many examples of this kind of derived complexity. In a previous column (February 1986, page 6), I discussed the example of diffusion-limited aggregation in the context of fractal descriptions of nature.[1] Since that column was written, many more physical examples of fractals have been studied, and much attention has been devoted to finding the physical basis of the produced structures. One situation, studied by Per Bak, Chao Tang and Kurt Weisenfeld[2] as well as others, involves model sand, which is added grain by grain on the top of a model sandpile. The sand is neatly stacked in columns, each of which lies upon the sites of a regular lattice. So long as the local slope—the difference in height between adjacent columns—is sufficiently small, the pile stands up. But as soon as the slope gets large enough, a local slide is started. This slide can add extra material downhill and thus trigger further events below it. The displacement of material can also undermine the sand above it, which then will also start in motion. Thus additions of a single grain will trigger cascades in which sand falls downhill in response to a too-large local slope of the pile. These avalanche events can be small or they can cover the entire system many times over.

Bak, Tang and Weisenfeld call the situation reached in this manner "self-organized criticality." Here "criticality" is a technical word that means that events of all sizes occur. In this case, it is easy to understand how the pile organizes itself. If the slope of the pile is initially very small, only a few small slides will occur, and so the pile will steepen. If the initial slope is very large, huge avalanches will sweep over the edges of the pile and the slope will then become less steep. The system hunts for a slope that is, in Goldilocks's words, "just right." At this slope, the pile is at its margin of stability against slide formation. There is some randomness in the stacking, and because the system is exactly at the stability margin, the randomness is magnified by the dynamical process so that avalanche events of all sizes occur. Thus the phrase "self-organized criticality" is intended to imply the operation of a feedback mechanism that ensures a steady state in which the system is marginally stable against a disturbance. One would like to know the laws that govern this state.

Notice that this almost trivial model contains three different levels of organization on which laws might operate: an individual grain, an avalanche and the composite effect of many slides. If one studies the individual grains of sand, a large avalanche is a very complex composite event, involving many displacements of grains. However, if one looks away from the individual grains and instead at the avalanche as a whole or, alternatively, at the results of many successive avalanches, one sees some understandable behavior. At each level, new "laws of nature" emerge. For example, Jean Carlson, Jennifer Chayes, E. R. Grannan and G. H. Swindle[3] have studied the dependence of the local slope of the pile upon space and time as the cumulative result of many avalanche events. In the steady state, this slope approaches a critical value; away from equilibrium, it obeys a simple diffusion equa-

Leo Kadanoff, a condensed matter physicist, is John D. and Catherine T. MacArthur Professor of Physics and Mathematics at the University of Chicago.

tion. The self-organized criticality makes the diffusion coefficient in this equation diverge as the slope approaches its critical value. (The observed behavior of the coefficient is that it varies as an inverse power of slope minus critical slope.)

As another example, consider the kind of turbulence that arises when a fluid is heated from below. Hotter fluid is less dense than colder, so heated regions tend to rise and colder ones to fall. The heating thus puts the fluid into motion, and if the induced flow velocities are sufficiently high the motion can show considerable irregularity in space and time. This process is called convective turbulence. However, "turbulent" does not mean random. The complex motion of the fluid contains characteristic patterns, events and structures that show through all the randomness. This example, as distinct from the sand-slide models, is a real physical system. The essential question is, Can we learn anything from the simple models that we can then apply to understanding turbulence? Since we do not understand turbulence on any very deep level, we cannot yet be sure what insights will really apply, but there may well be carryover from one of these nonlinear dynamical problems to another. Based on the discussion above, we might look for self-organization via a feedback mechanism and the production of characteristic structures and events.

Before considering the underlying mechanisms, let's look at the evident structures. Systems heated from below move a large amount of heat by producing thermal plumes. These mushroom-shaped objects (see the figure on page 9) each arise from a localized source of heat sitting at the bottom of a reservoir of fluid. The hot fluid rises in the stem of the mushroom, and when it reaches the cap expands outward to fill a larger region. Plumes burst into our attention at the end of World War II as the gross visible manifestation of that awful source of heat, the nuclear explosion. Plumes are also observed at the surface of the Sun. As upwellings in the molten part of the Earth, they are responsible for much localized volcanic and seismic activity. We also feel them as the thermals that can shake us up when we travel by airplane.

The figure shows a picture of such a plume caught at the surface of a vessel containing water undergoing convective turbulence. The plume is rising from a thin layer called a boundary layer; such layers exist at the surfaces where heat is being transferred. (The actual surface of the water is marked by the light brown line toward the bottom of the figure. The boundary layer is the dark brown region just above the line that marks the surface.)

In one possible view of this situation, the number of plumes and the thickness of the boundary layer are set by a feedback process. In this scenario the boundary layer is excited by plumes and then goes unstable, giving rise to more plumes. If the layer is too thick, then it will be highly unstable, and lots of plumes will come off, ripping with them parts of the hot layer. The layer will then thin down. Conversely, if the layer is too thin, few plumes will rise up and heat conduction will make the layer thicken. This argument implies that the thickness of the boundary layer should depend upon the motion of plumes across the entire cell. As a consequence, the boundary layer thickness should depend upon the height of the cell. This dependence is indeed confirmed by experiment,[4] supporting the idea of feedback.

This turbulent fluid also obeys different laws at different levels. At the lowest level, we have the partial differential equations that describe fluid flow. At the next, the temperature at a given point in the system goes up and down to form a richly complicated time history. This chaotic history results from the intricate patterns of motion of many different plumes. Nonetheless, an almost isolated plume obeys very simple laws. Finally, at the level of the entire system, there is yet another form of simple law: In the turbulent regime, the heat conducted through the system exhibits a simple power-law dependence on the temperature difference across it.

The lesson? I see many different laws at many levels of description. I see intricate and complicated behavior both resulting from and resulting in simple behavior at lower and higher levels of description. I see partially understood examples of connections among levels and among different problems, and interesting subject matter for continuing research.

References

1. B. Mandelbrot, *The Fractal Geometry of Nature*, Freeman, New York (1983).
2. P. Bak, C. Tang, K. Weisenfeld, Phys. Rev. Lett. **59**, 381 (1987).
3. J. Carlson, J. Chayes, E. R. Grannan, G. H. Swindle, Phys. Rev. Lett. **65**, 2547 (1990).
4. See A. Libchaber, in *Proc. Second Latin American Workshop on Non-Linear Phenomena*, Santiago, Chile, 1990, Elsevier, New York, to be published. ■

Bubble, bubble, boil and trouble

Daniel H. Rothman[1] and Leo P. Kadanoff
The James Franck Institute
The University of Chicago
5640 South Ellis Avenue
Chicago, Illinois 60637

SUMMARY

Multiphase fluid mixtures present not only fascinating examples of nonequilibrium pattern formation, but also illustrate rather general questions concerning new emergent levels of organization. To illustrate these issues, we use lattice-gas cellular automata to study a buoyant mixture of hot and cold bubbles. We find that, depending on the volume fraction of bubbles, the model can exhibit either a coarsening instability familiar from studies of sedimentation, or a convective instability similar to the large-scale flows of Rayleigh-Bénard convection.

[1]Permanent address: Department of Earth, Atmospheric, and Planetary Sciences, Massachusetts Institute of Technology, Cambridge, MA 02139

The derivation of hydrodynamic equations from microscopic physics typically follows from a coarse-graining of the complex many-body interactions of molecular dynamics [1]. Recent work, however, has shown that hydrodynamic equations may in fact be obtained from highly simplified, discrete microscopic models known as *lattice-gas cellular automata* [2, 3, 4, 5, 6]. Here we use the lattice-gas method to construct a buoyant mixture of hot and cold bubbles, and ask whether an analogous coarse-graining of the resulting many-bubble interactions can be applied to our bubble mixture to obtain a model of Rayleigh-Bénard convection. We consider this question to be interesting not only because of its relevance to the determination of constitutive equations for multiphase flows [7], but also because we believe that it serves to illustrate how interesting qualitative questions may be answered by simulations of hydrodynamic cellular automata. In particular, our simulations show that our naive formulation of constitutive equations can be incorrect, due at least in part to a symmetry-breaking coarsening instability known from studies of two-component sedimentation [8, 9, 10].

Our model is constructed from a variation of the lattice-gas models introduced in references [2, 11, 12]. In the simplest two-dimensional lattice gas [2], identical particles of equal mass hop from site to site on a triangular lattice. When particles meet at a site they scatter, thus undergoing a collision. The precise rule for collisions is unimportant, but collisions must conserve mass and momentum. The conservation of these quantities [13] in addition to the symmetries of the triangular lattice allow one to obtain the following constitutive equations for the coarse-grained dynamics [3, 4, 5, 6]:

$$\nabla \cdot \mathbf{v} = 0 \tag{1}$$

$$\partial_t \mathbf{v} + h(\rho)(\mathbf{v} \cdot \nabla)\mathbf{v} = -\frac{1}{\rho}\nabla p + \nu \nabla^2 \mathbf{v}. \tag{2}$$

Here ρ denotes the spatially-averaged density of particles, $\mathbf{v}$ the spatially-averaged particle velocity, p the pressure, and ν the kinematic viscosity. The first equation represents the conservation of mass while the second represents the conservation of momentum. The factor $h(\rho)$, equal to unity for real fluids, differs from unity in our simulations [14], a point we return to below.

Bubbles are created in our model by including additional species of fluid and by employing a collision rule at interfaces that creates surface tension. The surface-tension inducing collision is anti-diffusive: it maximizes the flux of a particular species in the direction of the gradient of the local density of that particular species [11]. For sufficiently large densities of both species in a two-fluid model, mixtures are unstable and phase separation occurs spontaneously, resulting in an equilibrium state in which two distinct phases are separated by an interface. In a three-fluid model much the same phase-separation behavior is observed; in this case, however, an appropriate choice of surface tensions and fluid concentrations can result in an equilibrium state in which two distinct bubbles are immersed in a sea of the third fluid [12]. For the present study we employ an N-fluid model [15], a generalization of the three-fluid model of reference [12]. This many-bubble model is simply a two-dimensional suspension of an arbitrary number of hydrodynamically-interacting bubbles which can never coalesce. An example of such a system is shown in Figure 1. The bubble shapes are approximately round, but they may deform due to interface fluctuations and the interaction of viscous stresses with surface tension. The capillary number $\rho\nu v/\sigma$, where σ is the surface tension, typically ranges from approximately 0.1 to 1.

In the simulations we discuss below there are two kinds of bubbles, "hot" and "cold," each of radius a. Heat cannot diffuse into or out of the bubbles. The hot, positively buoyant, bubbles are subjected to a constant upward acceleration, while the cold, negatively buoyant, bubbles are subjected to an equal and opposite downward acceleration. By designating ϕ_1 and ϕ_2 to be the average concentration of hot and cold bubbles, respectively, at a coarse-grained scale much larger than a, and defining a "temperature" $T = \phi_1 - \phi_2$, we assume that a simple coarse-graining of the Navier-Stokes equations (1) and (2),

$$\nabla \cdot \bar{\mathrm{v}} = 0 \tag{3}$$

$$\partial_t \bar{\mathbf{v}} + h(\rho)(\bar{\mathbf{v}} \cdot \nabla)\bar{\mathbf{v}} = -\frac{1}{\rho}\nabla p + \nu \nabla^2 \bar{\mathbf{v}} + T\mathbf{g}, \tag{4}$$

describes mass and momentum conservation in the bubble mixture. Here $\mathbf{g}$ represents the gravitational acceleration, $\bar{\mathbf{v}}$ represents the coarse-grained velocity, and the gradients are also taken at the same coarse-grained scale.

To complete the description of the bubble mixture, we assume that the temperature field evolves due to advection, Brownian motion, and dispersion. Neglecting any cross-diffusivities, we subsume the latter contributions into a self-diffusivity κ and write

$$\partial_t T + (\bar{\mathbf{v}} \cdot \nabla)T = \kappa \nabla^2 T. \tag{5}$$

Except for the factor $h(\rho)$, equations (3), (4), and (5) are identical to the Boussinesq equations of thermal convection [16]. Thus we are led to the following question: Can our bubble mixture simulate Rayleigh-Bénard convection?

To answer this question experimentally, we put the model in a box with rigid walls and approximately unit aspect ratio, and "heat it from below" by instantaneously converting cold bubbles to hot when they touch the bottom boundary, and likewise converting hot bubbles to cold at the top [17]. The usual non-dimensionalization of the Boussinesq equations yields (after dropping primes)

$$\nabla \cdot \bar{\mathbf{v}} = 0 \tag{6}$$

$$\mathrm{Pr}^{-1}[\partial_t \bar{\mathbf{v}} + h(\rho)(\bar{\mathbf{v}} \cdot \nabla)\bar{\mathbf{v}} + \nabla p] = \nabla^2 \bar{\mathbf{v}} + \theta \mathbf{k} \tag{7}$$

$$\partial_t \theta + (\bar{\mathbf{v}} \cdot \nabla)\theta = \nabla^2 \theta + \mathrm{Ra}\ \bar{\mathbf{v}} \cdot \mathbf{k}, \tag{8}$$

where θ is the dimensionless temperature perturbation from a uniform gradient, $\mathbf{k}$ is the unit vector in the vertical direction, the Prandtl number $\mathrm{Pr} = \nu/\kappa$, and the Rayleigh number $\mathrm{Ra} = g\Phi h^3/\kappa\nu$, where h is the height of the box and $\Phi = \phi_1 + \phi_2$ is the fraction of the box filled with bubbles. The box is initialized with an equal number of hot and cold bubbles, each with size $a \approx 5$. We choose Φ ranging from about 0.3 to 0.4 and g large enough such that Brownian motion is negligible compared to advection. Finally, we observe that since heat diffuses neither into nor out of bubbles, then $\kappa \approx 0$

(again neglecting Brownian motion) and therefore $\mathrm{Pr} \to \infty$. Thus we expect that any effects of the non-physical term $h(\rho)$ on the left-hand-side of (7) are negligible due to high Pr. Furthermore, since κ is small, only a negligible g is needed for Ra to exceed the critical value for convection.

However, the choice of an experimental criterion to establish the model's equivalence to the Boussinesq equations is not obvious. In principle, one could compute coarse-grained gradients and local time derivatives from simulations, and show to what degree of accuracy the simulations obey the Boussinesq equations. Here, however, we choose a simpler criterion. Laboratory experiments, numerical simulations, and theoretical studies all show that the salient feature of unit-aspect-ratio Rayleigh-Bénard convection just above the convective threshold is a large-scale circulation of a size comparable to the box size [18]. Thus we simply ask whether our convective bubble mixture can result in such a circulation. Although a positive answer does not guarantee adherence to the Boussinesq equations, a negative answer does result in immediate disqualification.

Surprisingly, immediate disqualification is precisely what we found. We observed transient blobs of bubbles, rising and falling throughout the box, never organizing themselves into large-scale flows. Thus we conclude that equations (3), (4), and (5) are *not* in general the correct constitutive equations for our model.

Where, then, is the fallacy in our reasoning? We believe our error lies in the simplifications implicit in the temperature equation (5). Specifically, our implicit assumption was that there were sufficiently large regions of almost uniform fluid motion so that $\bar{\mathbf{v}}$ and the hydrodynamic analysis made sense. But perhaps the properties of the fluid are much more nonuniform than would be compatible with our smoothed-out hydrodynamic description.

To understand why there might be additional structure in the system, we note that, in the absence of the adverse temperature gradient, our system is a special case of *two-component sedimentation* [8, 9, 10], in which the two types of sedimenting particles are identical except for their densities, each of which differs from the density of the suspending fluid by the same amount but with different sign. Batchelor and Janse van Renseburg [10] have derived criteria for the linear stability of a concentration wave in such a system. The essential result is that if the two species of sedimenting particles have a sufficiently strong hydrodynamic interaction, then the concentration wave is linearly unstable. In their analysis, which assumes $\kappa = 0$, a "sufficiently strong" interaction arises if the concentrations of both species are large enough. Precise predictions of the critical concentrations are not available, but experiments in quasi two-dimensional cells show that, for the case $\phi_1 = \phi_2$, one obtains $\Phi_c \approx 0.17 \pm 0.03$ [10].

In our context, however, the salient point is not just that concentration waves may be unstable, but that the most unstable wavelength itself grows with time. This is qualitatively evident from the experiments in Ref. [10] and also from the simulation of sedimentation shown in Figure 2. The simulation is initiated with randomly distributed hot and cold bubbles as before, with a bubble concentration of 0.40. There are no walls, and boundaries are periodic in both directions. The results show that the instability manifests itself as a coarsening phenomenon, in which regions of ascending red bubbles and descending blue bubbles grow, developing into large-scale fingers, some of which are similar to thermal plumes.

A direct consequence of this coarsening is the inability to define the coarse-grained scale implicit in the definition of $\bar{\mathbf{v}}$ in equations (3), (4), and (5). Any finite length scale eventually becomes small compared to the size of sedimenting fingers and plumes. Thus the symmetry between hot and cold bubbles is broken, and regions composed primarily of hot or cold bubbles grow. The phenomenon is loosely analogous to spinodal decomposition, in which thermodynamically unstable mixtures spontaneously separate into two distinct phases [19]. Here, however, the phase separation is not driven by a favorable free energy but is instead driven mechanically by hydrodynamic interactions.

The coarsening in Figure 2 can be quantified, allowing us to empirically determine a phase diagram for the stability of our two-component sedimentation. At discrete intervals in time, we compute the two-dimensional power spectrum

$$S(\mathbf{k},t) = \frac{1}{n_x n_y}\left|\sum_{\mathbf{x}} e^{-i2\pi\mathbf{k}\cdot\mathbf{x}}[\phi_1(\mathbf{x},t) - \phi_2(\mathbf{x},t)]\right|^2, \tag{9}$$

where $\phi_1(\mathbf{x},t)$ and $\phi_2(\mathbf{x},t)$ are the local concentrations of hot particles and cold particles, respectively, at time t at a site with coordinates given by $\mathbf{x}$, $\mathbf{k}$ is the discrete wavevector, and n_x and n_y are the number of lattice sites in the x-direction and y-direction, respectively. At the initial stages of the instability, the macroscopic structures are approximately isotropic, so we compute circular averages $\hat{S}(k,t) = \langle S(\mathbf{k},t)\rangle$, where the angle brackets indicate averaging over all wavevectors $\mathbf{k}$ of magnitude $k = |\mathbf{k}|$. We then define a characteristic length R from the inverse of the first moment of $\hat{S}$:

$$R(t) = \frac{\sum_k \hat{S}(k,t)}{\sum_k k\hat{S}(k,t)}. \tag{10}$$

Unstable mixtures are then defined to be those for which $R(t)$ grows with time; if $R(t)$ does not grow, the mixture is considered stable.

In our simulations, both the bubble concentration Φ and the mean bubble speed determine stability of the mixture. The latter parameter is significant here because, for sufficiently weak gravitational accelerations, Brownian motion can be significant. Thus we consider a phase plane in the space of Φ and the free-fall Peclet number $\mathrm{Pe}^* = av_f/\kappa$, where the latter parameter is controlled operationally by g, v_f is the free-fall velocity of a bubble of size a, and κ is estimated from simulations of Brownian motion.

We consider boxes of size $n_x = n_y = 128$ and particular combinations of Φ and Pe^* that allow us to experimentally determine the stability boundary. Figures 3, 4, and 5 show the results of a representative calculation for fixed $\mathrm{Pe}^* = 11$ at varying values of Φ. (Here and elsewhere the results are obtained by averaging over 5 independent simulations.) The spectra $\hat{S}(k,t)$ are plotted in Figure 3 for the case $\Phi = 0.20$; one sees that $\hat{S}(k,t)$ is invariant with time for dimensionless wavenumbers $ak > 0.5$, whereas for $ak < 0.5$ the wavenumber of the peak in $\hat{S}(k,t)$ decreases with time while the peak value itself increases with time. The high-wavenumber behavior reflects the constancy of the small-scale spatial fluctuations, which are determined entirely by the bubble size a. The low-wavenumber behavior, on the other hand, is a consequence of the coarsening. The time dependence of the coarsening is plotted in the graph of $R(t)$ in Figure 4; in the interval shown, $R(t)$ increases approximately linearly with time [20]. Figure 5 summarizes the results of similar computations of $R(t)$ for varying values of Φ. Here we plot only the value of the slope of $R(t)$, where the slope is obtained by a least squares fit of a straight line that intersects $R(0)$. Since the slope represents the best-fit linear rate of coarsening, we conclude that the stability boundary is at $\Phi = .125 \pm .025$.

By collecting many such data, we determine the phase diagram in Figure 6. Inside the curve (high Φ and Pe*), the sedimenting mixture is unstable and therefore coarsens; outside the curve the mixture is stable. In the limit of vanishing diffusivity (high Pe*), the results show stability for Φ less than about 0.1. On the other hand, as Φ increases, the smallest possible velocity in unstable sedimentation decreases. These results confirm simple physical intuition: a high bubble concentration (but still short of the close packing limit) and a large ratio of advection to diffusion creates favorable conditions for strong hydrodynamic interactions between bubbles, and thus instability of the mixture.

These results may now be brought to bear on the Rayleigh-Bénard problem. If indeed the coarsening in the sedimentation problem is prohibiting us from writing the constitutive equations (3), (4), and (5), then we should expect recovery of these same Boussinesq equations if we choose a *stable* combination of Φ and Pe*. This reasoning indeed appears to be correct. In Figure 7 we show an example of our numerical Rayleigh-Bénard experiment, where now we have chosen Pe$^* = 55$ and $\Phi = 0.1$, a combination that the results of Figure 6 show to yield approximately marginal stability for sedimentation. The results clearly show a large-scale circulation, thus indicating that a coarse-graining of the simulation at a scale much greater than the bubble size may indeed satisfy the Boussinesq equations. By using the Brownian diffusivity for κ, the Rayleigh number is estimated to be about 10^7.

In concluding, we note that our simulations have shown that 1) a naive coarse-graining of a bidispersed mixture of bubbles depends crucially on the stability of the mixture; 2) the stability of the mixture depends in turn on both the concentration of bubbles and a dimensionless measure of their mean falling speeds; and 3) if the mixture is in fact stable, then it can reproduce large-scale flows similar to those observed in Rayleigh-Bénard convection above the convective threshold. Finally, we note that our bubble model offers an interesting approach to the simulation of high Prandtl number convection and may perhaps offer an attractive alternative to the methods employed, for example, in Ref. [21]. Consistent with most other successful applications of lattice-gas cellular automata, we feel that our model is best suited for answering qualitative, rather than quantitative, questions. The precise limits of its applicability, however, remain to be determined.

This work was supported in part by the Materials Research Laboratory and the Department of Geophysical Sciences at The University of Chicago, by NSF Grant 9017062-EAR, and by the sponsors of the MIT Porous Flow Project. We thank R. Santos and J. Olson for help in the visualization of Figures 2 and 7 and F. Cattaneo for interesting discussions.

REFERENCES

[1] S. Chapman and T. G. Cowling, *The mathematical theory of non-uniform gases* (Cambridge University Press, Cambridge, 1970)

[2] U. Frisch, B. Hasslacher, and Y. Pomeau, Phys. Rev. Lett. **56**, 1505 (1986).

[3] U. Frisch, D. d'Humières, B. Hasslacher, P. Lallemand, Y. Pomeau, and J.-P. Rivet, Complex Systems **1**, 648 (1987).

[4] S. Wolfram, J. Stat. Phys. **45**, 471 (1986).

[5] L. Kadanoff, G. McNamara, and G. Zanetti, Phys. Rev. A **40**, 4527 (1989).

[6] G. Zanetti, Phys. Rev. A **40**, 1539 (1989).

[7] M. Ishii, *Thermo-fluid dynamic theory of two-phase flow* (Eyrolles, Paris, 1975).

[8] R.L. Whitmore, Brit. J. Appl. Phys. **6**, 239 (1955).

[9] R.H. Weiland, Y. P. Fessas, and B. V. Ramarao, J. Fluid Mech. **142**, 383 (1984).

[10] G. K. Batchelor and R. W. Janse van Rensburg, J. Fluid Mech. **166**, 379 (1986).

[11] D.H. Rothman and J.M. Keller, J. Stat. Phys. **52**, 1119 (1988).

[12] A. K. Gunstensen and D.H. Rothman, Physica D **47**, 47 (1991).

[13] Spurious invariants may also exist. See [6] and D. d'Humieres, Y.H. Qian, and P. Lallemand, *in Discrete kinetic theory, lattice gas dynamics, and foundations of hydrodynamics*, ed. R. Monaco (World Scientific, Singapore, 1989).

[14] A. K. Gunstensen and D.H. Rothman, Physica D **47**, 53 (1991).

[15] D.H. Rothman, in *Microscopic simulations of complex hydrodynamics*, M. Mareschal and B. Holian, eds., (Plenum Press, New York, in press).

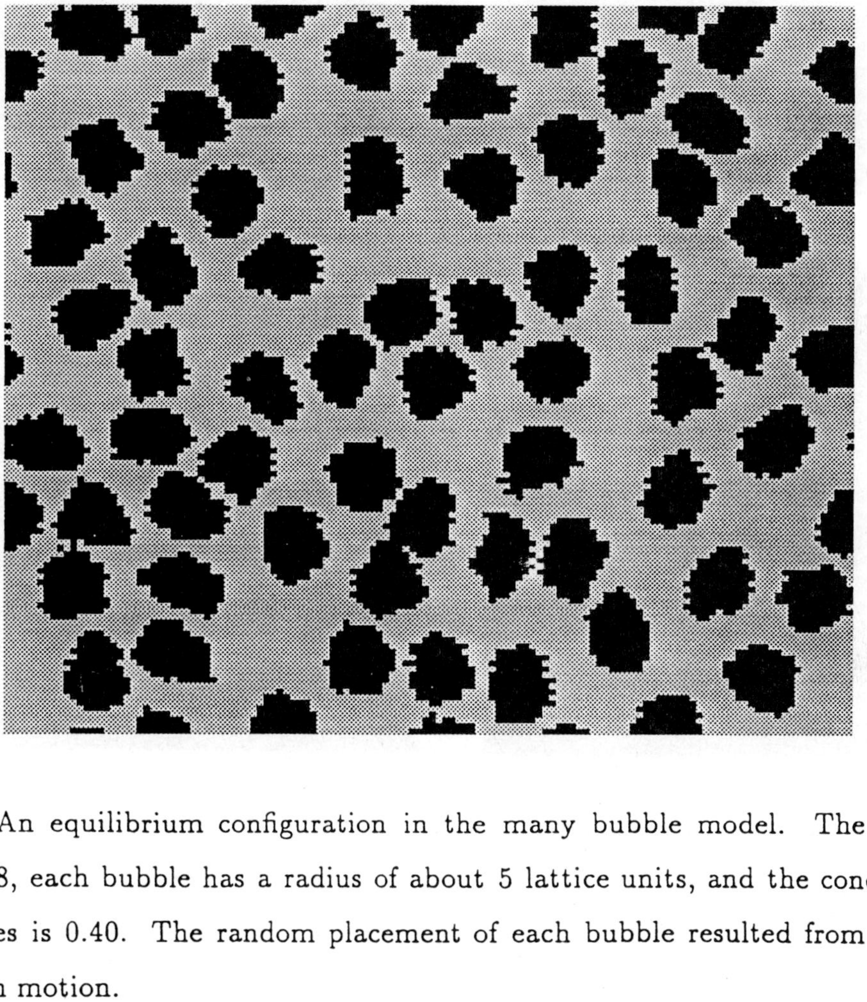

FIG. 1. An equilibrium configuration in the many bubble model. The lattice is 128×128, each bubble has a radius of about 5 lattice units, and the concentration of bubbles is 0.40. The random placement of each bubble resulted from collective Brownian motion.

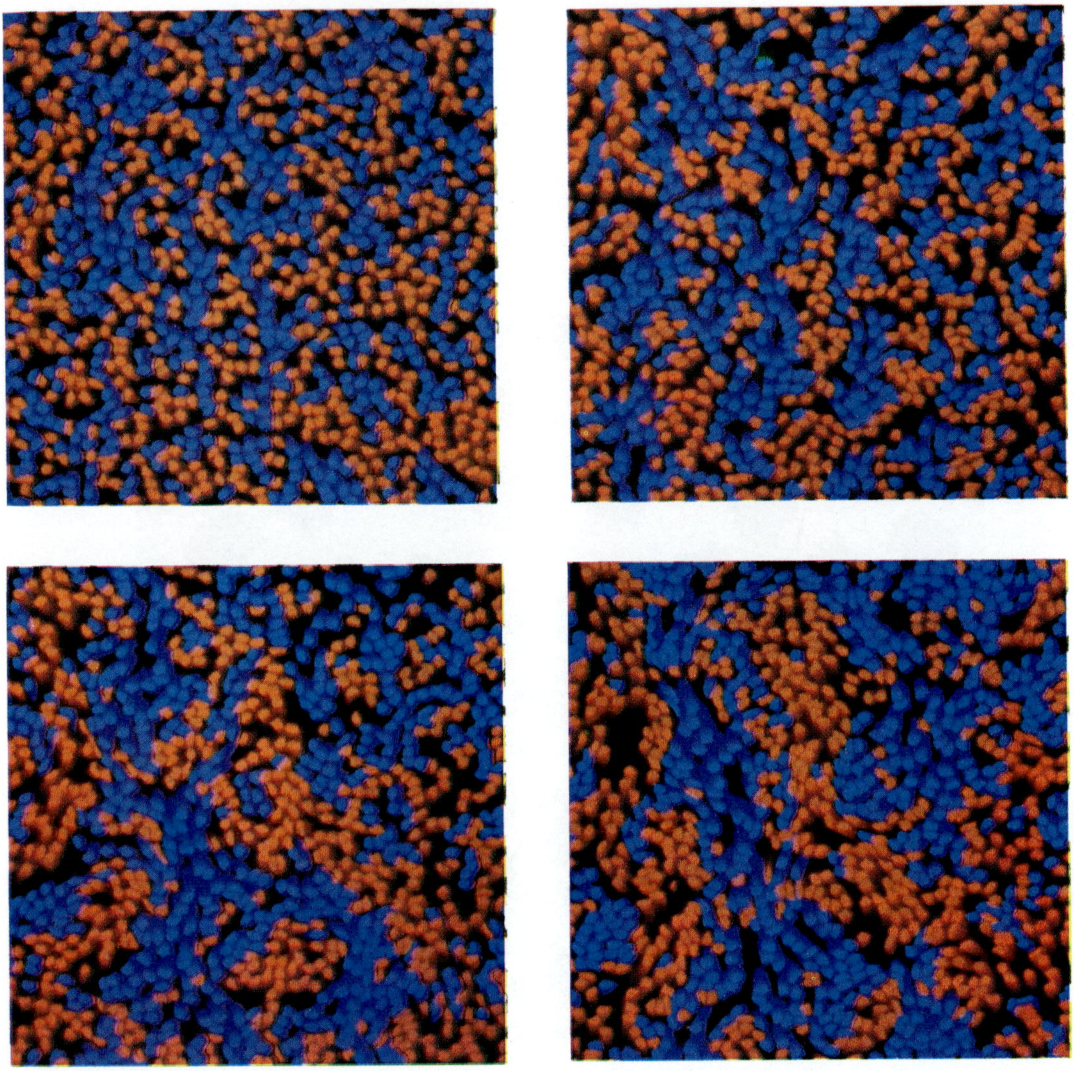

FIG. 2. A simulation of two-component sedimentation. The lattice contains 512×512 points; positively buoyant bubbles are red and negatively buoyant bubbles are blue; there are 1024 bubbles, each with a radius of about 5 lattice units, encompassing a total volume fraction $\Phi = 0.4$. Left to right, the snapshots are shown after 1000, 3500, 6000, and 8500 time steps, respectively. The recent motion of the individual bubbles prior to each snapshot is indicated by a reverse fade-out: the more distant in time prior to the present configuration, the more pale is the shade of red or blue. If bubble trajectories cross, the more recent trajectory takes precedence. Note the emergence of coherent large-scale structures as time evolves, some taking the shape of fingers or columns while others look more like the heads of plumes.

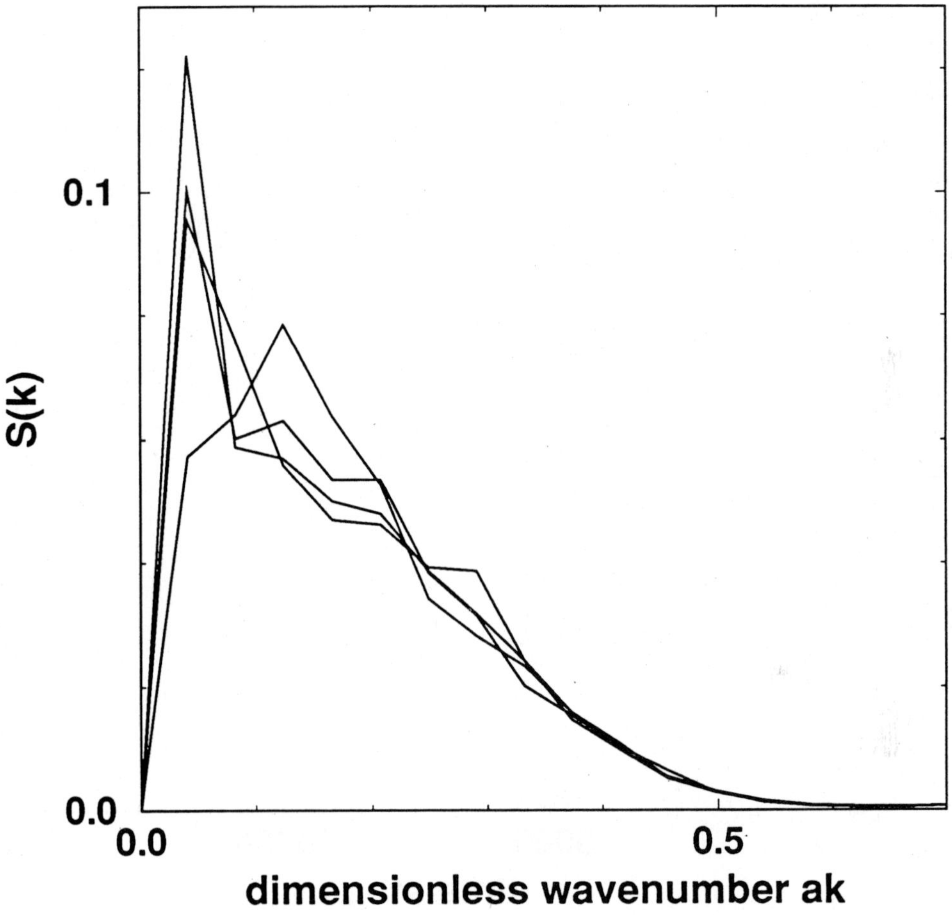

FIG. 3. The spectra $\hat{S}(k)$, computed at 0, 2500, 5000, and 7500 time steps. The earliest spectrum at time $t = 0$ was computed after the bubble mixture was brought to equilibrium but just prior to initiation of the gravitational force. As time increases the peak value of $\hat{S}(k)$ increases and shifts to lower wavenumbers. In this example, the volume fraction $\Phi = 0.20$ and $\text{Pe}^* = 11$.

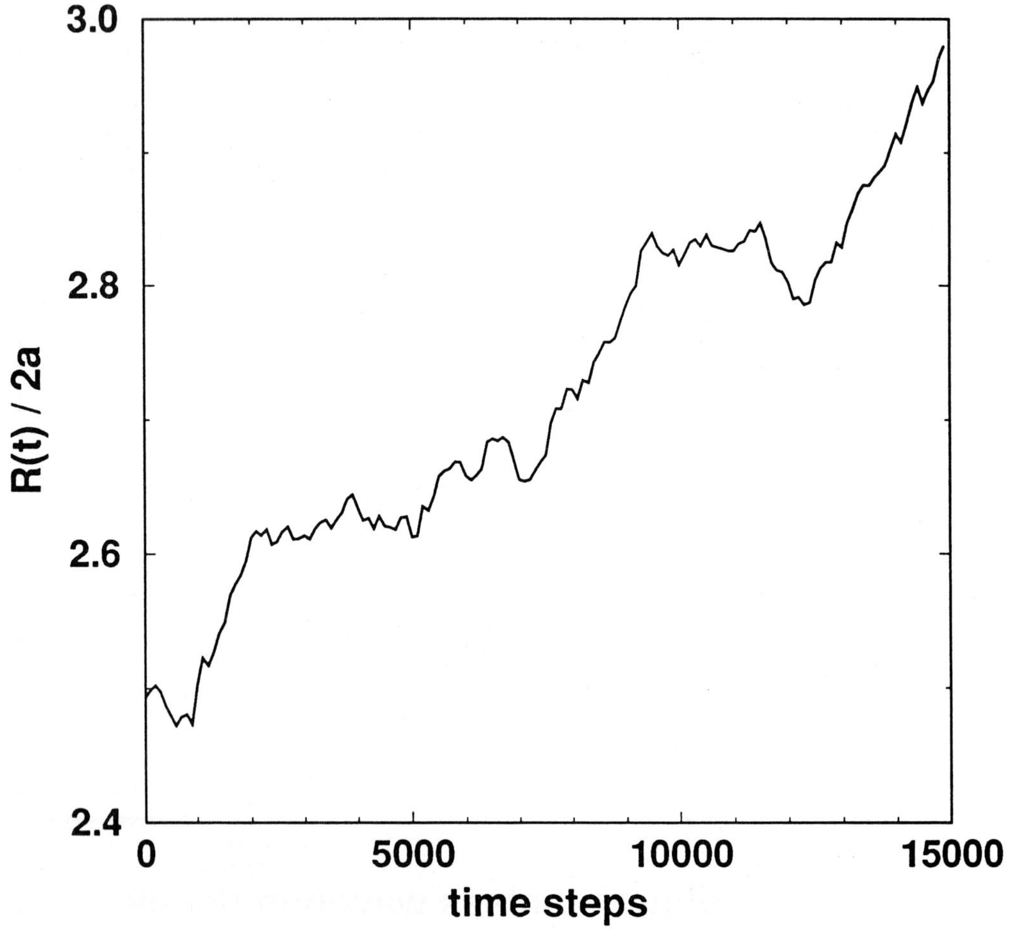

FIG. 4. Dimensionless blob size $R(t)/2a$, computed from spectra $\hat{S}(k,t)$ at time steps $0, 100, \ldots, 15000$, the last corresponding roughly to the length of time it takes bubbles to traverse the periodic box twice.

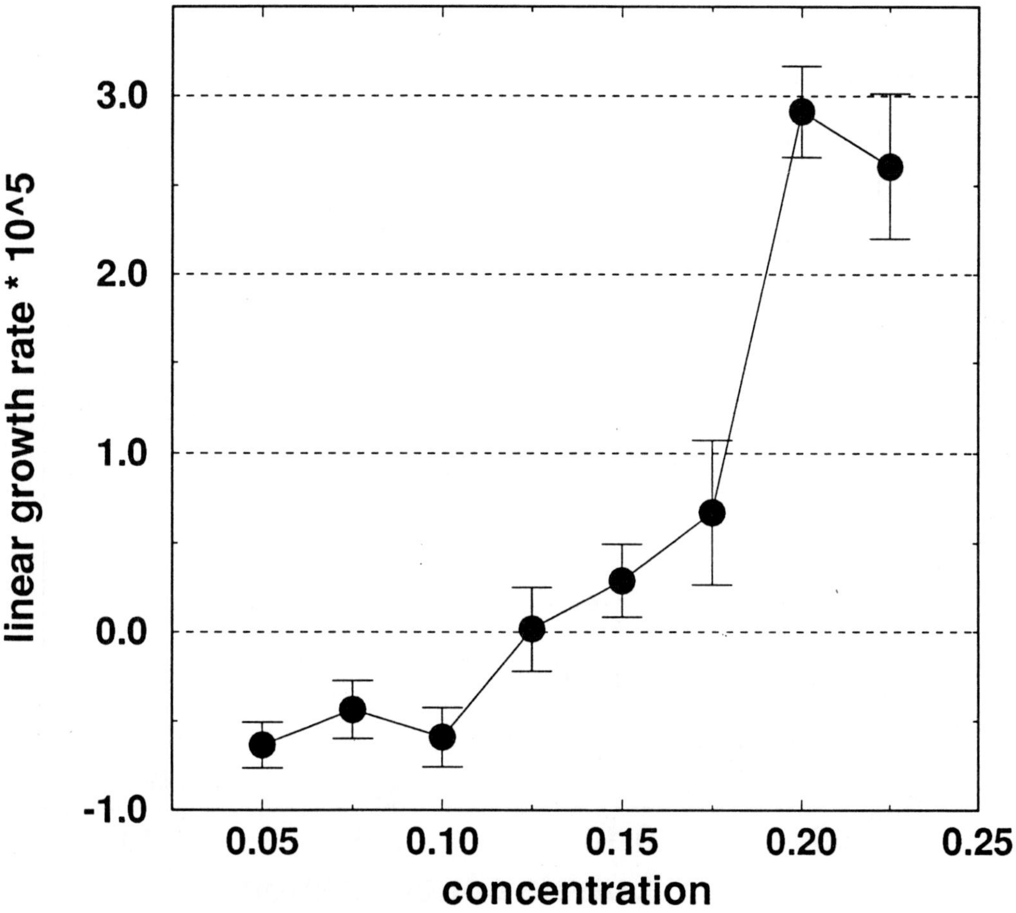

FIG. 5. A plot of the best fit linear growth rate, in units of dimensionless blob size $R(t)/2a$ per time step, as a function of bubble fraction Φ. Each point was computed with Pe^* fixed at 11.

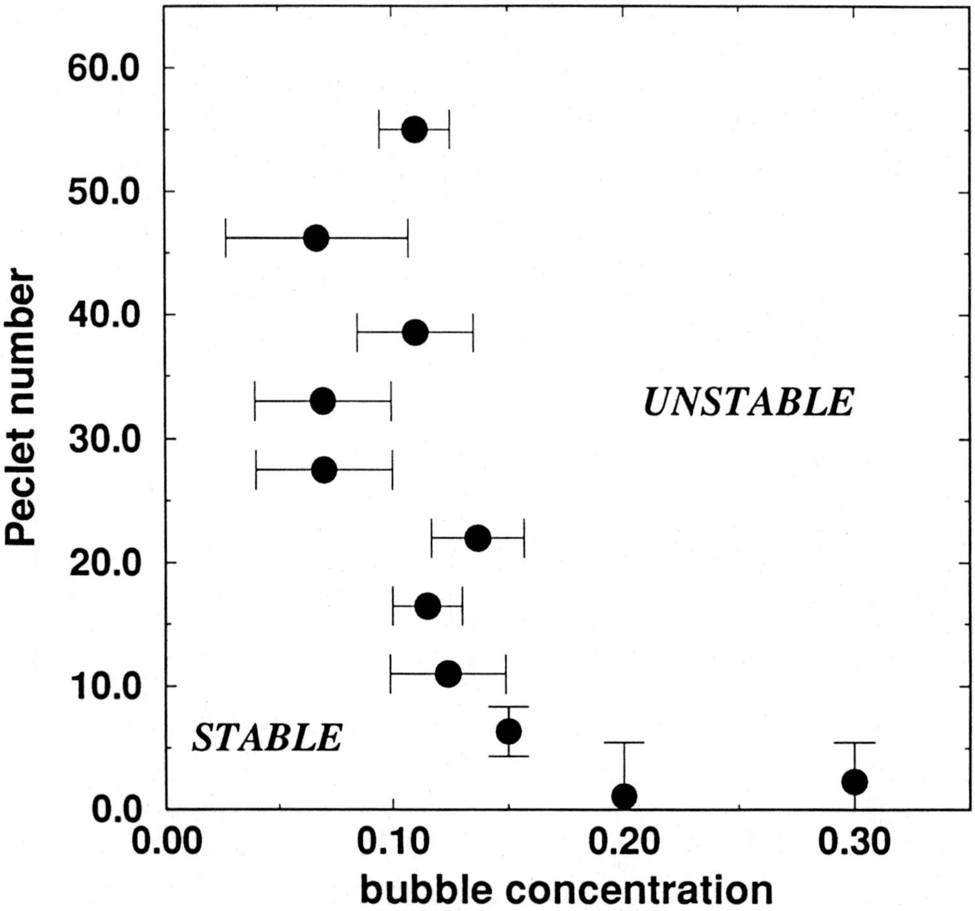

FIG. 6. Phase diagram for stability of two-component sedimentation. Error bars correspond roughly to the granularity of plots such as the one in Figure 5. Values of Pe* are approximate.

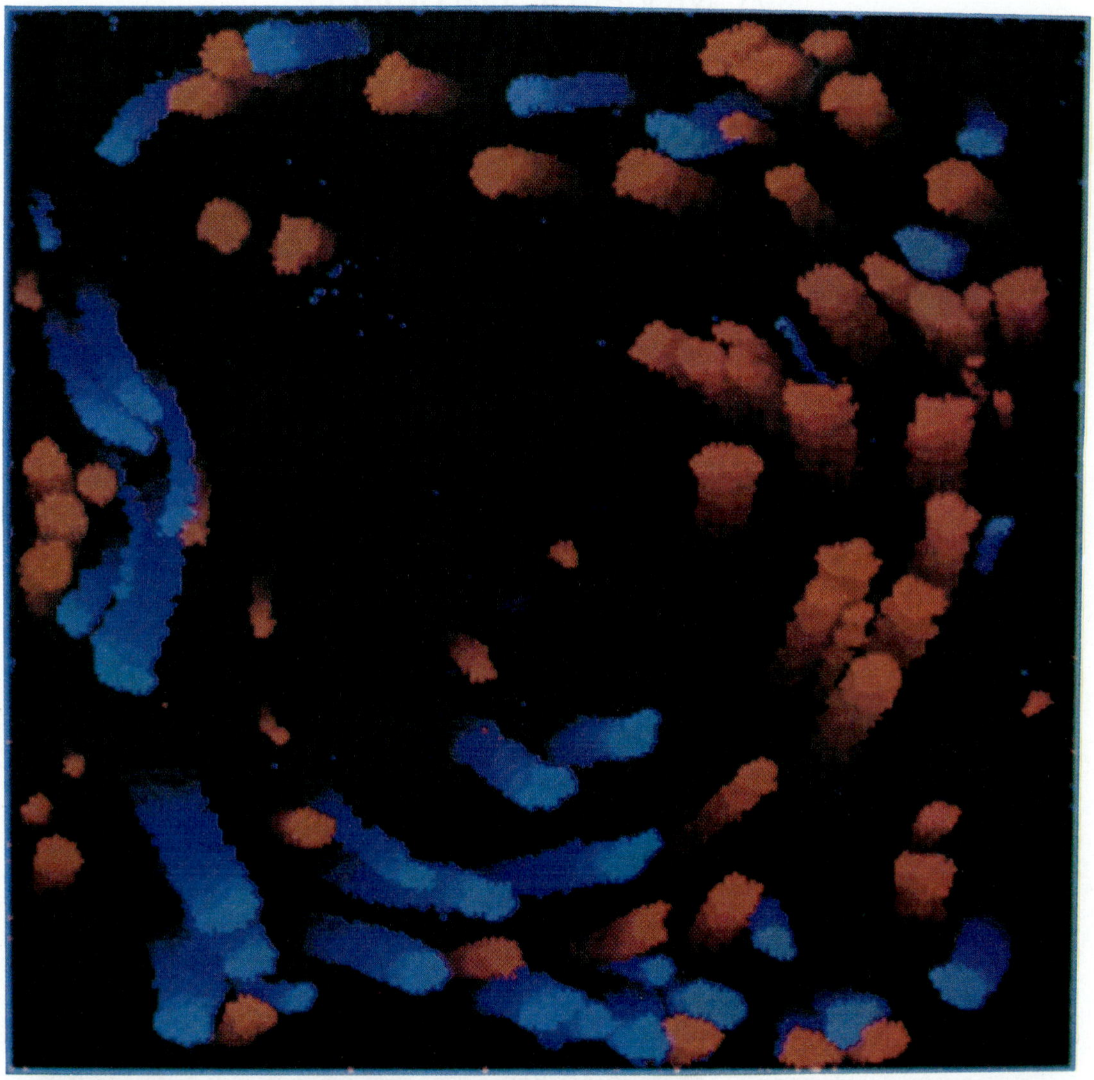

FIG. 7. A snapshot of a simulation of thermal convection on a 256×256 lattice, after 85000 time steps (approximately 6 circulation times). The volume fraction $\Phi = 0.10$ and $Pe^* = 55$. As in Figure 2, the recent motion of bubbles prior to the snapshot is indicated by a reverse fade-out.

REFERENCE FRAME

A Model of Turbulence

Leo P. Kadanoff

Turbulence is one of *those* problems. Interesting. Vexing. Longstanding. Unsolved.

The basic physics is easy to describe: Imagine a fluid that initially has some sort of simple flow with a nonuniform velocity. As time goes on, the nonlinear interactions among the fluid elements will tend to produce more and more complex structures, with more and more fine details. This complexity will continue to grow until, at a very fine scale, it is limited by viscous damping, which tends to smooth everything out. Note, for example, the extremely complex small structures in figure 1.

The parameter for describing the degree of complexity is the Reynolds number $\mathrm{Re} = LU/\nu$, where U is the typical nonuniformity in velocity, L is the size of a region in which the nonuniformity occurs and ν is the kinematic viscosity of the fluid in question. In everyday systems, fine scales are produced precisely because this viscosity is extremely small—of order 10^{-6} m^2/s for water.

Figure 2 (on page 13) shows a schematic of these turbulent processes, using concepts introduced over 70 years ago by L. F. Richardson. The dynamics is best described in Fourier-transform language in which the wavevector $\mathbf{k}$ serves to describe the inverse of the different spatial scales. Some large-scale process, perhaps stirring, is used to feed energy into the fluid at small $\mathbf{k}$. The range of scales at which this addition of energy occurs is called the integral range. Then the nonlinear coupling of velocities causes the energy to cascade toward larger wavevectors. The range of scales dominated by the cascade is called the inertial range. This range can be quite large, since it is proportional to a power of the Reynolds number. For even larger $\mathbf{k}$'s, one passes into the viscous range. Here, viscous processes dissipate the energy into heat and smooth out all details. The problem of turbulence is to predict and understand what happens when the inertial range is large.

LEO P. KADANOFF, *a condensed matter theorist, is John D. and Catherine T. MacArthur Professor of Physics and Mathematics at the University of Chicago.*

A quantitative analysis of energy conservation in turbulent flow was introduced by Andrei N. Kolmogorov in two papers written in 1941 and 1962 that are now generally called K41 and K62. The key idea is that all the energy that is added on the largest scale flows unchanged through the inertial range toward higher $\mathbf{k}$. The flow is described by a current. By dimensional analysis Kolmogorov found that the current of energy is proportional to $\mathbf{k}[\mathbf{u}(\mathbf{k})]^3$, where $\mathbf{u}(\mathbf{k})$ is the spatial Fourier transform of the velocity. The conservation of energy implies that in the inertial range this current must be independent of $\mathbf{k}$. The 1941 paper thus gives an order of magnitude statement

$$u(\mathbf{k}) \approx \left(\frac{\varepsilon}{k}\right)^{1/3}$$

where ε is the energy current.

There are two crucial assumptions in K41:

▷ The current ε is assumed to be roughly constant and thus essentially equal to the energy input per unit time.

▷ Further, since the dissipation never directly appears in the argument, the energy flow is assumed to depend

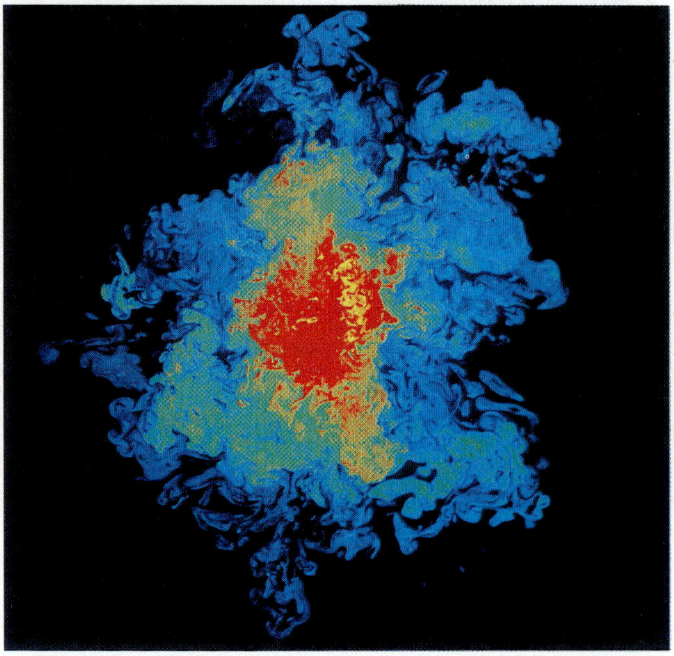

JET FLUID CONCENTRATION in a plane normal to the jet axis, 300 diameters downstream of the jet nozzle, measured by laser-induced fluorescence. Black indicates the reservoir fluid into which the jet is discharging. Pale yellow, red, green and blue indicate successively lower levels of jet fluid concentration. (Courtesy of Harry Catrakis and Paul Dimotakis, Caltech.) FIGURE 1

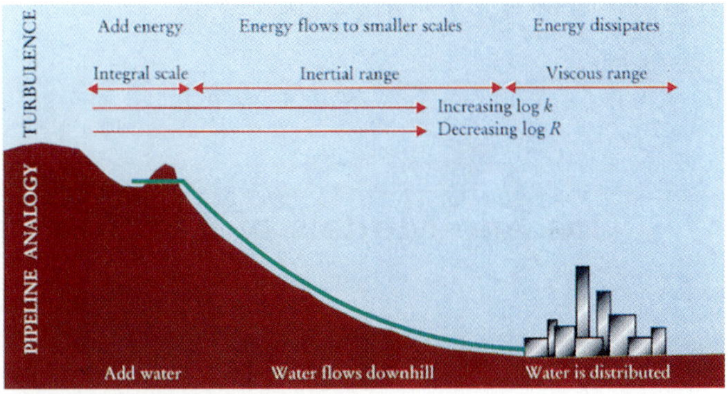

TURBULENCE AS A DOWNHILL FLOW. In this physical model energy is added on the integral scale and flows "downhill" toward shorter scales, R, and higher values of the wavevector $\mathbf{k}$. The process can be viewed as a kind of pipeline in which the energy flows downhill through a long inertial range and finally is dissipated in a viscously determined range of scales. FIGURE 2

only on what happens in the inertial and integral scales. Thus, for example, the value of ε should depend only on the nature of the external forcing and should be independent of the Reynolds number.

This argument, and many subsequent arguments, sees the energy flow as a kind of pipeline. (See again figure 2.) Energy is added to the pipe at the large scales, flows through the very long pipe and then flows out via dissipation at the far end. The 1941 analysis implies that the flow through the pipe at a given point depends only on processes upstream of that point. Thus, for example, inertial-range behavior would not be influenced by the value of the viscosity, since the viscosity appears only in the dissipative range, far downstream.

But the pipeline analogy casts doubt upon the assumption that no effects flow upstream. In a pipeline, a choking of the outlet would tend to diminish the flow.

Initially the field focused on the other basic assumption of the K41 theory, that of the constancy of energy dissipation. Partially in response to a criticism by Lev Landau, Kolmogorov changed his own point of view. In K62 he argued that because of fluctuations, the energy input should be a function of space and time that can vary over many, many orders of magnitude. These huge fluctuations are termed *intermittency*. While still assuming that effects only flow downstream, Benoit Mandelbrot and others developed richer cascade models that showed how a highly fluctuating behavior might arise and be reflected in a complex fractal structure of $\mathbf{u}(\mathbf{k})$.

Theory cannot for the moment distinguish between the weak fluctuations of K41 and the intermittency of K62. To settle questions of this kind we should look to experiment. The experimental evidence suggests the existence of intermittency but is not yet conclusive. We might also look to simulations. Turbulence problems can easily be set up on the computer, since a turbulent system may be correctly described by a well-known partial differential equation, the Navier–Stokes equation. Unfortunately computers are too slow to allow simulations to cover a sufficient range of scales. How can we learn more? One way is to set up models that correctly realize in specific form the previous theoretical ideas and then see how consistent these ideas might be.

Several one-dimensional models have been developed that include in a very rough way cascades from shell to shell in momentum space. One such model, which has received considerable attention lately, is connected with the names of E. B. Gledzer, K. Okitimi and M. Yamada (inelegantly referred to as GOY). This is the simplest model that can correctly realize a cascade through orders of magnitude of wavevector $\mathbf{k}$. The system is described by the Fourier coefficients $\mathbf{u}(\mathbf{k})$ of the velocity. Imagine a cascade that goes from a $\mathbf{k}$ of order the inverse size of the system, L^{-1}, to one in a shell that contains $\mathbf{k}$'s a factor of λ larger, and then to a higher shell with $\mathbf{k}$'s of order $\lambda^2 L$ and so forth until finally we reach $\lambda^N L$, a wavevector that sits far into the dissipative range. The real system involves many, many Fourier coefficients $\mathbf{u}(\mathbf{k})$, with the number per shell increasing sharply with the shell number. This model tries to describe the flow by keeping precisely one Fourier coefficient for each shell. The equations of motion are picked to mimic Navier–Stokes behavior, with nonlinearity, viscous damping and energy conservation all included. The model involves two parameters that have no direct counterpart in Navier–Stokes: the shell width λ and a parameter c. The latter determines the ratio of the energy flux toward higher wavenumber to the flux in the opposite sense.

Because the equations of this model involve many fewer variables than do the Navier–Stokes equations, its qualitative properties can be established with the aid of simulations. We can thus ask, Do the size of the fluctuations in the model agree with K41 or with K62? The question is sharply posed, and it may be sharply answered. The answer is yes. Yes? Yes! In one range of parameters (c close to unity) the system shows small fluctuations and K41 behavior, while in another range (smaller c) it shows very strong fluctuations and something more like K62 behavior. For this domain, the model permits independent fluctuations in the velocity ratio between neighboring shells. When one multiplies out a set of many independent random multipliers, the result can have truly huge fluctuations. Thus the model serves as a partial justification for both K41 and the later, more fluctuating theories.

However, K41 and some later theories say that the inertial-range behavior should be independent of ν. In response to some prompting from Zhen Su She (University of Arizona), Norbert Schörghofer, Jane Wang, Detlef Lohse, Roberto Benzi and I looked at the GOY model's inertial range. In one situation Schörghofer's numerics showed that these theoretical guesses were wrong: In the static solution, the energy flux down the pipe certainly does depend on ν, even in the limit of high Reynolds number. In contrast to the early theories (*both* K41 and K62), here the flow down the energy pipeline depends on an interaction between inlet and outlet conditions, induced by correlations carried up and down the pipe. (The detailed manifestation of the correlation in the GOY model depends on the shell thickness and is thus unphysical.) This correlation result points us toward the possibility that both inlet and outlet might play a role in determining the flow in real turbulence. If so, there will be some Reynolds-number dependence of inertial-range behavior. Thus a vastly oversimplified model led to insights that might well play back into a deeper understanding of a truly complex phenomenon.

Models are fun, and sometimes even instructive. ■

Tr. J. of Physics
21 (1997) , 1 – 14.

Cascade Models of Turbulence and Mixing *

Leo P. KADANOFF[†]
The Research Institutes
The University of Chicago
5640 South Ellis Ave.
Chicago, Illinois 60637, USA

Abstract

This note describes two kinds of work on turbulence. First it describes a simplified model of turbulent energy-cascades called the GOY model. Second it mentions work on a model of mixing in fluids. In addition to a brief historical discussion, I include some mention of our own work carried on at the University of Chicago by Jane Wang, Detlef Lohse, Roberto Benzi, Norbert Schörghofer, Scott Wunsch, Tong Zhou and myself. Our own studies are in large measure the outgrowth of a paper by M. H. Jensen, G. Paladin, and A. Vulpiani [1]. I mention this connection with some sadness because I recall Paladin's recent death in a mountain accident.

1. Turbulence and Mixing

Turbulence is one of *those* problems. Interesting. Vexing. Long-standing. Unsolved.

Mixing is also in much the same situation. In both cases, we know all the basic equations involved, but we really do not understand the physics in any full way. Instead we have models and metaphors which are believed to provide a partial understanding of what is going on, but both the relevance of the models, and the conclusions to be drawn from them, are quite controversial.

* This talk is based upon two recent articles: Leo P. Kadanoff, A model of turbulence, Physics Today, (September 1995) p 11. and Leo P. Kadanoff, Turbulent Excursions, Nature **382** p. 116 (1996). The talk itself was given at three meetings, at Cargese, Ankara, and Istanbul in the summer of 1996. The modeling work described here was done in collaboration with Jane Wang, Detlef Lohse, Roberto Benzi, Nobert Schörghofer, and Scott Wunsch. This publication is supported in part by the U.S. National Science Foundation.

† e-mail address: LeoP@Uchicago.edu

Today I shall focus on turbulence and the closely associated problem of mixing in a turbulent environment. I shall be interested in the conceptual and mathematical models which describe these phenomena, and most specifically on how one draws physical conclusions from such models.

The basic physics of turbulence is easy to describe: Imagine a fluid which initially has some sort of simple flow with a non-uniform velocity. In the situation in which velocities are small in comparison to the speed of sound the flow is approximately incompressible. It is then described by the Navier Stokes equation

$$\partial_t U(r,t) + (U \cdot \nabla)U \ = \ \nu \nabla^2 U - \nabla P(r,t) + F(r,t); \quad \nabla \cdot U = 0. \tag{1}$$

in units in which the density is equal to one. Here the velocity U is carried from place to place by the non-linear term $(U \cdot \nabla)U$. The diffusion term with its associated viscosity, ν, tends to smooth out the velocity field while F is an external force which describes the stirring process which puts the fluid into motion. The incompressibility condition is that the divergence of U is zero and is enforced by having the system pick an appropriate spatial dependence of the pressure, P . To get an alternative form of the Navier Stokes equation, one takes the spatial Fourier transform of equation (1). Use the symbols u, p and f for the spatial Fourier transforms of velocity, pressure, and force. The resulting equation reads

$$\partial_t u(k) + \int dq (u(k-q) \cdot iq) u(q) = -\nu k^2 u(k) - ikp(k) + f(k) \tag{2}$$

$$k \cdot u(k) = 0$$

To describe the mixing imagine some quantity suspended in the flow. This quantity might for example be heat. The density of the quantity, perhaps the temperature $T(r,t)$, obeys the flow equation

$$\left[\partial_t + U \cdot \nabla - D\nabla^2\right] T(r,t) = 0; \quad \nabla \cdot U = 0 \tag{3}$$

Here D is the diffusion coefficient for the passive scalar, while the mixing-term (the one proportional to U) describes how the velocity field moves the heat about.

The nonlinear processes built into the Navier Stokes equations produce the interesting structures in turbulent flow. As time goes on, the interactions among the fluid elements, represented by the $U \cdot \nabla U$ term, will tend to produce more and more complex structures, with more and more fine details. The complexity will continue to grow until, at a very fine scale, it is limited by viscous damping which tends to smooth everything out [2]. Of course, the $U \cdot \nabla T$ term will also produce complex structures. The difference is that this convective mixing equation, equation (3), has no non-linear character. Because of this lack of feedback, the structures in T can be quite different from the structures in U.

The parameter for describing the degree of complexity in velocity is the Reynolds number, $Re = LU/\nu$, where U is the typical non-uniformity in velocity, L is the size of a region in which the non-uniformity occurs and ν is the kinematic viscosity of the fluid in question. A similar parameter in the mixing equation is LU/D. In everyday systems, fine scales are produced precisely because these parameters tend to be quite large, as a consequence of the small size of the diffusion coefficients. For example the kinematic viscosity of water is of the order $10^{-6} m^2/sec$.

Conservation laws are always important. Since the diffusion parameters ν and D are small, it is reasonable to say that quantities are conserved in the absence of diffusion. For the Navier Stokes equation, the conserved quantities include

$$\text{the momentum} = \int dr\ \rho u(r,t) \tag{4}$$

$$\text{the kinetic energy} = \int dr\ \frac{\rho}{2} u(r,t)^2 , \quad \text{and} \tag{5}$$

$$\text{the helicity} = \int dr\ u(r,t) \cdot (\nabla \times u(r,t)) \tag{6}$$

The mixing in equation (3) moves temperature from place to place but, in the absence of diffusion, does not change the distribution of temperature-values over the entire system. For this reason, all moments of the temperature

$$\int dr\ [T(r,t)]^p \tag{7}$$

are equally conserved in the absence of diffusion.

It is usually considered that the conservation of kinetic energy is particularly crucial in defining the qualitative nature of turbulent processes. The figure shows a cartoon view of these processes, a view first introduced by Richardson[3] and then given more mathematical form by Kolmogorov in his wonderful 1941 papers[4]. The idea is best described in Fourier transform language, in which the wave vector k serves to describe the inverse of the different spatial scales. Some large scale process, perhaps stirring, is used to feed energy into the fluid at small k. The range of scales at which this addition of energy occurs is called the integral range. Then the non-linear coupling of velocities causes the energy to cascade toward larger wave vectors. The range of scales dominated by the cascade is called the inertial range. This range can be quite large since it is proportional to a power of the Reynolds number. For even larger k's, one passes into the viscous range. Here, viscous process dissipate the energy into heat and smooth out all details. The problem of turbulence is the problem of predicting and understanding what happens when the inertial range is large. The mixing process can be understood in a similar way as a question about what happens to some additive thrown into the pipe at the top.

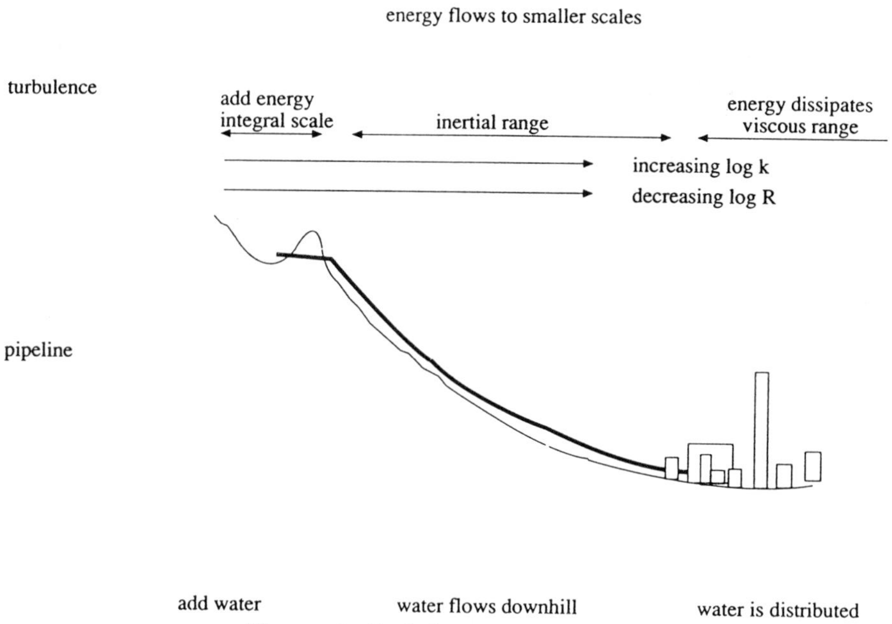

Figure 1. Turbulence as a Pipeline

2. Turbulent Models

A quantitative analysis of the energy flow was introduced by Kolmogorov in papers written in 1941 and 1962, and called K41 and K62. (See reference 4). The key idea is that all the energy which is added on the largest scale flows unchanged through the inertial range toward higher k. The flow is described by an energy current, ϵ. By dimensional analysis Kolmogorov found that the current of energy is proportional to $k[u(k)]^3$. The conservation of energy implies that in the inertial range, the time average of this current must be independent of k. In the 1941 work, it was assumed that the fluctuations in energy current are not very large. In that case, the velocity has a typical value which is given by the order of magnitude statement

$$u(k) \sim \left(\frac{\epsilon}{k}\right)^{1/3} \tag{8}$$

where ϵ is the time-averaged energy current of equivalently the average rate of energy dissipation. There are two crucial assumptions in K41.

(a) The energy flux, ϵ, is assumed to be roughly constant, and thus essentially equal to the energy input per unit time.

(b) Further, it is assumed that far into the inertial range, the system remembers as little as it possibly can about the conditions at which energy is added or dissipated.

It must remember the value of ϵ since that determines the total energy flow at all scales in the inertial range. In K41 it is assumed that this ϵ determines the flow and that the system does not know anything about either the value of L or of the dissipation length scale η.

This argument, and many subsequent arguments sees the energy flow as a kind of pipeline. (See again the figure.) Energy is added to the pipe at the very large scales, flows through the very long pipe toward smaller scales, and then flows out via dissipation at the far end.

The 1941 analysis assumes that the flow only depends upon the average ϵ. The 1962 analysis includes fluctuation in ϵ, but it assumes that the flow through the pipe at a given point only depends upon processes upstream of that point. Thus for example, inertial range behavior would not be influenced by the value of the viscosity since the viscosity appears only in the dissipative range, far downstream. Since the fluctuations in different regions of the pipe would be independent multiplicative effects one might expect to get fluctuations which are much larger than Gaussian.

But the pipeline analogy makes the assumption that influence only flows downstream seem doubtful. Certainly a choking of the outlet would tend to diminish the flow.

In the first decade after K41, scholars of turbulence focused upon the other basic assumption of the K41 theory, the assumption of the constancy of energy dissipation. Partially in response to a criticism by Landau, Kolmogorov changed his own point of view. In K62, he argued that, because of fluctuations, the energy input should be a function of space and time which can vary over many, many orders of magnitude. These huge fluctuations were termed *intermittency.* While still assuming that influence only flows downstream, Benoit Mandelbrot and others [5] developed richer cascade models which showed how a highly fluctuating behavior might arise and be reflected in a complex fractal structure of $u(k)$.

The mixing defined by equation (3) can be described in much the same terms. The K41 ideas suggest that there should be an entropy current proportional to $ku(k)[T(k)]^2$ and that this current should not be dependent upon k in the entire range of k^{-1} between the forcing scale and the dissipation scale. The consequence once again is a $k^{-1/3}$ behavior, but now for $T(k)$. However, once again, one might ask whether fluctuations will produce very important deviations from the simple estimated behavior.

Theory cannot for the moment distinguish between the weak fluctuations of K41 or the intermittency of K62. To settle questions of this kind we should look to experiment. The experimental evidence suggests the existence of intermittency, but is not yet conclusive. We might also look to simulations. Turbulence problems can easily be set up on the computer since the turbulent system may be correctly described by the Navier Stokes equation, our equation (1). Unfortunately, computers are too slow to allow simulations to cover a sufficient range of scales. How can we learn more? One approach is to set up models which correctly realize in specific form the previous theoretical ideas and then see how consistent these ideas might be. One effort in this direction has been the development of a one-dimensional model which induces in a very rough way cascades from shell to shell in momentum space.

3. GOY Model

The model is connected with the names of Gledzer, Ohkitani, and Yamada (inelegantly called GOY)[6,7]. It is among the simplest models which can correctly realize a cascade through orders of magnitude of wave vector, k. To derive it, look at the Fourier transform version of the Navier Stokes equation, as given by equation (2). Here, the system is described by the Fourier coefficients of the velocity, $u(k)$. Imagine a cascade which goes from a k of order the size of the system, to one in a shell which contains k's a factor of λ larger, and then a higher shell with k's of order $\lambda^2 L$, and so forth through $\lambda^N L$, a wave vector which sits far into the dissipative range. Thus we have k-values given by $k_n = k_0 \lambda^n$. The real system involves many, many Fourier coefficients, $u(k)$, with the number per shell sharply increasing with the shell-number, n. This model tries to describe the flow by keeping precisely one Fourier coefficient for each shell, so it is described by $u_n(t)$, for $n = 1, 2, \ldots, N$ with each u_n being a complex number. The equations of motion are picked to mimic Navier-Stokes, with non-linearity, viscous damping and energy conservation all included. Thus the model equation has the form

$$\frac{d}{dt} u_n(t) = C_n + D_n + F_n, \quad n = 1, 2, \ldots, N \tag{9}$$

where we will choose a number of shells, N to be some reasonable number like 25. The dissipation term is just like the one in equation (2), namely

$$D_n = -\nu k_n^2 u_n. \tag{10}$$

We will take the integral scale length, L, to be equal to one and then insure a large inertial range by picking the viscosity, ν, to be much smaller than one. The forcing term is taken to be at large scales, so we pick

$$F_n = \delta_{n,1} f. \tag{11}$$

Finally, one needs a term to represent the cascade which sends energy to higher and higher wave vectors. To mimic the structure of the Navier Stokes equation, this cascade term should be of first order in wave vector and second order in velocity. The Richardson-Kolmogorov picture sees a step by step process in which information is carried from one wave vector to neighboring ones, so we should couple nearby shells. One structure which will do this is the GOY form:

$$C_n = (ak_n u_{n+1} u_{n+2} + bk_{n-1} u_{n+1} u_{n-1} + ck_{n-2} u_{n-2} u_{n-1})^* \tag{12}$$

with a, b and c being undetermined coefficients.

There is a lot in the model-building which is undetermined. There is no particularly good reason for picking these specific three terms to make up the cascade term (12). We did it in our work[8,9] mostly because the previous authors had done so and we wished to build upon their ideas and results. Previous authors, K. Ohkitani and M. Yamada, (see reference 7) and M. H. Jensen, G. Paladin, and A. Vulpiani (see reference 1) and also R. Benzi, L. Biferale, and G. Parisi[10] used it because it gave a pattern of u_n versus time which looked rather like the fluctuating velocity fields $U(r,t)$ of turbulence.

Notice that in the absence of forcing and dissipation we have two conserved quantities of the form

$$C = \sum_n |u_n|^2 z^n \tag{13}$$

which are conserved so long as the coefficients in equation (10), (11), (12) obey

$$a + bz + cz^2 = 0. \tag{14}$$

A reasonable definition of a conserved kinetic energy is a sum of velocities squared of the form:

$$E = \frac{1}{2} \sum_n |u_n|^2 \tag{15}$$

which then requires the coefficients to obey

$$a + b + c = 0. \tag{16}$$

Given this condition, the energy obeys the conservation law

$$\frac{d}{dt} E_n = (J_n - J_{n-1}) + \delta_{n,1} \text{ Re } (f u_1^*) - \nu k_n^2 |u_n|^2 \tag{17}$$

The first term on the right hand side arises from the flow produced by the energy cascade. It is of the form of the divergence of an energy current, where the current has the form

$$J_n = a\Delta_n - c\Delta_{n-1} \quad \text{with} \quad \Delta_n = \text{ Re } (k_n u_n u_{n+1} u_{n+2})$$

The other terms in equation (17) represent the addition of energy at the forcing scale and its dissipation by viscosity.

We are now ready to start to understand the model. We can, without loss of generality pick

$$a = 1 = f$$

since these conditions only set the scale for time and for velocity. Then the model contains three parameters. The most physical one is ν, which is essentially the inverse Reynolds number and is taken to be very small. Next, λ sets the scale for discretization. If the model is a good one, results should be independent of λ. Finally the one remaining parameter is b, an artificial parameter which essentially defines different turbulence models. Given equation (12), the parameter b determines the ratio of the energy flux toward higher wave number to the flux in the opposite sense.

Thus b might be considered to be a free parameter in the theory. But, in reference [8] we argue that fixing a particular relation of b to λ,

$$b = -1 + \lambda^{-1} \tag{18}$$

makes the most sense physically because that makes the second conserved quantity in the GOY model scale with distance in the same way as the helicity defined in equation (4).

But the connection (18) between b and λ seems a bit contrived. If we put this connection aside, the model involves two parameters which have no direct counterpart in Navier Stokes: the shell width λ, and b. We study the model for b in the range $-1 < b < 0$. Because the model equations involve many fewer variables than those in the Navier Stokes equations, its qualitative properties can be established with the aid of simulations. We can thus ask, do the size of the fluctuations in the model agree with K41 or with K62? The question is sharply posed and it may be sharply answered. The answer is yes. Yes?...Yes!

In one range of parameters (b close to zero) the system shows a static behavior, no fluctuations, and exactly the K41 scaling of equation (7).

In contrast, in another range (more negative b) the static solution is unstable against fluctuations. The model solution then shows a time-dependent state with fluctuations in u_n which become stronger for higher values of n in the inertial range. The net result is something more like K62 behavior. (In fact if the relation between λ and b is chosen in accord with equation (18) then we are in this dynamic regime and there is a very close quantitative relation between the intermittency observed in experiments and simulations on turbulent systems and the intermittency of the GOY model. This point was developed in detail in reference 8.) In the GOY model, we can easily understand how we might find very large fluctuations in the magnitude of u_n for large n. The small n behavior is fixed by the forcing. To understand larger n, notice that the model permits independent fluctuations in the velocity ratio between neigboring shells. This idea was pointed out by R. Benzi, L. Biferale, and G. Parisi[10]. When one multiplies out a set of many independent random multipliers, the result can have truly huge fluctuations. Thus the model served as a partial justification for both K41 and the later, more fluctuating, theories.

However, K41 and K62 both suggest that only the value of ϵ matters. In K41 ϵ is constant throughout the inertial range; in K62 ϵ fluctuates and varies increasingly as one passes to smaller and smaller distances. But ϵ is still the only piece of information effectively passed along through the inertial range. As a consequence of this information flow from large to small distances, both theories suggest that the inertial range behavior should be independent of ν. The GOY model disagrees with this conclusion. In the static region of the model-behavior, not one but three pieces of information are passed through the inertial range. According to reference 9, these pieces of information include the values of

$$\frac{u_{3q}}{\lambda^q}, \qquad \frac{u_{3q+1}}{\lambda^q}, \qquad \text{and} \qquad \frac{u_{3q+2}}{\lambda^q}.$$

The product of these three numbers gives the energy flux. However the remaining two ratios represent information carried faithfully through the inertial range, but not predicted by the Kolmogorov arguments. (In fact there are four pieces of information carried, corresponding to the phases and the magnitudes of these quantities.). Because more information is passed along, there is more possibility for communication between the integral and the dissipative scales. In response to some prompting from Z.S. She, a group of us (Norbert Schörghofer, Jane Wang, Detlef Lose, Roberto Benzi, and I [8,9]) showed that in the static situation the energy flux down the pipe does certainly depend on ν, even in the limit of high Reynold's number. In contrast to the early theories (*both* K41 and K62) here the flow down the energy pipeline does depend upon an interaction between inlet and outlet conditions, induced by correlations carried up and down the pipe. The mechanism for the correlation in the GOY model is quite unphysical in that it makes direct use of the shell thickness, λ. However, this result points us toward the possibility that both inlet and outlet might play a role in determining the flow in real turbulence. If so, there will be some Reynolds number dependence of inertial range behavior. Thus, a vastly oversimplified model led to insights which might well play back to a deeper understanding of a truly complex phenomenon.

4. Mixing

The ν-dependence described above only holds in the static case. I would guess that when fluctuations are important, the extra pieces of information carried through the inertial range are destroyed by an averaging process. Nonetheless there are two issues raised by this argument which might indeed by relevant for real turbulence:

(a) How much information is carried through the inertial range?

(b) Can the ends of this range be replaced by simple boundary conditions? In particular, can the viscosity be replaced by a boundary condition in which the velocity is required to approach zero as k goes to infinity? If that is true, probably the exact form of the dissipation is irrelevant to the structure of inertial range behavior[11].

Somewhat similar questions hold for mixing. Turbulent stirring mixes the temperature distribution and produces small-scale structure from large-scale variations. There is a kind of inertial range in which one is far away from the full smoothing produced by the diffusion term in equation (3). To understand this range, it is reasonable to ask about how much information is transferred through this scaling regime. Maybe the existence of many conservations laws is relevant for this discussion. One could also ask whether the dissipation can effectively be replaced by a boundary condition at the start of the dissipative range. As part of the same discussion one could ask also whether the value of the diffusion coefficient matters.

Recent calculations have focused upon the special case in which fluctuations in the temperature field are produced by extremely rapid time-variations in a Gaussian velocity field. Many recent authors agree that the resulting temperature field shows extremely large fluctuations roughly like that of the K62 theory. In fact, Robert H. Kraichnan, Victor Yakhot, and Shiyi Chen[12] claim that they have found an exact solution to this multiscaling problem, based upon old theoretical work of Robert Kraichnan[13] It is further claimed that this result will be robust under small changes in the way the flow is set up. In contrast, other authors[14-16] look at the problem slightly differently and fail to find the claimed results and/or robustness. There is a real disagreement.

One possible approach to discussing this disagreement and also to gaining a better understanding of what makes the very large fluctuations is to deal with model problems which have roughly the right nature. Once can extend the GOY model to handle temperature mixing. In fact this was done in the early paper of M. H. Jensen, G. Paladin, and A. Vulpiani, reference 1. More recently, the calculation has been extended to the rapid-time-variation situation by A. Wirth and L. Biferale[17] who wrote down an equation of the form

$$\left(\frac{d}{dt} + Dk_n^2\right)\theta_n(t) = f(t)\delta_{n,1} + C_n^* \tag{19}$$

$$\begin{aligned} C_n = \; & k_n(\theta_{n+1}u_{n+2} + u_{n+1}^*\theta_{n+2}) \\ & + k_{n-1}(\theta_{n+1}u_{n-1} - u_{n+1}\theta_{n-1}) \\ & - k_{n-2}(\theta_{n-2}u_{n-1}^* + u_{n-2}\theta_{n-1}) \end{aligned}$$

where they take the u_n's to be a prescribed velocity field varying rapidly in time. They report a result more complex than simple Kolmogrov scaling, even though they have just one conservation law, for

$$\sum_n |\theta_n|^2.$$

It will be interesting to see if this model sheds any light on the nature of the mixing problem.

5. One Dimensional Incompressible Models

At Chicago, Scott Wunsch, Tong Zhou and I are working on a slightly different model involving a flow through two one-dimensional pipes. These pipes are labeled by an index σ which takes on the values plus or minus one. In one version of the model, the equations are:

$$\left[\frac{\partial}{\partial t} - D\frac{\partial^2}{\partial x^2} + \sigma u(x,t)\frac{\partial}{\partial x}\right] T_\sigma(x,t) \; + \; v(x,t)\left[\frac{T_+(x,t) - T_-(x,t)}{2}\right] = \lambda[T_{-\sigma}(x,t) - T_\sigma(x,t)] \tag{20}$$

In addition, we demand $\partial u/\partial x = v$. This relation is a kind of incompressibility condition. According to Scott Wunsch and Tong Zhou[18], this model has two conservation laws which hold in the absence of dissipation ($\lambda = D = 0$). These two conserved quatities are the 'energy ',

$$H = \sum_\sigma \int dx \; T_\sigma(x,t) \tag{21}$$

and the 'entropy '

$$S = \sum_\sigma \int dx \; [T_\sigma(x,t)]^2 \tag{22}$$

Unlike the case of real 'passive scalars' however higher powers of the temperature are not conserved. Nonetheless, the model might have considerable features in common with the standard models of passive scalars. To see this, we should ask: Does this model have large fluctuations in T? If we pick a simple scaling behavior for v, e.g. Gaussian with zero time correlation range and scaling properties in x, then what will we find for T? Will we have a simple scaling or a multiscaling?

We can also do a different version of the model which has an infinite number of conservation laws in the zero diffusion limit. Pick x to be a discrete index, $x = 1, 2, \ldots N$. The model should have diffusion and also advection. First, consider the equation for diffusion

$$T_\sigma(x, t+\tau') \; = \; T_\sigma(x,t) \tag{23}$$

$$
\begin{aligned}
&+D\tau'[T_\sigma(x+1,t)+T_\sigma(x-1,t)-2T_\sigma(x,t)] \\
&+\lambda\tau'[T_{-\sigma}(x,t)-T_\sigma(x,t)]
\end{aligned}
$$

Here, τ' represents a time step and D and λ are respectively measures of the diffusion rate in the x and 'y'directions.

The second process is advection which involves the composition of a series of swirls and vortices. The important thing is that the process be volume preserving. The simplest volume preserving process is one in which the temperature at two sites are interchanged:

$$
\begin{aligned}
&S_1(x) \; : \\
&\qquad T_\sigma(x,t+\tau)=T_{-\sigma}(x,t) \quad \text{for both values of } \sigma.
\end{aligned}
\tag{24}
$$

Another basic swirling process has a new temperature which obeys

$$
\begin{aligned}
T_\sigma(x,t+\tau) &= T_{-\sigma}(x,t) \\
T_\sigma(x+1,t+\tau) &= T_\sigma(x,t) \\
T_{-\sigma}(x+1,t+\tau) &= T_\sigma(x+1,t) \\
T_{-\sigma}(x,t+\tau) &= T_{-\sigma}(x+1,t)
\end{aligned}
\tag{25}
$$

for a single fixed value of σ. Notice how this process represents motion around a little square. We get a little more motion if we pick instead the result of doing (25) twice. We call the resulting process $S_2(x,x+1)$ and it has the effect

$$
\begin{aligned}
&S_2(x,x+1) \; : \\
&\qquad T_\sigma(x,t+\tau)=T_{-\sigma}(x+1,t) \\
&\qquad T_\sigma(x+1,t+\tau)=T_{-\sigma}(x,t)
\end{aligned}
\tag{26}
$$

for both values of σ. One gets larger swirls by putting together several of these smaller swirls. For example, one can define a process $S_n(x,x+n,\sigma)$ by defining a vector

$$
\begin{aligned}
W(t) &\equiv (T_\sigma(x,t),T_\sigma(x+1,t),\ldots, \\
&\qquad T_\sigma(x+n,t),T_{-\sigma}(x+n-1,t),\ldots T_{-\sigma}(x,t)) \\
&= (w_1,w_2,\ldots,w_{2n})
\end{aligned}
\tag{27}
$$

and then finding this same vector at a later time as a cyclic rearrangement of the original vector

$$
\begin{aligned}
S_n(x,x+n,\sigma) \;&: \\
W(t+\tau) &= (w_{n+1},w_{n+2},\ldots,w_{2n},w_1,w_2,\ldots,w_n) \\
&= (T_\sigma(x+n,t),T_{-\sigma}(x+n-1,t),\ldots T_{-\sigma}(x+n,t), \\
&\qquad T_{-\sigma}(x,t),T_\sigma(x,t),T_\sigma(x+1,t))
\end{aligned}
\tag{28}
$$

Notice that in the absence of diffusion, the probability distribution for temperature is invariant in time. Thus every sum of the form

$$S_p = \sum_\sigma \sum_x [T_\sigma(x,t)]^p \tag{29}$$

is independent of time. The combination of swirl plus diffusion can produce many of the characteristic effects of passive scalars in an incompressible flow.

We choose to have many processes like (27), (28) followed by some diffusion like (23). (Notice that the swirl process (27), (28) are picked so as to maximize the motion for a given size swirl. This gives an efficient calculation (I think) but may limit the range of scaling.) We should choose to have the probability of various value of n picked so as to get the right scaling.

That should give us a simple and effective passive scalar computation. we are now working on the construction, testing, and analysis of this kind of model.

Acknowledgement

I would like to thank Jane Wang, Detlef Lohse, Roberto Benzi, Norbert Schörghofer, Scott Wunsch, Tong Zhou, and Peter Constantin who were my collaborators on much of this work. Very helpful comments came from Emily Ching, Adrianne Fairhall, Z-S. She, L. Biferale, and Itamar Procaccia. This research was supported by the NSF's DMR. The writeup is in the public domain.

References

[1] M. H. Jensen, G. Paladin, and A.Vulpinai, *Phys. Rev. A* **43**, (1991) 798. See also T. Bohr, M.H. Jensen, G. Paladin, A. Vulpiani, *Dynamical Systems Approach to Turbulence*, Cambridge Nonlinear Science Series (In press)..

[2] See, for example, the extremely complex small structures in the picture by Harry Catrakis and Paul Dimotakis in Physics Today, (September 1995) p 11.

[3] See the writeup in Benoit Mandelbrot, *The Fractal Geometry of Nature*, Freeman and Company, New York, 1983 p. 401 ff.

[4] These papers are called K41 in the field. These and Kolmogorov's other work on turbulence are elegantly described in U. Frisch, *Turbulence*, Cambridge University Press, 1995.

[5] See the Frisch book, op. cit. , for details

[6] E. B. Gledzer, *Sov. Phys. Dokl.* **18**, (1973) 216.

[7] K. Ohkitani and M. Yamada, *Prog. Theor. Phys.* **81**, (1989) 329 *J. Phys. Soc. Jpn.* **56**, (1987) 4210. *Prog. Theor. Phys.*, **79**, (1988) 1265.

[8] L. Kadanoff, D. Lohse, J. Wang, R. Benzi, *Physics Fluids* **7**, (1995) 617 L. Kadanoff, *Physics Today* **48**, (1995) 11.

[9] N. Schörghofer, L. Kadanoff, D. Lohse, *Physics D* **88** (1995). 40-54. L. Kadanoff, N. Schöghofer, D. Lohse, *Physica D,* in press (1996).

[10] R. Benzi, L Biferale, and G. Parisi, *Physica D* **65**, 163 (1993)

[11] Gregory Eyink and Jack Xin "Universality of Inertial Convective Range in Kraichnan's model of a Passive Scalar," preprint chao-dyn 9605012 23 May 1996

[12] Robert H. Kraichnan, Victor Yakhot, and Shiyi Chen, *Phys Rev Lett.* **75,** (1995) 240-244.

[13] R. Kraichnan, *Phys. Fluids* **11**, (1968) 945-953.

[14] Boris I. Shraiman and Eric D. Siggia, C.R., *Acad. Sci. Paris,* **t.321 Serie II b,** p.279-284, 1995

[15] M. Chertkov, G. Falkovich, I. Kolokolov, and V. Lebedev, *Phys. Rev. E* **52**, (1995) 4924-4941,

[16] A. Gawedzki and A. Kupianen, *Phys Rev Lett.* **75**, (1995) 3608-3611.

[17] A. Wirth and L. Biferale, preprint 1996.

[18] Private communications

ELSEVIER

Physica D 100 (1997) 165–186

PHYSICA D

Scaling and linear response in the GOY turbulence model

Leo Kadanoff [a], Detlef Lohse [a,b,1], Norbert Schörghofer [a,*]

[a] The James Franck Institute, University of Chicago, 5640 South Ellis Avenue, Chicago, IL 60637, USA

[b] Fachbereich Physik, Universität Marburg, Renthof 6, 35032 Marburg, Germany

Received 28 March 1996; revised 20 June 1996; accepted 20 June 1996
Communicated by F.H. Busse

Abstract

The GOY model is a model for turbulence in which two conserved quantities cascade up and down a linear array of shells. When the viscosity parameter, ν, is small the model has a qualitative behavior which is similar to the Kolmogorov theories of turbulence. Here a static solution to the model is examined, and a linear stability analysis is performed to obtain response eigenvalues and eigenfunctions. Both the static behavior and the linear response show an inertial range with a relatively simple scaling structure. Our main results are: (i) The response frequencies cover a wide range of scales, with ratios which can be understood in terms of the frequency scaling properties of the model. (ii) Even small viscosities play a crucial role in determining the model's eigenvalue spectrum. (iii) As a parameter within the model is varied, it shows a "phase transition" in which there is an abrupt change in many eigenvalues from stable to unstable values. (iv) The abrupt change is determined by the model's conservation laws and symmetries.

This work is thus intended to add to our knowledge of the linear response of a stiff dynamical system and at the same time to help illuminate scaling within a class of turbulence models.

PACS: 47.27.–i; 47.27.Jv; 24.27.Eq; 05.45.+b; 02.10

1. Introduction

In turbulent flow a hydrodynamic system couples together many different length scales and thus shows in a single process a huge range of relaxation rates. There are a variety of simplified models of turbulence which are also intended to show this wide range of frequency and wave number scales. One such model goes under the inelegant acronym of "GOY". The model couples together a large number of shells, each with its characteristic scale of wave vectors and relaxation times. Shells are spaced logarithmically in wave vector. The nth shell is characterized by a single complex velocity, U_n, which then depends upon time, t. The model is a linked set of ordinary differential equations for all these U_n with equations picked to mimic those of real hydrodynamic flow.

* Corresponding author.

[1] E-mail: lohse@cs.uchicago.edu.

We do not know whether the model has much to do with turbulence. But it certainly illustrates the behavior of a stiff system. (Stiff systems are ones in which numerical simulations are made difficult by the effects of a huge range of relaxation rates.) In this paper, we describe the time dependence of the model in the simplest possible situation, the linear response to disturbances around a static solution. The response is described as an eigenvalue problem, with the response matrix being a large matrix which inherits the conservation laws, the scaling properties, and the symmetry principles of the GOY model. We look for the eigenvalues and eigenstates of the matrix. As we find them we see that the linear response, in turn, shows a considerable richness including scaling behavior and the analog of a phase transition. Much of this behavior can be understood in terms of the several symmetries and conservation laws of the original model.

The richness of behavior in this linear response theory serves to remind us that there is one area of scaling or similarity theory which has not been fully explored, the determination of eigenfunctions for matrices which have an underlying scaling structure. We also do not understand very much about turbulence, or even about the flow of information up and down dynamical linear chains. This paper is about these three not-fully-understood areas of applied mathematical science.

To start, we show the most interesting results of our study. We plot in Fig. 1, the eigenvalues of the linear stability in a sort of polar diagram in which the polar coordinates, θ and r, respectively, are the eigenvalues' phase and are proportional to the logarithm of the magnitude of the eigenvalue. (See Eq. (24) for a precise definition.) The two different parts of the plot show the response to a purely real disturbance (in part 1a) and to a purely imaginary disturbance (in part 1b). This distinction is meaningful because both the basic equations and the static solution are purely real. For this figure we have picked a particularly small value of the viscosity parameter so as to arrive at a simple scaling behavior. Indeed the simple spacing of the points in Fig. 1(a) shows that a simple multiplicative law generates the higher-order eigenvalues from the lower-order ones. That is why the points fall on a straight line with regular spacings. There is a tremendous range of scaling of the eigenvalues, and it all looks provocatively simple. This paper is mostly aimed at producing a partial explanation of these figures.

The paper starts with the introduction of the GOY model, and treats scaling for the static solution in Section 2. In Section 3 we discuss the linear stability analysis in general and apply it to the consideration of the scaling of the simpler, real component of the response. Also a relation between eigenvalues and conservation laws is derived. Section 4 discusses the linear stability in the imaginary response sector and the phase transition and scalings seen there. Section 5 summarizes the results. Footnotes and appendices discuss the findings that are not necessary for the main line of thought.

1.1. The model

The basic structure of GOY originated from Gledzer [1] was motivated by the cascade structure of turbulent eddies and conservation laws. Later, Ohkitani and Yamada [2] generalized the model and carried out numerical studies that revealed chaotic behavior and a dynamic scaling of velocity fluctuations. These studies have been extended in [3–6]. A popular introduction can be found in [7].

The basic ingredient is the hierarchical structure: The nth shell is characterized by a wave vector of length

$$k_n = k_0\lambda^n, \quad n = 1, 2, \ldots, N \tag{1}$$

with $\lambda > 1$ and by a complex velocity mode U_n. The Navier–Stokes dynamics is mimicked by the following set of ODEs:

$$\frac{\mathrm{d}}{\mathrm{d}t}U_n = F_n - C_n(U^*) - D_n, \tag{2}$$

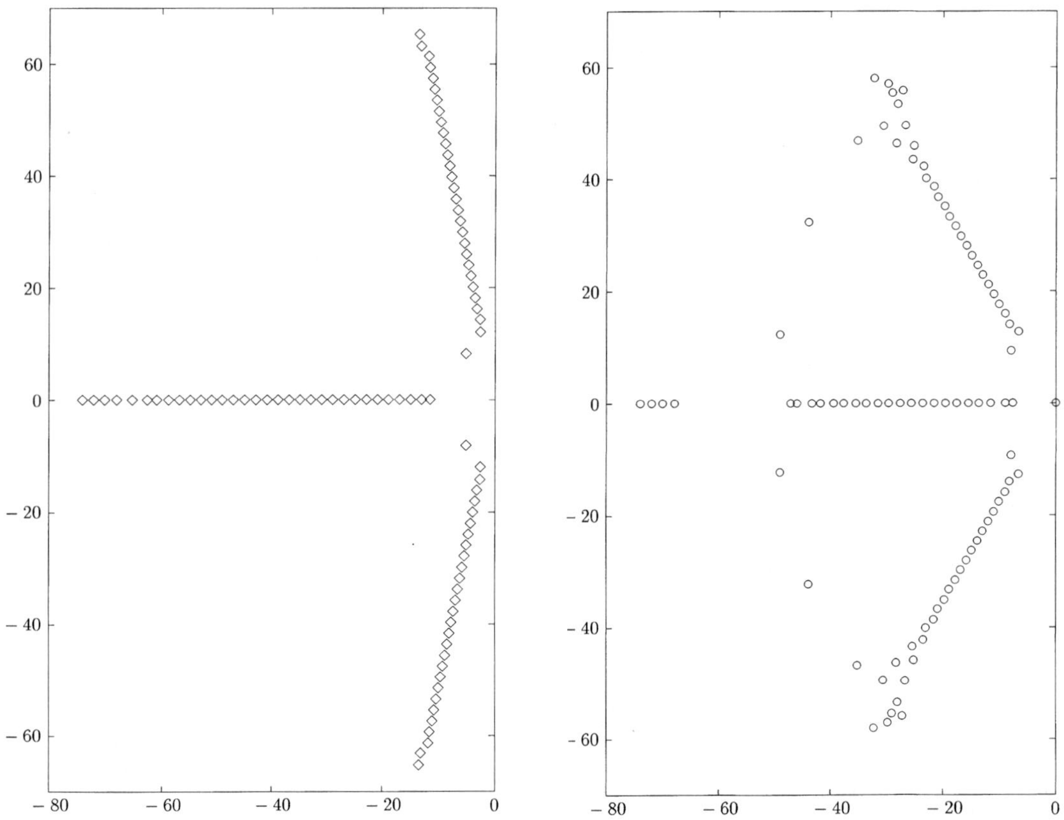

Fig. 1. Linear stability eigenvalues. The case considered is $\epsilon = 0.3$, $\nu = 10^{-3}16^{-26}$, $N = 90$. Here we plot the amplitude (diamonds) and phase (circles) of the eigenvalues in a kind of polar plot. The phase in this polar diagram is the phase of the eigenvalue. The radial coordinate is given by Eq. (24) which essentially produces the logarithm of the eigenvalue.

where the three terms represent, respectively, forcing, cascade processes and dissipation. (The $*$ indicates the complex conjugation.)

We pick a forcing on the first shell

$$F_n = \delta_{n,1} f. \tag{3}$$

Most previous studies [2,3,6,9] of the GOY model use a forcing on the fourth shell. Our choice of the first shell seems to give a simpler structure to the results. More details can be found in Appendix A. In our numerical work we shall choose $\lambda = 2$, $k_0 = \lambda^{-1}$, $f = 1$ unless otherwise stated.

The dissipation term is

$$-D_n = -\nu k_n^2 U_n \tag{4}$$

and is the k-space representation of the usual viscous dissipation process.

The cascade term couples the shell n to its nearest and next nearest neighbors,

$$C_n(U) = k_n U_{n+1} U_{n+2} - \epsilon k_{n-1} U_{n-1} U_{n+1} - (1-\epsilon) k_{n-2} U_{n-1} U_{n-2}. \tag{5}$$

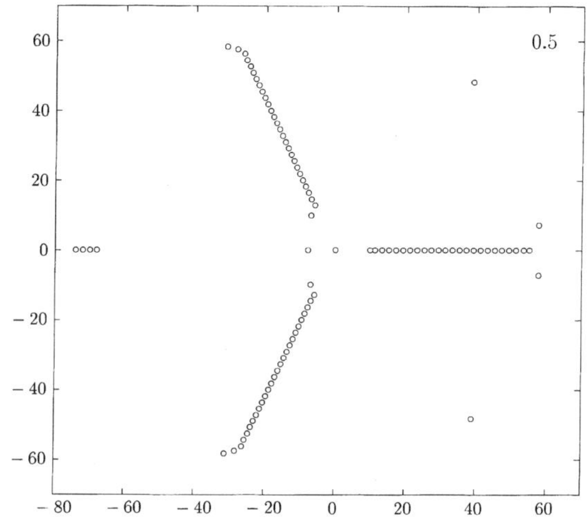

Fig. 2. The same as Fig. 1(b) except that now ϵ has the value 0.5. (The amplitude matrix becomes eventually unstable only at $\epsilon \approx 0.558$. For small enough ϵ the moduli eigenvalues will not change much in the dynamics. These modes are unstable only for certain ν and only very slightly. For larger $\epsilon > 0.56$ the moduli eigenvalue become increasingly unstable. We assume that the dimension of the chaotic attractor will increase correspondingly, but did not do a careful analysis on this.)

The boundary conditions are simply $U_{-1} = U_0 = 0$ and that the velocities go to zero as n becomes large. (In our numerics we represent that by cutting off at some large shell numbered N and then using the conditions $U_{N+1} = U_{N+2} = 0$.) Since the static equations couple n with the four immediately neighboring shells, we need four boundary conditions to define the problem. Two of the conditions are at the low-n end, and two at the high. The nature of the boundary conditions will become crucial later on.

The model parameter ϵ determines the ratio between upscale and downscale coupling. It gives us a convenient tool for varying the model and seeing qualitatively different ranges of behavior.

In the inviscid, unforced limit there are two conserved quantities: the energy

$$E = \frac{1}{2} \sum_n |U_n|^2 \tag{6a}$$

and a second conserved quantity of the form

$$H = \sum_n \frac{|U_n|^2}{(\epsilon - 1)^n} \tag{6b}$$

that can be roughly associated with the helicity in fluid motion [6]. Even though the association is not perfect, we shall call this quantity the helicity in this paper (for a GOY type model with an improved representation of the helicity, see [8]).

The GOY model shows dynamical as well as static behavior. There is a static solution [9] of the GOY model ($\mathrm{d}U/\mathrm{d}t = 0$) in which the phases are zero. When we wish to point particularly to the static solution, we shall write it as u_n, while we use U to refer to either the static or the dynamic case. In this paper, we study the static solution, u_n, and its linear stability properties.

The phases are not uniquely fixed for u_n [9] which is part of a one parameter family of solutions. (We pick the particular solution which has u_n real and positive.) The invariance in the phase will be important in our linear

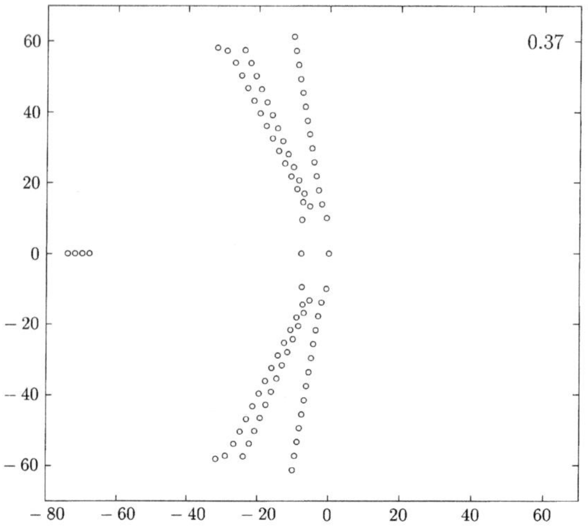

Fig. 3. The same as Fig. 1(b) except that now ϵ has its critical value, $\epsilon_{\rm bif}$ which is about 0.37. Notice that there are now six branches, all with complex eigenvalues. The equal spacing on each branch is evidence of scaling. But, evidently, the scaling is quite different from that shown in Figs. 1(b) and 2.

stability analysis. It gives rise to a zero eigenvalue. There are other approximate invariance properties which will be discussed when we get to the linear stability analysis.

2. Scaling (of static solution)

We are interested in understanding the behavior of the model in the limit as the viscosity becomes small. In that limit, there are three regions of k-space, or of n, called the stirring subrange (SSR), the inertial subrange (ISR), and the viscous subrange (VSR). The last is dominated by the viscosity term and has a velocity which decays very rapidly with k. (In the GOY model, the decay is exponential in a power [9] of k_n.) The SSR is naturally enough the range in which stirring is directly important. In our case this comprises only $n = 1$. The ISR is the intermediate range of wave vectors between these two. Here, the behavior is dominated by the cascade term. In the center of the ISR, the solution is best described by using the product of three velocities:

$$\Delta_n(u) = k_n U_n U_{n+1} U_{n+2}. \tag{7}$$

In the region in which only the cascade term matters, this product is

$$\Delta_n = A + B(\epsilon - 1)^n, \tag{8}$$

where A and B are adjustable constants of integration, representing, respectively, the energy and helicity flux through the inertial range. We shall work with ϵ in the unit interval so that, for large n, the A term dominates. In this region: [2]

$$U_n = W_n k_n^{-1/3}, \tag{9}$$

[2] At present it is hardly possible to analyze the structure of the second, "fluxless" solution with $A = 0$. It is unstable, so it is hard to obtain from the dynamical numerics. All solutions we have we essentially obtain from the dynamics which converged to a static solution (flux-solution K41). Then we continuously varied parameters and thus could also analyze the unstable solution. But they were continuously connected to stable ones.

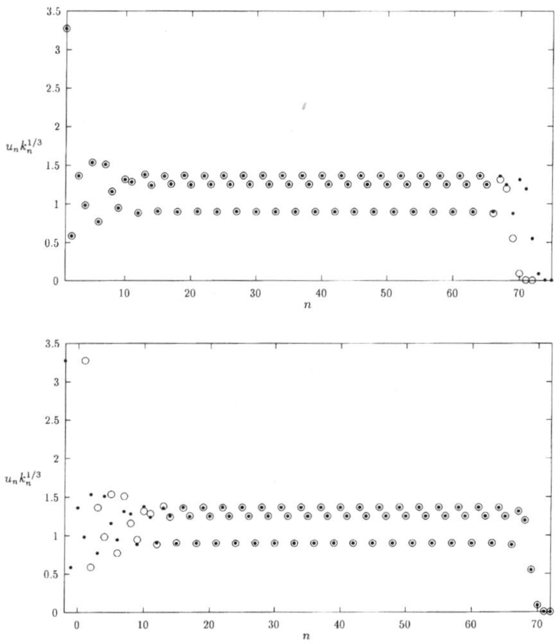

Fig. 4. We plot $W_n = k_n^{1/3} U_n$ against n. Notice how W_n oscillates in the inertial range and how it falls off quite rapidly in the dissipative range. (The latter is for $n > N_{\rm D} \approx 67$.) The actual calculation is done for $\epsilon = 0.3$, $\nu = 10^{-3}16^{-20}$ (circles) and $\nu = 10^{-3}16^{-21}$ (dots). The behavior of W in these two simulations is compared. In the second (dots), N is increased by three and ν is decreased by a factor of λ^{-4}. For all n smaller than 56 or so the two calculations agree. (b) A comparison like that in (a) except that now W_n is compared with W'_{n-3}. These agree top plotting accuracy for all n bigger than 20 or thereabouts.

where W_n is a periodic function with period three. This behavior continues as n increases until one enters the dissipative range, and U_n begins to fall off very rapidly.

Notice that we have, as expected, four undetermined parameters in the solution. In the ISR these parameters are A and B and the two parameters which define the period-three oscillations.

The scaling behavior is illustrated in Fig. 4 which plots $W_n = k_n^{1/3} U_n$ against n for a case in which ν has the value $10^{-3}16^{-20}$. Notice how U_n shows a simple scaling for large n up to n of about 67, and then it nosedives. The nosedive occurs as n increases above a dissipative threshold, which is achieved when the dissipation term and the cascade terms are roughly of equal size. Usually, W_n is of the order unity in the inertial range so that this dissipative cutoff can be computed as a function of a threshold value of wave vector, called $k_{\rm D}$, which obeys

$$k_{\rm D}^{-4/3} \sim \nu. \tag{10}$$

The condition is thus that the viscous effects should dominate for values of n larger than this dissipative threshold, that is,

$$n > N_{\rm D} \sim -\tfrac{3}{4} \log_\lambda \nu. \tag{11}$$

There is a simple scaling theory which we can apply to this case. Since the static solution has an asymptotic period three [6], the system should be almost unchanged by the transformation: [3]

[3] The change in N would be unimportant were we really working with very large values of N. For numerical convenience we work with N only seven or eight larger than $N_{\rm D}$. This change in N in Eq. (12b) eliminates effects produced by changing the cutoff. Throughout this

$$\nu \to \nu' = \nu/\lambda^4, \tag{12a}$$

$$N \to N' = N + 3. \tag{12b}$$

In particular, as one goes from the primed to the unprimed situation, the static solution should be basically the same at both ends, with the large, n end being only modified by having smaller values of u. Let u' be the solution for the velocity in the situation changed according to Eqs. (12a) and (12b). The two scaling symmetries can be written as

$$\text{(S)} \quad u'_n = u_n(1 + \mathrm{O}(k_n/k_{N_\mathrm{D}})^{2/3}) \quad \text{for } N_\mathrm{D} - n \gg 1, \tag{13a}$$

$$\text{(L)} \quad \lambda u'_{n+3} = u_n(1 + \mathrm{O}(\epsilon - 1)^n) \quad \text{for } n \gg 1. \tag{13b}$$

In these equations "S" stands for small k and "L" refers to large k. Note that the first of these equations remains valid in the SSR while the latter remains valid in the VSR.

In Eqs. (13a) and (13b) we have included estimates of the error resulting from the terms we have not taken into account in constructing the scaling. In the small k-range, we do not include viscous effects, so the error estimate is the relative size of the viscous term. For large k, we do not include the helicity flux represented by the B in Eq. (8), so this effect is put into the error. The global error is set by the sizes of each of these errors at the far ends of the VSR. One such error is the value of ν. The other, and usually larger effect, is the effect of the helicity flux term, B, at the large-n end of the ISR. This term has the order of magnitude,

$$\text{error} = \mathrm{O}(1 - \epsilon)^{N_\mathrm{D}}. \tag{14}$$

This scaling error must be at least as large as the maximum of this error and ν.

In terms of W, defined by Eq. (9), the two scaling equations imply, respectively, that $W'_n = W_n$ and $W'_n = W_{n+3}$. In the center of the ISR, both these scaling symmetries are valid. Then W has a period-three symmetry:

$$\text{(I)} \quad W'_n = W_n = W_{n+3} \quad \text{for } n \gg 1 \text{ and } N_\mathrm{D} - n \gg 1. \tag{15}$$

"I" refers to an intermediate range which occurs in the middle of the ISR.

To check our thinking, we calculate the static solution of the GOY model. Deviations from scaling can be seen by looking at the behavior of

$$\delta_{\mathrm{S},n} = 1 - W_n/W'_n, \tag{16a}$$

$$\delta_{\mathrm{L},n} = |1 - W_n/W'_{n+3}|. \tag{16b}$$

Both quantities are plotted in Fig. 5. Also shown in this figure are theoretical lines which show the errors defined in Eqs. (13a), (13b) and (14). Theory and experiment show excellent agreement.

3. Eigenvalue spectrum

3.1. Linear response

The next stage is to do linear stability analysis. We consider small deviations about the static solution u_n by writing

$$U_n(t) = u_n(1 + \delta\Phi_n \mathrm{e}^{\sigma t}). \tag{17}$$

---

paper we use a prime to denote quantities changed by the transformation of Eqs. (12a) and (12b).

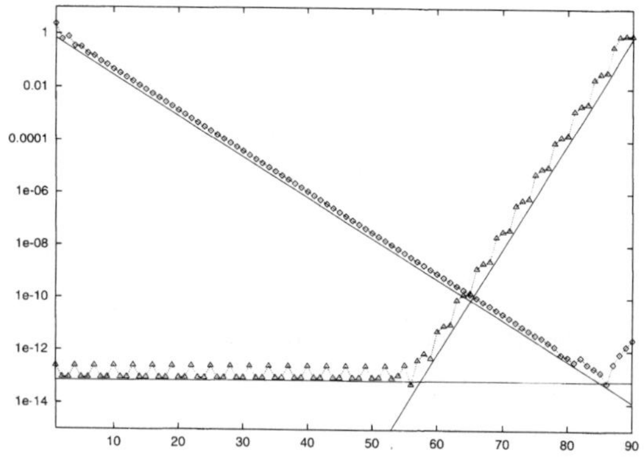

Fig. 5. The error in the scaling relations for the velocity. The deviation from unity of the ratio of two pieces of data in Eqs. (16a) (triangles) and (16b) (diamonds). The slope of the solid lines show the theoretical estimates of the error.

Since the static solution u is real, the eigensolutions split neatly into oscillations of the phase and the amplitude, corresponding, respectively, to $\delta\Phi_n$ being real and imaginary. Then we have an eigenvalue equation

$$\sigma\delta\Phi_n = \sum_m A_{nm}\delta\Phi_m. \tag{18}$$

We distinguish the two different cases with subscripts M for magnitude and ϕ for phase. We use a superscript j to denote which eigenvalue is being considered. In what follows we will always order the eigenvalues so that the magnitude of the eigenvalue increases with increasing values of the index j. Then the eigenstate will be very small for those shells which have n considerably smaller than j. We will then argue that the behavior of the response matrix for n and m of order j will play a large role in determining the eigenvalue.

Note that the response matrix A has two parts. The dissipative part is

$$D_{nm} = \nu\delta_{nm}k_n^2, \tag{19a}$$

and the cascade response is

$$C_{nm} = u_m \frac{\partial}{\partial u_m} C_n(u). \tag{19b}$$

Now the structure of A is different for the two kinds of response. For the magnitude response,

$$A_{\mathrm{M}} = -C - D, \tag{20a}$$

while for the phase response

$$A_\phi = C - D. \tag{20b}$$

3.2. *Eigenvalue spectra*

At the center of the ISR, u_n shows a simple scaling superposed on top of a period-three behavior. Thus, the response matrix also has a period-three scaling:

$$C_{n+3,m+3} = \lambda^2 C_{n,m} \quad \text{for } n, m \text{ in region I.} \tag{21}$$

Consequently, if C dominates the behavior of the matrices A in the determination of eigenvalues in some range of j, then the eigenvalues would obey the scaling property

$$\sigma_{\mathrm{M}}^{j+3} = \lambda^2 \sigma_{\mathrm{M}}^{j} \quad \text{for } j \text{ in region I}, \tag{22a}$$

$$\sigma_{\phi}^{j+3} = \lambda^2 \sigma_{\phi}^{j} \quad \text{for } j \text{ in region I}. \tag{22b}$$

In region I, then, the logarithms of the eigenvalues should be evenly spaced along straight lines with spacing of $\log \lambda^2$. Also, we might think that if C dominates the behavior of the matrices A, then the phase and magnitude eigenvalues should be the same except for a minus sign. (See Eqs. (20a) and (20b)). The viscosity is unimportant in the entire region S. Thus we expect

$$\sigma_{\mathrm{M}}^{j} = -\sigma_{\phi}^{j} \quad \text{for } j \text{ in region S}. \tag{23}$$

Fig. 1 shows the eigenvalue spectra for $\epsilon = 0.3$ for both modulus and phase stability matrices A_{M} and A_{ϕ}. As the distance between the eigenvalues σ^j in the hierarchical GOY model grows roughly exponentially, it is hard to visualize the spectra in the complex plane. To have some kind of visualization we use a kind of polar representation in which the phase of the plotted point is exactly the phase of σ while the distance from the center of the polar plot is given by

$$r = \log_{\lambda}(1 + 2^{10}|\sigma|). \tag{24}$$

The factor of 2^{10} is put in to enhance the visibility of eigenvalues with small values of $|\sigma|$. The plot has unstable eigenvalues showing up on the right-hand side of the origin while stable ones show up on the left. For large eigenvalues, even spacings on the plot mean that successive eigenvalues have ratios which are a constant. Thus even spacings are indicative of some kind of simple scaling.

In both the magnitude and the phase sector, the eigenvalues are arranged in several branches. Within each branch there are regions of even spacing, indicative of simple scaling. The magnitude eigenvalues seem particularly simple with three well-defined branches: A set of real eigenvalues and a pair of complex conjugate branches. All eigenvalues are stable. The phase eigenvalues show a more complex structure with what looks like more regions of simple scaling. Nonetheless both sets of eigenfunctions show the even spacing demanded by scaling.[4] As one might expect, the minimal modulus eigenvalue is about the size of the entries in the first row of the cascade matrix C, the modulus of the second one is of the order of the second row, and so on. Generally, the modulus of the nth eigenvalue is about the size of the nth row of the matrix. This rule holds roughly until the eigenvalue with j equal to N_{D} is reached whereupon the successive eigenvalues are real and have values of the order of the diagonal elements in the dissipation matrix, D. In the center of the inertial range the eigenvalues change from real to a complex conjugated pair and back to real and so on with a period three.

Thus, much of what we see is what we might expect. But not all Eq. (23) is completely inconsistent with the pictures we are seeing. This equation implies that if we have stable eigenvalues in the magnitude sector, we should have unstable ones in the phase sector. But all eigenvalues in Fig. 1 are stable. This result cannot be consistent with the notion that viscosity is unimportant for the eigenvalues in some region of the response. We have other difficulties. The pattern of phase response eigenvalues looks much more complex than the pattern of magnitude eigenvalues. Why should that be so?

[4] The existence of three branches is not due to the existence of a period three in the solution. One can assume an approximate solution $u_n = k_n^{-1/3}$, but the corresponding matrix still consists of three radial branches, although we have no period three in the "solution" any more.

Another difficulty is associated with the prediction of Biferale [11,12] that there would be a change in behavior at the special value of ϵ for which the contributions to the helicity sum in Eq. (6b) grow toward the high-n end of the inertial range. The value is

$$\epsilon_{\text{bif}} := 1 - \lambda^{-2/3}. \tag{25}$$

We have just seen that the model is stable for $\epsilon = 0.3$, which lies below the Biferale value ϵ approximately 0.37. Look at Fig. 2, which is the analog of Fig. 1(b) but now for $\epsilon = 0.5$. The magnitude spectrum looks much the same as before, but there is a qualitative change in the phase spectrum. Now, there are a large number of unstable eigenvalues in the phase spectrum. We interpret what we are seeing by saying that there is a qualitative change in properties of the asymptotic (ν goes to zero) model at $\epsilon = \epsilon_{\text{bif}}$. For $0 < \epsilon < \epsilon_{\text{bif}}$ there are at most a finite number of unstable eigenvalues. At ϵ just above ϵ_{bif} the system acquires an infinite number of unstable eigenvalues. Fig. 3 shows the behavior of the spectrum at the Bifarale point. Once again the phase spectrum has a qualitatively new character, while the magnitude spectrum remains much the same as before. We want to understand better how this behavior arises from the linear stability analysis.

3.3. Eigenvalues in the magnitude sector

The scaling behavior of the eigenvalues in the magnitude sector is given by the very same ideas which we already used in our analysis of u_n. As we shall see, some new ideas will be required for the phase sector. For this reason, we shall dispose of the magnitudes here and move on to a more extended discussion of phase eigenvalues in Section 4.

The general structure of the eigenstates is illustrated in Fig. 6. Here we look at right eigenvectors with $j = 26$. The eigenvalue equation in the magnitude sector is

$$\sigma^j \delta\Phi_n^j = -\sum_m (D_{nm} + C_{nm}) \delta\Phi_m^j. \tag{26}$$

For small n the eigenvector is very small. In fact, the eigenvector must decrease rapidly as n decreases to enable the eigenvalue term from swamping the right-hand side of the equation. This decrease is a falloff from a plateau which occurs at n about equal to j. At this point, the eigenvalue term and the cascade term in Eq. (26) are about equal in size. As n increases still further, the three parts of the cascade term each become much larger than the

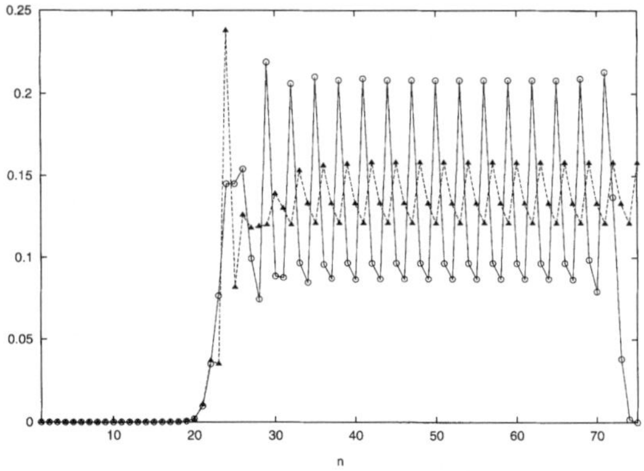

Fig. 6. The general structure of right eigenvectors for the the amplitude $\delta\Phi_n^{(26)} u_n k_n^{1/3}$ (solid) and phase $\delta\Phi_n^{(26)}$ (dashed line) ($N = 75, \epsilon = 0.3, \nu = 10^{-3}16^{-21}$). In both cases the modulus of the eigenvector is displayed. The amplitude eigenvector gets damped when it enters the viscous range, while the period three of the phase eigenvector continues into the viscous range.

eigenvalue term. To ensure the cancellation of the different parts of the cascade term the eigenvector settles down to an oscillation with period three, which holds throughout the entire ISR. Finally as n enters the VSR, the eigenstate once again falls off quite rapidly. The behavior of the phase eigenstate is much the same, except that there is no falloff in the VSR.

One can describe the same process in physical terms by saying that the eigenvalue term adds energy to the system which then cascades toward the VSR, where the energy is dissipated.

Now one can see why the dissipation plays such a large role in determining the behavior of the eigenvalues. The deviation $\delta\Phi_n^j$ produces a change in the conserved quantities, both the helicity and the energy. This change cascades down through the ISR toward the VSR. Because of the conservation of energy this cascade is not damped and remains constant in size until the VSR is reached. Since the eigenvector remains constant into the VSR, the VSR behavior serves as a kind of large n boundary condition on the eigenstate.

A simple count shows how this works. For large n in the ISR, $\delta\Phi_n$ is periodic with period three. Then there are two parameters which describe the wave function:

$$p_1^j(k) = \frac{\delta\Phi_{3k+1}^j}{\delta\Phi_{3k}^j}, \tag{27a}$$

$$p_2^j(k) = \frac{\delta\Phi_{3k+2}^j}{\delta\Phi_{3k}^j}. \tag{27b}$$

In our previous work [9] we found that there were two conditions upon the large-n velocities, required to keep these velocities from blowing up deep into the VSR: Two parameters, two conditions. Everything is determined. Thus we might expect that all eigenstates with sufficiently small j would have the very same values of the parameters for large enough n in the ISR. Table 1 serves to check this point. In this table we have shown values of the ratio of these parameters for various different eigenstates. According to the theory, the ratio should be unity. Clearly the theory works well for the magnitude eigenstates. (Table 1 also shows that the constancy of the parameters does not work for the phase eigenstates, but that story will be told later.)

Table 1
Behavior of parameters for eigenstates

j	n	Magnitude		Phase	
		$\mathrm{Re}(p_1^j(n)/p_1^{j+1}(n))$	$\mathrm{Re}(p_2^j(n)/p_2^{j+1}(n))$	$\mathrm{Re}(p_1^j(n)/p_1^{j+1}(n))$	$\mathrm{Re}(p_2^j(n)/p_2^{j+1}(n))$
18	35	1.02897	0.99597	0.02867	0.81927
18	45	1.00132	1.00041	−4.83889	22.2840
18	50	0.99986	1.00002	0.02867	0.81891
18	55	1.00004	1.00002	0.31876	−0.00242
18	62	1.00000	1.00000	0.02867	0.81891
18	70	1.00000	1.00000	0.31876	−0.00242
13	16	75.57590	2.08638	−0.58804	−51.81228
13	35	0.99648	1.00048	40.53829	−0.11043
13	45	0.99984	0.99995	−0.00677	0.02073
13	55	1.00000	1.00000	−0.42772	−51.26321
13	71	1.00000	1.00000	40.53836	−0.11044

Note. Here j refers to the jth eigenfunction while n describes the components of that eigenfunction. These two parameters describe the large-n behavior in the ISR. They are fixed by the value of $\nu = 10^{-3}16^{-21}$ in the magnitude sector but vary considerably in the phase sector. The ratios are essentially the same for the n values in between the given values. (Correlations between phase values with n different by a multiple of 3 stem from the period three in the eigenvectors.)

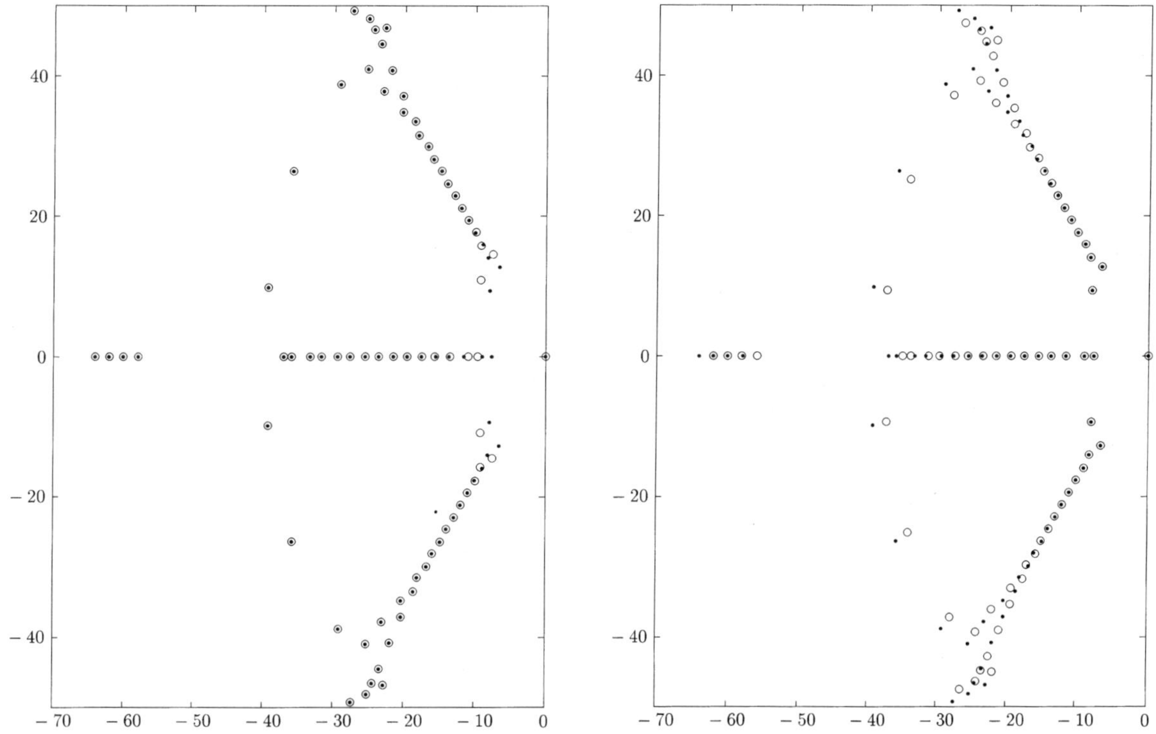

Fig. 7. (a) Eigenvalues of the phase matrix for $\epsilon = 0.3$, $N = 90$, and $N = 93$. One can see three main branches and the validity of scaling law S. (b) The same data, but the eigenvalues for $N = 90$ are multiplied by 4. Scaling law L is at work.

Thus, the matching into the VSR uses two of the four boundary conditions on our linear chain problem. A very similar mechanism sets the other two boundary conditions in the regions in which n is of order j. There are two quantities which can determine the behavior of the eigenfunction in this region. The first is the variation in helicity coefficient B produced by the disturbance. This variation produces a term in $\delta\Phi_n^j$ which varies as $(\epsilon - 1)^n$ and which then grows to be of relative order unity when n is of order j. By setting this coefficient, and also the value of the σ^j, one has two coefficients at ones command. These two are just enough to ensure that the eigenfunction decays, rather than grows, as n goes to one.

There is only one important difference between the problem of determining the static solution u_n and the eigenfunctions. For the former, we have forcing on the first shell. In the latter, the important forcing is produced by the eigenvalue term, and occurs on the jth shell. The scaling analysis for the jth eigenvalue is modified because the range of the cascade becomes j to $N_{\rm D}$ rather than the 1 to $N_{\rm D}$ that we used in the analysis of the velocity. Thus the scaling rule, with corrections, becomes on the small-j end

$$\sigma'^{(j)} = \sigma^{(j)}(1 + \mathrm{O}(k_j/k_{N_{\rm D}})^{4/3})) \quad \text{for } j \text{ in region S.} \tag{28a}$$

Here, as before, the prime indicates a decreasing in the viscosity by a factor of λ^4 together with a shift in the cutoff, N. The corresponding result on the large-j end is the statement

$$\sigma'^{(j+3)} = \lambda^2\sigma^{(j)}(1 + \mathrm{O}(\epsilon - 1)^j) \quad \text{for } j \text{ in region L.} \tag{28b}$$

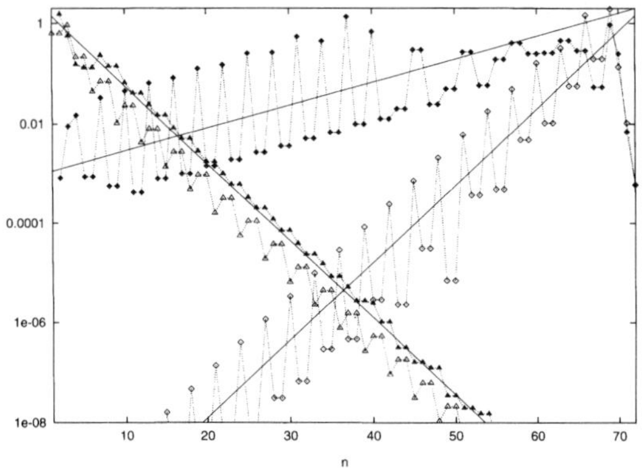

Fig. 8. Errors in the scaling laws for the eigenvalues, Eqs. (28a), (43a) and (28b), (43b); $\epsilon = 0.3$. Here we plot $|R| - 1$ where R is the ratio of the two sides of the equation. The diamonds describe the ratio in Eqs. (28a) and (43a) while the triangles do the same for Eqs. (28b) and (43b). The unfilled symbols are for the amplitude matrix and the filled ones stand for the phase matrix. Lines are theoretical estimates of the errors. (Above ϵ_{bif} the errors undergo some modifications.)

Both together give the scaling law:

$$\sigma'^{(j+3)} = \lambda^2 \sigma^{(j)} \quad \text{for } j \text{ in region I.} \tag{29}$$

To check our analysis, we should check these rules. The first check, done in Fig. 7(a), is to plot overlays of the primed and unprimed eigenvalues on polar plots like that in Fig. 1. We show the phase eigenvalues here, but the scaling fits are even better for the magnitude eigenvalues. The two spectra overlay precisely at small j, as expected, and fit badly for large j. In the second check, one plots overlays of σ and σ'/λ^2 as shown in Fig. 7(b). This overlay shows agreement between the two for large eigenvalues but not for small. A more careful check is to take the ratio of the nearby eigenvalues in these two figures. Call this ratio R. The quantity $|R| - 1$ is a quantitative measure of the errors in our statements (28a) and (28b). Fig. 8 plots this ratio versus j and shows that the order errors vary as stated in these equations.

Thus, one can feel that the magnitude response is understood reasonably well.

3.4. Eigenvalues and conservation laws

One basic principle about this model is that the dissipation can play a large role in determining the ISR behavior. To see this fact in more detail, we examine the effect of the conservation laws for energy and helicity upon the eigenvalue analysis.

Let us recollect that the conservation laws for the system can be expressed as

$$\sum_n U_n C_n(U) h^n = 0. \tag{30}$$

Here h is a quantity which defines the two conservations. For energy conservation $h = 1$; for helicity

$$h = (\epsilon - 1)^{-1}. \tag{31}$$

For the jth right eigenvector $\delta\Phi^j$ and its eigenvalue σ^j, we have an eigenvalue equation of the form

$$\sigma^j \delta\Phi_n^j = \sum_m (\pm C_{nm} - D_{nm}) \delta\Phi_m^j. \tag{32}$$

Here the minus sign corresponds to the magnitude eigenvalues and the plus to the phase eigenvalues. Multiply by $(u_n)^2 h^n$ and sum over all n to obtain

$$\sigma^j \sum_n (u_n)^2 h^n \delta\Phi_n^j = \pm \sum_{nm} h^n (u_n)^2 C_{nm}(u) \delta\Phi_m^j - \nu \sum_n h^n (u_n)^2 k_n^2 \delta\Phi_n^j. \tag{33}$$

The conservation identity (30) is true for any U and hence also for $U = u(1+\delta\Phi)$. Since $\delta\Phi$ is a small perturbation we can expand around u.

$$0 = \sum_n h^n U_n C_n(U) = \sum_n h^n \left[u_n C_n(u)[1+\delta\Phi_n^j] + \sum_{nm} (u_n)^2 h^n C_{nm}(u) \delta\Phi_m^j \right]. \tag{34}$$

In this way we have expressed the Jacobian (19b) in terms of the cascade itself. Eqs. (30), (34), and (2) yield

$$-\sum_{nm} h^n (u_n)^2 C_{nm} \delta\Phi_m^j = \sum_n h^n u_n C_n \delta\Phi_n^j = \sum_n (-D_n + F_n) u_n h^n \delta\Phi_n^j. \tag{35}$$

Now we can rearrange the eigenvalue equation. In the case of magnitude response the dissipation term in Eq. (35) adds to the dissipation term in Eq. (32) to give us an identity for the eigenvalue:

$$\sigma_{\mathrm{M}}^j \sum_n h^n (u_n)^2 \delta\Phi_n^j = -2\nu \sum_n h^n (u_n k_n)^2 \delta\Phi_n^j + f h u_1 \delta\Phi_1^j. \tag{36a}$$

The left-hand side of this equation is the rate of decay of the conserved quantity, as determined by the eigenvalue. On the right-hand side we see that the decay is (naturally enough) not produced by the cascade but only by the dissipation through viscous damping and also by the addition through the external force, f. Thus we understand once more that the dissipation must have a crucial role to play.

If we go through the same calculation for the phase response, the result is totally different. Instead of adding to one another, the dissipation terms cancel out (!), leaving us with the identity

$$\sigma_\phi^j \sum_n h^n (u_n)^2 \delta\Phi_n^j = -f h u_1 \delta\Phi_1^j. \tag{36b}$$

Compared to the magnitude sector, the phase sector shows a quite different form for the conservation law identities. We can no longer say that dissipation produces decay. Instead we say that the relevant quantity is added through the force term and then changes in time because of this addition. We now turn to a more detailed consideration of the phase sector.

4. Phase response

4.1. Establishment of phase transitions

From what we have seen, both u_n and the entire magnitude sector of the linear response vary smoothly as ϵ passes through the Biferale value given by Eq. (25). When $\lambda = 2$, this transitional value is 0.37. The story is different for the phase response. Whenever ϵ passes through the critical value, then there is a quite apparent change in the structure of the eigenvalues. As the viscosity goes to zero, this change involves having a large number of eigenvalues pass from being stable to unstable. In the asymptotic limit, one third of the entire set of ISR eigenvalues undergo such a passage at this point.

To see the evidence for this proposition, return to Figs. 1(b), 2, and 3. These pictures, respectively, apply below, above, and at the phase transition in ϵ. Away from the phase transition, the phase eigenvalues fall into two classes:

The eigenvalues fall onto three lines of constant phase: one real and two with opposite phases. Others, the "deviating eigenvalues" do not seem to fall into the simple pattern, and have phases which change from eigenvalue to eigenvalue. Since we are interested in scaling properties, we want to know something about the pattern which arises when the viscosity is taken to zero. An examination of the spectrum shows that its three main branches grow longer as ν becomes smaller, but that the number of deviating eigenvalues does not grow. Fig. 9 presents a counting of eigenvalues in the phase response sector. They are divided into the following categories: real, "constant phase" complex, and deviating eigenvalues. They are then counted at different values of N_D. We see that the number of deviating eigenvalues remains the same as we change the number of shells in the ISR. The same analysis – and result – applies above the transition. Consequently, for $\nu \to 0$ the number of deviant eigenvalues becomes negligible compared to the number in the three main branches. In the asymptotic limit, each of the main branches has one third of the total eigenvalues. Thus, the deviating eigenvalues are not a scaling limit phenomenon, but rather a transient which defines the approach to scaling at the ends of the ISR.

There is a main branch on the real axis both above and below the transition. Above the transition, this branch is unstable; below it is stable. At the critical point, we have a change involving, in the $\nu \to 0$ limit, an infinite number of eigenvalues.

Fig. 10, together with Figs. 1(b), 2, and 3, presents the flow of phase eigenvalues as ϵ goes from below to above the transition. We start at Fig. 10(a) with the familiar spectrum at $\epsilon = 0.2$. Real eigenvalues collide and turn to complex ones and these (so created) deviating group of eigenvalues wanders towards the imaginary axis. If we had chosen a larger system, the number of deviating eigenvalues would still be the same, and only the straight branches would gain in members. The smaller the system the more is the discontinuous transition smeared out by finite size effects. (For finite ν also the critical point ϵ_c weakly depends on ν and slightly deviates from ϵ_{bif}, see Fig. 11. These finite size effects spoil the regularity of the phase transition. A related transition with less disturbing finite size effects is discussed in Appendix B.)

As the deviating group is close to the phase transition the two conjugate straight branches of eigenvalues split into two (Fig. 3), causing a breakdown of scaling law S. A scaling of the form S^2 (i.e., make use of the symmetry $n \to n+6$) however, is still valid. The scaling of the solution and of the magnitude eigenvalues does not break down at this point.

Above the transition Fig. 10(d) the deviating group has curved in the other direction and the eigenvalues return to the now unstable side of the real axis.

There is a hidden structure of the deviating group not easily seen in our representation (24); these eigenvalues fall on a straight line in a $\log(|\sigma|) - \arg(\sigma)$ plot. The inset in Fig. 9 shows (the upper half of) the deviating group and the real eigenvalues for $\epsilon = 0.33, 0.37$, and 0.4. A dotted line marks the border of instability at $\frac{1}{2}\pi$. Remember again that only branches with constant phase (here horizontal) will contribute in the $N \to \infty$ limit. From Fig. 9 (inset) it is evident that with $N \to \infty$ instability will occur immediately above ϵ_{bif}, since an arbitrary small tilting suffices to produce unstable real eigenvalues. Moreover, for $\nu \to 0$ there is not even a finite number of unstable eigenvalues for $\epsilon < \epsilon_{bif}$ (see Fig. 11). This means that ϵ_{bif} coalesces with ϵ_c:

$$\epsilon_{bif} = \epsilon_c. \tag{37}$$

Although the instability mechanisms consists of one pair of complex conjugate eigenvalues after another crossing the imaginary axis, at $N \to \infty$ an infinite number of such pairs coalesce and an infinite number of oscillatory instabilities is unleashed within an arbitrarily small change in ϵ.

For $\nu \to 0$ the destabilization scenario of the fixed point is thus different than has been suggested in the light of "finite-size simulations" [5,9]. On one side of the phase transition we have totally stable behavior, while on the other side we have immediately an infinite number of unstable modes on all scales. These instabilities are purely exponential and not oscillatory.

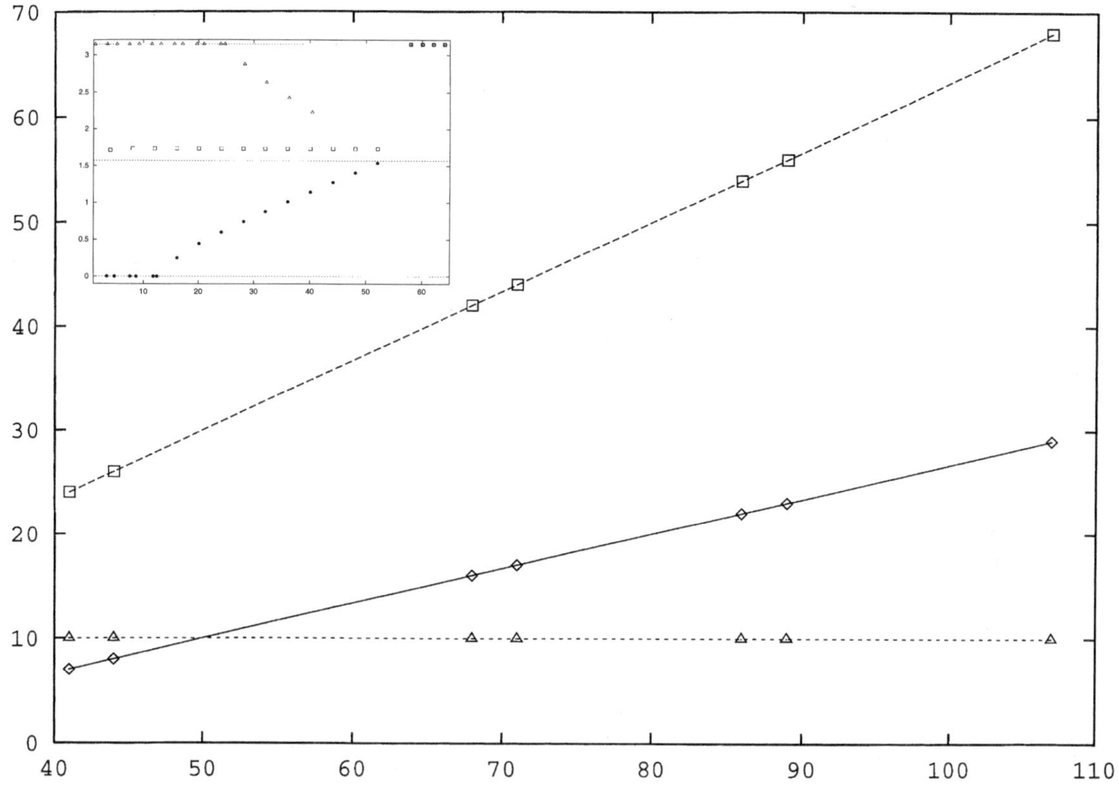

Fig. 9. Number of real (solid line), "straight" complex (dashed), and deviating eigenvalues (dots) with increasing shell number N_D. The slope of the lines is exactly $\frac{2}{3}$, $\frac{1}{3}$ and 0, respectively. Inset: Main branch of "deviating" eigenvalues and real eigenvalues in a $\log_\lambda |\sigma| - \arg(\sigma)$ plot, for $\epsilon = 0.33$ (triangles), $\epsilon = 0.37$ (squares), and $\epsilon = 0.4$ (circles). The tilted branch rotates with changing ϵ and is horizontal at the phase transition and unstable above (black). It can be seen that in the $N \to \infty$ limit an infinite number of eigenvalues turns unstable immediately above the phase transition.

4.2. Asymptotics

The phase transition is a change in the behavior of a very large number of eigenfunctions all at once. Biferale [11] has explained this phase transition as a blockage in the energy flow caused by a flow of helicity. Since the blockage can occur anywhere in the ISR it seems reasonable to assume that, when the conditions are right, blockages can occur at many places affecting many eigenfunctions and eigenvalues at once.

Another way of asking the same question is more mathematical in character. To see this phase transition we must have some kind of change in the small-ν asymptotics of the system. That is we must have some sort of change which can be seen by an eigenfunction with j much larger than unity and much smaller than N_D. So we need an asymptotic theory of eigenfunctions for this system.

To understand the phase transition, we must first understand how the phase sector can be so unstable. The key is given in Table 1. In that table we see that in contrast to the magnitude sector, the phase sector shows far more flexibility in the large-n limit of the eigenfunction. The magnitude sector has but one value of p_1 and p_2 for all eigenfunctions in this region. The phase sector can actually have two different linear combinations. The two

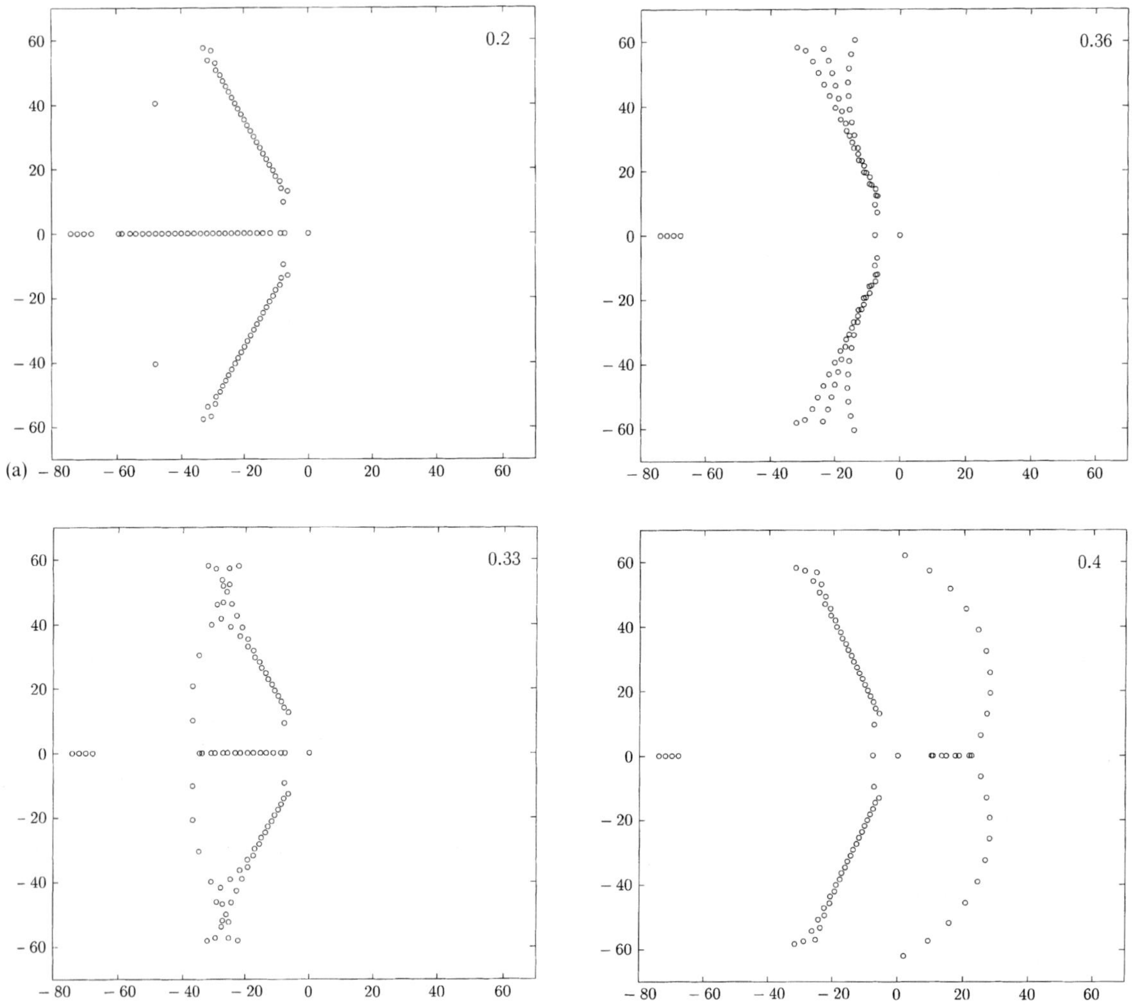

Fig. 10. This sequence of pictures shows the phase eigenvalues with changing ϵ. We start with three completely stable branches in (a). Real eigenvalues turn complex and move towards the imaginary axis (b) until fewer and fewer real eigenvalues are left (c). Notice, all the "deviating" eigenvalues that do not lie on a straight line are finite size effects, i.e., they constitute only a negligible part for a sufficiently large matrix. These deviating eigenvalues form a straight line at $\epsilon_{\text{bif}} = 0.37$ (Fig. 3). Now the spectrum consists of six complex branches and the scaling law (S) is no longer valid and has to be replaced by scaling law (S^2). Above ϵ_{bif} (d) eigenvalues return to the real axis, but are now unstable. Essentially there are two complex branches and one real branch again as it was below ϵ_{bif}. An even higher value of $\epsilon = 0.5$ is shown in Fig. 2.

permitted combinations are:

$$\delta\Phi_{3k+1} = 0, \quad \delta\Phi_{3k+2} = -\delta\Phi_{3k} \tag{38}$$

and also, for example,

$$-\tfrac{1}{2}\delta\Phi_{3k+1} = \delta\Phi_{3k+2} = \delta\Phi_{3k} \tag{39}$$

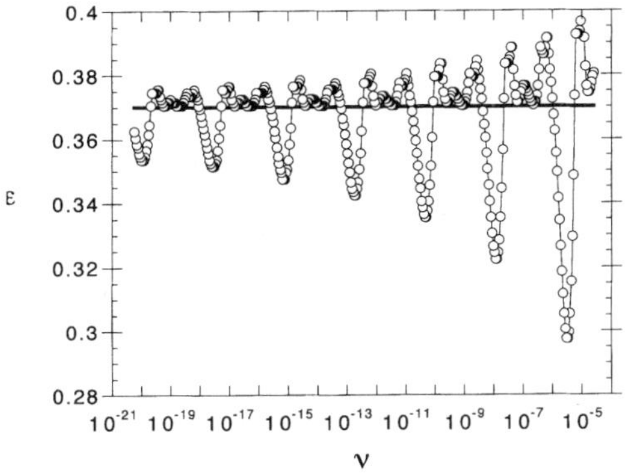

Fig. 11. Border of stability ϵ_c as a function of ν for $\lambda = 2$. ϵ_c approaches the Biferale prediction (thick horizontal line) for small ν. Some of the points are due to a complex pair of eigenvalues passing through the imaginary axis, others are due to a single real eigenvalue going through zero.

for integer values of k. The changes in expression (38) are an exact symmetry of the GOY system and produce a phase sector eigenfunction which has eigenvalue zero. The changes in expression (39) satisfy the eigenvalue equation for all n values except $n = 1$. Here, the eigenvalue equation fails because of the forcing term. However, this combination also forms a possible high-n behavior. All eigenfunctions have one or another linear combination of these two behaviors as the high-n limit. But there are two such linear combinations possible. In comparison to the magnitude case, we seem to have lost a boundary condition at the high-n end.

Thus, we state the first contrast between the magnitude sector and the phase sector. The magnitude sector can be described by giving two boundary conditions at the high-n side of the ISR; the phase sector can be described by giving only one.

We must have, someplace, an extra boundary condition for the phase sector. Bifarale directed our attention to the conservation of helicity. Instead of doing a full analysis of the possible extra conditions, we just follow his direction and look at the helicity conservation law, Eq. (36b), in the phase sector. (The reader will recognize that some leap of faith is required around this point in the argument.) This equation is

$$\sum_n h^n (u_n)^2 \delta\Phi_n^j = -\frac{f h u_1 \delta\Phi_1^j}{\sigma_\phi^j}. \tag{40}$$

But, for an eigenvalue with j in the middle of the ISR, the first component of the eigenfunction is very, very small. This component goes to zero more rapidly than an exponential in j. For this reason, it is quite reasonable to neglect the right-hand side of the helicity identity and write instead:

$$\sum_n h^n (u_n)^2 \delta\Phi_n^j = 0 \quad \text{for } j \text{ in region I.} \tag{41}$$

We propose that this identity replaces the lost high-n boundary condition for the asymptotics of the phase sector.

This proposition has several quite attractive features. We expect the eigenfunction to be of roughly the same order of magnitude over the whole range in which n is greater than j. Then for $\epsilon < \epsilon_{\rm bif}$ the summation has its main contribution at the lower-n end and then falls off as

$$\delta\Phi_n^j \left(-\frac{1-\epsilon_{\rm bif}}{1-\epsilon}\right)^n. \tag{42a}$$

When we are at the critical point, the summation does not fall at all but the summand looks like

$$\delta\Phi_n^j(-1)^n. \tag{42b}$$

Finally, above the critical point the major contributions come at the high-n end and then fall off toward lower n with the same law as shown in Eq. (42a). Near the phase transition, these behaviors produce long-range correlations between the different parts of the ISR. It is these correlations which form the key to the phase transition.

From these assumptions, one can find scaling laws for the phase eigenvalues, namely on the small-j end

$$\sigma'^j = \sigma^j\left(1 + \mathrm{O}\left(\left(\frac{1-\epsilon_{\mathrm{bif}}}{1-\epsilon}\right)^{N_\mathrm{D}-j}\right)\right) \quad \text{for } j \text{ in region S} \tag{43a}$$

Here, as before, the prime indicates a decrease in the viscosity by a factor of λ^4 together with a shift in the cutoff, N. The corresponding result on the large-j end is the statement, which we have used before.

$$\sigma'^{(j+3)} = \lambda^2\sigma^{(j)}(1 + \mathrm{O}(\epsilon - 1)^j) \quad \text{for } j \text{ in region L.} \tag{43b}$$

Both together give the scaling law:

$$\sigma'^{(j+3)} = \lambda^2\sigma^{(j)} \quad \text{for } j \text{ in region I.} \tag{44}$$

To check the accuracy of our thinking we show in Fig. 7 polar plots in which we overlay the eigenvalues σ_j with, respectively, the eigenvalues σ_j' and $\lambda^{-2}\sigma_{j+3}'$. According to the theory, the first of these should agree very well for small magnitudes of the eigenvalue, while the second should agree on the large magnitude end. The figures bear this out. For a more accurate check, we construct errors as, for example,

$$\delta = |\sigma_j/\sigma_j'| - 1. \tag{45}$$

These errors are shown in Fig. 8 and compared with the theoretical estimates taken from Eqs. (43a) and (43b). Since the estimated and the actual errors agree rather well, we can argue that we have the essence of the phase eigenvalues.

5. Summary and conclusions

The GOY model has a linear response behavior about a static solution in which the stability eigenvalues extend over a huge range of frequencies. In the ISR, the eigenvalues show a scaling behavior limited by disturbances from stirring or from viscous effects. The scaling of the eigenvalues is more complex in the phase sector than in the magnitude response. The phase sector shows a phase transition connected with a boundary condition which moves from one end of the ISR to the other. This boundary condition is derived from the conservation law for the model's version of helicity. Eigenvalues change quite abruptly as the model's parameter passes through its phase-transition value.

Much of the scaling mirrors the scaling properties of the static GOY model. However, there is scaling associated with the phase transition which we have not investigated in any detail. We have also not fully established the scaling behavior of the linear-stability eigenvectors. These studies are left for the future.

For now, we have demonstrated scaling in the linear response of a still system. We have seen a new phase transition and gained a qualitative understanding of its source.

Acknowledgements

It is our pleasure to thank Luca Biferale, Jean-Philippe Brunet, Peter Constantin, Greg Huber, and Norman Lebovitz for helpful discussion. We are also indebted to Peko Hosoi, who programmed our IDL animations. This work has been supported by the ONR, the DOE contract number DE-FG02-92ER25119, the MRSEC Program of the National Science Foundation under Award Number DMR-9400379, and by the Deutsche Forschungsgemeinschaft (DFG) through its SFB185.

Appendix A. Large scale forcing

Most studies of the GOY model employed $F_n = f\delta_{n,4}$ as large scale forcing. Though this forcing seems innocent, it has some disadvantages. The velocity component u_3 is much smaller than $u_{1,2,4,5,6,\ldots}$ [9], which is clearly unphysical. We force the system on level $n = 1$. As can be seen from Eqs. (2) and (5), with $u_3 = 0$ and $n \to n - 3$, the shift in the forcing by three shells just shifts the solution by three shells as well. [5]

With the different forcing the borderline of stability becomes only slightly different. For instance, ϵ_c beyond which (2) becomes unstable is in the standard case ($\lambda = 2$, $\nu = 10^{-7}$, $f = \sqrt{2} \times 5 \times 10^{-3}$) shifted from 0.370 to 0.377. More significant is the change in the amplitude of the ϵ_c oscillations with viscosity. This viscosity dependence is shown in Fig. 11 for $\lambda = 2$, and looks similar for other values of λ. In contrast to the case with forcing on the fourth shell (see [9], Fig. 12) the stability border ϵ_c now approaches the constant ϵ_{bif} for vanishing viscosity $\nu \to 0$,

$$\lim_{\nu \to 0} \epsilon_c = \epsilon_{\text{bif}} := 1 - \lambda^{-2/3}. \tag{A.1}$$

Some of the transitions to instability in Fig. 11 are due to a complex pair of eigenvalues passing through the imaginary axis, others are due to a single real eigenvalue going through zero. [6]

In the main text we have studied the behavior of the overall eigenvalue spectrum. The stability border, on the other hand, is determined only by the first eigenvalue crossing the imaginary axis. Nevertheless, in the main text we have two strong evidences for a dependence of the stability border on the particular forcing. As we have seen in Section 3.2 the modulus of the eigenvalues is set by the cascade term. The eigenvalues that decide the stability are the small ones close to zero. Their size should depend on how the solution looks like at the very first shells, and this is obviously influenced by the kind of forcing. Also formula (36b) shows that the stability border should depend on the particular forcing.

---

[5] With the forcing $F_n \propto \delta_{n,1}$ a static solution of Eq. (2) exists *for all* $0 < \epsilon < 1$, i.e., the saddle node bifurcation found in [9] is an artifact of the forcing $F_n \propto \delta_{n,4}$. In particular, the overall existence of the static solution now allows for an analysis of the eigenvalue spectrum for the parameter values $\lambda = 2$ and $\epsilon = 0.5$ [2–4].
The artificially small u_3 (for $F_n \propto \delta_{n,4}$-forcing) also leads to strongly non-normal eigenvectors. This was first discovered by Brunet [10], who also calculated the corresponding pseudoresonance. The reason for that can be seen from (19b) and (20b): $(A_\phi)_{3,m}$ with $m = 1, 2, 4, 5$ is small as $u_3 \sim \nu$ is so small. All the eigenvectors of A_ϕ then have a huge 3rd component and are thus nearly parallel to each other. This makes A_ϕ very non-normal, the more, the smaller ν is.

[6] The model with the forcing on the fourth shell shows a transition from stable to unstable behavior only via two complex conjugated eigenvalues going through the imaginary axis. With the forcing on the first shell instability for some ν occurs via a real eigenvalue going through zero, followed by a series of Hopf bifurcations (take e.g. $\nu = 10^{-8}$, $f = \sqrt{2} \times 0.005$, $k_0 = 0.0625$). However, from the viewpoint of the scenario of instability in the limit $\nu \to 0$, treated in Section 4.1, this additional type of transition is not fundamentally different from the other one.

Appendix B. A related phase transition

In this appendix we analyze the related phase transition in the spectrum of the matrix

$$A_\xi = \xi \cdot C - D. \tag{B.1}$$

The matrices C and D contain the solution of (2). For general ξ the matrix A_ξ is *not* the stability matrix of the system. Only for $\xi = -1$ we have $A_{-1} = A_{\mathrm{M}}$ and for $\xi = 1$ we have $A_1 = A_\phi$. The generalized amplitude matrix A_ξ with $\xi < 0$ shows the same features as A_{M} itself. So we only study the interesting case of A_ξ with $\xi > 0$ which displays a *phase transition* at $\xi = 1$. The advantage of studying this phase transition rather than the one in ϵ discussed in the body of the paper is that here the critical point $\xi = 1$ does not depend on the viscosity. Thus finite size effects are less complicated than for the transition in the spectrum of A_ϕ at the ν dependent ϵ_{c}. Taking $\xi \approx 1$ (but $\xi \neq 1$) can be thought of as equivalent to slightly varying the dynamical equation (2) and thus changing the phase symmetry of that equation, which leads to a different solution. We perform the analysis for $\epsilon = 0.3$ as it is the spurious stabilization of the phase matrix for $\epsilon < \epsilon_{\mathrm{c}} \approx 0.37$ which we want to understand.

We start with $\xi = 0$. All eigenvalues $-\nu k_n^2$ of $A_0 = -D$ are of course stable. However, with increasing ξ, more and more eigenvalues of A_ξ turn *positive* as soon as $\xi \gtrsim \nu$. This feature holds for $|\lg \nu| \sim 30$ orders of magnitude in ξ. From comparing the viscous and the inertial contribution to A_ξ we obtain that the largest eigenvalue increases as

$$\sigma_{\max} \sim \xi^{3/2} \nu^{-1/2}. \tag{B.2}$$

$\sigma_{\max}$ becomes as large as 10^{12} for $\xi \approx 0.63$ and the standard parameters $\lambda = 2$, $\epsilon = 0.3$, $N = 81$, $\nu \sim 10^{-30}$. This is the behavior we see in the spectrum of $A_\phi = A_1$ for $\epsilon > \epsilon_{\mathrm{c}}$ and what we had expected also for smaller ϵ.

Why and how is stability achieved for $\xi = 1$, i.e., why does $A_1 = A_\phi$ not have any positive eigenvalue for $\epsilon < \epsilon_{\mathrm{c}}$?

For ξ growing beyond 0.63 towards 1, two real eigenvalues merge, form a complex pair which moves in the complex plane on a circle and finally turns stable through an inverse Hopf bifurcation. This happens again and again through a self-similar cascade of bifurcations towards phase symmetry at $\xi = 1$ – no unstable eigenvalue is left. This phase symmetry of $A_1 = A_\phi$, discovered in [9], reflects itself in a zero eigenvalue, as discussed above.

We now analyze the singularity in detail. We introduce the distance τ from the singularity at $\xi = 1$,

$$\xi = 1 + \tau \tag{B.3}$$

and increase $|\tau|$ on a log scale from $\tau = 0$. In this way we break the phase symmetry in a stronger and stronger way.

We describe the features for $\tau \leq 0$. For $\tau = 0$ we only have negative eigenvalues and one eigenvalue equals zero. This center manifold eigenvalue turns *positive* for $|\tau| > 0$, signaling instability. Moreover, two different small (modulus wise) real eigenvalues (< 0) merge on the real axis and form a complex pair which wanders towards the $\mathbb{R}\sigma = 0$ axis and turns unstable via a Hopf bifurcation. This happens as early as for $\tau \approx 10^{-10}$, see Fig. 12. In the right complex half plane they continue to wander on a kind of circle, finally merging again on the real axis, now forming two positive real eigenvalues. But meanwhile a second pair of complex eigenvalues has formed in the left half plane which again turns unstable via a Hopf bifurcation and then finally merges. This happens again and again, leading to 3, 5, 7, 9, 11, ... positive eigenvalues. The real parts of the positive eigenvalues are plotted in Fig. 12. The self-similarity of the cascade of bifurcations can clearly be observed. The maximal eigenvalue $\sigma_{\max} \propto \tau^{4/3}$. The scaling law mirrors the classical K41 scaling of u_n.

For $\tau > 0$ the behavior is very similar to the $\tau < 0$ case. The only difference is that the center manifold (zero) eigenvalue turns stable first and we now have an *even* number 2, 4, 6, 8, ... of positive eigenvalues.

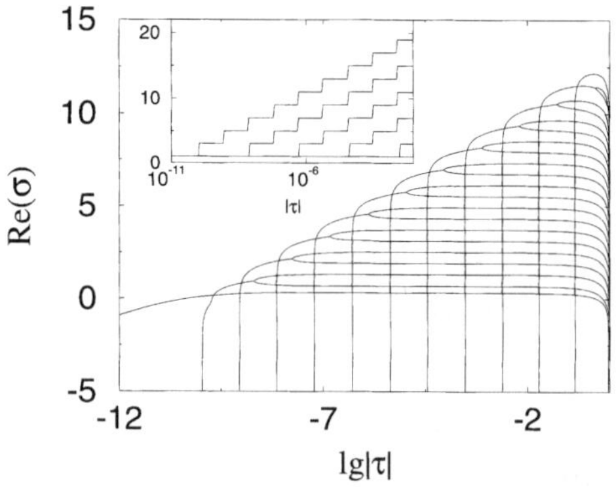

Fig. 12. All positive eigenvalues of A_ξ for $\tau = \xi - 1$ between 0 and -1 on a log–log scale in $|\tau|$, corresponding to a lg versus lg lg ξ plot. The self-similar cascade of bifurcations towards phase symmetry has its origin in the self-similarity of the matrix itself. The parameters are $N = 80$, $\nu \sim 2 \times 10^{-29}$, $\epsilon = 0.3$, $\lambda = 2$. For $\tau = -1$ (right edge of the plot) we have $A_\xi = -D$ and all eigenvalues are again stable. The inset shows the number of positive eigenvalues of $A_{1-\tau}$ as a function of $|\tau|$ for various system sizes $N = 80, 68, 56, 44, 32, 20$, left to right, corresponding to viscosities $\nu = \nu_0 \times 16^i$, $i = 0, 4, 8, 12, 16, 20$, respectively. $\nu_0 \sim 2 \times 10^{-29}$, $\epsilon = 0.3$.

To demonstrate the finite size effects of the transition we plot the number of positive eigenvalues as a function of lg τ for various system sizes N, i.e., viscosities ν, see the inset of Fig. 12. In the $N \to \infty$ limit, even for an arbitrarily small $|\tau|$, we have an infinite number of unstable eigenvalues.

References

[1] E.B. Gledzer, Sov. Phys. Dokl. 18 (1973) 216.
[2] M. Yamada and K. Ohkitani, J. Phys. Soc. Jpn. 56 (1987) 4210; Prog. Theor. Phys. 79 (1988) 1265; K. Ohkitani and M. Yamada, Prog. Theor. Phys. 81 (1989) 329.
[3] M.H. Jensen, G. Paladin and A. Vulpiani, Phys. Rev. A 43 (1991) 798; D. Pisarenko et al., Phys. Fluids A 5 (1993) 2533.
[4] R. Benzi, L. Biferale and G. Parisi, Physica D 65 (1993) 163.
[5] L. Biferale, A. Lambert, R. Lima and G. Paladin, Physica D 80 (1995) 105.
[6] L. Kadanoff, D. Lohse, J. Wang and R. Benzi, Phys. Fluids 7 (1995) 617.
[7] L. Kadanoff, Physics Today 48 (1995) 11.
[8] R. Benzi, L. Biferale, R. Kerr and E. Trovatore, Phys. Rev. E, (1996), in press.
[9] N. Schörghofer, L. Kadanoff and D. Lohse, Physica D 88 (1995) 40.
[10] J.P. Brunet, private communication (1995).
[11] L. Biferale, private communication (1995).
[12] L. Biferale and R. Kerr, Phys. Rev. E 52 (1995) 6113.

6

Singularities and Similarities in Interface Flows

Andrea L. Bertozzi, Michael P. Brenner, Todd F. Dupont, and Leo P. Kadanoff

6.1 Introduction

The onset of singularities in systems of nonlinear partial differential equations is an important issue in fields ranging from general relativity [27], to thermodynamic phase transitions [10], to fluid dynamics [13]. The development of a mathematical singularity, when some quantity associated with the PDE "blows up," reflects the creation of a new structure in the physical system which in turn forces the mathematical formulation to change. Whether or not such singularities are possible for a given system can be a difficult question. A famous problem from the theory of homogeneous incompressible fluids is the question of finite time singularity development in the three-dimensional Navier–Stokes equation: It is unknown if an initially smooth solution can develop a finite time singularity in which the vorticity becomes unbounded [23]. To date, no rigorous proof or counterexample exists; neither numerical nor physical experiments have produced definitive answers [22, 25]. When a particular system allows finite time singularities, many related questions become relevant. For example, do all singularities have universal characteristics, or are there many possible behaviors? Which quantities are unbounded at the singular time?

In this chapter we study these questions for a model equation describing a simple hydrodynamic system that is both easily accessible to experiments and well known to develop singularities in finite time. Consider two different fluids separated by an interface that evolves with dynamics including a pressure jump determined by the Gibbs–Thomson relation

$$\Delta p = \gamma\kappa. \tag{1.1}$$

Here, κ is the mean curvature of the surface, Δp denotes the pressure jump across the interface, and γ is the surface tension of the interface. Whenever the topology of the interface changes, the mean curvature and hence the pressure field develops a singularity. In typical situations, the singularity can also cause the flow velocity to diverge. This type of singularity happens whenever a mass of fluid separates into two pieces. Examples in nature range from dripping faucets to nuclear fission.

The mathematical description of such surface tension driven flows includes the Navier–Stokes equation [1] for the fluid velocity $\mathbf{v}$

$$\begin{aligned} \mathbf{v}_t + (\mathbf{v} \cdot \nabla)\mathbf{v} &= -\nabla p/\rho + (\mu/\rho)\nabla^2 \mathbf{v}, \\ \nabla \cdot \mathbf{v} &= 0 \end{aligned} \tag{1.2}$$

coupled with boundary conditions at the interface. Here ρ is the density, μ the viscosity, and p the pressure. At the interface we impose equation (1.1) along with the condition that the normal component of the fluid velocity is continuous across the interface and that the interface moves with the fluid velocity.

In this chapter, we consider singularity formation in a 1D model equation that follows from the above equations using the lubrication approximation and specific geometric constraints. The physical systems are flow in a Hele–Shaw cell and the flow of a thin film of viscous liquid on a solid surface. By reducing the hydrodynamic equations to a partial differential equation in only one spatial dimension we can more easily obtain numerical and analytical results.

The chapter proceeds as follows: After this section we derive the equations of motion in the two examples above. In Section 6.1.3, we present the model equation and discuss its mathematical features. In Section 6.1.4 we summarize our observations of singularity formation in the system. An important feature of the simulations is the ubiquitous presence of "similarity" (i.e., self-similar) solutions in the development of singularities. In Section 6.2, we describe in detail the types of similarity solutions associated with the model system and describe their characteristics. In Section 6.3, we present a detailed analysis of the numerically observed singularities for the model equation and demonstrate that most singularities are well described by similarity solutions. Finally, in Section 6.4 we point out unsolved problems.

6.1.1 Hele–Shaw Flow

In this case, two fluids (usually air and a viscous liquid) are trapped between two closely separated flat surfaces. For small Reynolds number, the evolution equation is the Navier–Stokes equation (1.2) without the inertial terms. The equation is simplified because the pressure is only a weak function of the z coordinate perpendicular to the flat surfaces. Thus, the essential dynamics follow from averaging the fluid velocity field over the perpendicular z direction to obtain Darcy's law [2]

$$v = -\frac{b^2}{12\mu} \nabla p(x, y). \tag{1.3}$$

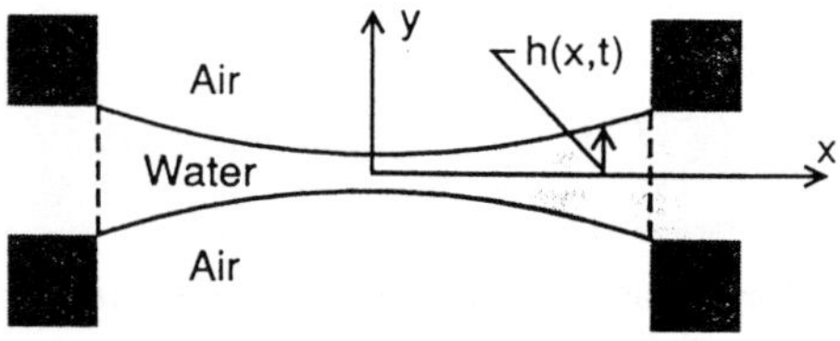

FIGURE 1.1. Picture of the thin neck in a Hele–Shaw environment. This picture is motivated by an experiment [16]. Here, $2h(x,t)$ represents the width of the neck at position x and time t.

Here b is the plate separation, μ is the viscosity, and v is the average horizontal fluid velocity. The xy plane is parallel to the flat surfaces.

We further simplify the problem by considering the evolution of a very thin neck of liquid, such as that of Figure 1.1. We assume the neck is symmetric about the center line $y = 0$ and has a small width $2h(x,t)$ which varies slowly with x. Using the lubrication approximation [7], we assume that the pressure field is independent of y, so that the x component of the velocity, u, is

$$u = -\frac{b^2}{12\mu}\partial_x p \simeq \frac{b^2\gamma}{12\mu}h_{xxx}, \tag{1.4}$$

since in this approximation, $\kappa \simeq h_{xx}$. Finally, the conservation of mass implies that

$$h_t + l_x = 0, \tag{1.5}$$

where $l = uh$ is the current of matter in the x direction. Combining (1.4) and (1.5) we obtain [8, 14, 15]

$$h_t + \frac{b^2\gamma}{12\mu}(hh_{xxx})_x = 0. \tag{1.6}$$

Equation (1.6) involves two approximations: First it assumes that the Reynolds number of the flow is small. This condition might be violated near a singularity. The derivation also assumes that the width of the thin neck $2h$ is much greater than the plate spacing b. When $2h < b$, the dynamics is intrinsically three dimensional and Darcy's law (1.3) does not apply. We point out, however, that in favorable circumstances, equation (1.6) might apply until $h \sim b$. In such a situation the Reynolds number of the flow around the singularity does not become large until the flow is fully three dimensional. For more information see [4].

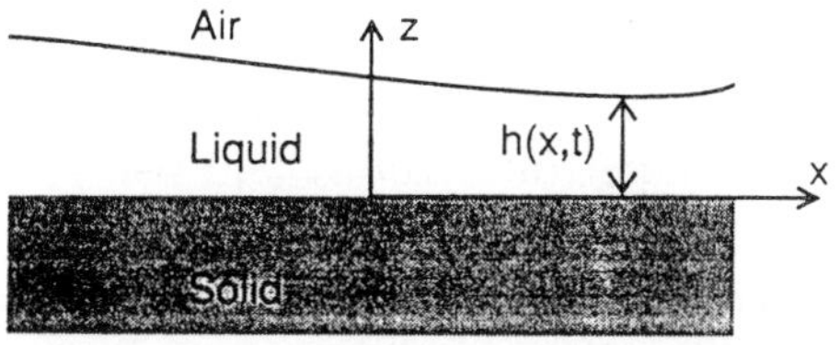

FIGURE 1.2. Picture of thin film of viscous liquid on a solid surface. Here, h represents the local height of the liquid at position x and time t.

6.1.2 Flow of a Thin Film

In a similar fashion, we derive approximate equations for the evolution of a one-dimensional thin layer of viscous liquid on a solid surface. We consider a thin film of local thickness $h(x, t)$ (see Figure 1.2). Here, there are two different boundary conditions: the tangential component of the liquid stress tensor vanishes at $z = h(x,t)$, the air–liquid interface. In addition, we must prescribe a boundary condition at the liquid–solid interface. The classical "no-slip" boundary condition, $u_{\|}|_{z=0} = 0$ ($u_{\|}$ is the component of $\mathbf{v}$ parallel to the solid surface), is a valid approximation as long as the microscopic solid–liquid interactions do not significantly affect the macroscopic flow field [9]. This condition is reasonable for films with a thickness much larger than the range $\tilde{h}$ of the solid–liquid interactions. However, the no-slip boundary condition breaks down near contact lines (where the liquid thickness approaches zero) [20, 26] and loses applicability when the film thickness is of order $\tilde{h}$, a length scale that depends on the particular physical problem at hand. In all cases, the solid exerts an enormous frictional force on the liquid, so that $\partial_z u(z,t)|_{z=0}$ is very large. Thus, the boundary conditions imply that the gradient of $\partial_z u_{\|}$ is large across the liquid film, so that the dominant viscous stress in the Navier-Stokes equation is $\partial_z^2 u_{\|}$. For low Reynolds number flows, this viscous stress balances the horizontal pressure gradient, $\nabla_{\|} p$. Again using the lubrication approximation, so the pressure field is independent of the coordinate normal to the solid surface, we obtain an equation for the horizontal component of the liquid velocity

$$u_{\|} = \frac{\nabla_{\|} p(x,y)}{\mu}\left(\frac{z^2}{2} - hz - B(x,y)\right). \tag{1.7}$$

The interactions at the solid interface determine the function B. Averaging (1.7) over the z direction, we obtain

$$\bar{u}_{\|} = -\frac{\nabla_{\|} p}{3\mu}(h^2 + 3B). \tag{1.8}$$

If we assume that h is independent of y, then the equation depends on only one spatial dimension. We are interested in the case where surface tension dominates the pressure; in particular, we neglect the effect of gravity. The evolution equation for h becomes [17]

$$h_t + \left(\frac{\gamma}{3\mu}(h^3 + Bh)h_{xxx}\right)_x = 0. \tag{1.9}$$

The choice of B depends on conditions at the solid surface. When $h \gg \tilde{h}$, we can take $B = 0$. When the surface is porous, $B = \alpha$, where α is the porosity of the surface [24]. Also of interest are polymeric liquids, which obey the boundary condition $u|_{z=0} = c\partial_z u|_{z=0}$, which yields $B = ch$. Here c is a length scale associated with the polymer [9]. The reader should note that for film thicknesses $h \ll \tilde{h}$, the term Bhh_{xxx} dominates the flux in equation (1.9).

6.1.3 The Model Equation

The examples considered above have a common feature: When h is sufficiently small, the dominant term in the flux has the same functional form, other than the exponent of h. Mathematically, these cases differ only in that the liquid velocity is proportional to h_{xxx} (Hele–Shaw and liquid on porous medium) , hh_{xxx} (polymeric liquids), and $h^2 h_{xxx}$ (macroscopic thin film). The velocities all have the general form

$$u \sim h^{n-1} h_{xxx}. \tag{1.10}$$

In nondimensional form, neglecting the terms with higher powers of h, the evolution equation is

$$h_t + (h^n h_{xxx})_x = 0. \tag{1.11}$$

We note that equations of this general form with $1 \leq n \leq 2$ also appear in the literature as slip models for modeling motion near contact lines [19, 26].

We now ask the following questions: Does equation (1.11) allow h to go to zero in finite time? Does this answer depend on on the exponent n? We will show that the value of n determines whether such singularities are possible. There are several transitions in allowable behavior as n varies, some of which are in the range corresponding to the physical models.

Mathematical Background of the Model Equation

The model equation is a fourth-order parabolic equation. The degeneracy of the equation as $h \to 0$ requires h to be bounded away from zero for standard parabolic theory to ensure well-posedness. In this work we consider equation (1.11) as an initial value problem on the bounded domain $[-1, 1]$. In general, one can use either periodic boundary conditions on the interval or two fixed conditions for h or higher derivatives at each endpoint. We specifically consider the fixed conditions

$$h(\pm 1) = 1, \qquad h_{xx}(\pm 1) = p \qquad \text{(pressure)}, \qquad (1.12)$$

$$h(\pm 1) = 1, \qquad h_{xxx}(\pm 1) = \pm c, \quad c > 0 \qquad \text{(current)}. \qquad (1.13)$$

The boundary condition (1.12) comes from the work of Constantin et al. [8] and Dupont et al. [11] in analyzing the Hele–Shaw problem, where p has the physical interpretation of the external *pressure*. The boundary condition (1.13) corresponds a fixed *current*, drawing liquid out of the cell at a finite rate (a situation in which a singularity must occur). Another choice introduced by Bernis and Friedman [3] is

$$h_x(\pm a) = 0, \quad h_{xxx}(\pm a) = 0. \qquad (1.14)$$

In the remainder of this section we combine both theoretical results and numerical data to illustrate when singularities are allowed in (1.11). We begin with some mathematical aspects of the solutions and present rigorous results. Then, we summarize the results of numerical simulations.

Rigorous Results

Existence of Solutions

If we consider smooth initial data $h_0(x)$ satisfying the chosen boundary conditions *and* $h_0(x) > 0$, then, in general, there exists a solution to (1.11) on a finite time interval. This solution is infinitely differentiable and can be continued in time as long as $h(x,t) > 0$. The reader will find a detailed proof of this fact with the boundary conditions (1.14) in [3]. Here we summarize the main ideas as well as differences resulting from different boundary conditions. First we note that h describes a height, a non-negative quantity. Therefore, we demand that the solution satisfies $h \geq 0$ and replace (1.11) by

$$h_t + (|h|^n h_{xxx})_x = 0. \qquad (1.15)$$

Proof of local existence of solutions to (1.15) follows from first proving local existence of solutions to the regularized equation

$$h_t + (f_\epsilon(h) h_{xxx})_x = 0, \qquad (1.16)$$

where $f_\epsilon(h) \to |h|^n$ as $\epsilon \to 0$ and $f_\epsilon(h) > \epsilon$ [an example is $f_\epsilon(h) = (h^2 + \epsilon^{2/n})^{n/2}$]. For a given ϵ, the equation is uniformly parabolic and, thus, by classical theory, has a solution on a short time interval. To prove the existence of solutions to equation (1.15) we pass to the limit in ϵ. As long as $h_\epsilon(x,t)$ is bounded away from zero uniformly in ϵ, all derivatives converge uniformly to a C^∞ solution $h(x,t)$ of the original equation. If $h(x,t) > 0$, the limit is unique. Such a limiting process is crucial for proving the existence of weak solutions even when h is not bounded away from zero

[3]. We continue this process in time until $h(x,t)$ hits zero for some value of x. Notice that finite time $h \to \infty$ type singularities are impossible for many of the boundary conditions. On any finite time interval, with either the boundary conditions (1.12), (1.14), or periodic boundary conditions, $\|h_x(\cdot,t)\|_{L^2[-1,1]}$ is an a priori bounded quantity on any time interval $[0,T]$ which forbids h to become unbounded in finite time.

Existence and Behavior of Singularities

We ask whether it is possible for $h \to 0$ either in finite or infinite time. Clearly boundary conditions play an important role. For example, in the case of the "pressure" boundary conditions (1.12), h definitely goes to zero if $p > 2$; the issue is whether it happens in finite or infinite time (see below). Moreover, for the current boundary conditions, a finite time singularity *always* occurs; the issue is whether it occurs on the boundary or in the interior.

We begin by noting that for $n \geq 1$ if $h \to 0$, then necessarily $\int_0^{t_c} |h_{xxxx}|_{L^\infty} dt$ must blow up. We present a formal argument that can be made rigorous:

$$\begin{aligned}
\frac{d}{dt}(h_{min}(t)) &= \frac{d}{dt}(min_x(h(x,t))) = \frac{d}{dt}(h(x_{min},t)) \\
&= h_t(x_{min},t) + h_x(x_{min},t)\dot{x}_{min} \\
&= h_t(x_{min},t) = -h^n h_{xxxx}(x_{min},t) - nh^{n-1}h_{xxx}h_x(x_{min,t}) \\
&= -h_{min}(t)^n h_{xxxx}(x_{min},t).
\end{aligned}$$

Hence, for $n \geq 1$, $h_{min}(t)$ can only go to zero in finite time if $\int_0^{t_c} h_{xxxx}\,(x_{min},t)dt$, and, hence, h_{xxxx}, diverges. Here we use the fact that $h_x(x_{min},t) = 0$.

We now show a result that rules out finite time singularities for some values of n. Furthermore, we include additional results that rule out finite time singularities when h_{xx} is bounded; we observe that this bound holds in many (but not all) of our simulations. We remark that our numerical simulations indicate that the bounds on n presented here may not be sharp. See Section 6.3 for more details.

Theorem 6.1.1 *Let h be a solution to (1.11) with either periodic boundary conditions or (1.14) and smooth initial data $h_0(x) > 0$. Then (1) if $n \geq 3.5$, then there exists a unique smooth solution $h(x,t)$ for all time that satisfies $h(x,t) > 0$. (2) If $n \geq 2$, then the above is true, provided that $h_{xx}(x,t)$ remains bounded.*

The proof is an extension of the one found in [3] showing (1) to be true for $n \geq 4$. We omit some details that are identical to those found in [3]. First consider the case of periodic or (1.14) boundary conditions. We first note that $\int h_x^2$ is an energy function because integration by parts yields $(d/dt)\int h_x^2 = -\int h^n h_{xxx}^2$. The Cauchy inequality then implies an a priori bound on the $C^{1/2}$ norm of h. There is also an a priori bound on $G(h,n) = \int h^{3/2-n}$ by integration by parts:

$$\begin{aligned} dG/dt &= (\tfrac{3}{2}-n)\int h^{\frac{1}{2}-n}(-h^n h_{xxx})_x dx \\ &= (\tfrac{3}{2}-n)(\tfrac{1}{2}-n)\int h^{-\frac{1}{2}} h_{xxx} h_x dx \\ &= -(\tfrac{3}{2}-n)(\tfrac{1}{2}-n)\left(\int h^{-\frac{1}{2}} h_{xx}^2 - \tfrac{1}{2}\int h^{-3/2} h_x^2 h_{xx} dx\right) \\ &= -(\tfrac{3}{2}-n)(\tfrac{1}{2}-n)\left(\int h^{-1/2} h_{xx}^2 - \tfrac{1}{4}\int h^{-5/2} h_x^4 dx\right). \end{aligned}$$

Using the fact that

$$\left|\int h_x^4 h^{-5/2}\right| = \left|2\int h_x^2 h_{xx} h^{-3/2}\right| \le 2\left[\int h^{-1/2} h_{xx}^2\right]^{1/2}\left[\int h_x^4 h^{-5/2}\right]^{1/2},$$

we see that for $n > 3/2$

$$dG/dt \le 0.$$

Hence, G is a priori bounded. For $n \ge 3.5$ this gives an a priori bound on $\int 1/h^2$ which in conjunction with the a priori bound in the $C^{1/2}$ norm of h makes it impossible for $h(x,t)$ to reach zero.

Part (2) follows similarly, by noting that for $n \ge 2$, $\int h^{-1/2}$ is a priori bounded. If h_{xx} is bounded, then $h \to 0$ implies $\int h^{-1/2}$ must become unbounded.

We now consider the constant pressure boundary conditions (1.12). In this case the energy function is $\int \varphi_x^2$, where $\varphi(x,t) = h(x,t) - h_\infty(x)$ and $h_\infty(x)$ is the minimum energy solution satisfying the boundary conditions

$$h_\infty(x) = 1 - \frac{P}{2} + \frac{P}{2}x^2. \tag{1.17}$$

For the linear equation, with $n = 0$, the solution tends toward h_∞ at large times. Note, however, that $h_\infty < 0$ on $(-x_{cr}, x_{cr})$, $x_{cr} = \sqrt{2/p - 1}$ for $p > 2$. For $p > 2$, the solution will not attain the "least energy" state $h_\infty(x)$ because h cannot change sign. We assume that the solution is non-negative for all time[1]; hence, we expect an alternative "least energy" state that satisfies this constraint. For $p > 2$, the correct choice is the weak solution

[1] $n \ge 1$ guarantees existence of a non-negative (possibly weak) solution for all time. The proof is a direct extension of a similar result described in [3].

$$\begin{aligned} w_\infty(x) &= 0, \quad |x| < x_c, \\ w_\infty(x) &= \frac{p}{2}(|x| - x_c)^2, \quad 1 \geq |x| \geq x_c, \\ x_c &= 1 - \sqrt{2/p}. \end{aligned} \tag{1.18}$$

This weak solution has a jump in its second derivative at $\pm x_c$ and hence will produce a singularity in h if $h \to w_\infty(x)$ in infinite time. Thus, when $p > 2$, we always expect a singularity to form, either in finite or infinite time. The details of the infinite time case with $n = 1$ are presented in [8, 11, 28].

When $p < 2$, there need not be a singularity, since the solution can tend toward the positive solution $h_\infty(x)$. However, we cannot rigorously rule out the possibility of finite time singularities for the constant pressure boundary conditions with $n < 4$. This result follows from an extension of Theorem 1.1. First we integrate by parts to see that $\int \varphi_x^2$ is a priori bounded where $\varphi = h - h_\infty$, which, in turn, gives an a priori bound for $\int h_x^2$ and from the boundary conditions, a bound for $\int h$. A calculation shows that

$$\frac{d}{dt}\left[\int h^{2-n} + (n-2)\int h\right] = \frac{d}{dt}[H_n(h)] = -(n-2)(n-1)\int (h_{xx}^2 - p h_{xx})dx.$$

For $n \geq 2$ this time derivative is bounded from above by the constant $(n-2)(n-1)p^2/2$. Hence, finite time singularities are impossible for $n \geq 4$. Note that this bound is higher than the bound of $n \geq 3.5$ for the boundary conditions of Theorem 1.1. The boundary conditions (1.12) provide extra boundary terms that we cannot control as in the a priori estimates used for proving Theorem 1.1. However, we can show part (2) of Theorem 1.1. The bound on $\int h$ gives a bound on $\int_0^T (h_{xxx}(1) - h_{xxx}(-1))dt$ and knowing that h_{xx} is a priori bounded gives definite bounds on $\int_0^T [h_x^3(1) - h_x^3(-1)]$ and $\int_0^T [h_x(1) - h_x(-1)]$. Using these facts, we can bound all of the boundary terms in the dG/dt integral to show that G is a priori bounded on any time interval and produce the following:

Corollary 6.1.1 *Let h be a solution to (1.11) with boundary conditions (1.12) and smooth initial data $h_0(x) > 0$. If $p > 2$, then h will go to zero in either finite or infinite time. If (1) $n \geq 4$, then finite time singularities are impossible and h hits zero in infinite time. If (2) $n \geq 2$, the above is true providing that h_{xx} remains bounded.*

If we consider the "current" (1.13) boundary conditions, we force a finite time singularity to happen and the question becomes one of whether or not it happens on the boundary or in the interior. We have the following:

Theorem 6.1.2 *Let h be a solution to (1.11) with boundary conditions (1.13) and smooth initial data $h_0(x) > 0$. (1) If $n \geq 3.5$ then $h(x,t)$ goes to zero in finite time with the singularity occurring on the boundary of* *the domain, that is at least one of* $h_x(1,t)$, $h_{xx}(1,t)$ $h_x(-1,t)$, $h_{xx}(-1,t)$ *becomes unbounded as* $t \to t_c$, *and* $min_I(h(x,t)) \to 0$ *as* $t \to t_c$ *for all arbitrarily small neighborhoods* I *of the boundary. (2) If* $n \geq 2$, *then the above is true provided that on every compact set contained in the interior of* $[-1,1]$ $h_{xx}(x,t)$ *remains bounded.*

We note that the bound on h_{xx} at the end of this theorem will depend sharply on the chosen set. Numerically we observe that the minimum of h goes to zero by moving to the boundary as it touches down for $n \gtrsim 2.0$. The proof of Theorem 1.2 follows in the same fashion as that of Theorem 1.1. We integrate by parts while keeping the boundary terms. By previous arguments, the only way that the results of Theorem 1.1 can fail is if one of the boundary terms blows up. Since h_{xxx} is constant on the boundary, at least one of $h_{xx}(\pm 1)$ or $h_x(\pm 1)$ must become unbounded. In the case where $n \gtrsim 2$, we cannot require that h_{xx} be bounded on all of I since typically its boundary value will blow up as $t \to t_c$. However, we can rule out interior singularities with the assumption that on *every* set contained in the interior of I, h_{xx} remains bounded (the bound is not uniform). The result that h goes to zero in any arbitrarily small open neighborhood of the boundary comes from the fact that one can show as in [3] that h_{xx} and h_x (and all higher derivatives) will remain bounded in a region where h remains bounded away from zero.

6.1.4 Simulations and Similarity Solutions

We summarize numerical results for the model system. Some of the results extend previous research conducted for the $n = 1$ case in [8, 11, 28]. We classify the singularities as follows:

(a) Solutions in which $h \to 0$ in finite time versus solutions in which $h \to 0$ at infinite time. In the latter situation, the infinite time singularity is a result of the solution trying to reach a "least energy" equilibrium state, $w_\infty(x)$, that has $w_\infty(x) = 0$ on a set of positive measure and a jump discontinuity in $(w_\infty)_{xx}$.

(b) Solutions which have reflection symmetry about the singular point versus solutions which do not.

(c) Solutions in which $h \to 0$ at an interior point $[x \in (-1,1)]$ versus $h \to 0$ at a boundary point $(x = \pm 1)$. Boundary singularities are necessarily asymmetric, in the sense of (b).

There are six different cases described by this list. Table 1 shows the ones we observe in our simulations. The cases listed in Table 1 are those for which we have both numerical and analytical evidence for the singularity.

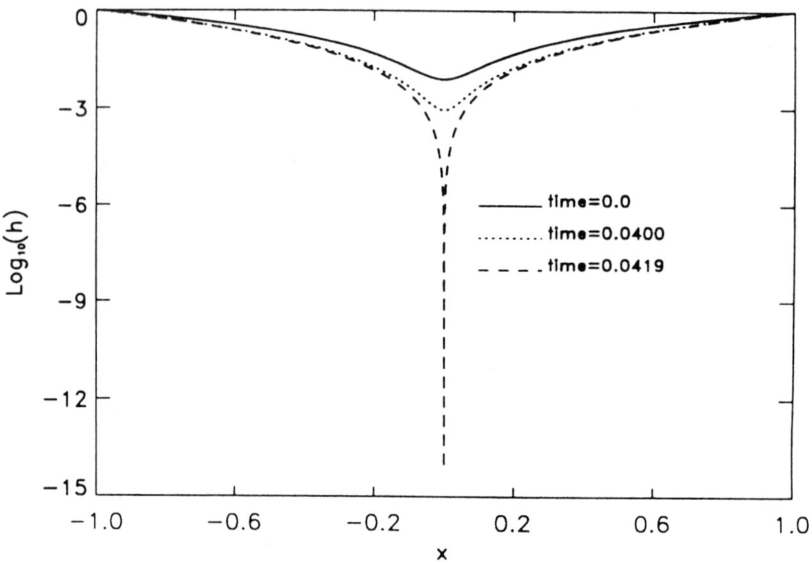

FIGURE 1.3. Finite time singularities at the center of the computational domain for $n = 0.75$. We show a log-linear plot of $h(x,t)$ for successive times before the singularity. Initial conditions and boundary conditions for this solution are detailed in Section 6.3.

TABLE 1. Singularities seen in simulations.

case	time	type of touchdown	boundary condition	observed for
i	finite	symmetrical	(1.12)	$0 \leq n < 1$
ii	finite	asymmetrical	(1.12) and (1.13)	$0.75 \lesssim n \lesssim 1.2$
iii	infinite	asymmetrical	(1.12)	$0.75 \lesssim n \leq \infty$
iv	infinite	symmetrical	(1.12), $p = 2$	$0.75 \lesssim n \leq \infty$
v	finite	boundary	(1.13)	$n \gtrsim 2$
vi	infinite	boundary	(1.13)	not observed

In Figures 1.3–1.6, we show typical simulation results illustrating the behaviors of Table 1. Figure 1.3 shows case (i). Here, h goes to zero at $x = 0$ ($n = 0.75$). Figure 1.4 shows case (ii). Here, h becomes small at two nonzero values of x ($n = 1.1$). Figure 1.5 shows case (v), in which a singularity occurs at the boundary of the computational domain ($n = 2.5$). Finally, Figure 1.6 shows case (iii). Here, h remains positive for all time but develops a zero at infinite time ($n = 2.0$). The zero actually develops in the entire interval $[-x_c, x_c]$. Case (iv) is a special case of (iii). Here the fact that $p = 2$ causes the singularity to be symmetric about the minimum,

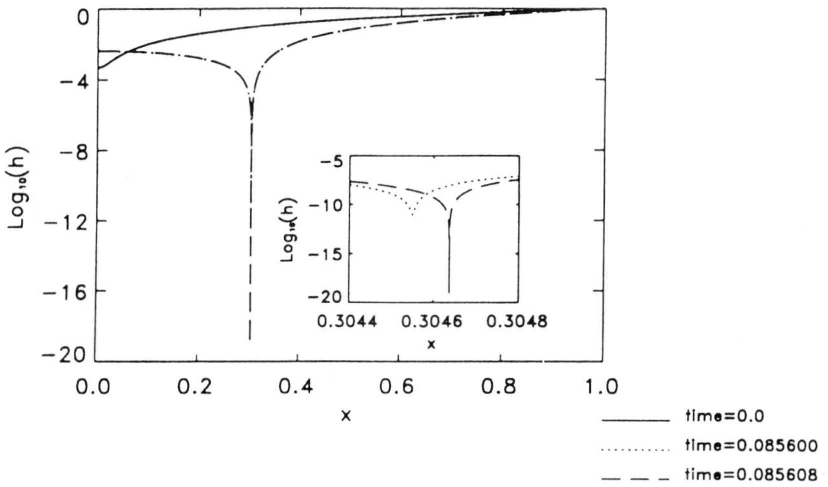

FIGURE 1.4. Finite time singularity at $\pm x_0$ for $n = 1.1$. We show a log-linear plot of $h(x,t)$ versus x for successive times before the singularity. Initial conditions and boundary conditions for this solution are detailed in Section 6.3. The inset shows a blow-up near the singularity.

since in this case $x_c = 0$.

In general, a change in the initial condition will obviously affect the nature of the solution at a later time, especially with regard to the possibility and location of singularity formation. We use the above classification because we believe that for a fixed value of n, there exists an open, possibly "large," set (in the appropriate function space) of initial conditions which lead to the same type of singularity, with many universal characteristics independent of the specific properties of the initial conditions. In our simulations, we consider a two-parameter family of initial conditions, described in Section 6.3.1. The characterizations presented in Table 1 delineate the different classes of observed singularities.

In all of the above cases, we would ideally like to *prove* that the observed behavior is possible for certain initial data and develop a detailed analytical understanding of the mathematical nature of the singularity. We lack rigorous proofs for these examples, but we can make arguments for the observed behaviors based on analytical properties of similarity solutions of the model equation. In most simulations, the region around the singularity is described by a time-independent function, apart from changes in scale;

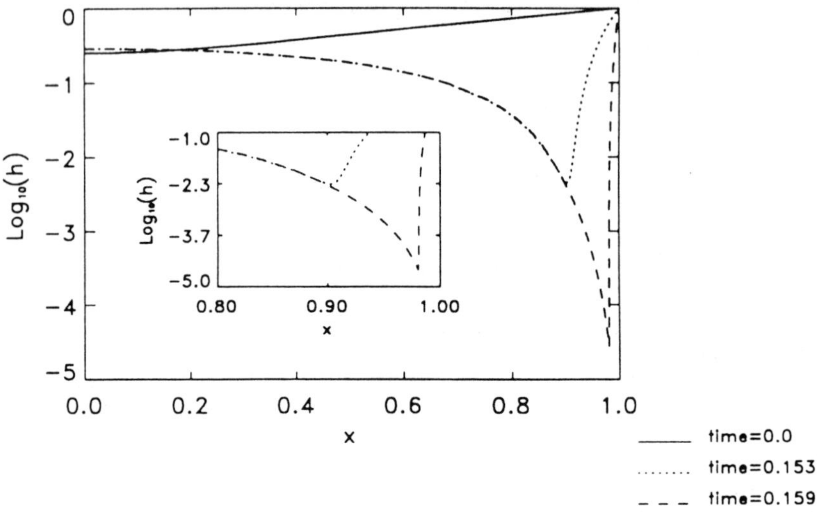

FIGURE 1.5. Finite time singularity at the edge of the simulation region for $n = 2.5$. We show a log-linear plot of $h(x,t)$ versus x for successive times before the singularity. Initial conditions and boundary conditions for this solution are detailed in Section 6.3. The inset shows a blow-up near the singularity.

that is, h has the form

$$h(x,t) = \tau(t)H(\eta), \quad \eta = \frac{x - x_p(t)}{\tau(t)^q}. \tag{1.19}$$

Here, $x_p(t)$ is the position of the minimum of h, and $\tau(t)$ is a time-dependent length scale which describes the size of the region over which the singularity occurs. A singularity happens when $\tau \to 0$. We can obtain an ODE for the shape of H by substituting (1.19) into the evolution equation and making certain assumptions about the relative size of the time-dependent terms in the equation, based on the observed behavior. We must then be able to match the similarity solution, H, at large values of η to a solution in an outer region or to boundary conditions. The matching condition puts a rather severe constraint on the types of similarity solutions that are admissible. In many of the cases described in Table 1, the matching conditions break down at specific values of n, explaining the transitions in the numerical solutions of the PDE.

In the rest of this chapter we explore the link between similarity solutions of the type (1.19) and formation of singularities in the model equation. In the next section, we discuss several classes of similarity solutions, delineated

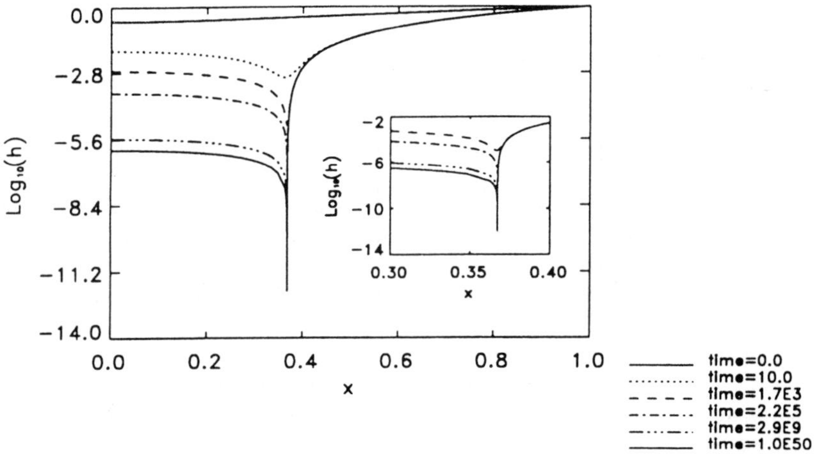

FIGURE 1.6. Infinite time singularity at $\pm x_c$ for $n = 2.0$. We show a log–linear plot of h versus x for large times. Initial conditions and boundary conditions are detailed in Section 6.3. The inset shows a blow-up near the singularity.

by the relative dominance of the different terms in the time-dependent equation. For each class, we study the admissible behavior of the functional form of H. In Section 6.3 we use the classification scheme developed in Section 2 to explain many of the observed characteristics of the numerical solutions.

6.2 The Similarity Solutions

We recall the model equation from Section 6.1,

$$h_t + (h^n h_{xxx})_x = 0. \tag{2.1}$$

We consider similarity solutions of the form

$$h(x,t) = \tau(t) H\Big(\frac{x - x_p(t)}{\tau(t)^q}\Big). \tag{2.2}$$

Solutions of this type arise in the numerical simulations both in the region around a singularity (the *pinch region*) and in other nonsingular parts of the solution. Typically, the scaling (2.2) represents the *dominant* behavior around a singularity; there are corrections to scaling which are of smaller order in $t_c - t$.

In order for a similarity solution to describe the region around a singularity, the solution must satisfy the following conditions:

(a) $\tau(t) \to 0$ as $t \to t_c$, the singular time.

(b) $H(\eta)$ is well behaved at large values of η, in order to match H to the outer solution or boundary conditions.

(c) $H(\eta) > 0$.

There are not always solutions which satisfy all of these conditions simultaneously. In particular, Theorem 1.1 indicates that solutions which satisfy (a) through (c) for finite t_c and $x_p \in (-1, 1)$ do not exist when $n > 3.5$. In this section, we analyze the types of similarity solutions of equation (2.1) and determine the values of n for which solutions satisfying the above conditions exist.

Moreover, in our simulations, for certain boundary conditions and values of n, we observe finite time singularities which do not have the simple scaling form (2.2). By determining when a self-similar singularity is *not* possible for a system that allows for finite time singularities, we hope to gain some insight into the development of such complex singularities. We address this issue in Section 6.4.

Similarity solutions are useful for describing more than just the region around a finite time singularity. We observe self-similar solutions, typically with compact support, in other parts of the solution. Another goal of this section is to analyze the existence of these "soliton" solutions. These solutions play an important role in some of the singularity mechanisms described in Section 6.3.

6.2.1 Derivation of Similarity Solutions

Substituting the form (2.2) into the basic equation (2.1) gives

$$\frac{\tau_t}{\tau}(1 - q\eta\partial_\eta)H - \frac{\dot{x}_p}{\tau^q}H_\eta + \tau^{n-4q}(H^n H_{\eta\eta\eta})_\eta = 0. \tag{2.3}$$

We classify self-similar solutions by assuming that, to leading order, τ and x_p have power law behavior in $t_c - t$ and then choosing relative sizes of $\dot{\tau}$, $\dot{x}_p$, and τ^{n-4q}. We emphasize that by "similarity solution" we refer merely to solutions which exhibit scaling. In some cases this scaling will yield an exact solution to (2.1). However, in most instances, the scaling gives an approximate solution that only solves (2.1) to leading order in a small variable (usually τ or a power of τ). We classify these different types of similarity solutions by the relative sizes of the time-dependent coefficients in equation (2.3).

q Equation

One possibility is that all the terms in (2.3) are of the same order of magnitude when η is of order unity. In that case, $x_p = \alpha\tau^q$, so that x_p can be absorbed into a redefinition of the variable η.[2] The similarity equation is

$$\lambda(1 - q\eta\partial_\eta)H = (H^n H_{\eta\eta\eta})_\eta \tag{2.4}$$

with τ given by

$$\frac{\tau_t}{\tau} = -\lambda\tau^{n-4q}. \tag{2.5}$$

The minus sign appears because τ necessarily decreases in time. There are two unknown parameters in this equation, q which defines the scaling of the solution, and α, which determines the location of the self-similar solution in x. There are two cases: If $4q > n$, then $\tau \to 0$ in finite time, as $t \to t_c$ with the power law

$$\tau \sim (t_c - t)^{\frac{1}{4q-n}}. \tag{2.6}$$

In the other case, $4q < n$, $\tau \to 0$ in infinite time

$$\tau \sim t^{\frac{-1}{n-4q}}. \tag{2.7}$$

Henceforth, the similarity equation (2.4) is called the q equation. So far, we have not found a situation in which the q equation describes the region around the minimum of h. However, solutions to the $q = 0$ equation do arise in the central region $[-x_c, x_c]$ for singularities of type (iii) of Table 1. Note that solutions to the q equation are exact self-similar solutions to (2.1).

Velocity Equation

Another possibility is that the motion or *velocity* of the singular point $x_p(t)$ dominates the first term in (2.3), that is,

$$|\tau^{q-1}\tau_t| \ll \left|\frac{dx_p(t)}{dt}\right|. \tag{2.8}$$

Then assume the $\dot{x}_p(t)$ term balances the third or "current" term in (2.3) so that

$$\frac{dx_p(t)}{dt} = a\tau^{n-3q}, \tag{2.9}$$

where a is an undetermined constant of order unity. These assumptions lead to an approximate solution to (2.3) with the self-similar form (2.2), where H satisfies

$$aH + b = H^n H_{\eta\eta\eta}. \tag{2.10}$$

Here b is a constant of integration of order unity. Since the velocity of the minimum point, $\dot{x}_p$ dominates the time derivative of the pinch region, we call equation (2.10) the velocity equation. Notice that this solution has

[2] If $q = 0$, then x_p satisfies $\dot{x}_c = \tau^n$.

two undefined "critical indi(es": We know neither the value of q nor the time dependence of τ. In general, matching conditions fixes these indices. In Section 6.3, we show that the velocity equation is relevant for understanding case (v) of Table 1, touchdown on the boundary.

Current Equation

A third possibility is that the time derivative h_t [the first two terms in (2.3)] is negligible, so that

$$\begin{aligned} |\dot{x}_p| &\ll \tau^{n-3q}, \\ |\tau_t| &\ll \tau^{n-3q}. \end{aligned} \tag{2.11}$$

In this case, H obeys

$$(H^n H_{\eta\eta\eta})_\eta = 0. \tag{2.12}$$

If the constant of integration, denoted by λ, is positive, we have

$$H^n H_{\eta\eta\eta} = \lambda. \tag{2.13}$$

This equation expresses the fact that the *current* is independent of position near the singularity. Note that again the similarity solution does not fix either q, or the time dependence of τ. In Section 6.3, we show that this similarity equation applies near the singularity in several different situations, including cases (ii) and (iii) of Table 1.

Parabolic Equation

We can also consider a case where the time dependences obey (2.11) but the constant of integration in (2.13) is zero. Then, H obeys

$$H_{\eta\eta\eta} = 0. \tag{2.14}$$

We call this the parabolic equation, for, in general, H is a quadratic function of η. In Section 6.3 we show that this solution is applicable in the central and outer regions of case (iii), and in the pinch region of case (i).

Table 2 summarizes similarity solutions we observe in the simulations. Here x_0 is the position in $[0, 1]$ where the singularity occurs. Since our simulations are symmetric about 0, we see an identical singularity at $-x_0$. Section 6.3 contains detailed analysis of the connections between the simulations and the similarity solutions.

TABLE 2. Similarity solutions seen in simulations.

location of singularity	time	observed for	pinch region equation	center region equation
$x_0 = 0$	finite	$0 \le n \le 1$	parabolic	same
$0 < x_0 < 1$	finite	$0.75 \lesssim n \lesssim 1.2$	current	not self-similar
$0 < x_0 < 1$	infinite	$0.75 \lesssim n \le 2$	current	q equation, $q = 0$
$0 < x_0 < 1$	infinite	$2 < n < \infty$	current	parabolic
boundary	finite	$n \gtrsim 2$	velocity	not self-similar

6.2.2 Qualitative Properties of the Similarity Solutions

First we investigate the properties of solutions relevant in the neighborhood of a singularity. These solutions must satisfy conditions (a)–(c) listed at the beginning of this section. In particular, we need to determine the values of n for which the current equation and the velocity equation admit solutions which satisfy conditions (b) and (c) above. To satisfy (b), the solution must be well behaved as $\eta \to \pm\infty$. To satisfy (c), it is necessary that the solution never crosses $H = 0$. We call the solution which satisfies both of these properties a global solution.

First we consider the current equation (2.13). As we will see in Section 6.3, this is the correct equation to describe both finite and infinite time singularities of types (ii) and (iii). We give a heuristic discussion of the types of solutions that are allowed. Reference [5] contains rigorous arguments. There are three ways in which a solution H to the current equation can behave as $|\eta| \to \infty$. The behaviors are

$$\begin{array}{lcl} (L)\ H & = & A|\eta| + B|\eta|^{3-n} + \ldots, \\ (Q)\ H & = & A\eta^2 + B|\eta|^{3-2n} + \ldots, \\ (S)\ H & = & A|\eta|^{\frac{3}{n+1}}, \end{array}$$

where A is positive and B and subsequent coefficients in the expansion depend on the sign of η. For the last behavior (S), A must satisfy the condition

$$A^{n+1}\frac{3}{n+1}\left(\frac{3}{n+1} - 1\right)\left(\frac{3}{n+1} - 2\right) = \mathrm{sgn}(\eta). \qquad (2.15)$$

In order for solution (L) to be valid as $|\eta| \to \infty$ we need $|\eta|^{3-n} \ll |\eta|$ for large $|\eta|$, or $n > 2$. Likewise for solution (Q) to work at large $|\eta|$, we need $n > 1/2$. For solution (S) to work, A is positive; hence, (2.15) applies for $0 < n < 1/2$ or $n > 2$ if $\eta > 0$ and $n < 0$ or $1/2 < n < 2$ if $\eta < 0$.

A similar analysis holds for the compactly supported solutions. Here it is necessary to determine which solutions are possible close to the location, η_0, a zero of H; for simplicity we assume that $\eta_0 = 0$. Again there are three possible behaviors: (L), (Q), and (S), with small η expansions identical to the large $|\eta|$ expansions displayed above. Hence, for (L) to work at small $|\eta|$, we need $|\eta| \gg |\eta|^{3-n}$ or $n < 2$. Solution (Q) works if $n < 1/2$, and (S) is possible for the values of n that guarantee the positivity of A.

In Table 3 we summarize the possible behaviors of the current equation. The first two rows summarize the possible behaviors as $H \to 0$ or $H \to \infty$. The boldface symbols **L**, **Q**, and **S** denote solutions to the ODE that are parametrically stable. The last four rows summarize the possible forms of the solution. Here, a global solution is a solution that never hits zero. A soliton is a solution for which $H(\eta) = 0$ at two values of η. Other solutions satisfy $H(\eta) = 0$ at one value of η and can be continued to a positive solution either as $\eta \to +\infty$ (zero on left) or as $\eta \to -\infty$ (zero on right).

A similar analysis holds for the velocity equation (2.10). Tables 4 through 6 summarize the results of the three possible cases $b = 0$, $b < 0$, and $b > 0$.
In Table 5, we present the possible behaviors for the velocity equation with $b > 0$. In Table 6, we present the possible behavior of the velocity equation with $b < 0$.

TABLE 3. Current equation (2.13) behavior.

	$0 \leq n \leq 1/2$	$1/2 < n < 2$	$2 \leq n$
possible zeros	S ($\eta > 0$),**L**,Q	S ($\eta < 0$),**L**	**S** ($\eta > 0$)
possible infinities	**S** ($\eta > 0$)	S ($\eta < 0$),**Q**	S ($\eta > 0$),L,**Q**
soliton	possible	possible	not possible
global	not possible	possible	possible
zero only on left	possible	possible	possible
zero only on right	not possible	possible	not possible

TABLE 4. Velocity equation (2.10) with $b = 0$.

	$0 < n < 3/2$	$3/ < n < 3$	$3 \leq n$
possible zeros	S ($\eta > 0$),**L**,Q	S ($\eta < 0$),**L**	**S** ($\eta > 0$)
possible infinities	**S** ($\eta > 0$)	S ($\eta < 0$),**Q**	S ($\eta > 0$),L,**Q**
soliton	possible	possible	not possible
global	not possible	possible	possible
zero only on left	possible	possible	possible
zero only on right	not possible	possible	not possible

6.3 Simulation Results Compared with Similarity Solutions

6.3.1 Numerical Method

Our numerical scheme is an adaptation of a code described previously in [11]. Thus, some of the language here comes, with permission of the authors, from this source. The simulations use a conventional finite-difference method. The code is an implicit, two-level scheme based on central differences. We also use a dynamically adaptive mesh composed of a fixed macrogrid and adaptive microgrid for higher resolution of singularities. In certain instances, we use a multilevel microgrid for extremely fine resolution of singularities. The finite-difference scheme is essentially identical to the scheme used in [8, 11, 28]. They compared their results to simulation results obtained from a finite element method and found excellent agreement. The new features in the code are the incorporation of "current" boundary conditions (1.13), and a dynamically adaptive multilevel mesh for resolution of moving singularities.

TABLE 5. Velocity equation (2.10) with $b > 0$.

$0 < n < 1/2$	$1/2 < n < 3/2$	$3/2 < n < 2$	$2 < n < 3$	$3 \leq n$	
possible zeros	S($\eta > 0$),**L**,Q	**L**,S ($\eta < 0$)	S ($\eta < 0$),**L**	S ($\eta < 0$),**L**	**S** ($\eta > 0$)
possible infinities	**S**($\eta > 0$)	**S** ($\eta > 0$)	S ($\eta < 0$)**Q**	S ($\eta < 0$),**Q**	S ($\eta > 0$),L,**Q**
soliton	possible	possible	possible	possible	not possible
global	not possible	not possible	possible	possible	possible
zero only on left	possible	possible	possible	possible	possible
zero only on right	not possible	possible	possible	possible	not possible

TABLE 6. Velocity equation (2.10) with $b < 0$.

	$0 \leq n < 1/2$	$1/2 < n < 3/2$	$3/2 < n < 2$	$2 < n < 3$	$3 \leq n$
possible zeros	S($\eta < 0$),**L**,Q	**L**,S ($\eta > 0$)	S ($\eta > 0$),**L**	S ($\eta > 0$),**L**	**S** ($\eta < 0$)
possible infinities	**S** ($\eta > 0$)	**S** ($\eta > 0$)	S ($\eta < 0$)**Q**	S ($\eta < 0$),**Q**	S ($\eta > 0$),L,**Q**
soliton	possible	possible	possible	possible	not possible
global	not possible	not possible	possible	possible	possible
zero only on left	not possible	not possible	possible	possible	not possible
zero only on right	possible	possible	possible	possible	possible

We consider solutions to (1.11) that are symmetric about $x = 0$. Given an initial condition satisfying $h(x, 0) = h(-x, 0)$, the solution retains this symmetry. Thus, we can solve the equation on the interval $[0, 1]$, discretized by the N mesh points,

$$0 = x_1 < x_2 < \cdots < x_N = 1.$$

At each computational time, the arrays h_i and p_i, $i \in [1, ..., N]$, approximate $h(x,t)$ and $-h_{xx}(x,t)$, and v_j, $j \in [1, ..., N-1]$ approximates $h_{xxx}(x,t)$. The h_i and p_i values exist at the point x_i, and v_i is the computed third derivative at the center of the interval, $(x_i + x_{i+1})/2$. The following picture depicts these associations:

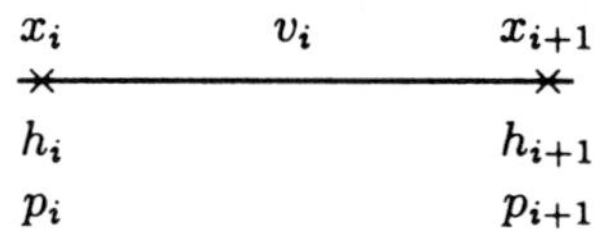

We use the notation

$$\begin{aligned}
\Delta x_{i+1/2} &= x_{i+1} - x_i, \\
x_{i+1/2} &= \tfrac{1}{2}(x_{i+1} + x_i), \\
\Delta x_i &= x_{i+1/2} - x_{i-1/2}, \\
h_{i+1/2} &= \tfrac{1}{2}(h_{i+1} + h_i), \\
\partial h_{i+1/2} &= (h_{i+1} - h_i)/\Delta x_{i+1/2}, \\
\delta^2 h_i &= (\partial h_{i+1/2} - \partial h_{i-1/2})/\Delta x_i.
\end{aligned} \tag{3.1}$$

For simplicity we describe the difference scheme in space first and later indicate the time step process. We replace the equation in (1.11) by:

$$(h_i)_t + (h^n_{i+1/2} v_i - h^n_{i-1/2} v_{i-1})/\Delta x_i = 0, \tag{3.2}$$

$$v_i + \partial p_{i+1/2} = 0, \tag{3.3}$$

$$p_i + \delta^2 h_i = 0. \tag{3.4}$$

We impose the "pressure" boundary conditions by setting $h_N = 1$ and $p_N = -p$ and using the symmetry at $x = 0$. We impose the "current" boundary conditions by setting $v_N = c$, which actually fixes h_{xxx} half of a mesh point away from the boundary instead of on the boundary. This can potentially lead to numerical errors when the solution touches down on the boundary. We cut down the error by dynamically refining the mesh on the boundary as the singularity progresses. This adjustment is both effective and computationally inexpensive.

The time discretization of the above set of differential-algebraic relations uses a simple two-level scheme. In advancing from time t to time $t + dt$ we replace the time derivative terms by difference quotients involving the solution at the old time level (time t) and the as yet unknown solution at the new time level (time $t + dt$). We evaluate the other terms using a weighted average of the solution at the two time levels; a typical weight is $\theta = 0.55$ on the advanced time level and $1 - \theta = 0.45$ on the old time level.

At each time level, the fully discrete system is a set of nonlinear equations which we solve using Newton's method. If one chooses an appropriate order for the computational unknowns, the Jacobi matrix has all of its nonzero entries very close to the diagonal. Thus, the solution of the linear equations in Newton's method is not a prohibitive expense.

We dynamically choose the time steps to control several aspects of the simulation. If the result of the time step violates any of a list of constraints, it rejects the step and tries again with a smaller step size. To avoid using unnecessarily small time step sizes, if we easily meet all the constraints for several steps, we increase the step size by about 20% on the next step. The first constraint comes from local time truncation. Another constraint rejects a step for which the minimum of h decreases by more than 10%. We also assure that the correction on the first iteration of Newton's method is a very small fraction of the change over the step, where the initial guess at the change was the change over the previous step, corrected for any difference in dt's. This last constraint allows us to solve the equation using only one Newton iteration per time step.

We use highly graded spatial grids that are very fine near the singular points and less fine in other parts of the region. At any given time, the mesh has locally constant Δx_i's that increase or decrease by a factor of 2 at any point where they change. (In fact, all the Δx values are negative powers of 2.) The location of the fine grid moves as the solution evolves based on a set of rules that we vary depending on the particular simulation. For example, when we wish to resolve a finite time singularity occurring at a point $x_p(t)$ that moves with some speed (for instance, n close to 1 with current boundary conditions), we choose the degree of resolution to dynamically depend on the minimum of h and the location of the resolution to depend on the location of $x_p(t)$. When either this location moves with respect to a higher-level mesh *or* when the minimum goes beyond a certain threshold, we adapt the mesh. Regardless of the particular remeshing rule, in locations where we add mesh points, we compute h at these points by cubic spline interpolation. We compute h_{xxx} by linear interpolation between nearest neighbors. When we remove mesh points, we merely keep the values of h and h_{xxx} fixed at the remaining points. After the remeshing takes place, we decrease the time step by a factor of 20 for a single step and then resume with the former step size.

We consider a two-parameter family of "initial" conditions for both sets of boundary conditions. The parameters are chosen to be experimentally

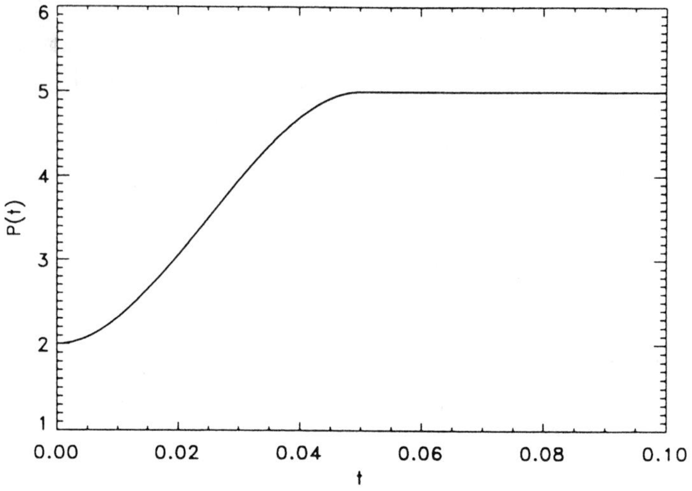

FIGURE 3.1. Typical profile of the boundary pressure as a function of time used in the pressure boundary conditions. We vary the value of the pressure at $t = 0$, and also the rate of increase.

realizable. In the case of the "pressure" boundary conditions, we start with a boundary pressure $p = 2 - \epsilon$, and choose the initial data to be the static parabola corresponding to this pressure

$$h(x, 0) = \frac{\epsilon}{2} + \frac{(2-\epsilon)}{2}x^2. \tag{3.5}$$

From this initial state we increase the pressure to some fixed value p. We can then consider the "initial condition" to be the state at the instant $h_{xx}(\pm 1)$ reaches its final fixed value. Figure 3.1 shows a typical graph of the boundary pressure as a function of time. We usually set the maximum p to be 5. The initial conditions thus have essentially two parameters: ϵ and the rate r at which the pressure on the boundary increases. Both of these parameters are relevant; varying either can influence the type of singularity produced.

For the current boundary conditions, we prepare the initial state by starting out with a static parabola at a pressure $p = 2 - \epsilon$ and a boundary current of zero. We then gradually increase the boundary current to its constant final value.

6.3.2 Infinite Time Singularities

Throughout this section and the rest of this chapter we use $x_p(t)$ to denote the (time-dependent) location of a minimum in h. We use x_0 [$= \lim_{t\to t_c} x_p(t)$] to denote, in general, the static location of the point where $h(x,t)$ goes to zero, and x_c denotes this location for the special case of infinite time singularities.

For $p < 2$, the constant pressure (1.12) boundary condition need not lead to a singularity. Instead, as t goes to infinity, $h(x,t)$ may approach the parabolic solution

$$h_\infty(x) = 1 - \frac{p}{2} + \frac{p}{2}x^2 \quad \text{(parabolic solution)}. \tag{3.6}$$

Expression (3.6) is the lowest-energy non-negative solution. [Recall from Section 6.1 that the "energy" for these boundary conditions is $\int \varphi_x^2 dx$, $\varphi = h(x,t) - h_\infty(x)$]. However, for $p > 2$, the parabolic solution (3.6) goes through zero. Instead, the lowest-energy function is the weak solution

$$w_\infty(x) = \begin{cases} \frac{p}{2}(|x| - x_c)^2 & \text{for } |x| > x_c \\ 0 & \text{otherwise} \end{cases} \tag{3.7}$$

with $x_c = 1 - \sqrt{2/p}$. This solution is not a classical solution to the equation as there is a discontinuity in $(w_\infty)_{xx}$ at x_c. If a finite time singularity does not intervene, the numerical solutions converge to the lowest-energy non-negative function in infinite time. When $p < 2$, h_∞ is strictly positive, so that as $h(x,t) \to h_\infty$, there is no singularity. However, if $p \geq 2$, $w_\infty(x) = 0$ for $x \in [-x_c, x_c]$, so that as $t \to \infty$ the solution must go to zero on this interval. In this particular case, we must also have a singularity in h_{xxx} since $(w_\infty)_{xx}$ has a jump discontinuity at $\pm x_c$.

We have initial conditions which lead to such infinite time singularities for all $n \gtrsim 0.75$. These initial conditions typically have $(r, \epsilon) = (100, 1)$ (see the previous section for a discussion of the two-parameter family of initial conditions), although there is some variation as a function of n. For $n < 0.75$, we have not found an initial conditions which leads to such a solution. In all cases, a finite time singularity intervenes. We conjecture that infinite time singularities exist for $n > 0.5$, although such solutions become much harder to find as $n \to 0.5$.

The infinite time singularities have the following generic structure: At large times, $h(x,t)$ develops two minima near $x = \pm x_c$. As $t \to \infty$, $h(\pm x_c, t) \to 0$ and $h_{xxx}(\pm x_c, t) \to \infty$. We call the region near $\pm x_c$ the "pinch region." The "central region" $[-x_c, x_c]$ contains a local maximum in h at $x = 0$. The height at the center (and in the entire central region) approaches zero as $t \to \infty$. The rate of decrease of the central region is slower than that in the pinch region. Finally, the outer region comprises $[x_c, 1]$ and $[-x_c, -1]$. Here, the solution approaches the static parabolas $\frac{p}{2}(|x| - x_c)^2$.

We now present self-similar solutions which accurately describe the approach to zero in the central region and the pinch region. Constantin et al. [8] carried out this analysis in great detail for the case $n = 1$; we extend their results to general n. In the central region, h approaches a similarity solution of the form $h(x,t) = h_0(t)C(x)$, where $C(\pm x_c) = 0$. There are two possible solutions of this type. The first possibility is that h is a solution to the q equation of Section 6.2. The only q equation that can describe the central region has $q = 0$, for when $q \neq 0$, the solutions do not have fixed support. The $q = 0$ solution satisfies

$$h_0(t) = \frac{\lambda}{n} t^{-1/n}$$

with

$$\lambda C = (C^n C_{xxx})_x \tag{3.8}$$

and

$$C(\pm x_c) = 0.$$

In the language of Section 6.2, C must be a soliton solution. As shown there, this type of solution only exists for $n \leq 2$. An alternative in the central region is the parabolic solution

$$h(x,t) = h_0(t)\left(1 - \frac{x^2}{x_c^2}\right). \tag{3.9}$$

This type of solution potentially applies to all values of n. However, it will turn out that it is only possible to match this central region solution onto the pinch region for $n > 2$. Note that the time dependence $h_0(t)$ is not fixed in this case but is determined by the matching. In both cases, $h_0(t)$ goes to zero as t goes to infinity. The solution thus asymptotically approaches the weak solution $w_\infty(x)$ in the central region.

Before proceeding to the pinch region, we check that the numerical solutions agree with the central region solutions. In Figure 3.2 we plot $h(x,t)/h(0,t)$ versus x for $n = 0.9$. We show data for five different times. The above theory predicts that the data should collapse onto a single curve. The solid line in the figure is a solution to equation (3.8) with the initial conditions $C(0) = 1$, $C_\eta(0) = 0$, $C_{\eta\eta}(0) = -2.95$, and $\lambda = 8.40$. The agreement is excellent. In Figure 3.3 we show a similar plot for $n = 3$. Again the data collapse and agree with the parabola (3.9).

Now we consider the solution in the pinch region. We focus on the pinch region near $+x_c$. (The same analysis holds near $-x_c$.) Here, we argue that h approaches a similarity solution

$$h(x,t) = h_{min}(t)H(\eta), \qquad \eta = \frac{x - x_c}{\xi(t)}, \tag{3.10}$$

where H obeys the current equation. The solution matches onto the central region as $\eta \to -\infty$ and onto the outer region as $\eta \to \infty$. The match onto

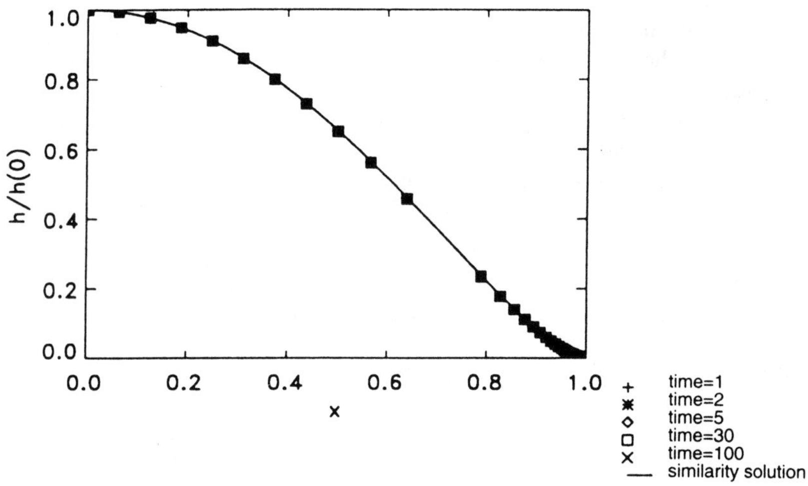

FIGURE 3.2. Rescaled profiles for the central region for $n = 0.9$. Each different symbol represents a numerical solution for a different time. The solid line is a solution to the similarity equation (3.8), computed using the conditions $C(0) = 1$, $C_\eta(0) = 0$,$C_{\eta\eta}(0) = -2.95$, $C_{\eta\eta\eta}(0) = 0$, and $\lambda = 8.40$.

the outer region requires $H(\eta) \sim A\eta^2$ at large η. The results of Section 6.2 imply this behavior is possible for $n > 1/2$ (see Table 3). Furthermore, since h_{xx} on the boundary is time independent, we need

$$h_{min} \sim \xi^2. \tag{3.11}$$

In Figure 3.4, we analyze the behavior in the pinch region by plotting $h(x,t)/h_{min}(t)$ against $(x - x_c)/h_{min}(t)^{1/2}$ for $n = 0.9$ for five different times. The similarity solution predicts that the data for different times should collapse onto a single curve. Indeed, the data collapse quite well. The solid line is a solution to the current equation with the initial conditions $H(0) = 1$, $H_\eta(0) = 0$, $H_{\eta\eta}(0) = 3.05$, and $\lambda = 1.5$.

Next, we turn to the calculation of the time-dependent coefficient $\xi(t)$ [or, alternatively, $h_{min}(t)$]. The current J from the central region, given by

$$J = \frac{d}{dt}\int_0^{x_c} h(x,t)dx,$$

fixes the time dependence of $\xi(t)$. On the other hand, the pinch region solution implies that

$$J \sim \xi^{2n-1}.$$

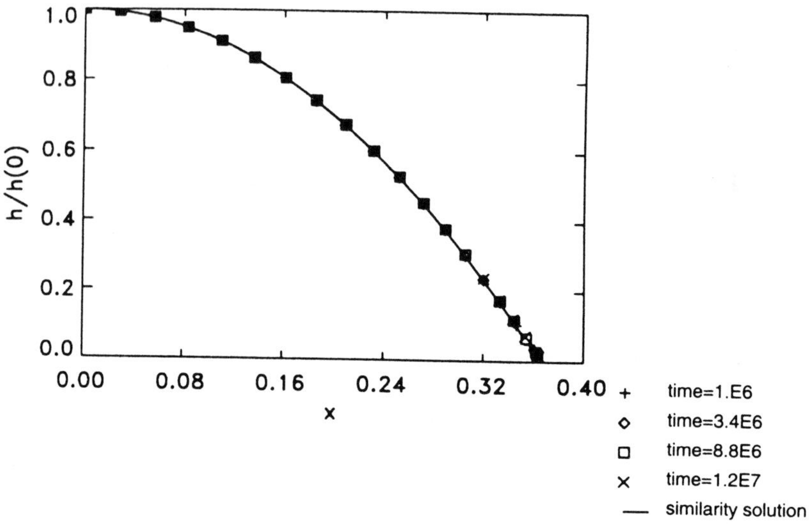

FIGURE 3.3. Rescaled profiles for the central region for $n = 3.0$. Each different symbol represents a numerical solution for a different time. The solid line is a solution to the similarity equation (3.9).

Combining these two results yields

$$h_0(t)^n \sim t\xi^{2n-1}. \tag{3.12}$$

Thus, $h_0(t)$ determines $\xi(t)$. For $1/2 < n < 2$, (3.8) gives

$$h_0 \sim t^{-1/n} \qquad (1/2 < n < 2).$$

For the parabolic solution, the time dependence is fixed by the matching: Near x_c, the edge of the central region, $h(x,t) \approx h_0(t)(x_c - x)$. This means that the solution in the pinch region must have the asymptotic behavior $H(\eta) \sim A\eta$ as $\eta \to -\infty$. From Table 3 of Section 6.2 this asymptotic behavior is only possible for $n > 2$. The match also requires the time dependences agree, so that

$$h_0(t) \sim \xi(t) \qquad (n > 2). \tag{3.13}$$

Combining these results, we find that $h_0(t)$ and $h_{min}(t)$ have the time dependences

$$h_0(t) \sim t^{-p(n)}, \quad h_{min}(t) \sim t^{-q(n)}. \tag{3.14}$$

For $1/2 < n < 2$, the exponents are

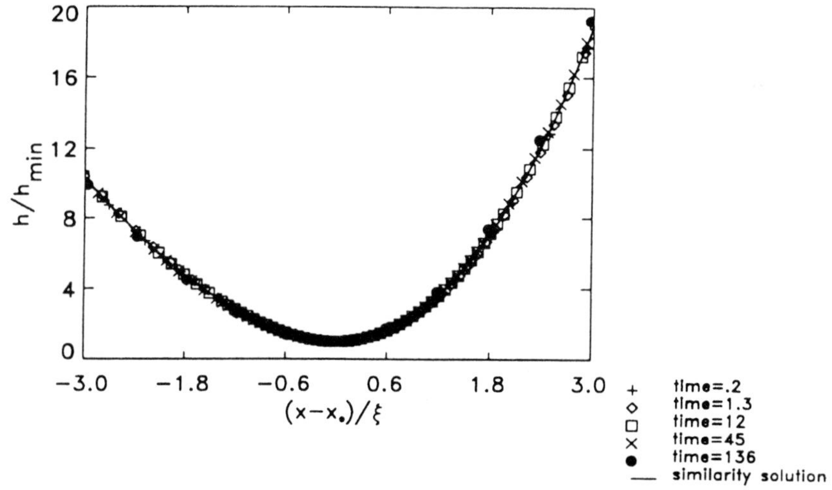

FIGURE 3.4. Rescaled profiles in the pinch region for $n = 0.9$. Each different symbol represents a numerical solution for a different time. The profiles are rescaled by $h_{min}(t)$ and $\xi(t)$, as explained in the text. The solid line is a solution to the current equation (2.13), computed using the conditions $H(0) = 1$,$H_\eta(0) = 0$, $H_{\eta\eta}(0) = 3.05$, and $\lambda = 1.5$.

$$p(n) = \frac{1}{n}, \qquad q(n) = \frac{2(1+n)}{n(2n-1)}, \tag{3.15a}$$

whereas for $n > 2$, we have

$$p(n) = \frac{1}{2(n-1)}, \qquad q(n) = \frac{1}{n-1}. \tag{3.15b}$$

Relations (3.11) and (3.15a)–(3.15b) compare very well with our numerical data. For example, in Figure 3.5 we compare log–log plots of numerical data against t for $n = 0.9$ with the slopes described by (3.15a) and find excellent agreement. In Figure 3.6 we compare log–log plots for numerical data against t for $n = 3.0$ with the slopes described in (3.15b), again with excellent agreement.

We perform similar analyses for many values of n. As a summary of these results in Figure 3.7 we show our empirically determined values of $p(n)$ (from least squares fit on the data) plotted against theoretical values for a wide range of n. In Figure 3.8 we show a similar plot for $q(n)$. No discrepancies seem to exist.

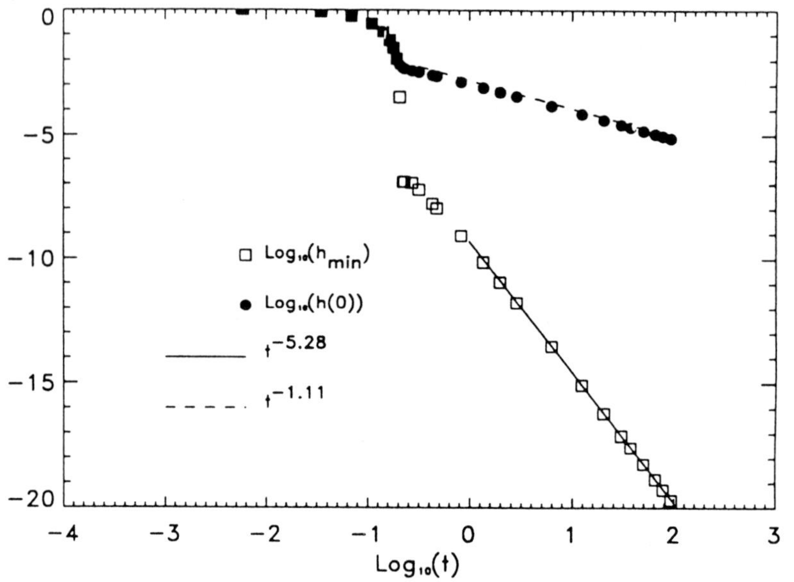

FIGURE 3.5. Time dependences of the minimum height, h_{min}, and the height $h(0,t)$ for $n = 0.9$. The solid and dashed lines are the predictions of the theory.

6.3.3 Finite Time Singularities at $\pm x_0$, $0 < |x_0| < 1$

The analysis of the preceding section is simplified by the fact that the singular point does not propagate. In this section we describe a *finite time* singularity in which the singular point does propagate. Dupont et al. [11] and Zhou [28] first discovered and analyzed this singularity in the case $n = 1$, using initial data with $(r, \epsilon) \approx (100, 1/64)$.

The singularity has the same overall structure as the previous infinite time case: There is a pinch region, a central region and an outer region. As before, we model the solution in the pinch region by a similarity solution of the form

$$h(x,t) \approx h_{min}(t)H(\eta), \qquad \eta = \frac{x - x_p(t)}{\xi(t)}, \tag{3.16}$$

where H obeys the current equation with current

$$J(t) = \frac{h_{min}(t)^{n+1}}{\xi(t)^3}, \tag{3.17}$$

and the pinch point $x_p(t)$ is a linear function of time as $t \to t_c$. However, in contrast to the infinite time singularities, the central region and outer

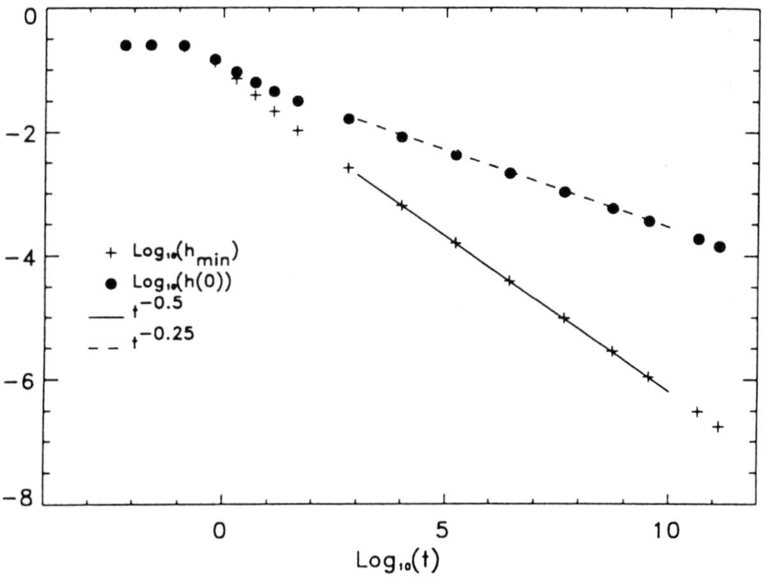

FIGURE 3.6. Time dependences of the minimum height, h_{min}, and the height $h(0,t)$ at $x = 0$ for $n = 3.0$. The solid and dashed lines show the predictions of the theory.

regions do not converge as $t \to t_c$ to simple similarity solutions. Hence, the matching of the pinch region to the rest of the solution is more complicated than in the previous case. In this section we focus only on the leading order behavior (3.16) in the pinch region.

We have numerical evidence for this finite time singularity for $0.75 \lesssim n \lesssim 1.25$ with both pressure and current boundary conditions.[3] The outer region is roughly parabolic. This suggests the pinch region scaling

$$h_{min}(t) \sim \xi(t)^2,$$

and, moreover, as $\eta \to \infty$,

$$H(\eta) \sim c\eta^2. \tag{3.18}$$

Table 3 of Section 6.2 states that this asymptotic behavior is only possible for $n > 1/2$. To illustrate the characteristics of this type of singularity we

[3]It is more difficult to find initial conditions which access this solution with the pressure boundary conditions. Here, we only find this type of singularity for $0.80 < n < 1.20$.

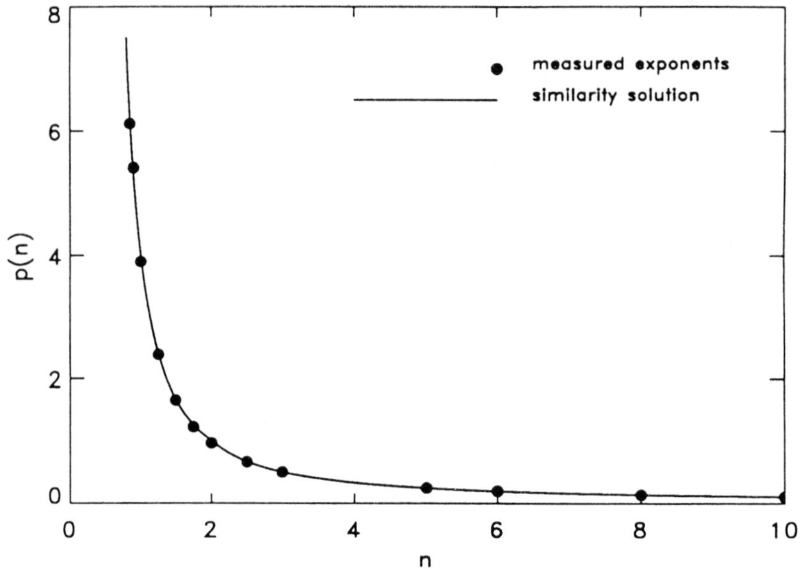

FIGURE 3.7. Exponent $p(n)$ of $h_{min} \sim t^{-p(n)}$ as a function of n. The measured exponents are compared with those of the similarity solution, equations (3.15a) and (3.15b). Error bars, as determined by the least squares fit, are typically ± 0.01. There is also a significant error which depends on the rate of convergence of the similarity solution.

show numerical results for the case $n = 1.1$. Figure 3.9 shows $h_{min}(t)$ as a function of $t_c - t$ and indicates that the singularity occurs in finite time. To illustrate that the pinch region obeys (3.18), in Figure 3.10 we show a plot of h_{xx} versus x at several different times close to the singular time for $n = 1.1$. Indeed, on the edges of the pinch region, h_{xx} approaches a constant value. Figure 3.11 shows a plot of $h(x,t)/h_{min}(t)$ as a function of $(x - x_p(t))/h_{min}^{1/2}$ for five different times, where we numerically compute $x_p(t)$ to satisfy $h(x_p(t),t) = h_{min}(t)$. The collapse of the data verifies the self-similar behavior of the solution. The solid line is a solution to the current equation with $H(0) = 1$, $H_\eta(0) = 0$, $H_{\eta\eta}(0) = 1.7$, and $\lambda = 0.5$. The agreement is excellent. We see roughly the same scaling behavior for solutions with $1.25 \gtrsim n \gtrsim 0.75$.

In order to completely understand this singularity we need to determine $h_{min}(t)$. This requires a complete match to the outer and central region. Dupont et al. [11] accomplished this for the $n = 1$ case. The corrections

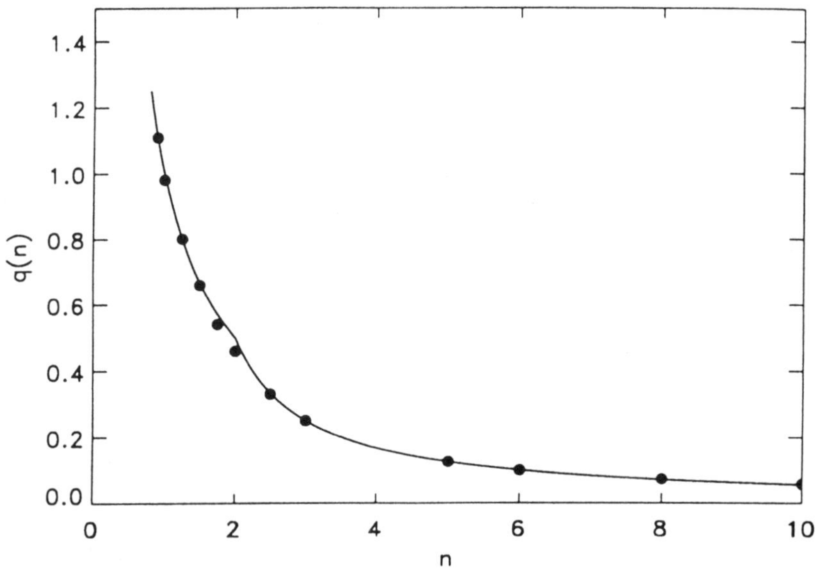

FIGURE 3.8. Exponent $q(n)$ of $h(0) \sim t^{-q(n)}$ as a function of n. The measured exponents are compared with those of the similarity solution, equations (3.15a) and (3.15b).

involve logarithmic terms and predict that for $n = 1$,

$$J(t) \sim (t_c - t)/|\ln(t_c - t)|. \tag{3.19}$$

This agrees quite well with the $n = 1$ numerical data. However, we do not know how to extend the result (3.19) beyond the special case $n = 1$. So far, we do not find completely convincing results from either the numerics or the analytics.

6.3.4 Edge Singularities

For large values of n, finite time singularities do not occur in the interior of the spatial domain.[4] However, one can force a finite time singularity at any value of n by using the current boundary conditions (1.13) which specify the constant rate at which $\int h$ decreases. In this specific case, h

[4]This fact is forced upon us for $n > 3.5$ by the theorems of Section 6.1 (see, in particular, Theorem 1.2 of Section 6.1).

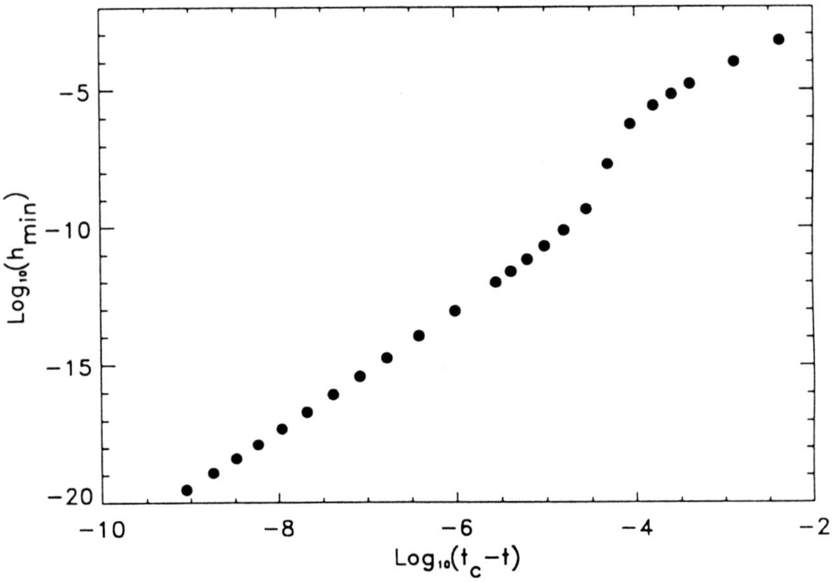

FIGURE 3.9. Time dependence of asymmetrical finite time singularities for $n = 1.1$. We show $\log(h_{min})$ versus $\log(t_c - t)$.

must go to zero in finite time, causing a singularity. For large n, the system forms singularities at the edge of the domain ($x = \pm 1$), as in the case of Figure 1.5. In our simulations, we observe that singularities form on the edge of the computational domain for $n > 2$. For $n < 1.5$, we never observe such an edge singularity. Our simulations are inconclusive as to whether the singularity forms in the interior or the edge for $1.5 < n < 2$.

The $n > 2$ edge singularities have a characteristic form. The minima of h, at $\pm x_p(t)$, progress to the boundary. For simplicity, we again consider the side close to the $x = 1$ boundary. Near the boundary but far from $x_p(t)$, $h(x,t)$ is a parabola:

$$h(x,t) = \frac{(x - x_p(t))^2}{(1 - x_p(t))^2} \quad \text{for} \quad 0 < \frac{x - x_p(t)}{1 - x_p(t)} \sim 1. \tag{3.20}$$

The current in the pinch [near $x_p(t)$] is quite small, and in fact goes to zero as $t \to t_c$. Hence, the current at the boundary, which is fixed by the boundary condition, controls the flow out of the region $[x_p, 1]$. For $t < t_c$,

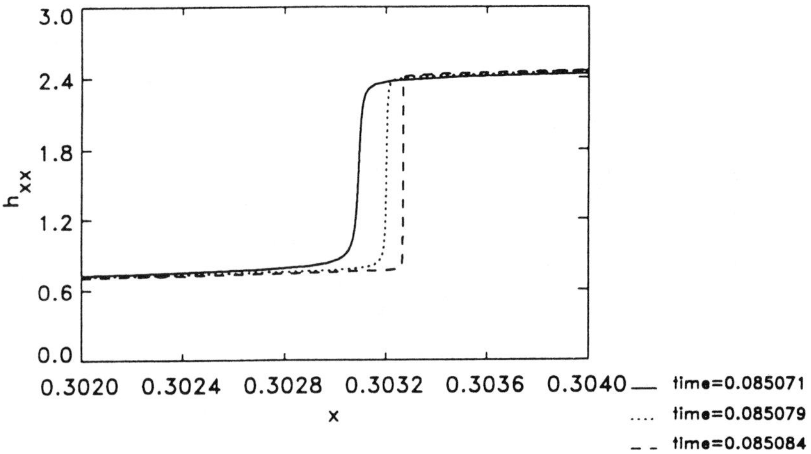

FIGURE 3.10. Plot of h_{xx} versus x in the region around the singularity near the singular time for $n = 1.1$.

the constant current boundary condition fixes the time derivative of x_p as

$$J(x=1,t) \approx -\partial_t \int_{x_p}^{1} h(x,t) \sim \frac{dx_p(t)}{dt}, \tag{3.21}$$

using equation (3.20). Since $J(1,t) \sim 1$ we find $x_p(t) \sim 1 - c(t_c - t)$.

In the pinch region, h is described by a similarity solution

$$h(x,t) = \xi^s H(\eta), \quad \eta = \frac{x - x_p(t)}{\xi(t)}. \tag{3.22}$$

Here, s is an as yet undetermined parameter. Since the time derivative of ξ^s is small compared with the change in $h(x,t)$ due to the propagation of the singular point, H obeys the velocity equation of Section 6.2 and hence satisfies

$$aH + b = H^n H_{\eta\eta\eta}. \tag{3.23}$$

Also ξ satisfies

$$\xi^{sn-3} \sim \frac{dx_p(t)}{dt} \sim \text{constant} \tag{3.24}$$

so that $s = 3/n$. In order for H to fit onto the quadratic form (3.20), we need both $H(\eta) \sim \eta^2$ as $\eta \to \infty$, and that ξ^{sn-2} be proportional to $(1 - x_p(t))^{-2}$. We thus find the time dependence

$$\xi(t) \sim (t_c - t)^{2n/(2n-3)}. \tag{3.25}$$

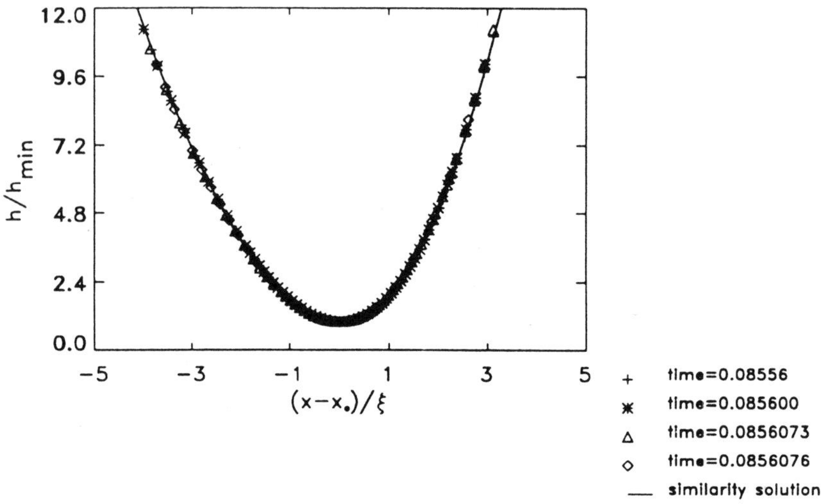

FIGURE 3.11. Rescaled profiles for the pinch region for $n = 1.1$. Each symbol represents a numerical solution at a different time. The solid line is a solution to the current equation, with $H(0) = 1$, $H_\eta(0) = 0$, and $H_{\eta\eta}(0) = 1.70$. The parameter $\lambda = 0.5$.

The matching to the outer region breaks down for $n < 3/2$, since equation (3.23) does not admit the asymptotic behavior $H \sim A\eta^2$ in this regime. This similarity solution can only describe singularity formation on the boundary for $n > 3/2$.

In order to complete the solution we must match to the central region. Recall in the case of infinite time singularities, the flux or current of fluid leaving the central region determines time dependences in the pinch region. Here the time dependence (3.25) follows from only the match to the boundary conditions. However, it is still true that the flux of fluid from the central region affects the solution. The neglect of this flux is only valid near the boundary, where the total flux is of order one [much larger than the flux from the central region, which is $O(\xi^s)$].

We verify that near the boundary, the similarity solution (3.23)–(3.24) holds. To check this we look at the specific case $n = 7.5$. We study the solution near $x^*(t)$, the maximum of h_{xxx}. The maximum x^* occurs to the right of x_p, the minimum of h. As $t \to t_c$, $h_{xxx}(x^*(t), t)$ diverges. Figure 3.12 shows the relation between $h_{xxx}(x^*(t), t)$ and $\xi(t)$. Figure 3.13 shows the dependence of $x^*(t)$ on $t_c - t$. Figure 3.14 shows the dependence of $h_{xxx}(x^*(t), t)$ on $t_c - t$. The solid line in each case shows the prediction of the similarity solution. The agreement is excellent. We also need to check

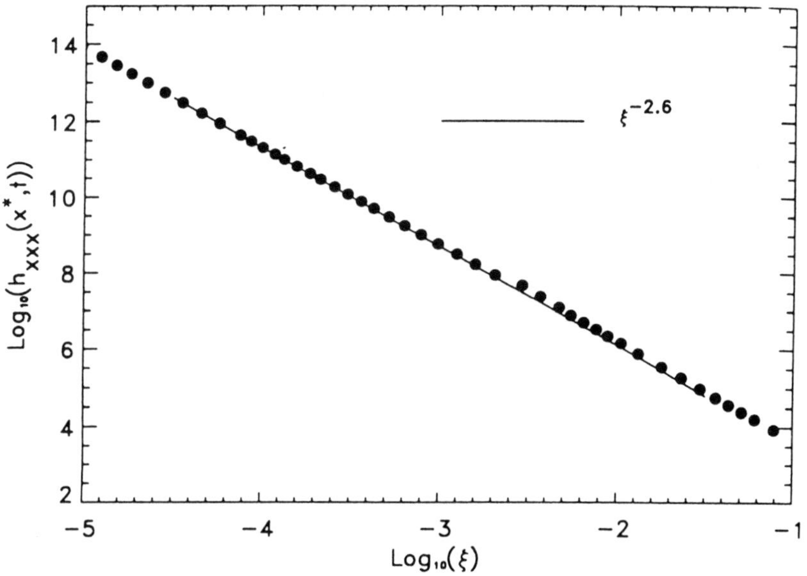

FIGURE 3.12. Dependence of $h_{xxx}(x^*,t)$ on ξ in for touchdown on the boundary, $n = 7.5$. The solid line represents the prediction of the theory.

that the functional form around the minimum is described by a solution to equation (3.23). It is *not* possible to fit the data with a single solution to the velocity equation. Upon rescaling, equation (3.23) becomes

$$H_{\eta\eta\eta} = c\frac{\text{sgn}(b) + \text{sgn}(a)H}{H^n} \tag{3.26}$$

where $\text{sgn}(a)$ denotes the sign of a. For simplicity we take $\eta = 0$ to correspond to x^*. The sign of a is necessarily positive since from above $\text{sgn}(a) = \text{sgn}(\dot{x}_p)$. The fact that $x^* \neq x_p$ in the simulations means that $\text{sgn}(b) = -\text{sgn}(a)$. It follows from this that $(H_{\eta\eta\eta})_\eta = 0$ both at the maximum of $H_{\eta\eta\eta}$ and the minimum of H. Although $(H_{\eta\eta\eta})_{\eta\eta} < 0$ at the maximum of $H_{\eta\eta\eta}$, $(H_{\eta\eta\eta})_{\eta\eta} > 0$ at the minimum of H, indicating that the similarity solution veers from the data at the minimum of H. This disagreement is a natural consequence of neglecting of the flux from the central region. As emphasized above we do not expect the similarity solution to hold for $x < x_p$.

However, a solution of (3.26) fits the data up to the minimum of h. This solution satisfies the conditions $H_{\eta\eta\eta}(0) = 1$ (an arbitrary choice) and also $H(\eta) \to 1$ as $\eta \to -\infty$. These two boundary conditions *completely*

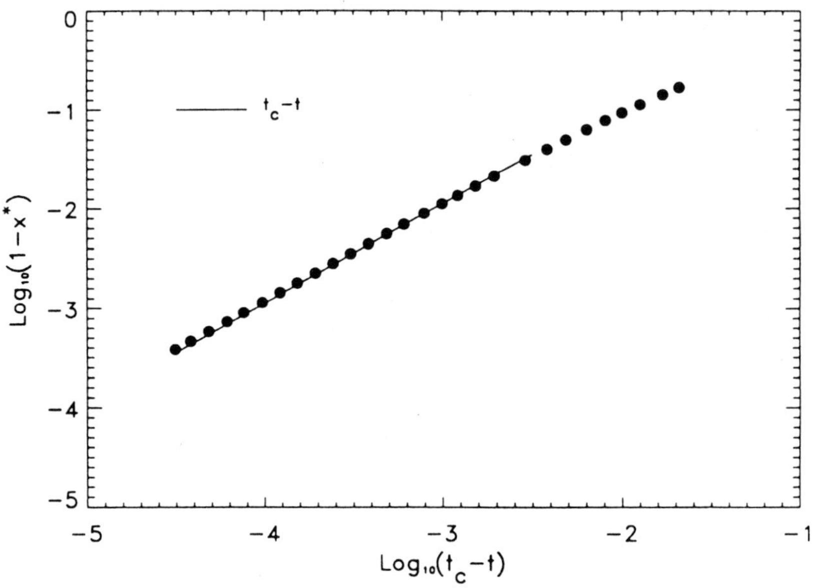

FIGURE 3.13. Dependence of $x^*(t)$ on $t_c - t$ for touchdown on the boundary, $n = 7.5$. The solid line gives the prediction of the theory.

determine the solution, for there are two exponentially growing solutions of the linearized (3.26) as $\eta \to -\infty$. The solution satisfying these conditions for $n = 7.5$ has $H(0) = 7.5/6.5$, $H_\eta(0) = 0.3542$, and $H_{\eta\eta}(0) = 0.7077$. In Figure 3.15 we compare this solution with results of a numerical simulation. In the upper half of the figure we show $h(x,t)/h_{xxx}(x^*(t))^{-1/6.5}$ versus $(x - x^*(t))/h_{xxx}(x^*(t))^{-7.5/19.5}$ for four different times. In the lower half of the figure we show $h_{xxx}(x,t)/h_{xxx}(x^*(t))$ versus $(x - x^*)/h_{xxx}(x^*(t))^{-7.5/19.5}$. The solid lines are the solutions described above. Indeed, the agreement between the similarity solution and the data is excellent up to the minimum of h, where the numerical data clearly deviate from the similarity solution. Beyond this point, the numerical data do not even collapse. Furthermore, there is an interesting dynamic structure in this region (not visible in Figure 3.15) that we defer until Section 6.4.

The similarity solution seems to agree with the numerics for a wide range of n. As an indication, in Figure 3.16 we show the scaling exponent $q(n)$ of $h_{xxx}(x^*,t) \sim \xi^{q(n)}$ as a function of n. In Figure 3.17 we show the scaling exponent $p(n)$ of $h_{xxx}(x^*,t) \sim (t_c - t)^{p(n)}$. The points represent the result of least squares fits to the data. The error in the points depends on how close the simulation is to t_c, the singular time. In each case, the solid line

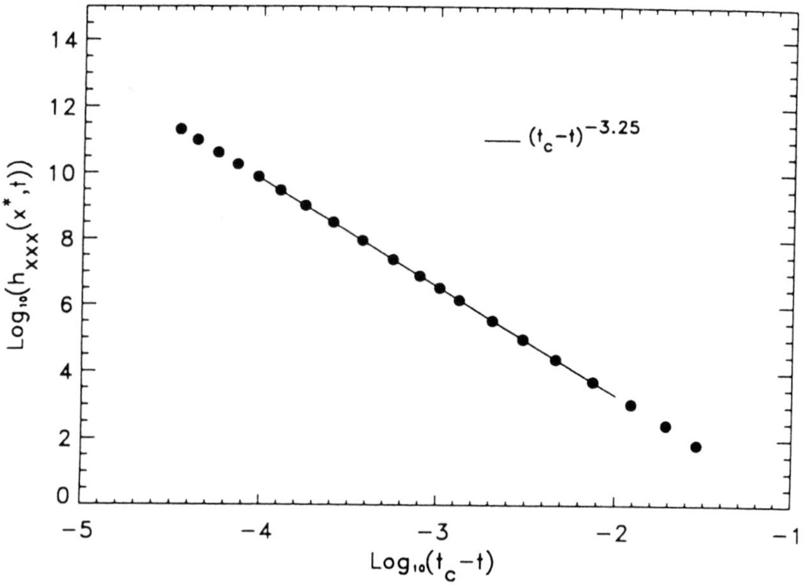

FIGURE 3.14. Dependence of $h_{xxx}(x^*,t)$ on $t_c - t$ for touchdown on the boundary with $n = 7.5$. The solid line shows the prediction of the theory.

represents the prediction of the similarity solution

$$p(n) = \frac{6(1-n)}{2n-3}, \qquad q(n) = \frac{3}{n} - 3. \tag{3.27}$$

The agreement is quite reasonable. However, we caution that Figures 3.16 and 3.17 do *not* indicate that the similarity solution describes the data down to $n = 1.5$. Without a match to the central region we cannot accurately predict the range of n over which the solution is valid. Recall that in the previous sections the crucial factor in determining when a similarity solution breaks down is the matching to the other regions. In Section 6.4 we present some interesting features of the matching region, illustrating its nontrivial nature.

6.3.5 *Finite Time Singularities at* $x_0 = 0$

All of the finite time singularity mechanisms so far considered have pinch regions which are asymmetric under reflection about $x_p(t)$. We also observe

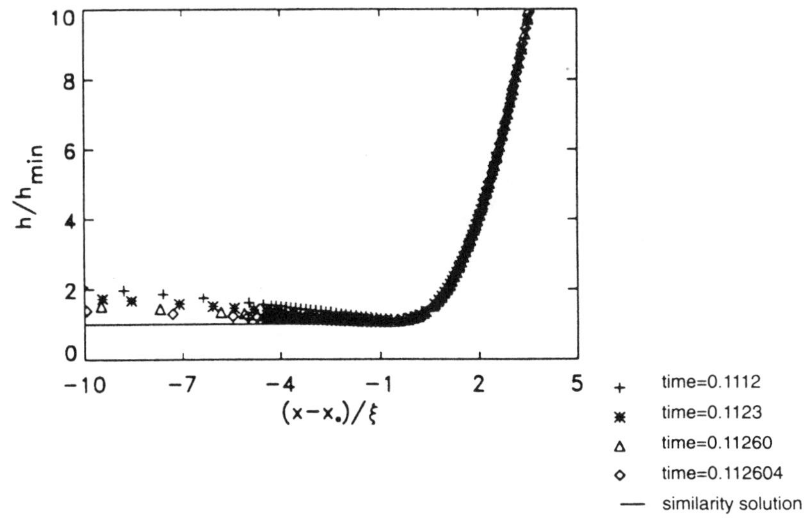

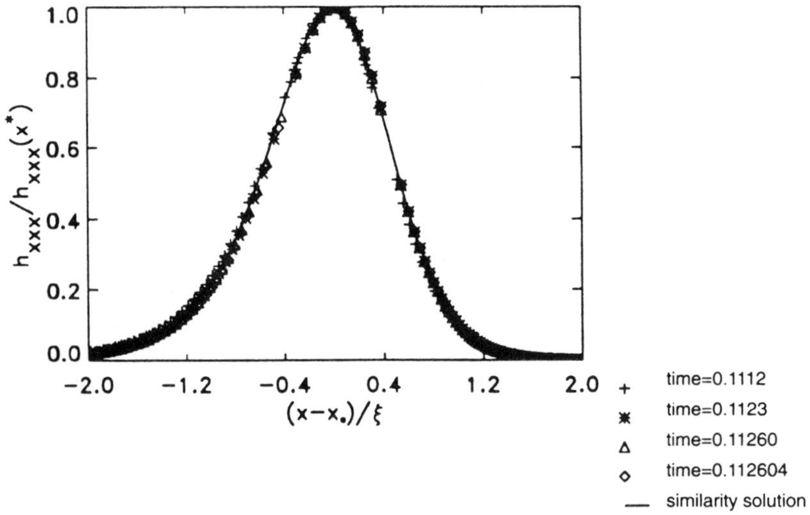

FIGURE 3.15. Rescaled profiles near the singularity for $n = 7.5$. The upper figure shows rescaled height profiles, and the lower figure shows rescaled h_{xxx} profiles. Each different symbol represents a numerical solution at a different time. The solid line is a solution to the velocity equation with initial conditions as described in the text. Notice that the scaling breaks down to the right of the minimum.

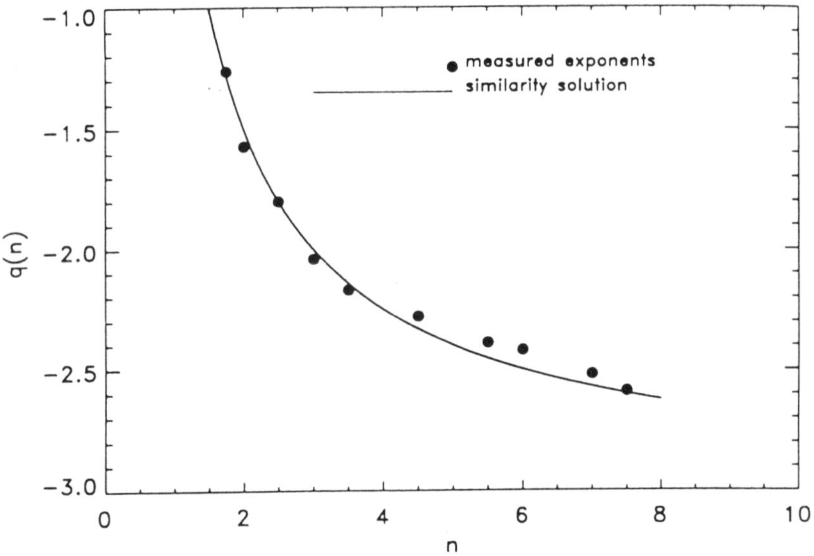

FIGURE 3.16. Exponent $q(n)$ of $h_{xxx}(x^*, t) \sim \xi^{q(n)}$ for boundary singularities with $n \geq 1.5$. The measured exponents are compared with those of the similarity solution.

finite time singularities with symmetric pinch regions for $n < 1$.[5] We also have numerical evidence for these singularities up to $n \approx 1.4$, although at present we have little theory for $n \geq 1$.

For small n ($n < 0.5$) and either "current" or "pressure" boundary conditions with $p > 2$, every initial condition we attempt gives a singularity of this type. For $n > 0.5$, some initial conditions converge to the other singularities discussed above. Typically, this singularity corresponds to choosing $\epsilon \sim 1/64$ and r very small. For each value of n and ϵ there is a critical rate r_c below which solutions generically converge to this type of singularity. For $n = 1$ and $\epsilon = 1/128$, the critical r_c is about 30. The critical rate r_c changes as a function of n.

We construct a similarity solution for these solutions with $n < 1$ as follows: To lowest order, the solution solves the parabolic equation

$$h(x,t) \approx (t_c - t) + Bx^2 \equiv H_0(x,t). \tag{3.28}$$

[5] We note that it is easy to find infinite time singularities with symmetric pinch points; these occur generically when $p = 2$.

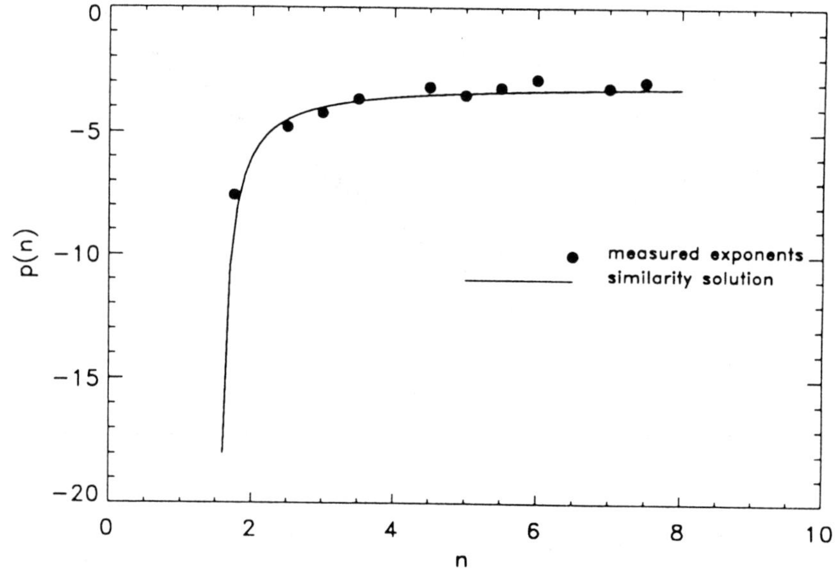

FIGURE 3.17. Exponent $p(n)$ of $h_{xxx}(x^*, t) \sim (t_c - t)^{p(n)}$ for boundary singularities with $n \geq 1.5$. The measured exponents are compared with those of the similarity solution.

Writing $h = H_0 + g$, g must satisfy

$$((H_0 + g)^n g_{xxx})_x = 1 \tag{3.29}$$

so that

$$g_{xxx} = \frac{x}{(g + H_0)^n}. \tag{3.30}$$

We formally expand equation (3.30) in powers of g. If $n < 1$, the successive terms in the expansion decrease for small $|x|$ and $|t_c - t|$.

The first-order correction to H_0 satisfies

$$g_{xxx} = \frac{x}{H_0^n}. \tag{3.31}$$

In Figure 3.18 we show the dependence of the minimum height, h_{min} on ξ, the characteristic width, for $n = 0.75$. In Figure 3.19 we show the dependence of h_{min} on $t_c - t$ for $n = 0.75$. In both cases, the solid lines represents the prediction of the similarity solution. The agreement is excellent. We also check that the the correction to H_0, which satisfies equation

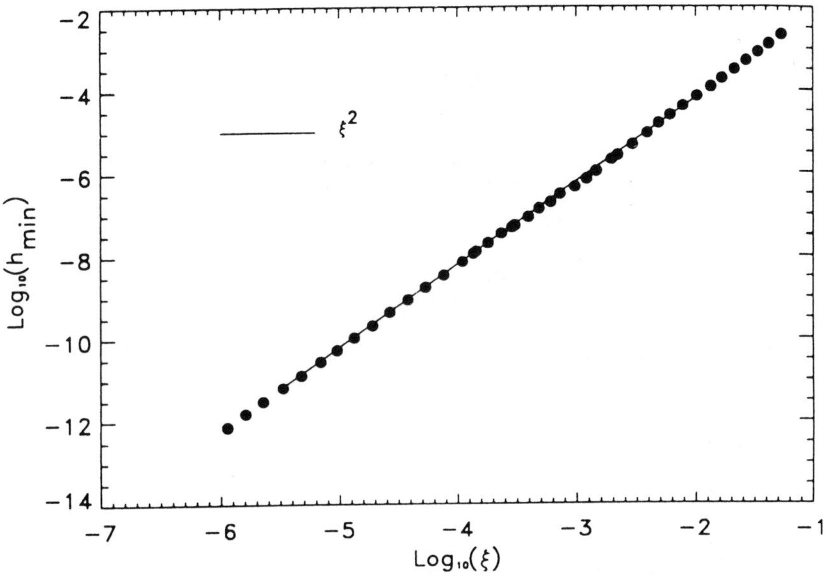

FIGURE 3.18. Dependence of h_{min} on ξ in the pinch region for finite time singularity at $n = 0.75$. We show $\log(h_{min})$ versus $\log(\xi)$. The solid line represents the prediction of the theory.

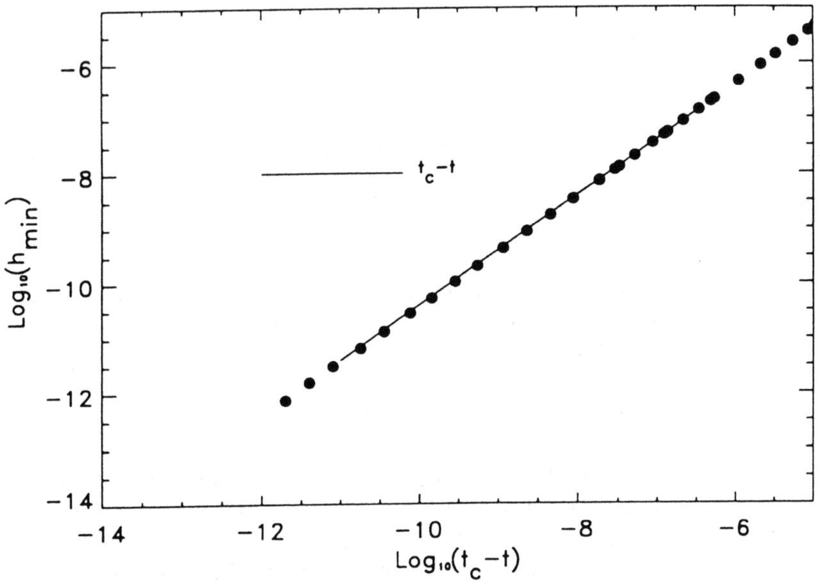

FIGURE 3.19. Time dependences for finite time singularity at $x = 0$ for $n = 0.75$. We show $\log(h_{min})$ versus $\log(t_c - t)$. The solid line represents the prediction of the theory.

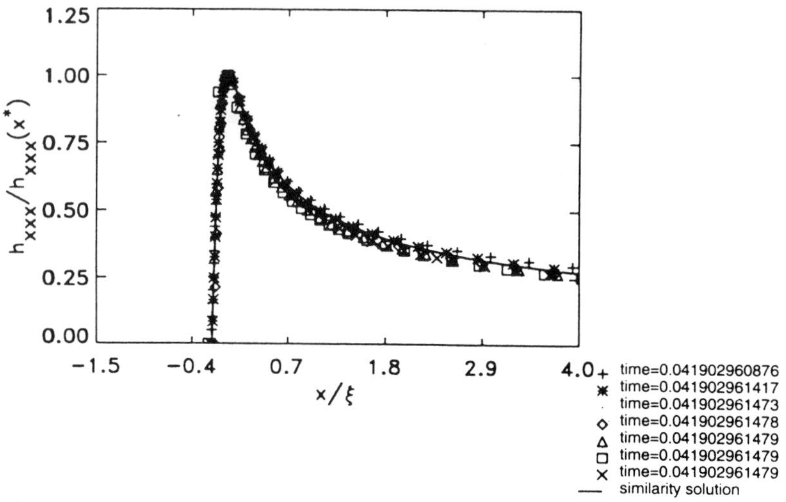

FIGURE 3.20. Rescaled h_{xxx} versus x profiles near the singularity for $n = 0.75$. Each symbol represents a numerical solution at a different time. The solid line is the first-order correction to the parabolic solution, as determined by equation (3.31).

(3.31), is present in the data. We can isolate the correction by examining h_{xxx}. For $n > 0.5$, h_{xxx} has a maximum at

$$x^* = \left(\frac{t_c - t}{B(2n-1)}\right)^{1/2}. \tag{3.32}$$

In Figure 3.20 we plot $h_{xxx}/h_{xxx}(x^*)$ versus $x/(h_{xxx}(x^*))^{-2}$ for seven different times at $n = 0.75$. The data collapse quite well. The solid line represents a solution (3.31) with $B = 69.2$ In Figure 3.21 we show a similar plot for $n = 0.25$. Here we plot $h_{xxx}(x,t)/h_{min}^{0.25}$ versus $x/h_{min}^{0.5}$ for seven different times. Again the data collapse; the agreement with the similarity solution $B = 2.5$ is excellent. We repeat this analysis for many values of n less than one, with similar agreement. As an example, in Figure 3.22 we show the exponent of $h_{min} \sim \xi^{q(n)}$ as a function of n. The solid line represents the prediction of the similarity solution, and the points are the results of simulations. In Figure 3.23 we show the exponent of $h_{min} \sim (t_c - t)^{p(n)}$ as a function of n, for both theory and simulations. Again the agreement is excellent.

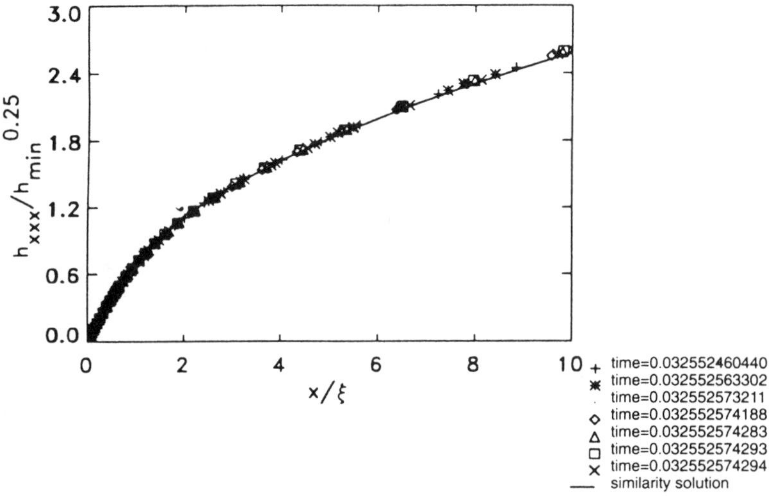

FIGURE 3.21. Rescaled h_{xxx} versus x profiles near the singularity for $n = 0.25$. Each symbol represents a numerical solution for a different time. The solid line is the first-order correction to the parabolic solution, as determined by equation (3.31).

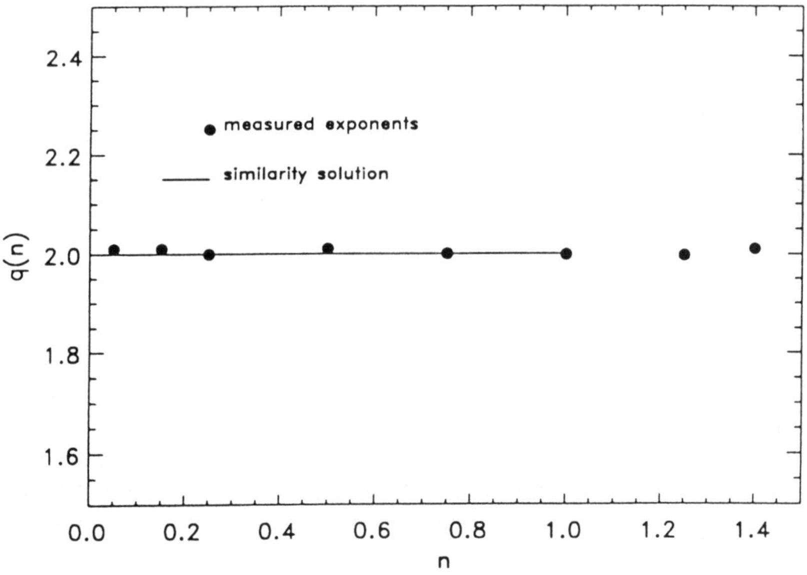

FIGURE 3.22. Exponent $q(n)$ of $h_{min} \sim \xi^{q(n)}$ for finite time touchdown at $x = 0$. The measured values are compared with the exponents of the similarity solution.

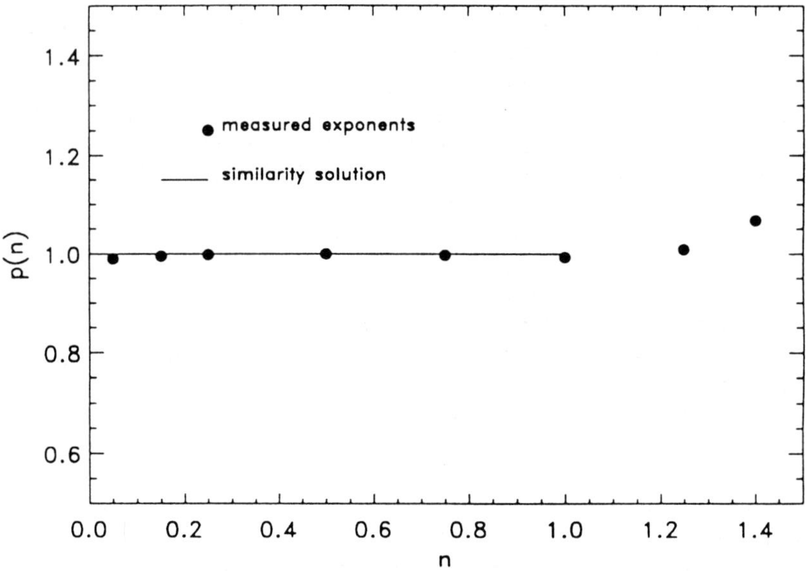

FIGURE 3.23. Exponent $p(n)$ of $h_{min} \sim (t_c - t)^{p(n)}$ for finite time touchdown at $x = 0$. The measured values are compared with the exponents of the similarity solution.

We also include in Figure 3.22 and 3.23 points for $n > 1$. Although it is possible (with a slight modification) to construct a similar expansion for $1 \leq n < 2$, such solutions do not agree with the measured exponents. Numerically we do not seem to observe this singularity all the way up to $n = 2$.

6.4 Unsolved Problems

6.4.1 Singularities and Similarity Solutions

In Section 6.3, we describe many singularities of the model equation (1.11) which exhibit self-similar structure. We compare the characteristics of these singularities to those of various similarity solutions that approximate the model equation. Our analysis is in the spirit of matched asymptotics, formulated for analyzing the solutions of ordinary differential equations [12]. In order for a similarity solution to describe the region around the singularity (the inner region) we must be able to match the solution to boundary conditions (the outer region). We discover in Section 6.3 that in many instances, the matching is sufficient to qualitatively predict transitions in the model PDE. The transitions obtained via the matching principle are consistent with the transitions observed in the numerical solutions. However, matching conditions for PDEs are much more complicated than those in traditional problems of matched asymptotics; in particular, we must include the time dependence of the solutions. We are completely successful in only two cases: the infinite time singularities and the finite time singularities at $x = 0$, for $n < 1$. These cases both have "singular" minimum points that do not propagate. In the cases where the singular points do propagate, that of edge singularities and finite time singularities at $0 < x_0 < 1$, we do not have a complete matching analysis. In both of these cases, we have a consistent theory for the leading order behavior of the singularity but we do not know how to match this behavior to the rest of the solution. A third case with incomplete analysis is that of finite time singularities at $x_0 = 0$ for $n \geq 1$.

The construction of a singular solution using matched asymptotics is important because it indicates that the numerical solution reflects properties of the PDE. Whether this solution is actually realized in practice depends on its stability. Brenner and Bertozzi [6] proved linear stability for a two-dimensional variant of the $q = -1$ solution. Their analysis also shows that the $q = -1$ solution and the parabolic solution of Section 6.3.2 are linearly stable to perturbations with support inside the support of the similarity solution. Our simulations indicate that many other similarity solutions are stable, although at present we have no proof.

In the cases where there is not a similarity solution we exercise extreme caution when interpreting numerical results on singularity formation. We must keep in mind that the simulations only track the solutions to a minimum height, typically $10^{-15} - 10^{-20}$. We cannot rule of the possibility that the numerical solutions do not converge on a particular self-similar singularity uniformly as $h \to 0$; instead a completely different type of behavior could set in at a small height beyond our resolution. When a similarity solution exists, we know that the numerical singularity is a true singularity of the PDE. However, in cases where there is no theory, simulations alone do *not* provide ample evidence for the existence of a singularity. As an indication of the subtlety of this issue, we describe a situation that arises for $n = 1.6$, with initial conditions described in Section 6.3.1. The solution initially appears to be the same type of singularity as the finite time singularities at $x = 0$. However, when the minimum height reaches below 10^{-15}, the nature of the solution changes drastically. The minimum begins to propagate toward the boundary, carrying with it an extraordinarily complicated structure. Figure 4.1 shows height profiles of the singularity at early times, and Figure 4.2 shows height profiles at slightly later times. Without rigorous theory to support the data, we can never be certain that the equation will not "fool" us in this way.

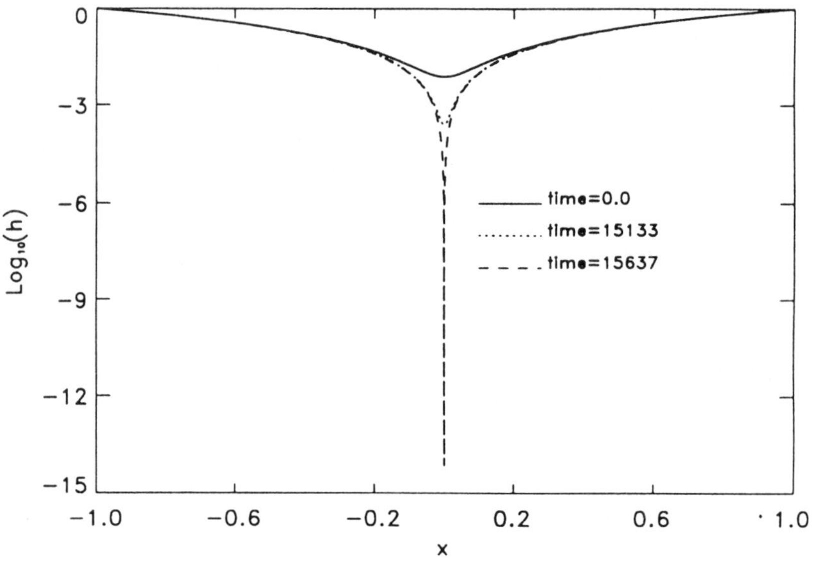

FIGURE 4.1. Early time structure of solution for $n = 1.6$. It appears that there will be a singularity at $x = 0$.

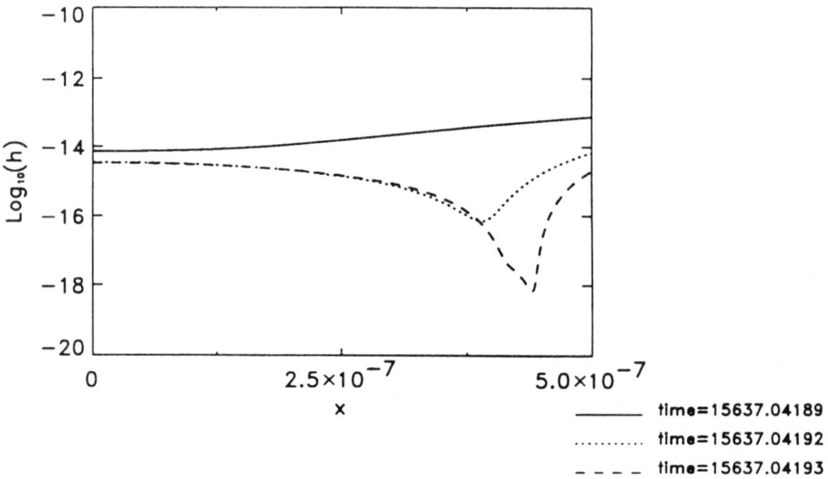

FIGURE 4.2. Blow-up of the preceding $n = 1.6$ solution at a slightly later time. The minimum is no longer at $x = 0$.

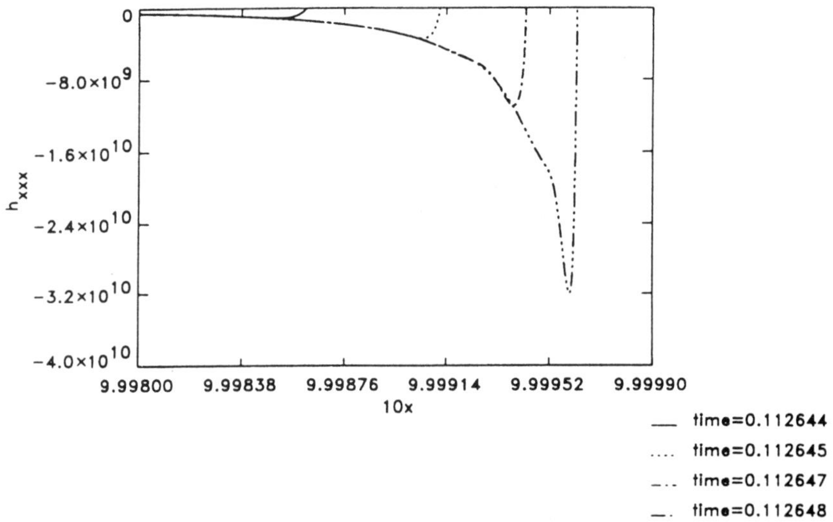

FIGURE 4.3. Numerical solutions at the foot of the maximum of the h_{xxx} profile in the case of edge singularity for $n = 7.5$.

6.4.2 Complex Singularities

This chapter focuses on similarity solutions which are extremely prevalent in the data. However, not all of the observed singularities are self-similar. As an illustration of the complex behavior, we discuss two examples: The first arises near the minimum of h in the edge singularity. As we discuss in Section 6.3, the similarity solution does not account for the small flux traveling across the pinch region from the central region. This flux causes an interesting dynamic structure near the minimum of h. Figure 4.3 shows the typical behavior we observe for higher values of n. It depicts successive profiles of h_{xxx} versus x, to the left of the minimum for $n = 7.5$. (Recall that the theory of Section 6.3 applies only to the right of the minimum.) We see the formation of a pronounced "dip," which has scaling structure with *different* exponents than the scaling theory presented in Section 6.3. This singularity thus has two different scaling regions. Values of n closer to the critical value 1.5 produce solutions which have an even more complex structure. Figure 4.4 shows the analogous profiles for $n = 2.5$. As in the $n = 7.5$ case, a "dip" begins to form. However, as depicted in Figure 4.5, at later times the profile has irregular oscillations with a frequency that increases as $t \to t_c$. We greatly resolve these oscillations to ensure that they are not produced by the numerics.

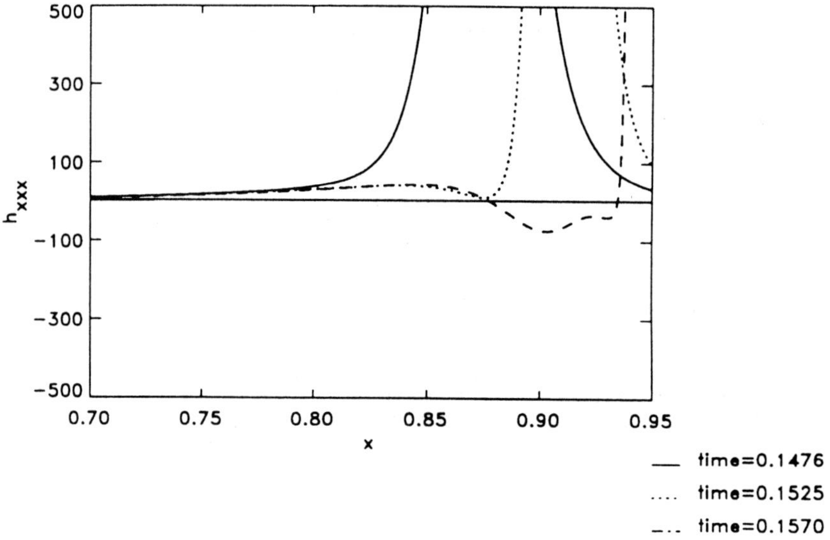

FIGURE 4.4. Numerical solution at the foot of the maximum of the h_{xxx} profile in the case of edge singularity for $n = 2.5$. Note the formation of a foot-like structure, as in the $n = 7.5$ case.

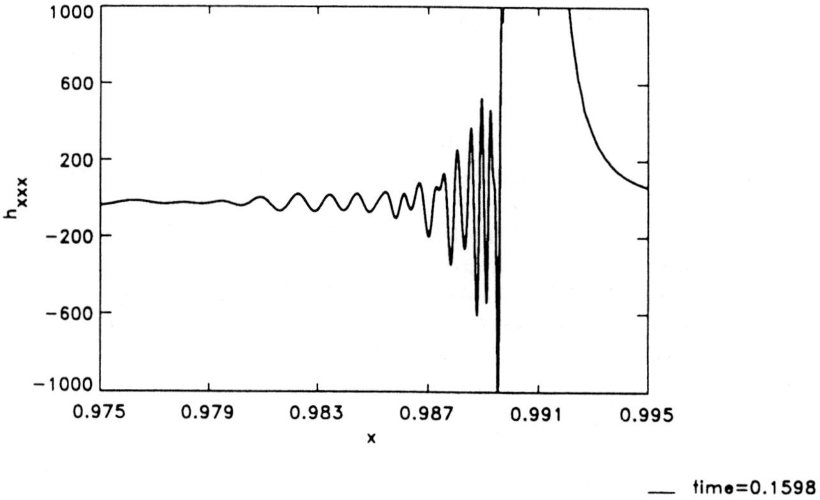

FIGURE 4.5. Numerical solution near the foot for $n = 2.5$ at a later time.

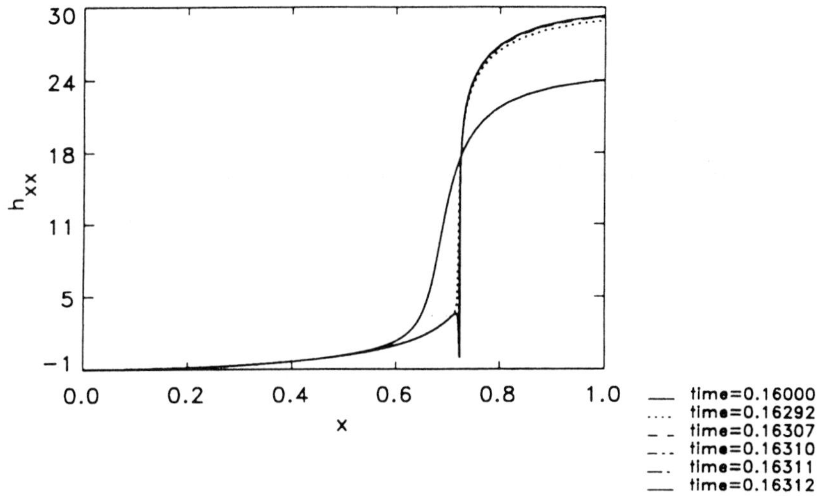

FIGURE 4.6. Successive time profiles of h_{xx} versus x for $n = 1.4$ with pumping boundary conditions. The gross feature of the plot is that as the singularity is reached, the profile is mainly monotonic with a pronounced tip that was not present for the cases $0.75 < n < 1.25$.

Another complex singularity arises generically for current boundary conditions with $1.5 > n > 1.25$. In the previous section, we argue that in the pinch region simulations with $0.75 \leq n \leq 1.25$ the current equation represents the leading order behavior for these solutions. This solution has the feature that h_{xx} is monotonic in x in the pinch region. In contrast, Figure 4.6 shows successive time profiles of h_{xx} versus x for $n = 1.4$. Here, although the global picture looks somewhat monotonic, it is actually considerably more complicated. To see this, in Figure 4.7 we blow up the area around the pinch which contains a local minimum in h_{xx} in the shape of a "tip." Blowing up the "tip" region again in Figure 4.8 we see that this is not a simple minimum, but in fact it posseses another tip, which, in turn, has yet another tip upon finer resolution. We do not know whether this type of structure persists until the singularity.

6.4.3 Mathematical Questions

A number of important mathematical questions still remain. For instance, what is the critical value of n, above which finite time singularities are impossible? Is it the same for all boundary conditions? The theorems of

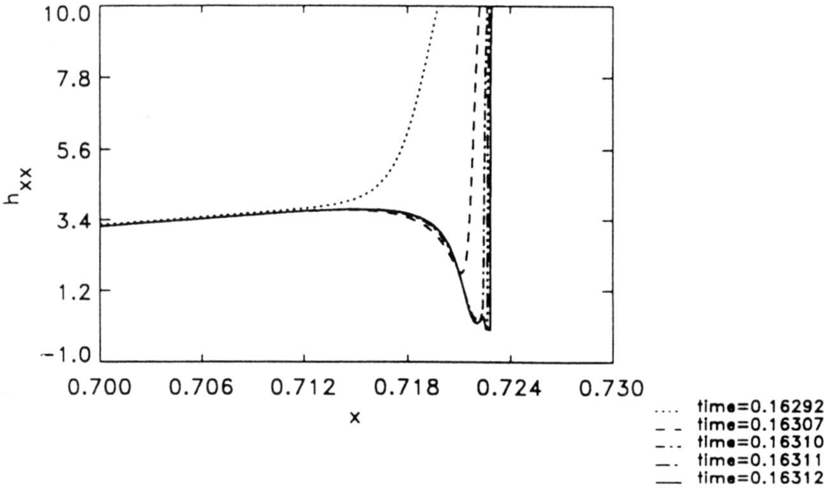

FIGURE 4.7. Blow-up of the previous figure. Notice the "tip" has the added structure of a smaller tip.

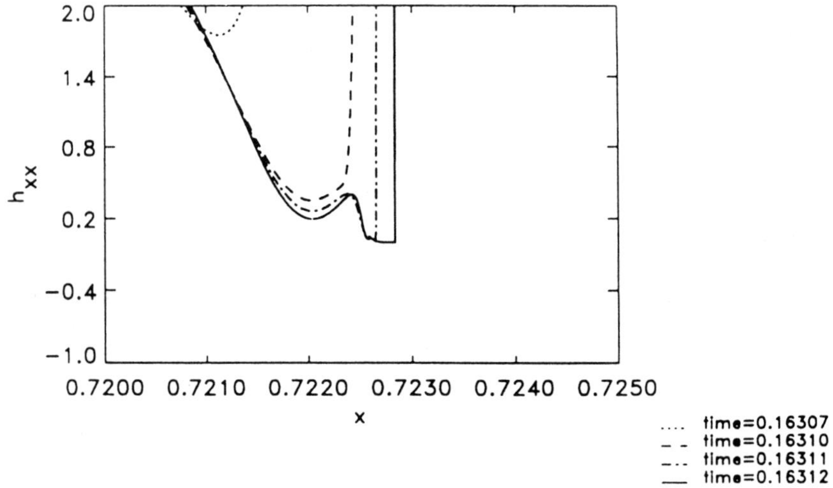

FIGURE 4.8. Blow-up of the "tip" region in the previous figure. The "tip" region also has a "tip".

Section 6.1 place an upper bound on this critical value of n. However, our simulations indicate that the critical value is probably lower. Another open problem is a rigorous proof for the existence of finite time singularities at *any* nonzero value of n larger than zero. Our numerical simulations are convincing evidence for the possibility of such a proof. A related problem not addressed in this chapter is the continuation of solutions beyond the singularity. Bernis and Friedman [3] prove the global in time existence of non-negative weak solutions for $n \geq 1$ with the boundary conditions (1.14). However, uniqueness is unknown. An open problem is whether or not additional boundary conditions (on the edge of the support of the solution) are needed after the singularity. If such conditions are needed, how do we pick out the physically relevant solutions? Perhaps similarity solutions again play a role, as in the work of Keller and Miksis describing a related but different surface tension problem [21].

Another issue not discussed in this chapter is comparison with experiment. The only practically realizable systems correspond to $n = 1$ (Hele–Shaw cell) and $n = 3$ (thin film on solid surface). There is reason to believe that our singularity results apply to Hele–Shaw experiments with sufficiently viscous fluids (see [4]). Experiments on singularities in Hele–Shaw systems are in progress (see [16, 18]). Hopefully the results will stimulate further theoretical developments.

Acknowledgments This research was supported by the Department of Energy and also by the Materials Research Laboratory of the University of Chicago through NSF grant number DMR-8819860. AB is partly supported by an NSF postdoctoral fellowship. MB acknowledges the support of a GANN fellowship. In addition, we would like to thank the following people for their helpful discussions: Stephanella Boatto, Peter Constantin, Ray Goldstein, Mike Shelley, and Sumin Zhou. We are particularly grateful to Piero Olla for critically reading Section 6.2.

References

[1] G.K. Batchelor, *An Introduction to Fluid Dynamics.* Cambridge University Press, Cambridge, 1967.

[2] D. Bensimon, L.P. Kadanoff, S. Liang, B.I. Schraiman, and C. Tang, Viscous flows in two dimensions. *Rev. Mod. Phys.* **58**, 977 (1986).

[3] F. Bernis and A. Friedman, Higher order nonlinear degenerate parabolic equations. *J. Diff. Equations* **83**, 179–206 (1990).

[4] A. Bertozzi, M. Brenner, T. Dupont, and L. Kadanoff. Unpublished results.

[5] Stephanella Boatto, Leo Kadanoff, and Piero Olla, *Phys. Rev. E* **48**, 4423 (1993).

[6] M. Brenner and A. Bertozzi, On the spreading of droplets on a solid surface. *Phys. Rev. Lett.* **71**(4), 593–596 (1993).

[7] A. Cameron, *Principles of Lubrication.* Longmans, London, 1966.

[8] P. Constantin, T. Dupont, R. Goldstein, L. Kadanoff, M. Shelley, and S. Zhou, Droplet breakup in a model of the Hele–Shaw cell. *Phys. Rev. E* **47**(6), 4169–4181 (1993).

[9] P.G. de Gennes, Wetting: Statics and dynamics. *Rev. Mod. Phys.* **57**, 827–863 (1985).

[10] C. Domb and M.S. Green (eds.), *Phase Transitions and Critical Phenomena.* Academic Press, London, 1972.

[11] T. Dupont, R. Goldstein, L. Kadanoff, and S. Zhou, Finite-time singularity formation in Hele–Shaw systems. *Phys. Rev E* **47**(6), 4182–4196 (1993).

[12] M. Van Dyke, *Perturbation Methods in Fluid Mechanics.* Parabolic Press, Stanford, CA, 1975.

[13] J. Eggers and T.F. Dupont, Drop formation in a one-dimensional approximation of the Navier–Stokes equation. To appear, *J. Fluid Mech.*

[14] M.J. Shelley, R.E. Goldstein, and A.I. Pesci, Topological transitions in Hele–Shaw flow. In *Singularities in Fluids, Plasma, and Optics*, R.E. Caflisch and G.C. Papanicolou (eds.), pp. 167–188.

[15] R.E. Goldstein, A.I. Pesci, and M.J. Shelley, Topology transitions and singularities in viscous flows. *Phys. Rev. Lett.* **70**(20), 3043–3046 (1993).

[16] R.E. Goldstein, T.G. Mason, and E. Shyamsunder. Private communication.

[17] H.P. Greenspan, On the motion of a small viscous droplet that wets a surface. *J. Fluid Mech.* **84**, 125–143 (1978).

[18] D. Grier and N. Morgan. Private communication.

[19] L.M. Hocking, Sliding and spreading of this two-dimensional drops. *Quart. J. Mech. Appl. Math.* **34**, 37–55 (1981).

[20] Chun Huh and L.E. Scriven, Hydrodynamic model of steady movement of a solid/liquid/fluid contact line. *J. Colloid Interface Sci.* **35**, 85–101 (1971).

[21] J.B. Keller and M.J. Miksis, Surface tension driven flows. *SIAM J. Appl. Math.* **43**(2), 268–277 (1983).

[22] R.M. Kerr, Evidence for a singularity of the three dimensional incompressible Euler equations. *Phys. Fluids A* **5**, 1725 (1993).

[23] A.J. Majda, Vorticity and the mathematical theory of incompressible fluid flow. *Comm. Pure Appl. Math.* **39**, 5187–5220 (1986).

[24] P. Neogi and C.A. Miller, *J. Colloid Interface Sci.* **92**, 338 (1984).

[25] A. Pumir and E.D. Siggia, Development of singular solutions to the axisymmetric Euler equations. *Phys. Rev. Lett.* **68**, 1511–1514 (1992).

[26] E.B. Dussan V and S. Davis, On the motion of a fluid-fluid interface along a solid surface. *J. Fluid Mech.* **65**, 71–95 (1974).

[27] Robert M. Wald, *General Relativity.* University of Chicago Press, Chicago, 1984.

[28] S. Zhou, *Interface Dynamics: Bubble Growth and Droplet Breakup in the Hele-Shaw Cell.* Ph.D. thesis, University of Chicago, 1992.

REFERENCE FRAME

Singularities and Blowups

Leo P. Kadanoff

We have gotten spoiled in physics. Our mathematics works so very well. Most of our problems involve differential equations, and most of our fundamental differential equations continue to make sense in a quite global fashion. Maxwell's equations or the Schrödinger equation have the property that, if you start out with a physically reasonable situation, the equation's time development gives you another physically reasonable solution and so on to eternity, or at least to the eternity defined by the equation.

But these equations are quite special. They are linear, and wonderfully robust. In recent years, physicists, mathematicians and others have turned their attention to equations that develop mathematical singularities from nothing. You start from a very smooth initial situation and just wait. After a time, an infinity shows up in the solution or in one of its derivatives. Sometimes you can continue the solution past the singularity. Sometimes you need new physics (perhaps an additional boundary condition) to see what happens next. Sometimes you can say nothing at all beyond the singularity time.

The simplest example of singularity formation is drawn from the elementary study of ordinary differential equations. Look at the equation $dx / dt = ax$, with a being positive and constant, and an x that is positive at time zero. This could be the growth equation for something—perhaps the concentration of some compound in the atmosphere. Note that the solution grows exponentially in time, but x remains positive and well-behaved for all finite times. Perhaps this outcome is bad, but the worst takes an infinitely long time to arise. In contrast, imagine that the growth rate, a, is itself linear in the concentration of the contaminant: $a = cx$, with c being constant. A quick calculation shows that the concentration obeys: $x(t) = 1 / c(t^* - t)$, with ct^* being an abbreviation for the positive quantity $1 / x(0)$. Notice that the concentration blows up at t^* and subsequently has the senseless property of being negative.

This example, drawn from the theory of ordinary differential equations, is too trivial for research-style investigation. However, examples of singularity formation based upon partial differential equations are hot research topics. One example is provided by black hole formation in general relativity. Black holes are singularities. In an elegant numerical study, Matt Choptuik[1] showed that in contrast to the generally held view, holes of mass smaller than the Chandrasekhar limit could be formed and that, in the process of formation, the solutions would oscillate periodically in $\ln(t^* - t)$. Many others published papers that agreed with and extended this conclusion.

Choptuik's study is one of a series of recent works that study how partial differential equations develop singularities that are universal and scale-invariant. These properties go together. For example, a group of us (Michael Brenner, Peter Constantin, Leonid Levitov, Alain Schenkel, Shankar C. Venkataramani and I) are studying a problem in which bacteria are attracted by something that they produce. This attractant undergoes a rapid diffusion process.[2] As the bacteria produce a more concentrated region of attractant, they are bound together into a tight little clump in which their

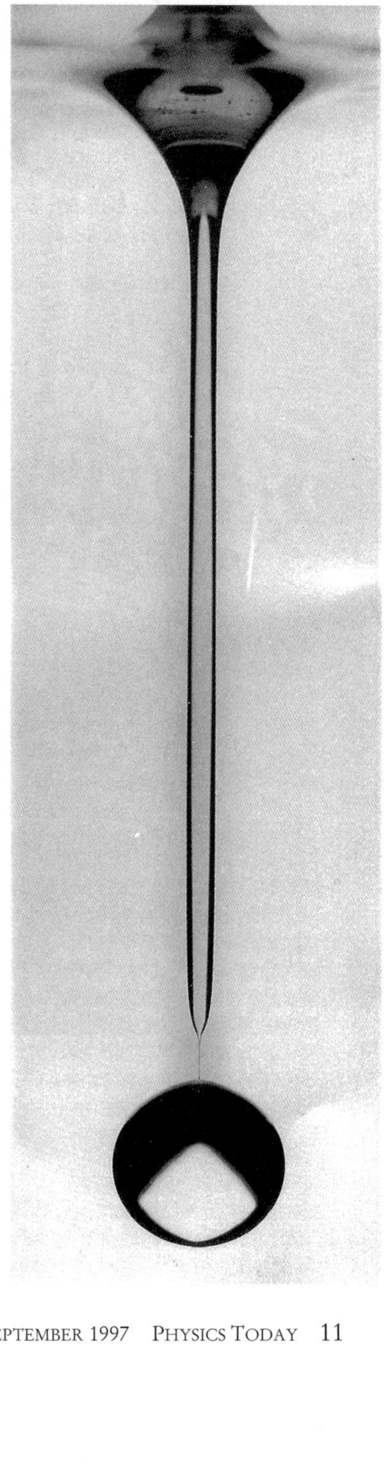

A DROP IS PRODUCED by the flow through a faucet of a glycerol/water mixture with a viscosity of 1 poise. The singularities occur at the precise space/time points at which the thread breaks. (Photo from S. D. Shi, M. P. Brenner, S. R. Nagel, Science **265**, 219 (1994).)

LEO P. KADANOFF, *a condensed matter theorist, is John D. and Catherine T. MacArthur Professor of Physics and Mathematics at the University of Chicago.*

density, ρ, goes to infinity near a singular point as

$$\rho(\mathbf{r}, t) = (t* - t)^{-1} F\left(\frac{|\mathbf{r} - \mathbf{r}_0|}{(t* - t)^{1/2}}\right) + \text{(smaller terms)}$$

Note that the singular term always has the same basic shape, but its size and extension vary in time. Hence we call it scale-invariant. We use the word "universal" to suggest that the result will remain the same if the initial situation is varied slightly, or even perhaps if the differential equation is changed a little bit. The equation describes the simplest way in which a solution to a partial differential equation can "go bad."

This same sort of mathematical structure appears time and time again. For example, scientists have long studied situations in which a mass of fluid forms a thin neck and that neck breaks so that the fluid separates into two pieces. (See the figure on page 11.) A group of theorists here at the University of Chicago looked for similarity solutions in these situations. Using simulations, we[3] found a whole zoo of such solutions, one of which was, in parallel, achieved experimentally.[4] In this solution, the derivative of the pressure blew up at the breaking point. This situation of broken necks is part of a broader problem, discussed by Pierre-Gilles de Gennes[5] and others, in which one must understand fluid flow on a surface that is partially wet and partially dry. The first time a dry spot appears on an initially wet surface, there is a mathematical singularity. After that time, one must somehow deal with the new boundary conditions required to describe the motion of the interface. The number and type of those boundary conditions can, in principle, be determined by studying another kind of similarity solution, the one that describes the wet–dry edge.

There is a classical, and unsolved, problem related to these singularity issues. The most fundamental equations for fluid flow are the Navier–Stokes and Euler equations. These equations[5] have a large number of near-singular behaviors that interfere strongly with the construction of economical, accurate and reliable simulations. These difficulties, in turn, limit advances in such fields as weather prediction and the design of explosive devices.

Accurate simulation of singular or near-singular behavior is both difficult to achieve and hard to assess. In the bacterial example, we ran into a particularly favorable situation in which we had theorems and stability analysis that told us what to expect. In one region of behavior, the theorems ruled out any singularity, yet each single simulation that we did showed some apparent singularity. Only the most careful comparisons of solutions obtained by different methods sufficed to show that the computed singularities were artifacts of the computational technique.

Choptuik concluded that the best approach to computational understanding of singularity formation was to have a maximum opportunity for "on line" human intervention in the simulation process. He designed techniques for achieving interactivity in the context of the Cray supercomputer at the University of Texas at Austin. We had a simpler computational problem so we could get where we wanted with workstations. For us too, interaction with the computer and extensive tests of computational technique were necessary before we could feel any confidence in our results.

Perhaps this experience gives some practical lessons. The Department of Energy is now in the process of learning how to use another generation of supercomputers, bigger and faster than the previous ones but, for the moment, more weakly supplied with software. These computers will be used in part to replace understanding that might have been obtained from testing of nuclear weapons. If the experience of the singularity work is any guide, in this new design and testing process, human understanding will remain essential. It is not sufficient to set up the code and let the computer zip along. It zips alright, but to where? It will remain crucially important to monitor the computational process in detail, to see and understand the steps involved in each stage of the computation. The weapons designers should require many internal tests of the consistency of the numerical technique. To enhance reliability, they should test their computer results against previous experiment and observation as much as possible. Even then, one cannot have absolute confidence that the computer will produce truth. Computer simulation is, at the edge, an art as much as a science.

References

1. M. Choptuik, Phys. Rev. Lett. **70**, 9 (1993).
2. E. O. Budrene, H. C. Berg, Nature **376**, 49 (1995).
3. A. Bertozzi, M. Brenner, T. F. Dupont, L. P. Kadanoff, in *Trends and Perspectives in Applied Mathematics*, vol. 100, L. Sirovitch, ed. (1994), page 155.
4. M. Brenner, J. Eggers, K. Joseph, S. Nagel, X. D. Shi, Phys. Fluids **9**, 1573 (1997).
5. P.-G. de Gennes, Rev. Mod. Phys. **57**, 827 (1985). ■

VOLUME 74, NUMBER 8 PHYSICAL REVIEW LETTERS 20 FEBRUARY 1995

Breakdown of Hydrodynamics in a One-Dimensional System of Inelastic Particles

Yunson Du, Hao Li, and Leo P. Kadanoff

The James Franck Institute, The University of Chicago, Chicago, Illinois 60637

(Received 15 August 1994)

We study dynamics of nearly elastic particles constrained to move on a line with energy input from the boundaries. We find that for typical initial conditions, the system evolves to an "extraordinary" state with particles separated to two groups: The majority of the particles get clamped into a small region of space and move with very slow velocities; a few remaining particles travel between the boundaries at much higher speeds. Such a state clearly violates equipartition of energy. The simplest hydrodynamic approach fails to give a correct description of the system.

PACS numbers: 05.20.Dd, 47.50.+d, 81.35.+k

In the study of a many-particle system with interactions, a hydrodynamic approach is naturally used when the quantity of interest is of a macroscopic nature with length and time scale much larger than the typical microscopic scales. Instead of focusing on the detailed dynamics of individual particles, one usually writes down a set of equations that describe the evolution of some macroscopic quantities such as local density, temperature, and flow velocity, with transport coefficients derived from statistical considerations [1]. The derivation of the hydrodynamic equations [2] relies on the assumptions that the particles reach local equilibrium with equipartition of energy, and the statistical properties can be simply described by local macroscopic quantities, i.e., the temperature. However, as we shall demonstrate that, with the introduction of a very small dissipation in the microscopic dynamics, a many-particle system can reach "extraordinary" states where local equilibrium is destroyed. As this happens, the hydrodynamic approach fails to give a correct picture for the system.

The specific example we shall show in this Letter is a one-dimensional many-particle system in which particles interact via *inelastic* collisions which conserve momentum but dissipate kinetic energy. Such a model was originally motivated by the studies of granular materials [3-7]. Consider a horizontal column of N sizeless inelastic particles with identical mass confined by two walls of infinite masses $L = 1$ apart. When two particles collide, the velocities after collision, v_1' and v_2', are expressed in terms of the velocities before collision, v_1 and v_2, as

$$\begin{aligned} v_1' &= \epsilon v_1 + (1-\epsilon)v_2 , \\ v_2' &= (1-\epsilon)v_1 + \epsilon v_2 . \end{aligned} \qquad (1)$$

Here $\epsilon = (1 - r)/2$, with r the coefficient of restitution defined by $v_1' - v_2' = -r(v_1 - v_2)$. If $r = 1$, the collision is perfectly elastic, and if $r = 0$, the collision is completely sticky. If there is no energy input, all the particles will come to rest after the initial kinetic energy is dissipated through collisions. To see nontrivial dynamics, one has to drive the system. We choose to drive the system by pumping energy from the left side wall. The rule is as follows. When the leftmost particle hits the left wall, it will be returned with a random velocity v_0 of a Gaussian distribution, $\sim \exp(-v_0^2/2T_0)$. The collision between the rightmost particle and right wall is perfectly elastic, resulting in no energy change. The above boundary conditions therefore mimic a left wall held at constant temperature and a right wall thermally insulated in a thermodynamic sense.

In a conventional hydrodynamic approach, one describes the system by a set of macroscopic quantities: particle number density $\rho(x,t) \equiv \langle \sum_i \delta(x - x_i(t)) \rangle$, macroscopic flow velocity $u(x,t) \equiv \langle \sum_i v_i(t)\delta[x_i(t) - x] \rangle$, and temperature $T(x,t) \equiv \langle \sum_i [v_i(t) - u(x,t)]^2 \delta(x_i(t) - x) \rangle$, where the $\langle\,\rangle$ represents a coarse-grained average over a small region of space or time. Assuming local equilibrium, one can derive the following set of equations based on mass, momentum, and energy balances [8,9]:

$$\begin{aligned} \partial_t \rho &= -\partial_x(\rho u) , \\ \rho \partial_t u &= -\rho u \partial_x u - \partial_x(C_1 \rho T) , \\ \rho \partial_t T &= -\rho u \partial_x T - C_1 \rho T \partial_x u + \partial_x^2 (C_2 T^{3/2}) - C_3 \epsilon \rho^2 T^{3/2} , \end{aligned} \qquad (2)$$

with C_1, C_2, and C_3 numerical constants. The above equations were derived in the context of granular sand flow. They are similar to the Navier-Stokes equation for an ideal gas except that inelastic collisions between particles lead to an additional energy dissipation term $C_3 \epsilon \rho^2 T^{3/2}$ in the energy balance equation.

What do the hydrodynamic equations tell us about our system? Given the boundary conditions $T(0,t) = T_0$, $\partial T/\partial x(1,t) = 0$, and $\rho u = 0$ at $x = 0, 1$ (no mass flow through the walls), the above equations predict that, regardless of the initial condition, the final state is a steady state with no macroscopic flow. This steady state can be solved analytically. The solution is smooth and stable (see Fig. 1 for a typical solution). We note that as $\epsilon \to 0$ with N fixed, $T(x)$ and $\rho(x)$ become uniform.

We now contrast the prediction of the hydrodynamic equations with our numerical simulation. We simulate this system by using an event-driven code which searches out where and when the next collision happens [5,6,10,11]. The configuration is updated after each collision. We use a number of particles N and a coefficient of restitution r such that $N(1 - r) < 1$, while N is large.

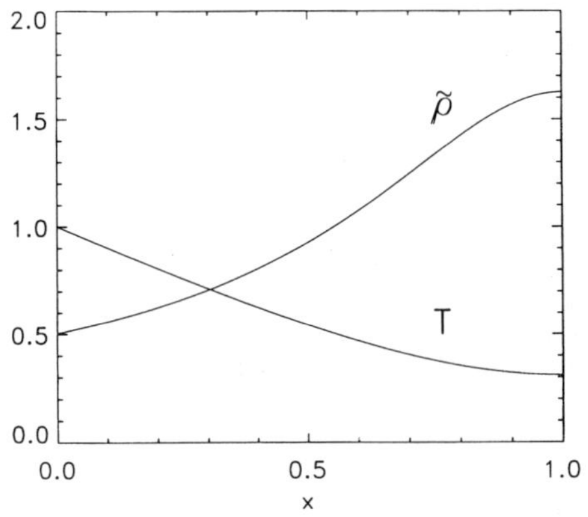

FIG. 1. Density ($\tilde{\rho} = \rho/N$) and temperature (T) profiles from the steady state solution of the hydrodynamic equations. Here $N = 100$, $\epsilon = 0.005$, and $C_2/C_3 = 11.8$.

Such a restriction is necessary since it is known [5,12] that, for sufficiently large $N(1 - r)$, one encounters a singularity where a group of particles become so sticky that they collide with each other infinitely often in a finite amount of time.

We start the simulation by generating an arbitrary initial configuration. Figure 2 shows the result of a typical simulation. At the beginning the particles are uniformly distributed. Only the leftmost particle has a nonzero velocity. Soon all the particles except the leftmost one move toward the right wall. This motion continues until all the $N - 1$ particles are squeezed at the right wall and get "clamped" into a small region. The size of the clamped cluster is very small compared to the dimension of the box, and the particles in the cluster move with a very small velocity. The leftmost particle travels at a relatively much faster speed, $v \sim \sqrt{T_0} = 1$, between the left wall and the cluster, delivering the energy it gained from the collision with the left wall to the cluster. The collisions between the fast particle and the cluster also provide the necessary pressure to keep the cluster clamped. When the leftmost particle obtains from the left side wall a velocity far less than its typical velocity, $\sqrt{T_0} = 1$, the cluster then may have time to burst out [Fig. 2(d)], and after a transient time, the $N - 1$ particles will again get clamped at the right wall [Fig. 2(e)]. Figure 3 shows the motion of the center of mass. In this case, the typical size of the cluster is $\sim 10^{-3}$ and the typical velocity of the particles in the cluster is $\sim 10^{-3}$. Notice that there are frequent bursts of the cluster when the velocity of the incident particle is very small (due to the random Gaussian distribution). These bursts are then brought back to the clamped states after tens of collisions with the fast particle, indicating that the $N - 1$ particles tend to stick to the right wall. We have checked various initial conditions to make sure that the realization of this final state is independent of the initial condition. For example, we tried an initial condition with particles having randomly chosen velocity between $-\sqrt{T_0}$ and $\sqrt{T_0}$.

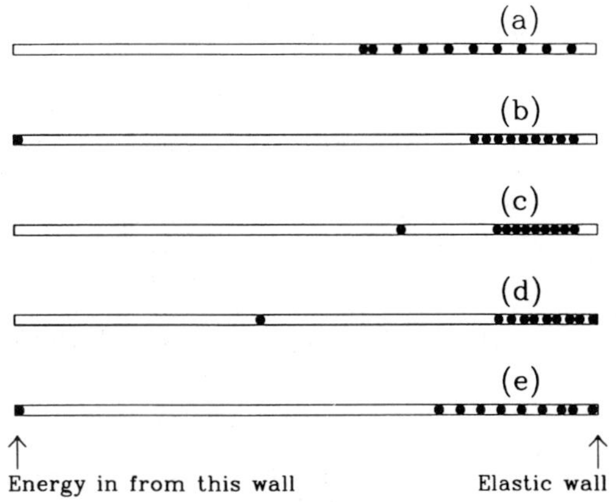

FIG. 2. Snapshots of a ten-particle system. Here $\epsilon = 0.05$ and $T_0 = 1$. (a) Snapshots of particles at a transient time. (b), (c) The "extraordinary state" where nine particles are squeezed into a small space an one particle runs fast between these nine particles and the left side walls. (d) A bursting cluster. (e) The burst is brought back to the clamped state. Note that in order to make particles visible, we assume each particle is a sphere of radius $r = 0.01$, thus the plotted particles center at $x' = x + (2i - 1)r$, where i is the particle index counting from the left to the right, i.e., the leftmost particle has an index $i = 1$.

We observe that the particles collide with each other and the initial kinetic energy gets dissipated after a transient period so that eventually the system ends up to the same final state as described above. We have simulated systems with various N and ϵ (with $N\epsilon < 1$), the above picture does not change. We should emphasize that the formation of the clump does not disappear as $\epsilon \rightarrow 0$, contrary to the prediction of the hydrodynamic equations. In fact, with fixed N and decreasing ϵ, the particles in the clump get squeezed into a smaller space and move with slower speeds.

Note that if we replace the inelastic particles by perfectly elastic ones, the situation is completely different. The

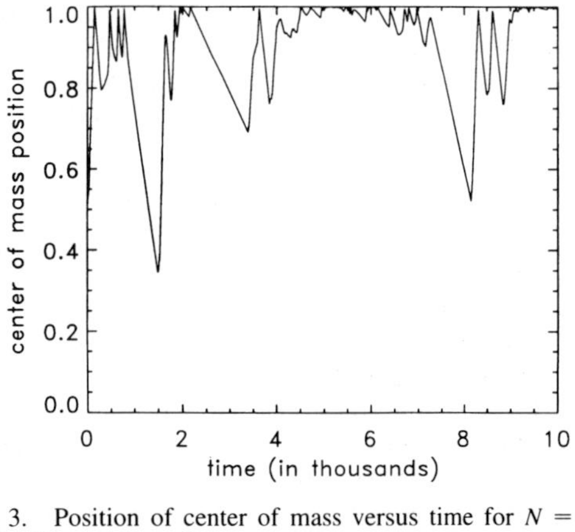

FIG. 3. Position of center of mass versus time for $N = 100$, $\epsilon = 0.005$, and $T_0 = 1$.

collision between two particles simply causes them to exchange their velocities. After a while, the initial velocity of any particle will be given to the leftmost one and get randomalized by the collision with the wall. Thus, the system will eventually reach a state where the particles distribute themselves uniformly in space, with velocities assuming the same Gaussian distribution as v_0. Therefore, the inelasticity plays an important role in destroying the local equilibrium.

The above inelastic result is independent of how the energy is pumped in at the boundary. For example, similar phenomenon occur when the left wall is rapidly oscillating with a small amplitude or when the wall simply kicks the leftmost particle back with a constant velocity. (In the latter case, we note that the number of fast particles between the left wall and the "clump" depends on the initial condition.) We also simulated a case where both side walls kick back the incoming particle with a constant velocity. In this case, almost all particles are squeezed to form a cluster and move slowly between the two side walls with two groups of fast particles running between the cluster and two side walls.

The case where the left wall kicks the particle back with constant velocity $v_0 = 1$ and initially $N - 1$ rightmost particles are at rest is quite illustrative, since the final state is periodic and detailed dynamics can be well understood. In this case, $N - 1$ particles are squeezed into a very small cluster, colliding with the fast particle and the wall as if they were just one big particle. Figure 4 plots the center of mass motion of the cluster as a function of time. We see that initially the center of mass accelerates toward the right wall until the cluster hits the wall with a large velocity and bounces back. After several collisions with the right wall and the leftmost particle, the cluster begins a periodic motion. In one period, the cluster first moves with a constant velocity V_{in} towards the wall. After colliding with the wall, it comes back with a constant velocity V_{out}. This velocity is changed back to V_{in} after the collision with the fast particle. This process repeats with the period equal to the traveling time of the fast particle.

The above state can be understood analytically by using the following simple picture. Consider $N - 1$ particles initially placed close to the wall with zero velocity, label them in order as particle $1, 2, \ldots, N - 1$ with particle 1 closest to the wall. Let particle N incident in from left with unit velocity. Since the collision is nearly elastic, after first collision between particle N and $N - 1$, particle $N - 1$ acquires a large velocity $1 - \epsilon$, and particle N nearly comes to rest with a small velocity ϵ. Then the next collision will be between particle $N - 1$ and $N - 2$, so on so forth, until particle 2 collides with particle 1, giving particle 1 a velocity $(1 - \epsilon)^{N-1}$. This velocity is reversed after the elastic collision of particle 1 with the wall, and the collisions propagate back to particle N and knock it out with a velocity $(1 - \epsilon)^{2(N-1)}$. This completes the collision of fast particle with the cluster, giving the cluster a center of mass velocity, $V_d = [1 + (1 - \epsilon)^{2(N-1)} - 2(1 - \epsilon)^{N-1}]/(N - 1) \approx [1 - \exp(-N\epsilon)]^2/(N - 1)$, in the limit $N \gg 1$ and $\epsilon \gg 1$. Therefore, the cluster acquires a small drifting velocity V_d towards the wall. For a small $N\epsilon$, $V_d \sim N\epsilon^2$.

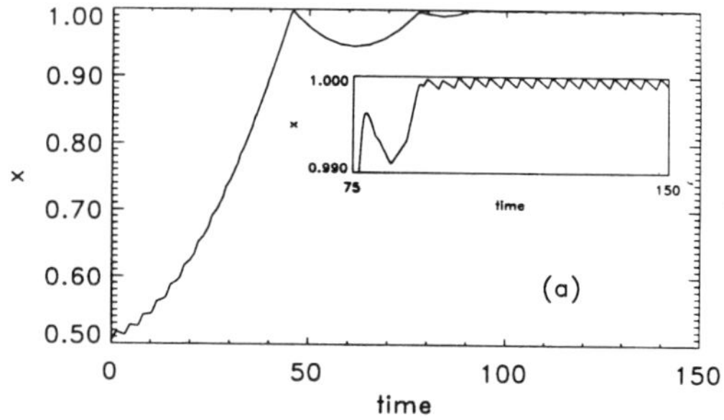

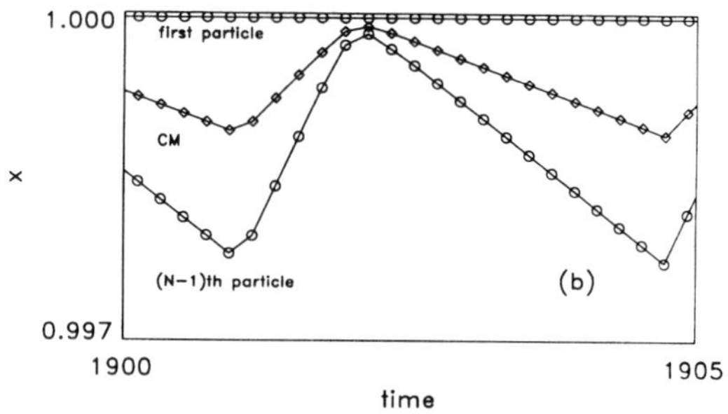

FIG. 4. Position of center of mass versus time for $N = 100$, $\epsilon = 0.005$. The left side wall kicks the incident particle with a velocity $v_0 = 1$. (a) The transient period. The inset is a blowup showing that after the transient, the c.m. moves periodically forever. (b) The periodic motion of the center of mass for one period. In (b) we also plot the positions of the first (closest to the right wall) and $(N - 1)$th particles to show that the clump synchronously undergoes periodic size changes. The plotted solid lines in (b) simply connect data. From (b), we obtain $V_{\text{in}} = 1.11 \times 10^{-3}$ and $V_{\text{out}} = 4.08 \times 10^{-4}$.

To figure out how the center of mass velocity of the drifting cluster is changed due to collisions with the wall, consider simply that the $N - 1$ particles move with some velocity V_{in} towards the wall. First, particle 1 reverses its velocity by a collision with the wall, then particle 1 collides with 2. The collision propagates until particle $N - 2$ collides with $N - 1$, so that $N - 1$ gets a velocity away from the wall. The same sequence of collision repeats itself, starting with a collision between particle 1 and the wall. After $N - 1$ such collision wave, the velocities of all the $N - 1$ particles change direction, and the whole cluster moves away from the wall with a center of mass velocity $V_{\text{out}} = \alpha V_{\text{in}}$, where α is a constant which depends on N and ϵ. For $N\epsilon \ll 1$, $\alpha = 1 - 2(N - 2)\epsilon$. There are another $N - 1$ waves of collisions between the particles (not with the wall) which rearrange the velocities (but do not affect V_{out}). We find after the rearrangement the velocity differences between the consecutive particles are constant, $\Delta V / V_{\text{in}} \sim 2(N - 3)\epsilon^2$ to lowest order, leading to a uniformly expanding column.

We now use the above collision pictures to calculate several quantities in the final state. Since $V_{in} = V_d - V_{out}$ and $\alpha V_{in} = V_{out}$, we obtain $V_{in} = V_d/(1 + \alpha)$ and $V_{out} = \alpha V_d/(1 + \alpha)$. The width of the cluster l can be roughly estimated by the distance the center of mass moves away from the wall, giving $l \approx \alpha V_d/(1 + \alpha)^2$, assuming the traveling period of the fast particle is 1. For a small $N\epsilon$, $l \sim N\epsilon^2$, and $V_{in} \approx V_{out} \sim N\epsilon^2$. This analysis shows that with N fixed and $\epsilon \to 0$, the clamped state become even singular. For $N = 100$ and $\epsilon = 0.005$, we find $\alpha = 0.34$, $V_{in} = 1.17 \times 10^{-3}$ and $V_{out} = 3.97 \times 10^{-4}$, which agrees quite well with the simulation. Numerically, we also find that the cluster is uniformly expanding while it is moving away from the wall.

The above scalings can also be understood via a Boltzmann equation for one-dimensional inelastic collisions [13]. Let $f(v,x,t)$ be the phase-space distribution density function such that $f(v,x,t)dxdv$ is the number of particles located between x and $x + dx$ with velocities between v and $v + dv$. Then the Boltzmann equation for this system is

$$\left(\frac{\partial}{\partial t} + v\frac{\partial}{\partial x}\right)f(v,x,t) = -\int du\,dU\,dV|u - v|f(v,x,t)f(u,x,t)\delta(V - v + \epsilon(v - u))\delta(U - u + \epsilon(u - v))$$
$$+ \int du\,dU\,dV|U - V|f(V,x,t)f(U,x,t)\delta(v - V + \epsilon(V - U))\delta(u - U + \epsilon(U - V)), \quad (3)$$

where the two terms on the right hand side are the usual scattering out and scattering into beam terms of the Boltzmann equation. For small ϵ, the equation can be simplified as

$$\left(\frac{\partial}{\partial t} + v\frac{\partial}{\partial x}\right)f(v,x,t) = \epsilon\frac{\partial}{\partial v}\int du\,(v - u)f(v,x,t)f(u,x,t)|v - u|. \quad (4)$$

When $N \to \infty$ and $\epsilon \to 0$ with $N\epsilon$ fixed, the $N - 1$ particles clumped into a small region of space $\sim N\epsilon^2$ and moved at a typical velocity $\sim N\epsilon^2$. We then have that $f \sim N/(N\epsilon^2)^2$. The two left hand side terms are $\sim 1/N\epsilon^4$, which are balanced by the right hand side term which is $\sim N\epsilon/(N\epsilon^4)$. Note that this approach implies that in the limit $\epsilon \to 0$ with $N\epsilon =$ const the system converges to a well-defined distribution function.

We now give a rough estimate of the transient time. For a system with no energy input, it takes certain amount of time to dissipate its initial energy, call such a time τ_d. Consider also a system driven by the boundary energy input, with an initial condition that all the particles are uniformly distributed with negligible velocity. Call τ_s the time it takes to squeeze the particles to the end. We assume that the transient time is the larger of τ_d and $\tau_s \cdot \tau_s$ can be estimated using the drift velocity the cluster acquired after each collision with the fast particle. This leads to $\int_0^{\tau_s} t\,dt\,V_d \sim 1$, which yields $\tau_s \sim \sqrt{2/V_d}$. For a system where particles have random initial velocities V_0 and uniform density n, a simple energy balance leads to $\tau_d \sim 1/\epsilon n V_f$, where V_f is the typical velocity of clamped particles. For $N = 100$ and $\epsilon = 0.005$, we find $\tau_s \sim 36$, which roughly agrees with the simulation that starts with particles having zero initial velocities.

A similar phenomenon has been previously reported in the cooling of inelastic particles [12–14]. In that case, a clump forms as particles gradually lose their kinetic energy. Here we have seen that, driven by boundary energy sources, a one-dimensional system of many inelastic particles may as well collapse into an extraordinary state where particles move with velocities of totally different magnitudes (i.e., no equipartition of energy) and therefore hydrodynamics fails to give a correct description. It remains to be seen whether such a behavior persists in a driven system in higher dimensions.

We thank Peter Constantin, Sidney Nagel, Heinrich Jaeger, Tom Witten Jonathan Miller, Elizabeth Grossman, Muhittin Mungan Gianluigi Zanetti, and William Young for helpful discussions. This research was supported in part by the University of Chicago Material Laboratory by the DOE and the NSF.

[1] L. D. Landau and E. M. Lifshitz, *Fluid Mechanics* (Pergamon Press, New York, 1959).

[2] S. Chapman and T. G. Cowling, *The Mathematical Theory of Non-uniform Gases* (Cambridge University Press, Cambridge, Great Britain, 1970).

[3] H. M. Jaeger and S. R. Nagel, Science **255**, 1523 (1992), and references therein.

[4] S. F. Shandarin and Ya. B. Zeldovich, Rev. Mod. Phys. **61**, 185 (1989).

[5] S. McNamara and W. R. Young, Phys. Fluids A **4**, 496 (1992).

[6] B. Bernu and R. Mazighi, J. Phys. A Math. Gen. **23**, 5745 (1990).

[7] C. S. Campbell, Annu. Rev. Fluid Mech. **22**, 57 (1990).

[8] P. K. Haff, J. Fluid Mech. **134**, 401 (1983).

[9] J. T. Jenkins and S. B. Savage, J. Fluid Mech. **130**, 187 (1983).

[10] D. C. Rapaport, J. Comp. Phys. **34**, 184 (1980).

[11] S. Luding, E. Clement, A. Blumen, J. Rajchenbach, and J. Duran (to be published).

[12] I. Goldhirsch and G. Zanetti, Phys. Rev. Lett. **70**, 1619 (1993).

[13] S. McNamara and W. R. Young, Phys. Fluids A **5**, 34 (1993).

[14] S. McNamara and W. R. Young, Phys. Rev. E **50**, R28 (1994).

PHYSICAL REVIEW LETTERS

VOLUME 78 | 26 MAY 1997 | NUMBER 21

Self-Organized Short-Term Memories

S. N. Coppersmith,[1] T. C. Jones,[2] L. P. Kadanoff,[1] A. Levine,[3] J. P. McCarten,[2] S. R. Nagel,[1] S. C. Venkataramani,[1] and Xinlei Wu[2]

[1]*The James Franck Institute, The University of Chicago, 5640 Ellis Avenue, Chicago, Illinois 60637*
[2]*Department of Physics and Astronomy, Clemson University, Clemson, South Carolina 29694-1911*
[3]*Exxon Research & Engineering Company, Route 22 East, Annandale, New Jersey 08801*

(Received 17 December 1996)

We report short-term memory formation in a nonlinear dynamical system with many degrees of freedom. The system "remembers" a sequence of impulses for a transient period, but it coarsens and eventually "forgets" nearly all of them. The memory duration increases as the number of degrees of freedom in the system increases. We demonstrate the existence of these transient memories in a laboratory experiment. [S0031-9007(97)03273-0]

PACS numbers: 03.20.+i, 46.10.+z, 72.15.Nj

We present a deterministic nonlinear dynamical system with many degrees of freedom which self-organizes to store memories, in that a configuration-dependent quantity "learns" preselected values. The system, a simple discretized diffusion equation, encodes multiple memories during an extended transient period, but, in the limit of long times, retains no more than two of them. This system thus displays a mechanism by which memories are forgotten as well as learned.

We demonstrate: (1) Short-term memories are exhibited by a system with two degrees of freedom $N = 2$, and become more pronounced as N is increased. (2) The interval in which multiple memories are encoded typically grows as the square of the system's linear extent. (3) Many features of the dynamics, including their duration, can be understood analytically. (4) The mechanism is robust and is manifest in experiments on a sliding charge-density wave solid.

Consider a system of coupled maps

$$x_j(\tau + 1) = x_j(\tau) + \mathrm{int}\left[k \sum_{i(nn)} [x_i(\tau) - x_j(\tau)] + (1 - A_\tau) \right], \tag{1}$$

where i, j are the particle indices, the sum is over nearest neighbors, τ is the time index, and $\mathrm{int}[z]$ is the largest integer less than or equal to z. These equations describe the evolution of the positions x_j of N particles in a deep periodic potential, with nearest neighbor particles connected by springs of spring constant $k \ll 1$, in the presence of force impulses $(1 - A_\tau)$. They describe the dynamics of sliding charge-density waves (CDW's) [1–3], and are closely related to models of a variety of dynamical systems [4]. The one-dimensional, τ-independent version of this system ($A_\tau = A$) has been studied previously [2,5] (see also Ref. [3]). Here, we consider A's which repeatedly cycle through M different values.

The self-organization that occurs as these maps evolve is manifest in the discrete curvature variables [5], $c_j(\tau) = k \sum_{i(nn)} [x_i(\tau) - x_j(\tau)]$, which obey

$$c_j(\tau + 1) - c_j(\tau) = k \sum_{i(nn)} \{\mathrm{int}[c_i(\tau) + (1 - A_\tau)] - \mathrm{int}[c_j(\tau) + (1 - A_\tau)]\}. \tag{2}$$

Figure 1 shows normalized histograms of $\mathrm{frac}(c) = c - \mathrm{int}(c)$ for a two-dimensional system with $M = 5$, periodic boundary conditions, and a random initial configuration of x's. Memory encoding is shown by the accumulation of c's with $\mathrm{frac}(c) = \mathrm{frac}(A_\tau)$. For a

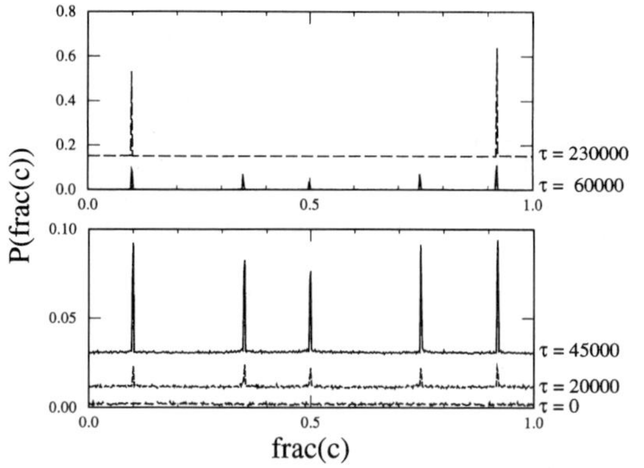

FIG. 1. Memory formation of Eq. (1) in a two-dimensional 100×100 system with periodic boundary conditions, $k = 0.0001$, initial condition of x's chosen randomly from the interval $[0, 1\,000\,000]$, and $M = 5$ ($A = [0.1, 0.35, 0.5, 0.75, 0.92]$). P is the proportion of c's with fractional parts frac(c) within a bin of width 0.002. For clarity, successive curves are offset vertically. The lower panel illustrates the short-term accumulation of c's at each value of A; the upper panel demonstrates that at long times only two peaks persist.

while all M memories are encoded to a similar degree; eventually all are forgotten except for two values of A [6]. No evolution occurs after the last trace, a fixed point of the map (2).

Figure 2 shows the curvature variables $c_j(\tau)$ versus time τ for one-dimensional chains with one free and one fixed end:

$$x_0(\tau) = 0; \quad x_{N+1}(\tau) = x_N(\tau) \qquad \text{for all } \tau\,. \qquad (3)$$

During the evolution, each c_j sticks at values corresponding to each A_τ. This tendency is more pronounced for

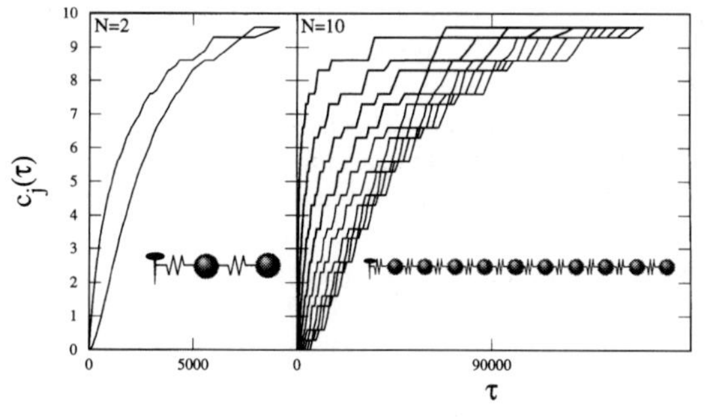

FIG. 2 Plot of curvatures $c_j(\tau)$ versus τ for N-particle chains with fixed-end boundary conditions Eq. (3), $k = 0.001$ and $A = [9.3, 9.6]$, and initial condition $x_j(\tau = 0) = 0$ for all j. The memories are manifest in the plateaus (more pronounced for $N = 10$ than for $N = 2$) when the c_j have values with fractional part of 0.3 and 0.6. Only one memory is retained at long times.

$N = 10$ than for $N = 2$, indicating that larger systems encode transient memories more effectively. At the fixed point, only one memory (rather than two as in the model with periodic boundary conditions) is encoded.

In CDW experiments, memory encoding is manifest as synchronization of the response to a repeated train of driving pulses so that V/I (V = voltage, I = CDW current, which is proportional to the CDW velocity v_{CDW}) decreases just as each pulse ends. The correspondence between V/I and the c's is discussed in detail in Refs. [2,3]. Heuristically, it follows because $x_j(\tau)$ can be thought of as the position of particle j after pulse τ, and the int functions in Eqs. (1) arise because after each pulse every particle falls into the nearest potential minimum. The memory values are at the discontinuities of the int functions, which for the highly overdamped dynamics relevant to CDW's [1] means that many particles are at potential maxima at the end of a pulse. Since particles mounting the potential go slower than those descending it (again, implied by overdamped dynamics), when many particles have c's on memory values, then a preponderance of particles are at potential maxima, which in turn implies that the ratio v_{CDW}/V is increasing at the end of each pulse. *Single* memory retention using identical pulses has been seen previously [7,8]. Here we report *multiple* memory encoding [9]. Figure 3 shows the successful training of a sample using five different four-pulse sequences (current pulses). For this sample, we investigated 25

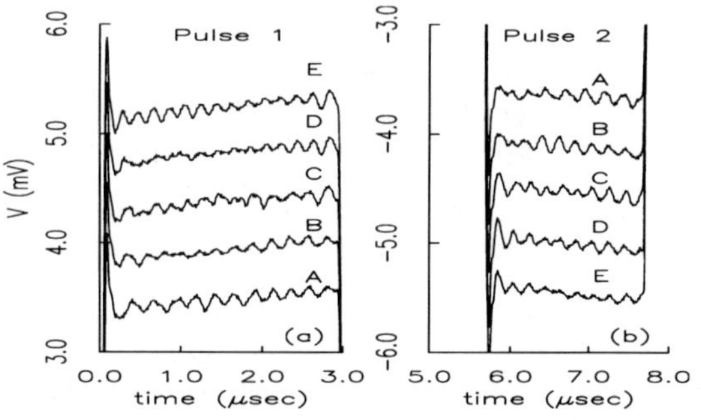

FIG. 3. Trained voltage response curves of $NbSe_3$ for five different four-pulse sequences. The evidence for multiple-memory encoding is the decreasing magnitude of the voltage at the end of each pulse. (Note the two pulses have opposite signs. The voltage response for pulses 3 and 4 are not shown.) The drive is a sequence of four current pulses with magnitudes $[I, -I, I, -I]$ and durations 2.95, 2.05, 1.75, and 2.75 μs. The drive magnitude I, is 20.76, 21.16, 21.55, 21.94, and 22.33 μA for curves A, B, C, D, and E, respectively. Training consisted of over one million pulse repetitions. Measurements were performed using a two-wire, silver-paint contact configuration. Sample dimensions are 5.2 $\mu m^2 \times 980$ μm. Additional silver paint strips of 43 μm and 100 μm in width are attached to the sample, centered at distances 13% and 58% between the probe contacts, respectively. The curves are offset for clarity, and averaged 200 times to reduce noise. $T = 50$ K, and $E_T = 47$ mV/cm.

different four-pulse sequences and observed that the voltage response at the end of each pulse had a negative slope (indicating the retention of a memory) 85% and a positive slope only 5% of the time. Thus, multiple memories are observed in experiments as well as simulations.

To understand why the memories form, consider the "nailed" case of Fig. 2. Initially, each impulse causes every x_j to increment by the same amount, so that only the spring near the nail stretches. As time progresses, this spring stretches more and more, until the force it exerts becomes large enough to keep the first particle from going into the next well under application of A_1, the impulse with the smaller fractional part. (At this point, the spring force is insufficient to change the action of A_2, the impulse with the larger fractional part.) The second spring then starts to stretch, which, on the next iteration, gives just enough added force to restore the first particle to its initial motion for *both* impulses. So now the first spring stretches on alternate applications of the impulse A_1, and the total spring force on particle 1 increments alternately by $+k$ and $-k$. Therefore, c_1 oscillates around the memory value A_1, leading to a plateau on the plot of Fig. 2. Eventually, the second spring is stretched to the point that the second particle also hangs up at the impulse A_1. This, in turn, starts the stretching of the third spring, etc. A memory for A_1 is created whenever the local curvature just cancels the fractional part of A_1. A similar analysis holds for all the other impulses A_τ that are applied. As time progresses, the c's get stuck at all the different possible memory values.

Another way to understand why the curvatures stick at the values of A_τ is to note that when c_j passes through a memory value, then the right hand side of Eq. (2) changes discontinuously, and in particular can change sign if the neighboring c's have the appropriate values. If this happens, then c_j oscillates with amplitude $\propto k$, and sticks at the memory value [10]. This sticking can take the form of either a fixed point or a cycle in the local c values.

For $k \to 0$, the dynamics separates into three regimes. The smallest motions are the $O(k)$ back-and-forth motions at the memory values, which lead to minute serrations in the plateaus that are not visible on the scale of our figures. The largest motions, involving changes in c_j which are larger than of order unity, and changes in τ which are much larger than $1/k$, are described by a linear discrete diffusion equation

$$\tilde{c}_j(\tau + 1) - \tilde{c}_j(\tau) = k \sum_{i(nn)} [\tilde{c}_i(\tau) - \tilde{c}_j(\tau)], \quad (4)$$

obtained by linearizing the int functions in Eq. (2). Fig. 4, which shows snapshots of configurations at two different times, demonstrates the accuracy of the linearized equation in reproducing the evolution on large scales. Numerically, the maximum deviation between the solutions to Eqs. (2) and (4) for identical initial conditions, $\sup_{j,\tau} |c_j(\tau) - \tilde{c}_j(\tau)|$, is less than unity for all system sizes, parameter values, boundary conditions, and initial

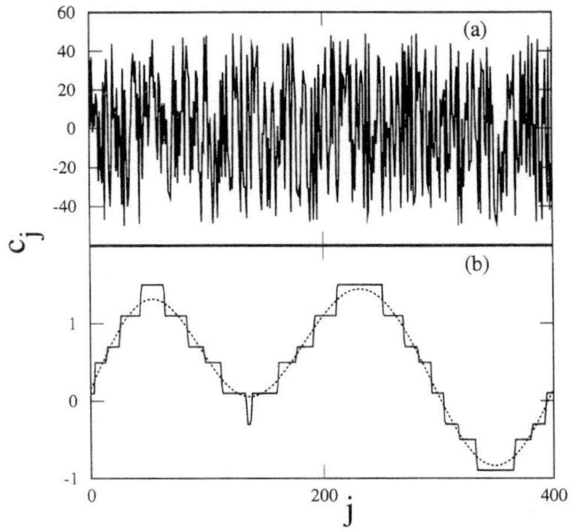

FIG. 4. Snapshots of curvature c_j versus position j of a one-dimensional chain of length $L = 400$ with periodic boundary conditions at (a) $\tau = 0$ and (b) $\tau = 1\,200\,000$. (The fixed point is reached at $\tau \sim 9\,300\,000$.) Parameter values are $A = [1.1, 1.5, 1.7]$, $k = 0.001$. The dashed line in (b) is the configuration obtained using Eq. (4) with the same initial conditions, demonstrating that this linear diffusion equation captures accurately the large scale evolution of the system.

conditions investigated. Using the nature of the nonlinearity in Eq. (2) together with the fact that $|z - \text{int}(z)| \leq 1$ for all z, one can obtain a rigorous analytic bound on this difference that grows as $\log(L)$, where L is the system's linear extent [11], which is sufficient to insure the applicability of the memory duration estimates given below.

Of course, neither of the regions just discussed produces the memory effects. The memories come from an intermediate region involving variations in c_j much larger than k but smaller than unity. Figure 2 shows that, on this intermediate scale, all the c_j's show a very simple behavior. They are either (1) stuck at one of the memory values, or (2) between memory values, with the c_j varying linearly in time. This sequence of stepwise linear motions progressively reduces the variation in the discrete curvature of c_j. Thus, the intermediate and large scale motions of c_j can be described as different kinds of diffusive smoothing.

To characterize how the memory durations depend on system parameters, first consider the case when the number of memories $M = 1$. The onset time τ_{onset} is determined by the condition that the *curvature of the curvature* $\nabla^2 c$ is ~ 1. This condition follows because the c's stick only when the discontinuity in the integer function is large enough to cause the right hand side (rhs) of Eq. (2) to change sign. For multiple memories, we sum the rhs of Eq. (2) over the M terms in the cycle and note that the discontinuity occurs in only one of M terms in the sum, leading to the condition $\nabla^2 c \sim 1/M$. For any M, the transient memories disappear when the *range* of curvatures becomes too small, $\delta c \equiv c_{\max} - c_{\min} \sim 1$. To relate the onset and forgetting times to these conditions on the curvatures, we use the linearized map Eq. (4), whose evolution obeys (when $\tau \gg 1/k$)

$$c_j(\tau) = \sum_q e^{i\vec{q}\cdot\vec{j}} c(q, \tau = 0) e^{-kq^2\tau}, \quad (5)$$

where $c(q, \tau = 0)$ is the spatial Fourier transform of the initial condition $c_j(\tau = 0)$. For the random initial conditions for c shown in Fig. 3, $\nabla^2 c$ is dominated by $q \sim \sqrt{k\tau}$ and the onset time $\tau_{\text{onset}} \sim c_0 M/k$, where c_0 is a typical value of $c(q, \tau = 0)$, independent of the linear extent L. At long times δc is dominated by the longest wavelength mode, so that $\tau_{\text{forget}} \sim \frac{L^2}{4\pi^2 k} \ln(c_0)$. Thus, larger systems remember longer.

The system of Eqs. (2) is deterministically driven towards a fixed point. Once this point is reached, it is impossible to retrieve the short-term memories. However, it is possible to keep the transient memories from decaying by adding noise to the system [12]. For example, in the nailed case of Fig. 2, moving the nail slowly but randomly with time creates the possibility of continuously encoding new memories. That noise can lead to the *retention* of memories is important for understanding our experiments. Permanently encoded multiple memories are observed, but only in samples of $NbSe_3$ with additional conducting strips attached to the crystal between the contacts, an arrangement known to induce noise [13].

Our memory mechanism can be compared to the "Hopfield memory" [14], which is a dynamical system with parameters adjusted so that particular configurations, which encode the desired patterns, minimize an energy functional. There, the memory is encoded in the long-time dynamics, and there is no intrinsic "forgetting" mechanism. Moreover, changing the remembered value requires nontrivial adjustment of the microscopic couplings of the model. In the CDW system studied here, the information is encoded in an evolving system, the control parameter (pulse size) is easily varied in the laboratory, and the self-organization is exhibited via standard transport measurements.

One avenue for further investigation is to characterize better the effects of noise, which, as discussed above, plays an important role in the CDW experiments. We would also like to identify other experimental systems which exhibit this short-term memory. Because the present model is just a discretized diffusion equation, having properties which seem quite robust, we are optimistic that physical embodiments exist that could be made into useful devices.

We thank S.E. Brown, P.B. Littlewood, M.L. Povinelli, and R. Thorne for fruitful discussions. This work was supported in part by the MRSEC program of the National Science Foundation under Award No. DMR-9400379. A.L. acknowledges support by an AT&T Graduate Fellowship.

[1] See, e.g., G. Grüner, Rev. Mod. Phys. **60**, 1129 (1988); R.E. Thorne, Phys. Today **49**, 42 (1996), and references therein.

[2] S.N. Coppersmith and P.B. Littlewood, Phys. Rev. B **36**, 311 (1987).

[3] S.N. Coppersmith, Phys. Rev. A **36**, 3375 (1987).

[4] P. Bak, C. Tang, and K. Wiesenfeld, Phys. Rev. Lett. **59**, 381 (1987); J. Carlson and J. Langer, Phys. Rev. Lett. **62**, 2632 (1989); A.V.M. Herz and J.J. Hopfield, Phys. Rev. Lett. **75**, 1222 (1995).

[5] C. Tang *et al.*, Phys. Rev. Lett. **58**, 1161 (1987).

[6] Which memories are retained at long times can be determined by using the observation that at the fixed point the c_j's do not change under application of *any* A_τ, so that [from Eqs. (1) and (2)] $P' + \text{frac}(A_\tau) < c_j < P' + 1 + \text{frac}(A_\tau)$ for some fixed integer P'. The left and right inequalities are most restrictive for those A_τ with the largest and smallest values of $\text{frac}(A_\tau)$, respectively.

[7] R.M. Fleming and L.F. Schneemeyer, Phys. Rev. B **33**, 2930 (1986).

[8] S.E. Brown, G. Grüner, and L. Mihaly, Solid State Commun. **57**, 165 (1986).

[9] Multiple memory encoding has been observed in two out of three samples of $NbSe_3$ which have additional conducting strips attached to the crystal between the contacts. The role of these extra strips is discussed below.

[10] Reference [5] notes the importance of the oscillations induced by the nonlinearity in the context of the fixed point selection when a single memory is encoded.

[11] S.N. Coppersmith, L.P. Kadanoff, and S.C. Venkataramani (unpublished).

[12] M. Povinelli, S.N. Coppersmith, and S.R. Nagel (unpublished).

[13] M.P. Maher *et al.*, Synth. Met. **43**, 4031 (1991).

[14] See, e.g., J.J. Hopfield, Proc. Natl. Acad. Sci. U.S.A. **79**, 2554 (1982); J.J. Hopfield, Proc. Natl. Acad. Sci. U.S.A. **81**, 3088 (1984); D.J. Amit, H. Gutfreund, and H. Sompolinsky, Phys. Rev. A **32**, 1007 (1985); D.J. Amit, H. Gutfreund, and H. Sompolinsky, Ann. Phys. (N.Y.) **173**, 30 (1987).

Complex Systems **1** (1987) 791–803

A Poiseuille Viscometer for Lattice Gas Automata

Leo P. Kadanoff
Guy R. McNamara
Gianluigi Zanetti
The Enrico Fermi and James Franck Institutes, The University of Chicago, 5640 South Ellis Avenue, Chicago, IL 60637, USA

Abstract. Lattice gas automata have been recently proposed as a new technique for the numerical integration of the two-dimensional Navier-Stokes equation. We have accurately tested a straightforward variant of the original model, due to Frisch, Hasslacher, and Pomeau, in a simple geometry equivalent to two-dimensional Poiseuille (Channel) flow driven by a uniform body force.

The momentum density profile produced by this simulation agrees well with the parabolic profile predicted by the macroscopic description of the gas given by Frisch et al. We have used the simulated flow to compute the shear viscosity of the lattice gas and have found agreement with the results obtained by d'Humiéres et al. [10] using shear wave relaxation measurements, and, in the low density limit, with theoretical predictions obtained from the Boltzmann description of the gas [17].

1. Introduction

In a now classic paper, Frisch, Hasslacher, and Pomeau [1] proposed a new technique for solving the two-dimensional Navier-Stokes equation based on the implementation of a lattice gas automaton. Their original idea has recently been extended to two-dimensional binary fluids, two-dimensional magnetohydrodynamics, three-dimensional Navier-Stokes, and other interesting problems [4].

Two-dimensional lattice gas automata have been described in great detail in reference 3. We will therefore give only a very short description of the model in order to define the nomenclature used.

Lattice gas automata are based on the construction of an idealized microscopic world of particles living on a lattice. The particles can move on the lattice by "hopping" from site to site. In the specific examples considered in this paper, we allow only hops from a site to its nearest neighbors (a particle may also remain stationary at its current site) and we indicate these motions with the vectors $\vec{C}^\alpha$. The $\vec{C}^\alpha$ are traditionally interpreted as the momenta of the particles. (We are using the lattice spacing, the "mass" of a particle, and the simulation time step as fundamental units.) To simplify even further, we suppose that there cannot be more than one particle with a given momentum at a given site. The population at each site can then be represented by an $l+1$ element binary vector, $\{f^\alpha(\vec{x})\}$, where l is the number of nearest neighbors and $\vec{x}$ is the label of a lattice site. We can now define the microscopic number density

$$\hat{\rho}(\vec{x}) = \sum_\alpha f^\alpha(\vec{x}) \tag{1.1}$$

and the microscopic momentum density

$$\vec{\hat{g}}(\vec{x}) = \sum_\alpha \vec{C^\alpha} f^\alpha(\vec{x}). \tag{1.2}$$

The time evolution of the gas is produced by the effect of two alternating steps: the "hopping" phase we described above and a collision phase. In the latter, the $\{f^\alpha\}$ of each site are transformed according to a set of collision rules. The rules can change from site to site or from time step to time step, but in any case they will conserve the microscopic densities $\hat{\rho}$ and $\vec{\hat{g}}$ on each site.

It is possible to construct macroscopic densities from $\hat{\rho}$ and $\vec{\hat{g}}$ by averaging in space and time over appropriate regions. The time evolution of the macroscopic number and momentum densities, ρ and $\vec{g}$, can be expressed, in the appropriate limit, in terms of the conservation laws

$$\partial_t \rho + \partial_k g_k = 0, \tag{1.3}$$
$$\partial_t g_i + \partial_j T_{ij} = 0, \tag{1.4}$$

(where Latin indices now denote Cartesian coordinates). It should be noted that we express the above densities in units of mass and momentum per unit area rather than units of mass and momentum per lattice site, as used by other authors (for instance, [9,10]). We completely ignore all the mathematical difficulties implied in the derivation of equations (1.3) and (1.4) [3], but we note that ρ and $\vec{g}$ in equations (1.3) and (1.4) are intended to be small perturbations from the equilibrium state, $\vec{g}=0$ and $\rho=$ constant.

The structure of the stress tensor T_{ij} reflects the symmetries of the underlying lattice. Frisch et al. have shown that a hexagonal lattice possesses sufficient symmetry to obtain the right structure for T_{ij}. By this we mean that up to higher derivatives and $O(g^4)$, it is possible to write

$$T_{ij} = \lambda(n) g_i g_j + p(n, g^2)\delta_{ij} - \nu(n)(\partial_j g_i + \partial_i g_j - \delta_{ij}\partial_k g_k) - \xi(n)\delta_{ij}\partial_k g_k, \tag{1.5}$$

where the quantities ν and ξ can be interpreted as transport coefficients while λ (which equals 1 for standard Navier-Stokes) arises from the absence of Galilean invariance for the lattice gas [3,13]. In the limit of incompressible flow, equations (1.3) and (1.4) together with the constitutive relation

(1.5), can be rescaled to the incompressible Navier-Stokes equation [3]. Thus, we can interpret this lattice gas as an analog computer capable of solving the two-dimensional incompressible Navier-Stokes equation.

Note that nowhere is there an attempt to simulate the microscopic behavior of a real fluid. Lattice gas automata are quite distinct from molecular dynamical simulations [20]. While both kinds of simulations seem to produce the expected macroscopic behavior for the fluid (in the sense of giving the expected constitutive relations for the macroscopic currents), they represent two completely different approaches to the problem. Molecular dynamical simulations attempt to faithfully model the microscopic behavior of a real fluid, while lattice gas automata extract only the minimal microscopic properties required to obtain the desired macroscopic behavior [5–7].

This suggests two interesting paths of research. The first, more technically oriented, concerns how well the results obtained from this new technique agree with real fluids, while the second concerns the more profound question of the connection between the microscopic and macroscopic aspects of many body systems [21]. In this paper, we principally address technical questions: the quantitative accuracy of the constitutive relation, equation (1.5), in a particular simple example, and the comparison of the effective kinematic viscosity measured in our steady non-equilibrium simulation with the values obtained by shear wave relaxation methods.

2. The simulation model

The object of our simulation is a steady forced flow between two walls with no-slip boundary conditions. We are simulating a steady flow because it allows us to obtain good accuracy in the measurements of ρ and $\vec{g}$ by extensive time averaging. We are simulating a channel with null velocity at the walls because for weak forcing (low Reynolds number), the g profile is expected to be a parabola and there is a simple relation between the maximum g, the forcing level, and kinematic viscosity ν. The actual simulation setup described below is conceptually very different from a direct implementation of a no-slip boundary channel flow but, as we will show, gives the same parabolic momentum profile.

The simulation system we have employed is a model of forced two-dimensional Poiseuille flow [13,22–24]. The system is a hexagonal lattice with an equal number of rows and columns (figure 1). Note that the system width, W, is $\sqrt{3}/2$ times the length, L, due to the unequal row and column spacings. The flow is forced by adding momentum in the positive x direction to the system at a constant rate: After each time step, we randomly select a lattice site and, if possible, apply one of the microscopic forcing rules described in figure 2. Each successful application of a forcing rule adds one unit of momentum to the system. The forcing process is repeated until the desired amount of momentum has been transferred to the gas; fractional amounts of momentum to be added to the system are accumulated across

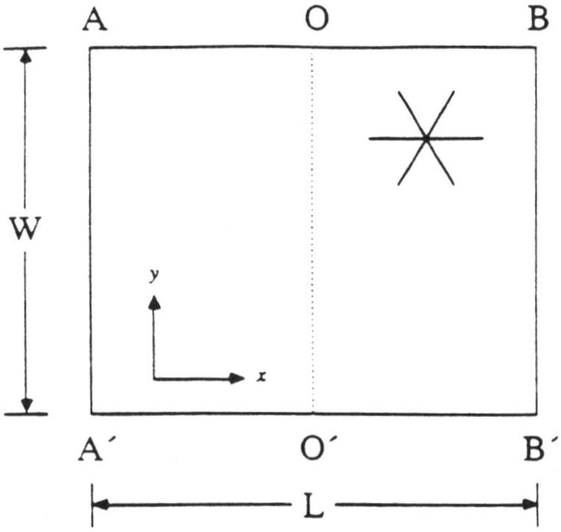

Figure 1: The simulation model. The walls AA′ and BB′ are joined by periodic boundary conditions while "Möbius strip" boundary conditions (see text) are used to connect the AB and A′B′ walls. The representative lattice site shown in the upper right-hand corner illustrates the orientation of the underlying hexagonal lattice.

time steps until they sum to an amount greater than 1, at which time one additional unit of momentum is added to the gas. The result of this process is a constant body force applied to the gas uniformly across the width and length of the channel.[1]

The forcing level employed in the present work varies from 0.3 to 2.8 units of momentum per time step. Within this range, the resulting flow is steady when averaged over a period of the order of a few diffusion times, L^2/ν. For a steady flow, the equations for the forced flow [22,3] become

$$0 = \partial_k g_k \tag{2.1}$$

$$\partial_k(\lambda g_k g_l) = -\partial_l p + \partial_k(\nu \partial_k g_l) + f_l \tag{2.2}$$

where $\vec{f} = (f, 0)$ is the average force per unit area.

The two walls perpendicular to the flow, AA′ and BB′ in figure 1, are mapped onto each other by periodic boundary conditions. The walls parallel to the flow, AB and A′B′, are mapped onto each other by "Möbius strip" boundary conditions. This boundary condition can be described as a two-step process whereby particles crossing the boundary have their position

[1]The actual forcing scheme is slightly more complicated since it must compensate for inhomogeneity in the momentum and number densities due to the macroscopic flow (see [11]). The forcing algorithm randomly selects a lattice row and column and then searches along the row until it finds a site where a forcing rule may be successfully applied. The program terminates if no forcing operation can be performed on a selected row. This guarantees that forcing operations will be uniformly distributed across the width of the channel, despite variations in the mass and momentum densities.

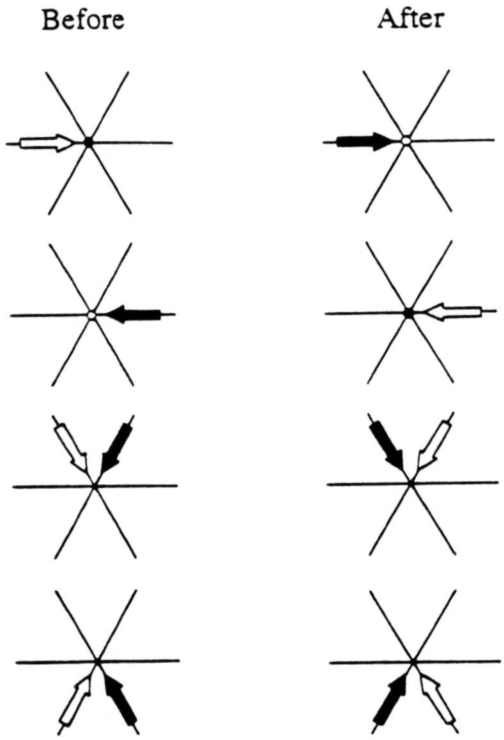

Figure 2: Forcing rules. The four pairs of diagrams represent the microscopic forcing rules used in the simulation. The black symbols indicate occupied states while the outlined symbols indicate vacant states. States not indicated in a diagram may be either filled or vacant. Each forcing operation adds one unit of momentum in the x direction.

and velocity reflected with respect to the line OO′ and then standard periodic boundary conditions are applied. The alternative to the Möbius strip boundary would be the use of no-slip boundary conditions, for instance, random scattering of the particle impinging on the walls. Both boundary conditions dissipate the momentum injected into the gas by the uniform body force, but the no-slip condition creates a layer at the boundary (a Knudsen layer [14]) whose thickness is of the order of a mean-free path.[2] Since the mean-free path for our model is typically about 5 lattice spacings and the system is only 32 lattice rows in width, Knudsen layers along both the upper and lower boundaries would significantly distort the Poiseuille flow momentum profile.

The combination of a uniform body force directed in the positive x direction and vanishing fluid velocity along the upper and lower boundaries gives rise (through equations (2.1) and (2.2)) to a parabolic momentum density profile

$$g_x(y) = \frac{g_{max}}{(\frac{W}{2})^2}(y^2 - (\frac{W}{2})^2), \tag{2.3}$$

with $g_y = 0$ and

$$g_{max} = \frac{1}{8}\frac{FW}{L\nu} \tag{2.4}$$

where we have neglected the corrections $O(g_x(\partial_y g_x)^2)$ due to variation of ρ across the width of the system (see [11]) and y is measured from the axis of the channel. We extract this momentum profile from the simulation by averaging the microscopic momentum density in time and along the lattice rows (lines of constant y).

At this point, we note that the flow which develops in the channel is equivalent to that obtained by applying to a system of length L and width $2W$, with periodic boundary conditions in both directions, the "square wave" force field

$$\vec{f}^*(x,y) \quad = (f,0) \quad \text{for } 0 \leq y < W, \tag{2.5}$$

$$\vec{f}^*(x,y) \quad = (-f,0) \quad \text{for } W \leq y < 2W. \tag{2.6}$$

We have verified this correspondence by comparing simulation results from runs employing Möbius strip boundary conditions with runs using square wave forcing (see figure 3). Both types of flow exhibit long wavelength instabilities related to the existence of inflection points in the momentum profile at $y = 0$ and $y = W$. The critical Reynolds number given by linear stability analysis for infinitely long channels (Kolmogorov flow [15,16]) is quite small, $Re_{cr} \approx 1.11$, but a finite length-to-width ratio increases Re_{cr}. The particular length-to-width ratio used in our simulation, $1/\sqrt{3}$,

[2]The Knudsen layer is caused by the matching between the artificial particle distribution imposed by this kind of boundary condition and the non-equilibrium particle distributions imposed by the macroscopic flow in the bulk.

appears to be stable even for the largest Reynolds number obtainable in our simulation (≈ 50), as was suggested by the linear stability analysis of the problem.

3. Results

In this section, we will occasionally refer to mass density in number of particles per site $n, n = (\sqrt{3}/2)\rho$, instead of number of particles per unit area, ρ; this is done for notational convenience. Figure 4 shows a typical momentum profile obtained from our simulation. The average number of particles per site in this run is $n = 2.1$, the system dimensions are $W = 16\sqrt{3}$ and $L = 32$ (corresponding to a 32×32 lattice), and we have used the model-II collision rules described in reference 3. The profile was obtained by averaging the microscopic momentum density $\vec{g}$ in the direction parallel to the flow and on one million iterations. The g_y component appears to be due entirely to statistical noise; it is small: $\max |g_y(y)/g_{max}| < 0.01$, where g_{max} is g_x at the center of the channel.

The solid line represent a parabola fitted to the simulation results which are shown as symbols. The fit is very good; if we define

$$e(y) = \frac{|g_x(y) - h(y)|}{g_x(y)}$$

where $h(y)$ is the fitted parabola, then $\max |e(y)| < 0.01$ over the central region of the parabola (roughly the 26 centermost rows).

The result quoted above can be improved by increasing the number of time steps on which the simulation in figure 4 is averaged. However, improvements obtained by averaging are limited by systematic deviations from a parabolic profile which can be reduced only by decreasing the amplitude of $\vec{g}$. These systematic deviations are due to higher-order terms, $O(g^4)$, neglected in equation (1.5), and to the presence of a term proportional to g^2 in the expression, derived in [1,2,3]; for the pressure p, see [11].

We can use equation (2.4) to relate the maximum measured velocity to the applied force. This permits us to define an effective kinematic viscosity

$$\nu = \frac{1}{8}\frac{FW}{Lg_{max}}. \tag{3.1}$$

In figure 5, we plot the measured ν as a function of the reduced density for a set of simulations using the model II collision rules of reference 3. The system used for the measurement was 32×32, and the forcing level was very weak, so that the typical Mach number, defined as the ratio between the speed of sound and the hydrodynamical velocity, $\vec{g}/\rho$, is approximately ≈ 0.1.

In the same figure, we compare our viscosity measurements with data obtained by relaxation measurements [10]. The errors bars for the latter set of data are set at about 3 percent of the actual measured viscosity [12], while the errors on our data set are about 1 percent and are not indicated.

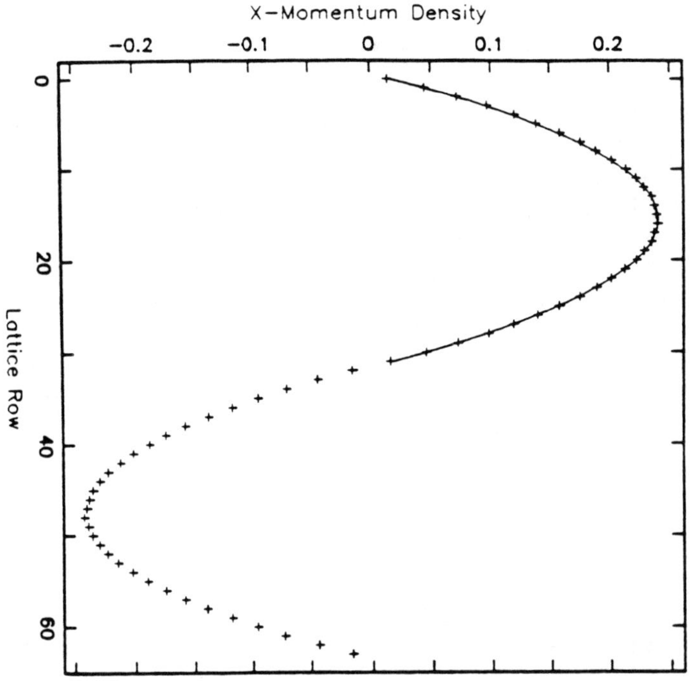

Figure 3: Möbius strip boundary conditions versus square wave forcing. The symbols represent the x momentum density profile for square wave forcing (see text) with periodic boundary conditions in both directions. The solid line is a parabolic fit to the momentum density profile obtained using Möbius strip boundary conditions on a system half as wide. Both simulations were run at a density of 2.1 particles per site and the profiles were averaged over one million time steps.

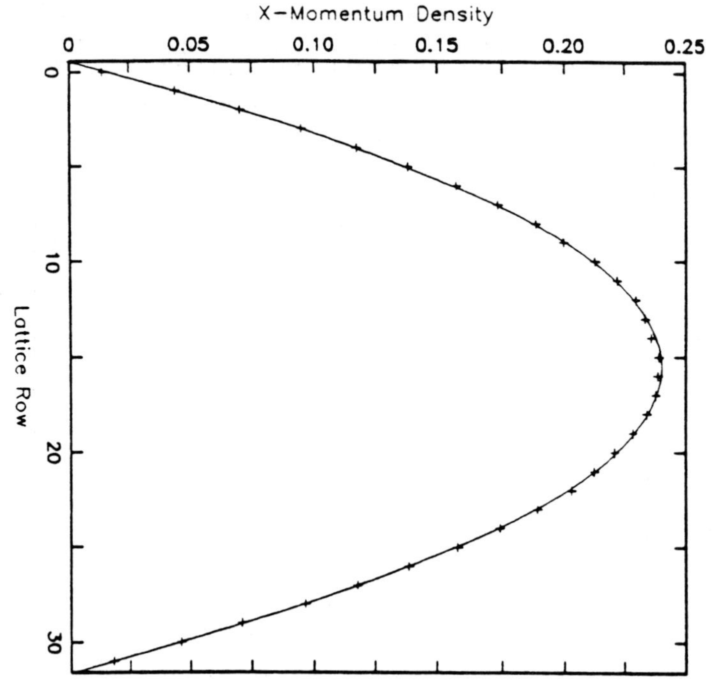

Figure 4: Typical momentum profile. The x momentum density profile for a 32×32 system run at $n = 2.1$ particles per site and a forcing level $F = 0.76$ momentum units per time step. The profile was averaged over one million time steps. The solid line is a parabolic fit to the simulation data points (symbols).

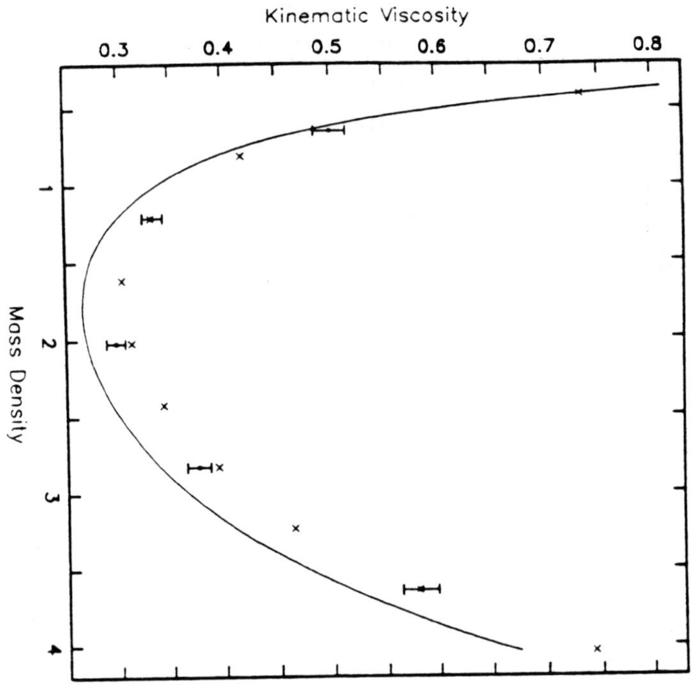

Figure 5: Kinematic viscosity versus mass density. Viscosity values derived from the present work are shown as crosses. Symbols with error bars are the results obtained by d'Humières et al. The solid line is the theoretical value obtained using Chapman-Enskog techniques.

In both cases, the errors are only rough estimates since they were not computed from first principles but were estimated by comparing similar runs with different initial conditions.

The two sets of viscosity data appear to be consistent, except in the range $2 < \rho < 3$, where our data are consistently greater than the results of reference 10. It would be tempting to relate the discrepancy between the two viscosities to viscosity renormalization effects [18], but we presently do not have any conclusive evidence.

The solid line is the shear viscosity as calculated by the technique of Michel Hénon [25] and by other authors using Chapman-Enskog techniques [17]. Both methods are based on an approximate theoretical description of the gas in which the correlations between particles are completely neglected.

In the low density limit, there is a very good agreement with the theory, as is expected since the relative importance of particle correlation becomes negligible in that limit. Note that we do not quote results for $\rho < 0.4$ because for these densities the mean-free path (see [22]) of the particles becames comparable to the width of the channel.

4. Conclusion

We have given some precise simulation evidence that LGAs are correctly represented by the constitutive relation, equation (1.5). We have also shown that the simulation of channel flow gives the expected parabolic profile to a good degree of accuracy and that the effective kinematic viscosity obtained by these steady non-equilibrium flows is in reasonable agreement with that obtained by d'Humières et al. using shear waves relaxation measurements.

The technique used for these simulations is capable of providing reasonably accurate measurement of viscosity; the particular kind of boundary conditions used allow a wide range of lattice size width and makes feasible the study of the dependence of the kinematic viscosity on the width of the simulation channel. For two-dimensional fluids, there are rather precise predictions, based on renormalization group arguments and other techniques, for this dependence [18]. In first-order perturbation theory, the kinematic viscosity diverges as the logarithm of the box size. We have some preliminary results, to be published elsewhere, which indicate the presence of this effect even within the range of channel widths accessible by our method of simulation.

Acknowledgments

We would like to thank B. Hasslacher and U. Frisch for interesting and helpful discussions, and G. Doolen and T. Shimomura for the original version of the FORTRAN program we have used. This work was supported by ONR and NSF-MRL, and we gratefully acknowledge the hospitality of the Los Alamos National Laboratory, where a portion of this work was done.

References

[1] U. Frisch, B. Hasslacher, and Y. Pomeau, *Phys. Rev. Lett.*, **56** (1986) 1505.

[2] S. Wolfram, *J. Stat. Phys.*, **45** (1986) 471.

[3] U. Frisch, D. d'Humières, B. Hasslacher, P. Lallemand, Y. Pomeau, and J. Rivet, "Lattice Gas Hydrodynamics in Two and Three Dimensions", *Complex Systems*, **1** (1987) 648.

[4] See the bibliography of [3].

[5] L. P. Kadanoff and J. Swift, *Phys. Rev.*, **165** (1968) 310.

[6] J. Hardy and Y. Pomeau, *J. of Math. Phys.*, **13** (1972) 1042.

[7] J. Hardy, O. de Pazzis, and Y. Pomeau, *J. of Math. Phys.*, **14** (1973) 1746; *Phys. Rev.*, **A13** (1976) 1949.

[8] D. d'Humières, P. Lallemand, and U. Frisch, *Europhys. Lett.*, **2** (1986) 297.

[9] D. d'Humières, Y. Pomeau, and P. Lallemand, *C. R. Acad. Sci. Paris II*, **301** (1985) 1391.

[10] D. d'Humières, P. Lallemand, and T. Shimomura, "Computer simulations of lattice gas hydrodynamics", LANL preprint.

[11] J. Dahlburg, D. Montgomery, and G. Doolen, "Noise and Compressibility in Lattice-Gas Fluids", *Physical Review*, in press.

[12] D. d'Humières, personal communication.

[13] D. d'Humières and P. Lallemand, *C. R. Acad. Sci. Paris II*, **302** (1985) 983.

[14] C. Cercignani, *Theory and Application of the Boltzmann Equation*, (New York, Elsevier, 1975).

[15] L. D. Meshalkin and Ya. G. Sinai, *J. Appl. Math. (PMM)*, **25** (1979) 1700.

[16] Z. S. She, "Large scale Dynamics and Transition to Turbulence in the Two-dimensional Kolmogorov Flow", *Proceed. Fifth Intern*, Beer-Sheva Seminar on MHD Flows and Turbulence, Israel, March 2–6, 1987.

[17] J. P. Rivet and U. Frisch, *C. R. Acad. Sci. Paris II*, **302** (1986) 267.

[18] Y. Pomeau and P. Résibois, *Phys. Rep.*, **19** (1975) 63.
K. Kawasaki and J. D. Gunton, *Phys. Rev.*, **A 8** (1973) 2048;
D. Forster, D. Nelson, and M. Stephen, *Phys. Rev.*, **A 16** (1977) 732;
T. Yamada and K. Kawasaki, *Prog. of Theor. Phys.*, **53** (1975) 111; Y. Pomeau and P. Résibois, *Phys. Rep.*, **19** (1975) 63.

[19] B. Alder and T. Wainwright, *Phys. Rev. Lett.*, **18** (1970) 968.

[20] D. J. Evans and G. P. Morriss, *Phys. Rev. Lett*, **51** (1983) 1776; D. H. Heyes, G. P. Morriss, and D. J. Evans, *J. of Chem. Phys.*, **83** (1985) 4760.

[21] L. P. Kadanoff, *Physics Today*, **39** (1986) 7.

[22] C. Burges and S. Zaleski, "Buoyant Mixtures of cellular Automaton Gases", *Complex Systems*, **1** (1987) 31.

[23] K. Balasubramanian, F. Hayot, and W. F. Saam, "Darcy's law for lattice gas hydrodynamics", Ohio State University preprint (1987).

[24] D. H. Rothman, MIT preprint (1987), submitted to *Geophysics*.

[25] M. Hénon, *Complex Systems*, **1** (1987) 762.